MOLECULAR BIOLOGY

structure and dynamics of genomes and proteomes

MOLECULAR BIOLOGY

structure and dynamics of genomes and proteomes

Jordanka Zlatanova

Kensal E. van Holde

Garland Science
Taylor & Francis Group

Garland Science
Vice President: Denise Schanck
Senior Editor: Summers Scholl
Editorial Assistants: Michael Roberts and William Sudry
Development Editor: Sharon Whitehead
Production Editor: Natasha Wolfe
Illustrator and Cover Design: Patrick Lane
Copyeditor: Heather Whirlow-Cammarn
Typesetter: Thomson Digital
Proofreader: Jo Clayton
Indexer: Indexing Specialists (UK) Ltd
Permissions Coordinator: Casey Mein

About the Authors
Jordanka Zlatanova is Professor Emeritus in the Department of Molecular Biology at the University of Wyoming. She earned her PhD and DSc degrees in Cellular and Molecular Biology from the Bulgarian Academy of Sciences.

Kensal E. van Holde is Distinguished Professor Emeritus in the Department of Biochemistry and Biophysics at Oregon State University. He earned his PhD in Physical Chemistry at the University of Wisconsin—Madison.

Library of Congress Cataloging-in-Publication Data

Zlatanova, J., author.
Molecular biology : structure and dynamics of genomes and proteomes / Jordanka Zlatanova, Kensal E. van Holde.
p. ; cm.
Includes bibliographical references.
ISBN 978-0-8153-4504-6
I. Van Holde, K. E. (Kensal Edward), 1928-, author. II. Title.
[DNLM: 1. Molecular Biology—methods. 2. Genome—physiology.
3. Proteome—physiology. 4. Transcription, Genetic—genetics.
QU 34]
QH506
572.8—dc23
2015024184 Published by Garland Science, Taylor & Francis Group, LLC, an Informa business,
711 Third Avenue, New York, NY 10017, USA, and 3 Park Square, Milton Park, Abingdon OX14 4RN, UK.
Printed in the United States of America
ISBN 978-0-8153-4504-6

15 14 13 12 11 10 9 8 7 6 5 4 3 2 1

Visit our Website at http://www.garlandscience.com

Dedication

We dedicate this work to the memory of Dr. E. Morton Bradbury, a physicist who taught us how the most sophisticated physical techniques could be applied to solve problems in biology.

Preface

Molecular biology is the study of biology taken to the molecular level. It reveals the essential principles behind the transmission and expression of genetic information in terms of DNA, RNA, and proteins.

Having been long engaged in collaborative research, we realized that the field of molecular biology has undergone great changes in recent years, becoming a highly structure-based science. Whether it is the elegant, dynamic structure of the ribosome, and how that structure explains the details of protein synthesis, or the complex organization of the human genome and its transcription, structural information is the key to understanding molecular mechanisms. For over a decade, one of us has taught an advanced molecular biology course, both in the classroom and online, and simply could not find a suitable text. A new textbook with contemporary coverage was needed.

Molecular Biology: Structure and Dynamics of Genomes and Proteomes elucidates the exquisite relationship between molecular structure and dynamics and the transmission and expression of genetic information. New techniques, ranging from single-molecule methods to whole-genome sequencing, have deepened our understanding of molecular processes and widened the scope of our vision of their interrelationships. These important advances present a new paradigm and must be considered in context of the dramatic history of the field—one of the most remarkable in modern sciences. An important example is our frequent reference to the ENCODE project and its implications. At the other end of the scale are the numerous illustrations of molecular interactions by cryo-electron microscopy and single-molecule dynamics.

From a pedagogical point of view, molecular biology is becoming more dependent on visual representations, and our goal was to create a dynamic and engaging illustration program. Accordingly, nearly 700 illustrations comprise a substantial portion of the book. We feel that legends should fully complement the figures so that figure and legend together can be essentially self-standing. Therefore, we have provided comprehensive explanatory detail within the figures themselves.

The text is written primarily for students at the upper undergraduate to first-year graduate level. The book can be adapted for students with or without an introductory course in biochemistry. A number of features make it flexible in approach: the first four introductory chapters could be covered very quickly as a review by an advanced audience, but they provide essential background for students with less experience in the biological sciences. In a similar vein, we have organized contextual and practical material (history, techniques, general cell biology background, and medical applications) into supplemental boxes. So that this book is useful to as wide a spectrum of students as possible, certain boxes represent a deeper analysis of a topic than will be needed for every course—labeled as "A Closer Look." These insights can be bypassed for a more direct, compact course, but they are available to those instructors or students who wish to explore more deeply into the topic. An additional feature is our focus on references that cover cutting-edge techniques, like single-molecule methods, that allow true molecular insights. Every effort has been made to consult and refer to the most current work in every area.

With these devices, the text offers readable, essential coverage of the field. The book is organized as follows. We begin with two introductory chapters, first presenting the basic ideas and development of molecular biology and elements of genetics. There follow two more chapters detailing the substances under study: proteins and nucleic acids. These four chapters could be used for a brief review in an advanced course. The processes by which genetic information is expressed, regulated, and maintained compose the core of the book. These are taken in the order of transcription, translation, and replication, with each topic requiring several chapters to cover structures, mechanisms, and regulation. Special chapters in the core are devoted to recombinant DNA techniques and the structure of the genome. The book concludes with chapters on genetic recombination and DNA repair.

Finally, a special note to the student: you are entering one of the most dynamic and exciting areas of science. Enjoy it!

Jordanka Zlatanova **Kensal E. van Holde**

RESOURCES FOR INSTRUCTORS AND STUDENTS

Teaching and learning resources for instructors and students are available online. The homework platform is available to everyone, though instructors will need to set up student access in order to use the dashboard to track student progress on assignments. The Instructor's Resources on the Garland Science Website are password-protected and available only to adopting instructors. The Student Resources on the Garland Science Website are available to everyone. We hope these resources will enhance student learning and make it easier for instructors to prepare dynamic lectures and activities for the classroom.

Online Homework Platform with Instructor Dashboard

Instructors can obtain access to the online homework platform from their sales representative or by emailing science@garland.com. Students who wish to use the platform must purchase access and, if required for class, obtain a course link from their instructor.

The online homework platform is designed to improve and track student performance. It allows instructors to select homework assignments on specific topics and review the performance of the entire class, as well as individual students, via the instructor dashboard. The user-friendly system provides a convenient way to gauge student progress and tailor classroom discussion, activities, and lectures to areas that require specific remediation.

Instructor Dashboard displays data on student performance, such as responses to individual questions and length of time required to complete assignments.

Tutorials explain essential or difficult concepts and are integrated with a variety of questions that assess student engagement and mastery of the material. The tutorials were created by Kristopher Koudelka, Point Loma Nazarene University.

Instructor Resources

Instructor Resources are available on the Garland Science Instructor Resource Center, located at www.garlandscience.com and selecting the "instructor" tab. The Website provides access not only to the teaching resources for this book but also to all other Garland Science textbooks. Adopting instructors can obtain access to the site from their sales representative or by emailing science@garland.com.

Art of Molecular Biology: The images from the book are available in two convenient formats: PowerPoint® and JPEG. They have been optimized for display on a computer. Figures are searchable by figure number, by figure name, or by keywords used in the figure legend from the book.

Student Resources

The resources for students are available on the *Molecular Biology* Student Website, located at www.garlandscience.com and selecting the "student" tab.

Flashcards: Each chapter contains flashcards, built into the student Website, that allow students to review key terms from the text.

Glossary: The comprehensive glossary of key terms from the book is online and can be searched or browsed.

Acknowledgments

The authors and publisher of *Molecular Biology: Structure and Dynamics of Genomes and Proteomes* specially acknowledge and thank Kristopher J. Koudelka (Point Loma Nazarene University) for creating the tutorials. The contributions of the following people are also gratefully acknowledged:

Ivan Dimitrov (The University of Texas Southwestern Medical Center) for Box 3.7; Aleksandra Kuzmanov (University of Wyoming) for the gene therapy section of Chapter 5; William Bonner (National Cancer Institute) for Box 22.6; Jean-Marc Victor (Université Pierre-et-Marie-Curie) and Thomas Bishop (Louisiana Tech University) for help with Chapter 8. We also thank the numerous researchers that provided useful resources in terms of figures and expert advice.

The following scientists and instructors provided valuable commentary as readers, reviewers, and advisors during the development of the book:

Steven Ackerman (University of Massachusetts Boston); Paul Babitzke (The Pennsylvania State University); Aaron Cassill (The University of Texas at San Antonio); Scott Cooper (University of Wisconsin—La Crosse); Raymond Deshaies (California Institute of Technology); Martin Edwards (Newcastle University [UK]); Yiwen Fang (Loyola Marymount University); Errol C. Friedberg (The University of Texas Southwestern Medical Center); Fátima Gebauer Hernández (Centre for Genomic Regulation [Spain]); Paul D. Gollnick (State University of New York at Buffalo); Paul Gooley (The University of Melbourne [Australia]); Leslie A. Gregg-Jolly (Grinnell College); Andrew W. Grimson (Cornell University); David Hess (Santa Clara University); Walter E. Hill (The University of Montana); Peter L. Jones (University of Massachusetts Medical School); Nemat O. Keyhani (University of Florida); Raida Wajih Khalil (Philadelphia University [Jordan]); Hannah Klein (New York University); Kristopher J. Koudelka (Point Loma Nazarene University); Stephen Kowalczykowski (University of California, Davis); Krzysztof Kuczera (The University of Kansas); Gary R. Kunkel (Texas A&M University); Richard LeBaron (The University of Texas at San Antonio); Boris Lenhard (Imperial College London [UK]); Diego Loayza (Hunter College of The City University of New York); William F. Marzluff (The University of North Carolina at Chapel Hill); Mitch McVey (Tufts University); Marcel Mechali (National Centre for Scientific Research [France]); Corinne A. Michels (Queens College, City University of New York); Peter B. Moore (Yale University); Daniel Moriarty (Siena College); Greg Odorizzi (University of Colorado Boulder); Wilma K. Olson (Rutgers, The State University of New Jersey); Wade H. Powell (Kenyon College); Susan A. Rotenberg (Queens College, City University of New York); Wilma Saffran (Queens College, City University of New York); Michael J. Smerdon (Washington State University); Kathryn Leigh Stoeber (Anglia Ruskin University [UK]); Francesca Storici (Georgia Institute of Technology); Andrew Arthur Travers (University of Cambridge [UK]); Edward N. Trifonov (University of Haifa [Israel]); Peter H. von Hippel (University of Oregon); Hengbin Wang (The University of Alabama at Birmingham); Carol Wilusz (Colorado State University); Xuewu Zhang (The University of Texas Southwestern Medical Center); Zhaolan Zhou (University of Pennsylvania).

Brief Contents

Chapter 1:	To the Cell and Beyond: The Realm of Molecular Biology	1
Chapter 2:	From Classical Genetics to Molecular Genetics	15
Chapter 3:	Proteins	31
Chapter 4:	Nucleic Acids	65
Chapter 5:	Recombinant DNA: Principles and Applications	93
Chapter 6:	Protein–Nucleic Acid Interactions	127
Chapter 7:	The Genetic Code, Genes, and Genomes	145
Chapter 8:	Physical Structure of the Genomic Material	163
Chapter 9:	Transcription in Bacteria	193
Chapter 10:	Transcription in Eukaryotes	215
Chapter 11:	Regulation of Transcription in Bacteria	241
Chapter 12:	Regulation of Transcription in Eukaryotes	261
Chapter 13:	Transcription Regulation in the Human Genome	297
Chapter 14:	RNA Processing	315
Chapter 15:	Translation: The Players	347
Chapter 16:	Translation: The Process	371
Chapter 17:	Regulation of Translation	395
Chapter 18:	Protein Processing and Modification	421
Chapter 19:	DNA Replication in Bacteria	457
Chapter 20:	DNA Replication in Eukaryotes	485
Chapter 21:	DNA Recombination	515
Chapter 22:	DNA Repair	549

Detailed Contents

Chapter 1: To the Cell and Beyond: The Realm of Molecular Biology 1

1.1 INTRODUCTION 1

1.2 THE VITAL ROLE OF MICROSCOPY IN BIOLOGY 2

The light microscope led to the first revolution in biology 2

Biochemistry led to the discovery of the importance of macromolecules in life's structure and processes 5

The electron microscope provided another order of resolution 6

1.3 FINE STRUCTURE OF CELLS AND VIRUSES AS REVEALED BY MICROSCOPY 7

1.4 ULTRAHIGH RESOLUTION: BIOLOGY AT THE MOLECULAR LEVEL 8

Fluorescence techniques allow for one approach to ultraresolution 8

Confocal fluorescence microscopy allows observation of the fluorescence emitted by a particular substance in a cell 9

FIONA provides ultimate optical resolution by use of fluorescence 10

FRET allows distance measurements at the molecular level 10

Single-molecule cryo-electron microscopy is a powerful new technique 10

The atomic force microscope feels molecular structure 11

X-ray diffraction and NMR provide resolution to the atomic level 11

1.5 MOLECULAR GENETICS: ANOTHER FACE OF MOLECULAR BIOLOGY 12

Key concepts 13

Further reading 14

Chapter 2: From Classical Genetics to Molecular Genetics 15

2.1 INTRODUCTION 15

2.2 CLASSICAL GENETICS AND THE RULES OF TRAIT INHERITANCE 15

Gregor Mendel developed the formal rules of genetics 15

Mendel's laws have extensions and exceptions 20

Genes are arranged linearly on chromosomes and can be mapped 21

The nature of genes and how they determine phenotypes was long a mystery 22

2.3 THE GREAT BREAKTHROUGH TO MOLECULAR GENETICS 23

Bacteria and bacteriophage exhibit genetic behavior and serve as model systems 23

Transformation and transduction allow transfer of genetic information 25

The Watson–Crick model of DNA structure provided the final key to molecular genetics 25

2.4 MODEL ORGANISMS 26

Key concepts 28

Further reading 29

Chapter 3: Proteins 31

3.1 INTRODUCTION 31

Proteins are macromolecules with enormous variety in size, structure, and function 31

Proteins are essential for the structure and functioning of all organisms 31

3.2 PROTEIN STRUCTURE 33

Amino acids are the building blocks of proteins 33

In proteins, amino acids are covalently connected to form polypeptides 34

3.3 LEVELS OF STRUCTURE IN THE POLYPEPTIDE CHAIN 35

The primary structure of a protein is a unique sequence of amino acids 35

A protein's secondary structure involves regions of regular folding stabilized by hydrogen bonds 39

Each protein has a unique three-dimensional tertiary structure 42

The tertiary structure of most proteins is divided into distinguishable folded domains 44

Algorithms are now used to identify and classify domains in proteins of known sequence 45

Some domains or proteins are intrinsically disordered 49

Quaternary structure involves associations between protein molecules to form aggregated structures 51

3.4 HOW DO PROTEINS FOLD? 53

Folding can be a problem 53

Chaperones help or allow proteins to fold 55

3.5 HOW ARE PROTEINS DESTROYED? 58

The proteasome is the general protein destruction system 58

3.6 THE PROTEOME AND PROTEIN INTERACTION NETWORKS 60

New technologies allow a census of an organism's proteins and their interactions 60

Key concepts 63

Further reading 64

Chapter 4: Nucleic Acids 65

4.1 INTRODUCTION 65

Protein sequences are dictated by nucleic acids 65

4.2 CHEMICAL STRUCTURE OF NUCLEIC ACIDS 65

DNA and RNA have similar but different chemical structures 65

Nucleic acids (polynucleotides) are polymers of nucleotides 68

4.3 PHYSICAL STRUCTURES OF DNA 69

Discovery of the B-DNA structure was a breakthrough in molecular biology 69

A number of alternative DNA structures exist 71

Although the double helix is quite rigid, it can be bent by bound proteins 74

DNA can also form folded tertiary structures 75

Closed DNA circles can be twisted into supercoils 76

4.4 PHYSICAL STRUCTURES OF RNA 82

RNA can adopt a variety of complex structures but not the B-form helix 82

4.5 ONE-WAY FLOW OF GENETIC INFORMATION 84

4.6 METHODS USED TO STUDY NUCLEIC ACIDS 84

Key concepts 91

Further reading 92

Chapter 5: Recombinant DNA: Principles and Applications 93

5.1 INTRODUCTION 93

Cloning of DNA involves several fundamental steps 94

5.2 CONSTRUCTION OF RECOMBINANT DNA MOLECULES 95

Restriction endonucleases and ligases are essential tools in cloning 95

5.3 VECTORS FOR CLONING 100

Genes coding for selectable markers are inserted into vectors during their construction 100

Bacterial plasmids were the first cloning vectors 102

Recombinant bacteriophages can serve as bacterial vectors 103

Cosmids and phagemids expand the repertoire of cloning vectors 105

5.4 ARTIFICIAL CHROMOSOMES AS VECTORS 106

Bacterial artificial chromosomes meet the need for cloning very large DNA fragments in bacteria 106

Eukaryotic artificial chromosomes provide proper maintenance and expression of very large DNA fragments in eukaryotic cells 107

5.5 EXPRESSION OF RECOMBINANT GENES 108

Expression vectors allow regulated and efficient expression of cloned genes 108

Shuttle vectors can replicate in more than one organism 109

5.6 INTRODUCING RECOMBINANT DNA INTO HOST CELLS 109

Numerous host-specific techniques are used to introduce recombinant DNA molecules into living cells 109

5.7 POLYMERASE CHAIN REACTION AND SITE-DIRECTED MUTAGENESIS 110

5.8 SEQUENCING OF ENTIRE GENOMES 112

Genomic libraries contain the entire genome of an organism as a collection of recombinant DNA molecules 112

There are two approaches for sequencing large genomes 113

5.9 MANIPULATING THE GENETIC CONTENT OF EUKARYOTIC ORGANISMS 115

Making a transgenic mouse involves numerous steps 115

To inactivate, replace, or otherwise modify a particular gene, the vector must be targeted for homologous recombination at that particular site 115

5.10 PRACTICAL APPLICATIONS OF RECOMBINANT DNA TECHNOLOGIES 116

Hundreds of pharmaceutical compounds are produced in recombinant bacteria 116

Plant genetic engineering is a huge but controversial industry 118

Gene therapy is a complex multistep process aiming to correct defective genes or gene functions that are responsible for disease 122

Delivering a gene into sufficient cells within a specific tissue and ensuring its subsequent long-term expression is a challenge 122

Whole animals can be cloned by nuclear transfer 124

Key concepts 125

Further reading 125

Chapter 6: Protein–Nucleic Acid Interactions 127

6.1 INTRODUCTION 127

6.2 DNA–PROTEIN INTERACTIONS 128

DNA–protein binding occurs by many modes and mechanisms 128

Site-specific binding is the most widely used mode 129

Most recognition sites fall into a limited number of classes 131

Most specific binding requires the insertion of protein into a DNA groove 132

Some proteins cause DNA looping 133

There are a few major protein motifs of DNA-binding domains 134

Helix–turn–helix motif interacts with the major groove 134

Zinc fingers also probe the major groove 135

Leucine zippers are especially suited for dimeric sites 135

6.3 RNA–PROTEIN INTERACTIONS 136

6.4 STUDYING PROTEIN–NUCLEIC ACID INTERACTIONS 139

Key concepts 144

Further reading 144

Chapter 7: The Genetic Code, Genes, and Genomes 145

7.1 INTRODUCTION 145

7.2 GENES AS NUCLEIC ACID REPOSITORIES OF GENETIC INFORMATION 145

Our understanding of the nature of genes is constantly evolving 145

The central dogma states that information flows from DNA to protein 147

It was necessary to separate cellular RNAs to seek the adaptors 147
Messenger RNA, tRNA, and ribosomes constitute the protein factories of the cell 148

7.3 RELATING PROTEIN SEQUENCE TO DNA SEQUENCE IN THE GENETIC CODE 149
The first task was to define the nature of the code 149

7.4 SURPRISES FROM THE EUKARYOTIC CELL: INTRONS AND SPLICING 152
Eukaryotic genes usually contain interspersed noncoding sequences 152

7.5 GENES FROM A NEW AND BROADER PERSPECTIVE 153
Protein-coding genes are complex 153
Genome sequencing has revolutionized the gene concept 153
Mutations, pseudogenes, and alternative splicing all contribute to gene diversity 154

7.6 COMPARING WHOLE GENOMES AND NEW PERSPECTIVES ON EVOLUTION 156
Genome sequencing reveals puzzling features of genomes 156
How are DNA sequence types and functions distributed in eukaryotes? 158
Key concepts 161
Further reading 161

Chapter 8: Physical Structure of the Genomic Material 163

8.1 INTRODUCTION 163

8.2 CHROMOSOMES OF VIRUSES AND BACTERIA 164
Viruses are packages for minimal genomes 164
Bacterial chromosomes are organized structures in the cytoplasm 165
DNA-bending proteins and DNA-bridging proteins help to pack bacterial DNA 166

8.3 EUKARYOTIC CHROMATIN 168
Eukaryotic chromosomes are highly condensed DNA–protein complexes segregated into a nucleus 168
The nucleosome is the basic repeating unit of eukaryotic chromatin 168
Histone nonallelic variants and postsynthetic modifications create a heterogeneous population of nucleosomes 171
The nucleosome family is dynamic 176
Nucleosome assembly *in vivo* uses histone chaperones 177

8.4 HIGHER-ORDER CHROMATIN STRUCTURE 178
Nucleosomes along the DNA form a chromatin fiber 178
The chromatin fiber is folded, but its structure remains controversial 180
The organization of chromosomes in the interphase nucleus is still obscure 181

8.5 MITOTIC CHROMOSOMES 182
Chromosomes condense and separate in mitosis 182
A number of proteins are needed to form and maintain mitotic chromosomes 183
Centromeres and telomeres are chromosome regions with special functions 184
There are a number of models of mitotic chromosome structure 187
Key concepts 190
Further reading 190

Chapter 9: Transcription in Bacteria 193

9.1 INTRODUCTION 193

9.2 OVERVIEW OF TRANSCRIPTION 193
There are aspects of transcription common to all organisms 193
Transcription requires the participation of many proteins 195
Transcription is rapid but is often interrupted by pauses 197
Transcription can be visualized by electron microscopy 198

9.3 RNA POLYMERASES AND TRANSCRIPTION CATALYSIS 201
RNA polymerases are a large family of enzymes that produce RNA transcripts of polynucleotide templates 201

9.4 MECHANICS OF TRANSCRIPTION IN BACTERIA 202
Initiation requires a multisubunit polymerase complex, termed the holoenzyme 202
The initiation phase of bacterial transcription is frequently aborted 206
Elongation in bacteria must overcome topological problems 209
There are two mechanisms for transcription termination in bacteria 210
Understanding transcription in bacteria is useful in clinical practice 211
Key concepts 212
Further reading 213

Chapter 10: Transcription in Eukaryotes 215

10.1 INTRODUCTION 215
Transcription in eukaryotes is a complex, highly regulated process 215
Eukaryotic cells contain multiple RNA polymerases, each specific for distinct functional subsets of genes 216

10.2 TRANSCRIPTION BY RNA POLYMERASE II 217
The yeast Pol II structure provides insights into transcriptional mechanisms 217
The structure of Pol II is more evolutionarily conserved than its sequence 219
Nucleotide addition during transcription elongation is cyclic 219
Transcription initiation depends on multisubunit protein complexes that assemble at core promoters 222
An additional protein complex is needed to connect Pol II to regulatory proteins 224
Termination of eukaryotic transcription is coupled to polyadenylation of the RNA transcript 224

10.3 TRANSCRIPTION BY RNA POLYMERASE I 225

10.4 TRANSCRIPTION BY RNA POLYMERASE III 227
RNA polymerase III specializes in transcription of small genes 227

10.5 TRANSCRIPTION IN EUKARYOTES: PERVASIVE AND SPATIALLY ORGANIZED 228
Most of the eukaryotic genome is transcribed 228
Transcription in eukaryotes is not uniform within the nucleus 232
Active and inactive genes are spatially separated in the nucleus 233
10.6 METHODS FOR STUDYING EUKARYOTIC TRANSCRIPTION 234
A battery of methods is available for the study of transcription 234
Key concepts 238
Further reading 239
Chapter 11: Regulation of Transcription in Bacteria 241
11.1 INTRODUCTION 241
11.2 GENERAL MODELS FOR REGULATION OF TRANSCRIPTION 242
Regulation can occur via differences in promoter strength or use of alternative σ factors 242
Regulation through ligand binding to RNA polymerase is called stringent control 243
11.3 SPECIFIC REGULATION OF TRANSCRIPTION 244
Regulation of specific genes occurs through *cis–trans* interactions with transcription factors 244
Transcription factors are activators and repressors whose own activity is regulated in a number of ways 244
Several transcription factors can act synergistically or in opposition to activate or repress transcription 244
11.4 TRANSCRIPTIONAL REGULATION OF OPERONS IMPORTANT TO BACTERIAL PHYSIOLOGY 246
The *lac* operon is controlled by a dissociable repressor and an activator 246
Control of the *trp* operon involves both repression and attenuation 250
The same protein can serve as an activator or a repressor: the *ara* operon 252
11.5 OTHER MODES OF GENE REGULATION IN BACTERIA 255
DNA supercoiling is involved in both global and local regulation of transcription 255
DNA methylation can provide specific regulation 256
11.6 COORDINATION OF GENE EXPRESSION IN BACTERIA 256
Networks of transcription factors form the basis of coordinated gene expression 257
Key concepts 258
Further reading 259
Chapter 12: Regulation of Transcription in Eukaryotes 261
12.1 INTRODUCTION 261
12.2 REGULATION OF TRANSCRIPTION INITIATION: REGULATORY REGIONS AND TRANSCRIPTION FACTORS 262
Core and proximal promoters are needed for basal and regulated transcription 262
Enhancers, silencers, insulators, and locus control regions are all distal regulatory elements 263
Some eukaryotic transcription factors are activators, others are repressors, and still others can be either, depending on context 264
Regulation can use alternative components of the basal transcriptional machinery 267
Mutations in gene regulatory regions and in transcriptional machinery components lead to human diseases 267
12.3 REGULATION OF TRANSCRIPTIONAL ELONGATION 267
The polymerase may stall close to the promoter 267
Transcription elongation rate can be regulated by elongation factors 268
12.4 TRANSCRIPTION REGULATION AND CHROMATIN STRUCTURE 268
What happens to nucleosomes during transcription? 268
12.5 REGULATION OF TRANSCRIPTION BY HISTONE MODIFICATIONS AND VARIANTS 271
Modification of histones provides epigenetic control of transcription 271
Gene expression is often regulated by histone post-translational modifications 272
Readout of histone post-translational modification marks involves specialized protein molecules 272
Post-translational histone marks distinguish transcriptionally active and inactive chromatin regions 275
Some genes are specifically silenced by post-translational modification in some cell lines 275
Polycomb protein complexes silence genes through H3K27 trimethylation and H2AK119 ubiquitylation 276
Heterochromatin formation at telomeres in yeast silences genes through H4K16 deacetylation 277
HP1-mediated gene repression in the majority of eukaryotic organisms involves H3K9 methylation 277
Poly(ADP)ribosylation of proteins is involved in transcriptional regulation 279
Histone variants H2A.Z, H3.3, and H2A.Bbd are present in active chromatin 280
MacroH2A is a histone variant prevalent in inactive chromatin 282
Problems caused by chromatin structure can be fixed by remodeling 282
Endogenous metabolites can exert rheostat control of transcription 284
12.6 DNA METHYLATION 285
DNA methylation patterns in genomic DNA may participate in regulation of transcription 286
Carcinogenesis alters the pattern of CpG methylation 287
DNA methylation changes during embryonic development 288
DNA methylation is governed by complex enzymatic machinery 288
There are proteins that read the DNA methylation mark 289
12.7 LONG NONCODING RNAS IN TRANSCRIPTIONAL REGULATION 289
Noncoding RNAs play surprising roles in regulating transcription 289
The sizes and genomic locations of noncoding transcripts are remarkably diverse 290

12.8 METHODS FOR MEASURING THE ACTIVITY OF TRANSCRIPTIONAL REGULATORY ELEMENTS 293
Key concepts 294
Further reading 295

Chapter 13: Transcription Regulation in the Human Genome 297

13.1 INTRODUCTION 297
Rapid full-genome sequencing allows deep analysis 298

13.2 BASIC CONCEPTS OF ENCODE 298
ENCODE depends on high-throughput, massively processive sequencing and sophisticated computer algorithms for analysis 298
The ENCODE project integrates diverse data relevant to transcription in the human genome 299

13.3 REGULATORY DNA SEQUENCE ELEMENTS 300
Seven classes of regulatory DNA sequence elements make up the transcriptional landscape 300

13.4 SPECIFIC FINDINGS CONCERNING CHROMATIN STRUCTURE FROM ENCODE 301
Millions of DNase I hypersensitive sites mark regions of accessible chromatin 301
DNase I signatures at promoters are asymmetric and stereotypic 302
Nucleosome positioning at promoters and around TF-binding sites is highly heterogeneous 303
The chromatin environment at regulatory elements and in gene bodies is also heterogeneous and asymmetric 303

13.5 ENCODE INSIGHTS INTO GENE REGULATION 305
Distal control elements are connected to promoters in a complex network 305
Transcription factor binding defines the structure and function of regulatory regions 305
Transcription factors interact in a huge network 307
TF-binding sites and TF structure co-evolve 309
DNA methylation patterns show a complex relationship with transcription 311

13.6 ENCODE OVERVIEW 312
What have we learned from ENCODE, and where is it leading? 312
Certain methods are essential to ENCODE project studies 312
Key concepts 313
Further reading 314

Chapter 14: RNA Processing 315

14.1 INTRODUCTION 315
Most RNA molecules undergo post-transcriptional processing 315
There are four general categories of processing 315
Eukaryotic RNAs exhibit much more processing than bacterial RNAs 316

14.2 PROCESSING OF tRNAS AND rRNAS 316
tRNA processing is similar in all organisms 316
All three mature ribosomal RNA molecules are cleaved from a single long precursor RNA 316

14.3 PROCESSING OF EUKARYOTIC mRNA: END MODIFICATIONS 318
Eukaryotic mRNA capping is co-transcriptional 319
Polyadenylation at the 3′-end serves a number of functions 320

14.4 PROCESSING OF EUKARYOTIC mRNA: SPLICING 321
The splicing process is complex and requires great precision 321
Splicing is carried out by spliceosomes 322
Splicing can produce alternative mRNAs 322
Tandem chimerism links exons from separate genes 324
Trans-splicing combines exons residing in the two complementary DNA strands 328

14.5 REGULATION OF SPLICING AND ALTERNATIVE SPLICING 329
Splice sites differ in strength 329
Exon–intron architecture affects splice-site usage 329
Cis–trans interactions may stimulate or inhibit splicing 330
RNA secondary structure can regulate alternative splicing 332
Sometimes alternative splicing regulation needs no auxiliary regulators 332
The rate of transcription and chromatin structure may help regulate splicing 332

14.6 SELF-SPLICING: INTRONS AND RIBOZYMES 334
A fraction of introns is excised by self-splicing RNA 334
There are two classes of self-splicing introns 334

14.7 OVERVIEW: THE HISTORY OF AN mRNA MOLECULE 335
Proceeding from the primary transcript to a functioning mRNA requires a number of steps 335
mRNA is exported from the nucleus to the cytoplasm through nuclear pore complexes 335
RNA sequence can be edited by enzymatic modification even after transcription 336

14.8 RNA QUALITY CONTROL AND DEGRADATION 339
Bacteria, archaea, and eukaryotes all have mechanisms for RNA quality control 339
Archaea and eukaryotes utilize specific pathways to deal with different RNA defects 341

14.9 BIOGENESIS AND FUNCTIONS OF SMALL SILENCING RNAS 342
All ssRNAs are produced by processing from larger precursors 342
Key concepts 344
Further reading 345

Chapter 15: Translation: The Players 347

15.1 INTRODUCTION 347

15.2 A BRIEF OVERVIEW OF TRANSLATION 347
Three participants are needed for translation to occur 347

15.3 TRANSFER RNA 349
tRNA molecules fold into four-arm cloverleaf structures 350
tRNAs are aminoacylated by a set of specific enzymes, aminoacyl-tRNA synthetases 351
Aminoacylation of tRNA is a two-step process 352
Quality control or proofreading occurs during the aminoacylation reaction 353

Insertion of noncanonical amino acids into polypeptide chains is guided by stop codons 354

15.4 MESSENGER RNA **356**

The Shine–Dalgarno sequence in bacterial mRNAs aligns the message on the ribosome 357

Eukaryotic mRNAs do not have Shine–Dalgarno sequences but more complex 5′- and 3′-untranslated regions 358

Overall translation efficiency depends on a number of factors 360

15.5 RIBOSOMES **361**

The ribosome is a two-subunit structure comprising rRNAs and numerous ribosomal proteins 361

Functional ribosomes require both subunits, with specific complements of RNA and protein molecules 362

The small subunit can accept mRNA but must join with the large subunit for peptide synthesis to occur 364

Ribosome assembly has been studied both *in vivo* and *in vitro* 365

Key concepts 368

Further reading 369

Chapter 16: Translation: The Process **371**

16.1 INTRODUCTION **371**

16.2 AN OVERVIEW OF TRANSLATION: HOW FAST AND HOW ACCURATE? **371**

16.3 ADVANCED METHODOLOGY FOR THE ANALYSIS OF TRANSLATION **373**

Cryo-EM allows visualization of discrete kinetic states of ribosomes 373

X-ray crystallography provides the highest resolution 374

Single-pair fluorescence resonance energy transfer allows dynamic studies at the single-particle level 376

16.4 INITIATION OF TRANSLATION **377**

Initiation of translation begins on a free small ribosomal subunit 377

Cryo-EM provides details of initiation complexes 377

Start site selection in eukaryotes is complex 378

16.5 TRANSLATIONAL ELONGATION **379**

Decoding means matching the codon to the anticodon-carrying aminoacyl-tRNA 380

Accommodation denotes a relaxation of distorted tRNA to allow peptide bond formation 381

Peptide bond formation is accelerated by the ribosome 381

The formation of hybrid states is an essential part of translocation 383

Structural information on bacterial elongation factors provides insights into mechanisms 385

There is an exit tunnel for the peptide chain in the ribosome 387

Translation elongation in eukaryotes involves even more factors 388

16.6 TERMINATION OF TRANSLATION **388**

RF3 aids in removing RF1 and RF2 390

Ribosomes are recycled after termination 390

Our views of translation continue to evolve 391

Key concepts 292

Further reading 293

Chapter 17: Regulation of Translation **395**

17.1 INTRODUCTION **395**

17.2 REGULATION OF TRANSLATION BY CONTROLLING RIBOSOME NUMBER **395**

Ribosome numbers in bacteria are responsive to the environment 395

Synthesis of ribosomal components in bacteria is coordinated 396

Regulation of the synthesis of ribosomal components in eukaryotes involves chromatin structure 397

17.3 REGULATION OF TRANSLATION INITIATION **400**

Regulation of translation initiation is ubiquitous and remarkably varied 400

Regulation may depend on protein factors binding to the 5′- or 3′-ends of mRNA 400

Cap-dependent regulation is the major pathway for controlling initiation 400

Initiation may utilize internal ribosome entry sites 401

5′–3′-UTR interactions provide a novel mechanism that regulates initiation in eukaryotes 402

Riboswitches are RNA sequence elements that regulate initiation in response to stimuli 404

MicroRNAs can bind to mRNA, thereby regulating translation 405

17.4 mRNA STABILITY AND DECAY IN EUKARYOTES **407**

The two major pathways of decay for nonfaulty mRNA molecules start with mRNA deadenylation 407

The 5′ → 3′ pathway is initiated by the activities of the decapping enzyme Dcp2 408

The 3′ → 5′ pathway uses the exosome, followed by a different decapping enzyme, DcpS 410

There are additional pathways for mRNA degradation 412

Unused mRNA is sequestered in P bodies and stress granules 412

Cells have several mechanisms that destroy faulty mRNA molecules 416

mRNA molecules that contain premature stop codons are degraded through nonsense-mediated decay or NMD 416

No-go decay or NGD functions when the ribosome stalls during elongation 417

Non-stop decay or NSD functions when mRNA does not contain a stop codon 417

17.5 MECHANISMS OF TRANSLATION **418**

Key concepts 419

Further reading 419

Chapter 18: Protein Processing and Modification **421**

18.1 INTRODUCTION **421**

18.2 STRUCTURE OF BIOLOGICAL MEMBRANES **422**

Biological membranes are protein-rich lipid bilayers 422

Numerous proteins are associated with biomembranes 422

18.3 PROTEIN TRANSLOCATION THROUGH BIOLOGICAL MEMBRANES **423**

Protein translocation can occur during or after translation 423

Membrane translocation in bacteria and archaea primarily functions for secretion 424

Membrane translocation in eukaryotes serves a multitude of functions 424
Integral membrane proteins have special mechanisms for membrane insertion 426
Vesicles transport proteins between compartments in eukaryotic cells 428
18.4 PROTEOLYTIC PROTEIN PROCESSING: CUTTING, SPLICING, AND DEGRADATION 429
Proteolytic cleavage is sometimes used to produce mature proteins from precursors 430
Some proteases can catalyze protein splicing 431
Controlled proteolysis is also used to destroy proteins no longer needed 433
18.5 POST-TRANSLATIONAL CHEMICAL MODIFICATIONS OF SIDE CHAINS 433
Modification of side chains can affect protein structure and function 433
Phosphorylation plays a major role in signaling 435
Acetylation mainly modifies interactions 436
Several classes of glycosylated proteins contain added sugar moieties 436
Mechanisms of glycosylation depend on the type of modification 433
Ubiquitylation adds single or multiple ubiquitin molecules to proteins through an enzymatic cascade 445
Specificity of ubiquitin targeting is determined by a special class of enzymes 447
The structure of protein–ubiquitin conjugates determines the biological role of the modification 451
Polyubiquitin marks proteins for degradation by the proteasome 451
Sumoylation adds single or multiple SUMO molecules to proteins 452
18.6 THE GENOMIC ORIGIN OF PROTEINS 455
Key concepts 455
Further reading 456

Chapter 19: DNA Replication in Bacteria 457
19.1 INTRODUCTION 457
19.2 FEATURES OF DNA REPLICATION SHARED BY ALL ORGANISMS 457
Replication on both strands creates a replication fork 457
Mechanistically, synthesis of new DNA chains requires a template, a polymerase, and a primer 459
DNA replication requires the simultaneous action of two DNA polymerases 459
Other protein factors are obligatory at the replication fork 460
19.3 DNA REPLICATION IN BACTERIA 461
Bacterial chromosome replication is bidirectional, from a single origin of replication 461
DNA polymerase III catalyzes replication in bacteria 462
Sliding clamp β, or processivity factor, is essential for processivity 462
The clamp loader organizes the replisome 464
The full complement of proteins in the replisome is organized in a complex and dynamic way 465
DNA polymerase I is necessary for maturation of Okazaki fragments 468
19.4 THE PROCESS OF BACTERIAL REPLICATION 470
The replisome is a dynamic structure during elongation 470
19.5 INITIATION AND TERMINATION OF BACTERIAL REPLICATION 472
Initiation involves both specific DNA sequence elements and numerous proteins 473
Termination of replication also employs specific DNA sequences and protein factors that bind to them 476
19.6 BACTERIOPHAGE AND PLASMID REPLICATION 478
Rolling-circle replication is an alternative mechanism 481
Phage replication can involve both bidirectional and rolling-circle mechanisms 481
Key concepts 482
Further reading 482

Chapter 20: DNA Replication in Eukaryotes 485
20.1 INTRODUCTION 485
20.2 REPLICATION INITIATION IN EUKARYOTES 485
Replication initiation in eukaryotes proceeds from multiple origins 485
Eukaryotic origins of replication have diverse DNA and chromatin structure depending on the biological species 488
There is a defined scenario for formation of initiation complexes 493
Re-replication must be prevented 495
Histone methylation regulates onset of licensing 495
20.3 REPLICATION ELONGATION IN EUKARYOTES 495
Eukaryotic replisomes both resemble and significantly differ from those of bacteria 495
Other components of the bacterial replisome have functional counterparts in eukaryotes 499
Eukaryotic elongation has some special dynamic features 499
20.4 REPLICATION OF CHROMATIN 499
Chromatin structure is dynamic during replication 499
Histone chaperones may play multiple roles in replication 500
Both old and newly synthesized histones are required in replication 501
Epigenetic information in chromatin must also be replicated 503
20.5 THE DNA END-REPLICATION PROBLEM AND ITS RESOLUTION 504
Telomerase solves the end-replication problem 504
Alternative lengthening of telomeres pathway is active in telomerase-deficient cells 506
20.6 MITOCHONDRIAL DNA REPLICATION 507
Are circular mitochondrial genomes myth or reality? 507
Models of mitochondrial genome replication are contentious 508

20.7 REPLICATION IN VIRUSES THAT INFECT EUKARYOTES **509**
Retroviruses use reverse transcriptase to copy RNA into DNA 509
Key concepts 512
Further reading 513

Chapter 21: DNA Recombination 515

21.1 INTRODUCTION **515**
21.2 HOMOLOGOUS RECOMBINATION **515**
Homologous recombination plays a number of roles in bacteria 516
Homologous recombination has multiple roles in mitotic cells 517
Meiotic exchange is essential to eukaryotic evolution 517
21.3 HOMOLOGOUS RECOMBINATION IN BACTERIA **517**
End resection requires the RecBCD complex 518
Strand invasion and strand exchange both depend on RecA 520
Much concerning homologous recombination is still not understood 520
Holliday junctions are the essential intermediary structures in HR 523
21.4 HOMOLOGOUS RECOMBINATION IN EUKARYOTES **525**
Proteins involved in eukaryotic recombination resemble their bacterial counterparts 525
HR malfunction is connected with many human diseases 527
Meiotic recombination allows exchange of genetic information between homologous chromosomes in meiosis 529
21.5 NONHOMOLOGOUS RECOMBINATION **532**
Transposable elements or transposons are mobile DNA sequences that change positions in the genome 532
Many transposons are transcribed but only a few have known functions 533
There are several types of transposons 535
DNA class II transposons can use either of two mechanisms to transpose themselves 538
Retrotransposons, or class I transposons, require an RNA intermediate 539
21.6 SITE-SPECIFIC RECOMBINATION **540**
Bacteriophage λ integrates into the bacterial genome by site-specific recombination 540
Immunoglobulin gene rearrangements also occur through site-specific recombination 542
Key concepts 547
Further reading 548

Chapter 22: DNA Repair 549

22.1 INTRODUCTION **549**
22.2 TYPES OF LESIONS IN DNA **551**
Natural agents, from both within and outside a cell, can change the information content of DNA 551
22.3 PATHWAYS AND MECHANISMS OF DNA REPAIR **551**
DNA lesions are countered by a number of mechanisms of repair 551
Thymine dimers are directly repaired by DNA photolyase 555
The enzyme O^6-alkylguanine alkyltransferase is involved in the repair of alkylated bases 556
Nucleotide excision repair is active on helix-distorting lesions 556
Base excision repair corrects damaged bases 557
Mismatch repair corrects errors in base pairing 558
Methyl-directed mismatch repair in bacteria uses methylation on adenines as a guide 559
Mismatch repair pathways in eukaryotes may be directed by strand breaks during DNA replication 560
Repair of double-strand breaks can be error-free or error-prone 560
Homologous recombination repairs double-strand breaks faithfully 560
Nonhomologous end-joining restores the continuity of the DNA double helix in an error-prone process 561
22.4 TRANSLESION SYNTHESIS **566**
Many repair pathways utilize RecQ helicases 568
22.5 CHROMATIN AS AN ACTIVE PLAYER IN DNA REPAIR **568**
Histone variants and their post-translational modifications are specifically involved in DNA repair 568
22.6 OVERVIEW: THE ROLE OF DNA REPAIR IN LIFE **576**
Key concepts 576
Further reading 577

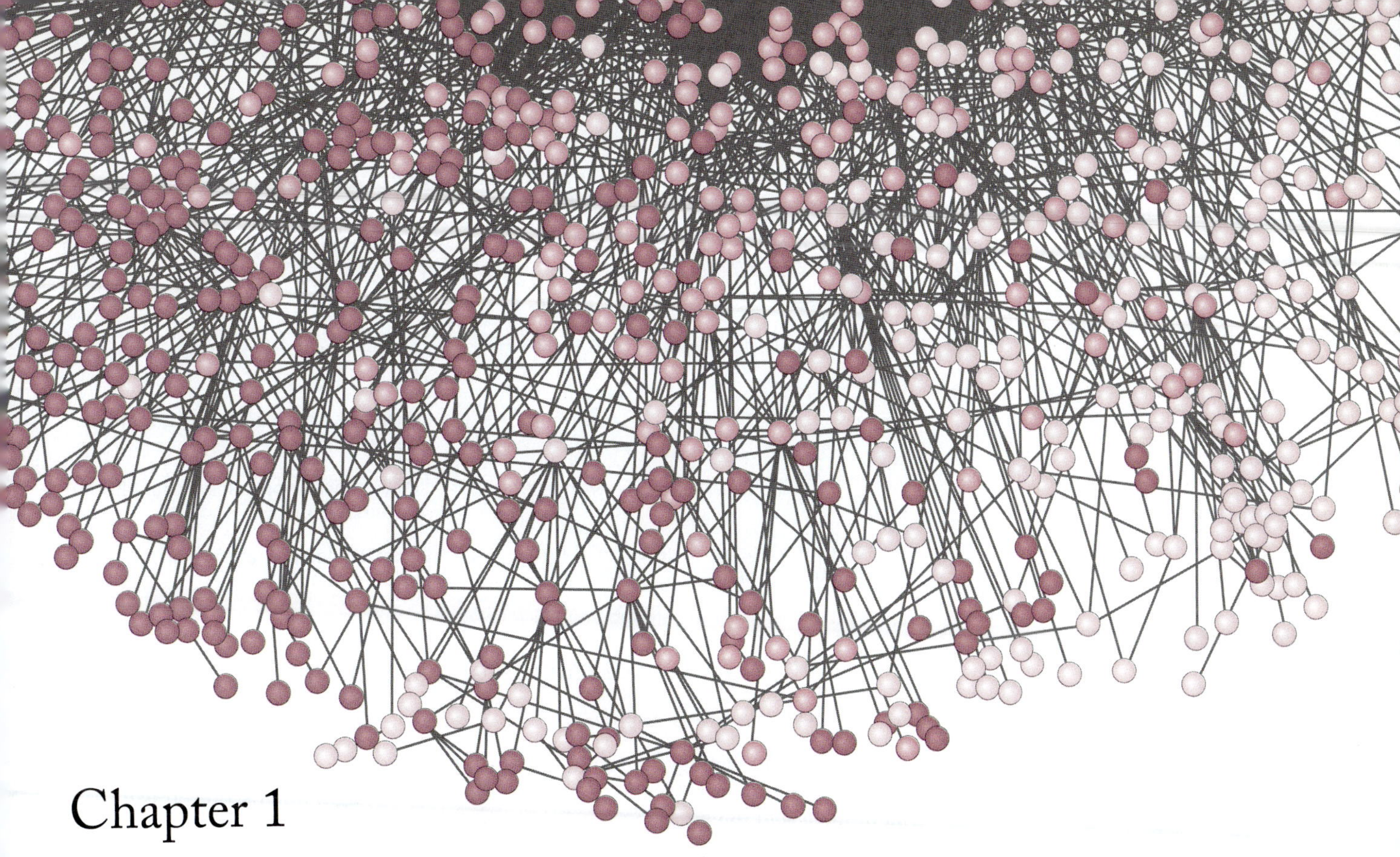

Chapter 1

To the Cell and Beyond: The Realm of Molecular Biology

1.1 Introduction

Every science has its own direction. Just as astronomy can be seen as a quest for understanding our universe on an ever larger scale, biology can be understood as a search for ever more minute structure in living things (**Table 1.1**). The earliest biologists must, of necessity, have depended on what the human eye could see to acquire their data. Thus biologists were initially concerned with whole animals and plants: their growth, their behavior, and their gross anatomy. Even so, the biology of antiquity was riddled with superstition and fancies, imagined monsters and absurd internal structures. The first person to begin a truly scientific study of animal and human anatomy was Leonardo da Vinci. He began these studies in about 1487 and combined his artistic talent with acute observation of actual dissections to produce anatomical data that were not rivaled for centuries. Indeed, the next great advances in biology had to wait for the scientific renaissance of the seventeenth century, typified by William Harvey's studies of the circulation of the blood in 1628.

In this chapter, we follow the development of ever more sophisticated techniques that have allowed examination of the structures of living things in greater and greater detail. The ultimate goal has always been to understand the fundamental processes of life and the structure that underlies them. This has now, to a considerable extent, been achieved even at the molecular level; the fine structure of life is accessible. We shall be referring to information derived from these methods again and again in chapters to come.

Table 1.1 Ways of studying biology at different scales.

Instrument	Approximate resolution limit	Comparable biological entities
human eye	0.3 mm	a mite
light microscope	0.3 μm	bacteria; eukaryotic cells; some organelles
electron microscope	0.3 nm	large macromolecules; organelle details; viruses
atomic force microscope	0.3–1 nm	large macromolecules; organelle details; viruses
X-ray diffraction or nuclear magnetic resonance	0.1 nm	small molecules; atomic groups in large molecules

1.2 The vital role of microscopy in biology

The light microscope led to the first revolution in biology

A revolution in biology occurred in the latter years of the seventeenth century, following Antonie van Leeuwenhoek's invention of the light microscope. This instrument increased the power of the eye a thousandfold and revealed a hitherto unsuspected world of microorganisms, such as **bacteria** and protozoa, and the multicellular nature of all higher organisms, the metazoa (see Table 1.1). Nevertheless, the traditional light microscope has its limits. It cannot resolve details much smaller than the wavelength of the light used because of the diffraction, or interference, of waves by an entity of such small size. Visible light has wavelengths of about 0.4–0.7 micrometers or μm (**Box 1.1**). In practice, this means that the resolving power of light microscopy, which is

Box 1.1 Radiation There are a number of places in this book where we must consider aspects of electromagnetic radiation. In this first chapter, for example, we are concerned with what can be learned about the structure of living matter by using different kinds of radiation as probes. Later in the book, we will discuss the damage that certain types of radiation can produce in tissues, cells, and molecules.

You are probably familiar, from earlier science classes, with the dual nature of radiation: particles and waves. Quantum mechanics shows that electromagnetic radiation can be visualized equally well as a stream of particles, known as photons, or as a train of electromagnetic waves. There is a fundamental relationship between the energy of a photon and the frequency ν, nu, of the wave that corresponds to it:

$$E = h\nu$$

Here h is Planck's constant, 6.6×10^{-34} J·s, and E is in joules, J. We will talk more often in terms of the wavelength λ, lambda, instead of the frequency of radiation. Since $\nu = c/\lambda$, where c is the velocity of light, we have

$$E = hc/\lambda$$

This tells us that longer wavelength radiation delivers lower-energy photons and vice versa. We can now look at the whole electromagnetic spectrum (**Figure 1**) from that point of view.

The first thing to notice is that the visible region of the spectrum, the part we can observe with our eyes, is an extremely narrow window onto the world. Note that the wavelength scale here is logarithmic; there is actually important radiation of wavelengths orders of magnitude longer or shorter than what we can see. Until about a hundred years ago, scientists were limited to that very narrow view of the world. Until very recently, it was believed that there was an inherent limitation to the resolving power of any radiation. Resolving power is the ability to perceive two closely situated objects as distinct and was believed to be irrevocably limited to a value of about half the wavelength of the radiation used. By this rule, the best resolution that could be obtained with the best microscope operating in the visible spectrum would be about 200 nm. Thus, for a long time, the search for more detail in living structures depended on the use of ever shorter wavelengths, X-rays for example. Recently, ways have been discovered to somewhat evade this diffraction limit, so microscopy is having a rebirth.

The other fundamentally important feature of the spectrum is how photon energy varies with wavelength. This means that what electromagnetic waves/particles can do to matter, including living matter, varies enormously over the spectrum. To make this clear, Figure 1 gives the energy of a mole of photons, or 6.02×10^{23} photons, at various wavelengths. This quantity is appropriate because we want to compare these values to the molar energies for various molecular processes: the energies required to break various chemical bonds, for example, which are conventionally expressed in kilojoules per mol.

Such comparisons provide interesting observations: for example, the human eye has evolved to use just a small region in the whole spectrum that offers the most advantages to the mechanisms of sight. The energies of photons in the visible region are barely on the threshold of promoting the chemical changes needed to induce a neural signal. Much shorter radiation would produce much more severe and damaging chemical changes. Compare the energy values for the visible light range given in Figure 1 with the energies necessary to break organic chemical bonds, which are typically in the range 300–600 kJ/mol. Fortunately, we are protected by our atmosphere from the more chemically active far-ultraviolet (UV) region of the spectrum. The atmospheric absorption of short-wavelength UV is why UV of wavelengths shorter than those of visible light is sometimes referred to as vacuum UV.

When we are exposed to radiation ranging from the near UV to the vacuum UV, or to even shorter X-rays and cosmic rays, serious chemical processes do occur. Longer UV produces skin burns and can do DNA damage that leads to skin cancer, but it cannot penetrate far enough into the body to damage internal cells. On the other hand, radiation such as X-rays or of even shorter wavelengths is strongly penetrating and can do harm throughout the body. Such damage is of two types: first, high-energy photons can directly damage DNA or proteins; second and more frequently, the damage is done by ion pairs that are produced by fragmentation of molecules, including water, by the radiation. This is why very high-energy radiation is often referred to as ionizing radiation.

As we look to types of radiation whose wavelengths are increasingly longer than those in the visible region, we find the energetic capabilities steadily diminishing. Infrared photons are especially effective in exciting intramolecular vibrations. As the wavelength increases toward the microwave region, molecular rotation changes and general heating, a reflection of increased overall molecular motions, become the rule. Finally, as we approach radio waves, very little direct energetic interaction with matter is seen, except with resonating antennae systems. This is why radio works: radio waves pass through material objects such as landscape features and buildings without significant absorption.

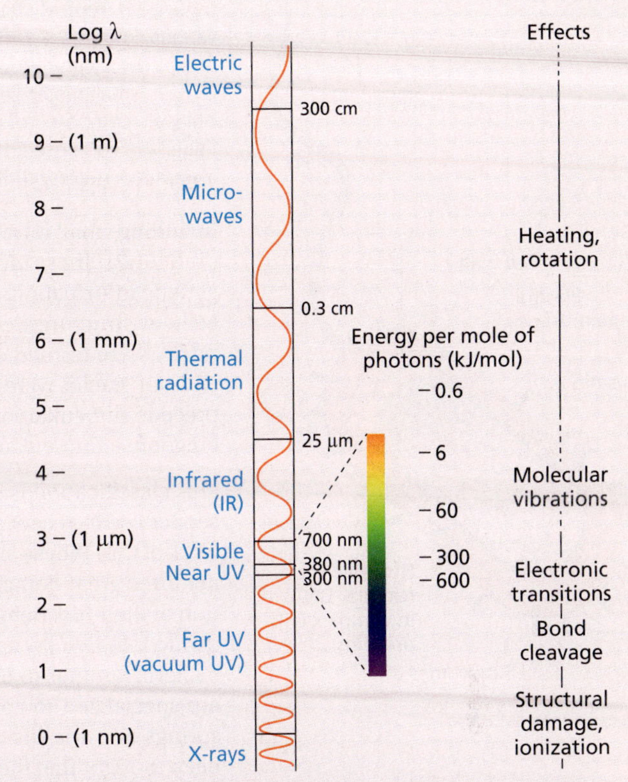

Figure 1 Range of electromagnetic radiation of interest to chemists and biologists. This schematic view is shown on a logarithmic scale. There are still longer and shorter waves than these. The borders between the different domains are somewhat arbitrary, except that the visual region is defined by the capability of the human eye.

the smallest distance at which two distinct objects can be resolved, is about 0.3 μm. Any objects closer than this will blur together. It is possible to extend these limits by the use of light emission from particles, either by **fluorescence** or by light scattering. If an object is illuminated from the side, to provide a dark field of view, even very small particles will show up as bright dots because of light scattering, but no internal detail can be seen.

The light microscope clearly revealed, however, that some kinds of cells—those from animals, plants, fungi, and even protozoa—have complex internal structures,

Table 1.2 Major cellular organelles.

Organelle	Found in	Functions
nucleus	all eukaryotes	repository of DNA as cellular information store; gene expression by transcription and processing of RNA
mitochondria	most eukaryotes	generation of useful energy for the cell by production of ATP through oxidative metabolism
chloroplasts	plants; eukaryotic algae	photosynthesis; production of ATP and fixation of carbon by use of light energy
peroxisomes	most eukaryotes	oxidative chemistry; heat production
endoplasmic reticulum	most eukaryotes	many functions, especially modification of newly synthesized proteins
Golgi apparatus	most eukaryotes	sorting of proteins and lipids destined for specific cellular locations

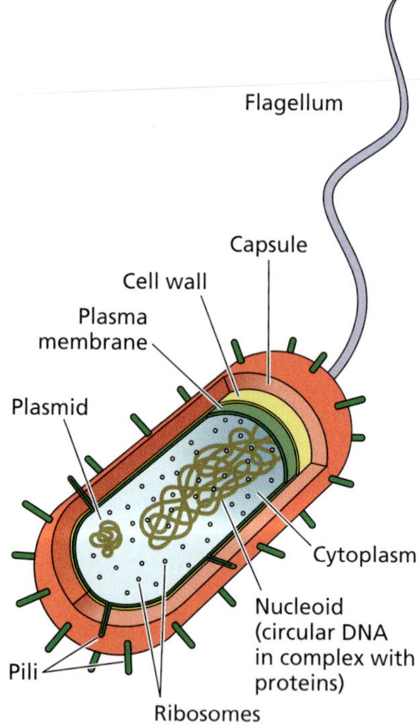

Figure 1.1 Typical structure of a bacterial cell. The major and minor components of the various entities are as follows: pili, composed of protein; plasmid, composed of lipid, protein, and nucleic acid; ribosomes, composed of nucleic acid and protein; cytoplasm, composed of protein, small molecules, and nucleic acid; plasma membrane, composed of lipid and protein; cell wall, composed of polysaccharide and protein; capsule, composed of polysaccharide; flagellum, composed of protein; and nucleoid, composed of nucleic acid and protein. (Adapted, courtesy of Mariana Ruiz Villarreal, Wikimedia.)

including small intracellular compartments of specialized functions called organelles (**Table 1.2**). These included the nucleus, which holds the chromosomes, the carriers of genetic information. Other creatures, bacteria in particular, did not have these intracellular structures; compare **Figure 1.1** and **Figure 1.2**. Organisms whose cells have nuclei were termed eukaryotes, from the Greek karyo for nut, kernel, nucleus; those without nuclei were called **prokaryotes**. The latter were believed to evolutionarily precede the eukaryotes, hence the name. All organisms were long divided into five kingdoms: four eukaryotic and one prokaryotic (**Box 1.2**).

This classification is still favored by some biologists, but studies of gene and protein sequences in recent years have suggested a different picture. Especially through the work of Carl Woese and his collaborators (see Box 1.2), it has been proposed that there are three great domains of living things (**Figure 1.3**, **Table 1.3**). Upon close examination of their biochemistry and certain nucleic acid sequences, the prokaryotes clearly divide into two domains. First, there are the Eubacteria, which comprise most of the bacteria familiar to us. Then there are the **Archaea**, consisting mostly of species that are specialized to conditions of extreme temperature and chemistry: sulfurous hot springs and the like. The archaea were originally termed archaebacteria, as it had been thought that these are bacterial species adapted to the harsh conditions on early Earth. They have some features in common with the eubacteria, and some with the eukaryotes, and some that are uniquely their own. As with all radical proposals, there are those who disagree. In this book, however, we do not use the term prokaryote but refer separately to Archaea or Eubacteria, often calling the latter by the familiar term, Bacteria. This is because, in terms of their biochemistry and molecular biology, the eubacteria and archaea are different enough that they need to be discussed separately. The eukaryotes are more consistently called Eukaryotae, but we shall generally use the more familiar term.

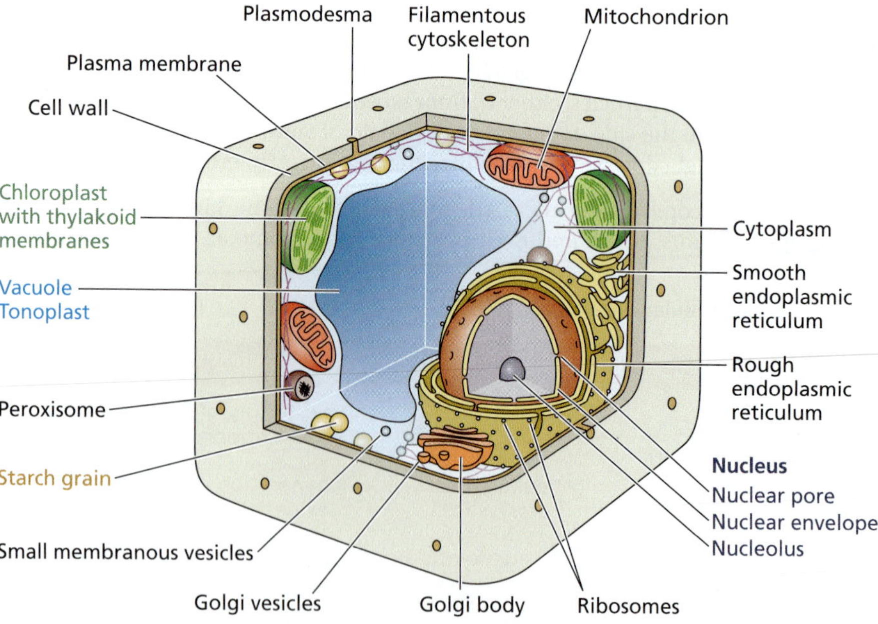

Figure 1.2 Typical structure of a eukaryotic plant cell. Note the presence of organelles, absent in bacteria. Most of those shown here are composed of both lipid and protein, but the nucleus, nucleolus, chloroplast, and mitochondrion also contain nucleic acid. (Adapted, courtesy of Mariana Ruiz Villarreal, Wikimedia.)

Box 1.2 Kingdoms and domains The rise of molecular biology, with its ability to look at organismal characteristics at the molecular level, has called into question some of the older ideas about phylogeny, the science of classification of organisms. Until about 1975, the older division of all organisms into five kingdoms—animals, plants, fungi, protists, and bacteria—was seldom questioned. A super grouping, with the first four kingdoms labeled eukaryotes and all bacteria as prokaryotes, was made primarily on the fact that all of the former had nucleated cells while the latter did not (**Figure 1A**). It seemed to be based on solid anatomical, developmental, and physiological distinctions. However, the emergence in the 1960s and 1970s of methods that allowed determination of the sequences of proteins and nucleic acids pointed to a different, and perhaps more fundamental, data set. A strong exponent of this approach was Carl Woese, of the University of Illinois. In the 1970s, Woese set out to determine the sequences of portions of ribosomal RNA from a wide variety of organisms. As we see in detail in Chapter 15, ribosomes are RNA-protein particles that serve as protein factories in every kind of cell, from bacterium to human. Their uniform, critical function would seem to make them excellent candidates for such an analysis.

It soon became obvious from this study that the ribosomal RNA sequences fell into just three distinct categories or domains (**Figure 1B**). One group was identical with the previously defined eukaryotes, but the bacteria seemed to be split into two quite distinct domains. One domain, termed Eubacteria, constituted the bacteria most familiar to microbiologists, including the workhorse of molecular biology, *Escherichia coli*. The other domain consisted largely of microorganisms that live in extreme environments, such as high temperature or high salinity, or that have unusual metabolisms, such as methane production. Because these features corresponded to what many have thought might be the conditions on the primitive Earth, these creatures were called Archaebacteria or Archaea. Subsequent studies showed that Archaea possess some other distinct and unifying features, such as unique kinds of membrane lipids. In other respects, such as protein composition, they often seem to lie between Eukaryotes and Eubacteria.

The scheme proposed by Woese leaves no place for the prokaryotes as a group, which leaves some well-respected biologists unhappy. It can be pointed out that, whatever the sequences say, there is a large group of organisms whose cells do not possess nuclei, and another large group that does. The controversy does not seem likely to be wholly resolved in the near future. In a sense, the argument is larger than one over names; it hinges on whether one feels the story of life is best read from sequences or from phenotypes.

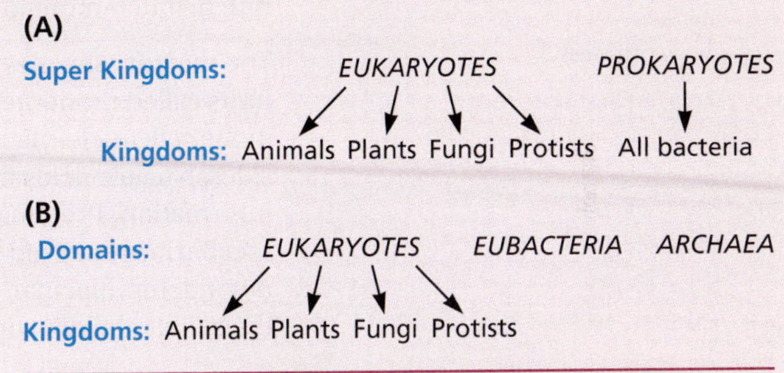

Figure 1 Two major schemes for classification of all organisms. (A) Five-kingdom model. (B) Three-domain model.

Biochemistry led to the discovery of the importance of macromolecules in life's structure and processes

Until about 1950, scientists were very limited in their ability to probe the structure of life to a finer level. The internal structure of cellular organelles, for example, remained beyond the resolution limits of the microscope. Nevertheless, the development of biochemistry during the late nineteenth and early twentieth centuries revealed much about organelle functions and the chemistry of life in general. Using selective techniques to isolate substances from cells and even purified organelles, biochemists were able to elucidate many of the chemical pathways in the cell and to localize these processes. During the course of this work, especially in the first half of the

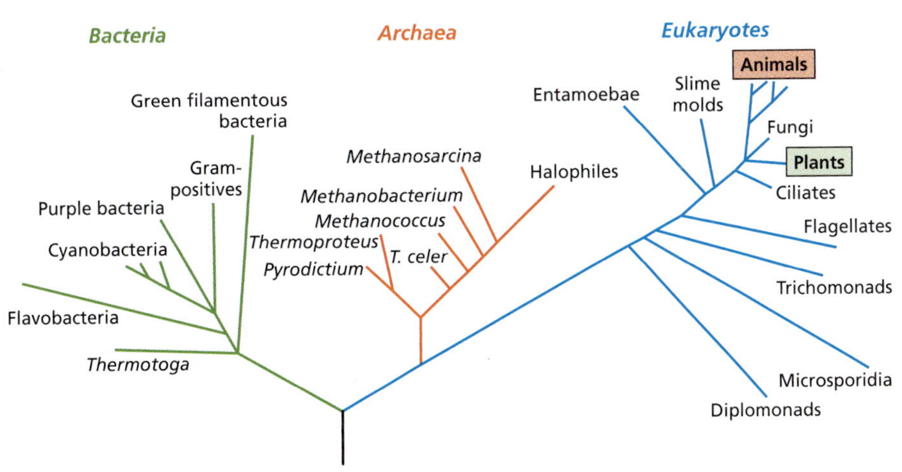

Figure 1.3 Tree of life as deduced from molecular data. A common unknown ancestor at the base of the tree gave rise to three different branches known as the domains of life: Bacteria, Archaea, and Eukaryotes. The lack of a known ancestor is designated by evolutionary biologists as a rootless tree. The lengths of the branches reflect how much the DNA of each lineage has diverged from their common ancestor. The tree demonstrates that most of life's genetic diversity is in the microbial domain; the entire animal kingdom in the old classification is represented by just a few twigs at one end of the tree. Notably, multicellular organisms such as fungi, plants, and animals have all evolved from unicellular organisms further down the tree. (Adapted from Woese CR [2000] *Proc Natl Acad Sci USA* 97:8392–8396. With permission from National Academy of Sciences.)

Table 1.3 The three domains of life.

Domain	Cell multiplicity	Major organelles	Examples	Characteristics of proteins, information transfer
Eubacteria or Bacteria	unicellular	none	*E. coli*; most common bacteria	different from Eukaryota; some similarities to Archaea
Archaea	unicellular	none	methanogens; heat- and salt-tolerant microorganisms	characteristics of both Eubacteria and Eukaryota; some unique lipids
Eukaryotae (eukaryotes)	unicellular, Protista, and multicellular (Metazoa)	nucleus, mitochondria, chloroplasts in plants	amoeba, human, apple tree	generally different from Eubacteria; some similarities to Archaea

twentieth century, it became clear that certain kinds of very large molecules, called **macromolecules**, played important roles in cellular structure and processes. A brief overview of the major classes of cellular constituents is given in Table 1.4. We will have much, much more to learn about some of these substances in later chapters.

The macromolecules of the cell are all polymers, formed by chemical linking of small units called monomers. As Table 1.4 shows, there are three classes of macromolecules in the cell: nucleic acids or polynucleotides, proteins or polypeptides, and polysaccharides. Nucleic acids are primarily involved in the storage and transmission of genetic information. Proteins play both structural and a myriad of functional roles, and polysaccharides are used both structurally and for storage of nutrient sugars. The lipids are not macromolecular, but they do aggregate to form cellular membranes. Examples of the cellular localization of these substances are shown in Figure 1.1.

The electron microscope provided another order of resolution

For over 200 years, the resolution limit of the light microscope seemed an insurmountable barrier. But in recent years, a number of ways have been devised to pass this limit. The first of these was the **electron microscope** or **EM**. The fundamental idea is simple: if resolution is limited by the wavelength of radiation, use shorter waves. Shorter-wavelength radiation is available in the electromagnetic spectrum (see Box 1.1), but it is not practical to simply go into the far-ultraviolet: such radiation is absorbed strongly by all organic molecules and even by water. One solution was found in the fact that electrons have both particle and wave character. A beam of electrons is a beam of charged particles and can be focused by electrostatic lenses. However, because of the dual character, electrons in the beam also have a characteristic wavelength. The wavelength will be determined by the energy given to the electrons and can be as small as 1 nanometer or nm. This would suggest a possible resolution of less than 1 nm;

Table 1.4 Molecular composition of a cell.

Class of substance	Role in cell	Common localization	Approximate dimension	Structural type
small molecules	nutrients, metabolic intermediates, signals	everywhere, especially in cytoplasm	a few nanometers or less	compact molecules
lipids	membrane formation, metabolites	most in membranes	nanometers, but aggregate	membrane formers
polysaccharides	nutrient storage, structural	cell wall, cytoplasm	micrometers to millimeters	fibrous
proteins or polypeptides	structural, enzymes, regulators, transporters	cytoplasm, nucleus, some in membranes	nanometers to micrometers	both fibrous and globular
nucleic acids or polynucleotides	information storage and transmission	most, but not all, in nucleus	micrometers to meters	mostly fibrous, some compact

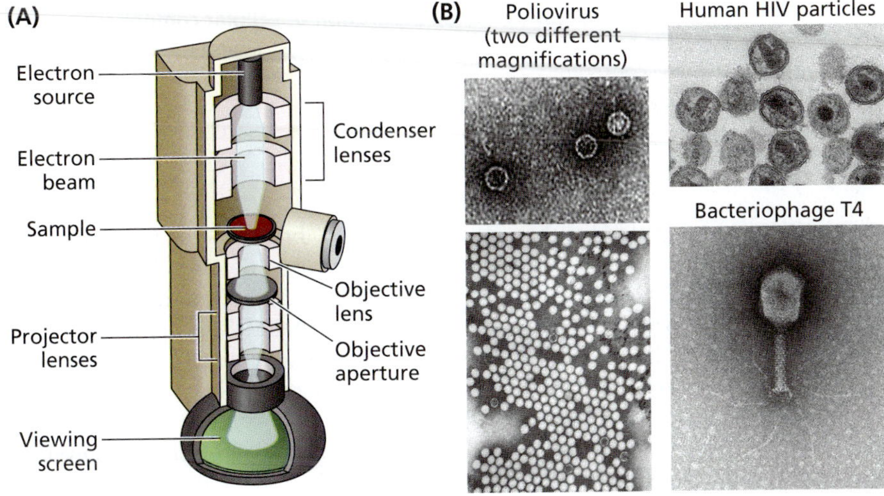

Figure 1.4 Transmission electron microscope. (A) Typical transmission EM, illustrating the principle. Electrons are emitted from a heated tungsten filament at the top, and the beam is focused by magnetic lenses onto the sample. The sample plane is then greatly magnified onto the detector by a second set of magnetic lenses. The detector can be either a fluorescent screen for visual observation, a photographic plate, or, more commonly today, a two-dimensional electronic detector. (B) Sample images of three different viruses. HIV, human immunodeficiency virus. (A, adapted, courtesy of Kevin G. Yager and Christopher J. Barrett, McGill University. B, top left, courtesy of FP Williams, United States Environmental Protection Agency. B, bottom left, courtesy of Frederick A. Murphy and Sylvia Whitfield, Centers for Disease Control and Prevention. B, top right, courtesy of Maureen Metcalfe and Tom Hodge, Centers for Disease Control and Prevention. B, bottom right, courtesy of Biophoto Associates.)

practical considerations lead to a true resolution of about 3 nm, which is still about 100 times the resolution of the best light microscope.

There are several techniques of electron microscopy, but the instrument most used for studies of cellular ultrastructure is the transmission electron microscope (Figure 1.4A). Modern electron microscopes have resolution capable of examining the overall structure of viruses (Figure 1.4B) and large protein and nucleic acid molecules. The entire path of the electron beam must be in high vacuum, and so must be the sample. This is a disadvantage of EM: with most variants of the technique, the sample will be wholly desiccated, and living samples cannot be examined. This limitation is partly overcome in **cryo-electron microscopy** or **cryo-EM**, in which the sample is embedded in quick-frozen ice or vitrified water (Figure 1.5). Cryo-EM provides a very clever way in which to envision the three-dimensional structure of complex multiunit molecular assemblies. But the primary use of EM over the past 70 years has been in defining the ultrastructure of cells.

1.3 Fine structure of cells and viruses as revealed by microscopy

A typical bacterial cell is shown in Figure 1.1. An outer lipid membrane, the plasma membrane, is in turn surrounded by a tougher cell wall, and sometimes by a capsule as well. These two latter structures are primarily carbohydrate or polysaccharide in composition. Many bacteria have short, hairlike pili on their surface; these aid in attachment to other bacteria or surfaces. Some bacteria also have long flagella, used for propulsion. Within the container of the plasma membrane lies the cytoplasm, where all metabolic processes occur and the protein enzymes needed to catalyze them are found. Also within the cytoplasm is the bacterial chromosome made of DNA with some attached proteins, and all of the molecular machinery needed to transmit the DNA's information for the synthesis of proteins. In the bacterial cell, there is little compartmentalization of processes, although some regions appear to be devoted to specific functions.

When looking at a eukaryotic cell, such as an animal cell or the plant cell depicted in Figure 1.2, we find a very different, much more complex structure, with extensive compartmentalization into organelles. For example, the genetic material, the chromosomes and the DNA they contain, is sequestered in the nucleus. There it is transcribed to

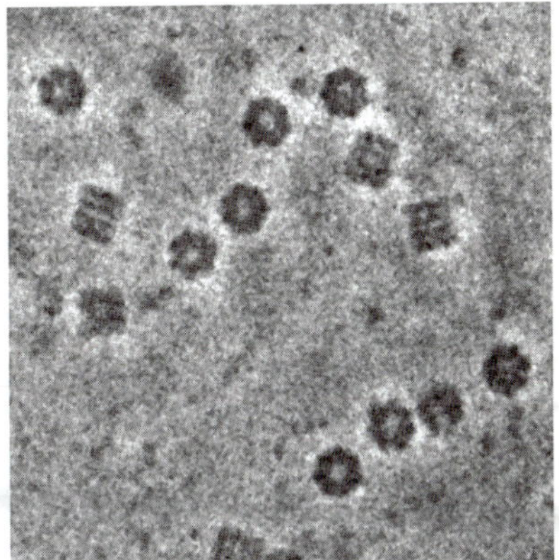

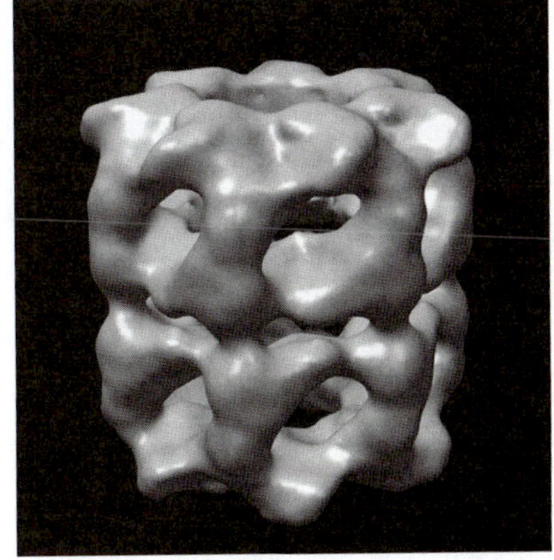

Figure 1.5 Example of cryo-EM image and three-dimensional (3D) reconstruction of the observed structure. (Top) Cryo-EM image of GroEL taken at 60,000× magnification. GroEL is a multisubunit chaperone that functions in protein folding in bacterial cells (see Chapter 3). (Bottom) 3D reconstruction of a side view at 1.2 nm resolution. (From Danev R & Nagayama K [2008] *J Struct Biol* 161:211–218. With permission from Elsevier.)

LYSE CELLS BY MECHANICAL BREAKAGE, OR ENZYMES, OR DETERGENT

Broken cells

CENTRIFUGE

Membranes, lipids

Cytoplasmic components

Organelles, DNA

Figure 1.6 A widely used technique for crude separation of cellular components. Cells are lysed or broken in buffer by a variety of techniques. After centrifugation in buffer, membranes and lipids will concentrate at the top because their density is less than that of buffer. DNA and some organelles, being large and denser, will concentrate at the bottom of the tube. Cytoplasmic components and low-molecular-weight molecules will distribute throughout the tube.

produce RNA copies, which are then exported to the cytoplasm to be translated into proteins. We will treat these processes and their control in detail in later chapters. Important organelles are listed and described in Table 1.2; the internal structure of each of these has been revealed, mostly by EM. Each organelle is surrounded by a lipid membrane, as is the whole cell by the plasma membrane. Plant cells, and some protozoa, also have tougher cell walls outside the plasma membrane. Within the plasma membrane, surrounding the organelles, is the cytoplasm, a concentrated mixture of small molecules, enzymes, regulatory proteins, and the RNA molecules involved in protein synthesis. If cells are broken open or lysed, and the lysate is diluted with appropriate buffer to approximate the cytoplasmic environment, differential sedimentation will separate cytoplasmic contents, lipid membranes, and organelles (**Figure 1.6**). Much of what we have learned of biochemistry has depended on just such separations.

The electron microscope also provided the first understanding of the structure, and to some extent the mode of functioning, of viruses. Most viruses are on the order of 50–100 nm in size. This is below the resolution limit of the light microscope. Viruses could be detected as dots in the dark-field ultramicroscope, but this revealed nothing of their structure. The development of the EM made it possible to view the shapes, and even structural details, of viruses (see Figure 1.4B). We now know a great deal about the internal and surface structure of many viruses, which has been of much importance in developing strategies against viral diseases such as AIDS (acquired immunodeficiency syndrome) and influenza.

1.4 Ultrahigh resolution: Biology at the molecular level

In recent years, several techniques have been developed that allow us to study biological structures down to the molecular level. These have been immensely important in establishing a true molecular biology. We say more about all of these methods in later chapters, but a brief overview will be helpful at this point.

Fluorescence techniques allow for one approach to ultraresolution

There are several ways in which the phenomenon of fluorescence can be used to greatly enhance either the compositional analysis or the effective resolving power of the light microscope. Recall that **fluorescence** involves the absorption of a photon of light at a given wavelength, depending on the molecule or atomic group illuminated, and its re-emission at a longer wavelength after dissipation of part of its energy (**Figure 1.7**). Thus, if a particular kind of molecule in the cell has a distinctive fluorescence excitation or emission spectrum, it can be distinguished from other substances. Sometimes this is accomplished by attaching a particular dye, known as a fluorescent label, specifically to the molecules of interest.

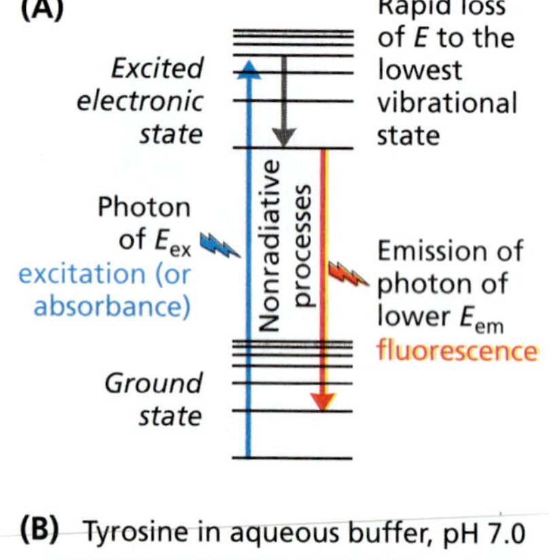

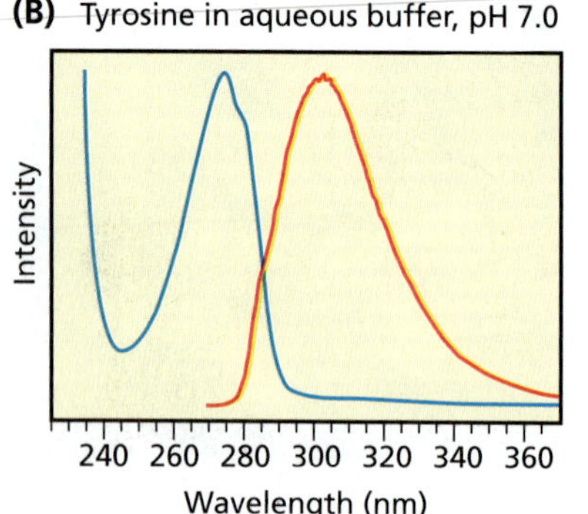

Figure 1.7 Fluorescence. (A) Principle. When a molecule absorbs light in the visible or near-UV spectral region, it is excited from its ground or lowest electronic state to a higher electronic state. It will often start from the lowest vibrational energy level in the ground state and move to a higher vibrational level in the excited state. The photon energy to do this is called the excitation energy, E_{ex}. The excited molecule can fritter away some of this energy, as heat, for example, falling to a lower vibrational level. If it then re-emits a photon as fluorescence, that photon will have a lower energy E_{em} and thus be of longer wavelength than the exciting photon. (B) Excitation spectrum, shown in blue, and fluorescence emission spectrum, shown in red, of the amino acid tyrosine, recorded in aqueous buffer at pH 7.0. Each electronic state has multiple sublevels corresponding to different vibration energy states, which accounts for the breadth of the excitation and emission bands.

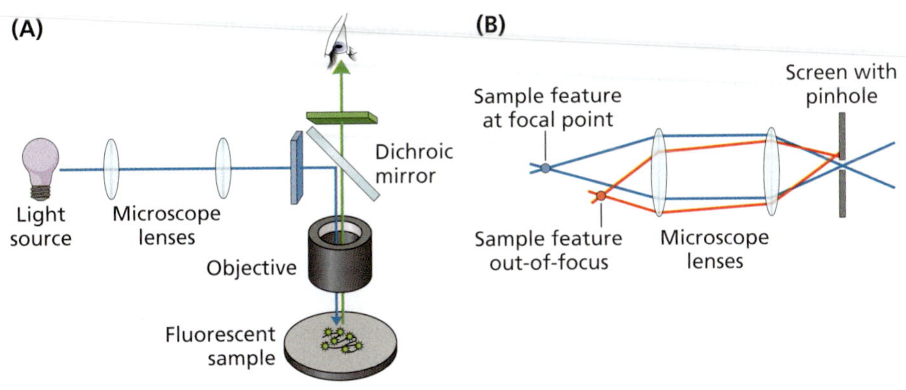

Figure 1.8 Confocal microscopy. (A) Basic principle. Exciting light from a source, today usually a laser, is reflected by a dichroic mirror onto the sample. Fluorescent light, of longer wavelength, coming from the sample is not reflected but transmitted by this mirror and goes to the observer or detector. Filters, shown as blue and green plates in the figure, improve the discrimination between exciting and fluorescent light. (B) Advantage of pinhole optics. Stray fluorescent light coming from parts of the sample not at the focal point will be mostly blocked by the screen around the pinhole and not reach the detector. (Adapted, courtesy of Eric Weeks, Emory University.)

Confocal fluorescence microscopy allows observation of the fluorescence emitted by a particular substance in a cell

How can one observe specifically, using a microscope, the fluorescence emitted by a particular substance in a cell? The simplest way is illustrated in **Figure 1.8A**: a simple confocal microscope. It uses a dichroic mirror or similar device that will reflect light of the exciting wavelength but transmit the fluorescent emitted light. Thus, the observer sees only light from fluorescent molecules, and ideally none of the other scattered light. The microscope is called confocal because both the exciting light and emitted light are focused on the same point in the sample, using the same set of lenses.

Even this simple technique can be very powerful. Suppose, for example, that a researcher wishes to know whether two different materials in the cell, say DNA and a particular protein, are associated with one another. By use of fluorescent dyes specific for each component, it is possible to record separately the localization of DNA and the protein. If the images superimpose, the components are co-localized in the cell.

The technique as described above suffers from one drawback: although the observer or detector is focused on the focal point in the sample, it will also receive some fluorescent light from other parts of the sample. This fuzzes the fluorescent image. The problem can be minimized by inserting a screen with a pinhole at the focal point before the detector (**Figure 1.8B**). Now the detector will primarily see light coming from one spot in the sample, giving much clearer resolution. This means, of course, that to investigate the whole sample, the focal point must be moved or scanned across and through the sample. This can be achieved by mirrors that are rotated in a precisely defined program. An example of a modern instrument that operates on this principle is described in **Figure 1.9A**. At present, lasers are almost always employed

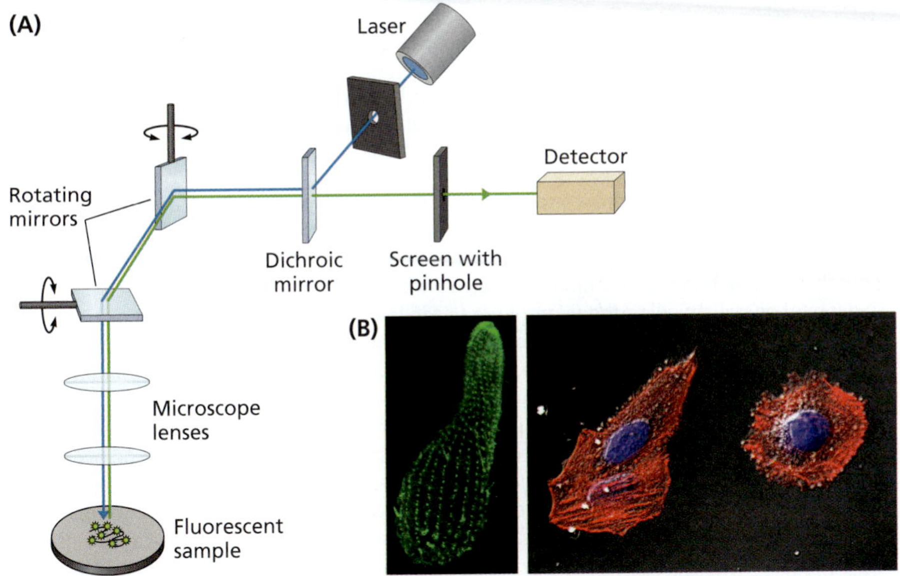

Figure 1.9 Schematic of conventional confocal microscope and some sample confocal images. (A) The system is essentially the same as described in Figure 1.8, but rotating mirrors have been added to allow the focal point to be scanned across and through the sample. (B) (Left) β-Tubulin in *Tetrahymena*, a ciliated protozoan; for more on tubulin structure, see Figure 3.15. The protein is stained with a green fluorescent dye. (Right) Cultured C6R cells, stained with a red fluorophore for actin and blue DAPI, 4′,6-diamidino-2-phenylindole, for DNA. (A, adapted from Prasad V, Semwogerere D & Weeks ER [2007] *J Phys: Condens Matter* 19:113102. With permission from Institute of Physics. B, left, courtesy of Pawel Jasnos, Wikimedia. B, right, courtesy of Noriko Kane-Goldsmith, Rutgers University.)

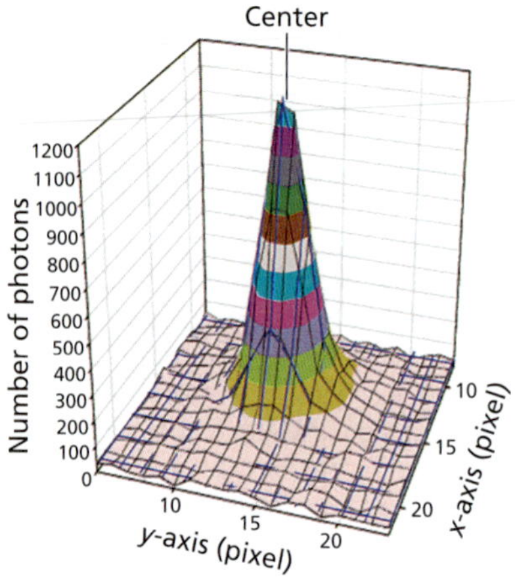

Figure 1.10 Fluorescence imaging with one-nanometer accuracy, FIONA. Location of a single fluorescent dye molecule on a surface by FIONA is illustrated. From the wavelength of the light used, one would expect location to within 250 nm. This is the traditional diffraction limit. However, FIONA locates the molecule to within 2.5 nm. Photons of fluorescent light are individually detected and counted; the more photons collected, the higher the accuracy of determination of the center position of the dye. The center of this distribution can be located to within ±1.25 nm when 10,000 photons have been counted. (Adapted from Yildiz A & Selvin PR [2005] *Acc Chem Res* 38:574–582. With permission from American Chemical Society.)

as light sources. The confocal scanning microscope can provide high resolution and precise spatial positioning of specific kinds of molecules in the cell (**Figure 1.9B**).

FIONA provides ultimate optical resolution by use of fluorescence

Photon-counting devices can be used to observe the emission of single photons from single molecules. In one direct application, called **fluorescence imaging with one-nanometer accuracy** or **FIONA**, many thousands of photons from one molecule or group are statistically analyzed on a two-dimensional detector. The result is a map of apparent emitter position (**Figure 1.10**). With enough photons counted, the maximum or position can be located within 1 nm. Two emitters, separated by less than 10 nm, can be resolved: about 30 times the resolution of the best conventional light microscope. Thus, it is possible to escape the diffraction limit long held to be an inviolable principle of microscopy. Many variants of these techniques are being devised.

FRET allows distance measurements at the molecular level

An even more widely used technique is **fluorescence resonance energy transfer** or **FRET** (**Figure 1.11A**). This method depends upon the fact that if two different fluorescent chromophores are sufficiently close together, excitation of one chromophore, the donor, may lead to a transfer of energy to another, the acceptor, which will then emit at its characteristic wavelength. The efficiency of transfer depends strongly on the distance between donor and acceptor (**Figure 1.11B**). This allows the use of FRET as a molecular ruler, which is especially good for distances of a few nanometers. The great advantage of FRET is that it can be applied to molecules in solution or even in living cells. Furthermore, it can be used to study dynamic internal motions in or between molecules.

Additional variants of fluorescence microscopy are continually being invented, and the light microscope has come back into a prominent role in biology. A major reason for this is that many light microscopic techniques can be used to follow the behavior of specific substances in living cells, by coupling these molecules to fluorescent probes.

Single-molecule cryo-electron microscopy is a powerful new technique

A recent advance in EM allows a more detailed examination of single molecules and particles. If many samples of a macromolecular particle are embedded in a film of

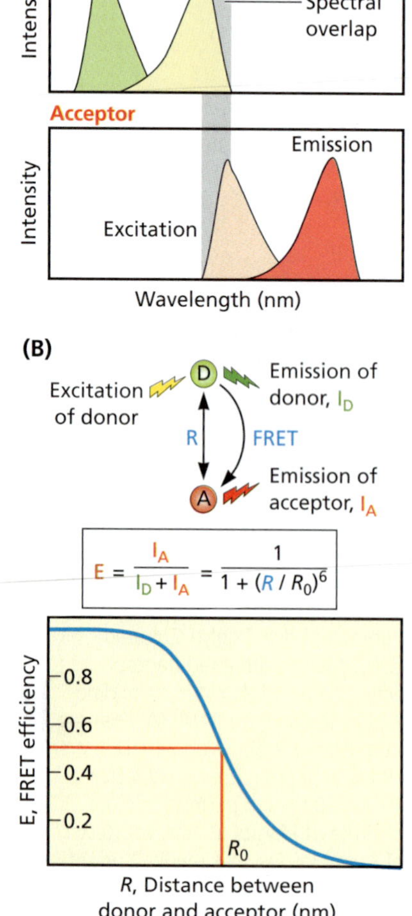

Figure 1.11 Principle of fluorescence resonance energy transfer, FRET, and its efficiency as a function of interdye distance. (A) Photophysics behind FRET. Recall from Figure 1.7 that absorption of a photon stimulates a transition from the ground state of the donor fluorophore to a higher vibrational state of the excited state. This is followed by a rapid loss of energy, bringing the fluorophore to the lowest vibrational state. From here, the dye can revert to the ground state by emission of a photon of lower energy or by a variety of nonradiative processes. If an acceptor fluorophore is in close proximity, energy can be transferred directly to it; the return to the ground state of the acceptor can occur through emission of photons of less energy. The panel shows the spectra of a donor–acceptor fluorophore pair. Note the condition necessary for transfer to occur: there should be overlap between the emission spectrum of the donor and the excitation spectrum of the acceptor. (B) Dependence of FRET efficiency on the distance between donor and acceptor fluorophores. The terms in the equation are schematically explained in the drawing above it. The distance R_0, which will correspond to 50% efficiency of energy transfer, is usually between 20 and 60 Å.

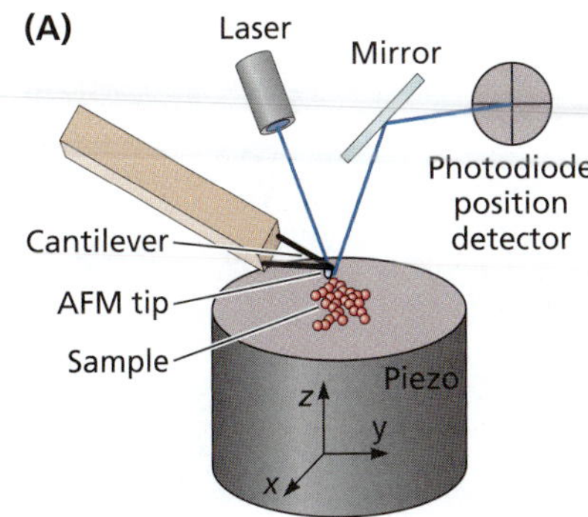

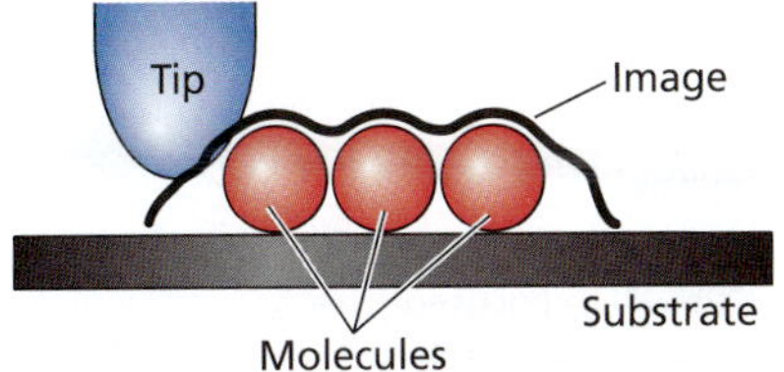

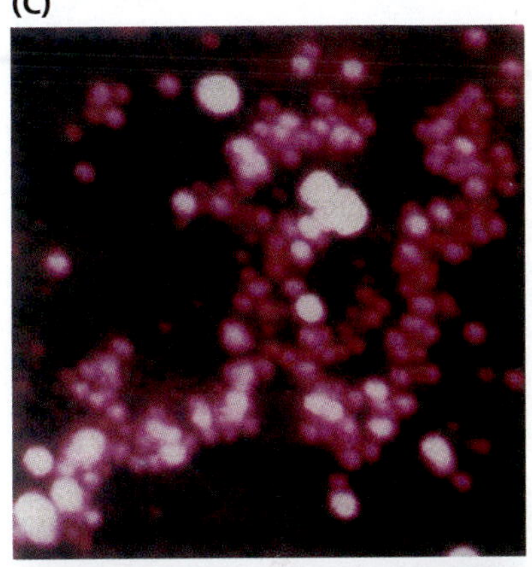

Figure 1.12 Principle of operation of the atomic force microscope, AFM. (A) Schematic of the instrument. A sharp tip is mounted at the end of a flexible cantilever; the tip makes repeated parallel scans across the biological sample deposited on an atomically flat surface such as mica, glass, or gold. A topographic image is created from changes in the laser signal caused by deflections of the cantilever. The deflections are caused by atomic-level tip/sample interactions: if the interactions are attractive, the cantilever bends toward the sample; if they are repulsive, the cantilever bends away from the sample. The AFM generally has a nanometer lateral resolution when imaging soft biological samples; the resolution is truly atomic with hard materials. (B) Illustration of the broadening effect of the tip. All structures appear somewhat wider than they actually are due to the fact that the tip, no matter how sharp, cannot exactly follow the molecules on the surface. These effects are usually small and do not change the overall perception of the structure. (C) Sample images of eukaryotic chromatin fibers. Chromatin is a complex between DNA and certain basic proteins, histones (see Chapter 8). The fiber was extracted from chicken erythrocytes and imaged unfixed on a glass surface in air. Even when imaging is done in air, there is a thin layer of liquid water above the sample, thus allowing the imaged molecules to preserve the structurally important water molecules, and thus, their native structure. Artificial coloring schemes are used to facilitate perception of structure heights. Finally, the digital image can be viewed at different angles, which is helpful in perceiving the three-dimensionality of the structure. (A, adapted from Zlatanova J, Lindsay SM & Leuba SH [2000] *Prog Biophys Mol Biol* 74:37–61. With permission from Elsevier. C, top, from Leuba SH, Yang G, Robert C et al. [1994] *Proc Natl Acad Sci USA* 91:11621–11625. With permission from National Academy of Sciences. C, bottom, courtesy of Sanford Leuba, University of Pittsburgh.)

vitrified water, they will generally take random orientations. This means that the EM is looking at a single kind of molecule from many different angles. If each molecule has the same conformation, there is in principle only one solution that will account for all views, and this can be found by sophisticated computer programs. Even if several conformations of the same molecule are present, it may be possible to find a solution that describes each conformation and determines how many molecules are in each category. Since the freezing techniques of immersing samples in liquid nitrogen are very rapid, a snapshot of the distribution of conformations present at the moment before freezing is obtained; for an example, see Figure 1.5.

The atomic force microscope feels molecular structure

Suppose that instead of looking at ultrastructure, it were possible to feel it. That is exactly what **atomic force microscopy** or **AFM** does. The principle is almost absurdly simple (**Figure 1.12**). A springy cantilevered arm, with a very fine tip at one end, is dragged, or more often taps its way, across a sample mounted on a very smooth surface. Light reflected off the cantilever is used to measure its deflection. The deflection is upward when atoms on the sample and on the tip of the arm experience mutual repulsion; the cantilever deflects or bends toward the sample if the atomic interactions are attractive. Under optimal conditions, AFM can attain a resolution comparable to that of EM. One great advantage of AFM, however, is that it can be applied under aqueous solutions, avoiding the desiccation artifacts that may accompany EM techniques. In a quite different application, the AFM tip can be used to stretch individual molecules with known force, opening a whole field of ultramicromechanics.

X-ray diffraction and NMR provide resolution to the atomic level

Perhaps the physical technique that has done most to create a true molecular biology is X-ray diffraction (see Box 3.6). The basic principles can be illustrated by diffraction experiments with visible light (**Figure 1.13**). Imagine passing a well-defined beam of light, such as that from a laser, through a very fine screen, with a blank wall some distance behind it. On the wall, you will see not only the spot produced by the direct beam but also, around it, a diffraction pattern, a rectilinear array of spots caused by reinforcement of rays from different holes in the screen. With experimentation, you will find that a finer screen gives wider spacing in the pattern, and vice versa. If the

(A)

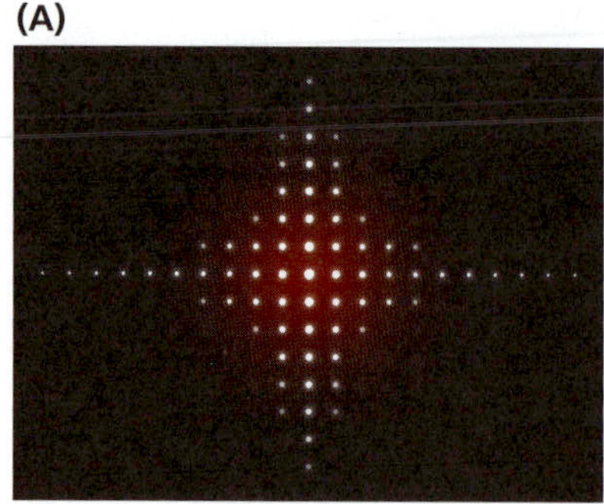

(B)

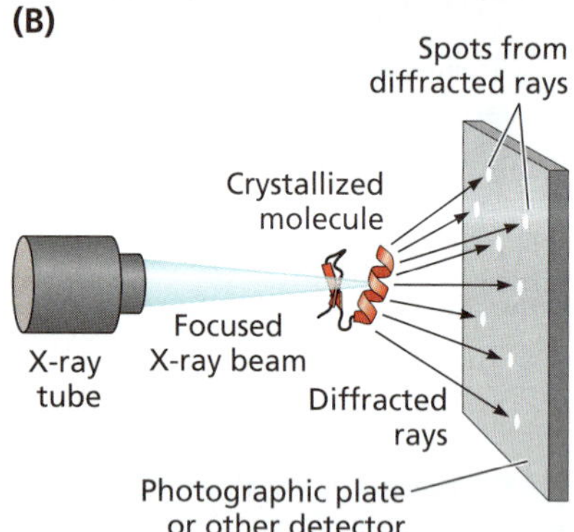

Figure 1.13 X-ray diffraction. (A) Basic principle of diffraction. If a two-dimensional screen with regularly spaced wires is placed in front of a laser light source, the kind of pattern shown will be obtained. (B) The same principle applies to X-rays diffracted through a crystal, except now the pattern is more complex and depends on the orientation of the crystal planes with respect to the beam direction.

screen has different spacing in the *x*- and *y*-directions, the same reciprocal relationship will hold. With a little mathematics, you can determine the lattice spacing in the screen from the spacing in the pattern. If the screen were made of helical wires, parallel in one direction, you could observe a cross-like diffraction pattern; this is what tipped Watson and Crick off to the helical structure of DNA (**Figure 1.14**). Finally, the shape of the holes will influence the relative intensities of spots.

Now, translate all of this to the molecular level, using a crystal or fiber of some macromolecule (see Figure 1.13B). As the spacings within a molecule are much smaller than those of a screen, we must use shorter-wavelength radiation to examine them. X-rays with wavelengths of about 0.1 nm will do nicely. Scientists performing X-ray diffraction studies sometimes still use the older unit, the angstrom, Å, where 1 Å = 0.1 nm. We usually employ the MKS or meter, kilogram, second unit, nanometers, but in some figures you will see angstroms. Analysis of the X-ray diffraction patterns generated by crystals is complicated by the fact that a crystal is a three-dimensional object, and the individual molecules whose structure we seek can be awesomely complex. It is important to note that the spatial arrangement of spots tells us only the lattice spacings of the macromolecular array in the crystal. The internal structure of the molecules determines the relative intensity of different spots. To go from the intensity data to the structure is not easy for large molecules, and this has been a major stumbling block for macromolecular crystallography.

Nevertheless, the technique has been brought to the point where structures can be determined, almost routinely, to a resolution of a few tenths of a nanometer. The resolution achievable depends on the quality of the crystals, which in turn determines the number of spots utilized in the analysis. There are many, many such structures in this book; examples of a protein structure at different levels of resolution are shown in **Figure 1.15**. It must always be remembered that the picture provided by X-ray diffraction is a static one, of molecules fixed into a crystal lattice, and thus may not correspond exactly to the molecule in solution or in the dynamic environment of the living cell.

Nuclear magnetic resonance, NMR, is a high-resolution technique that can be applied to macromolecules in solution. Utilizing a variety of isotopes with differing nuclear spins and analyzing spin–spin interactions, NMR is able to attain a structural resolution approaching that obtainable by X-ray diffraction while avoiding potential modification of structure from interactions within crystals. X-ray diffraction and NMR are discussed in more detail in Box 3.6 and Box 3.7.

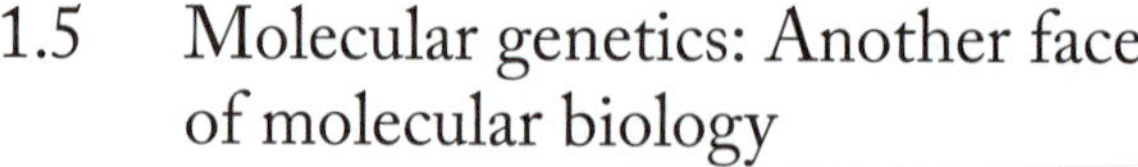

1.5 Molecular genetics: Another face of molecular biology

A major factor in the development of molecular biology was the discovery of the molecular basis of genetics. Although Mendel's laws were formulated in the 1860s (see Chapter 2), nearly 100 years passed before a physical basis for these laws could be proposed. A crucial step was the discovery that bacteria and viruses shared these laws

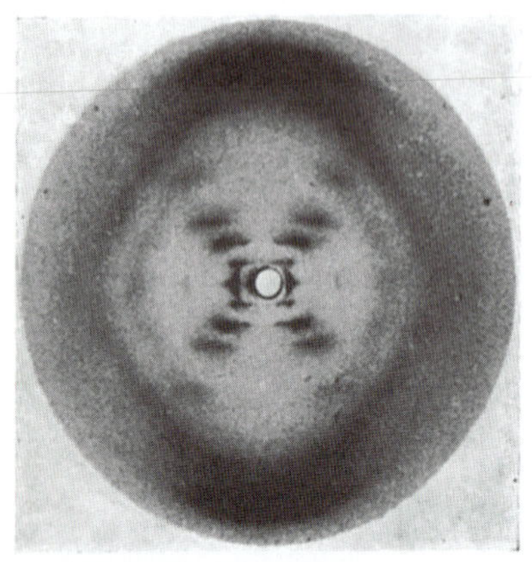

Figure 1.14 X-ray diffraction pattern from B-form DNA. This pattern was obtained by Rosalind Franklin and used by Watson and Crick to deduce the DNA structure. Since the pattern is from a fiber in which the DNA molecules are not perfectly aligned, rather than from a crystal, only a few, somewhat blurry spots are observed. However, the X pattern indicates a helical structure, and the spacing of spots allows deduction of the helical repeat distance along the molecule. (From Franklin RE & Gosling RG [1953] *Nature* 171:740–741. With permission from Macmillan Publishers, Ltd.)

(A)

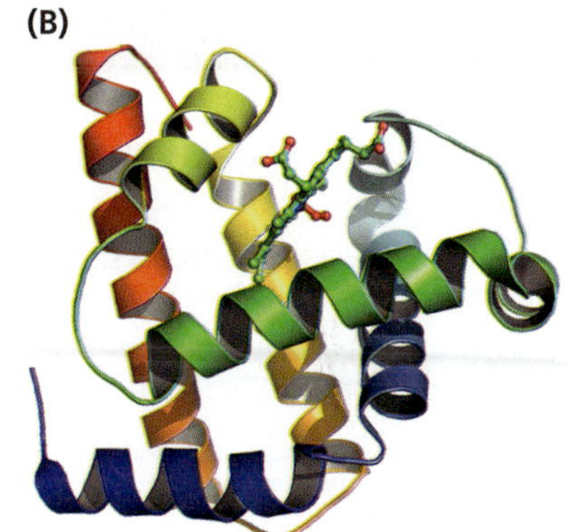
(B)

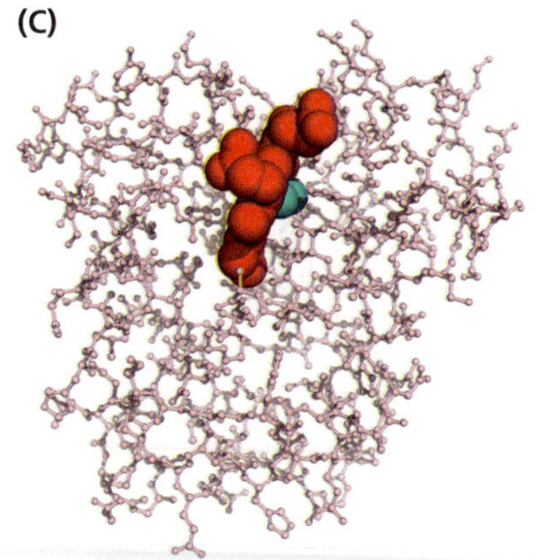
(C)

Figure 1.15 Structure of myoglobin at different resolutions as deduced from X-ray diffraction data. (A) Myoglobin at 6 Å resolution: this was the first protein structure ever determined by X-ray crystallography. (B) Myoglobin at 2 Å resolution; this was determined in the same laboratory only two years later. It shows a ribbon diagram for the protein and a ball-and-stick model for the heme group. (C) All atoms of the protein, with the heme group in bright red and a bound oxygen molecule in turquoise. The three models are viewed from slightly different angles. (A, from Dill KA & MacCallum JL [2012] *Science* 338:1042–1046. With permission from MRC Laboratory of Molecular Biology.)

with higher organisms and therefore provided simple and rapidly reproducing systems for study. Further, the monumental elucidation of the structure of DNA immediately pointed the way to a physical understanding of genetics. At about the same time, the nature and structure of proteins was becoming clear. We present an overview of these advances in genetics in Chapter 2.

Essential to all of these efforts was the discovery of ways in which polymers such as nucleic acids and proteins could be dissected, sequenced, and manipulated. In a few decades, analysis of sequences went from a laborious, year-long effort to determine the sequence of a small nucleic acid fragment to the ability to determine whole-genome sequences in a few days. We have learned how to make multiple copies of any particular gene sequence, change it at will, and insert it into an organism. Much of this can be done with relatively simple techniques, available in every laboratory. We will describe these where appropriate in later chapters. In addition, Chapter 5 is entirely devoted to this topic.

It is fair to say that, in a few decades of the mid-twentieth century, a whole new field of science was born. This discipline, which we call molecular biology, allowed the explanation of both genetics and biochemistry on a truly molecular basis. It has undergone explosive expansion over the following decades and is still evolving. This is what this book is about.

Key concepts

- Biology has advanced as a search for ever more minute details of life's processes and structure, using instruments that provide higher and higher resolution.
- At first, biology relied on the human eye and thus was restricted to the study of whole animals and plants and their gross anatomy and physiology.
- The light microscope, with resolution on the order of 0.3 μm, permitted discovery of microorganisms and the cellular structure of eukaryotes.
- Biochemistry has revealed not only much of the working principles of cells but also the existence of three great domains of life: Eubacteria, Archaea, and Eukaryoteae, or Eukaryotes.
- The electron microscope, with potential resolution to 0.3 nm, allowed investigation of the fine structure of cells and large macromolecules.
- Confocal fluorescence microscopy can localize specific kinds of macromolecules and their interactions in the cell.
- Other fluorescence techniques, such as FIONA and FRET, extend the resolution of light microscopy to the nanometer level.
- Atomic force microscopy permits high-resolution studies of macromolecular samples under aqueous conditions and their mechanical manipulation.
- X-ray diffraction and high-resolution NMR extend the range of possible studies to the atomic level and have helped to create a true molecular biology.
- Another major input came from the development of molecular genetics, beginning with the genetics of bacteria and viruses.

Further reading

Books

Alberts B, Johnson A, Lewis J et al. (2015) Molecular Biology of the Cell, 6th ed. Garland Science.

Egerton RF (2005) Physical Principles of Electron Microscopy: An Introduction to TEM, SEM, and AEM. Springer.

Frank J (2006) Three-Dimensional Electron Microscopy of Macromolecular Assemblies. Oxford University Press.

Goldman RD, Swedlow JR & Spector DL (2010) Live Cell Imaging: A Laboratory Manual. Cold Spring Harbor Laboratory Press.

Kuriyan J, Konforti B & Wemmer D (2013) The Molecules of Life: Physical and Chemical Principles. Garland Science.

Margulis L & Chapman MJ (2009) Kingdoms and Domains: An Illustrated Guide to the Phyla of Life on Earth, 4th ed. Academic Press.

van Holde KE, Johnson WC & Ho PS (2006) Principles of Physical Biochemistry, 2nd ed. Pearson Prentice Hall.

Reviews

Glaeser RM (2008) Macromolecular structures without crystals. *Proc Natl Acad Sci USA* 105:1779–1780.

Huang B, Babcock H & Zhuang X (2010) Breaking the diffraction barrier: Super-resolution imaging of cells. *Cell* 143:1047–1058.

Pace NR (2009) Mapping the tree of life: Progress and prospects. *Microbiol Mol Biol Rev* 73:565–576.

Prasad V, Semwogerere D & Weeks ER (2007) Confocal microscopy of colloids. *J Phys: Condens Matter* 19:113102.

Taylor KA & Glaeser RM (2008) Retrospective on the early development of cryoelectron microscopy of macromolecules and a prospective on opportunities for the future. *J Struct Biol* 163:214–223.

Vogel SS, Thaler C & Koushik SV (2006) Fanciful FRET. *Sci STKE* 2006:re2.

Woese CR (2000) Interpreting the universal phylogenetic tree. *Proc Natl Acad Sci USA* 97:8392–8396.

Zlatanova J & Leuba SH (2003) Chromatin fibers, one-at-a-time. *J Mol Biol* 331:1–19.

Zlatanova J, Lindsay SM & Leuba SH (2000) Single molecule force spectroscopy in biology using the atomic force microscope. *Prog Biophys Mol Biol* 74:37–61.

Zlatanova J & van Holde K (2006) Single-molecule biology: What is it and how does it work? *Mol Cell* 24:317–329.

Experimental papers

Kendrew JC, Bodo G, Dintzis HM et al. (1958) A three-dimensional model of the myoglobin molecule obtained by X-ray analysis. *Nature* 181:662–666.

Kendrew JC, Dickerson RE, Strandberg BE et al. (1960) Structure of myoglobin: A three-dimensional Fourier synthesis at 2 Å resolution. *Nature* 185:422–427.

Yildiz A, Forkey JN, McKinney SA et al. (2003) Myosin V walks hand-over-hand: Single fluorophore imaging with 1.5-nm localization. *Science* 300:2061–2065.

Yildiz A & Selvin PR (2005) Fluorescence imaging with one nanometer accuracy: Application to molecular motors. *Acc Chem Res* 38:574–582.

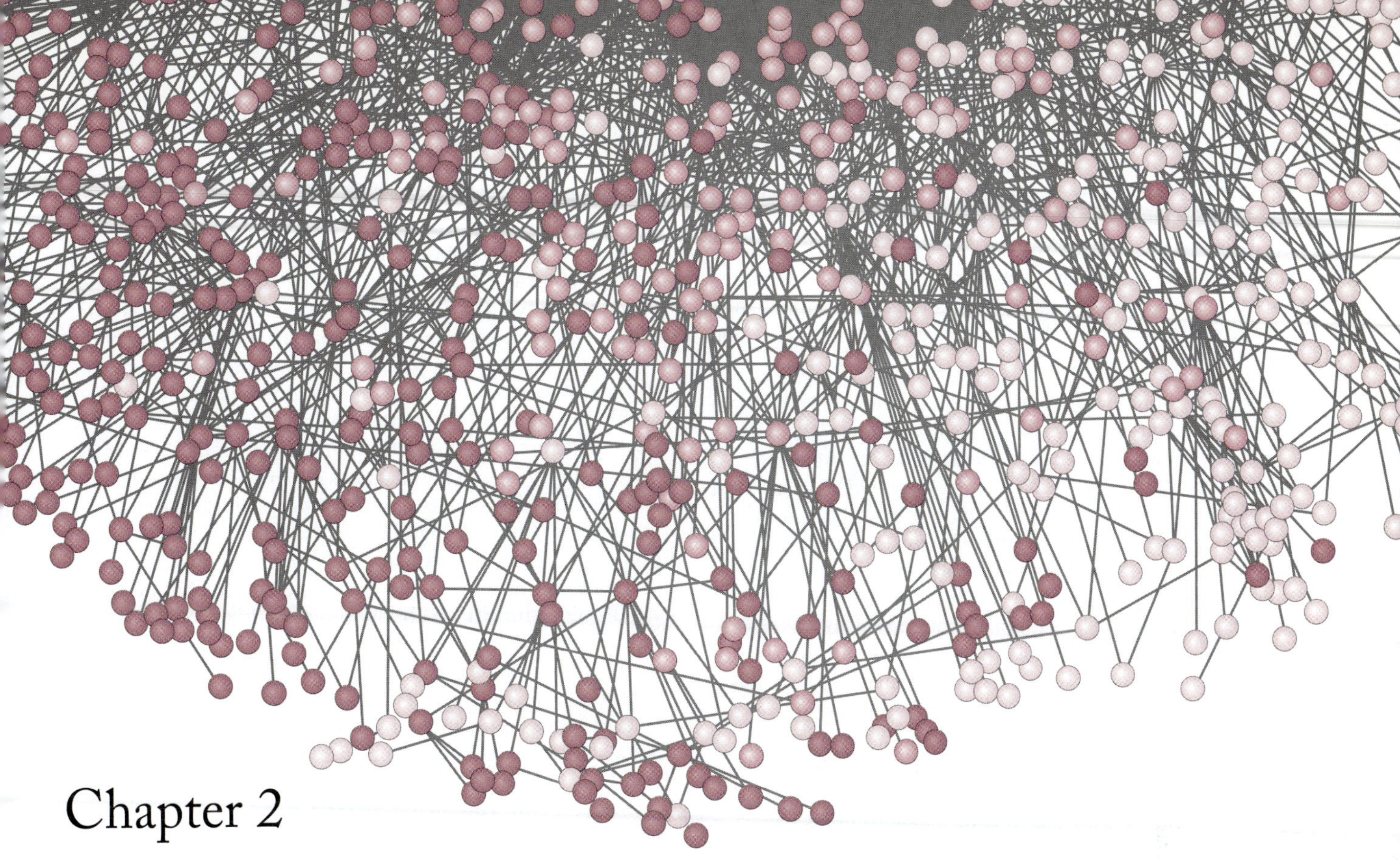

Chapter 2

From Classical Genetics to Molecular Genetics

2.1 Introduction

In Chapter 1, we showed how biology proceeded from macrostructure to microstructure through the development of ever more powerful techniques. We can now visualize much of the machinery of life at the molecular or atomic level and use that ability to explore the fundamental mechanisms and processes of life. There is, however, more to biology than structure and biochemistry. A true molecular biology must also explain how the information needed to produce biological structures and processes is stored, expressed, and transmitted from one generation to the next. This last task lies in the province of genetics. Much of genetics, or what we now call classical genetics, was developed before anything was known of the molecular processes involved, yet it provided a vital and important impetus and direction for the new science of molecular genetics.

2.2 Classical genetics and the rules of trait inheritance

Genetics is the study of inheritance. From earliest times, philosophers and plant and animal breeders have recognized that there are similarities between parents and their offspring, but any quantitative explanation for how traits are transmitted and which are favored was long in coming. It was not until the middle of the nineteenth century that the first rigorous analysis was carried out. Remarkably, this was not done by any of the eminent biologists of the day but by a little-known monk, Gregor Mendel.

Gregor Mendel developed the formal rules of genetics

As **Box 2.1** shows, Mendel was an unusual person to be doing such scientific research. As an Augustinian friar, he tended gardens at the monastery at Brno and studied, over a

Box 2.1 Two unusual scientists who defined classical genetics
Although many scientists contributed to the formulation of classical genetics, there are two names that stand out: Gregor Mendel and Thomas Hunt Morgan. Each was, in a sense, unusual among the scientists of his day and deserves special recognition. Their backgrounds and modes of doing science were entirely different.

Gregor Mendel
Mendel was born to a poor family in 1822 as Johann Mendel, in an area then part of the Austro-Hungarian Empire. He took the name Gregor when he entered the Augustinian abbey at Brno, presently in the Czech Republic, in 1843, where he spent most of his life. Mendel's scientific training appears to have been limited to two years at the University of Vienna, where he was sent by the abbey to study physics. Most of his genetic research was conducted in the gardens of the abbey, during the period 1856–1863. During that time, he is said to have raised and personally examined about 28,000 pea plants and kept careful quantitative records of their propagation. The results of this work, essentially Mendel's laws of genetics, were published in an obscure journal in 1866. Although Mendel sent reprints to most of the prominent biologists of the time, including Charles Darwin, he seems to have had little response. In fact, in the 35 years following publication, there were exactly three citations of his paper. In the present, frenzied climate of Science Citation Index, Mendel would not have fared well. Essentially, the work was neglected and forgotten and was only rediscovered around 1900.

Thomas Hunt Morgan
Morgan was born in 1866 to a prominent Kentucky family and, in contrast to Mendel, enjoyed the best of education. He obtained his Ph.D. in developmental biology from Johns Hopkins University in 1890 and spent the next decade on the faculty of Bryn Mawr College. In 1904, he accepted a professorship at Columbia University, where he spent his most productive years. Rather than working alone, as did Mendel, Morgan gathered about him an unusually talented group of students and postdoctoral researchers, many of whom also went on to make major contributions to genetics. These included Alfred Sturtevant, who constructed the first genetic map of a chromosome; George Beadle, who with Edward Tatum first related genes to proteins; Theodosius Dobzhansky, who showed how mutations could drive evolution; and Hermann Muller, who discovered the mutational effects of short-wavelength radiation. The list could be extended for further academic generations. Like Mendel, Morgan was distinguished by an ability to see the important general law in a mass of complex data.

According to some, Thomas Morgan can be credited with revolutionizing the way in which research groups operate. Up to Morgan's time, the European model, with a rigid, hierarchical structure dominated by an almost inaccessible professor, was emulated throughout the scientific world. In stark contrast, Morgan's fly lab had an informal, relaxed ambience; visiting scientists were often shocked to find all participants on a first-name basis. Gradually, this became the norm in American universities and to some extent elsewhere. If you find the lab in which you do graduate studies is a socially pleasant, relaxed place, you may owe a debt to Thomas Hunt Morgan.

period of about a decade, the genetics of garden peas. These were a good choice because peas can be raised rapidly and exhibit clearly recognizable traits that can breed true for many generations. Also, they are capable of either self-fertilization or cross-fertilization (**Figure 2.1** and **Table 2.1**). The unusual aspect of Mendel's work, for this field at the time, was his careful quantitation of the outcome of every breeding experiment.

In his experiments, Mendel would first choose a pair of stocks exhibiting contrasting traits in a particular character, yellow versus green seeds, for example, each of which he knew to breed true in self-fertilization. These are referred to as the parental phenotypes, the P1 generation. When these were cross-fertilized, it was always observed that only one of the two alternate traits was expressed in the progeny, called the F1 or filial 1 generation. In our example, the trait expressed in F1 is yellow (**Figure 2.2**; see Table 2.1). This showed Mendel that one trait, yellow seeds, was dominant over the other, green seeds. Why this occurred was shown by the next experiments, in which members of the F1 generation were self-hybridized. Now the other trait, the recessive trait, reappeared but in only one-fourth of the F2 progeny. This meant that there must be

Table 2.1 Examples of pairs of contrasting traits in peas studied by Mendel.

Part of plant	Traits	Phenotype of F1	Phenotypic ratios observed in F2
seeds	round/wrinkled	round	2.96 round/wrinkled
pods	green/yellow	green	2.86 green/yellow
flowers	violet/white	violet	3.15 violet/white
height of plant	tall/dwarf	tall	2.84 tall/dwarf

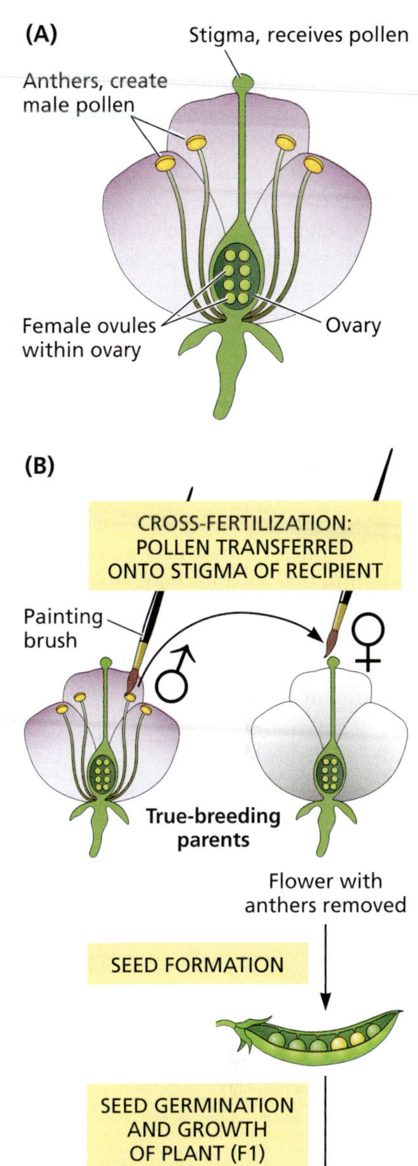

Figure 2.1 Mendel's experiments with garden peas. (A) Anatomy of the pea flower. Peas are self-fertilizing but can also be cross-fertilized. Clear-cut alternative, or antagonistic, forms of particular traits exist: seed color and shape, flower color, pod color and shape, stem length, and flower position. Crossing of plants that have different forms of these traits allowed the formulation of Mendel's laws of inheritance. (B) Mendel isolated and perpetuated true-breeding, or pure-breeding, pea lines, in which a trait that he was studying had remained constant from generation to generation for eight generations. He cross-fertilized antagonistic forms to produce hybrids; for each experiment he also did reciprocal crosses. The first filial or F1 generation was hybrid; it was then allowed to self-fertilize to produce second, third, and further filial generations: F2, F3, etc. Mendel then followed the inheritance patterns of the traits over several generations and quantified the data, leading to the most profound scientific understanding of inheritance laws.

units of heredity, now called genes, that could be distributed according to fixed rules and dictated the dominant and recessive states. Most eukaryotes are **diploid**, carrying two copies of each gene, now referred to as **alleles**, and therefore two alleles for each trait in their somatic cells (Box 2.2). In the case of seed color, the dominant allele is denoted *Y* and the recessive allele *y*. We can now understand the whole process by recalling the fact that, at fertilization, each parent donates one gamete, either sperm or ovum, to the fertilized egg. Because the gametes are **haploid**, this means that each parent donates one or the other of the two alleles, chosen at random. The F1 generation, obtained from cross-fertilization of two opposite homozygotes, each carrying two copies of just one of the alternate alleles, must be heterozygous, and only the dominant trait is expressed. In the F2 generation, however, random combination of gametes will lead to the possibilities *YY*, *Yy*, *yY*, and *yy*. Because *Y* is dominant, this yields the 3:1 ratio of yellow/green phenotypes observed. Mendel's experimental results, which are based on thousands of crosses, are summarized for several traits in Table 2.1.

From these experiments, Mendel formulated the two laws, which are really hypotheses, that constitute the basis of classical genetics. Mendel's first law, the law of segregation, states that the two alleles for each trait separate or segregate during gamete formation and then unite at random, one from each parent, at fertilization. The first law can be expressed in a number of ways. Here we choose to break it into several statements, in modern nomenclature.

- Variation in **phenotype** is explained by the existence of alternate versions of genes. These versions are called alleles.
- The alleles of each gene segregate, independently, one to each gamete.
- Every individual inherits two alleles of each gene, in one gamete from each parent.
- If the alleles differ, one will be dominant and one recessive. If the individual is heterozygous for an allele, only the dominant allele and trait will be expressed in the first generation.

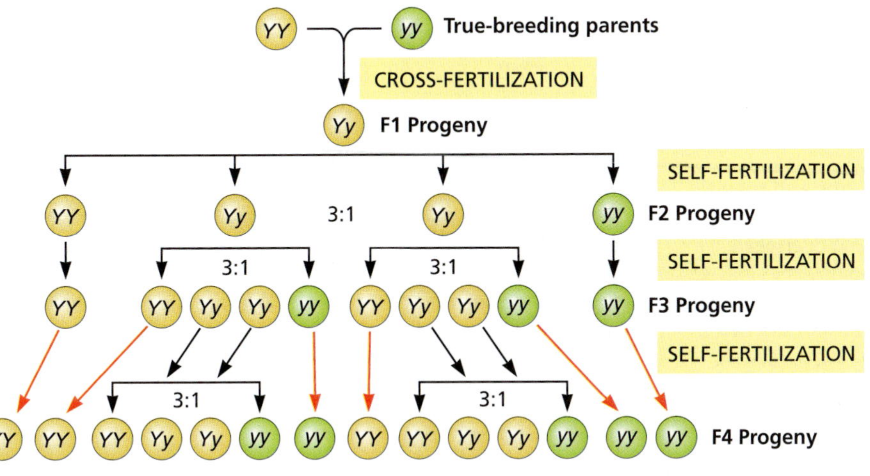

Figure 2.2 Mendel's results showing the distribution of a pair of contrasting traits. Seed color after four rounds of fertilization is illustrated. Capital *Y* stands for yellow color, whereas lowercase *y* stands for green color. For each trait, the plant carries two copies of a unit of inheritance, or two alleles of a gene in contemporary understanding. The trait that appears in all F1 hybrids is dominant; the antagonistic trait that remains hidden in F1 but reappears in F2 is recessive. In this specific example, yellow color is dominant and green is recessive. The schematic also illustrates the difference between homozygous and heterozygous individuals. Homozygous individuals of each generation breed true, as shown by red arrows, whereas heterozygous do not, as shown by black arrows.

Box 2.2 Sexual reproduction, mitosis, and meiosis "Birds do it, and bees do it. Indeed, researchers estimate that over 99.99% of eukaryotes do it, meaning that these organisms reproduce sexually, at least on occasion."—Sarah Otto

One of the oldest arguments for why sexual reproduction is so commonplace goes back to biologist August Weismann, who suggested, in the late 1880s, that sex generated variable offspring upon which natural selection can act. We now know that this notion may be an oversimplification, but its general tenets still hold.

A simple schematic of sexual reproduction, as it relates to humans, is presented in **Figure 1**. As illustrated in the schematic, two major types of cell division take place during the reproduction cycle: **mitosis** produces new diploid somatic cells, that is, cells of the body, whereas **meiosis**, which occurs in germ cells, gives rise to haploid gametes, sperm and ova (**Figure 2**). Both mitosis and meiosis begin by duplicating the genetic material, DNA (see Chapter 4), within the cell and then compacting it into condensed chromosomes, visible structures that are recognizable under the microscope. Chromosome structure is discussed in more detail in Chapter 8. Then mitosis and meiosis take different paths in the way the chromosomes behave, which leads to the production of either diploid cells or haploid gametes, respectively.

Cells that undergo mitosis go through different phases, each characterized by its own biochemical processes and structural reorganizations. It is beyond the scope of this text to describe them in detail, but the most important information that we will need on numerous occasions throughout the book is presented in **Figure 3**. The **cell cycle** is also described in more detail in Chapter 20, where we discuss eukaryotic DNA replication.

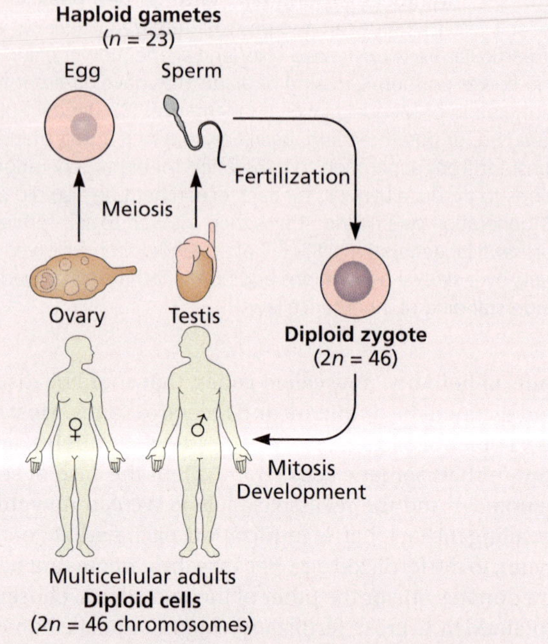

Figure 1 Conceptualized schematic of the sexual reproduction cycle in higher eukaryotes. Note the transitions between cells containing the diploid number of chromosomes, 46 in humans, and those with the haploid number of chromosomes, 23 in humans. The first transition occurs during the formation of the female and male gametes, egg and sperm, respectively, and involves a reduction division known as meiosis. The haploid gametes then reunite during fertilization to form the diploid zygote or fertilized egg. All cells of an adult organism are formed from further divisions known as mitosis.

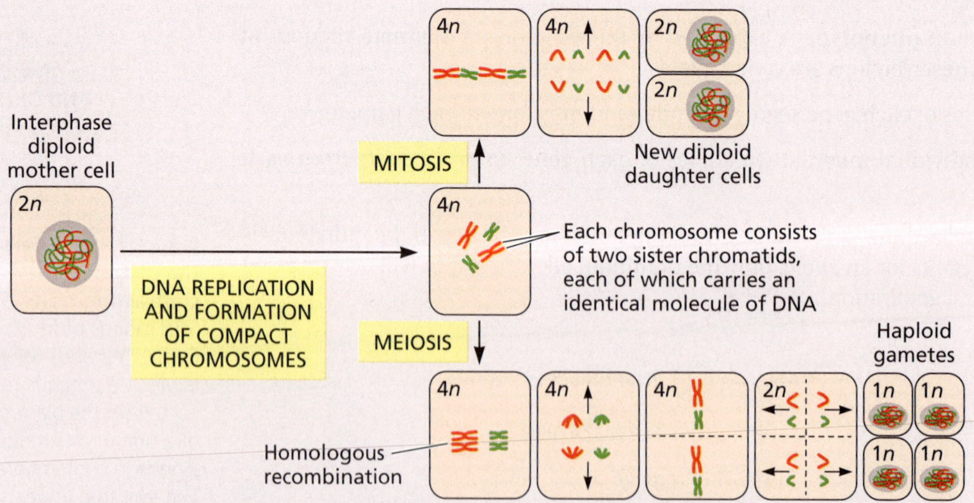

Figure 2 Conceptualized schematic of mitotic and meiotic cell divisions. Both mitosis and meiosis start with replication of DNA, formation of compact chromosomes, and disappearance of the nuclear membrane. In mitosis, the chromosomes line up individually at the equatorial plate and the two sister chromatids that make up each chromosome are moved toward the two opposite poles of the mitotic spindle. By contrast, in meiosis, the homologous chromosomes pair and the pairs align at the equatorial plate; two divisions then follow. During the first division in meiosis, each chromosome of a pair moves toward one of the opposite poles of the spindle; during a second division, the two sister chromatids of a chromosome move toward the opposite poles of new spindles formed in a direction perpendicular to the first one. As a result, four cells are formed, each with a haploid chromosome number.

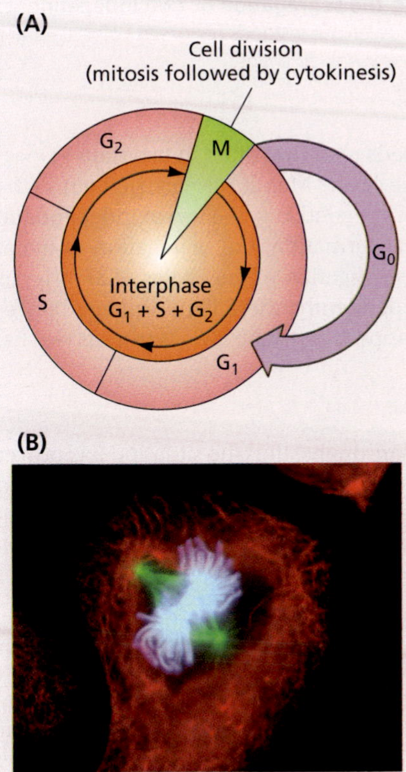

Figure 3 Mitotic cell cycle. (A) Schematic showing the traditional phases characteristic of cycling cells: G_1, cell growth and synthesis of components required for DNA synthesis; S, synthetic phase, DNA replication; G_2, preparation for mitosis; M, mitosis followed by cytokinesis or cell division. The G_0 detour is the special phase following mitosis during which differentiated cells perform their special function(s). This phase is not part of the normal cycle of continuously proliferating cells, in which M is immediately followed by G_1, the phase in which the cell prepares for DNA replication. Note that this rule holds true for most differentiated cells, but there are exceptions: some differentiated cells can perform their specialized function while actively proliferating. Note also that nonproliferating cells can re-enter the cell cycle under certain conditions: cancer cells are an *in vivo* example. (B) Fluorescent micrograph of a newt lung epithelial cell in the metaphase stage of mitosis. The cell was fixed and stained for immunofluorescence localization of microtubules, shown in green/yellow, and keratin filaments, shown in red; condensed mitotic chromosomes are in blue. By metaphase, the bipolar mitotic apparatus is fully formed and shaped like a spindle of thread. This structure supports the production of mechanical forces required for segregating the replicated chromosomes into daughter nuclei. In epithelial cells the spindle and its associated chromosomes are surrounded by a cage of keratin filaments, which prevents the motion of other cell organelles into the region containing the chromosomes. (B, courtesy of Conly L. Rieder, New York State Department of Health.)

Mitosis permits development of a wide variety of differentiated somatic cells from a few types of stem cells

Stem cells are cells that can both self-renew and generate progeny capable of following more than a single differentiation pathway. Of all of the stem cells in an organism, **embryonic stem cells** or **ES cells** have drawn the most attention. ES cells arise from the inner cell mass of the mammalian blastocyst (Figure 4) and can be maintained in culture in a pluripotent state. **Pluripotency** is defined as the ability to differentiate into every single cell type that is found in adult organisms. By contrast, multipotency is characteristic of lineage-restricted stem cells that are already committed to a certain pathway. For example, hematopoietic stem cells give rise to all kinds of blood cells but cannot become neurons or liver cells; neural stem cells can generate only neuronal and glial cells.

ES cells can be induced to differentiate *ex vivo* by culture conditions and can, under certain circumstances, develop an entire organism. This latter property has created considerable controversy in the public domain. The potential of ES cells to generate differentiated cells, tissues, and organs for use in clinical practice for the treatment of disease is deemed useful, but the ability to create whole human beings is seen as dangerous human cloning and is prohibited by law in most developed countries. By contrast, cloning of animals and plants has been accepted and provides the potential to improve food production.

Meiosis follows a path somewhat similar to mitosis; however, following DNA replication there occur two successive cell divisions. Therefore, a single cell will yield four haploid gametes, each carrying only one copy of each gene.

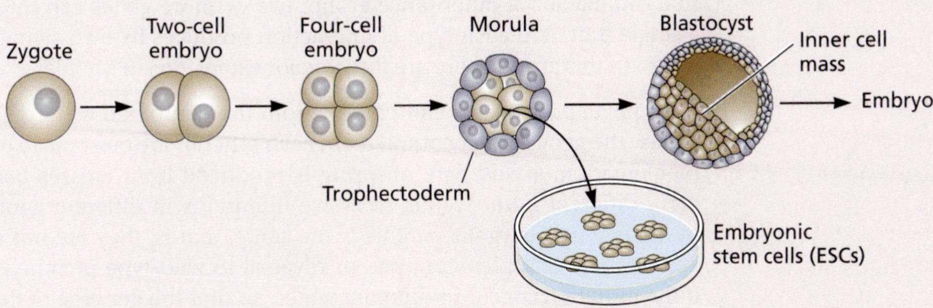

Figure 4 Schematic depicting how embryonic stem cells are derived.

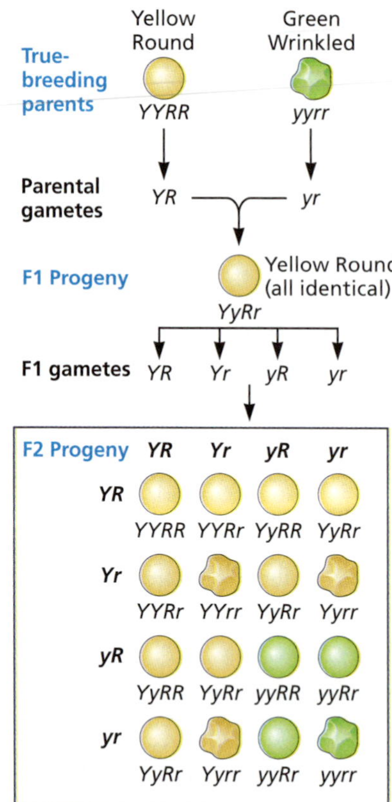

Figure 2.3 Mendel's results showing the segregation of two independent traits. Capital *Y* stands for yellow color, and lowercase *y* stands for green color; capital *R* stands for round shape, and lowercase *r* stands for wrinkled shape.

But what happens if one cross-breeds peas that differ in two traits? Does the segregation of alleles for one trait affect the other? Mendel also carried out such experiments and derived what is called Mendel's second law, the law of independent assortment. This law states that, during gamete formation, the segregation of the alleles of one allelic pair is independent of the segregation of the alleles of another allelic pair. In other words, traits segregate independently; there is no **linkage** between genes for different traits (**Figure 2.3**). This happens not to be always true, as we shall see.

Mendel's laws have extensions and exceptions

As with many great scientific breakthroughs, the true situation has proved to be more complicated than the initial insight suggested. There are many examples of exceptions to simple **Mendelian genetics**. In general, the extensions to Mendel's laws can be classified in two groups, depending on whether a trait is encoded by a single gene or by many genes, known as multifactorial inheritance.

In the single-gene inheritance group, there are three major extensions.

First, dominance is not always complete. In **incomplete dominance**, the hybrid resembles neither parent. In **co-dominance**, neither allele is dominant, with the F1 hybrid showing traits from both true-breeding parents. These relationships are schematically depicted in **Figure 2.4**.

Second, a gene may have more than two alleles. There are numerous examples of such genes, including those that determine human blood groups and the genes that code for human histocompatibility antigens, which are cell surface proteins that participate in proper immune responses. The latter proteins are encoded by three genes, each with between 20 and 100 alleles; each allele is co-dominant with every other allele at the molecular level. The most extreme example known to date is the olfactory genes, which have ~1300 alleles, only one of which is expressed. This **monoallelic gene expression** is strictly regulated by mechanisms that remain largely unknown. Usually cells express both alleles of a gene.

Third, one gene may contribute to several visible characteristics; this phenomenon is known as **pleiotropy**, from the Greek pleion, meaning more, and tropi, meaning to turn or convert. A classic example of pleiotropy is found in sterile males among the aboriginal Maori people of New Zealand. These men are sterile and have respiratory problems. The gene's normal dominant allele specifies a protein needed in both cilia and flagella; in men who are homozygous for the recessive allele, cilia and flagella do not function properly, affecting their abilities to both clear mucus from their respiratory tract and produce motile sperm.

In a multifactorial inheritance group, two or more genes can interact to determine a single trait, and each type of interaction produces its own signature of phenotypic ratios. In this group, there are three major extensions to Mendel's laws.

First, novel phenotypes can emerge from the combined action of the alleles of two genes. The genes either complement each other or are epistatic to each other. In complementation, a wild-type offspring is produced from crosses between strains that carry different homozygous recessive mutations in different genes; the phenotype of the two homozygous parents is the same, that is, they are not distinguishable by appearance. Complementation, or reversal to wild-type phenotype, can occur only if the mutations occur in different genes, so that the genome of the offspring carries one wild-type allele that complements the mutated allele of the same gene. In **epistasis**, one gene's alleles mask the effects of another gene's alleles. Biochemically, this situation arises when several genes participate in succession in a single biochemical pathway. The inactivity of a gene at the beginning of the pathway will hide the fact that subsequent genes may be expressed.

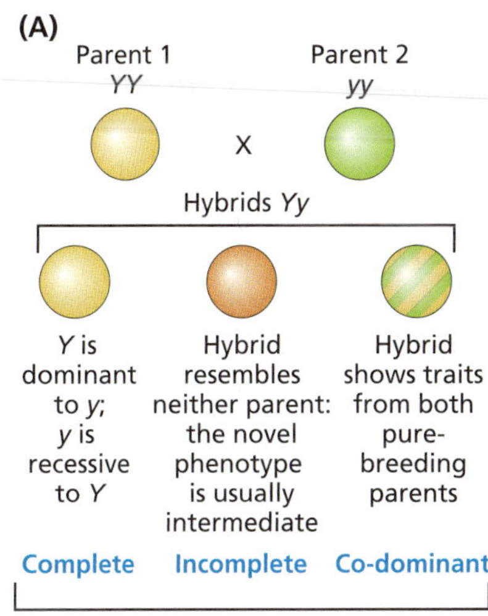

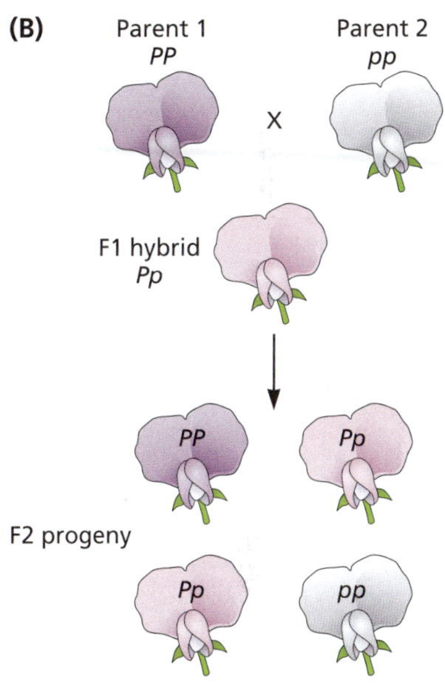

Figure 2.4 Dominance relationships between alleles. (A) Different dominance relationships between pairs of alleles are revealed by the phenotype of the heterozygote. Variations in dominance relationships do not detract from Mendel's law of segregation; rather, they reflect differences in the way in which gene products control the production of phenotype. (B) In this example of incomplete dominance, the petal color of the F1 heterozygote *Pp* is unlike that of either homozygote parent. The phenotypic ratios in F2 are an exact reflection of the genotypic ratios. The explanation of this behavior lies in the biochemistry of pigment production. *p* does not encode a functional enzyme for pigment production, hence a white color of the petals results. Conversely, the *PP* homozygote produces a double dose of the enzyme and hence a bright pink color is seen.

Second, a given genotype does not always produce the same phenotype: phenotype often depends on penetrance and expressivity. **Penetrance** describes how many members of a population with a particular genotype show the expected phenotype. Penetrance can be complete or incomplete. A frequently cited example of incomplete penetrance concerns the disease retinoblastoma: only 75% of people carrying a mutant allele for the retinoblastoma protein develop the disease. In addition, in some people who have retinoblastoma, only one eye is diseased: **expressivity** refers to the degree or intensity with which a particular genotype is expressed in a phenotype. It is important to understand that chance can affect penetrance and expression. For example, in the case of retinoblastoma, every cell carries the inherited mutation in one allele of the retinoblastoma gene, but a second chance event is needed for the disease phenotype to appear. Damaging radiation or errors in DNA replication in retinal cells provide the second hit, creating a mutation in the second copy of the retinoblastoma gene within one or more cells. Such situations gave rise to the two-hit hypothesis for the origin of cancer proposed by Alfred Knudson in 1971.

Third, there are also quantitative traits that vary continuously over a range of values. A good example is height and skin color in humans. These traits are polygenic and show the additive effects of a large number of genes and their alleles.

Here we mention the concept of **modifier genes**, which have secondary, more subtle effects on a trait. We also introduce the concept of **allele frequency** within a population. This is the percentage of the total number of gene copies in a population comprised of any one allele. The most prevalent allele, the one with the highest frequency in a population, is defined as the wild-type allele. In evolution, new alleles appear as a result of mutations.

Finally, it is now clear that the environment can affect the phenotypic expression of a genotype. When environmental agents cause a change in phenotype that mimics the effects of a mutation in a gene, this is known as **phenocopying**. A painful example of this phenomenon was the effect of the sedative drug thalidomide. If taken by pregnant women, this drug produced a phenocopy of a rare dominant trait called phocomelia, which disrupts limb development in the fetus.

Perhaps the most significant exception is the fact that Mendel's second law is not generally correct. There are many cases in which genes are linked, and discovery of this led to the second great advance in classical genetics.

Genes are arranged linearly on chromosomes and can be mapped

The next major step in genetic research was accomplished by Thomas Hunt Morgan, who spent most of his career at Columbia University (see Box 2.1).

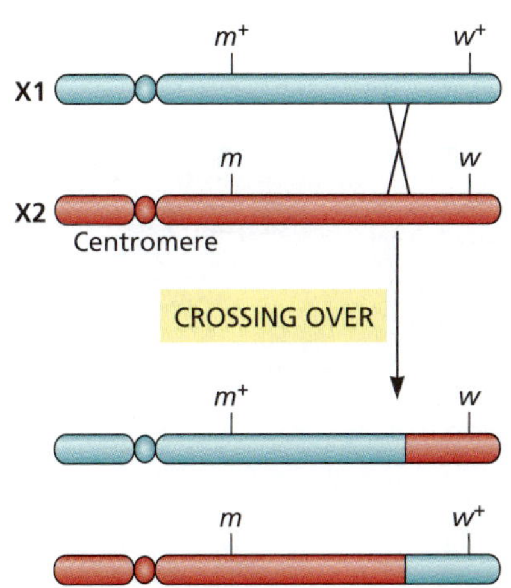

Figure 2.5 Recombination between the two X chromosomes of the female *Drosophila* fly. Chromosome X1 carries two wild-type alleles: m^+ codes for normal wings and w^+ determines red eyes. Chromosome X2 carries two mutant alleles, *m* coding for miniature wings and *w* coding for white eyes. During egg formation, a crossing-over or recombination event occurs somewhere between these two genes on the two chromosomes, resulting in two recombinant chromosomes, each of which carries a mixture of the parental alleles. This process creates a new combination of alleles, hence the name recombination.

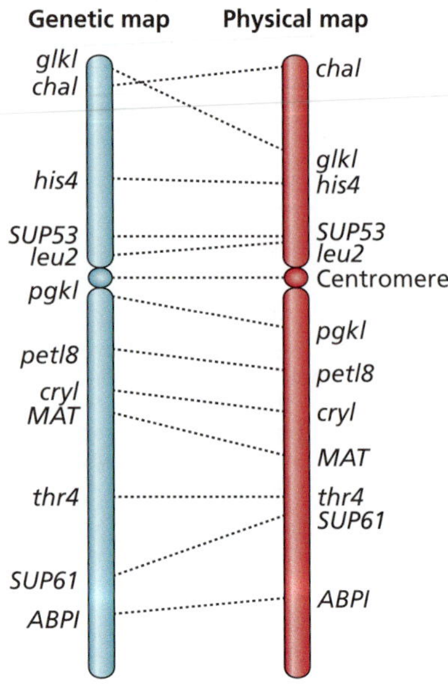

Figure 2.6 Genetic and physical maps of *S. cerevisiae* chromosome III. The genetic map was constructed by determining the frequency of recombination in genetic crosses; the physical map was determined by DNA sequencing. Despite some discrepancies between the two maps, their overall similarity is impressive. Note that the order of the upper two markers or identifiable genes has been incorrectly assigned on the genetic map; the relative positions of some markers are also somewhat different on the two maps. (Adapted from Oliver SG, van der Aart QJM, Agostoni-Carbone ML et al. [1992] *Nature* 357:38–46. With permission from Macmillan Publishers, Ltd.)

In the early part of the twentieth century, Morgan began working with fruit flies, *Drosophila melanogaster*. This was an even better choice than Mendel's peas because the flies breed very rapidly and could be raised in large numbers in very little space. This was important because Morgan was dependent, for much of his work, on the occurrence of rare, spontaneous mutations. These were the source of the changes in the genes that produced modified alleles and hence phenotypes. Thus the discovery, by Hermann Joseph Muller in Morgan's lab, that mutations could be induced by X-rays or other damaging radiation was very helpful. Muller was awarded the 1946 Nobel Prize in Physiology or Medicine "for the discovery of the production of mutations by means of X-ray irradiation."

Contrary to Mendel's observations, Morgan found that a number of traits of the flies appeared to be genetically linked. The difference with Mendel's conclusion may have lain in the fact that many of the pea traits studied by Mendel corresponded to genes on different chromosomes, which would be expected to segregate independently, whereas Morgan was initially concentrating on genes on the same chromosome, the female X chromosome. In any event, Morgan observed linked transmission of a number of genes. He hypothesized that the lack of linkage in some cases must result from recombination of alleles (**Figure 2.5**). Furthermore, he noted that the probability of such recombination must increase with the distance between the two genes in the chromosome. Thus, the degree of linkage must measure gene separation. Then something wonderful happened. Alfred Sturtevant, a student working in Morgan's laboratory, realized that this fact allowed mapping of genes on chromosomes. He skipped his assigned homework one night to produce the first genetic map. Soon this was extended to many *Drosophila* genes, and a new paradigm emerged: genes are arranged linearly on chromosomes. A sample genetic map is presented in **Figure 2.6**. In 1933, Morgan was awarded the Nobel Prize in Physiology or Medicine "for his discoveries concerning the role played by the chromosome in heredity."

The nature of genes and how they determine phenotypes was long a mystery

Despite rapid advances in classical genetics in the early years of the twentieth century, the nature of genes and their mode of function remained obscure. Indeed many geneticists preferred to think of them as abstract entities. Genes gained some substance through the examination of **polytene chromosomes**. These are parallel aggregates of a number of identical chromosomes, found in the salivary glands of some insects, including *Drosophila* (**Figure 2.7** and **Figure 2.8**). Upon proper staining and under the light microscope, they show a banded pattern, and some bands can be correlated with genes mapped by the Sturtevant–Morgan technique. Such studies reinforced the image of genes as physical objects, but of what were they made?

This same period saw the beginnings of protein studies, and the function of proteins as enzymes was well established. Probably for this reason, and because of misunderstanding of the nature of nucleic acids, most researchers before about 1940 assumed

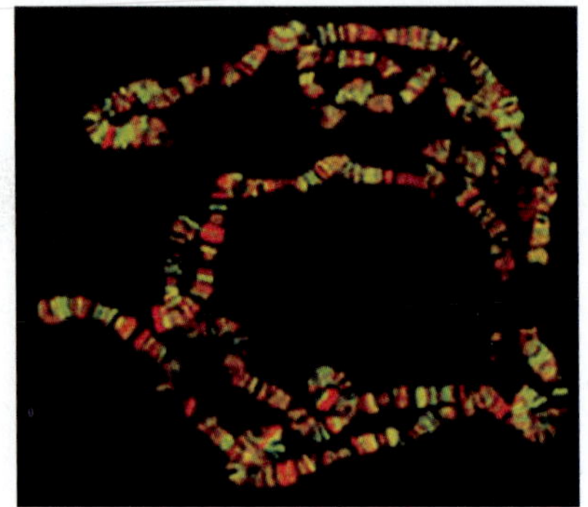

Figure 2.7 Polytene chromosomes with the typical banding patterns. The banding patterns of *Drosophila* polytene chromosomes were depicted by Calvin Bridges as early as 1935 and are still widely used to identify chromosomal rearrangements and deletions. A fluorescent image of the *Drosophila* salivary gland chromosomes stained for two different proteins, BRAMA in green and Pol II in red, is presented. The two proteins show a high degree of overlap. (From Armstrong JA, Papoulas O, Daubresse G et al. [2002] *EMBO J* 21:5245–5254. With permission from John Wiley & Sons, Inc.)

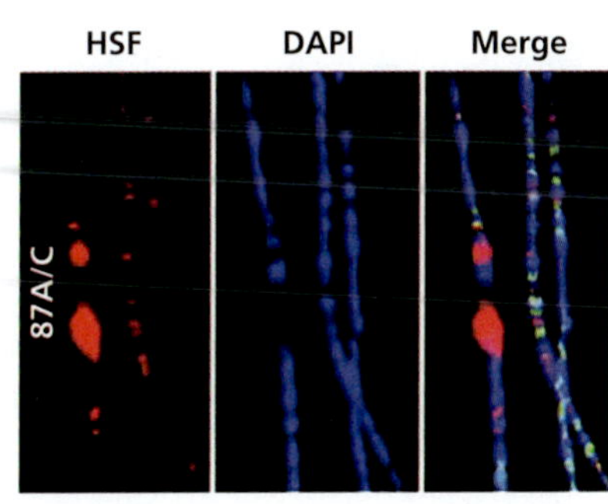

Figure 2.8 Banding patterns of polytene chromosomes reflect the transcriptional activity of chromosome regions. A fluorescent image of heat-shock-induced puffs, loci 87A/C, in *Drosophila melanogaster* chromosomes is shown. Inactive genes that are located in compact chromatin form condensed bands. In puffing, the material of the bands loosens and becomes decompacted, and local swelling of the chromosome region occurs. DNA is stained with 4′,6-diamidino-2-phenylindole or DAPI, shown in blue, while the heat-shock factor (HSF), a transcriptional activator of genes induced by heat shock, is shown in red. The merged image on the right indicates that the transcriptional activator is located in the puffed, decondensed regions. (From Armstrong JA, Papoulas O, Daubresse G et al. [2002] *EMBO J* 21:5245–5254. With permission from John Wiley & Sons, Inc.)

that genes were proteinaceous in composition. It was also becoming evident, however, that genes must dictate protein structure. Certain disease states that depended on the loss of one enzyme function exhibited Mendelian inheritance; this early observation led to the dictum one gene, one enzyme. After the brilliant work in several laboratories on sickle cell anemia (**Box 2.3**), this dictum had to be modified to one gene, one polypeptide chain. Today we realize that not only protein sequences but also non-protein-coding RNA sequences are dictated by genes, leading to further attempts to properly define a gene. As we see in Chapter 7, the definition of the term gene is still evolving, with a new definition recently proposed in 2012 as a result of studies on the human genome.

2.3 The great breakthrough to molecular genetics

Bacteria and bacteriophage exhibit genetic behavior and serve as model systems

In Chapter 4, we describe in detail the remarkable experiments that established once and for all that DNA is the genetic material. This recognition, together with the Watson–Crick structure of DNA, meant that by the mid-1950s a molecular theory of genetics had begun to develop. Much of the early work in this direction depended on the use of bacteria and viruses, particularly those viruses that infect bacteria, the bacteriophages or phage in scientific jargon.

The genetics of phage and bacteria had long been neglected, for neither follows the laws of classical Mendelian genetics. Both are haploid under most circumstances. However, in 1943, microbiologist Salvador Luria and physicist Max Delbrück provided convincing evidence of mutations in bacteria. In the same year, a major technical breakthrough was accomplished by Joshua Lederberg and Edward Tatum in the discovery of bacterial conjugation (**Figure 2.9**). In this process, one bacterium inserts all

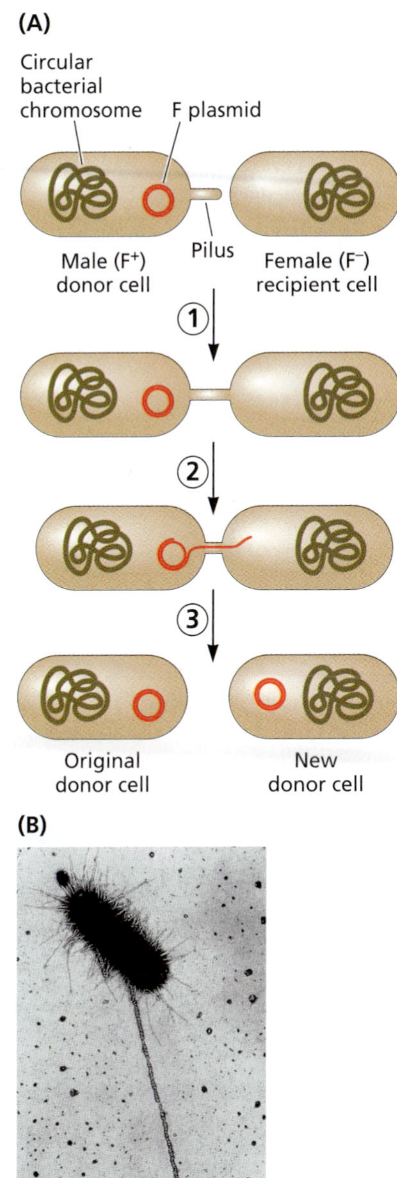

Figure 2.9 Bacterial conjugation. Conjugation between two bacterial cells can occur only when one of the partners carries the F or fertility plasmid; these cells are known as F-positive or F$^+$ cells. The F plasmid exists as an episome, that is, independently of the main bacterial chromosome. It carries its own origin of replication, an origin of transfer where nicking occurs to initiate transfer to a recipient F$^-$ cell, and a whole battery of genes responsible for formation of the pilus and attachment to the recipient cell. (A) Steps in the process of conjugation are as follows: (Step 1) Pilus attaches to recipient cell and brings the two cells together. Most probably the pilus is not directly used as a transfer channel. The channel is formed through the action of a specific enzyme at the base of the pilus, which initiates membrane fusion. (Step 2) The F plasmid is nicked and the nicked strand is unwound from the intact strand; transfer to the recipient cell begins. (Step 3) Single-stranded DNA is transferred to the recipient and copied to produce a double-stranded F plasmid; the single-stranded F plasmid in the donor cell is simultaneously copied to produce a double-stranded F plasmid. Sometimes the F plasmid is integrated into the genome of the donor; these strains are known as high frequency of recombination or *Hfr* strains. In such cases, the entire bacterial chromosome or a part of it can be transferred into the recipient cell. The amount of chromosomal DNA transferred depends on how long the two conjugating bacteria stay in contact: transfer of the entire chromosome normally requires ~100 minutes. Homologous recombination allows for integration of the transferred chromosome into the genome of the recipient cell. (B) Electron microscopic view of two bacterial cells in the process of conjugation. (B, courtesy of Charles Brinton and Judith Carnahan, University of Pittsburgh.)

Box 2.3 Sickle cell anemia: a key to molecular genetics A genetically transmitted disease called sickle cell anemia or SCA affects millions of people around the world, especially those of African origin or ancestry. It can be deadly and is often debilitating. Red blood cells, known as erythrocytes, are normally discoid in shape, but those in patients with SCA tend to be elongated or sickle-shaped, especially under conditions of oxygen deprivation (**Figure 1A**). Such distorted cells have two deleterious effects. First, they tend to block venous capillaries, causing pain and tissue damage. Second, they are more fragile than normal erythrocytes, being easily broken and releasing their hemoglobin; this leads to anemia.

It was first shown in 1949 by James Neel and E. A. Beet that sickle cell anemia is inherited as a Mendelian trait with two alleles. In individuals homozygous for the trait, the disease is fully expressed, with significant impairment of life quality or even early mortality. Heterozygous individuals, carrying one copy of the SCA allele, are often able to function quite normally, except in situations of oxygen deprivation or stress, where they will exhibit some sickling. Such individuals are often called carriers of the trait: they are not very sick, but their progeny will develop the disease if they are born homozygous for the SCA allele.

In the same year that Neel and Beet demonstrated the genetics of SCA, Linus Pauling and co-workers made a remarkable discovery. They found that the hemoglobin of people with SCA differed in electrophoretic mobility from that of healthy individuals. Furthermore, heterozygous individuals showed both electrophoretic bands (**Figure 1B**). This led Pauling to the conclusion that this was a molecular, genetic disease, the first clearly identified.

This observation was followed, in the 1950s, by protein-sequencing studies in the laboratory of Vernon Ingram. It was known by then that hemoglobin contains two types of protein subunits, called α and β. Ingram found that a single amino acid substitution in the β chain constituted the sole difference between normal and sickle cell hemoglobin. Thus, it could be deduced that a single mutation was responsible for the trait. This and Pauling's work strengthened the one gene, one protein hypothesis.

But why does this mutation cause sickling? Numerous studies had shown that the mutant protein induces the formation of long filaments of hemoglobin, which pack side by side within the erythrocyte (**Figure 1C**). Because hemoglobin is highly concentrated in these cells, this aggregation deforms the cells in the manner observed. The way in which the disease was expressed in homozygous and heterozygous individuals now became clear. In the former, no normal protein can be made; all of the hemoglobin can aggregate. In a heterozygous individual, both normal and aggregating hemoglobin would be mixed. Sickling could occur, but not so easily. Why did low oxygen have an effect? It was eventually shown that hemoglobin undergoes a conformational change as oxygen levels change. The low-oxygen form is more prone to aggregation.

There is one further twist to the story. Being heterozygous for the sickle cell trait can actually be advantageous in tropical regions where malaria is common. The malarial parasite must spend part of its life cycle within an erythrocyte. Sickled cells are inhospitable to the parasite because they are so fragile.

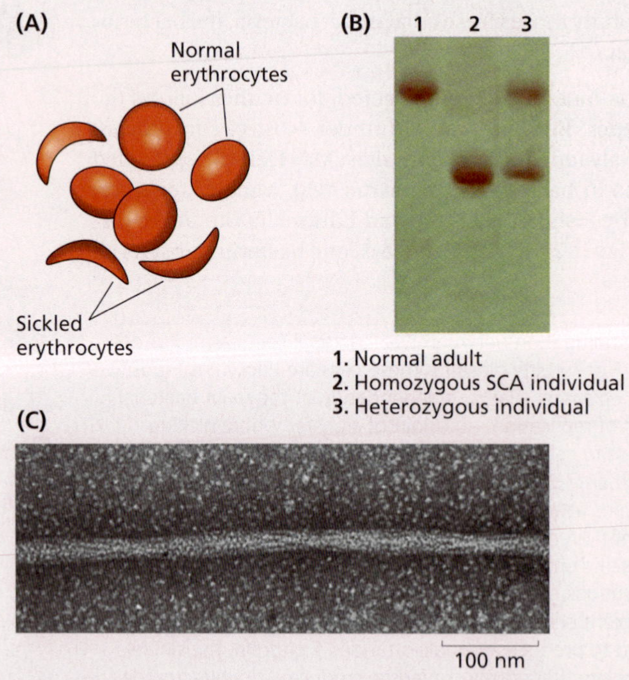

Figure 1 Sickle cell anemia, a molecular disease. (A) Shapes of normal and sickled erythrocytes. (B) Electrophoretic behavior of hemoglobin from normal individuals and individuals homozygous and heterozygous for the mutation in the β chain of hemoglobin. (C) Electron microscopic image of an aggregate of hemoglobin molecules in sickle cell anemia. The amino acid that is mutated in the β chain forms a protrusion that accidently fits into a complementary site on the β chain of other hemoglobin molecules in the cell. Instead of remaining in solution, these mutated molecules aggregate and become rigid, precipitating out of solution. (A, adapted, courtesy of Darryl Leja, National Human Genome Research Institute. B, courtesy of Michael W. King, Indiana University. C, from Dykes G, Crepeau RH & Edelstein SJ [1978] *Nature* 272:506–510. With permission from Macmillan Publishers, Ltd.)

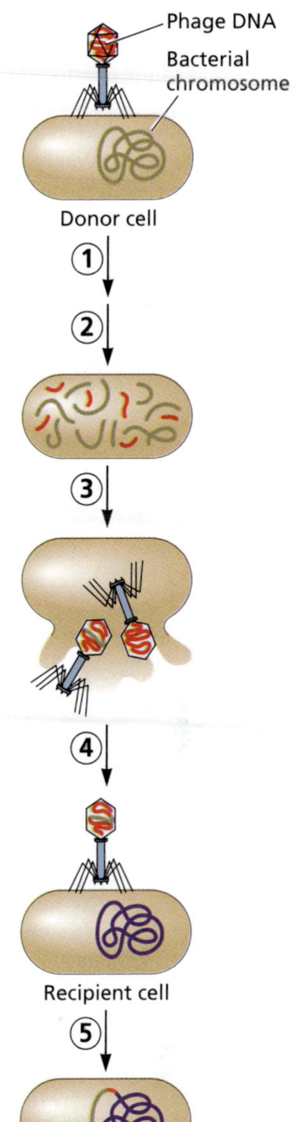

Figure 2.10 Conceptual schematic of transduction. Steps involved in transduction are as follows: (Step 1) A phage infects a bacterial cell. (Step 2) Phage DNA enters cell and replicates; phage proteins are made; bacterial chromosome is broken down. (Step 3) Occasionally, pieces of bacterial DNA are packaged into the phage head; some viral particles released during cell lysis contain bacterial DNA. (Step 4) A phage carrying bacterial DNA infects a new bacterial cell. (Step 5) Recombination between the donor bacterial DNA and the recipient bacterial DNA can occur; the recombinant cell is different from both the donor and the recipient cell.

or part of its DNA into another, followed by recombination of the two DNA molecules. With some strains, called high frequency of recombination strains, practically all of the donors are active, and conjugation can be synchronized. If conjugation is halted at a series of different times, different amounts of DNA will be transferred, permitting recombination of only those genes that have been transferred. This provided a convenient way to map genes on the bacterial chromosome before powerful sequencing methods were available.

Transformation and transduction allow transfer of genetic information

Conjugation is by no means the only way in which bacteria can acquire foreign DNA. As early as 1928, Frederick Griffith had demonstrated the phenomenon of **transformation** of bacterial strains by the transfer of some factor. However, he did not know the nature of that substance. It was not until 1944 that Oswald Avery, Colin MacLeod, and Maclyn McCarty showed that genetic transformation was caused by the transport of DNA, through solution, into the transformed bacterium. These experiments are at the very basis of molecular biology and are discussed in detail in Box 4.1.

Finally, DNA can be transferred between bacteria via bacteriophage, by a process called **transduction** (Figure 2.10). This usually involves **temperate phage**, which can exhibit two alternative life cycles: lytic and lysogenic. When in the lytic cycle, the virus can enter the bacterial cell, replicate, lyse the host bacterium, and go on to infect other bacteria. In the alternative lysogenic cycle, it can integrate its DNA into the bacterial chromosome. There it may remain for many bacterial generations in a dormant state, until some stimulus, such as radiation or chemical insult, causes it to be released from the host genome. It will then form new viruses and kill the bacterium. Packaging of the viral DNA into viral particles is a low-fidelity process: small pieces of bacterial DNA may become packed, alongside the phage genome, and thus transferred to the newly infected cell. At the same time, phage genes can be left behind in the bacterial chromosome into which they had been integrated. The phage can be modified by addition of foreign DNA, and can act as a **vector** to insert this into bacteria. We discuss the role played by transformation and transduction in genetic engineering in Chapter 5.

The Watson–Crick model of DNA structure provided the final key to molecular genetics

By 1955, two years after Watson and Crick's publication of the DNA double-helix model, the molecular basis of genetics was clear. Genes are made of DNA, which is carried in chromosomes. Bacteria and phage have haploid chromosomes, constituted of one or a few double-helical DNA molecules. These simple systems provided the entry into molecular genetics and key ideas for molecular biology. Most eukaryotes have two copies of double-helical DNA in each somatic cell but only one in each gamete (see Box 2.2). When cells replicate, DNA is duplicated by copying each of the strands of the double helix. As DNA replication occurs independently in each cell of a cellular population, at any point in time some cells will have DNA content characteristic of the G_1 phase, before replication begins; other cells will be in the process of replicating their DNA, in S phase; and still others will have their DNA replicated, in G_2 phase. The distribution of cells in the different stages of the cell cycle can be monitored by flow cytometry (**Box 2.4**). Mutations occur by modification of DNA sequences, and exchange of alleles occurs by recombination. Almost none of the details or control of

Box 2.4 Flow cytometry for cell analysis or sorting How can biologists determine the distribution of cell states in a collection of cells? In earlier years the only method was laborious microscopic analysis, and fractionation was not practical. The method most frequently used today is **flow cytometry** (**Figure 1**). A preparative variant of this method is **cell sorting**, which allows the fractionation of cellular populations into fractions highly enriched in cells of the same phase. Having such fractions at hand has been instrumental in experiments aimed at understanding the biochemical processes occurring in each phase. Flow cytometry has numerous other applications, both in the laboratory and in clinical practice. In addition to measuring total DNA content of cells, numerous other cellular molecules, as well as cellular volume and morphological complexities, can be assessed. The list of measurable parameters is very long and ever expanding, making flow cytometry one of the most important tools in molecular and cell biology.

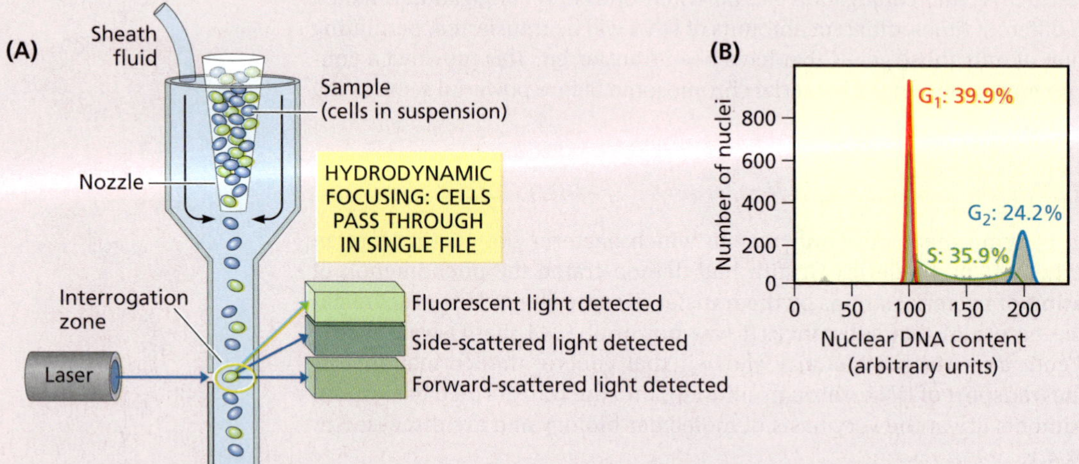

Figure 1 Flow cytometry, multiparameter analysis of individual particles in a population. (A) Principle of flow cytometry. A sample of cells randomly dispersed throughout the suspending liquid is injected into the cytometer. A fluidic system aligns the individual particles in single file, in a process known as hydrodynamic focusing. As a cell passes through the interrogation zone, it is hit by the light beam, which results in light scatter and possibly fluorescence, depending on whether molecules in or on the cell have been labeled with a fluorescent dye and on the spectral properties of the laser and the dye. A set of light detectors are situated at 90° with respect to each other to catch and quantify the strengths of the respective signals. The detection and analysis of light scatter and fluorescence gives information about each cell. When a cell passes through the interrogation zone, light is scattered in all directions. The forward-scatter detector registers light scattered up to a 20° angle from the excitation beam. The intensity of this signal is proportional to the cell size. The side scatter is detected at 90° from the beam and provides information on granularity and the internal complexity of the cells: the more granular a cell, the greater the side scatter. (B) Nuclear DNA content reflects the position of a cell within a cell cycle: thus, flow cytometric analysis of nuclear DNA content can be used for cell cycle analysis. Distribution of nuclear DNA content of nuclei isolated from broad bean meristem root tip cells prestained with the fluorescent stain DAPI is shown. (A, adapted, courtesy of Abcam, Inc. B, adapted, courtesy of Jaroslav Dolezel, Czech Academy of Sciences.)

these processes were understood in 1955. Much has been learned since then, as we show in following chapters.

2.4 Model organisms

Throughout this chapter, we have several times noted that the use of a particular organism was especially appropriate for a given study. In addition, you will find throughout the book that certain organisms have been used again and again as convenient models for whole categories of organisms (**Figure 2.11**). We describe some of these model organisms briefly below, with notes as to why they have been so often chosen. The genomic sequences of all of these organisms are now available.

Today bacteriophage λ is employed largely as a cloning vector, but it played an important part in the early development of genetics, especially because it has two alternative life cycles, lytic and lysogenic. In other words, the phage can either destroy or lyse the host bacterium or become integrated into its genome, existing in a dormant state that is propagated from cell generation to cell generation without any signs of the

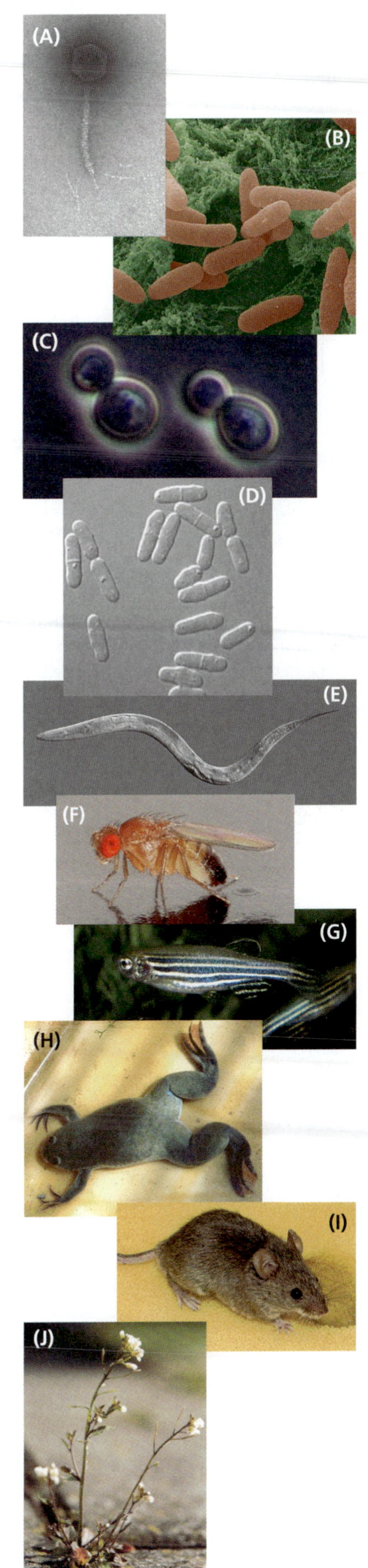

Figure 2.11 Picture gallery of some of the model organisms used most frequently in genetics research. (A) λ phage. (B) *Escherichia coli*. (C) *Saccharomyces cerevisiae*. (D) *Schizosaccharomyces pombe*. (E) *Caenorhabditis elegans*. (F) *Drosophila melanogaster*. (G) *Danio rerio*. (H) *Xenopus laevis*. (I) *Mus musculus*. (J) *Arabidopsis thaliana*. (A, courtesy of Bob Duda, University of Pittsburgh. B, courtesy of Peter Cooke and Stephen Ausmus, United States Department of Agriculture. C, courtesy of Maxim Zakhartsev and Doris Petroi, International University Bremen. D, from Gutterman JU, Lai HT, Yang P et al. [2005] *Proc Natl Acad Sci USA* 102:12771–12776. With permission from National Academy of Sciences. E, courtesy of Judith Kimble, University of Wisconsin. F, courtesy of André Karwath, Wikimedia. G, from Wikimedia. H, courtesy of Michael Linnenbach, Wikimedia. I, courtesy of George Shuklin, Wikimedia. J, courtesy of Brona Brejova, Wikimedia.)

viral DNA's presence, known as the lysogenic pathway. See Chapter 21 for a detailed description of this integration process and its control.

The bacterium *Escherichia coli* has been labeled the workhorse of molecular biology. There is practically no fundamental biochemical process, from DNA replication to protein synthesis, which was not first elucidated in *E. coli*. It is extremely easy to grow, in liquid culture or on solid agar plates, and metabolically very versatile, which has made it useful for studies of metabolic regulation.

The common budding or bakers' yeast *Saccharomyces cerevisiae* is among the simplest eukaryotes. Unicellular and easy to grow in large quantities, it provides a bridge between bacteria and the more complex eukaryotes. Its genetics has been very thoroughly studied, with many knockout strains available. In a knockout strain, a particular gene has been inactivated by recombinant DNA techniques (see Chapter 5). Studying such knockouts helps in elucidating the biological functions of genes. One difficulty in working with *S. cerevisiae* is the tough outer cell wall, which makes it difficult to insert substances. The fission yeast *Schizosaccharomyces pombe* is genetically similar to *S. cerevisiae* but lacks the tough outer layer. It divides, rather than buds, which is an advantage for some studies.

The free-living, primitive, unsegmented, and bilaterally symmetrical worm *Caenorhabditis elegans*, which was introduced to the field by Sidney Brenner, is a remarkably simple creature. The adult worm has only 1090 cells, and the lineage of each is precisely known (**Figure 2.12**). This makes it an outstanding candidate for developmental studies. Sydney Brenner, Robert Horvitz, and John Sulston were awarded the 2002 Nobel Prize in Physiology or Medicine "for their discoveries concerning genetic regulation of organ development and programmed cell death." Andrew Fire and Craig Mello investigated the regulation of gene expression in *C. elegans* and identified RNA interference or RNAi, a novel mechanism of gene silencing by double-stranded RNA (see Chapter 12). This work was awarded the 2006 Nobel Prize in Physiology or Medicine.

Drosophila melanogaster was the organism that provided the seminal studies in modern genetics. Morgan's fruit fly is easy to grow, in enormous numbers, in a very short time. This, plus the availability of a great many mutant strains, including many with mutations that affect general developmental patterns, make it still a useful model. The embryos are also used, especially for biochemical studies.

The small zebrafish *Danio rerio* is easy to grow and is very fecund, so it provides a convenient vertebrate model. Its special attraction lies in the fact that the embryos are transparent so that development of internal organs can be followed in live embryonic fish.

The African frog *Xenopus laevis* is useful because of its large and abundant eggs, which can easily be manipulated for injection studies and the like. *X. laevis* can rapidly produce thousands of embryos. A disadvantage of this model is that it is tetraploid and it takes years for the frog to reach sexual maturity. Another frog species, *Xenopus tropicalis,* is diploid and matures in less than 3 months; these two properties make it very attractive for genetic research.

The house mouse *Mus musculus* is the easiest mammal to study and has been used by generations of researchers. Despite a large evolutionary separation between mice and humans, ~85% of the mouse genome is very similar to that of humans. Figure 7.13 shows

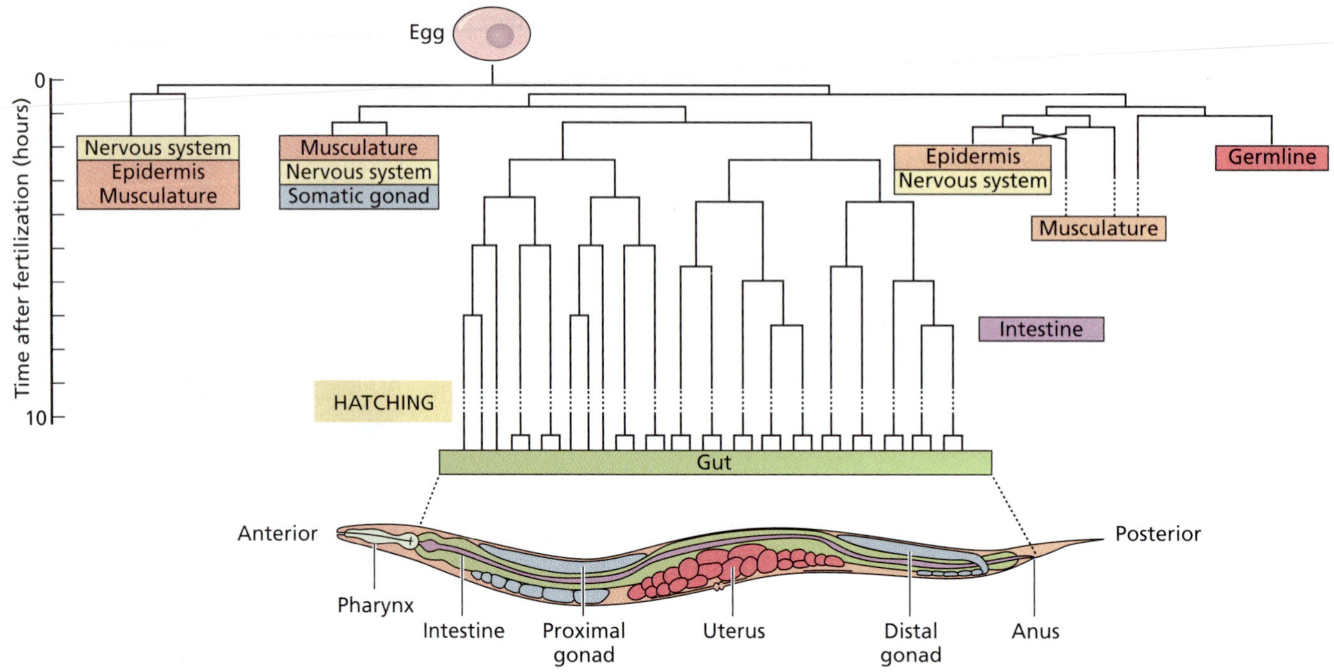

Figure 2.12 Lineage tree for cells that form the gut or intestine of *C. elegans*. The intestinal cells form a single clone, as do the germline cells; note that the cells of most other tissues do not. The schematic shows only a small portion of the entire lineage tree that gives rise to all cells in the adult organism. (Top, adapted from Alberts B, Johnson A, Lewis J et al. [2008] Molecular Biology of the Cell, 5th ed. With permission from Garland Science. Bottom, adapted from Sulston JE & Horvitz HR [1977] *Dev Biol* 56:110–156. With permission from Elsevier.)

a detailed comparison between a mouse and a human chromosome. By now, many purebred strains including those with specific genetic modifications are readily available.

Arabidopsis thaliana is a weed commonly known as thale cress. The plant is easy to grow and sexually matures in less than 6 weeks, producing ~5000 seeds per plant. This is the most commonly used plant model. First, it has a small genome, five pairs of chromosomes, whose entire sequence has been reported. Second, a large number of mutant lines and genomic resources or databases are available. Third, it is easy to transform by use of recombinant DNA technology techniques, including *Agrobacterium tumefaciens* (see Chapter 5).

Key concepts

- Gregor Mendel formulated the basic laws of genetics from experiments with garden peas.
- Mendel's first law states that alleles of a gene segregate independently.
- If two alleles are different, one dominates and determines the phenotype in the first hybrid progeny.
- Mendel's second law states that two traits, as dictated by pairs of alleles, will assort independently; genes are not linked.
- Thomas Hunt Morgan, in experiments on fruit flies, showed that genes are in fact sometimes linked.
- Linkage is proportional to the proximity of genes on a chromosome. This observation allows chromosome mapping.
- Experiments with bacteriophage and bacteria, plus the discovery of DNA structure, led to modern molecular genetics.
- In the development of molecular biology, certain model organisms have proved to be extremely useful.

Further reading

Books

Russell PJ (2005) iGenetics: A Molecular Approach, 2nd ed. Benjamin Cummings.

Sturtevant AH (2001) A History of Genetics. Cold Spring Harbor Laboratory Press.

Reviews

Amaya E, Offield MF & Grainger RM (1998) Frog genetics: *Xenopus tropicalis* jumps into the future. *Trends Genet* 14:253–255.

Botstein D, Chervitz SA & Cherry M (1997) Yeast as a model organism. *Science* 277:1259–1260.

Bradley A (2002) Mining the mouse genome. *Nature* 420:512–514.

Brenner S (1974) The genetics of *Caenorhabditis elegans*. *Genetics* 77:71–94.

Brenner S (1974) New directions in molecular biology. *Nature* 248:785–787.

Delbrück M (1945) Experiments with bacterial viruses (bacteriophages). *Harvey Lect.* 41:161–187.

Goffeau A, Barrell BG, Bussey H et al. (1996) Life with 6000 genes. *Science* 274:546–567.

Harland RM & Grainger RM (2011) *Xenopus* research: Metamorphosed by genetics and genomics. *Trends Genet* 27:507–515.

Herskowitz I (1988) Life cycle of the budding yeast *Saccharomyces cerevisiae*. *Microbiol Rev* 52:536–553.

Lederberg J (1946) Studies in bacterial genetics. *J Bacteriol* 52:503.

Lederberg J (1948) Problems in microbial genetics. *Heredity* 2:145–198.

Luria SE (1966) The comparative anatomy of a gene. *Harvey Lect.* 60:155–171.

Meinke DW, Cherry JM, Dean C et al. (1998) *Arabidopsis thaliana*: A model plant for genome analysis. *Science* 282:662–682.

Mitchison JM (1990) My favourite cell: The fission yeast, *Schizosaccharomyces pombe*. *Bioessays* 12:189–191.

Otto SP (2008) Sexual reproduction and the evolution of sex. *Nature Education* 1:182.

Sawin KE (2009) Cell cycle: Cell division brought down to size. *Nature* 459:782–783.

Visconti N & Delbrück M (1953) The mechanism of genetic recombination in phage. *Genetics* 38:5–33.

Wallingford JB, Liu KJ & Zheng Y (2010) *Xenopus*. *Curr Biol* 20:R263–R264.

Zhaxybayeva O & Doolittle WF (2011) Lateral gene transfer. *Curr Biol* 21:R242–R246.

Zhimulev IF, Belyaeva ES, Semeshin VF et al. (2004) Polytene chromosomes: 70 years of genetic research. *Int Rev Cytol* 241:203–275.

Experimental papers

Adams MD, Celniker SE, Holt RA et al. (2000) The genome sequence of *Drosophila melanogaster*. *Science* 287:2185–2195.

Blattner FR, Plunkett G III, Bloch CA et al. (1997) The complete genome sequence of *Escherichia coli* K-12. *Science* 277:1453–1462.

Evans MJ & Kaufman MH (1981) Establishment in culture of pluripotential cells from mouse embryos. *Nature* 292:154–156.

Hellsten U, Harland RM, Gilchrist MJ et al. (2010) The genome of the Western clawed frog *Xenopus tropicalis*. *Science* 328:633–636.

Lederberg J & Tatum EL (1946) Gene recombination in *Escherichia coli*. *Nature* 158:558.

Longo VD, Shadel GS, Kaeberlein M & Kennedy B (2012) Replicative and chronological aging in *Saccharomyces cerevisiae*. *Cell Metab* 16:18–31.

Morgan TH (1910) Sex limited inheritance in *Drosophila*. *Science* 32:120–122.

Mouse Genome Sequencing Consortium (2002) Initial sequencing and comparative analysis of the mouse genome. *Nature* 420:520–562.

Sturtevant AH (1913) The linear arrangement of six sex-linked factors in *Drosophila*, as shown by their mode of association. *J Exp Zool* 14:43–59.

Timmons L, Tabara H, Mello CC & Fire AZ (2003) Inducible systemic RNA silencing in *Caenorhabditis elegans*. *Mol Biol Cell* 14:2972–2983.

White RM, Sessa A, Burke C et al. (2008) Transparent adult zebrafish as a tool for *in vivo* transplantation analysis. *Cell Stem Cell* 2:183–189.

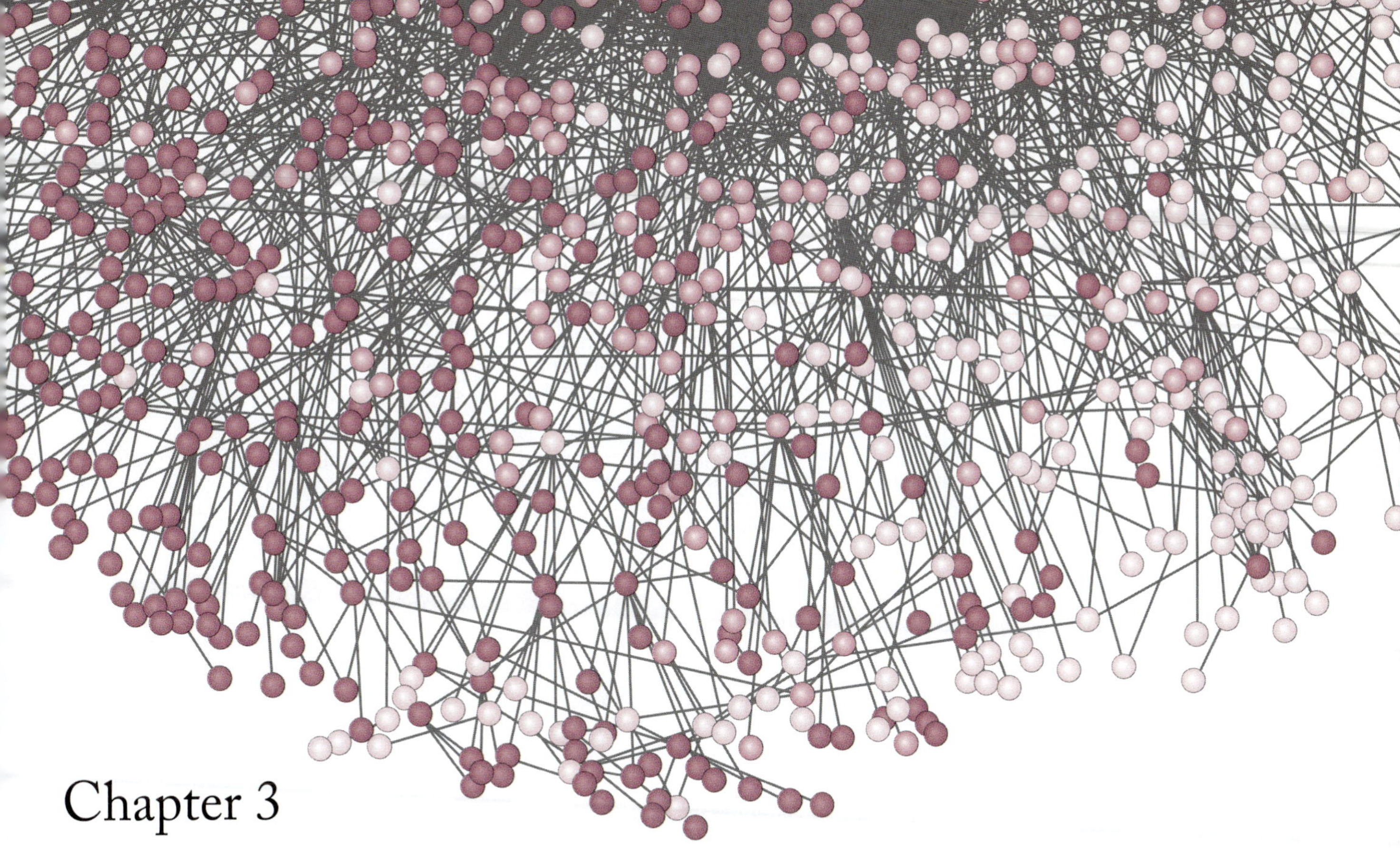

Chapter 3

Proteins

3.1 Introduction

Proteins are macromolecules with enormous variety in size, structure, and function

In the image of a cell shown in Figure 1.1, one class of macromolecules, proteins, is almost everywhere. As we see in this chapter, proteins are so widespread in biology because they are exceedingly versatile components, exhibiting enormous variety in both structure and function (**Table 3.1**). They are the most versatile and important working molecules in the cell.

All proteins are large molecules: some are very large, with masses in the millions of Daltons. Some are fibrous and highly extended and thus have mainly structural roles. Similarly, scaffold proteins connect and hold in place proteins whose functions are interconnected. A vast class of more compact proteins includes molecules that act as signals within and between cells, as transporters of small molecules, as regulators of cellular processes, and as enzymes. Enzymes are the catalysts responsible for facilitating the myriad chemical reactions that a living cell or organism utilizes in its metabolism and growth.

Proteins are essential for the structure and functioning of all organisms

It has been estimated that a typical human cell contains about a billion protein molecules, of over 20,000 different kinds. Some are present as only a few molecules per cell, while others are present in the millions. The directions for producing this vast array of complicated molecules are encoded in every cell of an organism, in its DNA. The pathways through which DNA communicates this information to direct the formation of protein molecules, how it selects which proteins are to be produced in each cell, and how that information is preserved from cell to cell and generation to generation is a major focus of molecular biology, the topic of this book. Proteins are one, but not the only, end product of information stored in DNA, but they are vital to the cell. Thus, it is important that we begin by finding out how proteins are structured, and how they use that structure.

Table 3.1 Major functions of proteins.

Group of proteins	Function	Characteristic structural features	Examples
enzymes	catalyze more than 4000 biochemical reactions, often with high specificity and enormous efficiency	the active site may have only a few amino acid residues in direct contact with the substrate, and usually only 3–4 residues are involved in actual catalysis	enzymes of the energy-producing cycles; enzymes involved in maintenance and flow of genetic information in the cell; digestive enzymes
structural proteins	organize intracellular structures; extracellularly, provide mechanical support to cells and tissues	often multisubunit proteins, in which individual subunits interact with each other to form fibers	inside cells, tubulin forms microtubules; actin forms actin filaments that support the plasma membrane; in the nucleus, histones form octameric protein cores around which DNA wraps to form structures known as nucleosomes; collagen and elastin are extracellular proteins that form fibers in tendons and ligaments
scaffold proteins	hold together proteins that are part of a signaling or catalytic pathway	multidomain; many apparently evolutionarily distinct types	scaffold proteins are essential to the signaling pathway that allows yeast to sense pheromones; Hsp70 and Hsp90 regulate sequential protein folding
transport proteins	carry small molecules or ions; those embedded in membranes carry substances through the membranes	changes in conformation of the binding site usually accompany binding of the transported molecule or ion	albumin in the bloodstream carries lipids; hemoglobin in red blood cells carries oxygen; transferrin carries iron; protein calcium pumps transport Ca^{2+} into muscle cells to trigger muscle contraction
motor proteins	move along fibers of proteins or molecules of nucleic acids to transport organelles or substances or to synthesize macromolecules	molecular motors are usually ATPases, enzymes that use energy from the hydrolysis of ATP or other nucleoside triphosphates to move or carry cargo along molecular tracks	myosin in muscle cells participates in muscle contraction by sliding on actin filaments; kinesin moves along microtubules to transport organelles or substances around the cell; dynein makes cilia and flagella beat or rotate; DNA and RNA polymerases move along DNA strands during synthesis of DNA or RNA
storage proteins	serve as depot for storage of small molecules and ions in certain cells or as repositories of amino acids for synthesis of other proteins	often resemble enzymes, having multiple binding sites, but lack catalytic function	ferritin stores iron in liver cells; ovalbumin in egg white of birds and casein in the milk of mammals are sources of amino acids for embryos or newborns; endosperm proteins in plant seeds feed germinating and developing embryos
signaling proteins	provide communication between and within cells	often highly specific in their interactions with cell surfaces or intracellular structures; belong to the class of proteins that are intrinsically disordered	hormones and growth factors circulate in the bloodstream to coordinate functions of individual cell types and tissues: insulin controls glucose levels in blood; epidermal growth factor stimulates cell growth and division in epithelial cells
receptor proteins	usually membrane proteins that detect signals (environmental or developmental) and transmit them to the cell interior to elicit appropriate cellular responses	usually dimerize in the membrane in response to signals	rhodopsin in the retina detects light; insulin receptor mediates cellular response to glucose by interacting with insulin

regulatory proteins	regulate multiple reactions such as transcription, by binding to DNA or proteins of transcriptional machinery, and translation, by binding to components of translational machinery	usually contain both sites for binding to target, such as DNA, RNA, or proteins, and sites that bind regulators	lactose repressor in bacteria binds to DNA regions encoding enzymes that participate in utilization of lactose; numerous transcription factors stimulate transcription of specific genes in eukaryotes in response to environmental and developmental cues; enzymes that introduce or remove postsynthetic modifications of proteins, thereby controlling their function
highly specialized proteins with various functions	highly variable functions depending on the organism and environment	often have unusual amino acids or modifications	antifreeze proteins in organisms that live in freezing environments; stress proteins that allow organisms to handle high temperatures or salinity; glue proteins that attach marine organisms to rocks; proteins with protective functions against foreign organisms, for example, antibodies

3.2 Protein structure

Amino acids are the building blocks of proteins

Proteins are polymers of α-amino acid monomers. The general structure of an α-amino acid is shown in **Figure 3.1**. These are **amino acids** in which the amine group is attached to the α-carbon or Cα, the one next to the carboxyl group. All amino acids contain this core, but they differ in the side chain R that is also attached to the Cα. Proline is technically an imino acid as a result of its cyclic side-chain structure. Glycine is an unusual amino acid because its side chain R is a single hydrogen atom and thus its Cα is a center of symmetry. The Cα of all other amino acids is an asymmetric center and therefore they have stereoisomers, designated D and L (see Figure 3.1). Only the L-isomer is found in native proteins, although some D-isomer amino acids do exist in living cells. Why nature chose only the L-isomer to make proteins remains a mystery.

The structures of the 20 classical amino acids that DNA codes for and specifies in the proteins of all living organisms are shown in **Figure 3.2**. Recently it has been shown that selenocysteine and pyrrolysine are also DNA-coded amino acids that are included in proteins, albeit rarely. These two unusual amino acids and their incorporation into protein chains are discussed in Chapter 15. The variety of side chains in the different amino acids allows a great variety of interactions with the solvent environment, other protein molecules, or other groups within the same protein molecule. Some side chains are aliphatic or aromatic; these will tend to be hydrophobic and so are typically packed into the interior of protein molecules. Others, such as those with acidic or basic groups, or their carboxamides such as glutamine, will be hydrophilic, preferring to associate with the water surrounding a protein. Other amino acids contain specialized groups, such as hydroxyl or sulfhydryl. Altogether, the variety in the side chains of

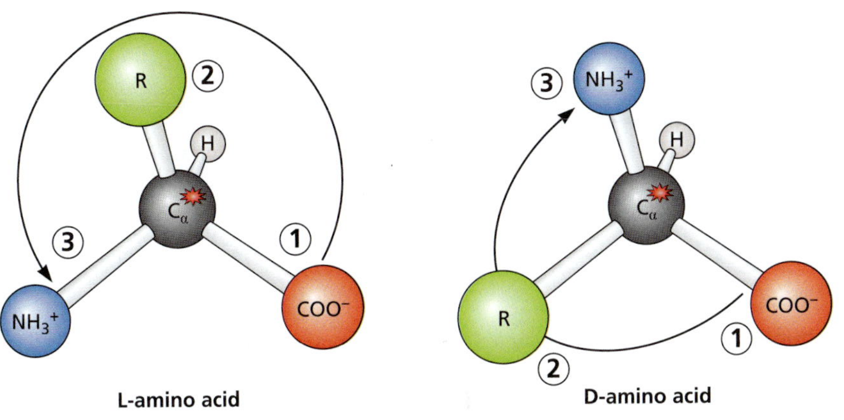

Figure 3.1 Chemical structure of α-amino acids. R denotes a chemical group that is unique to each amino acid. The asymmetric C atom satisfied its four valencies by the attachment of four different groups. Molecules that contain such an atom, known as a chiral center and designated here by a star, are optically active as they rotate the plane of polarized light in different directions. C-O-R-N law: Place the hydrogen behind the chiral center and count around from the COOH group to the R group to the NH_2 group. If these groups are arranged counterclockwise, the molecule is the L-stereoisomer; if clockwise, it is the D-stereoisomer. Proteins consist exclusively of L-amino acids, but D-amino acids are also found in living organisms. For example, D-alanine is a component of bacterial cell walls.

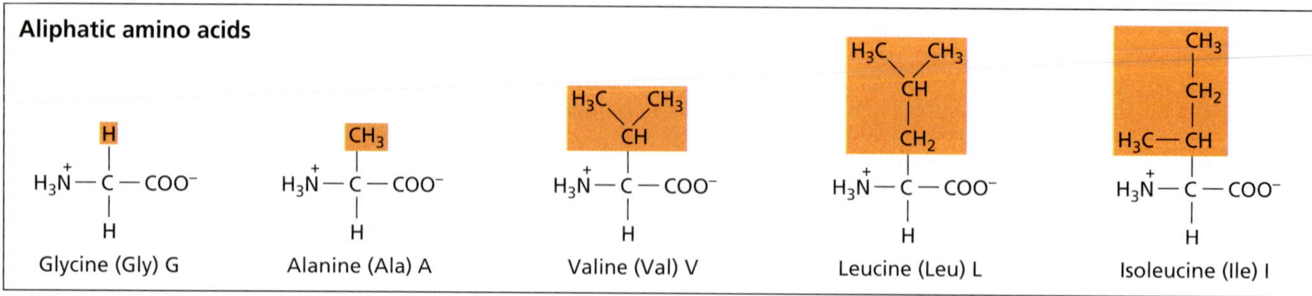

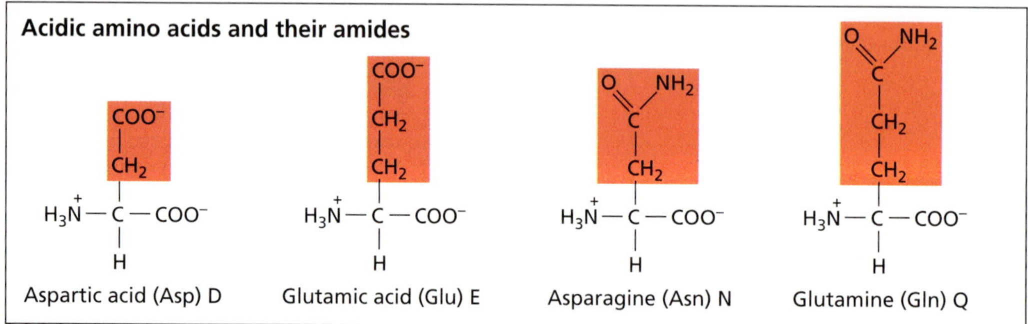

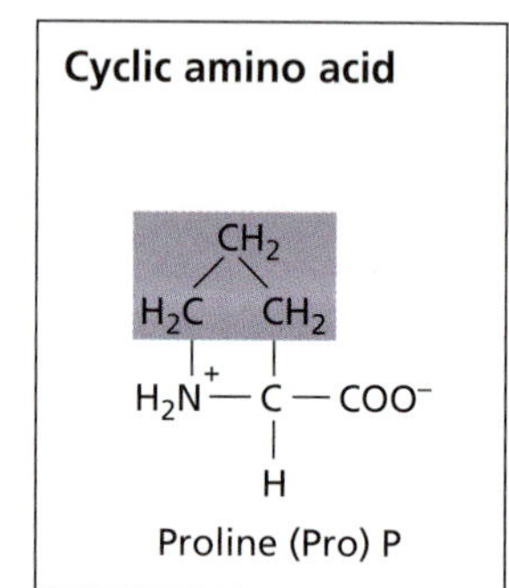

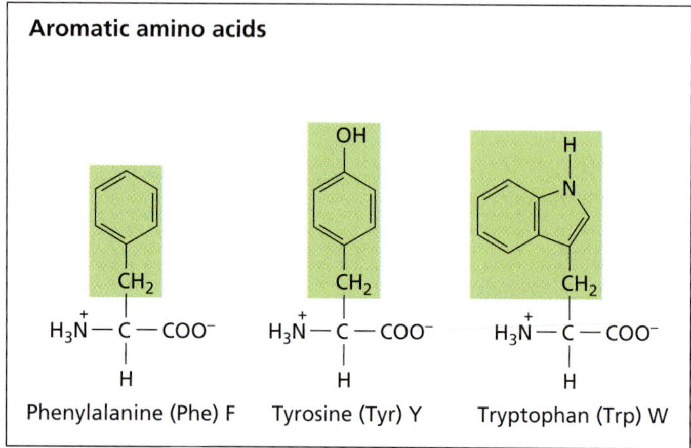

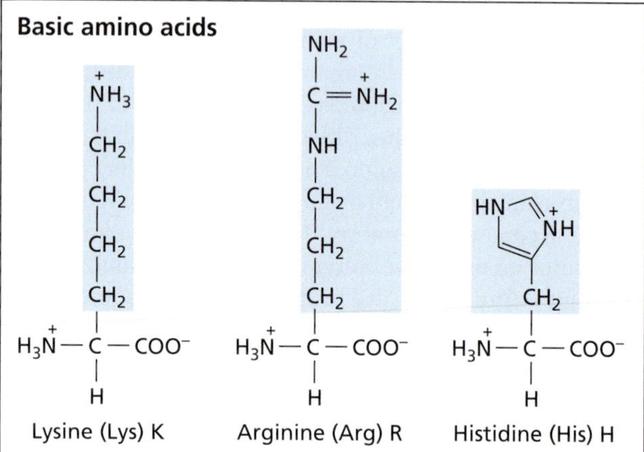

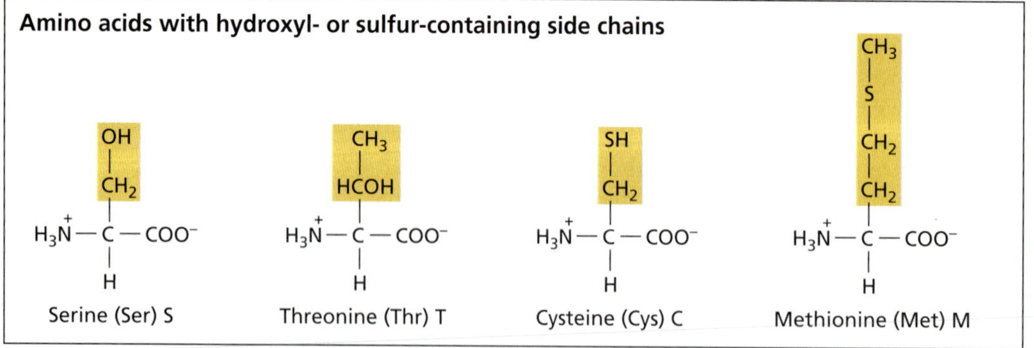

Figure 3.2 Classical amino acids found in proteins. The amino acids are grouped according to the chemical properties of their side chains. Each amino acid is presented by its full name, its three-letter abbreviation, and its one-letter code.

their amino acids provides protein molecules with a formidable tool kit for molecular interactions.

In proteins, amino acids are covalently connected to form polypeptides

The bonds that connect amino acids together into a protein polymer are called **peptide bonds**, and the polymer is a **polypeptide**. These bonds are formed between two monomers by a process that effectively eliminates a water molecule (Figure 3.3A),

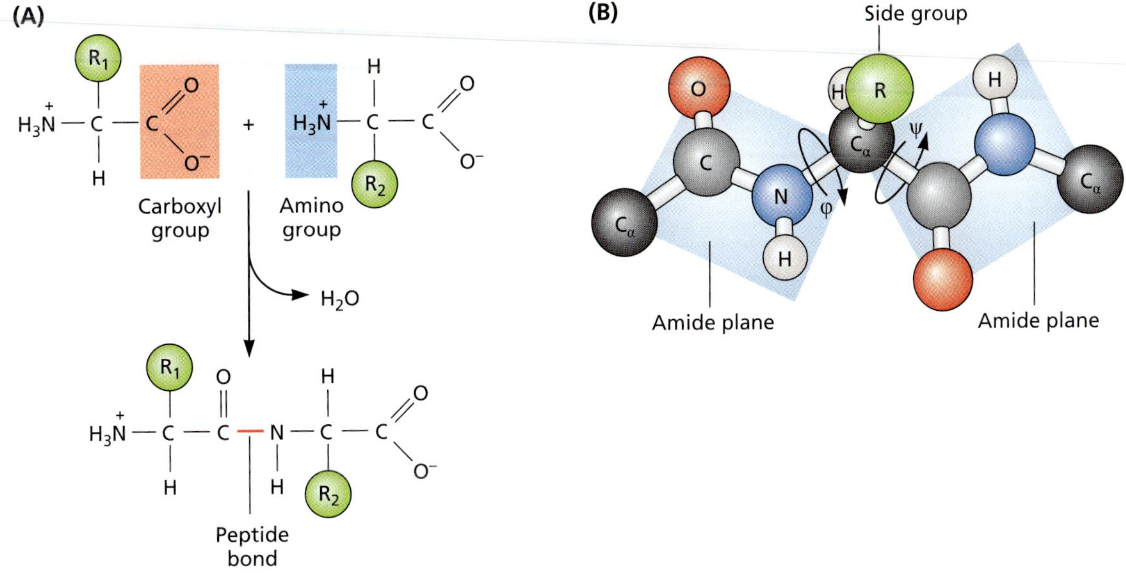

Figure 3.3 The peptide bond. (A) A peptide bond is formed between two amino acids by elimination of a molecule of water. (B) The peptide bond is planar and rigid. Rotation is possible only around the N–Cα bond (angle φ), and around the Cα–C bond (angle ψ). The group of atoms about the peptide bond could exist in two possible configurations, *cis* and *trans*. The *trans* configuration, in which the adjacent Cα atoms are placed further apart, is favored in proteins, as it avoids steric clashes between bulky R groups.

although in cells the reaction is actually much more complex and indirect. The remainder of the amino acid that remains within the polypeptide chain is called an **amino acid residue**. The reverse of this process, cleaving the peptide bond by adding a water molecule, is called **proteolysis** and is a specialized form of **hydrolysis**, or splitting by adding water. In aqueous solution, the hydrolysis of polypeptides occurs spontaneously, although the process is slow unless it is catalyzed by proteolytic enzymes. Consequently, the cell must use a special energy source to make proteins from amino acids. Spontaneous hydrolysis also explains why no protein lasts forever.

The peptide bond is rigid and planar because of the directionality of the N- and C-carboxyl orbitals (Figure 3.3B). However, rotation is relatively free about the C-carboxyl-Cα and Cα-N bonds. This allows a long polypeptide chain to have great flexibility and many possible conformations.

3.3 Levels of structure in the polypeptide chain

The primary structure of a protein is a unique sequence of amino acids

Proteins are linear polypeptides that have distinguishable ends (Figure 3.4). On one end, called the **N-terminus**, is an unreacted NH_2 group. The other end, with unreacted COOH, is the **C-terminus**. At physiological pH, these groups are usually charged: NH_3^+ and COO^-. The remarkable, and extremely important, characteristic

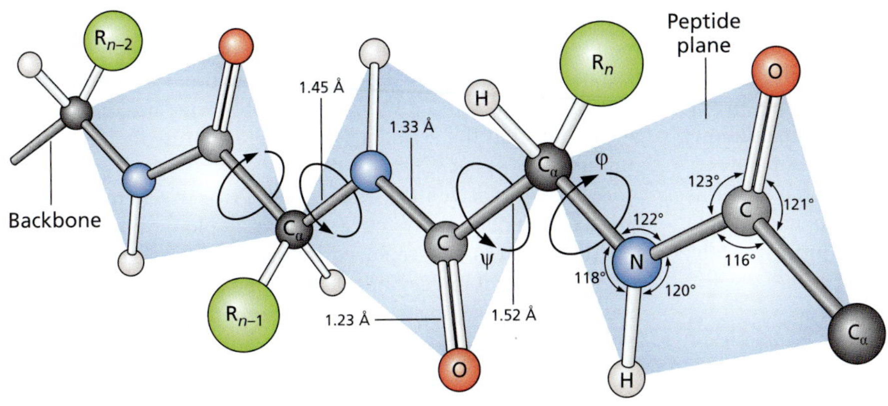

Figure 3.4 Extended polypeptide chain showing typical backbone bond lengths and angles. R_{n-2}, R_{n-1}, and R_n represent different side chains. (Adapted from Petsko GA & Ringe D [2004] Protein Structure and Function. With permission from Oxford University Press.)

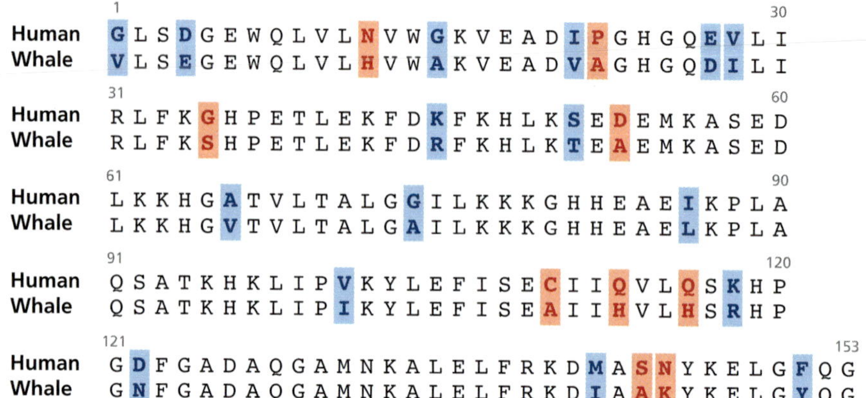

Figure 3.5 Amino acid sequences of myoglobin from two sources: human and sperm whale. Single-letter abbreviations are used. Numbering of amino acid residues along the polypeptide chain conventionally starts at the N-terminus. Myoglobin is highly conserved between these two species: 84% of the residues are identical. Another 16 of the amino acid substitutions are conservative, meaning an amino acid is replaced by an amino acid of the same chemical type; these are shown by blue boxes. Only nine substitutions are nonconservative, designated by red letters. There are no insertions or deletions.

of proteins is that each has a defined and unique sequence of amino acid residues. This sequence is, by convention, written from the N-terminus to the C-terminus, using either the three-letter or one-letter abbreviated codes given in Figure 3.2. This sequence of amino acid residues constitutes the **primary structure** of the protein. The primary sequences of myoglobin from the muscles of humans and sperm whales are shown as an example in Figure 3.5. If myoglobins from millions of people are sampled, most will have exactly the same sequence, although an occasional mutant variety, differing in one or more **residues**, might turn up. On the other hand, myoglobins from other animal species will have significantly different sequences. Other proteins, with completely different functions, will have very different sequences. There are examples, however, of proteins that have quite different sequences carrying out the same function in different organisms. An example is myohemerythrin

Box 3.1 Svedberg, the ultracentrifuge, and giant protein molecules In the 1920s, biochemists had begun to realize that proteins were polypeptides, but there was no clear agreement as to their size. In order to investigate this problem, biophysicist Theodor (The) Svedberg developed an entirely new kind of instrument, the analytical ultracentrifuge. This is simply a high-speed centrifuge equipped with an optical system that allows observation of the sedimenting substance. To investigate some smaller proteins, very high rotational speeds are needed, up to 100,000 revolutions per minute, and observation requires very clever design.

Even early studies with this instrument revealed that some proteins were *very* large, much larger than anyone had expected. Sedimentation velocity is expressed in terms of the sedimentation coefficient, where s = velocity of sedimentation/centrifugal field strength. The s value has the dimension of seconds, and 10^{-13} s is called one **Svedberg**, abbreviated as S. This gives an approximate measure of molecular size, and we still use the unit today; for example, 20S particles. Approximately, for spherical molecules:

$$s = M(1 - \bar{v}\rho)/6\pi\eta r N \qquad \text{(Equation 3.1)}$$

where M is molecular mass ρ is solution density, $\bar{v}$ is the partial specific volume (≈1/protein density), η is solvent viscosity, r is radius of the particle, and N is Avogadro's number (6.02×10^{23} molecules/mole).

Svedberg found that some proteins had sedimentation coefficients as large as 100S, which corresponded to molecular masses in the millions of Daltons. More exact values were later obtained by a variation of the technique called sedimentation equilibrium, in which sedimentation at lower speeds was allowed to continue until the protein came to equilibrium against back-diffusion. These studies proved beyond a doubt the immense size of many proteins. They also showed unequivocally that many proteins were homogeneous in molecular mass, all molecules of a given protein having the same size. This was a first hint that the sequences of different proteins might be unique.

Svedberg received the 1926 Nobel Prize in Chemistry for "his work on disperse systems." Svedberg's pioneering method has now been largely replaced by mass spectrometry for the measurement of molecular mass (see Box 3.9), but it still has many uses in studying macromolecular interactions and quaternary structure.

Box 3.2 Electrophoresis: General principles The use of **electrophoresis** is a powerful and versatile method for the analytical separation of biopolymers. It is fair to say that modern molecular biology could not have developed without this technique, and it remains probably the most widely used method. It is extensively employed, in a wide variety of variants, to separate both proteins and nucleic acids.

The basic idea is very simple. If a protein carrying a net electrical charge is placed in an electric field, there will be a net force on it that causes it to move toward the electrode of opposite charge. The molecule will accelerate until, very soon, its velocity is large enough that the drag force from the surrounding medium just balances the electrical driving force: at this point, the molecule is at steady-state velocity. The driving force is given by ZeE, where Z is the number of units of electron charge e, + or −, on the molecule and E is the electrical field strength. The drag force is fv, where v is velocity and f is the frictional coefficient, which depends on the size and shape of the protein molecule.

Thus, at steady-state velocity:

$$fv = ZeE \text{ or } v/E = Ze/f \text{ or } U = Ze/f \quad \text{(Equation 3.2)}$$

The velocity per unit field is termed the electrophoretic mobility U. This definition is much like that of the sedimentation coefficient (see Box 3.1): each involves the ratio of driving force to frictional resisting force. Nowadays, we do not generally measure absolute mobilities but carry out electrophoresis in some kind of gel matrix and measure the mobility (U_i) relative to that of some small, rapidly moving dye molecule (U_d). The relative mobility of component i is defined as

$$U_{ri} = U_i / U_d \quad \text{(Equation 3.3)}$$

If protein molecules are migrating in a gel matrix, in an apparatus like that shown in Figure 1, they will be separated into discrete bands and their relative mobilities can be calculated from the distance d_i each has traveled after some time of electrophoresis relative to the distance traveled by the tracking dye band:

$$U_{ri} = d_i / d_d \quad \text{(Equation 3.4)}$$

The fact that electrophoresis is usually carried out in some kind of gel matrix introduces another complicating but useful factor into the analysis. A gel will have a sieving effect, depending on its concentration and the size of the protein molecules. Small molecules can slip through the gel almost as easily as they move through pure solvent, while large or asymmetric molecules have more impediments. If we experimentally graph the log of relative mobility versus the concentration of the gel, we usually observe a straight line. Such graphs, known as **Ferguson plots**, are shown in Box 3.3 for several different kinds of molecules. The limit at zero gel concentration corresponds to what would have been observed in pure buffer, where mobility is determined only by Z and f. The slope of the line depends on the protein's molecular size and shape, being steeper for large and asymmetric molecules. These general principles can be used to devise different kinds of separations for different biopolymers or different problems.

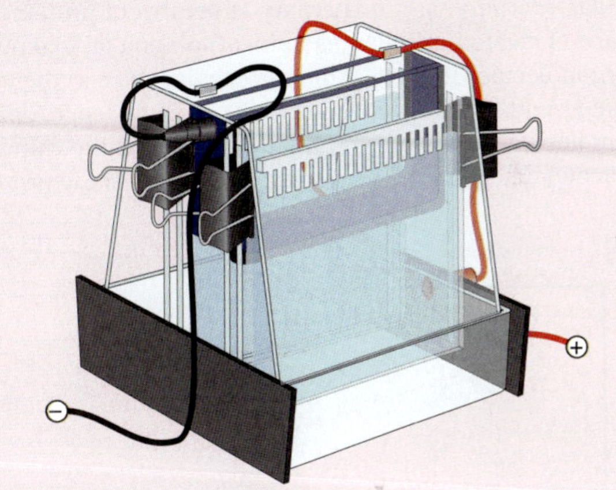

Figure 1 Typical apparatus for vertical polyacrylamide gel electrophoresis. The gel is cast between the two glass plates, which are separated by thin plastic spacers and held together by the paper clips. The comb at the top will be removed, leaving wells into which the samples will be loaded.

from certain primitive invertebrates. Like myoglobin, myohemerythrin stores oxygen in tissues, but its sequence is altogether different from that of myoglobin. Evolution has sometimes found different routes to the same function. Conversely, there are also many examples of proteins that have similar sequences but carry out quite different functions.

Myoglobin, with only 153 residues, is a small protein. Most proteins contain about 200–300 residues, and some are as large as 4000 residues. The number of possible proteins is almost infinite; theoretically, 200 residues can be arranged into 20^{200} different polypeptides. In the 1920s, the discovery that proteins were indeed large, covalently linked molecules (Box 3.1) came as a shock to many scientists, who were accustomed to the small molecules of organic chemistry. The later discovery that proteins have unique sequences was of equal or even greater significance. A number of methods were originally used to sequence proteins directly, whereas today, almost all protein sequences are deduced from their encoding gene sequences. One technique that remains of importance in analyzing proteins is **gel electrophoresis** (Box 3.2 and Box 3.3). Box 3.4 illustrates another set of widely used techniques, those that rely on specific interactions between proteins and antibody molecules.

Box 3.3 Electrophoresis: Techniques for protein electrophoresis

Native gels

The simplest technique is to simply load a solution containing a mixture of proteins onto a gel like that shown in Box 3.2. The gel is saturated with a buffer that maintains the proteins in their native, undenatured state, providing the possibility of detecting associations between molecules that might be of significance *in vivo*. By staining the gels with a quantitative stain, relative protein concentrations can be measured by the intensity of the stained bands. The disadvantage of native gels is that they tell virtually nothing about the molecule. Ferguson plots can be used to obtain a protein's free mobility (**Figure 1**), but that parameter depends on both charge and size. Even the order in which proteins appear after migration may depend upon gel concentration.

SDS gels

Detergents such as sodium dodecyl sulfate, SDS, destroy the secondary, tertiary, and quaternary structures of proteins to produce elongated micelles, each incorporating an extended protein molecule. The size of the micelle depends on the molecular weight of the protein. Because each attached detergent molecule carries one negative charge, and the number of attached detergent molecules is proportional to the protein chain length, the charge will also be proportional to protein molecular weight, as the residual charge on the protein is usually negligible compared to the micellar charge. Thus, SDS separates particles whose charge and size, and hence frictional coefficient, are both proportional to the molecular weight of the protein in the micelle. So according to Equation 3.2, all such particles should have about the same free mobility at a gel concentration of zero. Nevertheless, their Ferguson plots will be very different, with slope increasing with molecular weight (see Figure 1C). Thus, at any given gel concentration, we can sort proteins by molecular weight (see Figure 1D). Furthermore, if we include on the gel a lane containing a mixture of known proteins, we can calibrate the scale and thus measure the approximate molecular weights of the unknown proteins.

Note, however, that some information has been lost by denaturing the proteins. In particular, the multichain quaternary structure of proteins—other than multisubunit proteins that are held together by covalent bonds, such as disulfide bonds—will have been destroyed. This can be checked by repeating the experiment in the presence of a reducing agent, which will cleave these bonds.

SDS gel electrophoresis has been an important tool in molecular biology but is now being supplanted rapidly for many purposes by mass spectrometry, MS (see Box 3.9), which is much more accurate and versatile. Nevertheless, gel

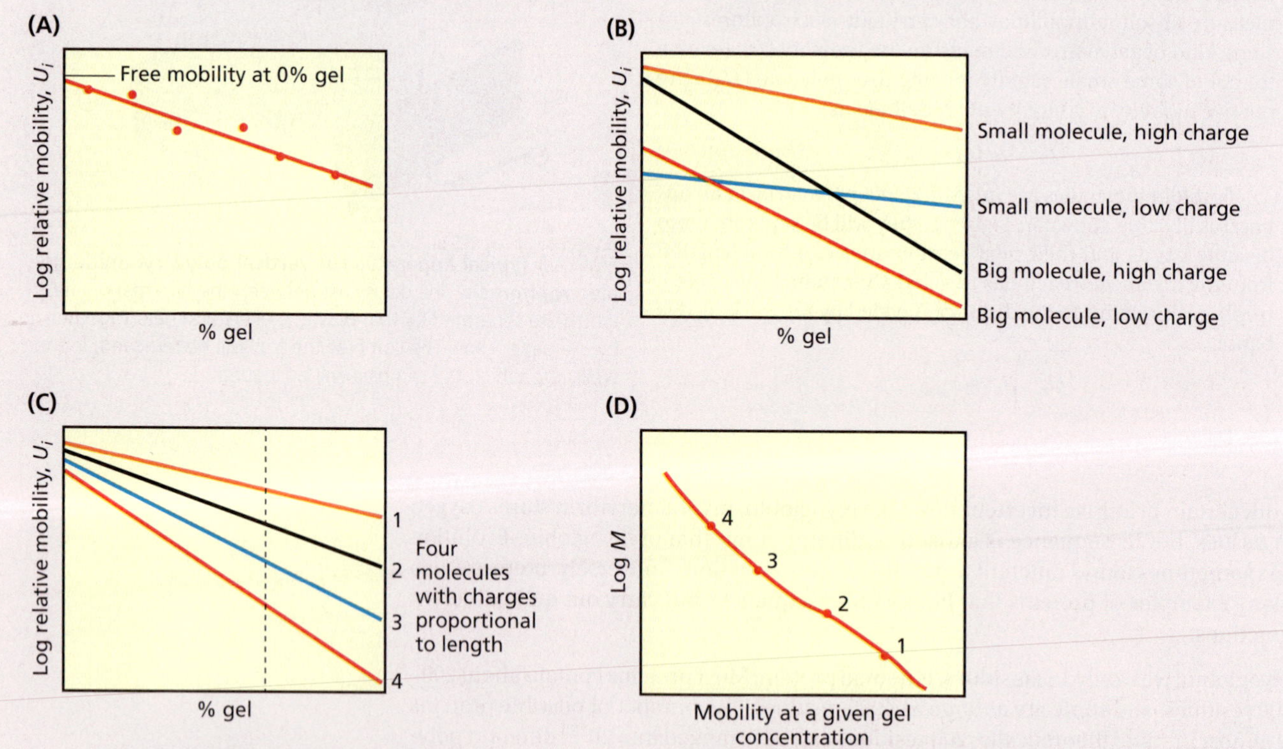

Figure 1 Mobility of molecules in gel electrophoresis. A Ferguson plot shows the log of relative mobility U_i as a function of concentration of the gel matrix. (A) Ferguson plot for a single protein: extending the plot to 0% gel gives the theoretical free mobility of that molecule. (B) Ferguson plots for four proteins that differ in size and charge: the free mobility depends mainly on charge, but the slope of the lines depends mainly on size. (C) When the molecular charge is proportional to the length of the molecule, the free mobilities are almost the same but the sieving effect of the gel is more pronounced for longer molecules (the molecules are numbered in order of increasing length and charge). (D) Relationship between molecular weight M and mobility, at a given gel concentration. If log M is plotted against mobility at a given gel concentration for molecules of known M, the standard curve obtained can be used to determine the molecular weight of a molecule of interest. The graph shown is constructed from gel concentration values from part C, black broken line. Points are numbered to correspond to the lines in part C.

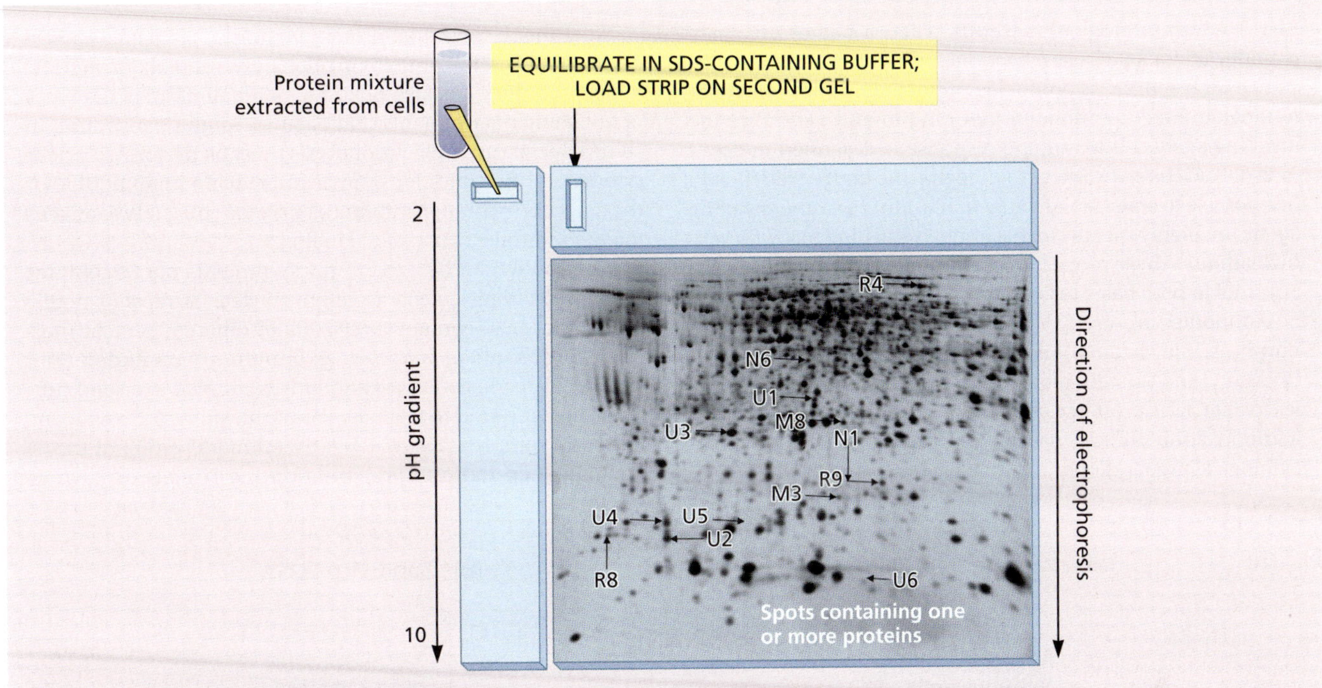

Figure 2 Two-dimensional gel electrophoresis for separation of complex protein mixtures. The isoelectric-focusing first-dimension gel is not stained for proteins because the stain interferes with the second-dimension gel electrophoresis. (Adapted from Jagadish SVK, Muthurajan R, Oane R et al. [2010] *J Exp Bot* 61:143–156. With permission from Oxford University Press.)

electrophoresis is a simple, inexpensive method that can be employed in any lab, whereas MS requires expensive complex equipment.

Isoelectric focusing and two-dimensional gels

Each protein contains a unique constellation of acidic and basic side chains. The charged state of each of these, +, –, or 0, will depend on the solution pH and will change as the pH is adjusted. There will be some pH at which the + and – charges exactly balance, so the net charge is zero. This is known as the **isoelectric point** of the protein. It is possible to fractionate proteins according to their isoelectric points by creating a gel across which there is a gradient in pH. In such a gel, each protein will concentrate at its own isoelectric pH where no electrical force is acting on that molecule.

Today, such fractionation is generally used as a part of a more powerful scheme, termed **two-dimensional gel electrophoresis**. As shown in Figure 2, a mixture of proteins is first separated by **isoelectric focusing** in a tube or narrow strip. This is then attached to a second-dimension gel slab that contains SDS, and electrophoresis is conducted in the perpendicular direction. Here, each of the proteins will be separated on the basis of molecular weight. Thus, each protein ends up on the slab at a point determined by both size and relative number of negatively and positively charged groups. It is possible, in some cases, to display the whole proteome of a bacterium in this way. Furthermore, points can be picked from such a display for high-resolution mass spectrometry (MS) (see Box 3.9).

A protein's secondary structure involves regions of regular folding stabilized by hydrogen bonds

Although the polypeptide chain is capable of great flexibility (see Figure 3.4), it also exhibits levels of defined folding, which are known as secondary and tertiary structure. Essential to the stability of these structures are noncovalent interactions between portions of the polypeptide chain. These interactions are summarized in Table 3.2. All are weaker than covalent bonds, with the hydrogen bond being the strongest. Electrostatic interactions between positive and negative charges on side chains can also be important. At physiological pH, arginines, lysines, and most histidines are positively charged, whereas aspartic and glutamic acids carry negative charges. Van der Waals interactions include a number of ways in which neutral molecules can attract one another, for example, via dipoles or charge fluctuations.

The term **secondary structure** is reserved for regions of regular folding within the protein chain that are stabilized by hydrogen bonds between amide hydrogens and

Box 3.4 Immunological methods When a higher organism is under attack by viruses, bacteria, or other extraneous substances, it responds by producing protein molecules called immunoglobulins or antibodies that bind to and neutralize the invading agents. The immune response is described in detail in Box 21.6. Here, it suffices to note that the antibodies elicited in response to a particular antigen, introduced by the invading agent, are highly specific to that antigen and bind to it with very high affinity. These properties of antibodies make them a valuable tool in both basic research and clinical practice.

Antibodies are used in a variety of ways and for different purposes. The enzyme-linked immunosorbent assay or ELISA (**Figure 1**) and Western blot (**Figure 2**) techniques use antibodies to detect the presence of a particular antigen in a sample. In addition, antibodies are used to localize a particular antigen in cytological preparations.

Note that if the initial protein mixture is subjected to SDS electrophoresis, the proteins are denatured and may lose some antigen binding; see also Box 21.6 for a definition of the types of antigenic determinants that elicit an immune response. In some special cases, electrophoresis must be performed under nondenaturing conditions. An example of an immunoblot for the detection of human immunodeficiency virus (HIV) antigens in blood samples is presented in Box 5.6.

Antibodies can be used to purify proteins out of complex biological samples, such as cell or nuclear extracts or bodily fluids, that may contain thousands of different protein molecules. The method is known as **immunoprecipitation** (**IP**). In recent versions of IP, the antibody is coupled to a solid substrate of some kind to facilitate the purification of the antigen–antibody complex (**Figure 3**). A useful modification of the IP technique, **co-immunoprecipitation** (**co-IP**), also known as

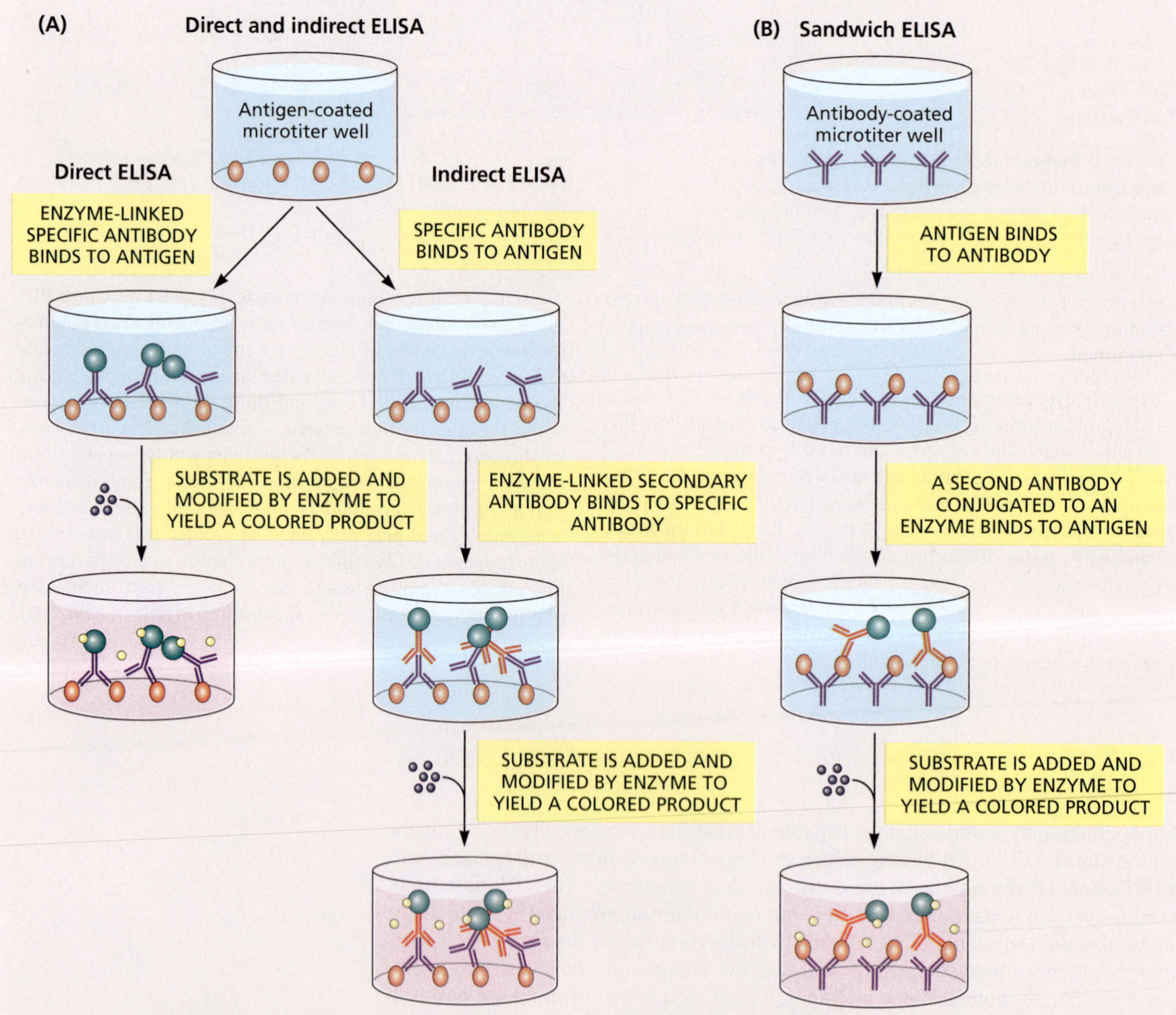

Figure 1 Enzyme-linked immunosorbent assay (ELISA). The ELISA technique uses microtiter plastic plates that contain 96 small wells, so that 96 reactions can be carried out simultaneously. The final readout of the assay is visible color, which is automatically quantified by special ELISA reader instruments. Depending on whether the first molecule immobilized on the walls of the microtiter wells is an antigen or an antibody, we discern two major types of assays: (A) direct and indirect ELISA or (B) sandwich.

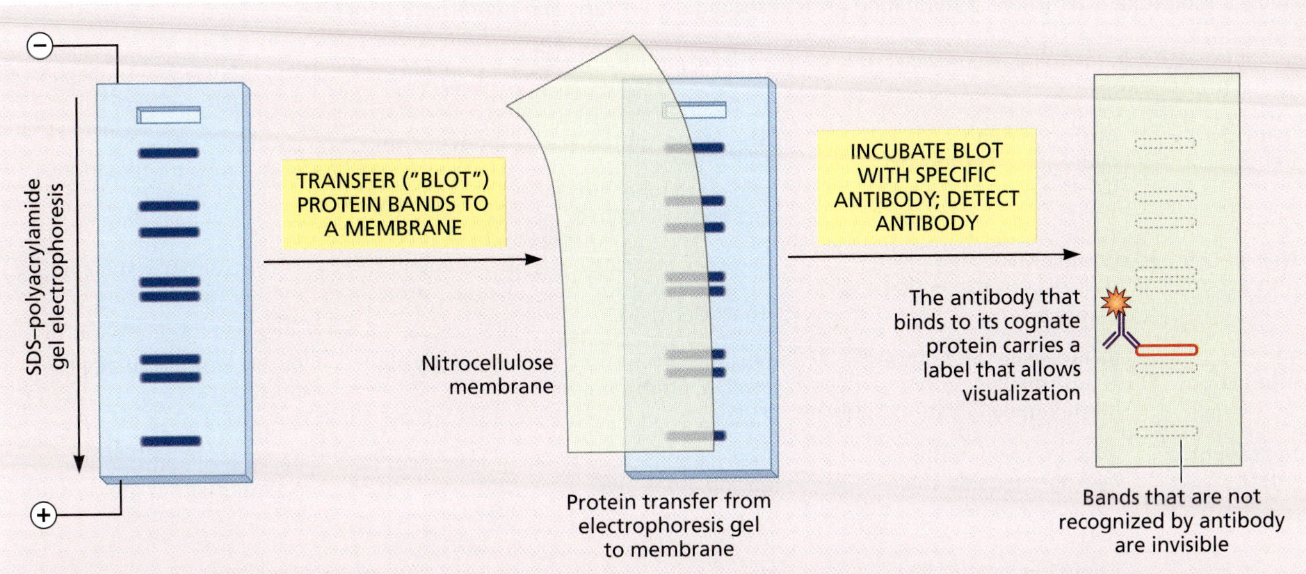

Figure 2 Immunoblotting, also known as Western blotting. The name Western blot mimics the name Southern blot, a technique for transfer of nucleic acids from electrophoretic gels to membranes introduced by Edwin Southern and named after him. Note that if the initial protein mixture is subjected to SDS electrophoresis, the proteins are denatured and may lose some antigen binding. See Box 21.6 for a definition of the types of antigenic determinants that elicit immune response. In some special cases, electrophoresis must be performed under nondenaturing conditions. An example of an immunoblot for detection of HIV antigens in blood samples is presented in Box 5.6.

pulldown), can identify interacting proteins or protein complexes present in complex samples: by immunoprecipitating one protein member of a complex, additional members of the complex may be captured and can be identified. Identification of proteins in the pellet can be done by Western blots, if respective antibodies to the presumed protein partners exist, or by sequencing protein bands purified from electrophoretic gels. The advantages of co-IP include the fact that the proteins are in their native state and in the physiologically relevant stoichiometry. The main disadvantages are that (i) only stable interactions can be detected; (ii) it is unclear whether the interactions are direct or through other members of the complex; and (iii) mixing of cellular compartments during cell lysis may lead to identification of interactions that do not occur *in vivo* due to protein compartmentalization.

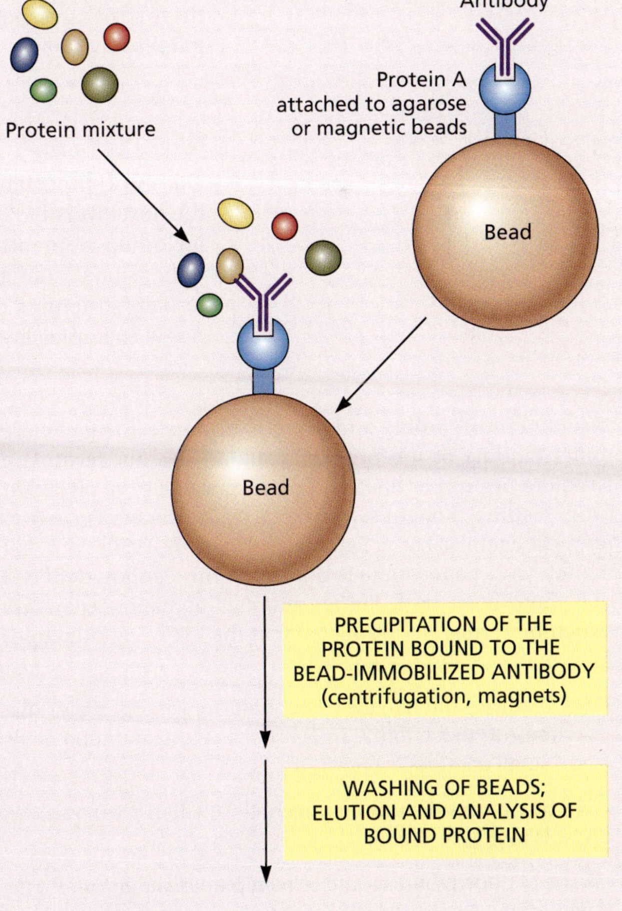

Figure 3 Immunoprecipitation (IP). The complex mixture of proteins in a cell lysate or bodily fluid is incubated with antibodies that are specific for the protein of interest. The antibody is attached to protein A immobilized on some solid support, usually resin or magnetic beads. Protein A is found in the cell wall of the bacterium *Staphylococcus aureus* and has the ability to bind immunoglobulins through their common Fc domains (see Box 21.6). The protein of interest interacts with the antibody, and the entire bead–protein A–antibody–antigen complex can be conveniently separated from the unreacted proteins. This is done by simple low-speed centrifugation in the case of agarose beads or by applying magnets to the tube wall in the case of magnetic beads. The precipitated complex is washed with buffer to remove any contaminating proteins, and the proteins attached to the antibody are eluted and analyzed.

Table 3.2 Molecular interactions determining protein secondary, tertiary, and quaternary structure.

Type of interaction	Participants in interaction	Typical amino acid side chains	Range of interaction (nm)	Comment
hydrogen bonds	hydrogen donors (D) and hydrogen acceptors (A): (D)–H···(A)	asparagine, glutamine, serine, threonine, lysine, arginine; also backbone H–N, O=C	about 0.3 nm (D–A)	strongest of the noncovalent interactions
charge–charge interactions	positively and negatively charged groups on side chains: ...(+)...(–)...	lysine (+), arginine (+), glutamic acid (–), aspartic acid (–)	0.5–2.0 nm "long" range	may be either stabilizing (+/–) or destabilizing (–/–, +/+)
van der Waals interactions	all molecules and groups; arises from electron inhomogeneity in molecules	most important for small, aliphatic amino acids	about 0.3 nm	become repulsive at very short distances
hydrophobic effect	all amino acids with hydrophobic side chains	phenylalanine, leucine, isoleucine	not defined; close packing of residues	packing of hydrophobic residues within the protein, away from water

carbonyl oxygens in the polypeptide backbone. As described in **Box 3.5**, Linus Pauling proposed a number of possible secondary structures that both satisfied the planarity of the **peptide** group (see Figure 3.3B) and allowed a maximum of hydrogen bonding. These have since been confirmed repeatedly in structures of real proteins. The most important secondary structures are the **α-helix** and **β-sheet**, as shown in **Figure 3.6**. In the α-helical conformation, hydrogen bonds are made between residues along the α-helix; in the β-sheet structure, hydrogen bonds are between parallel or antiparallel folds of the chain. The side chains of the residues do not play a direct role in forming these structures, but by steric effects they may favor one or the other. This can be made graphically clear in Ramachandran plots, as shown in Box 3.5.

Each protein has a unique three-dimensional tertiary structure

The most important component of structure in determining a protein's function is the next level of folding, termed the **tertiary structure**. The whole polypeptide chain, including elements of defined secondary structure, is folded into a unique three-dimensional configuration. There would seem to be an almost limitless variety of tertiary structures that can be formed by polypeptide chains of different sequence. A few examples are shown in **Figure 3.7**. In recent years, it has become possible to

Box 3.5 Linus Pauling and protein structure The first clear understanding of the principles underlying the secondary structure of polypeptides and proteins came from remarkably perceptive thought experiments by the famous American physical chemist Linus Pauling. In the early 1950s, working mainly with paper models and intuition, Pauling deduced that there must be a limited number of stable secondary structural forms. He postulated that the following principles must hold:

- Bond angles and lengths must conform, without major distortions, to those found in X-ray structures of amino acids and small peptides.
- No two atoms should approach closer than their van der Waals radii (see Table 3.2).
- Each peptide group should remain planar and in *trans* configuration (see Figure 3.3B), as had been observed in studies of small peptides. This permits rotation only about the C-Cα and N-Cα bonds.
- The overall structure must allow a maximum amount of hydrogen bonding, for stabilization.

Pauling and his collaborators found only a small number of chain configurations that satisfied all of these requirements. The two most important were the α-helix and the β-sheet (see Figure 3.6). Most remarkably, Pauling's completely theoretical predictions were accurately verified when the first X-ray diffraction studies of globular proteins were obtained (see Box 3.6). This can be demonstrated quite directly by using a graphical presentation devised by physicist/biochemist Gopalasamudram Ramachandran. Because the pair of phi/psi (φ/ψ) angles at any residue define the conformation at that residue (see Figure 3.3B), one may map these conformations on a graph with coordinates φ and ψ, as shown in **Figure 1**. Because of steric interference, not all parts of the map are available, and this depends somewhat on side chains. When one maps the individual φ/ψ pairs found from the X-ray structure of a particular protein, most of the points fall close to that expected for a right-handed α-helix or for β-sheet structures, thus

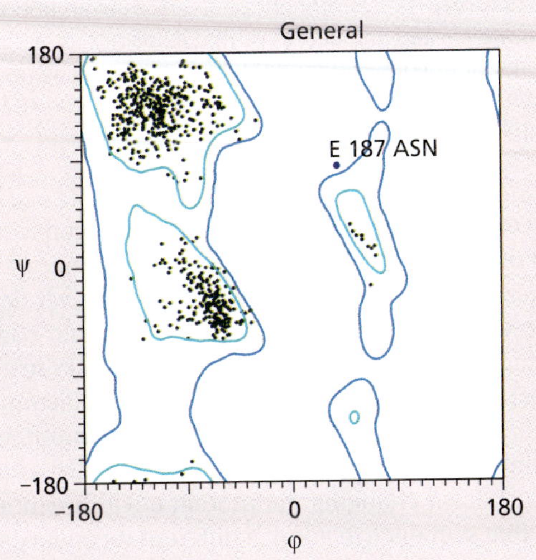

Figure 1 Ramachandran plot. This plot was generated from X-ray diffraction study of human PCNA, a clamp protein involved in DNA replication. The regions on the graph defined by the inner and outer contour lines respectively denote the ranges of conformations favored or simply allowed by steric restrictions. The dots correspond to actual φ/ψ value pairs for different residues in the protein. Note that they cluster in two regions: near φ = −57°, ψ = −47° and near φ = −129°, ψ = +125°. The first set of coordinates is as predicted by Pauling for a right-handed α-helix, and the second lies between two predicted β-sheet structures. (Courtesy of Jane Richardson, Duke University.)

confirming Pauling's vision. Pauling was awarded the Nobel Prize in Chemistry in 1954 for "his research into the nature of the chemical bond and its application to the elucidation of the structure of complex substances" (**Figure 2**).

Figure 2 Julian Voss-Andreae's α-helix for Linus Pauling. Powder-coated steel, 2004, height 10 ft (3 m). The sculpture stands in front of Pauling's childhood home on 3945 SE Hawthorne Boulevard in Portland, Oregon. (Courtesy of Julian Voss-Andreae, Wikimedia.)

determine such structures, often to atomic detail, by the techniques of X-ray diffraction (**Box 3.6**) and multidimensional nuclear magnetic resonance (NMR) (**Box 3.7**). To date, thousands of structures have been determined and can be found in publicly available protein data banks. The question that must be asked is: if so many structures are possible, why does each protein have its unique one?

The specific tertiary structure that a protein will adopt is dictated by noncovalent interactions, primarily between side chains but also with the surrounding water molecules. Characteristics of these interactions are given in Table 3.2. Of equal importance is the fact that hydrophobic residues tend to pack into the core of the molecule, away from water; this is known as the hydrophobic effect (**Figure 3.8**). The hydrophobic effect is not truly a force, like van der Waals interactions, but a consequence of the effect of hydrophobic groups on water structure. If inserted into water, they tend to arrange surrounding water molecules into rigid cages. Taking hydrophobic groups out of water into a protein interior releases these water molecules into their natural freedom. The increase in entropy associated with such freedom drives hydrophobic residues into the interiors of proteins. Hydrophilic residues, on the other hand, tend to be located on the surface of the protein molecule, in contact with surrounding water molecules (Figure 3.8). A very important consequence of these interactions is that the primary structure—that is, the amino acid

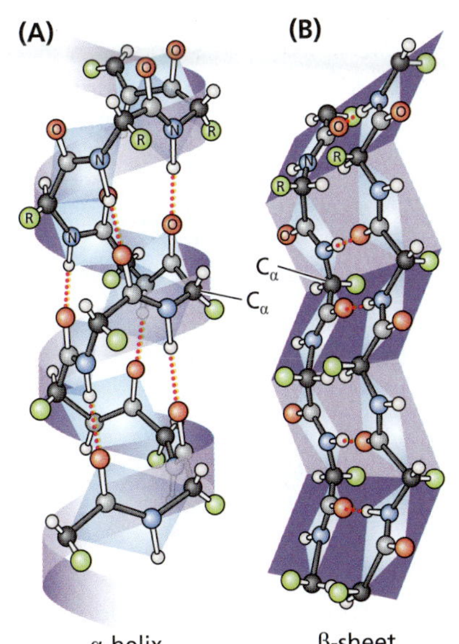

Figure 3.6 α-Helix and β-sheet: The two most common regular secondary structures in proteins. These structures differ mainly in how their stabilizing hydrogen bonds are formed. (A) In the α-helix, the hydrogen bonds are within a single length of coiled polypeptide chain and are oriented almost parallel to the helix axis. The number of residues in this helix is 3.6/turn of the helix, which corresponds to 0.54 nm/turn. (B) In the β-sheet, the hydrogen bonds occur between spatially adjacent chains. These could belong to the same polypeptide chain folded back upon itself or to two individual chains of proteins that have multisubunit structures, as in quaternary structure. In the β-sheet, the hydrogen bonds are almost perpendicular with respect to the chain direction. (Adapted courtesy of Geis Archives, Howard Hughes Medical Institute Art Collection.)

(A)

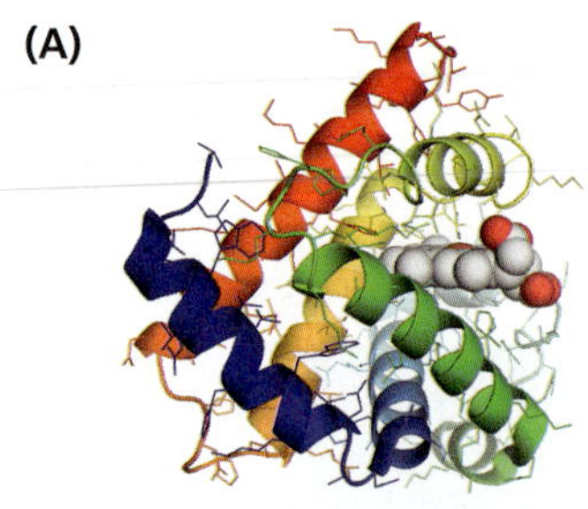

Sperm whale myoglobin

(B)

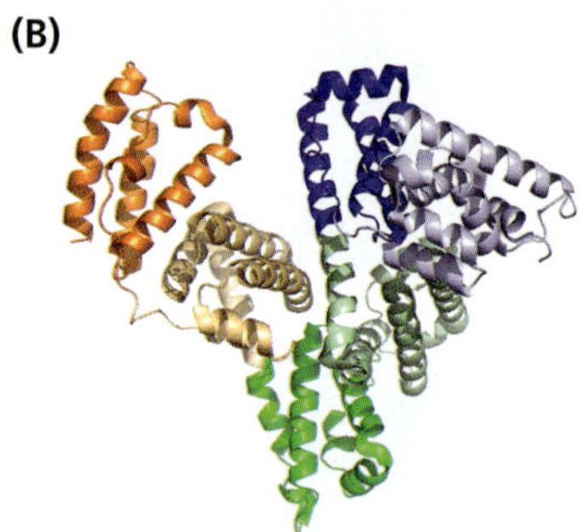

Human serum albumin

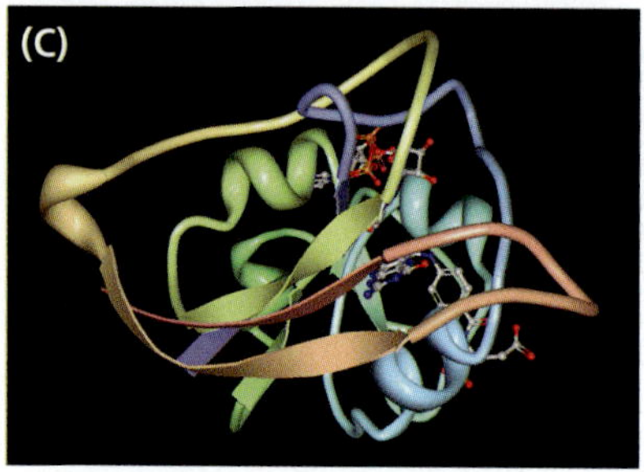

Dihydrofolate reductase

(D)

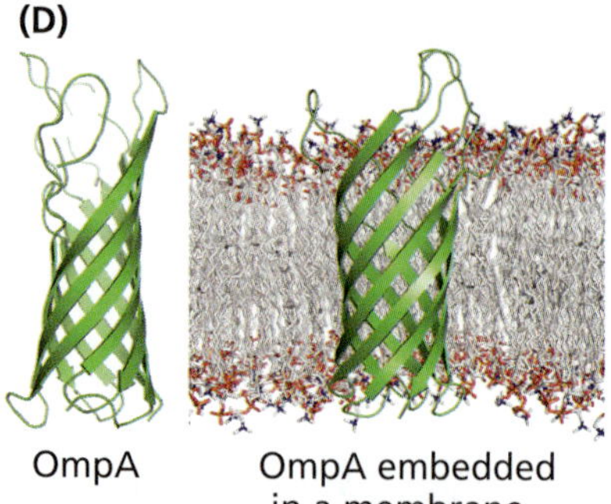

OmpA OmpA embedded in a membrane

(E)

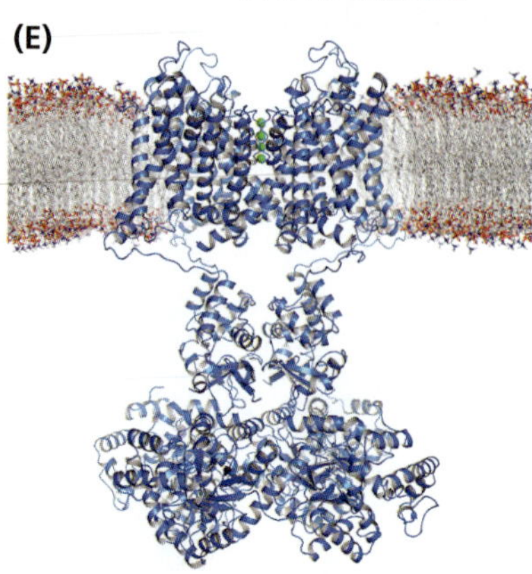

Potassium channel embedded in a membrane

Figure 3.7 Gallery of protein structures. The convention used here employs ribbons to represent secondary structures that are twisted in α-helices or that lie parallel or antiparallel in β-sheet structures; thin lines represent regions that do not have a simply defined secondary structure. (B, courtesy of Chung-Eun, University of Hawaii at Manoa. C, courtesy of the Research Collaboratory for Structural Bioinformatics.)

sequence—dictates both the secondary and tertiary structure of each protein molecule. Thus the enormous diversity of protein structures is a direct consequence of the enormous number of possible sequences and will be determined by whatever determines those sequences. Today, we know that the DNA sequence of each protein-coding gene determines the protein sequence. In some cases, the secondary or tertiary structure of a protein may be modified by interactions with small molecules or other macromolecules. For example, we note that the ionic composition of the medium may promote conformational changes, as can the binding of certain small molecules, known as **allosteric modulators (effectors)** (Box 3.8). Even changing the protein environment around a particular sequence may cause that sequence to adopt a different secondary structure, as for so-called **chameleon sequences in proteins**. Finally, chemical modification of the protein side chains can affect conformation. However, because the structure of the protein defines and allows these interactions, the genetic information is still dominant.

The tertiary structure of most proteins is divided into distinguishable folded domains

If the tertiary structures of many proteins are examined, it soon becomes evident that they can often be divided into folded **domains**. Although the precise definition of a domain is still a matter of contention, the common opinion is that they are units of protein structure that could fold autonomously, that is, independently of other domains in the protein. Domains are often quite compact and are connected by relatively unstructured segments of the polypeptide chain. Their connectivity may be enhanced by metal ion binding or occasionally by covalent disulfide bonds formed by oxidation of sulfhydryl side chains. Domains range from 25 to ~500 residues in length. It seems likely that many domains have evolved independently and have been connected to form specifically functional proteins later in evolution. Domain structure is common; about 90% of all known human proteins contain distinguishable domains.

Protein evolution seems to have proceeded from a relatively small number of domain types, which then further evolved and connected into the multidomain proteins we see today. This may account for the fact that features of protein tertiary structure seem to be even more conserved than sequence and can be used as clues to recognizing distant protein relatives, whose relatedness would not be evident from their sequences alone. A gallery of some common domain types is shown in Figure 3.9. Note that domains can be largely classified by their combinations of α and β secondary structure elements.

Very often, the combination of several domain types within a protein molecule allows it to carry out multitasking, with each domain having a specific function. An example is pyruvate kinase (Figure 3.10A), a key enzyme in the utilization of glucose as an energy source. This protein has three domains: an all-β regulatory domain and two α/β domains. One of these α/β domains catalyzes the dephosphorylation of phosphoenolpyruvate to pyruvate, and the other accepts the adenosine diphosphate, ADP, that is converted to adenosine triphosphate, ATP, in the reaction. The regulation in this case is allosteric, mediated by a number of metabolites that bind to the regulatory domain and modify the catalysis. See Box 3.8 for a detailed discussion of allosteric regulation.

There are also proteins in which one domain type is repeated. Often, this is an economical way to build a very large protein, such as the giant muscle protein titin (Figure 3.10B), which consists of hundreds of repetitions of immunoglobulin or Ig-like domains (see Figure 3.9). This also demonstrates the versatility of domains; the Ig domains were originally noted in the structure of antibodies, which have no common function with titin. Evolution makes do with whatever useful building blocks it can find.

Box 3.6 Max Perutz, John Kendrew, and the birth of protein crystallography Today we know the detailed structures of an enormous number of proteins, a level of understanding unimaginable half a century ago. Most of this information has come from studies of the diffraction of X-rays from protein crystals. In 1950, the technique of X-ray diffraction had been known for decades and used widely to determine the structure of small molecules. Indeed, much of the information about amino acid configurations that Pauling used to model protein secondary structures came from such studies. Yet Pauling himself had expressed doubts that a molecule as complex as a protein could ever be elucidated by this technique.

To get a simple idea of what is involved, consider the following experiment: if a very fine-mesh wire screen is placed in the path of a collimated light beam, like that from a laser, a rectangular pattern of bright spots can be observed on the wall behind the screen. This is the **diffraction pattern** of the screen, produced by reinforcement of in-phase waves coming to the wall from different apertures in the screen. The use of different screens can demonstrate much about diffraction. For example, a finer screen gives more widely spaced spots; there is a reciprocal relationship between object and its diffraction pattern. If a screen with wider horizontal spacings than vertical spacings is used, the pattern will show wider vertical spacings than horizontal spacings. In fact, the spacings of patterns on the wall can be used to deduce the periodic structure of the screen (see Figure 1.13).

Any crystal is a periodic structure. But there are two ways in which a protein crystal differs from the screens in our example: the spacings are very small, in the nanometer range, and the structure is three-dimensional, not two-dimensional like a screen. The small spacings necessitate the use of radiation of very short wavelength, X-rays, to get a diffraction pattern. The three-dimensional structure means that there are periodicities in three dimensions, so the crystal must be viewed from different angles. Finally, protein molecules are very complex. To obtain details of the molecular structure, both the positions and intensities of many, sometimes hundreds, of spots in the diffraction pattern must be measured. Unfortunately, this is not enough. Because the diffraction patterns are from a three-dimensional lattice, different spots will correspond to diffracted rays in different phases. The intensities do not tell this, and it was this phase problem that held up progress for many years. For small molecules, known constraints on the structure could lead to a resolved structure, but such methods were hopeless for proteins.

Such was the conundrum faced by researchers Max Perutz and John Kendrew at Cambridge University in the early 1950s. The breakthrough came in 1953, when Perutz discovered a way through the phase problem. If one could insert a heavy metal atom at the same place in every protein molecule in a crystal, it would perturb the intensities of different spots to various extents. With a series of such isomorphous replacements, all phases could be deduced and the structure could be solved.

By 1958 the first results were published when Kendrew's group presented a low-resolution 0.6 nm structure of myoglobin (Figure 1). At this resolution, little detail could be resolved, but it was clear that the structure was very complex and much less regular than previous guesses had suggested. The work was pushed to higher resolution, and by 1960, a 0.25 nm structure could be published. This showed, for the first time, the α-helices predicted by Pauling in a real protein.

In the same year, Perutz and collaborators produced a 0.55 nm structure of hemoglobin. This was a much more ambitious project because each hemoglobin molecule is a noncovalent assembly of four myoglobin-like chains. The results showed, for the first time, how proteins could associate to yield quaternary structure.

Perutz and Kendrew were awarded the 1962 Nobel Prize in Chemistry "for their studies of the structures of globular proteins."

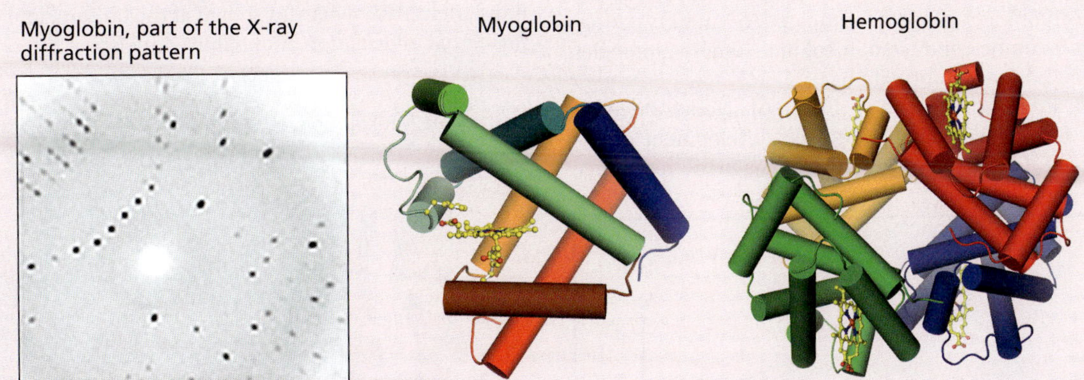

Figure 1 X-ray diffraction pattern and structures of oxygen-binding proteins. The diffraction image is from Francis Crick's work in Max Perutz's laboratory, 1949–1950. (Left, courtesy of Thomas Splettstoesser, Wikimedia.)

Algorithms are now used to identify and classify domains in proteins of known sequence

Only a few decades ago, the elucidation of a single protein structure was a huge challenge (see Box 3.5). Today, the number of known structures is so large that the challenge has shifted to making sense of the great variety of tertiary structures in terms of

(A)

VLSEGEWQLV LHVWAKVEAD VAGHGQDILI RLFKSHPETL EKFDRFKHLK
TEAEMKASED LKKHGVTVLT ALGAILKKKG HHEAELKPLA QSHATKHKIP
IKYLEFISEA IIHVLHSRHP GDFGADAQGA MNKALELFRK DIAAKYKELG
YQG

(B)

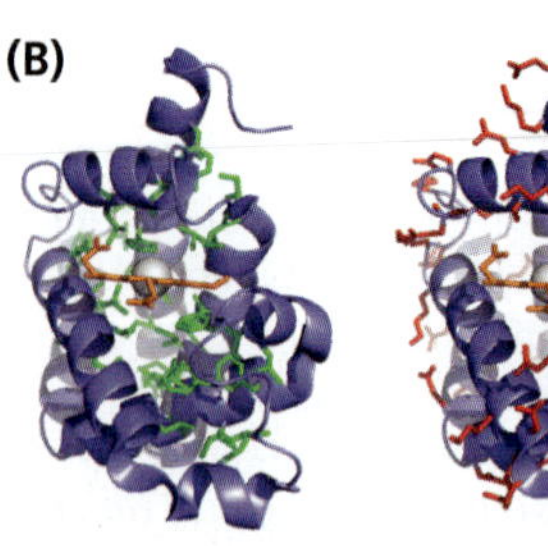

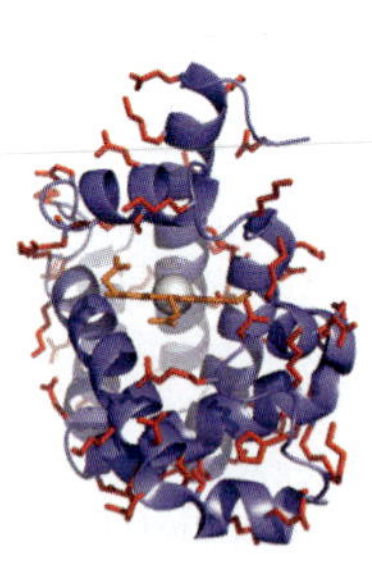

Figure 3.8 Distribution of hydrophilic and hydrophobic residues in sperm whale myoglobin. (A) Distribution in sequence: note the two types seem to be randomly scattered. (B) Distribution in tertiary structure. Hydrophilic residues, shown in red, are mainly on the surface, in contact with water, while hydrophobic residues, shown in green, are largely buried. (from Mathews CK, van Holde KE, Appling DR & Anthony-Cahill SJ [2013] Biochemistry, 4th ed. With permission from Pearson Prentice Hall.)

Box 3.7 Nuclear magnetic resonance studies of biopolymers

Nuclear magnetic resonance or NMR is a second technique, in addition to X-ray crystallography, that can provide detailed information about macromolecular structure. It is based on the fact that certain atomic nuclei, such as ^{1}H, ^{13}C, ^{15}N, and ^{31}P, possess a feature called spin that makes them behave like tiny magnets and hence take up different preferred orientations in an external magnetic field. These different orientations differ slightly in energy, so the appropriate radio frequency or rf energy can manipulate the orientations. When the radio frequency is just right, the nuclei resonate and emit radio waves of their own. For example, almost every proton in a protein or nucleic acid has, by virtue of its local environment in the molecule, a different resonance frequency. Thus, scanning the rf spectrum of a sample while it is in an intense magnetic field gives a proton spectrum (Figure 1). This can be useful in many ways, but the real power of NMR came with the discovery of spin–spin interactions.

The spin state of one nucleus can be perturbed by another spin that is nearby. This allows mapping of internuclear distances in a folded molecule. By use of such techniques, it is now possible to deduce, with fair accuracy, the three-dimensional structures of proteins and oligonucleotides. As Figure 2 shows, the results are close to those from X-ray crystallography. NMR has the enormous advantage, as compared to X-ray crystallography, that it can be conducted in solution, where molecules assume their natural conformations.

Richard Ernst was awarded the 1991 Nobel Prize in Chemistry for "his contributions to the development of the methodology of high resolution nuclear magnetic resonance (NMR) spectroscopy." Kurt Wüthrich was awarded a share of the 2002 Nobel Prize in Chemistry for "his development of nuclear magnetic resonance spectroscopy for determining the three-dimensional structure of biological macromolecules in solution."

NMR in medicine: MRI and MRS

Two powerful variations of NMR, magnetic resonance imaging (MRI) and magnetic resonance spectroscopy (MRS), have found widespread application in medicine. The advantage of MRI is that it is completely noninvasive and does not use harmful radiation. MRI uses magnetic gradients to spatially paint the position of the water proton spins. Submillimeter resolution images can be obtained in a matter of seconds from most soft tissues. MRI can aid in tumor localization, detection of Alzheimer's disease, stroke followup, sports injury evaluation, assessment of functional heart defects, etc. Figure 3A shows an example of the use of MRI to detect a tumor in the brain.

MRS uses magnetic gradients to localize one or several regions or voxels for investigation, but it also keeps the spectral information from the voxel so as to get a metabolic profile of the investigated region. The bone MRS profile shown in Figure 3B illustrates the connection between the known chemical structure of lipids, shown as dots of different colors connected with lines, their MRS spectrum shown at the bottom, and the compositional information obtainable with MRS.

Figure 1 Sample proton NMR spectrum. The solution contains a simple organic compound, ethylbenzene. Each group of signals corresponds to protons in a different part of the molecule.

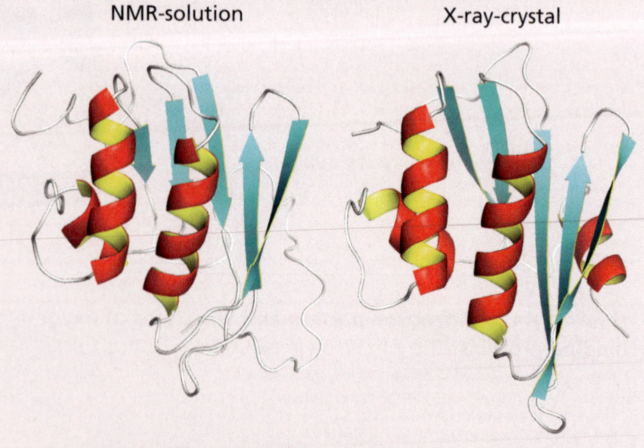

Figure 2 Comparison between NMR solution structure and crystal structure. The protein domain used here as an example is the RNase H domain of HIV-1 reverse transcriptase. Note that there are slight differences between the structures near the N- and C-termini; these may well result from the effects of packing the molecules into crystals for the X-ray study.

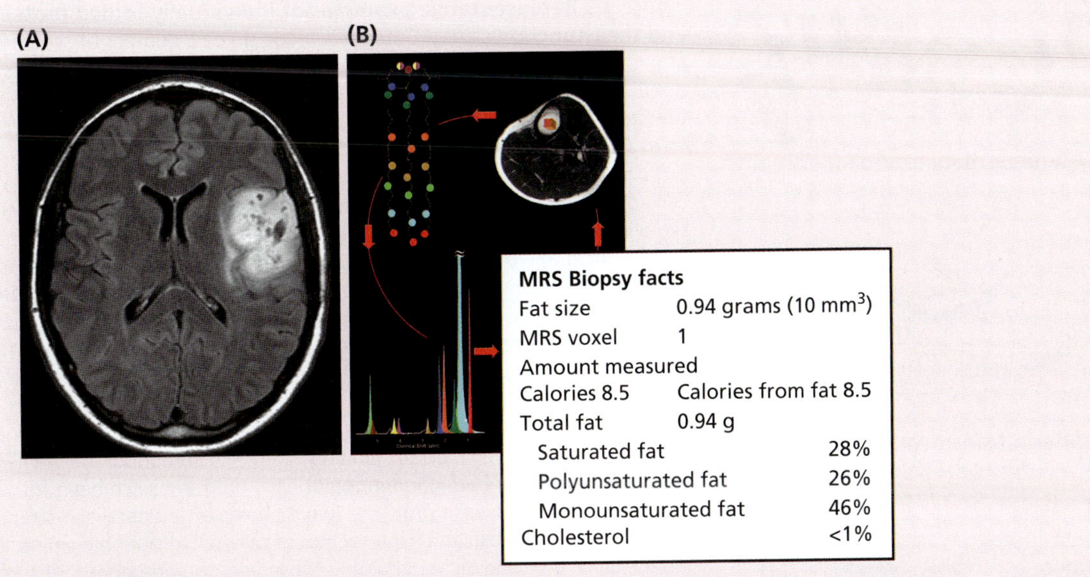

(A)
(B)
MRS Biopsy facts
Fat size 0.94 grams (10 mm3)
MRS voxel 1
Amount measured
Calories 8.5 Calories from fat 8.5
Total fat 0.94 g
Saturated fat 28%
Polyunsaturated fat 26%
Monounsaturated fat 46%
Cholesterol <1%

Figure 3 NMR in medicine. (A) MRI of a brain tumor; (B) MRS analysis of the lipid content of a tissue biopsy. (Courtesy of Ivan Dimitrov, University of Texas Southwestern.)

their function and evolution. Recognition of the importance of domains as structural building blocks and potential elements of evolutionary change has led to the development of automated and semiautomated methods for the identification of domains and their classification. None of these methods is perfect because of enormous difficulties in identifying discontinuous domains (see Figure 3.10) or domains that are

Box 3.8 Allostery Most proteins and other macromolecules are capable of specifically binding one or more kinds of small molecules, called **ligands**. There are many cases in which the binding of one ligand influences, either positively or negatively, the strength with which another ligand is bound. This phenomenon is called **allostery**, from the Greek *allos*, other, and *stereos*, space, and is a very common mode of biochemical regulation. Two classes of allostery can be distinguished. In homeotropic allostery, all ligands are the same. An example is hemoglobin, which has four binding sites on each molecule for the ligand oxygen. Binding even one oxygen molecule facilitates the binding of more oxygen molecules to the same hemoglobin molecule. By contrast, in heterotropic allostery, the ligands are different. For example, the affinity of many enzyme molecules for their substrates is modified by the binding of effector molecules. In some cases, the final product of a long metabolic pathway may allosterically inhibit the first enzyme in the pathway. Thus, a cell that already has a surplus of that product is saved from the wasteful process of making still more. Both positive and negative allosteric controls are important regulators of cell processes. Sometimes both occur on the same protein; for example, some transcription factors that regulate the expression of genes are subject to both positive and negative controls.

How does allostery work at the molecular level? For simplicity, imagine an enzyme that has a binding site for its substrate and a second site for an effector. It is tempting to assume that the binding of the effector molecule simply forces the protein into a conformation that favors binding of the substrate. But the most successful explanations have come from a slightly different model (Figure 1). It is postulated that the allosteric protein has two different conformations, called R and T, and can oscillate between these two states. The R-state has higher affinity for substrate and thus better enzyme function. Suppose effector binding favors formation of the R-state. Then, in the presence of effector, the protein will be more likely to be found in the R-state and thus enzymatic activity will be higher. Of course, in other cases, effector binding might favor the T-state; the effector is then an allosteric inhibitor.

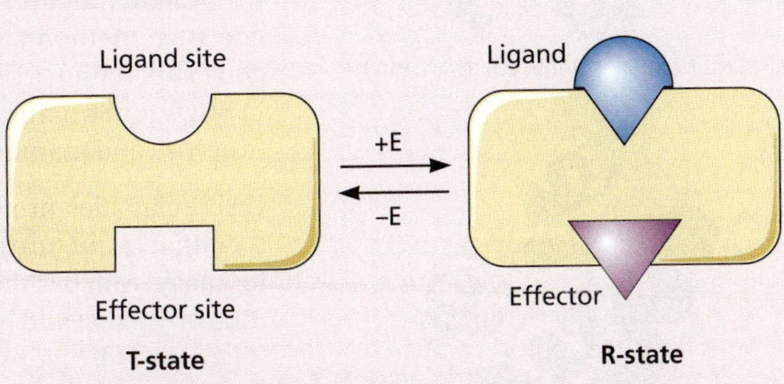

Figure 1 Simple model for allostery. Note that in this case both the effector and the ligand site have greater binding affinity when the protein is in the R-state. This results in positive allostery. If the T-state had greater affinity for the ligand, the effector (E) would diminish affinity for ligand by favoring formation of the R-state; this is the case with the Lac repressor, for example.

(A)

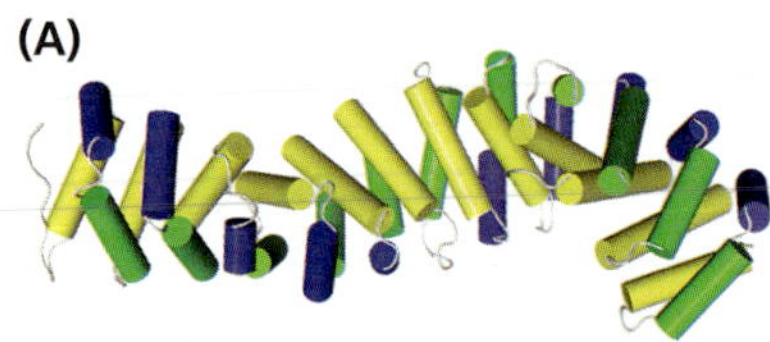

β-Catenin armadillo-repeat domains

(B)

Cadherin transmembrane domains

(C)

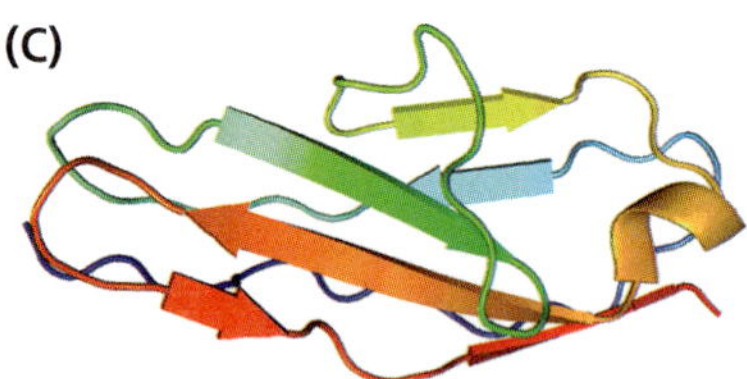

Immunoglobulin-like domains

(D)

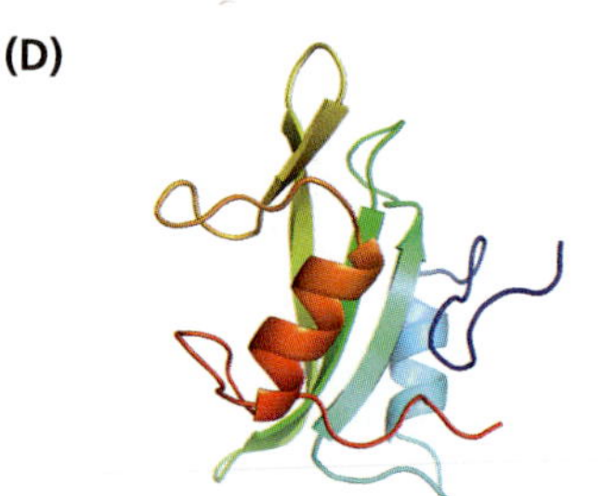

Src homology 2 domain (SH2)

(E)

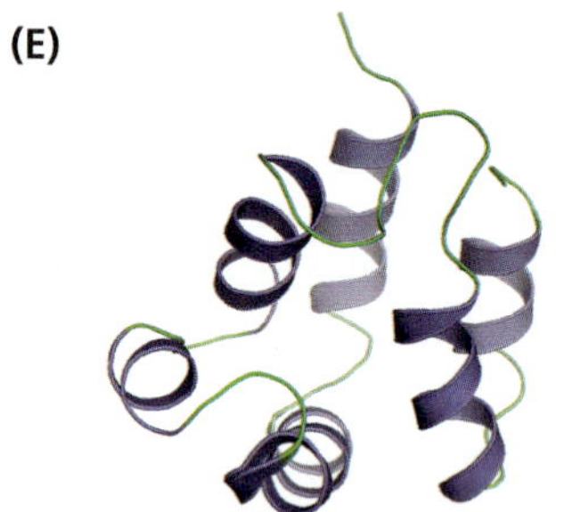

DED (death effector domain)

(F)

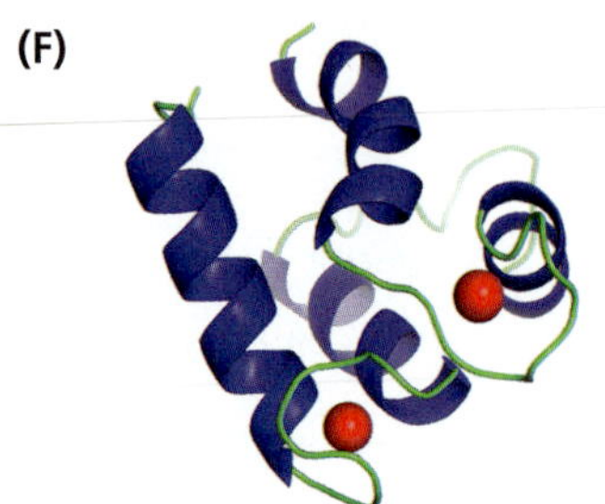

EF-hand motif

Figure 3.9 Representative examples of individually folded protein domains and their functions. (A) β-Catenin is made up of a number of repeats of the armadillo domain. Each armadillo repeat contains approximately 42 residues, forming three helices shown in yellow, blue, and green, arranged in a shape that has a triangular cross section. The 12 contiguous repeats of the armadillo repeat domain form a superhelix that features a long, positively charged groove. Many different protein partners bind to portions of the repeat domain. β-Catenin is involved in cell adhesions, which link the adhesion protein cadherin to the cytoskeleton, or in signaling pathways that change gene expression: in the presence of appropriate signals, β-catenin moves to the nucleus and interacts with transcription factors to activate target genes. (B) Cadherin repeats are named after cadherin proteins that play a fundamental role in calcium-dependent cell–cell adhesion. Adjacent cells adhere to each other via homophilic binding between cadherin molecules on their surfaces. Cadherins are typically formed by a single-pass transmembrane domain and multiple extracellular repeats. Each cadherin repeat consists of about 110 amino acid residues forming seven antiparallel β-strands, which are tightly linked by hydrogen bonds, and a highly conserved calcium-binding motif. (C) Immunoglobulin-like domains were originally described in immunoglobulins but have since been found widely distributed among diverse proteins. Two β-sheets of antiparallel β-strands form a sandwich that is stabilized by interaction between hydrophobic amino acids on the inner side of the sandwich and highly conserved disulfide bonds. The immunoglobulin superfamily includes proteins with roles in the immune system, such as cell surface antigen receptors, antibodies, and major histocompatibility complex; in muscle contraction, such as titin; and in cell adhesion. The immunoglobulin superfamily is the most populous known family of proteins, with more than 765 members in humans. (D) The SH2 domain contains a large β-sheet flanked by two α-helices, shown in orange and blue. A total of 120 SH2 domains, contained within 115 proteins, are encoded in the human genome. The SH2 domain binds to phosphorylated tyrosine and is found in signal transduction proteins. Extracellular signals sensed by receptors in membranes are converted to the chemical signal phosphorylated tyrosine; this initiates a series of events that eventually results in altered patterns of gene expression. (E) The death effector domain, DED, is a protein interaction domain that allows protein–protein binding by DED–DED interactions. DED is composed of a bundle of six α-helices. The dimerization of DED domains is mediated primarily by electrostatic interactions. DED proteins are involved in apoptosis (programmed cell death) which occurs through the activation of proteolytic cascades. The major players in these cascades are caspase proteases, whose inactive precursors are activated by DED–DED interactions with specific adaptor molecules. The structure shown is that of the DED domain of FADD. (F) The EF-hand motif consists of two perpendicular 10–12-residue α-helices with a 12-residue loop region between, forming a single calcium-binding site as a helix–loop–helix. Calcium ions interact with residues contained within the loop region. EF-hand domains are often found in single or multiple pairs; two pairs are shown here. A typical representative of EF-hand proteins is calmodulin, which can bind to and regulate a number of different protein targets in response to calcium binding.

tightly associated. Nevertheless, the potential value of such methods for understanding protein function and evolution is undeniable. In the following, we shall describe one such method called **CATH**: Class, Architecture, Topology or fold, and Homologous superfamily. CATH is a semiautomated method; it has one strong competitor in SCOP, or Structural Classification of Proteins, which relies more heavily on human expertise and manual classification of domains.

CATH classifies protein domains at four levels: classes, which are the predominant secondary structure organization, α, β, or α/β; architecture, which is the overall shape; topology, which is the fold, spatial arrangement, and connectivity of secondary structural elements; and homology, which is the highest level of families and superfamilies (**Figure 3.11**). **Figure 3.12** illustrates the distribution of domain structures among the three major protein classes. How are domains distributed overall among proteins? A comprehensive study in 2005 examined this question on a database consisting of the total genome sequences of 150 organisms. Approximately 1 million protein-coding genes were identified, of which 850,000 could be classified into 50,000 families, each of unique domain organization. The remainder were mainly singletons and small proteins. By use of CATH and similar algorithms, 80% of the domains could be grouped into only 5000 families, with the remainder falling into a category of new families.

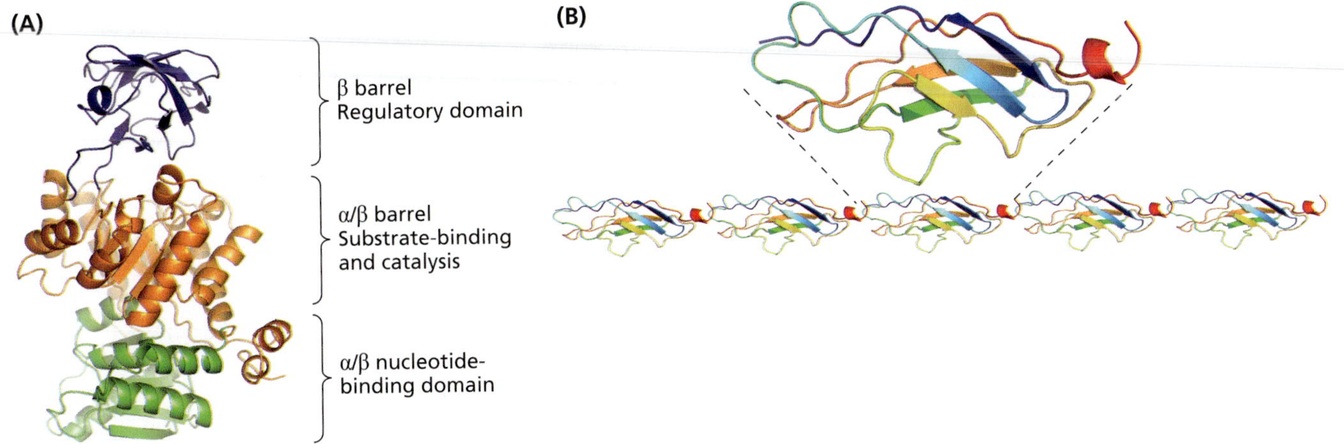

Figure 3.10 Two examples of multidomain proteins containing either distinct domains or closely related repetitive domains. (A) Pyruvate kinase, a three-domain protein. The central α/β barrel domain is present in numerous different enzymes that catalyze completely unrelated reactions. It is formed by different noncontiguous segments of the polypeptide chain coming in close spatial proximity, hence the name discontinuous domain. The β-barrel regulatory domain is continuous, formed by a single stretch of the polypeptide chain. (B) A portion of the giant muscle protein titin that consists of repeating immunoglobulin-like domains. These domains are believed to confer elasticity to the muscle fibers. (B, adapted, courtesy of Wikimedia.)

Thus, a relatively small number of domains has been used to construct an enormous variety of proteins.

The next step is to predict protein function from the many sequences that we recognize in the genome but have not yet isolated as proteins. Unfortunately, the protein sequence does not in itself allow us, as yet, to predict the tertiary structure, except by analogy to proteins of very similar sequence. However, similarities between known protein tertiary structures provide better hints to function and evolution than do the sequences themselves.

Some domains or proteins are intrinsically disordered

After the remarkable early successes in X-ray diffraction studies of easily crystallizable proteins, the idea became fixed that every protein probably had well-defined secondary and tertiary structures. As more and more proteins of diverse functions were investigated, however, it became clear that the situation is more complicated. Researchers began to find regions in polypeptide chains that simply did not show up in the crystallographic maps. These regions differed from those portions of the chain that do not have any standard type of secondary structure but are still well defined in the map, such as loops and turns. The sections that are not found on crystallographic maps instead correspond to chain segments that have such mobility that they differ in conformation in different molecules in the crystal or at different times. These are referred to as **intrinsically disordered proteins or regions**. Evidence for such structural features of proteins in solution also comes from other techniques, notably high-resolution NMR (see Box 3.7). It became clear that many parts of many proteins have intrinsic flexibility. Indeed, we now know some cases in which an entire protein molecule behaves very much like a random coil, devoid of regular secondary and tertiary structure. These are often referred to as intrinsically disordered proteins or ID proteins. It is important to remember that intrinsic disorder is just as much a consequence of amino acid sequence as is order. ID proteins are often characterized by a combination of low hydrophobicity and high charge, both of which favor their opening up to solvent.

It is believed that the high flexibility of ID proteins facilitates their interaction with multiple other proteins; their conformation can fit to different partners. Figure 3.13 presents one well-studied example, the p53 tumor suppressor protein. Mutations in p53 are associated with a number of human malignancies. p53 participates in a number of biological pathways, interacting with a different partner in each case. A key to this lies in the disordered N- and C-terminal portions of p53, which can adopt many

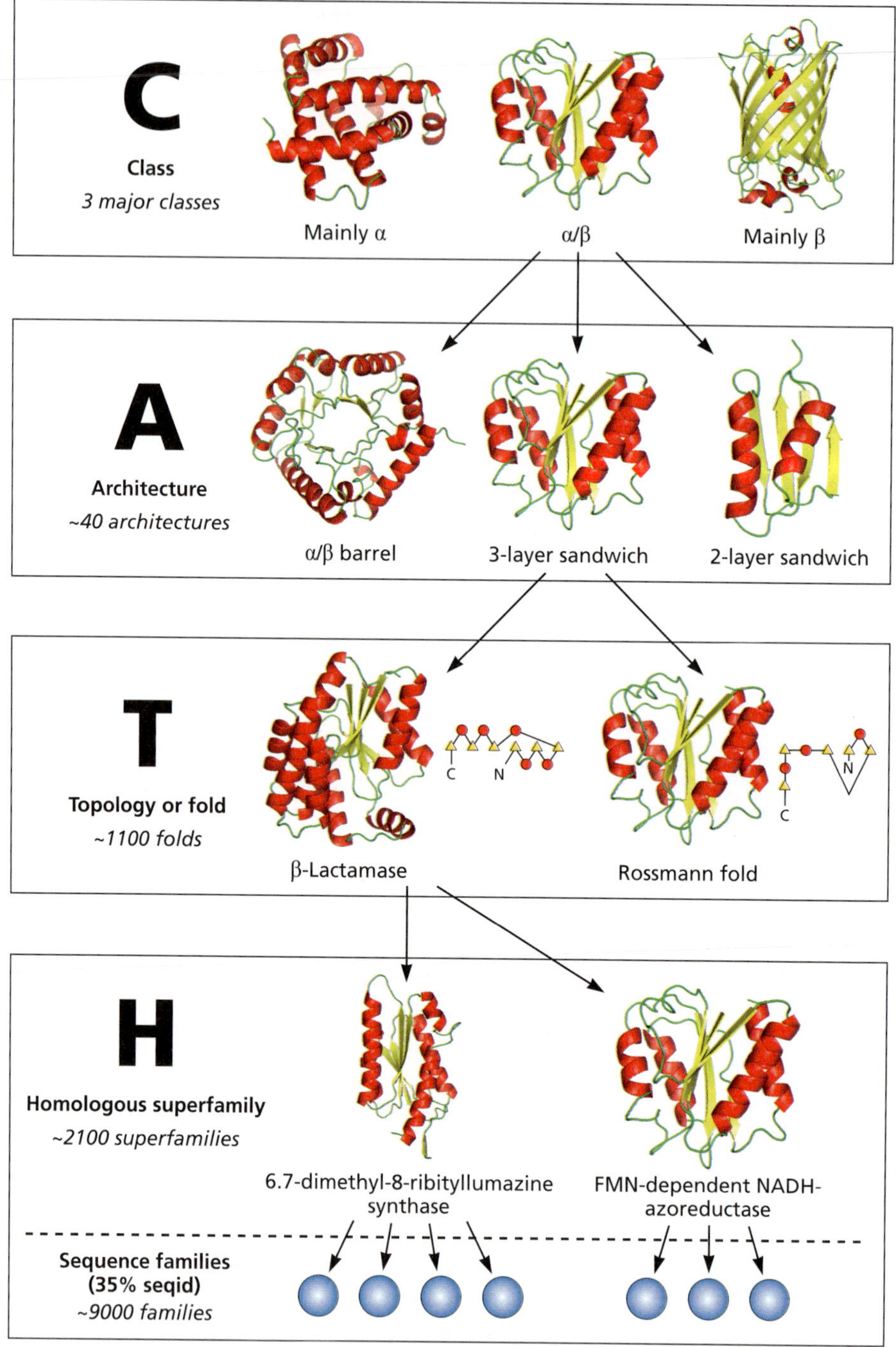

Figure 3.11 Structural classification of proteins by CATH. CATH is one of several existing databases that use algorithms to automatically assign domains in proteins with known sequences. The four major hierarchical levels for classifying proteins according to their structural features are Class, Architecture, Topology or fold, and Homologous superfamily. Seqid refers to sequence identity between the domains. (Adapted, courtesy of Christine Orengo and Ian Sillitoe, University College London.)

conformations. The p53 example illustrates another point: most post-translational protein modifications occur in these disordered regions, presumably to tailor the protein to specific partners. Figure 3.14 presents an example of a protein known to recognize and react with methylated lysines and with other groups. Again, flexible peptides can adapt to the geometry of the binding site, promoting strong interactions.

Functional uses for disordered protein regions have also been found in enzymes, where such regions may promote promiscuity, in the sense that a given enzyme may accommodate a variety of similar substrates. This may have facilitated enzyme evolution. The disordered regions correspond more frequently to regulatory sites rather than to the catalytic site. This is believed to be because catalysis requires a fairly rigid molecular structure, complementary to the transition state in the reaction, whereas regulation may occur through binding of a number of different effectors, which a disordered region can accomplish.

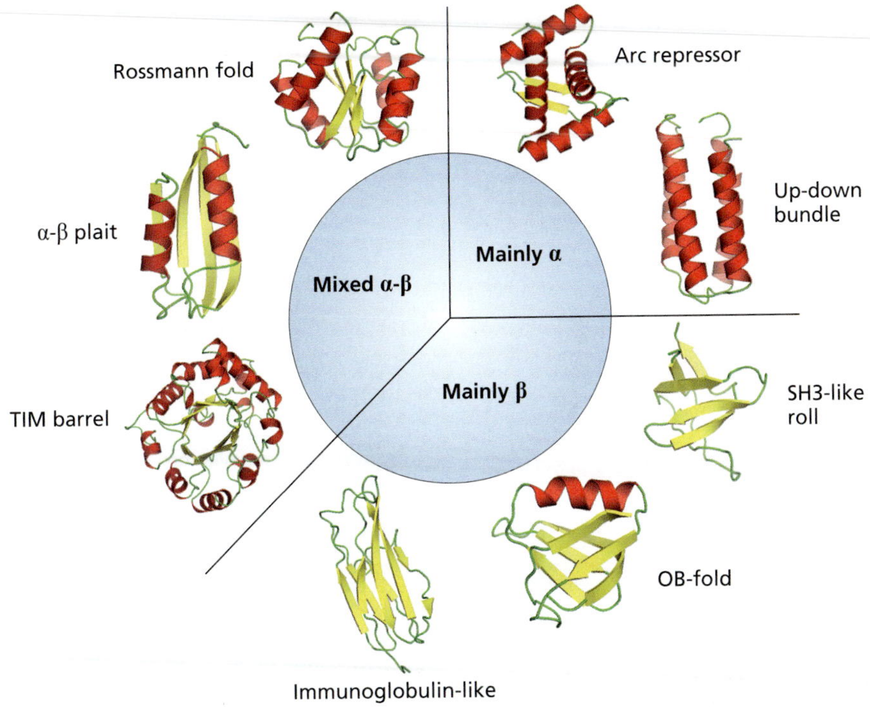

Figure 3.12 Structural classification of proteins by CATH. Distribution of domain structures among the three major classes: mainly α, mainly β, or mixed α-β, with representative examples in each class. α-Helices are colored in red; β-sheets are shown in yellow. (Adapted from Cuff AL, Sillitoe I, Lewis T et al. [2011] *Nucleic Acids Res* 39:D420–D426. With permission from Oxford University Press.)

Perhaps the most striking utilization of unstructured protein regions is in signaling proteins, which must interact with a variety of other proteins as part of a signaling pathway. To date, no signaling protein–protein interactions that involve structured domains have been identified. Thus, there is a vast class of proteins with functions in signaling and regulation that utilize a lack of defined protein structure.

Quaternary structure involves associations between protein molecules to form aggregated structures

Many proteins exhibit a fourth level of organization, referred to as **quaternary structure**. This involves noncovalent association between individual protein molecules, which may themselves have multiple domains, to form defined aggregated structures. These may be long chains of globular proteins, as in the muscle protein actin (**Figure 3.15**). More usual are symmetrical arrays of a small number of monomer units, often called **protomer** units: arrays of 2, 4, and 6 protomers are common, but cases of up to 48 units are known. Some examples are shown in **Figure 3.16**. Quaternary structures can be considered special cases of the very general phenomenon of protein–protein interactions. Such interactions may be quite stable, resulting in what we call quaternary assemblies or multisubunit proteins. But they may, in other cases, be transitory but nonetheless vital to the dynamics of the cell. How complex such patterns can be is only now being recognized, in a discipline called bioinformatics.

The protomers comprising a quaternary assembly may be the same or of several kinds; if only one type is involved, they are usually arranged in some variety of point-group symmetry, with one or more axes of symmetry. In Figure 3.16, topoisomerase shows twofold symmetry; rotation of the molecule 180° about an axis in the plane of the page will yield exactly the same image. Proliferating cell nuclear antigen, PCNA, has a threefold symmetry axis, perpendicular to the plane of the page. Hemoglobin is said to have pseudo-twofold symmetry, because the two kinds of subunits are almost, but not quite, identical.

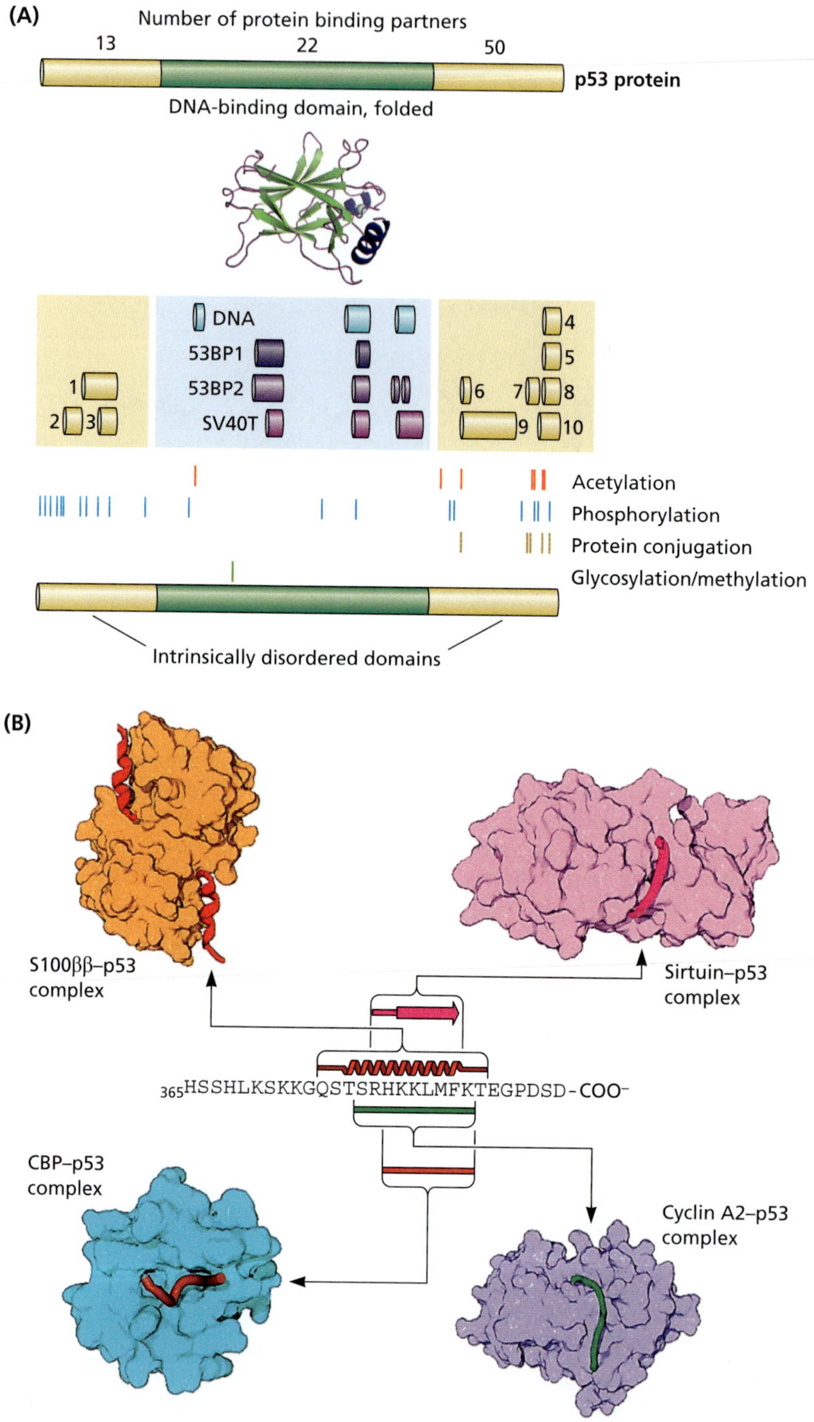

Figure 3.13 Tumor suppressor protein p53: A protein that contains both a structured region and two intrinsically disordered domains. (A) p53 domain structure and protein partners that bind to the different domains. More than 70% of the protein partners interact with the nonfolded regions of p53, which comprise only 29% of the length of the polypeptide. Some proteins, as well as DNA, interact with the structured DNA-binding domain. The exact portions of the p53 polypeptide chain that interact with the respective partners are depicted by boxes of different colors. Note that some partners interact with more than one region in p53. The boxed, numbered proteins have been co-crystallized with p53; the crystal structures with cyclin A (4), sirtuin (5), CBP (8), and S100ββ (10) are presented in part B. Post-translational modification sites are shown by colored vertical bars. Note that the majority of these sites are located in disordered regions. (B) Comparison of experimentally determined X-ray structures of the same C-terminal disordered region, in shades of red or in green, which adopts different structures upon interacting with different partners. (A, adapted from Dunker AK, Silman I, Uversky VN & Sussman JL [2008] *Curr Opin Struct Biol* 18:756–764. With permission from Elsevier. B, adapted from Oldfield CJ, Meng J, Yang JY et al. [2008] *BMC Genomics* 9(Suppl 1):S1. With permission from BioMed Central.)

Only rarely is molecular size itself a reason for a protein to have a defined quaternary structure. More often than not, such arrangements facilitate allosteric regulation of protein function. Consider the difference between hemoglobin (see Figure 3.16) and myoglobin, which is very much like a protomer of hemoglobin. Myoglobin

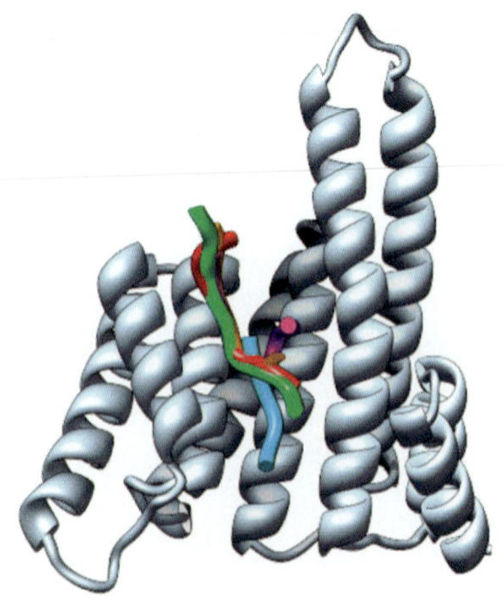

Figure 3.14 Protein segments with different disordered sequences use their flexibility to adapt to a common binding site. Crystal structures of five intrinsically disordered peptides bound to the highly structured binding pocket of protein 14-3-3ζ. The protein, shown in light blue, binds to various protein partners by interacting with their disordered segments: peptides from serotonin *N*-acetyltransferase, shown in green, and histone H3, shown in purple, as well as three other distinct peptides, shown in orange, red, and cyan. Each peptide has a distinct sequence and modifications and follows a different path through the 14-3-3ζ peptide-binding pocket. (From Oldfield CJ, Meng J, Yang JY et al. [2008] *BMC Genomics* 9(Suppl 1):S1. With permission from BioMed Central.)

Figure 3.15 Proteins can form filaments of identical or nonidentical subunits. (A) Actin, the monomer globular actin, G-actin, and the helical polymer of actin filaments (F-actin). G-actin is globular in structure; the polymer is a two-stranded helical arrangement of G-actin. Polymerization is induced upon binding of ATP to G-actin; ATP hydrolysis occurs, but ADP stays bound to the filament. The asymmetry of the individual subunits leads to polarity of the filament, defined as + and – ends; the two ends grow at different rates. (B) Tropocollagen, the basic unit of collagen fibers, is a helix of three polypeptide chains, each ~1000 residues in length. Each chain is a left-handed helix with ~3.3 residues/turn. The three helices wrap around each other in a right-handed sense; the structure is stabilized by hydrogen bonds. The structure imposes constraints on the primary sequence, with glycine-proline and hydroxyproline forming a repetitive motif. The image of collagen fibers on the right is taken by atomic force microscopy, AFM. (C) Microtubule, a hollow cylinder. Along the microtubule axis, α- and β-tubulin heterodimers are joined end-to-end to form protofilaments with alternating α- and β-subunits. Staggered assembly of 13 protofilaments yields a helical arrangement of tubulin heterodimers in the cylinder wall. (B, from Mathews CK, van Holde KE, Appling DR & Anthony-Cahill SR [2012] Biochemistry, 4th ed. With permission from Pearson Prentice Hall. C, bottom, from Li H, DeRosier DJ, Nicholson WV et al. [2002] *Structure* 10:1317–1328. With permission from Elsevier.)

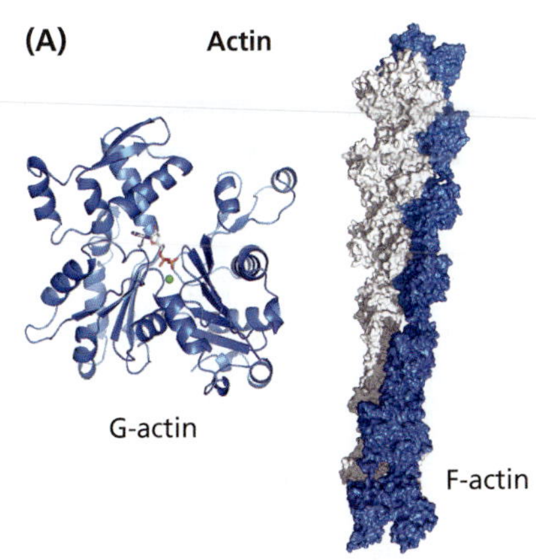

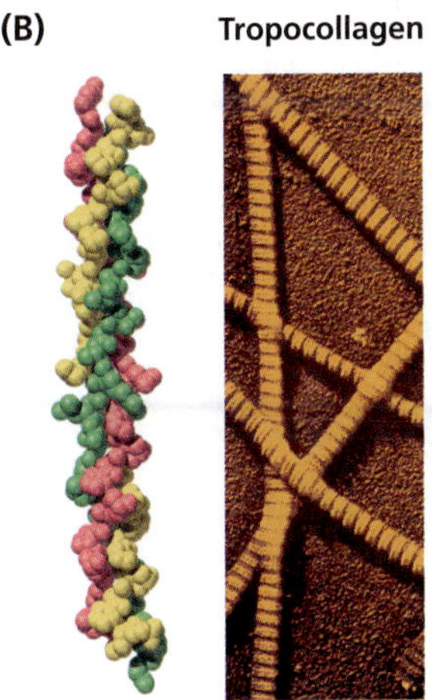

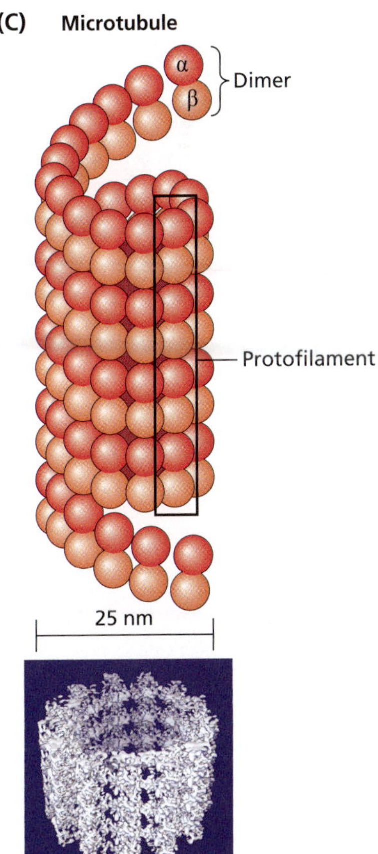

will bind oxygen, and does so in tissues. But hemoglobin is much better suited to the delivery of oxygen because it binds and releases oxygen cooperatively: the presence of oxygen on one protomer facilitates its binding to another. Furthermore, the binding of other allosteric effectors to hemoglobin modifies oxygen affinity, allowing adaption to varied circumstances. For a deeper understanding of allostery, see Box 3.8.

In some cases, quaternary structure becomes so elaborate and extensive that a molecule is described as a **multiprotein complex**. Later chapters feature many examples of quaternary complexes involving as many as a dozen or more different kinds of protomers, grouped together to fulfill some complex function and its regulation. Such complexes are often associated with fundamental cell processes, like replication of DNA, transcription of DNA into RNA, or synthesis of proteins. Many of the protomer types in such complexes bind allosteric effectors, which in turn modulate the function of the whole complex. More and more, it is becoming clear that the cell is a network of interacting molecules. In many respects, quaternary structure and domain structure are complementary ways of achieving the same goals. Quaternary assembly, however, has the advantage that functional units can be exchanged for varied needs.

Quaternary structure is stabilized by the same kinds of noncovalent interactions that account for secondary and tertiary structure in proteins (see Table 3.2). As these involve specific groups on the surfaces of the protomers, we know that the quaternary arrangement is dictated by the protomer tertiary structure, and ultimately by the sequence of the polypeptide chain or chains.

3.4 How do proteins fold?

Folding can be a problem

As we see in Chapter 16, proteins are synthesized as unstructured polypeptide chains. The protein-synthesizing machinery of the cell does not itself impose structure. To become fully functional, the chain must fold into its appropriate secondary and tertiary conformations and, in some cases, associate into quaternary structures. How this happens has long intrigued molecular biologists. Surely it cannot happen by a random search through possible conformations: even if there are only two orientations at each residue, a 100-residue polypeptide would have 2^{100} conformations. Searching only a fraction of these, at nanoseconds per search, would take billions of years. Does each kind of protein have a template protein that shows it how to fold? Then what folds the template?

In reality, it appears that protein folding often can proceed in a stepwise, orderly manner dictated by the primary structure, the information the protein is provided with as

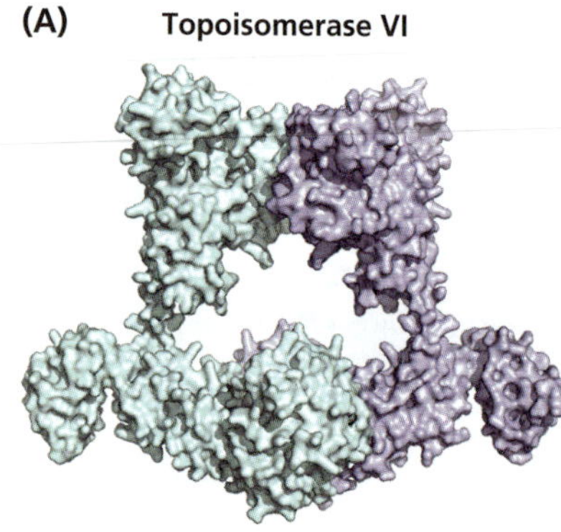

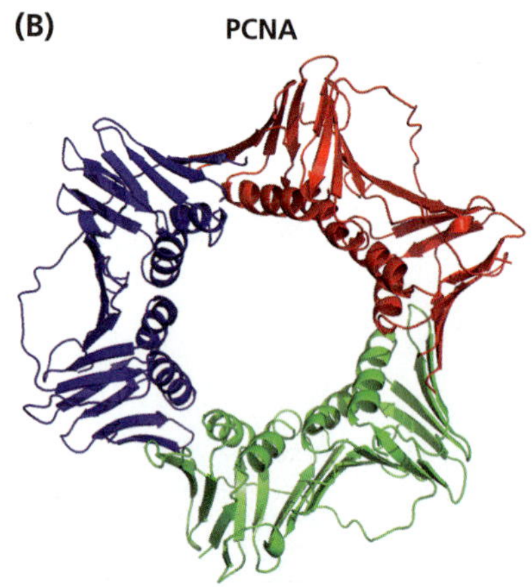

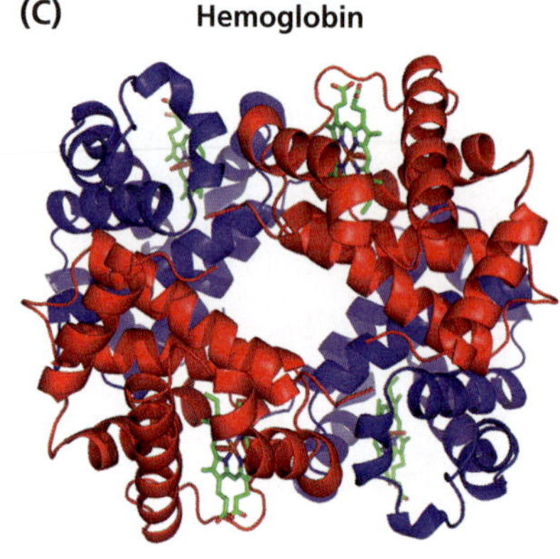

Figure 3.16 A gallery of proteins whose quaternary structures contain identical or highly similar protein subunits. (A) Topoisomerase VI is an example of proteins that contain two identical subunits. (B) PCNA, proliferating cell nuclear antigen, is formed by three identical subunits. (C) Hemoglobin contains two α- and two β-subunits.

it is made. This is referred to as the **self-assembly principle**. Evidence for this comes from many experiments that show unfolded proteins spontaneously attaining their functional conformation after having been denatured; that is, having had their native structure unraveled. A classic example is shown in **Figure 3.17A**, which depicts the thermal denaturation of ribonuclease, a small globular protein. If ribonuclease in solution is heated above 40°C, all evidence for defined secondary and tertiary structure, as well as enzymatic activity, is lost. Upon recooling, all of these features are quantitatively regained. Furthermore, comparison of data from various techniques indicates that this is an all-or-none transition: at any intermediate temperature, a mix of native and denatured molecules is present. The argument is as follows: different techniques for studying protein structure are sensitive to different aspects of the structure, which relate to various parts of the protein. If proteins are folded piecewise, then tracking the folding by a range of methods should lead to distinguishable curves in different experiments. Instead, the curves obtained by different methods are superimposable (**Figure 3.17B**). Thus, we know that folding, at least in this case, is a fast all-or-none process, occurring in seconds or minutes.

Such experiments have shown that for many proteins the amino acid sequence, and thus the DNA gene sequence that dictates it, can define the conformation and functionality without external aid. The native structure represents a free energy minimum, the state of optimum stability in physiological conditions (**Figure 3.17C**). This is a very important principle, one of the cornerstones of molecular biology. However, having stated it, we must add some important qualifications. First, after a protein is synthesized in the cell, it may be subject to chemical modification. Such post-translational modifications, PTMs, are of two general types: a part of the protein may be cut off by specific proteolysis, or side-chain residues may be chemically modified. A list of the more important modifications is given in **Table 3.3**. Details of the modifications, how they are made, and the functions they serve are presented in Chapter 18. A powerful and extraordinarily precise tool for locating and identifying such modifications is mass spectrometry, MS (**Box 3.9**). Although post-translational modifications are not directly dictated by the DNA sequence, they are certainly influenced by it; the correct amino acid residue must be at the correct position for modification to occur. Furthermore, modification always requires specific enzymes, which themselves are specified in the DNA. So DNA still has the final say.

Second, the folding of many proteins, especially in the crowded environment of the cell, is undoubtedly more complicated and difficult than in the test tube. Interactions with other cellular molecules, including proteins, may block correct folding. Or there may exist other states, distinct from the native state, in which the protein can be trapped. A way of looking at this is shown in Figure 3.17C. The deepest point in the free energy landscape is the native conformation. In the complex kinetic process of becoming folded, the protein may be temporarily trapped in an adjacent, shallow energy well.

It has recently been proposed that multiple conformations close in energy to the native conformation may actually serve biological purposes. A protein that exhibits such behavior might be promiscuous in function, serving more than one purpose or allowing easy evolvability into new functions. But such alternative available states may also lead to misfolding, which can have serious consequences.

Misfolded proteins are generally useless and may even be harmful to the cell, especially through their tendency to aggregate: they must be dealt with. There are two major ways to deal with misfolded proteins. First, they can be given a chance to avoid misfolding or they can be allowed to refold to their native, functionally active state. Macromolecular structures that can help to accomplish these activities are called chaperones. Second, misfolded proteins may be sent to specialized cellular machineries for

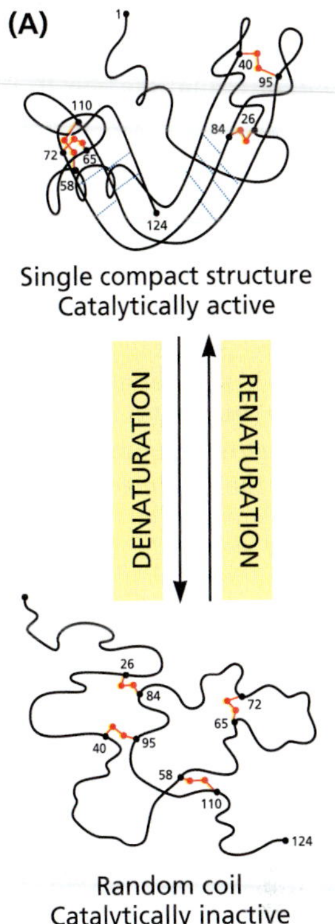

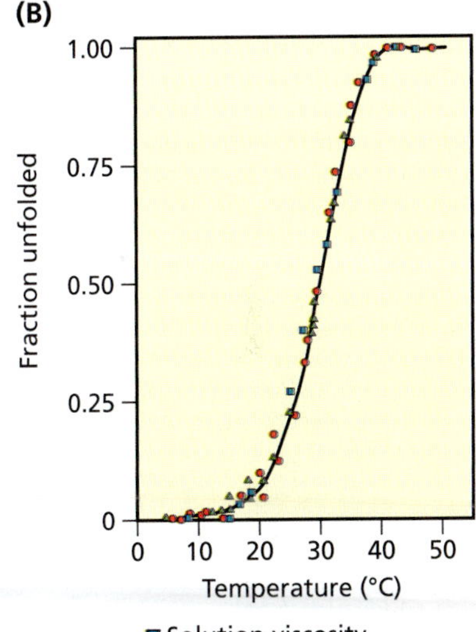

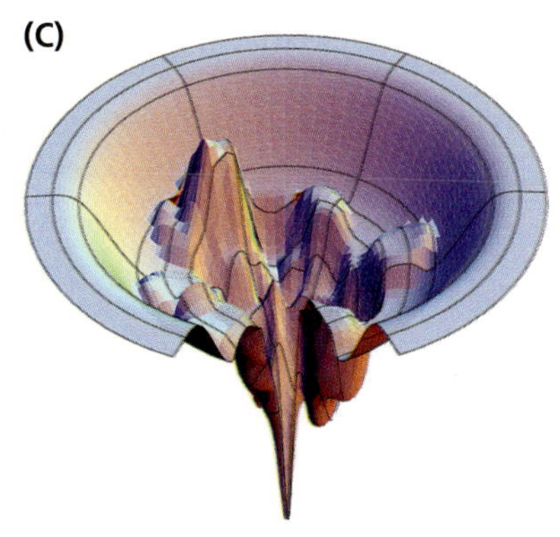

Figure 3.17 Protein folding. (A) Thermal denaturation of ribonuclease A. When ribonuclease is heated above a certain temperature, it undergoes a conformational transition from a highly ordered, catalytically active structure into a catalytically inactive random coil that experiences continuous fluctuation among a large number of extended conformations. When the enzyme is inactivated by elevated temperature, its disulfide bonds remain intact, allowing for faster renaturation upon reduction of the temperature; other denaturation treatments may lead to rupture of these bonds and hence to slower renaturation. If disulfide bonds are reduced concomitant with denaturation, reoxidation will lead to scrambled structure with incorrect bonds. However, removal of denaturant previous to reoxidation yields correct structure. (B) A range of physical methods has been used to study the unfolding (denaturation) and the refolding (renaturation) of proteins. (C) Hypothetical free energy map for a protein. The vertical axis represents free energy, and the *x*- and *y*-axes represent different conformations; actually there should be about 2^{100} such dimensions, but that is hard to draw. The map illustrates that there is one true free energy minimum, the most stable conformation or native state, and many pseudominima at slightly different conformations in which the molecule could, theoretically, become temporarily trapped. (A, B, adapted from Mathews CK, van Holde KE, Appling DR & Anthony-Cahill SJ [2012] Biochemistry, 4th ed. With permission from Pearson Prentice Hall. C, from Dill KA & Chan HS [1997] *Nat Struct Biol* 4:10–19. With permission from Macmillan Publishers, Ltd.)

degradation (**Figure 3.18**). Both refolding and specific degradation require energy expenditure through ATP hydrolysis, which is believed to lead to conformational transitions in the respective proteins.

Chaperones help or allow proteins to fold

Chaperones are huge protein complexes usually considered as protein folding machines. Chaperones enclose the protein, protect it from the cellular environment, and give it a chance to fold properly. One might say they serve as protein incubators. More than 50 chaperone families have been described to date. We must note, though, that chaperones have a number of additional cellular functions, sometimes seemingly opposite to their role in folding. For example, proteins destined to function in or be transported through membranes should not fold before being inserted into the membrane, and their premature folding is prevented by special chaperones. Chaperones also interact with nascent polypeptide chains emerging from their site of synthesis on ribosomes, so that they do not associate (aggregate) with chains synthesized nearby.

As a specific example, we consider the best-studied bacterial chaperone system, **GroEL/GroES**, also known as chaperonin 60/chaperonin 10. The GroEL/GroES system is indispensable for the survival and growth of the bacterial cell as numerous proteins, both small and large, depend on it for proper folding. The subunit composition

Table 3.3 Most important postsynthetic modifications of proteins.

Name of modification	Adduct	Site	Example functions of modification
acetylation	acetyl	lysine, N-terminal amino acid	protein recognition in gene regulation
phosphorylation	phosphate	serine, threonine, tyrosine, histidine	activation of enzymes, signaling
methylation	methyl	lysine, arginine	protein recognition in gene regulation
glycosylation	polysaccharides	asparagine, serine, threonine	cell recognition, protein recognition
ubiquitylation	small protein ubiquitin	lysine	marks proteins for degradation

Box 3.9 Mass spectrometry Mass spectrometry (MS) was once considered to be applicable only to small molecules, but in recent years it has proven to be a powerful method of analyzing proteins. This technique separates ionized molecules by the velocity they gain in an electric field and/or their deflection in a magnetic field. The basic principle is quite simple (Figure 1). First, there must be a way to produce the molecules in an ionized state or set of states. The ionized molecules are then accelerated through a vacuum by an electric field. Different molecules or ions will gain different velocities, depending on their mass-to-charge (m/z) ratio. They can then be separated either by time-of-flight through the vacuum or by applying a magnetic field to bend their trajectories. In the simple case shown in Figure 1, the molecule CO_2 has been ionized so as to produce the positive ion CO_2^+ and fragments C^+, O^+, and CO^+, which have different m/z ratios. The ions are detected, and a spectrum is produced. The method is capable of extraordinary precision.

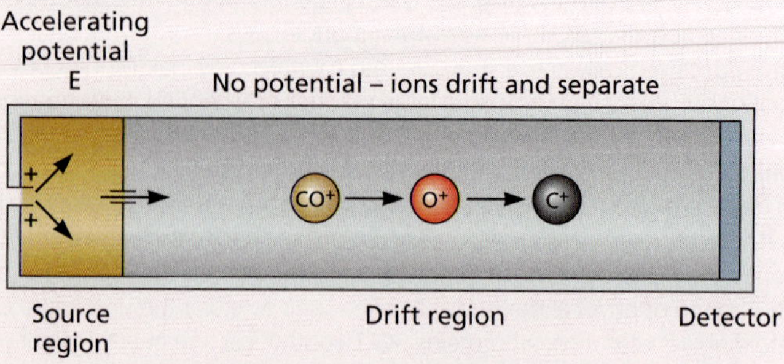

Figure 1 Simplified cartoon of a time-of-flight mass spectrometer. The sample is introduced into the source chamber, where ions are accelerated to a velocity inversely proportional to $(m/z)^{1/2}$, where z is their charge and m is their mass. They then drift through the drift chamber, arriving at the detector at times proportional to $(m/z)^{1/2}$.

The application of MS to large molecules such as proteins required the development of ways to produce individual protein molecular ions and to get them into the gas phase for injection into the vacuum. The two principal methods used today are **matrix-assisted laser desorption and ionization (MALDI)** (Figure 2) and **electrospray ionization (ESI)**. In MALDI, the protein molecules are literally blasted out of an inert matrix material by a laser pulse. In ESI, microdroplets, each containing one or a few protein molecules, are sprayed into the spectrometer through a charged nozzle. This gentler technique shows promise of broader applications in analyzing biological materials.

The exquisite sensitivity and resolving power of modern mass spectrometers make remarkable feats possible. The molecular weights of even large proteins can often be determined to within a few atomic mass units. This makes it possible to analyze small protein modifications. Peptides produced by fragmentation of the protein—by specific proteolytic enzymes, for example—can be separated and often identified, leading to rapid sequencing. Even information about a protein's three-dimensional structure can be gained by using the mass spectrometer to follow hydrogen/deuterium exchange, as exchange is more rapid for residues near or on the protein surface, which are in contact with solvent.

Several Nobel Prizes in Physics and Chemistry have been awarded throughout the years for different developments in the field of mass spectrometry. A share of the 2002 Nobel Prize in Chemistry went to John B. Fenn and Koichi Tanaka "for their development of soft desorption ionisation methods for mass spectrometric analyses of biological macromolecules."

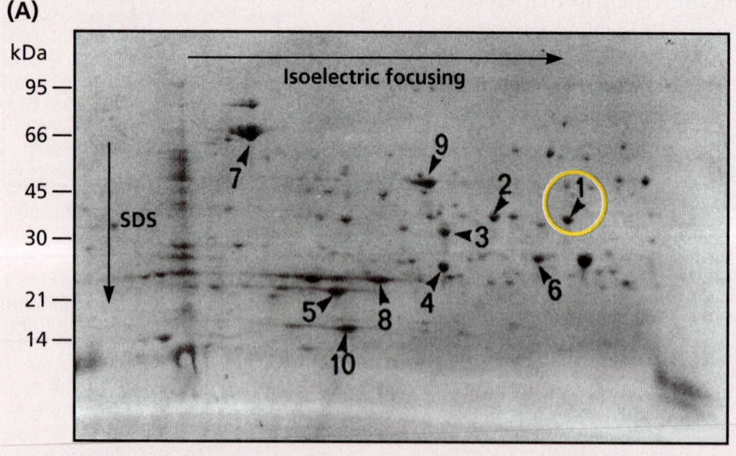

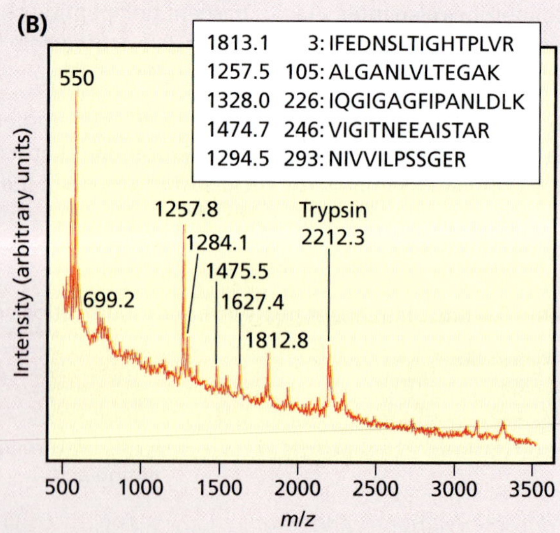

Figure 2 Mass spectrometric identification of a protein. (A) Blot on a poly(vinylidene difluoride) membrane of a two-dimensional electrophoretic gel of an *Escherichia coli* cell lysate. The proteins on the blot were stained with Coomassie blue. Spot 1 was excised with a razor blade and trypsin-digested on the membrane for MALDI analysis, resulting in the profile shown in B. (B) All five masses determined from the spectrum matched five peptides for cysteine synthase from protein databases, thus positively identifying spot 1 as cysteine synthase. Only one protein matches the five masses in a protein database of over 100,000 proteins. (Adapted from Henzel WJ, Watanabe C & Stults JT [2003] *J Am Soc Mass Spectrom* 14:931–942. With permission from Springer Science and Business Media.)

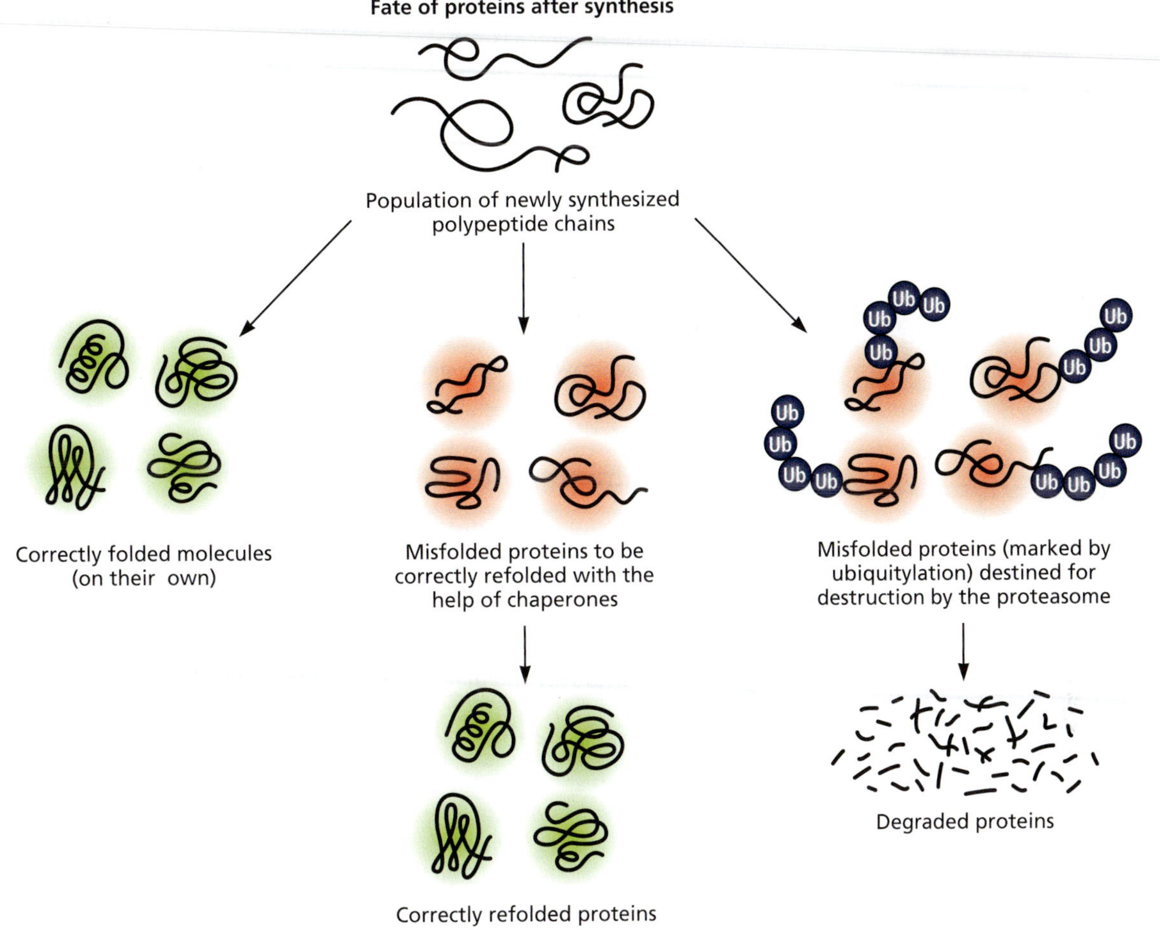

Figure 3.18 Three possible fates of proteins after synthesis. Some proteins can correctly fold by themselves. The cell has evolved two alternative pathways to take care of misfolded proteins: they are either refolded with the help of specialized folding machines, known as chaperones, or sent for degradation by a destruction machine, the proteasome.

and structure of this large protein complex are shown in Figure 3.19, and the reaction cycle is schematically presented in Figure 3.20.

Another important member of the chaperone family is **Hsp90** (heat-shock protein 90). This protein, originally identified as one of the proteins induced by stress conditions including heat, is the most abundant heat-shock protein under normal physiological conditions. It has roles in both protein folding and protein degradation and has been implicated in multiple biochemical pathways, including signal transduction. Although we know quite a bit about Hsp90 (Figure 3.21), its mode of function is still not fully understood, nor do we fully understand why it can exhibit such diverse functions.

ATP hydrolysis by Hsp90 is rather slow: in humans, Hsp90 hydrolyzes one ATP molecule in about 20 min. The slow hydrolysis indicates that the complex conformational rearrangements that occur in the Hsp90 dimer during its action are coupled to ATP hydrolysis (see Figure 3.21). Such coupling is a recurring theme in many reactions that involve ATP hydrolysis. The current view is that ATP binding and hydrolysis are instrumental in shifting the protein from one conformation to another; these conformational transitions, in turn, determine the activity of the protein. The equilibrium between different protein conformations is usually regulated in the cell; in the case of Hsp90, there are a large number of co-chaperones that bind to the C-terminus of Hsp90 and regulate its ATPase activity.

If the conformational transitions are inhibited one way or another, the protein is inactivated. Drugs that inactivate Hsp90 by blocking the ATP-binding site, and thus preclude the ATP-mediated conformational transitions, are widely used in

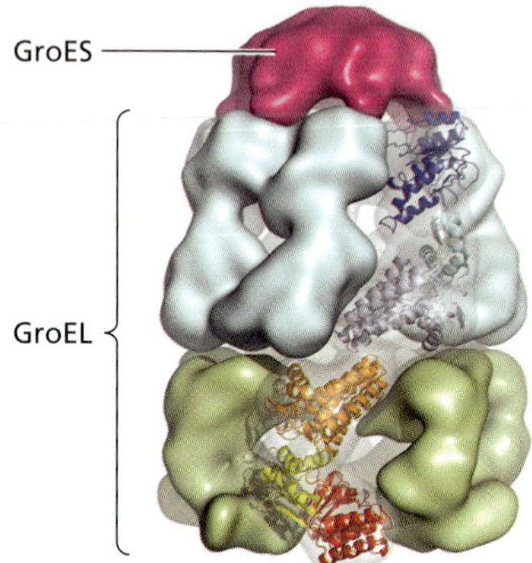

Figure 3.19 Structure of GroEL/GroES. The two heptameric rings of GroEL have a characteristic double-doughnut structure, as modeled from electron microscopic maps. Ribbon representations of the atomic structure of two GroEL protomers are superimposed on the model. Each GroEL subunit consists of an equatorial domain shown in light blue in the top molecule, which contains the nucleotide-binding site, an intermediate hinge domain shown in jade, and an apical domain shown in royal blue. The apical domain is located at the opening of the GroEL cylinder and contains the binding site for both GroES and the chaperoned polypeptide. The polypeptide binds at a hydrophobic groove in the apical domain facing the central channel. GroES, also termed the lid, is a dome-shaped heptamer that consists almost exclusively of β sheets.

clinical practice as anti-tumor agents. Presumably, these drugs induce apoptosis through inhibiting growth-factor signaling pathways; Hsp90 stabilizes growth-factor receptors.

3.5 How are proteins destroyed?

All proteins are eventually destroyed by hydrolysis; this accounts for the fact that proteins in fossils cannot last forever. But hydrolysis is slow unless it is catalyzed by proteases, and in general proteases cannot discriminate between proteins that are damaged or no longer needed in the cell and those that are vital. Therefore, there must be pathways that can mark and then destroy some proteins while not affecting others. Such pathways are known and play vital roles in the cell.

The proteasome is the general protein destruction system

The **proteasome** is responsible for the degradation of hundreds, possibly thousands, of badly misfolded or otherwise aberrant (mutated) proteins. It also destroys regulatory proteins, such as transcription factors or cell-cycle regulators, that must be available within a cell for strictly regulated brief periods and need to be destroyed once they have performed their function. Thus, transcription factors that regulate the expression of genes in response to temporary stimuli are needed only over a short period of time, and cyclins, the proteins that regulate the progression of the cell through the cell cycle, need to be present only during specific phases of the cycle and then must be degraded to allow the cell to move to the next phase.

Proteins that are destined for degradation by the proteasome will be marked by the covalent attachment of chains of the small protein molecule **ubiquitin**. The structural feature recognized by the ubiquitylation system is thought to be the presence of stable patches of hydrophobic amino acid residues exposed on the surface of the misfolded

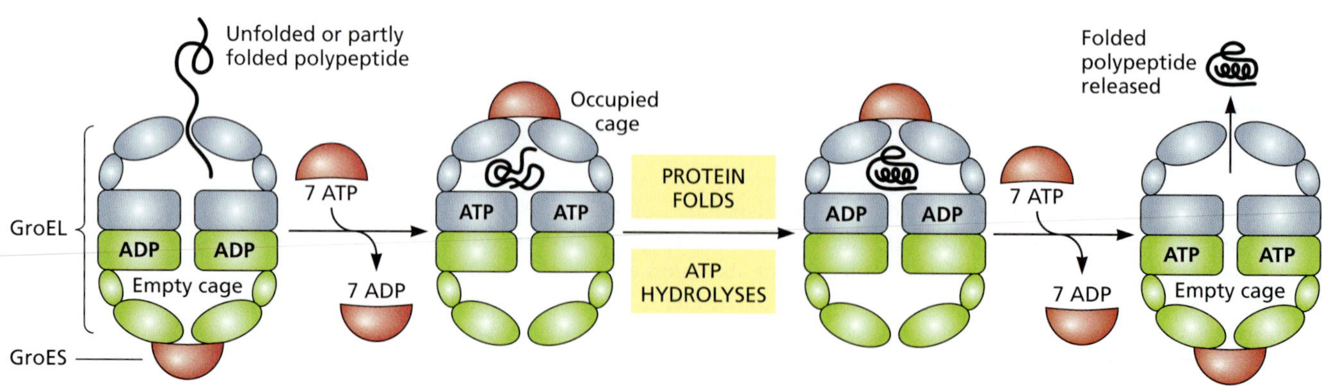

Figure 3.20 Reaction cycle of the bacterial GroEL/GroES chaperonin system. The nucleotide-binding abilities of the two GroEL rings are mutually exclusive: thus ATP or ADP can be bound to only one ring at a time. The GroES lid binds to the ring that has the nucleotide bound, creating a cage. The binding of unfolded or partly folded substrate protein to the open GroEL ring is followed by binding of ATP and GroES. The ring cavity now closes and the substrate is released into the cage, where it can fold. Following ATP hydrolysis, the binding of ATP to the opposite ring triggers the dissociation of GroES and the dissociation of folded substrate protein from the complex. The chaperonin complex is now ready to accept another polypeptide. (Adapted from Ellis RJ [2006] *Nature* 442:360–362. With permission from Macmillan Publishers, Ltd.)

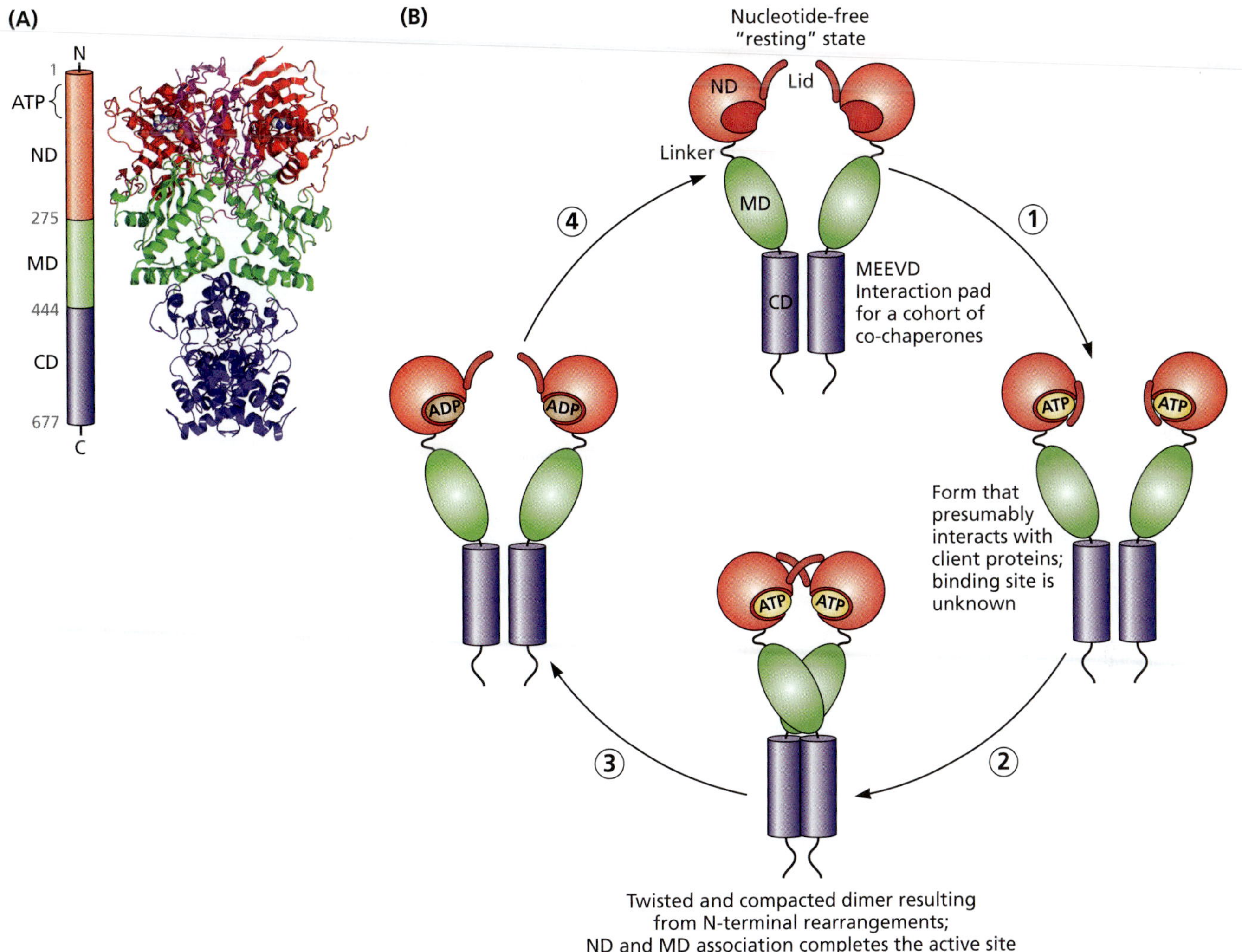

Figure 3.21 Chaperone Hsp90 and its ATPase cycle. (A) Structure of chaperone Hsp90 from yeast. The active protein is a dimer, with each subunit consisting of three distinct domains. The N-terminal domain ND, shown in red, possesses a deep ATP-binding pocket, where ATP binds in an unusual kinked conformation. The N-terminal domain in eukaryotes is connected to the middle domain MD, shown in green, by a long flexible linker sequence. The C-terminal domain CD, shown in royal blue, is the dimerization domain and contains, in eukaryotes, an amino acid motif MEEVD that binds to a variety of co-chaperones whose function is to regulate the ATPase activity of the N-terminus. (B) ATPase cycle of Hsp90. (Step 1) Following ATP binding, the lid flaps over the ATP-binding pocket. (Step 2) Lid binds to N-domain of the other subunit, producing a strand-swapped, transiently dimerized conformation. (Step 3) ATP hydrolysis; NDs dissociate from MDs; monomers separate N-terminally; lid opens. (Step 4) ADP and inorganic phosphate are released. The exact mode of client-protein binding remains unknown, as does the actual mechanism of protein folding. (A, adapted from Wikimedia. B, adapted from Mayer MP, Prodromou C & Frydman J [2009] *Nat Struct Mol Biol* 16:2–6. With permission from Macmillan Publishers, Ltd.)

chain. These are not present on proteins with normal tertiary structure; rather, such residues are usually buried within the protein. Their presence on the surface indicates that something is wrong. This ubiquitylation "kiss of death" leads the protein to a multisubunit structure called the proteasome.

The proteasome degradation process is shown schematically in **Figure 3.22A**. The eukaryotic proteasome consists of a 20S core particle (see Box 3.1 for the definition of S) formed by four heptameric rings, two outer α-rings and two inner β-rings (**Figure 3.22B** and **Figure 3.22C**), and one or two 19S regulatory particles of very complex composition. Some of the proteins in the core particle possess activities that hydrolytically degrade the client proteins. The main function of the 19S regulatory particle is to first recognize the ubiquitylated protein and then regulate access of the client protein to the 20S degradation chamber. The opening to this chamber is tiny, only 1.3 nm in diameter, and only unfolded polypeptide chains can pass through it. Thus,

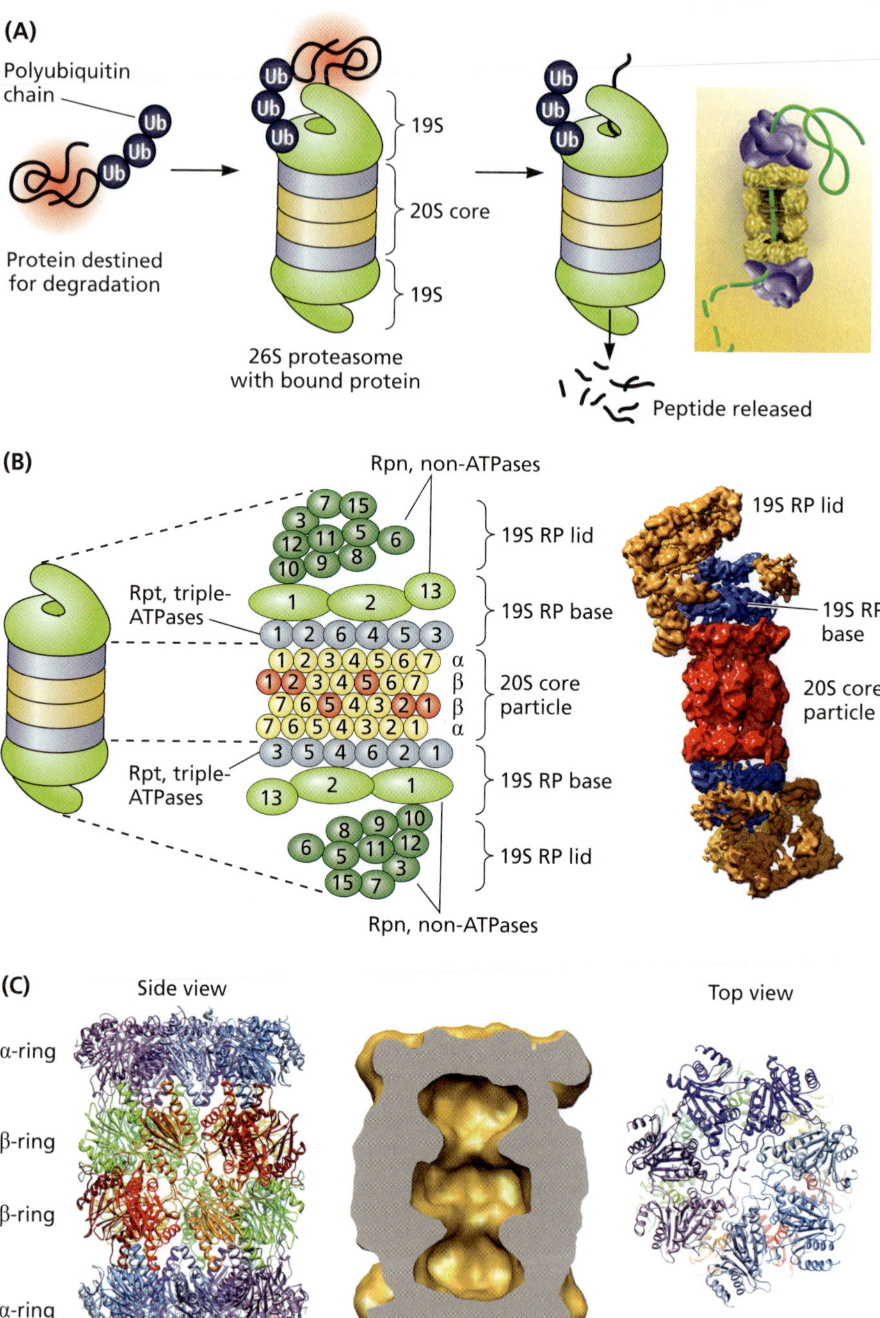

Figure 3.22 The proteasome: Structure and protein degradation. (A) Left, proteins to be degraded by the proteasome are usually modified by polyubiquitin chains, which are recognized and bound by components of the 19S regulatory particle. The misfolded protein needs to be unfolded to pass through the narrow gate of the 20S core particle. Right, cross section of the proteasome in action. (B) Left, schematic of subunit structure of the 26S proteasome. RP, regulatory particle, consisting of base and lid components; Rpn, regulatory particle non-ATPases; Rpt, regulatory particle ATPases. Right, three-dimensional reconstitution of the proteasome from single particles visualized by electron microscopy. (C) Left, architecture of the 20S core particle from yeast, ribbon diagram. Middle, volume of the particle calculated from atomic coordinates, cut in half to show three chambers within the particle. Right, top surface of the α-ring: the N-termini of the individual subunits form a closed gate that blocks access to the inner chamber. (A, left, adapted from Hochstrasser M [2009] *Nature* 458:422–429. With permission from Macmillan Publishers, Ltd. A, right, from Wikimedia. B, left, adapted from Tanaka K [2009] *Proc Jpn Acad Ser B* 85:12–36. With permission from The Japan Academy. B, right, from Lasker K, Förster F, Bohn S et al. [2012] *Proc Natl Acad Sci USA* 109:1380–1387. With permission from National Academy of Sciences. C, from Cheng Y [2009] *Curr Opin Struct Biol* 19:203–208. With permission from Elsevier.)

access to the degradation chamber is regulated by the 19S particle first unfolding the proteins. In addition, specific proteins in the 19S complex act as ubiquitin receptors, whereas others function to remove the polyubiquitin chains once the client protein is recognized and bound.

3.6 The proteome and protein interaction networks

New technologies allow a census of an organism's proteins and their interactions

Recent advances in high-throughput technologies have allowed unprecedented insights into the entire protein content of cells and the interactions among proteins. The library of all proteins encoded in the genome of a cell is termed the **proteome**. Many proteins are presently recognized only on the basis of portions of sequenced genomes (see Chapter 7); they have not yet been isolated as biochemical entities, and practically nothing is known about their possible functions. A huge

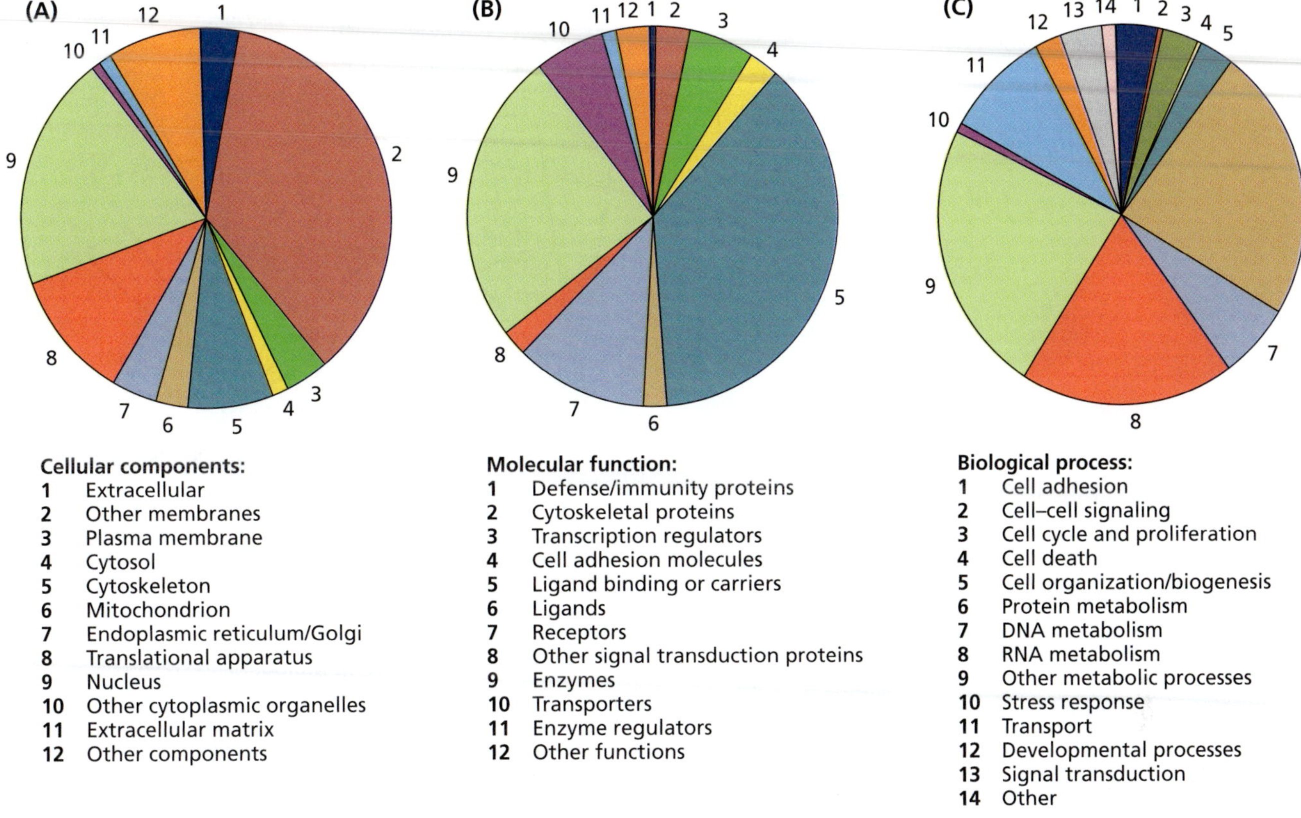

Figure 3.23 Distribution of the entire human proteome into three distinct large categories of proteins. The categories, as defined by the Gene Ontology database, are (A) cellular components, (B) molecular function, and (C) biological process.

amount of effort now goes into using available sequence information, at the gene and protein levels, and structural data to predict functions of proteins with closely related sequences. These predictions then have to be verified by actual biochemical experimentation.

Another significant challenge is to classify the proteome of a given species into different protein classes. Huge databases are created by the international scientific community to create orderly classifications and derive biologically relevant insights into these seemingly chaotic data. Figure 3.23 illustrates the classification of the entire human proteome into different classes related to the subcellular localization of the proteins, their molecular function, and the biological processes in which the proteins participate. The known proteomes of some other model species, such as the yeast *Saccharomyces cerevisiae*, the worm *Caenorhabditis elegans*, and the fruit fly *Drosophila melanogaster*, have also been classified into these same categories. It is important to realize that these achievements, as impressive as they are, constitute only the first steps in understanding the way the proteome functions.

Another state-of-the-art development in the proteome field is the creation of **interactome maps**. These are databases that identify, through high-throughput experimental methods, all of the possible binary interactions, that is, interactions between any two proteins, in the entire proteome. Other databases identify interactions in protein complexes, involving many protein partners. Some interactome maps focus on particular proteins or pathways—Figure 3.24 shows an example—while others are more ambitious, attempting to cover large portions of proteomes. Figure 3.25 presents a recent interactome of a large portion of the yeast proteome; Figure 3.26 shows different representations of a portion of the human interactome. Clearly, the range of interactions is wonderfully complex, even with our presently limited knowledge.

Again, it must be noted that creating these beautiful-looking maps is only the first step in a process that tries to make sense of all this complexity. In fact, attempts to

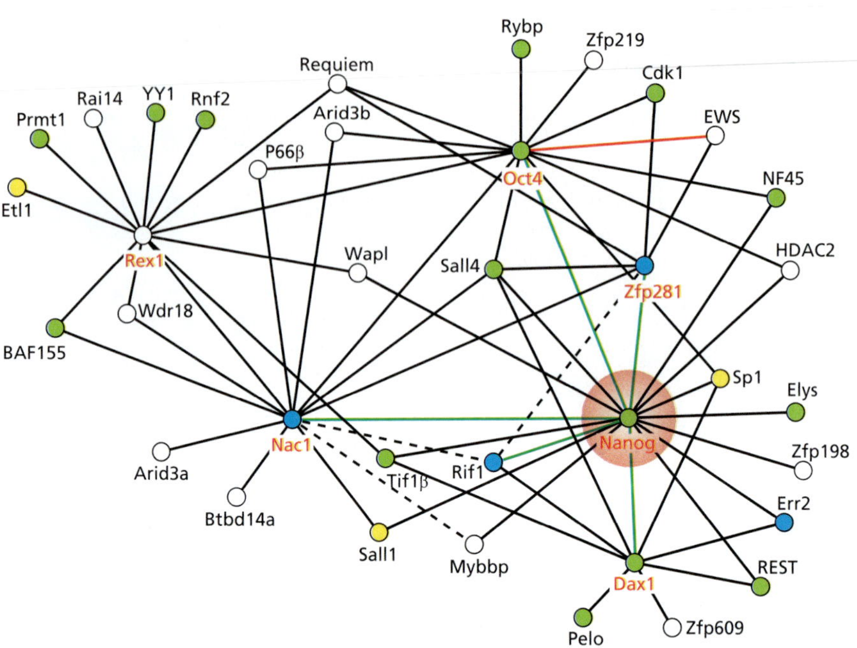

Figure 3.24 Partial interaction network for the protein Nanog. Nanog is known to participate in defining the pluripotent state of embryonic stem (ES) cells, that is, their ability to differentiate into numerous different cell types, depending on the applied stimuli. Affinity-purified Nanog partners were identified by mass spectrometry; in an iterative fashion, protein partners of these primary partners were then identified. The new protein partners were further verified experimentally by alternative techniques such as co-immunoprecipitation. The network is highly enriched in nuclear factors that are individually critical for the pluripotent state of the ES cell; the presence and levels of these factors are coordinated during further differentiation of the stem cells into different cell types. The tight protein network seems to function as a cellular module dedicated to pluripotency. The lines of different color that connect the proteins in the network indicate the methods by which the interaction has been verified experimentally. The circles of different colors denote proteins whose presence is essential for a particular stage of development, with the white circles designating proteins whose absence has not been correlated with developmental defects. (From Wang J, Rao S, Chu J et al. [2006] *Nature* 444:364–368. With permission from Macmillan Publishers. Ltd.)

represent large, complex databases in a meaningful way, in both science and technology, have given rise to an entirely new discipline called visual analytics. Scientists not only seek ways to graphically represent an interactome (see Figure 3.25 and Figure 3.26) but also try to detect "what's in there," to extract the basic principles that

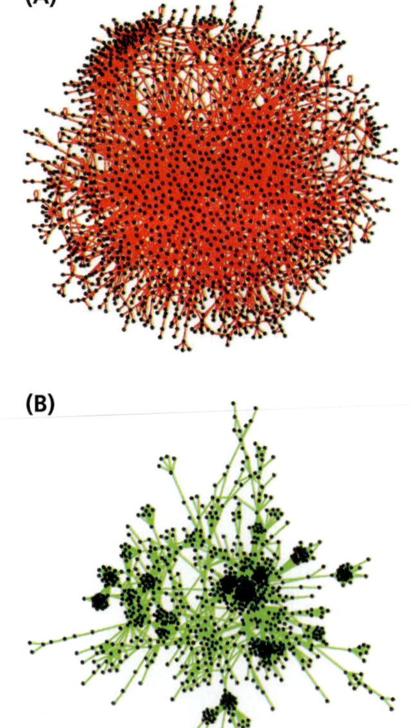

Figure 3.25 Yeast protein interaction maps representing binary protein–protein interactions. Binary protein–protein interactions are those between two proteins. These maps are derived from two different high-throughput experimental methods. Individual proteins are represented by dots, and interactions are shown by lines connecting the dots. (A) This interaction map from October 2008 combines three high-quality proteomewide data sets, obtained by yeast two-hybrid screening. The map covers ~20% of the entire binary yeast interactome and contains 2930 interactions among 2018 proteins. (B) Map of binary interaction extracted from co-complex interactome maps, which are obtained by high-throughput co-affinity purification followed by mass spectrometry, a principle similar to that presented in Box 3.9. The map is composed of 9070 associations between 1622 proteins. The two maps constructed by use of different system biology tools are fundamentally different but complementary in nature: they reflect different connections and biological properties of the interacting partners. Certain proteins are highly connected, with numerous partners; these are termed hub proteins. The absence of such proteins leads to a number of different phenotypes. Interestingly, hub proteins are characterized by intrinsic disorder, which may form the molecular basis for their promiscuous interaction with multiple partners. Interacting with different partners in different conditions constitutes the molecular basis for pleiotropic effects, that is, different roles of the same protein in multiple biochemical pathways. (From Yu H, Braun P, Yildirim MA et al. [2008] *Science* 322:104–110. With permission from American Association for the Advancement of Science.)

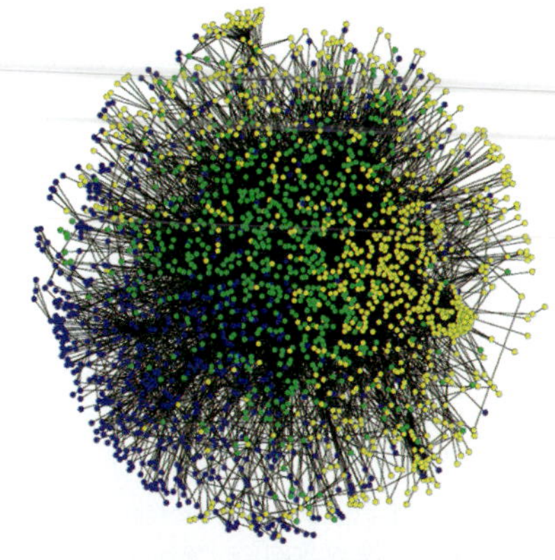

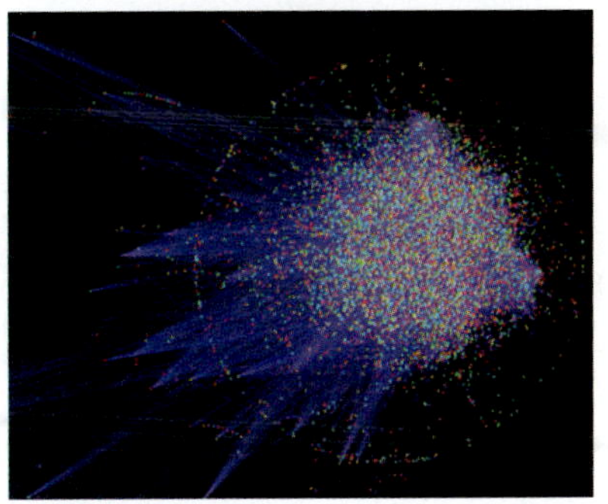

Figure 3.26 Different graphic representations of portions of the human interactome. These styles of representation are frequently used, but there are numerous ways to graph experimental interactome data. The estimated size of the binary interactome ranges from 130,000 to ~650,000 interactions; only ~8% of them have been identified. Ambiguity about the size of the human interactome stems from difficulties in identifying which of the possible biophysical interactions actually occur in cells. With the presently estimated number of human proteins, nearly 250 million protein pairs need to be tested to map the whole human interactome: such a huge undertaking is possible only with application of high-throughput approaches. The worm *Caenorhabditis elegans* appears to have a protein number similar to that in humans, and the sizes of the two interactomes appear to be of the same order. Then, what makes humans different from worms? We do not have the answer yet. (A, courtesy of Seth Berger, Mount Sinai School of Medicine. B, from Wikimedia.)

regulate the interactions of individual components. The insights that interactome mapping is expected to provide into the functioning of living organisms cannot be overstated.

It is obvious that the enormous diversity of proteins that exist in living organisms demands that every organism, in fact every cell, must have some repository of information to specify the primary structures of many proteins. The expression of this information must be under strict control as different cell types, even within a single organism, need different proteins to function, and these needs will change with growth and development. There must also be a way to preserve and transmit this information from generation to generation. That is the focus of this book.

Key concepts

- Proteins are polymers of L-α-amino acids, joined by peptide bonds. Thus, they are polypeptides.
- There are millions of kinds of proteins in living organisms, and each has a unique amino acid sequence or primary structure, which dictates its function.
- Certain regular foldings of the polypeptide chain are preferred: the most important are the α-helix and β-sheet. These are referred to as the secondary structure of the protein.
- Most proteins have, in addition, a preferred three-dimensional folding, referred to as the tertiary structure.
- The tertiary structure of the polypeptide chain can often be divided into recognizable domains. Some proteins have only one domain, while others have many. Occasionally, domains or even whole proteins lack any regular secondary or tertiary structure. These are called intrinsically disordered domains or proteins.
- Sometimes folded polypeptide chains associate noncovalently with one another to form more complex structures. These associated complexes are referred to as the quaternary level of structure.
- In many cases, protein folding and association processes are spontaneous and dictated by the primary structure.
- In the complex milieu of the cell, proteins can sometimes misfold. Chaperone complexes may engulf a misfolded protein and allow it to refold correctly. Alternatively, protease complexes may digest the aberrant protein back to its constituent amino acids.
- Recent proteomic studies are beginning to unravel the very complex interactions among proteins in the cell.

Further reading

Books

Branden C & Tooze J (1999) Introduction to Protein Structure, 2nd ed. Garland Publishing.

Creighton TE (1993) Proteins: Structures and Molecular Properties. WH Freeman.

Kyte J (1995) Structure in Protein Chemistry. Garland Publishing.

Pauling L (1960) The Nature of the Chemical Bond, 3rd ed. Cornell University Press.

Petsko GA & Ringe D (2003) Protein Structure and Function. Primers in Biology series. New Science Press.

van Holde KE, Johnson WC & Ho PS (2006) Principles of Physical Biochemistry, 2nd ed. Pearson Prentice Hall.

Reviews

Anfinsen CB & Scheraga HA (1975) Experimental and theoretical aspects of protein folding. *Adv Protein Chem* 29:205–300.

Clore GM & Gronenborn AM (1989) Determination of three-dimensional structures of proteins and nucleic acids in solution by nuclear magnetic resonance spectroscopy. *Crit Rev Biochem Mol Biol* 24:479–564.

Dunker AK, Oldfield CJ, Meng J et al. (2008) The unfoldomics decade: An update on intrinsically disordered proteins. *BMC Genomics* 9(Suppl 2):S1.

Ellis RJ (2006) Protein folding: Inside the cage. *Nature* 442:360–362.

Orengo CA & Thornton JM (2005) Protein families and their evolution: A structural perspective. *Annu Rev Biochem* 74:867–900.

Tanaka K (2009) The proteasome: Overview of structure and functions. *Proc Jpn Acad Ser B* 85:12–36.

Wandinger SK, Richter K & Buchner J (2008) The Hsp90 chaperone machinery. *J Biol Chem* 283:18473–18477.

Experimental papers

Lander ES, Linton LM, Birren B et al. (2001) Initial sequencing and analysis of the human genome. *Nature* 409:860–921.

Waterston RH, Lindblad-Toh K, Birney E et al. (2002) Initial sequencing and comparative analysis of the mouse genome. *Nature* 420:520–562.

Yu H, Braun P, Yildirim MA et al. (2008) High-quality binary protein interaction map of the yeast interactome network. *Science* 322:104–110.

Web sites

CATH: Protein Structure Classification Database at UCL.http://www.cathdb.info

Disprot: Database of Protein Disorder.http://www.disprot.org

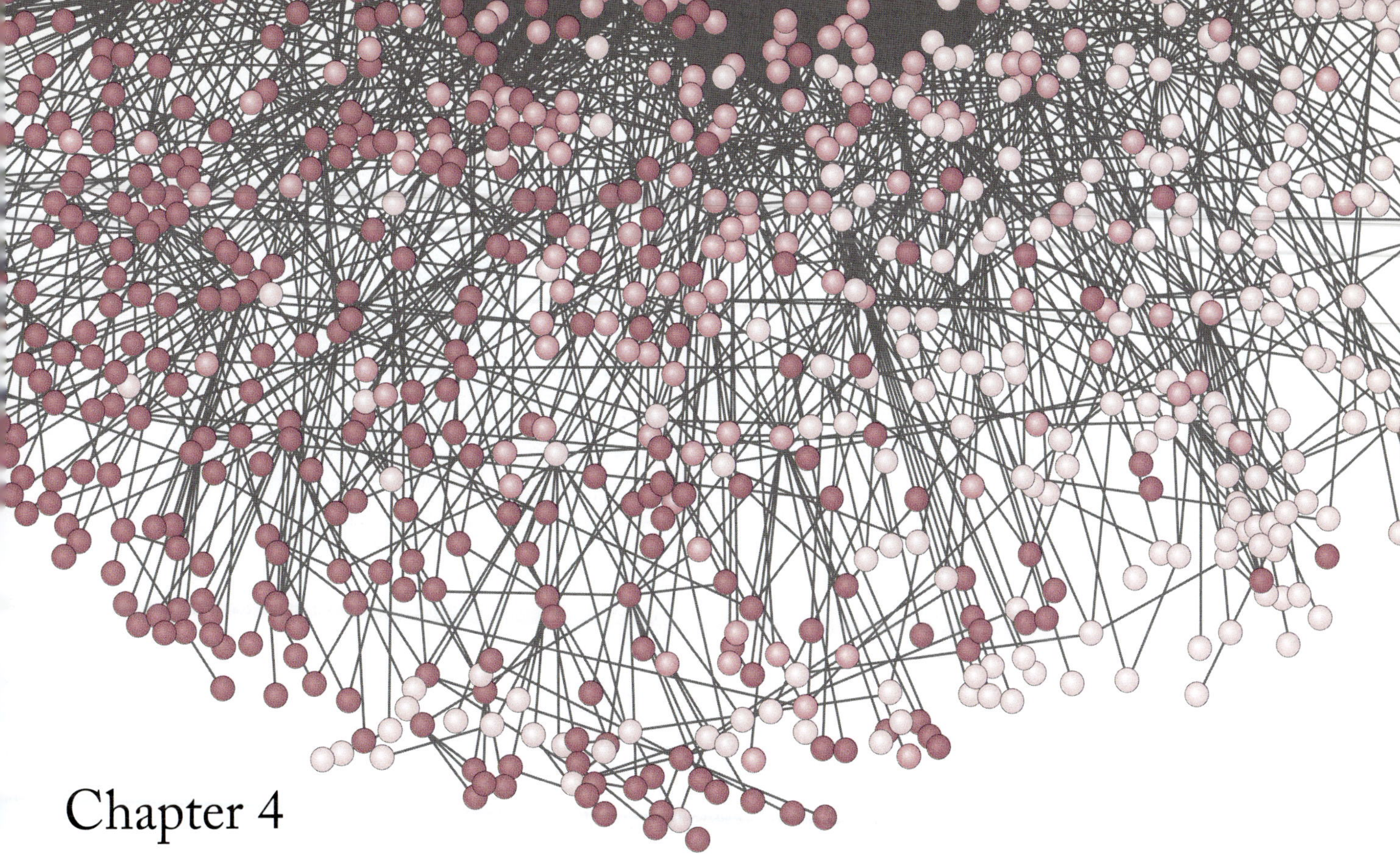

Chapter 4

Nucleic Acids

4.1 Introduction

Protein sequences are dictated by nucleic acids

We have seen that proteins, in their enormous variety, can play a host of roles in the cell, both structural and functional. Each protein accomplishes this by having a unique amino acid sequence, which determines its secondary, tertiary, and quaternary structures. The information that dictates these sequences must somehow be stored in the cell, expressed in proteins, and transmitted through generations of cells and organisms. These vital functions are provided by biopolymers called **nucleic acids**, or **polynucleotides**, of which there are two kinds: **ribonucleic acid** (**RNA**) and **deoxyribonucleic acid** (**DNA**). This chapter will be devoted to describing these nucleic acids, their structures and possible conformations, and the multiple ways in which they can store and transmit information in the cell.

4.2 Chemical structure of nucleic acids

DNA and RNA have similar but different chemical structures

The general structure for the monomeric units for DNA and RNA—the nucleoside monophosphates, or **nucleotides**—involves a five-carbon sugar attached to a base unit and a phosphate (**Figure 4.1**). The mononucleotides are members of a whole class of nucleoside phosphates, many of which have other biological roles; for example, **adenosine triphosphate (ATP)** and guanosine triphosphate (GTP) are important energy currencies in the cell. In each case, the phosphate is attached to the 5′-carbon of the sugar, and in the polynucleotide, there is a phosphodiester link

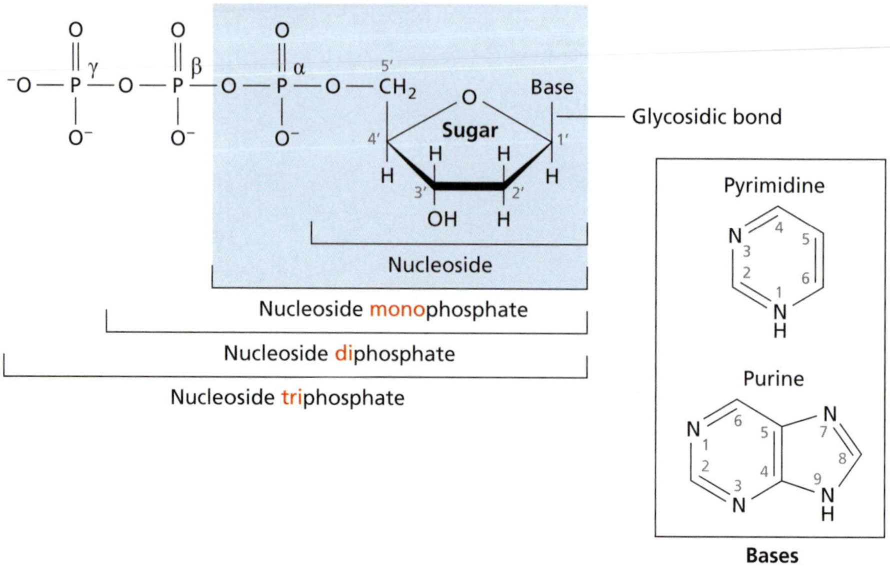

Figure 4.1 Chemical composition of nucleotides. Nucleotides contain a five-carbon (pentose) sugar—either deoxyribose in DNA, as shown here, or ribose in RNA (see Figure 4.2)—as well as a nitrogenous base and phosphate group(s). The base and sugar together form a nucleoside; depending on the number of phosphate groups attached to a nucleoside, it may be designated a nucleoside mono-, di-, or triphosphate. Nucleic acids are made of nucleoside monophosphates, shaded in blue, which contain nitrogenous bases that are derivatives of either pyrimidine or purine.

to the 3′-carbon of the next residue (see Figure 4.4). The sugars involved are ribose in RNA and 2′-deoxyribose in DNA.

The bases are all derived from either purine or pyrimidine, shown in Figure 4.1. The kinds of bases found in nucleic acids are shown in **Figure 4.2A**. The basic groups contained in nucleic acids are always attached to the 1′-carbon of the sugar. Four different kinds are found in DNA: **adenine (A)**, **guanine (G)**, **cytosine (C)**, and **thymine (T)**. RNA contains the same bases except that **uracil (U)** substitutes for thymine. A ribose or deoxyribose with a base attached is called a **nucleoside** (**Figure 4.2B**). When

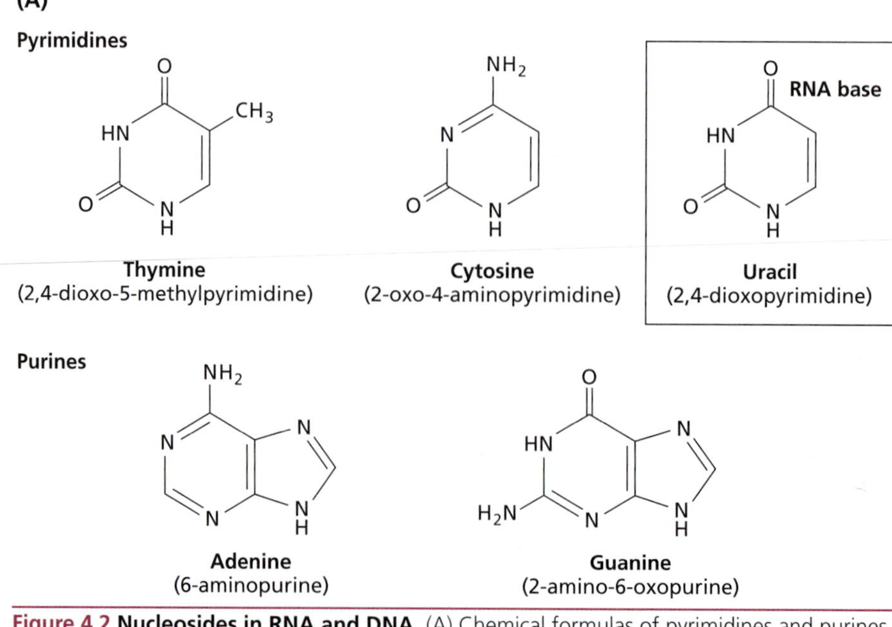

Figure 4.2 Nucleosides in RNA and DNA. (A) Chemical formulas of pyrimidines and purines. Uracil, shown in the box, is present only in RNA.

(B)

Deoxyadenosine (in DNA)

Adenosine (in RNA)

Base	DNA nucleoside	RNA nucleoside
Adenine	Deoxyadenosine	Adenosine
Guanine	Deoxyguanosine	Guanosine
Cytosine	Deoxycytidine	Cytidine
Thymine	Deoxythymidine (thymidine)	—
Uracil	—	Uridine

Figure 4.2 (B) Differences between the nucleosides in DNA and RNA are highlighted in red. The table shows the nomenclature of nucleosides formed by the addition of each base to the respective pentose.

a phosphate group is attached to the 5′-carbon of the sugar, a nucleotide, also called a nucleoside 5′-phosphate, is obtained (see Figure 4.1).

As depicted in **Figure 4.3A**, the sugar moiety of nucleic acids is in the ring or β-furanose form. To a first approximation, we may say that four of the five atoms of the ring lie in

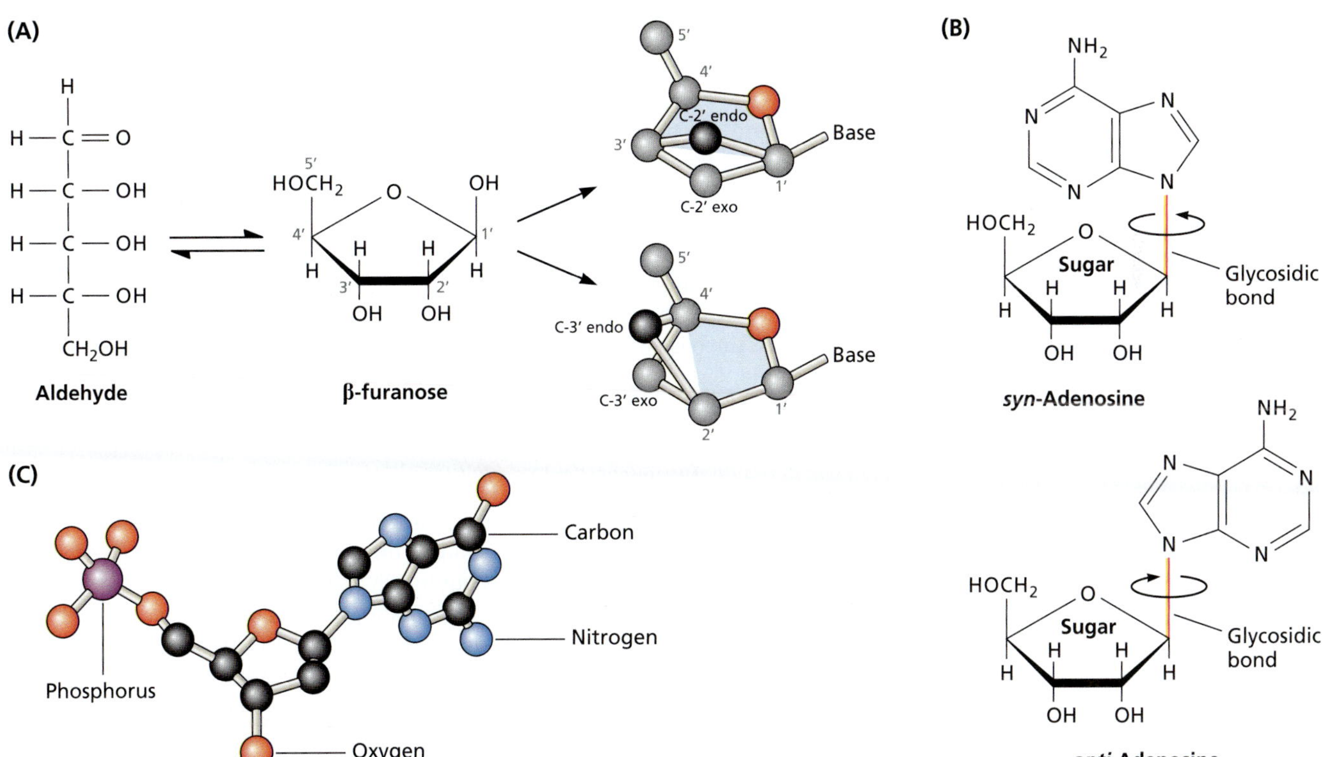

Figure 4.3 Conformations of nucleotide components. (A) Conformation of the ribose ring. In solution, there is equilibrium between the straight-chain or aldehyde form and the ring or β-furanose form of ribose; RNA contains only the ring form. The ribofuranose rings in nucleotides exist in two different puckered conformations: four of the five atoms in the ring lie in a single plane, while the fifth atom, C-2′ or C-3′, is either on the same side of the plane as the C-5′ atom, in *endo*nucleosides, or on the opposite side, in *exo*nucleosides. (B) *Syn* and *anti* conformations of nucleosides. Rotation around the glycosidic bonds allows the existence of two nucleoside conformations: *syn* and *anti*. In solution, purines rapidly equilibrate between the two conformers, while in pyrimidines the *anti* predominates. In nucleic acids, the preferred conformation is *anti*. (C) The three-dimensional structure of deoxyguanosine 5′-monophosphate (dGMP); hydrogen atoms have been omitted for clarity. The plane of the purine ring is almost perpendicular to that of the furanose ring. The purine base is in the *anti* conformation. The sugar ring is puckered in the 3′-*exo* conformation. The phosphoryl group attached to the C-5′ atom is positioned well above the sugar and far away from the base.

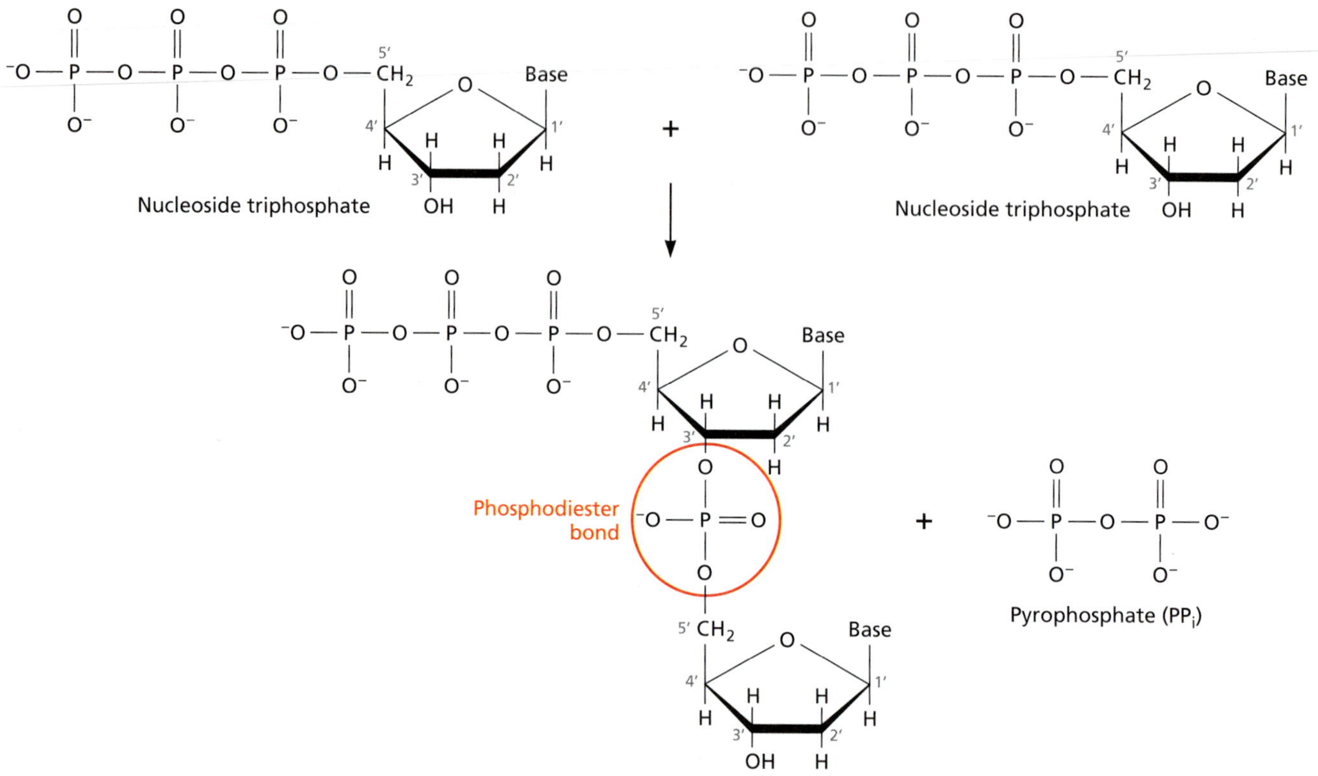

Figure 4.4 The phosphodiester bond. In DNA and RNA, the phosphodiester bond is the linkage between the C-3′ atom of one sugar molecule and the C-5′ atom of another, through the 5′-phosphate and a 3′-OH group.

the same plane; the position of the fifth atom, either C-2′ or C-3′, with respect to the plane determines whether the conformation is that of an ***endo*nucleoside**, with the fifth atom on the same side of the plane as the C-5′ atom, or an ***exo*nucleoside**, with the fifth atom on the opposite side of the plane. Actually the situation is more complex, with a variety of slightly different sugar conformations found. Nucleosides also have two conformations determined by the non-free rotation around the glycosidic bond (**Figure 4.3B**). The preferred conformation in nucleic acids is the *anti* conformation, in which the base and the sugar ring are as far away from each other as possible. As an example, **Figure 4.3C** shows the structure of the entire nucleoside monophosphate unit of deoxyguanosine monophosphate (dGMP).

Nucleic acids (polynucleotides) are polymers of nucleotides

Polynucleotides, or nucleic acids, can be visualized as being formed by the removal of a water molecule between the 5′-phosphate of one nucleotide and the 3′-hydroxyl of another. As in protein formation, this is not how it actually happens in the cell. Addition of a nucleoside triphosphate with elimination of pyrophosphate more closely describes the process; see **Figure 4.4** and Chapters 9 and 19. The chains can be very long, involving millions or even billions of nucleotide residues. Nucleic acids are acidic, because the linking phosphate groups carry negative charges. Like proteins, most nucleic acids have unique sequences. Note that there is an unreacted phosphate on one end of the chain, called the 5′-end, and an unreacted hydroxyl at the other end known as the 3′-end, except in cases where the molecule is circular. Thus, each polynucleotide chain has polarity (**Figure 4.5**).

The sugar–phosphate backbone of any nucleic acid is repetitious and uniform; the sequence uniqueness is carried by the order of the attached basic groups. Thus, we can compactly write the unique sequence of a DNA molecule as in this example, AGTCCTAAGCCTT, starting with the base at the 5′-end on the left according to convention. In RNA, uracil substitutes for thymine and so the corresponding RNA polynucleotide chain would read AGUCCUAAGCCUU.

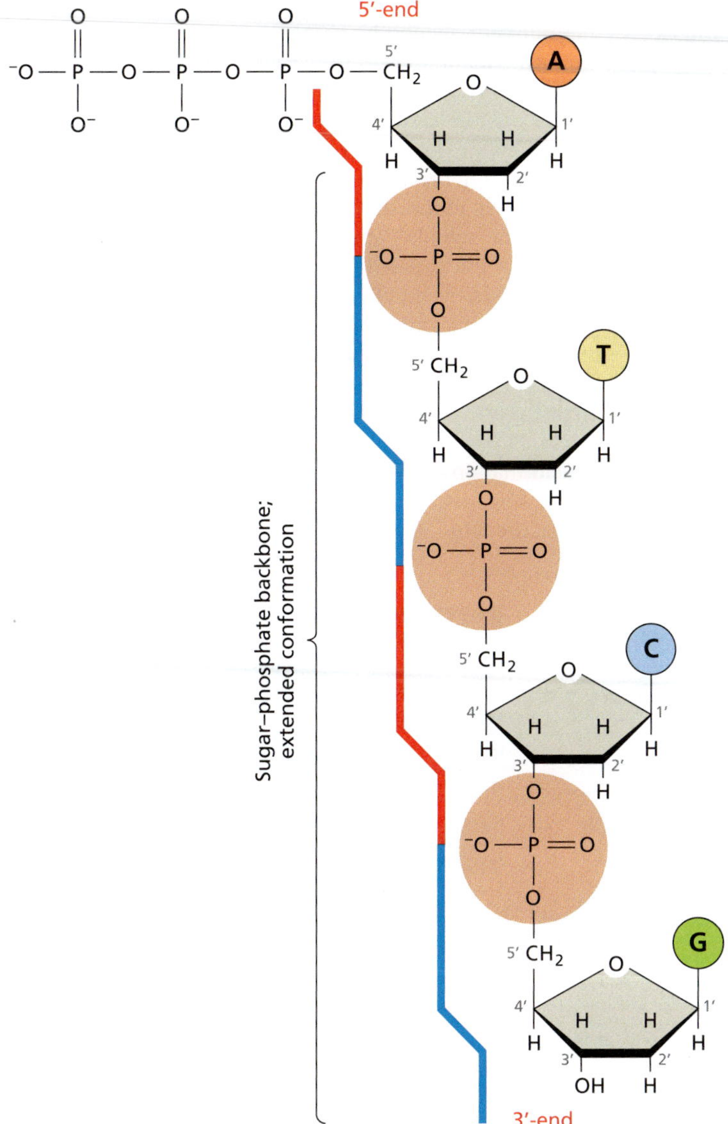

Figure 4.5 A tetranucleotide. The backbone of the chain is formed by the phosphates and sugars, marked in alternating blue and red for the successive nucleotide residues; bases are indicated by the circled letters. Note that the chain possesses polarity: the 5′-end and the 3′-end are chemically different. Reading the sequence from the top, in the 5′ → 3′ direction, will give ATCG; reading from the bottom, in the 3′ → 5′ direction, will give GCTA. By convention, nucleotide sequences in nucleic acids are written and read from 5′ to 3′.

4.3 Physical structures of DNA

Discovery of the B-DNA structure was a breakthrough in molecular biology

By the early 1950s, interest in DNA as a candidate for the genetic material had reached a high level of excitement. Critical experiments, described in Box 4.1 and Box 4.2, had strongly indicated this to be the long-sought genetic substance. Scientists still had no idea about DNA's structure, although there were already some provocative clues. For example, it had been shown that a peculiar relationship existed between the stoichiometry of bases in natural DNA: the amount of A was always about the same as the amount of T, with a similar result for G and C. These general rules concerning the base composition of any DNA were deduced by Edwin Chargaff from analysis of the nucleotide composition of DNA samples from different organisms (Table 4.1). The first of **Chargaff's rules** states the equal representation of A and T and of C and G in any DNA molecule; in other words, the sum of purines, A + G, equals the sum of pyrimidines, C + T. The second rule concerns the overall composition of DNA in different organisms: it varies from one species to another. Any proposed structure must explain Chargaff's rules and clarify the even more mysterious problem of how the genetic material could be accurately replicated when cells divide or organisms reproduce.

It is now widely known that the solution came from the brilliant insights of James Watson and Francis Crick, guided by the experimental work of others, including that of

Box 4.1 DNA is the carrier of genetic information: the experiments of Griffith and of Avery, MacLeod, and McCarty
Frederick Griffith was a British health officer working on developing a vaccine against pneumococcus infections. He was interested in understanding why multiple types of *Streptococcus pneumoniae*, some virulent and some nonvirulent, were often present over the course of the disease. He entertained the idea that one bacterial type might somehow change into another, rather than patients being infected simultaneously by multiple types at the onset of disease. He did his experiments with two strains of *S. pneumoniae*, rough (nonvirulent) and smooth (virulent), which he injected into mice either separately or in combination (**Figure 1**). The interesting and totally unexpected result came when he injected a mixture of heat-inactivated virulent strain and live nonvirulent strain: the mice unexpectedly died despite the virulent strain having been inactivated. Moreover, Griffith was able to recover live virulent strain from the dead mouse. How could this happen? Evidently, some "principle" had transformed the nonvirulent strain into a killer. The question was, what was that principle?

The answer came in 1944, almost 20 years later, when Oswald Avery, Colin MacLeod, and Maclyn McCarty, working at the Rockefeller Institute for Medical Research, identified the chemical nature of the substance that caused transformation. Saline-soluble components of heat-killed bacteria were chemically fractionated by the simple procedures available at the time, and eventually the active portion was identified as a substance whose physical properties and chemical composition were consistent with those of DNA. Final proof that the transforming principle was DNA came from enzymatic treatments: proteases and ribonucleases did not affect the principle, but a crude preparation of DNase destroyed the extract's transformation ability. It became clear beyond doubt that the bearer of genetic information was DNA. However, few took note at the time.

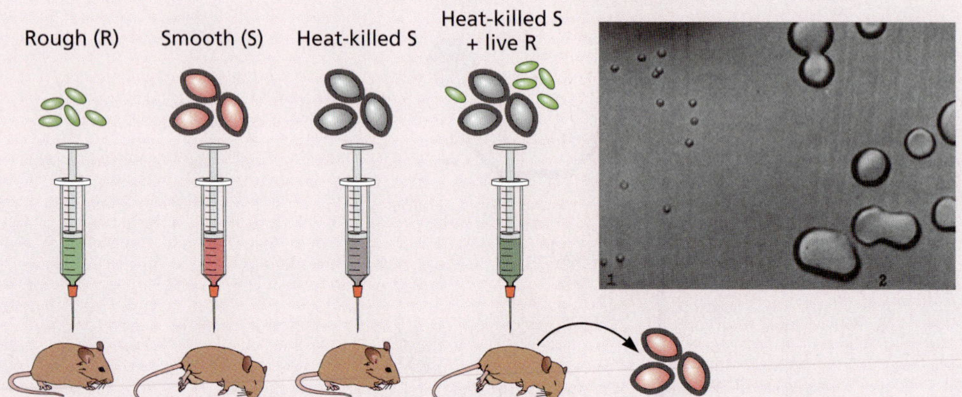

Figure 1 The Griffith experiment. After the mouse is injected with heat-killed S and live R, live S bacteria are isolated from the tissues of the dead mouse. Something must have transformed the live R into live S. The micrograph at right shows the appearance of *Streptococcus pneumoniae*: colonies of rough nonvirulent strains on the left and smooth virulent strains on the right. The smooth colony appearance is due to the presence, on the cell surface, of a polysaccharide capsule that makes the strain resistant to phagocytosis by the cells of the host immune system. (Right, from Avery OT, MacLeod CM & McCarty M [1944] *J Exp Med* 79:137–158. With permission from Rockefeller University Press.)

Chargaff and the crystallographic results of Rosalind Franklin and Maurice Wilkins (**Box 4.3**). The right-handed double-helix model proposed by Watson and Crick (**Figure 4.6**) fitted all the requirements of a genetic material so perfectly that it just had to be correct. First, the A–T and G–C base pairings allowed strong hydrogen-bonding between the two DNA strands (Figure 4.6) and accounted for the peculiar base stoichiometry, A = T and G = C, noted above. This structure also put the 1′-carbons in either pair exactly the same distance apart, 1.08 nm, allowing for a uniform, unstrained double helix. Finally, and most importantly, the model provided a strong hint, noted by Watson and Crick, as to how DNA could be replicated. Note that the two strands in the double helix, which are antiparallel, meaning they run in opposite directions, are exact complements of one another. If the strands were separated and each was used as a template to specify a new strand, two new double-stranded molecules would be obtained, each an exact copy of the original **duplex** (**Figure 4.7**). This mode of copying, called semiconservative replication, was demonstrated to be correct by Matthew Meselson and Frank Stahl (see Chapter 19).

The original Watson–Crick model was an inspired guess, based on fiber X-ray diffraction patterns. Such a pattern does not provide the kind of unambiguous structure that can be obtained from X-ray diffraction patterns from single crystals (see Box 3.6). Many

Box 4.2 The Hershey–Chase experiment Evidence that the genetic material must be DNA goes back to at least the 1944 experiments of Avery and collaborators (see Box 4.1). Even up to the 1950s, however, many scientists believed that the more complex proteins must play this role. This idea was disproved, once and for all, by a simple experiment performed by Alfred Hershey and Martha Chase (Figure 1). They made use of a virus of bacteria, the bacteriophage T2, which infects *Escherichia coli*, producing more phage inside the bacterium and lysing it to release the progeny phage. Electron microscopy had already shown that the T2 phage has a protein head, which is a container for DNA, and a tail that attaches the phage to the surface of the bacterium, but there was no evidence that the whole phage entered the cell. Something must be delivered into the cell to cause the infection, but what?

Hershey and Chase made use of the fact that DNA contains phosphorus, whereas proteins have very little. Conversely, proteins contain sulfur but DNA has none. Two batches of phage were grown: one batch was grown on radioactive phosphorus, which would selectively label the DNA, while the other was grown on radiolabeled sulfur compounds, to selectively label proteins. Each batch of phage was used to infect a separate culture of *E. coli*. After the phage had been attached for a while, they were shaken off the bacteria by use of a Waring blender, and the infected bacteria were separated from the empty phage "ghosts" by centrifugation. Most of the phosphorus label, and hence the DNA, was found in the cell pellet containing the bacteria, whereas virtually all of the sulfur label, and hence the protein, remained in the supernatant with the phage ghosts. Thus it was shown that DNA, not protein, had been inserted into the bacteria. Furthermore, when the bacteria were lysed by the phage, the emerging phage progeny contained a portion of the DNA label. The phage progeny contained protein, but this was NOT the protein that had been in the infecting phage because it did not contain the radioactive sulfur label. Thus the DNA must somehow have directed the formation of new protein within the bacterium.

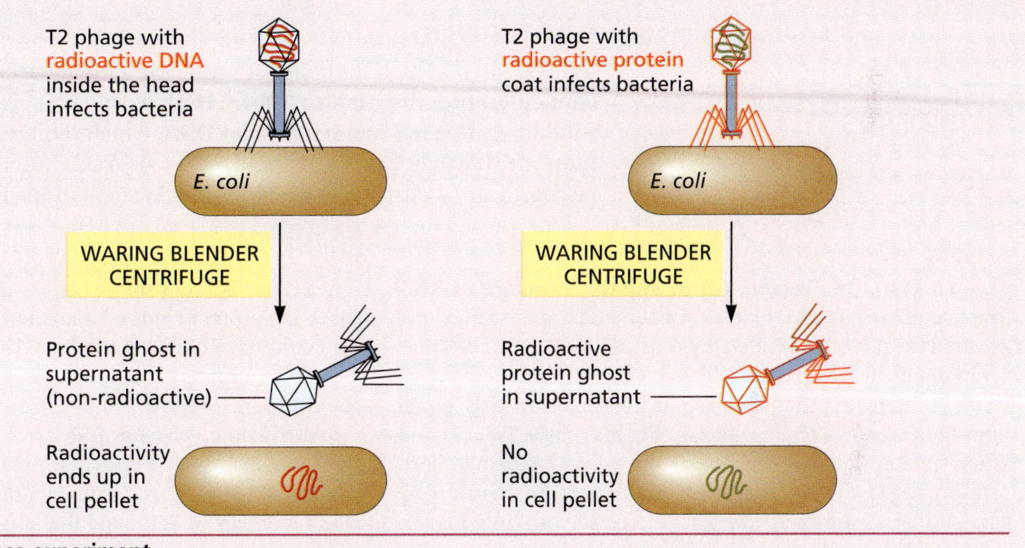

Figure 1 The Hershey–Chase experiment.

years passed before true crystals of small DNA molecules were obtained and studied by this technique. The results turned out to be remarkably close to those predicted by the Watson–Crick model (Figure 4.8). There are in fact nearly, but not exactly, 10 base pairs for each turn of the helix, and the bases are almost, but not exactly, perpendicular to the helix axis (Table 4.2).

At elevated temperature, DNA will **denature** in the sense that the duplex will dissociate into single strands (Figure 4.9). This process often occurs at a sharply defined temperature and is therefore also called melting. The melting point is increased by high G/C content or high salt, facts that are often useful in experimental protocols. Denatured DNA can reconstitute double-stranded base-paired duplexes upon slow cooling or annealing. The ability of single strands of DNA of complementary base sequence to form DNA duplexes forms the basis for a number of experimental hybridization techniques for labeling DNA fragments.

A number of alternative DNA structures exist

DNA is, in fact, capable of adopting more than one type of duplex structure, depending on the base sequence in certain regions of the polynucleotide chain and the environmental conditions. The one that Watson and Crick studied is termed **B-form DNA**

Table 4.1 Base composition of DNA (mole %) and ratios of bases that led to the formulation of Chargaff's rules. The slight deviations in the A/T, G/C, and purines/pyrimidines ratios from 1.00 reflect uncertainties in experimental measurements. Bacteriophage φX174 genome is single-stranded and thus does not follow Chargaff's rules. (Data reported in Sober HR [ed] [1970] Handbook of Biochemistry: Selected Data for Molecular Biology, 2nd ed. With permission from CRC Press.)

Organism	A	G	C	T	(C+G)	A/T	G/C	Purines/ Pyrimidines
bacteriophage φX174	24.0	23.3	21.5	31.2	44.8	0.77	1.08	0.89
Escherichia coli	23.8	26.8	26.3	23.1	53.2	1.03	1.02	1.02
Mycobacterium tuberculosis	15.1	34.9	35.4	14.6	70.3	1.03	0.99	1.00
yeast	31.7	18.3	17.4	32.6	35.7	0.97	1.05	1.00
Drosophila	30.7	19.6	20.2	29.5	39.8	1.03	0.97	1.01
corn/maize	26.8	22.8	23.2	27.2	46.1	0.99	0.98	0.98
calf	27.3	22.5	22.5	27.7	45.0	0.99	1.00	0.99
pig	29.8	20.7	20.7	29.1	41.4	1.02	1.00	1.01
human	29.3	20.7	20.0	30.0	40.7	0.98	1.04	1.00

and is favored at high humidity; it is therefore the most common form *in vivo* and in solution *in vitro*. There is also **A-form DNA**, existing especially under conditions of low humidity, and a left-handed **Z-form DNA**, which requires special base composition. These are shown in **Figure 4.10**, and the detailed structural parameters characteristic of each form are listed in Table 4.2. Occasionally single, unpaired DNA strands are found in

Box 4.3 Franklin, Watson, Crick, and the structure of DNA The tale of how, in 1953, James Watson and Francis Crick discovered the double-stranded helical structure of DNA is familiar to many. But the way in which it actually happened is a bit more complicated than the popular version and tells much about how science really works. In the first place, Watson and Crick were an unlikely pair for such a groundbreaking effort. Watson came to the Cavendish Laboratory in Cambridge, England, as a young postdoctoral biologist, with little knowledge of DNA. Crick was still working on his Ph.D.; his studies had been interrupted by the Second World War.

Both shared the strong belief that the structure of DNA was important and could be solved with the aid of X-ray diffraction studies of DNA fibers. But Watson had no experience with the method, and Crick's experience was largely theoretical. Furthermore, they lacked the appropriate equipment to obtain such data and the director of the Cavendish Laboratory, Sir Lawrence Bragg, was unenthusiastic about their embarking on such a project. On the other hand, excellent equipment and long experience in diffraction studies of fibrous polymers existed in the laboratory headed by Maurice Wilkins at King's College, London. There, a young researcher named Rosalind Franklin had, for several years, been attempting to get better and better diffraction patterns from DNA fibers prepared under various circumstances. Franklin was a very careful worker, unwilling to publish or share results until she was convinced of their validity. By 1952, however, she was obtaining good patterns from two forms of DNA, an A-form found at low humidity and the B-form found in wet samples. The patterns strongly indicated that the B-form was helical, with about 10 residues per turn of the spiral and a spacing of 0.34 nm between residues.

These data still left the overall structure unclear: there could have been one or more chains in the molecule, and the bases could have been on the outside or the inside of the chain. Franklin apparently wished to wait for more data before publication.

Meanwhile, Watson and Crick were attempting to build reasonable models based on the skimpy older data and stereochemical constraints. Linus Pauling had proceeded in much the same way in deducing the structure of the α-helix of proteins and was by now also trying to do the same with DNA. Both came up with first models that were soon realized to be unlikely, putting the bases on the outside of the helix and crowding the negatively charged phosphates in the center.

January 30, 1953, was a pivotal day in the history of molecular biology. Watson went to London and stopped to see Rosalind Franklin and Maurice Wilkins. The meeting with Franklin did not go well, but the subsequent chat with Wilkins was quite another matter. Wilkins showed Watson the very fine diffraction patterns that Franklin had obtained for B-DNA. Watson was astounded and immediately realized the clear implications: that B-DNA was indeed helical, with 10 units per turn. This provided a framework on which models could be built. Franklin had come to the same conclusion but was opposed to model building.

Watson and Crick were not in the least inhibited and immediately began building three-dimensional models using Franklin's findings. Available data on the density and water content of DNA fibers convinced them that the structure involved two DNA strands. Furthermore, models showed that the phosphate backbones could not lie crowded at the center of the double helix but must be at the periphery with the bases inside. This initially caused problems: the bases were of different dimensions, the purines bigger than the pyrimidines. Fitting them together would yield an irregular bumpy helix. A further difficulty came from the fact that Watson and Crick were initially using the wrong tautomers of G and T; Jerry Donohue, a postdoctoral associate, corrected this. Watson soon realized that there was one way in which a smooth helix could be produced: if a purine was always across from a pyrimidine and vice versa. Furthermore, it became evident that only pairing A with T and G with C allowed the formation of a number of hydrogen bonds between the conjugate pairs (see Figure 4.6). But then A = T and G = C, exactly following Chargaff's rules. Suddenly everything fell into place, even a mechanism for transmitting genetic information between generations: if the chains must be complementary, then separating and copying each of the separated strands would lead to two double helices identical to the original.

In ascribing credit for this remarkable discovery, one must remember that the critical data were obtained by Rosalind Franklin during years of hard, careful work. But the inspiration that put all the pieces of the puzzle together was clearly that of Watson and Crick. When the Nobel Prize in Physiology or Medicine was awarded in 1962, it was divided between Watson, Crick, and Wilkins. Franklin had died a few years earlier without comparable honors: the Nobel Prize is never awarded posthumously.

nature, but for most purposes, we can consider the B-form to be the biologically most important form, and the Watson–Crick model is a good description of its predominant helical structure.

Certain DNA viruses contain a radically different form of DNA, in which the phosphate backbone is inside and the bases project outward. This form is sometimes called the Pauling form, as it corresponds closely to an incorrect early proposal by Pauling for B-DNA structure (see Box 4.3).

Because the sequence of one DNA strand dictates the complementary sequence of the other, we do not usually bother to write down both sequences in describing a duplex DNA; by convention, one strand is written from its 5′-end and the sequence of the complementary strand is understood.

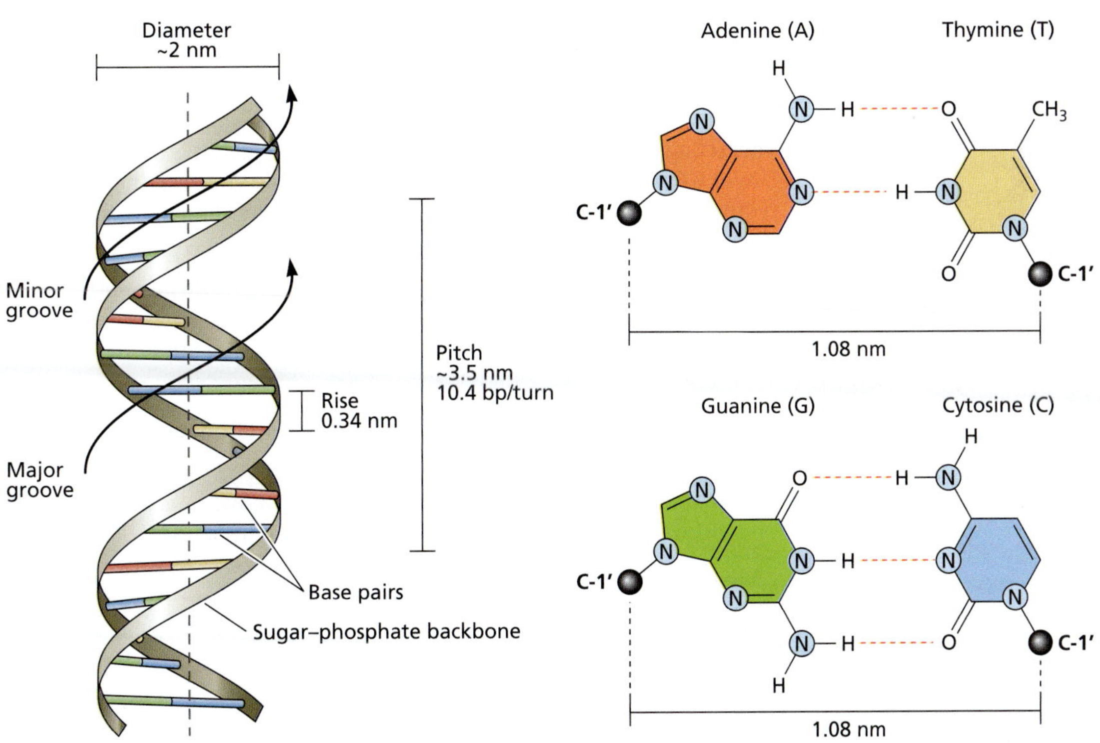

Figure 4.6 Schematic of the double-helix structure of B-form DNA. Base pairing occurs between a purine on one strand and the corresponding pyrimidine on the other strand; the pairing between these complementary bases allows the distances between the C-1′ on the sugar moieties to be exactly the same for both adenine–thymine and guanine–cytosine base pairings. Hydrogen-bonding between the complementary bases and stacking interactions between successive pairs are both important in stabilizing the helix. The original Watson–Crick model closely resembles this but with 10.0 base pairs/turn. (Adapted, courtesy of the U.S. National Library of Medicine.)

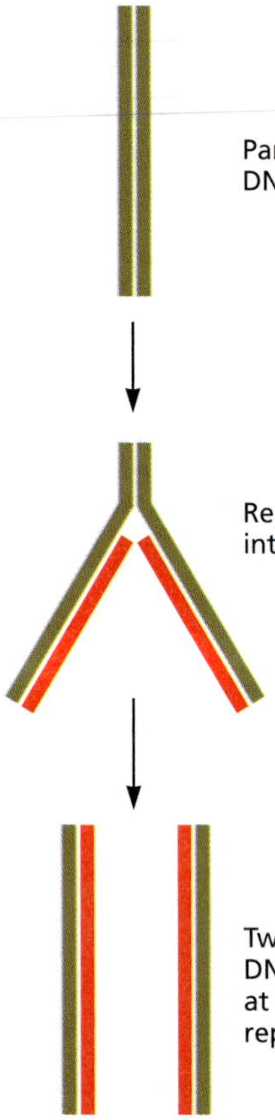

Parental DNA duplex

Replication intermediate

Two daughter DNA duplexes at the end of replication

Figure 4.7 The double-helical structure of DNA allows for a simple semiconservative mechanism of DNA replication. The unwinding of the two strands of the parental duplex, shown in brown, is accompanied by synthesis of two complementary daughter strands, shown in red, each using one parental strand as a template. This model of replication of the genetic material was envisaged by Watson and Crick in 1953, as soon as they deciphered the structure of DNA.

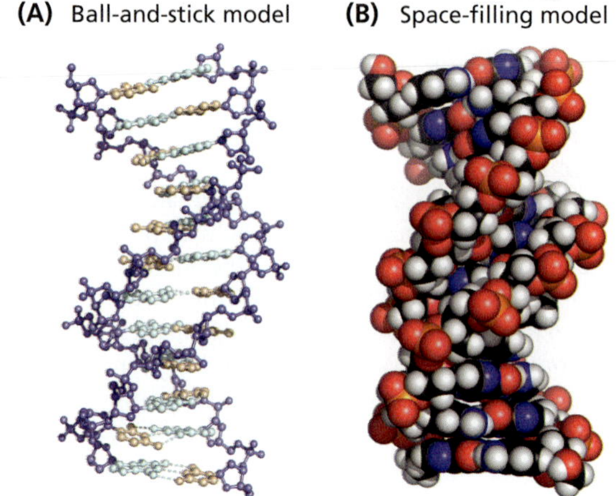

Figure 4.8 Structure of B-DNA as derived from crystallographic studies. (A) Ball-and-stick model. (B) Space-filling model. The base stacking is especially clear in this model, in which each atom is represented by a sphere of its van der Waals radius.

Although the double helix is quite rigid, it can be bent by bound proteins

Most depictions of B-DNA suggest that it is a rigid rodlike molecule, but this is not an exact description. DNA duplexes are bendable, and in some cases they contain built-in bends. The flexibility of any polymer molecule can be expressed in terms of a quantity called the **persistence length**. This may be understood from the following thought experiment: Imagine that you are holding one end of a molecule rigidly in your hand. How accurately can you predict the position of the other end? If the molecule is extremely rigid, you can make an exact prediction. Such a molecule is said to have a very large persistence length. On the other hand, if the molecule is extremely flexible and floppy, you can make no good prediction as to the position of the other end. This molecule has a very small persistence length. Many studies of B-DNA in solution have resulted in a value of about 45 nm for the persistence length, corresponding to ~130 bp. This limited bendability places constraints on the structures that DNA can form by interacting with proteins. Thus, for example, although it is possible to wrap 147 bp of DNA 1.67 turns about the protein core of a nucleosome (see Chapter 8), this wrapping requires significant bending energy, as well as some changes in **twist** and some dislocations in the helix.

In addition to this basic flexibility, DNA can have intrinsic bends, dictated by base sequence. The stacking together of base pairs can be asymmetric; some pairs will stack more tightly on one side of the helix than on the other. To take one example, a

Table 4.2 Structural parameters of the three forms of double-helical DNA.

	B-form	A-form	Z-form
handedness (helical sense)	right-handed	right-handed	left-handed
diameter (nm)	~2.0	~2.6	~1.8
base pairs per helical turn	10.4	11	12
helix rise per base pair (nm)	0.34	0.26	0.37
base tilt normal to the helix axis (deg)	6	20	7
sugar pucker conformation	2′-*endo*	3′-*endo*	2′-*endo* for pyrimidines; 3′-*endo* for purines
glycosidic bond conformation	*anti*	*anti*	*anti* for pyrimidines; *syn* for purines

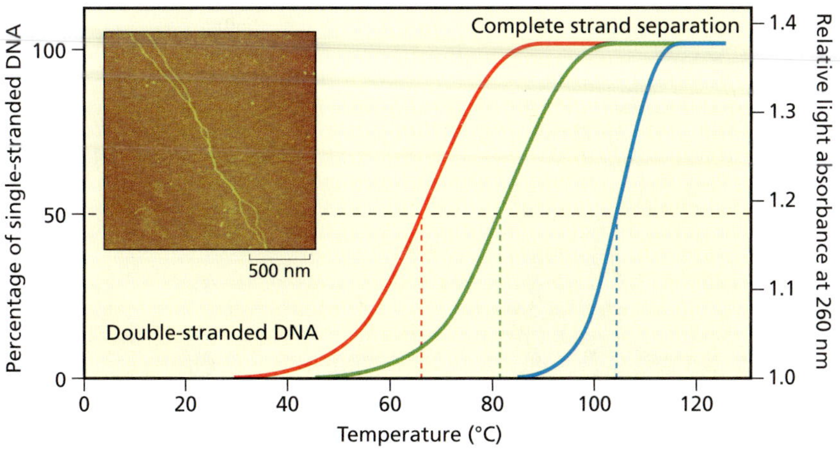

Figure 4.9 Melting of double-stranded DNA. Absorbance at 260 nm is usually measured in high-salt buffers, 100 mM Na^+, as the temperature of the solution is slowly raised at a rate of 1°C/min. The melting temperature is defined as the temperature at which 50% of the DNA is still in the form of a duplex. The three different curves illustrate the dependence of melting behavior on the base composition of the DNA. Poly[d(AT)] is represented by the red line, naturally occurring DNA by black, and poly[d(GC)] by blue. The melting temperature, T_m, is indicated by the corresponding dotted lines. (Inset) Electron micrograph of partially denatured DNA. (Inset, from Liu Y-Y, Wang P-Y, Dou S-X et al. [2005] *Sci. Technol. Adv. Mater.* 6:842–847. With permission from Elsevier.)

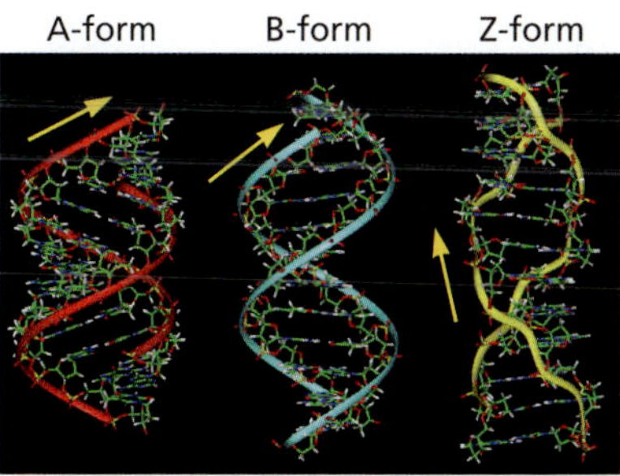

Figure 4.10 Alternative forms of DNA double helices. The three major forms of the DNA double helix, A, B, and Z, are shown. Note the handedness of the sugar–phosphate backbone, shown by yellow arrows: it is right-handed in the A and B structures but left-handed in Z. Note that the B-form has a wider major groove and a narrower minor groove. These grooves are important in DNA–protein interactions (see Chapter 6). In the A-form, both grooves are nearly the same width. (From Wikimedia.)

series of regularly spaced dA/dT pentamers [$d(A/T)_5$ tracts] will produce significant bending. The existence of intrinsically bent DNA may be important in facilitating the binding of certain proteins to DNA (see Chapter 6). Intrinsic DNA bending can be detected by its effect on the electrophoretic mobility of DNA in gels: bent DNA will migrate more slowly than linear DNA of the same length. An additional method for estimating the degree of bending involves the circularization of small pieces of DNA by use of a ligase, an enzyme that is capable of joining two pieces of DNA into a single uninterrupted molecule. The experiments are done with pieces of DNA that are so short, and hence rigid, that they would not be able to form circles on their own if they do not contain built-in bends. Circularization will be greatly enhanced if the molecule is bent, with the probability of circularization going up with larger or sharper or more numerous bends. Gel electrophoresis is a convenient method to estimate the probability of circularization, as it readily separates linear molecules from circular ones.

DNA can also form folded tertiary structures

Special DNA sequences can engender three-dimensional DNA structures. One example is found in palindromic sequences. In linguistics, a **palindrome** is a sequence of letters or words that reads the same way in either direction: "Able was I ere I saw Elba," Napoleon might have said. Consider the sequence shown in Figure 4.11. Because of its palindromic nature, it can exist either in the normal linear duplex or in the **cruciform** structure. Cruciforms are usually extruded when a topologically constrained DNA molecule is subjected to high levels of negative supercoiling stress. A single-stranded palindrome can form a **hairpin** structure.

A more exotic **triple helix** structure, **H-DNA**, results from the formation of triple-strand regions that are stabilized in part by unusual hydrogen-bonding, known as Hoogsteen base pairing (Figure 4.12). This structure requires strand regions that are either all-purine or all-pyrimidine and was dubbed H-DNA because it requires high concentrations of protons to form the triple helix. Triple helices are also important in the interaction of single-stranded RNA with duplex DNA.

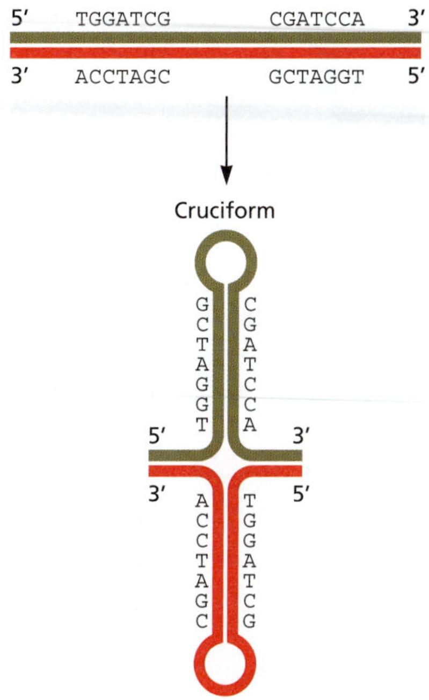

Figure 4.11 Cruciforms (double hairpins). These structures are extruded when palindromic sequences of double-stranded DNA are subjected to high negative superhelical stress.

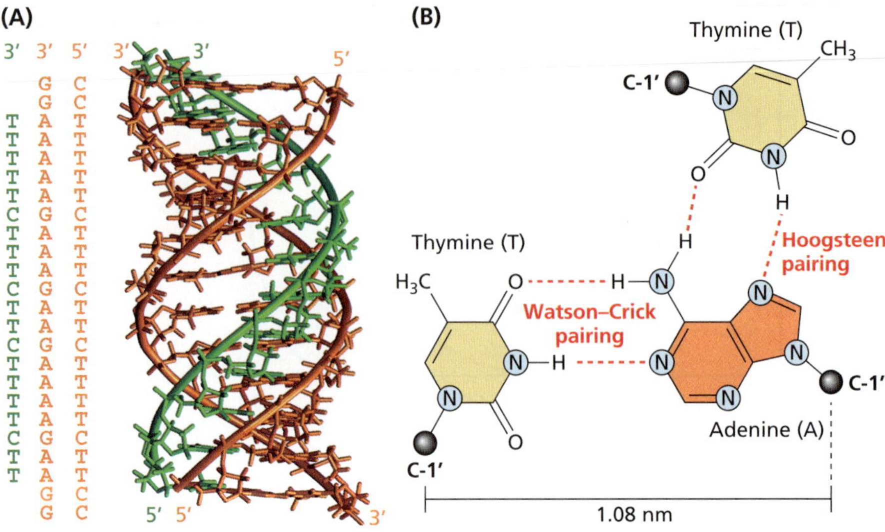

Figure 4.12 Triplex DNA. (A) A third pyrimidine strand, shown in green, binds to the major groove of pyrimidine–purine duplex DNA to form a triple-stranded DNA. (B) Hoogsteen-type base pairing occurs in addition to the normal Watson–Crick base pairing in these triplexes. (A, courtesy of Taejin Kim and Tamar Schlick, New York University.)

A great variety of complicated structures can, in fact, be generated by synthesizing specific single-stranded oligonucleotide sequences and letting them interact or anneal with each other to form double-helical structures in their complementary regions. The ability to design such artificial structures is now being utilized in the construction of complex DNA-based nanostructures for applications in nanotechnology (**Figure 4.13**).

Closed DNA circles can be twisted into supercoils

The range of possible DNA structures becomes even more remarkable when we consider closed circular duplexes. Many DNA molecules, especially those found in bacterial cells, are closed circles. In addition, most of the DNA of higher organisms exists as linear molecules that form loops through attachment to insoluble protein networks in the eukaryotic nucleus. Such constrained loops formally behave as circles; that is, they are **topologically constrained**. All such molecules have a unique property: they can be supercoiled, which means that the axis of the coiled double helix is coiled about itself. A simple example of supercoiling can be seen in a telephone cord. The cord is coiled, but chances are very good that the coil has become twisted and supercoiled about itself.

To explain more fully what this signifies, a few thought experiments are helpful. First, suppose that we have a linear DNA, containing 105 base pairs, with exactly 10.5 base pairs/turn. Assume it is lying on a flat surface, and with ultramicrotweezers we push it into a circle, with the ends touching (**Figure 4.14A**). We now, by some clever enzyme, join the ends, forming a closed circular DNA. It will be a *relaxed* circle that remains flat on the surface because it had an integral number of

(A)

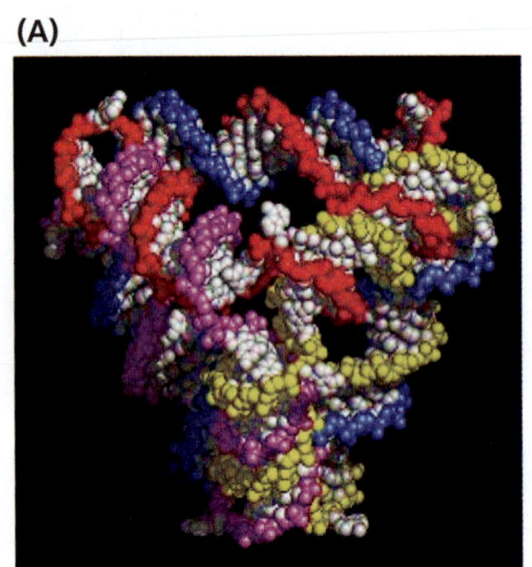

(B)

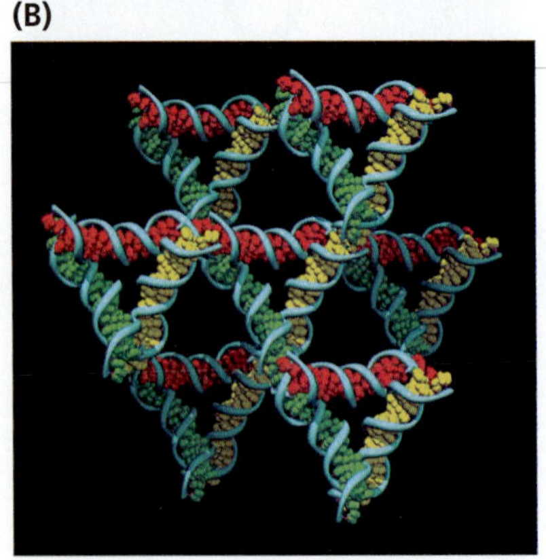

Figure 4.13 Rational design of self-assembled three-dimensional DNA crystals. A number of complicated artificial structures have been created by annealing single-stranded oligonucleotides that possess complementary sticky ends. These ends associate with each other preferentially, assume B-DNA structure, and guide the formation of the three-dimensional crystal. (A) DNA tetrahydron, manufactured in the laboratory of Andrew Turberfield. (B) A structure known as the tensegrity triangle has been manufactured, crystallized, and resolved to 4 Å. (A, from Goodman RP, Schaap IAT, Tardin CF et al. [2005] *Science* 310:1661–1665. With permission from American Association for the Advancement of Science. B, from Zheng J, Birktoft JJ, Chen Y et al. [2009] *Nature* 461:74–77. With permission from Macmillan Publishers, Ltd.)

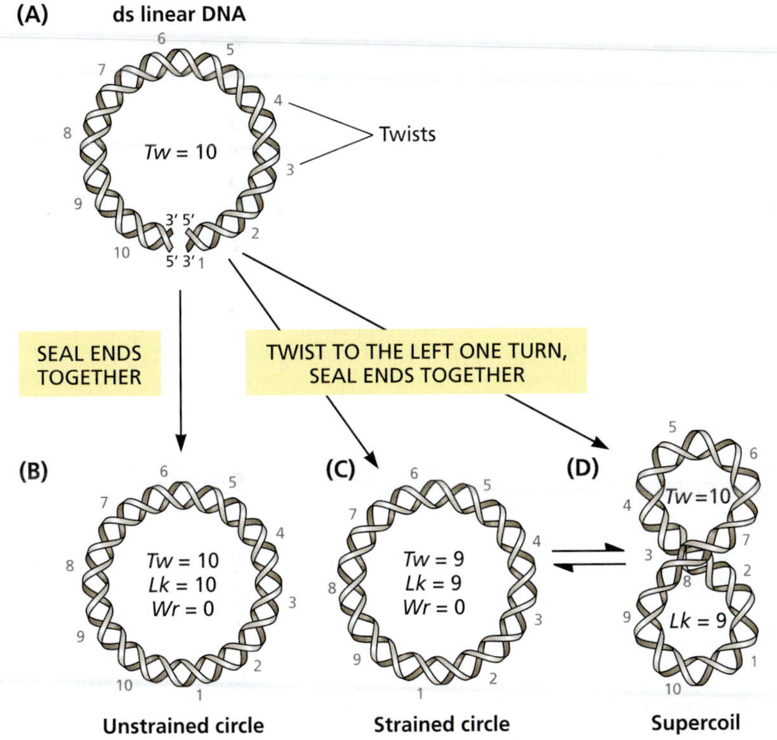

Figure 4.14 Forming a DNA supercoil. (A) The double-stranded (ds) linear DNA molecule depicted here has 105 bp and a pitch of 10.5 bp/turn, giving Tw (twist) = 10. (B) Because the number of turns is an integer, that is, the last turn is completed, the 5′- and 3′-ends are oriented such that the molecule can be closed into a circle that is relaxed and lies flat on the surface. In an unstrained circle, *Lk* is 10, *Tw* is 10, *Wr* (writhe) is 0, and the pitch is 10.5 bp/turn. (C) If the number of turns is reduced by one before the ends are joined, the newly formed circle will have only nine turns. For this circle to lie flat, the pitch must change from 10.5 to 11.67 bp/turn. *Lk* is 9, *Tw* is 9, and *Wr* is 0. This molecule is said to be underwound, as more base pairs are needed to form each turn of the helix than is normal for B-DNA. The molecule is strained because its helix is untwisted with respect to the thermodynamically stable B-form. (D) Rather than changing its twist, a strained molecule may writhe or supercoil. Here, the pitch is 10.5 bp/turn, *Lk* is 9, *Tw* is 10, and *Wr* is –1. A molecule that is supercoiled cannot lie flat on the surface. When a DNA molecule is strained, it initially absorbs the stress by changing its twist, but soon after that it undergoes a transition to a supercoiled structure. This transition has been dubbed the buckling transition. (Adapted from Mathews CK, van Holde KE & Ahern KG [2000] Biochemistry, 3rd ed. With permission from Pearson Prentice Hall.)

turns: 105 bp divided by 10.5 bp/turn = 10 turns. In such a molecule the 5′-end of each strand will exactly meet with the 3′-end of the same strand and could be easily joined into a circle by our smart enzyme (Figure 4.14B). But now suppose we had unwound the DNA by one turn, by rotating one end counterclockwise while holding the other end stationary, before joining the ends. Now we have a circular DNA containing 105 bp in 9 turns rather than in 10. It is said to be underwound, and its **linking number**, the number of times one strand crosses the other, has been decreased by one. This DNA is not relaxed and is under strain. It can accommodate that strain in various ways. As in Figure 4.14C, it could spread the strain over the whole circle, making a DNA molecule with 11.67 bp/turn, that is, 105 bp with 9 turns. This DNA circle can also lie flat on the surface. Another more likely possibility is that the DNA axis would make one turn about itself, producing a negatively supercoiled structure; note that the actual crossing of the DNA duplexes is right-handed or positive; this can be seen in Figure 4.14D. This writhing allows the DNA to gain back one right-hand twist, which is compensated by the supercoil. The DNA is now back to 10 turns, at 10.5 bp/turn. The supercoiled DNA, because it crosses itself, cannot lie flat on a planar surface.

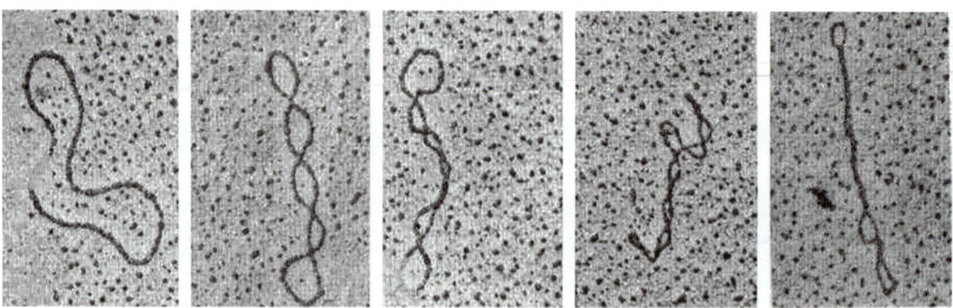

Figure 4.15 Visualizing DNA supercoiling. Electron micrographs of a circular DNA molecule with increasing degrees of supercoiling from left to right. (From Kornberg A & Baker TA [1992] DNA Replication, 2nd ed. With permission from University Science Books.)

DNA supercoiling can be modeled by using a piece of elastic belt or rubber band. A relaxed band joined to form a circle corresponds to a relaxed circular double-stranded DNA molecule, with edges marking the two strands. We can model supercoiling by holding one end of the band, twisting the other end, and then joining the two ends. If the starting model is a closed circle, like a rubber band, it must first be cut before twisting and resealing. As the band represents a B-form helix, the coiling we see is a secondary level of coiling. When the band is twisted even more before joining, the supercoiled structure becomes more complex and more compact at the same time, and we can observe branches forming. This principle of supercoiling is clearly evident in the series of lightly supercoiled DNA molecules shown in the micrographs of **Figure 4.15**. The relationship between linking number *Lk*, twist *Tw*, and writhe *Wr* can be put on a quantitative basis as described in **Box 4.4**.

At high levels of supercoiling, the DNA can become quite contorted (Figure 4.15). Two general kinds of supercoils may be formed: plectonemic and toroidal (or solenoidal) structures (**Figure 4.16**). Note that the level of overall compactness of the

Box 4.4 A Closer Look: Supercoiling, linking number, and superhelical density It is possible to describe the configurations of closed circular duplex DNA in a quantitative manner. First, note that once a DNA molecule has been closed into a circle, it has an invariant topological quantity that can only be changed by cutting, twisting, and resealing. This is the linking number, *Lk*, which is defined as the number of times two circles are interlinked. In the case of DNA, the two circles are the two individual DNA strands in a closed circular molecule. It must be an integer, and its sign is positive for right-handed crossings and negative for left-handed ones (**Figure 1**).

The linking number is distributed between twist, *Tw*, the number of times each strand twists around the DNA axis, and **writhe**, *Wr*, the number of times the DNA axis crosses itself.

$$Lk = Tw + Wr \qquad \text{(Equation 4.1)}$$

These relationships are shown in Figure 4.14. Note that the forms shown in Figure 4.14 parts C and D both have a linking number of 9 but it is distributed differently between twist and writhe.

The amount of supercoiling present in a circular DNA molecule is measured in terms of the supercoil density, σ:

$$\sigma = (Lk - Lk_0)/Lk_0 \qquad \text{(Equation 4.2)}$$

where Lk_0 is the linking number of the molecule if relaxed, which equals Tw_0 because a relaxed molecule has zero writhe. A typical σ value for DNA molecules *in vivo* is about −0.05, which means that there are about five negative supercoils for every 100 turns, or about 1000 bp, of DNA. Negative supercoiling is most common; most naturally occurring circular DNAs are underwound.

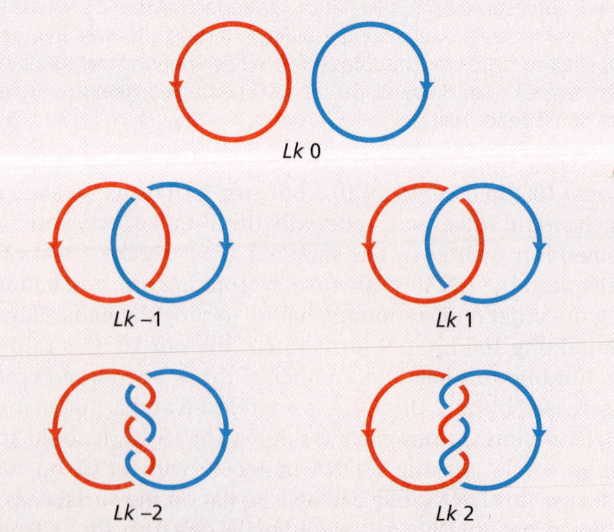

Figure 1 Linking number. Any two closed curves in space can be moved, cut, and resealed to be linked in a number of ways; a few simple examples are shown. The way the two curves interlace determines the linking number and its sign: *Lk* is positive for right-handed crossings and negative for left-handed ones.

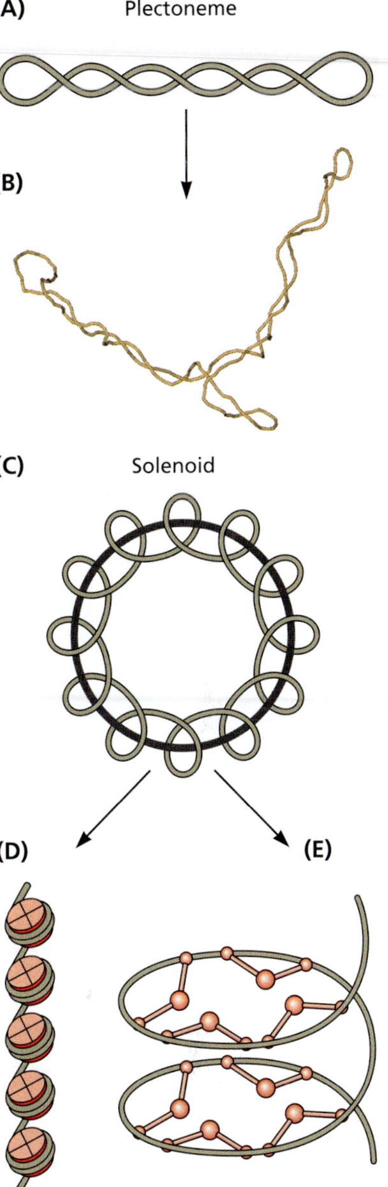

Figure 4.16 Three-dimensional trajectories of superhelical DNA. Schematic presentation of the two forms adopted by superhelical DNA: (A) plectoneme and (C) solenoid, also called toroid. (B) Typical conformation of supercoiled plectonemic DNA: computer simulation of a supercoiled molecule of 3500 base pairs in length with a supercoil density σ of −0.06 under physiological ionic conditions. Note the similarity of the computational model to the electron microscopy images presented in Figure 4.15. Two forms of solenoidal supercoil are shown: (D) Wrapping of DNA around histone octamers, shown as red cylinders, compacts the DNA, which forms a left-handed solenoid (see Chapter 8). (E) Condensin proteins, shown as red ball-and-stick structures, that participate in forming mitotic chromosomes affect global DNA writhe by forming large positive solenoids. (A, C, adapted from Travers A & Muskhelishvili G [2007] *EMBO Rep* 8:147–151. With permission from John Wiley & Sons, Inc. B, courtesy of Alex Vologodskii, New York University. D, E, adapted from Holmes VF & Cozzarelli NR [2000] *Proc Natl Acad Sci USA* 97:1322–1324. With permission from National Academy of Sciences.)

same molecule differs between the two forms, with toroids forming more compact structures than plectonemes. Negative values of writhe correspond to left-hand coiling in a toroid but right-hand crossing in plectonemic structures. Although difficult to visualize, this can be demonstrated by wrapping a piece of tubing in several toroidal turns about a cylinder and then removing the cylinder while the ends of the tubing are held. Toroidal supercoiling is common in eukaryotic DNA, which is often packed in very large amounts into a small nucleus. In eukaryotic chromosomes in the interphase state, DNA toroids are supported by proteins that are associated with chromatin (see Chapter 8). However, cells usually use both forms of supercoiling to compact their DNA, as exemplified by the *Escherichia coli* chromosome (Figure 4.17). Although bacteria usually possess only one giant circular DNA, this circle is partitioned into individual domains that possess different structures and different, regulated degrees of supercoiling.

Cells possess enzyme molecules that can regulate the superhelicity of their DNA. To do this, the linking number must be changed, and this requires cutting and resealing the duplex. These enzymes are called **topoisomerases**, and there are two general classes: type I and type II. Type I topoisomerases change the linking number in steps of one, whereas type II topoisomerases change it in steps of two. This difference mechanistically stems from the number of DNA strands being broken and resealed in individual enzymatic steps: one DNA strand for type I topoisomerases and both strands for type II enzymes. Table 4.3 provides more information on the chemistry of the reactions and the actual mechanisms that change the *Lk*. Even within one topoisomerase class, the mechanisms regulated by individual enzymes can differ. The number of topoisomerases described to date in bacteria, archaea, and eukaryotes is very large and expected to grow.

Type IA topoisomerases act according to the enzyme-bridging model, as shown in Figure 4.18: they nick one DNA strand and allow the unbroken strand to pass to relieve the superhelical stress. Members of type IB subfamily use the controlled rotation mechanism (Figure 4.19). In this mechanism the enzyme first nicks one strand and then covalently connects its free 5′-phosphate to an active-site tyrosine. The free 3′-OH end then rotates by one turn before attacking the phosphate residue. This reseals the DNA duplex, with the linking number changed by one unit. Type II topoisomerases (Figure 4.20 and Figure 4.21) use a two-gate mechanism: they make a double-strand break and pass an intact double helix through the gate created by the broken helix. Although most topoisomerases act to relax superhelical

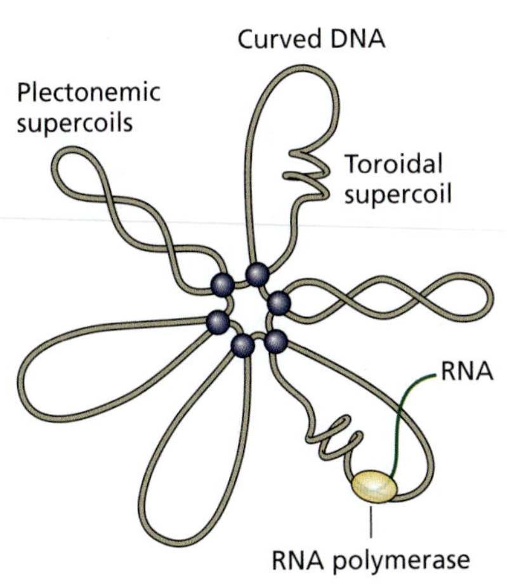

Figure 4.17 Supercoiled DNA domains in the *E. coli* chromosome. The drawing shows the presence of both plectonemic loops and loops containing toroidal or solenoidal structures; the proteins involved are not depicted. This is a simplified cartoon of the chromosome; in real life, there are perhaps as many as 400 different domains. The loops exist due to the presence of specific structural proteins bound at the bases of the loops. Intrinsically curved DNA that bends spontaneously because of a specific sequences of bases tends to be localized at the tips of supercoils. (Adapted from Willenbrock H & Ussery DW [2004] *Genome Biol* 5:252–256. With permission from BioMed Central.)

Table 4.3 Subfamilies of DNA topoisomerases.

Topoisomerase subfamily[a]	Breakage of DNA[b]	Mechanism of changing *Lk*	Representative members
IA	DNA strands transiently break one at a time; active-site tyrosyl becomes covalently linked to a 5′-phosphoryl group	enzyme-bridging: broken DNA ends are bridged by the enzyme and the single strands pass through one another	bacterial topo I and III, yeast topo III, *Drosophila* topo IIIα and IIIβ, mammalian topo IIIα and IIIβ
IB	DNA strands transiently break one at a time; tyrosyl becomes covalently linked to a 5′-phosphate group	controlled rotation: the 3′-end of the broken strand remains attached to tyrosine; ionic interactions with the other end permit rotation of the DNA segment relative to duplex held by the enzyme	eukaryotic topo I, mammalian mitochondrial topo I, poxvirus topo
IIA	a pair of strands in the double helix is transiently broken in concert by dimeric enzymes; strand-breakage reaction same as IA	gate: the enzyme uses the energy of ATP hydrolysis to transport one intact DNA duplex through a double-strand break in a second duplex	bacterial gyrase, topo IV, phage T4 topo, yeast topo II, *Drosophila* topo II, mammalian topo IIα and IIβ
IIB	same as IIA	same as IIA	*Sulfolobus shibatae* topo VI

[a]The mechanisms of action of the different subfamilies IA, IB, IIA, and IIB are depicted in Figures 4.18–4.21, respectively.
[b]Transesterification between the active-site tyrosyl and a DNA phosphate leads to breaking of the DNA backbone and formation of a covalent enzyme–DNA intermediate. The final step of re-formation of intact DNA happens by reversal of this reaction.

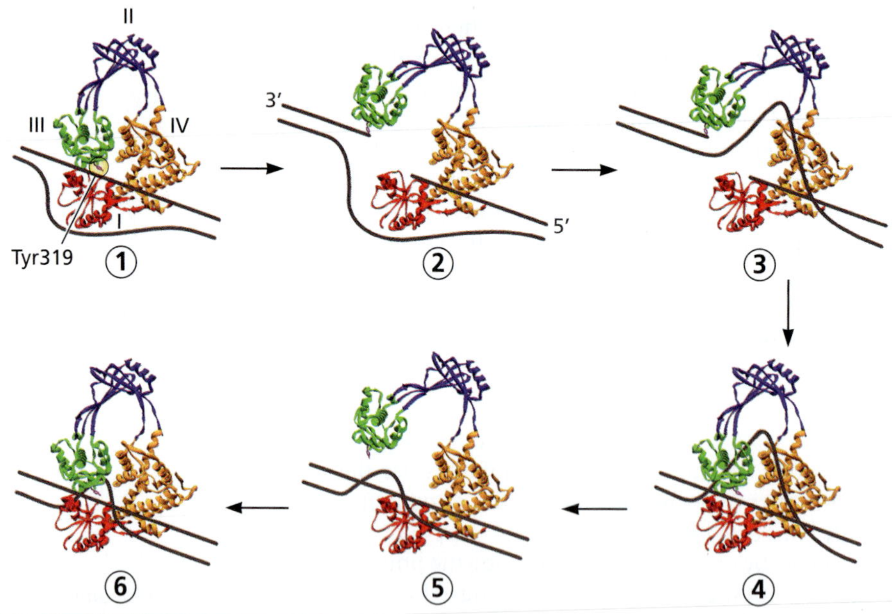

Figure 4.18 Enzyme-bridging model. Shown is the relaxation of one turn of a negatively supercoiled plasmid DNA by *E. coli* topo I, from the IA subfamily. The two strands of DNA are shown as dark gray lines and are not to scale. The four domains of the protein are colored differently and designated by Roman numerals. The C-terminal domain is missing from the crystal structure and from this schematic. The active-site tyrosine, Tyr319, is located in domain III. The protein assumes the configuration of a fattened torus or doughnut shape, with a positively charged hole large enough to accommodate double-stranded DNA. The arrangement of the four domains allows for the conformational plasticity needed for DNA binding and relaxation. The DNA strand to be cleaved is bound to the surface of the protein. The length of the intact strand that is passed through the open gate is exaggerated to simplify the drawing. The protein is proposed to oscillate between a closed conformation, shown in steps 1, 4, and 6, and an open conformation, shown in steps 2, 3, and 5, that provides DNA access to the central hole. The steps are as follows: (Step 1) Binding of a single-stranded portion of the DNA to the enzyme. (Step 2) After cleavage, Tyr319 is covalently bound to the 5′-phosphate of the cleaved strand; the other end occupies a nucleotide-binding site in domain I; domain III lifts away from domain I to allow passage of the intact strand. (Step 3) The intact strand passes into the hole of the torus. (Step 4) The clamp closes and the cleaved strand is religated. (Steps 5 and 6) The protein opens and closes a second time to release the passed strand: the cycle is complete. (From Champoux JJ [2001] *Annu Rev Biochem* 70:369–413. With permission from Annual Reviews.)

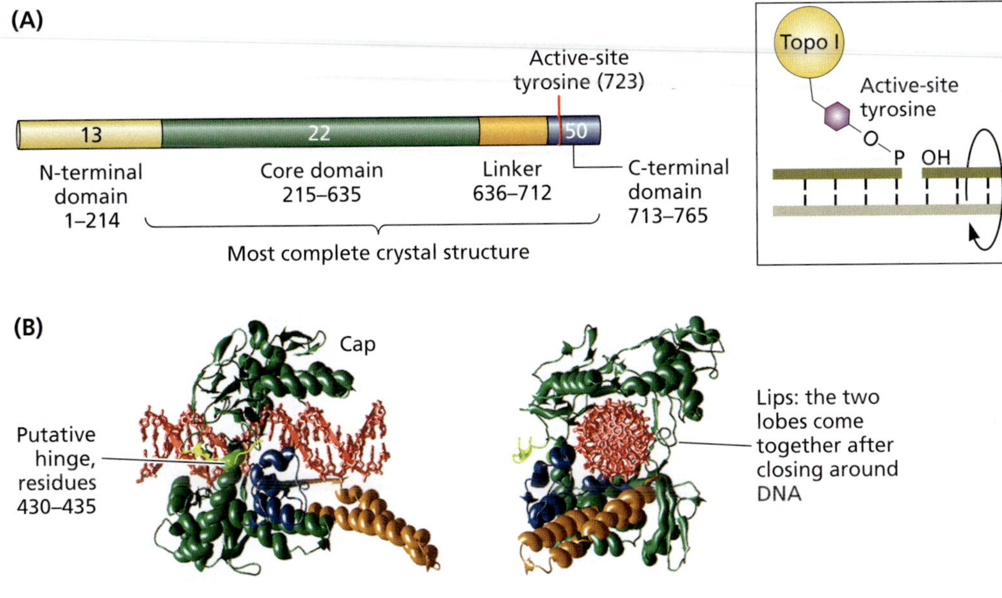

Figure 4.19 Schematic representation of human topo I from the IB subfamily. (A) Domain organization, with the active-site tyrosine marked. This enzyme acts through the controlled rotation mechanism, shown in the box. (B) Crystal structure of the enzyme bound to DNA, with individual domains colored according to the top schematic. Note that the protein clamps around DNA, shown in red. (B, from Leppard JB & Champoux JJ [2005] *Chromosoma* 114:75–85. With permission from Springer Science and Business Media.)

tension, others, such as the bacterial **DNA gyrase**, can introduce negative supercoils. This requires energy from ATP hydrolysis to work against the superhelical stress.

Topoisomerases are important targets for anti-cancer agents that are widely used in clinical practice (Box 4.5). We discuss the importance of supercoiling in DNA packaging and in the regulation of gene activity in much more detail in other chapters.

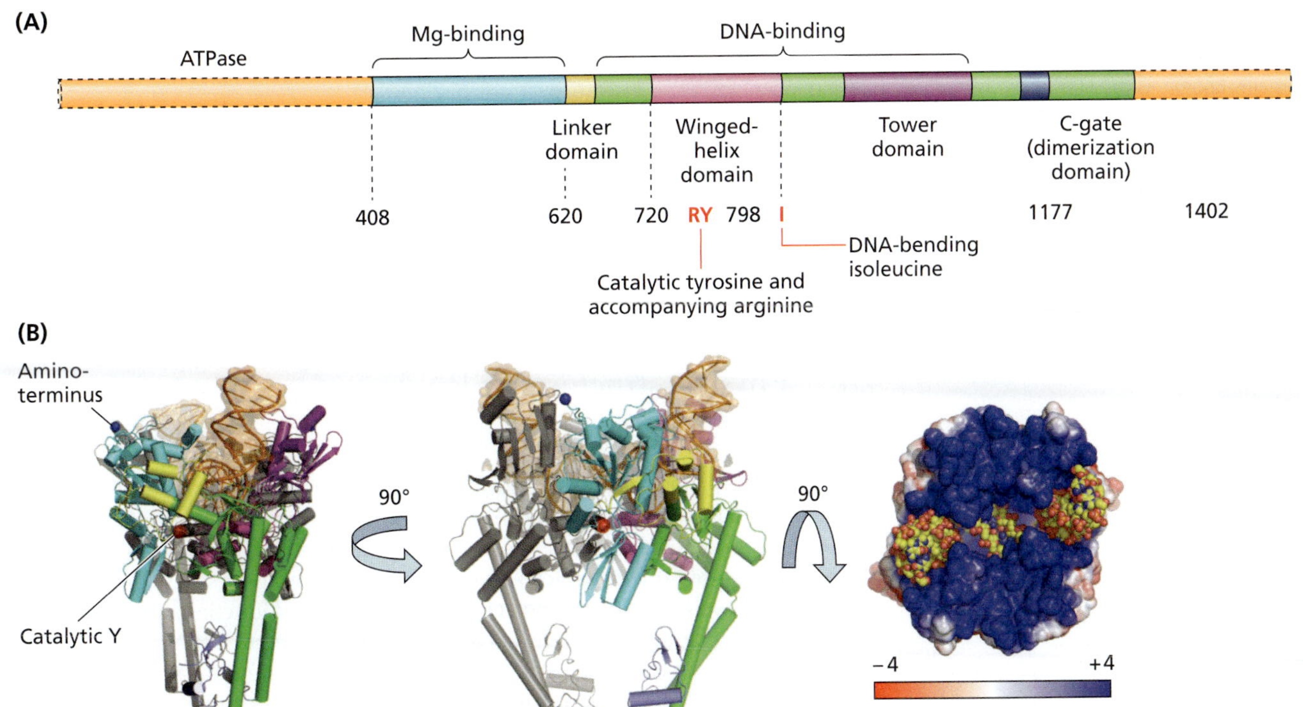

Figure 4.20 *Saccharomyces cerevisiae* topo II structure: DNA binding and cleavage core. (A) Domain organization. Auxiliary scaffolding domains are shown in green. (B) Orthogonal views of the enzyme bound to DNA. One protomer is colored gray, while the other one follows the color scheme from the top diagram. DNA is shown as a double-stranded oligonucleotide, in orange with a transparent surface; it contains complementary four-base 5′-overhangs to mimic the intermediate complex after the double helix is cut in a staggered way. The surface representation at right is colored by electrostatic potential, with red for negative charge and blue for positive. Note how the negatively charged DNA fits snugly into the positively charged surface. (Adapted from Dong KC & Berger JM [2007] *Nature* 450:1201–1205. With permission from Macmillan Publishers, Ltd.)

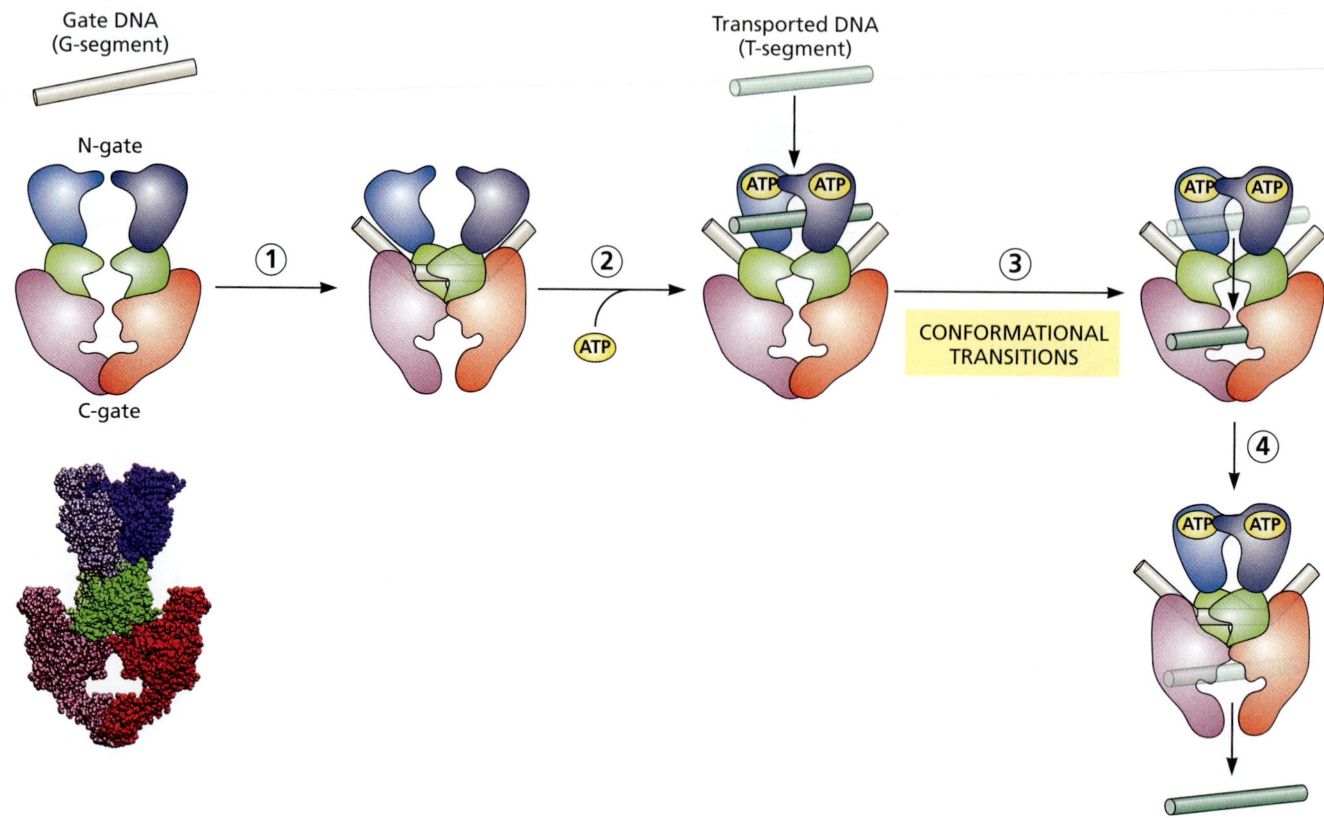

Figure 4.21 Two-gate model for DNA transport during the catalytic cycle of topo II. Topo II utilizes the energy from ATP to transport one intact DNA duplex through an enzyme-mediated 4-bp staggered double-strand break in a second duplex. The reaction proceeds in the following steps: (Step 1) The N-terminal ATPase domains known as the N-gate are separated, allowing topo II to bind a DNA duplex, shown as a gray tube, and cause significant DNA bending. This DNA has been termed the gate or G-segment because it serves as a gate through which another strand must pass. (Steps 2 and 3) ATP binding to the N-terminal domains triggers the dimerization and capture of a second DNA duplex, known as T-segment DNA and shown as a green tube. Closure of this N-gate induces a succession of conformational changes that promote the cleavage and opening of the G-segment, followed by transport of the T-segment DNA through the break. (Step 4) The T-segment then exits through the opened C-terminal dimerization interface or C-gate. The DNA gate closes to allow G-segment religation. Inset: Crystal structure of yeast topo II is shown below the first structure. ATPase N-domains are in blue, the DNA gate is in green, and the C-terminal portions that enclose a hole large enough to accommodate the T-segment DNA are in red. (Adapted from Collins TRL, Hammes GG & Hsieh T [2009] *Nucleic Acids Res* 37:712–720. With permission from Oxford University Press.)

4.4 Physical structures of RNA

RNA can adopt a variety of complex structures but not the B-form helix

Ribonucleic acid, RNA, differs chemically from DNA in only two respects. First, RNA contains the sugar moiety ribose instead of the 2′-deoxyribose in DNA (see Figure 4.2B). Second, RNA contains the base uracil, U, instead of thymine, which has the chemical structure 5-methyluracil. These differences may seem small, but they serve to qualify the two kinds of nucleic acids for very different functions in the cell. One direct consequence is that RNA cannot form the double-stranded B-form found in cellular DNA: the hydroxyl group in ribose gets in the way, whereas the smaller hydrogen atom in deoxyribose does not. Complementary RNA strands can form double helices, but these are invariably in the A-form.

In fact, most of the RNA found in cells is generated as single-stranded copies of one strand of DNA. In such copying, which is called **transcription**, the base A in DNA pairs with U in the newly forming RNA strand; other pairings are as described for DNA. However, chemical modification of bases is much more common in RNA than in DNA. We have much more to say about transcription in other chapters; here we present only a brief overview that will define the roles of RNA in biology.

Box 4.5 Topoisomerase inhibitors as anti-cancer agents Inhibitors of topoisomerases are now widely used in clinical practice as anti-cancer agents. All topoisomerase reactions proceed through multistep mechanisms: initial cleavage of the DNA, relaxation, and then religation of the phosphodiester backbone(s). Certain drugs trap the enzyme following the cleavage step and thus prevent the religation step.

Topo I is a ubiquitous and essential enzyme in higher eukaryotes, including humans. *Top1*-knockout mice lacking this enzyme die early during embryogenesis. The severity of this mutant phenotype is linked to the need to relax the negative supercoiling generated during replication and transcription.

How do drugs trap the enzyme–DNA complex and prevent religation? The religation step requires a nucleophilic attack on the tyrosyl–DNA-phosphodiester bond by the 3′-OH end of the broken DNA stand (see Table 4.3 and Figure 4.19). For this reaction to occur, the participating partners must be properly aligned. Any misalignment, whether it is caused by preexisting lesions in the DNA or by drug binding, inhibits religation. Topo I action is reinitiated only when the proper alignment is restored by repairing the DNA lesion or removing the drug.

The only drugs in clinical use to trap human topo I–DNA complexes are camptothecin and its water-soluble derivatives (**Figure 1**). Camptothecin is a plant alkaloid first isolated from the bark of the Chinese tree *Camptotheca acuminata*. The natural alkaloid, although it possesses anti-cancer activity, is very toxic and had to be derivatized to reduce its side effects. The major advantage of this drug is its extreme specificity: topo I is its only target. In addition, camptothecin penetrates vertebrate cells readily and quickly, and its binding to the topo I–DNA complex is reversible and of low affinity, which allows precise control over its action. The crystal structure of topo I with bound camptothecin (Figure 1) reveals the structural basis for its action: the drug intercalates between the ends of the broken DNA strand, thus precluding their proper alignment, which is needed for religation.

Recent efforts have been directed toward finding a small molecule that could inhibit bacterial topo I in the same way, leading to bacterial death. Research with natural mutants of the enzyme has identified amino acid residues close to the active-site region whose mutation inhibits the normal Mg^{2+} binding and thus the religation step of the reaction.

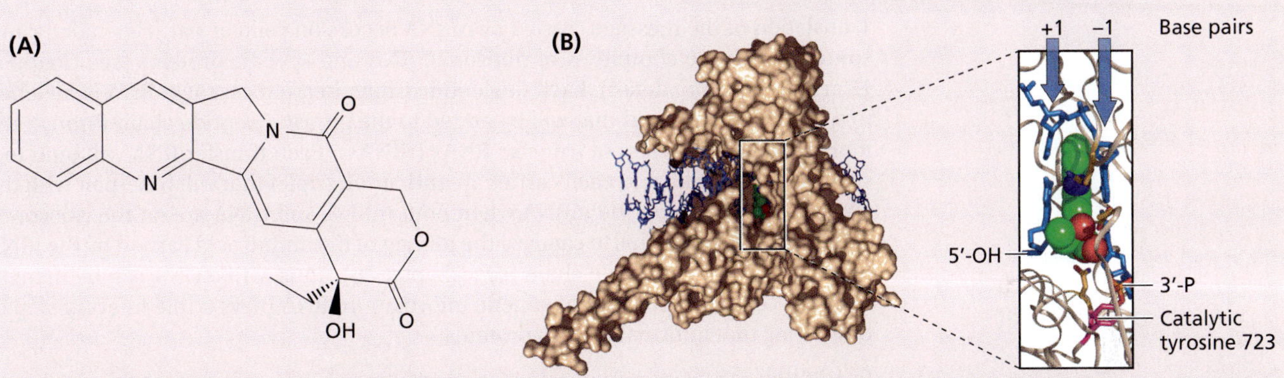

Figure 1 Mechanism of action of the anti-cancer drug camptothecin. (A) Structure of camptothecin and (B) trapping of the topo I DNA-cleavage site by camptothecin. In the enlarged view, the drug molecule, with carbons in green, oxygens in red, and nitrogens in blue, is shown stacked between the DNA base pairs, shown in blue, that flank the cleavage site, shown as a brown ribbon. The nick is on the lower DNA strand, and religation of the 5′-end of the nicked DNA is precluded by the presence of the intercalated drug. The N-terminal 200 amino acid residues of the enzyme are not shown, as their atomic structure is not known. (B, from Pommier Y [2006] *Nat Rev Cancer* 6:789–802. With permission from Macmillan Publishers, Ltd.)

In terms of three-dimensional structure, some of any single-stranded RNA will be in a disordered, **random-coil** conformation. However, interactions between complementary RNA regions that may be in distant parts of the chain promote specific and often complex folds of many RNA molecules. The simplest folds are hairpin structures or half-cruciforms formed by palindromes (see Figure 4.11), but much more complicated structures are common. In fact, the structural potential of RNA may be almost as diverse as that of proteins. Extended regions of defined secondary structure, such as α-helices and β-sheets in proteins and A-form duplexes in RNA, are found in both types of macromolecule, and in both cases these regions can further fold to form tertiary structure. These structures appear, in many cases, to have evolutionarily adapted to specific functions. Some RNA molecules bind to proteins and modify their function; others, called **ribozymes** (see Box 14.1), have catalytic functions like protein enzymes. Thus, the information passed from DNA to RNA can have two kinds of function: to code for proteins or to produce functional RNA molecules. It is becoming evident that the latter expresses the largest fraction of the genome.

4.5 One-way flow of genetic information

In the years following the elucidation of the Watson–Crick DNA structure, a series of studies revealed how the transfer of genetic information from DNA actually occurs. Actually, the exploration of this huge and complex problem is still a major effort in molecular biological research and the subject of much of this book. But to provide a perspective and clarify the roles of DNA and RNA, a brief overview is useful at this point. The essentials were termed by some the central dogma of molecular biology:

$$\text{DNA} \xrightarrow{\text{Transcription}} \text{RNA} \xrightarrow{\text{Translation}} \text{Protein}$$

RNA molecules were at first seen to function primarily as intermediates in the transfer of information from DNA sequence to protein sequence. The amino acid sequence information for any protein, coded in the DNA sequence, is transcribed into a **messenger RNA (mRNA)** during the copying of one of the two DNA strands of a protein-coding gene. This mRNA must be translated, via a code, from the four-letter language of the four bases in polynucleotides to the 20-letter language of the 20 amino acids in proteins. It is clear that a single nucleotide or even a doublet would not work; 4 or even $4 \times 4 = 16$ does not provide enough combinations to code for 20 amino acids. It was found that each amino acid corresponds to one or more nucleotide triplet(s). The genetic code is very nearly universal over all of life. The code and its properties are discussed in detail in Chapter 7.

Translation of the message carried by mRNA occurs on cellular particles called **ribosomes**, which are complexes of ribosomal RNA and specific proteins (see Chapters 15 and 16 for more detail). Each nucleotide triplet in the messenger RNA is matched to a specific amino acid that will be added to the growing peptide chain through the function of a small class of **transfer RNAs (tRNAs)**. Each transfer RNA will bind to a specific amino acid, and each carries an **anticodon** triplet that will base-pair with the appropriate **codon** on the mRNA when both mRNA and tRNA are on the ribosome. The ribosome then helps to catalyze the joining of the amino acid carried by the tRNA to the growing polypeptide chain (see Chapter 16). Thus, while DNA can be thought of as the cellular repository of genetic information, RNA plays a role in every step of translating that information into proteins.

In addition, it becomes increasingly clear that many RNA molecules have regulatory roles in many of the processes of information transfer and that RNA can serve a wide variety of cellular functions. This and our present recognition that information can sometimes pass from RNA back to DNA requires a revision of the simple scheme written above:

$$\text{DNA} \underset{\substack{\text{Reverse} \\ \text{transcription}}}{\overset{\text{Transcription}}{\rightleftarrows}} \begin{matrix}\text{mRNA} \\ \text{ncRNA}\end{matrix} \xrightarrow{\text{Translation}} \text{Protein}$$

In this new diagram, ncRNA stands for all of those noncoding RNAs that do not code for proteins but do other things. In higher organisms, this utilizes by far the larger part of the genome, and we have only glimpses as to what functions these ncRNAs may be performing. The change reminds us that the dogma was never a dogma anyway, for that term connotes an unchanging truth. Nothing in science is unchanging.

4.6 Methods used to study nucleic acids

Over the years, numerous methods have been developed to study nucleic acids. Some of these are rather simple in terms of instrumentation and execution; others require sophisticated and expensive machines. The most common methods—gel electrophoresis (**Box 4.6** and **Box 4.7**), gradient centrifugation (**Box 4.8**), nucleic acid hybridization (**Box 4.9**), and DNA sequencing (**Box 4.10**)—are presented here in a series of boxes. Additional methods are detailed in relevant chapters, mainly in Chapter 5.

Box 4.6 Gel electrophoresis of linear nucleic acid molecules Gel electrophoretic analysis is central to almost every experimental study of the molecular biology of nucleic acids. It provides a simple, but highly discriminating, way to separate mixtures of DNA or RNA according to molecule size. The principle is the same as that of SDS gel electrophoresis of proteins (see Box 3.3) but is much more direct. A series of DNA molecules of different lengths will have charge proportional to length, and the friction factor is also roughly proportional to length. Thus, according to Equation 3.2 in Box 3.2, a wide range of DNA and RNA molecules will have about the same free mobility, but their retardation by the gel will increase with increasing length, so that at any finite gel concentration the biggest molecules will move most slowly (**Figure 1A**).

A graph of the distance migrated or relative mobility against the logarithm of the molecular weight, which can be expressed in base pairs, will give nearly a straight line (**Figure 1B**). A set of DNA fragments of exactly known sequence, and hence length, can be run on the same gel as calibrating markers. Using different gel concentrations makes it possible to study nucleic acids in different molecular weight ranges: concentrated gels can give resolution down to the base-pair range, whereas very dilute gels allow migration of molecules up to 100,000 bp or more. Special techniques, such as pulsed-field electrophoresis, permit even greater ranges.

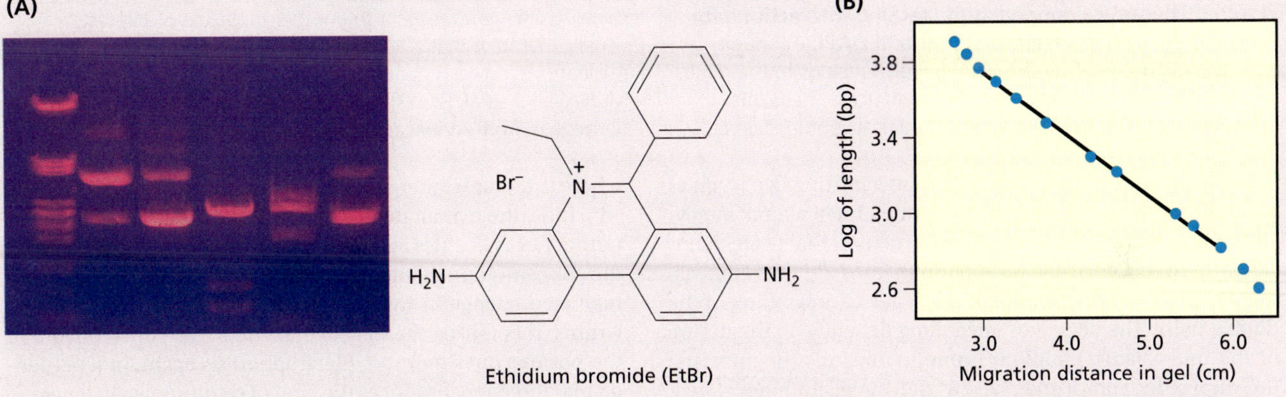

Figure 1 Agarose gel electrophoresis of DNA. (A) Example of an agarose gel stained with ethidium bromide (EtBr). EtBr intercalates between the DNA bases and, when exposed to UV light, fluoresces with an orange color. Although EtBr has been used for years, the mutagenic and carcinogenic properties of the intercalator made researchers look for alternatives for visualization of DNA bands on gels, for example, SYBR Green. Some of the new dyes have higher sensitivity. (B) Linear dependence of electrophoretic mobility on the logarithm of the length of DNA. (A, left, courtesy of Markus Nolf, Wikimedia. B, based on data from Seminars on Science. With permission from American Museum of Natural History.)

Box 4.7 Gel electrophoresis of circular and supercoiled nucleic acid molecules Because the sieving effect in gel electrophoresis depends on molecular dimensions, it should not be surprising that the technique can be used to separate linear from closed circular molecules of the same length. Moreover, **topoisomers** of a given circular DNA molecule can be resolved. Recall that the more supercoiled the DNA molecule is, the more compact it is (see Figure 4.15). As one might expect, the mobility of the same circular molecule increases with the absolute value of the linking number, until a limit is approached for highly supercoiled molecules (**Figure 1**).

This very simple analysis has one drawback, however: it cannot distinguish positive from negative supercoils. But there is a clever trick to accomplish this. First, the mixture of molecules is electrophoresed in one dimension. This gives a series of bands, each of which may contain both positive and negative supercoils of a given absolute *Lk*. The one-dimensional gel is then

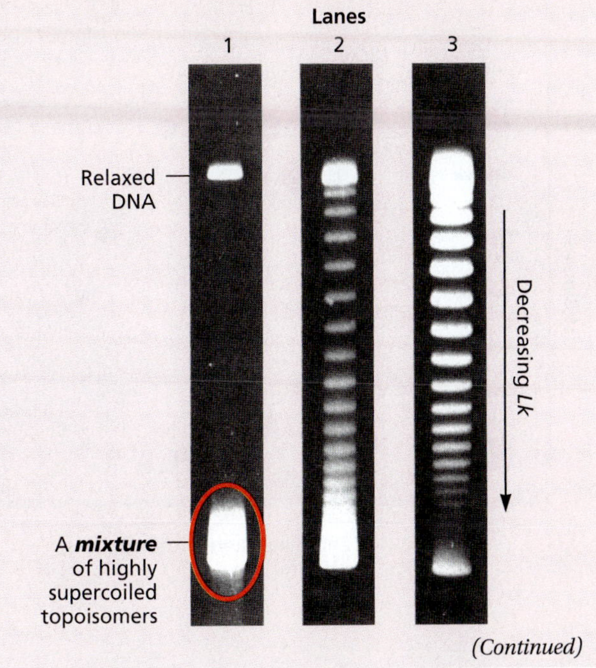

Figure 1 Separation of individual topoisomers of supercoiled DNA by one-dimensional agarose gel electrophoresis. Lane 1, supercoiled DNA isolated from bacterial cells; lanes 2 and 3, same DNA treated with type I topoisomerase, which partially relaxes the supercoils, for 15 and 30 min, respectively. The individual bands between relaxed and supercoiled DNA bands represent topoisomers of decreasing *Lk*; the *Lk* of the DNA molecules in neighboring bands differs by one. (From Keller W [1975] *Proc Natl Acad Sci USA* 72:2550–2554. With permission from National Academy of Sciences.)

(Continued)

Box 4.7 (Continued)

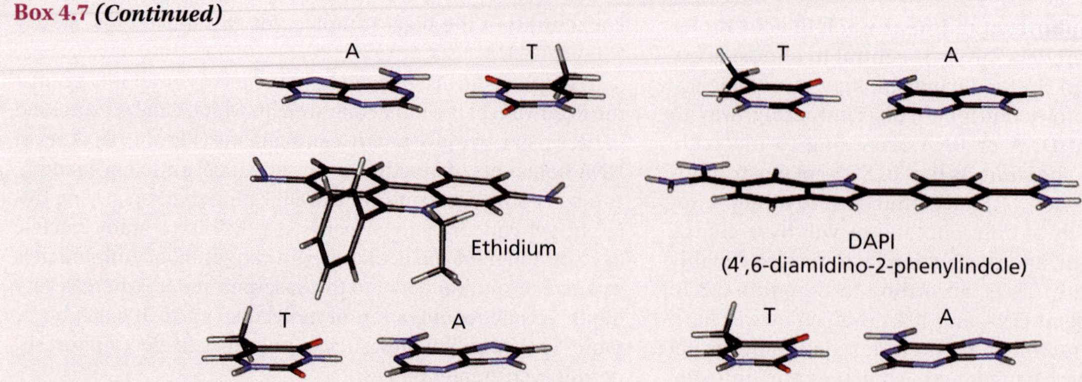

Figure 2 Theoretical modeling of stacking interactions for two widely used intercalators: EtBr and DAPI. Although 4′,6-diamidino-2-phenylindole (DAPI) typically binds to the minor groove, it can also, under certain circumstances, intercalate between DNA bases. DAPI is widely used as a counterstain to visualize DNA with respect to another molecule of interest, usually a protein that is stained with a different dye in cellular/nuclear cytological preparations. (Adapted from Řeha D, Kabeláč M, Ryjáček F et al. [2002] *J Am Chem Soc* 124:3366–3376. With permission from American Chemical Society.)

soaked in a buffer containing a DNA intercalating dye [such as ethidium bromide (EtBr) or chloroquine]. Intercalators are cyclic organic compounds with planar structures that place themselves between adjacent base pairs (**Figure 2**), thus increasing the distance between them: this leads to unwinding of the double helix. The degree of unwinding depends on the nature of the intercalator. Ethidium bromide unwinds (or untwists) the duplex by approximately -26° per molecule intercalated, whereas the unwinding angle of chloroquine is approximately -8°. Thus, the overall degree of unwinding will depend on the number of molecules intercalated, which in turn is a function of the concentration of intercalator in the buffer. The intercalator-mediated change in twist must be compensated by changing writhe. This makes the negative isomers less supercoiled and the positive ones more so. Electrophoresis again, in a perpendicular direction, produces the kind of result shown in **Figure 3**.

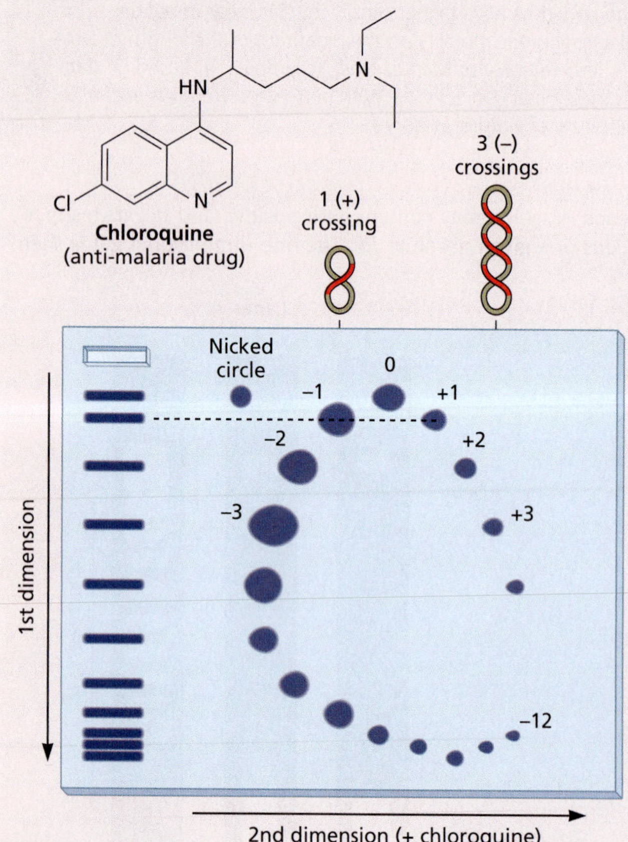

Figure 3 Discrimination between positively and negatively supercoiled DNA by two-dimensional electrophoresis. The dashed line indicates that a single band in the first dimension, if it happens to contain both positive and negative supercoils, now produces two spots in the second dimension. The spot numbers indicate the writhe of the molecule in the second-dimension gel. Remember, the crossing of the two DNA helices in an overwound molecule with positive writhe is actually negative (see Figure 4.14). The magnitude of change depends on the concentration of the intercalator in the buffer: In the specific example illustrated here, the chloroquine concentration, and hence the number of molecules intercalated, is such that it unwinds two helical turns in the DNA molecule, which is compensated by creation of two positive writhes. Thus, the intercalation of chloroquine into the +1 topoisomer in the original topoisomer mixture has changed its geometry; it now contains three rather than one negative crossings. At this chloroquine concentration, the –1 topoisomer in the original topoisomer mixture has changed its geometry from one positive crossing to one negative crossing. As a result of these changes, the two topoisomers that were indistinguishable in the first dimension can now be separated from each other in the second dimension because of the difference in compaction imposed by chloroquine intercalation. Recall that the *Lk* of a given molecule is invariant and cannot change without breaking and resealing of the DNA backbone, which does not occur here. Thus, what changes following chloroquine intercalation is the geometry of the molecule, not its *Lk*.

Box 4.8 Gradient centrifugation Two techniques that are based on sedimentation in a density gradient are commonly used for the separation of macromolecules, especially nucleic acids. The principles and applications of the two techniques are somewhat different.

Density gradient velocity sedimentation

This technique is often referred to as **sucrose gradient sedimentation** because sucrose is most commonly used to prepare the gradient of solution density, although other substances such as glycerol can be used. The principle is very simple (Figure 1). Centrifuge tubes are filled with sucrose solutions from a mixing device, which places the most dense solution at the bottom of each tube with a smoothly decreasing gradient of sucrose concentration, and hence density, toward the top. A thin layer of the macromolecule mixture in a less dense solution is then placed on top, and the tubes are spun in a swinging-bucket centrifuge rotor. The purpose of the gradient is to allow sedimentation to proceed without convective mixing. The mixture will then separate into bands of different sedimentation velocity, which roughly reflects molecular weight, especially for a series of homologous polymers like nucleic acids. The bands can be individually collected by dripping the contents from the tubes, as shown, or by careful pumping. The technique can be used for either analytical or preparative purposes.

Density gradient equilibrium centrifugation

In this method, a gradient is developed in a solution of a dense salt—often cesium chloride, but many other dense salts can be used—just by spinning a centrifuge rotor at a very high speed. This is an equilibrium gradient that will remain stable indefinitely as long as the centrifugal field is maintained. If the mixture of macromolecules to be separated is dissolved in the original salt solution, each component will band at a point in the gradient corresponding to its own density. Thus the method separates on the basis of molecular density, not size. It can be sensitive to extremely small differences in density. An example of density gradient equilibrium centrifugation is seen in the Meselson-Stahl experiment (see Box 19.1).

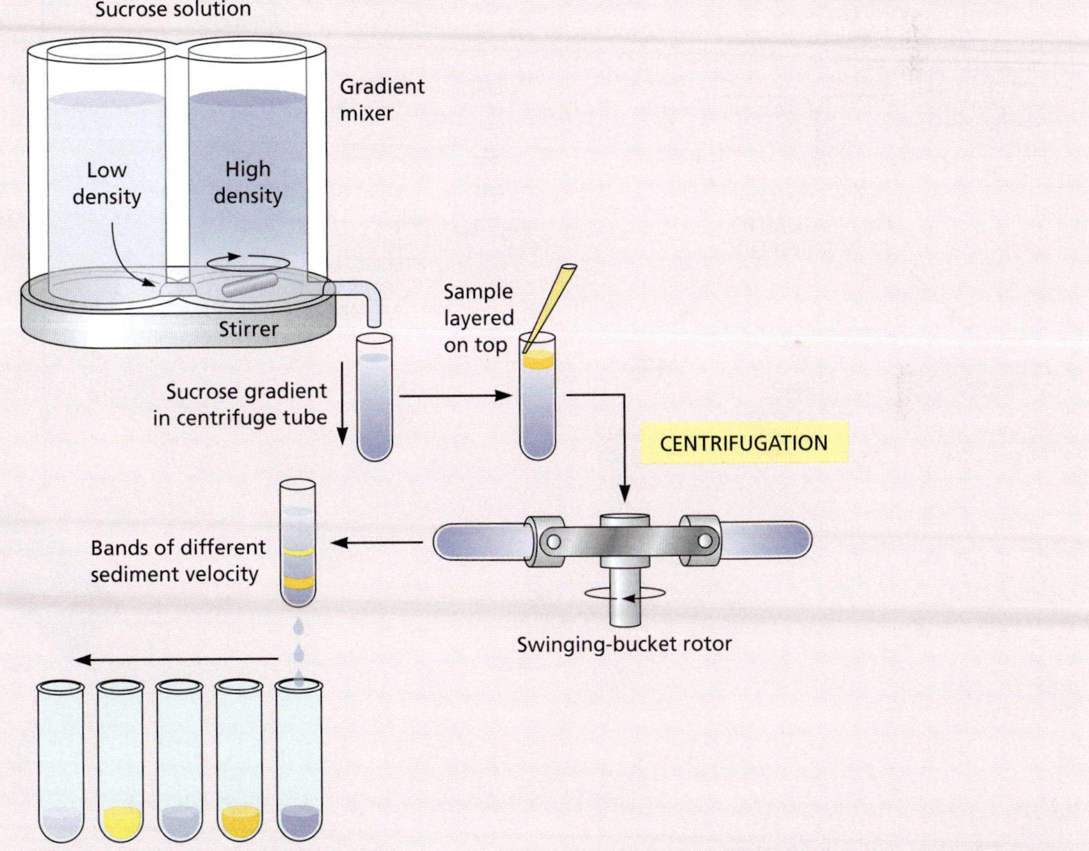

Figure 1 Sucrose gradient centrifugation. Steps in fractionating mixtures of DNA molecules by sucrose gradient centrifugation are shown. A sucrose gradient is pre-formed in a centrifuge tube by use of a special mixer device. The mixture of molecules is layered on the top of the gradient in a low-density solution. The tube is placed in a swinging-bucket rotor and centrifuged; the different molecules move through the gradient at different velocities that are dependent on the molecular mass and shape of the molecules. Following centrifugation, the content of a tube is collected into a series of tubes, through a hole punched in the bottom of the tube or by a pumping device.

Box 4.9 Nucleic acid hybridization Hybridization is a very useful technique with a myriad of applications in molecular and cellular biology. Hybridization detects sequence complementarity between labeled short probes and DNA target molecules present in a mixture (**Figure 1**). The probe can be either a purified restriction fragment of DNA or a synthetic oligonucleotide that encompasses the sequence of interest and is labeled with radioactivity (**Figure 2**); with digoxigenin or biotin, which are detected by chemiluminescence (**Figure 3**); or with fluorescent dyes that are visualized by fluorimetry (**Figure 4**).

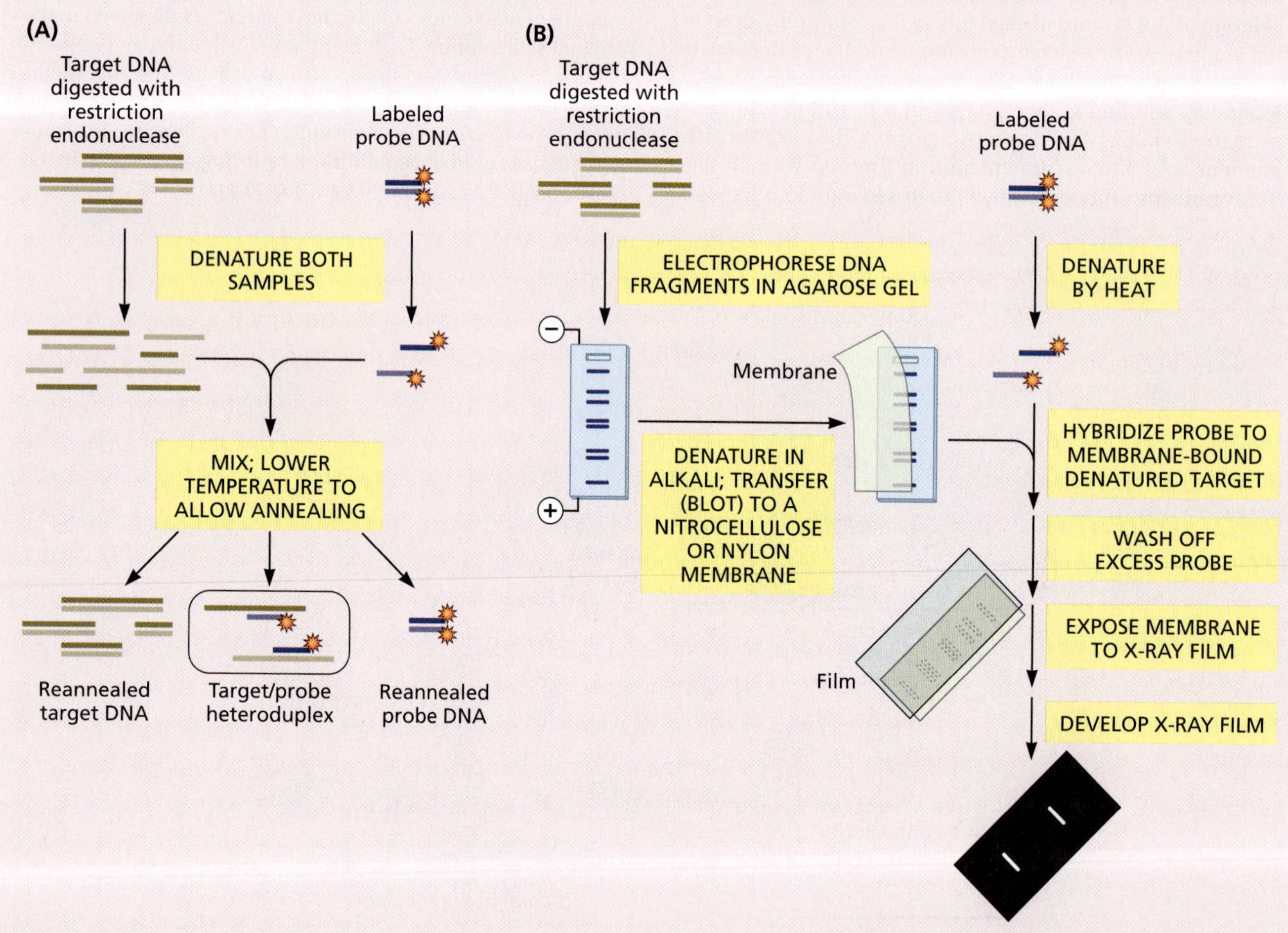

Figure 1 Nucleic acid hybridization. Hybridization detects sequence complementarity between short labeled probes and DNA target molecules present in a mixture. The probe can be either a purified restriction fragment of DNA or a synthetic oligonucleotide that encompasses the sequence of interest. It must be labeled in some way (see Figure 2). (A) If the reaction is performed in solution, three possible DNA duplexes will be present after the annealing step. (B) The Southern blot hybridization protocol involves size separation of the initial DNA mixture by agarose gel electrophoresis, in-gel denaturation, transfer (blotting) of the material onto filters, on-filter hybridization, removal of excess labeled probe, exposure to X-ray film if the probe is radioactively labeled, and film development. Only the DNA fragment(s) that contain sequences complementary to those in the probe will appear as bands on the film. If the probe is labeled by digoxigenin or biotin, the detection method is chemiluminescence, whereas fluorimetry is used to visualize probes labeled with fluorescent dyes. (Adapted from Strachan T & Read A [1999] Human Molecular Genetics, 2nd ed. With permission from Garland Science.)

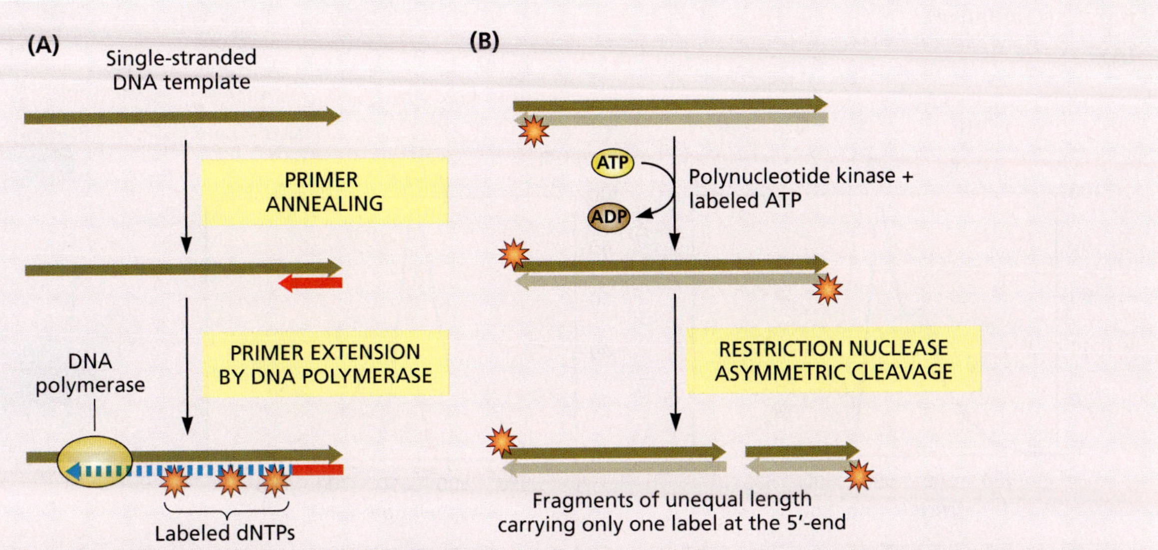

Figure 2 Two common methods for labeling DNA fragments. (A) Enzymatic copying of DNA single-stranded template in the presence of labeled deoxynucleoside triphosphates (dNTPs); alternatively, normal dNTPs can be utilized in the presence of labeled primer. (B) End-labeling uses the bacteriophage enzyme polynucleotide kinase to transfer a single labeled phosphate from ATP to the 5′-end of each DNA chain. Another method, nick translation, is also widely used (see Chapter 19). The first method produces DNA fragments that are uniformly labeled along the entire chain; these chains are appropriate for use in hybridization reactions, as they provide easily detectable signals of high intensity. The end-labeling procedure, on the other hand, produces probes that contain smaller amounts of label, as only one labeled atom is incorporated by the kinase into each DNA strand; these probes are invaluable for other applications, such as DNA footprinting (see Chapter 6). For such applications, an additional, restriction nuclease digestion step is required to produce fragments that carry just one label at one of the 5′-ends.

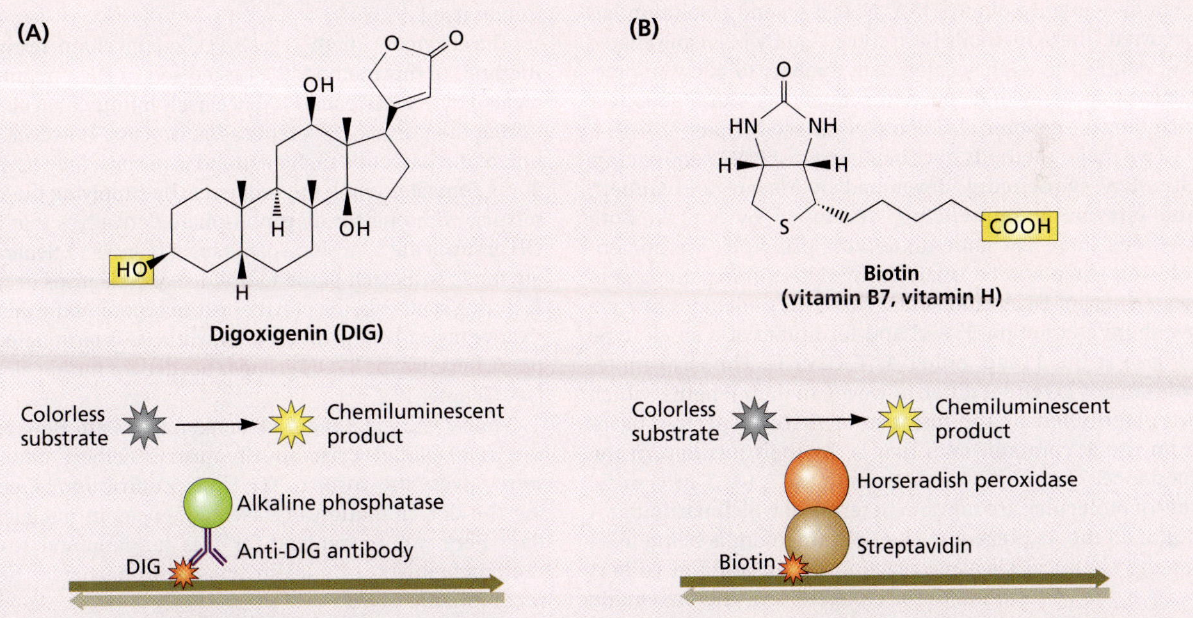

Figure 3 Chemiluminescent detection of digoxigenin- or biotin-labeled DNA samples. (A) Detection of digoxigenin (DIG)-labeled nucleotides is done by use of anti-DIG antibodies conjugated to alkaline phosphatase. The DIG OH group used for attachment to bases in dNTPs, through linkers, is boxed in yellow; the modified dNTPs still have their bases available for base-pairing. When an appropriate substrate is added, alkaline phosphatase breaks it down to a compound that emits a photon of light. (B) Detection of biotinylated DNA is typically done by use of streptavidin or its relatives, such as avidin, proteins with high affinity for biotin. Streptavidin is preconjugated to horseradish peroxidase. The COOH group of biotin used for attachment to bases in dNTPs, through linkers, is boxed in yellow. Breakdown of substrate, such as luminol peroxide, results in the release of a photon of light.

(Continued)

Box 4.9 ***(Continued)***

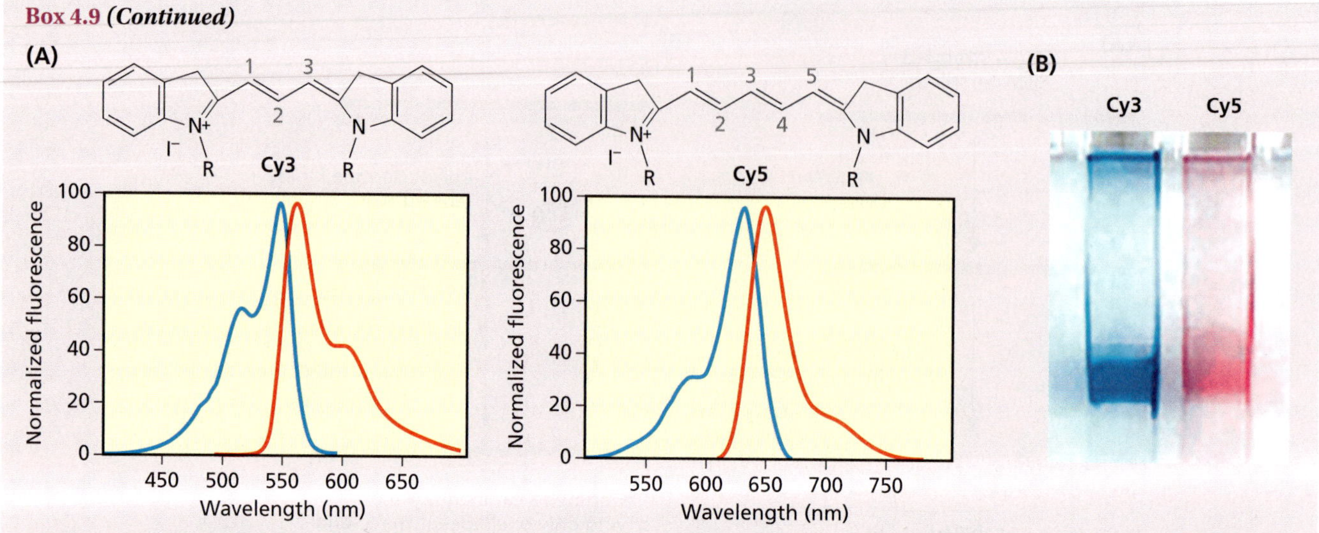

Figure 4 Detection of fluorescently labeled DNA molecules. (A) Absorption (blue lines) and emission spectra (red lines) of unconjugated *N*-hydroxysuccinimide (NHS) esters of the two fluorescent dyes, Cy3 and Cy5, most commonly used to label nucleotides. Many other fluorophores are also on the market. Cy3 is excited maximally at 550 nm and emits maximally at 570 nm, in the green part of the spectrum; Cy5 is excited maximally at 649 nm and emits maximally at 670 nm, in the far red part of the spectrum. The spectral overlap of Cy3 and Cy5 makes them a good choice of dyes for fluorescence resonance energy transfer (FRET) experiments (see Figure 1.11). (B) Agarose gel with DNA samples labeled with Cy3 and Cy5; the fluorescence is excited by lasers of different wavelengths. (A, adapted courtesy of Shanghai Open Biotech, Ltd. B, from Ramsay N, Jemth AS, Brown A et al. [2010] *J Am Chem Soc* 132:5096–5104. With permission from American Chemical Society.)

Box 4.10 DNA sequencing Methods developed in the late 1970s led to the revolutionary capability to determine the nucleotide sequence of any DNA. Now a second revolution has occurred. These methods have subsequently been automated and coupled to sophisticated data analysis to allow unprecedented accuracy and speed in sequencing entire genomes, including the genome of *Homo sapiens* (see Chapter 7).

Two major methods have been used for DNA sequencing: **chemical sequencing**, developed by Maxam and Gilbert, and **enzymatic sequencing**, developed by Sanger. Both methods share the same underlying principle: the nucleotide sequence is reconstructed by determining the size of nested sets of DNA molecules. The DNA molecules in each set share a common 5′-end and terminate at a single type of base at the 3′-end, either A, T, C, or G. The DNA molecules in any given set differ, however, in their lengths, which are determined by the distance of these respective bases from the 5′-common end; that is, by their position in the sequence. The processes that create the A (or T or C or G) sets of molecules are random in terms of which particular A out of all the As present in the DNA fragment is being used for the chemical cleavage reaction of Maxam and Gilbert or as a site of termination of elongation in the enzymatic method of Sanger. We will illustrate how the method works for the enzymatic method, as the chemical method is no longer used.

The enzymatic method is also called the **chain-termination method**. In this method, the nested sets of DNA fragments are created by *in vitro* reactions that closely mimic chain elongation during the normal DNA replication process. However, elongation of the nascent daughter strand is purposefully terminated, at random sites along the sequence, by supplying the reaction mixture with nucleoside triphosphate derivatives that lack the OH group at the 3′-position of the sugar (Figure 1). Separate runs are made with each of the four dideoxy derivatives or ddNTPs. If, during synthesis, the polymerase incorporates the ddNTP derivative instead of the dNTP, then synthesis is terminated at that point, because the 3′-OH group is needed for further elongation (see Chapter 19).

Walter Gilbert, Frederick Sanger, and Paul Berg received the 1980 Nobel Prize in Chemistry. Gilbert and Sanger were given the award "for their contributions concerning the determination of base sequences in nucleic acids." Paul Berg was recognized for "his fundamental studies of the biochemistry of nucleic acids, with particular regard to recombinant-DNA."

(A)

2′-deoxy-NTP

2′,3′-dideoxy-NTP

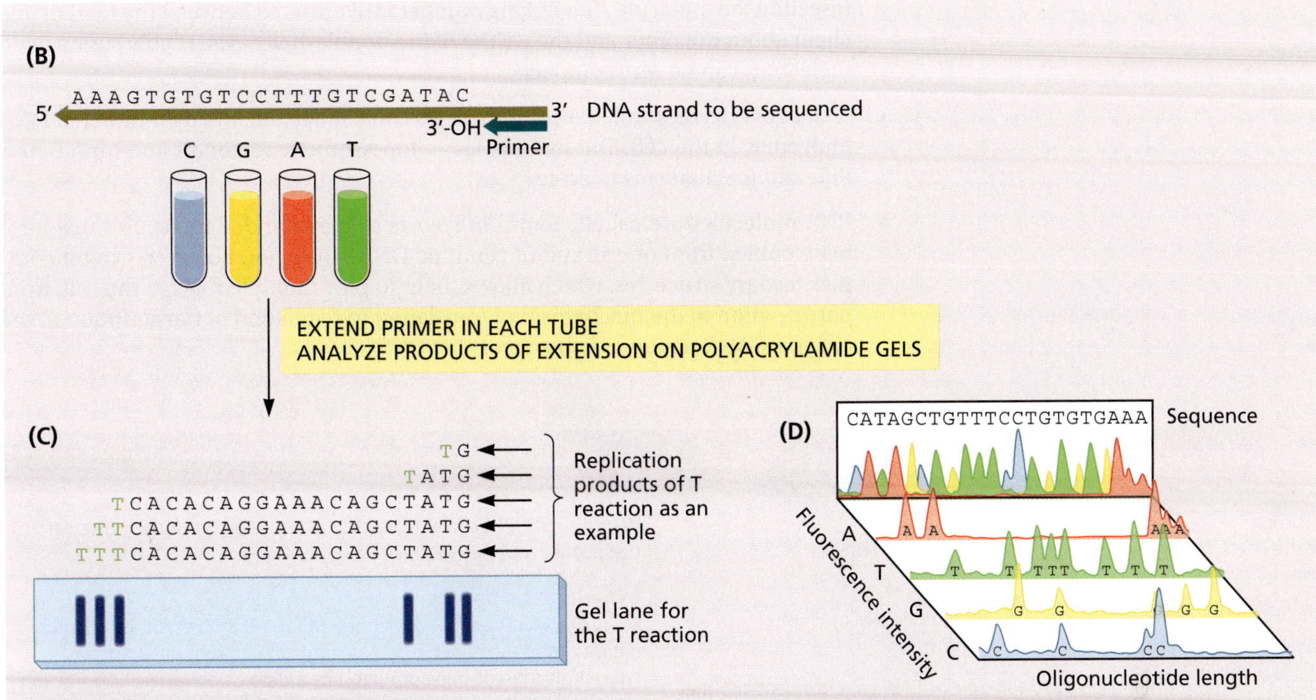

Figure 1 Sanger sequencing reaction. (A) Single-stranded DNA is amplified in the presence of fluorescently labeled dideoxy-NTPs (ddNTPs) in a reaction that mimics the normal replication reaction (see Chapter 19). The presence of dideoxythymidine triphosphate (ddTTP) will cause DNA synthesis to abort when the polymerase encounters an A in the DNA template strand. (B) Four reaction mixtures containing DNA template, primer, DNA polymerase, all four dNTPs, and one of the four ddNTPs at a ratio of one ddNTP to 100 dNTP. Each ddNTP is labeled with a different fluorophore. When the ratio of ddNTP to dNTP in the reaction mixture is 1:100, the frequency of termination will be, on average, once per 400 nucleotides. (C) The fragments of DNA are then separated via polyacrylamide gel electrophoresis and visualized with four different laser beams to excite the fluorescence of each individual fluorophore. Each reaction mixture is in a separate lane. (D) Finally, the data are compiled into a sequence with the aid of a computer. An electrophoretogram of a finished sequencing reaction is shown. (D, adapted from *The Science Creative Quarterly* [2009] 4. With permission from Fan Sozzi.)

Key concepts

- Two kinds of biopolymers are utilized for information storage and transmission in the cell: ribonucleic acid, RNA, and deoxyribonucleic acid, DNA. Both polymers are built on a repetitive backbone of sugar moieties linked by phosphodiester bonds. In RNA the sugar is ribose; in DNA the sugar is 2′-deoxyribose. Attached to the 1′-atom of each sugar is a basic unit, either a purine or a pyrimidine. In DNA, the purines are adenine (A) and guanine (G) and the pyrimidines are cytosine (C) and thymine (T). The same bases are found in RNA except that uracil (U) substitutes for T. The sequence of bases along the polynucleotide chain provides unique identity to each DNA or RNA molecule.
- DNA and RNA have significantly different structures and functions *in vivo*. The B-form DNA structure first proposed by Watson and Crick consists of antiparallel, double-stranded helices, with base pairing of A with T and G with C in the complementary strands. The helix is right-handed with ~10 bp/turn. Other structures are possible under special circumstances.
- The B-form DNA structure is quite stable, allowing it to serve as a repository of genetic information, and its complementary double-stranded nature allows for accurate copying and thus information transfer from generation to generation.
- Some DNA molecules are circular, and these allow further conformational variation, especially supercoiling. A supercoiled DNA is characterized by its linking number, *Lk*, the total number of times one strand is interlinked with the other. The linking number is a topological invariant and can be changed only by cutting and

resealing the molecule. The linking number is distributed between the twist of one chain about the other and the writhe of the double-helix axis. Highly supercoiled molecules can be very compact.

- A class of enzymes called topoisomerases can change the linking number of DNA molecules in the cell. The two classes of topoisomerases, topo I and topo II, use different mechanisms to do so.
- RNA molecules are usually found *in vivo* as single-stranded molecules that have been copied from one strand of genomic DNA. They can, however, exhibit complex tertiary structures, which allows them to play multiple roles in the cell, from participation in the mechanism of translation to regulation of transcription to the enzyme-like functions of ribozymes.

Further reading

Books

Bates AD & Maxwell A (2005) DNA Topology, 2nd ed. Oxford University Press.

van Holde KE, Johnson WC & Ho PS (2006) Principles of Physical Biochemistry, 2nd ed. Pearson Prentice Hall.

Wang JC (2009) Untangling the Double Helix: DNA Entanglement and the Action of the DNA Topoisomerases. Cold Spring Harbor Laboratory Press.

Watson JD (1981) The Double Helix: A Personal Account of the Discovery of the Structure of DNA. Norton Critical Editions.

Reviews

Champoux JJ (2001) DNA topoisomerases: Structure, function, and mechanism. *Annu Rev Biochem* 70:369–413.

Crick FHC, White JH & Bauer WR (1980) Supercoiled DNA. *Sci Am* 243:100–113.

Frank-Kamenetskii MD & Mirkin SM (1995) Triplex DNA structures. *Annu Rev Biochem* 64:65–95.

Paleček E (1991) Local supercoil-stabilized DNA structures. *Crit Rev Biochem Mol Biol* 26:151–226.

Rich A & Zhang S (2003) Z-DNA: The long road to biological function. *Nat Rev Genet* 4:566–572.

Stellwagen NC (2009) Electrophoresis of DNA in agarose gels, polyacrylamide gels and in free solution. *Electrophoresis* 30 (Suppl. 1):S188–S195.

Wang JC (2002) Cellular roles of DNA topoisomerases: A molecular perspective. *Nat Rev Mol Cell Biol* 3:430–440.

Experimental papers

Avery OT, MacLeod CM & McCarty M (1944) Studies on the chemical nature of the substance inducing transformation of pneumococcal types: Induction of transformation by a desoxyribonucleic acid fraction isolated from *Pneumococcus* type III. *J Exp Med* 79:137–158.

Griffith F (1928) The significance of pneumococcal types. *J Hyg* 27:113–159.

Maxam AM & Gilbert W (1977) A new method for sequencing DNA. *Proc Natl Acad Sci USA 74*:560–564.

Meselson M & Stahl FW (1958) The replication of DNA in *Escherichia coli*. *Proc Natl Acad Sci USA* 44:671–682.

Sanger F, Nicklen S & Coulson AR (1977) DNA sequencing with chain-terminating inhibitors. *Proc Natl Acad Sci USA* 74:5463–5467.

Watson JD & Crick FHC (1953) Molecular structure of nucleic acids: A structure for deoxyribose nucleic acid. *Nature* 171:737–738.

Zheng J, Birktoft JJ, Chen Y et al. (2009) From molecular to macroscopic via the rational design of a self-assembled 3D DNA crystal. *Nature* 461:74–77.

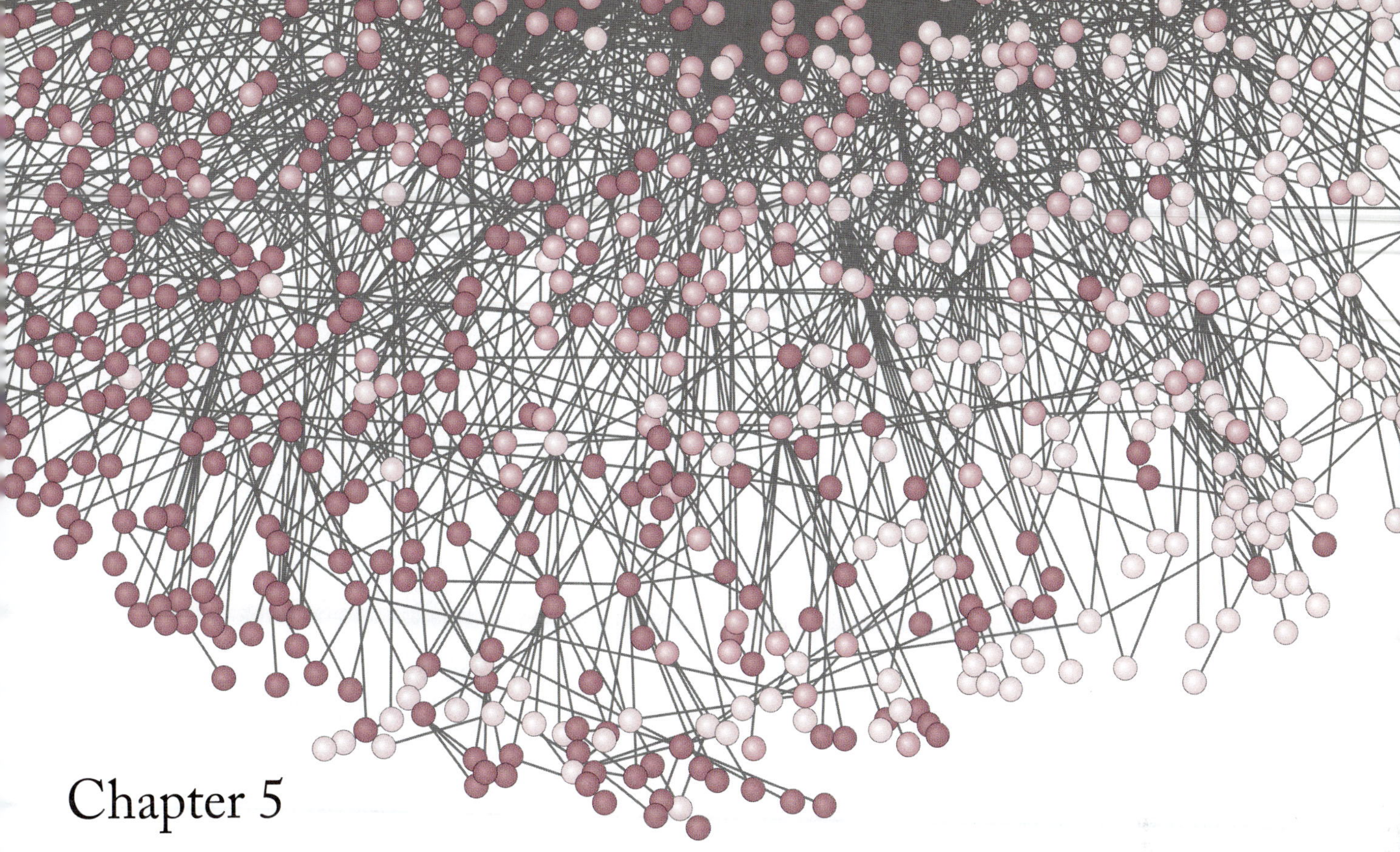

Chapter 5

Recombinant DNA: Principles and Applications

5.1 Introduction

Molecular biology made a giant leap forward with the introduction of **recombinant DNA technology**. These techniques, introduced in the early 1970s, allow the isolation and copying of any gene, thus generating the amount and purity of DNA necessary to determine the gene's nucleotide sequence. Moreover, they provide a way to express genes at the high level needed for thorough investigation of their products, most importantly proteins. Finally, recombinant DNA technology makes it possible to introduce any desired mutation at a predetermined position in any gene, thus allowing the production of proteins with desired properties that can be rather different from the properties of the original native proteins. Modified genes can be introduced into living cells as a way of studying their biochemical and physiological roles. For the first time, we are able to modify at will the genetic content of cells and whole organisms.

From a purely practical viewpoint, recombinant DNA technology has given rise to the contemporary biotechnology industry that produces and markets hundreds of new drugs and vaccines. It also underpins the production of genetically modified crops with desirable characteristics, such as improved nutritional value or resistance to insects, herbicides, drought, and salinity. Finally, it has given rise to gene therapy projects that aim to cure genetic diseases. Although these projects are not yet mature enough for general application in clinical practice, they are making rapid and significant progress.

The importance of these technologies for both basic research and other human activities is enormous and will be referred to again and again in the coming chapters. Therefore, in this chapter, we introduce the benchmark techniques that collectively form the basis for the creation and use of recombinant DNA molecules. We begin with a general introduction to the concept and the individual steps of the fundamental technique

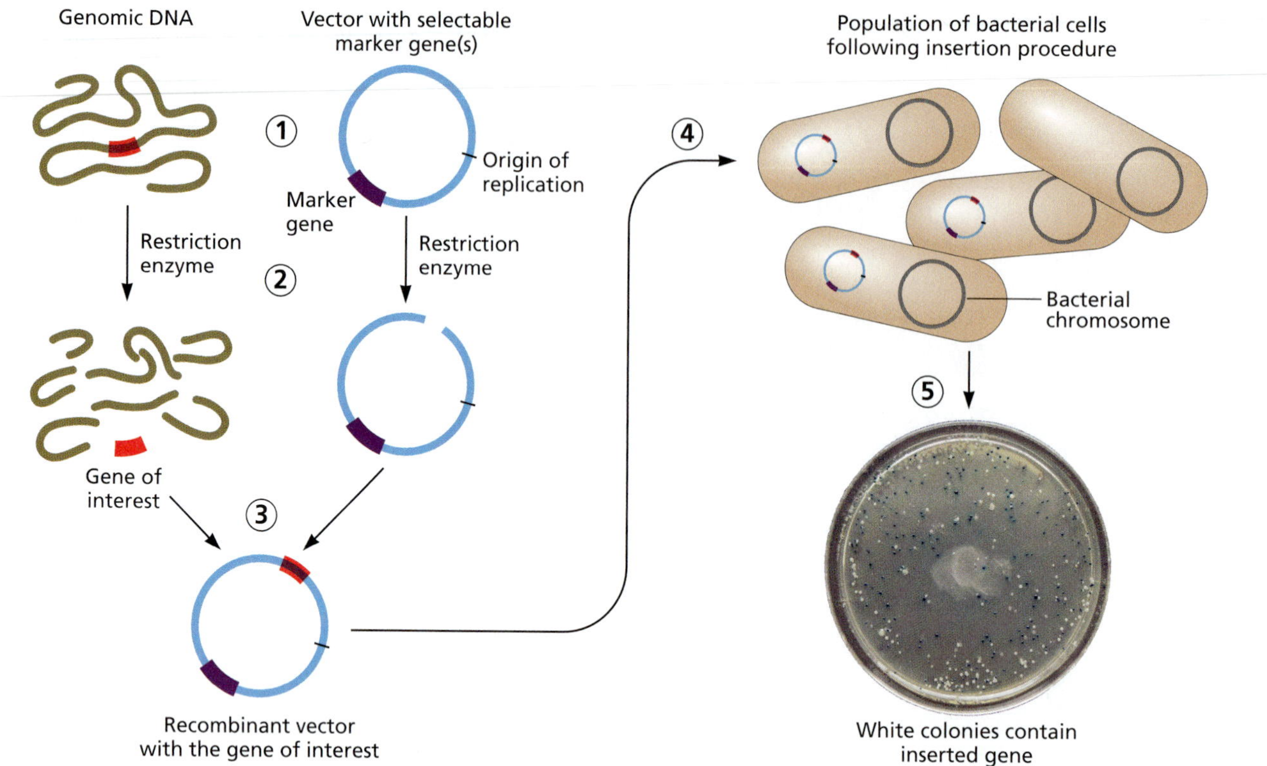

Figure 5.1 General scheme for DNA cloning in bacteria. DNA cloning can be divided into several steps: (Step 1) Isolate DNA from an organism and prepare cloning vector with marker gene. (Step 2) Produce recombinant DNA molecules via treatment of both donor and vector DNA with a restriction endonuclease and a ligase. The restriction endonuclease treatment will digest the organismal DNA into a large set of linear DNA fragments of heterogeneous length. Note that the type of fragment ends and the length of the fragments will depend on the restriction endonuclease used. (Step 3) Ligate the DNA fragments produced into the vector linearized by restrictase treatment. This step results in three different kinds of vectors: those that carry the gene of interest, those that have other DNA fragments from the original mixture inserted, and those that self-ligate without incorporating any foreign DNA. For simplicity, only a recombinant vector carrying the gene of interest is depicted. (Step 4) Introduce the population of vector molecules into bacterial cells by transformation or transduction, plate the cell population onto solid agar plates, and allow individual cells to proliferate and form colonies on the plate. It is important to dilute the bacterial cultures before plating so the individual colonies can be observed and further tested. Four different kinds of bacterial cells are obtained following the vector insertion procedure, reflecting the three different kinds of vectors described above as well as cells that have not taken up any vector. (Step 5) Screen for clones that contain the gene of interest. The step is performed in three stages. First, cells that did not take up the plasmid are eliminated by growing the cell population on solid medium, such as agar, that contains an antibiotic. Cells that do not contain a vector will die, because the antibiotic resistance gene is carried only by the vector. Second, one should be able to discriminate between cells that carry recombinant vectors and those that carry original vector, self-ligated with no insert. This is usually done by monitoring a gene on the vector that becomes inactivated when a foreign fragment is inserted into it. If the insertion is into a gene that confers resistance to a second antibiotic, these cells will become sensitive to this drug; these are the desired cells. Other genes, like the one carrying the enzyme β-galactosidase, can be used for insertional inactivation. Third, there should be a way to find the bacterial clone that carries the gene of interest against the background of all other recombinant bacteria. This is done by colony hybridization with a labeled probe that contains the gene of interest or a portion of it. (Petri dish, courtesy of Stefan Walkowski, Wikimedia.)

in recombinant technology: the making of multiple copies of a DNA sequence by the technique called **cloning**.

Cloning of DNA involves several fundamental steps

A **clone** is a group of identical copies of some biological entity; a colony of bacteria grown from a single bacterium is a clone. When we speak of the cloning of DNA, we mean the production, in some organism, of multiple copies of a particular sequence. The development of such techniques in the 1960–1970 period solved the problems noted above.

As illustrated in **Figure 5.1**, all cloning projects begin with two essential components: the piece of DNA to be cloned and the vector molecule that will be used to introduce the sequence of interest into a living host cell. By definition, vectors are relatively

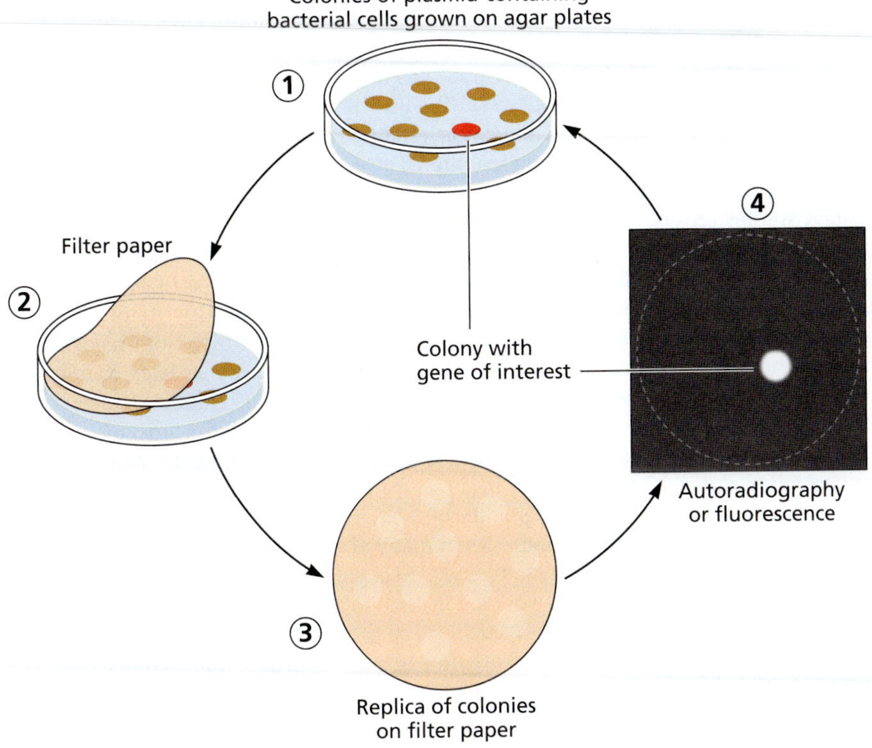

Figure 5.2 Nucleic acid hybridization identifies the colony that carries the gene of interest. (Step 1) Grow colonies of plasmid-containing bacterial cells on agar plates. The colony containing the gene of interest is shown in red. (Step 2) Press a filter against colonies to create a replica of the colonies where some cells from each colony adhere to the filter. (Step 3) Wash the filter with an alkali solution to denature the DNA. The resulting ssDNA is then hybridized with the gene of interest labeled in some way; only the colonies that contain the gene of interest will hybridize to the probe and become labeled. (Step 4) Detect the label in an appropriate way: by autoradiography if the probe is radioactively labeled or by fluorescence if the probe carries a fluorophore. Then compare the pattern with probe to the pattern on the original plate to identify the colony containing the gene.

small DNA molecules that can incorporate the foreign nucleotide sequence, be introduced into a cell, and then be maintained stably in that cell.

The next step is insertion of the donor DNA segments into the vector. This is essentially a cut-and-paste process that requires several enzymes, including restriction endonucleases for cutting and DNA ligases for resealing. Once the donor sequences are inserted into a vector, the resulting recombinant vectors are introduced into living cells. There are a variety of ways to do this, each optimized for the specific host, be it a bacterium or a eukaryotic cell. This step results in the production of several different types of host cells: those that did not take up any vector, those that contain the unaltered self-ligated vector, and those that contain recombinant vectors. The majority of this last group of host cells may contain sequences in which the investigator is not interested; a relatively small proportion will have the recombinant of interest. In order to retrieve the clone of interest from this heterogeneous cell population, various selection or screening methods are applied, as illustrated in Figure 5.1 and Figure 5.2. The reader will learn about these methods along the way, when we describe the properties of the cloning vectors by using specific examples.

5.2 Construction of recombinant DNA molecules

Restriction endonucleases and ligases are essential tools in cloning

To insert a given DNA sequence into a vector, one must first cleave the vector, allow recombination, and then ligate the recombinant vector. Two classes of DNA enzymes

Box 5.1 The discovery of restriction endonucleases: How basic research has implications far beyond science The existence of enzymes that cleave nucleic acids was recognized as early as 1903, but the fact that some could make specific cleavages came only half a century later. The first hint of such activity came from seemingly unrelated genetic studies in the laboratory of the pioneering molecular geneticist Salvatore Luria. In 1952, Luria and Mary Human published studies on the comparative susceptibility of various bacterial strains to several bacteriophages. The results were curious: some strains were almost completely resistant to some phage but vulnerable to others. The word almost here is significant; even in resistant strains, there were a few bacteria in which the phage could grow, and these phages, when harvested, were fully potent against the previously resistant strain of bacteria. Similar results, with other phage, were soon obtained in other laboratories. The phenomenon was first termed host-controlled variation and later **host restriction**.

The puzzling result waited a decade for explanation. In 1962, Werner Arber and associates at the University of Geneva proposed that the resistant bacterial strains contained an enzyme that degraded the incoming phage DNA. But why did such an enzyme not degrade host DNA? The answer must be that the host DNA had somehow been modified by another enzyme to resist this degradation. DNA methylation by a methylase was suspected as the source of host resistance by Arber as early as 1965 but was not experimentally demonstrated until 1972. The postulate of a protecting enzyme also explained the occasional observation of acquired infectivity: in rare cases, the DNA of invading phage could itself be methylated.

Early studies on the restriction process centered largely on type I restriction endonucleases, which do not cleave at a specific site. It was a major breakthrough when, in 1970, Hamilton Smith and colleagues at Johns Hopkins University discovered type II restriction endonucleases, which both recognize and cleave at a sequence-specific site. A year later, Kathleen Danna and Daniel Nathans, also at Johns Hopkins University, used such an enzyme to digest the DNA of the virus SV40 into discrete fragments, which they separated by gel electrophoresis. This led in turn to the routine isolation and cloning of specific fragments of DNA. Restriction endonucleases play fundamental roles in many of the most important techniques in molecular biology. Recognition of the enormous potential of this work led in 1978 to the award of the Nobel Prize in Physiology or Medicine to Arber, Smith, and Nathans "for the discovery of restriction enzymes and their application to problems of molecular genetics."

are essential for creating any recombinant DNA molecule: restrictases and ligases. DNA polymerases constitute another very important class of DNA enzymes used to manipulate nucleic acids, but as they encompass a wide variety of enzymes, each with its own role and mechanism of action, we describe them in relevant chapters elsewhere in the book.

The discovery of **restriction endonucleases**, also called **restrictases**, is one of the most important developments in molecular biology; for a historic account, see Box 5.1. The majority of these enzymes catalyze the cleavage of DNA at specific nucleotide sequences, producing defined double-stranded DNA fragments suitable for numerous applications in basic research and biotechnology.

More than 3000 restrictases have been described to date. They are all DNA **endonucleases**, that is, they hydrolyze internal phosphodiester bonds within the polynucleotide chain (Figure 5.3). **Exonucleases**, by contrast, hydrolyze the chain from either its 3′- or 5′-end, removing one nucleotide at a time. The name of each restrictase carries information about the bacterial species and strain in which it was originally discovered; a Roman numeral is used to differentiate between several enzymes from the same strain. For example, ***Eco*RI** designates one of the two known restrictases in *Escherichia coli* strain R; ***Hae*III** (see Figure 5.3) denotes the third enzyme present in *Haemophilus aegyptius*. There are three general classes of restrictases. Type II enzymes are of primary interest to us because their ability both to recognize a particular DNA site and to cut at a defined point within that site makes them ideal tools for recombinant DNA technology.

Each restriction enzyme is characterized mainly by the specific sequence it recognizes and the site where it cuts the double helix (Table 5.1). Most recognition sequences are 4, 5, or 6 bp in length, although longer sites have also been described. The longer the recognition site, the less frequently it occurs in DNA. For instance, enzymes that target 4-bp sites cut about once every 256 or 4^4 nucleotide pairs, whereas 6-bp sites will be found on average once in 4096 or 4^6 nucleotide pairs. As illustrated in Figure 5.3A, the enzyme can attack either one of the two ester bonds in the phosphodiester linkage, producing different products. In addition, different enzymes can cut the opposite strands of the DNA helix, either in a staggered way to produce **overhangs**, also known as **sticky ends**, or exactly opposite each other to

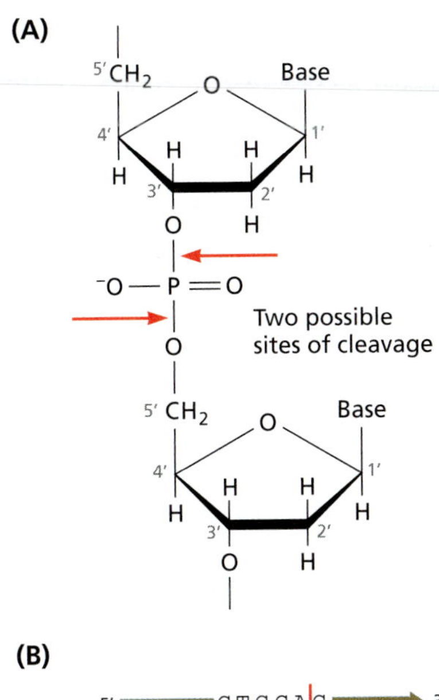

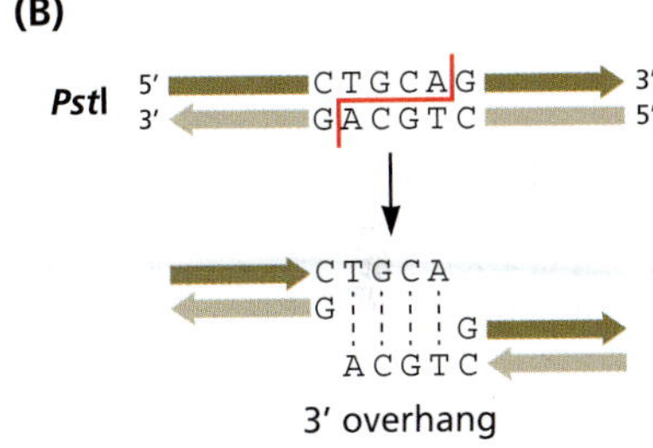

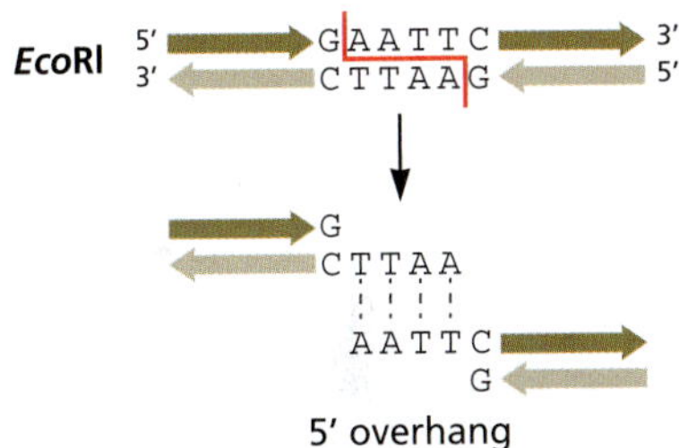

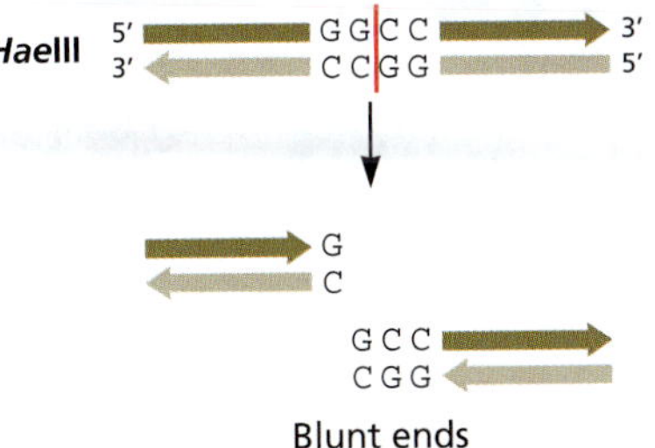

Figure 5.3 Cleavage action of restriction endonucleases. (A) In general, endonucleases that cleave polynucleotide chains internally can be specific with respect to which ester bond in the phosphodiester linkage they cleave, thus producing DNA fragments of different termini, as depicted. Type II restriction enzymes produce 5′-phosphates and 3′-OH groups only. (B) Cleavage sites on the two DNA strands can be offset, by as many as four nucleotides, or occasionally even more, to produce overhangs, also known as sticky ends. When there is no offset of the cutting positions, blunt-end products are created.

produce **blunt ends**, also known as flush ends (see Figure 5.3B). Both kinds of products can be ligated to form uninterrupted molecules, but the ligation efficiency of blunt-ended fragments is much lower than that of fragments containing overhangs, because the latter can actually base-pair with each other, increasing the probability of the two ends staying in close proximity and in proper orientation for ligation to occur.

Type II restriction endonucleases have proven to be of great value in recombinant technologies, as they possess well-defined specificity for the cleavage sequence, and thus the products of their action are very well defined. These enzymes usually recognize palindromic sequences, which read the same on both complementary DNA strands, 5′ → 3′ and 3′ → 5′, and bind as homodimers. The cleavage patterns of type II restrictases can be rather diverse. Figure 5.4 illustrates some frequent cleavage patterns and introduces two important concepts: **isoschizomers** and methylation-sensitive isoschizomers.

Isoschizomers have the same recognition site but come from different bacterial species; they may cut the sequence at different places, thus adding to the diversity of cleavage patterns and DNA ends produced. **Methylation-sensitive isoschizomers** may or may not cleave a sequence, depending on whether the sequence carries methyl groups. These pairs of enzymes have proven instrumental in studying the methylation patterns of specific sequences of interest.

The initial binding of type II restrictases occurs at random sites, and the subsequent search for the recognition sequence involves one-dimensional translocation along the DNA. Numerous restrictases have been crystallized and their structures have been determined in their free forms, bound to noncognate (nonspecific) DNA fragments, and bound to cognate sequences. Figure 5.5 compares the crystal structures of *Eco*RV when bound to a random sequence and when bound to its specific recognition site. Note that interaction with the cognate nucleotide sequence leads to conformational changes in the enzyme that allow it to embrace the DNA more tightly.

Table 5.1 Classification and properties for restriction/modification systems. All enzymes require Mg^{2+} for cleavage; only type I enzymes require ATP for functioning of the motor. The letter N stands for any nucleotide.

	Type I	Type II	Type III
example	*Eco*B	*Eco*RI	*Eco*PI
recognition site	TGAN$_8$TGCT	GAATTC	AGACC
cleavage site	random, up to 10 kb away from recognition site	between G and A on both strands	24–26 bp 3′ to recognition site
methylation site	TG<u>A</u>N$_8$TGCT ACTN$_8$<u>A</u>CGA	GA<u>A</u>TTC CTT<u>A</u>AG	AG<u>A</u>CC TCTGG only one strand is methylated
nuclease activity	yes	yes	yes
methylase activity	yes	no; methylation of recognition site by a separate enzyme	yes

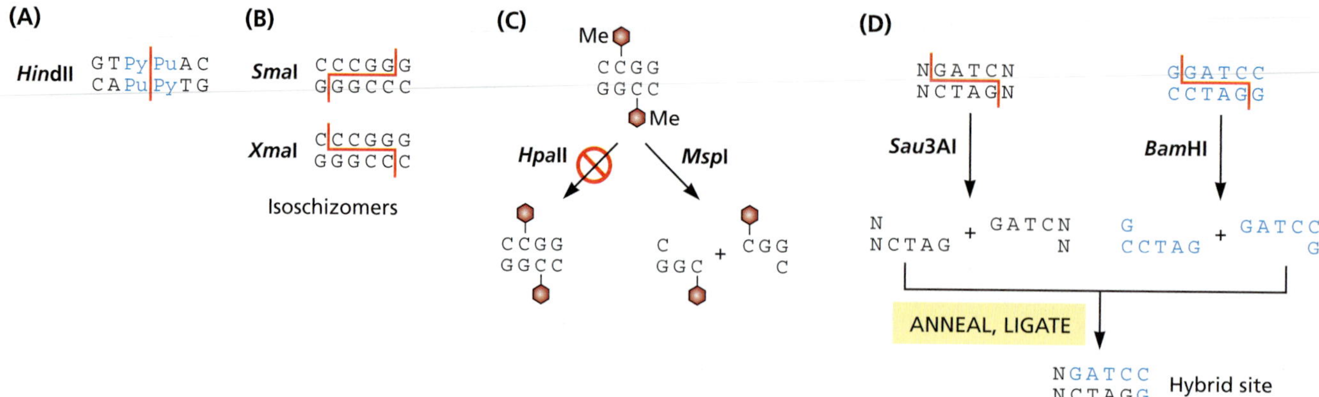

Figure 5.4 Diversity of restriction endonuclease cleavage patterns. (A) Ambiguity of recognition sequences, as illustrated for the example of *Hind*II, the first type II enzyme to be described. Py stands for either pyrimidine, C or T; Pu stands for either purine, A or G. All possible sequences of this kind are cleaved by *Hind*II. (B) Isoschizomers are enzymes from different sources that recognize the same target. Some pairs of isoschizomers cut their targets at different places, as in the example shown. (C) Methylation-sensitive isoschizomers: if the recognition sequence is methylated, in this case on the C, one enzyme of the isoschizomer pair will cut it while the other one will not. In the specific example shown, *Hpa*II is methylation-sensitive and will not cut when C is methylated, while *Msp*I will cut whether C is methylated or not; that is, *Msp*I is methylation-indifferent. Such pairs of enzymes are extensively used in research to probe the methylation status of a particular sequence. (D) Production of a hybrid site by cohesion of complementary sticky ends generated by two different enzymes. One of these enzymes (*Sau*3AI) recognizes a 4-bp sequence that is embedded in the hexanucleotide sequence recognized by a different enzyme, *Bam*HI in this case. The sticky ends produced by these enzymes can base-pair to produce a hybrid site. The new site is sensitive to *Sau*3AI but may not constitute a target for *Bam*HI, depending on the nucleotides adjacent to the original *Sau*3AI site.

Type I restrictases are multifunctional enzymes that combine both nuclease and methylase activity in a single protein trimeric complex (**Figure 5.6**). One subunit, R, carries the restriction function, cleaving the DNA at random locations, sometimes very far, 10 kilobase pairs or more, from the enzyme's recognition sequence; it also has an ATP-dependent motor activity that is involved in translocating the DNA. The second subunit, M, carries out the DNA methylation reaction, which occurs in the recognition site itself. The third subunit, S, recognizes the specific sequence to which the complex binds. For cleavage to occur, the recognition/binding site has to be brought into proximity with the cleavage site; that is, the intervening DNA has to loop out (at the same time it is supercoiled) (see Figure 5.6). Therefore, these enzymes are unusual molecular motors that bind specifically to DNA and then move the rest of the DNA through this bound complex.

Type III restriction endonucleases are very similar to type I, but they do not require ATP, they methylate just one of the strands, and their cleavage site is relatively close to the recognition sequence.

DNA ligases are also among the most important enzymes in the cell. They are indispensable for numerous cellular processes, including DNA replication, recombination, and repair, whenever there is a need to seal nicks in the DNA double helix or join broken helices. Of interest to us here is their essential role in joining fragments of double-helical DNA to form intact molecules in recombinant DNA technologies. Ligases form phosphodiester bonds between the 3′-OH and 5′-phosphate termini of DNA fragments (**Figure 5.7A**). The reaction involves several steps. Adenylation of the active-site lysine uses ATP in phages and eukaryotes but uses NAD^+,

(A) Complex with noncognate DNA

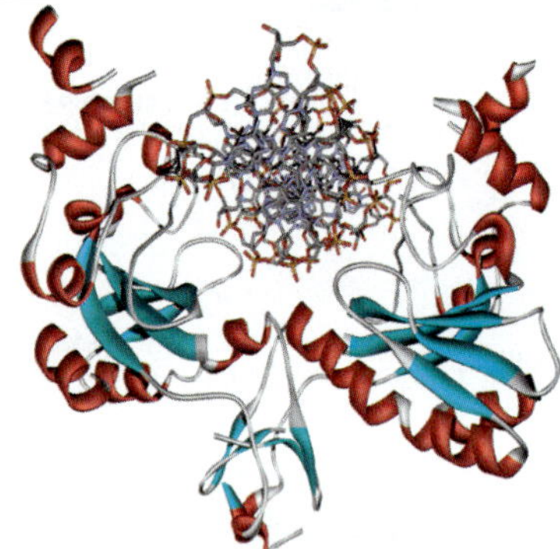

(B) Complex with cognate DNA

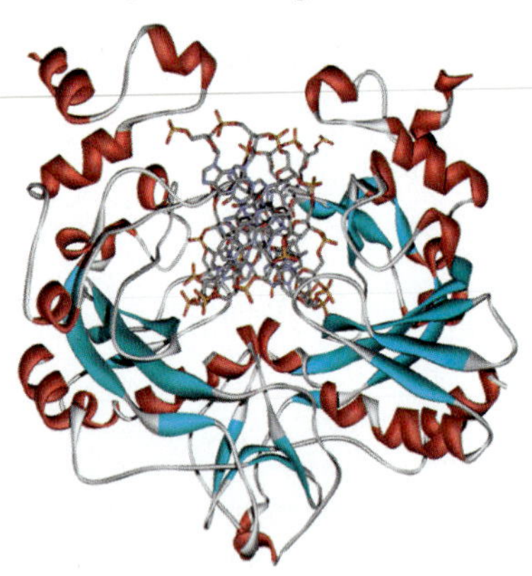

Figure 5.5 Comparison of crystal structures of nonspecific and specific enzyme–DNA complexes for *Eco*RV. (A) The nonspecific enzyme–DNA complex presumably closely reflects the structure of the complex that slides along DNA during the initial one-dimensional search for the recognition sequence. (B) The specific enzyme–DNA complex crystallized in the presence of Ca^{2+} can be considered to resemble the enzyme–substrate complex. Note the conformational transitions in some of the helices that embrace the DNA once the enzyme recognizes the sequence to be cleaved. (From Pingoud A, Fuxreiter M, Pingoud V & Wende W [2005] *Cell Mol Life Sci* 62:685–707. With permission from Springer Science and Business Media.)

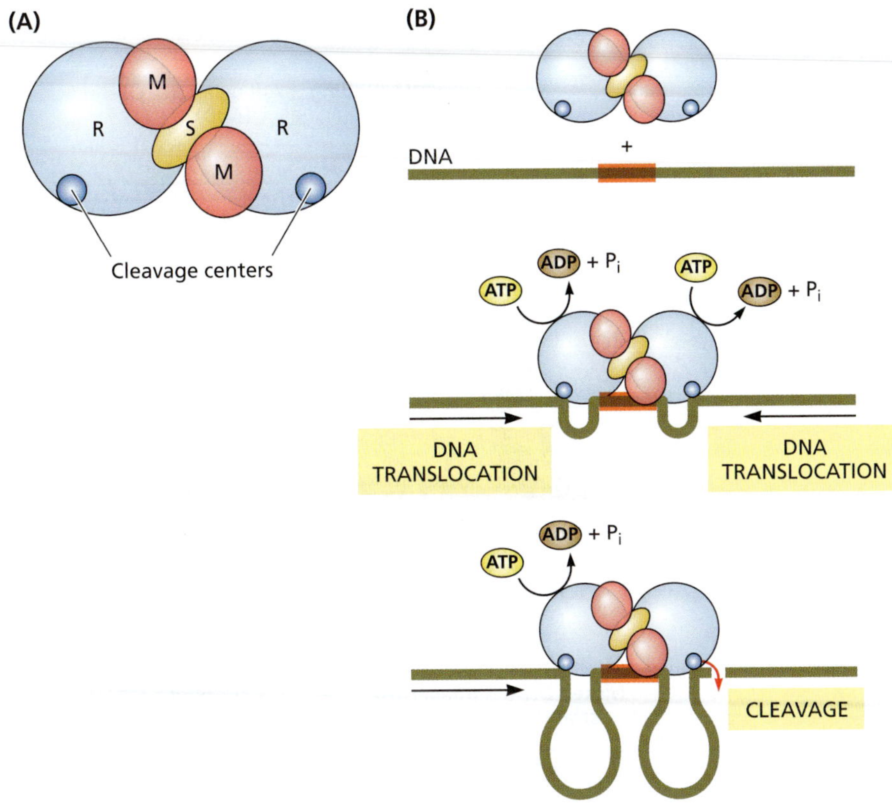

Figure 5.6 Structure and activity of *Eco*R124I, a type I restrictase. (A) The enzyme, with subunit composition R_2M_2S, possesses three functionally distinct subunits. The S subunit is responsible for DNA binding specificity, while M is required for DNA methylation, that is, modification activity. R, along with the core M, is absolutely required for DNA cleavage, that is, restriction activity. It is also responsible for ATP binding, hydrolysis, and DNA translocation; that is, it is a molecular motor. The R subunit also possesses helicase activity and domains responsible for assembly of the complex. (B) Successive steps in the motor activity of *Eco*R124I. The orange block represents the DNA binding/recognition site. When the motor complex binds to the DNA at the recognition site, it also attaches to adjacent DNA sequences. Then the motor begins to translocate these adjacent sequences through itself, following the helical path of the DNA in the complex. This leads to the creation of an expanding loop of positively supercoiled DNA. During translocation, the entire complex remains tightly bound to the recognition sequence. The process is bidirectional, with each R subunit acting as a molecular motor, and proceeds until blocked by some external event, usually another enzyme also translocating the DNA; other blockages include DNA topology. Cleavage follows blockage of the translocation; the site of cleavage is random because translocation blocking is a random process. (Adapted from Pennadam SS, Firman K, Alexander C & Górecki DC [2004] *J Nanobiotechnol* 2 [10.1186/1477-3155-2-8]. With permission from BioMed Central.)

nicotinamide adenine dinucleotide, in bacteria. Figure 5.7B and Figure 5.7C depict the domain structure of human ligase I and the crystal structure of this enzyme complexed with DNA.

For the purposes of recombinant DNA technology, a distinction should be made between ligases that can only join DNA ends on overhangs and those that can ligate blunt-end DNA fragments. Most use is made of the first type: it is mechanistically easier to join the phosphodiester backbones of fragments that are kept in close proximity by base pairing of complementary overhangs. The *E. coli* **DNA ligase** is a representative of this type. Joining together blunt ends is much more challenging, and the enzymes that are capable of doing this are much less efficient. An example of such an enzyme is the DNA ligase from bacteriophage T4. The efficiency of blunt-end ligation can be increased by using small oligonucleotides that carry restriction sites for sticky-end production, known as **DNA linkers**. The strategy for using linkers is depicted in Figure 5.8.

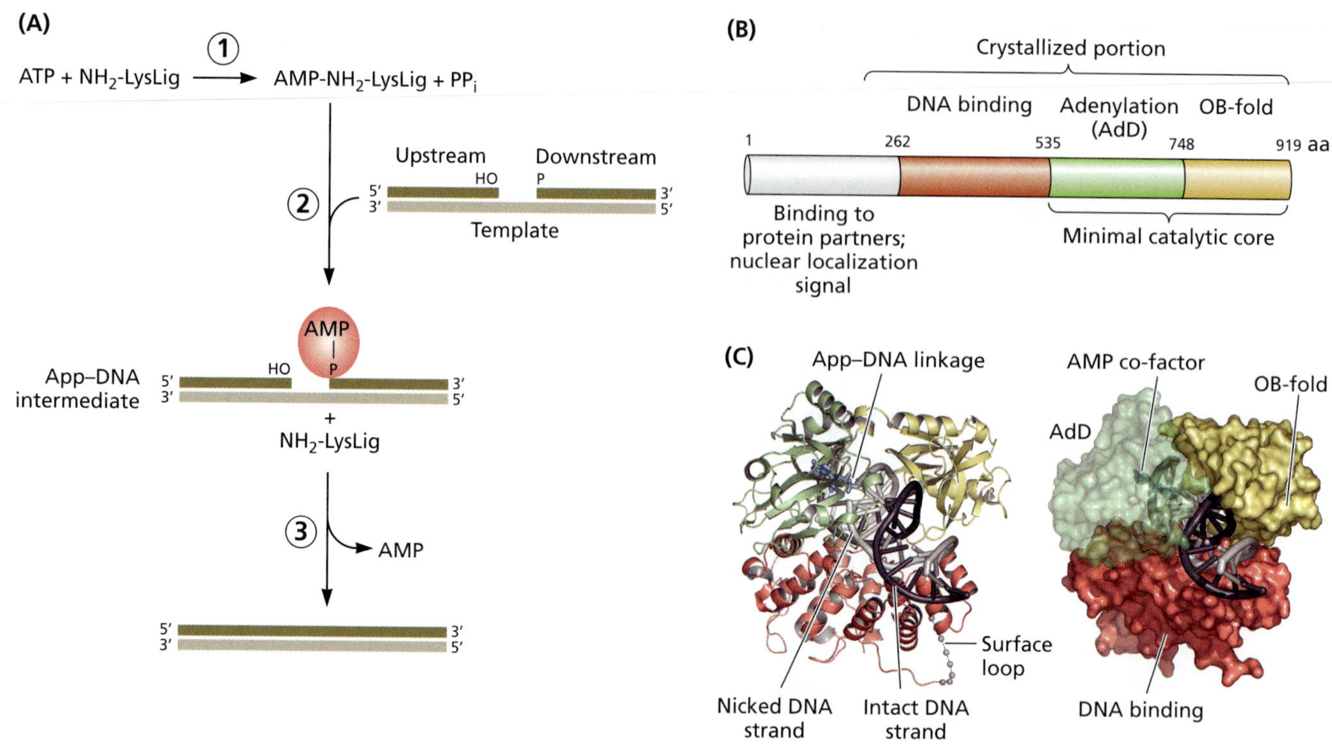

Figure 5.7 DNA ligation. (A) Steps in the reaction catalyzed by DNA ligase. (Step 1) Enzyme–AMP is formed by attack of lysine on the α-phosphate of ATP, releasing inorganic pyrophosphate. (Step 2) The 5′-phosphate of the nicked DNA strand, downstream side, attacks the Lys–AMP intermediate to form an App–DNA intermediate, where pp indicates a pyrophosphate linkage, 5′-P to the 5′-phosphate of AMP. (Step 3) The 3′-OH end of the nicked strand, upstream side, attacks the 5′-P of App–DNA, covalently joining the DNA strands and liberating AMP. Thus, ligation pays an energy price of one ATP. (B) Domain organization of human ligase I. OB-fold is the oligonucleotide/oligosaccharide-binding domain. (C) Left, the three domains, colored as in part B, fully encompass the App–DNA reaction intermediate. The intact DNA strand is shown in black, the nicked strand is shown in gray, and the App–DNA linkage is shown in blue. Gray spheres indicate a poorly ordered surface loop at residues 385–392. Right, molecular surface: the adenylation domain, AdD, is semitransparent to highlight the AMP cofactor held within its active site. (Adapted from Pascal JM, O'Brien PJ, Tomkinson AE & Ellenberger T [2004] *Nature* 432:473–478. With permission from Macmillan Publishers, Ltd.)

The availability of site-specific endonucleases and DNA ligases made it possible to create the first recombinant DNA molecules in the early 1970s (**Box 5.2**).

5.3 Vectors for cloning

Genes coding for selectable markers are inserted into vectors during their construction

As indicated in Figure 5.1, the construction of vectors involves an important step: the introduction of genes that can be used for the selection of cells that carry the vector. The most commonly used selectable markers are antibiotic resistance genes, conferring resistance to antibiotics such as tetracycline, ampicillin, chloramphenicol, and kanamycin. The protein products of these genes act by a variety of pathways, some preventing the antibiotic from entering the cell and others destroying or enzymatically modifying the antibiotic to inactive forms.

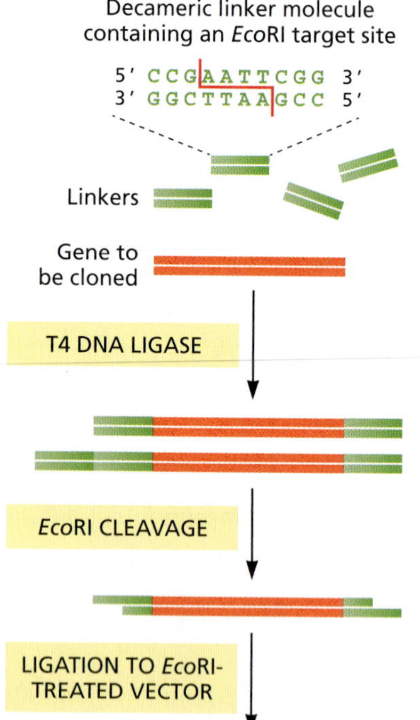

Figure 5.8 DNA linkers for joining of blunt-end DNA fragments. Linkers can be chemically synthesized to contain any restriction target site. Blunt-end ligation is performed by the DNA ligase from phage T4, as the *E. coli* DNA ligase will not catalyze blunt-end ligation except under special reaction conditions of macromolecular crowding. Linkers can be ligated to both ends of the blunt-end DNA fragment to be cloned; upon treatment with the restriction endonuclease that recognizes the linker sequence, sticky ends are produced. This treatment is then followed by standard cloning techniques that use sticky ends for efficient ligation of the foreign gene into a vector treated with the same enzyme.

Box 5.2 The first recombinant DNA molecules The discovery of type II restriction endonucleases in 1970 (see Box 5.1) had profound repercussions. In 1972, Paul Berg and his group at Stanford University carried out the first experiments to modify natural DNA specifically. They used the *Eco*RI restrictase, newly discovered by Herbert Boyer at the same institution, to cleave the DNA chromosome of the mammalian virus SV40 at one specific point, opening the closed circle into a linear form. They then used the enzyme terminal transferase to add 3′ extensions of either poly(dA) or poly(dT) onto such molecules, after resection of the 5′-ends with exonuclease (**Figure 1A**). Such molecules could then be annealed together through A–T base pairing.

Single-stranded regions were filled in by DNA polymerase and ligated by ligase. In this way, Berg and colleagues were able to construct dimers and higher oligomers of SV40 DNA and to insert foreign DNA fragments into the viral DNA. Because SV40 can infect eukaryotic cells, they envisioned their procedure as a gateway for mammalian DNA modification.

This was a powerful and groundbreaking result: it was the first time that a natural DNA molecule had been specifically modified in a premeditated way. The technique was somewhat awkward, however, especially in the requirement for poly(dA) and poly(dT) tails, which necessarily involved inserting some extraneous dA/dT tracts into the product.

Shortly thereafter, it was recognized that the very nature of the cut made by *Eco*RI, as well as many other restriction endonucleases, necessarily left overlapping, complementary ends on the cleaved product (**Figure 1B**). In 1973, Stanley Cohen and Herbert Boyer took advantage of this in a seminal paper that laid the groundwork for recombinant DNA research.

Cohen and Boyer worked with **plasmids** that could be introduced into bacteria and contained all of the elements needed for their replication in the bacterial cell. They also contained expressible genes for resistance to various antibiotics, which allowed rapid and efficient screening for colonies of bacteria containing each plasmid. Finally, each contained an *Eco*RI site, placed so as not to interfere with plasmid replication or gene expression. Cohen and Boyer were able, by taking advantage of the self-complementarity of the DNA ends produced by *Eco*RI cleavage, to make recombinant combinations of antibiotic-resistant plasmids and then "clone" these in the bacterial host.

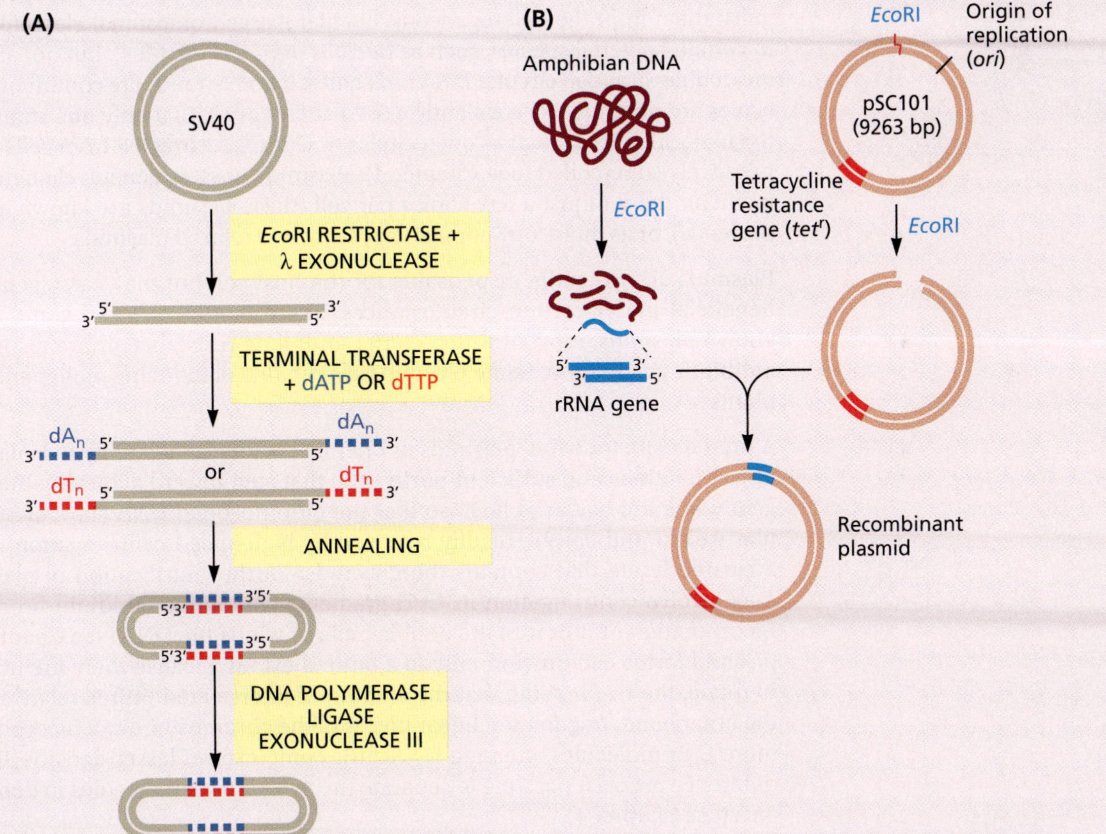

Figure 1 The first experiments to produce recombinant DNA molecules. (A) The Berg experiment to produce covalently closed SV40 dimer circles. The procedure involved converting circular SV40 DNA to a linear form; adding single-stranded complementary homodeoxypolymeric extensions to the 3′-ends of the linearized molecules; annealing these complementary extensions to form noncovalently closed circular duplex structures; and filling the gaps with DNA polymerase and sealing the nicks with DNA ligase to form concatemeric closed circular molecules. A similar procedure produced circular SV40 DNA molecules containing λ phage genes and the galactose operon of *E. coli*. (B) The Cohen–Boyer experiment to construct recombinant biologically functional plasmids carrying the rRNA gene from an amphibian. Importantly, the recombinant plasmids were capable of replicating in *E. coli* and could be selected for because the antibiotic-resistant gene they carried was active. This allowed bacteria transformed with the recombinant plasmid to live and propagate in antibiotic-containing medium, whereas nontransformed bacteria died out.

(Continued)

Box 5.2 *(Continued)* Cohen and Boyer did not miss the much broader implications of their work. In the concluding paragraph of their 1973 paper, they stated: "The general procedure described here is potentially useful for insertion of specific sequences from prokaryotic or eukaryotic chromosomes or extrachromosomal DNA into independently replicating bacterial plasmids." When the enormous range that recombinant DNA methods have since enveloped is considered, the statement is remarkably modest.

In 1977, Boyer's laboratory and outside collaborators described the first-ever synthesis and expression of a peptide-coding gene, the somatostatin gene. In August 1978, they produced synthetic insulin, followed in 1979 by a growth hormone. In 1980, Paul Berg was awarded the Nobel Prize in Chemistry for "his fundamental studies of the biochemistry of nucleic acids, with particular regard to recombinant-DNA"). Cohen and Boyer have also won numerous prestigious awards for their major contributions to both basic research and the practical applications of recombinant DNA technology.

A very large repertoire of cloning vectors based on plasmids, viruses, or combinations thereof are used in recombinant DNA technology. Which vector one would choose depends on the target cell, the amount of DNA to be incorporated, and the purpose of incorporation.

Bacterial plasmids were the first cloning vectors

Plasmids are extrachromosomal nucleic acid molecules carried by bacteria that replicate independently of the main bacterial chromosome. All plasmids of a given type within a bacterial cell will have exactly the same size. These properties distinguish them from the heterogeneous, circular extrachromosomal elements that exist in certain bacterial species, such as *Bacillus megaterium*. Most, but not all, plasmids are double-stranded circular DNA molecules. If both strands are continuous, the molecules are described as covalently closed circles or CCC; if only one strand is intact, the molecules are known as open circles or OC. CCC extracted from cells are usually negatively supercoiled (see Chapter 4). Plasmids possess genetic elements that can maintain them in just a few copies per cell (these plasmids are known as stringent plasmids), or as multicopy populations, known as relaxed plasmids.

Plasmids are generally dispensable for the host cell but may carry genes that are beneficial under certain circumstances. There is a long list of genes carried by known plasmids, most of which confer antibiotic resistance or are responsible for antibiotic production. Some plasmids are capable of inducing malignant tumors in plants.

A prerequisite for using plasmids as cloning vectors is the ability to isolate them in large quantities and sufficient purity. The first step toward achieving this goal is to gently lyse the bacterial host, so that the chromosomal DNA stays of high molecular weight and can be readily removed by high-speed centrifugation, producing a cleared lysate that contains the plasmids. Further purification of plasmid DNA may involve centrifugation in CsCl gradients containing ethidium bromide, EtBr. Incorporation of EtBr into the double helix unwinds the DNA (see Chapter 4). CCC plasmid forms can unwind only to a limited extent because there are no free ends to rotate, the result being that the amount of intercalated EtBr is relatively low. On the other hand, fragmented linear pieces of the chromosome can incorporate many more EtBr molecules, creating DNA–EtBr complexes of lesser density. The density difference makes it possible to separate the two kinds of molecules in density gradients (see Chapter 4).

The alternative method depends on the different denaturation properties of linear and CCC DNA forms. There exists a narrow range of high pH, 12.0–12.5, in which linear molecules denature but circular molecules remain intact. Upon neutralization of the solution, chromosomal DNA partially renatures, forming a highly insoluble network that is easily removed by centrifugation.

Plasmids that are chosen to be cloning vectors have several desirable characteristics: low molecular weight, ability to confer selectable traits on the host cell, and availability of single sites for restrictase cleavage. The first requirement stems from the ease of purifying small plasmids that remain intact during the isolation procedure. In addition, smaller plasmids are usually present in multiple copies, allowing for stable

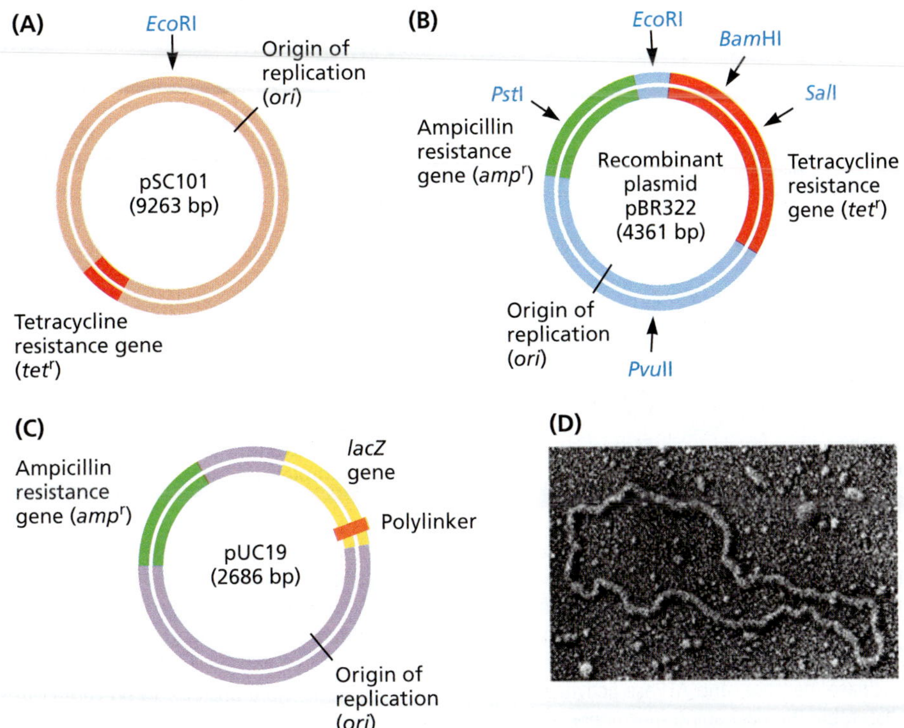

Figure 5.9 Some bacterial cloning vectors. (A) pSC101, the first bacterial plasmid used for cloning (see Box 5.2). p stands for plasmid, and SC stands for Stanley Cohen. (B) pBR322, in which BR stands for Bolivar Rodriguez, the postdoctoral researcher who constructed the plasmid, is an early cloning vector that is itself a recombinant DNA molecule. It contains an origin of replication sequence *ori* from a naturally occurring plasmid, ColE1, and two selectable marker genes from two other plasmids: an ampicillin resistance gene, shown in green, and a tetracycline resistance gene, shown in red. If the restriction site to be used for cloning lies within one of these genes, that specific gene is inactivated upon insertion of the foreign sequence. This is a useful property since it can be used for selection of the bacteria that carry the foreign gene. All bacteria that carry the vector, whether recombinant or not, will be resistant to the antibiotic specified by the intact resistance gene. Bacteria carrying the foreign gene will become sensitive to the antibiotic whose gene was inactivated due to the insertion. (C) pUC19, in which UC stands for University of California, where the vector was constructed by Joachim Messing and colleagues. It contains, in addition to the standard antibiotic selectable marker and origin of replication, the *lacZ* gene, shown in yellow, which encodes the enzyme β-galactosidase. The *lacZ* gene is split by a polylinker, an array of multiple restriction sites for cloning, also known as a multiple cloning site or MCS. Thus, transformed cells containing the plasmid with inserted gene can be distinguished from cells containing the original, nonrecombinant plasmid by the color of the colonies they produce in an appropriate medium. Nonrecombinant cells will turn blue because the presence of active β-galactosidase, from the unsplit *lacZ* gene, will transform a colorless substrate present in the medium to a blue compound. Recombinant cells will be white because the insert has inactivated the *lacZ* gene. (D) Electron microscopic image shows the typical appearance of an isolated plasmid. (D, from Cohen SN [2013] *Proc Natl Acad Sci USA* 110:15521–15529. With permission from National Academy of Sciences.)

high-level expression of the genes they carry; this becomes especially important for genes that confer selectable phenotypes.

The number of derivative plasmids that have been used for cloning is enormous. These are commercially available and allow the researcher to choose the appropriate vector for a specific project. **Figure 5.9** presents a few classical examples of plasmid cloning vectors and introduces some of the strategies used for each vector to easily select the transformed recombinant bacteria.

Recombinant bacteriophages can serve as bacterial vectors

Plasmid vectors are often used for cloning small DNA segments, up to 10 kbp. Cloning larger fragments, such as individual large genes or fragments to be used for the construction of genomic libraries, requires the use of alternative vectors, such as derivatives of bacteriophages. The most widely used phage for cloning is phage λ and its derivatives. There are two types of λ vectors: insertion and replacement. In **insertion vectors**, the linear phage DNA molecule is cut with a restriction enzyme that cleaves the sequence only once. The two fragments generated are joined to the foreign DNA fragment, and the resulting recombinant is packaged into infectious

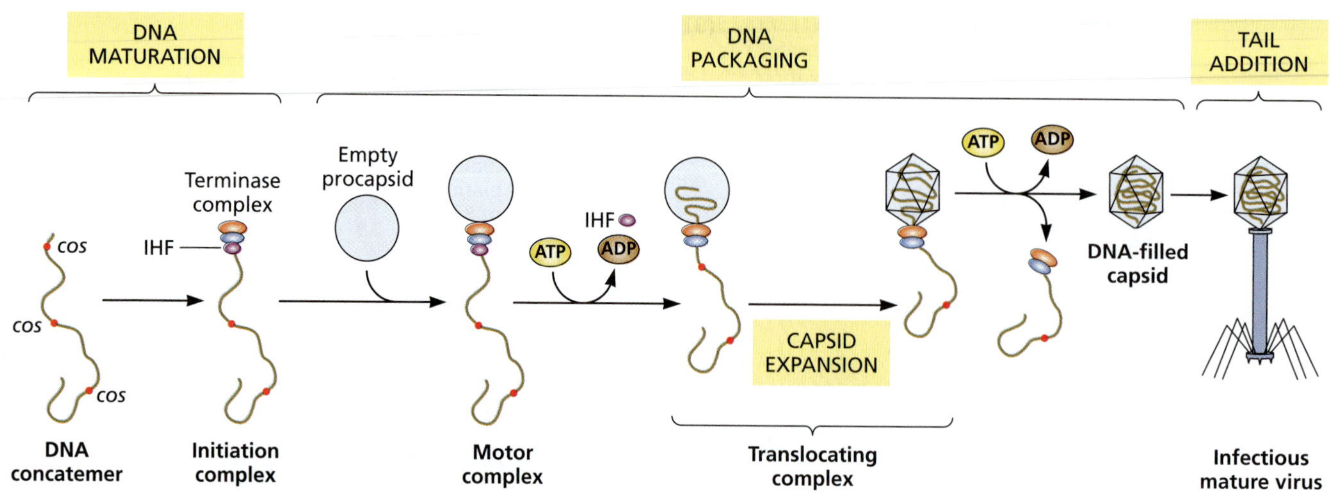

Figure 5.10 Packaging of λ phage DNA into infectious mature viral particles. The heterodimeric terminase enzyme complex and the *E. coli* integration host factor (IHF) bind cooperatively to a *cos* site in a DNA concatemer. This is the form of the viral genome that is produced during replication; it contains multimers of genome-size fragments connected by *cos* sequences. The terminase possesses *cos*-site cleavage endonuclease, helicase, and ATPase activities that work in concert to assemble the packaging machinery on the concatemer and prepare the end for packaging. The same enzyme also possesses a DNA- and ATP-dependent translocase activity that packages DNA into an empty procapsid, thus serving as a molecular motor. The procapsid is an empty protein shell composed of four different proteins; the tail is composed of 11 different viral glycoproteins. The terminase–concatemer complex, named the initiation complex, binds to the portal, a ringlike structure in the procapsid, to form the motor complex which pumps the viral DNA into the capsid. DNA packaging triggers a process of expanding the procapsid, with additional viral proteins binding to it. Once the packaging motor encounters the next *cos* site in the concatemer, terminase again cuts the duplex, and its helicase activity helps in packaging the genome to near-liquid-crystalline density. Following the addition of some more proteins to prevent DNA release through the portal, the preassembled tail is added to complete virion assembly. It is now possible to assemble infectious viral particles *in vitro* starting with seven individual proteins, purified procapsids and tails, and mature λ DNA. Importantly, the same system can be used to produce infectious viruses containing recombinant DNA. (Adapted from Gaussier H, Yang Q & Catalano CE [2006] *J Mol Biol* 357:1154–1166. With permission from Elsevier.)

mature viral particles (**Figure 5.10**). This approach, as easy as it sounds, imposes a strict limitation on the size of foreign DNA that can be cloned. This is because the capacity of the phage head is limited; it does not allow packaging of molecules whose size deviates from the wild-type λ genome by more than ±5%. In order to overcome this problem, **replacement vectors** were introduced. These vectors make use of the fact that about a third of the λ genome is not necessary for replication and can thus be replaced by any piece of foreign DNA without affecting the infectivity of the recombinant phage. The construction of recombinant replacement phages is illustrated in **Figure 5.11**. These were named Charon vectors by their creators, Fred Blattner and colleagues, after Charon, the ferryman in Greek mythology, who ferried the dead to the underworld across the rivers Styx and Acheron. The analogy may seem far-fetched but it does confer the idea of transfer: of dead souls in Greek mythology and of foreign DNA in cloning.

Another bacteriophage often used as a cloning vector for subsequent DNA sequencing through Sanger's method is the filamentous phage of *E. coli*, M13 (**Figure 5.12A**). The life cycle of M13 is depicted in **Figure 5.12B**. The M13 genome is a 6.4 kbp single-stranded circular DNA encapsulated in a coat consisting of ~2700 copies of a small coat protein. Additional minor proteins participate in the attachment of the phage to host *E. coli* cells. The coat is amazingly flexible, changing its size to accommodate DNA molecules of very different sizes, ranging from a few hundred nucleotides to around twice the size of the native genome. The genome of M13 has been modified to serve as a cloning vector as depicted in **Figure 5.12C**.

M13-derived vectors can be used in a technique called **phage display**. This method allows large peptide and protein libraries to be expressed on the surface of the filamentous phage and thus presented to potential interaction partners. It permits the selection of expressed peptides and proteins, including antibodies, that have high affinity and specificity for almost any target, including specific proteins or DNA

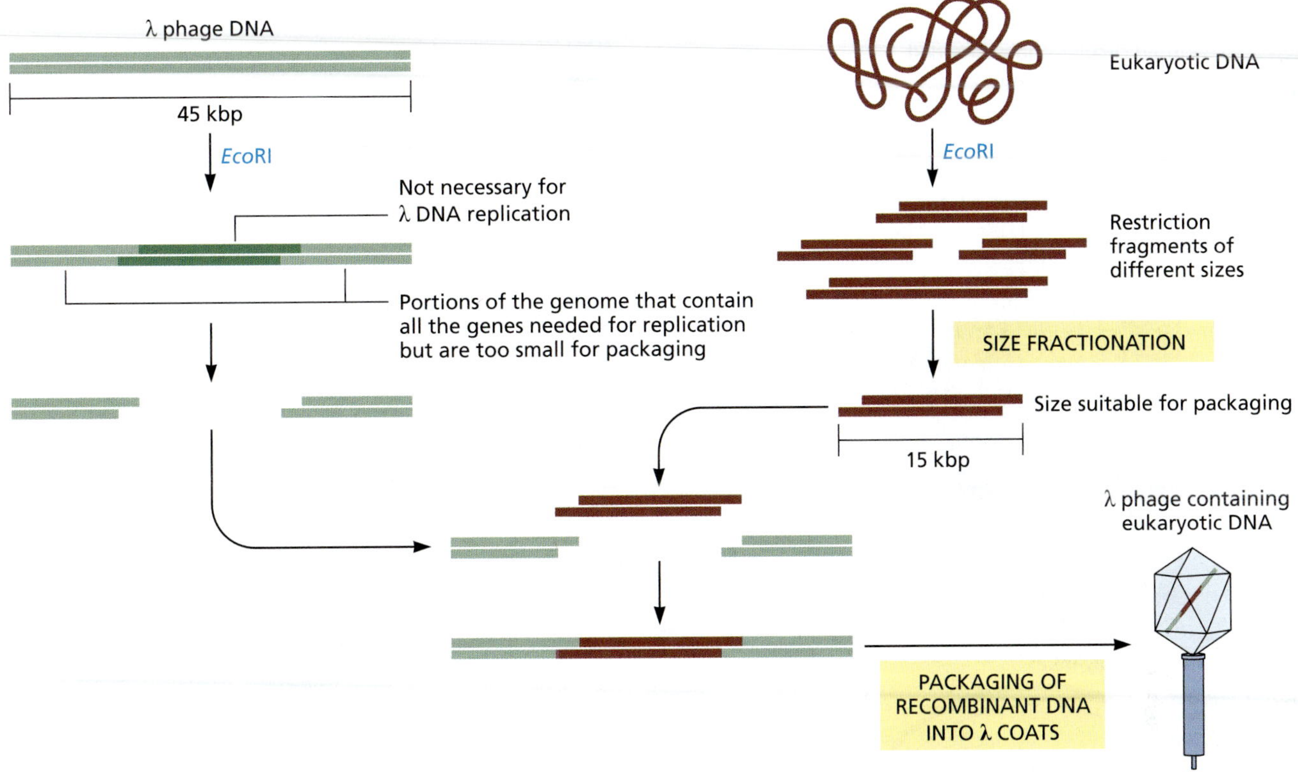

Figure 5.11 Use of λ phage as a vector. The λ phage genome contains a large middle region that does not participate in DNA replication and therefore can be replaced by a piece of foreign DNA. The size of the insert should be ~15 kbp to allow efficient packaging of the recombinant DNA into infectious viral particles.

sequences. The direct link between the experimental phenotype, that is, the properties of the peptides or proteins on the surface, and the genotype encapsulated within the vector allows the selection of molecules that are optimized for binding. In an important application, phage display facilitates the engineering of antibodies to amend their size, affinity, and effector functions.

How does phage display work? DNA fragments are inserted into the middle of gene 3, the protein product of which is crucial for attachment of the phage to the bacterial pilus through which it enters the cell (see Figure 5.12B). When the p3 gene is expressed, it produces p3 fusion proteins that contain the amino acid sequence encoded by the insert. The fusion proteins become incorporated into the phage coat in the same way as wild-type p3. In this way, the fusion proteins are accessible and can be readily probed for interactions with desired targets. The method is outlined in **Figure 5.13**.

Cosmids and phagemids expand the repertoire of cloning vectors

Cosmids are modified plasmids that carry *cos* sites from phage λ in addition to the standard origin of replication and selectable marker genes. They are capable of cloning large, >40 kbp fragments and are introduced into bacterial cells by standard protocols. *Cos* sites enable the DNA to be packed into infectious viral particles *in vitro* (see Figure 5.10), thus facilitating their purification. Like plasmids, cosmids perpetuate in bacteria because they do not carry the genes for lytic infection and thus do not kill their bacterial hosts.

Phagemids are constructs that combine the features of phages and plasmids. One commonly used phagemid is **pBluescript II**, which is derived from the plasmid pUC19 and contains two origins of replication, an ampicillin resistance gene, and a *lacZ* gene. The multiple cloning site splits the *lacZ* gene, allowing for white-blue screening (see Figure 5.9). **pBluescript** is an expression vector, as the cloning site is flanked by promoter sequences derived from phages T3 and T7.

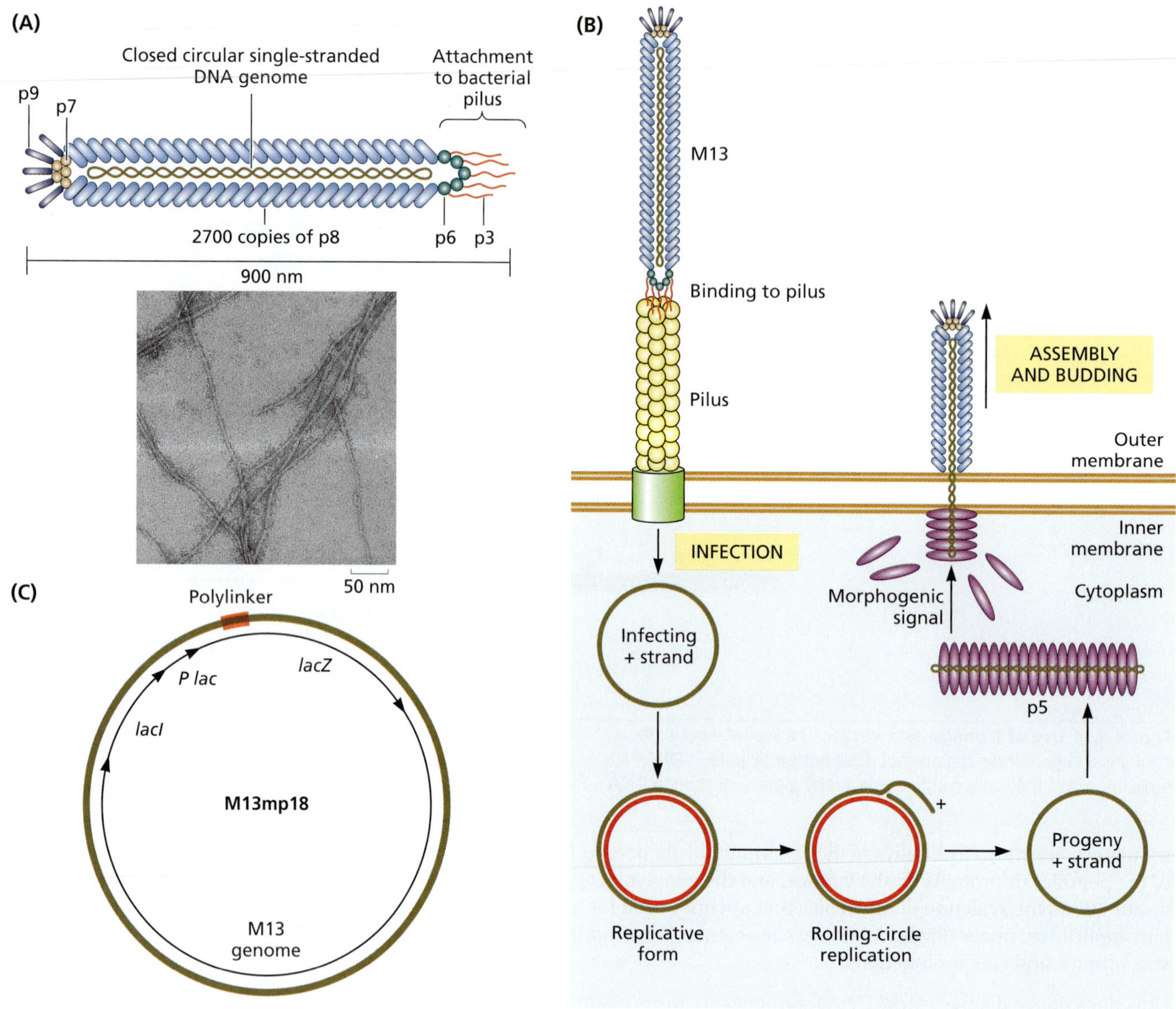

Figure 5.12 Bacteriophage M13, its life cycle, and its derivative constructed as a cloning vector. (A) Structure of M13. The single-stranded DNA (or ssDNA) circular genome is encapsulated by a flexible capsid consisting of ~2700 molecules of one of the phage-encoded proteins, p8. (B) M13 life cycle. M13 enters bacterial cells through pili; following uncoating, the ssDNA circle is replicated to the so-called replicative form, which is double-stranded. This form is then replicated by the rolling-circle mechanism (see Chapter 19), producing single-stranded DNA, the form that is incorporated into the viral particle. The ssDNA is temporarily packaged, with the help of p5, into a form that is finally transformed into a mature particle. The viral proteins taking part in this process are all membrane-attached. Assembly is followed by budding of the phage from the infected cell, without lysis of the host cell. (C) M13 as a cloning vector. The following elements have been inserted into the genome: the *lacZ* gene, to allow white–blue selection for plasmids carrying foreign DNA fragments (see Figure 5.9); a *lac* promoter upstream of the *lacZ* gene; and the gene for the Lac repressor, *lacI*, to allow regulation of the *lac* promoter (see Chapter 11). The derivatives of M13 used as cloning vectors are named after the specific polylinker region they contain. (A, electron microscope image, from Murugesan M, Abbineni G, Nimmo SL et al. [2013] *Sci Rep* 3 [10.1038/srep01820]. With permission from Macmillan Publishers, Ltd.)

5.4 Artificial chromosomes as vectors

Bacterial artificial chromosomes meet the need for cloning very large DNA fragments in bacteria

Bacterial artificial chromosomes or **BACs** are based on functional fertility plasmids, called F factors. These are naturally occurring plasmids that control their own replication and copy number. The plasmids are maintained at one or two copies in each *E. coli* cell, which is advantageous for the cloning of large fragments that would undergo

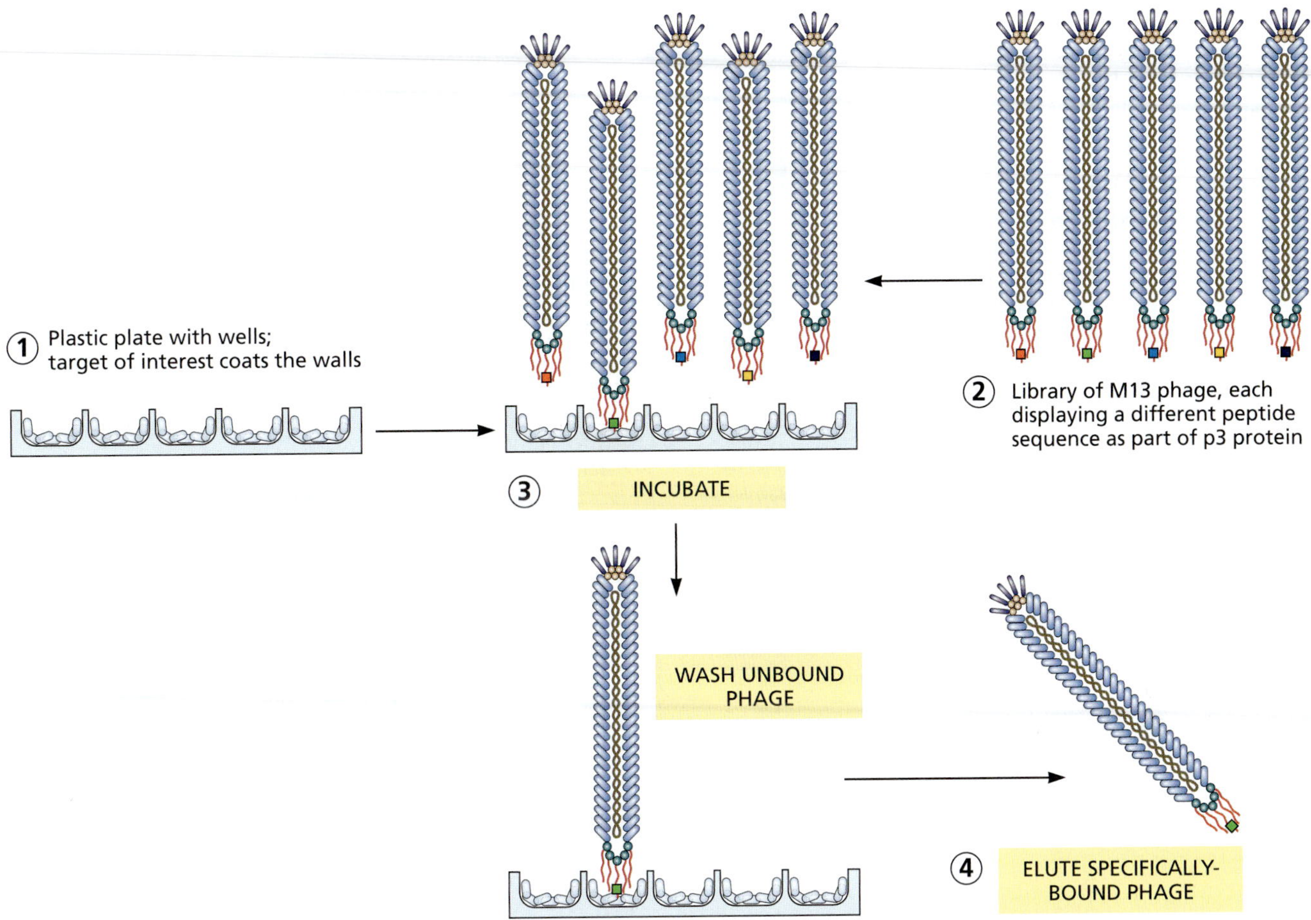

Figure 5.13 Phage display identifies DNA sequences that encode peptides or proteins of desired characteristics. As an example, we present the sequence of steps in phage display screening to identify polypeptides that bind with high affinity to desired target protein or DNA sequence. (Step 1) Target proteins or DNA sequences are immobilized to the wells of a plastic microtiter plate. (Step 2) A library of DNA sequences that code for peptides/proteins is prepared by inserting these sequences into the gene encoding coat protein 3. Other coat proteins can also be used, including the major protein p8 (see Figure 5.12). Thus, the products expressed by the recombinant genes are fusion proteins between p3 and the peptide/protein encoded by the inserted DNA sequence. The fusion proteins end up on the phage surface. (Step 3) This phage-display library is added to the plate and allowed time to bind. Then a buffer wash removes the unbound phages, while those carrying the fusion proteins that interact with the immobilized target remain bound. (Step 4) The specifically attached phages are eluted with excess known ligand for the target or by lowering pH. The eluted phages are used to generate more phages by infection of suitable bacterial hosts. The new phage mixture is enriched, containing considerably fewer irrelevant (nonbinding) phages than were present in the initial mixture. (Step 5) Steps 1–4 are repeated to achieve further enrichment in phages carrying binding proteins. Following further bacterial-based amplification, the DNA in the interacting phage is sequenced to identify the interacting proteins or protein fragments. This sequence can be further mutagenized *in vitro* by site-specific mutagenesis (see Box 5.4) to optimize binding characteristics.

significant sequence rearrangements if cloned in high-copy-number plasmids. In addition to the sequences and genes involved in plasmid replication and its control, the original BAC vector contains a chloramphenicol resistance marker and a cloning site. More recent versions also contain the β-galactosidase gene for easy screening.

Eukaryotic artificial chromosomes provide proper maintenance and expression of very large DNA fragments in eukaryotic cells

Eukaryotic artificial chromosomes are large artificial constructs carrying all the important structural and functional elements that define eukaryotic chromosomes: telomeres, centromeres, and replication origins. A **yeast artificial chromosome** or **YAC** (**Figure 5.14**) allows cloning of fragments larger than 100 kbp and up to 3000 kbp. In addition to the cloning of large genes, YACs are useful for physical mapping of complex genomes and for constructing genomic libraries. YACs and other yeast expression vectors have an advantage over BACs in that they can be used to express eukaryotic

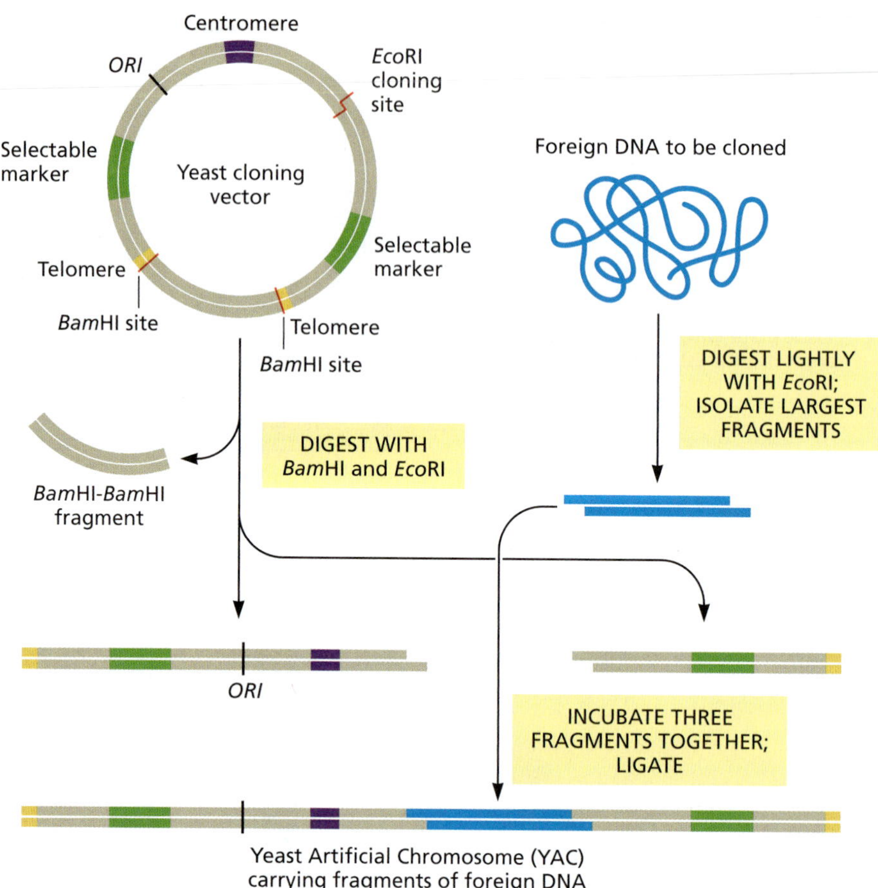

Figure 5.14 Yeast artificial chromosomes allow cloning and expression of large DNA fragments in yeast. The yeast cloning vector constructed for these purposes carries the important elements that would allow the vector to multiply in the yeast host cell: a centromeric and two telomeric regions and, most importantly, a yeast origin of replication (autonomous replication sequence). As with all cloning vectors, it also carries specific restriction sites and selective markers: genes whose protein products would allow the host cells to grow under selective conditions so that only the cells carrying the markers could survive.

proteins that require post-translational modifications. The disadvantage of YACs is that they are less stable than BACs, allowing significant gene rearrangements due to high copy numbers.

5.5 Expression of recombinant genes

Expression vectors allow regulated and efficient expression of cloned genes

If the aim of a project is to produce large amounts of protein, whether native or mutated, specialized vectors called **expression vectors** must be used. As their name implies, these vectors are capable not only of carrying an inserted gene but also of expressing the protein product. They must ensure that the gene will be transcribed, and in a form that can be translated into protein, by use of the host cell's machinery. Expression vectors ideally have the following general characteristics:

- They should be present and maintained in the host cells at high copy number.
- They should be able to integrate into the host genome to ensure stable propagation of the gene of interest from cell generation to cell generation.
- They should carry a promoter that is functional within the given host. Promoters are elements where RNA polymerases bind to initiate gene transcription (see Chapters 11 and 12).
- It should be possible to regulate the promoter to ensure that the gene is expressed only when desired (see Chapters 11 and 12). This is especially important if the protein happens to be toxic to the host cell.

There are also desirable characteristics for the cloned gene itself. As redundancy exists within the genetic code (see Chapter 7), the same amino acid can be encoded by more than one codon. Some organisms make more frequent use of some of these codons than the others specifying the same amino acid. In order to optimize the expression of a foreign protein, it is desirable for its gene to be optimized for codon usage in the host cell. Current technologies make this possible by using codon exchange rather routinely.

It is also mandatory to use intron-free genes. Introns are noncoding regions found within most eukaryotic genes. In the eukaryotic nucleus, introns are spliced out, but bacteria cannot accomplish this. Even if a eukaryotic gene is expressed in a eukaryotic host such as yeast, the use of intron-free constructs is preferable, as it avoids complications that could arise from alternative splicing (see Chapter 14).

Other potential problems exist in expression projects: the protein may be unstable or may be secreted into inclusion bodies, which are sequestered intracellular vesicles that accumulate insoluble or faultily folded proteins. Solubilizing such proteins can present a huge problem. Some of these difficulties can be overcome by the construction of genes that encode **fusion proteins**, in which the desired protein is fused to a normal host protein. These are usually more stable than naked foreign proteins and their purification is simpler. If it is preferable for the protein of interest to be secreted outside the cell, special signal sequences can be added, usually to the N-terminus of the protein, that ensure such secretion.

Shuttle vectors can replicate in more than one organism

Cloning projects sometimes require the same vector to multiply in two different organisms, most frequently in *E. coli* and yeast. The need for two different hosts arises, for example, when one is interested in expressing yeast genes. These can be cloned and multiplied as parts of bacterial vectors, but they cannot be expressed in the bacterial host because appropriate yeast sequences that are needed for gene expression are missing. Such expression sequences can be included in the bacterial vector, but this is often impractical as it results in inserts that are too large. In addition, bacterial cells do not have the ability to introduce the postsynthetic modifications that are often necessary for proper functioning of a eukaryotic protein (see Chapter 18). Thus, when the aim of a project is to express and properly modify recombinant eukaryotic proteins, the respective genes are first multiplied in bacterial cells and then expressed as part of the same vector in the appropriate eukaryotic host. Shuttle vectors must thus contain origins of replication that will function in both bacterial and eukaryotic hosts.

The **yeast episomal plasmid vector**, **Yep**, is an example of a shuttle vector. It contains the *E. coli ori*, origin of replication sequence and *amp*^{r}, ampicillin resistance gene and also sequences derived from the large yeast plasmid known as the 2-μm plasmid. The 2-μm plasmid portion contains the yeast *ORI* and two genes, *REP1* and *REP2*, that are responsible for replication and stable maintenance of about 50 copies of the plasmid in the yeast cell. Yep also carries a yeast marker gene, usually *His3* or *LEU2*, both involved in the biosynthesis of amino acids. To be able to use these for selection, the host cell must be deficient in the biosynthesis of these amino acids; the deficiency will be corrected by expression of the vector-borne genes.

5.6 Introducing recombinant DNA into host cells

Numerous host-specific techniques are used to introduce recombinant DNA molecules into living cells

The techniques for introducing recombinant DNA molecules into living cells for further propagation are highly specific for both the type of host cell, bacteria or eukaryote, and the type of cloning vector used. Plasmid vectors are introduced into bacterial cells by transformation techniques that involve prior treatment of bacteria with calcium chloride at low temperature followed by a brief heat shock. It is believed that calcium chloride affects the structure of the cell wall and may help in the binding of DNA to the cell surface. The heat shock stimulates the actual DNA uptake.

The efficiency of introducing recombinant DNA molecules into bacterial cells is greater with phage-based cloning vectors. The development of *in vitro* packaging techniques to pack the modified DNA in the phage head represented a major improvement in this stage of the process.

Introduction of DNA into eukaryotic cells is more complicated, especially for yeast and plant cells, which possess, in addition to the plasma membrane, outer protective polysaccharide layers. These layers must be removed, usually by enzymatic treatments, to produce **protoplasts**, which survive and are competent for transformation. Transformed plant protoplasts are grown on selective medium to produce large masses of undifferentiated cells known as calli, plural for callus. These can then be transferred to nutritional medium containing specific growth hormones that allow regeneration of whole fertile transformed plants.

The search for efficient and more universal methods of introducing cloned genes into a wide variety of microbial, plant, and animal cells led to the development of purely mechanical methods. One such technique is **electroporation**. When subjected to an electric shock—for example, brief exposure to a voltage gradient of 4000–8000 V/cm—cells take up exogenous DNA through holes created in the plasma membrane. Drugs such as colcemid that arrest cells at metaphase increase transformation efficiency. This increase is attributed to the lack of a nuclear membrane during this stage of the cell cycle, as the nuclear membrane also needs to be crossed for stable integration of the gene into the genome.

Another method for transformation of animal cells or the protoplasts of yeast, plant, and bacterial cells is liposome-mediated gene transfer. **Liposomes** are unilamellar (single-bilayer) vesicles prepared from cationic lipids that readily and spontaneously form complexes with DNA in solution. The positively charged liposomes also bind to cultured cells and presumably fuse with the plasma membranes. The use of liposomes for transformation or transfection is known as **lipofection**.

Researchers have developed **gene guns** to facilitate gene transfer into cells that are notoriously difficult to transform. The method can be applied not only to unicellular organisms but also to plant leaves or entire animals, such as *Drosophila* and mice. It has been particularly useful for chloroplast transformation, as there are no other known methods to introduce foreign DNA into these organelles. The efficiency of the gun system varies, with skin cells being the most susceptible to transformation. It has already been used for delivering hepatitis B DNA-based vaccine to mice and humans.

The gene gun method relies on the ability of nucleic acids to adhere to biologically inert metal particles, such as gold or tungsten. The DNA–particle complex is accelerated under partial vacuum and hits the target tissue in its acceleration path. There are numerous ways to create acceleration, such as the use of pneumatic devices, magnetic or electrostatic forces, and sprayers. In a variant of the method, uncoated metal particles are shot through a solution containing the cells and the DNA. Presumably, they pick up DNA in transit and carry it into the cell.

5.7 Polymerase chain reaction and site-directed mutagenesis

In the late 1970s and early 1980s, two methods were introduced that became indispensable for the further development of recombinant DNA technology. Both methods won Nobel Prizes soon after they were introduced, in recognition of their immediate and broad impact. The first method, **polymerase chain reaction** or **PCR**, makes it possible to obtain large amounts of homogeneous DNA fragments *in vitro* by using repetitive cycles of denaturation, polynucleotide synthesis, and renaturation or annealing. Synthesis proceeds from small synthetic primers complementary to the ends of the sequence of interest. Because of the high temperature needed for the denaturation and annealing reactions, the method requires the use of a thermostable form of DNA polymerase that is capable of synthesizing DNA strands at high temperature (**Box 5.3**). The PCR technique makes it possible to prepare substantial quantities of specific DNA fragments from

Box 5.3 Polymerase chain reaction The polymerase chain reaction, PCR, is a method for obtaining large amounts of highly purified DNA molecules suitable for analysis or manipulation. It was introduced in 1987 by Kary Mullis. Mullis shared the 1993 Nobel Prize in Chemistry with Michael Smith "for contributions to the developments of methods within DNA-based chemistry."

The PCR reaction usually consists of a series of 20–40 repeated cycles, where each cycle consists of three discrete temperature steps: denaturation, annealing, and elongation (Figure 1). During the denaturation step, the DNA sample is heated to 94–98°C for 20–30 s: the high temperature disrupts the hydrogen bonds between complementary bases and separates the two DNA strands. During the annealing step, the temperature is lowered to 50–65°C for 20–40 s to allow annealing of the primers, 15–30-nucleotide synthetic DNA fragments, to the single-stranded DNA template. The primers are complementary in sequence to the ends of the region of interest and possess a free 3′-OH group, which is needed for the action of DNA polymerase at the third step. Typically, the annealing temperature is set to be slightly below the melting temperature of the primer–DNA hybrid duplex, by 3–5°C. Stable DNA–DNA hydrogen bonds are formed only when the primer sequence very closely matches the template sequence. A couple of bases that are not complementary to each other are usually tolerated by the duplex. The ability to form such nonperfect, but still stable, hybrid duplexes is helpful when one tries to amplify a sequence from an organism using knowledge about similar, but probably not identical, sequences from other organisms to design the primers. The exact conditions—temperature, ion concentration, and so on—under which the annealing step is performed determines what is known as the **stringency** of hybridization. The more stringent the conditions, for example, temperature closer to the melting temperature of the hybrid, the lower the probability that an imperfect, mismatched duplex will be stable enough to allow the next step of the reaction cycle to occur, and vice versa.

The Primer extension/elongation step involves the extension of the primer by a thermostable form of DNA polymerase (see Chapter 19). The temperature at this step depends on the DNA polymerase used. The commonly used polymerase is **Taq polymerase**, a naturally thermostable enzyme from the thermophilic bacterium *Thermus aquaticus* that works optimally at 75–80°C, but other thermostable DNA polymerases are used for specialized PCR assays. The chemistry of elongation does not differ from that of the normal *in vivo* process. The extension time depends both on the DNA polymerase used and on the length of the DNA fragment to be amplified. The entire cycle is then repeated. Under optimum conditions, the amount of DNA target is doubled at each extension step, leading to exponential amplification of the specific DNA fragment.

The products of PCR can be detected in one of several ways. Following the reaction, the product can be analyzed by electrophoretic techniques. Hybridization to microarrays, described in Chapter 6, can be used to detect and analyze more complex products.

Real-time PCR makes the process quantitative

An important recent development is **real-time PCR**, in which the products of the reaction are monitored during the course of amplification, usually through fluorescence methods. Real-time PCR shows good correlation between that cycle in the entire process during which unambiguous product signals

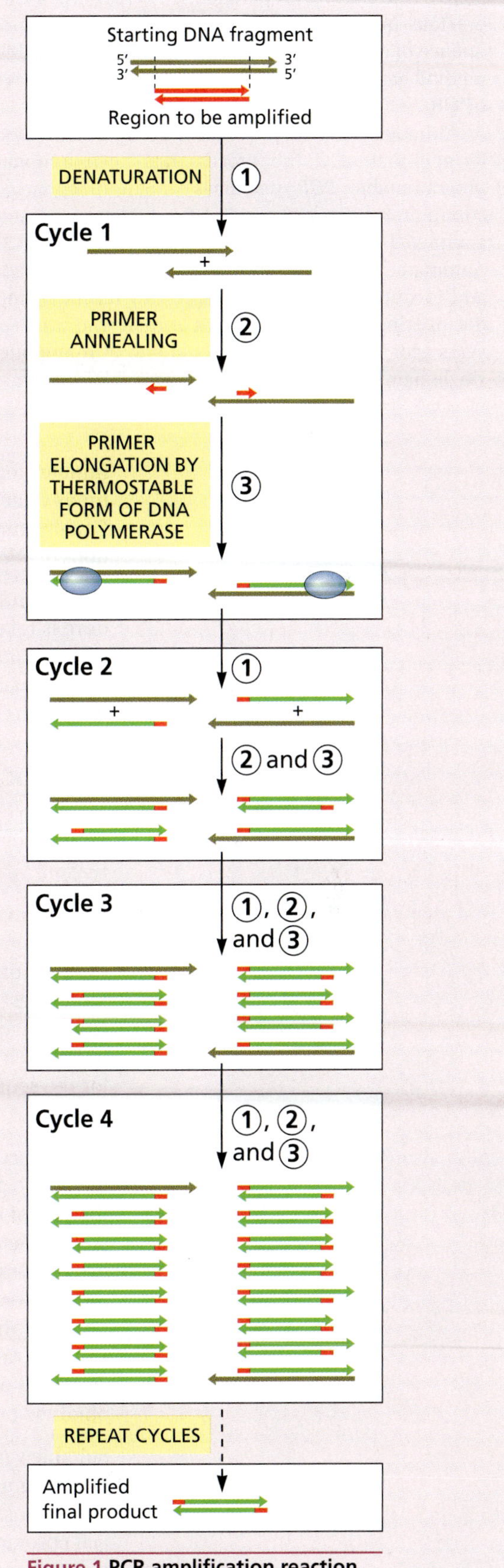

Figure 1 PCR amplification reaction.

(Continued)

Box 5.3 ***(Continued)*** become measurable, usually referred to as the **threshold cycle** or C_T, and the concentration of template in the starting preparation. Thus real-time PCR provides precious quantitative information on the amount of the sequence of interest in the sample. This unique capability of the method has earned its alternative name, **quantitative PCR** or **qPCR**.

Although the PCR process would be quite tedious if performed manually, the process has been automated, and programmable PCR machines or thermocyclers are now commonplace in most molecular biology laboratories. The complete PCR reaction can be readily performed in 3–4 hours. Automated PCR is nowadays widely used in forensic cases and to study DNA from fossils. A very recent example is the deciphering of the genome of a human who lived some 4000 years ago: the DNA was recovered and PCR-amplified from a single hair found in a permafrost area in North America. The initial sequencing of four short gene sequences from fossils of Neanderthals reported in 2007 and 2008 was also PCR-based; high-throughput sequencing techniques were later used to compile draft sequences of the entire Neanderthal genome.

Reverse transcription PCR can amplify RNA sequences into DNA

In some cases, researchers are interested in amplifying only genes that are actually transcribed in the cell; that is, those that participate in the flow of information from genes to proteins or that perform different kinds of structural, enzymatic, or regulatory functions. To that end, a reverse transcription step is performed before the PCR reaction. During this step, pools of RNA molecules extracted from cells and a special kind of polymerase, **RNA-dependent DNA polymerase** or **reverse transcriptase** (see Box 20.6), is used to produce a double-stranded DNA copy of the RNA molecules. Once the copy DNA or cDNA is at hand, the standard PCR protocol is used.

minute amounts of starting material, without the need for cloning. It has proved of enormous value in extracting sequence information from minute amounts of DNA in fossils. Equally importantly, PCR is indispensable for the analysis of complex natural populations of microorganisms, most of which cannot be grown under laboratory conditions.

The second method, of no less importance, is **site-directed mutagenesis**. The procedure is relatively simple and allows the creation of any desired mutation at any preselected site (**Box 5.4**). Introducing defined mutations is critical to our understanding of the biochemistry and physiology of genes, their control regions, and the proteins they encode.

5.8 Sequencing of entire genomes

Genomic libraries contain the entire genome of an organism as a collection of recombinant DNA molecules

The cloning and sequencing of individual genes of interest was the first application of recombinant DNA technology in research. It was soon realized that it would be possible, by use of recombinant technology, to prepare large sets of recombinant DNA molecules—vectors plus inserts—that can contain entire genomes in the form of overlapping sequences. These **genomic libraries** can be multiplied in bacteria and stored indefinitely. The availability of such libraries was the most important prerequisite for the sequencing of whole genomes, first for viruses, then bacteria and yeast, then higher eukaryotes, and culminating in 2001 with the complete human genome.

Creating complete libraries is often technically challenging. The size of the library needed to guarantee representation of all the sequences of a genome is a function of the size of DNA fragments that can be accommodated in the cloning vector and the size of the genome itself. Simple calculations indicate that, in order to include any desired sequence with a probability of 99%, the number of recombinant clones in a human library should be close to 700,000. These calculations are based on the assumption that the genome has been fragmented randomly, for example, by mechanical shearing, as in sonication. The cloning of sheared fragments is relatively inefficient, so usually an acceptable, albeit not perfectly random, population of DNA fragments is obtained by limited digestion with restrictases.

Suitable vectors for creating genomic DNA libraries should allow the cloning of large DNA fragments, to reduce the requisite number of clones for full representation, and should possess high cloning efficiency. λ phage vectors, cosmids, and bacterial artificial chromosomes (BACs) fulfill these requirements. A major advantage of such vectors is that they can be efficiently packaged *in vitro* to produce highly infectious viral particles.

Box 5.4 Site-directed mutagenesis: A universal method to mutate any DNA sequence in a desired way For many years, scientists have been on the hunt for methods to mutate DNA sequences, mainly to study the effect of mutated genes, and hence their altered protein products, on protein properties and function. The initial approaches used chemicals to modify an amino acid in the polypeptide chain directly, without modifying the gene sequence, or used physical or chemical agents to mutagenize the DNA coding for the protein *in vivo*. Both methods suffered from a major drawback: they were unable to change one specific amino acid at one predetermined position within the molecule. Both methods produced complex sets of molecules that had to be further fractionated and characterized in order to isolate the molecule of interest.

A major breakthrough occurred in 1978, when Michael Smith and colleagues at the University of British Columbia in Vancouver, Canada, came up with a relatively simple *in vitro* procedure to create any mutation at any preselected site. The procedure underwent many modifications, but the principle stays the same (**Figure 1**):

- Synthesize a short, ~20-nt-long oligonucleotide sequence that complements the DNA sequence of the intended mutation site in the gene of interest but that carries deliberate mistakes, such as single-base substitutions, short deletions, or insertions at its center.
- Anneal the oligonucleotide to a single-stranded cloned sequence of interest; the product will contain a base or several bases that cannot base-pair with the gene and, depending on the change introduced in the oligonucleotide, may form a loop. The correctly matched bases on both sides of the mutation will keep the annealed complex stable.
- DNA polymerase will use the annealed oligonucleotide as a primer to synthesize an entire strand complementary to the template, but retaining the modified oligonucleotide; DNA ligase will ligate the nick.
- The new double-stranded circle is then introduced into bacterial cells and the bacteria are allowed to multiply. Various techniques can be used to get rid of the original molecule.

Michael Smith won a share of the 1993 Nobel Prize in Chemistry for "his fundamental contributions to the establishment of oligonucleotide-based, site-directed mutagenesis and its development for protein studies." Smith understood what it took to be successful in science: "In research you really have to love and be committed to your work because things have more of a chance of going wrong than right. But when things go right, there is nothing more exciting." The prize was shared with Kary Mullis "for his invention of the polymerase chain reaction (PCR) method" (see Box 5.3).

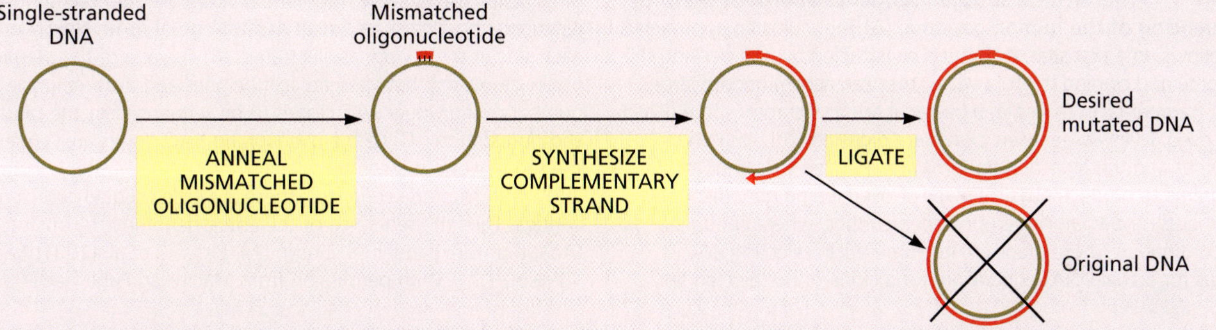

Figure 1 Principle of site-directed mutagenesis. The single-stranded DNA could be that of the single-strand DNA phage M13 containing the cloned gene; alternatively, if the gene had been cloned in a double-stranded vector, the DNA would first need to be denatured, to allow hybridization with the synthetic oligonucleotide. Several rounds of replication of the ligated product introduced into a bacterial cell will result in the formation of a mutated sequence of the original DNA. Convenient methods exist to get rid of the latter. (Adapted from Kunkel TA, Roberts JD & Zakour RA [1989] In Recombinant DNA Methodology Wu R, Grossman L & Moldave K eds, 587–601. With permission from Academic Press.)

There are two approaches for sequencing large genomes

In the whole-genome shotgun approach, the entire genome is cloned as a series of recombinants, and each clone is sequenced. Computational methods are then used to reconstruct the entire genome, through alignment of clones that contain overlapping sequences. Such overlapping clones always exist, as the initial random fragmentation of the genome is performed on a population of cells, each of which will produce its own random set of fragments. The second strategy for genome sequencing is the hierarchical **shotgun sequencing approach**, also called BAC-based or clone-by-clone sequencing. This approach involves generating and organizing, through restriction nuclease mapping, a set of large-insert clones, typically 100–200 kbp each, and then separately performing shotgun sequencing on the appropriately chosen clones. The latter approach has been selected, following "lively scientific debate," for sequencing the human genome. **Figure 5.15** illustrates the approach, as used by the International Human Genome Sequencing Consortium.

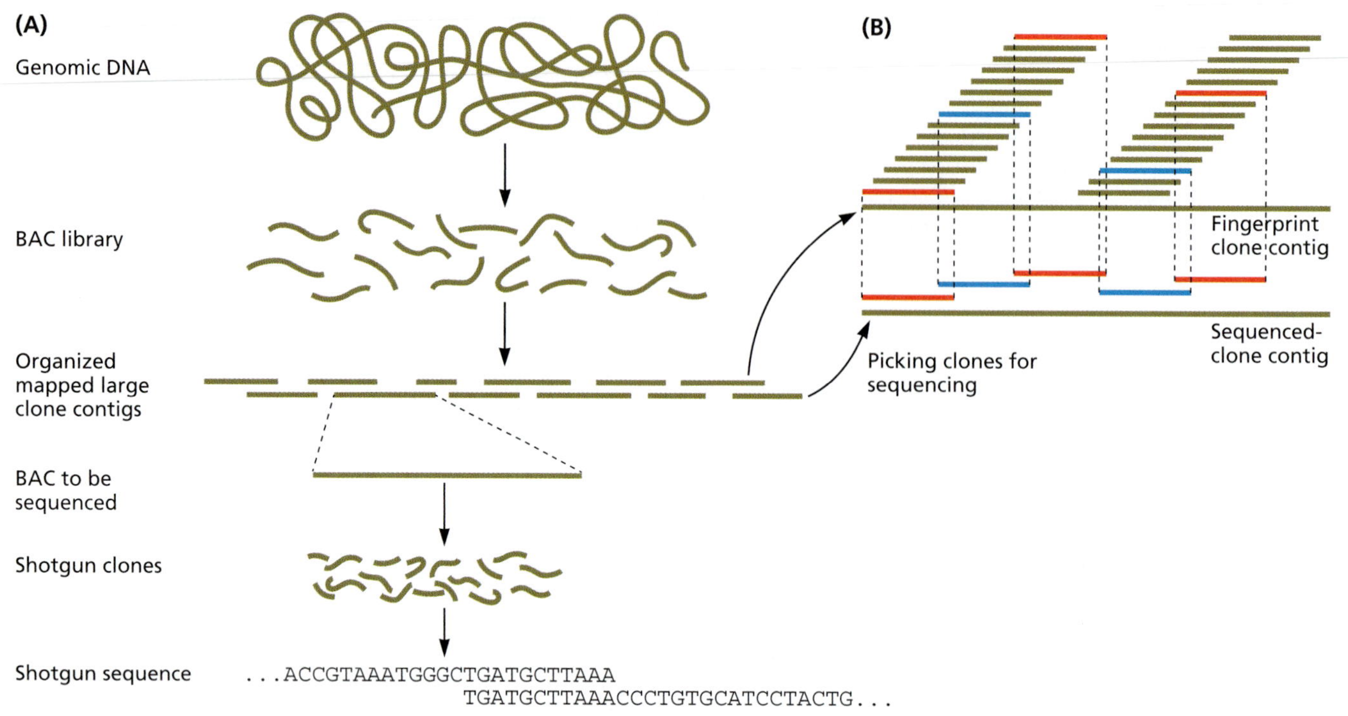

Figure 5.15 Hierarchical shotgun sequence approach used for sequencing of the human genome. (A) The multistep approach is as follows. In a first step, a library is constructed by fragmenting the genome and cloning it into a vector that can accommodate large DNA fragments: in this case, a bacterial artificial chromosome (BAC) was used. Many large-insert clones are digested with restriction enzymes and the digestion patterns are analyzed by computer programs to produce physical maps of overlapping genomic DNA fragments, known as fingerprint clone contigs, where contig stands for contiguous. These maps are instrumental in selecting BAC clones for actual DNA sequencing; for a clone to be selected, all of its restriction enzyme fragments, except the two vector–insert junctions, must be shared with at least one of its neighbors on each side in the contig, as illustrated. (B) The aim of this step is to minimize overlap between adjacent clones and thus minimize the number of individual BAC clones to be sequenced by the random shotgun strategy. Once the overlapping BAC clones are sequenced, the sequences are assembled, again through computer programs, into uninterrupted sequences; these are called sequenced-clone contigs. Thus the sequence of the entire genome is reconstructed. (Adapted from Lander ES, Linton LM, Birren B et al. [2001] *Nature* 409:860–921. With permission from Macmillan Publishers, Ltd.)

Genomic libraries are not the only type of library important to research and biotechnology. Very useful libraries have been created to represent the messenger RNA (mRNA) contents of eukaryotic cells. The genes that encode proteins are first transcribed into precursor mRNA, which then undergoes a series of processing reactions to create mature mRNA molecules. A major step in processing is removal of the intervening noncoding sequences (introns) from the original transcript, to produce the uninterrupted nucleotide sequences that will be translated on ribosomes to give rise to proteins (see Chapter 14). Each cell type in an adult eukaryotic organism is characterized by a different set of mRNA molecules; some of these are shared among all cells, and some are unique to that specific cell type. The unique mRNA molecules are the ones responsible for the differentiated phenotype of that cell type: they code for proteins that perform highly specialized functions.

The best way to study which protein genes are transcribed, and later translated, in a given cell type is to isolate the population of mature mRNA from the cytoplasm and then, by use of *in vitro* methods such as reverse transcription, to produce double-stranded DNA copies of this population. These are known as **cDNAs** or **copy DNAs**. Cloning this entire population of cDNAs into bacterial vectors produces **cDNA libraries**. These can be maintained stably in bacteria and stored indefinitely, exactly as genomic libraries are. The existence of such libraries has practical applications: because they lack introns, cDNA clones are a convenient source of sequences to be directly expressed in bacterial cells.

5.9 Manipulating the genetic content of eukaryotic organisms

The development of recombinant DNA techniques has given us the power to change, at will and in specific ways, the genomes of plants and animals. The gene transfer techniques described in this chapter allow us to produce what are referred to as transgenic organisms. These are in fact new organisms that will breed true if they are homozygous for the genetic change.

Making a transgenic mouse involves numerous steps

In the early 1980s, a number of laboratories developed a technique for introducing new genetic material into eukaryotic genomes in a way that would be transmitted from generation to generation. An egg is removed from a female mouse and fertilized *in vitro*. The vector containing the gene of interest, as well as sequences necessary for proper transcription of the gene and markers to allow screening for proper incorporation, is injected into the male pronucleus, the haploid nucleus of the sperm, after the sperm enters the egg. The male pronucleus then fuses with the pronucleus of the egg to form the diploid nucleus of the embryo. The timing of the microinjection is critical because the introduced DNA must integrate into the genome prior to the duplication of genetic material that takes place before the first embryo cell division. If integration occurs following that division, the mouse will be mosaic, with some wild-type and some transgenic cells. Integration of the vector into the host genome is random with respect to the underlying genomic DNA sequence.

At the next stage, the egg is transplanted to the uterus of a foster mouse, and a pup will be born. If integration is successful, the resulting mouse will be heterozygous in the region of integration. This can be detected by the markers employed, which are often chosen to regulate fur color. If two such mice breed together, some of the progeny, 25%, will be homozygous for the new trait (see Chapter 2). These will breed true. Images of such transgenic mice are presented in **Figure 5.16**.

Site-directed mutagenesis makes it possible, in principle, to insert a protein with a new or modified sequence into an organism and to study its effect on the whole organism. However, it must be remembered that this is the addition of a function; the original variant of the gene is still present and, in most cases, functional. If we want to take away a gene or replace it with an altered version, the challenge is somewhat more difficult.

To inactivate, replace, or otherwise modify a particular gene, the vector must be targeted for homologous recombination at that particular site

This goal, long thought impossible, was achieved by Mario Capecchi in the late 1980s. The technique will be illustrated by the widely used procedure to produce

(A)

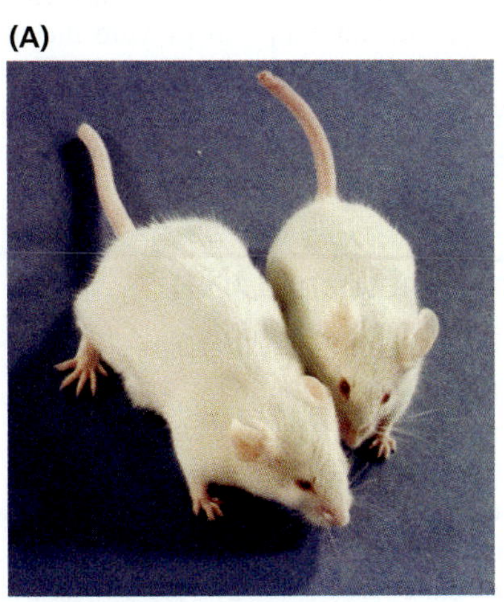

(B)

Figure 5.16 Transgenic mice. (A) Super mouse. Richard Palmiter from the University of Washington and Ralph Brinster from the University of Pennsylvania microinjected the human growth hormone gene into freshly fertilized mouse eggs. Expression was ensured by fusion of a mouse promoter, that of the metallothionein I gene, to the structural gene coding for the human growth hormone. A significant percentage of mice stably incorporated the fusion gene into their genome and exhibited high concentrations of the hormone in the blood. The transgenic mice grew to twice the size of normal mice. (B) Green mice. Masaru Okabe and colleagues from Osaka University in Japan engineered mice that exhibit green fluorescence by injecting into mouse embryos a jellyfish gene that codes for a protein called GFP, green fluorescent protein, that can fluoresce under UV illumination. The construct that was injected contained the GFP cDNA, an uninterrupted coding sequence from jellyfish, under the control of elements which ensure that the protein is expressed at high, easily detectable levels. The transgenic mice expressed GFP in every cell type tested—brain, liver, adrenal gland, etc.—except sperm, red blood cells, and hair. The newborn mice were all green under UV illumination until they grew fur, at which point only the naked body parts were fluorescent. The mice are not just a scientific curiosity; their green cells can be used in basic research to follow, for example, the fate of green cancer cells transplanted into normal mice. In addition, one can monitor the response of these cells to treatment. (A, courtesy of Ralph Brinster, University of Pennsylvania. B, from Okabe M, Ikawa M, Kominami K et al. [1997] *FEBS Lett* 407:313–319. With permission from Elsevier.)

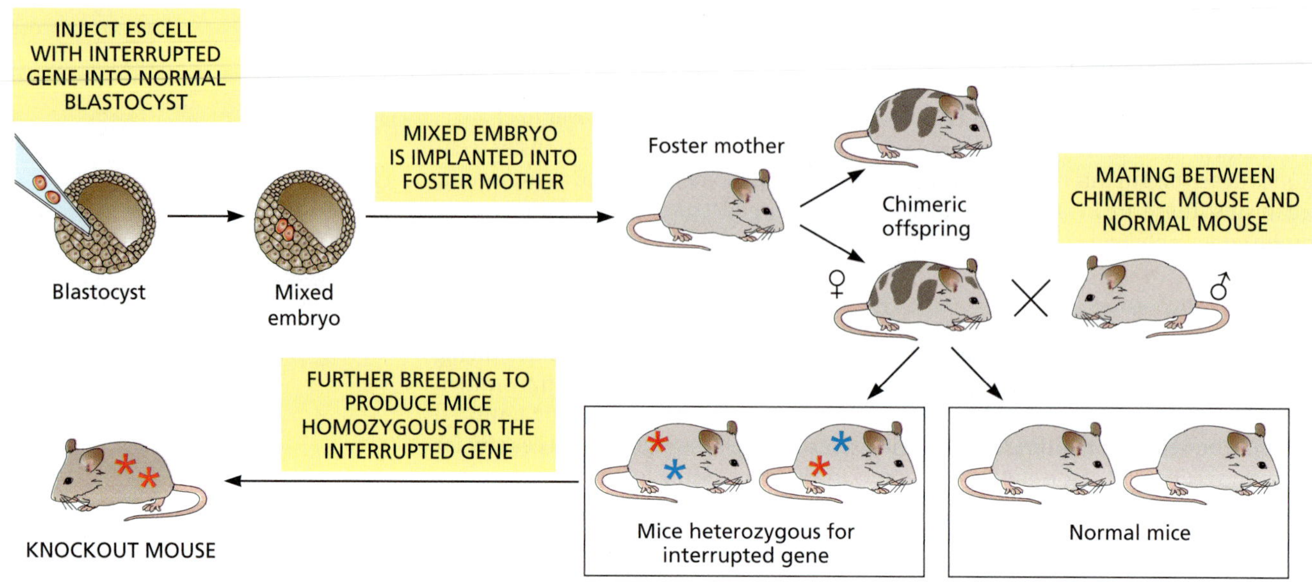

Figure 5.17 Making a knockout mouse by introducing an *in vitro* altered gene into mice. (Adapted from the Nobel Foundation.)

knockout mice, in which a specific gene has been inactivated by interrupting it with an extraneous sequence.

The procedure begins with the construction of a vector containing the interrupted gene, together with enough native flanking sequences to favor homologous recombination in the recipient cell (see Chapter 21). Markers for screening are also included, often within the interrupted gene. The vectors carrying the altered gene are inserted, by chemical means or by electroporation, into embryonic stem or ES cells (see Box 2.2) taken from a mouse embryo. These are pluripotent cells that have the potential to develop into any kind of cell of the organism if provided the right molecular cues from the environment. The cells are selected for recombination events during growth in a culture medium. Recombinant cells are then inserted into another blastocyst mouse embryo (**Figure 5.17**). The resulting mixed embryo that contains both normal cells and cells with the interrupted gene is implanted into the uterus of a foster mouse. The newborn mice will be chimeras, with a mosaic of altered heterozygous cells and wild-type cells. If such females are allowed to grow to maturity and bred to wild-type males, some of the progeny will be wild-type while the rest will be heterozygous for the interrupted gene. However, homozygous mice can be obtained by interbreeding the heterozygous mice. These mice will survive and breed further, if the lack of the gene is not lethal.

Modifications of the above procedure can be used to generate **knock-in** organisms, in which a modified gene is specifically substituted for the wild-type gene, or **knock-downs**, in which the regulation of a particular gene is modified. All of these techniques are very time-consuming and relatively inefficient, but they can provide the most definitive evidence for the *in vivo* function of a particular gene. Capecchi shared the 2007 Nobel Prize in Physiology or Medicine with Sir Martin Evans and Oliver Smithies for their "discoveries of principles for introducing specific gene modifications in mice by the use of embryonic stem cells."

5.10 Practical applications of recombinant DNA technologies

Hundreds of pharmaceutical compounds are produced in recombinant bacteria

This is the most developed field of application of recombinant DNA technology. The list of compounds produced and used in clinical practice already encompasses hundreds of products, most of which are proteins. These include life-saving products such as human insulin (**Box 5.5**), growth hormone, and blood clotting factor VIII

Box 5.5 Production of recombinant human insulin for therapeutic purposes The history of using insulin to treat diabetes is long and fascinating. Insulin is produced by a small portion of the pancreas, in the β-cells in the islets of Langerhans. In the first decade of the nineteenth century, physician George Zülzer showed that sugar levels in the blood of depancreatized dogs could be controlled by injecting a pancreatic extract. He then injected the extract into a dying diabetic patient, producing some initial improvement, but the patient died when the extract, manufactured by a local company in Berlin, was exhausted.

In 1922, physicians Frederick Banting and Charles Best purified insulin from pancreatic extract and successfully treated a 14-year-old patient with a severe form of the disease. Banting shared, with John Macleod, the 1923 Nobel Prize in Physiology or Medicine "for the discovery of insulin." This animal insulin was not ideal: pancreases were in short supply and the product itself was not very effective, as animal and human insulins are not exactly the same.

When recombinant DNA technologies became available, efforts at both pharmaceutical companies and universities were focused on developing recombinant insulin to be produced by bacteria. Two major strategies were employed: separate production and purification of the two chains, A and B, of the mature, biologically active insulin (Figure 1) and production of proinsulin, a precursor molecule that undergoes several steps of processing *in vivo* to give rise to mature insulin. These steps are discussed in detail in Chapter 18.

In 1978, David Goeddel, then at Genentech, in collaboration with Keiichi Itakura and Arthur Riggs at the City of Hope Medical Center in California, produced the first recombinant human insulin. The strategy was to clone and express the A and B chains of insulin in separate bacterial strains and then to purify the chains and join them together to form a functional product, by allowing the two S-S bridges that hold the chains together to form *in vitro*. The DNA fragments coding for the A and B chains were chemically synthesized, and then cloned in a derivative of pBR322 that carried a copy of the β-galactosidase gene under the control of *lac* operator (see Chapter 11), as shown in Figure 1A. Following *in vivo* synthesis of the chains, the β-galactosidase portion of the fusion proteins could be efficiently cleaved off by treatment with cyanogen bromide.

In the same year, Walter Gilbert's laboratory at Harvard reported the cloning of the rat proinsulin using pBR322 as a cloning vector (see Figure 1B). In this approach, the researchers first obtained double-stranded cDNA copies of mRNA from a rat pancreatic tumor and cloned the cDNA into the *Pst*I site of pBR322. Note that plasmid pBR322 is itself recombinant, containing antibiotic resistance genes from different sources. The cloning restriction site lies within the ampicillin resistance gene; the encoded fusion product is secreted into the periplasmic space, thus facilitating its purification. More recent efforts have focused on the production of numerous rapid-acting and/or long-acting insulin analogs.

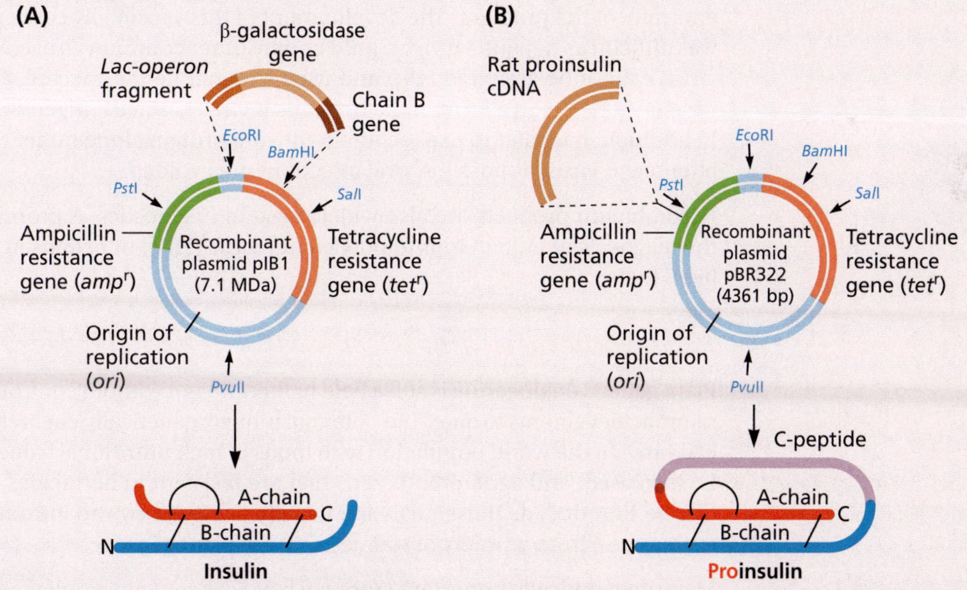

Figure 1 Two major strategies used for production of mammalian insulin in *E. coli*. (A) The two-strain approach, in which the two chains of mature insulin are produced by separate recombinant *E. coli* strains. The schematic presents the construct for the B-chain; the same strategy was used to create the strain that produced the A-chain. (B) In the first reported cloning of proinsulin as a single polypeptide, rat proinsulin cDNA was inserted into the *Pst*I site of pBR322. The connecting C-peptide can be removed from proinsulin enzymatically, as it is *in vivo*.

(see Box 21.4). Other important applications include the production of recombinant vaccines. A prominent example is the production of hepatitis B vaccine. The World Health Organization estimates the number of carriers of hepatitis B virus worldwide to be ~350 million, with 1 million new cases annually. An estimated 75–100 million infected individuals may die from liver cirrhosis or liver cancer, underlying the

Box 5.6 HIV diagnosis relies on methods that use recombinant products or procedures All three common methods for diagnosing HIV infection have been developed through DNA recombinant methods. Principles of the two immunological tests, ELISA or enzyme-linked immunosorbent assay and Western blots, are described in Chapter 3: they detect antibodies to specific recombinantly produced HIV proteins that are present in blood samples from patients (Figure 1). The test that directly detects the RNA genome of HIV uses reverse transcriptase polymerase chain reaction or RT-PCR. Development of the RT-PCR test was made possible by molecular cloning and sequence analysis of the HIV genome.

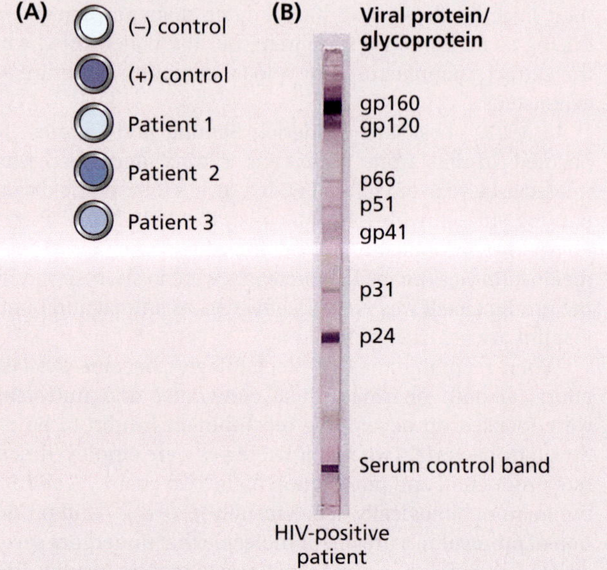

Figure 1 Immunological detection of HIV infection. (A) Enzyme-linked immunosorbent assay (ELISA) test. Blood samples from three patients were analyzed by ELISA. In the case of positive ELISA, a Western blot test is performed. (B) Western blot test. The interpretation of this test is not as straightforward as that of ELISA, as the viral protein bands that are detected by antibodies present in the blood sample are not always visible. The consensus among clinicians is that, to be conclusive, the profile must have at least five bands. (B, from Suthon V, Archawin R, Chanchai C et al. [2002] *BMC Infect Dis* 2 [10.1186/1471-2334-2-19]. With permission from BioMed Central.)

enormity of the problem. The development of this vaccine by conventional methods was difficult as hepatitis B virus, unlike some other common viruses such as the polio virus, cannot be grown *in vitro* and used to generate the vaccine. The recombinant vaccine is based on a form of the hepatitis B virus surface antigen produced in yeast. In addition, it is safer to use recombinant viral protein alone rather than intact albeit attenuated virus, as possible viral infections are avoided.

Recombinant products are also widely used in diagnostics. A prominent example is the diagnosis of human immunodeficiency virus (HIV) infections in humans, as outlined in Box 5.6.

Plant genetic engineering is a huge but controversial industry

Plant genetic engineering is still a field in flux and not without controversy, despite the many achievements to date. The potential is huge: genetically engineered or GE plants can provide the world population with foods of high nutritional value, crop yields can be increased, and agricultural crops that are resistant to herbicides and insecticides can be introduced. These goals are usually achieved by the introduction of one or more genes, from a variety of sources, into the plant genome.

Many economically important crops such as soybean, maize, alfalfa, and cotton have been engineered to show herbicide resistance by introducing a recombinant gene whose expression results in resistance to the commonly used herbicide glyphosate. Other crops are made resistant to insects by introducing a recombinant form of a bacterial protein that is selectively toxic to insects. The original gene from *Bacillus thuringiensis* encodes the Bt toxin, which kills insects. The traditional way of dealing with insect predators was to spray fields with the bacterium from aircraft. This method has been largely replaced by creating recombinant plants carrying the *Bacillus* gene in their genomes.

A more recent development concerns efforts to engineer plants genetically as biofactories for the production of vaccines. A case in point is the production of edible hepatitis B vaccine. Several food plants, including banana, potato, lettuce, and carrots, have been successfully engineered to induce immunological protection against hepatitis B virus. Similar projects will inevitably follow.

The controversy surrounding GE plants centers around two main issues: the safety of these crops for human consumption and the fear that such artificially created plants may disturb the delicate ecological balance in nature. While the first issue seems to be resolved in favor of GE food, the second is still of concern.

Before discussing one of the major achievements of plant genetic engineering, the development of golden rice, we should introduce the most commonly used plant vector. This vector is a derivative of a naturally existing large plasmid that inhabits the bacterium *Agrobacterium tumefaciens.* A portion of the plasmid is transferred to plant cells through wounds in the plant and causes tumor growth; hence the name Ti plasmid, tumor-inducing. The structure and biology of Ti plasmids, and their modification to form plant transformation vectors, are detailed in **Box 5.7**.

Box 5.7 A Closer Look: *Agrobacterium*, the natural vector for plant genetic engineering The genus *Agrobacterium* includes bacterial species that live in the soil. Four *Agrobacterium* species cause neoplastic transformation in a wide variety of plants, mainly dicotyledons. The tumor formed by infection with *A. tumefaciens,* new classification *Rhizobium radiobacter*, is known as **crown gall**.

The Ti plasmid resides in the bacterial cell (**Figure 1**) and transforms plant cells after *Agrobacterium* infection through wounds in the root or stem. When wounded, the plant gives off certain chemical signals, phenols and sugars. These signals are recognized by receptors in the bacterial cell membrane and activate a cascade of *Agrobacterium* virulence (*vir*) genes necessary for transmission of the transfer DNA (T-DNA) region from the Ti plasmid to the plant chromosome (**Figure 2**). The T-DNA becomes stably integrated into the plant nuclear genome. With the exception of *virA* and *virG*, the *vir* genes are essentially silent in the free-living bacterium. Different *vir* gene products participate in different stages of the process: some copy the T-DNA, while others prepare the T-DNA for passage into the host and then open a channel in the bacterial cell membrane, through which the T-DNA passes. The T-DNA in the cytoplasm interacts with still other Vir proteins to be able to enter the cell nucleus. Curiously, these proteins contain plant nuclear localization signals, indicating co-evolution of the bacterium and the respective plant hosts. The process of integration of T-DNA into the plant chromosome is less well understood. One theory is that the T-DNA waits until the plant DNA is being replicated or transcribed and then inserts itself into the exposed plant DNA. Note that proteins bound to DNA constitute an obstacle to processes that use DNA as a template; these are usually temporarily removed to allow access to the DNA.

The potential for using the Ti plasmid for introducing foreign genes into plants was recognized almost immediately after the plasmid was discovered. The two borders are the only *cis*-acting elements essential for DNA transfer. Thus, the native wild-type oncogenes and the opine synthase genes can be replaced by any sequence of interest, which will then be transferred to the plant cell and integrated into the genome. The *vir* genes that are needed for transfer and integration could be supplied in *trans*, on a second plasmid, thus forming a binary transformation system (**Figure 3**). Plant protoplasts, cells whose cell wall has been removed by enzymatic treatment, are transformed with the desired genes and then grown in culture to give rise to the callus, an undifferentiated mass of transformed cells. Whole plants are regenerated from calli and grown into mature seed-forming plants in the greenhouse.

Despite the achievements of *Agrobacterium*-mediated plant transformation, some problems remain, mainly concerning the inability of some economically important crops

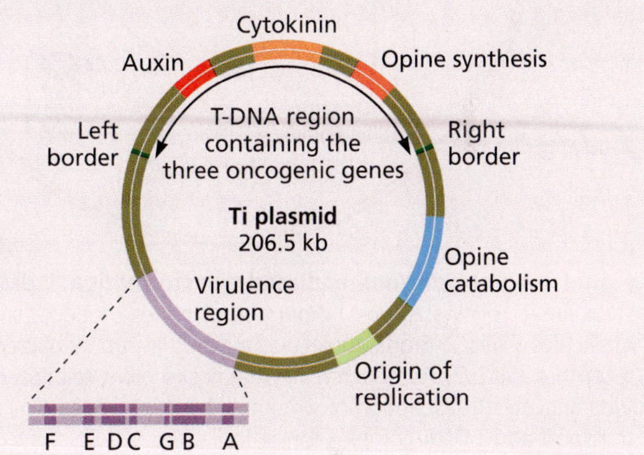

Figure 1 Ti or tumor-inducing plasmid of *A. tumefaciens.* The Ti plasmid has 196 genes densely packed along the genome. These genes form three major regions. The T-DNA region is the portion of the plasmid that is transferred to the plant cell and stably integrated into the plant nuclear genome. The amount of plasmid DNA transferred to the DNA of the plant cell represents 0.0011% of the DNA of the tumor cell. The T-DNA region contains two genes that encode plant hormones, auxin and cytokinin. The products of these genes stimulate proliferation of the transformed plant cell to produce the tumor mass. The third gene encodes an enzyme that is involved in the synthesis of opines, which are sugar–amino acid conjugates synthesized by the transformed cell and used solely by the free-living bacterium as a dedicated food source. The T-DNA region is delineated by highly homologous short, 25–28 bp direct repeats, known as borders. The left and right borders are the only *cis*-acting element essential for T-DNA transfer. The other two regions that are involved in bacteria–plant interactions act in *trans*; that is, they are not transferred to the plant cell but their protein products act in both the bacterium and the plant cell during the processes of infection and integration. The first region encompasses the 35 virulence genes organized in seven operons, A–G; they are absolutely required in several steps. Importantly, the *vir* genes can be removed from the plasmid without adversely affecting its ability to insert its own DNA into plant cells. This characteristic has been exploited for the creation of Ti plasmid-based vectors as tools for introducing foreign DNA in plant genetic engineering projects. The second region participates in opine uptake and metabolism.

Box 5.7 *(Continued)*

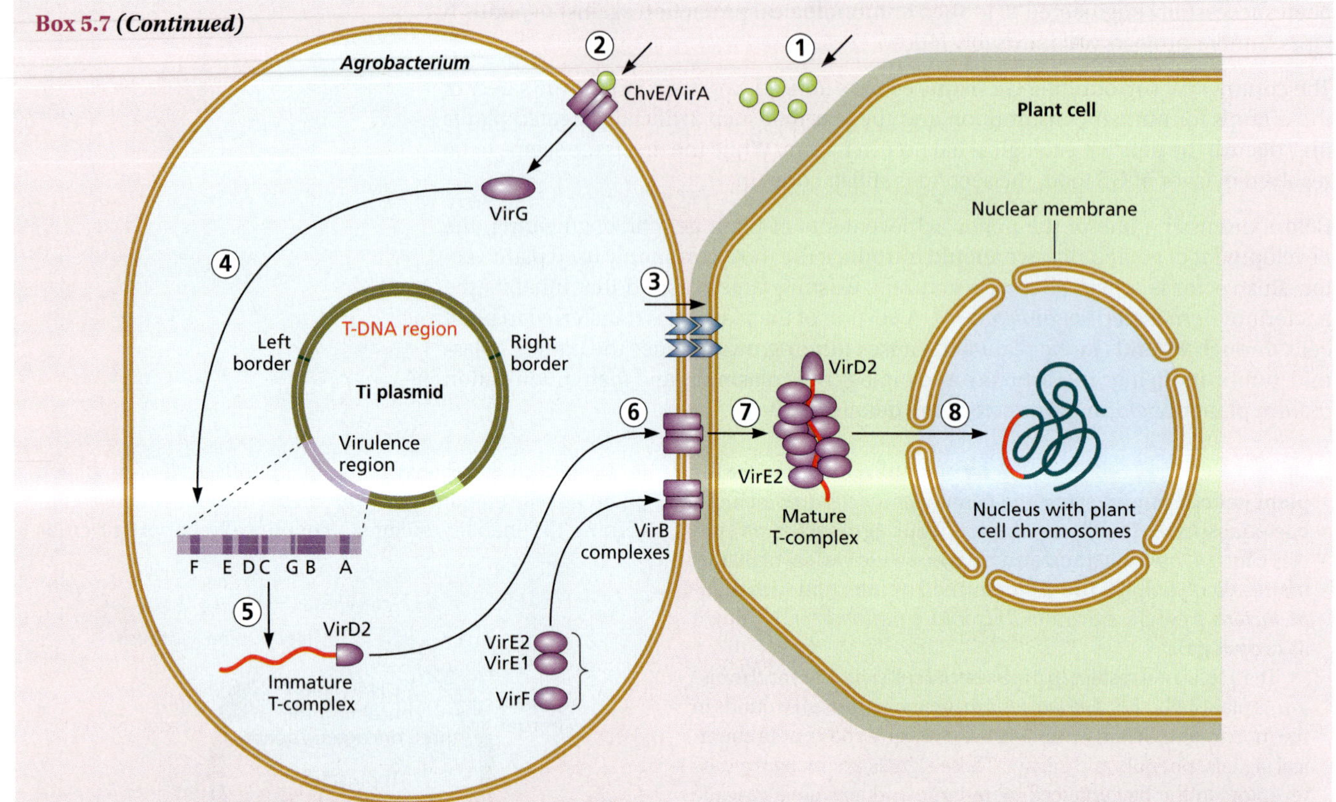

Figure 2 *Agrobacterium*-mediated transformation. Individual steps are as follows: (Step 1) Wounded plant cell produces signal molecules. (Step 2) Signal molecules are recognized by bacterial receptors. (Step 3) *Agrobacterium* attaches to plant cell. (Step 4) VirG activates transcription of *vir* genes, resulting in proteins for transport and T-DNA complex formation. (Step 5) T-DNA is excised from Ti plasmid to form immature T-complex. (Step 6) Immature T-complex is transported to the plant cell. (Step 7) T-complex matures by binding to VirE2 proteins and moves to the nucleus. (Step 8) T-DNA randomly integrates into a plant chromosome, and bacterial genes are expressed. Proteins encoded in the *vir* region of the Ti plasmid are as follows: VirA/VirG, a signal transduction system; VirD1/VirD2, endonucleases targeting the border-specific sequences and releasing single-stranded T-DNA; VirD2, covalently attaches to the 5′-end of the T-strand to form the immature T-complex; VirB and VirD4, form the secretion system for the transfer of immature T-complex and other virulence proteins into the plant cell; and VirE2, the binding of which transforms the immature T-complex into a mature one. Both VirE2 and VirD2 possess nuclear localization signals for invasion of the nucleus. Plant factors are involved in passing the nuclear membrane and integration, but their exact function is less clear. (Adapted from Păcurar DI, Thordal-Christensen H, Păcurar ML et al. [2011] *Physiol Mol Plant Pathol* 76:76–81. With permission from Elsevier.)

to be infected by *A. tumefaciens*. Studies of the host proteins involved in the process are underway; understanding their function will be instrumental in broadening the range of susceptible plant species.

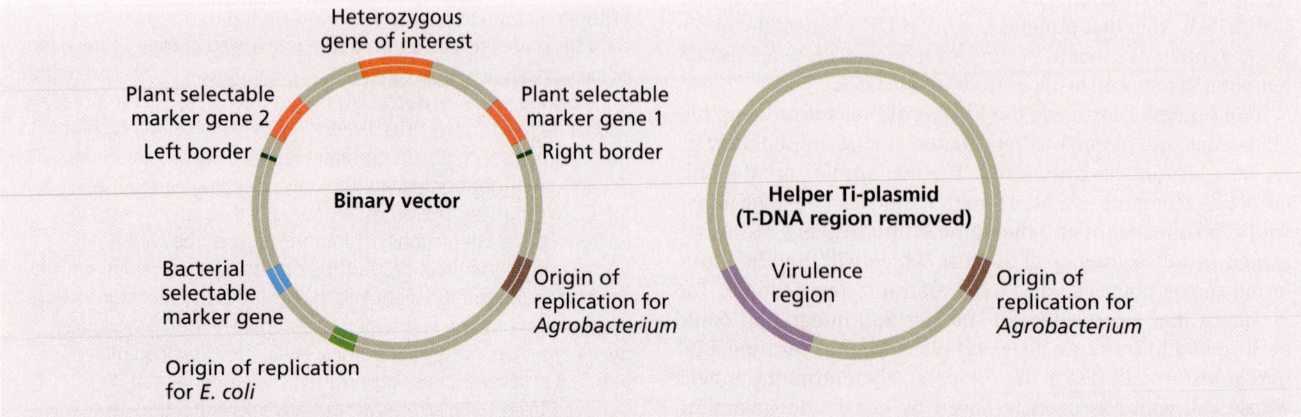

Figure 3 Typical binary vector and helper plasmid used for plant transformation. The bacterial selectable marker and origin of replication are needed to produce the vector in the necessary purity and quantities in *E. coli* before transformation of the plant. (Adapted from Păcurar DI, Thordal-Christensen H, Păcurar ML et al. [2011] *Physiol Mol Plant Pathol* 76:76–81. With permission from Elsevier.)

The engineering of golden rice, a rice cultivar with improved nutritional value, involved the introduction of several genes of different origins into the plant genome. The idea was to develop a crop that would significantly diminish two major nutritional deficiencies, especially in populous countries where rice is the main staple food. The engineering work was complemented by standard breeding practices that collectively resulted in this wonder food (**Box 5.8**). It has been predicted that the substitution of presently grown rice cultivars with golden rice will constitute a major step toward solving the world's nutrition problem.

Box 5.8 Golden rice, a solution to vitamin A and iron nutritional deficiencies Vitamin A deficiency affects some 400 million people worldwide, leaving them vulnerable to infections, dwarfism, and blindness. Another nutritional deficiency worldwide is that of iron, which can lead to anemia. It is estimated that most pre-school children and pregnant women in developing countries, and at least 30–40% in industrialized countries, are iron-deficient. As rice is the main staple food for more than half the world's population, efforts to improve its nutritional quality to overcome these deficiencies are very promising. **Golden rice** has been engineered to produce β-carotene (provitamin A) in its endosperm, the edible part of rice. Note that rice plants naturally produce β-carotene in the leaves but not in the endosperm. In addition, the plant was engineered to increase its iron content and to improve iron absorption.

Golden rice was first created through collaboration between two laboratories, those of Ingo Potrykus in Switzerland and Peter Beyer in Germany. These researchers transformed rice with two β-carotene biosynthesis genes: *psy*, which encodes phytoene synthase from daffodil, *Narcissus pseudonarcissus*, and *crtI*, which encodes phytoene desaturase from the soil bacterium *Erwinia uredovora*. **Figure 1A** shows the biochemical pathway involved in β-carotene synthesis. The *psy* and *crtI* genes were placed under the control of endosperm-specific promoters so they are expressed only in the endosperm. The end product of the engineered pathway is the red pigment lycopene, which is further processed to β-carotene by endogenous plant enzymes, giving the rice the distinctive yellow color for which it is named (**Figure 1B**).

Three recombinant plasmids were created that carry different combinations of these genes; these vectors were separately electroporated into *A. tumefaciens* and the bacterium was co-cultivated with calli from immature rice embryos. Transgenic plants were regenerated from calli resistant to the two antibiotics used for selection of transformants, rooted, and transferred to the greenhouse.

Subsequently, golden rice has been bred with local rice cultivars in Asia and the United States. Field-grown golden rice produces 4–5 times more β-carotene than golden rice grown under greenhouse conditions. Substituting the daffodil *psy* gene with the same gene from maize resulted in a transgenic variety that produces even more carotenoids. The recommended dietary allowance will readily be met by consuming about 150 g of the most high-yielding strain.

The iron supplementation project involved introducing three genes into rice plants (**Figure 2**). Rice actually has a lot of iron, but only in the seed coat. However, because unpeeled rice quickly becomes rancid in tropical and subtropical climates, the seed coat has to be removed for storage.

Once β-carotene- and iron-rich rice strains became available, the Potrykus team cross-bred them, producing hybrids that combined both improvements. In more recent developments, the iron content in polished rice was increased even further by using a somewhat different approach. The genes introduced into an existing rice variety were the ferritin gene, with which we are already familiar, and the gene encoding nicotianamine synthase. The latter enzyme is involved in a complex biochemical pathway that leads to enhanced uptake of iron from the soil: the product of nicotianamine synthase, nicotianamine, binds the iron temporarily and facilitates its transportation into the plant. In combination, the expression of these genes provides a synergistic effect on iron uptake and storage in the kernel: the transformed plant produces kernels containing sixfold more iron than the original variety.

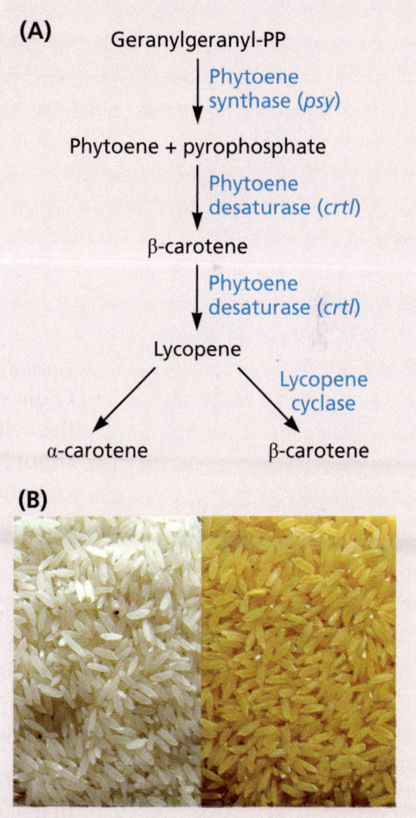

Figure 1 Golden rice. (A) Synthetic pathway for β-carotene synthesis and the enzymes involved. Geranylgeranyl-PP is naturally synthesized in immature rice endosperm. Phytoene synthase and phytoene desaturase were the enzymes used for transformation of immature rice endosperm. The increased β-carotene content gives the bright yellow color to the grains. (B) Wild-type and golden-rice grains photographed side by side. (B, courtesy of Golden Rice Humanitarian Board.)

Box 5.8 ***(Continued)***

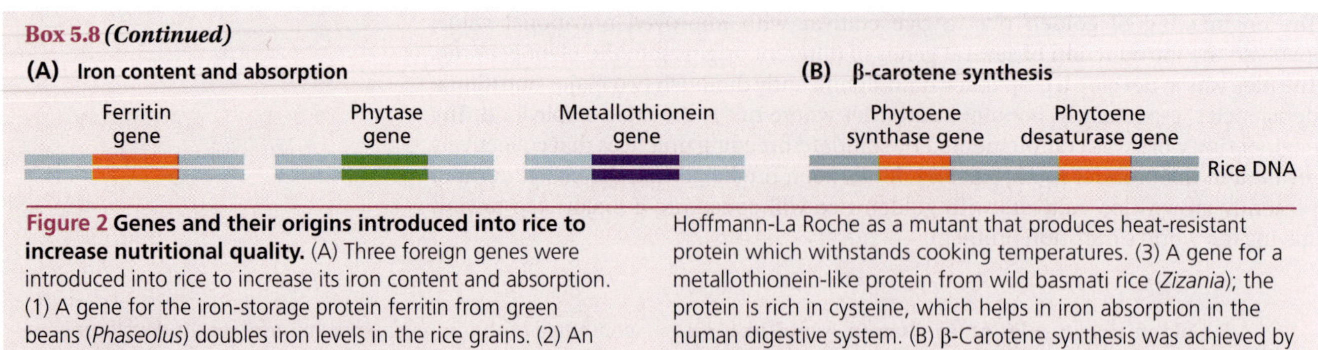

Figure 2 Genes and their origins introduced into rice to increase nutritional quality. (A) Three foreign genes were introduced into rice to increase its iron content and absorption. (1) A gene for the iron-storage protein ferritin from green beans (*Phaseolus*) doubles iron levels in the rice grains. (2) An *Aspergillus* fungal gene for an enzyme known as phytase was added. Phytase breaks down phytate, a sugarlike molecule that ties up 95% of dietary iron and prevents the human body from absorbing it. The gene was provided by pharmaceutical company Hoffmann-La Roche as a mutant that produces heat-resistant protein which withstands cooking temperatures. (3) A gene for a metallothionein-like protein from wild basmati rice (*Zizania*); the protein is rich in cysteine, which helps in iron absorption in the human digestive system. (B) β-Carotene synthesis was achieved by the introduction of two essential genes involved in the synthetic pathway: a daffodil (*Narcissus*) gene for phytoene synthase and a bacterial gene from *Erwinia* for phytoene desaturase.

Gene therapy is a complex multistep process aiming to correct defective genes or gene functions that are responsible for disease

Gene therapy combines recombinant techniques with the aim of correcting defective genes or gene functions responsible for disease development. Administration of the functional gene, instead of the protein itself, is performed because proteins are quickly degraded, whereas a properly integrated gene will continue to be expressed. There are two major approaches:

1. A normal gene may be inserted into a random, nonspecific location within the genome to supplement a nonfunctional gene. This approach is classified as **gene addition** therapy, since the nonfunctional gene stays in the genome. Examples might be the use of tumor suppressor genes in the treatment of a malignancy or immunostimulatory genes to treat an infectious disease. Gene addition therapy could also be used to provide the cell with transgenes whose expression can control the expression of a mutated gene.

2. A mutated gene can be swapped for a normal gene through homologous recombination (see Chapter 21). This approach is representative of **gene replacement** therapy.

Gene therapy is a complex multistep process involving several steps in the production of the vector carrying the gene of interest. Once the vector delivers the transgene into the cell, the gene must travel through the cytoplasm and enter the nucleus, where it should be stably integrated into the nuclear genome: only integrated gene copies can be consistently replicated as part of the genome. Finally, properly regulated expression must be achieved. This is not a trivial task, as the majority of vectors will insert the gene they carry at random locations. Such random insertion can create two kinds of problems. In most cases, the gene will end up in a heterochromatic environment that does not permit its transcription (see Chapter 12). In other cases, the gene may land within other genes or their regulatory sequences, which will lead to the inactivation of these host genes. This may be detrimental to the host cell, to an extent that depends on the importance of the inactivated gene. Researchers are constantly improving existing methods or coming up with new approaches to overcome these problems.

One of the biggest problems in the development of gene therapy is the immunogenicity of the vector. On one hand, the immune response can very effectively eliminate transduced cells. On the other hand, the development of adaptive immunity against viral vectors can prevent their readministration.

Delivering a gene into sufficient cells within a specific tissue and ensuring its subsequent long-term expression is a challenge

Because the anionic nature of DNA hinders transfer across the membrane, the focus of efficient gene therapy is on vector-mediated delivery. There are two types of vectors: viral and nonviral vectors.

Viruses possess the natural ability to enter the cell and nucleus efficiently. **Viral vectors** differ in several properties, all of which affect the success and safety of the entire process (Table 5.2).

Vector capacity is the size of the transgene(s) that could be packed into the vector. It is determined by the size of the viral genome itself and by the amount of viral nucleic acid, DNA or RNA, that could be excised and replaced by a transgene without affecting viral trafficking ability.

Efficiency of vector production in cell lines used as incubators becomes an issue as more and more viral genes are deleted and replaced with transgenes. The viral vector must be replication-incompetent: that is, it must carry only the essential elements necessary for transgene packaging and expression. The viral genes necessary for viral particle production are removed from the vector and provided in *trans* by another virus in the vector-producing cell lines.

Tropism, or host range, is the ability of the virus to infect different cell types. Viruses enter the cell through endocytosis, which is triggered by binding of the virus to specific receptor proteins on the cell surface. These may or may not be present on all cells. Modifications of the viral genes that encode the envelope and capsid proteins can overcome this problem. Once in the cell, viral DNA is released from the capsid/envelope structures and needs to enter the nucleus. Crossing the nuclear membrane may be an insurmountable obstacle for some viruses, especially retroviruses. This may be the reason retroviruses can infect only rapidly dividing cells; the elimination of the nuclear envelope during mitosis provides any exogenous DNA or RNA with a window of easy entry into the nucleus.

The overall transfection efficiency depends on the efficiency of the steps involved in viral uptake, entry into the nucleus, and escape from degradation. Viruses containing

Table 5.2 Viral vectors for gene therapy. (Adapted from Waehler R, Russell SJ & Curiel DT [2007] *Nat Rev Genet* 8:573–587. With permission from Macmillan Publishers, Ltd.)

Feature	Adenoviral vector	AAV vector	Retroviral vector	Lentiviral vector
particle size, nm	70–100	20–25	100	100
cloning capacity, kbp	8–10	4.9	8	9
chromosome integration	no	no; yes if *rep* gene is included	yes	yes
cell entry mechanism	receptor-mediated endocytosis; microtubule transport to nucleus		receptor binding, conformation change of Env, membrane fusion, internalization, uncoating, nuclear entry of reverse-transcribed DNA	
transgene expression	weeks to months; highly efficient short-term expression, appropriate for treatment of cancer or acute cardiovascular diseases	>1 year; medium- to long-term expression for nonacute diseases, onset of transgene expression after ~3 weeks	long-term correction of genetic defects	
emergence of replication-competent vector *in vivo*	possible but not a major concern		risk is a concern	
infects quiescent cells	yes	yes	no	yes
risk of oncogene activation by vector	no	no	yes	yes

Abbreviations: AAV, adenovirus-associated virus; Env, viral envelope protein.

a dsDNA genome have, in general, higher transfection efficiency than ssDNA and RNA viruses; the latter need to first convert their genomes into dsDNA.

Nonviral vectors have been developed in order to overcome the immunogenicity and safety issues linked to the use of viral vectors. Although nonviral vectors are much safer then viral vectors, they have an important limitation—an impaired ability to enter the nucleus—which has so far limited their use to proliferating cells only. The simplest option is the direct introduction of therapeutic DNA into target cells. The use of liposomes, artificial lipid spheres with an aqueous core containing the DNA, is another option. In still another approach, the DNA is chemically linked to molecules that are capable of binding to special cell-surface receptors or of facilitating nuclear transfer. The list of nonviral vectors and the ways to enhance their transformation efficiency is steadily growing. Prominent among these are human artificial chromosomes, with practically unlimited capacity and stability of gene expression and no immunogenicity.

The administration of gene therapy to humans has been and remains controversial. Following the first excitement around the promise of curing genetic diseases through modern biotechnology approaches, there has been growing skepticism about the efficiency and safety of the procedures. The community in general and clinicians in particular became wary after an 18-year-old boy died from multiple organ failure four days after treatment, and several patients in a Paris-based clinical trial developed leukemia. It became painfully clear that treatment is risky, although in many cases successful and life-saving. More recent efforts aimed at reducing these risks have been promising and there is little doubt that "gene therapy deserves a fresh chance."

Whole animals can be cloned by nuclear transfer

Although it is not, strictly speaking, a recombinant DNA technique, whole-animal cloning seems appropriate to conclude this section on applications of modern gene technology. A clone of an animal will, by definition, be an exact copy of that animal and must, if our current understanding is correct, carry exactly the same genetic information.

Such clones can be produced by the technique of nuclear transfer. This method involves taking a cell nucleus from the donor and injecting it into an enucleated egg. This is then fertilized and allowed to develop as an embryo, which is then transplanted into the uterus of a foster mother. Success in such experiments began in 1952, when Robert Briggs and Thomas King cloned leopard frogs, *Rana pipiens*, by transferring nuclei from early embryo cells into enucleated eggs. Briggs and King cloned 27 tadpoles from 104 successful nuclear transfers. Attempts to obtain normal tadpoles from nuclei extracted from differentiated cells failed. In the 1960s, John Gurdon successfully cloned *Xenopus laevis* frogs, starting from nuclei from intestinal epithelial cells of tadpoles. Sir John Gurdon shared the 2012 Nobel Prize in Physiology or Medicine with Shinya Yamanaka for "the discovery that mature cells can be reprogrammed to become pluripotent."

The experiments with amphibian cloning were fairly easy, because frog eggs are large, but manipulating the tiny eggs of mammals proved much more difficult. Nevertheless, over the next decade researchers were able to successfully clone a number of mammals, from mice to sheep. In 1997, the sheep Dolly was cloned from an epithelial cell of an adult donor, a major breakthrough that attracted public attention. The cloning of frogs, sheep, and other animals from differentiated cells proved beyond doubt the **principle of genetic equivalence**, which states that all cells of an adult organism contain the same genetic information and differ only in what portion of that information is being used in a specific cell type. Thus, nuclei from cells that are differentiated retain all the genetic information needed for the development of an entire organism; they just need to be reprogrammed, for example, by being transferred to the environment of an enucleated egg.

At first glance, the cloning of domestic animals and pets may sound attractive, but it has not yet proved so commercially. Cloning of an animal is a time-consuming and expensive business; often hundreds of trials are required before success. It can sometimes be efficient in producing lines of laboratory animals with specific attributes and has been proposed as a means of rescuing species on the verge of extinction, especially those that do not breed well in captivity. Advances in the technology of nuclear transplantation might change this picture.

Key concepts

- Recombinant DNA technology allows us to clone, and subsequently sequence, genes or whole genomes. It allows us to manipulate genes at will and to insert or delete them from cells or organisms. It has revolutionized molecular biology.
- The essential enzymes for this technology are restriction endonucleases, which specifically cut DNA sequences; ligases, which reconnect them; and polymerases, which extend or copy sequences.
- Cloning vectors are used to insert desired sequences into the genome of a cell or organism. Commonly used vectors include bacterial plasmids, bacteriophage, and artificial chromosomes.
- An expression vector is required for an inserted gene to produce a protein product in the host. If the gene is eukaryotic, introns should be first eliminated from the gene construct.
- Shuttle vectors can multiply in different hosts and are sometimes required for the cloning of eukaryotic genes.
- Once the desired gene has been incorporated into an appropriate vector, the vector must be delivered to the host cell and often to the nucleus. A wide variety of techniques, chosen according to the vector and host cell, can achieve this.
- In order to assure that the desired gene has in fact been taken up by the host, genes coding for detectable markers are usually incorporated into the vector.
- Polymerase chain reaction makes multiple copies of a sequence *in vitro*.
- Site-directed mutagenesis permits gene modification at specific sites.
- Cloning and sequencing allow the construction of genomic libraries. Ideally, each library will contain the entire genomic sequence. This has now been accomplished for a wide variety of genomes, from viral to human.
- Construction of transgene animals, such as knockout mice, is a major tool in biological research.
- Recombinant DNA techniques allow the industrial production of a number of proteins that are of direct therapeutic value to humans. They also allow the production of genetically modified plants, incorporating resistance to chemicals or insects or conferring higher nutritional value.
- Gene therapy promises to provide resistance to malignant or disease states in humans.
- Whole-animal cloning has now been accomplished with many mammal species but has been of limited practical use to date.

Further reading

Books

Glick BR, Pasternak JJ & Patten CL (2010) Molecular Biotechnology: Principles and Applications of Recombinant DNA, 4th ed. ASM Press.

Kumar A & Garg N (2005) Genetic Engineering. Nova Science Publishers, Inc.

Old RW & Primrose SB (1994) Principles of Gene Manipulation: An Introduction to Genetic Engineering. Blackwell Science Ltd.

Primrose SB & Twyman R (2009) Principles of Gene Manipulation and Genomics, 7th ed. Wiley.

Sambrook J & Russell D (2001) Molecular Cloning: A Laboratory Manual, 3rd ed. Cold Spring Harbor Laboratory Press.

Watson JD, Tooze J & Kurtz DT (1983) Recombinant DNA: A Short Course. WH Freeman.

Reviews

Beyer P, Al-Babili S, Ye X et al. (2002) Golden rice: Introducing the beta-carotene biosynthesis pathway into rice endosperm by genetic engineering to defeat vitamin A deficiency. *J Nutr* 132:506S–510S.

Capecchi MR (2005) Gene targeting in mice: Functional analysis of the mammalian genome for the twenty-first century. *Nat Rev Genet* 6:507–512.

Cohen SN (1975) The manipulation of genes. *Sci Am* 233:25–33.

Gurdon JB & Byrne JA (2003) The first half-century of nuclear transplantation. *Proc Natl Acad Sci USA* 100:8048–8052.

Mullis KB (1990) The unusual origin of the polymerase chain reaction. *Sci Am* 262:56–65.

Mullis KB & Faloona FA (1987) Specific synthesis of DNA *in vitro* via a polymerase-catalyzed chain reaction. *Methods Enzymol* 155:335–350.

Nathans D & Smith HO (1975) Restriction endonucleases in the analysis and restructuring of DNA molecules. *Annu Rev Biochem* 44:273–293.

Pingoud A, Fuxreiter M, Pingoud V & Wende W (2005) Type II restriction endonucleases: Structure and mechanism. *Cell Mol Life Sci* 62:685–707.

Schell J, Van Montagu M, De Beuckeleer M et al. (1979) Interactions and DNA transfer between *Agrobacterium tumefaciens*, the Ti-plasmid and the plant host. *Proc R Soc London, Ser B* 204:251–266.

Experimental papers

Briggs R & King TJ (1952) Transplantation of living nuclei from blastula cells into enucleated frogs' eggs. *Proc Natl Acad Sci USA* 38:455–463.

Campbell KH, McWhir J, Ritchie WA & Wilmut I (1996) Sheep cloned by nuclear transfer from a cultured cell line. *Nature* 380:64–66.

Capecchi MR (1980) High efficiency transformation by direct microinjection of DNA into cultured mammalian cells. *Cell* 22:479–488.

Chilton MD, Saiki RK, Yadav N et al. (1980) T-DNA from *Agrobacterium* Ti plasmid is in the nuclear DNA fraction of crown gall tumor cells. *Proc Natl Acad Sci USA* 77:4060–4064.

Cohen SN, Chang AC, Boyer HW & Helling RB (1973) Construction of biologically functional bacterial plasmids *in vitro*. *Proc Natl Acad Sci USA* 70:3240–3244.

Gaussier H, Yang Q & Catalano CE (2006) Building a virus from scratch: Assembly of an infectious virus using purified components in a rigorously defined biochemical assay system. *J Mol Biol* 357:1154–1166.

Goeddel DV, Kleid DG, Bolivar F et al. (1979) Expression in *Escherichia coli* of chemically synthesized genes for human insulin. *Proc Natl Acad Sci USA* 76:106–110.

Hutchison CA 3rd, Phillips S, Edgell MH et al. (1978) Mutagenesis at a specific position in a DNA sequence. *J Biol Chem* 253:6551–6560.

Jackson DA, Symons RH & Berg P (1972) Biochemical method for inserting new genetic information into DNA of simian virus 40: Circular SV40 DNA molecules containing lambda phage genes and the galactose operon of *Escherichia coli*. *Proc Natl Acad Sci USA* 69:2904–2909.

Kunkel TA (1985) Rapid and efficient site-specific mutagenesis without phenotypic selection. *Proc Natl Acad Sci USA* 82:488–492.

Kunkel TA, Roberts JD & Zakour RA (1989) Rapid and efficient site-specific mutagenesis without phenotypic selection. In Recombinant DNA Methodology (Wu R, Grossman L & Moldave K eds), pp 587–601. Academic Press.

Lander ES, Linton LM, Birren B et al. (2001) Initial sequencing and analysis of the human genome. *Nature* 409:860–921.

Palmiter RD, Brinster RL, Hammer RE et al. (1982) Dramatic growth of mice that develop from eggs microinjected with metallothionein-growth hormone fusion genes. *Nature* 300:611–615.

Pascal JM, O'Brien PJ, Tomkinson AE & Ellenberger T (2004) Human DNA ligase I completely encircles and partially unwinds nicked DNA. *Nature* 432:473–478.

Villa-Komaroff L, Efstratiadis A, Broome S et al. (1978) A bacterial clone synthesizing proinsulin. *Proc Natl Acad Sci USA* 75:3727–3731.

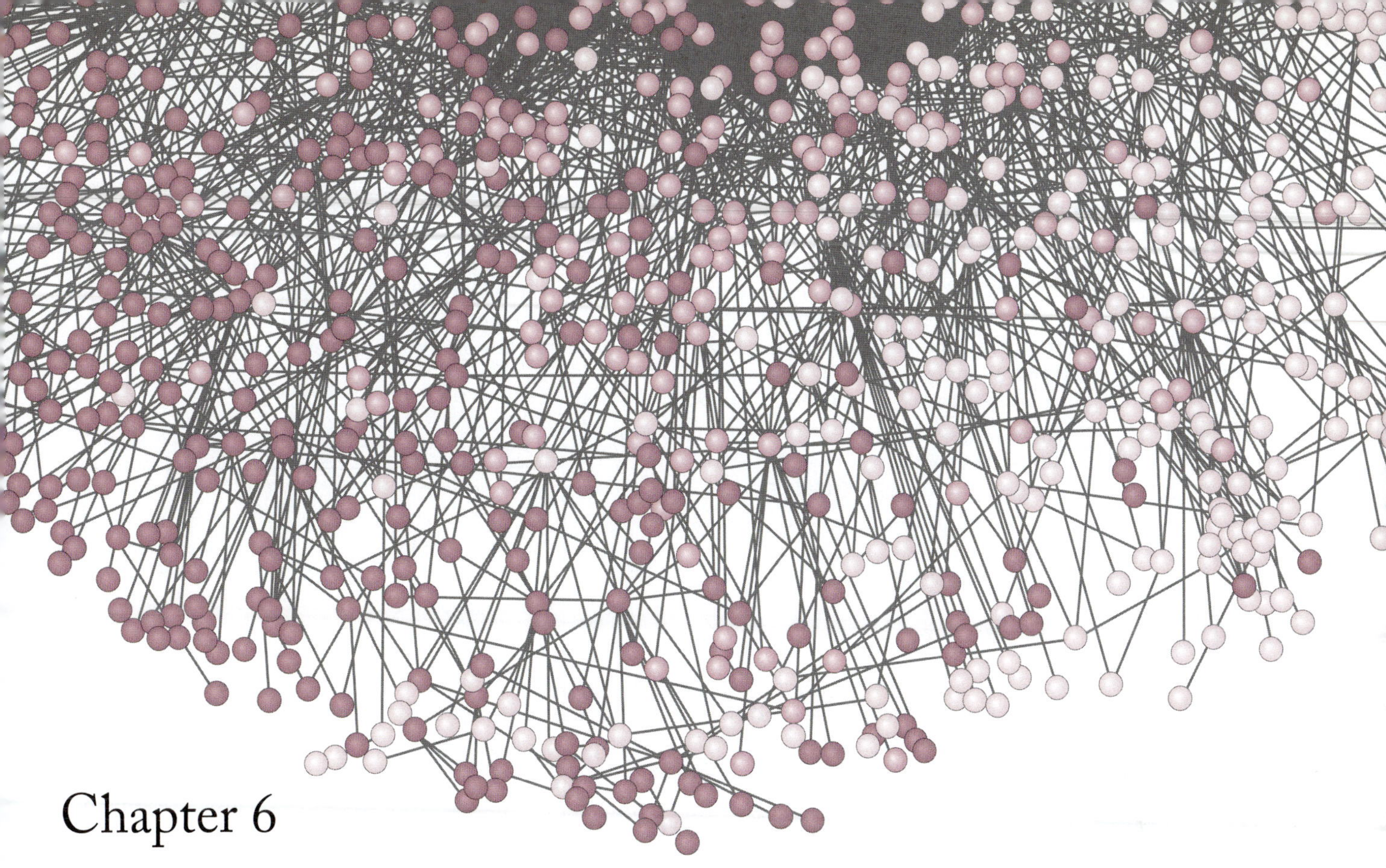

Chapter 6

Protein–Nucleic Acid Interactions

6.1 Introduction

In previous chapters, we have provided a structural, and to a small extent functional, overview of two major classes of biopolymers: proteins and nucleic acids. We have seen that proteins have a multitude of roles in cells and tissues; structural proteins and many enzymes usually act alone or with other protein partners. But virtually all processes involving nucleic acids also require the participation of proteins. These proteins are almost always bound to nucleic acids by noncovalent bonds, which might include hydrogen bonds, van der Waals interactions with bases, and electrostatic interactions between basic groups on proteins and nucleic acid phosphates. Protein–nucleic acid interactions can be specific, with the protein binding only to a particular nucleotide sequence, or nonspecific, in which case the protein can bind virtually anywhere on a particular kind of nucleic acid. Of course these definitions apply only to extremes: many specific proteins can also bind nonspecifically and then hunt along the DNA to find their specific sites. Many nonspecific binders also show weak sequence preferences. We find that there is a general, though not exact, distinction between the functions of these two classes of proteins. Nonspecific binders tend to play more structural roles—maintaining single-stranded DNA is an example—whereas specific binders often act as regulators of nucleic acid function. In this chapter, we look at modes and patterns of binding. We look at binding to DNA and to RNA separately, because both the mechanisms and the functions of protein binding to these two polynucleotides are different. We also briefly describe experimental methods for studying protein–nucleic acid binding. The oldest and simplest of these, which only allow detection of binding, are presented in Box 6.1 and Box 6.2.

Box 6.1 Filter binding This is the oldest method for detecting an interaction between a given protein and DNA (**Figure 1**). Although very simple, the method provides information on the strength of binding that cannot be accessed by many other methods. As with every other method, there are drawbacks too. A fundamental problem, especially when monitoring weak interactions, comes from the necessity to wash the filter after the sample is applied. If washing is insufficient, some DNA can be retained adventitiously, but too thorough washing may lose complexes because of dissociation. Thus the K_d values obtained must be regarded with some caution. The method is little employed today but played a major role in the early days of molecular biology.

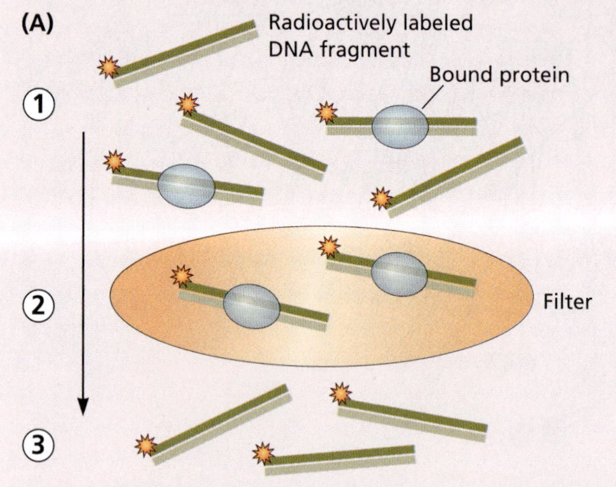

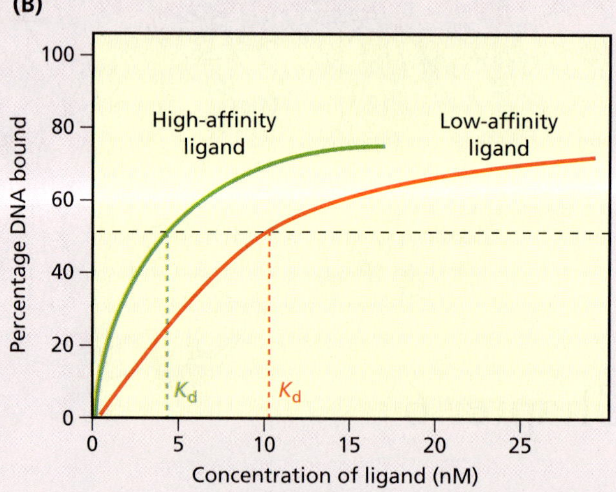

Figure 1 Filter binding assay. (A) The assay is performed in three steps: (Step 1) binding reactions, (Step 2) passing the incubation mixture through nitrocellulose paper filters, and (Step 3) detection of radioactively labeled DNA that is retained on the filter. The nitrocellulose filter is negatively charged and will not retain the negatively charged DNA unless it has been bound by protein; most proteins are positively charged. The exact amount of DNA retained on the filter is quantified by measuring the amount of radioactivity. (B) Titration curves, showing DNA fraction bound versus increasing amounts of protein present in the incubation mixture, are used to determine the affinity of protein binding.

6.2 DNA–protein interactions

DNA–protein binding occurs by many modes and mechanisms

Nonspecific binding of proteins to DNA can proceed very simply through electrostatic interaction between basic groups on the protein and the smooth, regular track of negatively charged phosphates that form the phosphodiester backbone (see Figure 4.5 and Figure 4.6). In fact, virtually any positively charged protein will tend to stick to DNA in solution, at least at low ionic strength. This is often also true for proteins that possess a recognition site for a particular nucleotide sequence: they can cling to the DNA elsewhere but more weakly. This is often a factor in facilitating the access to specific protein-binding sites. Sometimes nonspecific binding is used to coat the DNA, thereby protecting it against unwanted interactions, protecting it from degradation, or compacting it. Some proteins can do all of these things. An important example is found in **chromatin**, the protein–DNA complex in which DNA is compacted and sequestered in the eukaryotic nucleus. Much more is said about chromatin in Chapter 8; for now, suffice it to note that in chromatin, complexes of eight basic protein molecules form spools upon which nuclear DNA can be wound. To a large extent this binding is nonspecific, as must be expected for an interaction that involves most of the genome. Nevertheless some elements of nucleotide sequence preference or avoidance for the formation of such structure can be found.

There is a class of nonspecific DNA-binding proteins that preferentially interact with and stabilize single-stranded DNA (ssDNA). These **single-strand binding proteins, SSB**, or **helix-destabilizing proteins** as they are sometimes called, are involved in processes such as genetic recombination, DNA replication, and DNA repair, where

Box 6.2 DNA-affinity chromatography Another simple technique used to search for proteins that bind strongly to a particular nucleic acid sequence is to attach the polynucleotide fragments to silica beads in a chromatographic column and pour a mixture of suspected proteins through (**Figure 1**). Strongly binding proteins will be retained on the column even when dilute salt solutions are run through the column, which will elute weak binders. The investigator can then take advantage of the fact that even strongly binding proteins will dissociate from the nucleic acids under high salt conditions, allowing their isolation.

Neither filter binding nor DNA-affinity chromatography can tell us the exact sequence of the binding site, although some information can be obtained by using a variety of mutants in the presumed binding region. This is laborious, and there are now easier ways to determine binding sequences.

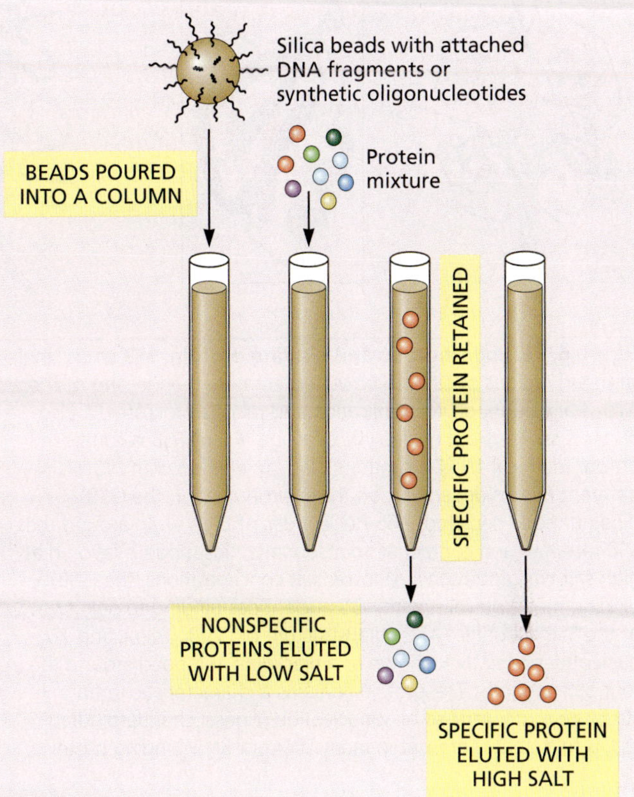

Figure 1 DNA-affinity chromatography allows purification of DNA-binding proteins. DNA fragments or synthetic oligonucleotides containing the protein-binding site are immobilized onto silica beads, which are then packed into a chromatography column. A mixture of proteins, usually a nuclear extract, is passed through the column: the binding protein is retained on the column, whereas proteins that do not bind or bind only weakly either pass through directly, without being retained, or can be eluted with low-salt concentration buffers. The specific binding protein can subsequently be recovered by eluting with a high-salt concentration buffer. More recent techniques use magnetic beads, which do not need to be packed into columns but can be easily manipulated by use of magnets external to the tube.

it is essential to maintain a region of denatured DNA for some time. SSB proteins are ubiquitous and can function as monomers, as seen in viruses; as homotetramers, as seen in bacteria; or as heterotrimers, such as replication protein A in eukaryotes. It was long believed that such proteins might bind to a transient opening in the helix and then bind cooperatively in a side-by-side fashion to expand the opening. However, recent work in the von Hippel laboratory indicates that this cannot be a universal mode; binding sites for the SSB are often so large that spontaneous openings of sufficient size would be very rare. This kinetic block means that a **helicase** molecule is needed to first unwind a suitable length of DNA. The SSBs can then bind and stabilize the single strand. A well-studied example, the SSB from *Escherichia coli*, is illustrated in **Figure 6.1**. *In vitro*, the protein binds to ssDNA either as a dimer or as a tetramer, depending on the conditions. In the cell, SSB binds as a tetramer, wrapping ~70 nucleotides of ssDNA around the protein subunits. This wrapping leads to an overall reduction in the contour length of the DNA, as seen in the electron microscope images presented in Figure 6.1C.

Site-specific binding is the most widely used mode

The binding of proteins to specific sites on DNA is fundamental to a vast range of functions and structures. It has been estimated that the human proteome contains several thousand site-specific DNA-binding proteins, each with a specific base sequence that it recognizes and binds to. Some binding proteins recognize only a single site in the whole genome, whereas others bind at multiple locations. The most intriguing question is, how do protein molecules distinguish particular DNA sequences? Even before we had specific knowledge, it was clear that the B-form double helix of DNA contained information that could be read by other molecules. The surface of the DNA duplex is defined by two deep grooves, known as the major and minor grooves (see Figure 4.6). The edges of the bases are exposed in each of these grooves in a way that presents a unique combination of chemical groups, which can interact with a protein that

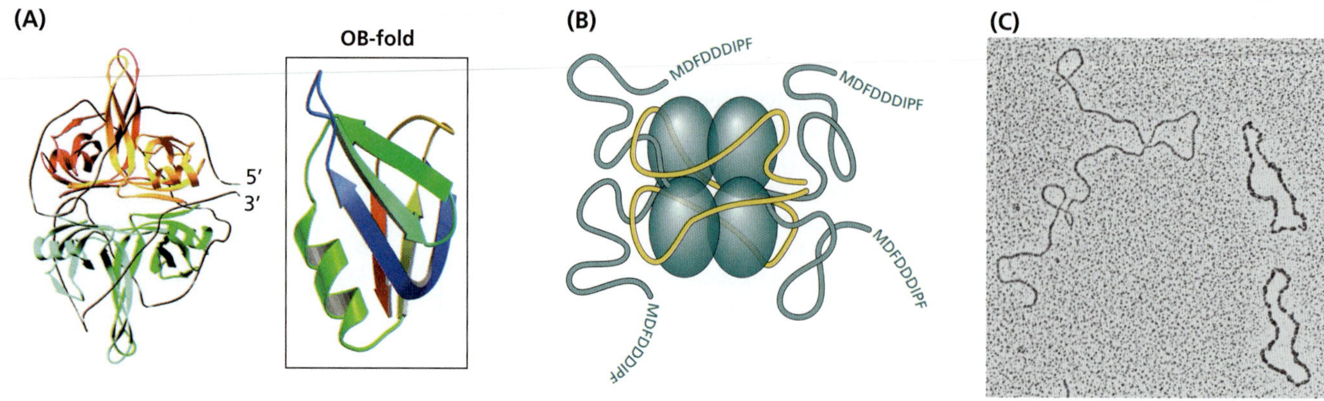

Figure 6.1 ***E. coli*** **single-strand binding protein.** SSB binds single-stranded regions of DNA to prevent premature reannealing of the two strands during DNA replication and related processes. It also protects ssDNA from nucleolytic cleavage. *E. coli* SSB is composed of four identical 19 kDa subunits that can interact with ssDNA in different modes depending on the environment. In the $(SSB)_{65}$ binding mode depicted, ~65 nucleotides of DNA wrap around the SSB tetramer and contact all four subunits; this mode is favored at high salt concentrations. At lower salt concentrations, the $(SSB)_{35}$ binding mode prevails, with ~35 nucleotides binding to only two of the SSB subunits. (A) Crystal structure of $(SSB)_{65}$ depicting 70 nucleotides of ssDNA, shown as a black line, wrapped around the four SSB subunits. The protein contains a characteristic tertiary structure motif, termed oligonucleotide/oligosaccharide-binding or OB-fold, that is present in many DNA- or RNA-binding proteins; its general topology is shown in the box. (B) Cartoon representing ssDNA as a yellow ribbon wrapped around the SSB core, corresponding to the structural model in part A, with the addition of the unstructured C-terminal tails, shown as gray lines, that are not observed in the crystal structure. The nine-amino-acid sequence, shown in single-letter amino acid code, at each C-terminus is responsible for the interaction of SSB with other metabolic proteins. (C) Electron microscopic image of naked DNA (left) and SSB-bound DNA (right). The reduction in contour length is due to the wrapping of the DNA around SSB. (A-B, adapted from Kozlov AG, Jezewska MJ, Bujalowski W & Lohman TM [2010] *Biochemistry* 49:3555–3566. With permission from American Chemical Society. A, inset, from Agrawal V & Kishan RKV [2001] *BMC Struct Biol* 1:5. With permission from Springer Science and Business Media. C, courtesy of Maria Schnos, University of Wisconsin—Madison.)

inserts into the groove. The combination includes, for example, both hydrogen-bond donors and acceptors, and these are presented in a different pattern for a GC base pair than for an AT base pair (**Figure 6.2**). Thus a protein, with its own hydrogen-bond acceptors and donors, can detect the difference between these two kinds of pairs. In addition to this, there are other kinds of interactions, such as van der Waals interactions between methyl groups on the DNA and nonpolar groups in the protein, that can convey the identity of a base pair.

It should be emphasized that many site-specific binders also exhibit a much weaker nonspecific binding mode, involving electrostatic interactions between positive amino acids on proteins and the featureless track of negative charges on the DNA surface. This can help a protein find its specific site, either by sliding along the DNA helix

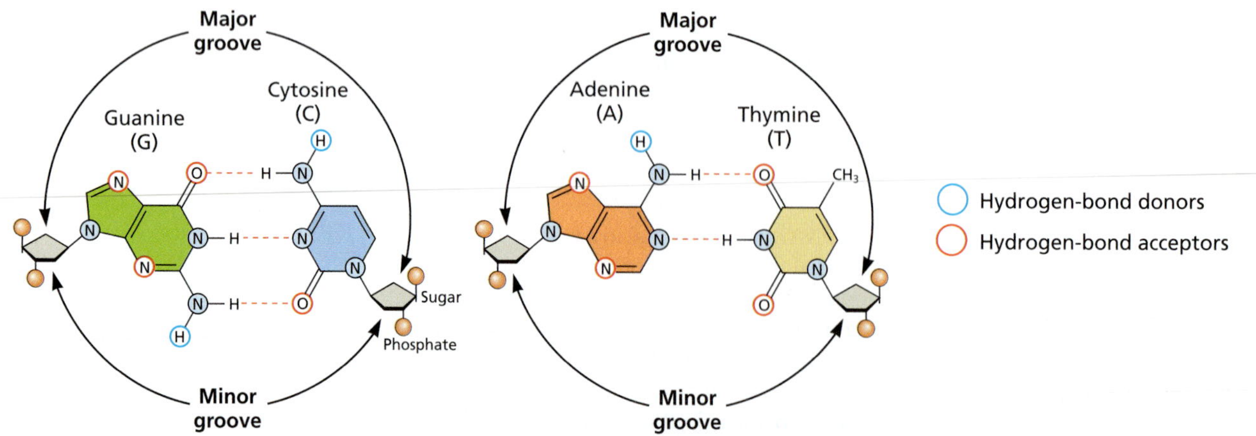

Figure 6.2 Proteins recognize specific sequences on the DNA by predominantly interacting with the major groove. Looking down the DNA helix, one can see the edges of the bases protruding into the major groove. Accessibility of these bases depends primarily on the width of the groove. Clearly, the minor groove is less accessible. A stack of several base pairs along the chain will create a unique constellation of chemical groups into the groove, which is recognized by DNA sequence-specific binding proteins.

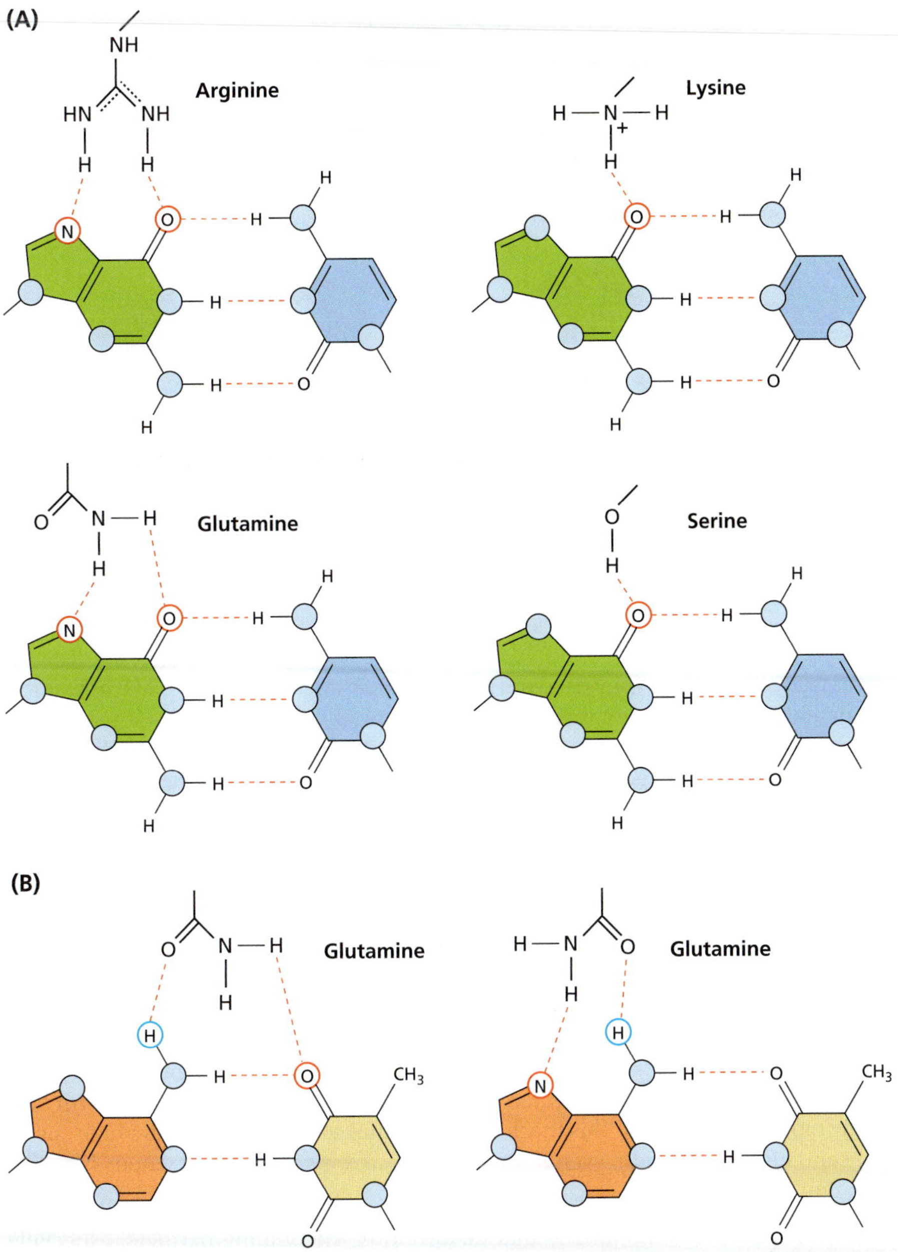

Figure 6.3 Possible patterns of hydrogen-bonding during recognition of base pairs by amino acid side chains. (A) Recognition of GC base pairs by arginine, lysine, glutamine, and serine side chains. (B) Recognition of AT base pairs by glutamine side chains. Note the two possible patterns of hydrogen-bonding.

or by hopping from one region to another across DNA loops. Such one-dimensional hunting can be more efficient than wandering through three-dimensional space.

Most recognition sites fall into a limited number of classes

There are many possible specific amino acid–nucleotide interactions that could promote protein recognition of a binding site (**Figure 6.3**). Analysis of many X-ray diffraction studies of protein–DNA complexes reveals no simple set of rules for such pairing. There is some preference for arginine to pair with G in GC pairs and for glutamine or asparagine to pair with AT pairs (Figure 6.3). Evolution appears to have adopted whatever works, often complicated by distortion of the DNA and/or protein to provide the best fit.

The number of base pairs involved in different DNA recognition sites varies widely, depending on the function of the protein. Some restriction endonucleases, for example, have relatively short recognition sites, 4–6 base pairs or bp, so that they can cleave genomic DNA in many places. Some transcription activator or repressor proteins, which are targeted against only one or a few sites in the genome, must have

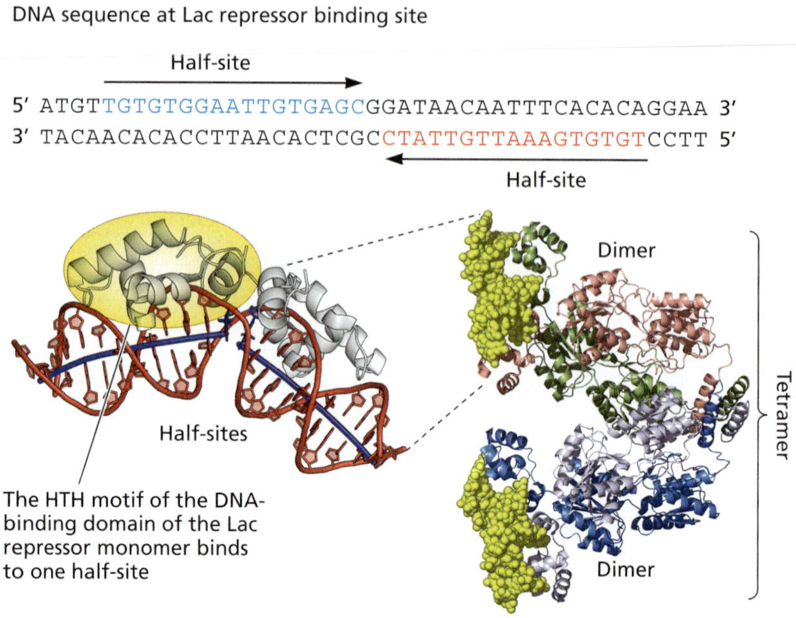

Figure 6.4 Binding of Lac repressor to DNA. The repressor binds to DNA via helix–turn–helix (HTH) motifs in the DNA-binding domains of two monomers that form a dimer. Note that the two half-sites in the DNA recognition sequence form a palindrome, which matches the symmetry of the dimer. The repressor binds to DNA as a tetramer, or dimer of dimers, formed by interactions of the C-termini; each dimer binds two half-sites. The monomers in the top dimer are presented in green and pink; the monomers in the bottom dimer are in violet. The Lac repressor kinks the DNA at a central CpG base pair step. The helix axes on both sides of the kink, shown in blue, show an abrupt change in the helix trajectory caused by the kink. (Right, from Wilson CJ, Zhan H, Swint-Kruse L & Matthews KS [2007] *Cell Mol Life Sci* 64:3–16. With permission from Springer Science and Business Media. Left, from Rohs R, Jin X, West SM et al. [2010] *Annu Rev Biochem* 79:233–269. With permission from Annual Reviews.)

larger recognition sites. They need not be excessively large in order to make unwanted accidental false recognition unlikely. Since the number of possible combinations of bases in a site of size n is 4^n, sites 10 bp long will occur only about every million base pairs, giving an expectancy of several thousand such sites in the whole human genome of 3.2 billion bp; some of these sites may be obscured by other proteins bound to or around them. When $n = 20$, an astronomical 10^{12} different sequences are possible; thus a particular, specific 20 bp site would be unlikely to be found anywhere within the genome.

Some recognition sites, such as those of the Lac repressor binding protein (**Figure 6.4**), are palindromic and can therefore accept a dimer of the protein, increasing both the specificity and binding affinity. As Figure 6.4 shows, Lac repressor actually takes this one step further by recognizing two adjacent palindromic pairs and binding as a tetramer. Indeed, as we shall see in further examples, site-specific binding proteins frequently exhibit tandemly repeated binding domains, which then interact with tandemly repeated DNA sites. This mechanism can make it possible to attain high specificity and binding strength, even if there are only a few specific interactions between each domain and the DNA.

Finally, it should be emphasized that distortion of the binding partners, especially bending of the DNA, is a common feature of site-specific binding. Because DNA bending exacts a free energy price, this must be returned by a favorable free energy gain from the interaction process itself. In other words, the binding must be strong to compensate for the bending.

Most specific binding requires the insertion of protein into a DNA groove

DNA binding can gain stability from protein interaction with the uniform phosphate backbone of DNA. The recognition of specific sites along the DNA chain, however, demands the insertion of a portion of the protein into one of the grooves in the DNA helix, where the specific base pairs can be sensed. Most DNA-binding proteins use the major groove. This is understandable because the major groove provides more recognition sites and in B-DNA it is significantly wider than the minor groove. The major groove is, in fact, wide enough to accommodate snugly a segment of α-helix, and this is a very common mode of binding. Although less common, there are examples in which two-stranded β-sheets, or the edges of even larger β-sheets, are inserted into the major groove. The good fit of α-helix or β-sheet into the major groove means that these protein domains can bind to recognition sites with little distortion of the DNA.

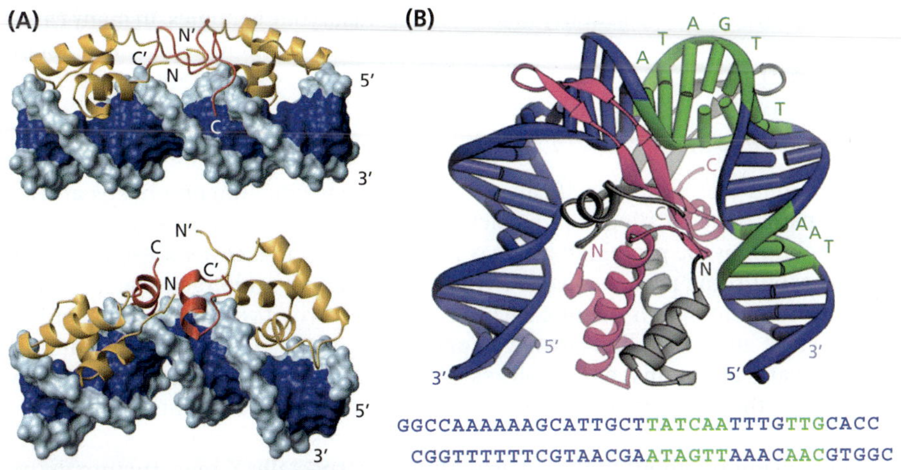

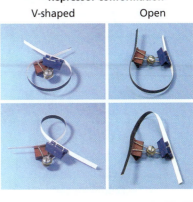

Figure 6.5 Minor-groove binders cause DNA bending. (A) DNA-binding domain of Lac repressor bound to nonspecific (top) and specific (bottom) nucleotide sequences. The hinge region, shown in red, is unstructured in both the free protein and the protein bound to nonspecific DNA; it folds into an α-helix that interacts with the minor groove in the specific complex. In the nonspecific complex the DNA adopts the canonical B-form conformation, whereas in the specific complex it is bent by ~36°. (B) Crystal structure of the integration host factor (IHF) complex with DNA: α subunit is shown in cyan, β subunit in purple, consensus sequence DNA in green, and less conserved DNA in blue. IHF is a small, 20 kDa heterodimeric protein that binds DNA in a sequence-specific manner and induces a large bend of >160°. This bending aids in the formation of higher-order structures in such processes as recombination, transposition, replication, and transcription. The protein contacts only the phosphodiester backbone and some bases in the minor groove; thus, it represents an example of indirect readout, where the protein relies on sequence-dependent structural features of the DNA, such as backbone conformation and flexibility. This sequence recognition is in contrast with direct readout, where sequences are distinguished through the unique functional groups of DNA bases in the major groove. (A, from Kalodimos CG, Biris N, Bonvin AMJJ et al. [2004] *Science* 305:386–389. With permission from American Association for the Advancement of Science. B, from Lynch TW, Read EK, Mattis AN et al. [2003] *J Mol Biol* 330:493–502. With permission from Elsevier.)

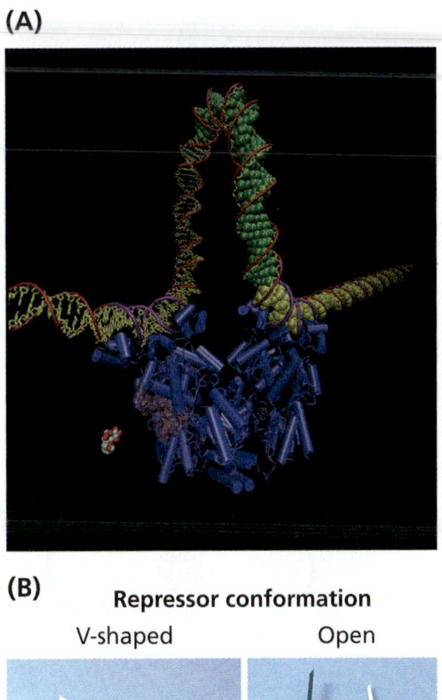

Figure 6.6 DNA looping. (A) Three-dimensional view of a loop formed by binding of a tetramer of the Lac repressor from *E. coli* to the two operators that control the expression of the *lac* operon (see Chapter 11). The tetramer is in dark blue; the DNA is represented by a combination of stick and space-filling models, with the backbone in red and purple. (B) Physical models corresponding to possible loop geometries. Photographs show loop models with the V-shaped or crystallographic repressor conformation (left column) and the corresponding configurations that result when the repressor is opened (right column). In each pair, the structure on the right was produced from that on the left by rotating the two half-tetramers, represented by a blue or a red clip, away from each other about the axis of the four-helix bundle, represented by a silver bolt. The paper strip representing the DNA is colored black on one side and white on the other to make any twist in the helix visible. (A, courtesy of Elizabeth Villa, University of Illinois at Urbana–Champaign Theoretical and Computational Biophysics Group. B, from Wong OK, Guthold M, Erie DA & Gelles J [2008] *PLoS Biol* 6:e232 [10.1371].)

Binding into the minor groove is a different matter. In the first place, the minor groove of B-DNA is occupied by a spine of precisely positioned water molecules. In consequence, binding to the minor groove involves the release of this structured water, with a large entropy increase. This appears to be a major free energy source for minor-groove binding. Second, the minor groove is just too narrow to accommodate an element such as an α-helix, so if such binding is to occur, the groove must be stretched open. There are a number of ways in which DNA can accommodate such stretching, but the most common is by compression of the adjacent major groove. This will have the consequence of bending the DNA helix at the site of protein binding (**Figure 6.5**). Thus, it is not surprising that minor-groove binders are almost synonymous with DNA benders. The energetic cost of such bending appears to be largely paid by the above-mentioned entropy increase resulting from the release of water. However, there may be an interesting synergy here, in that the water spine contributes to the rigidity of the double helix; so displacing it should make DNA more flexible.

It should be emphasized that major-groove binding can also give rise to bending. This can happen when a dimeric or oligomeric protein binds to adjacent sites in the DNA. There exist a number of ingenious ways to study DNA bending and relate it to protein binding.

Some proteins cause DNA looping

We have noted that many DNA-binding proteins exist as non-covalently-bound dimers or even larger oligomers. If such proteins have recognition sites on DNA that are some distance apart, they have the capability to produce DNA loops (**Figure 6.6**).

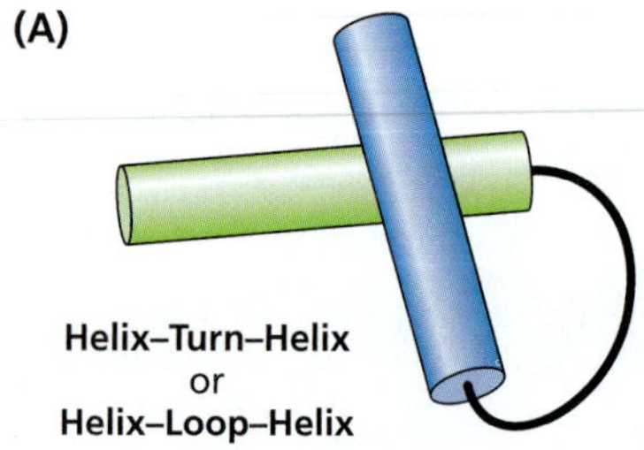

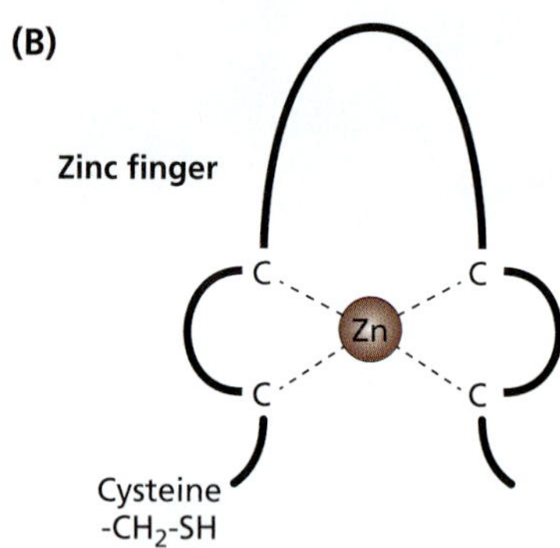

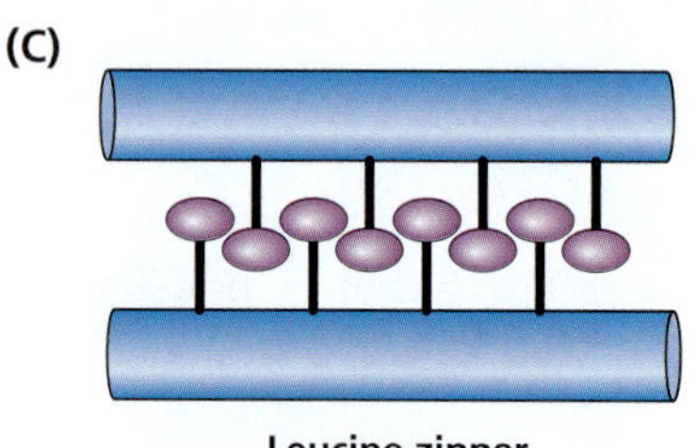

Figure 6.7 DNA-binding motifs. (A) Helix–turn–helix, HTH; (B) zinc finger; (C) leucine zipper.

DNA looping is observed in both prokaryotic and eukaryotic genomes. In many cases, it appears to isolate particular regions of the genome so as to regulate transcription therein (see Chapter 12).

There are a few major protein motifs of DNA-binding domains

Every protein that binds specifically to DNA for a specific function needs at least two domains. One is called the transactivating domain. It either senses an external signal to prompt the protein to bind, as in a transcription factor, for example, or it will carry out some process on the DNA, as a restriction endonuclease, for example. The other domain is the one we have been concerned with so far: the binding domain that attaches to the DNA at a specific site or sites. There may be multiple transactivating domains, or complexes of several proteins that are noncovalently attached to the binding domain. In many cases, the transactivating domains can cause conformational changes in the binding domains. This may suggest that there should be caution about the many experiments, especially X-ray structure determinations, in which only the purified or cloned DNA-binding domain, bound to DNA, is studied. This structure might not reflect the structure of the entire bound protein.

Surprisingly, a large fraction of binding domains appears to use only a small vocabulary of protein motifs, known as **recognition motifs**, for the actual binding, even if the binding serves very different functions. This may suggest that the DNA-binding domains of proteins evolved from only a few examples, ages ago. We describe here just three of the most frequently encountered motifs; they are depicted schematically in Figure 6.7.

Helix–turn–helix motif interacts with the major groove

Helix-turn-helix (**HTH**) motifs are common in both prokaryotic and eukaryotic transcription factors. The motif consists of a stretch of ~20 amino acid residues, divided into two α-helices, each about 7–8 residues long, separated by a turn or loop. The second of these two helices, from the N-terminus, is the recognition helix, which lies in the major groove of the DNA. A specific example, CRP activator, is shown in Figure 6.8. Here there are three α-helices. As the α-helix has 3.6 residues per turn, only two turns are being presented to the DNA, permitting the opportunity for only a few specific contacts. In many cases, HTH proteins bind as dimers or even tetramers, increasing the number of contacts and hence selectivity (see Figure 6.8). As mentioned above, the binding of more than one monomer can have another consequence: bending of the DNA. As Figure 6.8 illustrates, the geometry of

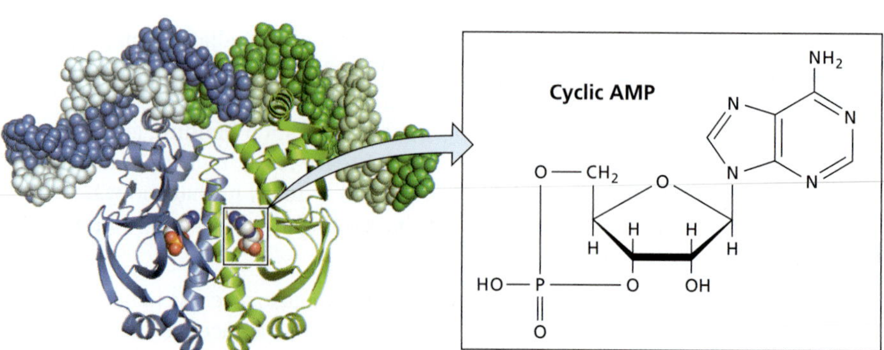

Figure 6.8 Helix–turn–helix protein bound to DNA. CRP activator is a cyclic AMP regulatory protein, also known as CAP or catabolite activator protein, that positively controls numerous operons in *E. coli*. Its activity is regulated by cyclic AMP binding. The protein binds as a dimer to a bipartite binding site, shown in green and blue, causing the DNA to bend ~90°. Each monomer has two structural domains: the N-terminal domain, amino acids 1–140, contains the cyclic AMP nucleotide binding site, whereas the C-terminal 50–60 amino acid domain contains a helix–turn–helix motif that interacts with the DNA.

protein–protein interactions can make it necessary for the DNA to bend in order to interact with all of the HTH motifs.

Zinc fingers also probe the major groove

Another motif frequently observed in DNA-binding domains is called the **zinc finger** because of the essential role of zinc in its structure. As shown in **Figure 6.9**, the most common variant has a short length of α-helix bound to a short β-sheet via the zinc atom. Usually, the zinc is coordinated to two histidine side chains on the α-helix and two sulfhydryl groups of cysteine residues on the β-sheet, but sometimes four sulfhydryls are used. In any event, the compact finger that is produced can be inserted into the major groove without distortion. The number of possible interactions is small, so specificity is often gained by having multiple fingers. These may be all contained in the sequence of one polypeptide chain, as in the TFIIIA transcription factor shown in **Figure 6.10**, or they may be contained in interacting proteins, as in the steroid receptor proteins in **Figure 6.11**.

Leucine zippers are especially suited for dimeric sites

Leucine zippers are protein–protein interaction motifs that help the stable formation of protein dimers to increase the specificity of interaction with DNA. Leucine zipper proteins always interact with DNA as homo- or heterodimers, held together by hydrophobic interactions of long α-helices (**Figure 6.12A**). These interacting protein tails each have leucine or isoleucine residues spaced about 3–4 residues apart. This means that all of these residues will lie on one side of the α-helix, creating a hydrophobic face (**Figure 6.12B**). This face can be buried away from solvent by having the helices make a gentle coiled coil about one another. The term zipper refers to the fact that isoleucine residues on the two α-helices often interdigitate like teeth of a zipper. At the end of this coil lie the recognition elements themselves, usually α-helix segments that insert into the major groove (**Figure 6.12C**). Zippers can be homodimers, which interact with a repeated DNA sequence, or heterodimers, which carry different DNA-recognition elements and thus interact with a pair of different DNA sites. This allows a sophisticated kind of control, in which two different protein factors must be present and interact to allow DNA binding.

The DNA-binding elements of proteins described in this section do not represent anything like the totality of modes of protein–DNA interactions. There are many variants on the motifs described above, and there are many proteins that seem to bind in unique ways. Important examples of protein–DNA complexes feature throughout this book.

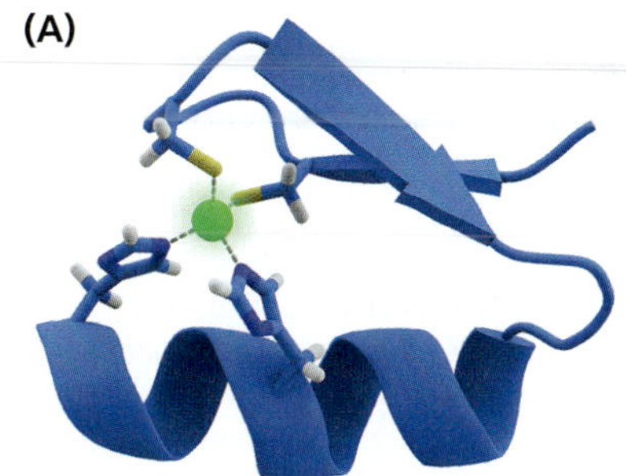

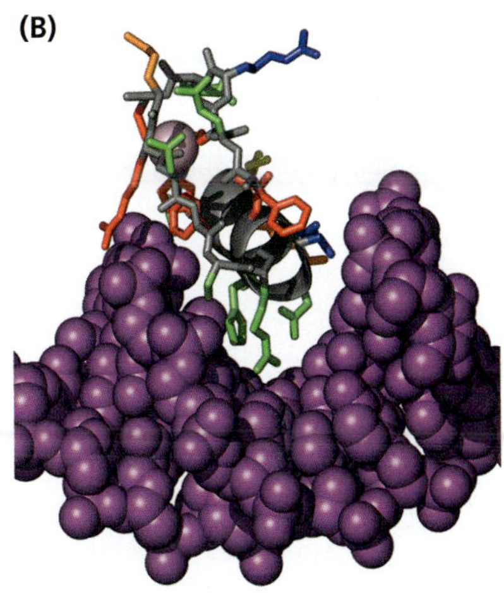

Figure 6.9 Some proteins bind DNA with multiple zinc fingers. (A) One of the three zinc fingers of transcription factor Zif268. In this case, the Zn atom is coordinated between two cysteine and two histidine residues. (B) The middle zinc finger of Zif268 bound to DNA, shown in purple, in the major groove, with the Zn(II) atom shown as a sphere. (A, courtesy of Thomas Splettstoesser, Wikimedia. B, from Magliery TJ & Regan L [2005] *BMC Bioinf* 6 [10.1186/1471-2105-6-240]. With permission from BioMed Central.)

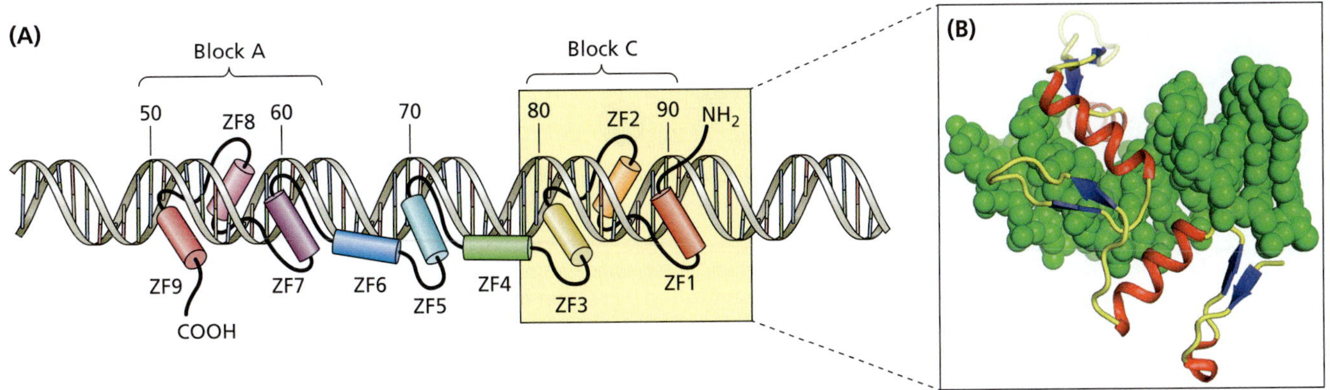

Figure 6.10 Arrangement of proteins containing multiple zinc fingers along the DNA. (A) Transcription factor TFIIIA binds to 5S rDNA, the gene that encodes the 5S RNA component of the ribosome, via multiple zinc fingers (ZF) that insert themselves into the major groove. The two major recognition regions, blocks A and C, are contacted by fingers 7–9 and 1–3, respectively. Fingers 4–6 are used when the protein interacts with RNA. (B) Crystal structure of fingers 1–3 bound to DNA. (A, adapted from Dyson HJ [2012] *Mol BioSyst* 8:97–104. With permission from Royal Society of Chemistry.)

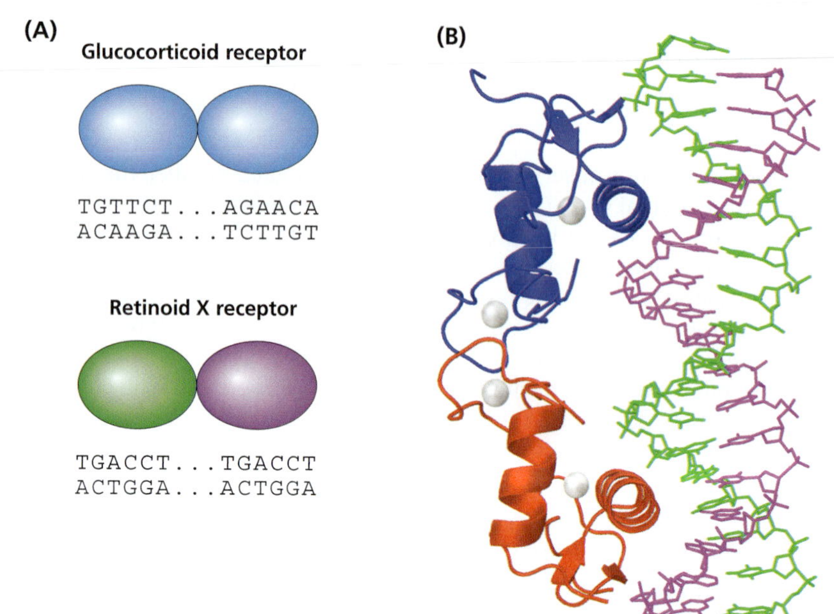

Figure 6.11 Zinc fingers in steroid receptors. (A) The sequences in DNA that bind steroid hormone receptors, called steroid response elements, consist of two short half-sites that may be palindromic or direct repeats. The receptor proteins bind palindromic sequences as head-to-head homodimers; receptor protein heterodimers bind direct repeats. The schematic shows examples of each binding mode. (B) Crystal structure of the estrogen receptor bound to DNA. The monomer receptors have two Zinc fingers each. The binding of the first finger determines the sequence-specific binding, while the second finger is responsible for dimerization; that is, Zinc fingers can also be protein–protein interaction domains.

6.3 RNA–protein interactions

Cellular RNA molecules interact with specific proteins, but in ways that are somewhat different than DNA-protein interactions. In the first place, there is a general difference in function between RNA-protein and DNA-protein complexes. The latter, as we have seen, often have regulatory functions, whereas RNA-protein complexes more often carry out catalytic processes. In some cases the protein moiety is the catalyst, as in RNA-processing complexes, whereas in other examples, the proteins appear to

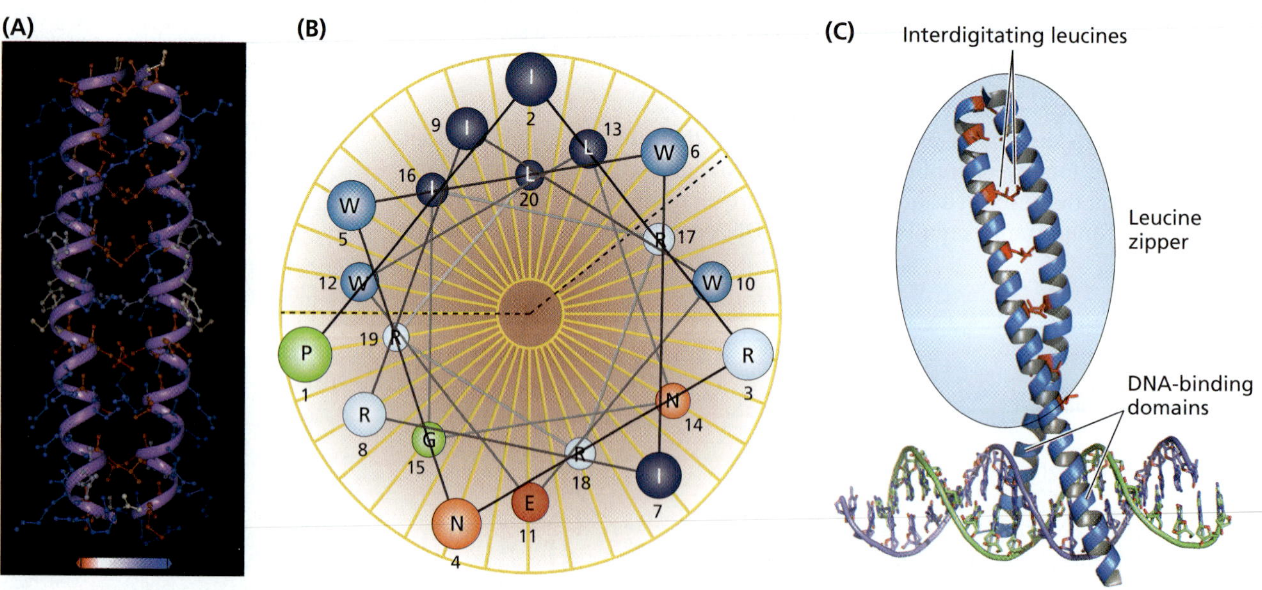

Figure 6.12 Leucine zippers. Leucine zippers are protein–protein interaction motifs that help the stable formation of protein dimers to increase the specificity of interaction with DNA. (A) Leucine zippers form via side-by-side interactions of amphipathic helices. These are characterized by the presence of hydrophobic, nonpolar residues on one side of the helical cylinder and hydrophilic and polar residues on the other side, resulting in different properties of the two sides. The colors of the side chains represent their hydrophobicity ranging from hydrophilic, shown in blue, to hydrophobic, shown in red. The leucine residues on one side of the helix interdigitate with the leucine residues on another helix from another protein molecule. (B) Amphipathic helices can be predicted by helical wheel analysis, in which a stretch of amino acid sequence is imaginarily arranged in a wheel that mimics the arrangement of amino acids in an α-helix. The example given is that of 20 N-terminal amino acid residues of chicken cathelicidin, chCATH-B1. Note that, in addition to leucines, the hydrophobic top side of the helix also contains isoleucine, another hydrophobic amino acid residue. (C) Structure of a leucine zipper protein heterodimer bound to DNA. The DNA-binding domains are usually basic. (A, courtesy of David E. Volk, University of Texas Health Science Center. B, adapted from Goitsuka R, Chen CH, Benyon L et al. [2007] *Proc Natl Acad Sci USA* 104:15063–15068. With permission from National Academy of Sciences.)

primarily function in determining the three-dimensional structure of a complex in which the RNA is the catalytic agent, known as a ribozyme (see Box 14.1). Another factor that makes RNA–protein interactions distinct from typical DNA–protein binding is the fact that cellular RNA is synthesized as single-stranded molecules. These will often, however, contain regions of self-complementarity that allow the formation of double-stranded structures, such as hairpins, interspersed with single-stranded loops. Such structures can be very complex (see Figure 15.17). Thus, RNA-binding proteins may have to recognize a variety of folded structures, not just sites in the grooves as in DNA. In addition, the numerous base modifications found in RNA provide additional opportunities for recognition. Probably because of this great variety of potential binding modes, we do not find RNA–protein recognition involving such a limited vocabulary as DNA–protein recognition.

The example, par excellence, of protein–RNA interaction is the ribosome, the particle that is the machine for protein synthesis in all cells (see Chapter 15). The bacterial ribosome, pictured in Figure 15.19 and Figure 15.20, contains three RNA molecules with a total length of 4566 bases, plus 55 different proteins. The structure of this immense, complex particle has now been solved by X-ray diffraction, for which the Nobel Prize in Chemistry was awarded in 2009 (see Box 15.5). Most of the proteins appear to play structural roles, complexly intertwined with the RNA, but some also have catalytic functions. The ribosome is not the only large protein–RNA complex, however. Much of the processing of messenger RNA that occurs in the eukaryotic cell nucleus involves such particles (see Chapter 14).

There does not seem to be, or perhaps we have just not yet recognized, a simple group of common protein motifs that are used in recognition of specific sites on RNA molecules. Rather, there are a variety of binding motifs, perhaps because there is little uniformity in RNA structure (**Figure 6.13**). One principle that seems to dominate RNA–protein binding is the use of a multiplicity of sites, in a modular fashion. In such a case,

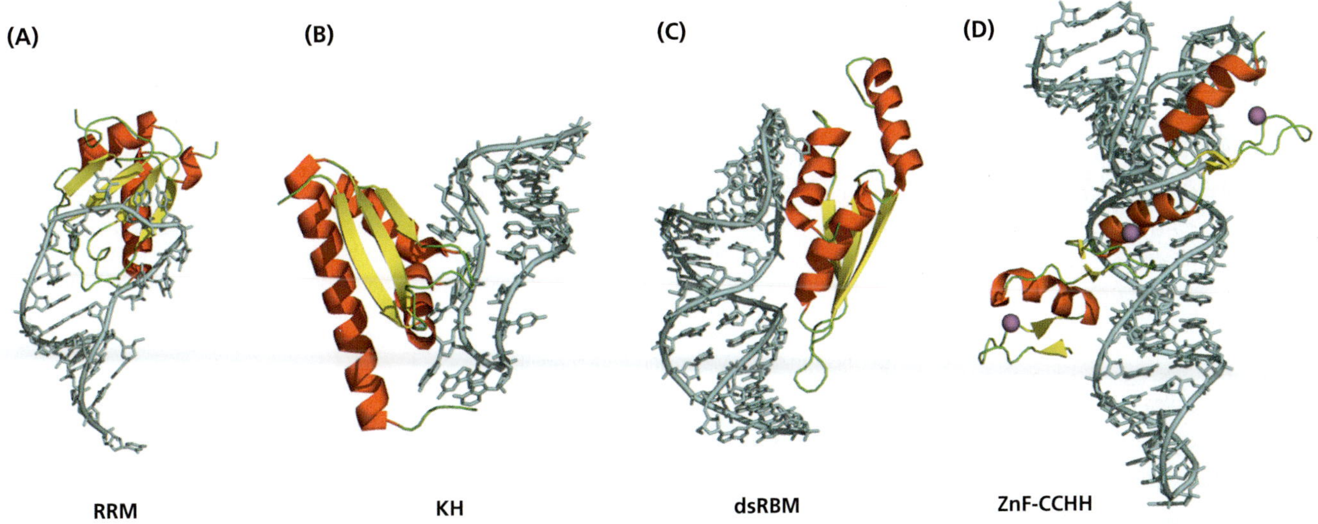

Figure 6.13 Crystal or solution structures of some common RNA-binding domains bound to RNA. (A–C) αβ domains; (D) zinc-finger domain (ZnF) domain of the CCHH motif type. (A) N-terminal RNA-recognition motif (RRM) of human U1A spliceosomal protein complexed with an RNA hairpin, in jade. Single-stranded bases in the RNA are recognized by the protein β-sheet, in yellow, and by two loops, shown in green, that connect the secondary structure elements. (B) KH (K-homology) domain of Nova bound to 5′-AUCAC-3′. KH domains bind to specific sequences, in both ssDNA and RNA, through a conserved GXXG sequence located in an exposed loop, shown in green. (C) Double-stranded RNA-binding domain (dsRBM) of Rnt1p RNase III bound to an RNA helix capped by an AGNN tetraloop. The protein loop, shown in green, interacts with the 2′-OH groups in the RNA minor groove, whereas Lys and Arg residues in the longer helix recognize the position of the phosphate atoms in the A-form helix. *In vitro*, most dsRBMs bind to dsRNA of any sequence provided it stretches for at least 12–13 base pairs and is not interrupted by too many bulges or internal loops; *in vivo*, however, proteins containing dsRBMs bind to and act on specific RNAs, and some domains even bind to other proteins. It is not clear how this non-sequence-specific domain is used to bind specific RNAs in the cell. (D) RNA-binding zinc-finger proteins: complex of three-finger peptide, fingers 4–6, from TIFIIIA and truncated 5S RNA. Recognition is primarily through residues in α-helices; in this specific case, the sequence-specific recognition is achieved through exposed bases in the loop regions of RNA. (From Chen Y & Varani G [2005] *FEBS J* 272:2088–2097. With permission from John Wiley & Sons, Inc.)

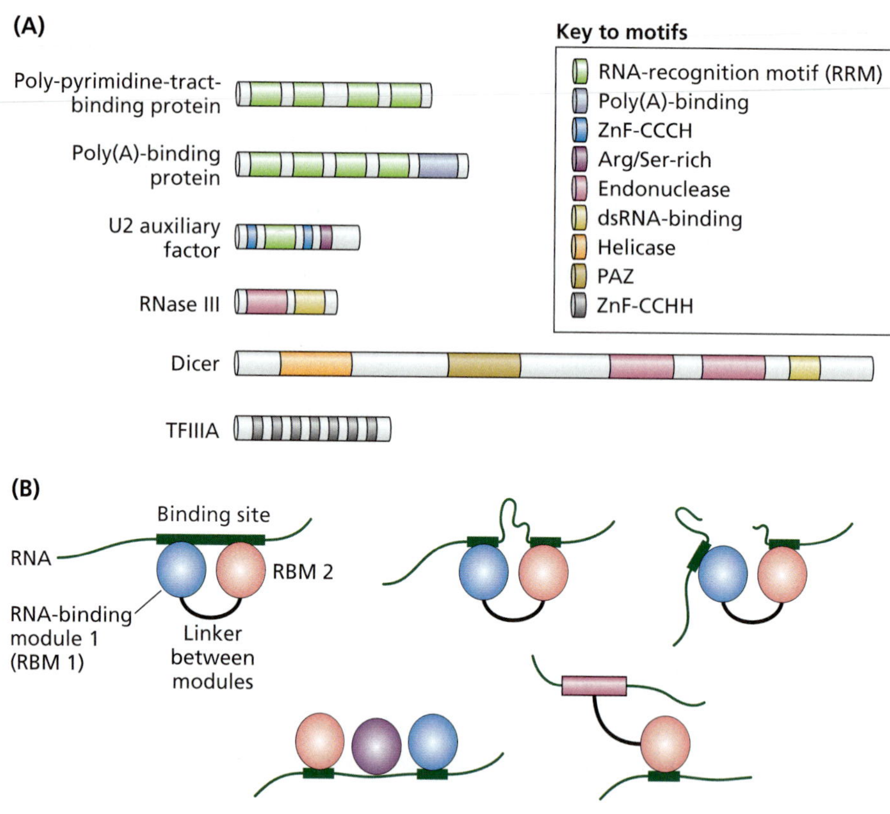

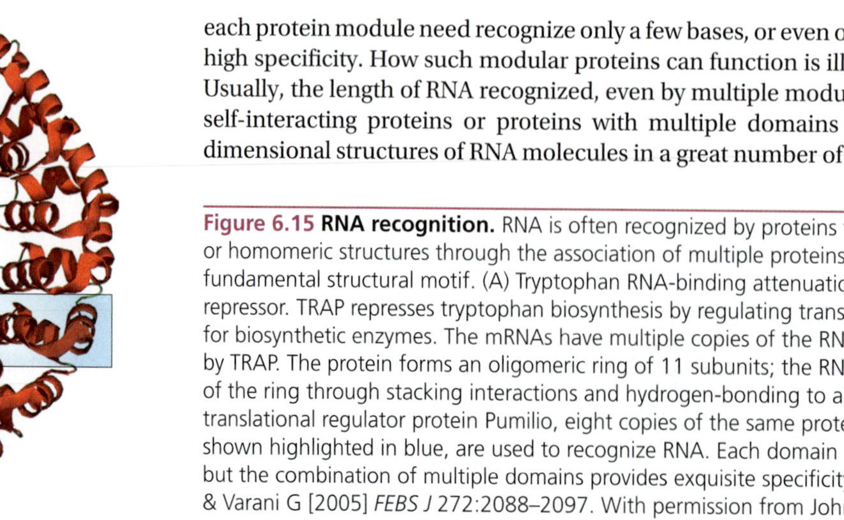

Figure 6.14 **Modular structures of RNA-binding proteins and their modes of action.** (A) Schematic representation of some RNA-binding protein families illustrating the arrangements of individual RNA-binding motifs along the polypeptide chain. Often, the proteins contain additional domains of varied functions. For example, Dicer and RNase III both contain an endonuclease catalytic domain followed by a double-stranded RNA-binding domain (dsRBM), so both proteins recognize dsRNA. In addition, Dicer has to specifically interact with stem–loop structures in microRNA precursors; it does this through additional domains that recognize the unique structural features of these RNAs. (B) The modular organization of RNA-binding proteins allows them to perform a variety of functional roles. The relatively low specificity and affinity of binding that is characteristic of individual domains is augmented significantly by the simultaneous binding of two or more domains. (Top) Multiple domains may combine to bind to long recognition sequences, or to sequences separated by long stretches of unrelated intervening sequences, or to sequences from different molecules. (Bottom) Two domains may function as spacers to properly position other modules; finally, the RNA-binding modules may help to define substrate specificity or regulate enzymatic activity. (Adapted from Lunde BM, Moore C & Varani G [2007] *Nat Rev Mol Cell Biol* 8:479–490. With permission from Macmillan Publishers, Ltd.)

each protein module need recognize only a few bases, or even only one base, to achieve high specificity. How such modular proteins can function is illustrated in Figure 6.14. Usually, the length of RNA recognized, even by multiple modules, is fairly short. Thus, self-interacting proteins or proteins with multiple domains can modify the three-dimensional structures of RNA molecules in a great number of ways (Figure 6.15).

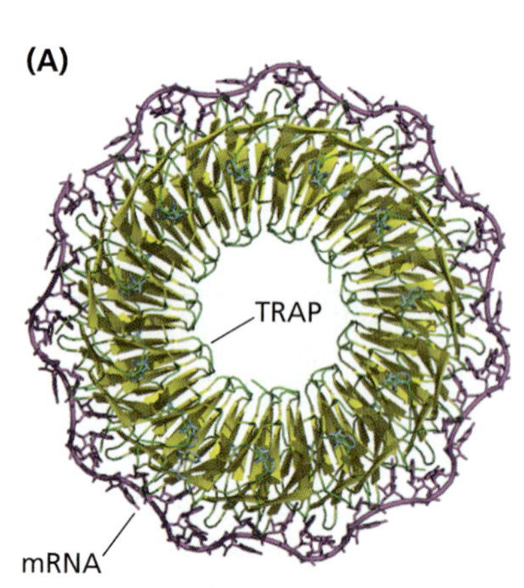

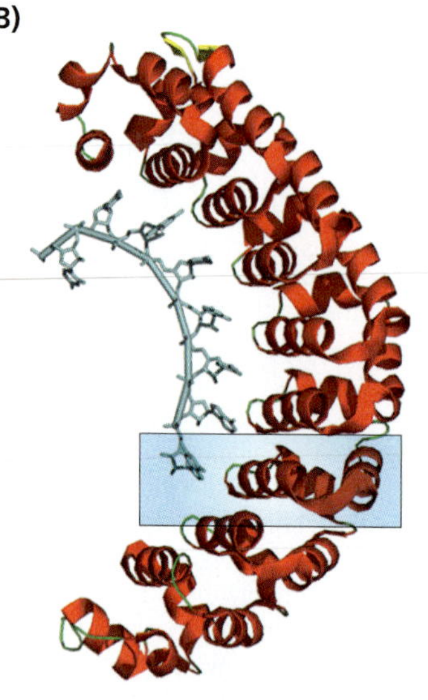

Figure 6.15 **RNA recognition.** RNA is often recognized by proteins that form heteromeric or homomeric structures through the association of multiple proteins or of the same fundamental structural motif. (A) Tryptophan RNA-binding attenuation protein (TRAP) repressor. TRAP represses tryptophan biosynthesis by regulating translation of mRNAs coding for biosynthetic enzymes. The mRNAs have multiple copies of the RNA sequence recognized by TRAP. The protein forms an oligomeric ring of 11 subunits; the RNA binds on the outside of the ring through stacking interactions and hydrogen-bonding to amino acids. (B) In the translational regulator protein Pumilio, eight copies of the same protein structural motif, shown highlighted in blue, are used to recognize RNA. Each domain binds a single nucleotide, but the combination of multiple domains provides exquisite specificity. (Adapted from Chen Y & Varani G [2005] *FEBS J* 272:2088–2097. With permission from John Wiley & Sons, Inc.)

Box 6.3 Band shift When a protein binds to DNA or RNA, the complex exhibits a lower electrophoretic mobility than the free polynucleotide, mainly because of charge neutralization. As Figure 1 shows, this can be used in a simple and fast technique that requires very little material. The DNA or RNA fragments studied, termed probes, are labeled either with radioactive isotopes or with fluorescent dyes, both of which can be easily detected. The method suffers from the disadvantage that unstable complexes may fall apart; methods such as chemical cross-linking may be used for added stability. The basic assay can be augmented by adding specific or nonspecific competitors. These competition reactions help to assess the degree of specificity of complexes between the labeled DNA fragment of interest and the protein. In another version of the assay, an antibody specific to the protein of interest is added to the incubation mixture: when the antibody binds to the protein–DNA complex, a supershift is produced. Caution should be taken not to overinterpret the relative intensity of bands, as there is always some possibility of dissociation of the complex during migration.

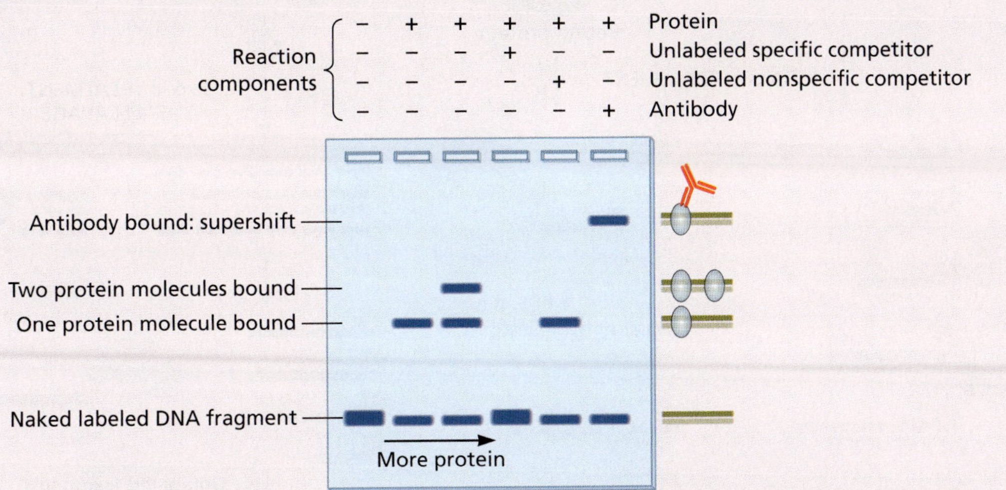

Figure 1 Gel shift assay. The gel shift assay, also known as electrophoretic mobility shift assay (EMSA) or as gel retardation assay, is performed in three steps: (1) binding reactions, (2) electrophoresis, and (3) probe detection. The DNA probe fragment is labeled either with radioactive isotope or with fluorescent dye. The labeled probe is incubated with a protein mixture to allow binding to occur. If proteins bind to the probe, its electrophoretic mobility is changed: the complex is retarded in comparison to the free DNA fragment. Following electrophoresis, the labeled probe is detected by the usual methods of autoradiography or fluorescence detection. The method suffers from the disadvantage that unstable complexes may fall apart during application of the electric field. The relatively low ionic strength of the electrophoresis buffer helps to stabilize transient interactions, and sometimes chemical cross-linking is used to create covalent bonds between the protein and DNA, thus stabilizing these complexes. The basic assay can be augmented by doing competition reactions in which an unlabeled probe—the DNA fragment itself, as a specific competitor, or a mutant or unrelated DNA fragment, as a nonspecific competitor—is added to the binding reaction mixture. These supplemented reactions help in assessing the degree of specificity of complexes between the labeled DNA fragment of interest and the protein. In another version of the assay, especially when it is performed with a mixture of proteins that presumably contains a protein of interest, an antibody specific to the protein of interest is added to the incubation mixture. When the antibody binds to the protein–DNA complex, a supershift is produced: that is, the protein–DNA complex is further retarded due to antibody binding.

6.4 Studying protein–nucleic acid interactions

Throughout this chapter we have described, via boxes, ways in which protein-nucleic acid complexes could be identified and characterized. However, it has required the application of a number of sophisticated techniques to provide the details of those interactions. Obviously, the most detailed way to examine the interactions between a given protein and DNA or RNA is to prepare the complex and then use either X-ray crystallography (see Box 3.6) or high-resolution NMR (see Box 3.7) to reveal the molecular details of the complex. This will show exactly what the binding site is and how the protein molecule recognizes it. A multitude of examples in this book illustrate how this is becoming an ever more realizable goal. But for this to be done, it is first necessary to define, at least partially, the binding site on the polynucleotide and the protein domain involved in binding. There are a number of less sophisticated, but no less important, methods to gain this and other information about the interaction. Several of these methods are described in Box 6.1, Box 6.2, Box 6.3, Box 6.4, and Box 6.5.

Box 6.4 DNA footprinting Many techniques that cleave nucleic acids can be used to locate bound proteins, sometimes with high precision. DNA footprinting depends on the fact that proteins bound to DNA will protect that DNA from cleavage in the region of binding. High-resolution gels can sometimes locate binding sites with base-pair precision. Figure 1 depicts two of the most common footprinting methods. In addition, these methods can be used to look for regions in the DNA that are protected by the bound protein from being methylated by DNA methyltransferases. If footprinting is carried out under varying salt conditions, some information can be gained about the relative affinities of adjacent binding sites.

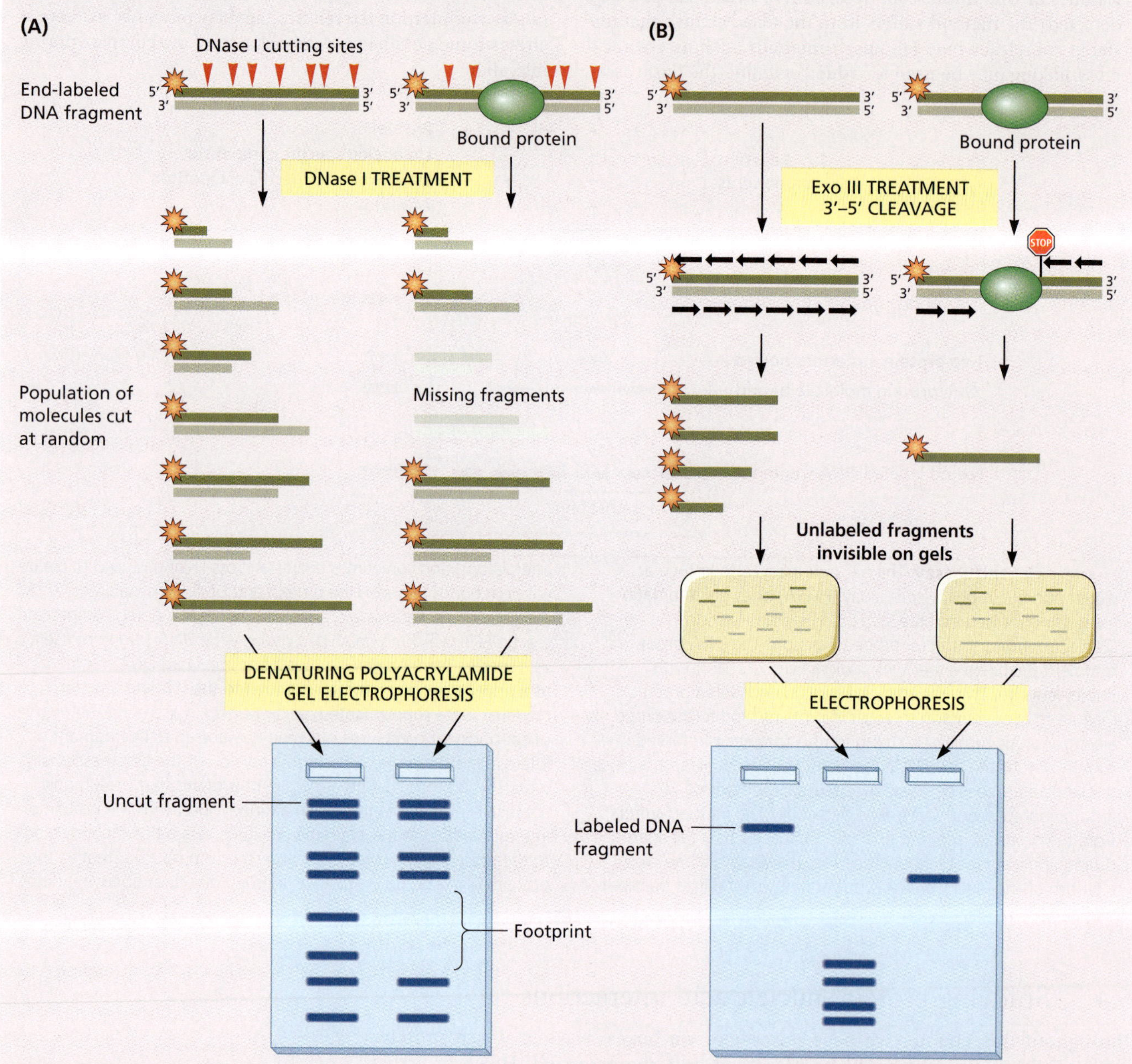

Figure 1 DNA footprinting methods. (A) DNase I footprinting. A protein bound to a specific site on the DNA should protect the DNA from enzymatic attack by pancreatic deoxyribonuclease, DNase I. The naked DNA and the protein–DNA complex are treated in parallel with DNase I under conditions in which each naked DNA fragment is cut once on average, at random. The two samples are analyzed side by side on high-resolution sequencing gels: the bands on such gels differ from their neighbors by one nucleotide. The protection by the bound protein gives rise to a footprint: absence of bands, or presence of bands of very low intensity, in the protected region. Since DNase I has some sequence specificity of cutting, alternative cleavage agents, such as the organic compound methidiumpropyl-EDTA–Fe^{2+} (MPE–Fe^{2+}), have been introduced. Another widely used agent produces hydroxyl radicals that also cleave DNA nonspecifically; the technique is known as hydroxyl-radical footprinting. (B) Exo III footprinting. Exonuclease III is an *E. coli* enzyme that cleaves the DNA exonucleolytically, from the ends, until it reaches the bound protein. The direction of cleavage is from 3′ to 5′. The small labeled DNA fragments resulting from the cleavage reaction are analyzed by high-resolution sequencing gels and are detected by autoradiography.

Box 6.5 Protein-induced DNA bending

Circular permutation assay

DNA molecules that are bent near their center move through an electrophoretic gel more rapidly than the same sequence bent near the ends. This makes it possible to locate sites of intrinsic curvature in DNA (Figure 1A). The DNA molecule is circularized by ligation, and then samples are relinearized by cleavage with different single-cutter restrictases. This places the bend nearest the end in some molecules and near the middle in others. The latter will electrophorese more rapidly. The same principle applies when the bending is produced by a bound protein (Figure 1B). This method can be used both to identify proteins that bend DNA and to roughly locate their binding sites.

DNA circularization assay

This assay detects the ability of a protein to bend DNA by estimating, from electrophoretic gels, the probability that short DNA will be circularized by DNA ligases (Figure 2). The DNA fragments are so short that, in the absence of protein-induced bending, they will only be able to link to each other to form linear dimers, trimers, and so on. If a protein induces DNA bending, then the probability of forming circles will increase in proportion to the degree of DNA bending.

DNA looping assay

DNA looping can reveal the locations of binding sites by direct microscopic observation, by either electron microscopy or atomic force microscopy.

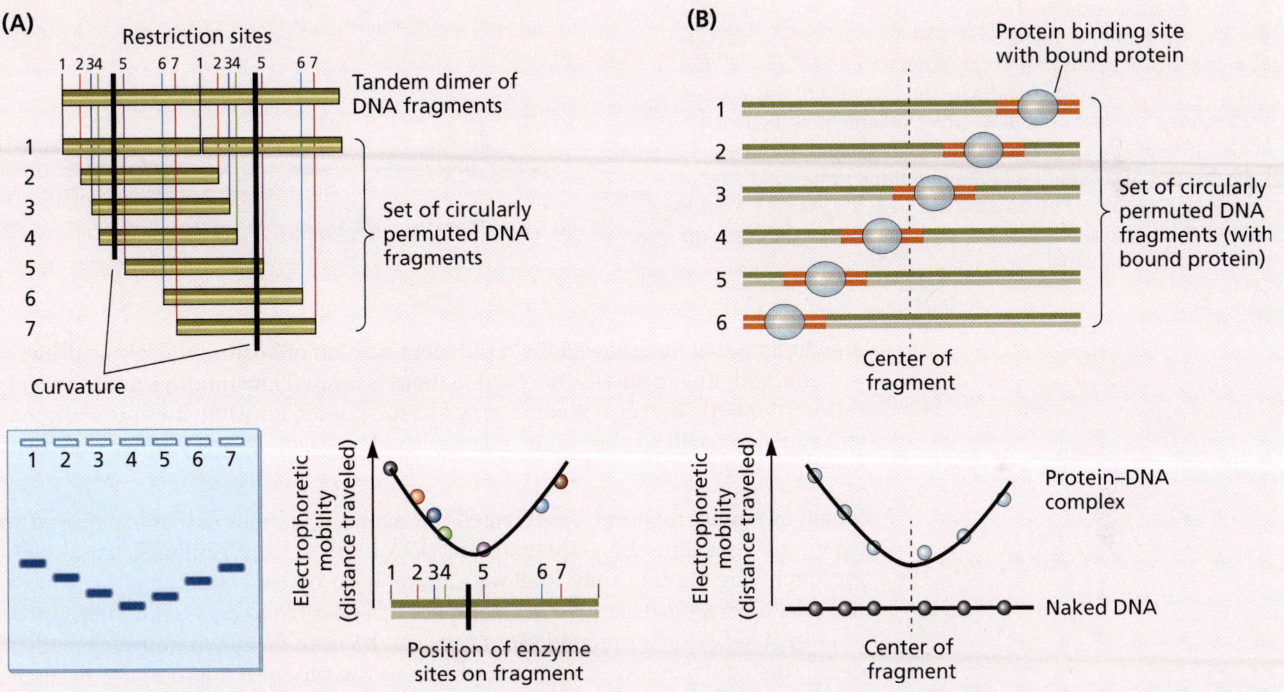

Figure 1 Circular permutation assay for localizing intrinsic curvatures in DNA fragments. (A) The DNA fragment of interest is cloned as a tandem dimer. Then separate samples of the dimer are each cut with a different restriction enzyme that cleaves only once within the fragment sequence. A set of restriction endonucleases will produce a set of cut fragments: for example, restriction enzyme 5 denoted on top of the dimer sequence will produce fragment 5, enzyme 6 will produce fragment 6, and so on. Importantly, because the nucleotide sequence is repeated within the dimer, these restriction fragments have the same nucleotide composition and length; they differ only in the position of the curvature relative to the ends. Each fragment is purified and run on nondenaturing polyacrylamide gel electrophoresis. The electrophoretic mobility of each fragment depends on the position of the curvature, with the fragment curved in the middle moving the fastest and the fragment with the curvature toward the end moving more slowly. Recall that the mobility through pores in the gel is a function of length and shape; when all fragments are of equal length, mobility depends only on the shape of the fragment. A fragment curved in the middle will have a more compact structure that moves more easily through the pores. Finally, the mobility is plotted as a function of the position of cutting, and the lowest portion of the curve defines the position of the curvature. (B) To detect protein-induced DNA bending, the naked DNA fragment is straight; that is, it does not contain intrinsic curvature. Each of the six DNA fragments is incubated with the protein of interest, and the protein–DNA complexes are run on nondenaturing polyacrylamide gel electrophoresis. The set of circularly permuted DNA fragments with the protein bound are identical in nucleotide composition and length, but they differ in the position of the protein binding site with respect to the fragment ends. Mapping to determine the bending center is performed as before.

Box 6.5 ***(Continued)***

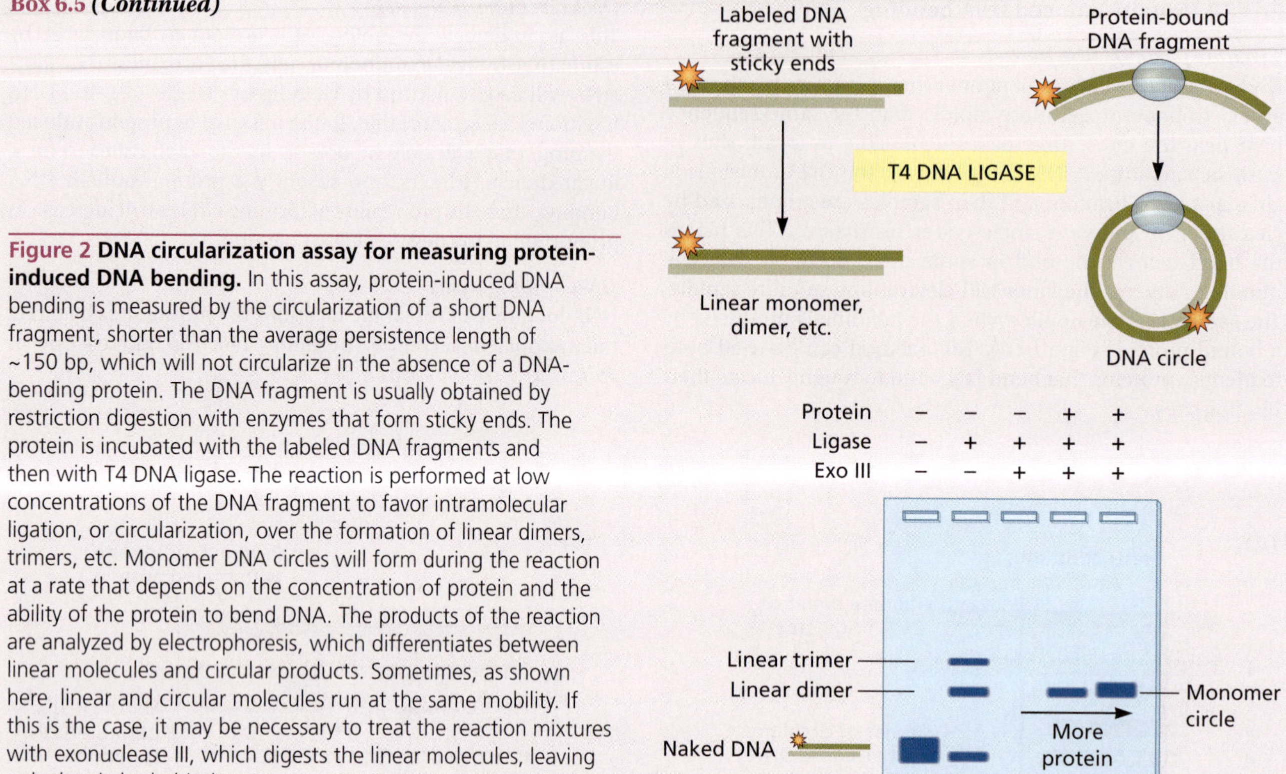

Figure 2 DNA circularization assay for measuring protein-induced DNA bending. In this assay, protein-induced DNA bending is measured by the circularization of a short DNA fragment, shorter than the average persistence length of ~150 bp, which will not circularize in the absence of a DNA-bending protein. The DNA fragment is usually obtained by restriction digestion with enzymes that form sticky ends. The protein is incubated with the labeled DNA fragments and then with T4 DNA ligase. The reaction is performed at low concentrations of the DNA fragment to favor intramolecular ligation, or circularization, over the formation of linear dimers, trimers, etc. Monomer DNA circles will form during the reaction at a rate that depends on the concentration of protein and the ability of the protein to bend DNA. The products of the reaction are analyzed by electrophoresis, which differentiates between linear molecules and circular products. Sometimes, as shown here, linear and circular molecules run at the same mobility. If this is the case, it may be necessary to treat the reaction mixtures with exonuclease III, which digests the linear molecules, leaving only the circles behind.

A technique that has allowed the rapid identification of hosts of nucleic acid-binding proteins and other proteins that bind to them is termed **chromatin immunoprecipitation** (Box 6.6). The term is something of a misnomer, for while it was initially and is still applied heavily to chromatin, it is applicable to virtually any protein–nucleic acid interactions.

Finally, we must mention site-directed mutagenesis, a method that has proved very useful for dissecting the architecture of DNA binding sites. Although not a method for identifying binding sites, it allows careful, base-by-base probing of the source of the interactions. As described in Chapter 5, it is possible to create mutations in DNA *in vitro*. A set of such mutated sequences can be used for *in vitro* binding studies to define the contribution of each nucleotide to the binding. Additionally, or alternatively, these mutated sequences can be introduced into living cells to observe changes in phenotypes and thus to understand the physiological significance of the specific protein–DNA binding in which the researcher is interested.

Box 6.6 Chromatin immunoprecipitation In recent years, a number of new methods have been developed that give an enormous amount of information about the genomic localization of protein binding sites and that can identify the proteins bound at these sites. The term chromatin immunoprecipitation (**ChIP**) is a bit of a misnomer: the technique does not depend on chromatin per se. Rather, the original idea was to fragment some large genomic region that carries bound cross-linked proteins and then use antibodies to precipitate or otherwise select those fragments that have a particular protein attached; the proteins can be bound to nucleosomal or nonnucleosomal portions of eukaryotic genomes. Actually, the same term is used for the method when it is applied to bacterial cells, and we know that chromatin is a characteristic feature of eukaryotic genomes only. After the immunofractionation of the material into protein-bound and unbound fractions, one can quickly assay the DNA sequences bound (Figure 1), frequently by very high-throughput methods. Furthermore, it is now becoming clear that many sites on the genome carry not only the protein with affinity to the site but also other proteins attached to it. So one can now hunt for clusters of proteins attached at particular sites. Unfortunately, such a vast array of data can now be quickly massed that the problem becomes one of data handling and interpretation.

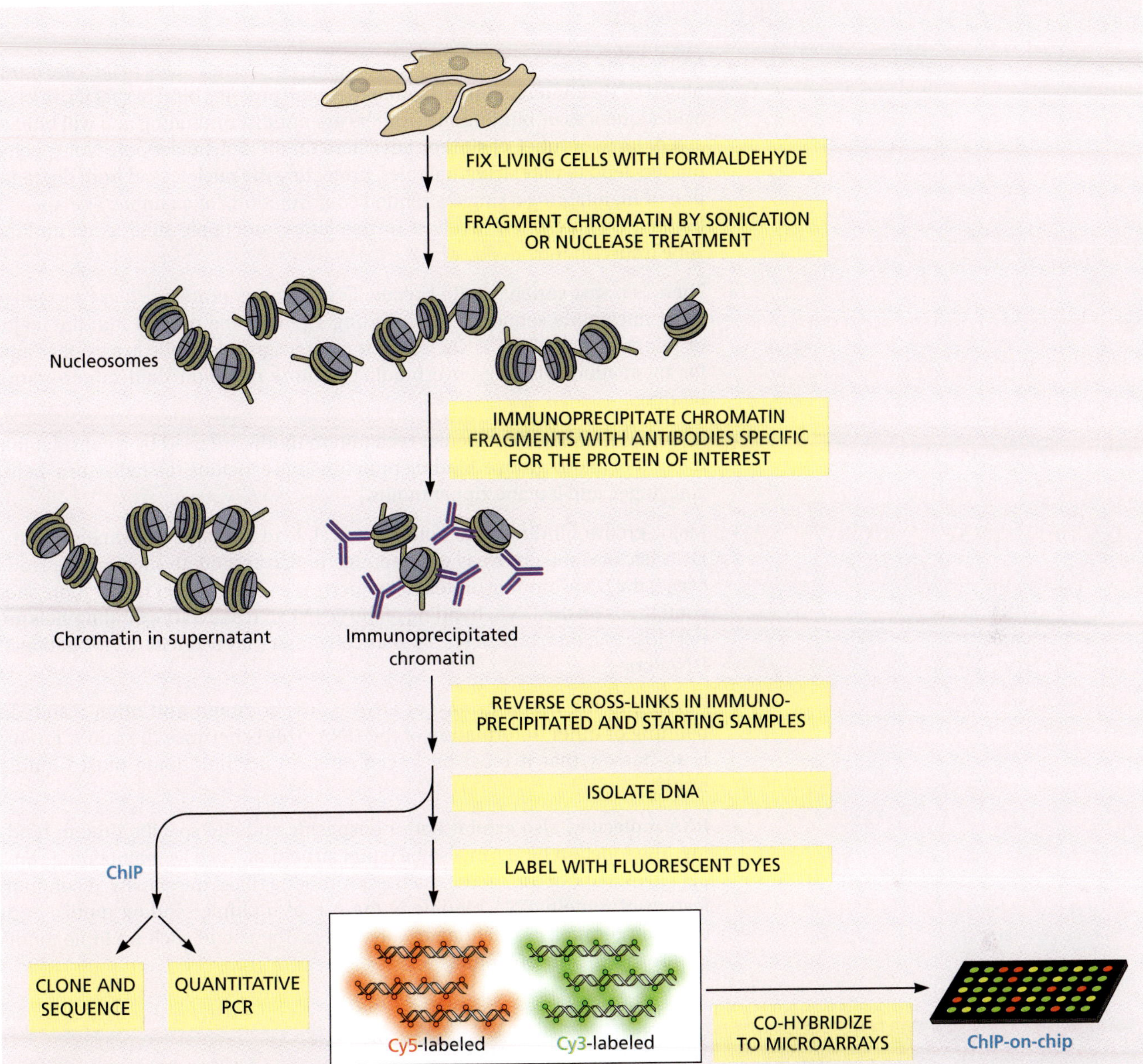

Figure 1 Chromatin immunoprecipitation and ChIP-on-chip. Genomewide localization analysis (GWLA) is a hybrid of two highly potent individual techniques that allows interrogation of protein–DNA interactions *in vivo*: chromatin immunoprecipitation (ChIP), and high-resolution, high-density DNA microarrays, known as chips. The combination of these two techniques has given GWLA its popular name: ChIP-on-chip or ChIP-chip. Living cells are subjected to a protein–DNA cross-linking agent, usually formaldehyde, to fix the protein of interest to its chromatin location in the genome (see Chapter 8). Chromatin is then fragmented by sonication or treatment with nucleases, either restriction endonucleases or micrococcal nuclease that cleaves predominantly in the linker regions between nucleosomes to produce core particles. The population of chromatin fragments is incubated with an antibody specific to the protein of interest, and the reacted fragments are immunoprecipitated. Typically, the antibodies are conjugated to beads to facilitate the separation of reacted from unreacted fragments. In classical ChIP, DNA isolated from the immunoprecipitated fraction is cloned and sequenced, or subjected to quantitative PCR (qPCR) amplification to identify the DNA fragments that had been cross-linked by the formaldehyde treatment. In ChIP-on-chip, the starting and immunoprecipitated DNA populations are amplified, labeled by two different fluorophores, mixed, and co-hybridized to microarrays. The microarray platforms are either commercially available or custom-made, depending on the specific project. The slides may contain a variety of probes arranged in a variety of ways; they may cover groups of genomic regions, such as open reading frames (ORFs) and intergenic sequences, entire chromosomes, or for organisms of relatively small genome sizes, entire genomes. In addition, the genome coverage can vary in the probe density, from one probe every ~300 bp to probes that are tiled in a way to ensure that each genome region or nucleosome is represented by several overlapping probes.

Key concepts

- Many of the structural and functional aspects of nucleic acids *in vivo* are modulated by their interaction with proteins. Some proteins bind to specific nucleic acid sequences or binding sites; others are nondiscriminating and will bind at a very large number of sites or anywhere on the polynucleotide. Nonspecific binders tend to play structural roles, protecting the nucleic acid from degradation or maintaining a single-stranded conformation, for example. Site-specific binding is more often involved in regulatory functions, such as controlling gene transcription.
- There is a vast variety of site-specific DNA-binding proteins. These recognize short nucleotide sequences by inserting a part of the protein into the major or minor groove of the DNA, where it can recognize specific arrays of groups for interaction. Major-groove binding is more common than minor-groove binding.
- A limited number of molecular recognition motifs are used by a considerable fraction of major-groove-binding proteins. These include the helix-turn-helix, zinc finger, and leucine zipper motifs.
- Major-groove binding does not, in general, lead to major deformation of the DNA because the groove is wide enough to accommodate such motifs. However, if the DNA-binding protein oligomerizes and recognizes two or more sites contiguous on the DNA, bending of the DNA can result. If the binding sites are not close together, binding to a protein oligomer may result in the formation of DNA loops.
- Binding to the minor groove of DNA is less common and often results in bending or other deformation of the DNA. This is because the minor groove is so narrow that it must be forced open to accommodate most binding motifs.
- RNA molecules also exhibit both nonspecific and site-specific protein binding. The function here can also be either structural, such as maintaining compact RNA folds, or functional, such as promoting ribozyme activity. A common feature of protein–RNA binding is the use of multiple binding motifs, each of which recognizes only one or a few sites. The use of such multiple motifs ensures specificity of binding.

Further reading

Books

Kneale GG (ed) (1994) DNA–Protein Interactions: Principles and Protocols. Methods in Molecular Biology, Vol. 30. Humana Press.

Neidle S (2002) Nucleic Acid Structure and Recognition. Oxford Press.

Rice PA & Correll CC (eds) (2008) Protein–Nucleic Acid Interactions: Structural Biology. RSC Publishing.

van Holde KE, Johnson WC & Ho P-S (2006) Principles of Physical Biochemistry, 2nd ed. Pearson Prentice Hall.

Reviews

Chen Y & Varani G (2005) Protein families and RNA recognition. *FEBS J* 272:2088–2097.

Lane D, Prentki P & Chandler M (1992) Use of gel retardation to analyze protein–nucleic acid interactions. *Microbiol Rev* 56:509–528.

Lunde BM, Moore C & Varani G (2007) RNA-binding proteins: Modular design for efficient function. *Nat Rev Mol Cell Biol* 8:479–490.

Privalov PL, Dragan AI & Crane-Robinson C (2009) The cost of DNA bending. *Trends Biochem Sci* 34:464–470.

Theobald DL, Mitton-Fry RM & Wuttke DS (2003) Nucleic acid recognition by OB-fold proteins. *Annu Rev Biophys Biomol Struct* 32:115–133.

von Hippel PH (2007) From "simple" DNA–protein interactions to the macromolecular machines of gene expression. *Annu Rev Biophys Biomol Struct* 36:79–105.

Experimental papers

Kalodimos CG, Biris N, Bonvin AM et al. (2004) Structure and flexibility adaptation in nonspecific and specific protein–DNA complexes. *Science* 305:386–389.

Kozlov AG, Jezewska MJ, Bujalowski W & Lohman TM (2010) Binding specificity of *Escherichia coli* single-stranded DNA binding protein for the χ subunit of DNA pol III holoenzyme and PriA helicase. *Biochemistry* 49:3555–3566.

Lynch TW, Read EK, Mattis AN et al. (2003) Integration host factor: Putting a twist on protein–DNA recognition. *J Mol Biol* 330:493–502.

O'Shea EK, Klemm JD, Kim PS & Alber T (1991) X-ray structure of the GCN4 leucine zipper, a two-stranded, parallel coiled coil. *Science* 254:539–544.

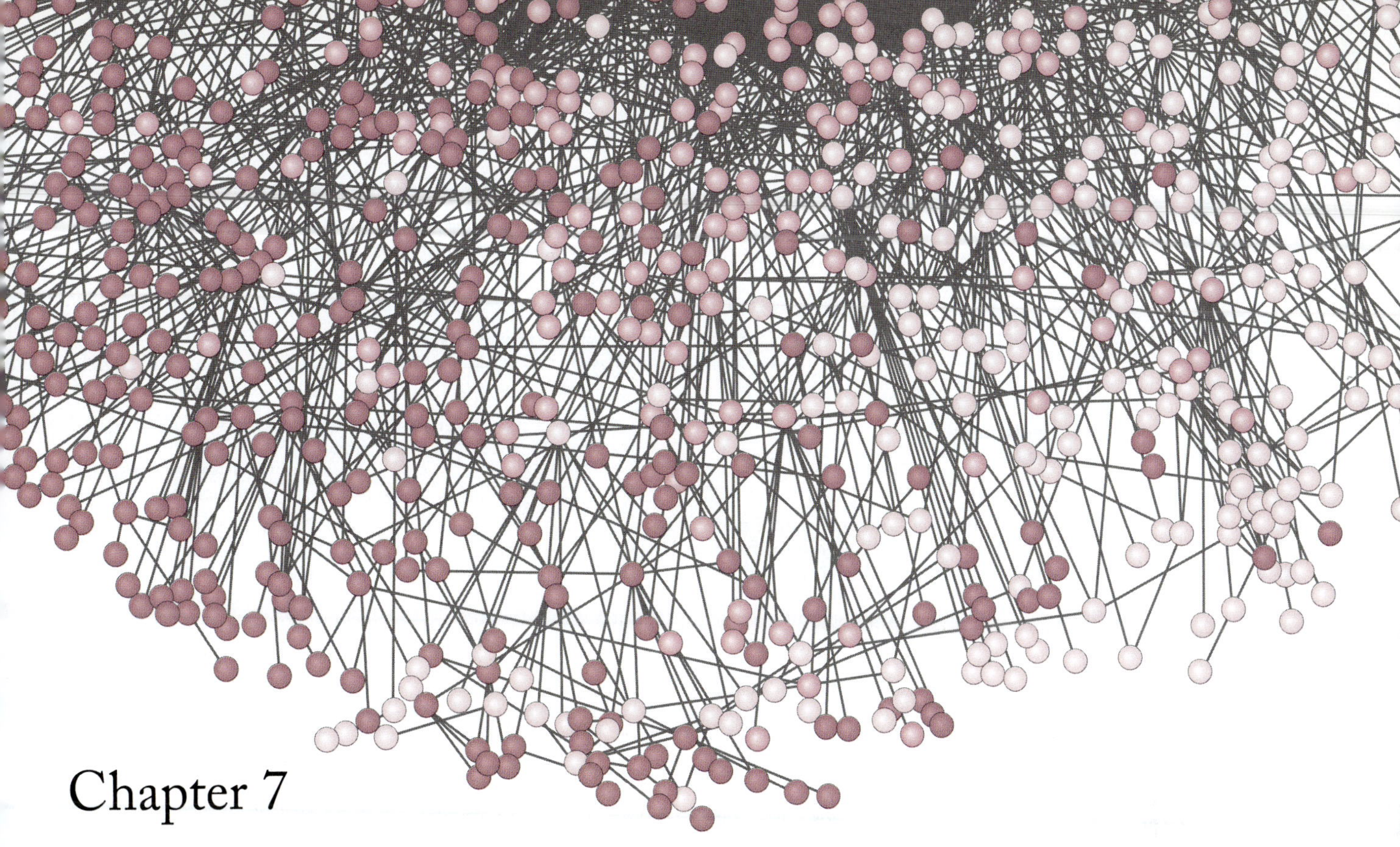

Chapter 7

The Genetic Code, Genes, and Genomes

7.1 Introduction

This chapter outlines the basic ideas of molecular biology, which are expanded in later chapters, and therefore takes a somewhat historical perspective. These ideas were largely developed in the remarkably short period of about 10 years between 1953 and 1963. Therefore, it seems most interesting to present these fundamental concepts by describing this remarkable burst of discovery. These developments are placed in a broader perspective in **Figure 7.1**, which also traces the emergence and elaboration of the concept of the gene.

7.2 Genes as nucleic acid repositories of genetic information

Our understanding of the nature of genes is constantly evolving

The golden age of classical genetics, described in Chapter 2, culminated with the outstanding studies of fruit flies by Thomas Hunt Morgan (see Box 2.1). From the work of Morgan and others, the formal rules for the behavior of genes were established, and by 1930 many gene locations were mapped on chromosomes (see Chapter 2). Yet the nature of the gene itself was then completely unknown and did not even seem of great importance to the geneticists of the time. Indeed, Morgan stated in his 1933 Nobel address, "it does not make the slightest difference whether the gene is a hypothetical unit, or whether the gene is a material particle."

That point of view began to change in the 1940s, as rapid advances in biochemistry made comprehending the workings of cells a foreseeable goal, and the role of proteins, particularly enzymes, became evident. For example, certain hereditary malfunctions

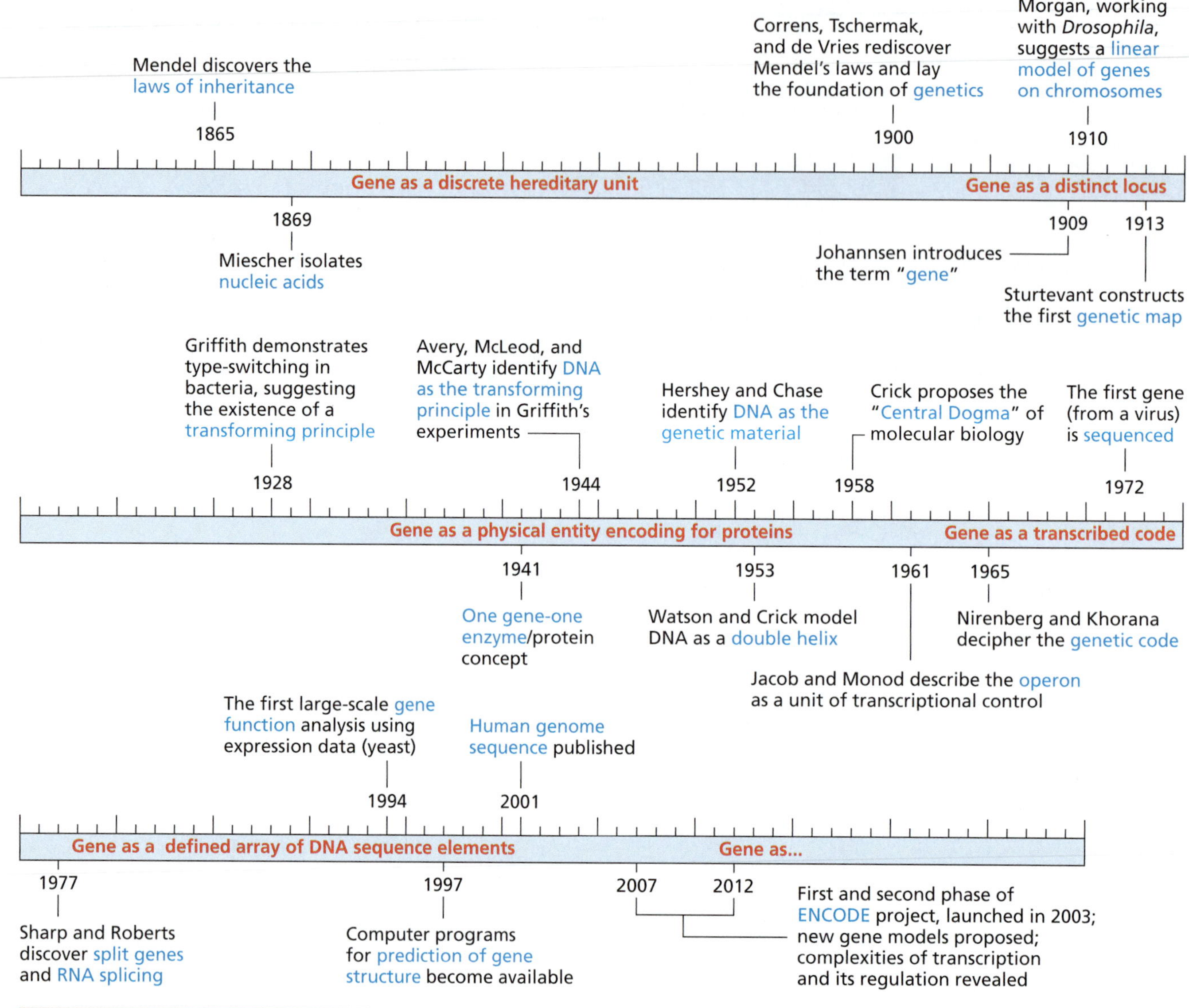

Figure 7.1 Evolution of the concept of the gene.

in metabolism, such as albinism, that mapped to specific gene locations also could be linked to the malfunction or lack of particular enzymes. This led to the idea that genes somehow dictated protein structure—one gene, one enzyme—and thus not only were heritable but also participated in the everyday workings of each cell. The concept was broadened to include proteins other than enzymes with the recognition of sickle-cell anemia as a genetic disease (see Box 2.3).

A critical breakthrough came with Sanger's sequencing of insulin, in the period 1949–1955 (see Chapter 3). Even the first results of this work showed that the amino acid sequence of each protein was defined and was unique for that protein. This demonstrated that the genetic material, whatever it was, was able to contain and transmit sequence information to direct the synthesis of proteins. Before Sanger's work, many biochemists expected that proteins might be built from repetitions of certain small patterns of amino acids. Such structures could be made by a small battery of enzymes, and this is in fact how long polysaccharide chains are put together. But the insulin molecule, and other proteins soon sequenced, showed no evidence of such periodic patterns of structure.

All this evidence showed that the genetic substance must be a template of specific sequence, which not only can be replicated in cell division but also carries a code of some kind that specifies protein sequences. With the Watson–Crick model of DNA structure of 1953, together with the experiments of Hershey and Chase (see Box 4.2), it finally became clear that the genetic substance must be DNA. At this moment in history, the gene became not just an abstract location on a chromosome but a part of a macromolecule that codes for a protein. We still find this to be a correct but far too limited definition. As we shall see, the information in DNA is indeed used to code for proteins but also has other very important functions. The issue of defining a gene is still a problem.

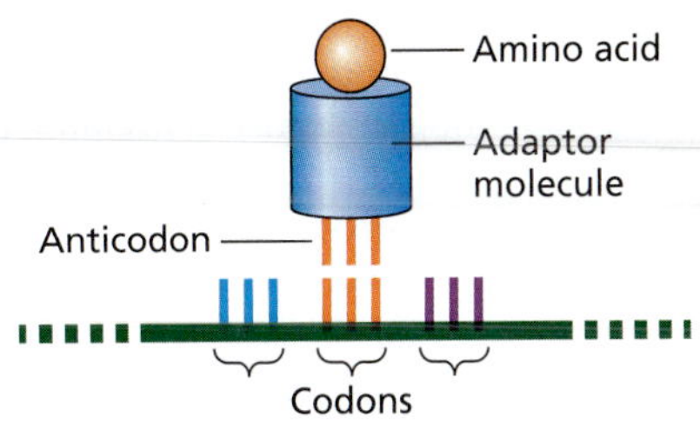

Figure 7.2 Concept of an adaptor molecule as suggested by Crick. The adaptor should have the ability to bind to just one specific amino acid; at the same time, it should carry a small oligonucleotide that should recognize and base-pair with bases forming the codon.

The central dogma states that information flows from DNA to protein

DNA and proteins contain the same information in two very different languages. Indeed, the two languages are so different that it was initially hard to see how translation from one to the other could occur. An early insight was expressed in what Francis Crick called the sequence hypothesis, which states that the information in a DNA sequence must be carried entirely in the sequence of bases, because the sugar–phosphate backbone is always the same, and hence devoid of information. Furthermore, he postulated the existence of a simple code that would translate the DNA sequence into a protein sequence, although at that time, 1958, the nature of the code or how it operated was wholly unknown. At the same time, Crick proposed what has come to be called the **central dogma** of molecular biology, stating that information flows only from polynucleotide to protein, never in the reverse direction. This was a remarkable, audacious hypothesis, made without a shred of experimental evidence. It is not really a dogma, either, as this is a theological term for something held to be an irrefutable truth. There is nothing irrefutable in science, but in fact the dogma has turned out to be correct, so far as it goes. Furthermore, it is an important idea because if information could leak back from cellular proteins into the genetic material, protein changes induced by environmental effects could be inherited. This would be consistent with Lamarckism, a long-discredited view of evolution. The picture of the flow of genetic information has been made more complicated by recent discoveries but still adheres to Crick's dogma; there is no evidence for information flowing from proteins to DNA.

This picture fails to answer the crucial question: how does information coded into a string of four kinds of bases get translated into a different string made from 20 kinds of amino acids? Once again Francis Crick, in a remarkable leap of intuition, proposed the correct answer. In 1955 he argued that there must be, in the cell, a collection of what he called **adaptor molecules**. Each of these must have, he said, the ability to bind just one kind of amino acid. But the adaptor must also have on it a small oligonucleotide that could make specific hydrogen bonds with a specific base sequence on a DNA or RNA template (Figure 7.2). This latter sequence we now call the codon; the complementary sequence on the adaptor molecule is termed the anticodon. The postulated adaptor molecules were not actually identified until about 5 years later, but they turned out to have just the properties that Crick had proposed. But meanwhile, thought and experimentation on these new problems proceeded apace. Many scientists in many countries were involved, but two at Cambridge, Francis Crick and Sydney Brenner, and two in Paris, François Jacob and Jacques Monod, stand out in importance. Above all, Francis Crick was the leader during this period (Box 7.1).

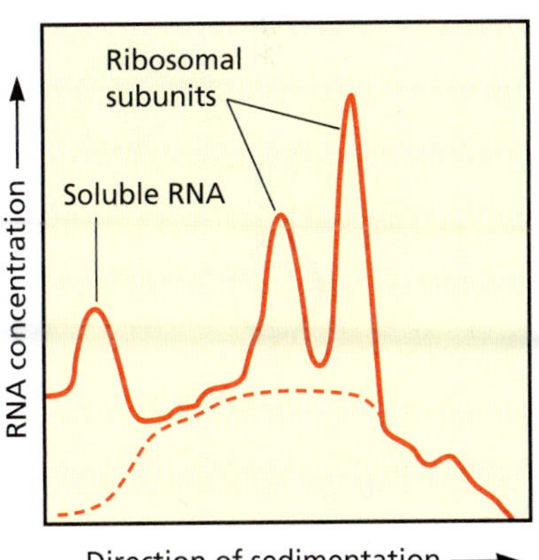

Figure 7.3 Schematic of the sedimentation velocity distribution commonly observed with preparations of whole cytoplasmic RNA. The dotted curve represents the RNA, including mRNA, that is not soluble tRNA or ribosomal subunits. The experiment uses density gradient sedimentation, which can accomplish the separation of the small soluble RNA fraction from ribosomal RNAs and all other large RNAs.

It was necessary to separate cellular RNAs to seek the adaptors

By the late 1950s, it became clear that whereas DNA in eukaryotic cells was confined to the nucleus, protein synthesis was occurring in the cytoplasm. It seemed obvious, then, to seek adaptor molecules in cytoplasmic RNA. There is, however, much RNA in the cytoplasm, in a wide range of molecular sizes. The most useful technique available in the 1950s to separate RNAs was ultracentrifugation to give preparations of cell contents free of nuclei, and hence free of DNA and nuclear RNA. RNA could be radiolabeled as cells were grown to aid in identification. The typical result is shown schematically in Figure 7.3. Note that gradient centrifugation is discussed in depth in Box 4.8. Most rapidly sedimenting is a group of compact, RNA-protein particles, originally called microsomes but now known as ribosomes. These were found to

Box 7.1 The genius of Francis Crick The name of Francis Crick is widely known because of his monumental intuition that led to the discovery of the double-helical structure of DNA. What is not so widely appreciated is the role he played in understanding the genetic code and the flow of information in the cell. Crick was not primarily known as an experimentalist. Rather, he displayed a remarkable ability to see to the heart of a scientific problem and an imagination freed of the constraints of tradition. A few of his most important contributions include the following:

- **1953** with James Watson, DNA structure
- **1955** adaptor hypothesis predicts the existence of small molecules, later identified as tRNAs, that match amino acid with codon; he also included the suggestion of stop codons
- **1957** sequence hypothesis predicts that the information in protein sequence is collinear with that in DNA
- **1958** central dogma states that information can flow into proteins but not back out
- **1961** with Sidney Brenner, François Jacob, and others, generation of the mRNA concept and genetic proof that the code is triplet and non-overlapping
- **1962** general features of the code: it is triplet, unpunctuated, and degenerate
- **1966** wobble hypothesis predicts that the third letter of a codon can take alternate positions to facilitate degeneracy.

Together, these contributions constitute a major portion of the origins of molecular biology and genetics.

be abundant, particularly in bacteria and other rapidly growing cells, and were soon, and correctly, suspected of being involved in some way in protein synthesis. A smaller amount of large but unstable RNA was long disregarded; this component turned out to be of critical importance. Finally, centrifugation revealed a substantial amount of low-molecular-weight, slowly sedimenting RNA, known for a long time as soluble RNA because it remained in solution after high-molecular-weight RNA was precipitated. It was among this low-molecular-weight RNA that Robert Holley, around 1960, found molecules that corresponded to the adaptors, which were henceforth referred to as **transfer RNA** or **tRNA**. Holley managed to isolate alanine-tRNA from yeast and, over a six-year period, determined its sequence (Figure 7.4). This was the first nucleic acid molecule ever to be sequenced, and the contrast in efficiency between Holley's laborious work and modern high-throughput methods is astounding. The possibility of internal hydrogen bonding within tRNA to form a folded structure was recognized immediately, as were the sites of amino acid attachment and the anticodon loop. The actual folded structure was later determined by X-ray diffraction studies (see Chapter 14).

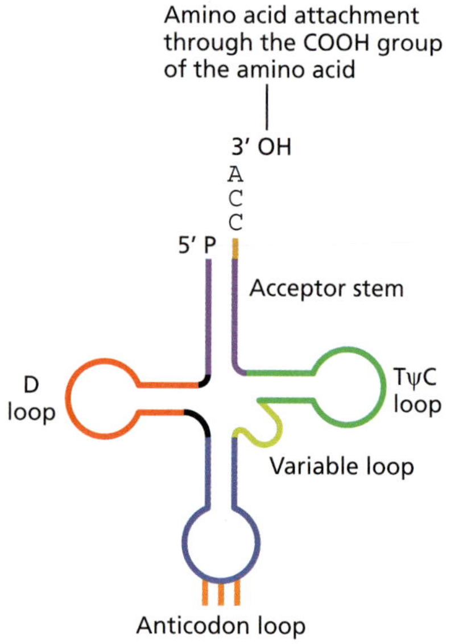

$GGGCGUGU^{m}GGCGUAGU_{(H2)}$
$CGGU_{(H2)}AGCGC^{m2}G$
$CUCCCUUUIGC^{m}I\psi GGGAGAGGU_{(H2)}C$
$UCCGGT\psi CGAUUCCGGA$
$CUCGUCCACCA$

Figure 7.4 Primary and secondary structure of alanine-tRNA. Alanine-tRNA was the first tRNA to be sequenced, by Robert Holley. The color scheme in the schematic of secondary structure corresponds to that in the letter code for primary structure. tRNA contains some base modifications that are discussed in Chapter 15.

Messenger RNA, tRNA, and ribosomes constitute the protein factories of the cell

The idea that genetic information passed through an RNA intermediate in its transfer from DNA to protein seems to have been held before it was suggested by any experimental evidence. The fact that DNA is in the nucleus and protein synthesis occurs in the cytoplasm argued strongly for a messenger. Certainly this concept is implicit in Crick's statement of the central dogma. And yet, until about 1961, there was no hypothesis as to what that intermediate might be. Or, rather, there was a wrong hypothesis, widely held, that the RNA present in ribosomes was the message. After all, ribosomal RNA was by far the most abundant RNA in the cytoplasm, where protein synthesis was known to occur. Furthermore, there was early evidence that ribosomes were somehow involved in protein synthesis. So it was assumed for some time that different ribosomes contained different RNA molecules, with each coding for a different protein. It was evident, however, that there were serious problems with this idea. First, whereas different proteins varied enormously in their polypeptide chain length, the RNA molecules in ribosomes seemed to be of very uniform size. Second, it was found that bacteriophage infection of bacteria rapidly gave rise to new protein synthesis, which could rapidly be turned off by destruction of the viral DNA. Whatever substance was coding for proteins in the cytoplasm was clearly quite unstable. Yet ribosomes were extremely stable structures, some even persisting through generations of bacteria.

The light seems to have dawned in a wholly informal discussion in Sydney Brenner's apartment at Cambridge in 1961. Only half a dozen scientists were present,

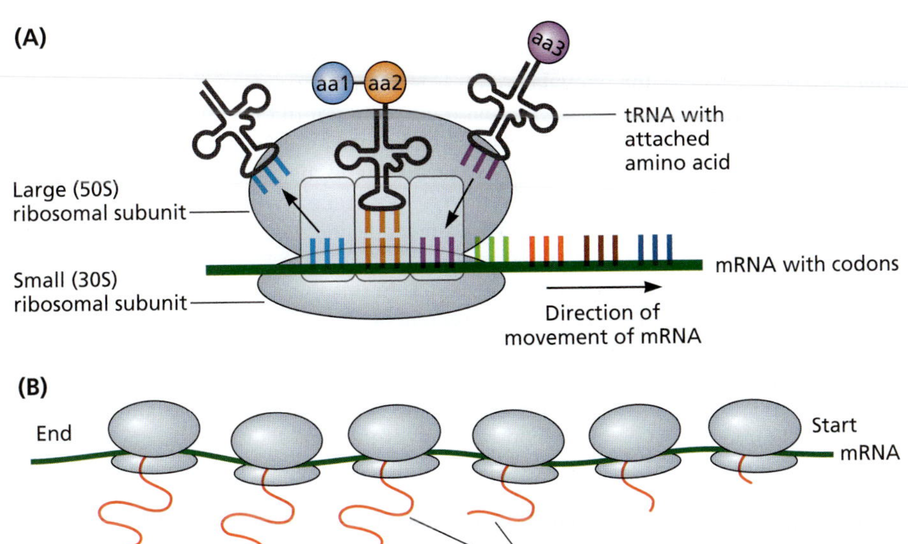

Figure 7.5 Schematic of a bacterial 70S ribosome translating a message. (A) The message can be visualized as moving from left to right, or the ribosomes from right to left. Details of this complicated process are given in Chapter 16. (B) A sequence of ribosomes, called a polyribosome, translating a single mRNA molecule. This schematic is based on electron microscopic images. As above, the ribosomes can be visualized as moving right to left.

and Crick, Brenner, and François Jacob seem to have led the discussion. It suddenly became clear that there might be an unstable, high-molecular-weight RNA fraction, copied from one strand of DNA, that carried the sequence message. This **messenger RNA (mRNA)** was postulated to associate with ribosomes. The function of the ribosomes then became clear: they allowed the reading of this message by permitting the adaptor molecules to pair with the codons on the message, and the ribosome could then facilitate peptide bond formation (Figure 7.5). This insight did not come all at once, and experimental support was needed. This soon came, for example, in work by Benjamin Hall and Solomon Spiegelman at the University of Illinois, who showed that new RNA synthesized following phage infection of bacteria was complementary to phage DNA.

7.3 Relating protein sequence to DNA sequence in the genetic code

The first task was to define the nature of the code

The major breakthroughs in understanding the roles of mRNA, tRNA, and ribosomes made the elucidation of the genetic code an even more important goal (Box 7.2). It was clear from the start that codons must be groups of at least three bases, known as triplets. One or two bases for each amino acid would not do, as they would code for only 4 or $4^2 = 16$ possible amino acids, respectively. A triplet code allows $4^3 = 64$ permutations, more than enough to code for all 20 amino acids. A number of models of possible codes had been suggested; examples are shown in Figure 7.6A and Figure 7.6B. It soon became clear that neither an overlapping code nor a punctuated code could survive the experimental tests. Crick and Brenner showed by inserting single or multiple base pairs into phage DNA that only a unpunctuated, non-overlapping triplet code could work. A consequence of a non-overlapping, unpunctuated triplet code is that there are three possible **reading frames** depending on where one begins to read the message (Figure 7.6C). Therefore, a specific start site must be specified to give a particular message. An insertion or deletion of any number of residues that is not a multiple of three will result in a **frameshift**, specifying a different protein sequence (Figure 7.6D).

But which codons corresponded to which amino acids? A first clue was given when Marshall Nirenberg showed that the homopolynucleotide poly(U) could direct the synthesis of polyphenylalanine. This was quickly confirmed and extended by the experiments of Gobind Khorana using repetitive-sequence synthetic RNAs as messages (see Box 7.2). Elucidation of the specific codons corresponding to each amino acid was a rather tedious undertaking but finally permitted clarification of the entire translational

Box 7.2 Cracking the genetic code The existence of some kind of code that could translate the language of nucleic acids into the language of proteins had been suspected by some even well before the revolution sparked by Watson and Crick's discovery of the structure of DNA. Indeed, there are hints of it in the 1946 book *What is Life?* by Erwin Schrödinger, an influential volume for early molecular biologists. Nevertheless, even as late as 1961, eight years after Watson and Crick presented their structure, there was not a shred of experimental evidence to suggest how DNA is translated. There were theories and models aplenty for various possible coding schemes, but these only generated debate among researchers.

Then, at the 1961 International Congress of Biochemistry held in Moscow, a paper was presented by Marshall Nirenberg. Because he was young and relatively unknown at the time, he was given about 15 minutes at the end of a session, and few of the better-known scientists were there. Among the exceptions was Matthew Meselson, who recognized the importance of Nirenberg's work and prevailed upon the organizers of the conference to grant Nirenberg another hearing. This time, the repercussions were intense.

Nirenberg, together with postdoctoral fellow Johann Matthaei, had been interested in the mechanism of translation of messenger RNAs, a concept just becoming formulated, into proteins and they had set up a cell-free system to study this process. Such systems, which had been devised by a number of other researchers, included ribosomes; a mix of low-molecular-weight RNAs, probably mostly tRNAs; and an ATP-driven energy-generating system. The low-molecular-weight RNAs could be shown, upon stimulation with viral or crude mRNA, to drive the incorporation of amino acids into protein.

Nirenberg and Matthaei simplified this experiment: they used synthetic polynucleotides containing only one base, added different radiolabeled amino acids to the reaction mixture, and asked which amino acids were incorporated into polypeptides for each polynucleotide message. Homopolyribonucleotides were available as a consequence of Severo Ochoa's development of their enzymatic synthesis in the late 1950s. Their first clear success was the demonstration that poly(U) led, overwhelmingly, to the formation of polyphenylalanine. Thus, codons composed only of U coded for phenylalanine. It is important to note that this experiment gave limited information: it did not say that the codon was UUU, because it had not been shown at this date that the code was triplet. It was silent on the question of punctuation. Rather, the Nirenberg–Matthaei experiment opened the first clear path as to how the code could be studied experimentally. From all accounts, the insiders in the field were intensely irritated at being beaten to this important discovery by relatively unknown researchers.

The Nirenberg–Matthaei approach became much more powerful when synthetic polynucleotides of more complex but regular sequence became available, mainly through the work of Har Gobind Khorana at the University of Wisconsin. As is shown in **Figure 1**, a homogeneous polynucleotide $(A)_n$ can code only for a polypeptide that is composed of a single amino acid, a repetition of a dinucleotide sequence $(AB)_n$ produces a product of alternating amino acids, and a repetition of a trimer $(ABC)_n$ yields three different homogeneous polypeptides simultaneously. Results of this kind firmly established that the code was triplet and unpunctuated. At the same time, Francis Crick and Sydney Brenner were answering the same question by purely genetic experiments with bacteriophage. In brief, they showed that making one- or two-base-pair insertions or deletions in the genome produced mutant phage because the reading frame for a gene was shifted, while making three closely spaced insertions or deletions led to something very close to the wild-type phenotype.

The genetic code was completely deciphered only in 1967, facilitated by another technical breakthrough by Nirenberg. He discovered that a single trinucleotide could attach to a ribosome and trap the appropriate tRNA. This tRNA would not pass an ultrafine filter and could thus be identified.

It was fitting that the 1968 Nobel Prize in Physiology or Medicine was awarded to Nirenberg, Khorana, and Robert Holley for breaking the genetic code. Holley had determined the sequence of the first tRNA (see Figure 7.4). It is interesting to note that none of the three could be considered members of the inner circle that had dominated the earliest years of molecular biology.

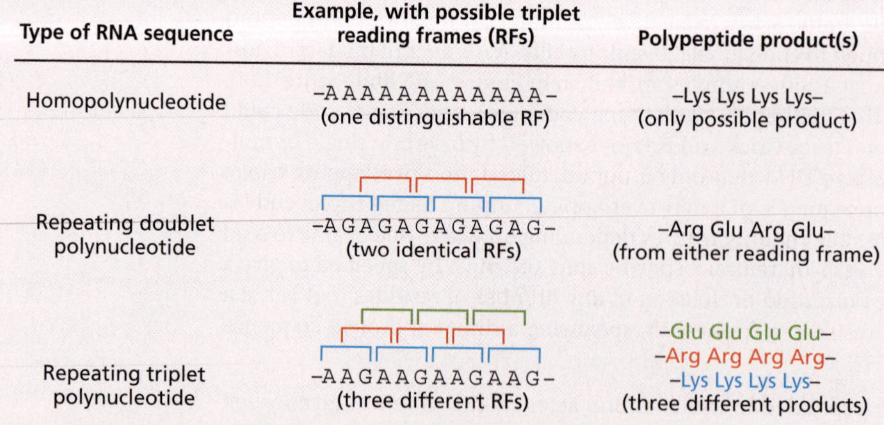

Type of RNA sequence
Example, with possible triplet reading frames (RFs)
Polypeptide product(s)
Homopolynucleotide
–AAAAAAAAAAAA–
(one distinguishable RF)
–Lys Lys Lys Lys–
(only possible product)
Repeating doublet polynucleotide
–AGAGAGAGAGAG–
(two identical RFs)
–Arg Glu Arg Glu–
(from either reading frame)
Repeating triplet polynucleotide
–AAGAAGAAGAAG–
(three different RFs)
–Glu Glu Glu Glu–
–Arg Arg Arg Arg–
–Lys Lys Lys Lys–
(three different products)

Figure 1 Properties of the genetic code. Experiments that used synthetic polynucleotides as templates for protein synthesis in cell-free systems demonstrated that the code was triplet, non-overlapping, and unpunctuated. They also allowed assignments of many codons to specific amino acids.

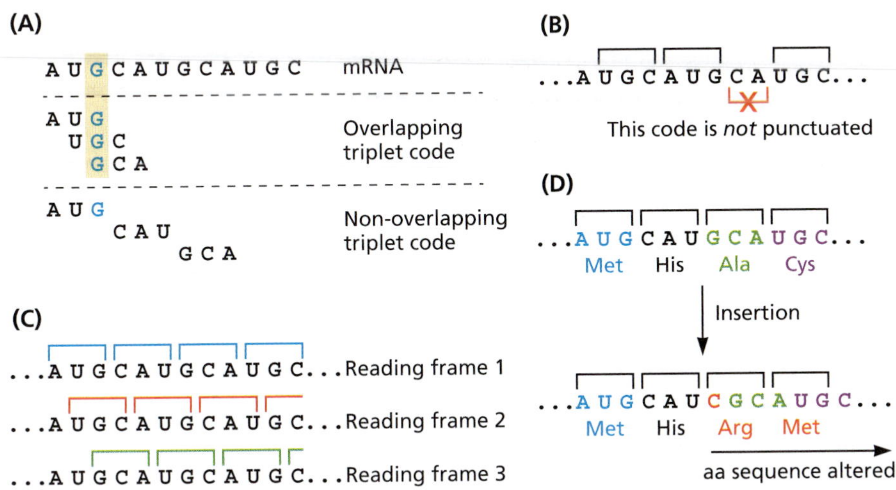

Figure 7.6 Basic features of the genetic code. (A) Schematic of the two possible ways a triplet code can be read: in the overlapping code, the same base can be part of three codons, whereas in the non-overlapping code, each base is part of only one codon. The genetic code is actually non-overlapping. (B) The code is unpunctuated: all nucleotides need to be read in succession and no gaps are allowed between successive codons. (C, D) The concepts of reading frame and frameshift mutations. (C) Each nucleotide sequence has to be read starting from a particular nucleotide. Specification of this first nucleotide defines the succession of codons and thus the primary structure of the polypeptide chain. If a different first nucleotide is specified, the same sequence can be read in a different frame, producing a different polypeptide. (D) Insertion of one nucleotide into the message would change the amino acid sequence in the downstream portion of the polypeptide chain. Such frameshift mutations usually lead to total loss of function of the encoded protein.

process. The code, as it is now known for most organisms, is summarized in Table 7.1. Several points become obvious:

- All 64 codons are used but three codons—UGA, UAA, and UAG—are commonly used as **stop codons** to tell the ribosome to stop reading the RNA message and release the polypeptide. In some organisms, these three codons can also code for amino acids (Table 7.2).
- There is a **start signal**, which always codes for methionine, but not all methionine codons indicate the start of a message that is to be translated into a polypeptide. The mechanism that distinguishes between an initiating and internal methionine is discussed in Chapter 16.
- The code is **degenerate**: most amino acids have more than one codon, with tryptophan and methionine being the only exceptions.
- Degeneracy lies mainly in the third letter of the codon; in other words, the first two letters of codons alone often specify the amino acid. Again, the insight of Francis Crick is important; he proposed the **wobble** hypothesis, which explains much of this. The wobble hypothesis is discussed in detail in Chapter 15.
- The code is almost, but not quite, universal. Exceptions to the general codon usage appear in mitochondria, which have their own translation systems, and also in a few primitive organisms (see Table 7.2). It is not surprising that the code has remained nearly constant over billions of years of evolution, as any

Table 7.1 The standard genetic code. Some characteristic features are as follows. (1) The code is unique. Each codon specifies only one amino acid in a particular organism. (2) The code is degenerate. One amino acid may be specified by multiple codons; for example, serine has six codons, glycine has four, etc. (3) The first two nucleotides are often enough to specify a given amino acid; for example, serine is specified by UC. (4) Codons with similar sequences specify amino acids of similar chemical properties. Codons for serine and threonine differ in the first letter; codons for aspartic acid and glutamic acid differ in the third letter. This ensures that many mutations will result in the incorporation of a similar amino acid that would not significantly affect the structure/function of the protein. (5) There are three stop codons that define the end of the polypeptide chain, and there is an initiation codon, AUG, that determines the point in the mRNA which codes for the first amino acid.

First position (5′)	Second position U	C	A	G	Third position (3′)
U	Phe	Ser	Tyr	Cys	U
	Phe	Ser	Tyr	Cys	C
	Leu	Ser	stop	stop	A
	Leu	Ser	stop	Trp	G
C	Leu	Pro	His	Arg	U
	Leu	Pro	His	Arg	C
	Leu	Pro	Gln	Arg	A
	Leu	Pro	Gln	Arg	G
A	Ile	Thr	Asn	Ser	U
	Ile	Thr	Asn	Ser	C
	Ile	Thr	Lys	Arg	A
	Met	Thr	Lys	Arg	G
G	Val	Ala	Asp	Gly	U
	Val	Ala	Asp	Gly	C
	Val	Ala	Glu	Gly	A
	Val	Ala	Glu	Gly	G

Table 7.2 Alternative codon usage.

Codon	Common use	Alternative use
CUU/C/G/A	Leu	Thr in yeast mitochondria
AGA, AGG	Arg	Stop in yeast and vertebrate mitochondria
AUA	Ile	Met in mitochondria of yeast, *Drosophila*, and vertebrates
CGG	Arg	Trp in some plant mitochondria
UGA	stop	Trp in mycoplasma and some plant mitochondria; alternatively selenocysteine, Sel, in a few taxa
UAA	stop	Gln in some protozoa
UAG	stop	Gln in some protozoa; pyrrolysine (Pyr) in some bacteria and archaea
CUG	Leu	Ser in *Candida albicans*

major change would have destroyed most cellular proteins. In fact, it seems remarkable that some changes have been tolerated.

7.4 Surprises from the eukaryotic cell: introns and splicing

Eukaryotic genes usually contain interspersed noncoding sequences

By 1967, the fundamentals of molecular biology seemed to be firmly established, at least insofar as bacteria and viruses were concerned. The roles of messenger RNA, transfer RNA, and ribosomes in transmitting information from DNA to protein were clear. The code had been completely worked out and seemed universal for viruses and bacteria. The essential enzymology for the transcription of mRNA from DNA by RNA polymerase was recognized.

Many researchers then turned their attention to the molecular biology of eukaryotes, confidently expecting more of the same, but surprises were in store. First, in eukaryotes, there turned out to be not one RNA polymerase, as in bacteria, but three, each with a different specialization (see Chapter 10). There were exceptions to the bacterial code in some lower eukaryotes and in mitochondria (see Table 7.2). But the great shock was the discovery, first by Phillip Sharp and Richard Roberts in 1977, that the mRNA found in the cytoplasm of eukaryotic cells need not necessarily be complete copies of the genes in the nuclear DNA. Specifically, it was found that within the genes there were stretches of DNA sequence that had no counterparts in the cytoplasmic mRNAs. These intervening sequences, later called **introns**, did not correspond to any part of the protein sequence. They lay between sequences that did code for protein, now called **exons**.

The first discovery of introns was in the genes of adenovirus, a virus that infects human respiratory cells, and so it was suspected at first that introns might be some idiosyncrasy of such viruses. This expectation was soon completely dashed when Pierre Chambon demonstrated that the chicken ovalbumin gene contained no less than seven intervening sequences, along with eight exons (Figure 7.7). Furthermore, the total length of introns was far greater than the total length of the exons that contained the cytoplasmic message. Soon, other eukaryotic genes were found to contain introns. We now know that eukaryotic genes without introns are in the minority.

Clearly, something special must take place in the eukaryotic nucleus to prepare mRNA from which the introns have been removed. In subsequent years, an elegant process of **splicing** was discovered. The gene, it turns out, is first completely transcribed, introns

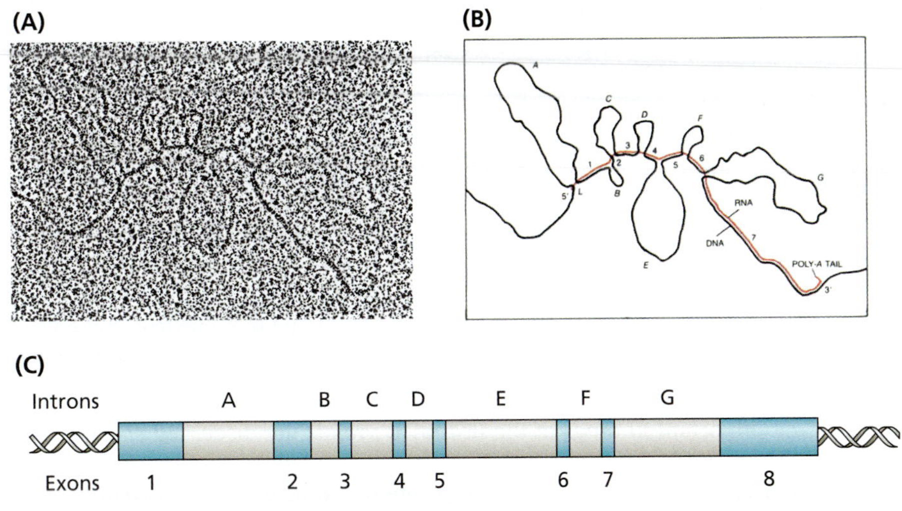

Figure 7.7 The ovalbumin gene and its mRNA illustrate the concept of split genes. The protein is 386 amino acids long and could have been encoded by a gene of 1158 base pairs, but the length of the gene is actually 7700 base pairs. (A) This anomaly was explained when ovalbumin mRNA was purified from the cytoplasm and allowed to hybridize with the ovalbumin gene, and the resulting hybrid was examined under the electron microscope or EM. (B) In this schematic representation of the EM image the black line shows the DNA; this is the complete gene including the introns (A to G), regions that are not present in the cytoplasmic mRNA. The mRNA is shown in red; it includes a 5' leader region (L) and seven exons (labeled 1 to 7). (C) A linear representation of the ovalbumin gene, with the leader plus 7 exons shown in red, and the introns labeled A to G shown in gray. (Adapted from Chambon P [1981] *Sci Am* 244:60–71. With permission from Scientific American.)

and all, to make a **pre-mRNA**. This is precisely cut and resealed to make a transcript containing only exon regions, in the proper order. After some further adjustments to the 5′- and 3′-ends, the mature message is delivered to the cytoplasm for translation. All of these processes are described in detail in Chapter 14.

7.5 Genes from a new and broader perspective

Protein-coding genes are complex

A generalized structure of a eukaryotic protein-coding gene is presented in Figure 7.8. In addition to the introns found in the **pre-messenger RNA**, there are regions remaining in the mature mRNA that do not code for amino acids. These are appropriately called **untranslated regions** (**UTR**), 5′ or 3′ depending on their location with respect to the coding sequence, and they play a role in the regulation of translation of the message (see Chapter 17). Each gene also contains a defined site where the actual process of transcription is initiated and has an associated **promoter** sequence, which is the binding site for RNA polymerases. As we see in Chapters 9 and 10, recognition of the promoter regions by the polymerases could be either direct, as in bacteria, or indirect, with a mechanism that involves general transcription factors, as in eukaryotes. Other regions that are associated with protein-coding genes are called **enhancers** and **silencers**; these are involved in the regulation of gene expression even though they may reside at considerable distance from the coding sequence. Thus, a gene cannot be considered as just the string of bases that codes for a protein sequence.

Genome sequencing has revolutionized the gene concept

The entire genetic content of an organism is referred to as its **genome**, just as the total list of proteins is its proteome (see Chapter 3). Modern, ultrafast techniques have allowed the sequencing of many whole genomes, in particular the human genome. This, in turn, has completely upset our established notions of what to call a gene. New discoveries of the past decade have made the old molecular definition of a gene obsolete.

The major contribution to analysis of the human genome and its functional elements came from the **Encyclopedia of DNA Elements** or **ENCODE** project consortium. This consortium was created in 2003 under the auspices of the National Human Genome

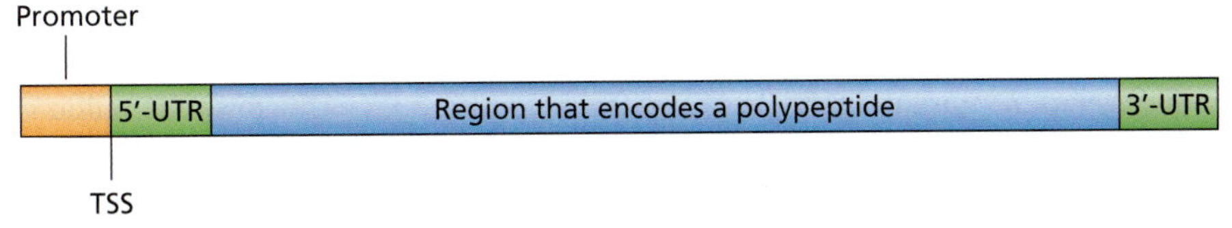

Figure 7.8 Generalized structure of a protein-coding gene. In addition to the portion of the gene that directs synthesis of the respective polypeptide, known as the coding region, each gene has a transcription start site, TSS, and usually two regions, one on each side of the coding region, that are present in the mRNA but are not translated into proteins; these are called untranslated regions or UTRs and perform regulatory functions during protein synthesis. Note that in eukaryotes the majority of protein-coding genes contain introns, as illustrated in Figure 7.7. Finally, each gene must have a promoter, a sequence upstream of the TSS that serves as a landing platform for the RNA polymerase.

Research Institute in the United States and consists of a large international group of scientists with broad expertise in genomewide experimental methods and data analyses by computational methods. The consortium operationally defines a **functional element of DNA** as a "discrete genome segment that encodes a defined product (for example, protein or non-coding RNA) or displays a reproducible biochemical signature (for example, protein binding or a specific chromatin structure)." The elements mapped across the human genome include all regions transcribed into RNA, protein-coding regions, transcription-factor binding sites, features of chromatin such as DNase I hypersensitivity and histone modifications (see Chapter 13), and DNA methylation sites. Data production efforts focused on two sets of human cell lines, designated tier 1 and tier 2. Tier 1 cell types were the highest priority set and comprised three widely studied cell lines, two malignant and one embryonic stem cell line; tier 2 included two further malignant cell lines and primary nontransformed cells from umbilical vein endothelium.

Many of the important findings of the ENCODE project are discussed elsewhere in this book. Here, we focus on the overall results that totally changed our long-held views on genes. ENCODE identified 20,687 protein-coding genes, with an indication that some additional protein-coding genes remain to be found. These genes constitute less than 3% of the entire genome. This proportion is much smaller than had been expected and made it impossible to continue to think that the major role of DNA is to code for proteins, especially since ENCODE data indicate that around 80% of the entire human genome is transcribed. A very large fraction of the human genome is, instead, being transcribed into a number of classes of RNA molecules, most of presently unknown function. The number of small, <200 nucleotide RNA molecules has been estimated at 8801 and the number of long noncoding RNAs (lncRNAs) at 9640. LncRNAs are generated through mechanisms and pathways similar to those that transcribe protein-coding genes, including use of the RNA polymerase specific to protein-coding genes (RNAP II, see Chapter 10). The project also annotated 11,224 **pseudogenes**. These are exact or nearly exact copies of genes that are not expressed as proteins. Unexpectedly, and contrary to our present understanding of pseudogenes, a substantial number, 863 or about 7% of these, are transcribed and associated with active chromatin features.

Given this new information, we no longer have a widely accepted definition for a gene. It is clear that the definition must include the ability of a gene to direct the production of either functional RNA or protein molecules. One of the main unresolved issues is whether or not to include in the definition of a gene the regulatory regions—promoters, enhancers, and so on—that accompany it (see Chapter 12). The problem is that most of these regions are ill-defined, and there is no clear-cut one-to-one correspondence between a single gene and a particular regulatory region. In other words, the same regulatory region can control the transcription of numerous genes; and vice versa, the same gene may be regulated by numerous regions, often at huge distances from each other and from the gene.

The new information also sheds a peculiar light on the history of molecular biology. Almost all of the very creative thinking that went on in the decade or so after 1953 was dedicated to solving the problem of how genes directed the synthesis of protein sequences (see Figure 7.1). We now find that this is a minor function of the genome. Yet it is hard to see how the field could have advanced so far and so rapidly along any other track.

Mutations, pseudogenes, and alternative splicing all contribute to gene diversity

The continued evolution of eukaryotes, with ever more diversified cell types and capabilities, has required the development of new protein-coding genes or modifications of old ones. As shown later (see Table 7.3), progression from the level of yeast to humans has involved an approximate four-fold increase in the number of protein-coding genes. A number of mechanisms can account for this. Obviously mutation is one, but in itself mutation is limited in potential. Frameshift mutations will almost invariably lead to a nonfunctional product. Point mutations in themselves may change

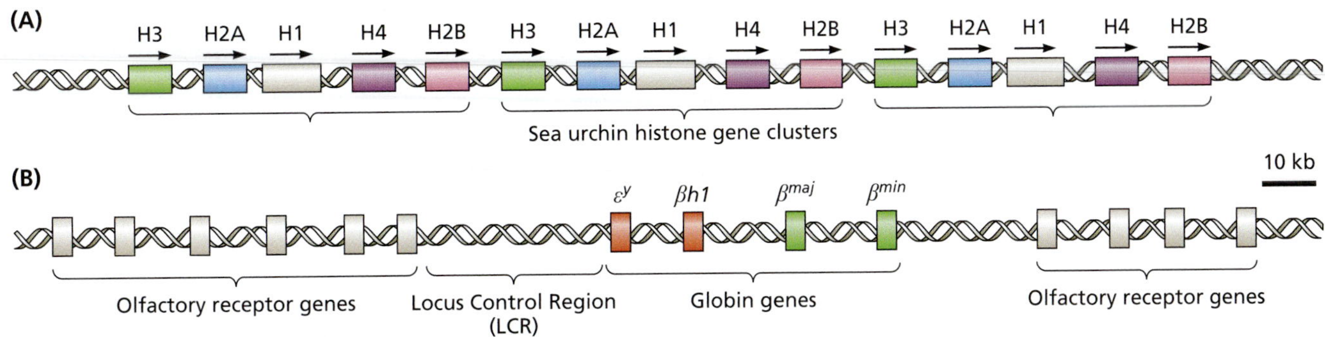

Figure 7.9 Multigene families consist of either identical or similar member genes. (A) Histone gene family in sea urchin. The genes for all five histones that participate in organizing chromatin structure are clustered, and the clusters are tandemly repeated numerous times; the number of repeats differs from species to species. (B) Organization of mouse globin genes. These genes differ slightly from each other; they encode slightly different globin proteins, which give rise to slightly different oxygen-bearing hemoglobin molecules. Expression of each of these genes is developmentally regulated during embryogenesis and in the adult organism, with the locus control region (LCR) playing a major role in this regulation.

the function or properties of a gene product, sometimes in ways that will be evolutionarily favored. However, point mutations will not increase the number of genes; they simply substitute one gene for another of similar or identical function.

Much effective evolution comes about through gene duplication. Mechanisms exist that can result in the duplication of a gene or even a contiguous set of genes (see Chapter 21). Initially, gene duplication may mean only that extra copies of the gene are present to produce more mRNA. This seems to be the case for the repeated copies of the sets of histone genes (Figure 7.9A), which may be an adaptation to the need for a great burst in histone synthesis to occur concomitantly with DNA replication.

Alternatively, the new copy might not be expressed, perhaps because it lacks a functional promoter or translational controls; it then becomes a pseudogene (Figure 7.10). Pseudogenes can also arise from reverse transcription of processed RNA by the enzyme reverse transcriptase (see Chapter 21). In either event, mutations can accumulate in the new DNA sequence. If this sequence is, or becomes, expressible, it may now serve a varied or even completely different function. An example of this is seen in Figure 7.9B, which shows the β-globin genes. Copies of an original globin gene have diverged, producing variants of hemoglobin that can function best at different developmental stages (see Figure 12.3).

In recent years, it has been discovered that considerable diversity in proteins is a consequence of **alternative splicing**. This process, whereby exons are spliced together in the nucleus to produce the mature mRNA, can be regulated so as to allow alternative

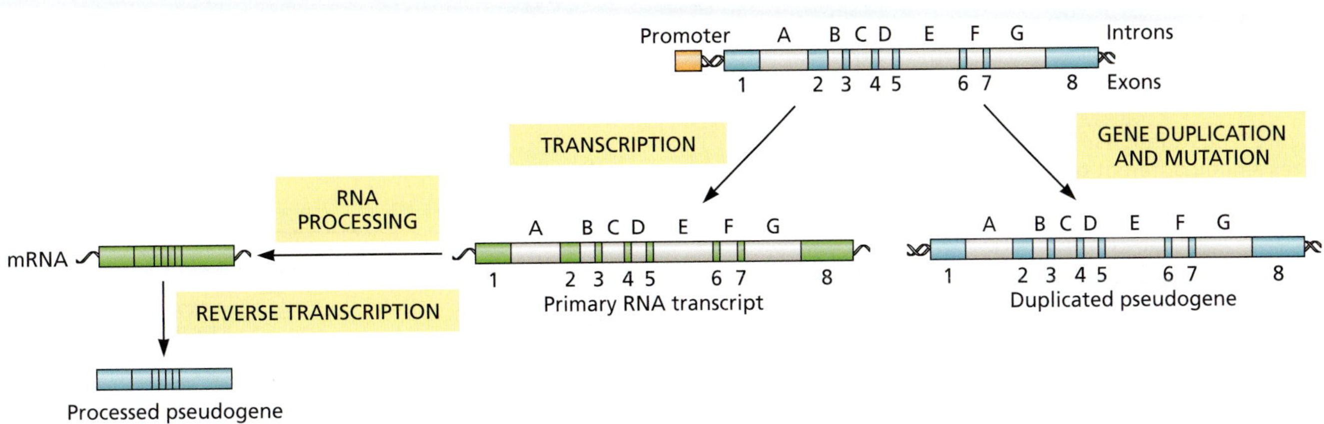

Figure 7.10 Pseudogenes. Pseudogenes are DNA sequences that are very similar to real, functional genes but lack the sequences necessary for gene expression. Duplicated pseudogenes arise by gene duplication events followed by mutations. Processed pseudogenes are formed by reverse transcription and mutation: the primary transcript is first processed to mature mRNA, which is then reverse-transcribed into double-stranded DNA.

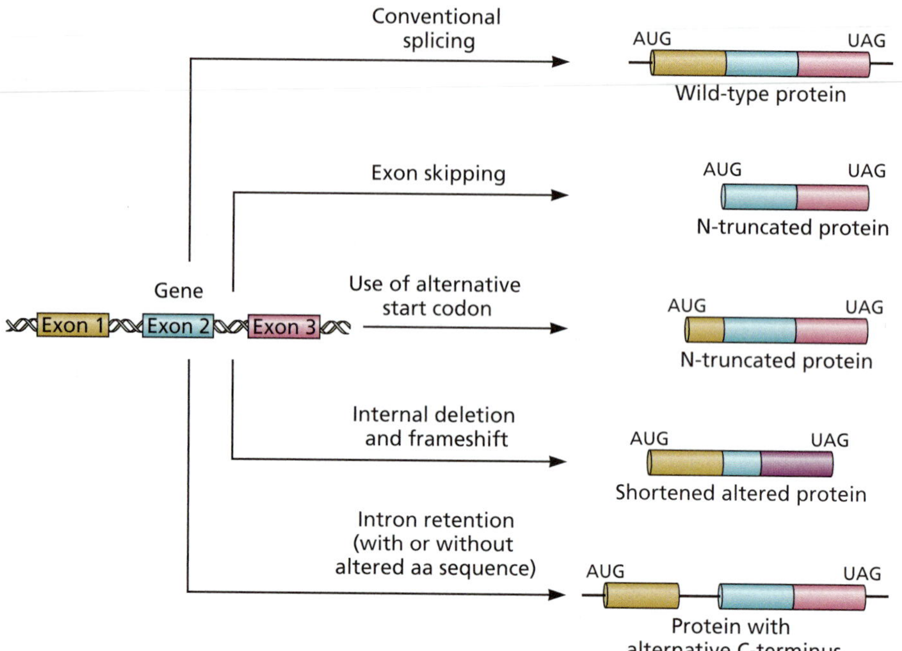

Figure 7.11 Possible pathways of alternative splicing. A hypothetical gene comprising three exons and two introns is shown. Note the different protein products of each pathway.

ways to put the exons together or to omit some entirely (**Figure 7.11**). An enormous variety of similar but functionally different proteins can be produced in this manner. For a more detailed discussion of alternative splicing, see Chapter 14.

7.6 Comparing whole genomes and new perspectives on evolution

Genome sequencing reveals puzzling features of genomes

With the development of ultrafast DNA sequencing, it has become possible to determine the entire genome sequences of many organisms. Comparison of even the gross features of these sequences has important implications with respect to evolution. The first genomes to be studied were those of viruses and bacteria, with the landmark publication of the *Escherichia coli* genome, sequenced by Ying Shao and collaborators, in 1997. Many other bacterial genomes have since been sequenced. Extension to the simpler eukaryotes, such as yeast, soon followed, and as of 2013, complete genome sequences are known for a few hundred eukaryotic species, among them at least 130 fungi, 30 insects, 7 nematodes, 21 plants, and 27 mammals, including humans (**Figure 7.12** and **Table 7.3**). The range in genome size is great, even among

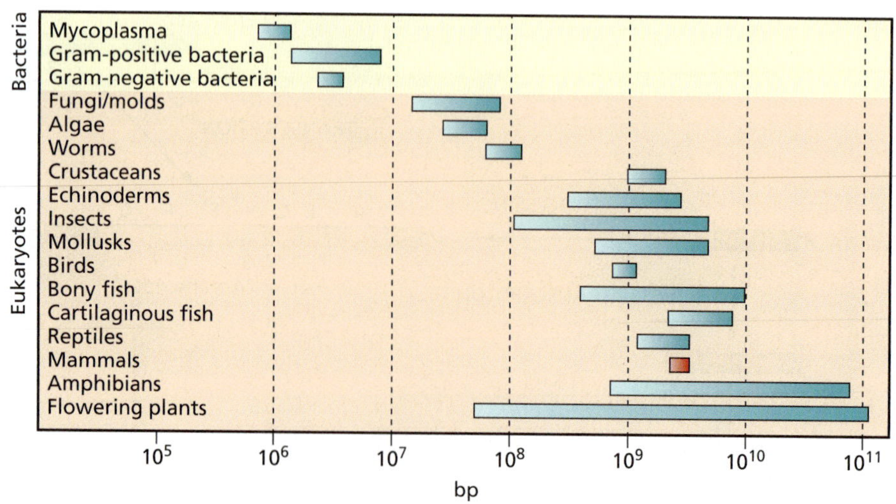

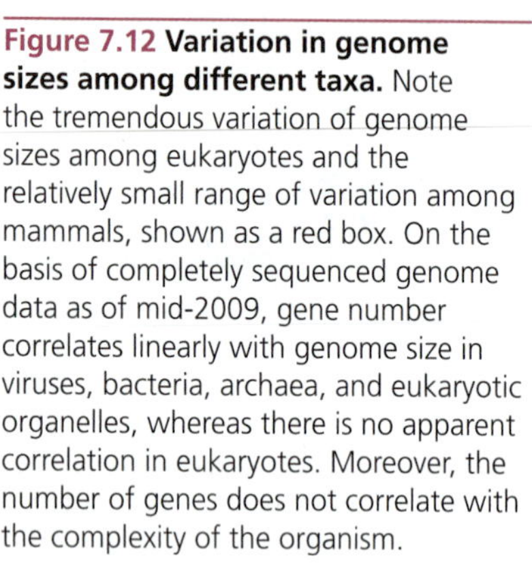

Figure 7.12 Variation in genome sizes among different taxa. Note the tremendous variation of genome sizes among eukaryotes and the relatively small range of variation among mammals, shown as a red box. On the basis of completely sequenced genome data as of mid-2009, gene number correlates linearly with genome size in viruses, bacteria, archaea, and eukaryotic organelles, whereas there is no apparent correlation in eukaryotes. Moreover, the number of genes does not correlate with the complexity of the organism.

Table 7.3 Genome sizes, gene numbers, and chromosome numbers for several selected organisms whose genome sequences have been completely determined.

Species	Genome size (× million base pairs)	Number of protein-coding genes	Number of chromosomes
phage λ	0.048	73	1[a]
human immunodeficiency virus	0.00975	9	2[b]
Haemophilus influenzae	1.83	1743	1
Bacillus subtilis	4.22	4422	1
Escherichia coli	4.6	4288	1
Saccharomyces cerevisiae	13.5	5882	16[c]
Saccharomyces pombe	12.5	4929	3[c]
Caenorhabditis elegans	100	18,424	12
Drosophila melanogaster	139	15,016	8
Arabidopsis thaliana	119	25,498	10
Oryza sativa, rice	390	37,544	24
Mus musculus, mouse	2500	23,786	40
Homo sapiens	3200	20,687	46

[a]Linear double-stranded DNA molecule.
[b]Single-stranded RNA molecules.
[c]Haploid.

eukaryotes, with the human genome being 200-fold larger than that of yeast. On the other hand, most metazoans seem to have roughly the same number of protein-coding genes, about 20,000–30,000.

Examination of Table 7.3 raises some curious questions. For example, humans have about the same number of protein-coding genes as the worm *Caenorhabditis elegans*. But humans have about 30 times more total DNA. What are we doing with all that extra DNA that the worm does not seem to need? One postulate is that extra information is needed to program the much more complex development and body plan of humans, and this information is carried in DNA that does not code for proteins. From this perspective, it seems naive to have thought that the major function of genes was to code for proteins. It is similar to thinking that the complete prescription of a factory is given by descriptions of the machines. How they are placed and interconnected, and how they are coordinately controlled, is essential to describing the whole and its functioning.

More peculiarities appear if we examine the distribution of protein-coding and non-protein-coding regions in vertebrates. Figure 7.13 depicts comparable regions in a human and three mouse chromosomes and shows how such regions tend to segregate (see Chapter 2). In general, there are chromosome regions that are gene-poor and others that are gene-rich. In addition, there are some significant differences in the density of genes on different chromosomes. For example, the male sex chromosome Y has only around 60 genes, which is far fewer than the average of about 1000. It is a tiny chromosome, with gene density far below the average. Furthermore, the Y chromosome is present in only one copy in the cells of a male, and so the

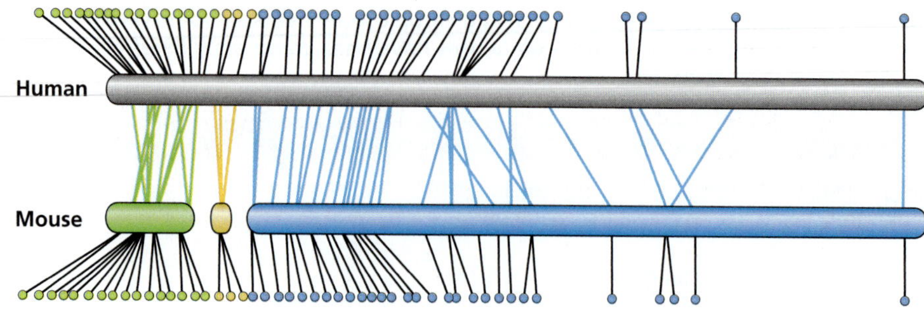

Figure 7.13 Schematic view of homologous regions in human and mouse chromosomes. Human chromosome 21 is shown in gray; a portion of mouse chromosome 16 is shown in blue, of 17 in orange, and of 10 in green. Protein-coding genes are presented as dots of the three colors corresponding to the three mouse chromosomes. Each pair of gene markers is joined with a line. Note the highly uneven distribution of genes along the chromosomes but the relatively high conservation of gene order. (Adapted from Hattori M, Fujiyama A, Taylor TD et al. [2000] *Nature* 405:311–319. With permission from Macmillan Publishers, Ltd.)

Y-linked loci are effectively haploid. Box 7.3 discusses the organization and function of Y-linked genes.

How are DNA sequence types and functions distributed in eukaryotes?

We conclude with an overview of the types of sequences found in eukaryotic DNA, using data from the human genome as an example. Some of the values given in Figure 7.14 must be regarded as approximate because it is not always possible to put sequence elements into well-defined categories. Furthermore, the task of systematizing and understanding the vast amount of data accumulated in the ENCODE project is far from complete.

Box 7.3 A Closer Look: Human Y chromosome and the SRY gene The human Y chromosome is the most extreme example of a gene-poor chromosome, as described in the text. It has the lowest gene density of all chromosomes. Y is evolving rapidly, especially in the regions that code for male sex-determining factors. The cells of a male are effectively haploid for the genes uniquely present in Y. These genes are marked in Figure 1; the most-studied gene involved in determining the anatomical sex of males is the ***SRY* gene**, whose protein product is shown in Figure 2.

Since its discovery, the importance of the *SRY* gene in sex determination has been extensively documented. For example:

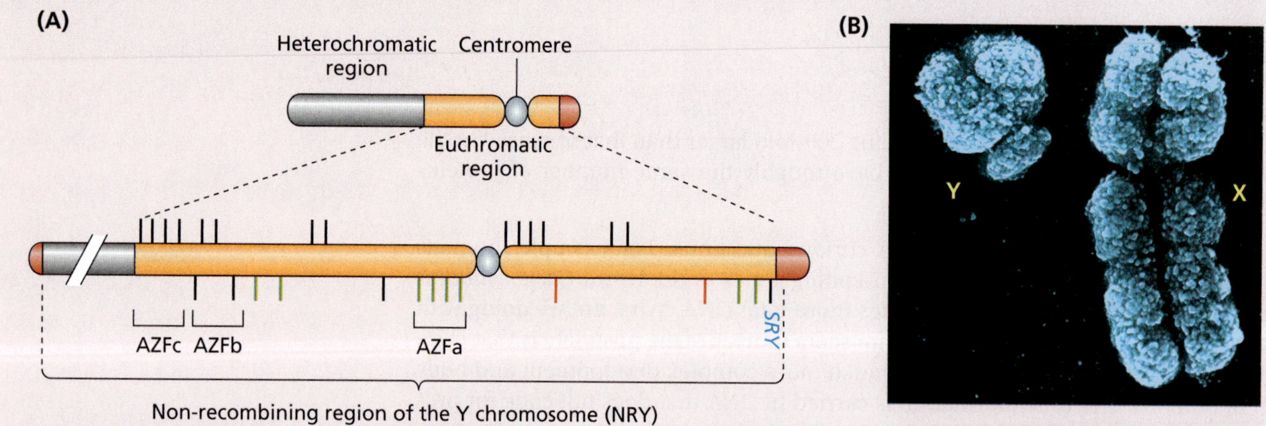

Figure 1 Structure of the human Y chromosome. (A) Yellow bar, expressed euchromatic portion of the non-recombining region of the Y chromosome, NRY; gray bar, heterochromatic portion of the NRY; light gray sphere, centromere; orange bars, pseudoautosomal regions that can recombine with X. Note that these recombining regions are thought to be evolutionary remnants from the time when the X and Y chromosomes shared long regions of homology. For clarity, genes in the pseudoautosomal regions are omitted. Genes are marked by vertical bars: those shown along the bottom have active X-chromosome homologs, whereas those shown along the top lack known X homologs. Genes marked with green bars are widely expressed housekeeping genes; genes with black bars are expressed in the testis only; and two genes with red bars are expressed only in a few unrelated tissues, tooth buds and brain. With the exception of the *SRY* or sex-determining region Y gene, all testis-specific Y genes are multicopy. *SRY* is an intronless gene that codes for a transcription factor, an HMG-box DNA-binding protein. Some multicopy gene families form dense clusters, the constituent loci of which are indistinguishable at the resolution of this map. Three regions that are often found to be deleted in infertile men, azoospermia factor regions AZFa, AZFb, and AZFc, are indicated. Males suffering from this infertility condition, which affects around 1% of the male population, have no measurable level of sperm in semen. (B) Micrograph illustrating the morphology and size of the two human sex chromosomes. (A, adapted from Lahn BT, Pearson NM & Jegalian K [2001] *Nat Rev Genet* 2:207–216. With permission from Macmillan Publishers, Ltd. B, courtesy of Indigo Instruments.)

- Humans with one Y chromosome and multiple X chromosomes—XXY, XXXY, etc.—are usually males.
- Individuals with a male phenotype and an XX, that is, female, karyotype have been observed; these males have the *SRY* gene in one or both X chromosomes, moved there by chromosomal translocation. These males are, however, infertile.
- Similarly, there are females with an XXY or XY karyotype. These females have no *SRY* gene in their Y chromosome, or the *SRY* gene exists but is defective or mutated, in a condition known as Swyer syndrome.

***SRY* and the Olympics**

The *SRY* gene was used as a means for gender verification at the Olympic Games held in 1992 and 1996. Athletes with a *SRY* gene were not permitted to participate as females, although all athletes in whom the gene was detected at the 1996 Summer Olympics were ruled false positives and were not disqualified. In the late 1990s, a number of relevant professional societies in the United States called for elimination of gender verification, stating that the *SRY*-based method was uncertain and ineffective. The screening was eliminated in time for the 2000 Summer Olympics.

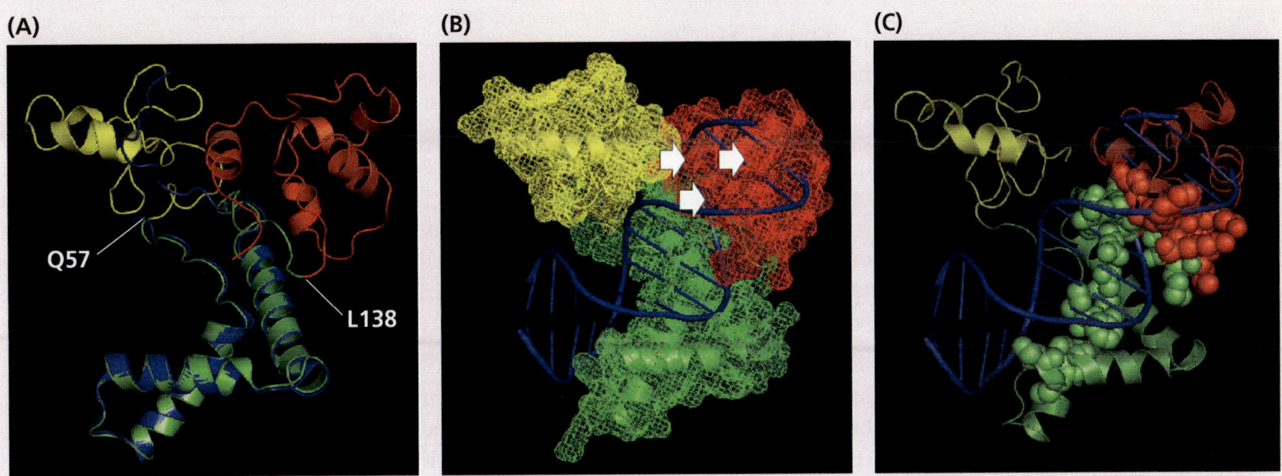

Figure 2 Computational model of the structure of the entire SRY protein bound to DNA. (A) The SRY protein computational model and the known structure of a representative HMG box, shown in blue, were structurally annealed by using the locations of Gln57 and Leu138 as points of alignment. SRY regions are shown as follows: N-terminal, yellow; HMG box, green; C-terminal, red. (B) In the models of SRY–DNA complexes, DNA passes through amino acid residues in the C-terminal region of SRY, shown by white arrows. The binding of the HMG box creates a bend in the DNA, which is a characteristic feature of all HMG boxes. (C) The DNA interacting cavity is formed by amino acid residues shown as green spheres in the HMG-box region and as red spheres in the C-terminal region of SRY. (From Sánchez-Moreno I, Coral-Vázquez R, Méndez JP & Canto P [2008] *Mol Hum Reprod* 14:325–330. With permission from Oxford University Press.)

We have already mentioned that the actual protein-coding sequences defined by the exons in transcribed genes represent a very small portion of the whole genome. At one time, much of the remainder was thought to be **junk DNA**, but we now know this is not true. If one sums together all sequences related to the production of proteins, the total is much larger, ~40%. These sequences would include not only exons but also

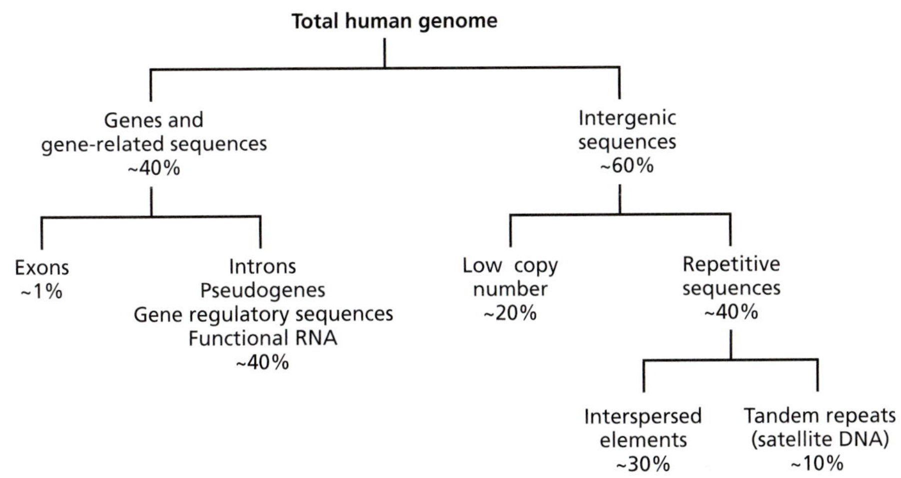

Figure 7.14 Approximate distribution of DNA functions in the human genome. Note that the DNA that actually codes for proteins is a very small fraction of the genome. The function of low-copy-number intergenic sequences is not well understood. Interspersed elements are mostly transposons. Satellite DNA can be further subdivided into mini- and microsatellites.

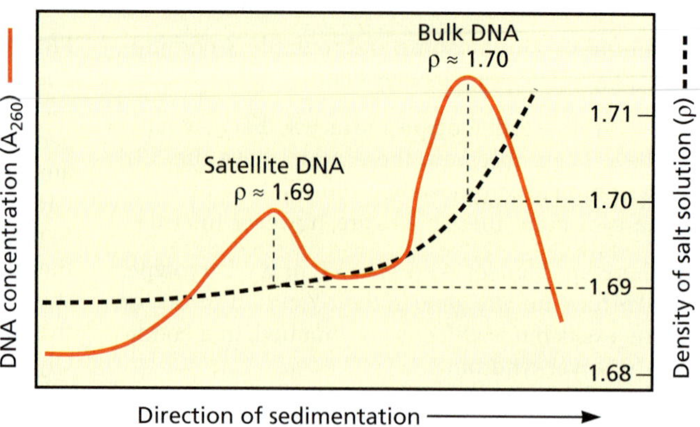

Figure 7.15 Density-gradient resolution of satellite DNA. The density gradient is created within the ultracentrifuge cell by the sedimentation–diffusion equilibrium of a dense salt such as CsCl. DNA will stably band at the position where its density matches that of the solution. In this case, an AT-rich satellite bands at lower density than bulk DNA.

introns, promoter regions, enhancers, silencers, pseudogenes, and RNA genes that function in the production of proteins: tRNAs, ribosomal RNAs, regulatory RNAs, and so on. The last group is ill-defined, as many of the RNA transcripts that are being discovered are of uncertain function (see Chapters 10, 12, and 13). The remaining 60% of the genome has been somewhat characterized, but its function is less clear. A substantial portion consists of transposons (see Chapter 21) which are scattered throughout the genome. About 10% of the total DNA is present as short tandem repeats, also referred to as **satellite DNAs**. These can be further subdivided into minisatellites, which are sequences of 10–15 base pairs repeated up to ~1000-fold, and microsatellites, which are segments of 2–6 bp present in up to 100 copies.

Satellite DNA is so named for the following reason: it can be identified as satellite bands, separated from the bulk DNA, in density-gradient equilibrium sedimentation (**Figure 7.15**). If whole DNA is cleaved with a restriction nuclease that does not cut in the repetitive sequence, the stretches of repeats produced will often have an overall base composition very different from the bulk: consider an $(AT)_n$ repeat, for example. DNA density depends on base composition, so such DNA will band separately from the bulk (see Figure 7.15). Analysis of microsatellites is being widely used in forensics (**Box 7.4**).

Box 7.4 Microsatellites and DNA Identification Certain features of microsatellite DNA have made this genome component especially useful in the identification of individuals from samples of DNA, known as DNA fingerprinting. A particular microsatellite will often differ from one individual to another, especially in the number of repeats. This is believed to result from slippage during DNA replication, because the DNA replication machinery has a hard time moving continuously on repetitive templates. Such satellites are often heterozygous, meaning that the two copies in a diploid individual differ in copy number. These differences are easily detected by gel electrophoresis as shown in **Figure 1**, which depicts a hypothetical forensic application of the method.

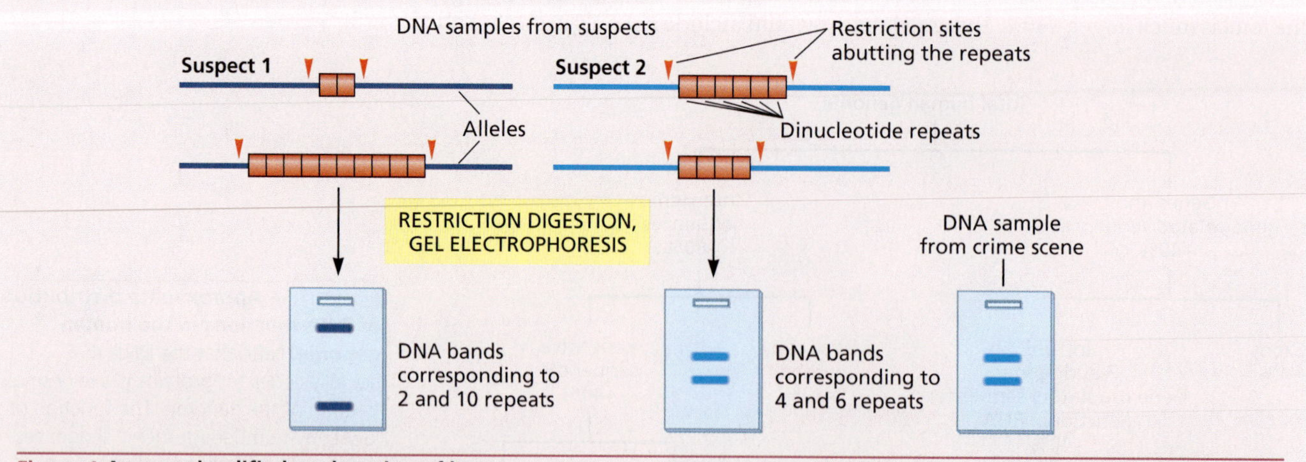

Figure 1 An oversimplified explanation of how measurements of dinucleotide-repeat lengths in microsatellites can be used in criminal investigations. Suspect 1 is very likely innocent; suspect 2 may be guilty. Similar tests at other loci are needed to provide definitive evidence.

Key concepts

- The decade between 1953 and 1963 saw the emergence of a whole new field, molecular biology, which also encompassed a new molecular genetics. We can summarize the developments as follows.
- The complementary double-stranded structure of DNA immediately suggests how genetic information might be duplicated during cell division.
- The central dogma proposes that there is a one-way flow of information from the nucleotide sequence of DNA to the polypeptide sequences of proteins, and there must exist a code to allow this translation.
- The intermediate in this transfer is messenger RNA, which copies one strand of DNA and is then translated into protein.
- The molecular machine on which translation occurs is the ribosome.
- In order to match amino acids to particular oligonucleotide codon sequences on the RNA, adaptor molecules known as transfer RNAs are required. These carry both an activated amino acid and the appropriate anticodon.
- The code is three-letter or triplet, non-overlapping, and unpunctuated.
- All 64 possible three-letter codons are used, which means the code is degenerate, with different amino acids having from one to six possible codons. Most variation is in the third place of the codon.
- One codon, AUG, is used as a start signal for translation, and the three codons UGA, UAA, and UAG are commonly employed as stops.
- The code is almost universal from microbes to humans, with a few exceptions, mostly in mitochondria.
- Bacterial mRNAs are faithful copies of the corresponding genes, but eukaryotic genes contain intervening sequences, known as introns, which must be removed so the functional or coding parts of the message, known as exons, can be spliced together.
- This splicing can sometimes be altered, in a process called alternative splicing, to yield different mRNAs and consequently rearranged proteins.
- Protein diversity in evolution seems to have resulted mainly from gene duplications and subsequent mutations and from alternative splicing.
- The entire genetic information of an organism is referred to as its genome; its complete protein library is known as its proteome.
- Genome size varies enormously over eukaryotes, but proteome size does not vary nearly so much. The reasons are obscure.
- Very little of the human genome is actually translated into protein. Much appears to be transcribed into various classes of functional but non-protein-coding RNA. A substantial fraction is distributed over several kinds of repetitive DNA.

Further reading

Books

Judson HF (1996) The Eighth Day of Creation: Makers of the Revolution in Biology. Cold Spring Harbor Laboratory Press.

Mount D (2004) Bioinformatics: Sequence and Genome Analysis, 2nd ed. Cold Spring Harbor Laboratory Press.

Ridley M (2006) Francis Crick: Discoverer of the Genetic Code. Harper Collins Publishers.

Watson JD (1968) The Double Helix: A Personal Account of the Discovery of the Structure of DNA. Atheneum.

Reviews

Balakirev ES & Ayala FJ (2003) Pseudogenes: are they "junk" or functional DNA? *Annu Rev Genet* 37:123–151.

Breathnach R & Chambon P (1981) Organization and expression of eucaryotic split genes coding for proteins. *Annu Rev Biochem* 50:349–383.

Chambon P (1981) Split genes. *Sci Am* 244:60–71.

Crick FH (1958) On protein synthesis. *Symp Soc Exp Biol* 12:138–163.

Frazer KA (2012) Decoding the human genome. *Genome Res* 22:1599–1601.

Nirenberg M (1965) Protein synthesis and the RNA code. *Harvey Lect* 59:155–185.

Sharp PA (1994) Split genes and RNA splicing. *Cell* 77:805–815.

Yanofsky C (2007) Establishing the triplet nature of the genetic code. *Cell* 128:815–818.

Experimental papers

Bailey JA, Gu Z, Clark RA et al. (2002) Recent segmental duplications in the human genome. *Science* 297:1003–1007.

Berget SM, Moore C & Sharp PA (1977) Spliced segments at the 5′ terminus of adenovirus 2 late mRNA. *Proc Natl Acad Sci USA* 74:3171–3175.

Chow LT, Gelinas RE, Broker TR & Roberts RJ (1977) An amazing sequence arrangement at the 5′ ends of adenovirus 2 messenger RNA. *Cell* 12:1–8.

Crick FHC, Barnett L, Brenner S & Watts-Tobin RJ (1961) General nature of the genetic code for proteins. *Nature* 192:1227–1232.

Dunham I, Kundaje A, Aldred SF et al. (2012) An integrated encyclopedia of DNA elements in the human genome. *Nature* 489:57–74.

Jeffreys AJ, Wilson V & Thein SL (1985) Individual-specific 'fingerprints' of human DNA. *Nature* 316:76–79.

Venter JC, Adams MD, Myers EW et al. (2001) The sequence of the human genome. *Science* 291:1304–1351.

Watson JD & Crick FHC (1953) Molecular structure of nucleic acids: a structure for deoxyribose nucleic acid. *Nature* 171:737–738.

Yanofsky C, Carlton BC, Guest JR et al. (1964) On the colinearity of gene structure and protein structure. *Proc Natl Acad Sci USA* 51:266–272.

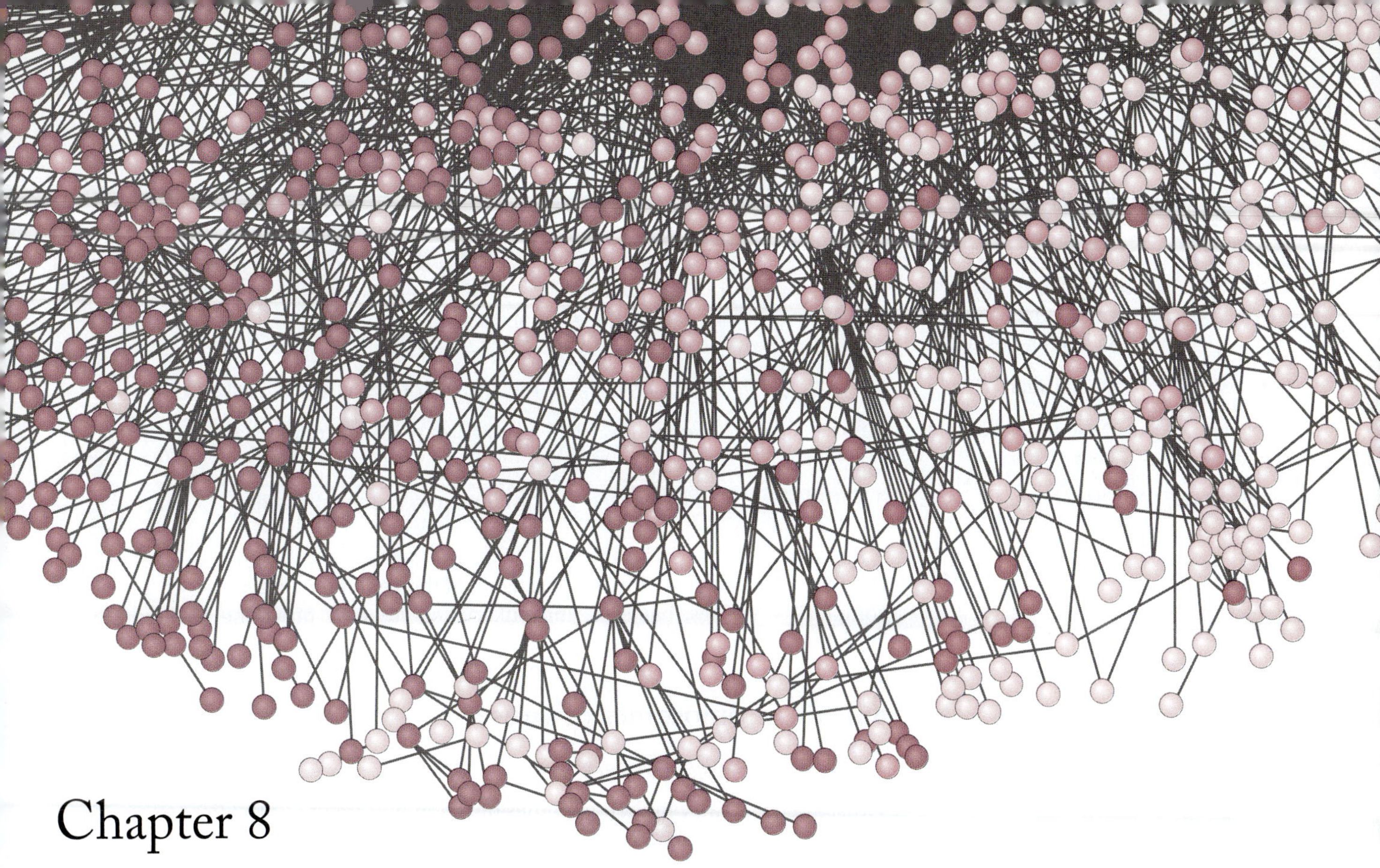

Chapter 8

Physical Structure of the Genomic Material

8.1 Introduction

The genome of any organism consists of one or more long polynucleotide chains. In viruses, the polynucleotide chain may be single-stranded or double-stranded, linear or circular, DNA or RNA. Bacteria usually have a single circular double-stranded DNA chromosome. In eukaryotes, each of several chromosomes carries one linear duplex DNA molecule. In each case, the size and informational complexity of all these genomes give rise to two problems that every organism must overcome. First, the extended length of these molecules far exceeds the dimensions of the capsids or cells that must contain them. If extended, the DNA in one human cell would be about 1 meter in length. Given that a human cell is a few micrometers in diameter, enormous compaction is required.

Second, at no time do all of the genes in a particular cell undergo transcription together. Even in viruses, there is programmed gene expression during the process by which these pathogens infect host cells. The problem becomes acute in higher organisms, or metazoa, which have many cell and tissue types, each carrying the same genome but each with its own requirements for specific gene expression. Switches in expression can be temporal as well as spatial. For example, bacteria must use alterations in gene expression to respond to environmental changes; the development of eukaryotes from embryo to adult involves an incredible series of gene expression switches.

These two problems, compaction and specific gene regulation, are dealt with in all organisms by proteins that are bound to the genome (see Chapter 6). In this chapter, we present an overview of the protein–nucleic acid complexes that are involved, focusing on those that determine the functioning genetic material. Specific details concerning the regulation of gene expression and replication are discussed in later

Table 8.1 Structural aspects of genomes.

Organismal level	Genome type	Typical genome size (bp)	Localization	Structural proteins
virus	dsDNA, dsRNA, ssDNA, ssRNA; linear or circular	10^3–10^5	viral envelope/ capsid	capsid/envelope proteins
bacteria	dsDNA; usually circular	10^6–10^7	cytoplasm	HU, IHF, minor proteins (see Table 8.2)
archaea	dsDNA; usually circular	10^5–10^6	cytoplasm	histone-like proteins (contain histone fold and interact through handshake motif; see main text)
eukaryotes	dsDNA; linear	10^7–10^{11}	nucleus	histones; non-histone proteins

chapters. Here, we describe the packaging of the genome in organisms of increasing complexity—viruses, bacteria, and eukaryotes—thereby providing background for later discussions of function (**Table 8.1**).

8.2 Chromosomes of viruses and bacteria

Viruses are packages for minimal genomes

Viruses need to carry in their genomes only a minimum of information: to construct particles that can enter a cell, replicate their nucleic acid and essential proteins, and then exit the cell. In many cases some information is taken from the host genome; for example, many viruses use host polymerases to replicate their genomes. Viral genomes come in a remarkable variety of forms. The polynucleotide may be DNA or RNA, and it may be single-stranded or double-stranded. The protein shell that surrounds the viral DNA or RNA, known as the **capsid**, is just sufficient to contain the genome, to permit the virus to get into a host cell, and to allow the newly replicated viruses within to escape to infect other cells. These requirements are met in a host of ways in different viruses, each using minimal machinery. As shown in the example of bacteriophage φ29 (**Figure 8.1**), some viruses are constructed in a way that allows

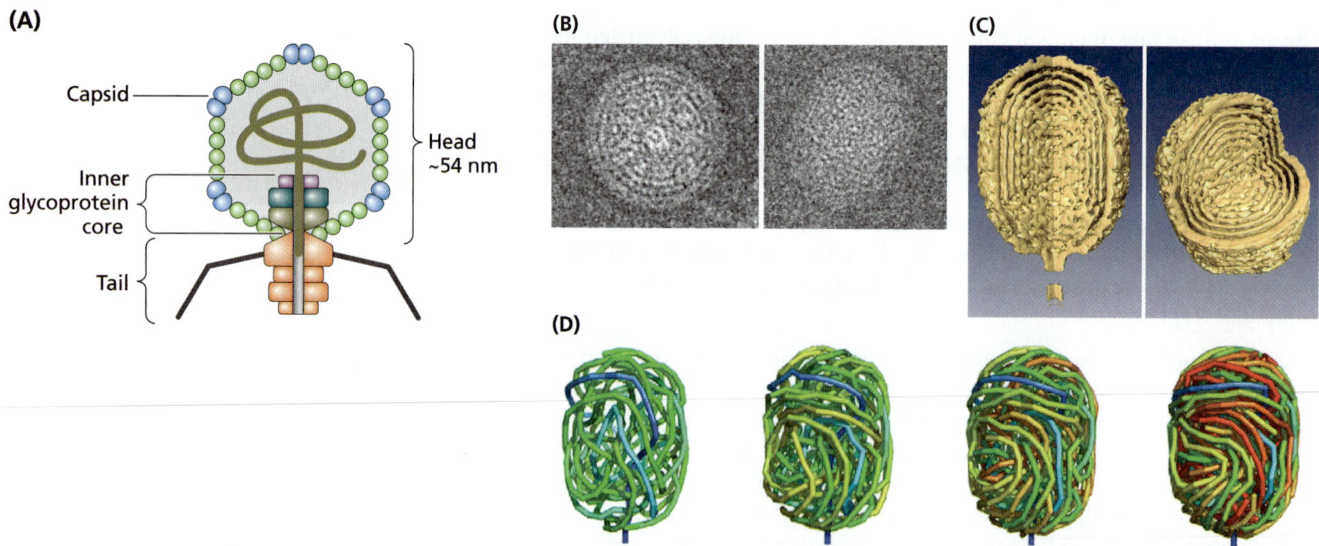

Figure 8.1 φ29 bacteriophage-like viruses. (A) Non-enveloped, head–tail structure. The tail is noncontractile and has a collar plus 12 tail fibers. The genome is double-stranded linear DNA of ~16–20 kb, encoding between 20 and 30 genes. (B) Typical cryo-EM images of mature φ29 particles at ~50,000-fold magnification. (C) Three-dimensional reconstruction of the full-length φ29 packaged genome. The degree of order increases from the center toward the capsid, with six concentric DNA shells visible in longitudinal and transverse sections of the volume. (D) Monte Carlo simulation of DNA packaging in φ29: consecutive images show stages at which 30%, 50%, 70%, and 100% of the genome has been packaged. The chain is colored blue at the end entering the confinement and red at the free end within the capsid at full packaging. The capsid is not shown. (A, adapted from ViralZone, Swiss Institute of Bioinformatics. B–D, from Comolli LR, Spakowitz AJ, Siegerist CE et al. [2008] *Virology* 371:267–277. With permission from Elsevier.)

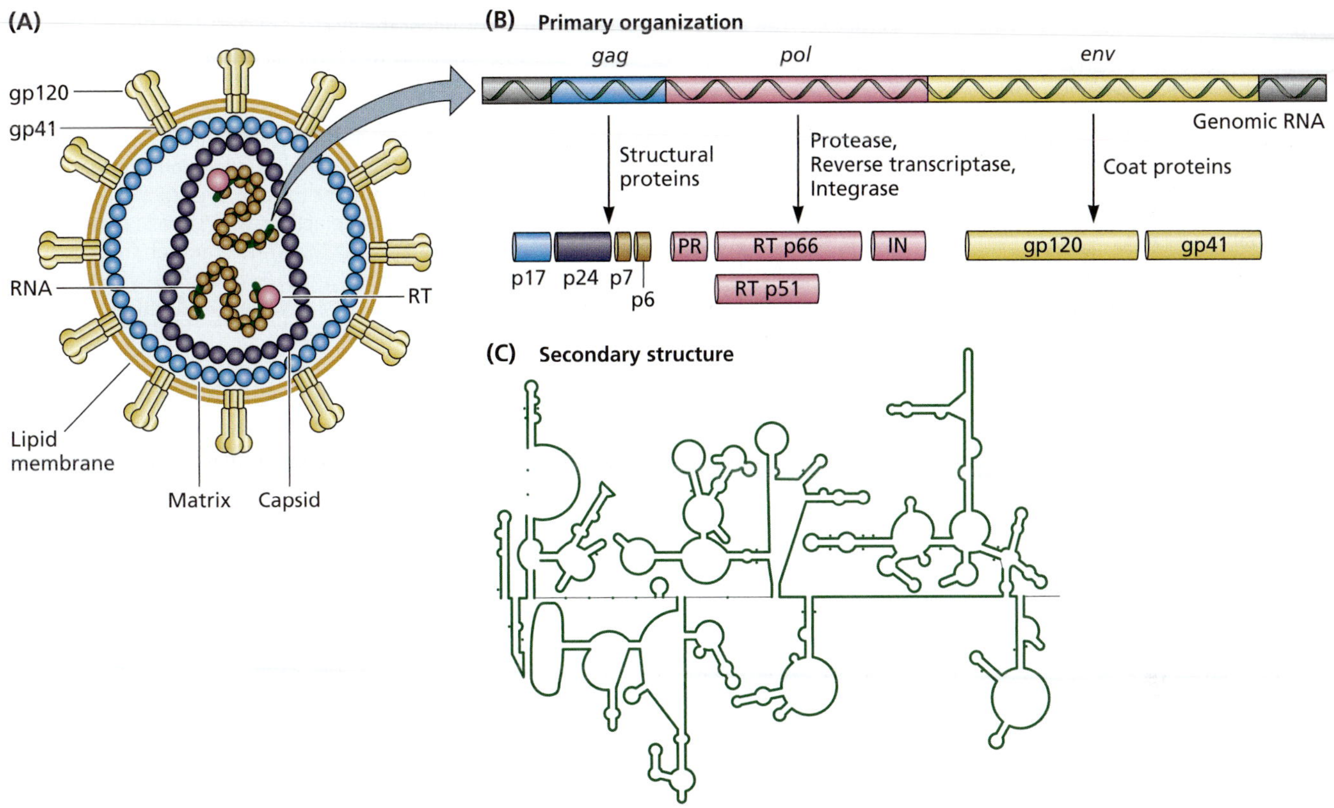

Figure 8.2 HIV and organization of its genome. (A) Anatomy of HIV. The outer coat of HIV is a layer of cell membrane stolen from the infected cell upon viral particle release. The coat is studded with protein spikes composed of two glycoproteins, gp120 and gp41, which serve to dock the virus onto receptor proteins on the surface of the host cell. This membrane layer envelops the matrix and capsid structures that encase the genetic material. The matrix proteins form a coat on the inner surface of the viral membrane and play a central role when new viruses bud from the surface of infected cells. The capsid proteins form a cone-shaped coat around the viral RNA, delivering it into the cell during infection. The proteins that form these structures are colored according to the color scheme representing the primary structure of the genome in part B. Three kinds of proteins that are essential for viral replication—reverse transcriptase (RT), integrase, and protease—are encoded by viral RNA and are present in the virions or mature viral particles. The HIV replication cycle is covered in more detail in Chapter 20. (B) Primary organization of the genetic material along the RNA molecule. (C) The secondary structure of the viral RNA is highly complex and is further coiled in three dimensions. The coiling is so dense that it was practically impossible to see any detail by electron microscopy. (C, adapted from Watts JM, Dang KK, Gorelick RJ et al. [2009] *Nature* 460:711–716. With permissionfrom Macmillan Publishers, Ltd.)

them to inject their DNA into a host cell; in these viruses, the DNA is tightly coiled within the capsid, and energy for injection is stored in the bending of the DNA and in electrostatic repulsion between DNA phosphates. This means that energy must be supplied to force the DNA into the capsid, for example, by ATP hydrolysis.

Other viruses, especially those that attack eukaryotic hosts, use a completely different method: they have an external membrane (envelope) that can fuse with the host's cell membrane to ensure internalization of the viral genome. The proteins or other molecules needed to accomplish this are synthesized by the host cell, using genetic instructions from the viral genome. The compaction of the nucleic acid to virus-compatible dimensions is in this case accomplished by the viral coat proteins, together with the presence of positive ions or small molecules bound to the DNA that help to neutralize the polynucleotide charge. The structural organization of the RNA genome of the human immunodeficiency virus, HIV, as a representative example of such enveloped viruses is presented in Figure 8.2. Many RNA viruses, including those that cause influenza and the common cold, are constructed in this fashion.

Bacterial chromosomes are organized structures in the cytoplasm

Most but not all bacteria possess a single, larger circular chromosome, often accompanied by one or more small circular plasmids. The large chromosome is

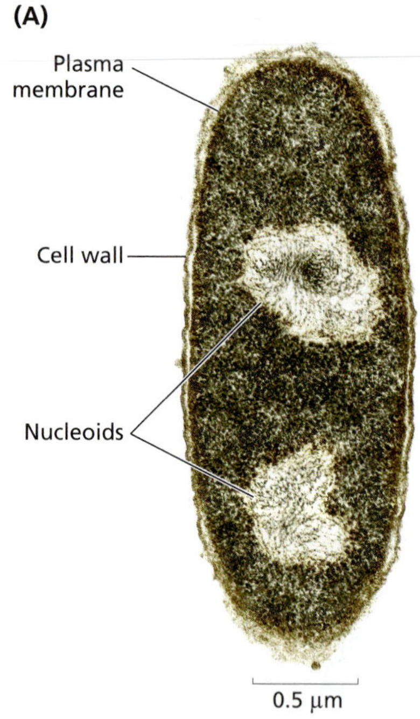

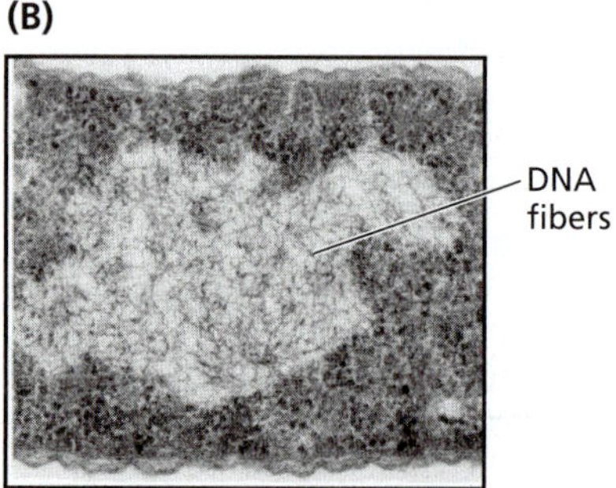

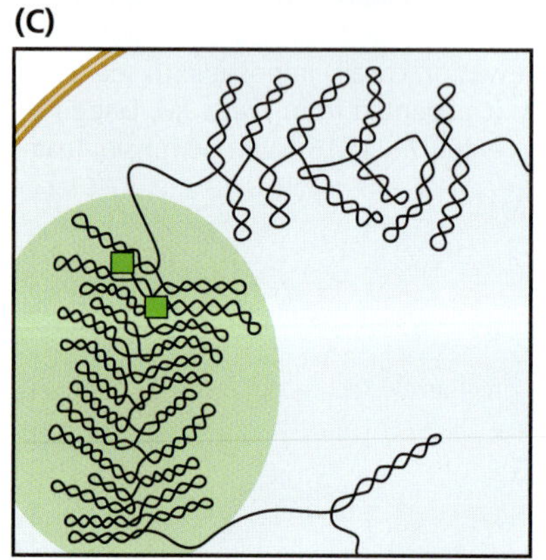

Figure 8.3 The bacterial nucleoid. (A) *E. coli* under the electron microscope. The cell has just undergone DNA replication and consequently has two nucleoids. (B) Transmission electron microscopy, TEM: osmium fixation of *E. coli* shows a confined nucleoid containing a network of randomly oriented DNA fibers. (C) Schematic representing one of the four DNA topological macrodomains in the nucleoid of *E. coli* and an adjacent less-structured region. Large portions of the circular chromosome occupy the same cellular space, forming macrodomains that are separated by less-structured regions; these regions prevent the domains from colliding. The DNA in the macrodomains is highly packed through interactions with specific protein factors, shown as green squares. (A, from Menge B & Wurtz M. With permission from Photo Researchers, Inc. B, from Eltsov M & Zuber B [2006] *J Struct Biol* 156:246–254. With permission from Elsevier. C, adapted from Boccard F, Esnault E & Valens M [2005] *Mol Microbiol* 57:9–16. With permission from John Wiley & Sons, Inc.)

free in the cell cytoplasm, as bacteria have no morphologically defined nucleus. The cytoplasmic complex of DNA and proteins is called the **nucleoid** (Figure 8.3). It must be tightly folded to fit within the bacterium. For *Escherichia coli*, the unfolded circumference of the duplex DNA circle would be ~1.6 mm, whereas the dimensions of the bacterium are around 2 μm. This corresponds to a packing ratio of about 1000.

A major factor contributing to compaction is supercoiling of the bacterial DNA. The bacterial DNA is highly negatively supercoiled. The equilibrium superhelical density is about –0.06; given a genome size of 4.6×10^6 bp, or 4.6×10^5 turns of DNA, this indicates about 3×10^4 superhelical twists (see Chapter 4). This can be seen *in vitro* as a highly branched plectonemic writhing of the DNA (see Figure 8.3). Whether this is the exact mode of supercoiling *in vivo* is unknown. The supercoiling is created and maintained by several factors. These factors include the binding of structural proteins that are a part of the bacterial nucleoid, processes such as transcription and replication (see Chapter 9), and the actions of gyrases, which are topoisomerases that introduce negative supercoiling into topologically constrained DNA molecules (see Chapter 4). Other topoisomerases regulate the level of superhelical stress.

DNA-bending proteins and DNA-bridging proteins help to pack bacterial DNA

A partial list of bacterial nucleoid-associated proteins is given in Table 8.2. The most extensively studied of these are the DNA-bending proteins **heat unstable (HU) protein**, and **integration host factor, IHF**, both of which exist in slightly postsynthetically modified forms and function as heterodimers. HU and IHF are structurally very similar, and as minor-groove binders they are capable of inducing very sharp bends and kinks in DNA (see Chapter 6). Recent evidence shows that both HU and IHF, in high concentrations, can induce helical rodlike structures in DNA. Examples from X-ray and atomic force microscopy (AFM) imaging studies are shown in Figure 8.4.

Both HU and IHF bind nonspecifically but preferentially to highly supercoiled regions. IHF possesses some tendency to bind to specific nucleotide sequences as well. It is probably appropriate to think of these proteins as primarily having a structural function, although it is clear from Table 8.2 that they do not constrain, even together, a large fraction of the genome. Alone, they could not yield the extensive, stable higher-order structure that we find in the eukaryotic genome. The level of stability that they provide is appropriate for the bacterial genome, which is continually involved in processes such as transcription, replication, recombination, and transposition. This high activity level would be inhibited by highly stable structures.

Another group of proteins, including **histone-like nucleoid structuring protein (H-NS)** and **structural maintenance of chromosome (SMC) proteins**, are known to bridge DNA. The action of these proteins in compacting DNA is illustrated in Figure 8.5. Other nucleoid proteins play predominantly functional roles. **Factor for inversion stimulation (Fis)**, for example, acts rather like a transcriptional activator in higher organisms: it can bind upstream from a promoter site and interact with the C-terminal tail of the RNA polymerase, so as to activate transcription from that promoter. In addition,

Table 8.2 Major structural proteins of the *E. coli* nucleoid. (Adapted from Johnson RC, Johnson LM, Schmidt JW & Gardner JF [2005]. In The Bacterial Chromosome [Higgins PN ed.], pp 65-132. With permission from ASM Press.)

					% DNA bound	
Abbr	Protein	Structure	Nature of binding site[a]	Site size (bp)	Exponential growth	Stationary phase
HU	heat unstable protein	Homodimer	kinked, NS	36	8	6
IHF	integration host factor	heterodimer	kinked, NS, also S	36	4	23
H-NS	histone-like nucleoid structuring protein	dimer, oligomer	curved, NS	10–15	1	1
SMC	structural maintenance of chromosomes	dimer, multimer (rosettelike)	NS			
StpA	suppressor of td mutant phenotype A	dimer, oligomer	curved, NS	10	1	1
Fis	factor for inversion stimulation	homodimer	curved, NS, some S	21–27	6	<1
Dps	DNA protection during starvation	dodecamer	NS	90?	<1	30

[a]NS, nonspecific binding; S, specific stronger sites.

it can bend DNA and help to compact it (Figure 8.6). **DNA protection during starvation protein (Dps)** is a peculiar, specialized protein that accumulates only when bacteria are starved and enter a stationary phase. Under these circumstances, it forms very extensive and regular liquid-crystal co-structures with the DNA, protecting the genome from disintegration. *In vitro*, the protein interacts with DNA and through self-association (aggregation compacts the DNA; see Figure 8.6). Overall, the bacterial nucleoid appears to be a very fluid, dynamic structure, well suited to the hectic metabolism of these organisms.

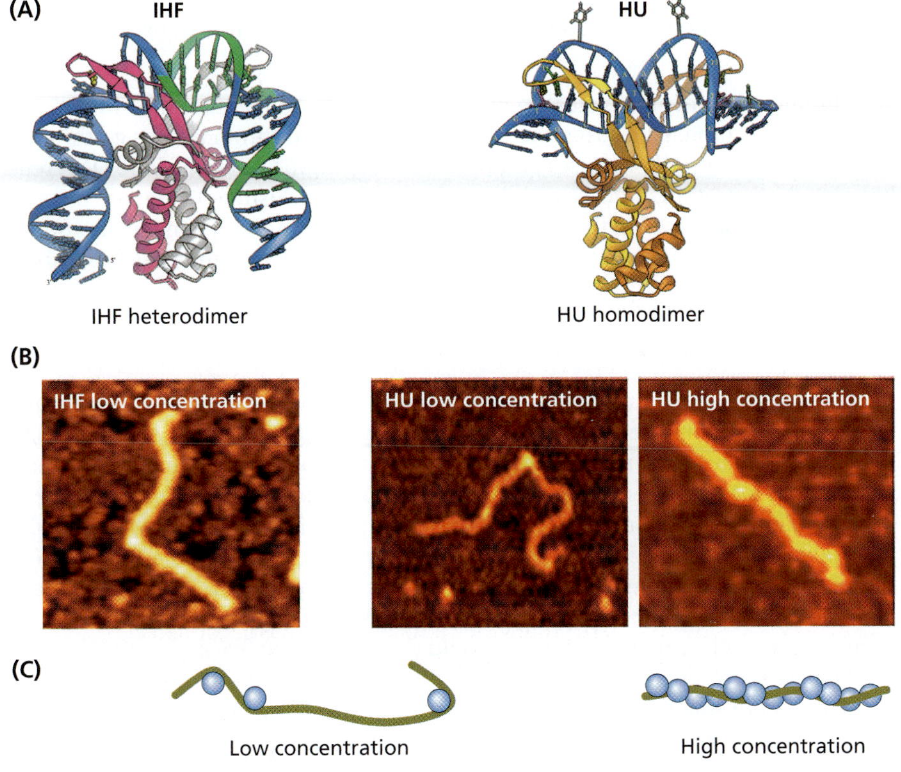

Figure 8.4 Interactions of DNA-bending proteins IHF and HU with DNA. (A) Protein–DNA co-crystal structures. IHF is bound sequence-specifically to the H′ site from phage λ (see Figure 6.5). (B) AFM images of protein–DNA complexes. (C) At low concentrations, both proteins, represented by blue spheres, bend the DNA. At high concentrations, the proteins induce the formation of rigid filaments, as illustrated in the right-hand image for HU; a similar structure has also been proposed for IHF–DNA filaments at high concentrations of the protein. (A, from Swinger KK, Lemberg KM, Zhang Y & Rice PA [2003] *EMBO J* 22:3749–3760. With permission from John Wiley & Sons, Inc. B, from Luijsterburg MS, Noom MC, Wuite GJL & Dame RT [2006] *J Struct Biol* 156:262–272. With permission from Elsevier.)

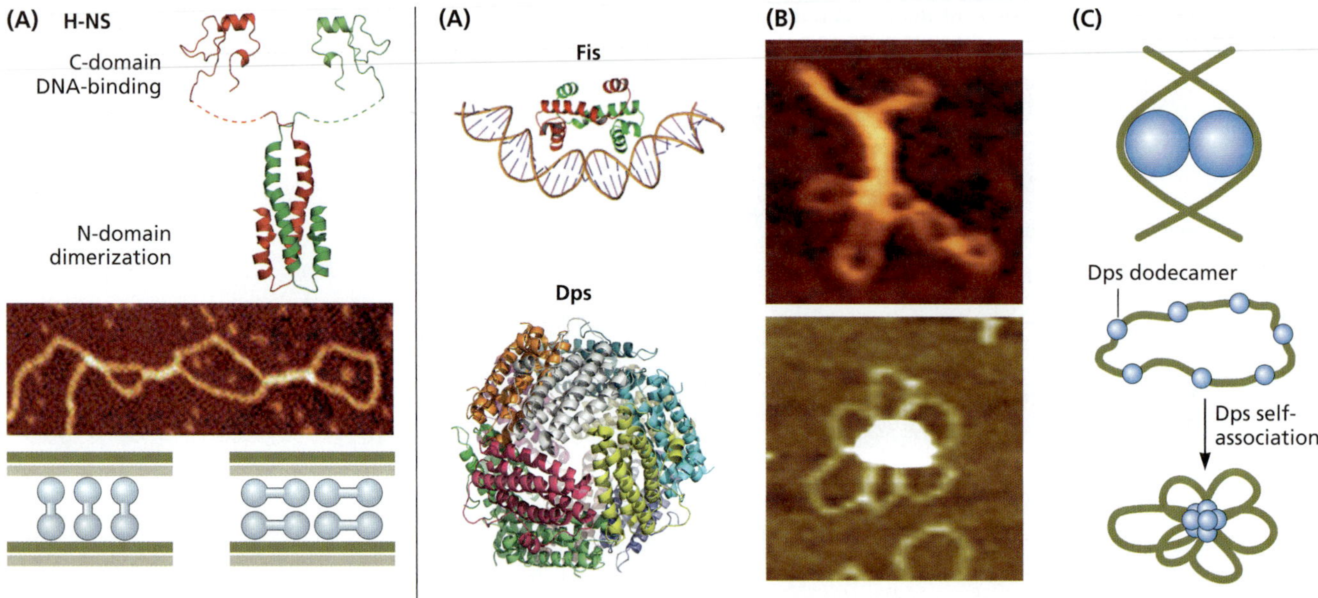

Figure 8.5 Interactions of DNA-bridging proteins H-NS and SMC with DNA. (A) H-NS. Top, structure of the protein dimer; middle, AFM image of DNA loops formed by duplex bridging by H-NS; bottom, two possible models of binding. (B) SMC. Top, AFM image of rosettelike SMC structures in solution. A typical field shows rosettelike structures. Bottom, speculative model of rosette formation by SMC dimers, in which SMC head domains form the central part of the structures and arms extend outward; DNA loops could be trapped by the arms. (A [top], adapted from Luijsterburg MS, Noom MC, Wuite GJL & Dame RT [2006] *J Struct Biol* 156:262–272. With permission from Elsevier. B, adapted from Mascarenhas J, Volkov AV, Rinn C et al. [2005] *BMC Cell Biol* 6:28. With permission from BioMed Central.)

Figure 8.6 Interactions of nucleoid-associated proteins Fis and Dps with DNA. (A) Crystal structures of Fis–DNA complex and pure Dps protein, a huge dodecamer. (B) AFM images of protein–DNA complexes created *in vitro*. (C) Perceived modes of binding to create the structures in the AFM images. (A–B, top, from Luijsterburg MS, Noom MC, Wuite GJL & Dame RT [2006] *J Struct Biol* 156:262–272. With permission from Elsevier. A, bottom, courtesy of Andrea Ilari, University of Rome. B, bottom, from Ceci P, Cellai S, Falvo E et al. [2004] *Nucleic Acids Res* 32:5935–5944. With permission from Oxford University Press.)

8.3 Eukaryotic chromatin

Eukaryotic chromosomes are highly condensed DNA–protein complexes segregated into a nucleus

In the great evolutionary leap from bacteria and/or archaea to eukaryotes, a major revision of genome structure occurred. The more complex lifestyles of higher organisms required controls that led to segregation of the genetic material in the cell nucleus and demanded more complex control of processes like transcription and replication.

The chromosomes of eukaryotes are highly condensed complexes of individual DNA molecules and proteins. During the cell cycle, eukaryotic chromosomes go through a series of biochemical processes and structural transitions. These include DNA replication, condensation, mitosis, and decondensation, as illustrated in **Figure 8.7** and described in Chapter 20. In this chapter, we are primarily interested in the interphase state of the chromosomes, because this is when gene expression and DNA replication occur.

The nucleosome is the basic repeating unit of eukaryotic chromatin

It has long been known that the nucleic acid in the interphase eukaryotic nucleus forms a highly compacted complex with basic proteins, now called chromatin. As early as 1869, chemist Friedrich Miescher extracted such a complex from human cells and, in a general sense, recognized its nature (**Box 8.1**). But understanding of the true nature of nucleic acids lagged, and their genetic significance was not fully and universally recognized until about 1950. Similarly, although the major DNA-associated proteins, **histones**, were recognized very early (see Box 8.1), their characterization had to await the surge in protein chemistry of the mid-twentieth century. It then became clear that the histones could be divided into five major classes, four of which were present in chromatin in stoichiometrically equal amounts (**Table 8.3**). All histones are small proteins and all are quite basic, being rich in lysine and arginine. We now know that a multitude of histone sequence variants exist, even in one

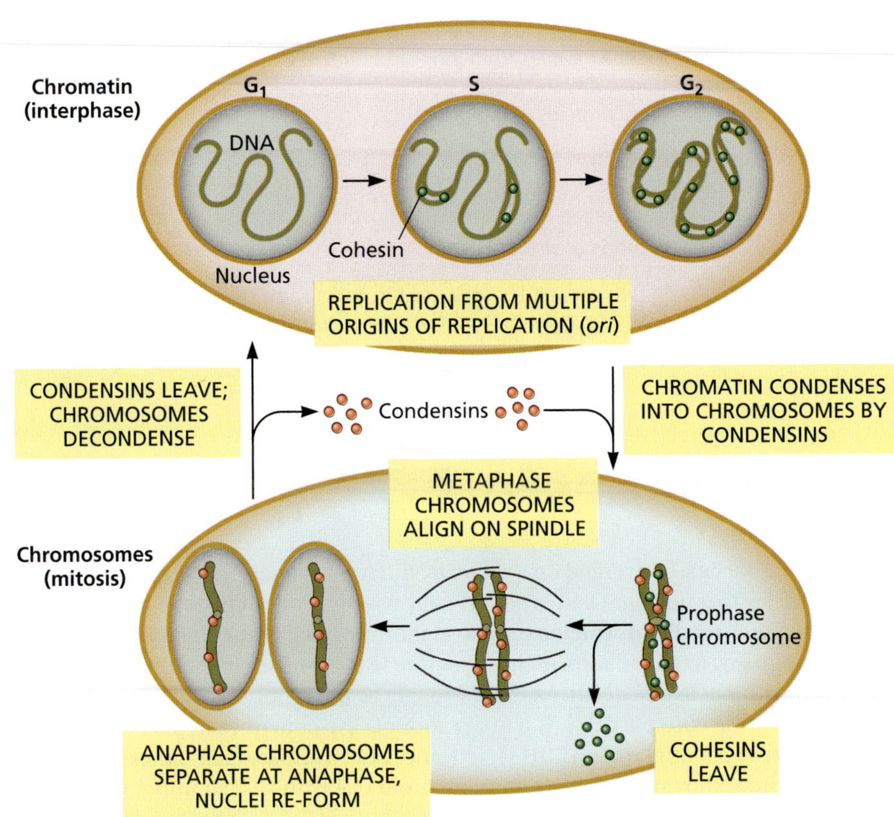

Figure 8.7 Dynamics of chromatin compaction and decompaction. Chromatin is compacted into mitotic chromosomes at the beginning of mitosis, and decompaction of mitotic chromosomes occurs at the end of mitosis. The schematic shows the timing of these events with respect to phases of the cell cycle and the association or dissociation behavior of the two major players, cohesin and condensin, during these transitions. In the S phase, sister chromatids are linked by cohesin molecules. Note that the nuclear membrane disintegrates during mitosis. The precise architecture of the prophase chromosome is unknown.

organism. Four of the histones were shown to be capable of specific heteroassociations *in vitro*: H3 and H4 form an $(H3/H4)_2$ tetramer, and H2A and H2B form a heterodimer. It soon became clear that the ubiquity of these histones among virtually all eukaryotes, from yeast to humans, and their abundance, about equal in mass to the nuclear DNA, meant that they must likely play a structural role in the organization of chromatin.

That chromatin was primarily a complex of DNA and histones was the general level of understanding that existed by about 1970, but the question as to how the histones and DNA were co-organized remained unanswered. A prevalent view held that the DNA was wrapped into some kind of superhelix, which the histones coated like insulation on a wire. But in this period, research in a number of laboratories, using very

Box 8.1 Discovery of chromatin The protein-DNA complex that we now call chromatin was discovered and partially understood at a remarkably early date in the history of biology. About 1870, in Germany, biochemist Friedrich Miescher isolated nuclei from pus cells taken from the bandages of hospital patients. This choice of material may sound odd to us, but it was actually very clever. Most cells from higher organisms exist in tough tissues, and it would have been difficult to extract nuclei by methods available at the time without destroying the nuclei and highly degrading the DNA. Miescher's choice of cell type avoided this difficulty. Miescher extracted a substance that was rich in phosphorus and protein, which he called nuclein. It probably was a crude chromatin preparation, enriched in DNA. In 1884, Albrecht Kossel, another student in the same laboratory at Tübingen, carried out an acid extraction of nuclei from goose red blood cells. This extraction yielded a mixture of basic proteins that Kossel called histon. By this date it was recognized that nuclei contained a complex of a phosphorus-rich acidic substance and basic proteins. One scientist, Otto Hertwig, went so far as to suggest in 1885 that this complex transmits the hereditary characteristics. It seemed that molecular biology was prepared to make giant leaps forward.

Regrettably, science sometimes proceeds in an erratic fashion, and it would be almost 70 years before Hertwig's prescient judgment was universally shared. One reason for the slow progress was the primitive state of biochemical techniques in the period. The necessary further experiments simply could not yet be done. Another problem came from misconceptions about the nature of DNA, which until well into the 1900s was thought to be a low-molecular-weight material of simple and regular composition. On the other hand, the complexity of proteins was becoming recognized, so it was widely felt that only they could serve as the material of genes. Only in the 1950s, with the Watson-Crick model, was the true role of DNA fully recognized. By then, separation and characterization of the histones had also advanced to a point where the true chemical nature of chromatin was understood. However, it took another 20 years for a correct structural model to appear.

Table 8.3 Histones: five major classes. The linker histones, also known as H1 histones, are a family of closely related members; they bind to the linker DNA between nucleosomal particles. The core histones derive their names from their structural role as organizers of the core particle. The sizes, amino acid compositions, and sequences of these histones vary somewhat from species to species; numbers in the table refer to bovine histones. Still, histones are among the most highly conserved proteins in evolution.

Histone	NCBI Gene ID	Number of amino acids	Number of Lys, Arg	Function
H1[a]	3005	194	56, 6	binds to linker DNA
H2A	92815	130	13, 13	two molecules take part in formation of histone octamer, the protein core of the nucleosome
H2B	8349	126	20, 8	same as H2A
H3	8350	136	13, 18	same as H2A
H4	121504	103	11, 14	same as H2A

[a]Member of histone H1 family.

different approaches, converged on a radically different model (Box 8.2). The essence of this model is that nuclear DNA is periodically wrapped about histone cores to form particles we now refer to as **nucleosomes** or, more correctly, **nucleosomal core particles**. A major key to elucidation of this structure was the observation that cleavage of chromatin with nucleases occurred most strongly at sites about 200 bp apart. Further digestion gave the core particles as a final product. The core of each particle consists of an octamer of histones: one $(H3\text{-}H4)_2$ tetramer plus two H2A-H2B dimers. The DNA wrapping is now known to be a left-handed toroid of 1.67 turns, comprising a stretch of 147 bp. A high-resolution X-ray diffraction structure of the core particle is presented in Figure 8.8.

Box 8.2 Discovery of the nucleosomal structure of chromatin Some major scientific breakthroughs come not through a burst of individual genius but from the accumulation of partial leads from a number of scientists. Such was the case for the evolution, over a few years, of the nucleosome model for chromatin. It all began, about 1970, when many researchers became discontented with the kind of uniform supercoil models that had been proposed on the basis of the available electron microscopy and X-ray scattering studies. The latter could be interpreted in various ways, and electron micrographs of condensed chromatin fibers looked, as one scientist quipped, "like an explosion in a spaghetti factory."

A partial breakthrough came when two groups, one led by Chris Woodcock at The University of Massachusetts and the other by Ada and Don Olins at Oak Ridge National Laboratory, began using a new spreading technique for the preparation of electron micrograph samples. Independently, these groups observed a distinctive beads-on-a-string structure, which fit no existing model. This work was presented by both groups at a cell biology meeting in November 1973. The Olins' work was published a few months later; Woodcock's was rejected by a reviewer. Meanwhile, others, working largely without knowledge of the above results or one another's advances, were following different trails. In Australia, Dean Hewish and Leigh Burgoyne were carrying out gentle nuclease digestions of chromatin, *in situ*, which yielded a peculiar pattern of ~200 bp repeats. The laboratory of one of this book's authors, Kensal van Holde, was using nuclease digestion and analytical ultracentrifugation to reveal compact particles of chromatin, containing something over 100 bp of DNA and a roughly equal mass of histone.

Meanwhile, Roger Kornberg and Jean Thomas, at the Medical Research Council Laboratories in Cambridge, had been investigating the interactions between histones using cross-linking methods. They demonstrated the existence of the $(H3/H4)_2$ tetramer. Building upon the Hewish–Burgoyne result and the known histone stoichiometry in chromatin, Kornberg was able to postulate in 1974 a chromatin structure in which stretches of ~200 bp of DNA were each associated with histone octamers to form discrete particles. These were later termed nucleosomes. With some adjustments, the model fitted all of the other data mentioned above and provided a new paradigm for chromatin research.

Further elucidation of the structure of the particle came quickly. Electron micrographs showed individual core particles to be roughly spherical, with a diameter of ~100 Å. Since each particle must contain DNA about 400 Å long, it became obvious that DNA must be coiled. That it was coiled on the outside of the particles was strongly indicated by the DNase I digestion experiments of Marcus Noll at Cambridge University. Definitive evidence that the DNA was outside the protein core was provided by low-angle neutron scattering. Altogether, many people in many laboratories, usually working independently, produced over a period of a few years a new, coherent picture of chromatin structure.

The final definition of the core particle came from X-ray diffraction studies of nucleosome crystals. The pioneer in this work was Aaron Klug at Cambridge University, who was awarded the Nobel Prize in Chemistry in 1982, partly for the first low-resolution 25 Å study, published in 1977. Since then, a series of crystallographic studies, by several laboratories, have steadily refined the structure until we now have the elegant picture shown in Figure 8.8.

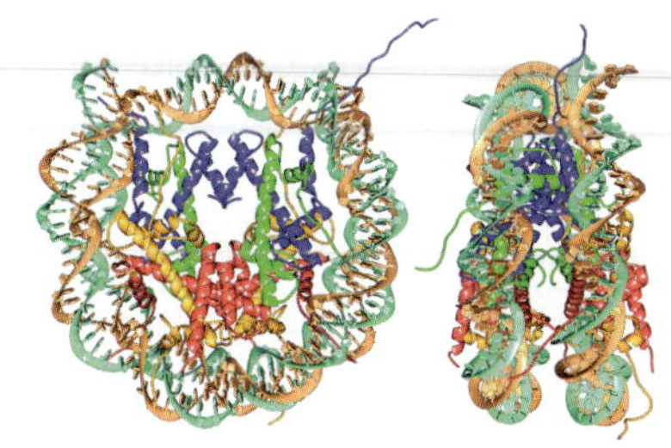

Figure 8.8 Crystal structure of the core nucleosome at 2.8 Å resolution. Two views perpendicular to the twofold axis of symmetry, known as the dyad, are shown. Color scheme: H3, blue; H4, green; H2A, yellow; H2B, red. The main technical difficulty in obtaining this high-resolution structure was getting good crystals. To reduce the inherent heterogeneity in particles isolated from nuclei, it was crucial to use recombinant histones that lack post-translational modifications and are devoid of the N-terminal tails, which seem highly disorganized. The tails in the structure show the presumptive locations of some of the tails. In addition, the DNA sequence used for reconstitution was constructed of two identical halves, connected head-to-tail. (From Luger K, Mäder AW, Richmond RK et al. [1997] *Nature* 389:251–260. With permission from Macmillan Publishers, Ltd.)

The structure of the histone octamer that forms the core of the nucleosome reflects the propensities of the histones to make specific heteroassociations. Thus, a $(H3\text{-}H4)_2$ tetramer forms the basis of the structure and is flanked by two H2A-H2B dimers. The overall structure has twofold symmetry; the axis of symmetry, known as the dyad, passes between the two H3-H4 dimers of the tetramer (**Figure 8.9**). The DNA appears to interact with the histones primarily through its phosphate groups, and this interaction is far from uniform along the length of the nucleosomal DNA. The strongest binding is to the H3-H4 tetramer, near the axis of symmetry, about 73 bp from either end of the DNA. Weaker binding is found at positions about 40 bp to either side of this point, positions where contacts with H2A-H2B dimers are made. This binding pattern has been demonstrated by single-particle experiments in which the DNA is mechanically peeled off the histone surface.

Although their sequences differ significantly, all of the histones that form the core of the nucleosome share a common structural motif termed the **histone fold** (**Figure 8.10**). This allows interaction of histones via the **handshake motif** to yield the dimers and tetramers essential for nucleosome formation. The major differences between these **core histones** lie in their termini; these extend beyond the nucleosome core, where they can make contact with other nucleosomes or other nuclear proteins. We shall also see that these histone termini contain the favored sites for many kinds of postsynthetic modifications.

The fifth type of histone, actually a family of related proteins, is referred to as the **linker histone** or the H1 family. These play a quite different role in chromatin structure than the core histones. H1s are associated with some but not all nucleosomes, and they are not part of the histone core. Instead, they associate primarily with the **linker DNA** (or **spacer DNA**) that lies between adjacent nucleosomes. The discovery and elucidation of the nucleosome and its role in chromatin structure has required the development and application of a number of techniques (**Box 8.3**).

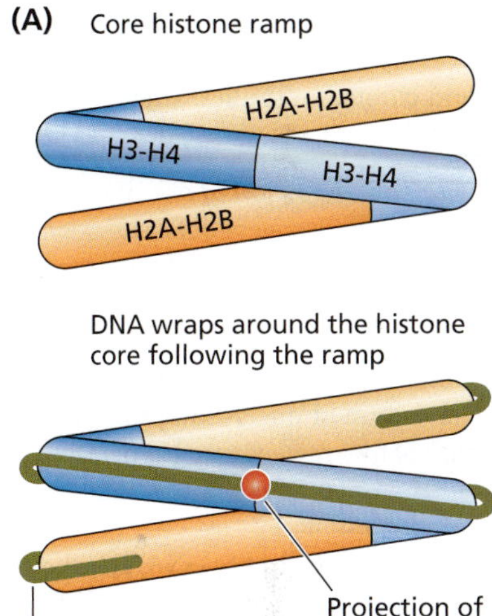

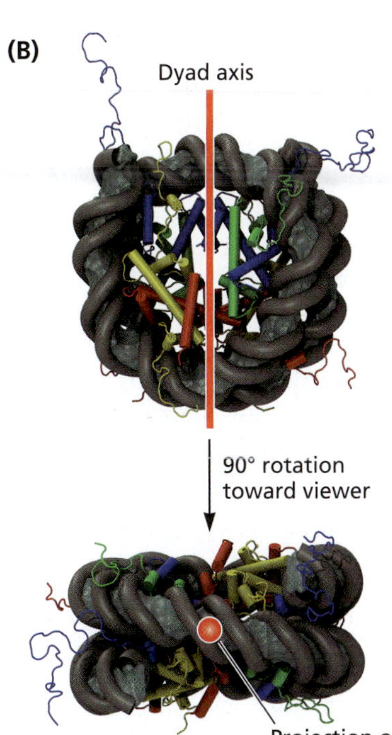

Histone nonallelic variants and postsynthetic modifications create a heterogeneous population of nucleosomes

Some of the eukaryotic histones, H1, H2A, and H3, exist in a number of nonallelic variants that differ somewhat in amino acid sequence. Recall from Chapter 2 that all genes in a eukaryotic cell exist in two copies or alleles, one coming from the mother and the other one from the father during fertilization. Some genes may have additional copies of slightly different sequences; these are termed nonallelic to discriminate them from the two conventional alleles of a single gene. Nonallelic gene variants also come in pairs, one inherited from the mother and the other one from the father. So for all practical purposes these are new gene pairs, similar to already existing gene pairs

Figure 8.9 Core particle symmetry about the dyad axis. (A) Core histone dimers interact with each other to form a spiral ramp that guides the wrapping of the DNA. The protein ramp is highly positively charged at the points of contact with the DNA. The schematic helps to visualize the symmetry of the particle and the location of the dyad axis. (B) All-atom simulation of core particle structure, based on the coordinates from the crystal structure. The dyad axis, the axis of twofold symmetry, is indicated in the face-on view of the nucleosome as a vertical red line. In the side view, it is coming directly toward the viewer and is indicated by the red circle. (B, courtesy of Jean-Marc Victor, Université Pierre et Marie Curie, and Thomas Bishop, Louisiana Tech University.)

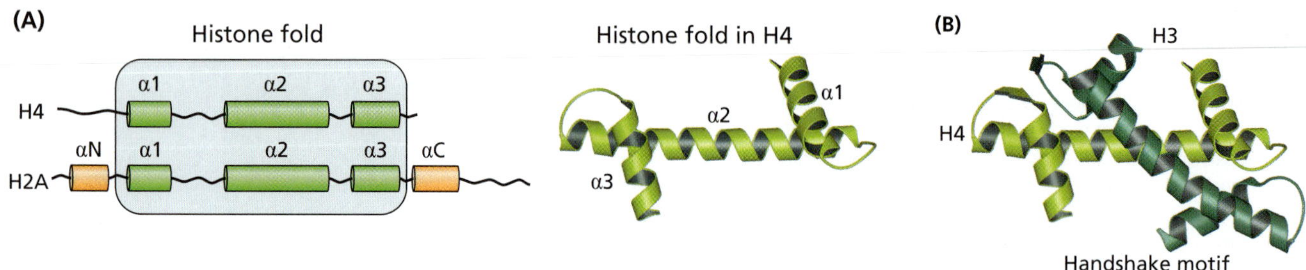

Figure 8.10 Histone fold and handshake motif. (A) All four core histones share a common structural motif called the histone fold. This fold consists of three α-helices arranged as shown in the crystal structure of histone H4, when H4 is part of the core particle. The individual histones differ in the presence or absence of additional α-helices at one or both ends of the polypeptide chain. Thus, histone H2A possesses additional α-helices at both ends of the molecule, αN and αC. (B) Individual histone molecules interact with each other along their long α2 helices in what is known as the handshake motif. The structures of the H2A-H2B dimers and H3-H4 dimers are almost superimposable. (Adapted from Harp JM, Hanson BL, Timm BL & Bunick GE [2000] *Acta Crystallogr* D56:1513–1534. With permission from International Union of Crystallography.)

Box 8.3 Studying the physical structure of eukaryotic chromatin uses a variety of methods Many of the methods we described in earlier chapters can be, and have been, applied to chromosome and chromatin studies. There are, however, two techniques that have been especially useful in this field, so we discuss them here.

Nuclease digestion of chromatin

The initial discovery of the nucleosome structure of chromatin involved, among other things, the use of endogenous endonucleases to digest chromatin. Later, the use of two exogenous nucleases, **micrococcal nuclease (MNase)** and deoxyribonuclease I (**DNase I**), provided a wealth of information on the organization of nucleosomes in fibers and on the internal structure of the nucleosome particle (Figure 1).

MNase cleaves both strands of DNA. It possesses both endonuclease and exonuclease activities. It first cuts chromatin in the exposed linker DNA, and then it trims the particles from the DNA ends. Unfortunately, MNase possesses some DNA-sequence preference, which necessitates including chromatin-free DNA samples as controls. Partial digestion of chromatin, followed by electrophoretic analysis of purified DNA, gives rise to the famous **DNA ladders**, DNA fragments that are multiples of a constant length (see Figure 1A); this length, which is specific to the cell and tissue type, has been named **repeat length**.

The use of DNase I provided a tentative answer to the question of where the DNA is located in the particle, inside or outside. DNase I makes single-strand nicks in double-stranded DNA. When DNA from DNase I-digested core particles is analyzed by electrophoresis under denaturing conditions, another type of ladder pattern is observed: the single-strand fragments on the gels are multiples of ~10 nucleotides (see Figure 1B). The simplest explanation of this pattern is that the nuclease cuts preferentially where the DNA is maximally exposed on the nucleosome surface. This early interpretation was soon supported by other studies. Later, high-resolution electrophoretic techniques provided information on subtle differences in the core particle structure that result from the incorporation of histone variants, post-translational histone modifications, or the activity of chromatin remodelers (see Chapter 12).

Mapping nucleosome positions in chromatin

The precise way in which nucleosomes are arranged on the chromatin fiber is important both for the higher-order structure of that fiber and for the regulation of transcription of

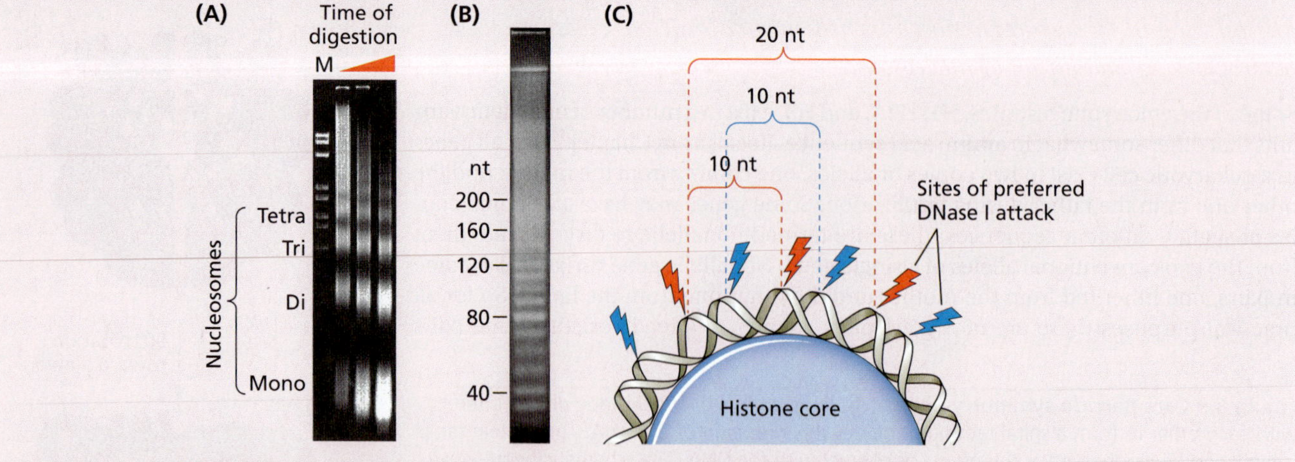

Figure 1 Nuclease digestion of chromatin. (A) MNase digestion pattern of chromatin from chicken erythrocytes: time course. The gel was stained with ethidium bromide and observed under UV light illumination. The lane marked M contains DNA restriction fragments used as size markers. (B) DNase I digestion pattern of core particles isolated from rat liver. (C) Schematic illustrating how the pattern is obtained: the nuclease cuts the DNA at sites of maximal exposure at the surface of the nucleosome. (A, from Mathews CK, van Holde KE, Appling DR & Anthony-Cahill SR [2012] Biochemistry, 4th ed. With permission from Pearson Prentice Hall. B, from Noll M [1974] *Nucleic Acids Res* 1:1573–1578. With permission from Oxford University Press.)

chromatin. A number of techniques have been developed for mapping nucleosome positions. As techniques have become more and more precise, it has become evident that many nucleosomes occupy well-defined locations.

The earliest studies of this kind utilized a technique called **indirect end-labeling** (Figure 2A). Suppose one wishes to examine a genomic region, of known DNA sequence. Whole chromatin is lightly digested—*in situ*, if desired—with MNase to produce a nucleosome ladder pattern. The purified DNA from the MNase digest is cleaved by a pair of restriction endonucleases that cut at defined sites bordering the region of interest. The DNA is then electrophoresed and blot-hybridized to a radioactive probe abutting one end of the DNA. This will reveal only bands corresponding in length to the several oligonucleosomes in the digest that hybridize to this probe sequence. The original method can give, at best, only approximate positions, because the nuclease may cut anywhere within the linker. The precision of locating nucleosomes can be improved by using cleavage reagents that cut naked DNA at almost every base pair, and then performing the electrophoresis on long, high-resolution sequencing gels. Appropriate reagents include DNase I, the reactive intercalator methidiumpropyl-EDTA-Fe^{2+} (MPE–Fe^{2+}), and hydroxyl radicals generated by an iron-catalyzed redox reaction. Under optimal conditions, these can give almost single-base resolution with short oligonucleosomes. However, there is always the possibility that a sample will contain a spectrum of closely related alternate positions, which will blur the results.

A quite different class of techniques involves **primer extension**, a procedure in which a synthetic oligonucleotide primer is extended by DNA polymerase, using a single-stranded DNA as a template (see Chapter 19). One such method is shown in Figure 2B.

Recent techniques allow the mapping of nucleosome positions over a whole genome. These rely on the use of microarrays covering wide genomic regions and ultrafast parallel sequencing techniques. In one application, a nuclease digest of whole chromatin is hybridized against such an array, and the captured nucleosomes are then sequenced. With current techniques, millions of nucleosomes can be sequenced and their locations can be mapped against the whole genome. Our knowledge of chromatin structure at the sequence level is suddenly becoming infinitely more comprehensive.

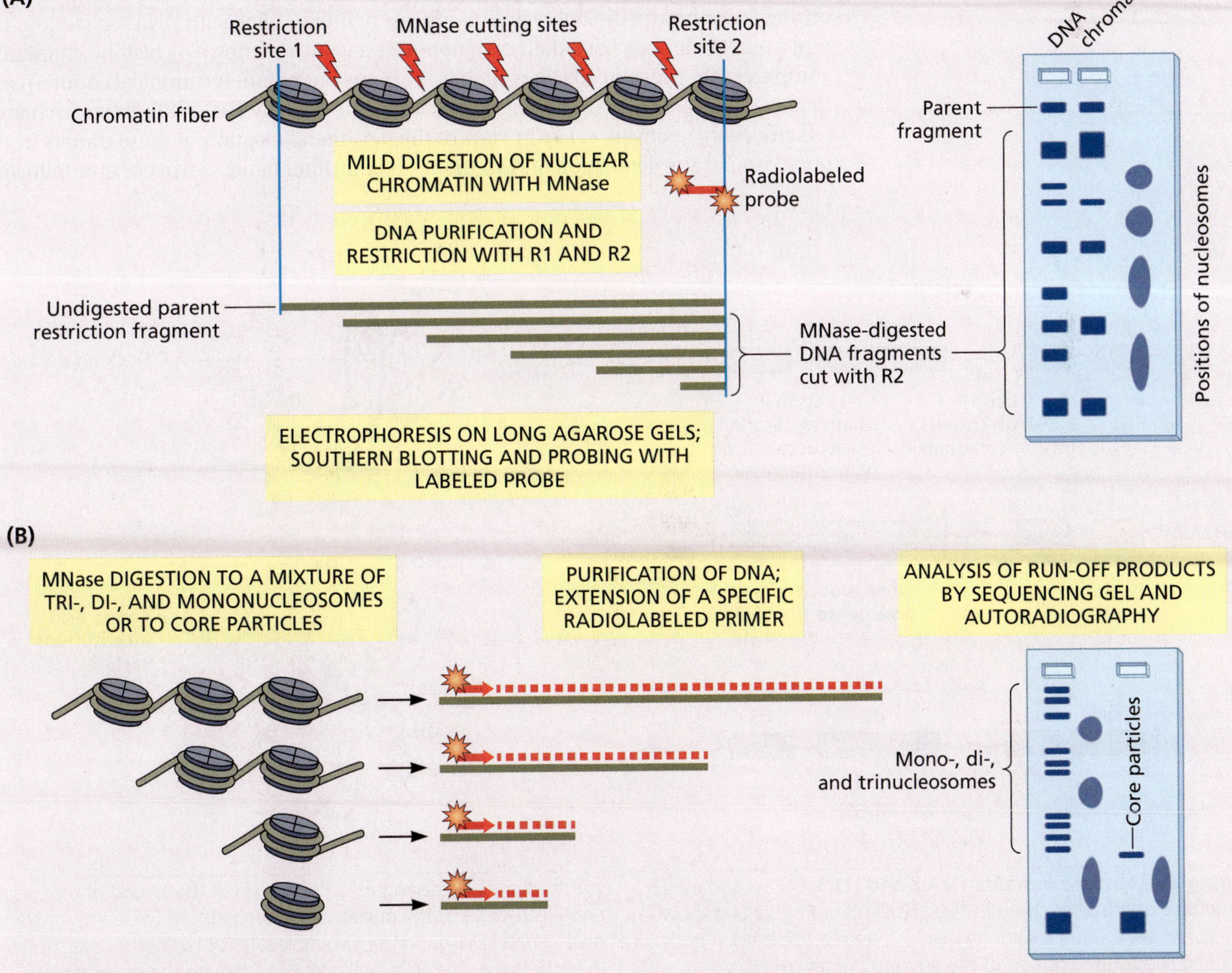

Figure 2 Two common methods for mapping nucleosome positions *in vivo*. (A) Indirect end-labeling; (B) primer extension. The autoradiograms of gels show idealized results that assume a single nucleosomal array on the gene of interest. Often, nucleosomes occupy slightly different positions that are spaced ~10 bp apart. (Adapted from Clark DJ [2010] *J Biomol Struct Dyn* 27:781–793. With permission from Taylor and Francis Group.)

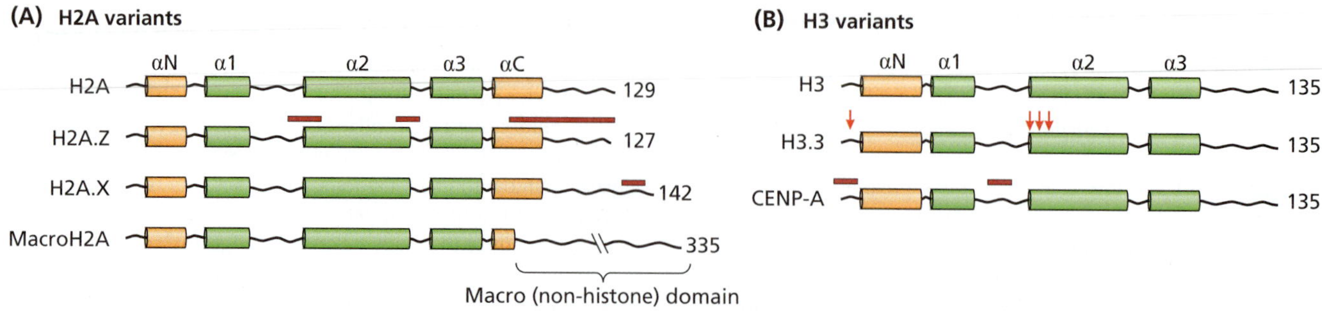

Figure 8.11 Nonallelic variants of core histones. The number to the right of each schematic denotes the length of the polypeptide chain. (A) H2A variants. The major regions of differences in the sequences and their position with respect to the histone fold, shown in green, are marked with red bars. MacroH2A contains a large non-histone domain fused to the N-terminal portion of the αC helix. (B) H3 variants. Red arrows indicate the four amino acid differences between H3.3 and the canonical H3 sequence. CENP-A is the variant present in chromosome centromeres; variant sequences are marked with red bars.

but possessing differences in the nucleotide sequence that give rise to slightly altered protein molecules. The histone nonallelic variants are also known as **replacement variants** because they are synthesized throughout the cell cycle and are deposited on nonreplicating chromatin, to replace the resident canonical variants. Examples are given in **Figure 8.11** and **Figure 8.12**.

Canonical histones are those that are synthesized and deposited onto chromatin only during the S phase of the cell cycle, in a process tightly coupled with DNA replication. The replacement variants have their own deposition vehicles, known as histone chaperone complexes, that specifically deposit these variants but not their canonical counterparts. Replacement of H2A with any of its variants requires only H2A-H2B dimer exchange, which occurs relatively easily in view of the peripheral location of these dimers in the nucleosome. The deposition of H3 variants, on the other hand, can occur only following

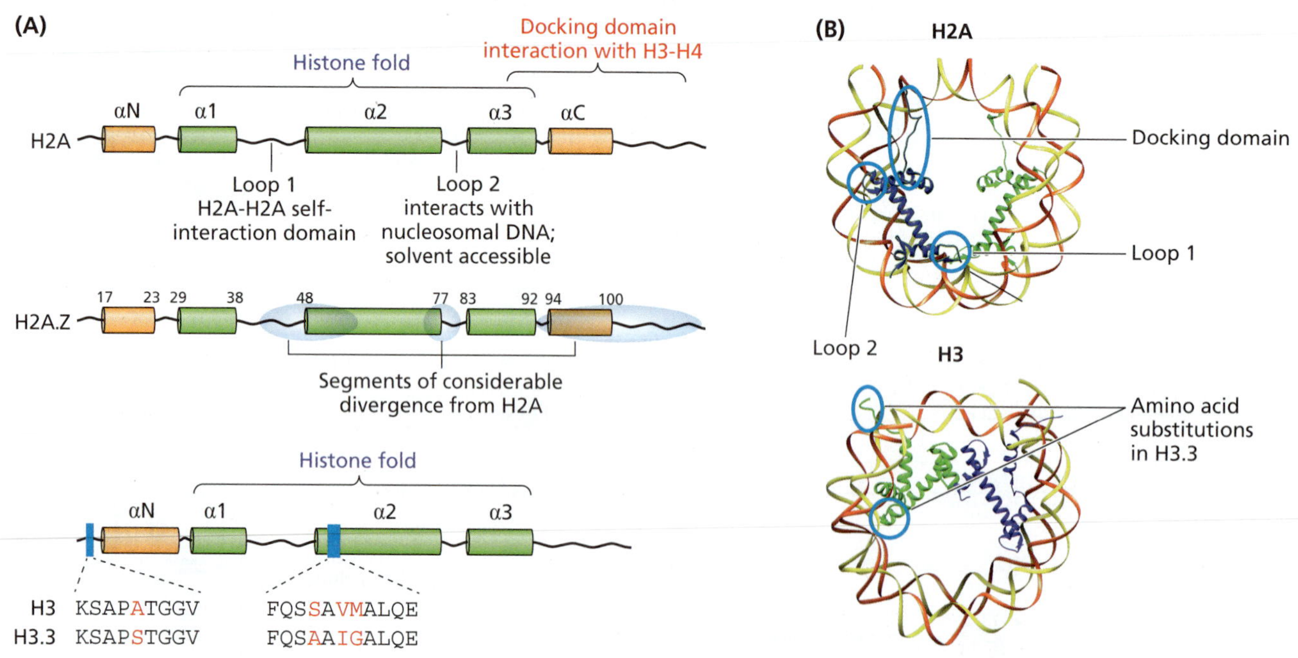

Figure 8.12 Histone variants H2A.Z and H3.3. (A) Secondary structures of the respective histones. Brackets with blue text show the canonical histone fold of helices 1, 2, and 3. The bracket with red text shows the position of the docking domain of H2A; this is the region of interaction between the H2A-H2B dimer and the H3-H4 dimer in this half of the nucleosome. Blue ovals denote the location of the H2A.Z segments with considerable divergence from H2A. Amino acid sequences in the H3 structure show the substitutions that distinguish H3 from H3.3. (B) Portion of the crystal structure of the nucleosome core particle. For clarity, only nucleosomal DNA and the two molecules of H2A or H3 are shown in each image. The blue circles or ovals denote the major regions of divergence between the canonical core nucleosome and the respective variant particles. (B, courtesy of Amit Thakar, University of Wyoming.)

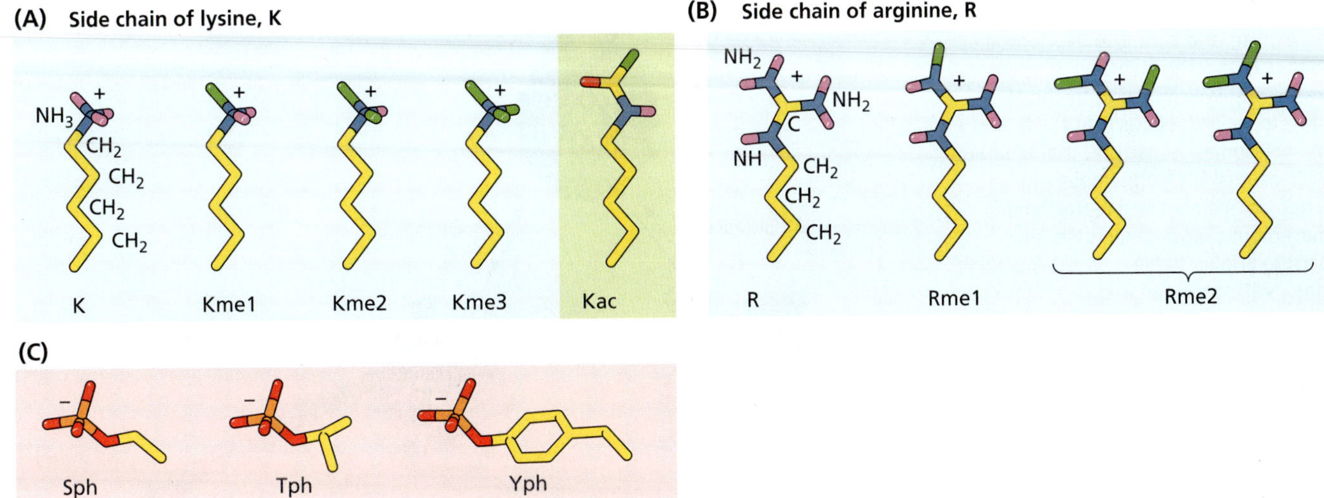

Figure 8.13 Known histone post-translational modifications and their effect on overall properties of the modified side chain. The different classes of modified amino acid residues are represented by stick models. Color code: yellow, carbon; blue, nitrogen; pink, polar hydrogen; red, oxygen; orange, phosphorus; green, methyl groups. Background shading denotes the charge of modified side chains at physiological pH: light blue, positive; pink, negative; light green, uncharged. (A) Lysine methylation and acetylation. Charge is ablated upon lysine acetylation, denoted as Kac, whereas all methylated forms of lysine, denoted as Kme, are positively charged or cationic at physiological pH. The incremental addition of methyl groups, shown in green, from K to Kme3 increases the hydrophobicity and the cation radius of the methylammonium group; the ability to donate hydrogen bonds concomitantly decreases. (B) Arginine methylation. (C) Phosphorylation of serine, threonine, and tyrosine introduces negative charge. (Adapted from Taverna SD, Li H, Ruthenburg AJ et al. [2007] *Nat Struct Mol Biol* 14:1025–1040. With permission from Macmillan Publishers, Ltd.)

total displacement of the resident histones from the DNA. As we find in Chapter 12, certain variants, H2A.Z and H3.3, for example, play important roles in the regulation of transcription. Others, like H2A.X, participate in DNA repair processes (see Chapter 22).

In addition to the possible combinations of **histone variants** that exist in chromatin fibers, there is also a very wide range of postsynthetic or post-translational modifications that can occur on each histone. These include acetylation of lysine residues, methylation of lysine or arginine, phosphorylation of serine, threonine, or tyrosine, and ubiquitylation of lysine as shown in Figure 8.13 and Figure 8.14. The majority of

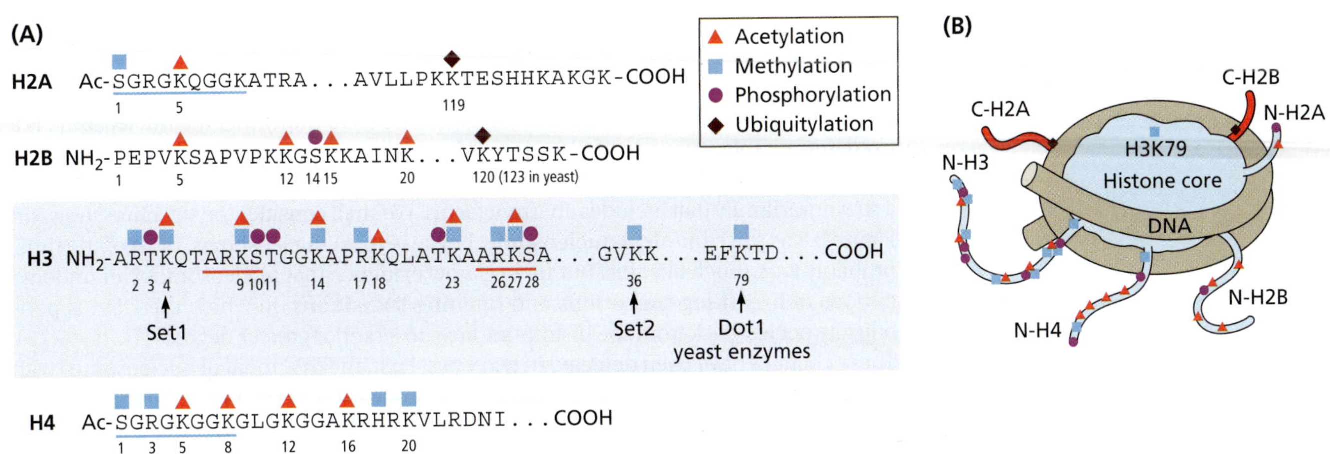

Figure 8.14 Distribution of known post-translational modifications along polypeptide chains of the four core histones. (A) Underlined sequences are motifs that are repeated either in two different molecules, underlined in blue, or within the same molecule, underlined in red. The significance of having such repeats is unclear. The best-studied modifications are those of histone H3. The enzymes that place the same modification on different residues are environment-specific: thus in yeast, different enzymes, Set1, Set2, and Dot1, modify lysines 4, 36, and 79, respectively. (B) The majority of post-translational modifications occur in the histone tails that protrude from the core particle structure, but recent studies have demonstrated the existence of modified residues within the histone fold, which is entirely inside the particle. A good example is the methylation of residue K79 in H3. Note that for clarity only half of the core histones in the particle are shown.

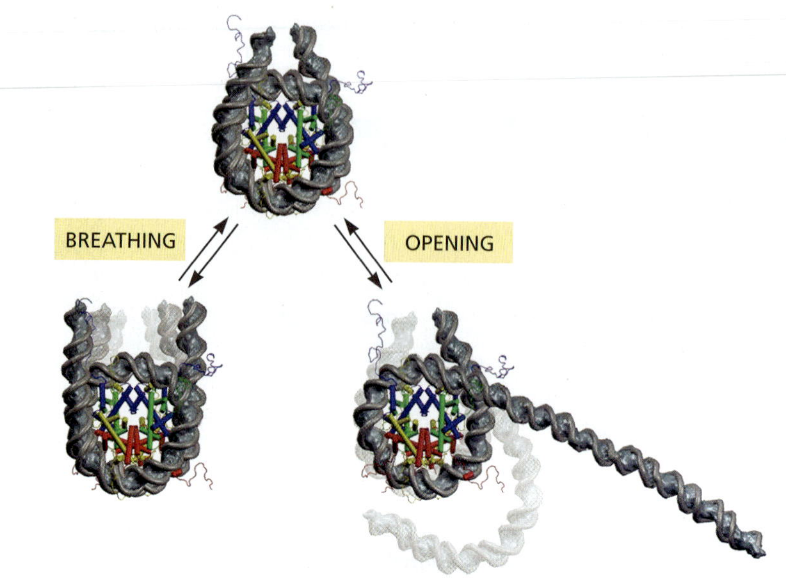

Figure 8.15 Nucleosome dynamics: breathing and opening. Nucleosomal DNA can undergo reversible spontaneous transitions, which involve the breaking of histone–DNA contacts at the end of the octasomes. In particles that breathe, nucleosomal DNA undergoes short-range detachment and reattachment movements; most of the time, the particle is in its stable or crystallographic state, with only brief excursions into a more open state. Opening occurs over a longer length of DNA and is much less frequent: the DNA that remains attached to the histones may be able to form just one full turn around the histone core. It is important to note that particles with such properties have been previously seen in biochemical experiments. (Courtesy of Jean-Marc Victor, Université Pierre et Marie Curie, and Thomas Bishop, Louisiana Tech University.)

post-translational modifications occur in the histone tails that protrude from the core particle structure, although modifications of residues within the histone fold, which is entirely inside the particle, have been recently recognized (see Figure 8.14).

Together with histone sequence variations, the modifications of histones in chromatin can potentially give rise to an enormous number of possible distinguishable nucleosome particles that differ in their stability and dynamic properties, and thus in their transcribability. We see in Chapter 12 that these modifications play vital roles in the specific regulation of the transcription of different genes.

The histone fold is not found in bacterial nucleoid proteins. Proteins with a very similar, almost identical, fold are present in Archaea, and seem to serve a very similar purpose: to compact the DNA and regulate gene activity. Thus, this kind of chromatin structure appears to be present in two of the three kingdoms of life, Archaea and Eukaryota. The cellular localization of chromatin is, however, different in each of these two domains of life. In archaea, as in bacteria, chromatin occupies the volume of cytoplasm, whereas in the eukaryotic cell it is confined to the nucleus, which appears to be a unique feature of eukaryotes.

The nucleosome family is dynamic

The level of detail shown in Figure 8.8 might be taken to imply that the nucleosome is a fixed, immutable structure. Nothing could be further from the truth. We now know that the nucleosome can undergo dynamic changes in conformation and even exists stably as a structural family that includes aberrant forms. We shall consider the structure shown in Figure 8.8 as the **canonical nucleosome**, because most nucleosomes, most of the time, probably look much like this. But there is now evidence that nucleosomes can undergo periods of **breathing transitions** and **opening transitions**, in which the DNA is periodically peeled back from the histone surface, to lesser or greater degrees (**Figure 8.15**). These changes have been detected in two ways. First, the treatment of nucleosomes with site-specific nucleases reveals cleavages of the nucleosomal DNA that should not occur if the DNA were tightly fixed around the histone octamer. These cleavages occur most easily near the periphery and are less frequent but still detectable near the dyad axis. Second, single-molecule fluorescence resonance energy transfer (FRET) experiments (see Chapter 1) yield direct evidence for both breathing and opening transitions.

There also exist, under some conditions, nucleosomal particles that have unusual histone cores, such as only a $(H3/H4)_2$ tetramer. This might be called a **tetrasome**, if we adopt the nomenclature in **Table 8.4**, in which the canonical particle becomes the **octasome**; more specifically, the L-octasome, because there are also conditions under which the DNA takes a right-handed twist about the histone core to form an R-octasome. All of

Table 8.4 Members of the nucleosome family with new and traditional names. Particles containing non-histone proteins in lieu of histones should have specific names to reflect the specific non-histone protein present. For example, the proposed $Scm3_2(H3/H4)_2$ particle might be called Scm3 hexasome.

New name	Traditional name	Histone composition/stoichiometry	Handedness of DNA superhelix
nucleosome (L-octasome or L-nucleosome)	nucleosome	$(H2A\text{-}H2B\text{-}H3\text{-}H4)_2$	left-handed
R-nucleosome (R-octasome)	reversome	$(H2A\text{-}H2B\text{-}H3\text{-}H4)_2$	right-handed
hexasome	hexasome	$H2A\text{-}H2B/(H3\text{-}H4)_2$	left-handed
L-tetrasome	tetrasome	$(H3\text{-}H4)_2$	left-handed
R-tetrasome	right-handed tetrasome	$(H3\text{-}H4)_2$	right-handed
hemisome	half-nucleosome	H2A-H2B-H3-H4 (could be variant-specific, such as H2A-H2B-CenH3-H4)	right-handed

the structures listed in Table 8.4 have been reported, but we do not yet know which may have physiological significance. A gallery of aberrant structures is shown in **Figure 8.16**.

Nucleosome assembly in vivo *uses histone chaperones*

Although nucleosomes can assemble spontaneously *in vitro* from DNA and histones, *in vivo* assembly and modification often utilize a number of carriers of histones, called **histone chaperones**. Histone chaperones have the ability to recognize specifically

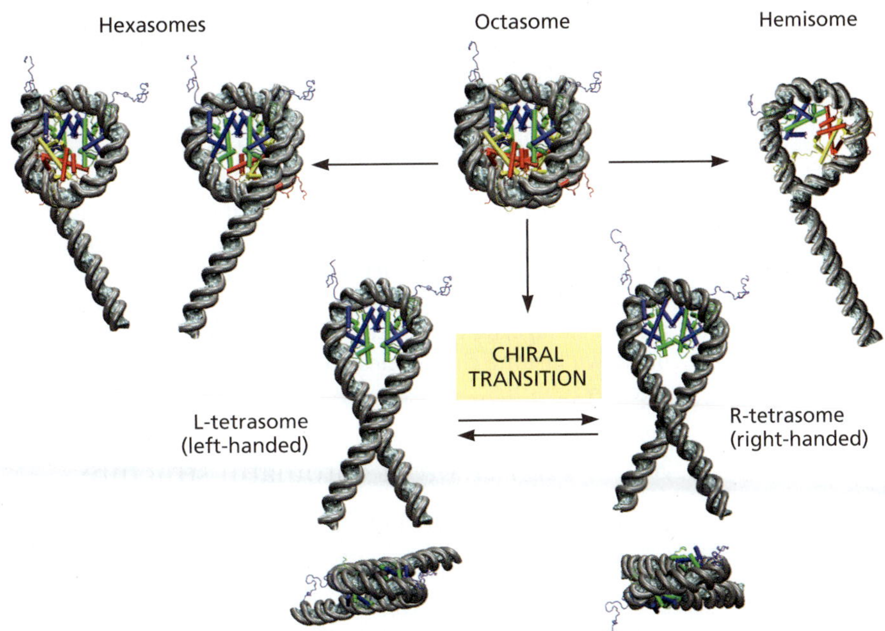

Figure 8.16 The nucleosome family. Numerous experiments have indicated the existence of nucleosome particles that deviate in histone composition and stoichiometry and in DNA trajectories from the canonical octasome. The models shown here are based on the atom coordinates in the highest-resolution, 1.9 Å X-ray structure available. They were created by deleting the appropriate histones from the octasome, allowing the released DNA segments to assume the traditional B-form structure. The models assume that histones are rigid bodies that can be pieced together like a jigsaw puzzle, even though this might not represent the physical reality. In the octasome, the entire 147-bp stretch of DNA interacts with the histones. In the tetrasome, removal of the H2A and H2B histones releases 48 bp from each end; thus, the tetrasome contains less than one DNA superhelical turn. Note that the tetrasome can undergo the so-called chiral transition from the canonical left-handed superhelix to a right-handed one, shown at two angles. There are two possibilities for the hexasome, depending on which H2A-H2B dimer, proximal or distal, has dissociated. The hexasomes maintain more than one DNA superhelical turn in contact with the histone core. The hemisome contains only one molecule each of the core histones rather than two. Formation of a hemisome from an octasome releases 48 bp from one end, as in the hexasome, and ~15 bp from the other end due to loss of H3-H4 contacts. Hemisomes have been described in centromeric chromatin. H3, blue; H4, green; H2A, yellow; H2B, red. (Courtesy of Jean-Marc Victor, Université Pierre et Marie Curie, and Thomas Bishop, Louisiana Tech University.)

Table 8.5 Representative examples of histone chaperones and their function. In addition to the functions listed here, some histone chaperones participate in histone transfer to enzymatic complexes that post-translationally modify histones.

Histone chaperone	Histone selectivity	Function
	Single Chaperones	
Asf1	H3-H4 and H3.3-H4	histone donor for CAF-1 and HIRA
HIRA	H3.3-H4	replication-independent deposition (of histone variants)
N1/N2	H3-H4	storage in *X. laevis* oocytes
nucleoplasmin (*X. laevis*), nucleophosmin (*H. sapiens*)	H2A-H2B	storage in *X. laevis* oocytes; cytosolic–nuclear transport; replication; transcription
Nap1	H2A-H2B	cytosolic–nuclear transport; replication; transcription
nucleolin	H2A-H2B	transcription elongation; assists chromatin remodeling
	Chaperone Complexes	
CAF-1 (p150, p60, and RbAp48)	H3-H4	deposition coupled to DNA synthesis (replication and repair)
FACT (Spt16, SSRP1)	H2A-H2B and H3-H4	transcription elongation

and bind to specific histone dimers and then to transfer them to other molecules. Importantly, chaperones are not part of the final product. Histone transfer processes include transfer from one chaperone to another; histone transfer onto DNA, known as deposition; histone removal from DNA, known as eviction; histone exchange reactions in already existing nucleosomal particles; and histone transfer to modification enzymes. Chaperones can also serve in chromatin remodeling (see Chapter 12), in histone storage, for example, in oocytes, and probably in other functions. A partial list of known histone chaperones, their histone specificities, and their main functions is given in **Table 8.5**.

8.4 Higher-order chromatin structure

Nucleosomes along the DNA form a chromatin fiber

Individual nucleosomes are organized into higher-level structures called chromatin fibers. There is convincing evidence that certain DNA sequences can provide strong, specific sites for **nucleosome positioning** (**Figure 8.17A**). What is perhaps most interesting is where nucleosomes are *not* found. The ENCODE project (see Chapter 13) has located millions of **nuclease-hypersensitive sites**; DNase I mapping (see Box 6.4 and Box 8.3) indicates that these sites are nucleosome-free. Some of these are at transcription start sites and are flanked by two strongly positioned nucleosomes. There is also evidence for what has been called **statistical positioning**, where one or more such strongly positioned nucleosome(s), often called index nucleosomes, provide boundary elements for the positioning of a series of spaced nucleosomes, with a regularity that decreases upon moving further from the index nucleosome (**Figure 8.17B**).

Despite some such specific positioning, the spacing of nucleosomes on the DNA is generally irregular. As shown by nuclease digestion experiments (see Box 8.3), the average spacing varies between species and cell types, from about 160 to 240 bp. In most cases, this is considerably larger than the 147 bp carried on the nucleosome core

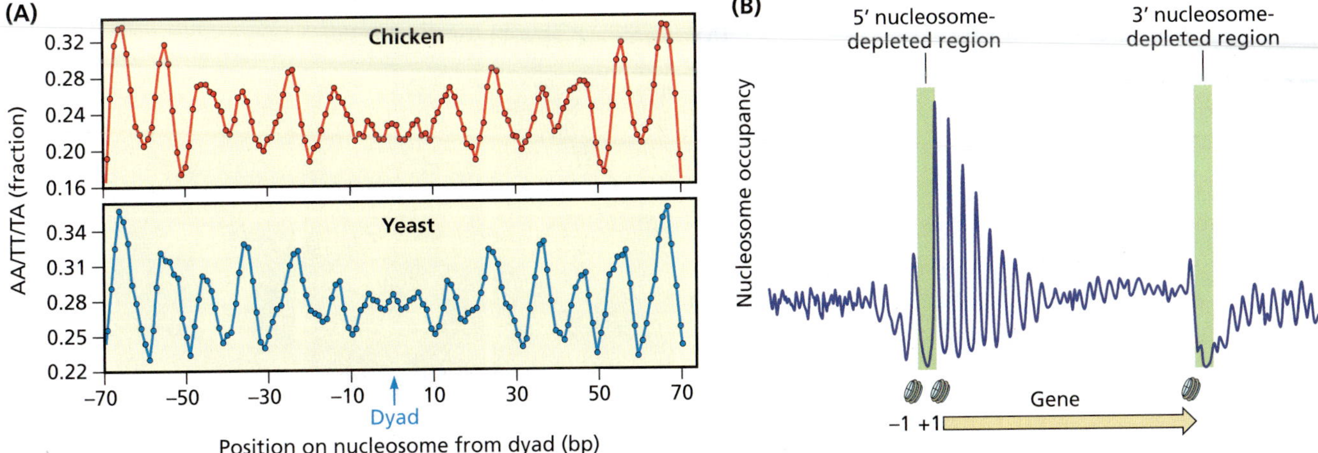

Figure 8.17 Nucleosome positioning. (A) Effect of nucleoside sequences on nucleosome positioning. Graphs show the fraction of AA, TT, and TA dinucleotides at each position of center-aligned chicken and yeast nucleosome-bound genomic sequences, indicating 10-bp periodicity of these dinucleotides in naturally occurring nucleosomes. For example, mononucleosomes were purified from log-phase yeast *Saccharomyces cerevisiae* cells following nuclease digestion, the DNA was extracted, and protected fragments of length ~147 bp were cloned and sequenced. (B) Typical nucleosomal occupancy map showing clear demarcation of the gene locus, identifiable by the presence of the 5′ and 3′ nucleosome-depleted regions or NDR. The 5′ NDR is flanked by rather tightly bound nucleosomes downstream of the transcription start site. (A, adapted from Segal E, Fondufe-Mittendorf Y, Chen L et al. [2006] *Nature* 442:772–778. With permission from Macmillan Publishers, Ltd. B, adapted, courtesy of Sergei Grigoryev, Penn State University, and Gaurav Arya, University of California, San Diego.)

particle, and so there must be linker DNA between particles. This is what produces the beads-on-a-string appearance noted by early electron microscopists (see Box 8.2). This linker DNA turns out to be the habitat of the fifth class of histones, the linker histones (see Table 8.3). Together with the core particle, they form what is called the **chromatosome**. Linker histones bind to the linker DNA entering and exiting the nucleosome in a manner that stabilizes the particle (Figure 8.18). In addition, these histones seem to play important roles in the formation and stabilization of higher-order folding of the

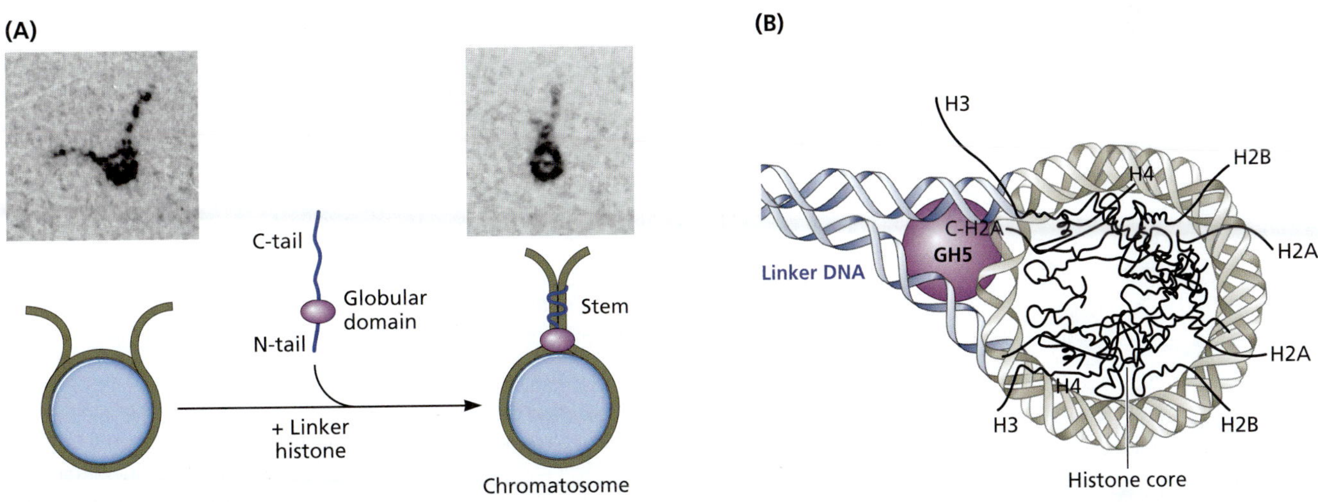

Figure 8.18 Chromatosome: a particle containing linker histones and more than 147 bp of DNA. (A) Linker histones, LH, bind to the nucleosome at the entry–exit point of nucleosomal DNA. Each member of the linker histone family possesses a short unstructured N-terminus, a globular domain (winged-helix motif), and a long unstructured C-tail. The binding of an isolated globular domain seals off two full turns of DNA around the histone octamer, compared to the 1.67 turn in the particle lacking LH. Interaction of the C-tail with the incoming and outgoing linker DNA creates the stem structure. The EM images above the two schematics are of nucleosomes reconstituted from naked DNA and histone octamers in the absence or presence of LHs. The stem structure has also been seen in situ. (B) To-scale schematic of the chromatosome based on the crystal structure of the core particle, the known dimensions of the globular domain GH5 of linker histones, and the expected trajectory of the two linker DNA segments. (A, EM images from Hamiche A, Schultz P, Ramakrishnan V et al. [1996] *J Mol Biol* 257:30–42. With permission from Elsevier. B, adapted from Leuba SH, Bustamante C, van Holde K & Zlatanova J [1998] *Biophys J* 74:2830–2839. With permission from Elsevier.)

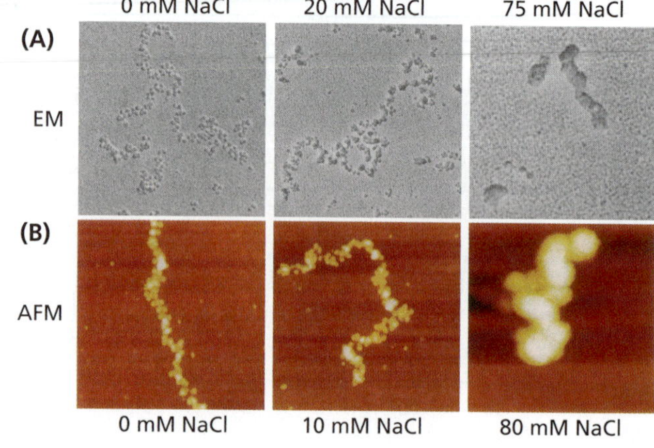

Figure 8.19 Micrographs of native chromatin fibers imaged by (A) EM or (B) AFM. In both cases, isolated fibers were fixed with glutaraldehyde under various salt concentrations, dialyzed against buffers that do not contain salt, and imaged. Compaction of the fiber with increasing ionic strength is clearly visible. (A, from Thoma F, Koller T & Klug A [1979] *J Cell Biol* 83:403–427. With permission from The Rockefeller University Press. B, from Zlatanova J, Leuba SH, Yang G et al. [1994] *Proc Natl Acad Sci USA* 91:5277–5280. With permission from National Academy of Sciences.)

chromatin fiber. Such folding is essential to give sufficient compaction of the chromatin in the nucleus. Note that formation of a nucleosome only compacts about 40 nm of DNA into a 10 nm particle. This compaction ratio of 4:1 is far from the values in the thousands necessary for proper compaction in the nucleus.

The chromatin fiber is folded, but its structure remains controversial

Does the chromatin fiber have a regular structure, and if so, what is it? That question has perplexed researchers for almost half a century. It has proved very difficult to visualize the chromatin fibers as they are packed into the interphase nucleus in any interpretable way. The crowding of macromolecules is just too great. Therefore, from the beginning of modern chromatin research, around 1970, investigators have utilized a variety of methods to extract, or to attempt to reconstruct, what appears to be at least a basic element in that structure. A host of studies using a wide variety of physical techniques, ranging from extraction of chromatin fiber from gently digested lysed nuclei to reconstruction of regularly spaced nucleosome arrays on repeated-sequence DNA, have given a superficially similar result: a more or less regular helical fiber of about 30 nm in diameter. This 30-nm fiber has been observed in a number of electron microscopic studies. Some typical views are shown in **Figure 8.19**, as are some AFM images taken under similar conditions.

We can say a few things about the structure of chromatin fibers with some confidence:

- It probably has physiological significance, since packed structures like this can be occasionally glimpsed in sectioned nuclei.
- It is stabilized by moderately high salt and especially by divalent ions.
- It is stabilized by the linker histones.

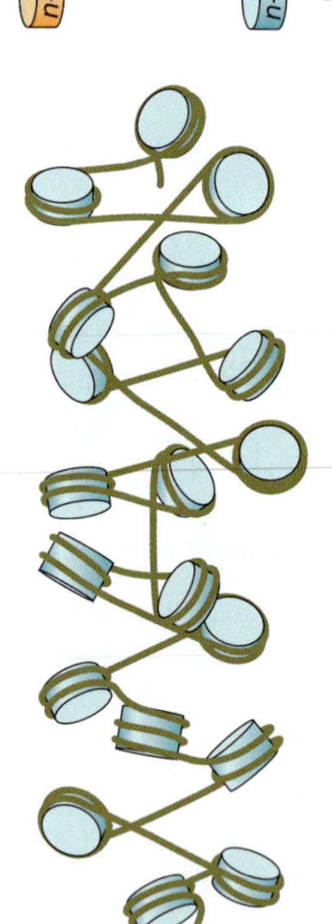

Figure 8.20 Models depicting the possible arrangement of nucleosomes in the 30-nm fiber. Note that none of the models has been universally accepted. (A) Side and top views of two major types of models: a one-start helix, or solenoid, and a two-start helix. A two-start helix can be visualized as follows: Imagine a zigzag ribbon of nucleosomes, where linkers pass back and forth between two rows of nucleosome particles. Now twist this ribbon into a helix. Nucleosomes are represented in both structures by cylinders. The linker DNA is not shown in the side view of the solenoid, since in most models it is located in the fiber interior; the linkers can be seen from the outside only in the two-start helical ribbon, where they help to create the zigzag ribbon of two rows of nucleosomes. Different color is used to indicate an example of nucleosomes, designated *n* and *n*+1, that are next to each other in the extended chromatin fiber; note that these nucleosomes have different spatial locations, and different neighbors, in the different models for the 30-nm fiber. The red lines are simply to emphasize the helical nature of both structures. (B) Straight-linker model: successive nucleosomes in the fiber are connected by straight linkers that criss-cross the fiber interior. The nucleosomes are situated at the periphery, as in the helical models. Further condensation will yield a lumpy, irregular fiber, like the one in AFM images at 80 mM NaCl (see Figure 8.19). (A, adapted from van Holde K & Zlatanova J [2007] *Semin Cell Dev Biol* 18:651–658. With permission from Elsevier. B, adapted, courtesy of Mikhail Karymov, California Institute of Technology.)

Beyond these areas of agreement, there is discord, even after decades of research. The main arguments are on a very fundamental point: what kind of a helix is this? The original proposal, by Mellema and Klug in 1976, was for a left-handed solenoidal helix with linker DNA passing directly from one nucleosome to the next along the spiral of the helix. But others have proposed two-start helices, or helices with complex internal connections. A few of the many models that have been proposed are shown in **Figure 8.20**. As one can see from these models, the proposed structures will be difficult to distinguish by simple electron microscopic observation: they look pretty much alike unless observed in fine detail.

There is also the fundamental question as to whether any of the models proposed, or indeed any uniform model, has any relevance to chromatin *in vivo*. Certainly, pieces of 30-nm fiber extracted from nuclei rarely show a smooth, regular helical structure; they are lumpy and bent. This may be what we should expect, as we know that linker lengths are locally heterogeneous, and this fact should argue against any wholly regular structure. Indeed, AFM images of chromatin being progressively compacted do not seem to converge on a single regular structure (see Figure 8.19B).

The organization of chromosomes in the interphase nucleus is still obscure

Early cytological studies of interphase nuclei revealed the existence of two readily distinguishable forms of chromatin structure: highly condensed **heterochromatin** and more dispersed **euchromatin**. These distinctions are still recognized and are thought to correlate with transcriptional activity, with transcription of the condensed heterochromatin being repressed. There also seem to be differences in the overall protein composition of these regions and in the presence of specific histone post-translational modifications (see Chapter 12). Newer techniques are able to show that the linear double-stranded DNA molecules that constitute individual interphase chromosomes occupy distinct portions of the nuclear volume, forming **chromosome territories** (**Figure 8.21**). The chromatin fiber within each territory can form individual loop domains, and domains that make temporal excursions out of their respective territories can be bridged together, presumably to allow their coordinated regulation through interactions with transcriptional factories (see Chapter 10). A significant portion of the chromatin fiber is associated with the lamina structure in the so-called **lamina-associated domains**, which are transcriptionally repressed. Intra- and interchromosomal interactions also form distinct **nucleoli** that contain both active and inactive copies of tandemly repeated ribosomal genes. A final recognizable structure that may have a role in the spatial and topological organization of the genome is the **nuclear matrix**, an insoluble meshwork of various skeletal proteins.

A curious, and still poorly understood, phenomenon is the distribution of gene-rich and gene-poor chromosomes (see Chapter 7) within the nuclear volume. The gene-poor chromosomes tend to locate at the nuclear periphery, whereas gene-rich chromosomes prefer the nuclear interior (**Figure 8.22**).

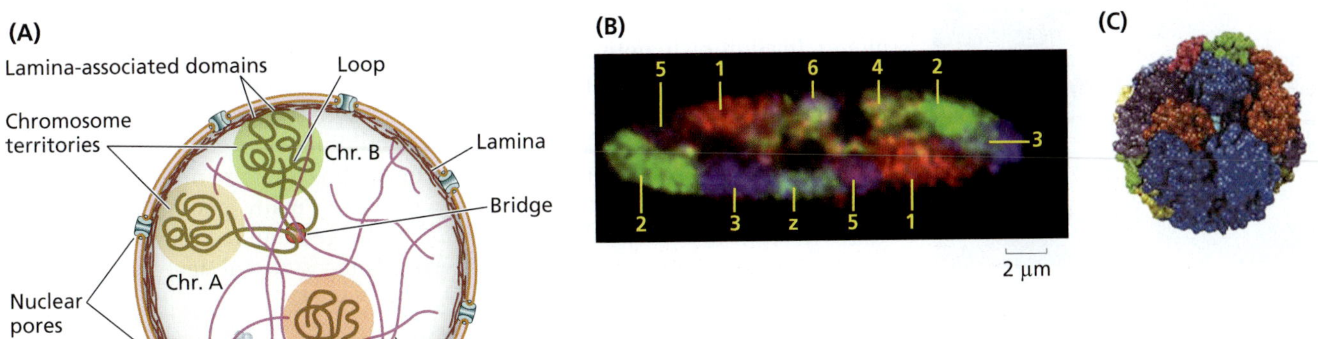

Figure 8.21 Chromosome territories in the interphase nucleus. (A) Organization of individual interphase chromosome territories (CTs) denoted here as Chr. A, Chr. B, and Chr. C, and shown in different colors. (B) Light optical section through a chicken fibroblast nucleus, showing mutually exclusive CTs. The individual chromosome territories were painted by a combinatorial scheme of several compounds, which were then detected by secondary antibodies labeled with the fluorescent dyes Cy3, fluorescein isothiocyanate (FITC), and Cy5. The two homologous chromosomes comprising each chromosome pair are seen in separate locations; note that only one of the two territories for each of chromosome 4 and chromosome 6 is displayed in this section. (C) Three-dimensional computational model of chromosome territories, visualized by use of pseudocolors, in a male diploid human cell nucleus. (B, from Cremer T & Cremer C [2001] *Nat Rev Genet* 2:292–301. With permission from Macmillan Publishers, Ltd. C, from Cremer T, Cremer M, Dietzel S et al. [2006] *Curr Opin Cell Biol* 18:307–316. With permission from Elsevier.)

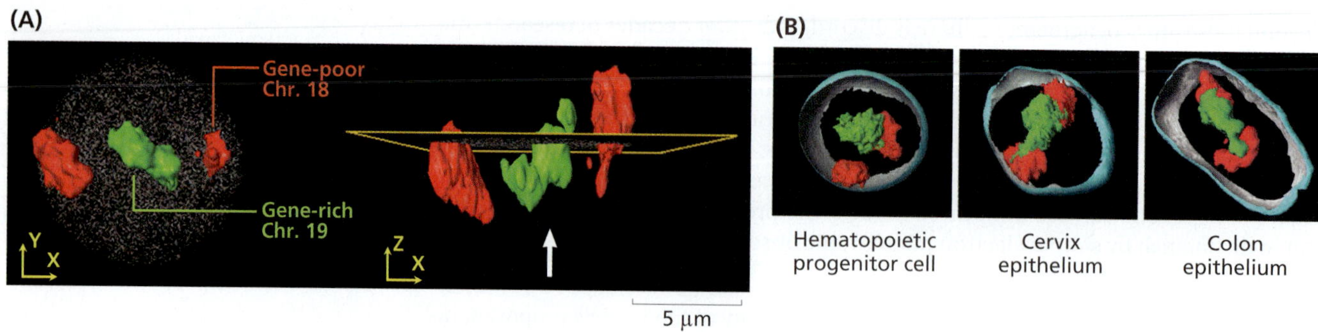

Figure 8.22 Gene-rich and gene-poor chromosome territories in human cells. (A) Lymphocyte nucleus: three-dimensional reconstructions of gene-poor chromosome 18 territory, shown in red, and gene-rich chromosome 19 territory, shown in green. Chromosome 18 territories were typically found at the nuclear periphery, whereas chromosome 19 territories were located in the nuclear interior. Left, an *X*, *Y* view: a section of the nucleus is shown in gray. Only the parts of the territories below this section can be seen. Right, an *X*, *Z* view: the arrow marks the side from which the section is viewed. (B) Three-dimensional reconstructions of the same chromosomes in three different types of cells. Part of the nuclear border is shown: outside, blue; inside, silver-gray. Usually CT19 is situated in the nuclear interior, whereas CT18 is found at the nuclear periphery, either side-by-side or at remote sites. Frequently, this distribution is lost in tumor cells. (A, from Cremer T & Cremer C [2001] *Nat Rev Genet* 2:292–301. With permission from Macmillan Publishers, Ltd. B, from Cremer M, Küpper K, Wagler B et al. [2003] *J Cell Biol* 162:809–820. With permission from The Rockefeller University Press.)

8.5 Mitotic chromosomes

Chromosomes condense and separate in mitosis

When cells enter mitosis after DNA replication, the nuclear envelope disintegrates and the paired sister chromatids undergo major condensation. The resulting **mitotic chromosomes** are easily isolated and can be observed by light microscopy (see Chapter 1). They have been known to scientists for more than 100 years. Mitotic chromosomes were first seen under the optical microscope when dividing cells were stained with specific stains. At certain stages of the division cycle, intensely stained bodies appeared and were named **chromosomes** because of their colored staining (**Figure 8.23**). Mitotic chromosomes are very compact and can be isolated from cells for further study. Gradually, it was recognized that chromosomes are the physical entities that carry the genetic information from cell generation to cell generation. The genetic information is first replicated in interphase and then evenly distributed between the two daughter cells during mitosis. Here, we focus on the molecular structure of chromosomes and the factors involved in the structural transitions from interphase chromatin to condensed chromosomes and back.

Despite years of research, our knowledge of the fine structure of chromosomes remains limited. Part of the reason is that the structure is so compact. Whereas the nucleosome achieves only 4–5-fold compaction of the DNA, and the 30-nm fiber no more than about 40-fold, the value for the mitotic chromosome is around 10,000-fold. The past 40 years have seen considerable efforts toward addressing this problem, mainly through the use of ever more sophisticated imaging techniques. In addition to optical microscopy, researchers have both transmission and scanning electron microscopy, fluorescence microscopy, and atomic force microscopy; the principles behind these techniques are briefly introduced in Chapter 1. Recently, proteomic analysis has identified a multitude of proteins, in addition to the already known topo II, cohesins, and condensins, that are present on mitotic chromosomes, but their functions remain obscure. Finally, genomewide studies have allowed the identification of sites along the genome to which some major chromosome proteins bind. However, these advances do not deal with the basic structural problem.

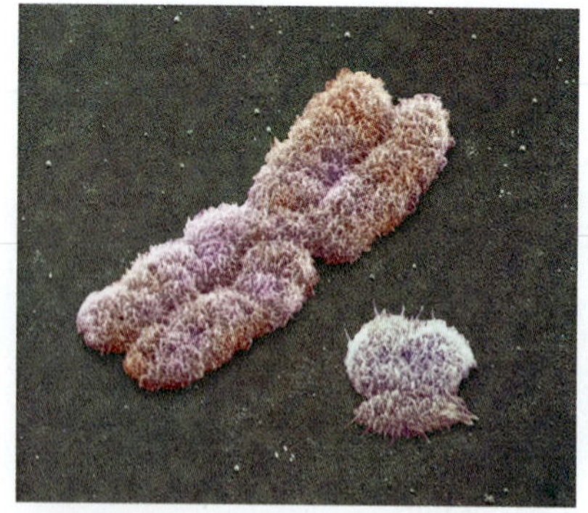

Figure 8.23 EM image of human chromosomes. Sex chromosomes X and Y are shown as viewed under scanning EM at magnification about 10,000-fold. (From Andrew Syred. With permission from Science Source.)

As mentioned above, most of our knowledge about chromosome structure comes from microscopic observations. A typical EM image is presented in Figure 8.23. **Figure 8.24** displays a gallery of AFM images that point to different structural characteristics within a single mitotic chromosome. A major advantage of AFM imaging

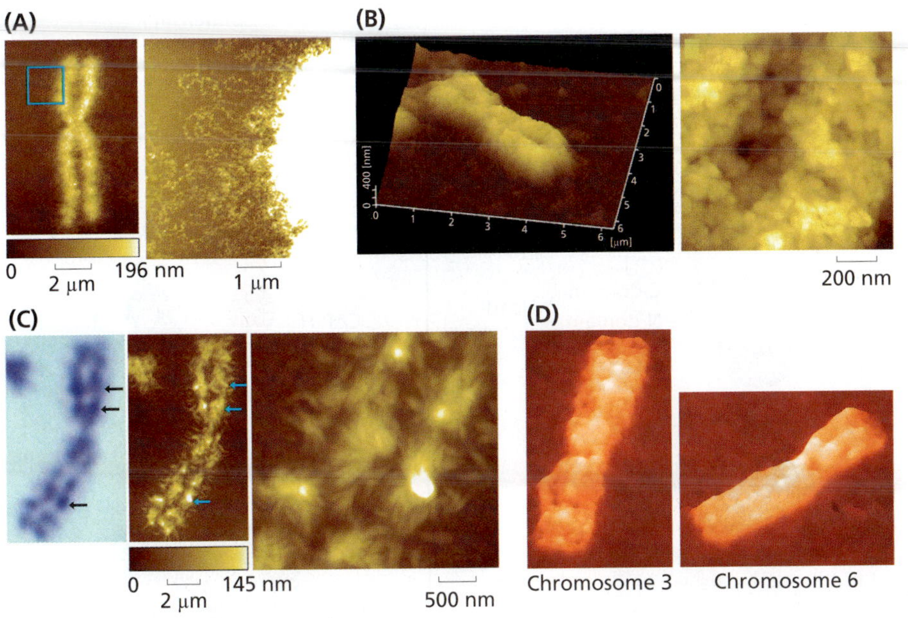

Figure 8.24 AFM images of human mitotic chromosomes. (A) Fixed chromosomes; some chromatin fibers accidentally unraveled during sample preparation, and these are visible as loops in the close-up view of the boxed region. The bar below indicates the height scale. (B) Unfixed chromosomes observed in buffer. The tilted view clearly shows the three-dimensional organization; the close-up on the right shows an aggregation of globular or fibrous structures (chromatin fibers). (C) Comparison between light-microscopic image, on the left, and AFM image, on the right, of a G-banded chromosome, produced by light trypsin treatment followed by Giemsa staining. The dark G-positive bands correspond to ridges in the AFM structure, whereas the pale G-negative bands correspond to grooves. The close-up shows that chromatin fibers are densely packed in the ridges but are rather sparse in the grooves. (D) AFM images of G-banded human chromosomes 3 and 6. Again, ridges and grooves are obvious along the entire chromosome length. (A–C, courtesy of Tatsuo Ushiki, Niigata University. D, courtesy of Stefan Thalhammer, Helmholtz Zentrum.)

is that it can be done on unperturbed, unfixed, and unstained specimens in liquid; thus, the resulting images are believed to reveal structures more closely resembling the physiologically relevant structures existing *in vivo*.

A number of proteins are needed to form and maintain mitotic chromosomes

Extensive biochemical studies have identified some of the most important proteins that play a role in forming and maintaining mitotic chromosome structure. One of these proteins was found, unexpectedly, to be topoisomerase II (see Chapter 4). Strangely, the enzymatic activity of topo II does not seem to be required, so the protein must play a purely structural role in this context.

A major step forward was the discovery of several protein complexes, **cohesins** and **condensins**, which have related structures. Their subunit composition and overall structure are depicted in **Figure 8.25**, which also illustrates the proposed modes of their interaction with chromatin fibers. The major protein subunits in both cohesins and condensins belong to the same protein family, structural maintenance of chromosomes. In addition, cohesins and condensins possess two or three non-SMC subunits; these differ between cohesins and condensins (see Figure 8.25), as well as between mitotic and meiotic cohesin and between condensin I and II, which are the two slightly different condensin complexes in vertebrates; they show different cellular localization in interphase but are both associated with mitotic chromosomes.

Cohesins, which attach to chromatin during DNA replication in S phase, embrace the two sister chromatids, keeping them close together until the prophase–metaphase transition in mitosis (see Figure 8.7). Condensins, on the other hand, bind to chromatin at the beginning of mitosis to compact the chromatin fiber in a manner not well understood. *In vitro* assays using purified components have shown that condensins introduce positive supercoiling in DNA (see Figure 8.25B), possibly indicating the importance of supercoiling in chromosome compaction. Cohesins are devoid of such activity.

If cohesin forms a ring to embrace sister chromatids, how do the two partners come together in this topologically linked structure? The process is driven by ATP hydrolysis as depicted in **Figure 8.26**. The next obvious question is how the ring is broken when the sister chromatids have to move to opposite poles of the mitotic spindle. The answer lies in the action of a specific protease, termed **separase**, which specifically cleaves one of the non-SMC subunits of the complex, Scc1, to open the ring.

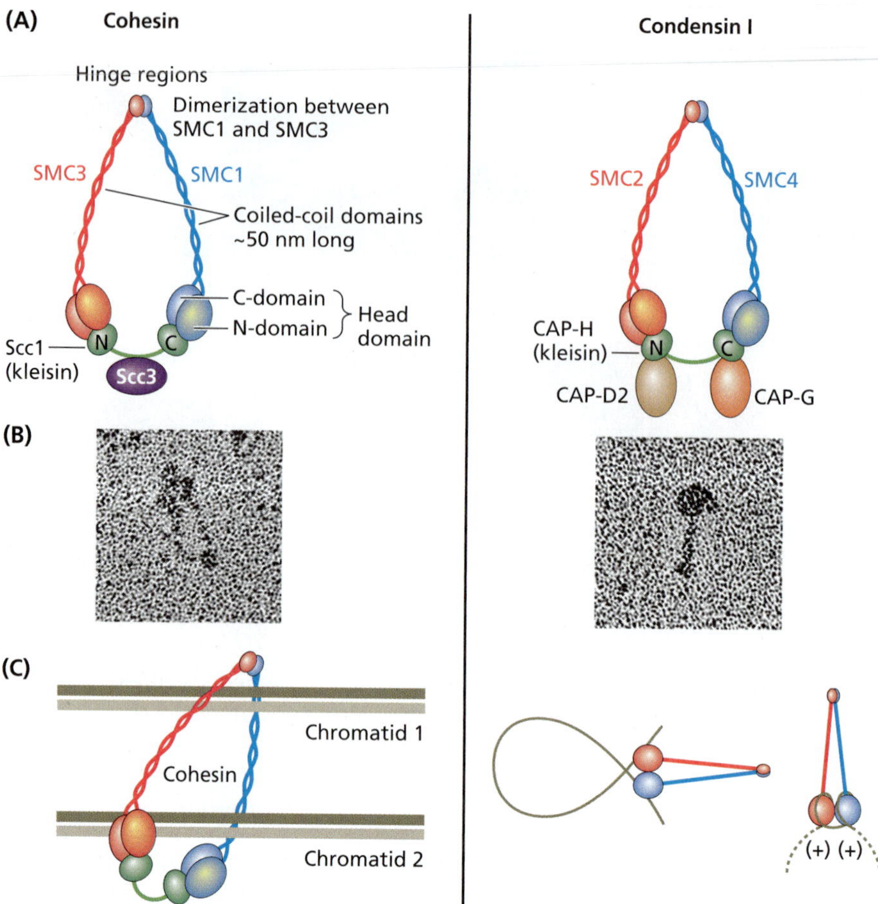

Figure 8.25 Cohesin and condensin structure and interaction with chromatin DNA. (A) Schematic representation of cohesin and condensin I. Both complexes contain a pair of SMC or structural maintenance of chromosomes proteins with additional non-SMC subunits. The five-domain structure of all SMCs consists of N- and C-termini that contain ATP-binding motifs and two long coiled coils that are separated by a hinge domain. Each SMC subunit folds back on itself to form a central region of two anti-parallel coiled coils, flanked on one side by the hinge region and on the other side by a head domain composed of the two N- and C-terminal ATP-binding domains. Two heterologous SMC subunits dimerize through their hinge domains, whereas the head domains interact with the non-SMC subunits. Scc1 and Scc3 are sister chromatin cohesion subunits in cohesin; CAP-H, CAP-D2, and CAP-G are the non-SMC subunits in condensin I. The second condensin complex, condensin II, contains alternative CAP subunits (not shown). Condensins I and II have different cellular localization in interphase, but both bind to chromosomes early in mitosis to play a role in chromosome condensation and segregation. (B) EM images of human cohesin and condensin. Note the differences in overall shape of the two complexes: ring shape versus rodlike. (C) Models of cohesin and condensin binding to chromatin DNA. Cohesin is believed to embrace both sister chromatids in a ring structure. For the dynamics of binding and dissociation of the two proteins during the cell cycle, see Figure 8.7. Condensin is known to introduce positive supercoils in DNA; this activity is believed to somehow contribute to condensation, although the precise mechanism remains obscure. The schematics illustrate two suggested possibilities. (B, from Anderson DE, Losada A, Erickson HP & Hirano T [2002] *J Cell Biol* 156:419–424. With permission from The Rockefeller University Press.)

Centromeres and telomeres are chromosome regions with special functions

There are two regions of the mitotic chromosome that have specialized structures and functions: centromeres and telomeres. Centromeres were initially recognized morphologically at major constrictions in mitotic chromosomes. Today, we know that centromeres secure faithful chromosome segregation during mitosis through their participation in the assembly of the kinetochore, the protein complex that mediates chromosome binding to the microtubules of the mitotic spindle and subsequently pulls chromosomes to the

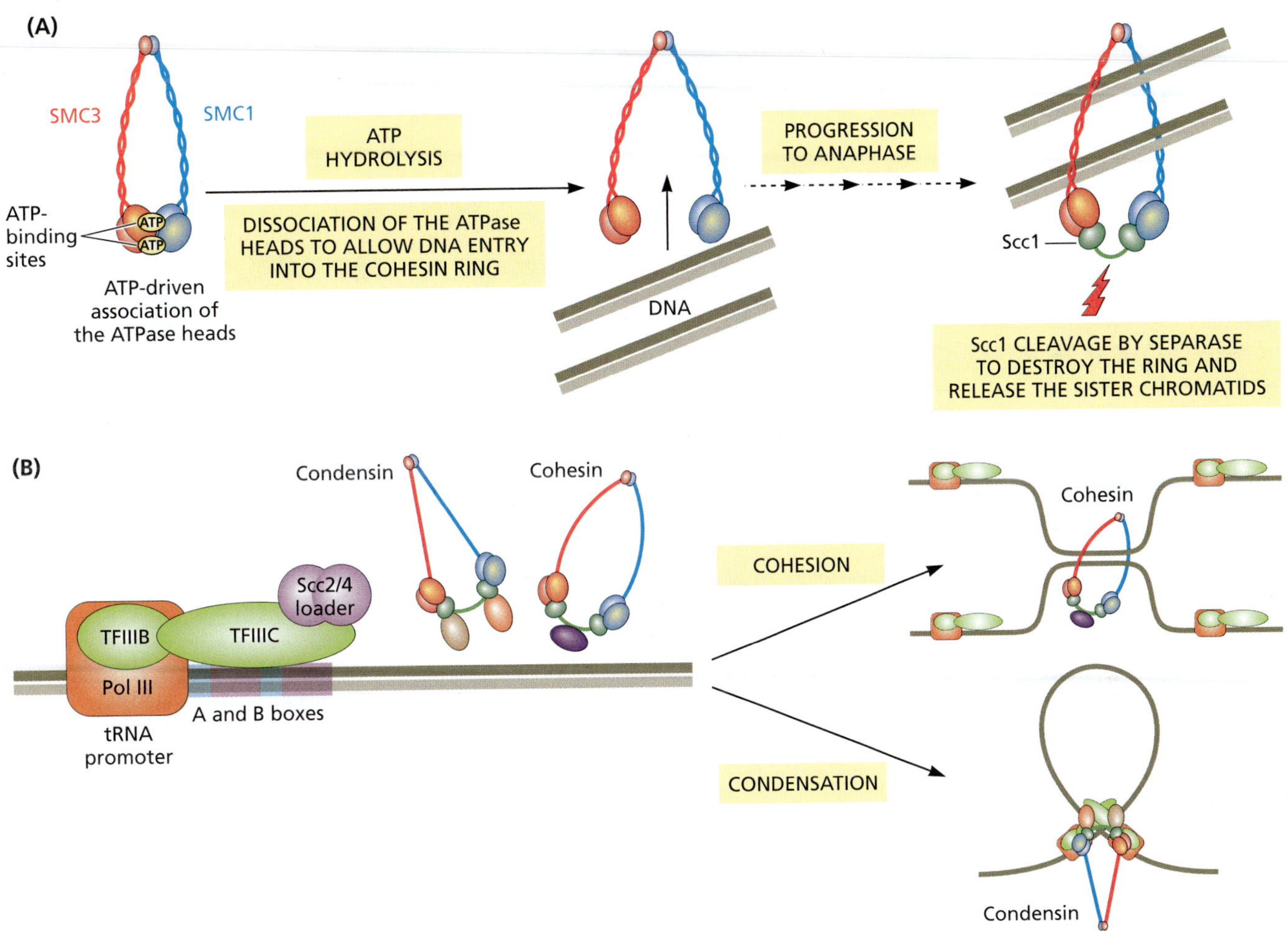

Figure 8.26 Loading of cohesins and condensins onto DNA. (A) Embracing sister chromatids during S phase, and their subsequent release from the embrace during anaphase, involves conformational transitions in the cohesin complex. The release also involves the action of a specific protease, separase, which cleaves the Scc1 non-SMC subunit of the complex. (B) Loading of both cohesin and condensin onto DNA occurs through the action of the same loader, a dimer of Scc2 and Scc4, labeled Scc2/4 loader. Genomewide distribution studies of cohesin and condensin in yeast have identified the sites of loading as the promoter regions of tRNA genes, either bound by the entire Pol III pre-initiation complex or by TFIIIC alone. There are close to 300 tRNA genes distributed across the genome that thus serve as loading platforms for both cohesins and condensins. Subsequent to loading, cohesin moves to distant locations, holding the two sister chromatids together from S phase until mitosis. Condensins, on the other hand, stay at the loading sites, forming DNA loops by bringing together distant TFIII sites. (Adapted from Gartenberg MR & Merkenschlager M [2008] *Genome Biol* 9:236. With permission from BioMed Central.)

poles during anaphase. Telomeres, on the other hand, are specialized nucleoprotein structures located at both ends of chromosomes. They function to protect the chromosome ends from nucleolytic degradation and/or recombination or repair events.

Centromeres come in different numbers and varieties. Most chromosomes have just one centromere, but some have more than one. Most organisms have localized centromeres, with centromere formation restricted to specific loci; other organisms, for example the nematode *Caenorhabditis elegans*, possess holocentric chromosomes, in which diffuse centromeres are present along the entire length of the chromosome. We focus our discussion here on localized centromeres, which are the most prevalent and best studied.

Localized centromeres are highly variable in size and sequence and belong to two general classes: **point centromeres**, found in budding yeast, and **regional centromeres**, found in fission yeast and most other eukaryotes. The point centromeres in budding yeast contain an essential 125-bp-long sequence that accommodates a single centromere nucleosome containing the CENP-A homolog Cse4; CENP-A is a special

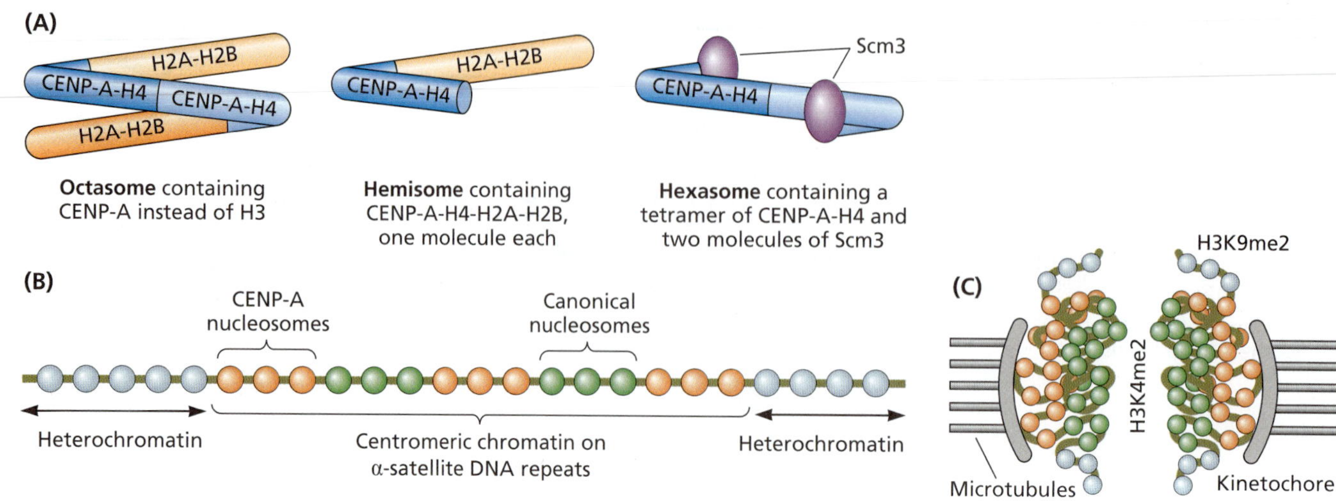

Figure 8.27 Unique chromatin organization of centromere regions in humans. (A) Possible compositions of the histone cores of centromeric nucleosomes. All particles contain the centromere-specific variant of histone H3, CENP-A. The CENP-A-H4 dimer can participate in the formation of an octamer, as in a canonical octasome. In addition, it can interact with a H2A-H2B dimer to form the core of the hemisome, a particle containing only one molecule each of the core histones. Finally, the CENP-A-H4 dimer has been proposed to form a hexamer, in which two CENP-A-H4 dimers interact with two molecules of non-histone kinetochore protein Scm3, the specific deposition vehicle for yeast CenH3. If proven, this will be the first example of a nucleosome particle that can accommodate non-histone proteins in lieu of histone H2A-H2B dimers. (B, C) Higher-order organization of centromeric chromatin. Two-dimensional chromatin fibers, with subdomains of nucleosomes containing centromeric histone CENP-A, shown in orange, interspersed with canonical H3-containing nucleosomes, shown in green. The canonical particles are dimethylated at lysine K4, denoted H3K4me2. This arrangement of nucleosomes occurs over a portion of the α-satellite DNA repeats; the remainder of the α-satellite DNA is assembled into heterochromatin, shown in light blue, that flanks one or both sides of the centromeric chromatin domain. Histone H3 in heterochromatin is dimethylated at lysine K9, a typical heterochromatin modification. At metaphase, when mitotic chromosomes condense, the interspersed domains coil in a way that stacks CENP-A nucleosomes on one face of the chromosome, where they can interact with kinetochore proteins. The H3-containing nucleosomes are oriented between sister kinetochores. (Adapted, courtesy of Michael Hendzel, University of Alberta.)

variant of histone H3. This single nucleosome interacts with the kinetochore, which attaches stably to a single microtubule during metaphase. The regional centromeres usually contain DNA repeats, whose sequence and organization is highly variable among species. The lack of conservation of centromeric DNA suggests that the DNA sequence does not determine centromere formation. In fact, we now know from numerous experiments that centromeric DNA is neither necessary nor sufficient for centromere formation. The only structural feature that unifies all regional centromeres is the presence of variant nucleosomes containing the CenH3 variant (CENP-A in human) in place of the canonical histone H3.

Figure 8.27A illustrates three different types of nucleosomal particles that contain CENP-A. The centromeric nucleosomes may contain (1) an octamer of histones; (2) an

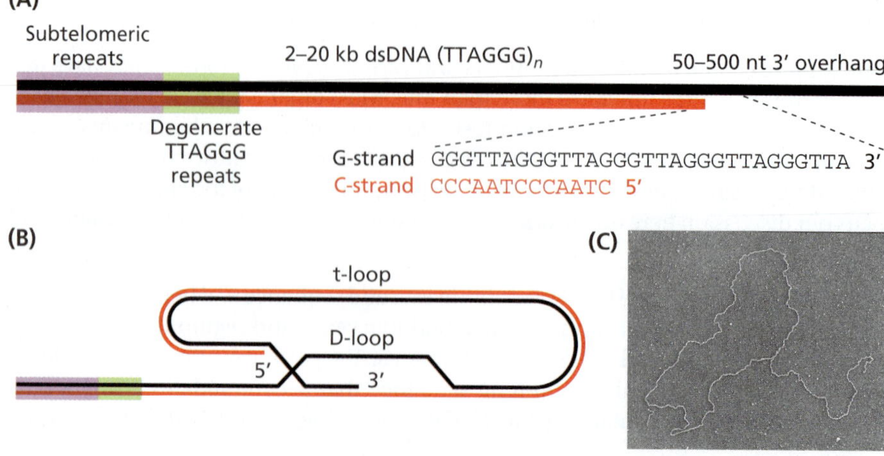

Figure 8.28 DNA structure of human telomeres. (A) The chromosomes end in an array of TTAGGG repeats that vary in length even in the same individual. The TTAGGG repeats are preceded by a segment of imperfect repeats and by subtelomeric repetitive elements. The long G-strand overhang favors the formation of G-quadruplexes. (B) Schematic of t-loop structure; the size of the loops is variable. The 3′ single-stranded end of the telomere displaces a strand from a homologous region in the dsDNA further upstream, leading to the formation of the D-loop; this displacement is catalyzed by TRF2. (C) Electron micrograph of a t-loop. (A–B, adapted from Palm W & de Lange T [2008] *Annu Rev Genet* 42:301–334. With permission from Annual Reviews. C, courtesy of Jack Griffith, University of North Carolina.)

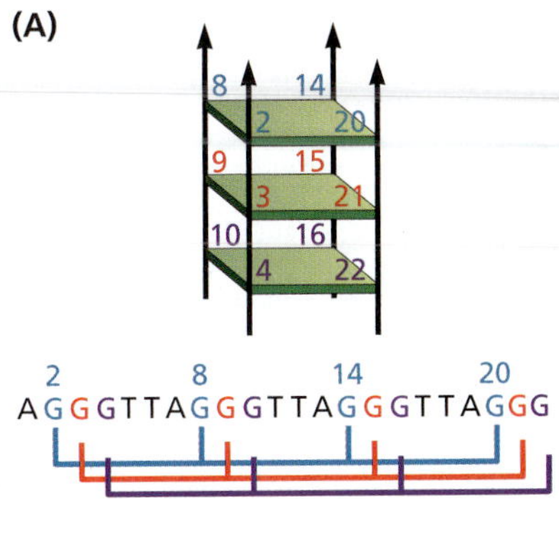

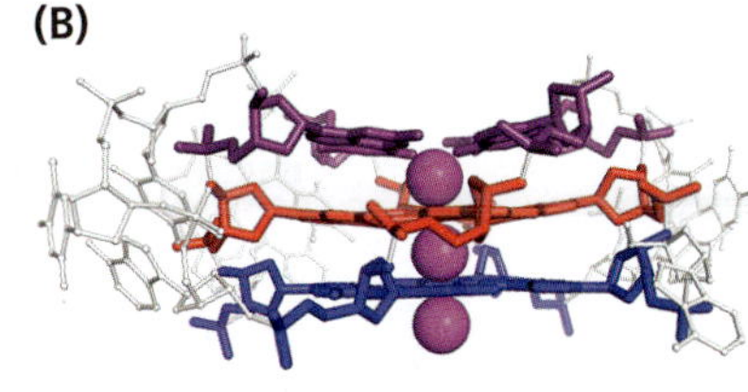

Figure 8.29 Short repeating sequence in the G-rich overhang from human telomere forms three stacked G quartets. (A) Schematic of the structure of the repeat sequence, with the guanine nucleotides forming the quartets colored the same way: 2, 8, 14, and 20, blue; 3, 9, 15, and 21, red; 4, 10, 16, and 22, purple. The top quartet contains the first G from each repeating unit; it is stacked above a quartet that contains the second G of each repeating unit, and so on. (B) The crystal structure shows that the DNA strand circles the G bases within the quartets; the G bases stack together in the center around coordinated metal ions.

unusual tetramer consisting of only one molecule each of CENP-A, H4, H2A, and H2B, with the particle containing this tetramer dubbed a **hemisome**; and (3) a regular CENP-A-H4 tetramer with some non-histone proteins in lieu of H2A-H2B dimers; this nucleosome is a hexasome. When the structure of regional centromeres was further elucidated by using fluorescent antibodies to canonical H3 or CENP-A, a very unusual interspersed distribution of canonical versus centromeric nucleosomes was detected (Figure 8.27B). This organization presumably allows the chromatin fiber to fold into a higher-order structure with two unequal faces: one interacts with the kinetochore, while the other one serves to bridge the two sister chromatids. An additional level of complexity involves the nonrandom distribution of certain histone post-translational modifications (Figure 8.27C).

Telomeres are essential because of a peculiar aspect of DNA replication: upon coming to the end of replicating a DNA duplex, the polymerase cannot quite finish one strand. This would result in shortening of the chromosome with each replication cycle and eventual loss of genetic information (see Chapter 20). Telomeres are special structures at the ends of chromosomes that avert this. The structure of telomeres is amazing. In most organisms, telomeric DNA consists of short, tandemly repeated sequences that end in single-stranded G-rich overhangs (Figure 8.28). These repeats have been added to the end of the DNA to protect against the deleterious effects of shortening: loss of these does not matter. However, there will always be 3′ overhangs. These could be mistaken by DNA repair enzymes for sites of DNA damage, which could lead, among other things, to fastening chromosomes together at their ends. This undesirable, unscheduled DNA repair is inhibited by the peculiar structures formed by these overhangs, known as **G-quadruplexes** (Figure 8.29). Alternatively, the overhang interacts with a homologous region further upstream to form **t-loops**. These t-loops interact with a protein complex termed **shelterin** (Figure 8.30), whose function is to shelter the structure from unwanted repair and degradation. In terms of chromatin structure, the relatively short telomeres of lower eukaryotes are nonnucleosomal, whereas the long telomeres of higher eukaryotes are mostly organized as tightly spaced nucleosomes that contribute to the overall telomere structure.

There are a number of models of mitotic chromosome structure

There are at least four major models for the structure of mitotic chromosomes (Table 8.6): folded fiber model, hierarchical helical coiling model, radial loop model,

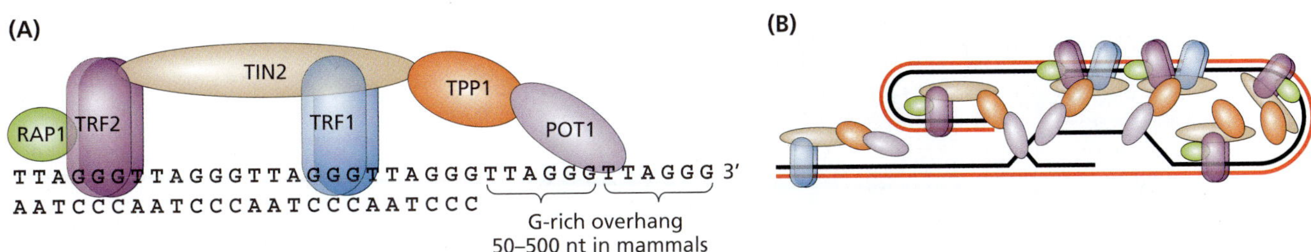

Figure 8.30 Shelterin complex. (A) The shelterin complex includes dimers of TRF1 and TRF2 proteins that bind to the double-stranded portion of the telomere, POT1 proteins that bind to the single-stranded G-rich overhang, and TIN2 and TPP1 proteins that serve as bridges. POT1 serves to suppress normal DNA repair processes that would recognize the single-stranded overhang as DNA damage. (B) Model of the shelterin complex bound to a telomere in a t-loop configuration; different subcomplexes can exist, such as those shown without the TRF1 dimer, because TRF1 and TRF2 bind independently of each other. (A, adapted from Palm W & de Lange T [2008] *Annu Rev Genet* 42:301–334. With permission from Annual Reviews. B, adapted from Denchi EL [2009] *DNA Repair* 8:1118–1126. With permission from Elsevier.)

Table 8.6 Four major models for mitotic chromosome structure.

Model	Methods	Features	Researcher(s), year
folded fiber model	light and electron microscopy	random tangle of chromatin fibers in all directions	Ernest DuPraw, 1965
hierarchical helical coiling model	light and electron microscopy	30-nm chromatin fiber is progressively coiled into 100-nm fibers, 200–250-nm fibers, and finally into 500–750-nm structures (the diameter of sister chromatids); a hollow center for the last coil is predicted	Francis Crick, 1977; John Sedat and Laura Manuelidis, 1978
radial loop model	electron microscopy of histone-depleted chromosomes	chromatin fiber loops extend radially from a proteinaceous scaffold containing topo II and condensin	Ulrich Laemmli, 1977
helical coiling of radial loop model	transmission and scanning electron microscopy	200-nm fibers consisting of radial loops of 30-nm fibers are helically coiled to form sister chromatids	Jerome Rattner, 1992

and helical coiling of the radial loop model. At present, the most popular competing models are the radial loop model and the hierarchical helical coiling model (**Figure 8.31**), both of which have substantial supportive evidence and may actually coexist. **Box 8.4** presents the radial loop model in detail.

It is evident that further work is needed to definitively prove or disprove any existing model. The advent of more and more sophisticated imaging technologies is expected to finally solve this long-standing mystery in molecular biology.

Box 8.4 The mitotic chromosome: the work of Ulrich Laemmli

It has long been known that DNA compaction in mitotic chromosomes is huge. Each mitotic chromosome contains a single molecule of DNA, which has to be shortened by ~10,000-fold

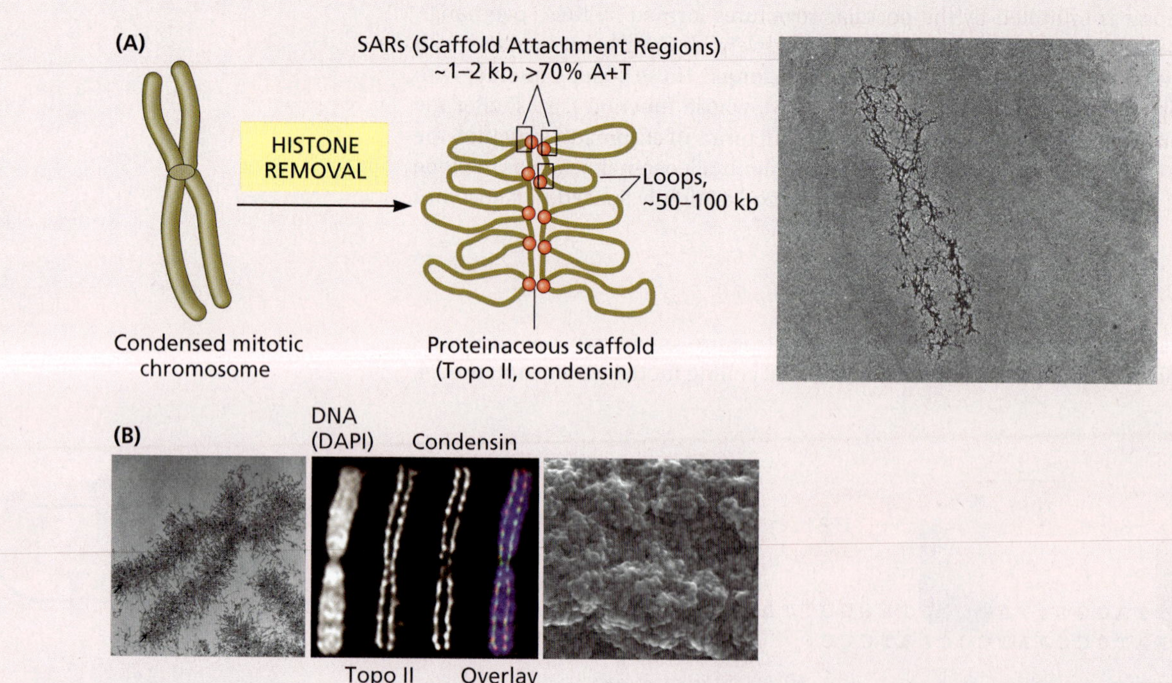

Figure 1 Mitotic chromosome. (A) Mitotic chromosomes following histone removal. (B) Left, swollen intact mitotic chromosomes imaged at low concentrations of divalent ions; center, fluorescence microscopy with fluorescently labeled antibodies against topo II and condensin; right, scanning EM image of a swollen chromosome. (A, right, from Paulson JR & Laemmli UK [1977] *Cell* 12:817–828. With permission from Elsevier. B, left and middle, from Maeshima K & Eltsov M [2008] *J Biochem* 143:145–153. With permission from Oxford University Press. B, right, from Shemilt LA, Estandarte AKC, Yusuf M & Robinson IK [2014] *Philos Trans R Soc Lond A* 372:20130144. With permission from Royal Society of Chemistry.)

prior to the phase in mitosis in which the genetic material is distributed evenly between the two daughter cells. The DNA compaction provided by the interphase chromatin fiber does not exceed ~40-fold, so significant additional compaction is needed.

Ulrich Laemmli wanted to understand both the physical organization of the fully compacted mitotic chromosome and the process of compaction itself. Laemmli started his work in the mid-1970s, when the prevalent models of chromatin compaction into chromosomes asserted some sort of self-assembly that was based on successive levels of coiling of the chromatin fiber. He proposed that certain non-histone proteins interact specifically with certain DNA regions to fold the chromatin fiber into loops, which extend radially from a central proteinaceous axis; hence the name of his model, the **radial loop model**. How did he come to this idea?

His early experiments relied on EM imaging of isolated mitotic chromosomes after removal of the histones (**Figure 1**). The resulting structure, shown in Figure 1A, was not completely unfolded; rather, it maintained the overall shape of the chromosomes. It also showed a conspicuous central structure, which was named the **scaffold**. Further studies identified two major protein components of the scaffold: topoisomerase II and protein SC2, one of the non-SMC subunits of condensin (see Figure 8.25). The DNA regions that attach to the scaffold were termed **scaffold attachment regions** or **SARs**; these turned out to be ~1–2 kb in length and highly enriched in AT base pairs. Regions of very similar length and composition were also found in interphase nuclei, as points of attachment of the chromatin fiber to components of the nuclear matrix. These were termed **matrix attachment regions** or **MARs**. It is believed that MARs and SARs are highly related, if not identical, and serve the same purpose in the interphase nucleus and in mitotic chromosomes: they attach dynamically to chromatin fibers and help to organize them into individual loop domains.

These early images of histone-depleted chromosomes showed DNA loops that spread out of the proteinaceous scaffold to form a dense halo. The challenge now was to prove that these DNA loops were also present in nonextracted chromosomes. But how can loops be seen in the highly compacted chromosome structure? The trick was to image slightly extended chromosomes obtained by exposing them to reduced concentrations of divalent ions, such that all proteins remained bound. EM images of such swollen chromosomes (see Figure 1B, left image) unequivocally confirmed the presence of loops. Scanning EM images (see Figure 1B, right image) were also consistent with chromatin loops: the knobs at the chromosome surface probably represent the tips of the loops. Of note, recent AFM images (see Figure 8.24) show a very similar morphology.

Further work focused on the role of the two scaffolding proteins, topo II and condensin, in the organization of the structure. Fluorescence microscopy using antibodies for these proteins gave the answer (see Figure 1B, center image). The two proteins exhibit axial distribution at the center of each chromatid, with alternating staining for topo II and condensin (barber-pole structure). The requirement for topo II for chromosome assembly was directly proven by biochemical experiments.

Years of work have led to the formulation of the radial loop model (**Figure 2**).

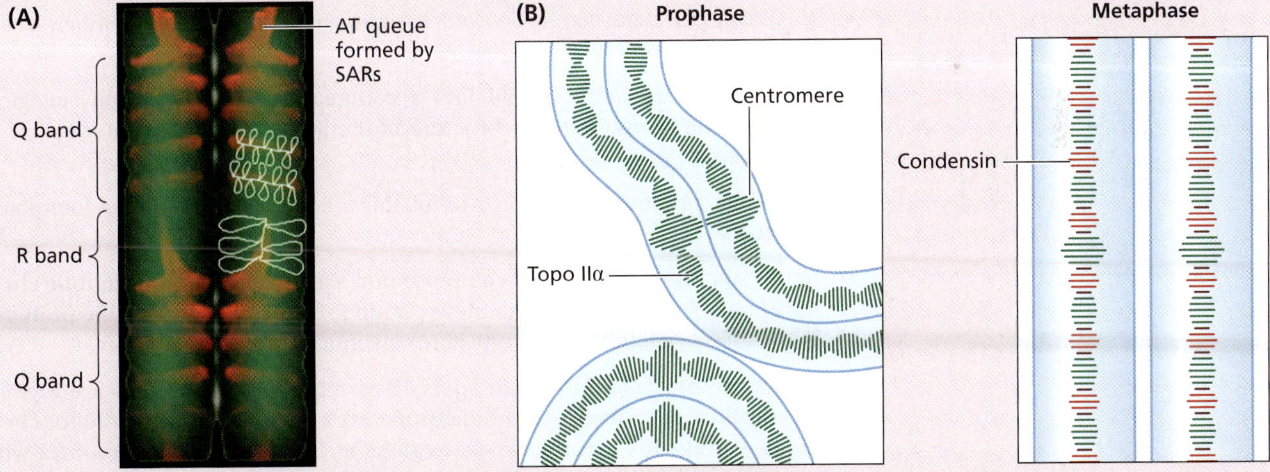

Figure 2 Laemmli's model of a condensed metaphase chromosome and two-step scaffolding model of chromosome assembly. (A) Laemmli's model of a condensed metaphase chromosome. Each chromatid is quite uniformly stuffed with DNA but contains an internal subregion called the AT queue, which is generated by juxtaposition of the SARs bound to the proteinaceous scaffold. The AT queue proceeds from telomere to telomere on an irregular coil-like path. The folding of the AT queue seems to correspond to cytologically recognized bands: Q bands from tight coiling and R bands from looser coiling. (B) Two-step scaffolding model of chromosome assembly: topo II first, then condensin. This model was proposed in view of the fact that prophase chromosomes do not contain condensin; condensin associates only during the pro-metaphase transition. During prophase, the bright foci of the topo II signal at centromeres and the topo II-marked chain are shown in green. The topo II chain may arise from mitotic extension of centric heterochromatin into the chromatid arms. This step of condensation yields prophasic chromatids that are longer and thinner than those of metaphase. Sister chromatids are shown. During metaphase, the chromosomes take on a characteristic barber-pole appearance. It is unclear whether coiling of the topo II- or condensin-marked chains, stacking, or some other structural transformation generates this appearance. (A, courtesy of Ulrich Laemmli, University of Geneva. B, adapted from Maeshima K & Laemmli UK [2003] *Dev Cell* 4:467–480. With permission from Elsevier.)

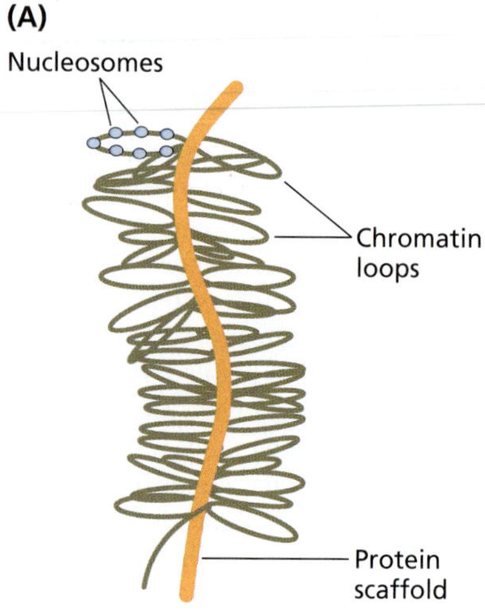

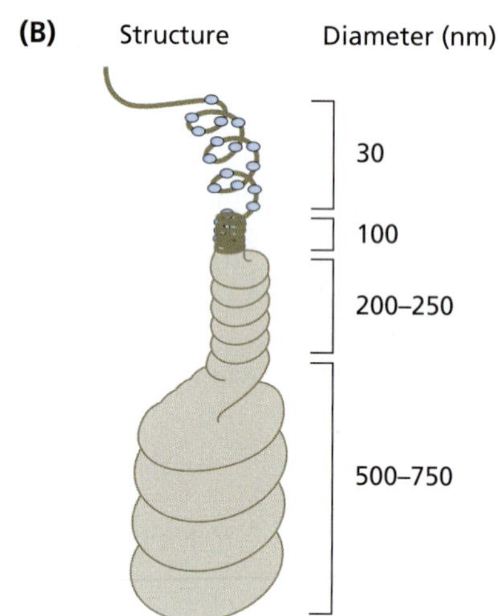

Figure 8.31 Two major competing models for chromosome structure. (A) Radial loop model; (B) hierarchical helical coiling model. This model proposes that the 30-nm fiber is further successively folded into helices of 100, ~250, and ~500 nm in diameter.

Key concepts

- The genomes of all organisms are compacted by interaction with specific proteins. These often also play roles in gene regulation.
- In viruses, the capsid constrains the DNA or RNA into highly compacted structures.
- In bacteria, a large circular chromosome is complexed by proteins in a dynamic manner that compacts it and yet makes it accessible for transcription.
- Eukaryotes utilize a set of structural proteins, the histones, to organize the chromatin into a repeating structure. Archaea have similar chromatin organization.
- The repeating elements of eukaryotic chromatin are called nucleosomes. Nucleosomes consist of core particles and linker DNA. In each core particle, about 147 bp of DNA is wrapped in 1.67 left-handed toroidal turns about an octamer of histones, two each of four types: H2A, H2B, H3, and H4. H2A and H2B form heterodimers, while H3 and H4 form a tetramer.
- Some of the histones exist in variant forms, and all are subject to a number of possible post-translational modifications. This allows an enormous range of variation in nucleosomes.
- Assembly of nucleosomes is facilitated by proteins called histone chaperones.
- The nucleosome is a dynamic structure and may exhibit a number of aberrant forms under special conditions.
- The chromatin fiber exhibits much specific positioning of nucleosomes, but their spacing is erratic and species-dependent.
- The linker DNA between nucleosomes is frequently occupied by members of the linker or H1 class of histones.
- Chromatin fibers tend to fold into a compact 30-nm fiber at physiological salt concentrations. The fine structure of the 30-nm fiber remains a matter of controversy.
- Individual chromosomes appear to inhabit unique territories in the interphase nucleus.
- In mitosis, sister chromatids condense into the long-recognized mitotic chromosomes. These are highly condensed and held together by specific proteins. The manner of folding of mitotic chromosomes is still under debate.
- Chromosomes contain certain specialized regions. Centromeres are regions for attachment to the microtubular apparatus of the mitotic spindle for chromatid segregation. Telomeres are regions at the ends of chromosomes with special structure to protect DNA from degradation or unscheduled DNA repair. Each of these regions has its own specific proteins that stabilize the structure.

Further reading

Books

Catalano CE (ed) (2005) Viral Genome Packaging: Genetics, Structure, and Mechanisms. Springer Verlag.

Higgins PN (ed) (2005) The Bacterial Chromosome. ASM Press.

Sumner AT (2003) Chromosomes. Organization and Function. Blackwell Science Ltd.

van Holde KE (1988) Chromatin. Springer Verlag.

Zlatanova J & Leuba SH (eds) (2004) Chromatin Structure and Dynamics: State-of-the-Art. Elsevier.

Reviews

Arya G, Maitra A & Grigoryev SA (2010) A structural perspective on the where, how, why, and what of nucleosome positioning. *J Biomol Struct Dyn* 27:803–820.

Cremer T & Cremer C (2001) Chromosome territories, nuclear architecture and gene regulation in mammalian cells. *Nat Rev Genet* 2:292–301.

De Koning L, Corpet A, Haber JE & Almouzni G (2007) Histone chaperones: an escort network regulating histone traffic. *Nat Struct Mol Biol* 14:997–1007.

Hudson DF, Marshall KM & Earnshaw WC (2009) Condensin: Architect of mitotic chromosomes. *Chromosome Res* 17:131–144.

Luijsterburg MS, Noom MC, Wuite GJ & Dame RT (2006) The architectural role of nucleoid-associated proteins in the organization of bacterial chromatin: a molecular perspective. *J Struct Biol* 156:262–272.

Maeshima K & Eltsov M (2008) Packaging the genome: the structure of mitotic chromosomes. *J Biochem* 143:145–153.

Palm W & de Lange T (2008) How shelterin protects mammalian telomeres. *Annu Rev Genet* 42:301–334.

Pisano S, Galati A & Cacchione S (2008) Telomeric nucleosomes: forgotten players at chromosome ends. *Cell Mol Life Sci* 65:3553–3563.

Torras-Llort M, Moreno-Moreno O & Azorin F (2009) Focus on the centre: the role of chromatin on the regulation of centromere identity and function. *EMBO J* 28:2337–2348.

van Holde K & Zlatanova J (1999) The nucleosome core particle: does it have structural and physiologic relevance? *BioEssays* 21:776–780.

van Holde K & Zlatanova J (2007) Chromatin fiber structure: Where is the problem now? *Semin Cell Dev Biol* 18:651–658.

Wood AJ, Severson AF & Meyer BJ (2010) Condensin and cohesin complexity: the expanding repertoire of functions. *Nat Rev Genet* 11:391–404.

Zlatanova J, Bishop TC, Victor J-M et al. (2009) The nucleosome family: dynamic and growing. *Structure* 17:160–171.

Experimental papers

Luger K, Mäder AW, Richmond RK et al. (1997) Crystal structure of the nucleosome core particle at 2.8 Å resolution. *Nature* 389:251–260.

Paulson JR & Laemmli UK (1977) The structure of histone-depleted metaphase chromosomes. *Cell* 12:817–828.

Segal E, Fondufe-Mittendorf Y, Chen L et al. (2006) A genomic code for nucleosome positioning. *Nature* 442:772–778.

Thoma F, Koller T & Klug A (1979) Involvement of histone H1 in the organization of the nucleosome and of the salt-dependent superstructures of chromatin. *J Cell Biol* 83:403–427.

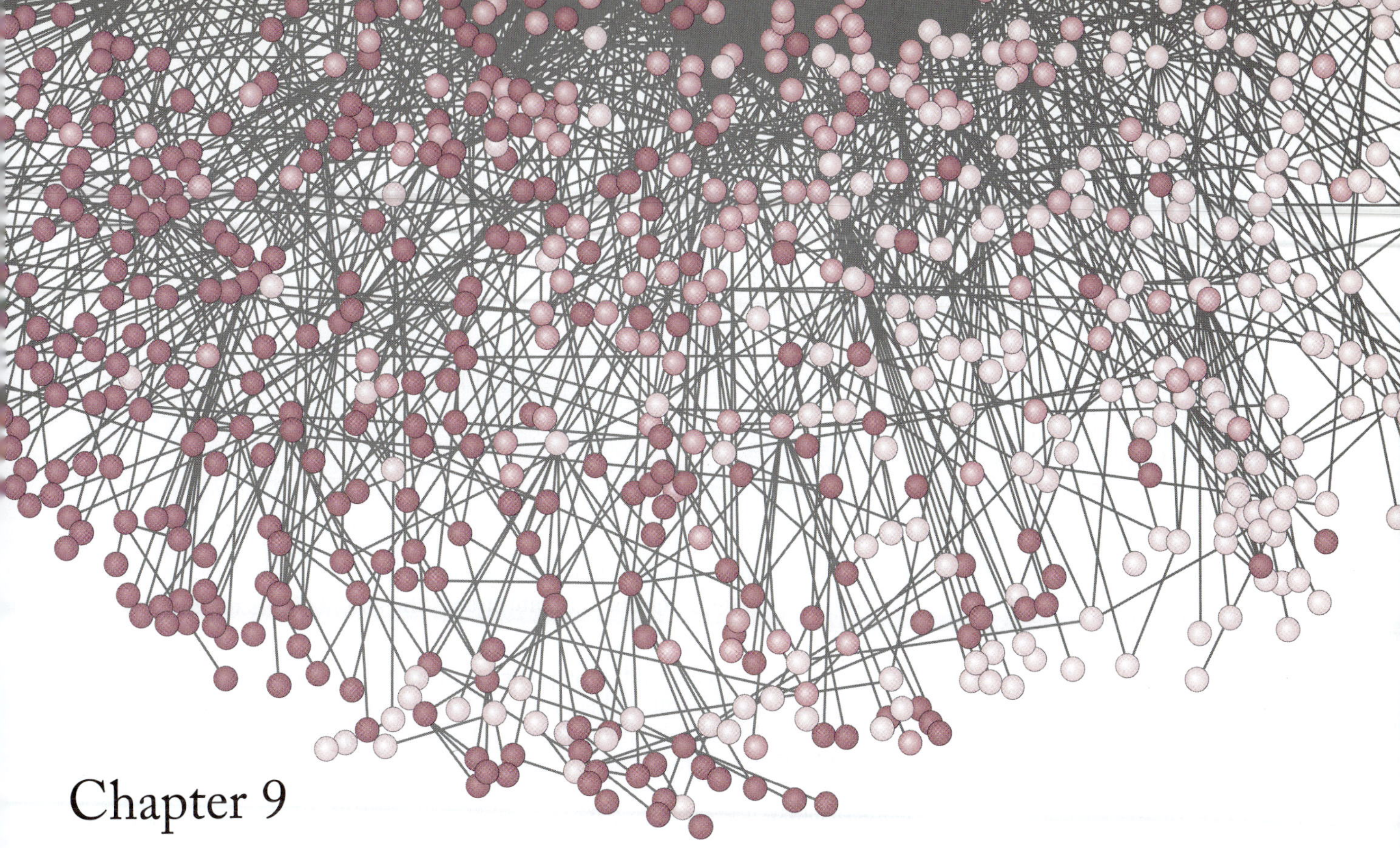

Chapter 9

Transcription in Bacteria

9.1 Introduction

As we have seen in previous chapters, the genetic information in the cell is stored in DNA. But this information is expressed in the form of specific RNA and protein molecules. In fact, all this expression requires the synthesis of RNA molecules that are complementary in sequence to one of the two strands of DNA. The process of RNA synthesis, or transcription, is an essential part of both production of mRNA templates for protein synthesis and generation of all the special RNA molecules needed in the cell. In this chapter, we first provide an overview of the basic mechanisms common to transcription in all organisms and then present a more detailed picture as to how transcription occurs in bacteria. Further aspects of the regulation of transcription in bacteria will be covered in Chapter 11.

9.2 Overview of transcription

There are aspects of transcription common to all organisms

To begin, we introduce a number of concepts and/or conventions used to describe transcription. First, the DNA strand whose sequence is being read to form RNA is called the **template strand**. Thus, the nucleotide sequence in RNA is a complementary copy of the sequence within this template strand. The nature of each nucleotide added to the nascent RNA chain is specified by base-pairing and complements the next unpaired nucleotide on the template strand. The pairing is exactly in accordance with the rules that define base pairing between the two complementary strands of a DNA molecule except that U, which pairs with A, substitutes for the T in DNA (**Figure 9.1**). Three different alternative designations are in use for the other DNA strand: **sense strand**, **coding strand**, or **nontemplate strand**. The first two designations reflect the fact that the sequence information in this DNA strand is exactly the same as in RNA, that is, that the two strands read exactly the same in the $5' \rightarrow 3'$ direction, with U in RNA substituting for T in DNA; compare the

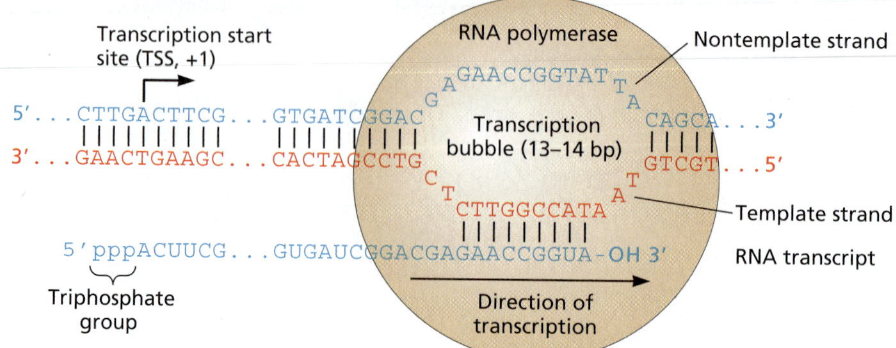

Figure 9.1 Orientation of a gene. By convention, the gene is said to be transcribed from its 5′- to its 3′-end; however, the template strand is actually copied from the 3′-end. This is required so that the new RNA strand complies with the normal rules for double-stranded nucleic acids: the DNA template strand and the nascent RNA are antiparallel and the bases are complementary. The nontemplate strand is also known as the coding strand or the sense strand, exactly the same as the codons in mRNA. In the process, a small, 13–14 bp portion of DNA is unwound, creating a transcription bubble. Moving along a gene in the 5′ to 3′ direction is described as moving downstream, whereas moving in the opposite 3′ to 5′ direction is known as moving upstream.

sequences in blue in Figure 9.1. These designations can be confusing, which is why we prefer to use the term nontemplate strand: the strand that is not directly used as the template.

Because the RNA corresponds in sequence to the nontemplate strand, we say, by convention, that a gene is transcribed from the 5′- to the 3′-end, despite the fact that the template strand is actually copied from the 3′-end (see Figure 9.1). Thus, the 5′-end of the RNA corresponds to the 5′-end of the nontemplate strand. The direction of transcription is defined by the requirement for a free 3′-OH group on the nascent RNA chain to which the next nucleotide is to be added. **Figure 9.2** shows the chemistry of the nucleotide addition reaction. In addition, the new RNA strand complies with the rules of complementarity and antiparallelism of polynucleotide strands that govern DNA structure. Moving along a gene in the 5′ → 3′ direction is described as moving

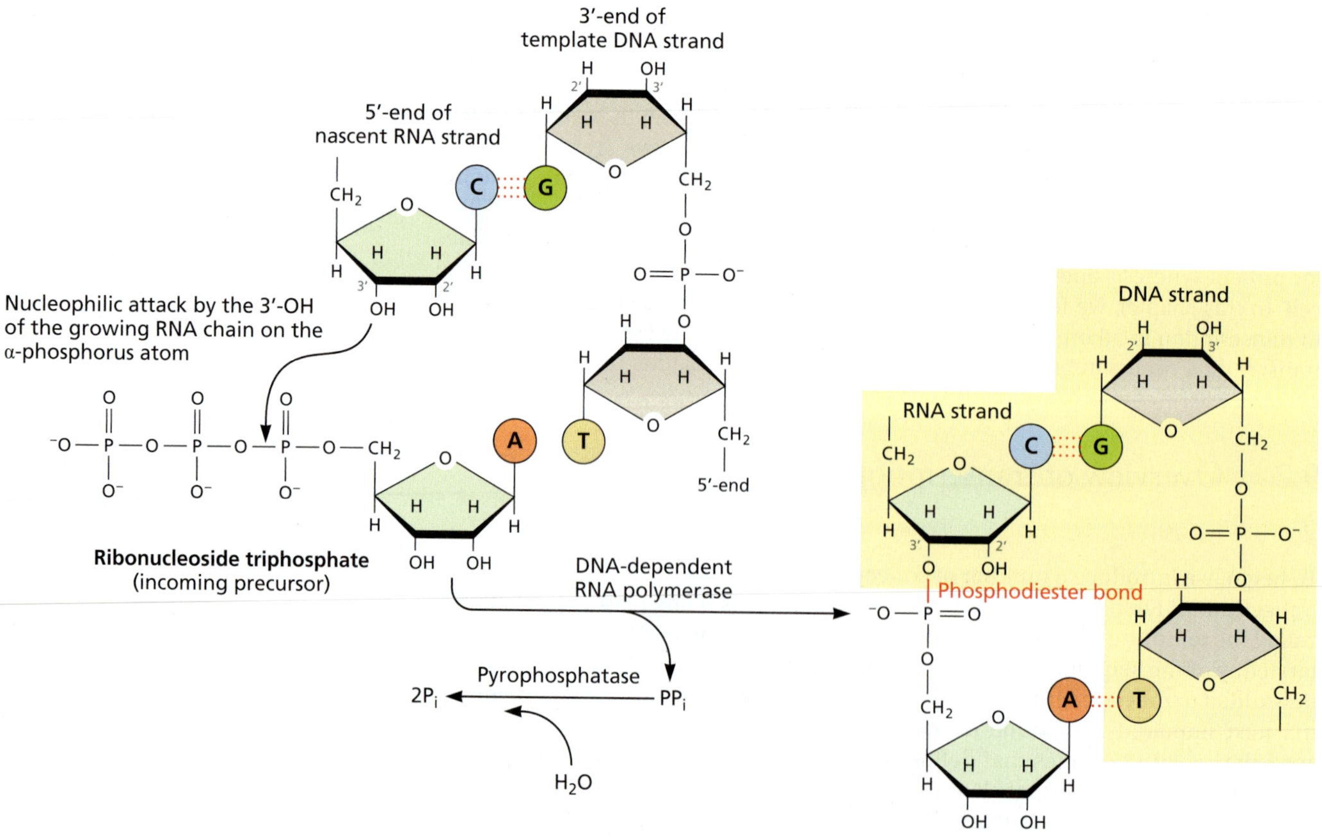

Figure 9.2 Chemistry of RNA synthesis. The new phosphodiester bond is formed between the 3′-OH group of the RNA and the 5′-phosphate of the incoming ribonucleoside triphosphate, with elimination of pyrophosphate. The pyrophosphate or PP_i that is released is quickly hydrolyzed by the abundant enzyme pyrophosphatase. This fast hydrolysis is needed to prevent the reversal of the polymerization reaction.

downstream, whereas moving in the opposite 3′ → 5′ direction is described as moving **upstream**.

In the classical view of transcription, only one strand in any given region of the genome is transcribed. We now know that this is not always the case; sometimes both can be transcribed, in opposite directions. The rules of nomenclature given above still apply to each transcript. Such **bidirectional transcription** is described in detail in later chapters.

Transcription requires the participation of many proteins

Template-dependent synthesis of RNA is catalyzed by a class of enzymes called **RNA polymerases** or **RNAPs**. The residues to be added to the transcript are presented to the enzyme as nucleoside triphosphates, and their hydrolysis with release of pyrophosphate (see Figure 9.2) provides the energy to drive the polymerase along the DNA in a unidirectional sense (Box 9.1). Nucleotide addition requires a magnesium-catalyzed

Box 9.1 Molecular motions and the second law of thermodynamics There are two different modes in which an RNAP molecule can move along DNA. After it has attached at random and is hunting for a promoter site, the molecule makes a random walk, also known as one-dimensional Brownian motion, back and forth on the DNA. By contrast, once it is caught by a promoter and begins transcription, the motion becomes almost exclusively one-directional, sometimes for tens of thousands of base pairs. What makes such a fundamental difference here?

As a clue, we see that each step of unidirectional motion happens only as a nucleotide residue is added to the RNA chain, with accompanying release of pyrophosphate. The need for an energy-releasing coupled reaction in order to achieve directional motion is general throughout biology and is a consequence of the second law of thermodynamics. One way of stating the second law is that you cannot get useful work out of a process without an expenditure of free energy. In fact, free energy change can be defined as that part of the total energy change available for useful work, given that some energy must always be wasted in heat.

Consider the single-molecule experiment shown in Figure 1. An RNA polymerase, with DNA passing through it, is fixed to a glass slide. One end of the DNA is attached to the cantilever spring of an atomic force microscope or AFM (see Chapter 1). We assume that the RNA polymerase is poised to clear a promoter and add nucleoside triphosphates to start transcription. As the DNA passes through the polymerase in one direction, it does work against the spring, bending it. On the other hand, if the polymerase were randomly hunting on the DNA, which it can do without nucleoside triphosphates, it might step back and forth, but no net displacement would occur over time and no net work would be done. However, a unidirectional motion can always be conceived as a way of doing work. Suppose a protein were proceeding around and around a circular DNA molecule. We could imagine building a nanomachine that made this motion turn an axle and do some work. This would be a perpetual-motion machine, which would violate the second law of thermodynamics. A source of energy coupled to the machine is needed to do work.

Where does the free energy come from? In the case of RNA polymerase, it comes from the reaction adding nucleotide residues. Cleavage of nucleoside triphosphates is highly exergonic, a good source of free energy for any process to which it is coupled. We find examples of such coupling in all sorts of one-directional processes in biology; examples include directed transport of ions through membranes, export of RNA from the eukaryotic nucleus (see Chapter 14), motion of vesicles on microtubules, and rotation of bacterial flagella. For many such processes, the hydrolysis of ATP provides the driving force.

There is another, more mechanistic way to look at unidirectional transport of RNAP. Each time the molecule takes the translocation step, there is a coupled nucleotide-addition reaction. Presumably, this requires the RNAP to undergo a conformational change so that it cannot step back. This ratchet mechanism is rather like that in a clock: as the spring drives the clock forward one second, the ratchet falls, preventing it from slipping back. Such a mechanism is easier to visualize, but the thermodynamic explanation is much more general; it really requires no model.

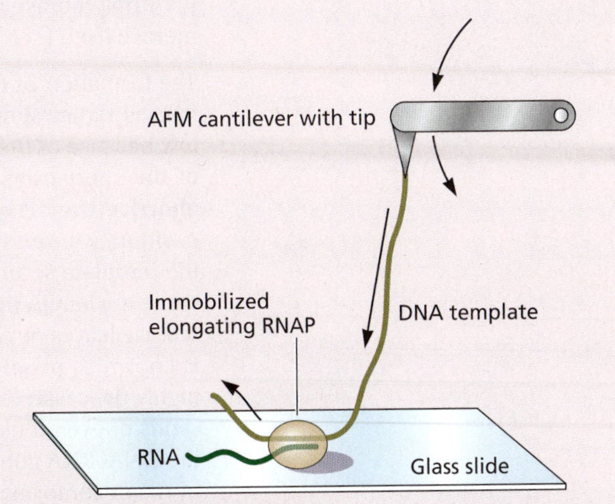

Figure 1 Single-molecule experiment demonstrating work done in transcription. The DNA molecule is being transcribed by RNA polymerase, shown as a tan oval, attached to a glass slide. The DNA pulls on the cantilever spring of an atomic force microscope or AFM, doing work against the spring. The energy for this work comes from nucleoside triphosphate hydrolysis.

nucleophilic attack by the 3′-OH at the end of the RNA chain on the α-phosphate of the incoming nucleoside triphosphate.

Transcription is a multistep process that involves **initiation**, **elongation**, and **termination**. Each of these steps is quite complex and can involve numerous protein factors in addition to the RNA polymerase itself. Moreover, despite the existence of similarities in mechanisms, each of several specific types of polymerase is characterized by its own features and functionality. In this chapter, we discuss the relatively simple mechanisms that operate in a bacterial cell. The more complex eukaryotic transcription is the focus of the next chapter.

The reason for this greater complexity in eukaryotes might be inferred from earlier chapters. Every cell of a eukaryote carries the entire genome of the organism, but each cell type and each tissue requires the expression of a specific subset of genes. This could be a functional reason for sequestering the eukaryotic genome in a nucleus, where it can be approached only by cell- or developmental-stage-specific signals. This is also one proposed function for compacting and protecting much of the DNA by chromatin structure as described in Chapter 8. The bacterial nucleoid, although also carrying many bound proteins, appears to be a less well-defined and stable structure. Almost all of the bacterial genome is potentially accessible all of the time, which seems appropriate to the simpler bacterial lifestyle.

We begin with an overview of the processes of initiation, elongation, and termination and then move on to consider details of their mechanism and regulation. As illustrated in the schematic in **Figure 9.3A**, the process of transcription begins with binding of the polymerase to DNA sequences in a region called the promoter. Immediately following the promoter is a signal, a few base pairs long, called the **transcription start site** or **TSS**. This marks the point, the specific nucleotide, at which transcription begins (see Figure 9.1). In simple terms, the first interaction of the RNAP with the promoter results in the formation of the **closed complex**. This is an oversimplification, as the polymerase actually binds at random sites along the DNA and then performs a one-dimensional search for the promoter; that is, it tracks the double helix until it finds the promoter sequence. This one-dimensional search is a common feature of protein–DNA interactions (see Box 9.1 and Chapter 6). In the next step, the polymerase melts the DNA at the promoter region, forming a **transcription bubble**, 13–14 base pairs long, in the **open complex**. Now DNA in the transcription bubble has its bases exposed and can be used as a template. RNA synthesis can begin. Note that DNA enters and exits the RNA polymerase as a double-stranded helix, so the bubble exists inside the polymerase itself (see Figure 9.3).

The formation of the first phosphodiester bond between the nucleotides of nascent RNA is rather slow. At the beginning, the stability of the RNA–DNA duplex is very low because of the few bases involved in base pairing. This leads to frequent release of the short transcripts from the polymerase in a process that has been dubbed **abortive transcription**. During this phase, the polymerase remains in its original position, without moving along the template, with the already transcribed portion of the template scrunching in its vicinity. Once the transcript has the chance to overcome the length threshold of ~15 bases, the process switches to elongation: the polymerase leaves or *clears* the promoter and moves into the body of the gene. This step is known as **promoter escape** or promoter clearance. Once elongation begins, it is highly **processive**, which means that the length of the DNA transcribed may be thousands or even tens of thousands of base pairs without the enzyme dissociating from the DNA–RNA complex. The transition from initiation to elongation is accompanied by major conformational changes in the polymerase that ensure a firm grip of the polymerase on the DNA–RNA complex. The final step in transcription usually involves recognition of **termination signals** or termination sequences on the DNA and occurs with the help of numerous termination factors. Termination results in dissociation of the triple or ternary DNA–RNA–polymerase complex: the transcript and the enzyme are released from the template, which regains its fully duplex character as the bubble closes.

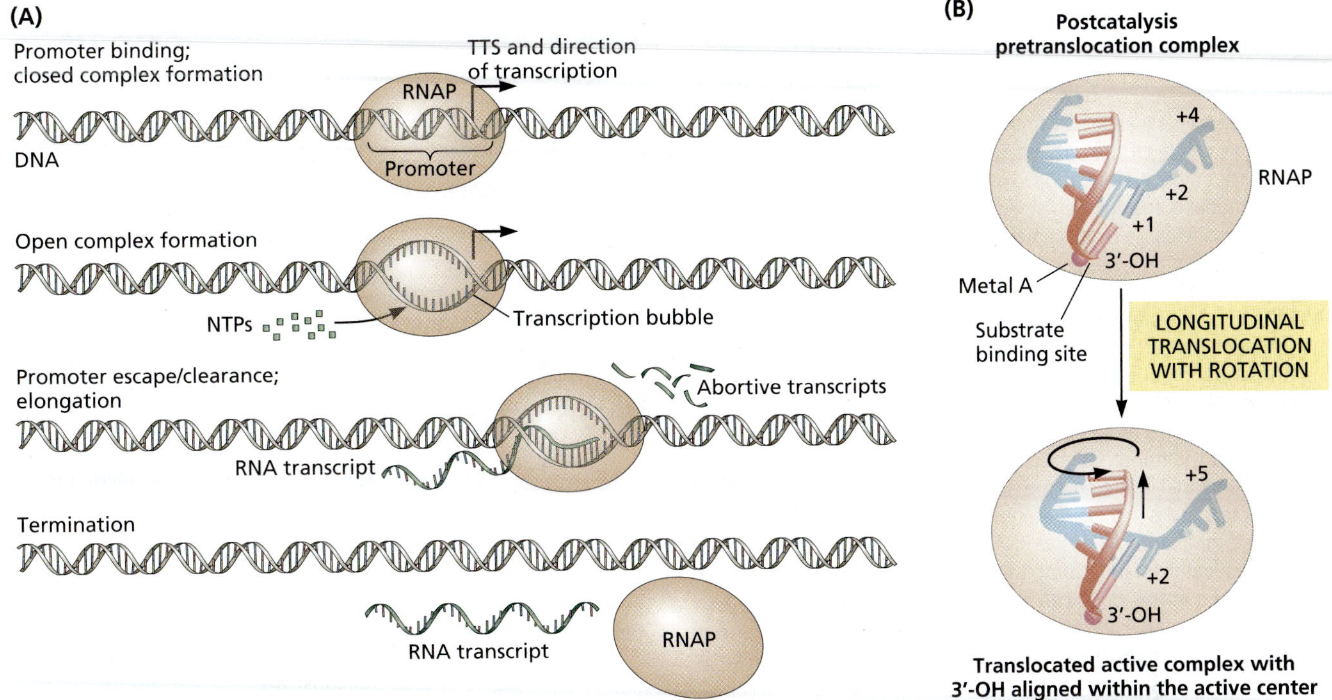

Figure 9.3 Basics of transcription. (A) Steps in transcription. RNAP binds to the promoter region on the DNA to form a closed complex. RNAP melts the DNA to form the open complex in which the DNA strands separate to form a transcription bubble of 13–14 bp. The incoming ribonucleotide triphosphates (NTPs) pair with the exposed DNA bases on the template strand and the first phosphodiester bonds are formed, linking nucleotides within nascent RNA molecules. At this stage, RNA synthesis is abortive, and most newly formed short transcripts are released from the polymerase. Once a transcript elongates beyond ~15 bases, the polymerase clears the promoter and enters the processive elongation phase, in which the DNA–RNA–polymerase ternary complex is very stable and the enzyme transcribes long stretches of DNA without dissociating. The characteristic high processivity of RNA polymerase distinguishes it from other DNA-tracking enzymes, such as DNA polymerases and most helicases. Finally, transcription is terminated, resulting in dissociation of the ternary complex. (B) Schematic of the DNA–RNA heteroduplex, based on the crystal structure of an elongating RNAP, illustrates the expected translocation movements of the DNA double helix with respect to the catalytic center of the enzyme, labeled here as metal A. DNA is shown in light blue; RNA transcript is shown in red. Top, postcatalysis pretranslocation complex; bottom, translocated active complex. Note that the transition involves two different kinds of motion, shown by black arrows: longitudinal translocation along the double helix and rotation, so that the 3′-OH group is aligned with the active center and the next nucleotide on the template strand, at position +2, is oriented properly to determine the chemical nature of the incoming ribonucleotide. (Adapted from Zlatanova J, McAllister WT, Borukhov S & Leuba SH [2006] *Structure* 14:953–966. With permission from Elsevier.)

Each nucleotide addition step during elongation is accompanied by two steps of the DNA with respect to the catalytic center of the polymerase, as illustrated in **Figure 9.3B**. The DNA template has to move longitudinally, so as to position the next template base in the catalytic center; simultaneously, the DNA has to rotate so that the 3′-OH group in the nascent RNA is properly aligned to attack the α-phosphorus atom of the next ribonucleoside triphosphate, thus powering the continuing chemical reaction. There are several contacts between amino acid residues in the enzyme and the incoming nucleotide to ensure that, first, it is a ribonucleotide rather than a deoxyribonucleotide, and second, it is the appropriate nucleotide to base-pair with the template base. Inappropriate nucleotides are simply rejected.

Transcription is rapid but is often interrupted by pauses

Transcription is a rapid process, especially in bacteria, where the average *in vivo* rate has been estimated at ~45 base pairs or bp per second. In eukaryotes, it is roughly tenfold slower. These values do not, of course, represent the actual rate of production of RNA in the cell, as many genes are being transcribed at the same time, and some of these have tandem polymerases traveling along them. The rate at which any one RNA polymerase moves along the DNA template is not uniform; the enzyme may temporarily pause on certain sequences before resuming elongation. There are

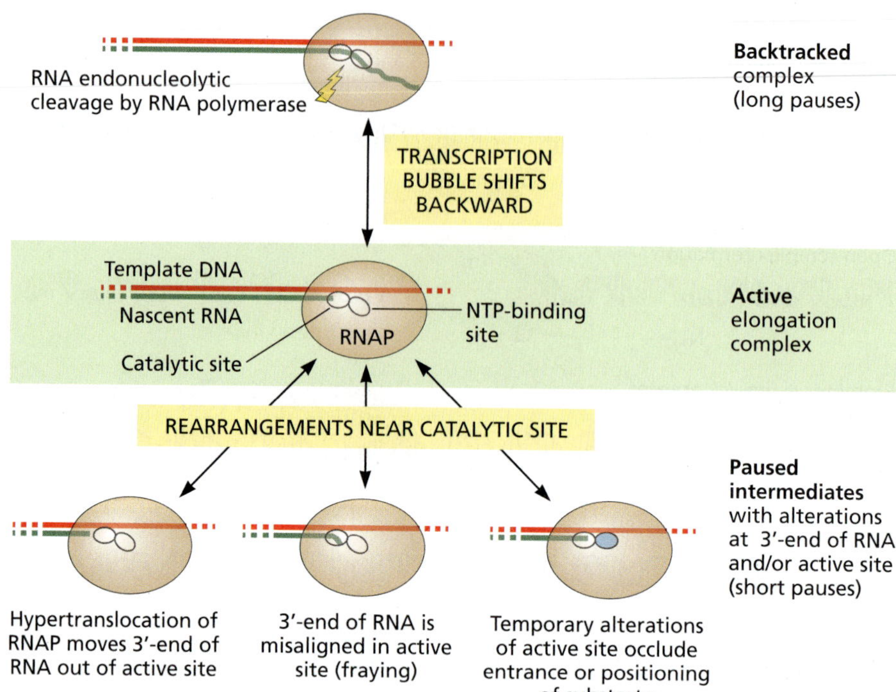

Figure 9.4 Conformational states of RNA polymerase elongation complexes. In the transcriptionally active complex, the 3′-end of the nascent transcript is properly aligned within the catalytic center of the enzyme, allowing for efficient elongation. Slight conformational rearrangements near the catalytic site cause displacement of the 3′-end from its proper position, slowing down transcription: this is thought to be the state inhabited by the frequently occurring short pauses. If the transcription bubble shifts backward along the DNA–RNA heteroduplex, in a movement known as backtracking, the 3′-end of the transcript is extruded through a channel in the enzyme; the backtracked state is inactive because of long-range misalignment of the RNA end and the catalytic site. Complexes that are halted at this stage can resume transcription only after the RNA tail is removed by the endonucleolytic action of the polymerase, a reaction stimulated by factors such as GreA and GreB from bacteria and TFIIS from eukaryotic cells. Recovery is slow, with the initial synthesis of the RNA portion that had been chopped off being many times slower than the normal continuous elongation process. (Adapted from Zlatanova J, McAllister WT, Borukhov S & Leuba SH [2006] *Structure* 14:953–966. With permission from Elsevier.)

two types of pauses, long and short (Figure 9.4 and Box 9.2). Short pauses arise from sporadic alterations in the position of the 3′-OH end of RNA and/or the active site. They occur randomly once every 100 bp on average. Recovery from these short pauses is spontaneous and fast. Long pauses are caused by **backtracking**, shifting the transcription bubble backward along the DNA–RNA heteroduplex, which leads to extrusion of the 3′-end of the transcript through a channel in the enzyme. The resulting drastic misalignment of the 3′-OH group with the catalytic site precludes further elongation. This pause cannot be spontaneously relieved; elongation can resume only after the extruded portion of the RNA chain is removed. This is accomplished by an endonucleolytic activity residing in the polymerase complex itself, in a reaction stimulated by specific protein elongation factors: GreA and GreB in bacteria, TFIIS in eukaryotes.

Recovery from long pauses is a slow process. It is fascinating, first, that such backtracking can happen, and second, that means have evolved to deal with it. Direct evidence of **pausing** has come from single-molecule studies (see Box 9.2). Conventional kinetic studies are not able to resolve such details: individual RNAP molecules may pause at different points in the process, whereas the rate of elongation between pauses is surprisingly uniform.

Termination of transcription in bacteria requires certain sequences on genes that give rise to particular sequences in the elongating transcripts. These RNA sequences are of two types, which can induce termination by two distinct mechanisms. We discuss these mechanisms later in the chapter.

Transcription can be visualized by electron microscopy

Visualization of transcription elongation by electron microscopy has produced instructive images (Figure 9.5). Spreading the contents of a bacterial cell onto microscopic grids showed DNA as a thin thread and polymerase complexes as small dots along the thread; RNA transcripts extend from the DNA in an almost perpendicular direction and have a knobby appearance resulting from the presence of proteins that bind to the nascent RNA chains (see Figure 9.5A). These proteins have different functions; most of them participate in RNA processing (see Chapter 14), which begins soon after a sufficient length of RNA is synthesized and is extruded from the exit channel of the polymerase. Figure 9.5 shows the **Christmas tree** structure seen

Box 9.2 Transcription studied by single-molecule techniques Development of methods that allow characterization of individual macromolecules is one of the most significant steps in contemporary biology. Single-molecule approaches provide direct real-time measurements with high spatial and temporal resolution: nanometer distances, millisecond time scales, and piconewton forces. Rather than measuring the population-averaged properties of molecules in solution, as in conventional biochemical and biophysical methods, these methods assess the properties of individual members of molecular populations. Thus, they provide insights into heterogeneity in the functional dynamics of molecules that were previously unattainable. Equally important, tracking the behavior of one molecule at a time circumvents the need for process synchronization. The significance of this feature of single-molecule methodology cannot be overstated, especially when multistep stochastic processes, such as transcription, are being investigated.

Here we illustrate the application of magnetic tweezers, MT, and optical tweezers, OT, to the study of initiation and promoter melting (Figure 1), elongation and pausing (Figure 2), and movements of the DNA template within the catalytic center of an RNAP molecule (Figure 3). In each case, information is obtained that could not be found by macroscopic averaging methods. Usually, the primary data are rather noisy and have to be computationally averaged or smoothed to produce interpretable results. Researchers are now in the position to obtain high-resolution data and reach conclusions that are beyond the capabilities of any other method.

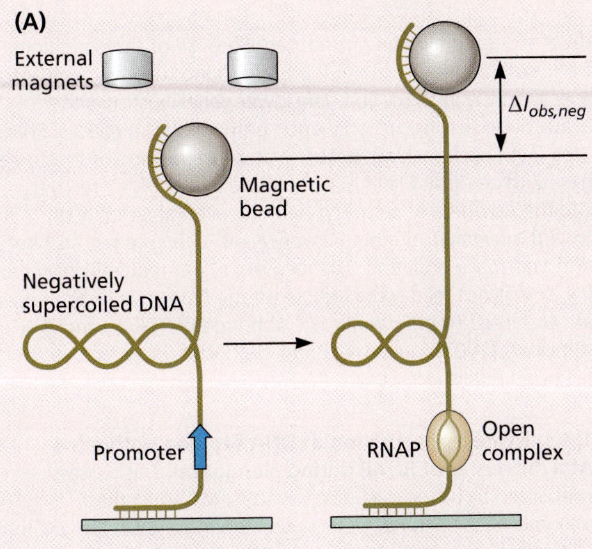

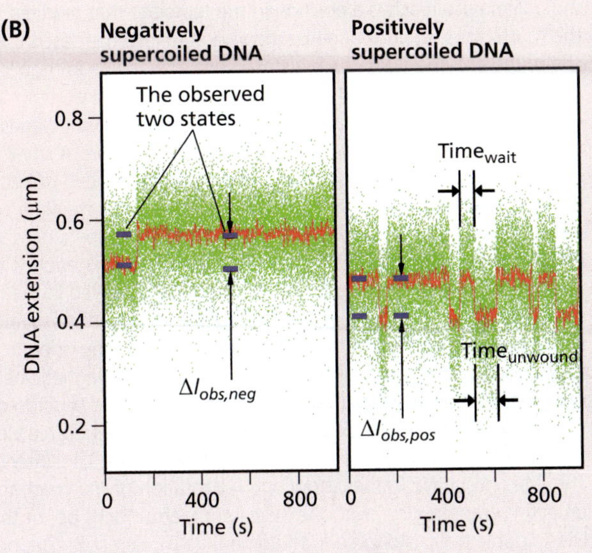

Figure 1 Promoter binding and melting by individual RNAP molecules. This process is visualized by use of magnetic tweezers, MT. (A) A double-stranded DNA molecule containing a single promoter is topologically constrained between a magnetic bead at one end and a glass surface at the other end. It is important that the DNA template be topologically constrained—that is, it should not contain nicks—and the attachment of the downstream end to the bead should not allow one strand of the DNA to swivel around the other one. If this is allowed to happen, then the motion of the DNA occurring as a result of melting could not be propagated to the bead, which is the only observable in this experiment. A stretching force is applied to the DNA tether by a pair of external magnets. The DNA end-to-end extension *l* is monitored in real time by video microscopy as the distance between the bead and the surface. Controlled rotation of the magnets rotates the bead, thus introducing supercoiling into the DNA: the formation of plectonemes leads to changes in *l*. In a torsionally constrained DNA molecule with a constant linking number *Lk*, a change in twist *Tw*, which is the number of times the two DNA strands cross each other in the double-helical structure, must be compensated by an equal but opposite change in writhe *Wr*, the number of supercoils. Now RNAP is added and the open complex is allowed to form. With negatively supercoiled DNA, unwinding of ~1 turn of promoter DNA by RNAP, that is, promoter opening, must result in a compensatory loss of ~1 negative supercoil and, correspondingly, in an increase in *l* as $\Delta l_{obs,neg}$. With positively supercoiled DNA, promoter opening will result in a compensatory gain of one positive supercoil, that is, a decrease in *l* will be observed, although not shown here. (B) DNA extension versus time plots indicating promoter opening by RNAP. (Left) Promoter opening on negatively supercoiled DNA is stable and effectively irreversible. (Right) Unstable and reversible promoter opening on positively supercoiled DNA. Green points represent raw data obtained at video rate 30 frames/s; red points show averaged data over a 1 s window; Time$_{wait}$ is the time interval between a rewinding event and the next unwinding event; and $T_{unwound}$ is the time interval between an unwinding event and the next rewinding event. (Adapted from Revyakin A, Ebright RH & Strick TR [2004] *Proc Natl Acad Sci USA* 101:4776–4780. With permission from National Academy of Sciences.)

(Continued)

Box 9.2 *(Continued)*

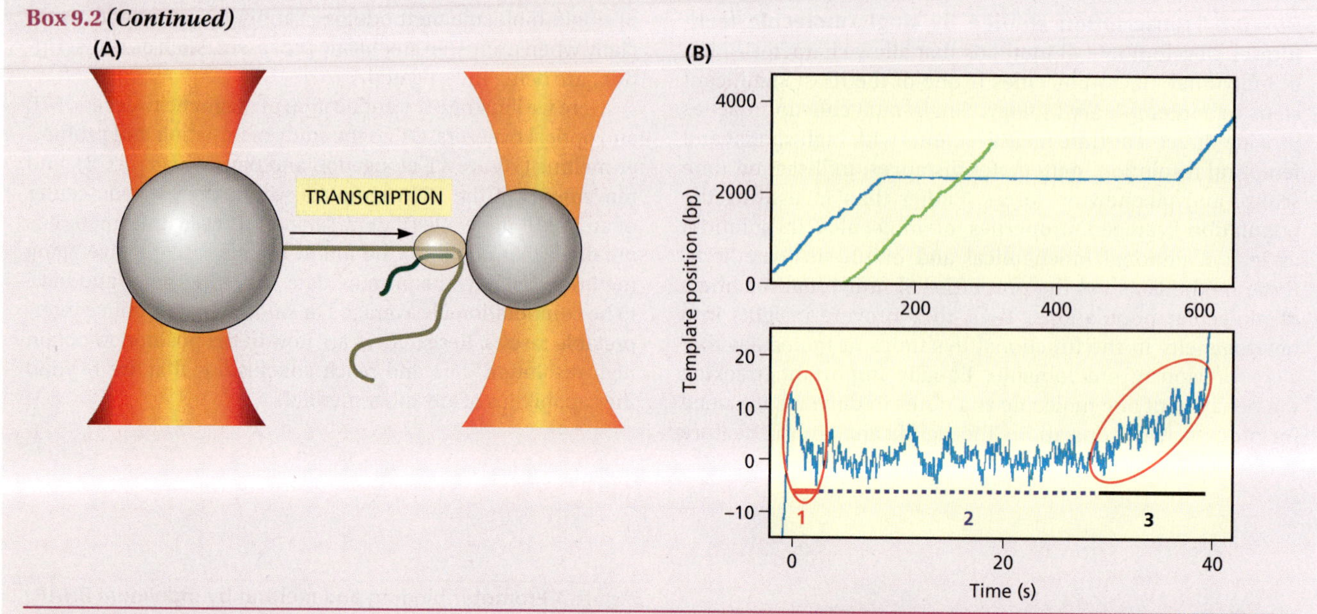

Figure 2 Transcription elongation and pausing. This process was studied by use of optical tweezers, OT. (A) Two beads are held in separate optical traps, with the right trap being 10 times weaker; that is, the force it applies to the bead to restore it to its equilibrium position in the trap is 10 times smaller. In the geometry depicted, the smaller right bead is attached to a single RNAP, and the larger left bead is attached to the downstream end of the DNA template. Transcript elongation pulls the beads together, with almost all motion appearing as displacement of the weakly trapped right bead. (B) Upper part shows records of two individual RNAP molecules, blue and green, transcribing the same template. A very long pause is seen in the blue trace. Note that the two molecules exhibit essentially the same, and constant, rate of elongation when not paused. Because of the conditions of the experiment, the elongation rate found here is significantly lower than the *in vivo* value. The lower panel illustrates the backtracking that occurs only upon entry into long pauses. Phase 1 (red line), backtracking: note the downward slope of the curve; phase 2 (blue dotted line), pause; phase 3 (solid line), recovery. Note the slowness of recovery: here the average velocity of forward movement during recovery is ~0.3 bp/s, versus ~13 bp/s during normal elongation. The recovery phase requires only about 10 s, so it is not readily discernible on the time scale of the upper part. (Adapted from Shaevitz JW, Abbondanzieri EA, Landick R & Block SA [2003] *Nature* 426:684–687. With permission from Macmillan Publishers, Ltd.)

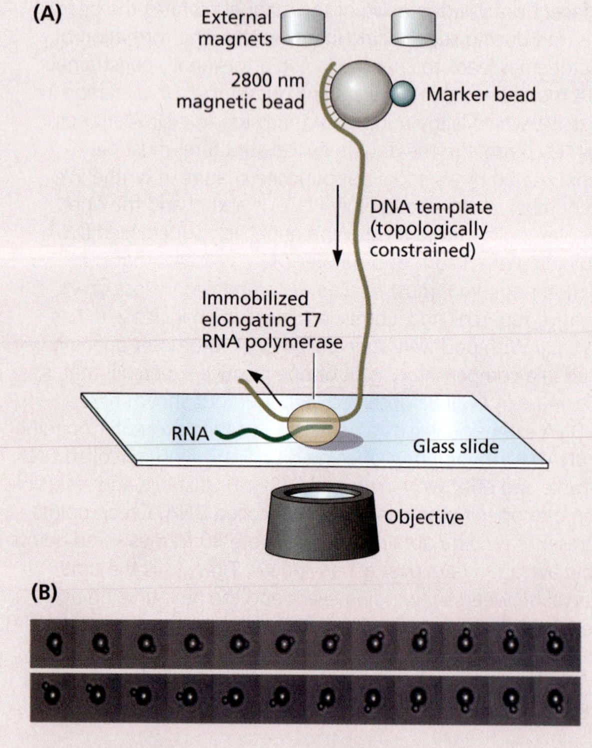

Figure 3 Direct observation of DNA rotation within the catalytic center of RNAP during elongation. This process is visualized by use of magnetic tweezers, MT. (A) In these experiments, initiation of transcription is performed in the test tube on a 4 kilobase-pair DNA template in the presence of only three out of the four precursor ribonucleoside triphosphates or NTPs. When the polymerase reaches a position in the template that requires that the fourth missing nucleotide be incorporated into the nascent RNA chain, the enzyme stalls. Such a stalled enzyme can be immobilized onto the glass slide of the MT setup. The downstream portion of the DNA template is attached to a magnetic bead that carries a smaller marker bead to allow visualization of rotation. A small magnetic force of 0.1 pN is applied to the magnetic bead to keep it above the surface. When all four NTPs are added into the flow cell, transcription resumes. During elongation, the DNA experiences two kinds of motion: longitudinal threading through and rotation within the catalytic center of the enzyme. Both can be observed in the MT setup: longitudinal motion will pull the DNA downward, resulting in the bead's becoming out of focus when visualized by video microscopy, and rotation of the DNA will rotate the magnetic bead, with the rotation visualized by video microscopy of the position of the smaller decorating bead. (B) Series of video frames from a video recorded at 20 frames/s; the frames shown represent 0.1 s intervals. The 24 consecutive frames show one full rotation of the bead, thus reflecting transcription over one helical turn, around 10 bp, of the DNA double helix. (Adapted from Pomerantz RT, Ramjit R, Gueroui Z et al. [2005] *Nano Lett* 5:1698–1703. With permission from American Chemical Society.)

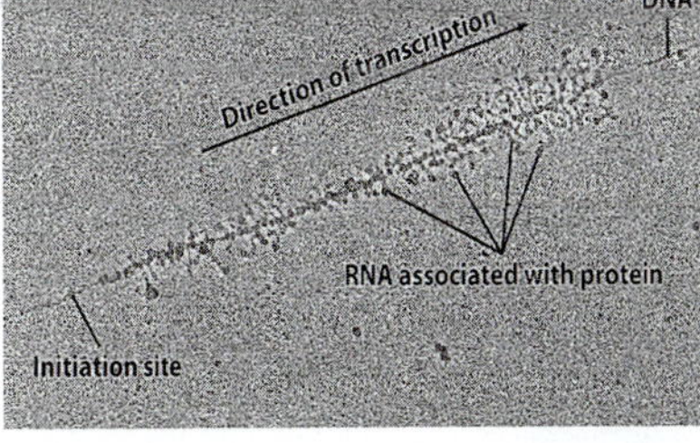

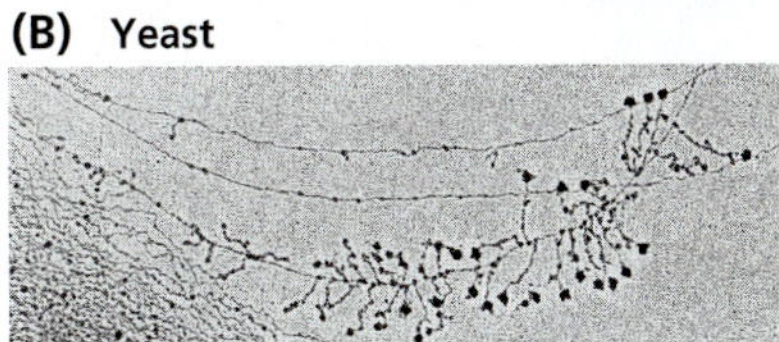

Figure 9.5 Transcription of ribosomal genes in (A) *E. coli* and (B) yeast. Both genes are transcribed from left to right, and the RNA products of each present the common Christmas tree appearance. The nascent rRNA chain associates with proteins that process it by nucleolytic cleavage before transcription is complete (see Chapter 14); these correspond to the knobs at the ends of transcripts. (A, from French SL & Miller OL [1989] *J Bacteriol* 171:4207–4216. With permission from American Society for Microbiology. B, from Dragon F, Gallagher JEG, Compagnone-Post PA et al. [2002] *Nature* 417:967–970. With permission from Macmillan Publishers, Ltd.)

at sites of ribosomal RNA transcription. These genes are transcribed at very high frequency, especially in growing bacterial cells, with numerous RNA polymerases moving along the template, one after the other. Thus the RNA chains close to the initiation site are very short, but their length gradually increases further downstream, creating structures that resemble the appearance of a tree. Interestingly, the transcription of ribosomal RNA in the eukaryotic nucleus produces very similar images (see Figure 9.5B).

We now turn to a more mechanistic description of the RNA polymerases and the processes occurring in each stage of transcription.

9.3 RNA polymerases and transcription catalysis

RNA polymerases are a large family of enzymes that produce RNA transcripts of polynucleotide templates

RNA polymerases are complicated cellular machines that perform the catalytic addition of nucleoside monophosphates to a growing RNA chain. This function can be performed with different templates. DNA is the template in most transcription, but RNA itself can serve as template in some specific processes. Thus, RNA polymerases are classified into DNA-dependent and RNA-dependent polymerases. If the nature of the template is not explicitly mentioned, it is assumed to be DNA. In addition, RNA polymerases can be single-subunit or multisubunit enzymes, with single-subunit enzymes being mainly of viral origin. The RNA polymerases in bacteria and archaea, as well as the nuclear enzymes in eukaryotic cells, are all multisubunit complexes.

Figure 9.6 presents the crystal structure of the single-subunit T7 phage RNA polymerase and the biological context in which it works. Figure 9.6B shows the two conformations of the protein–nucleic acid complex, referred to as the **initiation complex** and the **elongation complex**. There are major structural differences between these two complexes, indicating that some portions of the polypeptide chain undergo substantial movements or refolding during the transition from initiation to elongation. These transitions lead to the firm grip on the DNA that is needed for processivity during elongation.

Bacterial cells possess only one polymerase type that transcribes all genes, both those that directly participate in the process of information flow, such as ribosomal RNA, transfer RNA, and messenger RNA, and those that have other functions (Table 9.1). By contrast, eukaryotic cells have three different types of polymerases in the nucleus, with each type serving a distinct class of genes (see Chapter 10). The subunit structure of the *Escherichia coli* enzyme is given in Table 9.2, and a brief history of its discovery is presented in Box 9.3. Two α-subunits together with two structurally unrelated β-subunits are absolutely needed for catalytic function; they form the so-called **core RNA polymerase**. Sometimes the Ω subunit is included in the core, but it is not always needed. Another subunit, σ, is the **initiation factor** needed for promoter recognition during initiation.

A structural model of a free RNA polymerase from a bacterial source, *Thermus aquaticus*, is presented in Figure 9.7A, and Figure 9.7B shows the same enzyme now bound to DNA in the elongation complex. The DNA is sharply kinked within the enzyme; such DNA kinking is a conserved structural feature observed in all other polymerases examined.

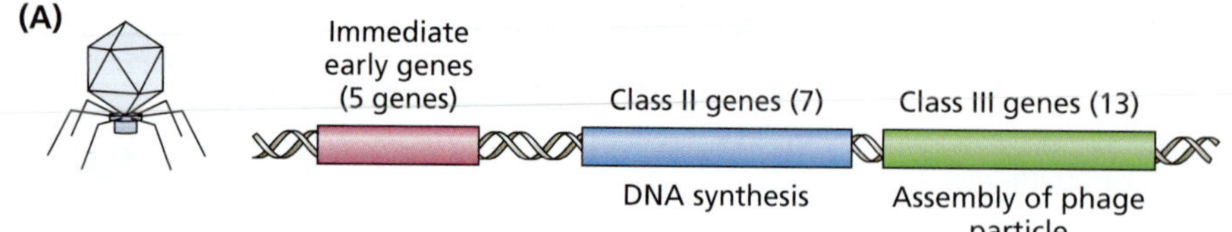

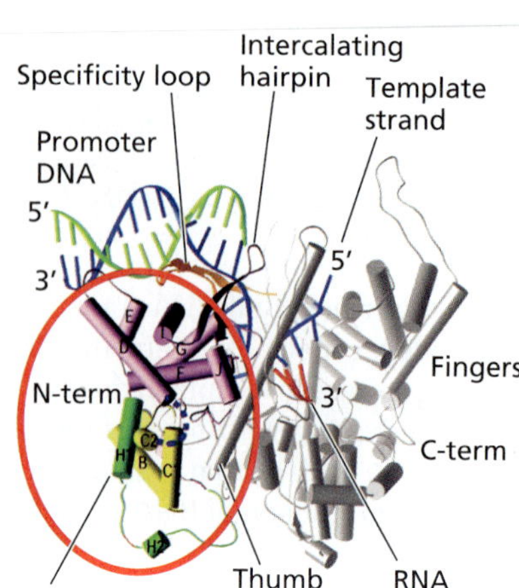

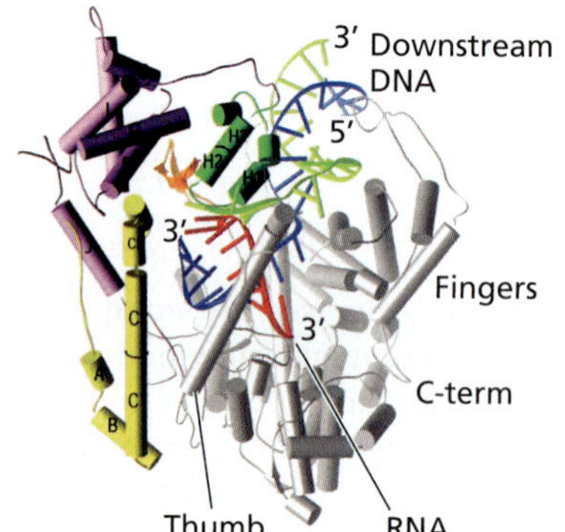

Figure 9.6 T7 phage RNA polymerase, a single-subunit, 98 kDa enzyme. (A) The linear duplex DNA genome of T7 phage is organized into three groups or classes of genes. Immediate early genes are transcribed by the host polymerase. One of these genes encodes the viral-specific T7 polymerase, responsible for all transcription later in infection; the products of the other early genes repress host transcription by inactivating *E. coli* RNAP. The other two classes of genes are responsible for DNA synthesis and assembly of the mature phage particle. Shown on the left is a model of the T7 phage particle. (B) Crystal structure of T7 RNA polymerase, illustrating the intricate choreography of the enzyme during the initiation–elongation transition. The corresponding residues in the N-terminal domains of the two complexes that undergo major refolding are colored in yellow, green, and purple, whereas the C-terminal domain, residues 300–883, is colored in gray. Template DNA is shown in blue, nontemplate DNA in green, and RNA in red. In the elongation complex, subdomain H, shown in green, has moved more than 70 Å from its location in the initiation complex. The specificity loop, shown in brown, recognizes the promoter during initiation and contacts the 5′-end of RNA during elongation, whereas the intercalating hairpin opens the upstream end of the bubble in the initiation phase but is not involved in elongation. The large conformational change in the N-terminal region of T7 RNAP facilitates promoter clearance. (B, from Steitz T [2006] *EMBO J* 25:3458–3468. With permission from John Wiley & Sons, Inc.)

9.4 Mechanics of transcription in bacteria

Initiation requires a multisubunit polymerase complex, termed the holoenzyme

Initiation in bacteria first requires binding of a specific initiation subunit, σ, to the core polymerase complex. The resulting complex, capable of recognizing promoter sequences and initiating transcription, is named the **holoenzyme**. The overall schematic of how initiation occurs is presented in **Figure 9.8**, and the conserved nucleotide sequences that form the promoter are shown in **Figure 9.9**. Analysis of numerous individual bacterial and viral, or bacteriophage, promoters recognized by σ^{70}, the most abundant σ factor, has led to the derivation of a **consensus sequence**. The consensus defines the base that occurs with the highest frequency at each position with respect to the TSS. Bacterial promoters turned out to be

Table 9.1 RNA types that constitute the protein-synthesizing machinery in rapidly growing *E. coli* cells.

Type of RNA	Steady-state level, %	Synthetic capacity,[a] %
ribosomal RNA	83	58
transfer RNA	14	10
messenger RNA	3	32[b]

[a]Relative amount of each type of RNA synthesized at any given point in time.
[b]Up to 60% in slowly growing cells.

Table 9.2 Subunit structure of RNA polymerase from *E. coli*. The core RNA polymerase complex comprises $\alpha_2\beta\beta'\Omega$. β and β′ are structurally unrelated. The holoenzyme consists of the core complex plus the initiation factor subunit σ.

Subunit	Function	MM, Daltons
β	active site	150,600
β′	active site	155,600
α	scaffold for assembly of other subunits, regulation platform	36,500
Ω	promotes RNAP assembly and stability	11,000
σ	initiation factor	70,300

bipartite: they contain two boxes of conserved sequences separated from each other by 16–18 nucleotides. The TSS itself is usually a purine. The consensus sequence of the box closest to the TSS is **TATAAAT**, and this is referred to as the **–10 box**. The consensus of the other box, referred to as the **–35 region**, is **TTGACA**. The relative strength of a promoter depends primarily on its affinity

Box 9.3 Discovery of RNA polymerase In the late 1950s, intense interest developed in finding the enzyme(s) involved in the synthesis of RNA. This was spurred in part by the spectacular success of Arthur Kornberg in the 1958 discovery of the first DNA polymerase and in part by the growing realization that RNA must play an important role in the transmission of information from DNA to protein. It must be remembered, though, that existence of messenger RNA was not recognized before 1961, so not only the hypothetical enzyme but also its very role was obscure. This may explain why some initial results were mostly confusing. For example, the first enzyme purported to be involved in RNA synthesis, polyribonucleoside phosphorylase, was soon suspect because it utilized only ribonucleoside diphosphates, only in high concentrations, and did not require a DNA template. Its biological role is now known to involve RNA degradation, and it participates in the trimming of pre-tRNAs (see Chapter 14). About this time, poly(A) polymerase was discovered. This enzyme used adenosine triphosphate, or ATP, as substrate but again did not require a template. We now know that this enzyme is involved in adding poly(A) tails to both bacterial and eukaryotic messenger RNAs.

The breakthrough came in 1960 thanks to results from four groups, working independently and publishing almost simultaneously. Two groups were working with bacteria, and the others were working with eukaryotic nuclei from liver and pea. In each case, they discovered an activity in crude extracts that could catalyze the incorporation of a radiolabeled nucleoside phosphate into RNA, using the triphosphate as donor. The critical observations were two requirements for significant incorporation of radioactivity: (1) All four ribonucleoside triphosphates—ATP; guanosine triphosphate, GTP; cytidine triphosphate, CTP; and uridine triphosphate, UTP—had to be present in the mixture, even though three of them were not labeled. (2) At least some DNA had to be present: destruction of DNA by DNase reduced incorporation. Point 1 shows that a true RNA synthesis, incorporating all four monomers, was taking place. Point 2 strongly suggests that a DNA template was directing this synthesis. The main results of the Hurwitz study are presented in graph form in Figure 1.

These results came at almost exactly the same time as the announcement by Jacob and Monod of the messenger RNA hypothesis (see Box 11.1); the broad outlines of molecular biology were now established. Within a few years, the *E. coli* polymerase had been purified and its subunit structure was established. By 1969, the role of the σ subunit was determined.

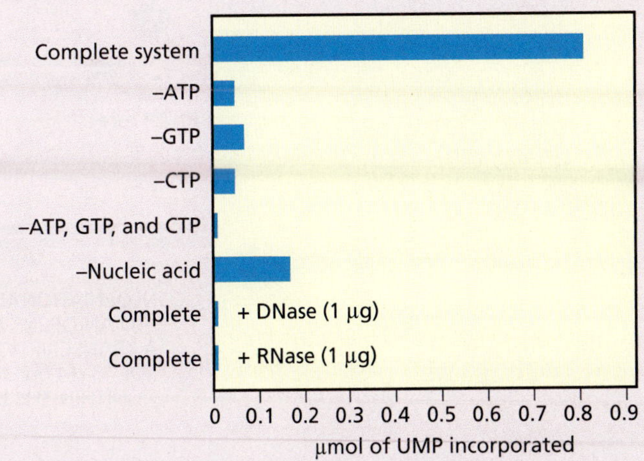

Figure 1 Requirements for uridine monophosphate incorporation into RNA. The complete system contained radiolabeled UTP, unlabeled ATP, GTP, and CTP, and *E. coli* DNA. After synthesis was allowed to proceed for 20 min, the reaction was stopped by addition of acid and the radioactivity of the acid-insoluble material was measured. (Based on data from Hurwitz J, Bresler A & Diringer R [1960] *Biochem Biophys Res Comm* 3:15–18. With permission from Elsevier.)

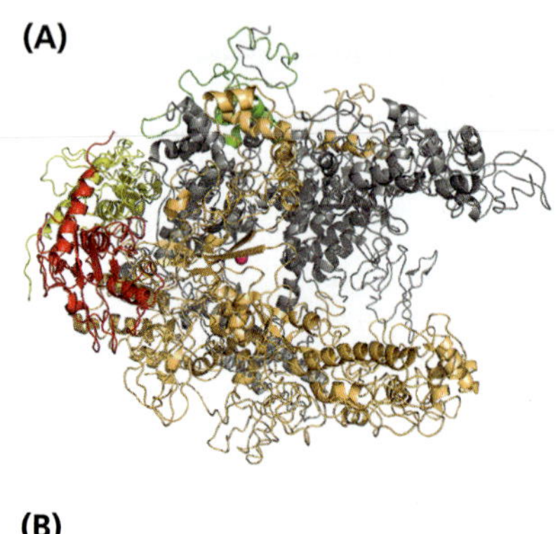

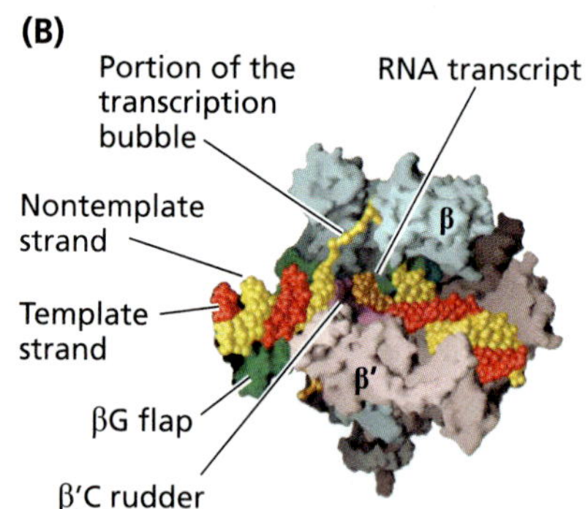

Figure 9.7 Bacterial *Taq* polymerase. (A) Backbone ribbon diagram of *Thermus aquaticus* or *Taq* core RNAP. The two α monomers are colored red and light green, and the Ω subunit is colored dark green. The large subunits are colored dark gray for the β′ and orange for the β subunit, respectively. The active site Mg^{2+} is shown as a red sphere. (B) Model of the ternary transcription elongation complex, RNA polymerase with nucleic acids, built on the *Taq* core RNAP crystal structure: β, cyan; β′, pink; α and Ω, light gray. DNA template strand is shown in red, DNA nontemplate strand in yellow, and RNA transcript in gold. Base-paired sections of the nucleic acids are also shown. Note that the upstream duplex is at a right angle to the downstream DNA. The nascent RNA exits the polymerase underneath a flexible element of β, called the flap domain or βG flap. A 9-bp RNA–DNA hybrid remarkably fits into this protein architecture. The hybrid extends from the enzyme active site to the βG flap and the β′C rudder. (A, from Werner F & Grohmann D [2011] *Nat Rev Microbiol* 9:85–98. With permission from Macmillan Publishers, Ltd. B, from Korzheva N, Mustaev A, Kozlov M et al. [2000] *Science* 289:619–625. With permission from American Association for the Advancement of Science.)

for the appropriate σ factor. Strong promoters, which ensure a high frequency of initiation, have sequences close to the consensus. The sequences of weak promoters deviate more from the consensus but the two boxes are still recognizable. New methods to map transcription start sites have greatly increased our understanding of transcription initiation.

Promoter recognition resides entirely in the σ subunit, of which there are several distinct representatives that recognize slightly different promoters in specific gene categories (**Table 9.3**). The structure of σ^{70} has been resolved (**Figure 9.10**). The protein has three well-conserved domains, all of which make extensive contacts with the core enzyme. The regions in domains σ2 and σ4 that recognize the DNA promoter sequences at –10 and –35 are solvent-exposed; thus, when the σ subunit interacts with the promoter, it is sandwiched between the polymerase and the DNA. Comparisons of binding parameters and half-lives of the core enzyme and the holoenzyme with promoter and random DNA sequences reveal that the presence of the σ subunit increases the affinity of polymerase for promoters and stabilizes the DNA–enzyme complex (**Table 9.4**).

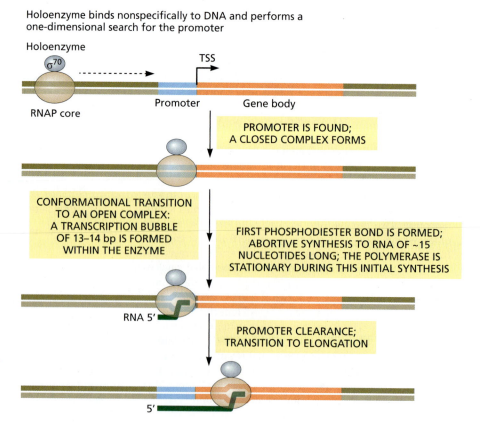

Figure 9.8 Initiation of transcription in *E. coli*. The transition from initiation to elongation is accompanied by conformational changes in the enzyme.

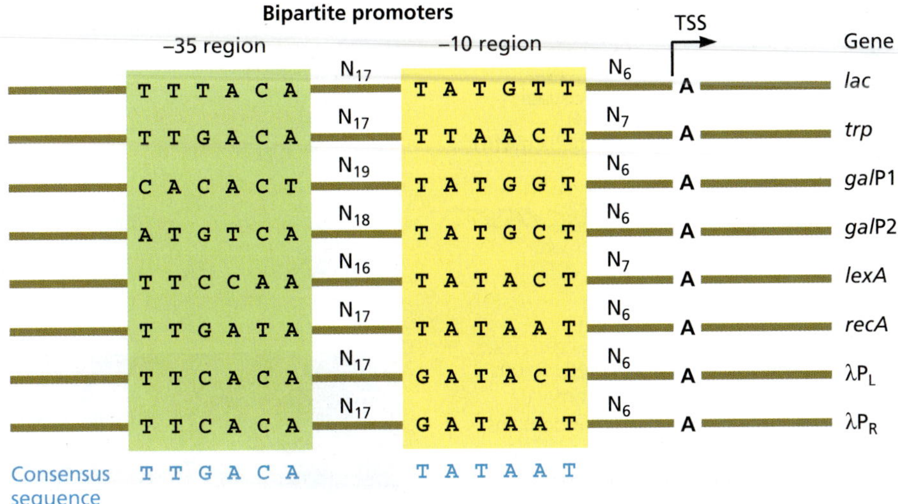

Figure 9.9 Transcription initiation at promoters in *E. coli*. Examples of individual promoters from eight phage and bacterial genes recognized by σ^{70} are shown, aligned to highlight common sequences. N stands for any nucleotide residue; the subscript denotes the number of intervening nucleotides between the two boxes in the promoter and between the –10 region and the transcription start site. The consensus sequence was derived from a much larger database of more than 300 well-characterized promoters; it reflects the highest frequency of occurrence of a base at a specific position. The frequency of transcription initiation for any given gene depends on the needs of the cell. The more closely the promoter resembles the consensus sequence, the stronger the promoter: that is, the more frequently transcription initiates from that promoter.

As Table 9.3 indicates, bacterial cells make use of slightly different σ factors to transcribe different groups of genes. Different bacterial species have different numbers of σ factors. The number seems to correlate with the complexity of the environment in which a given species lives: organisms with more varied lifestyles contain more σ subunits. Because all σ subunits employ the same core polymerase, there must be mechanisms that ensure the association of the appropriate factor under a given condition. Under certain circumstances, there is an orderly succession of alternative σ subunit usage. A common emergency encountered by bacteria is heat shock, a sudden exposure to an increase in temperature. A defensive response requires changes in protein, and hence RNA, synthesis. Figure 9.11 illustrates how bacteria respond to heat shock

Table 9.3 σ factors in *E. coli*. There are six factors in the σ^{70} family. σ^{70} and σ^{S} are closely related and recognize nearly identical core promoter sequences. Selectivity for specific genes may depend on protein factor binding. $\sigma^{54/N}$ constitutes a separate family on its own. N denotes any nucleotide. Housekeeping genes are defined as those that are constitutively expressed; they code for functions absolutely necessary for the life of the cell.

			Promoter sequence recognized	
Subunit	**Gene**	**Genes transcribed**	**−35**	**−10**
σ^{70}	*rpoD*	most housekeeping genes in growing cells	TTGACA	TATATT
$\sigma^{38/S}$	*rpoS*	starvation/stationary phase genes	TTGACA	TATAAT
$\sigma^{32/H}$	*rpoH*	genes whose expression is turned on when cells are exposed to heat	GTTGAA	CCCATNTA
$\sigma^{28/F}$	*rpoF*	genes involved in production of multiple flagella for rapid swimming	TAAA	GCCGATAA
$\sigma^{24/E}$	*rpoE*	genes involved in extreme heat-shock response	GAACCT	TCTAA
$\sigma^{19/F}$	*fecI*	*fec* (ferric citrate) gene for iron transport		
$\sigma^{54/N}$	*rpoN*	genes involved in nitrogen metabolism	none	CTGGCACNNNNNTTGCA

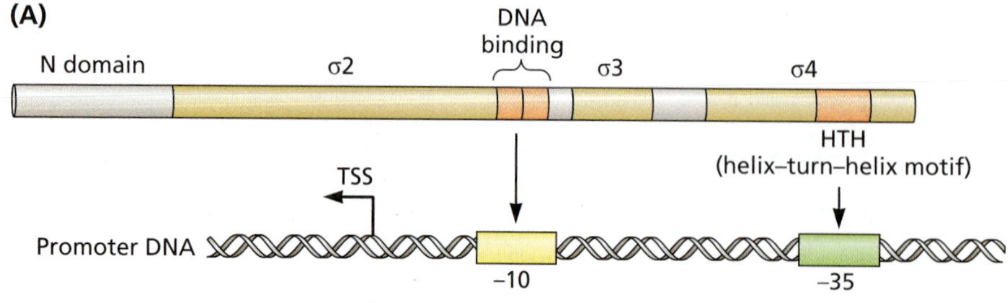

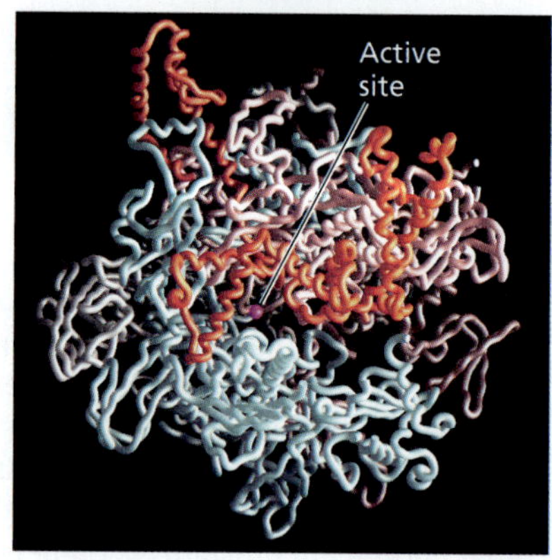

Figure 9.10 Structure of RNAP holoenzyme from *T. aquaticus*. (A) Conserved regions and functional assignment based on σ^{70}: arrows point to the regions of σ that contact conserved sequences in the promoter region. (B) Top, cartoon representation of σ domains, shown in yellow, bound to core RNAP. Only one α subunit, shown in gray, is visible in this view. The same color scheme is used in the high-resolution structure on the bottom, in which the pink sphere is the active site. The pincers of the crab-claw structure formed by the β and β′ subunits create a 27-Å-wide channel containing the active site. During elongation, downstream DNA reaches the active site via this channel. Each σ domain makes extensive contacts with the RNAP core; all promoter-recognition determinants in σ are solvent-exposed until DNA is bound, with a spacing that is consistent with the predicted separation of their target promoter elements. Thus, the σ domains bridge the promoter DNA and the polymerase. Conformational changes in both partners are integral to initiation. (Adapted from Young BA, Gruber TM & Gross CA [2002] *Cell* 109:417–420. With permission from Elsevier.)

through the consecutive use of two factors, $\sigma^{24/E}$ and $\sigma^{32/H}$. $\sigma^{24/E}$ is needed for transcription of the $\sigma^{32/H}$ gene; the $\sigma^{32/H}$ subunit then turns on an entire set of genes that participate in the heat-shock response. In general, it must be noted that transcription is a very tightly regulated process (see Chapter 11): numerous regulatory mechanisms are at play, in addition to the mechanisms involving σ factors.

In some cases, the α subunits play a role alongside σ in initiation. Each α subunit carries an extended C-terminal domain or CTD that is capable of interacting with sites upstream of the σ-sites described above. This provides another level of promoter discrimination in the initiation of bacterial transcription (see Chapter 11).

The initiation phase of bacterial transcription is frequently aborted

The initial steps of elongation are slow and the short RNA transcripts are frequently released from the ternary complex in what is known as abortive transcription. The reaction has been studied mainly *in vitro*, and only very recently *in vivo*, by using promoter sequences that produce unusually high yields of aborted transcripts and have

Table 9.4 DNA-binding properties of core RNA polymerase and holoenzyme containing σ^{70}. The role of σ is (i) to increase the affinity for promoter sequences while decreasing the affinity for nonpromoters and (ii) to stabilize the complex on promoter sequences while destabilizing it on random sequences. Promoter binding is 100 times faster than the maximum theoretical value for a diffusion-limited second-order reaction; this can be rationalized by a one-dimensional diffusion mechanism (scanning rate ~2000 bp in ~3 s).

	Dissociation constants, nM		Half-life of complex	
	Promoter	Random sequence	Promoter	Random sequence
core RNA polymerase	0.1	0.1	60 min	60 min
holoenzyme = core + σ subunit	0.02	1.0	2–3 h	3 s

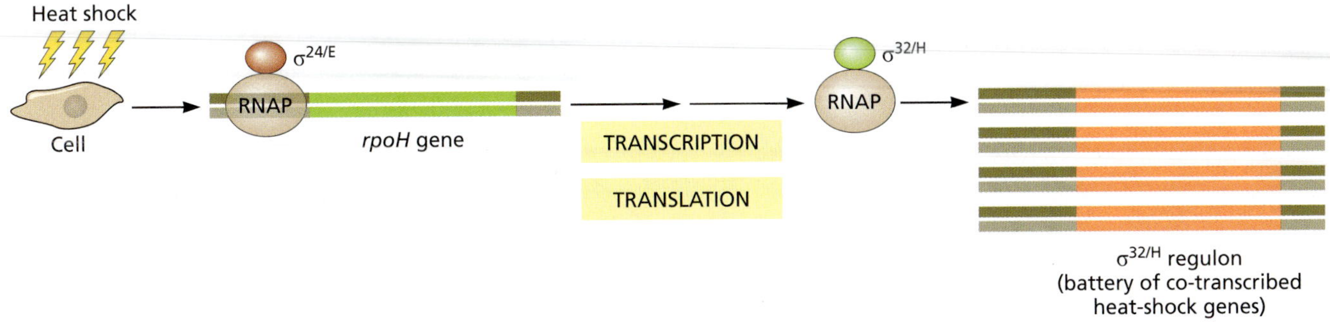

Figure 9.11 Alternative σ factors mediate *E. coli*'s global response to heat shock. The response of heat-shocked cells involves the successive use of two alternative σ factors: $\sigma^{24/E}$ is required for expression of the gene that encodes the second alternative factor $\sigma^{32/H}$, which in turn regulates the entire $\sigma^{32/H}$ regulon, a set of bacterial genes that share common regulatory elements for transcription (see Chapter 11). The $\sigma^{32/H}$ regulon is induced by excessive unfolding of proteins in the cell, caused by heat shock. The gene encoding chaperone GroEL, a protein involved in refolding misfolded or unfolded proteins, is a member of this regulon.

very slow rates of promoter escape. The use of one such promoter to visualize abortive transcripts on electrophoretic gels is illustrated in Figure 9.12. Note the extremely high frequency of abortion with this specific promoter: it takes the polymerase 300 trials before it can switch to stable elongation. With more conventional promoters, there are around 30–40 abortion events for each full-length transcript synthesized.

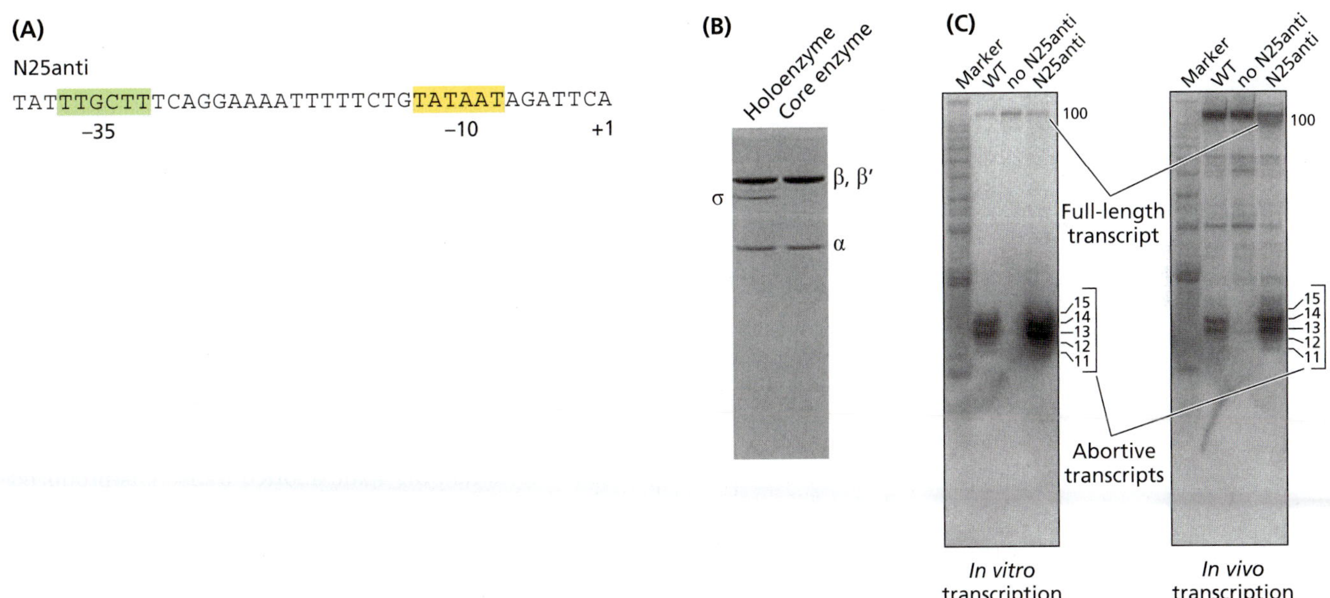

Figure 9.12 *E. coli* RNAP and detection of abortive transcription *in vitro* and *in vivo*. Transcription reactions were performed on DNA templates carrying the N25anti promoter, a 100-bp transcription unit, and the tR2 terminator. This entire construct is designated N25anti-100-tR2. (A) Phage T5 N25anti promoter is a classic model system for the study of abortive initiation and promoter escape, introduced in Michael Chamberlin's laboratory. The abortive/production ratio for this promoter is very high, ~300, as opposed to 40 for the closely related promoter N25. (B) SDS–polyacrylamide gel electrophoresis of purified commercially available preparations of *E. coli* RNA polymerase. The core enzyme contains α, β, and β′ subunits; the latter two co-migrate under these conditions. The holoenzyme also contains the σ subunit. This commercial RNAP preparation was used for the transcription experiments illustrated in part C. (C) Transcripts were analyzed following ethanol precipitation, electrophoresis on polyacrylamide gels containing urea, transfer to membranes, and hybridization with a ^{32}P-labeled probe complementary to the beginning of the transcribed region. The *in vivo* transcription reaction was performed on a plasmid carrying the template, which was introduced into live cells. Further experiments demonstrated that the *in vivo* products exhibited the hallmarks of abortive transcripts as defined *in vitro*: altering the strength of interactions between RNAP and promoter and between RNAP and σ factor alters the yields of 11–15-nucleotide transcripts. These interactions must not be too strong to allow ending of abortive initiation and enable promoter escape. (A, C, adapted from Goldman SR, Ebright RH & Nickels BE [2009] *Science* 324:927–928. With permission from American Association for the Advancement of Science. B, courtesy of Epicentre, an Illumina company.)

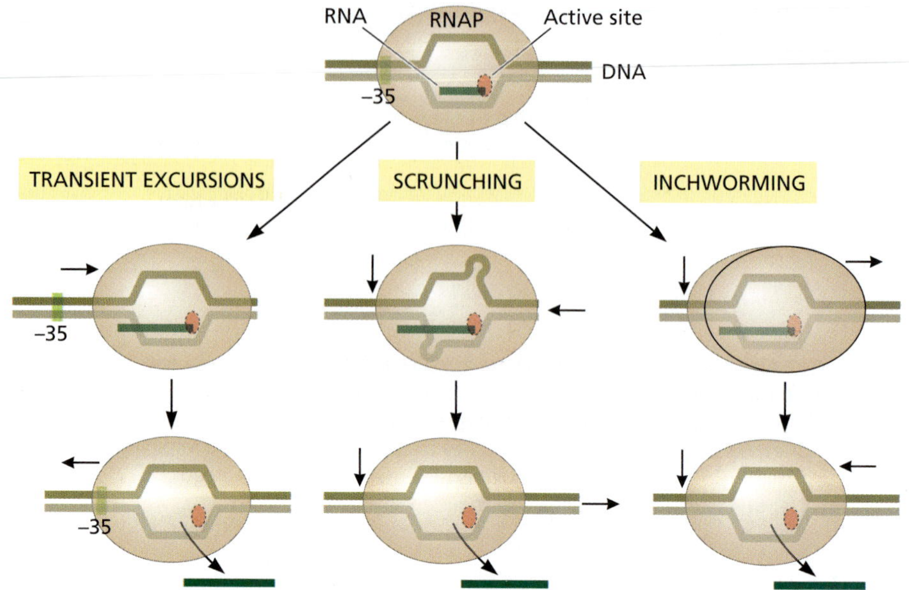

Figure 9.13 Competing models for abortive initiation. The models focus on the relative motions of RNAP and the DNA template, as indicated by arrows. The ability to synthesize numerous short transcripts that are released from the initially unstable transcribing complex, coupled with fast re-initiation of new transcripts, implies that the active site of the polymerase is moving along the DNA template in a forward direction but is still maintaining contacts with the promoter in order to allow re-initiation. *In vitro* experiments over the years have resulted in three models of how this might occur. (Left) In the transient excursion model, RNAP breaks its contacts with the promoter region while transcribing the initial portion of the RNA chain but quickly returns to the promoter once the short nascent transcript is aborted. (Right) In the inchworming model, the enzyme undergoes conformational changes that enlarge the footprint of the polymerase on the DNA, increasing the region over which the two molecules interact; note the slightly enlarged RNAP symbol. Thus the enzyme can continue transcribing while still maintaining its grip on the promoter. Release of the aborted transcript relaxes the RNAP to its normal dimensions. (Middle) In the most recent scrunching model, the RNAP does not change shape; its effective footprint is increased by pulling in a portion of the downstream template, resulting in a stressed DNA conformation. When the short transcript is aborted, the scrunched DNA is released, and the enzyme is ready for initiation of a new chain. The single-molecule experiments that led to the scrunching model are described in Figure 9.14. (Adapted from Herbert KM, Greenleaf WJ & Block SM [2008] *Annu Rev Biochem* 77:149–176. With permission from Annual Reviews.)

Importantly, Figure 9.12 demonstrates that the same aborted transcripts produced during the *in vitro* reaction are also observed in bacterial cells that contain introduced plasmids carrying the same template construct.

Understanding the mechanism of abortive transcription has been challenging because it was necessary to reconcile two seemingly opposing events. On one hand, the polymerase has to move along the DNA template to be able to synthesize a short RNA chain, a process that requires sequential reading of the sequence of bases in the template. On the other hand, the polymerase has to either preserve or quickly re-form contacts with the promoter to be able to re-initiate transcription once a short transcript is aborted. How can this be achieved?

Three models have been proposed to explain how abortive transcription may occur: transient excursions, inchworming, and scrunching (**Figure 9.13**). Single-molecule experiments were essential in deciding which of these models is correct. Single-pair fluorescence resonance energy transfer or FRET experiments are presented in **Figure 9.14**. Additional experiments using magnetic tweezers led to the same conclusion: abortive transcription occurs through a scrunching mechanism in which a stressed intermediate, with approximately one turn of additional DNA unwinding, is formed. Stress in this intermediate provides the driving force to break the existing interactions between RNAP and promoter and between RNAP and σ initiation factor, allowing promoter escape.

Elongation in bacteria must overcome topological problems

In bacteria, the process of elongation is simpler than in eukaryotes, principally because it does not, for example, have to deal with nucleosomes. The nucleoid structure in bacteria (see Chapter 8) appears to provide little resistance to the polymerase; there remains, however, the complication of **superhelical stress**.

It has been known for years that when an enzyme, such as RNA polymerase, tracks the DNA helix, superhelical stress is created in the DNA if it is topologically constrained. The conventional view of transcription elongation assumed that the polymerase spiraled around the DNA as it tracked the double helix. Such a movement would, of course, entangle the transcript around the DNA. We now realize that, in the cell, the polymerase is probably stationary and exerts both a linear pulling force and a rotary force on the DNA to thread it through its active center. The two kinds of motions—lateral translocation and rotation (see Figure 9.3B)—on the topologically constrained DNA create positive superhelical torsion ahead of the advancing polymerase and negative supercoiling in the wake of the enzyme (**Figure 9.15**).

As elongation proceeds along the topologically constrained loop domains in the bacterial nucleoid or in the eukaryotic nucleus, the buildup of positive supercoiling would be expected to stop transcription eventually, as the great energy cost of unwinding the overwound DNA would prohibit further elongation. Although it is true that topoisomerases will relieve the stress, a topoisomerase molecule must find the stressed region and then attach before relaxation can occur. Thus it is likely that, given the high rate of transcription, considerable stress will accumulate. The idea that high levels of transcriptional activity necessarily lead to high levels of superhelical stress is, however, a misconception. High levels of transcription usually reflect high frequency of initiation, which creates an array of polymerases tracking each other along the gene (see Figure 9.5). Such an array will partially neutralize the positive and negative stresses that are produced by each polymerase: that is, the negative stress in the wake of one polymerase will be partially neutralized by the positive stress created in front of the following polymerase, and so on. Nevertheless, transcription in a constrained system will always encounter at least transient effects of induced changes in supercoiling.

The topological consequences of transcription have been often overlooked, as most *in vitro* transcription systems use either short linear DNA templates, in which the torsion

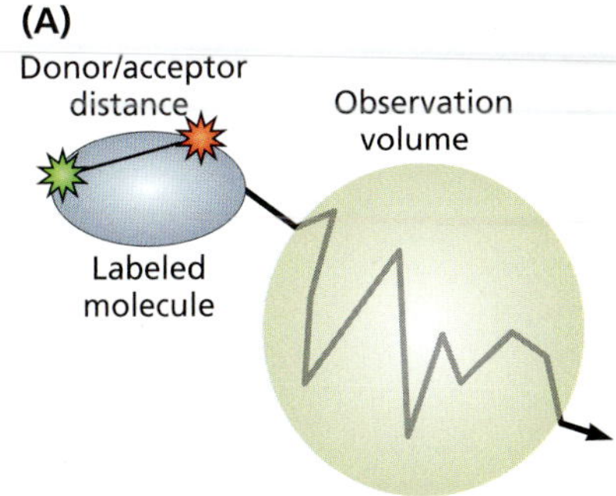

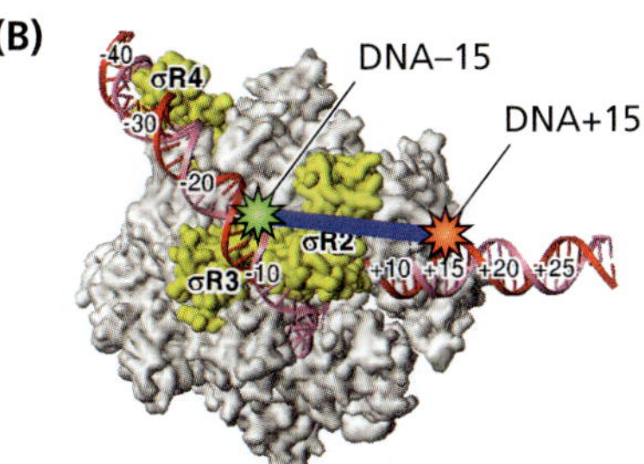

The open complex RP_o with positions of the donor (green) and acceptor (red) dyes

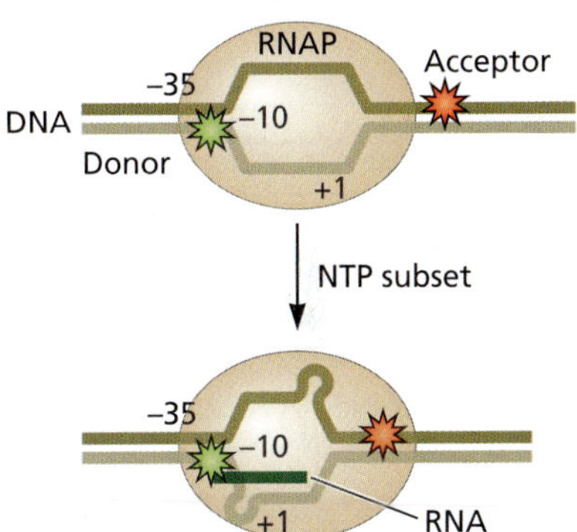

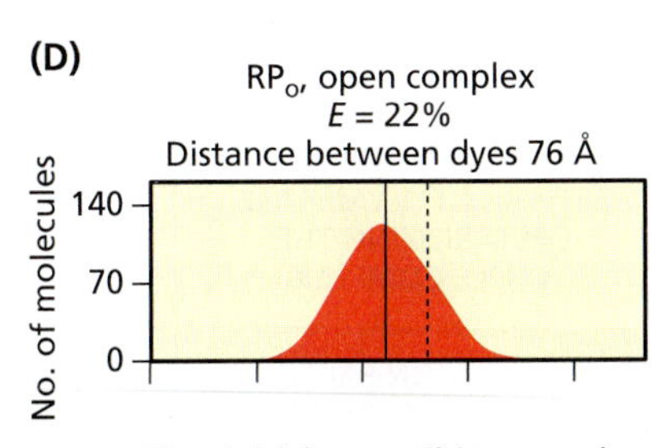

Figure 9.14 Single-pair FRET experiments demonstrating the DNA scrunching mechanism of abortive transcription. (A) Experimental approach: fluorescence resonance energy transfer or FRET between donor and acceptor dyes, attached at different positions within an open complex, was followed by confocal microscopy. Each molecule traverses the femtoliter-scale observation volume in about 1 ms; the zigzag line signifies the molecule diffusing in the observation volume. (B) Structural model of the open complex RP_O showing positions of the donor and acceptor dyes, shown as green and red starbursts, respectively, used in this specific experiment. Numbers on the DNA refer to the relative position of the given base with respect to the TSS, which is position +1. Both dyes are on the DNA. The RNAP core is shown in gray, and σ^{70} is in yellow. Other labeling sites were used in experiments addressing the validity of the transient excursion and inchworming models, neither of which was supported by the data. (C) Expected outcome of the experiment if the process of abortive transcription occurred through scrunching of the DNA. (D) FRET data for open complex and for initiation complex, which allows the formation of transcripts of up to 7 nucleotides; the maximal length of the aborted transcripts is determined by the presence of only two out of the four ribonucleoside triphosphates (NTPs). The data clearly indicate a decrease in the distance between fluorophores in the initiation complex: compare the positions of the two red peaks in the two complexes. Note that the scrunching model was also verified through the use of another single-molecule technique, magnetic tweezers. The experimental setup for the magnetic tweezers work was very similar to that presented in Box 9.2. (Adapted from Kapanidis AN, Margeat E, Ho SO et al. [2006] *Science* 314:1144–1147. With permission from American Association for the Advancement of Science.)

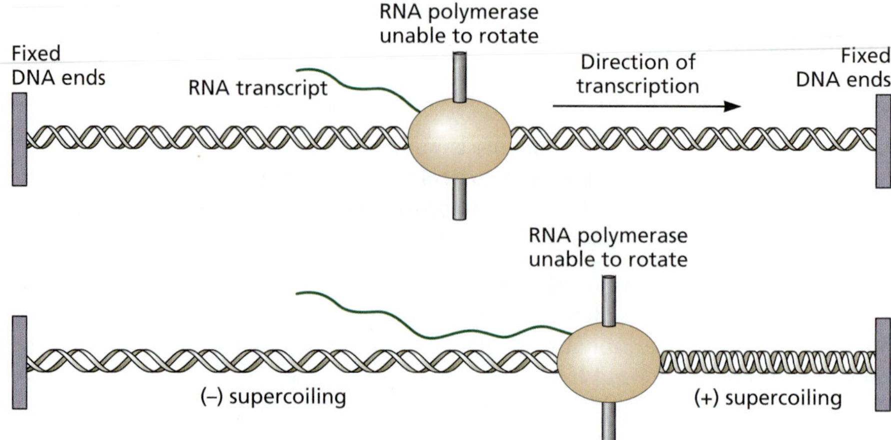

Figure 9.15 Topological consequences of transcription elongation in a topologically constrained DNA template. The DNA is represented as having fixed ends to satisfy the topological requirements for superhelicity. The transcription bubble does not rotate in the immobile RNA polymerase catalytic center, creating domains of positive and negative supercoiling, represented as DNA with different twist. Superhelical stress accumulates quickly because transcription of one helical turn, around 10 bp, leads to the creation of one superhelical turn. DNA structure cannot tolerate too much twist, so the stress can be relieved by the creation of writhe, which is coiling of the DNA axis.

dissipates from the free DNA ends, or circular plasmids, where the positive and negative supercoiling propagating from opposite directions around the circle eventually cancel each other out.

There are two mechanisms for transcription termination in bacteria

Bacteria use two very different mechanisms to terminate transcription: one intrinsic or sequence-dependent and the other protein-factor-dependent. In *Escherichia coli*, the number of genes using one mode or the other is about equally divided. However, in general, there is a wide range of relative usage of these two modes among bacterial species. Here, we will describe termination in *E. coli*, where the process is best understood:

1. Intrinsic or sequence-dependent termination: Intrinsic termination is entirely determined by the nucleotide sequence at the 3′-end of the gene and does not require any other factors. The sequence of the template near the termination site should give rise to an RNA transcript capable of forming a hairpin. This will be assured if the DNA contains an inverted repeat (**Figure 9.16**). An additional sequence feature is essential for intrinsic termination: the inverted repeat should be followed by a string of six or seven As. The formation of a hairpin in the RNA transcript causes the polymerase to pause and strips the RNA from the complex before further transcription can occur. The weak base pairing between the A string in the template and the complementary U string in the transcript contributes to dissociation of the complex: the pausing caused by the hairpin gives the string of A–U pairs many chances to dissociate. Proteins such as NusA stabilize the RNA hairpin, thereby lengthening the pause; this helps the dissociation process.

2. Rho-dependent termination: Factor-dependent termination makes use of **Rho factor**, a hexameric ATP-dependent DNA–RNA helicase. The steps in this

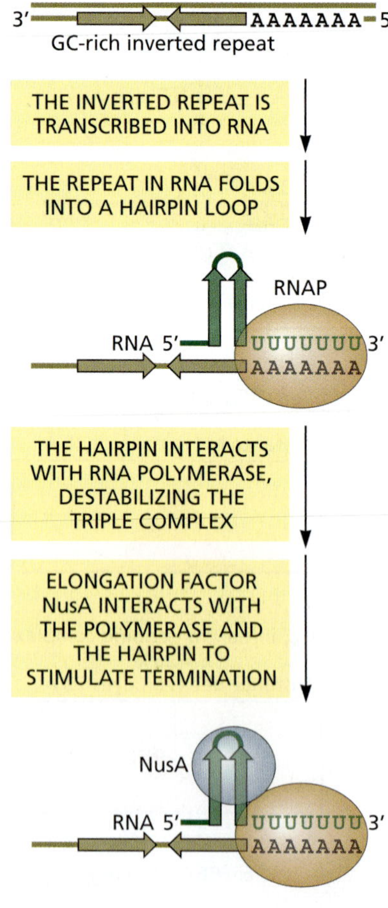

Figure 9.16 Intrinsic termination in *E. coli*. Transcription terminates when an inverted repeat in the RNA transcript forms a hairpin that is abutted by a string of uracils; the low stability of the RNA–DNA heteroduplex, due to A–U base pairing, contributes to the destabilization of the RNA–DNA hybrid in the transcription bubble. The RNA hairpin interacts directly with the RNA polymerase, destabilizing the triple DNA–RNA–polymerase complex. Additionally, the elongation protein factor NusA interacts with both the polymerase and the RNA hairpin structure to aid in dissociation of the complex.

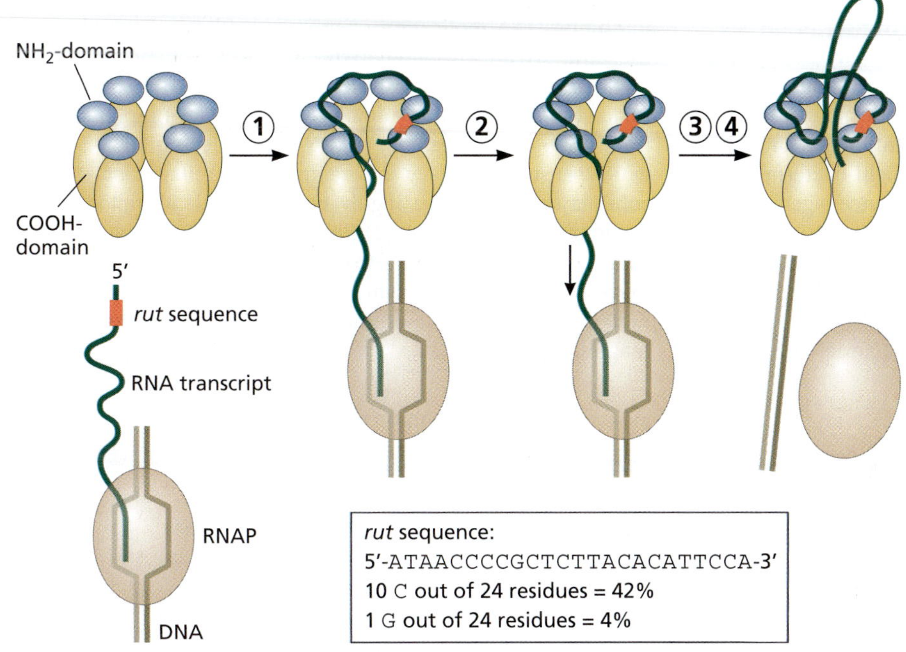

Figure 9.17 Rho-dependent termination of transcription in *E. coli*. Rho factor is a hexameric ATP-dependent DNA–RNA helicase that acts on DNA–RNA hybrids exclusively. Transcription termination occurs in four steps. (Step 1) The N-terminal domain of Rho, shown in blue, binds the Rho utilization or *rut* sequence on the transcript. (Step 2) The C-terminal domain of Rho, shown in yellow, binds the mRNA downstream of *rut*, and the ring closes. (Step 3) The C-terminus cyclically hydrolyzes ATP to propel itself along the mRNA in a 5′ to 3′ direction. (Step 4) Helicase action leads to disassembly of the transcription complex. (Inset) Sequence of Rho utilization or *rut* site of λ tR1 terminator. Note the highly skewed nucleotide content, with 42% of the residues being C and only 4% being G. This is a typical characteristic of the *rut* sequences: the efficiency of usage of these sites for termination increases with the length of the C-rich/G-poor region. *rut* sites are found at varying distances from the actual termination sites. (Adapted from Kaplan DL & O'Donnell M [2003] *Curr Biol* 13:R714–R716. With permission from Elsevier.)

termination process are schematically depicted in **Figure 9.17**. The process begins when Rho binds to a **Rho utilization site** or ***rut* site** on the RNA transcript. *rut* sites have highly skewed nucleotide content, being very rich in C and poor in G; otherwise, there is no recognizable sequence similarity among *rut* sequences in different transcripts. Once Rho binds to the *rut* sequence, it undergoes a conformational transition from an open ring to a closed ring that embraces the RNA (**Figure 9.18**). In an ATP-dependent process, Rho is propelled along the RNA transcript in a 5′ → 3′ direction; its DNA–RNA helicase activity pulls the RNA from the DNA in the transcription bubble, thus terminating transcription.

Understanding transcription in bacteria is useful in clinical practice

Today, the scientific community has acquired deep understanding of how transcription occurs in bacteria. This knowledge is absolutely necessary for the

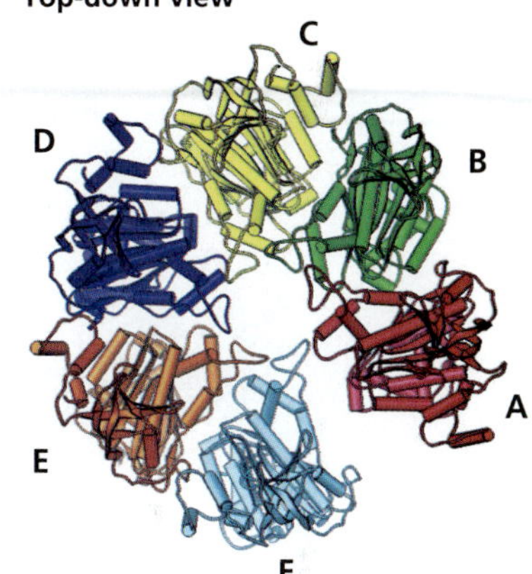

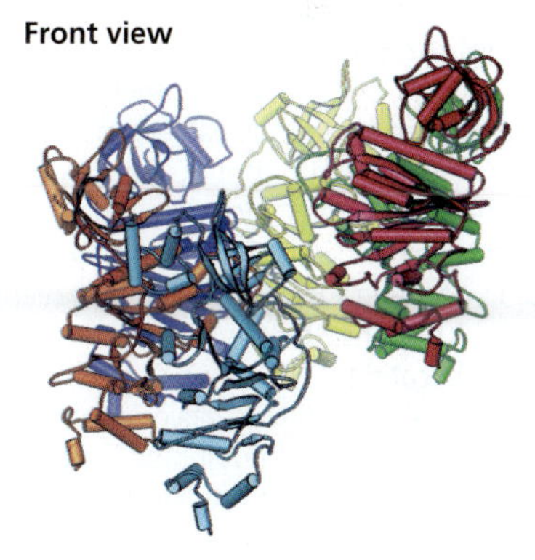

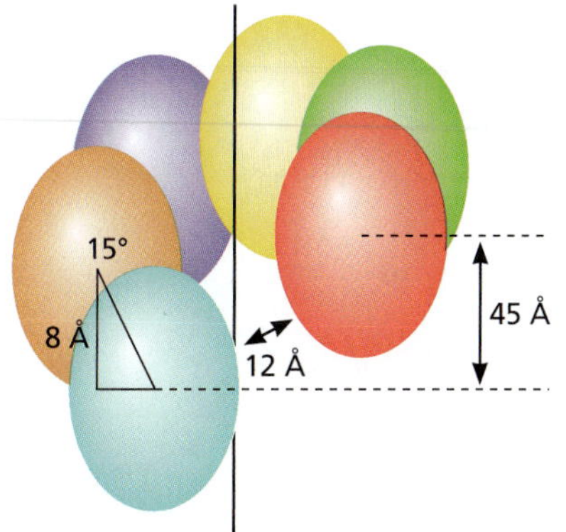

Figure 9.18 Crystal structure of Rho termination factor. Top-down view shows protomers labeled A–F. In the front view, the six subunits or protomers, represented in different colors, pack in an open hexameric ring. This is the conformation in which Rho binds to the *rut* sequence on the mRNA, after which the structure closes into a ring that embraces the RNA. The transition between open and closed ring conformations is accompanied by rotational shifts of subunits with respect to their neighbors. Shown at the bottom is a drawing of the relative rise and offset of adjacent Rho subunits as they wind about the axis of the ring, shown as a vertical line. Color scheme corresponds to the top-down and front views. Gap between monomers is 12 Å and helical pitch is 45 Å. (Adapted from Skordalakes E & Berger JM [2003] *Cell* 114:135–146. With permission from Elsevier.)

Box 9.4 Antibiotics: Inhibiting bacterial transcription

Antibiotics are a broad class of naturally occurring compounds, generally synthesized by bacteria or fungi. The role of these compounds in nature is to kill other bacteria in the competition for natural resources. Many antibiotics have been introduced into clinical practice to help fight bacterial infections; some of them are also used as anti-cancer agents. Here, we describe a couple of well-understood examples of widely used antibiotics that act by inhibiting bacterial RNAP through two entirely different mechanisms (Figure 1). The majority of antibiotics used in clinical practice act by other mechanisms, such as inhibiting the synthesis of the bacterial cell wall or inhibiting DNA replication or protein synthesis. Examples of the latter two are presented in the appropriate chapters.

Bacterial RNAP is a proven target for broad-spectrum antibacterial therapy. The suitability of RNAP as a target stems from three considerations: (1) RNAP is an essential enzyme, which ensures efficacy of treatment. (2) Bacterial RNAP subunit sequences are highly conserved, which allows for broad-spectrum activity. (3) Eukaryotic RNAP sequences diverge significantly from those of the bacterial enzymes, which makes the eukaryotic enzyme much less susceptible to inhibition: according to some estimates, by more than 3–4 orders of magnitude.

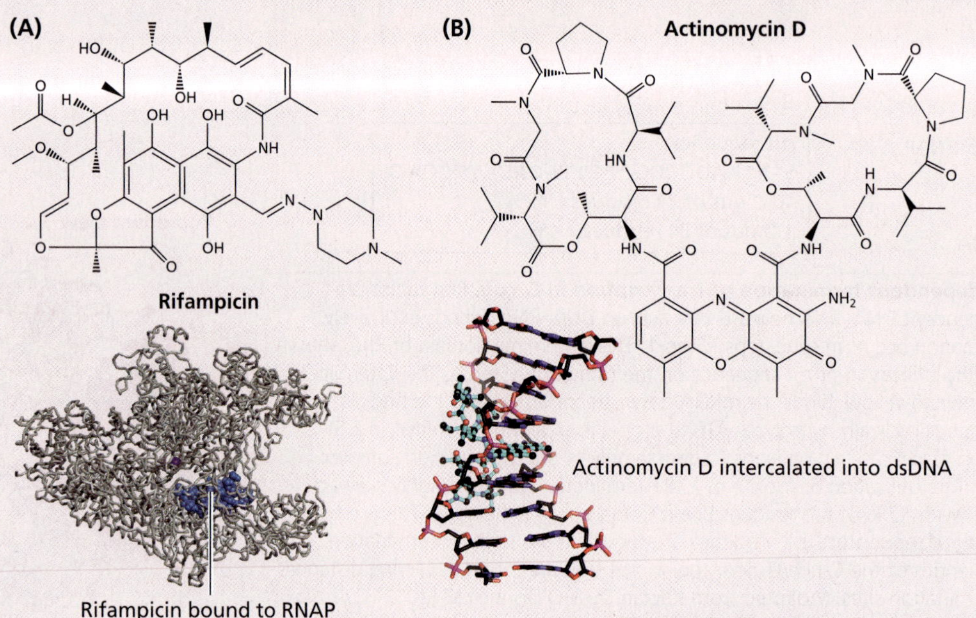

Figure 1 Examples of widely used antibiotics that act by inhibiting bacterial RNAP. (A) Rifampicin is a complex organic compound that is a semisynthetic derivative of rifamycin, which is produced by the soil bacterium *Streptomyces*. It is widely used for the treatment of both Gram-positive and Gram-negative bacterial infections, including tuberculosis. The introduction of this antibiotic in 1959 permitted a marked reduction of the time needed to treat tuberculosis patients from 18–24 months to 6–9 months, the effect being attributed to its ability to kill nonreplicating tuberculosis bacteria. Rifampicin acts by binding to a site within the RNAP active-center cleft, adjacent to the active center itself. It does not inhibit binding of the polymerase to the DNA template or the catalytic step of the reaction itself. Instead, it sterically interferes with binding of the newly formed short RNA–DNA hybrid after the addition of two or three nucleotides to the RNA chain: it blocks the channel into which the RNA–DNA hybrid must pass. (B) Actinomycin D is a polypeptide-containing antibiotic from a different strain of *Streptomyces*. It acts by intercalating its phenoxazone ring between neighboring base pairs in the DNA, thus affecting the ability of the DNA to serve as a template for transcription. At low concentrations, it selectively inhibits transcription without interfering with DNA replication. At higher concentrations, it inhibits the growth of all rapidly dividing cells and is used as an anti-cancer agent. (A, bottom, from Ho MX, Hudson BP, Das K et al. [2009] *Curr Opin Struct Biol* 19:715–723. With permission from Elsevier.)

development of drugs to combat bacterial infections. Box 9.4 presents examples of the mechanism of action of some common antibiotics that act by inhibiting bacterial transcription.

Key concepts

- Transcription is the process whereby the sequence of DNA is copied into an RNA sequence. A new RNA strand is assembled on DNA, complementary to one DNA strand and therefore identical in sequence to the other DNA strand.
- This process is catalyzed by enzymes called RNA polymerases or RNAPs and is driven by the hydrolysis of nucleoside triphosphates as nucleotide residues are

added to the RNA chain. The RNAP pulls DNA through itself to synthesize RNA chains from the 5′-end toward the 3′-end.

- There is a defined start site of transcription, contained within a promoter region. The RNAP binds strongly to the promoter and subsequently opens a bubble in the DNA duplex. This allows the initiation of transcription, although the process may abort before a critical length of the transcript is reached, ~15 nucleotides.
- In the elongation phase, the polymerase moves processively over long stretches of DNA but may pause from time to time. Short pauses do not create a problem but long pauses, in which proper placement of the 3′-end of the nascent RNA transcript in the bubble is lost, require special treatment.
- In bacteria, recognition of the promoter requires special σ subunits that are complexed with the RNAP core enzyme. There are a number of different σ subunits, which permits some selection in gene expression in response to environmental conditions.
- Although it might seem that accumulation of superhelical stress in the DNA template during elongation should inhibit transcription, the presence of multiple RNAPs on frequently transcribed regions should largely negate such effects.
- Topoisomerases also play a role in relaxing the DNA superhelical stress created by the polymerase as it tracks down the template.
- Termination of transcription in bacteria can occur via two kinds of mechanisms. Sequence-specific termination requires special sequences at the 3′-end of the transcript that favor hairpin formation followed by release of the transcript. In other cases, termination is produced by a helicase enzyme, which recognizes a site on the nascent transcript, moves along the RNA chain toward the polymerase, and literally pulls the RNA chain out of the polymerase.

Further reading

Books

Cooper GM & Hausman RE (2010) The Cell. A Molecular Approach, 3rd ed. Sinauer Associates Inc.

Wagner R (2000) Transcription Regulation in Prokaryotes. Oxford University Press.

Reviews

Darst SA (2001) Bacterial RNA polymerase. *Curr Opin Struct Biol* 11:155–162.

Dove SL & Hochschild A (2005) How transcription initiation can be regulated in bacteria. In The Bacterial Chromosome (Higgins NP ed), pp 297–310. ASM Press.

Greenleaf WJ, Woodside MT & Block SM (2007) High-resolution, single-molecule measurements of biomolecular motion. *Annu Rev Biophys Biomol Struct* 36:171–190.

Gruber TM & Gross CA (2003) Multiple sigma subunits and the partitioning of bacterial transcription space. *Annu Rev Microbiol* 57:441–466.

Hurwitz J (2005) The discovery of RNA polymerase. *J Biol Chem* 280:42477–42485.

Kaplan DL & O'Donnell M (2003) Rho factor: transcription termination in four steps. *Curr Biol* 13:R714–716.

Uptain SM, Kane CM & Chamberlin MJ (1997) Basic mechanisms of transcript elongation and its regulation. *Annu Rev Biochem* 66:117–172.

von Hippel PH & Pasman Z (2002) Reaction pathways in transcript elongation. *Biophys Chem* 101–102:401–423.

Zlatanova J, McAllister WT, Borukhov S & Leuba SH (2006) Single-molecule approaches reveal the idiosyncrasies of RNA polymerases. *Structure* 14:953–966.

Experimental papers

Harada Y, Ohara O, Takatsuki A et al. (2001) Direct observation of DNA rotation during transcription by *Escherichia coli* RNA polymerase. *Nature* 409:113–115.

Liu LF & Wang JC (1987) Supercoiling of the DNA template during transcription. *Proc Natl Acad Sci USA* 84:7024–7027.

Miller OL Jr & Beatty BR (1969) Visualization of nucleolar genes. *Science* 164:955–957.

Vassylyev DG, Sekine S, Laptenko O et al. (2002) Crystal structure of a bacterial RNA polymerase holoenzyme at 2.6 Å resolution. *Nature* 417:712–719.

Yin YW & Steitz TA (2002) Structural basis for the transition from initiation to elongation transcription in T7 RNA polymerase. *Science* 298:1387–1395.

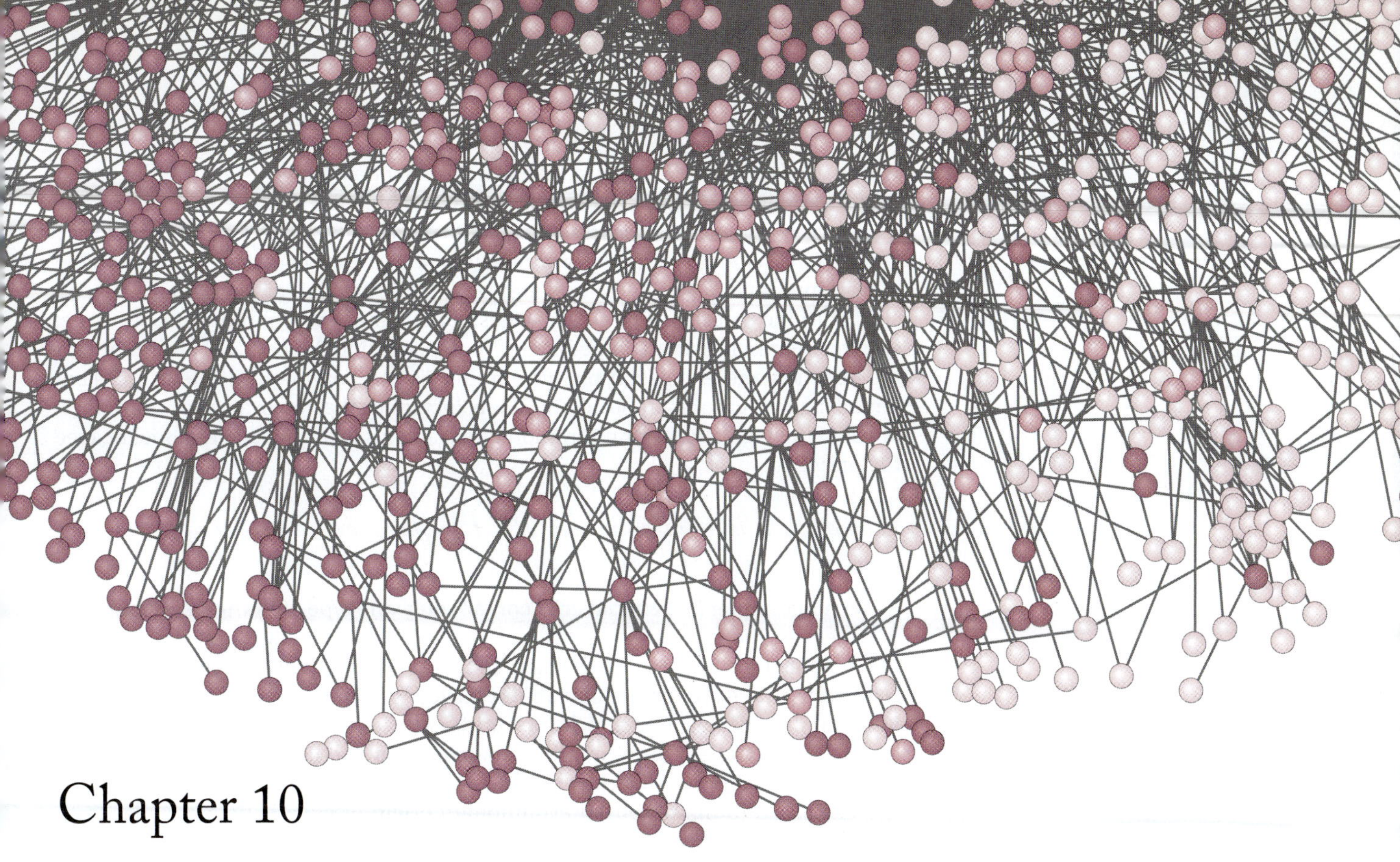

Chapter 10

Transcription in Eukaryotes

10.1 Introduction

Transcription in eukaryotes is a complex, highly regulated process

Eukaryotic transcription is more complex than that in bacteria, not only in terms of the mechanisms involved in the processes of initiation, elongation, and termination but also in terms of its regulation. This should not be surprising, for the typical eukaryote must tailor transcription to fit the needs of a host of cell types. In this chapter, we deal first, and mainly, with the transcription of mRNAs during the expression of protein-coding genes. This has been the major focus of research on eukaryotic transcription. We also briefly describe the expression of genes coding for other functional RNA molecules. However, we now recognize that mRNA production represents only a small fraction of total transcription in eukaryotes.

Gene expression, defined here as the production of a functional RNA transcript, is regulated at the levels of transcription and transcript processing to form a functional mature RNA. In the case of mRNA, its transport from the nucleus to the cytoplasm and its translation into a protein are also sites of regulation. Finally, the proteins themselves undergo post-translational modifications and different kinds of processing to give rise to a number of functionally different molecules. Protein stability or turnover is also a factor that determines the steady-state concentration of the protein in the cell. The complexity of gene expression is essential to the existence of eukaryotic metazoan organisms. A bacterium needs to respond only to environmental changes, whereas a metazoan must also preferentially express different parts of the same genome at different times in its many different cell and tissue types. Of these numerous and complex regulatory mechanisms that control gene expression, regulation of transcription is considered the most important: it is mainly through effects on transcription that cells respond to developmental signals and to the environment. In this chapter, we look at the basic mechanisms of transcription in the eukaryotic nucleus. Chapters 12 and 13 discuss the mechanisms by which transcription is regulated in eukaryotes.

Within a eukaryotic cell nucleus, the transcription apparatus must deal with a much more restrictive and complex milieu than that encountered by bacterial polymerases. As shown in Chapter 8, bacteria organize their genomes as nucleoids, with loosely associated proteins, allowing easy access to the DNA. The eukaryotic genome, on the other hand, is packaged into chromatin, and a large fraction of the DNA is wrapped about histone cores to form nucleosomes. Although not an entirely static structure, chromatin presents severe impediments to transcription. Indeed, it is now clear that the initiation and elongation phases of transcription both require at least transient remodeling of chromatin structure, often including the removal of nucleosomes. Much of the regulation of transcription in eukaryotes involves such remodeling (see Chapter 12).

Eukaryotic cells contain multiple RNA polymerases, each specific for distinct functional subsets of genes

As seen in Chapter 9, bacterial cells contain only one type of DNA-dependent RNA polymerase. The types of genes transcribed by this polymerase depend on which initiation factor σ is bound to the core enzyme. The situation in the eukaryotic cell is different: there are three multisubunit nuclear polymerases, abbreviated as Pol I, Pol II, and Pol III, each of which transcribes a certain set of genes. There are also organelle-specific enzymes found in mitochondria and chloroplasts (**Table 10.1**). It is interesting to note that the organelle polymerases are monomeric; this is usually attributed to the proposed evolutionary origin of eukaryotic organelles from bacteria symbiotically living in other bacterial cells, to eventually form an integral eukaryotic cell.

We first examine the structures of the three nuclear RNA polymerases. As shown in **Table 10.2**, all three enzymes are multisubunit, varying from 12 subunits in Pol II to 17 subunits in Pol III. Ten of the subunits, including those containing the catalytic center, are shared by the three enzymes. Other subunits differ but occupy similar peripheral positions in the enzyme complex, for example, the Rpb4/7 subcomplex. Pol III contains a three-subunit complex of C82, C34, and C31 that is unique to this polymerase. Combined data from crystallographic studies and cryo-electron microscopy (cryo-EM) or imaging, together with structural modeling, have produced good visualization of the respective structures, but the functions of some of the subunits and subcomplexes remain to be determined. From comparison of these structures, we may expect that the catalytic mechanism will be much the same for all three polymerases but that the targeting to different classes of genes, as determined by peripheral proteins, will be very different. We now consider each in turn. We begin with Pol II, which has been the most extensively studied because of its preference for protein-coding genes.

Table 10.1 Eukaryotic DNA-dependent RNA polymerases.

RNA polymerase	Number of subunits	Genes transcribed	Cellular localization	Molecules per cell
Pol I	14	pre-ribosomal RNA	nucleolus	40,000
Pol II	12	mRNA precursors; U1, U2, U4, and U5 snRNAs[a]; long non-coding RNAs	nucleoplasm	65,000
Pol III	17	5S and 7S rRNAs; tRNA; U6 snRNA[a]; other small RNAs	nucleoplasm	20,000
mitochondrial polymerase	monomeric; encoded by nuclear gene; similar to T7 RNAP	mitochondrial genes	mitochondria	unknown
chloroplast polymerase	monomeric; encoded by chloroplast gene; similar to polymerases in cyanobacteria	chloroplast genes in plants	chloroplasts	unknown

[a]Small nuclear RNAs, components of the spliceosome.

Table 10.2 Comparative structures of eukaryotic RNA polymerases I, II, and III. Pol I and Pol II structures have been fully resolved by crystallography. Pol III structure is based on cryo-EM, homology modeling of the core, and the X-ray structures of C17/25. All three enzymes contain a 10-subunit core to which additional subcomplexes are attached at the periphery. (Left, from Engel C, Sainsbury S, Cheung AC et al. [2013] *Nature* 502:650–655. With permission from Macmillan Publishers, Ltd. Middle and right, from Cramer P, Armache KJ, Baumli S et al. [2008] *Annu Rev Biophys* 37:337–352. With permission from Annual Reviews.)

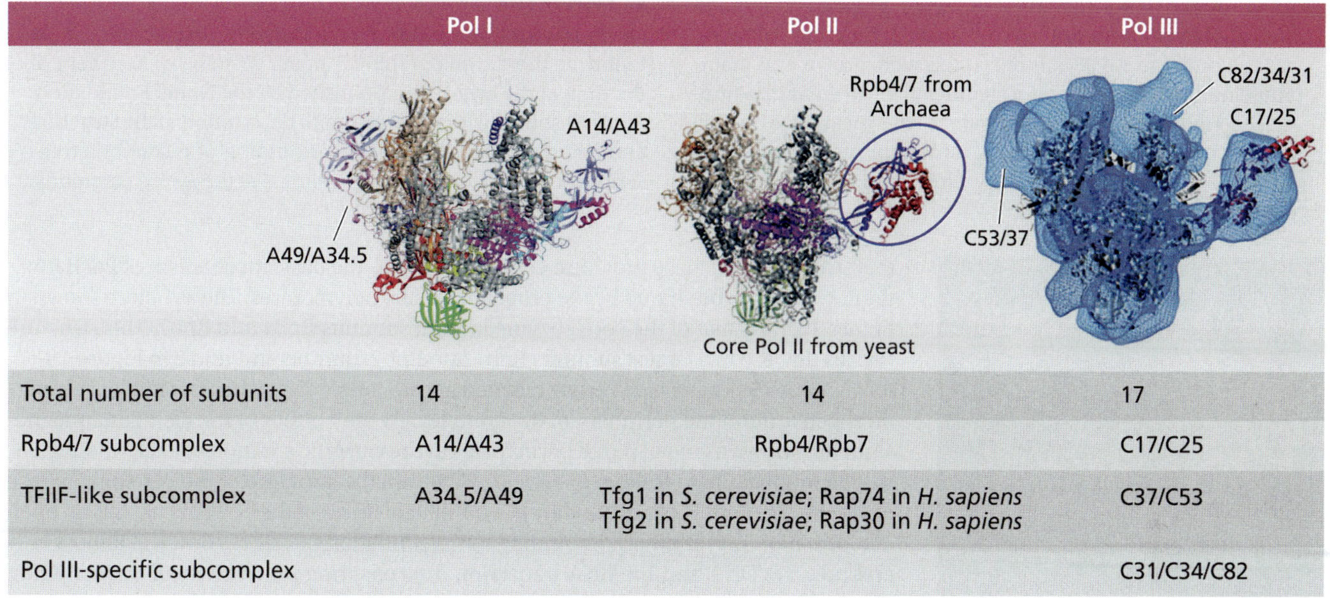

	Pol I	Pol II	Pol III
Total number of subunits	14	14	17
Rpb4/7 subcomplex	A14/A43	Rpb4/Rpb7	C17/C25
TFIIF-like subcomplex	A34.5/A49	Tfg1 in *S. cerevisiae*; Rap74 in *H. sapiens* Tfg2 in *S. cerevisiae*; Rap30 in *H. sapiens*	C37/C53
Pol III-specific subcomplex			C31/C34/C82

10.2 Transcription by RNA polymerase II

The yeast Pol II structure provides insights into transcriptional mechanisms

RNA polymerase II has been the most thoroughly studied, because it transcribes protein-coding genes to produce mRNA molecules; coding for proteins was long thought to be the major function of the genome. We now know, however, that there are many long noncoding RNA molecules produced in the nucleus, and Pol II appears to be responsible for these as well. Thus, Pol II is a very important enzyme.

Yeast Pol II is the only eukaryotic RNA polymerase whose structure has been resolved by crystallography (**Box 10.1**). Lessons about how the enzyme functions can be learned from

Box 10.1 Unraveling the molecular basis of eukaryotic transcription: The yeast Pol II story In 2006, the Nobel Prize in Chemistry was awarded to Roger Kornberg of Stanford University for "his studies of the molecular basis of eukaryotic transcription." The story of the research that led to atomic-level resolution of the structure of the multisubunit RNA polymerase II from yeast is fascinating and attests to the importance of innovation, persistence, and critical thinking in resolving difficult problems in science.

Kornberg was trained as a chemical physicist at Stanford, where he mastered NMR analyses of lipid membranes. This experience was useful later on, when he was learning how to crystallize Pol II (**Figure 1**). During his postdoctoral training, Kornberg worked with Aaron Klug at the Laboratory of Molecular Biology in Cambridge, a world-renowned center for X-ray diffraction studies of macromolecules. Attempts to understand how chromatin is transcribed led to Kornberg's work on the structure of Pol II and basal transcription factors.

The project was challenging and it took years to overcome major technical problems connected to growing crystals of sufficient quality for X-ray diffraction studies. The first Pol II crystals were two-dimensional (see Figure 1) and initially very poorly ordered. The problem lay in the heterogeneity of the protein preparation, resulting from substoichiometric

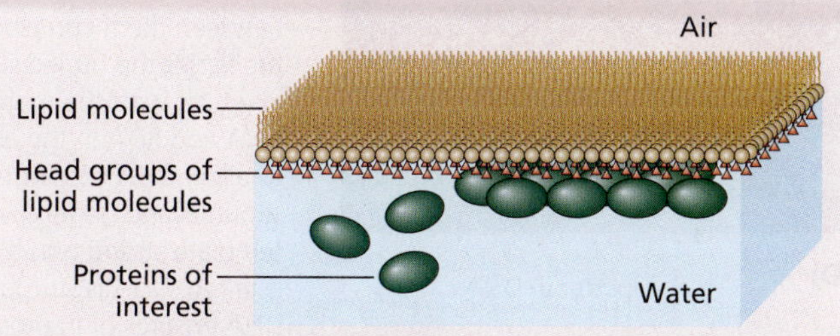

Figure 1 Two-dimensional protein crystallization on lipid layers. Rapid lateral diffusion of the lipids leads to the creation of single-layer-thick, two-dimensional or 2D crystals. The protein molecules that bind to the head groups of the lipid layer are constrained to two dimensions but are free to diffuse in the plane, thus leading to crystal formation. These 2D crystals can often add additional layers by epitaxial growth, built upon and in register with the first layer, forming thin three-dimensional or 3D crystals for X-ray analysis.

(Continued)

Box 10.1 *(Continued)* amounts of two small, nonessential subunits of the enzyme. A deletion mutant from yeast, lacking both of these subunits, solved this problem. The first ordered three-dimensional crystals obtained by epitaxial growth of two-dimensional crystals (see Figure 1) failed to diffract. Their slight yellowish tint led to the realization that oxidation was the problem, so all future crystals were grown and maintained under argon. Finally, a significant effort went into finding the right heavy-atom derivatives, the location of which in the structure is pivotal for any structure resolution. None of the 50 commonly used heavy-atom components worked, until iridium and rhenium compounds were eventually found to be effective.

The structure of yeast Pol II at 2.8 Å is key to understanding the molecular basis of eukaryotic transcription (see main text and Figure 10.1).

An account of the discovery of the Pol II structure can be found in the autobiographical notes written by Kornberg at the time of the award and published by the Nobel Foundation. Another source with more in-depth description of the structure and its importance for the understanding of eukaryotic transcription is the Nobel lecture presented at the award ceremony.

this structure, and these provide a general model, as the basic mechanism of Pol II functioning seems to be shared by the other eukaryotic polymerases. The structure shown in Figure 10.1A is that of the core enzyme lacking subunits Rpb4 and Rpb7. This structure reveals that the two largest subunits Rpb1 and Rpb2, numbered 1 and 2 in Figure 10.1B, form opposite sides of a positively charged active-center cleft, with the smaller subunits arranged around the periphery of the complex. At the bottom of the cleft lies a magnesium ion known as metal A that participates in the catalytic addition of each ribonucleoside monophosphate residue. Figure 10.1C shows the structure of the core polymerase during elongation. The **mobile clamp**, a domain of Rpb1, in the free enzyme has an open conformation that allows DNA binding during initiation but closes down to almost fully embrace the DNA and the RNA transcript, thus ensuring processivity of transcription. The structure also reveals the position of the **bridge helix** that connects the two large subunits; this helix is implicated in translocation motions of the enzyme with respect to the DNA .The conformation of the DNA–RNA hybrid at the catalytic center is intermediate between the A- and B-forms of DNA, as seen in Figure 10.1D.

The largest subunit of Pol II, Rpb1, contains a long **C-terminal domain** or **CTD** that is not present in the homologous subunits of Pol I and Pol III. CTD is characterized by the presence of numerous repeats—26 in yeast, 52 in humans—of a heptapeptide sequence, YSPTSPS. The CTD does not have a regular structure and so is not seen in crystallographic studies. The last ordered residue of Rpb1 seen in the crystal structure lies at the beginning of a linker that connects the structured body of Rpb1 to the unstructured

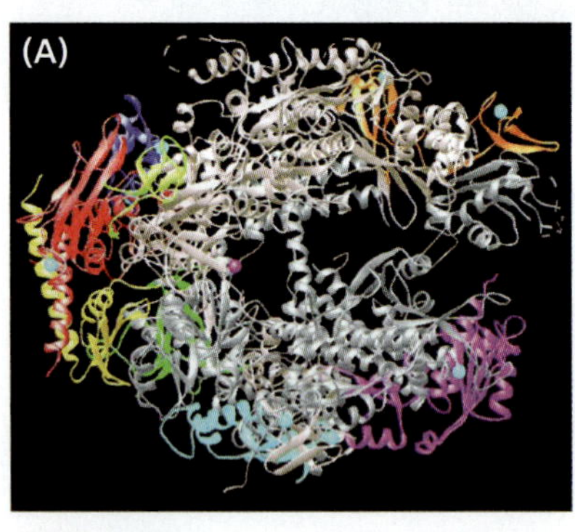

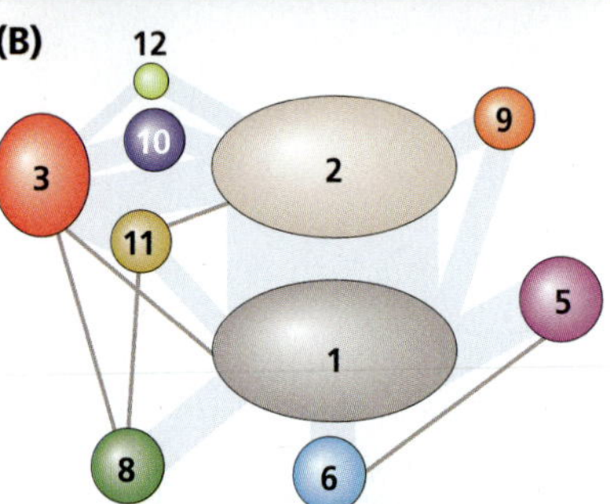

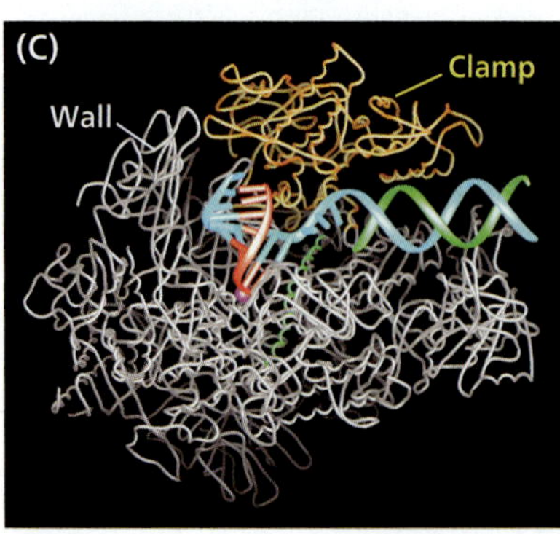

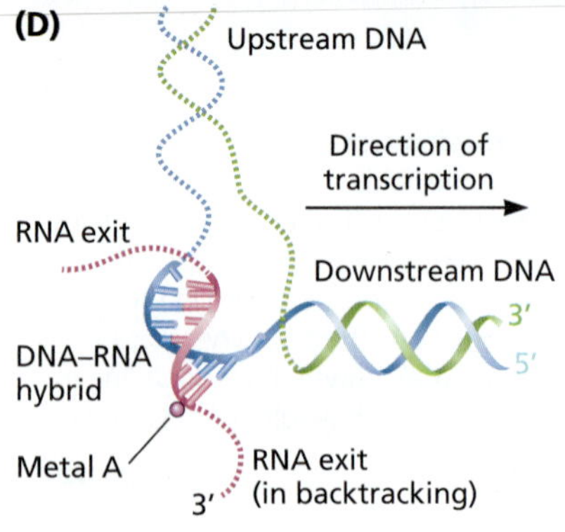

Figure 10.1 Yeast RNA Pol II structure and the enzyme in the act of transcription. (A) Structure of enzyme at 2.8 Å resolution; note that subunits 4 and 7 are missing from the crystal. The protein is shown in a ribbon representation, with the Mg^{2+} ion at the active center as a pink sphere. Light blue spheres are Zn^{2+} ions used in crystallization. The structure contains 3500 amino acid residues, with more than 28,000 non-hydrogen atoms. (B) This schematic provides the color code for individual subunits. The thickness of connecting bars between them corresponds to the surface area buried in the corresponding subunit interface: the larger the buried surface, the stronger the interaction. (C) Polypeptide chains are shown in white, with the mobile clamp, a domain in Rpb1, shown in orange. The clamp closes on DNA and RNA in the elongation complex but has a different orientation in the free enzyme to allow DNA binding at initiation. The bridge helix, shown in green, connects the two largest subunits and undergoes conformational transitions during the translocation steps. The DNA template strand is in blue, the nontemplate strand is in green, and the nascent RNA transcript is in red. DNA is surrounded by protein over ~270°. This tight grip of the polymerase on the DNA ensures high processivity of the enzyme. (D) This schematic represents the nucleic acids in the transcribing complex: the DNA template strand is in blue, the nontemplate strand is in green, and the nascent RNA chain is in red. Solid ribbons are from the crystal structure; dashed lines indicate possible paths of the nucleic acids not revealed in the structure. Note that the entering and exiting DNA duplexes are at an angle of nearly 90° with respect to each other. The conformation of the DNA–RNA hybrid is intermediate between canonical A- and B-forms of DNA. The schematic also reveals where the nascent transcript leaves the polymerase and where it ends up if backtracking occurs. (A–B, adapted from Cramer P, Bushnell DA & Kornberg RD [2001] *Science* 292:1863–1876. With permission from American Association for the Advancement of Science. C–D, adapted from Gnatt AL, Cramer P, Fu J et al. [2001] *Science* 292:1876–1882. With permission from American Association for the Advancement of Science.)

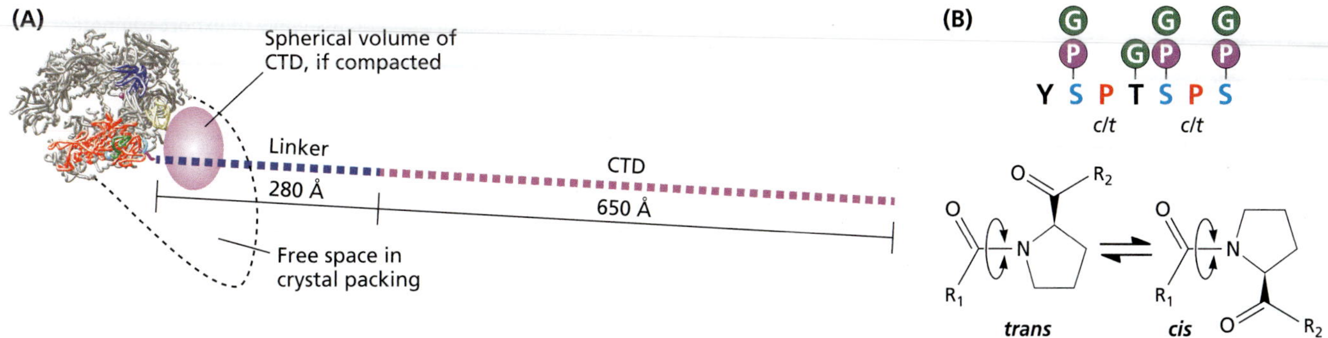

Figure 10.2 C-terminal repeat domain of Rpb1, the largest subunit of Pol II. Although this subunit is highly homologous in all three polymerases, only Rpb1 of Pol II contains a C-terminal repeat domain or CTD. (A) Size and location of CTD. Dashed lines represent the possible length of a fully extended linker and CTD. The absence of electron density corresponding to the linker and CTD provides evidence of disorder or motion, but even if disordered, these regions are unlikely to be in an extended conformation. The existence of free space in the crystal lattice for CTDs from four neighboring polymerases of Pol II suggests that the CTD may actually be compacted. (B) CTD contains the repeated heptapeptide sequence YSPTSPS, which can be phosphorylated, shown here as P, on all three serines, individually or in combination. These same residues can alternatively be glycosylated, shown here as G, with multiple glycosylated forms possible; Thr4 can also be glycosylated. Finally, both prolines can undergo a *cis–trans* isomerization reaction as illustrated. R_1 and R_2 represent continuations of the polypeptide chain toward the N- and C-terminus, respectively. Subsets of these modifications influence the function of Pol II by serving as binding platforms for specific proteins. The unit of recognition is believed to be a pair of heptapeptides. The number of repeats differs in different organisms, ranging from 26 in yeast to 52 in humans. (A, adapted from Cramer P, Bushnell DA & Kornberg RD [2001] *Science* 292:1863–1876. With permission from American Association for the Advancement of Science.)

CTD (**Figure 10.2A**). The unstructured CTD is probably not extended but compacts into a disordered state, close to the body of the enzyme. Individual amino acid residues within the repeat can undergo phosphorylation, glycosylation, and proline *cis–trans* isomerization (**Figure 10.2B**). The phosphorylation of serine residues and possibly also other modifications change during the transcription process (**Figure 10.3**). It is believed that these modifications, separately or in combination, determine the binding of various protein factors to the CTD, in turn regulating transcription elongation. This ability to interact with numerous partners is typical of disordered protein domains, as seen in Chapter 3. The CTD is also believed to play roles in RNA processing (see Chapter 14).

The structure of Pol II is more evolutionarily conserved than its sequence

Pol II structure is highly conserved across eukaryotic species: yeast and human Pol II sequences exhibit 53% overall identity, with the conserved residues distributed over the entire structure. As expected, sequence conservation between yeast and bacterial polymerases is less extensive, with the conserved regions clustering around the active center (**Figure 10.4A**). Despite the relatively weak sequence conservation, the conservation of structure is remarkable (**Figure 10.4B**). The eukaryotic enzymes share a core structure, and thus a conserved catalytic mechanism, with the bacterial enzyme (see Chapter 9). The differences in structure are at the enzymes' periphery and surface, where interactions with other proteins occur. The detailed structure of the Pol II enzyme complex is especially revealing with respect to its mechanism of action during RNA elongation. Therefore, we shall consider elongation first, leaving discussion of initiation and termination until later in the chapter.

Nucleotide addition during transcription elongation is cyclic

Transcription elongation passes through repetitive cycles of adding one new nucleotide to the growing RNA chain, followed by translocation steps that move the enzyme complex one base pair along the DNA and prepare the complex for the next nucleotide addition. Recent advances in deciphering the structure of triple Pol II–DNA–RNA complexes at different steps of the elongation process have made it possible to understand this nucleotide addition and translocation cycle. The entire cycle is illustrated in **Figure 10.5**. Most of the steps are highly evolutionarily conserved and seem to occur in a similar way in all three eukaryotic polymerases as well as in bacteria and archaea.

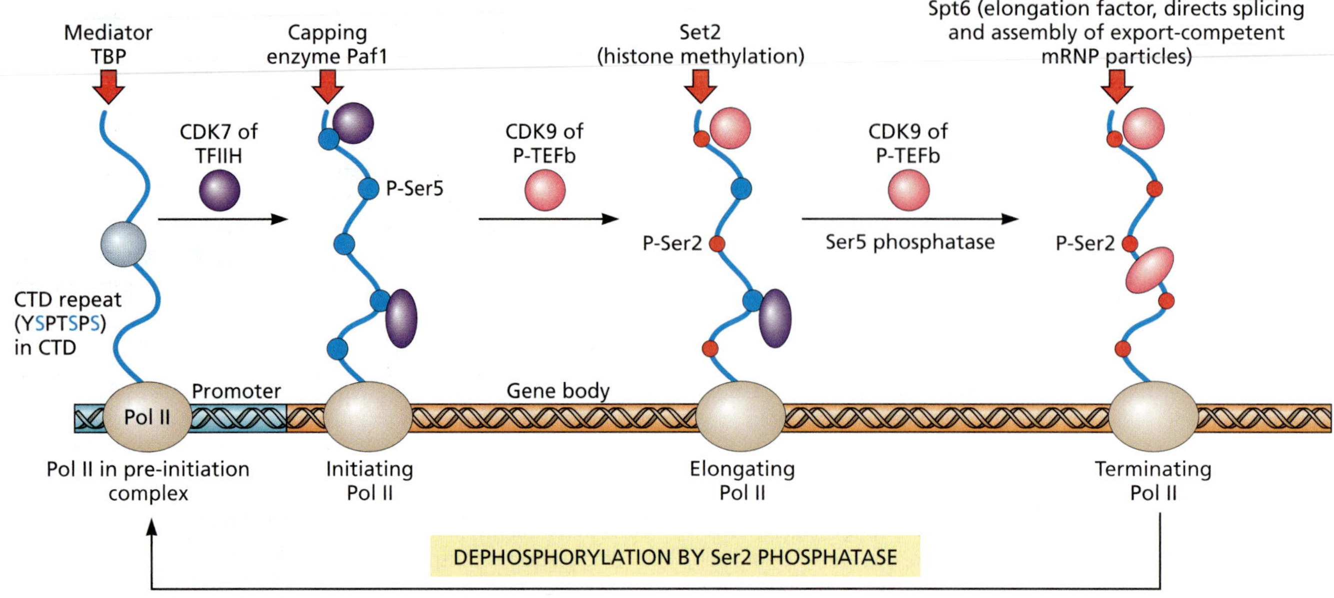

Figure 10.3 C-terminal repeat domain phosphorylation pattern. The phosphorylation pattern of CTD changes during the transcription process and determines which protein factors bind to it. The serines in positions 2 and 5, as well as 7, of the YSPTSPS repeat in CTD can undergo phosphorylation in a pattern that reflects the stage of transcription. In the pre-initiation complex, CTD is not modified. The transition to elongation is accompanied by phosphorylation on Ser5 residues, represented as blue spheres. Later in the elongation process the CTD becomes doubly phosphorylated, carrying modifications on both Ser5 and Ser2. Ser2 phosphorylation is depicted as red spheres. Finally, as the enzyme approaches the end of the gene, phosphorylation is exclusively on Ser2 residues. After termination of transcription, Pol II falls off the template and undergoes a dephosphorylation reaction, which allows it to re-initiate. The enzymes involved in these phosphorylation/dephosphorylation events are indicated above and below the arrows connecting the different stage polymerases. Exactly what controls these transitions, that is, the activities of the respective kinases, is still unclear. It is clear that the phosphorylation pattern of CTD determines what protein partners bind to the polymerase and control its activity. There are more than 100 different proteins that interact with CTD; some of them are depicted above the respective complex and are discussed throughout the book. (Adapted from Phatnani HP & Greenleaf AL [2006] *Genes Dev* 20:2922–2936, and Egloff S & Murphy S [2008] *Trends Genet* 24:280–288. With permission from Cold Spring Harbor Laboratory Press and Elsevier.)

Loading of the nucleoside triphosphate substrate occurs in two steps: pre-insertion and insertion. Recall from Chapter 9 that the DNA within the polymerase will be opened or denatured into a transcription bubble. The nucleoside triphosphate or NTP first binds to an open active center in the pre-insertion stage, when the selection of the appropriate NTP—that is, the one which base-pairs with the template base to be copied—takes place. Additional contacts help to discriminate between NTPs and deoxynucleoside triphosphates, dNTPs. In the second step, the correct NTP is delivered to the insertion site; this step is followed by the catalytic incorporation of the nucleoside monophosphate or NMP into the nascent RNA chain. The release of pyrophosphate completes this portion of the cycle.

Next, a complex series of motions in the polymerase allows the necessary translocation of the DNA with respect to the catalytic center to occur. Translocation also occurs in two steps: see Figure 9.3B for a depiction of the two kinds of motions involved in translocation. In step 1, the hybrid moves so that the downstream DNA translocates until the next DNA template base is properly located. In step 2, the DNA template base rotates by 90° to attain its correct orientation in the active center. Both of these steps

(A)

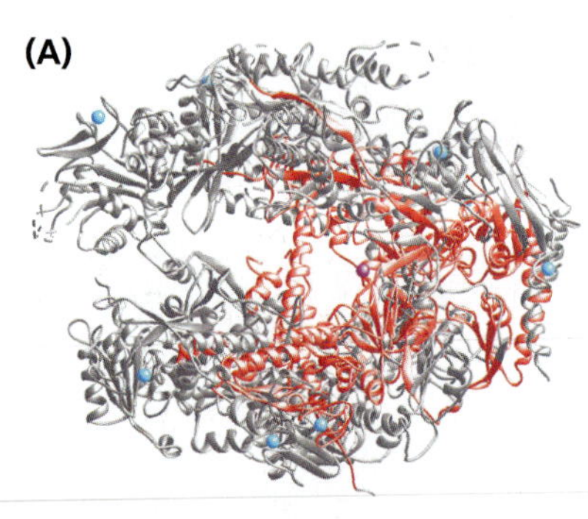

(B)

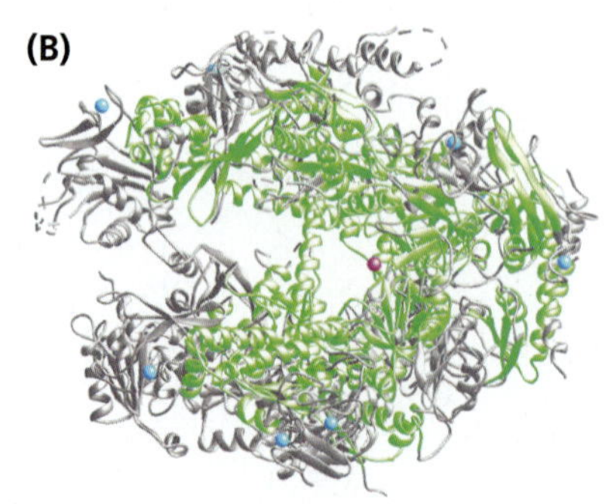

Figure 10.4 Comparison between the two largest subunits of bacterial and eukaryotic RNA polymerases. (A) Bacterial RNA polymerase from *Thermus aquaticus*; (B) eukaryotic RNA polymerase from *Saccharomyces cerevisiae*. The two structures are overlaid: blocks of sequence homology are shown in red, and regions of structural homology are shown in green. The magenta sphere is the Mg^{2+} ion in the active site. The structures show considerable conservation around the active center, suggesting a conserved catalytic mechanism. Note that structural homology is much higher than sequence homology. The differences are mainly located in the peripheral and surface features, where the polymerases interact with other proteins. (From Cramer P, Bushnell DA & Kornberg RD [2001] *Science* 292:1863–1876. With permission from American Association for the Advancement of Science.)

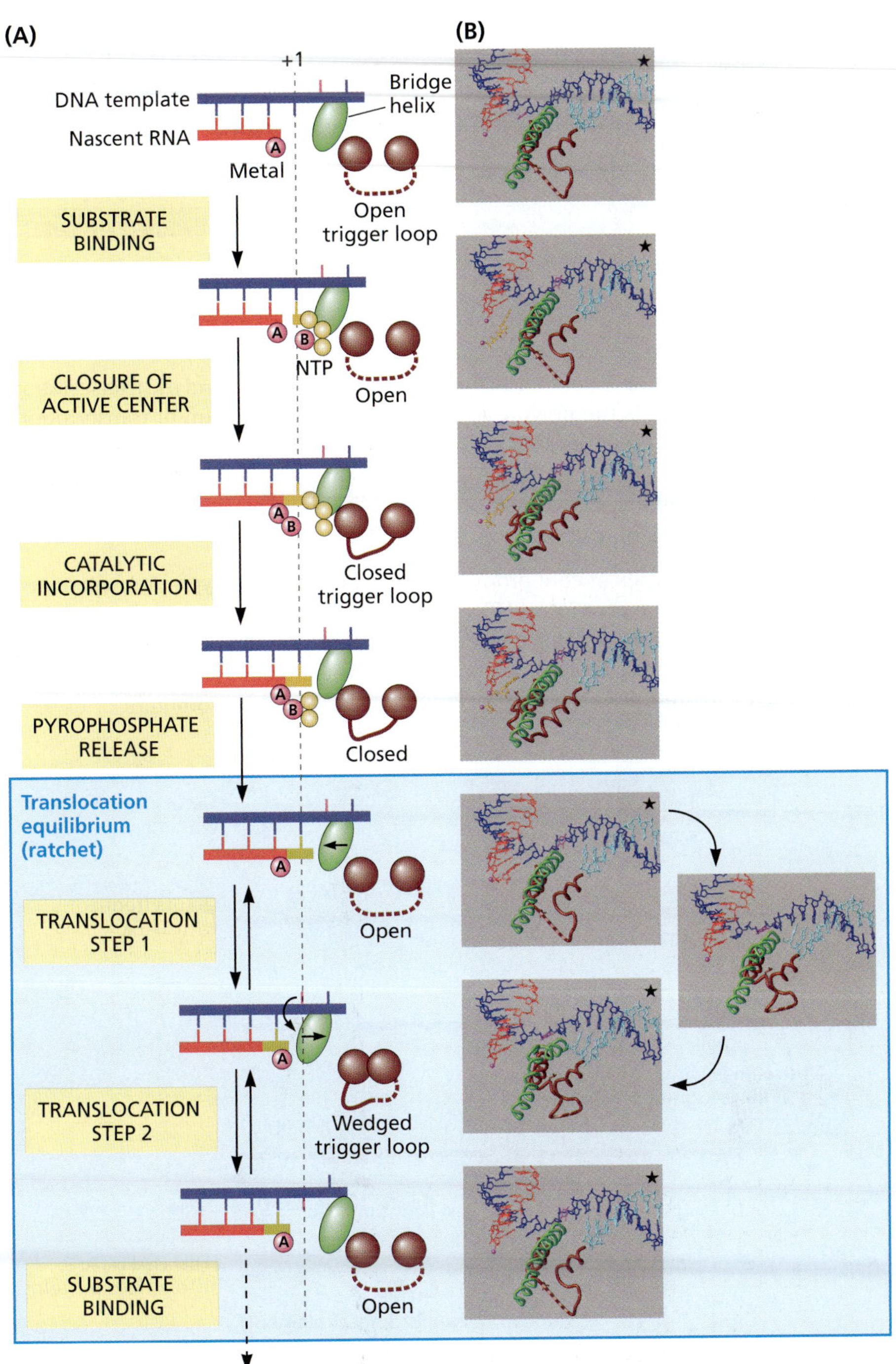

Figure 10.5 Model of the nucleotide addition cycle. (A) Schematic representation of RNA polymerase elongation complex showing the DNA template strand in blue, the nascent RNA chain in red, and two functional polymerase elements: bridge helix in green, Rpb1 residues 811–845, and trigger loop in brown, Rpb1 residues 1062–1106. The two catalytic magnesium ions, designated A and B, are shown in pink. The vertical dotted line indicates the position of DNA template base +1. The cycle is based on a Brownian ratchet model, which assumes that the ground state of the elongation complex is an equilibrium between interconverting pre- and post-translocation states, denoted by double arrows. This oscillation is temporarily stopped by substrate binding and resumes around the next template position after nucleotide addition. The cycle begins with the binding of an NTP substrate, shown in orange, to an open active center or open trigger loop. Folding of the trigger loop leads to closure of the active center. The growing 3′-end of the RNA chain becomes elongated by catalytic addition of the nucleotide, which results in formation of a pyrophosphate ion. The release of pyrophosphate destabilizes the closed conformation, leading to trigger loop opening. In the resulting pretranslocation state, the incorporated 3′-terminal nucleotide remains in the substrate site. Translocation possibly takes place in two steps. During step 1, the hybrid moves to the post-translocation position, the downstream DNA translocates until the next DNA template base, +2, reaches the pretemplating position above the bridge helix. During step 2, the DNA template base twists by 90° to reach its templating position in the active center, +1. Note the altered, flipped-out conformation of the bridge helix and the transition from an open to a wedged trigger loop conformation. After translocation of DNA and RNA, the elongation complex is in a post-translocation state, with a free substrate site to bind the next incoming NTP. The nucleotide addition cycle can then be repeated. (B) Structural models of various stages of the nucleotide addition cycle. Five of the seven different functional states of the elongation complex have been crystallographically resolved, as marked by a black star; two other states were modeled. (Adapted from Brueckner F, Ortiz J & Cramer P [2009] *Curr Opin Struct Biol* 19:294–299. With permission from Elsevier.)

are supported by conformational transitions in specific portions of Pol II. Once translocation is complete, the complex is ready for a new nucleotide addition cycle.

This one-directional translocation of the DNA through the enzyme requires an energy source. The only evident source is cleavage of the NTPs to add an NMP to the growing RNA chain and release pyrophosphate. This is an energetically highly favored reaction: the action of the polymerase is only catalytic. Thus, once NTPs are available, the process drives itself. The irreversibility of the process is further guaranteed by enzyme-catalyzed hydrolysis of the released pyrophosphate.

Fidelity of transcription, that is, avoiding incorporation of the wrong nucleotide or even a deoxyribonucleotide into the growing chain, is assured by the two-step nature of the elongation process: the incoming NTP is tested before the second step can occur. This testing is favored by the slow rate of transcriptional elongation in eukaryotes, which is about a tenth of the rate in bacteria.

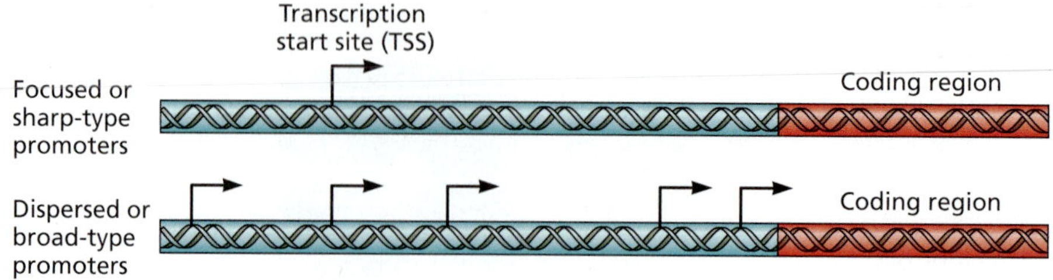

Figure 10.6 Focused versus dispersed core promoters. Focused or sharp-type promoters contain single or tightly clustered TSSs. They are more ancient and widespread in nature. Dispersed or broad-type promoters, on the other hand, contain multiple weak TSSs over a region of 50–100 nucleotides. They are usually found in CpG islands (see Chapter 12) and are more common than focused promoters in vertebrates. (Adapted from Juven-Gershon T, Hsu JY, Theisen JW & Kadonaga JT [2008] *Curr Opin Cell Biol* 20:253–259. With permission from Elsevier.)

Transcription initiation depends on multisubunit protein complexes that assemble at core promoters

The **Pol II core promoter** is defined as the DNA sequence(s) around the transcription start site, TSS, that binds the polymerase and directs the initiation of transcription. For many years, it was believed that a single motif upstream of TSS served as a universal promoter. This is known as the TATA box, similar to the –10 element in bacterial promoters (see Chapter 9). Now we know that promoter recognition in eukaryotes is much more varied and more complex.

There are two broad types of core promoters: **focused** or **sharp-type promoters** and **dispersed** or **broad-type promoters** (Figure 10.6). In vertebrates, less than a third of all promoters are focused. The motifs constituting focused promoters are relatively well-known (Figure 10.7), but less is known about the characteristic sequence motifs and factors involved in pre-initiation complex formation at dispersed promoters.

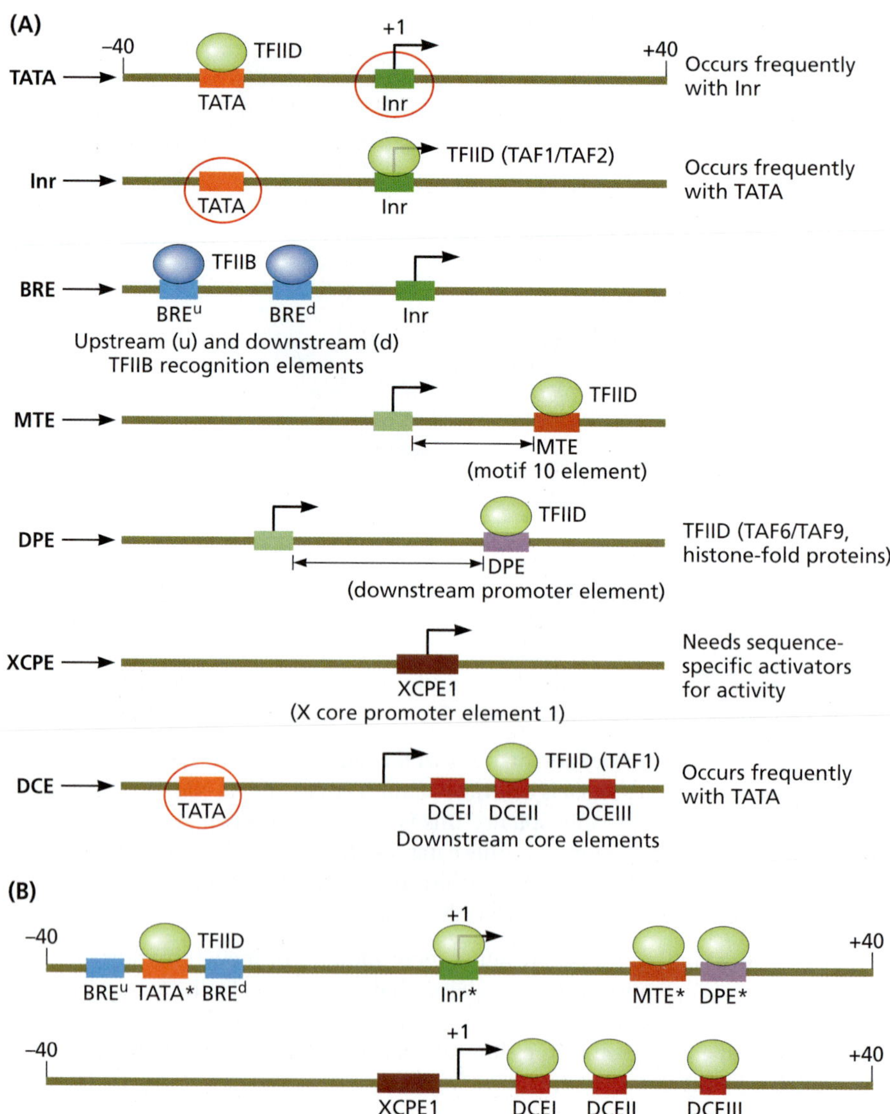

Figure 10.7 Known core promoter motifs for Pol II transcription. (A) Schematic representation of some individual promoters and the protein complexes that bind to them. Each individual element occurs in only a fraction of all promoters, with the Initiator element, Inr, being the most frequent. Double-headed arrows in some of the promoters indicate that the spacing between the two elements is critical. Often, promoter elements controlling specific genes work in combination. Elements that may or may not be present in a specific promoter are circled; for example, the TATA box is a secondary promoter element that is present in some Inr promoters. The synergy between motif ten element, MTE, and other promoter elements led to the creation of the super core promoter, SCP, which contains optimized versions of TATA, Inr, MTE, and downstream promoter element DPE. These elements are marked by asterisks in part B. SCP is the strongest existing promoter that functions both in *in vitro* transcription reactions and when introduced into cells in culture. SCP shows an unusually high affinity for TFIID. The green ovals represent TFIID that recognizes at least five of the seven promoters. That TFIID is a complex of TATA-binding protein (TBP), core subunits, and some of the TBP-associated factors or TAFs; thus the TFIID complexes that recognize specific promoter elements may be different. (B) To-scale diagram showing the mutual disposition of various promoter elements. The lower schematic illustrates elements that overlap the elements presented in the upper schematic. There is no promoter element that is universal, that is, present in all promoters. (A, adapted from Sandelin A, Carninci P, Lenhard B et al. [2007] *Nat Rev Genet* 8:424–436. With permission from Macmillan Publishers, Ltd. B, adapted from Juven-Gershon T, Hsu JY, Theisen JW & Kadonaga JT [2008] *Curr Opin Cell Biol* 20:253–259. With permission from Elsevier.)

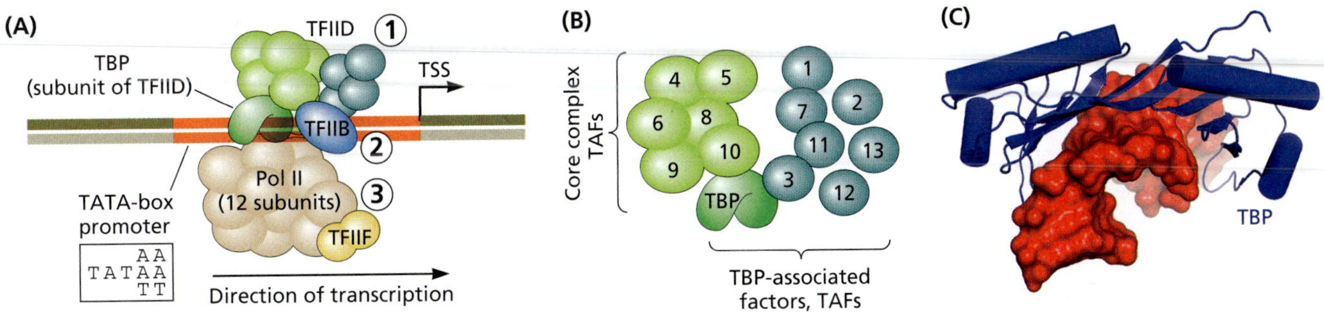

Figure 10.8 Assembly of minimal functional pre-initiation complex on the TATA box during transcription initiation. (A) Schematic of the partial PIC. TBP and the other components of TFIID bind to the TATA box and recruit all other general transcription factors, GTFs, and the polymerase itself. Additional general transcription factors E, H, and J are not shown. The numbers show the succession of binding *in vitro*. (B) TFIID: the core complex of six subunits, 4–6 and 8–10, depicted in light green, associates with combinations of other TAFs and TBP to form a large variety of different complexes. These different complexes bind to different types of promoters to initiate transcription. (C) Structure of TBP, shown in blue, bound to a promoter TATA box, shown in red. TBP induces a kink in the DNA strands and forces open the minor groove of the DNA double helix, where most of its contacts with DNA occur. The other GTFs, as well as the RNA polymerase II complex, assemble around it. The TBP–DNA complex is slightly asymmetrical, ensuring the directionality of transcription, that is, that transcription occurs on the correct DNA strand.

Transcription initiation in eukaryotes is exceedingly complex and depends upon assembly on the promoter of a huge multiprotein complex. The nuclear polymerases themselves are not even capable of recognizing promoter sequences; thus, they need the help of **basal transcription factors**, also called general **transcription factors** or GTFs, to bind to promoters and initiate the process. Basal TFs are absolutely required for transcription, independent of the frequency at which a particular gene is transcribed. The level of transcription is determined mainly by the frequency of transcription initiation and the stability of the minimal **pre-initiation complex** or **PIC**, which in turn depend on signals from regulatory proteins such as activators or repressors. These bind specific nucleotide sequences in regions distal to the basal promoter (see Chapter 12 for more detail).

A schematic of the minimal PIC assembled on a TATA-box promoter is presented in Figure 10.8A, and more detailed information about each of the PIC factors is given in Table 10.3. The first complex to bind to the promoter *in vitro* is

Table 10.3 Protein complexes involved in pre-initiation complex assembly. TFIIA, previously thought of as a general transcription factor, is actually a coactivator that interacts with activators and components of the basal initiation machinery to enhance transcription. (Adapted from Sikorski TW & Buratowski S [2009] *Curr Opin Cell Biol* 21:344–351.)

Protein complex	Number of subunits	Function
RNA Pol II	12	catalyzes transcription of all protein-coding genes and a subset of noncoding RNA, including RNA components of the spliceosome
TFIIB	1	stabilizes TFIID binding to promoter; aids in recruiting TFIIF/Pol II to promoter; directs accurate TSS selection
TFIID	14	subunits include TBP and TAFs (Figure 10.8); nucleates PIC assembly through either TBP binding to TATA box or TAF binding to alternative promoters (Figure 10.7); coactivates transcription through TAFs/activator direct interactions
TFIIE	2	helps to recruit TFIIH to promoters and stimulates its helicase and kinase activities; binds single-stranded DNA; essential for promoter melting
TFIIF	2–3	forms a tight complex with Pol II and enhances its affinity for TBP–TFIIB–promoter complex; necessary for TFIIE/TFIIH recruitment; aids in TSS selection and promoter escape; enhances elongation efficiency
TFIIH	10	ATPase/helicase activity for promoter opening and clearance (helicase activity for DNA repair that occurs on transcribed genes); kinase for Ser5 phosphorylation of CTD; facilitates transition from initiation to elongation
Mediator	24	regulates transcription by serving as a bridge between gene-specific activators (or repressors) and basal TFs; absolutely required for basal transcription from almost all Pol II promoters (can be regarded as a general TF and a signal processor)

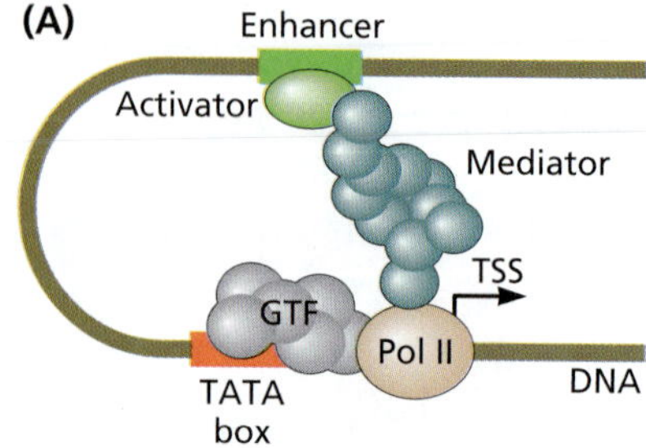

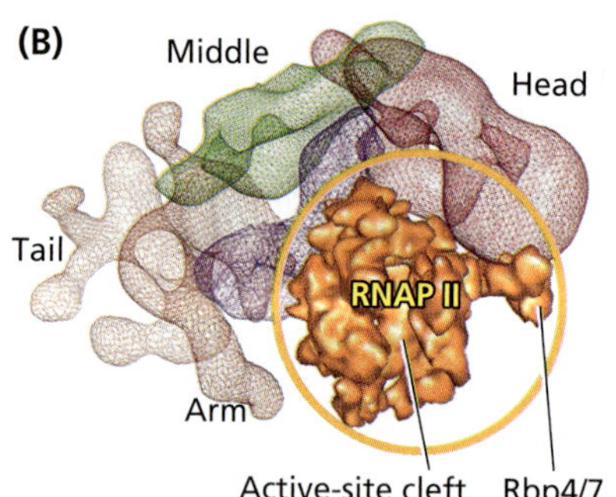

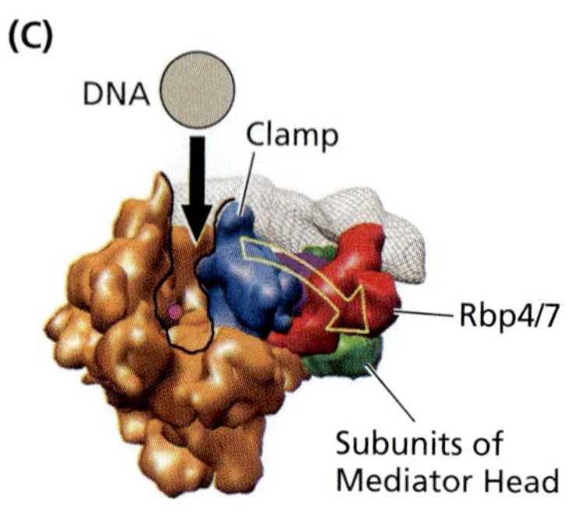

Figure 10.9 Mediator, a multiprotein complex of >20 subunits. Mediator is a co-activator, co-repressor, and general transcription factor all in one. (A) Schematic showing Mediator serving as a bridge between proteins that activate or repress transcription and the transcriptional machinery assembled at the promoter. Note the direct contacts with both the activator protein and Pol II. (B) Model of the Mediator–RNAP II complex, based on cryo-EM imaging, reveals contacts between the Head module of Mediator and the RNAP II subunits Rpb4/Rpb7. Interaction of RNAP II with Mediator is very extensive, with Mediator surrounding the face of the polymerase where TBP, TFIIB, and TFIIF presumably bind. All of these contacts are consistent with reported stabilization of the pre-initiation complex by Mediator. (C) Interaction between the Head module of Mediator and Rpb4/Rpb7 could affect the conformation of the polymerase clamp domain and possibly facilitate opening, shown by a white open arrow, of the RNA polymerase II active-site cleft, outlined in black, to allow access of double-stranded promoter DNA to the polymerase active site. (A, adapted from Kornberg RD [2007] *Proc Natl Acad Sci USA* 104:12955–12961. With permission from National Academy of Sciences. B, from Cai G, Imasaki T, Takagi Y & Asturias FJ [2009] *Structure* 17:559–567. With permission from Elsevier. C, from Cai G, Imasaki T, Yamada K et al. [2010] *Nat Struct Mol Biol* 17:273–279. With permission from Macmillan Publishers, Ltd.)

transcription factor IID, **TFIID**, which contains the **TATA-binding protein**, **TBP**, six core subunits, and a variable set of additional subunits, termed **TATA-binding protein associated factors** or **TAFs** (**Figure 10.8B**). TAFs discriminate among the alternative promoters recently recognized (see Figure 10.7). TBP binding to the TATA box induces a kink in the DNA and opens the minor groove of the double helix, wherein most of its contacts occur (**Figure 10.8C**). Remember that most proteins recognize nucleotide sequences in duplex DNA through the major groove (see Chapter 6). If a protein interacts with DNA through the minor groove, it usually must deform the double helix to ensure access to the base edges for sequence-specific recognition.

An additional protein complex is needed to connect Pol II to regulatory proteins

It was long believed that the six basal transcription factors TFIIB, D, E, F, H, and J were sufficient for transcription initiation. More recently, another large protein complex, **Mediator**, was discovered in yeast and subsequently also observed in mammalian cells (**Figure 10.9A**). As stated by Roger Kornberg, who discovered the complex, "Mediator is no less essential for transcription than Pol II itself." In addition to its role as a necessary basal factor, Mediator forms the basis for regulated transcription: it serves as a bridge between Pol II and the sequence-specific proteins that recognize regulatory elements surrounding the gene. Contacts between Mediator and Pol II are extensive and direct and involve Pol II subunits Rpb4 and Rpb7 (**Figure 10.9B** and **Figure 10.9C**). The binding between the two protein complexes stabilizes the pre-initiation complex for efficient initiation. Thus, according to Kornberg, "mediator is a co-activator, a co-repressor, and a general transcription factor, all in one."

The current model of how all these factors work together to promote initiation is presented in **Figure 10.10**. Importantly, once the polymerase clears the promoter to enter the processive elongation phase of transcription, a scaffold complex comprising Mediator, most of the basal TFs, and an activator protein stays behind; this scaffold can now be quickly repopulated by a free Pol II/TFIIF complex to form the closed promoter complex. This speeds up the next round of transcription initiation.

Termination of eukaryotic transcription is coupled to polyadenylation of the RNA transcript

Termination of eukaryotic transcription is strongly coupled to post-transcriptional processing of the RNA transcript (see Chapter 14). Each mRNA primary transcript, with a few exceptions, is **polyadenylated** at its 3′-end: a string of A residues is added by a special polymerase. The site of polyadenylation is defined by a particular sequence

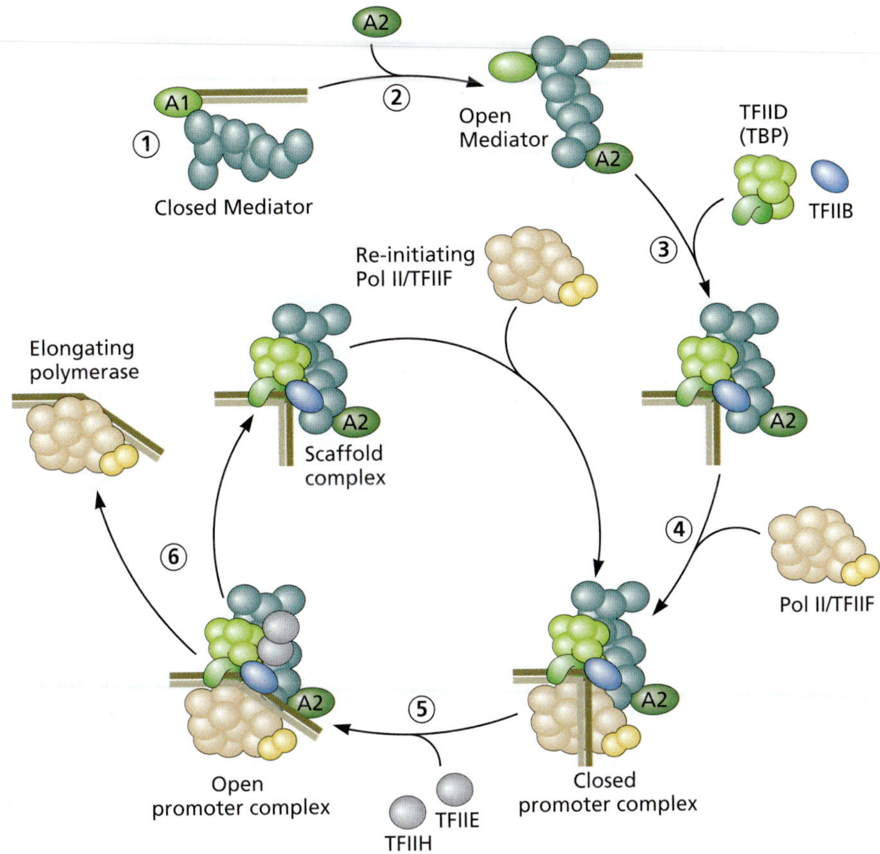

Figure 10.10 Structural model for Mediator initiation and re-initiation. The steps of the process are as follows: (Step 1) Mediator is recruited to a promoter through the action of activator A1. (Step 2) The action of a second activator, A2, leads to a conformational change of Mediator that now provides a binding site for components of the basal transcriptional machinery. (Step 3) Transcription factors TFIIB and TFIID position themselves at the promoter and form a landing platform for the Pol II/TFIIF complex. (Step 4) Pol II/TFIIF complex binds, or Pol II and TFIIF concomitantly bind, to the pre-formed Mediator/DNA/TBP/TFIIB complex. (Step 5) Through the action of TFIIE and TFIIH, the promoter opens and the template DNA strand reaches the polymerase active site. (Step 6) Following the abortive transcription phase, Pol II, or the Pol II/TFIIF complex, escapes the promoter, leaving behind a platform that facilitates rapid assembly of a new pre-initiation complex and re-initiation by new incoming Pol II/TFIIF complex. (Adapted from Asturias FJ [2004] *Curr Opin Struct Biol* 14:121–129. With permission from Elsevier.)

called the polyadenylation signal; the RNA transcript is cleaved at some point 3′ to the signal, and the transcribing polymerase is released. The CTD may be involved here. One function of polyadenylation seems to be to aid in exporting RNA from the nucleus to the cytoplasm. Thus, it is not surprising that polyadenylation is not found in bacteria.

10.3 Transcription by RNA polymerase I

As indicated in Table 10.1, Pol I transcribes the precursors of rRNA: the two large RNA molecules that participate in the formation of large and small ribosomal subunits and the small 5.8S rRNA (see Chapter 15). The direct product of transcription is a single 47S RNA, which is then processed to produce the above products. Thus, Pol I has only one kind of target gene. Transcription of the rRNA genes occurs in the nucleolus, where these genes are tandemly repeated multiple times (Figure 10.11). The transcriptional regulation of these genes is complex and poorly understood, as only a fraction of the genes are being transcribed at any given moment, even in rapidly proliferating cells where the demand for ribosomes, and hence rRNA, is huge.

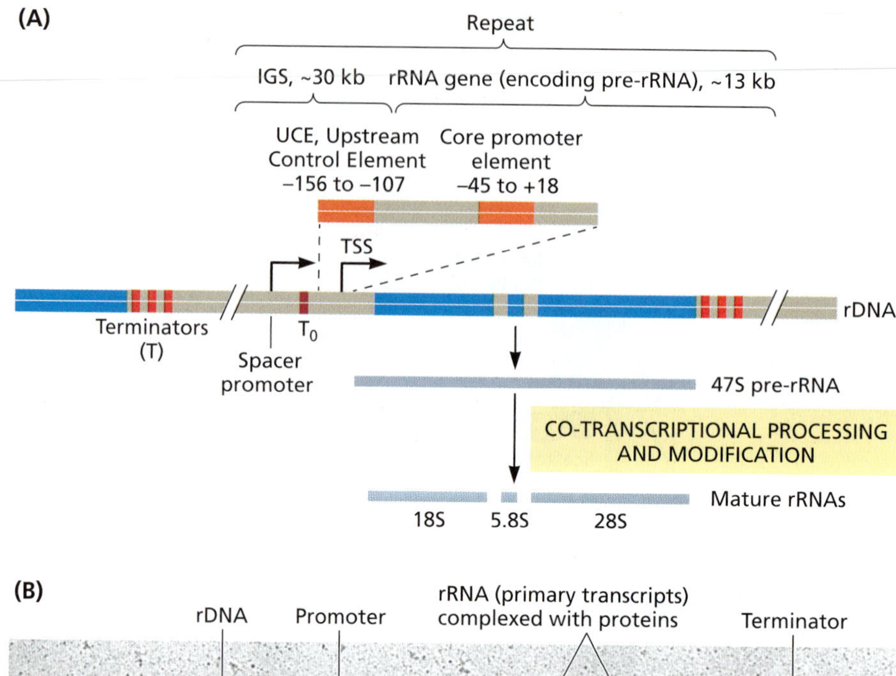

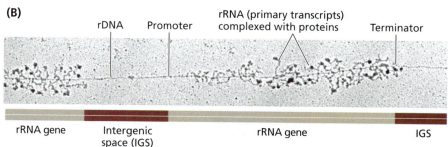

Figure 10.11 Repetitive nature of rRNA genes in eukaryotes. (A) Not-to-scale schematic of human rRNA repeats. The intergenic spacers, IGS, contain the transcription regulatory elements. A spacer promoter directs Pol I transcription of short-lived transcripts of yet unknown function. All rRNA sequences are transcribed as a large precursor that undergoes processing and base modifications to produce the mature RNA components of small and large ribosomal subunits,18S and 5.8S/28S rRNA, respectively. The promoter element is bipartite and consists of an essential core and an upstream control element. The 3′-end of the transcribed region contains several transcription-termination elements. (B) Electron microscopic image of yeast nucleolar chromatin. Progressively longer RNA transcripts associated with proteins are seen to originate from the Pol I complexes, visible as small dots on DNA. (Adapted from Russell J & Zomerdijk JCBM [2005] *Trends Biochem Sci* 30:87–96. With permission from Elsevier.)

The general transcription factors of the Pol I pre-initiation complex are schematically shown in Figure 10.12. The TBP is used in both Pol II and Pol I transcription, although the TAFs are specific for each polymerase. The TBP complex with TAFs contains the Pol I selectivity factor 1, SL1, which interacts with numerous other factors in the PIC (see Figure 10.12 and Figure 10.13). An important interactor is the upstream binding factor or UBF. Two UBF dimers interact with two DNA elements, the core promoter and the upstream control element or UCE, to form the highly specific architecture known as the **enhanceosome** (Figure 10.14). Two adjacent enhanceosomes bring the core promoter elements and the UCE together to help to recruit TBP.

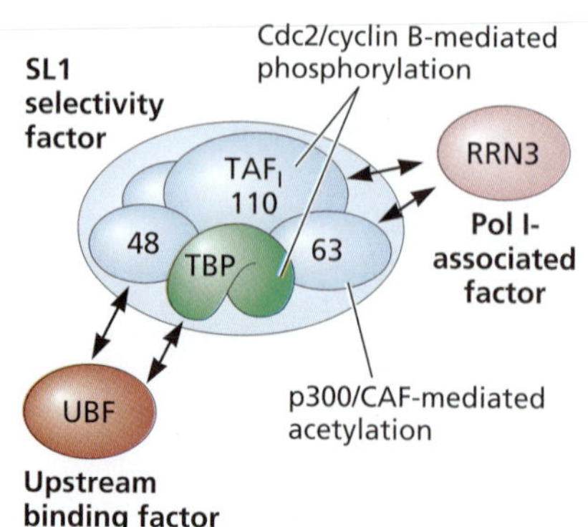

Figure 10.12 General transcription factors of eukaryotic Pol I pre-initiation complex. SL1 selectivity factor is a 300 kDa complex containing the TATA-binding protein, TBP, and at least three Pol I-specific TBP-associated factors or TAFs, of 110, 63, and 48 kDa in humans. It recruits Pol I to the promoter via interaction of its TAF_I63 and TAF_I110 subunits with the Pol I-associated factor RRN3. Some of these TAFs are known to be post-translationally modified: phosphorylated by cyclin-dependent protein kinases, which are regulators of the cell cycle, or acetylated by the abundant acetylase p300/CAF. SL1 interacts with the upstream binding factor, UBF, and with RRN3; these interactions are indicated as double-headed arrows connecting the proteins with the respective subunits of SL1. (Adapted from Russell J & Zomerdijk JCBM [2005] *Trends Biochem Sci* 30:87–96. With permission from Elsevier.)

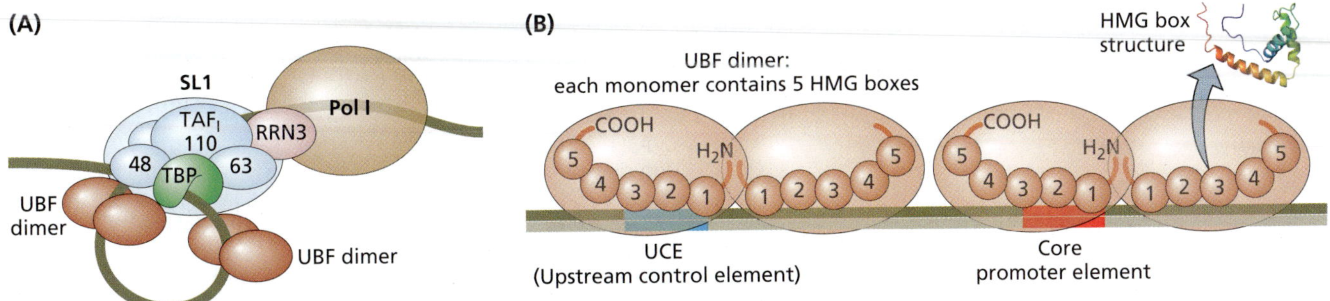

Figure 10.13 Molecular architecture of the mammalian Pol I pre-initiation complex. (A) Schematic reflecting recognized interactions in the Pol I PIC: SL1 selectivity factor interacts with DNA, hRRN3, and UBF; UBF interacts with rDNA and Pol I via RAF53, a homolog of Pol I A49 subunit, which is not shown. (B) Close-up of interactions of the two UBF dimers with the two DNA promoter elements: upstream control element, UCE, and core promoter element. At each of these two DNA sites, UBF binds as a dimer. Each UBF monomer has a domain structure with five HMG boxes. HMG boxes possess a characteristic L-shaped tertiary structure that is present in both sequence-specific transcription factors and nonspecific DNA binders. The box is named after the non-histone chromatin protein HMG, high-mobility group protein, where it was first identified. Binding of HMG boxes to DNA induces DNA bending. (A, adapted from Russell J & Zomerdijk JCBM [2005] *Trends Biochem Sci* 30:87–96. With permission from Elsevier. B, adapted from Stefanovsky VY, Pelletier G, Bazett-Jones DP et al. [2001] *Nucleic Acids Res* 29:3241–3247. With permission from Oxford University Press.)

The Pol I transcription cycle consists of several steps (**Figure 10.15**). An important step, termed re-initiation, allows for quick reassembly of the entire pre-initiation complex by recycling Pol I to the engaged promoter or enhanceosome structure. Recent X-ray diffraction studies of Pol I have clarified the structure and revealed elements involved in the transition from inactive to active polymerase. First, in the inactive form the active-site cleft is held open by an expander element. This must move out of the way before the cleft can close on the substrate. Second, the inactive form can dimerize via a connector element. This must dissociate for activation. Presumably, one or both are under allosteric control.

10.4 Transcription by RNA polymerase III

RNA polymerase III specializes in transcription of small genes

As indicated in Table 10.1, Pol III transcribes the small 5S component of the ribosome and tRNAs (see Chapter 15), one of the small nuclear RNAs that comprise the spliceosome (see Chapter 14), and other small RNAs, often of unknown function. Each of the genes encoding these RNAs possesses specific promoter sequences (**Figure 10.16**). Some of them resemble Pol II promoters and are located upstream of the TSS and contain TATA boxes. Others are totally internal, that is, they are located downstream of the TSS. In addition to the three major types of promoters illustrated in Figure 10.16, there are also mixed promoters that consist of elements from two different pure types.

The functional complexity of all these promoters is poorly understood. What is clear is that different promoter types require the assembly of different pre-initiation complexes (**Figure 10.17**).

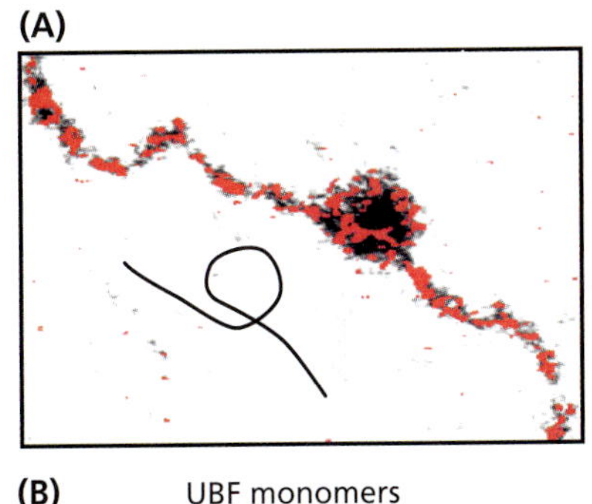

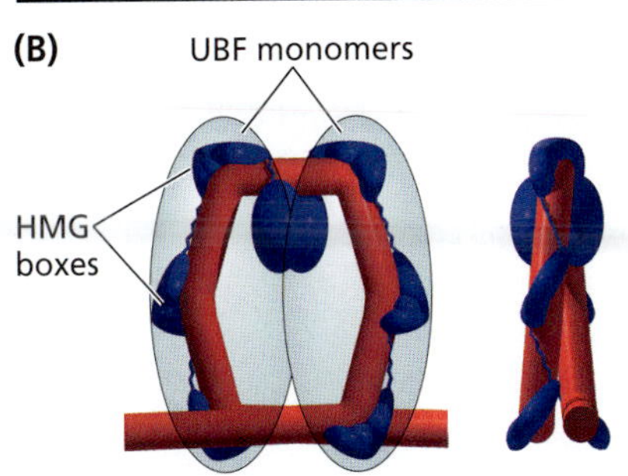

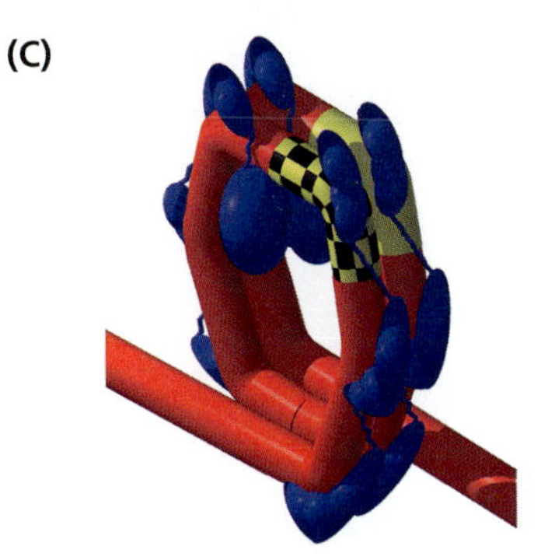

Figure 10.14 Establishment of molecular architecture of the mammalian Pol I pre-initiation complex. (A) Electron spectroscopic image of the enhanceosome, a dimer of UBF complexed with DNA. This imaging technique allows visualization of the location of phosphorus atoms from nucleic acids, shown in red, in the structure and superimposes that on the corresponding total mass image, shown in gray. (B) Low-resolution modeling of the enhanceosome. UBF monomers induce hemi-enhanceosomes, bending the DNA by 175°; two hemi-enhanceosomes are precisely phased by dimerization of UBF; thus, DNA looping in the enhanceosome is the result of noncooperative in-phase bending by two UBF molecules. Each UBF dimer organizes ~140 base pairs of DNA. (C) Two adjacent enhanceosomes bring together the UCE and core elements, shown in yellow; this structure helps to recruit TBP to initiate PIC formation. (A–B, from Stefanovsky VY, Pelletier G, Bazett-Jones DP et al. [2001] *Nucleic Acids Res* 29:3241–3247. With permission from Oxford University Press. C, courtesy of Victor Stefanovsky, Université Laval.)

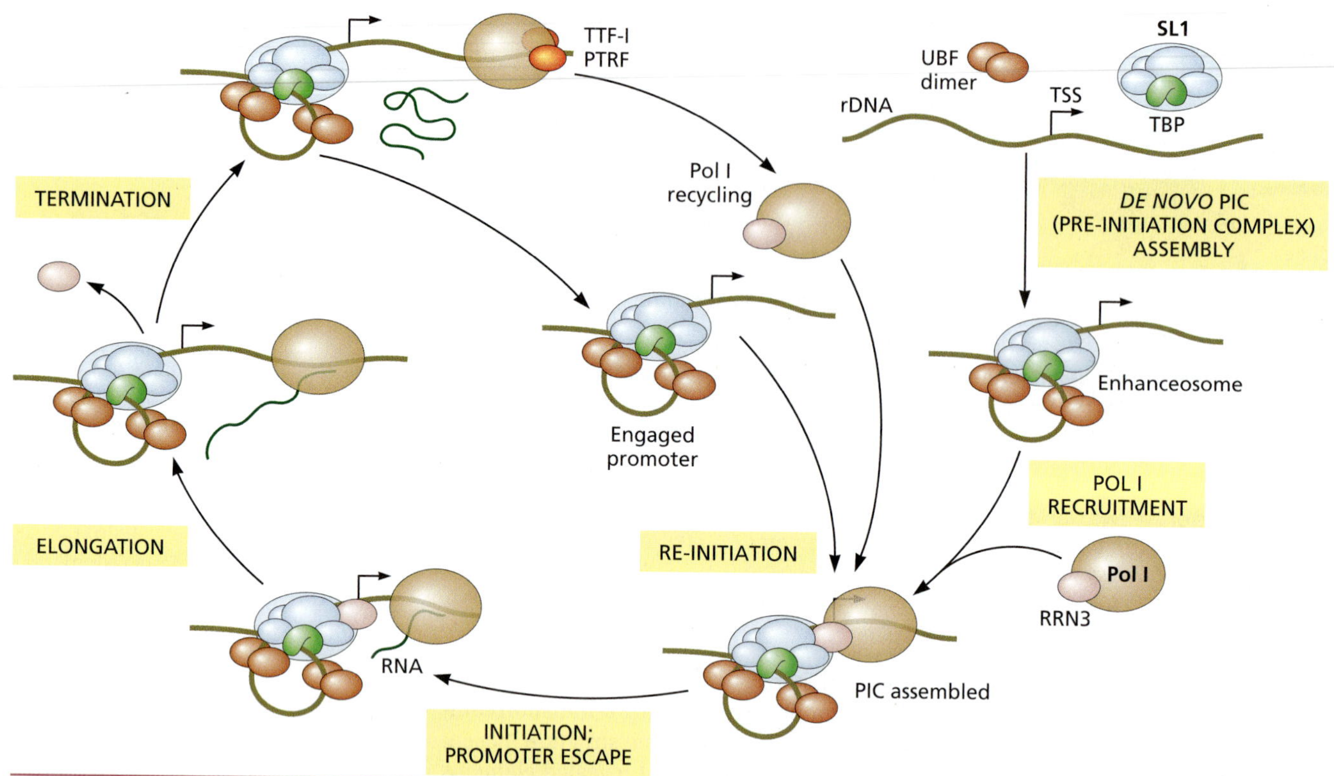

Figure 10.15 Pol I transcription cycle. The cycle consists of the following steps: pre-initiation complex assembly, Pol I recruitment, initiation, promoter escape, elongation, termination, and re-initiation. The first nucleotide addition steps are slow and the process goes through several abortive synthesis steps, during which short nascent RNA chains are released from the complex. Once the transcript reaches 10–12 nucleotides in length, the polymerase switches to a processive mode and can elongate, without dissociating from the template, for thousands of nucleotide addition cycles. Termination occurs when a specific termination factor, TTF-I, binds to the termination sequences at the 3′-end of the gene. TTF-I bends the termination site, forcing Pol I to pause. Another factor known as PTRF, polymerase I and transcript release factor, comes into play; PTRF acts in conjunction with a T-rich DNA sequence to release the polymerase and the RNA transcript. SL1 and UBF remain bound to the promoter the whole time, providing a scaffold upon which transcription can be quickly re-initiated. (Adapted from Russell J & Zomerdijk JCBM [2005] *Trends Biochem Sci* 30:87–96. With permission from Elsevier.)

10.5 Transcription in eukaryotes: pervasive and spatially organized

Most of the eukaryotic genome is transcribed

For decades, the eukaryotic genome has been viewed as a linear arrangement of genes that code either for proteins or for structural RNA molecules, such as rRNA and tRNA. Even the latter were involved in making proteins; the genome was therefore envisioned as dedicated to directing protein manufacture. This seemed reasonable so long as bacteria were the objects of study, but exploration of the eukaryotic genome raised disturbing questions. There is far more DNA in a human cell than is needed to account for any reasonable estimate of the number of proteins. Thus the concept of junk DNA, which coded for nothing, was put forward. Such an idea made many biologists unhappy. Why would the cell expend the energy to replicate worthless DNA? Why was it not selected against in evolution?

With the advent of high-throughput technologies for studying transcription on a genomewide scale (**Box 10.2**), this view has been drastically changed. Numerous labs all over the world participated in the Encyclopedia of DNA Elements or ENCODE project (see Chapter 13). The aim of the project was to provide a more biologically informative representation of the human genome by identifying and cataloging the functional elements it encodes. In the pilot phase of the project, which ended in 2007, 35 groups provided more than 200 experimental and computational data sets that examined in unprecedented detail a targeted 30 megabases, or 1%, of the human genome. The second phase of the project analyzed the entire human genome, with the

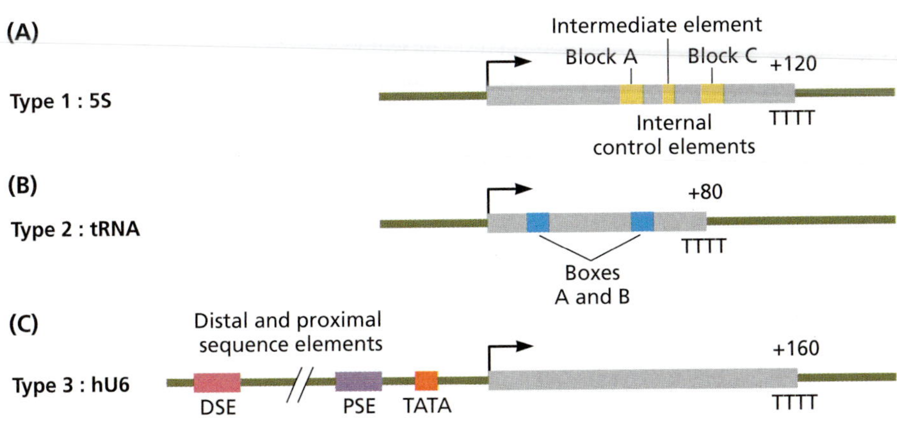

Figure 10.16 Three types of Pol III promoters. (A) Type 1 promoters are found in the 5S genes and consist of an internal control region, ICR, that is subdivided into A block, intermediate element, and C block. (B) Type 2 promoters are found in tRNA genes, adenovirus 2 VAI gene, and some other genes and consist of two gene-internal elements called the A and B boxes. (C) Type 3 promoters consist of a distal sequence element, DSE, that serves as an enhancer (see Chapter 12); a proximal sequence element, PSE; and a TATA box. Mixed promoters have also been described; they consist of elements from two different pure types. For example, the *S. cerevisiae* U6 gene promoter contains both a TATA box and A and B boxes; in *Schizosaccharomyces pombe*, nearly all tRNA and 5S genes contain a TATA box in addition to gene-internal elements, and the TATA box is required for transcription. (Adapted from Schramm L & Hernandez N [2002] *Genes Dev* 16:2593–2620. With permission from Cold Spring Harbor Laboratory Press.)

major findings released in 2012 (see Box 10.2). Importantly, these findings overturned accepted ideas and forced attempts at a new structural-functional definition of the gene (see Chapter 13).

A first surprise is that only a small percentage of the human genome codes for protein. Coding for proteins is clearly not its main function. The more evolved organisms use the smallest fraction of their genomes for this purpose.

Arguably the most important finding of the ENCODE project is that transcription in eukaryotes is pervasive: close to 75% of the sequences in the human genome give

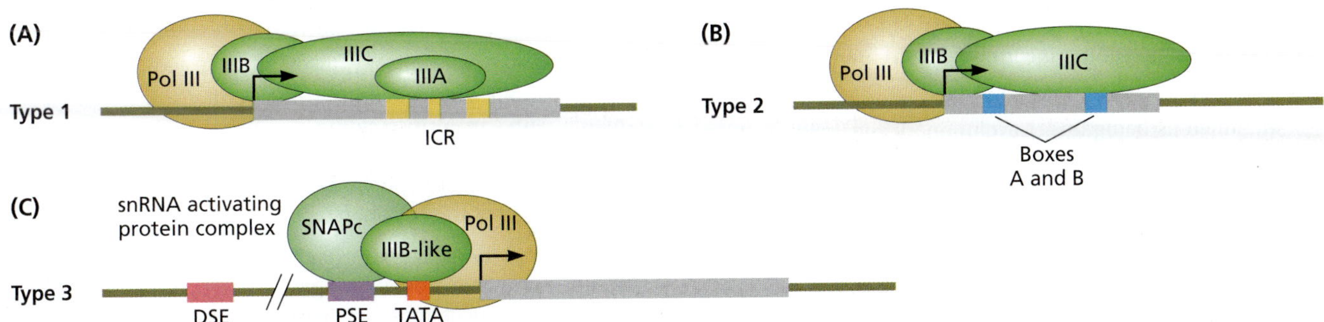

Figure 10.17 Transcription initiation complexes on Pol III promoters. (A) Type 1 promoters recruit TFIIIA, the founding member of the C_2H_2 zinc-finger family of DNA-binding proteins (see Chapter 6). The binding of TFIIIA then allows the binding of TFIIIC, which has five subunits in humans and *S. pombe*, six subunits in *S. cerevisiae*. Once the DNA/TFIIIA/TFIIIC complex is formed, another complex Brf1–TFIIIB joins, which in turn allows the recruitment of RNA Pol III. (B) Type 2 promoters can recruit TFIIIC without the help of TFIIIA because TFIIIC binds directly to the A and B boxes. This then allows the binding of Brf1–TFIIIB and RNA Pol III. In yeast, once Brf1–TFIIIB has been recruited to type 1 or 2 promoters, TFIIIA and/or TFIIIC can be stripped from the DNA with high salt or heparin treatment. Brf1–TFIIIB remains bound to the DNA and is sufficient to direct multiple rounds of transcription. (C) Metazoan-specific type 3 promoters recruit the small nuclear RNA activating protein or SNAP complex that binds to the PSE. Type 3 promoters also recruit Brf2–TFIIIB through a combination of protein–protein contacts with SNAPc and a direct association of the TBP component of Brf2–TFIIIB with the TATA box. Then RNA Pol III joins the complex. DSE functions as an enhancer: it contains specific sequences that bind transcription factors (see Chapter 12). (Adapted from Schramm L & Hernandez N [2002] *Genes Dev* 16:2593–2620. With permission from Cold Spring Harbor Laboratory Press.)

Box 10.2 What we learned about the functioning of the human genome from the ENCODE project The pilot phase of the ENCODE project examined a small selected portion, 1%, of the human genome; the second phase expanded this work to the entire genome. The major findings of the pilot project are given here as a direct quotation from the ENCODE Project Consortium in 2007. Some of the findings are directly relevant to transcription and its regulation, others to DNA replication, and still others look at evolutionary conservation and its relationship to function. We include all these findings here, since their importance for our overall understanding of the entire genome and its function is difficult to overstate. The findings of even the pilot phase of the ENCODE project are so revolutionary that they drastically change our views on the genes and genomes.

- The human genome is pervasively transcribed, such that the majority of its bases are associated with at least one primary transcript and many transcripts link distal regions to established protein-coding loci.
- Many novel non-protein-coding transcripts have been identified, with many of these overlapping protein-coding loci and others located in regions of the genome previously thought to be transcriptionally silent.
- Numerous previously unrecognized transcription start sites have been identified, many of which show chromatin structure and sequence-specific protein-binding properties similar to well-understood promoters.
- Regulatory sequences that surround transcription start sites are symmetrically distributed, with no bias towards upstream regions.
- Chromatin accessibility and histone modification patterns are highly predictive of both the presence and activity of transcription start sites.
- Distal DNase I hypersensitive sites have characteristic histone modification patterns that reliably distinguish them from promoters; some of these distal sites show marks consistent with insulator function.
- DNA replication timing is correlated with chromatin structure.
- A total of 5% of the bases in the genome can be confidently identified as being under evolutionary constraint in mammals; for approximately 60% of these constrained bases, there is evidence of function on the basis of the results of the experimental assays performed to date.
- Although there is general overlap between genomic regions identified as functional by experimental assays and those under evolutionary constraint, not all bases within these experimentally defined regions show evidence of constraint.
- Different functional elements vary greatly in their sequence variability across the human population and in their likelihood of residing within a structurally variable region of the genome.
- Surprisingly, many functional elements are seemingly unconstrained across mammalian evolution. This suggests the possibility of a large pool of neutral elements that are biochemically active but provide no specific benefit to the organism. This pool may serve as a 'warehouse' for natural selection, potentially acting as the source of lineage-specific elements and functionally conserved but nonorthologous elements between species.

The 2012 report emphasizes further aspects of the organization and functioning of the human genome. What follows is a direct quotation from that report. Some of these important facts will be useful when we consider regulation of transcription in Chapter 13.

- The vast majority (80.4%) of the human genome participates in at least one biochemical RNA- and/or chromatin-associated event in at least one cell type. Much of the genome lies close to a regulatory event: 95% of the genome lies within 8 kilobases (kb) of a DNA–protein interaction... and 99% is within 1.7 kb of at least one of the biochemical events measured by ENCODE.
- Classifying the genome into seven chromatin states indicates an initial set of 399,124 regions with enhancer-like features and 70,292 regions with promoter-like features, as well as hundreds of thousands of quiescent regions. High-resolution analyses further subdivide the genome into thousands of narrow states with distinct functional properties.
- It is possible to correlate quantitatively RNA sequence production and processing with both chromatin marks and transcription factor binding at promoters, indicating that promoter functionality can explain most of the variation in RNA expression.
- Many non-coding variants in individual genome sequences lie in ENCODE-annotated functional regions; this number is at least as large as those that lie in protein-coding genes.
- Single nucleotide polymorphisms (SNPs) associated with disease ... are enriched within non-coding functional elements, with a majority residing in or near ENCODE-defined regions that are outside of protein-coding genes. In many cases, the disease phenotypes can be associated with a specific cell type or transcription factor.

Annotation is the process through which scientists identify protein-coding and noncoding genes based on some common characteristic features of these elements. For example, a protein-coding gene should contain a recognizable promoter, a start codon, etc. Annotation can be performed either in an automated way or manually; the ENCODE project used both to create the GENCODE reference gene set.

rise to some kind of primary transcript. Note that this number combines all recognized transcripts in all kinds of human cells studied. No individual cell line shows transcription of more than ~57% of the total sequence transcribed across all cell lines. **Figure 10.18** presents the current view that transcription in eukaryotes is more than just a conversion of the information encoded in DNA into mRNA molecules. **Figure 10.19**

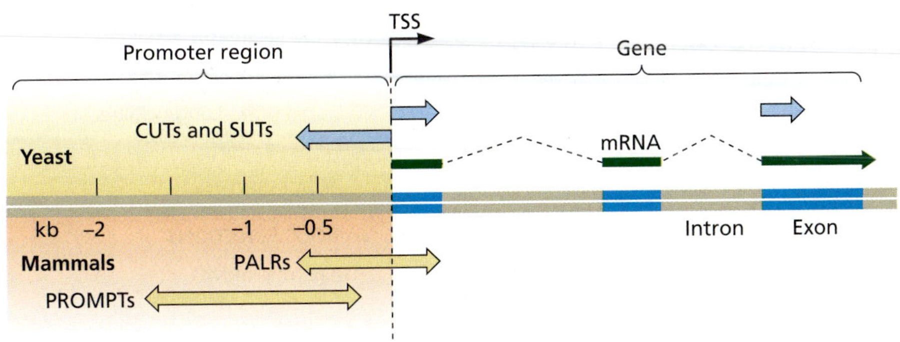

Figure 10.18 Transcription in eukaryotes is pervasive. In yeast, shown in the upper half of the schematic, genomic sequences around many transcription start sites, TSSs, are transcribed bidirectionally in long noncoding RNAs called cryptic unstable transcripts, CUTs, and stable unannotated transcripts, SUTs. Transcription of these noncoding RNAs is promoter-dependent, just like transcription of mRNA, and the transcripts are 5′-capped, as is mRNA (see Chapter 14). CUTs are easily degraded by RNA surveillance pathways and can be detected only in mutant cells where these pathways are disabled. They are 200–600 nucleotides or nt long and are usually heterogeneous at their 3′-ends. SUTs are more stable and can be detected in normal cells; average length is ~760 nt. Both CUTs and SUTs are proposed to be broad regulators of transcription in yeast. In mammals, shown in the lower half of the schematic, the complexity of transcription is even higher. Long noncoding transcripts such as promoter-associated long RNAs, PALRs, and promoter-upstream transcripts, PROMPTs, partially overlap the same DNA sequences; PROMPTs are less stable than PALRs. PROMPTs may be transcribed from unrecognized promoters. Mammals also contain short noncoding RNAs, <200 nt, as depicted in Figure 10.19. (Adapted from Carninci P [2009] *Nature* 457:974–975. With permission from Macmillan Publishers, Ltd.)

gives a much more detailed view of transcription occurring at and around a hypothetical mammalian gene. The complexity is enormous and presents a huge challenge in understanding how many of these transcripts represent just transcriptional noise and how many have actual function. If some of these mysterious transcripts have functions, what are they? In any event, the fact that so much DNA is transcribed makes the concept of junk DNA obsolete.

In general, the newly discovered transcripts are categorized into two major classes: **long noncoding RNAs**, known as **lncRNAs**, and **short noncoding RNAs**. Some of their functions are beginning to emerge, especially those of some of the short noncoding RNAs. We discuss these elsewhere, but in **Figure 10.20** we illustrate some notions about the possible functions of lncRNAs in order to provide an idea of the complexity of eukaryotic transcription.

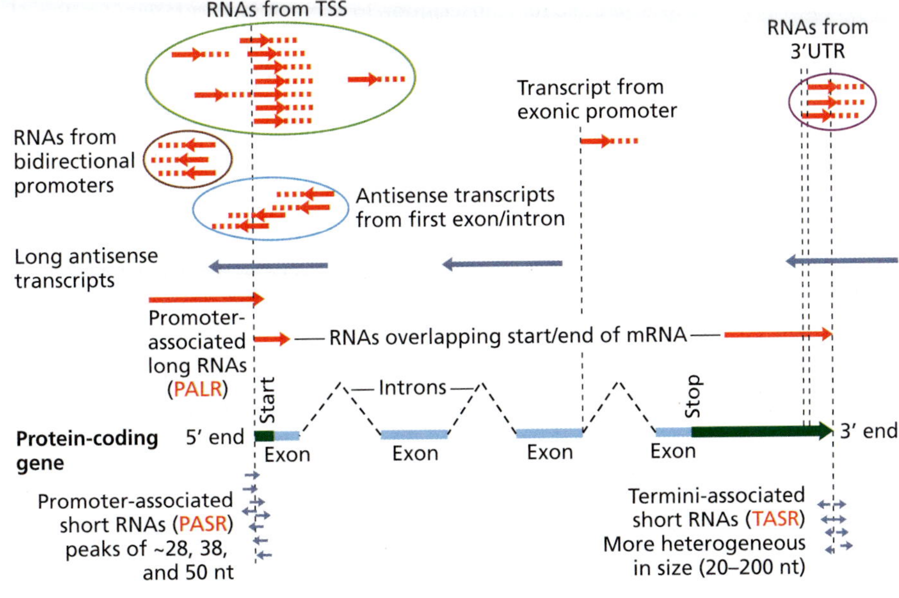

Figure 10.19 Hypothetical mammalian gene with a map of noncoding RNAs surrounding the mRNA portion. The map is not comprehensive. Small red arrows depict short stretches encompassing the first 20 nt of RNA. The method used for their identification does not identify the end of the transcript; hence they are marked with dotted lines. (Adapted from Carninci P, Yasuda J & Hayashizaki Y [2008] *Curr Opin Cell Biol* 20:274–280. With permission from Elsevier.)

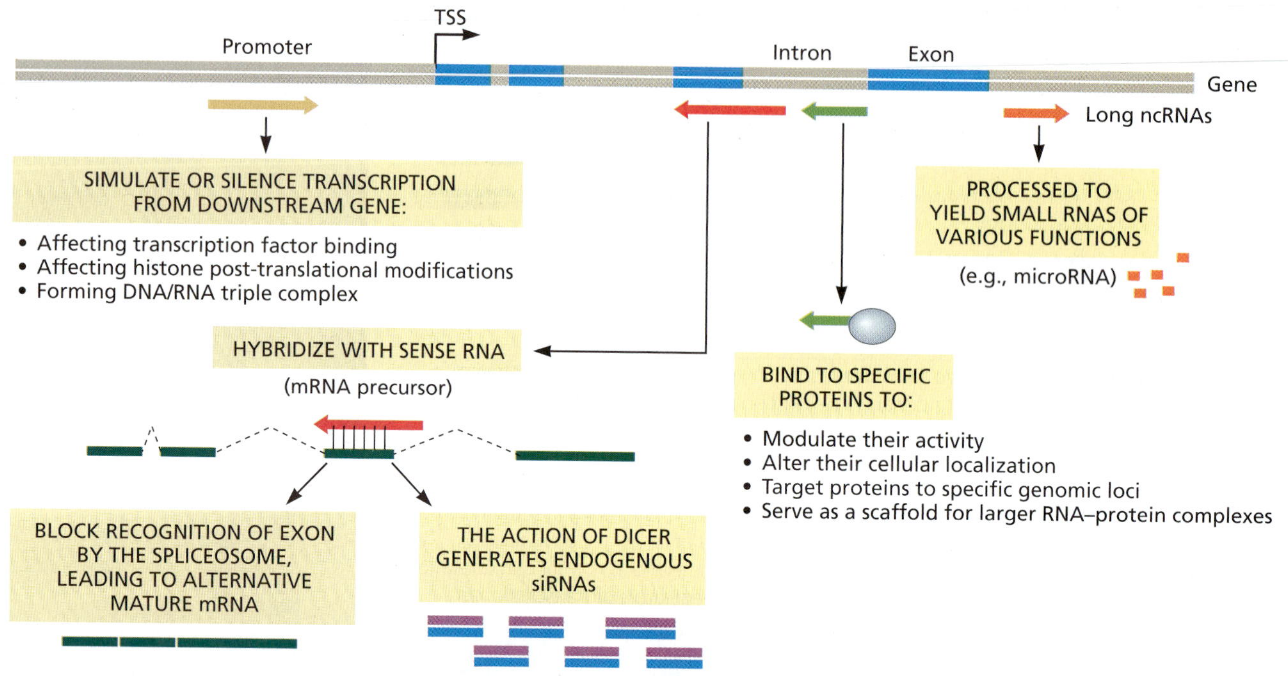

Figure 10.20 Possible functions of long noncoding RNAs. An interesting, recently recognized possibility is the post-transcriptional processing of mature long transcripts, both protein-coding mRNA and lncRNAs. These small RNAs may be further modified by addition of a cap structure (see Chapter 14). (Adapted from Wilusz JE, Sunwoo H & Spector DL [2009] *Genes Dev* 23:1494–1504. With permission from Cold Spring Harbor Laboratory Press.)

Transcription in eukaryotes is not uniform within the nucleus

Chromatin in the interphase nucleus is organized in chromosome territories. Following mitosis, chromosomes decompact to each occupy a distinct portion of the nuclear volume (see Figure 8.21 and Figure 8.22). In addition, there is considerable preference for gene-rich chromosomes to occupy the center of the nucleus, whereas chromosomes that are relatively poor in genes tend to be at the nuclear periphery. What governs the overall nuclear architecture is unclear.

Recent sophisticated imaging techniques have uncovered another unexpected feature of genome organization in terms of its activity. Transcription does not occur evenly throughout the nuclear volume; much of it occurs in specific sites termed transcription factories (**Figure 10.21**). A typical eukaryotic cell contains ~10,000 such factories, which are believed to harbor the full complement of molecules, such as polymerases and transcription factors, needed for transcription to occur. A factory is usually shared by a number of genes that may be located on different chromosomes. For such sharing to occur, genome regions that carry transcribed genes usually, but not always, move

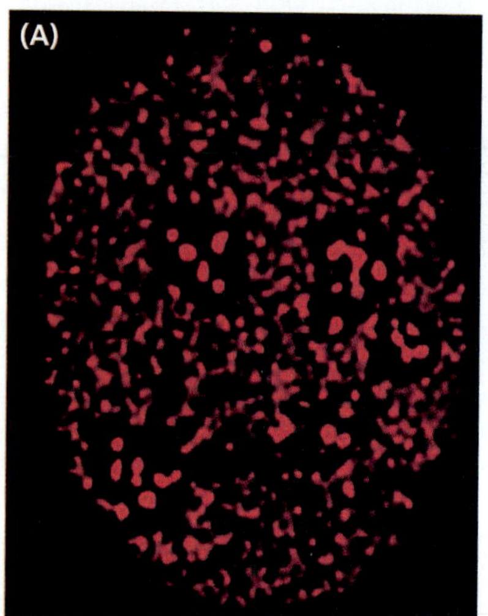

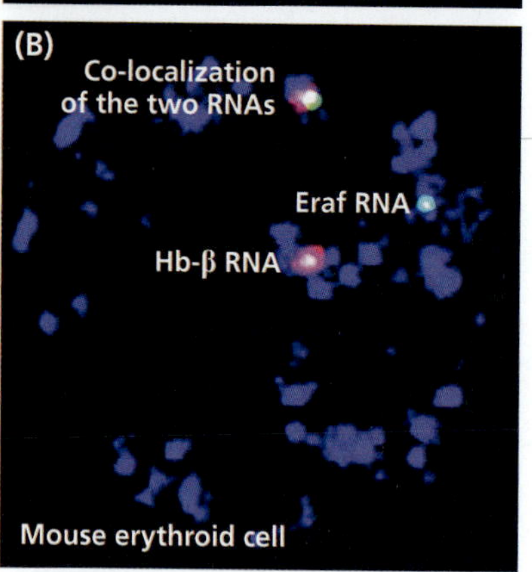

Figure 10.21 Spatial organization of transcription in the eukaryotic nucleus: transcription factories. Transcription in the nucleus is compartmentalized: it does not occur evenly throughout the nuclear volume but is mostly limited to specific, spatially confined sites termed transcription factories. The number of these factories per nucleus is estimated to be close to 10,000. These are subnuclear transcription centers that harbor transcription factors and polymerase molecules that can serve several genes. (A) In this image, the transcription sites are visualized by incorporation of bromo-dUTP, a derivative of uridine triphosphate, which can be detected by fluorescently labeled antibodies. (B) RNA immunofluorescence *in situ* hybridization, known as immuno-FISH, with a fluorescently labeled RNA probe. Hemoglobin β-chain RNA is shown in red; erythroid associated factor or Eraf RNA is shown in green. The two genes are located on the distal third of mouse chromosome 7, separated by 25 Mb; they co-localize in a factory in one of the chromosomes. The blue color indicates the localization of Pol II by immunofluorescence detection. (A, from Elbi C, Misteli T & Hager GL [2002] *Mol Biol Cell* 13:2001–2015. With permission from American Society for Cell Biology. B, from Chakalova L, Debrand E, Mitchell JA et al. [2005] *Nat Rev Genet* 6:669–677. With permission from Macmillan Publishers, Ltd.)

out of their respective chromosome territories to join a preexisting transcriptional factory. It is not clear what proteins or other macromolecules contribute to the formation of a factory. Suggested proteins include the polymerase molecules themselves as well as other protein factors, such as **CCCTC-binding factor** or **CTCF**. CTCF is a ubiquitous zinc-finger protein with numerous functions; considerable evidence puts it into the enviable position of a master organizer of the genome. It can form chromatin loops in *cis*, that is, between DNA sites on the same chromosome, or it can bridge sequences located on different chromosomes in *trans*. Consistent with a role as a genome organizer, thousands of CTCF binding sites have been identified in genomewide localization studies, with considerable enrichment around transcription start sites. The 2012 ENCODE data report 55,000 CTCF sites in each of the 19 tested cell lines, including normal primary cells and immortal lines. Contrary to previous ideas, the actual CTCF occupancy of these sites varies somewhat among cell lines. This variability may reflect the presumptive involvement of CTCF in the dynamics of the transcription factories.

Active and inactive genes are spatially separated in the nucleus

Another intriguing feature of eukaryotic transcription has been recently recognized. Using the **4C technology** depicted in **Figure 10.22A**, researchers have realized that active and inactive genes are spatially separated in the nucleus. The example shown in **Figure 10.22B** illustrates DNA contacts of the β-globin gene in fetal liver, where the gene is actively transcribed, and in fetal brain, where the gene is not expressed. It is immediately obvious that the β-globin gene makes contacts with very different regions in the chromosome depending on its expression status. Moreover, the active

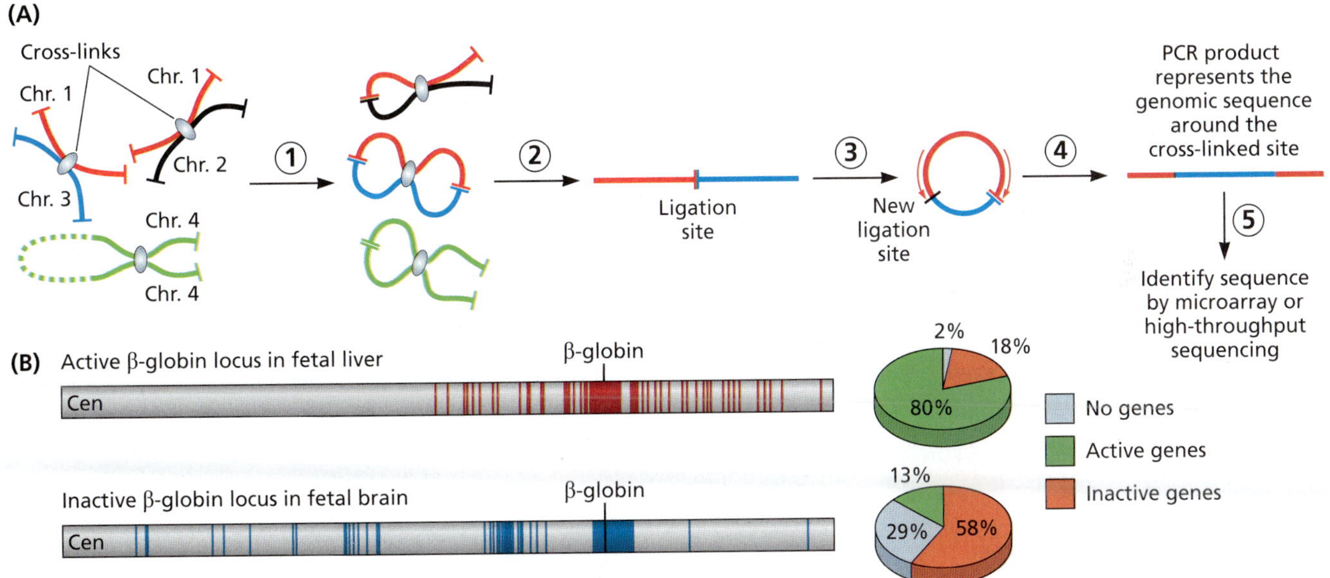

Figure 10.22 Active and inactive genes separate spatially in the nucleus. (A) Schematic of 4C technology used to identify DNA regions on the same or different chromosomes that interact *in vivo*. The 3C or chromosome conformation capture protocol consists of the following four steps: (Step 1) Cross-link chromatin *in vivo*; digest with restrictase, a six-cutter; and ligate cross-linked fragments. (Step 2) Reverse cross-links and digest with four-cutter to trim ligation junctions. (Steps 3 and 4) Ligate to create DNA circles and amplify circles by polymerase chain reaction, PCR, using oligonucleotide primers derived from the gene region of interest, shown in red. Steps 1–4 are followed by microarray or sequencing analysis of interacting regions (Step 5), where the identified sequences are mapped back to the genome. This step adds a fourth dimension, the fourth C, to the assay, allowing unbiased analysis of any sequence that contacts the gene of interest. (B) Long-range intrachromosomal interactions identified for the β-globin locus in its active state in mouse fetal liver and its inactive state in fetal brain. Chromosome 7, which hosts the β-globin locus, is presented as a gray horizontal bar. Vertical lines indicate the regions that interact with the β-globin locus: red for the active locus and blue for the inactive locus. The pie charts next to the chromosome characterize the distribution of interacting regions in terms of their gene content or gene activity. Note the difference in interactions between the active and inactive locus: the active locus preferentially interacts with other actively transcribed regions, whereas the inactive locus predominantly contacts silent regions. The constitutively active housekeeping gene *Rad23a* contacts essentially the same regions in the two tissues, the majority of which are transcriptionally active. (Adapted from de Laat W & Grosveld F [2007] *Curr Opin Genet Dev* 17:456–464. With permission from Elsevier.)

gene preferentially contacts other active genes, whereas the contacts of the inactive gene are mainly with other inactive genes. Presumably this organization reflects the organization of eukaryotic transcription in transcription factories.

10.6 Methods for studying eukaryotic transcription

A battery of methods is available for the study of transcription

Over the years, numerous methods have been introduced for the study of transcription. The majority of methods rely on radioactive or fluorescent labeling of transcripts and analysis on electrophoretic gels. More specific methods are used to map transcription start sites or TSS (**Box 10.3**) and to elucidate the structures near the 5′- and 3′-ends of genes (**Box 10.4**).

Box 10.3 Methods used to map transcription start sites
Mapping the exact position in a gene where transcription is initiated, the transcription start site or TSS, is important: it helps in the identification of promoter sequences and transcribed regions in the genome. Thus far, only a small portion of each sequenced genome has been annotated in terms of the genes it contains. Many of these genes code for proteins, but it is now clear that a significant portion of the genome codes for RNA as the final product.

There are two classical methods for mapping TSSs in known genes: **S1 nuclease protection** and primer extension. These methods are presented in schematic form in **Figure 1**.

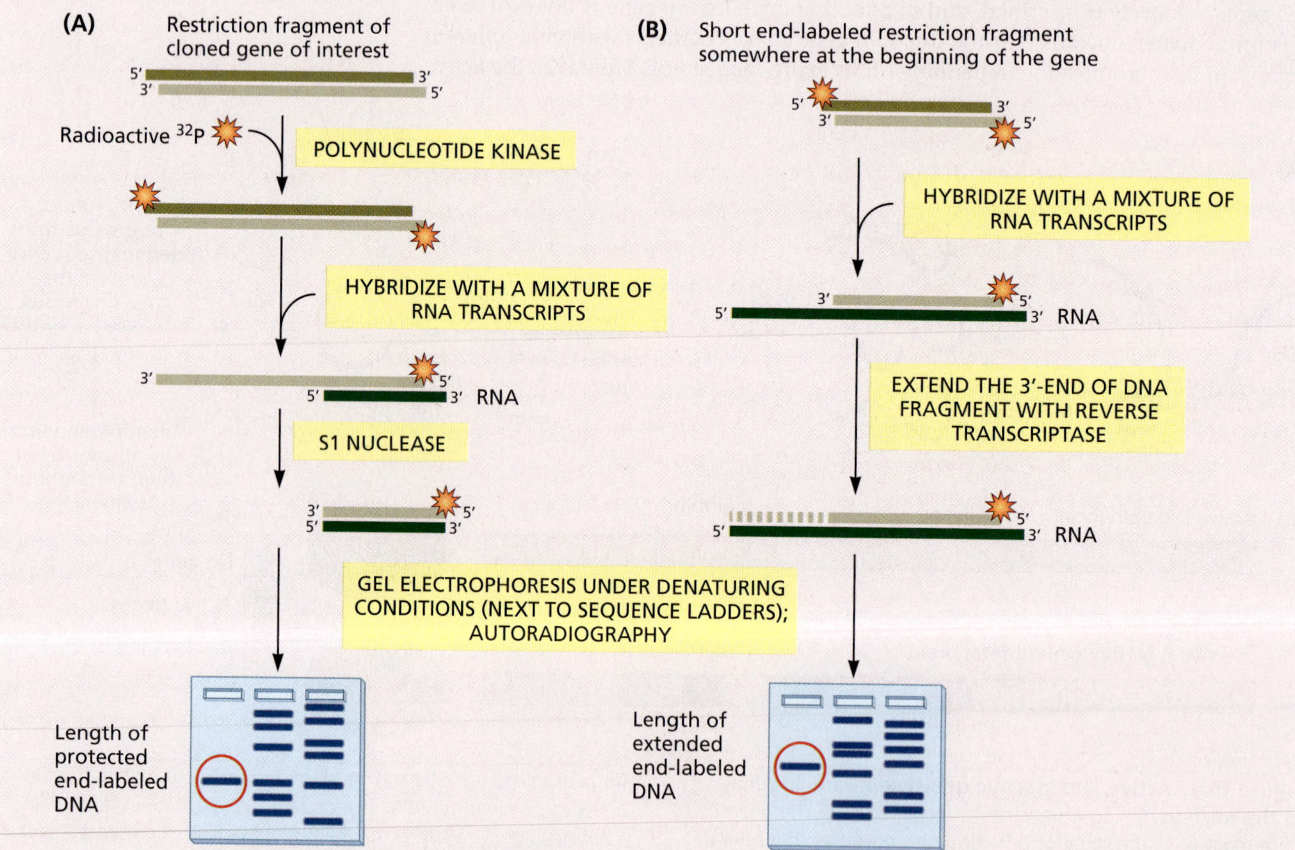

Figure 1 Mapping the position of transcription start sites. (A) S1 Nuclease protection assay. A restriction fragment from the 5′-end of a cloned gene is end-labeled at the 5′-ends, denatured, and hybridized to total RNA from cells expressing the gene. The DNA–RNA heteroduplexes are treated with the fungal enzyme S1 nuclease, which specifically cleaves single-stranded DNA or RNA, until it reaches the boundaries of the duplex; note that other enzymes with similar action can also be used. High-resolution electrophoresis is run to identify the size of the protected DNA fragment precisely; usually, the gel also contains sequencing lanes to determine the exact base at which transcription is initiated. (B) Primer extension assay. The restriction fragment expected to contain the TSS is deliberately chosen to be small. Hybridization with a cognate RNA will create an overhanging 5′-end, which can be used as a template for reverse transcription. The 3′-OH group of the DNA strand will serve as a primer in the reaction. The DNA will be extended until the 5′-end of the RNA is reached. The size of the extended DNA primer can be precisely determined as it was in the last step in the S1 protection assay.

Recently, it became possible to analyze the location of TSSs across entire genomes, with the human genome presenting the most complete and revealing data. In addition, high-throughput sequencing methods allow analysis of the entire **transcriptome**, which comprises the products of transcription of all genes active under specified conditions. **Transcriptome analysis** (Box 10.5) is the measurement of transcriptional activity of thousands of genes at once to create a global picture of cellular function. Recently, this analysis has been extended to cover entire sequenced genomes, including the human genome. In addition, transcriptome analysis can be used to provide a complete map of transcriptional activity of the genome in a certain bacterial species under a certain set of conditions. This analysis can be used to monitor the transcriptional response of the cell to changing environmental conditions or to drug treatment.

A simplified outline of the procedure for transcriptome analysis can be found in Box 10.5. The procedure is based on the use of **reverse transcriptase**, a specialized RNA-dependent DNA polymerase that is involved in the life cycle of viruses whose

Box 10.4 Methods used to identify 5′- and 3′-regions of eukaryotic genes In addition to mapping TSSs, identifying 5′- and 3′-regions of genes can significantly contribute to our overall understanding of functional elements of genomes. The need to identify the ends of genes led to the development of a variety of rather sophisticated methods. **RACE, rapid amplification of cDNA ends**, is diagrammed in Figure 1. **CAGE, cap analysis of gene expression**, is outlined in Figure 2. **PET, pair-end di-tag analysis**, is detailed in Figure 3.

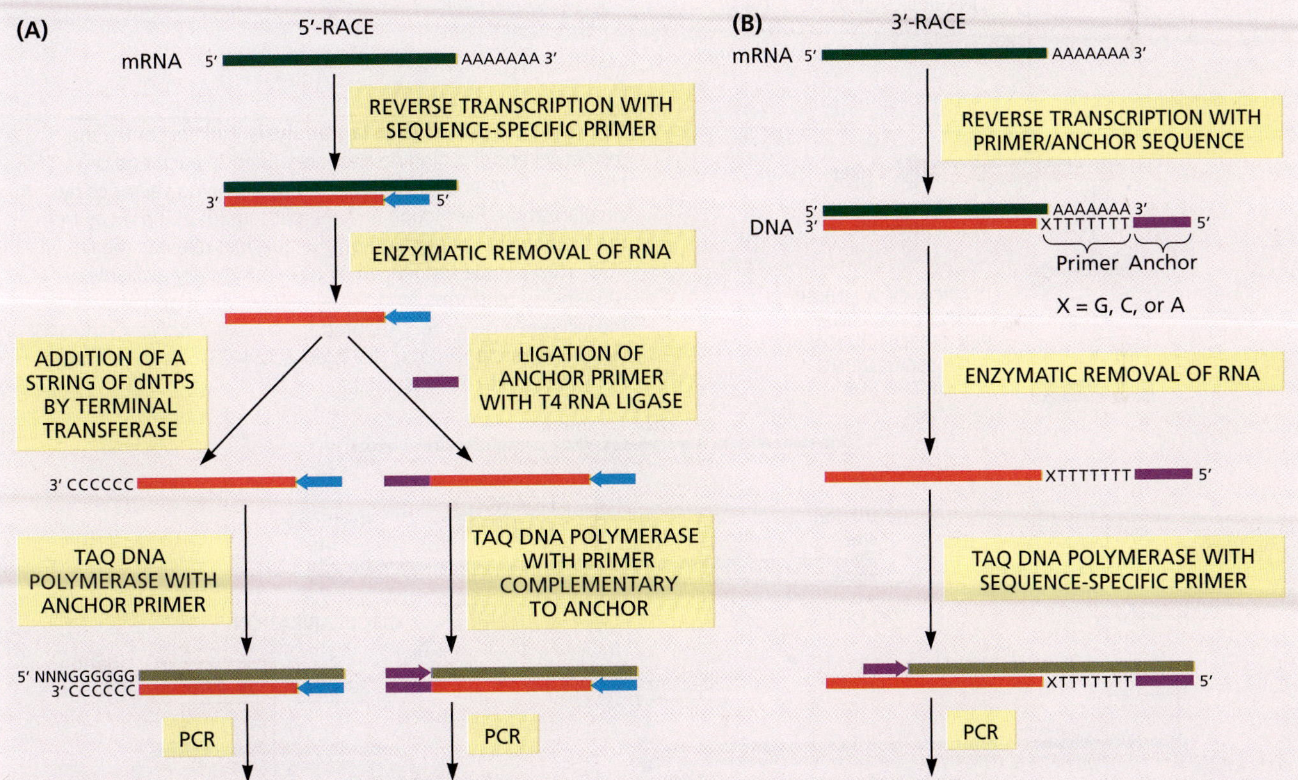

Figure 1 RACE: rapid amplification of cDNA ends. This method is a variation of reverse transcription polymerase chain reaction, RT-PCR, which amplifies unknown cDNA sequences corresponding to the 5′- or 3′-ends of the RNA. (A) In 5′-RACE, the first-strand cDNA synthesis reaction is primed by use of an oligonucleotide complementary to a known sequence in the gene. After removal of the RNA template, terminal deoxynucleotidyl transferase adds a nucleotide tail to create an anchor site at the 3′-end of the single-stranded cDNA. An anchor primer complementary to the newly added tail is used to synthesize the second cDNA strand. An alternative strategy uses an RNA ligase to add the necessary anchor. (B) In 3′-RACE, a modified oligo(dT) sequence serves as the reverse transcription primer. This oligo(dT) sequence is composed of a primer oligo(dT) sequence, which anneals to the poly(A)$^+$ tail of the mRNA, and an adaptor sequence at the 5′-end. A single G, C, or A residue at the 3′-end of the primer ensures that cDNA synthesis is initiated only when the primer/adaptor anneals immediately adjacent to the junction between the poly(A)$^+$ tail and 3′-end of the mRNA.

(Continued)

Box 10.4 *(Continued)*

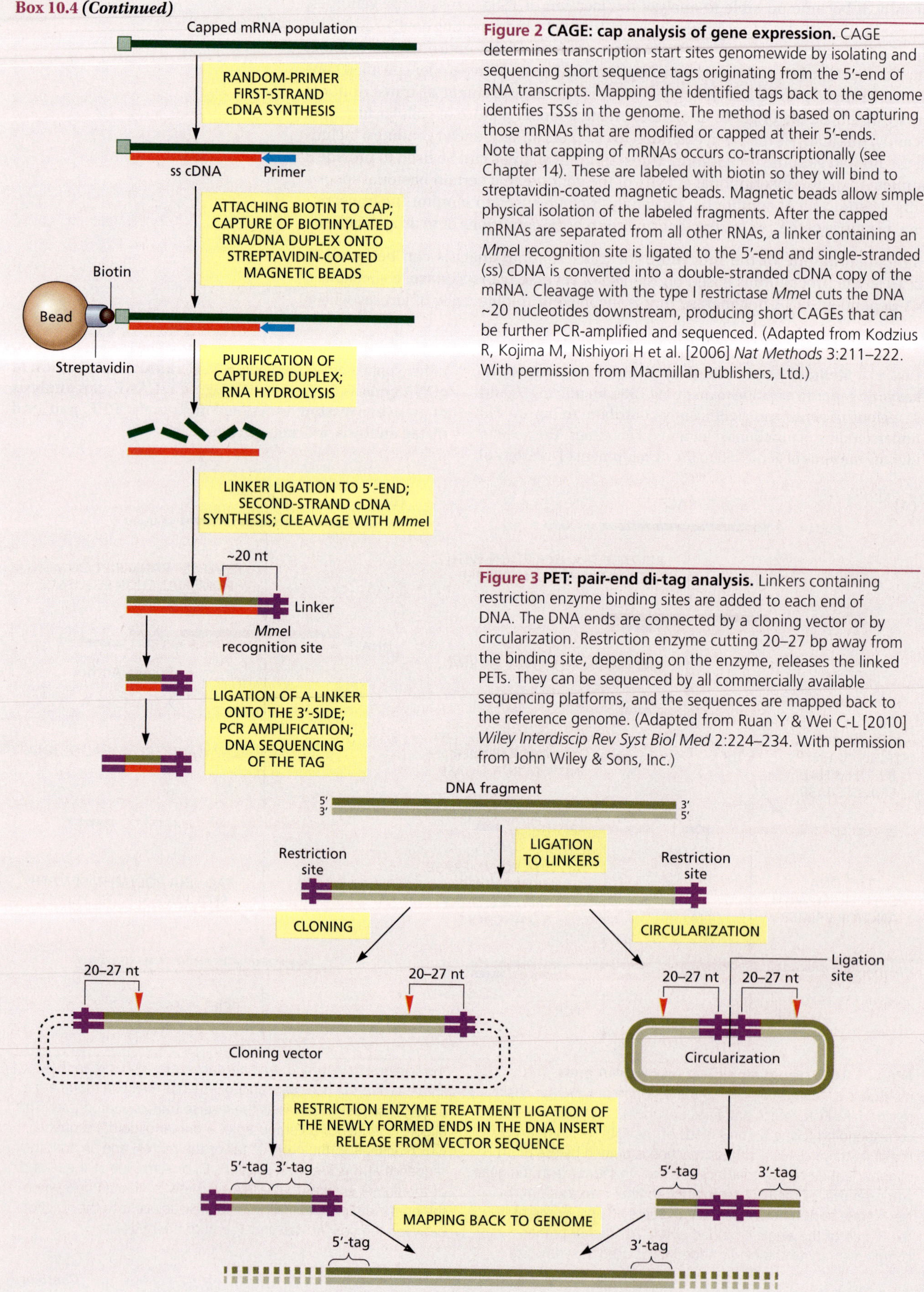

Figure 2 CAGE: cap analysis of gene expression. CAGE determines transcription start sites genomewide by isolating and sequencing short sequence tags originating from the 5′-end of RNA transcripts. Mapping the identified tags back to the genome identifies TSSs in the genome. The method is based on capturing those mRNAs that are modified or capped at their 5′-ends. Note that capping of mRNAs occurs co-transcriptionally (see Chapter 14). These are labeled with biotin so they will bind to streptavidin-coated magnetic beads. Magnetic beads allow simple physical isolation of the labeled fragments. After the capped mRNAs are separated from all other RNAs, a linker containing an *Mme*I recognition site is ligated to the 5′-end and single-stranded (ss) cDNA is converted into a double-stranded cDNA copy of the mRNA. Cleavage with the type I restrictase *Mme*I cuts the DNA ~20 nucleotides downstream, producing short CAGEs that can be further PCR-amplified and sequenced. (Adapted from Kodzius R, Kojima M, Nishiyori H et al. [2006] *Nat Methods* 3:211–222. With permission from Macmillan Publishers, Ltd.)

Figure 3 PET: pair-end di-tag analysis. Linkers containing restriction enzyme binding sites are added to each end of DNA. The DNA ends are connected by a cloning vector or by circularization. Restriction enzyme cutting 20–27 bp away from the binding site, depending on the enzyme, releases the linked PETs. They can be sequenced by all commercially available sequencing platforms, and the sequences are mapped back to the reference genome. (Adapted from Ruan Y & Wei C-L [2010] *Wiley Interdiscip Rev Syst Biol Med* 2:224–234. With permission from John Wiley & Sons, Inc.)

Box 10.5 Genomewide transcriptome analysis: mapping RNA transcripts on a genomewide scale Transcriptome analysis is the measurement of transcriptional activity of thousands of genes at once, to create a global picture of cellular function. Recently, this analysis has been extended to cover entire sequenced genomes, including the human genome (see Chapter 13). In addition to providing a complete map of transcriptional activity of the genome in a certain cell type under a certain set of conditions, such analyses can be used to compare different cell types or to monitor the transcriptional response of a cell to changing environmental conditions. The methodology has the potential to revolutionize medicine by comparing transcriptional profiles of normal versus diseased cells or by following the profile of cells during the course of pharmacological treatments, thereby monitoring success.

RNA from a whole cell, such as a bacterium, or a given cellular compartment, such as the cytoplasm or nucleus of a eukaryotic cell, is first converted into double-stranded (ds) copy DNA or cDNA with the help of an enzyme called reverse transcriptase. Reverse transcription is a process involved in the life cycle of viruses whose genomes are RNA-based. A specialized enzyme, RNA-dependent DNA polymerase or reverse transcriptase, copies the RNA genome into double-stranded DNA, which is then transcribed and translated as regular cellular DNA to produce mature viral particles (see Chapter 20). When reverse transcriptase is used in a laboratory setting to copy RNA molecules, the product is called cDNA. cDNA copies are further analyzed by either **DNA tiling arrays** (Figure 1A) or high-throughput sequencing (Figure 1B).

Microarrays are small, solid supports onto which the sequences from thousands of different genes are immobilized, or attached, at fixed locations. The supports themselves are usually glass microscope slides but can also be silicon chips or nylon membranes. The DNA, whether in the form of DNA, cDNA, or oligonucleotides, may be printed, spotted, or synthesized directly onto the support by use of robots. The RNA samples to be analyzed are fluorescently labeled and allowed to hybridize with the array, and the extent of hybridization is estimated by measuring the intensities of fluorescence in each individual spot (Figure 2). Tiling arrays contain millions

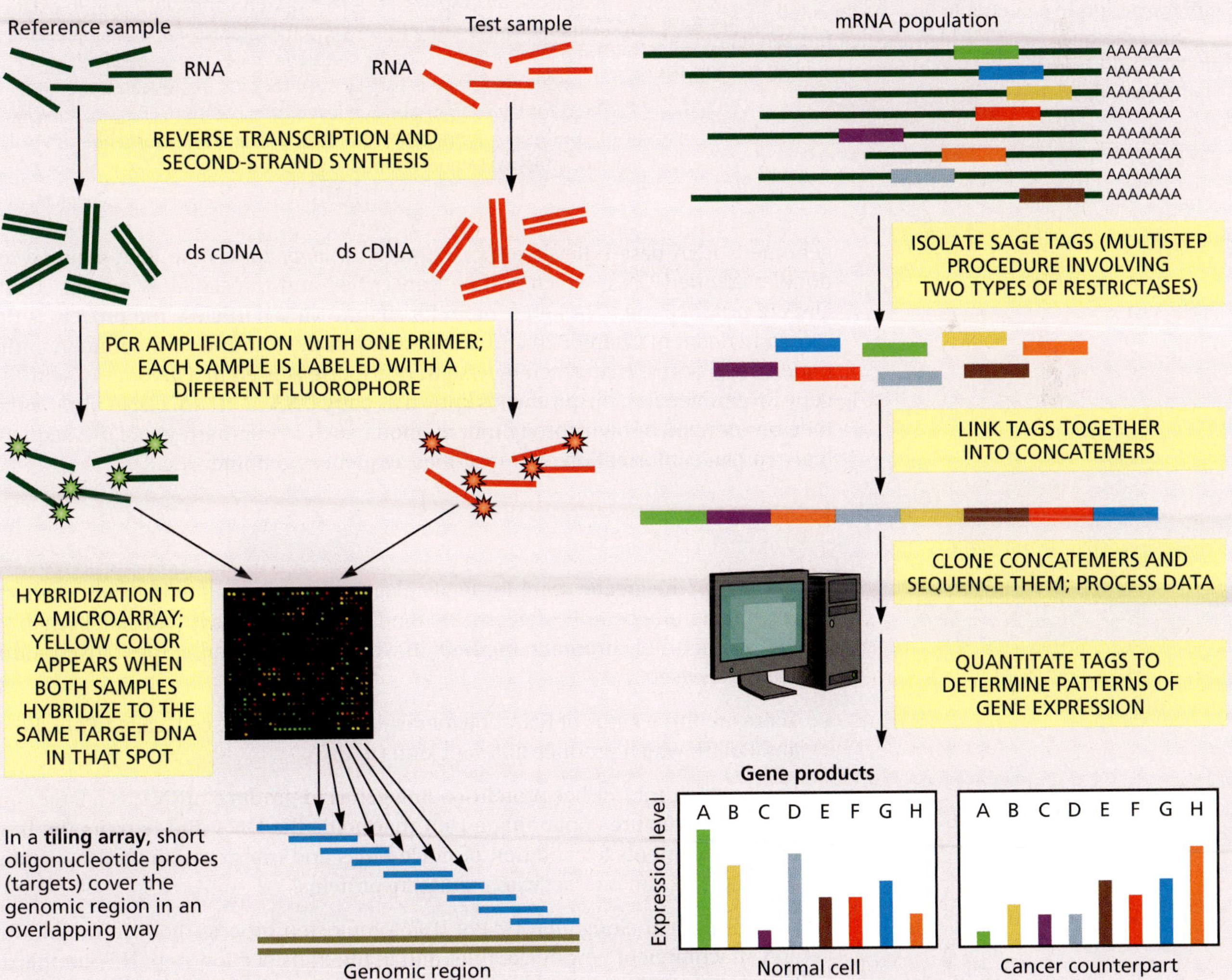

Figure 1 Genomewide transcriptome analysis. Principles behind the two major methods used for analysis of amplified populations of double-stranded cDNA molecules are illustrated: (A) DNA microarrays, usually tiling arrays, and (B) high-throughput sequencing, for example, SAGE, serial analysis of gene expression. (A, microarray, courtesy of John Coller, Stanford University.)

(Continued)

Box 10.5 ***(Continued)*** of spotted oligonucleotide sequences that cover large portions of a genome or entire genomes. For example, the Affymetrix tiling array provides 6.4 million spots, each containing oligonucleotides of length 25 nucleotides, spaced 10 base pairs apart, across the entire human genome. Such arrays allow unbiased analysis of the entire human transcriptome without any prior knowledge of genic and intergenic regions, introns and exons, etc. Microarrays provide a huge amount of noisy data. Interpretation of these data requires a combination of biological knowledge, statistics, machine learning, and the development of efficient algorithms that are able to select features above background noise.

The advent of next-generation sequencing platforms has made sequence-based expression analysis an increasingly popular, digital alternative to microarrays. Such platforms are now commercially available. In addition to providing exact sequences present in RNA populations, which are then mapped back to the genome, they give information about expression levels: one simply counts the number of times a gene is represented, or the number of hits, in the final data set. Thus, the more one counts—and these next-generation systems can count a lot—the better the measure of the copy number is; even rare transcripts in a population can be detected.

Once the microarray or high-throughput sequencing data are available, computer algorithms can be used to map these tags back to the reference genome.

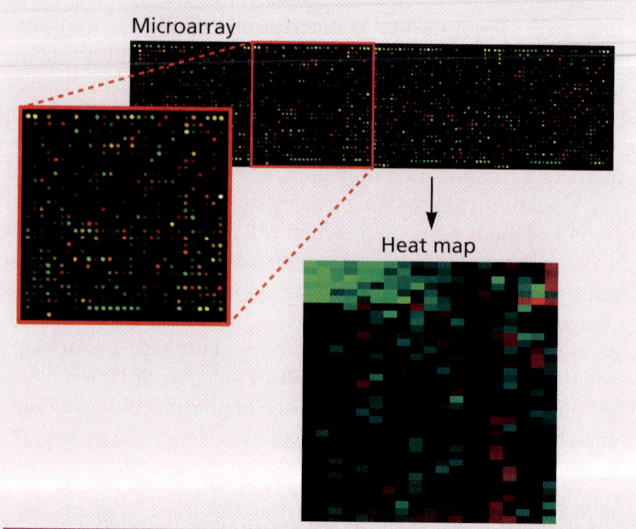

Figure 2 Representation of microarray data. Top image with enlarged inset is a photograph, taken under a fluorescence microscope, of an actual microarray following hybridization. The raw data are usually converted into a heat map, which is a two-dimensional display of the hybridization intensity values in a data matrix. In the case of transcriptome analysis, each cell represents the level of expression of one gene under some specified conditions. Green is usually used for high expression levels and red for lower-than-average levels of expression. (Top, microarray, courtesy of John Coller, Stanford University. Bottom, heat map, courtesy of Miguel Andrade, Wikimedia.)

genome is RNA-based. Reverse transcriptase, *in vivo*, copies the RNA genome into double-stranded DNA, which is then transcribed and translated as regular cellular DNA to produce the RNA and protein for mature viral particles. The process is presented in detail in Chapter 20 and illustrated with the example of the human immunodeficiency virus, HIV. When reverse transcriptase is used in a laboratory setting to copy RNA molecules, the product is known as **copy DNA** or **cDNA**. cDNA libraries can then be analyzed by high-throughput methods, such as microarrays or DNA sequencing, to provide information on transcribed sequences genomewide.

Key concepts

- Because of the greater complexity of eukaryotes, transcription in these organisms is a much more complicated process than in bacteria. This is reflected both in the more structured chromatin medium in which it occurs and in the transcriptional apparatus itself.
- There are three kinds of RNA polymerases in a eukaryotic cell, called Pol I, Pol II, and Pol III, which produce different kinds of RNA.
- Pol II, which transcribes protein-coding genes to produce mRNA, is a large multisubunit structure containing a cleft that holds the DNA and harbors the basic catalytic elements for addition of nucleotides and translocation. It also contains subunits that can interact with regulatory proteins.
- Transcriptional elongation by Pol II is a multistep process, both in the binding and attachment of new nucleotides and in the translocation step. This complexity assures selectivity and processivity in transcription.
- Nuclear RNA polymerases are not in themselves capable of recognizing promoters to initiate transcription. A set of basal transcription factors, TFs, associate with promoter DNA and the polymerase to form the minimal pre-initiation complex or PIC.

- For proper initiation of Pol II transcription, another huge protein complex, Mediator, is needed to allow the PIC to respond to regulating elements on the genome at some distance from the promoter.
- Termination of transcription in eukaryotes is closely linked to post-transcriptional processing of RNA.
- Pol I transcribes the genes for the major ribosomal RNAs. Somewhat like Pol II, it requires interactions with upstream elements via an upstream binding factor for proper initiation.
- Pol III transcribes certain small RNAs involved in ribosome function and the processing of mRNAs in the nucleus.
- Although it was formerly believed that only a small fraction of eukaryotic DNA is transcribed, the ENCODE project, which has exhaustively surveyed the human genome, revealed that the majority of DNA sequences are transcribed, much into noncoding RNA of presently unknown function.
- Recent research has also found that much Pol II transcription is localized in compact transcription factories in the nucleus and that active and inactive genes are spatially separated.

Further reading

Books

Conaway RC & Conaway JW (2004) Proteins in Eukaryotic Transcription. Elsevier Academic Press.

Goodbourn S (1996) Eukaryotic Gene Transcription. IRL Press.

Grandin K (ed) (2007) Les Prix Nobel. The Nobel Prizes, 2006. Nobel Foundation.

Reviews

Brueckner F, Ortiz J & Cramer P (2009) A movie of the RNA polymerase nucleotide addition cycle. *Curr Opin Struct Biol* 19:294–299.

Carninci P (2009) Molecular biology: The long and short of RNAs. *Nature* 457:974–975.

Carninci P, Yasuda J & Hayashizaki Y (2008) Multifaceted mammalian transcriptome. *Curr Opin Cell Biol* 20:274–280.

Chakalova L, Debrand E, Mitchell JA et al. (2005) Replication and transcription: Shaping the landscape of the genome. *Nat Rev Genet* 6:669–677.

Cramer P, Armache KJ, Baumli S et al. (2008) Structure of eukaryotic RNA polymerases. *Annu Rev Biophys* 37:337–352.

de Laat W & Grosveld F (2007) Inter-chromosomal gene regulation in the mammalian cell nucleus. *Curr Opin Genet Dev* 17:456–464.

Egloff S & Murphy S (2008) Cracking the RNA polymerase II CTD code. *Trends Genet* 24:280–288.

The ENCODE Project Consortium (2007) Identification and analysis of functional elements in 1% of the human genome by the ENCODE pilot project. *Nature* 447:799–816.

The ENCODE Project Consortium (2012) An integrated encyclopedia of DNA elements in the human genome. *Nature* 489:57–74.

Gerstein MB, Bruce C, Rozowsky JS et al. (2007) What is a gene, post-ENCODE? History and updated definition. *Genome Res* 17:669–681.

Juven-Gershon T, Hsu JY, Theisen JW & Kadonaga JT (2008) The RNA polymerase II core promoter: The gateway to transcription. *Curr Opin Cell Biol* 20:253–259.

Kornberg RD (2007) The molecular basis of eukaryotic transcription. *Proc Natl Acad Sci USA* 104:12955–12961.

Lenhard B, Sandelin A & Carninci P (2012) Metazoan promoters: Emerging characteristics and insights into transcriptional regulation. *Nat Rev Genet* 13:233–245.

Phatnani HP & Greenleaf AL (2006) Phosphorylation and functions of the RNA polymerase II CTD. *Genes Dev* 20:2922–2936.

Russell J & Zomerdijk JCBM (2005) RNA-polymerase-I-directed rDNA transcription, life and works. *Trends Biochem Sci* 30:87–96.

Schramm L & Hernandez N (2002) Recruitment of RNA polymerase III to its target promoters. *Genes Dev* 16:2593–2620.

Wilusz JE, Sunwoo H & Spector DL (2009) Long noncoding RNAs: Functional surprises from the RNA world. *Genes Dev* 23:1494–1504.

Experimental papers

Cai G, Imasaki T, Yamada K et al. (2010) Mediator head module structure and functional interactions. *Nat Struct Mol Biol* 17:273–279.

Cramer P, Bushnell DA & Kornberg RD (2001) Structural basis of transcription: RNA polymerase II at 2.8 angstrom resolution. *Science* 292:1863–1876.

Djebali S, Davis CA, Merkel A et al. (2012) Landscape of transcription in human cells. *Nature* 489:101–108.

Gnatt AL, Cramer P, Fu J et al. (2001) Structural basis of transcription: an RNA polymerase II elongation complex at 3.3 Å resolution. *Science* 292:1876–1882.

Kuhn CD, Geiger SR, Baumli S et al. (2007) Functional architecture of RNA polymerase I. *Cell* 131:1260–1272.

Nikolov DB, Chen H, Halay ED et al. (1995) Crystal structure of a TFIIB–TBP–TATA-element ternary complex. *Nature* 377:119–128.

Wang H, Maurano MT, Qu H et al. (2012) Widespread plasticity in CTCF occupancy linked to DNA methylation. *Genome Res* 22:1680–1688.

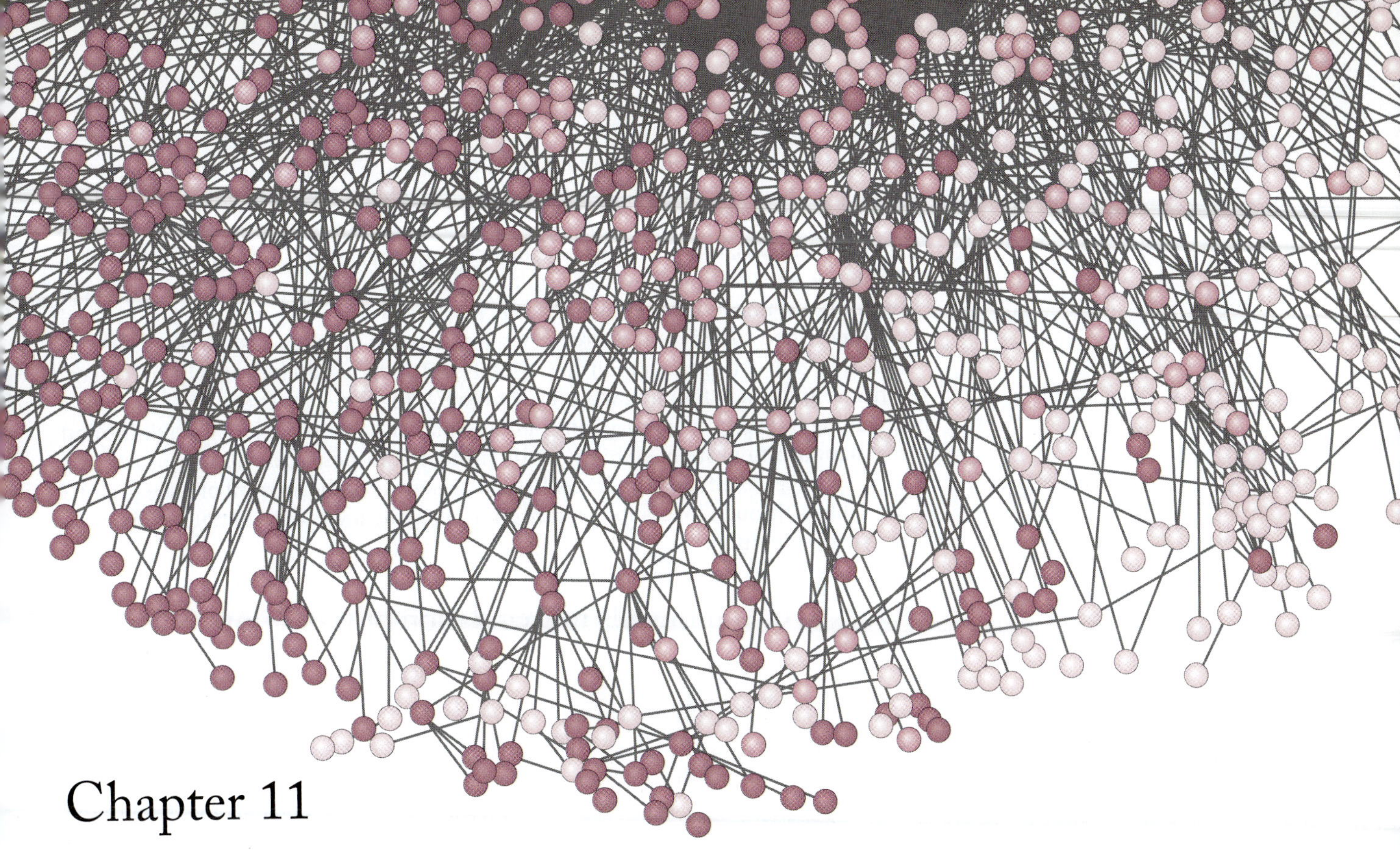

Chapter 11

Regulation of Transcription in Bacteria

11.1 Introduction

Traditionally, regulation of transcription in bacteria is viewed as a way for these single-celled organisms to adjust their activity to suit their immediate environment. However, the reasons for transcriptional regulation are more diverse. Some bacteria undergo major changes in lifestyle, as in sporulation. Some choose not to live alone; there are some special cases in which bacteria form colonies or biofilms, with some level of differentiation in the functions of individual members of the bacterial population. A well-studied example of colonial life is the formation of biofilms, called swarms. Gram-negative *Myxococcus xanthus* bacteria self-organize in response to starvation or other environmental cues, forming fruiting bodies, which are dome-shaped structures of approximately 100,000 cells. The individual bacteria within the fruiting bodies differentiate in terms of their overall metabolism and spore formation. But this behavior is unusual; most bacteria and archaea live solitary lives, at least at low population densities. In such cases, regulation of processes such as transcription is much simpler in bacteria than in metazoan eukaryotes. It must be noted, however, that archaea share aspects of transcriptional regulation with eukaryotes (see Chapter 12).

In this chapter, we focus on several distinct mechanisms used by the bacterial cell to adjust its metabolism to suit its immediate environment. Before discussing these mechanisms, it is important to make the distinction between **constitutive transcription** and **regulated transcription**. Genes that are essential to the vital functions of any cell, termed **housekeeping genes**, tend to be transcribed at all times in what is called a constitutive manner, although even housekeeping genes may be expressed at different levels under different growth conditions. Most other genes are transcribed only under certain conditions; thus, their transcriptional status is highly regulated. A regulated gene can be OFF, when its product is not needed, or ON, when the cell needs the product either for some vital function, under a specific set of conditions, or to increase its efficiency in the use of resources, such as scarce nutrients. An additional level of regulation ensures that the product is available in the amounts needed. In practice, this means that a regulated gene can be expressed or transcribed to different degrees. Perhaps one should regard such regulation as more like that of a dimmer than an ON/OFF switch. As we see here and in later chapters, the amount of functional protein product is regulated at several different levels including transcription, processing of the primary transcript, transcript stability and translation into protein, stability of the protein, and post-translational processing and modification of the protein. Still, primary control of the amount of gene product available is through regulation of transcription, at the levels of initiation, elongation, and/or termination.

11.2 General models for regulation of transcription

Regulation can occur via differences in promoter strength or use of alternative σ factors

The most direct way in which the transcription of different genes is controlled is from the structure of the promoter itself or from the composition of the holoenzyme complex. Supplies of both free RNA polymerase molecules and σ factors are limited in the cell, so there is intense competition between gene promoters for the holoenzyme. Because all promoters are more or less equally accessible to binding, promoter strength will be mainly defined by the comparative rates of dissociation of the holoenzyme from different promoters. The intrinsic strength of the promoter of each individual gene is a function of the extent to which the core promoter, the -10 and -35 boxes, conforms to the consensus sequences at these sites, because each base pair in the consensus sequence represents by definition the most favorable interaction with σ at that point in the sequence (see Chapter 9). The closer the -10 and -35 sequences are to the consensus, the stronger interactions will be and the stronger the promoter.

Two additional elements may be present in some promoters: the extended -10 element, or TG_n element, and the upstream element, or UP element. These are schematically depicted in **Figure 11.1**. The extended -10 element interacts with domain 3 of σ factor, whereas the UP element interacts with the C-terminal domain of the α subunit of RNA polymerase. These additional contacts between the promoter and the holoenzyme stabilize the initiation complex, contributing to higher levels of initiation.

Another level of regulation is brought about by the use of alternative σ factors. We discuss these factors in detail in Chapter 9, where we also present an example of their alternative usage for the induction of heat-shock genes (see Figure 9.11). The existence of alternative σ factors that recognize different promoters under different conditions contributes significantly to gene regulation within the bacterial cell.

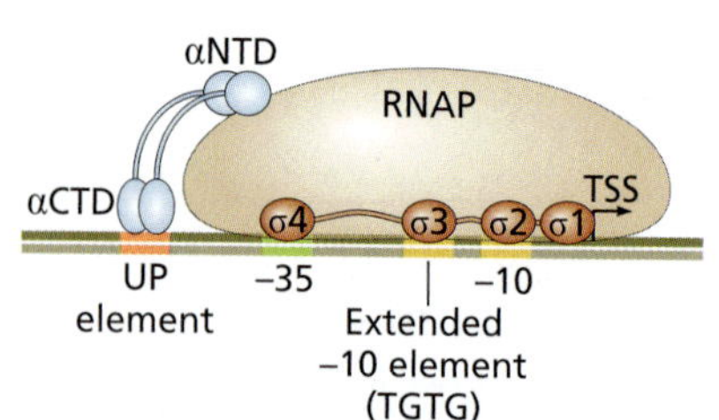

Figure 11.1 General promoter architecture. In addition to the −10 and −35 boxes, there are two additional gene elements: the extended −10 or TG_n element and the upstream or UP element. The extended −10 element is a 3–4 bp motif located immediately upstream of the −10 consensus element. It is rich in TG and interacts with domain 3 of the σ subunit of the RNA polymerase (RNAP) holoenzyme. The UP element is an AT-rich sequence of ~20 bp that is associated with strong promoters. It is located at somewhat varied distances from the −35 box and interacts with the C-terminal domains of the RNAP α subunit, known as αCTD. The variability in the exact location of the UP element is allowed because of the highly flexible nature of the linker region that connects the CTD and the N-terminal domains or NTD of the α subunit of RNAP. All four promoter elements participate in the initial binding of RNAP to the promoter, but the relative contribution of each element differs from promoter to promoter. Thus, the UP element is characteristically present in strong promoters that dock the RNA polymerase in a relatively stable pre-initiation complex.

Regulation through ligand binding to RNA polymerase is called stringent control

Bacteria have developed an amazing, highly controlled, broad-spectrum way to respond to stress conditions such as nutrient deficiency. When the cells experience amino acid starvation, they curtail synthesis of stable RNA molecules, such as ribosomal RNA or rRNA, in favor of comprehensive restructuring of metabolic gene expression, including some amino acid biosynthesis genes. This massive, coordinated metabolic response is known as the **stringent response** and is mediated by a small molecule, the **alarmone**, ppGpp.

ppGpp binds to the β and β′ subunits of RNA polymerase or RNAP, close to the active center, to inhibit the transcription of rRNA genes. Several non-mutually-exclusive models of the ppGpp-mediated negative regulation of transcription have been proposed: these include inhibiting open complex formation, reducing the stability of the open complex, inhibiting the process of promoter clearance, and increasing polymerase pausing. Although high-resolution structures of complexes of RNAP with both ppGpp and pppGpp are now available, the mechanism of polymerase inhibition is still not fully understood. The observed reduction in the stability of the open complex occurs on all promoters, but only promoters that have intrinsically unstable open complexes, such as those of the rRNA genes, are selectively inhibited.

The cell synthesizes ppGpp in response to stress by two different mechanisms (**Figure 11.2**). The lack of available amino acids leads to the accumulation of uncharged tRNA molecules, and these bind to the acceptor site in the ribosome, which is the site of entry of the incoming tRNA carrying its cognate amino acid (see Chapter 16). The binding of uncharged tRNAs to the ribosome signals the ribosome-associated enzyme RelA to synthesize pppGpp, which is then converted to the alarmone ppGpp. In a second, less-well-understood pathway, ppGpp is produced by the enzyme SpoT that senses most other stresses. SpoT is an enzyme of dual action, as it also hydrolyzes ppGpp. Mutant cells lacking both RelA and SpoT fail to inhibit the production of rRNA, and thus stringent control is relaxed in these mutant cells.

The second part of the stringent response is the positive regulation of a large number of genes that are induced during starvation. This positive regulation is aided by a protein factor, DskA, which is also bound to the polymerase. In addition, there is evidence that the release of the RNAP from the inhibited rRNA genes passively contributes to the up-regulation of these starvation-induced genes by raising the amount of free enzyme available.

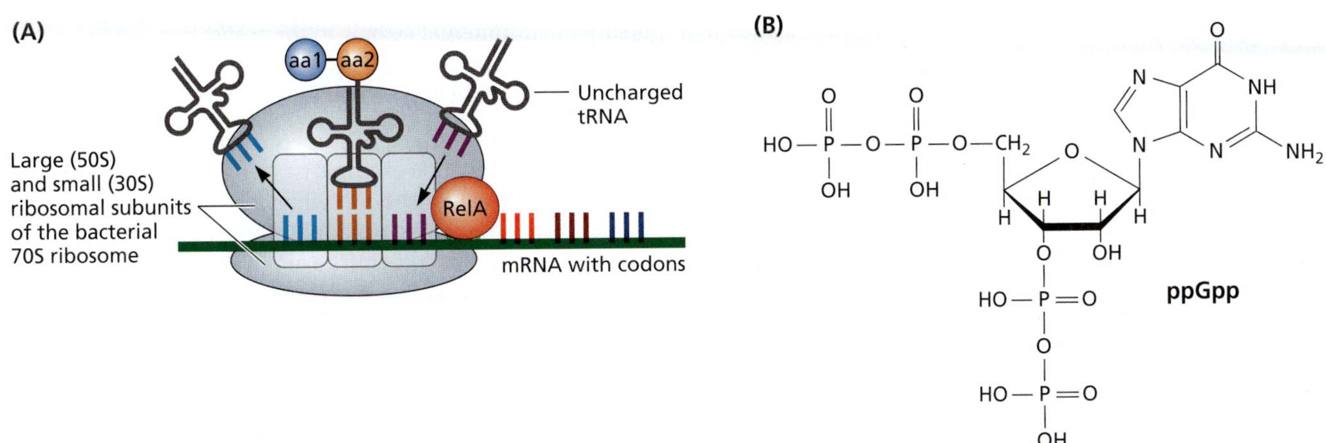

Figure 11.2 Stringent response to stress in bacteria. (A) The response is mediated by the alarmone ppGpp, whose synthesis is induced in response to nutrient stress. The absence of amino acids in the cell is sensed from large quantities of uncharged tRNA molecules. Uncharged tRNAs bind to the ribosome, signaling the ribosome-associated enzyme RelA to synthesize pppGpp, which is then converted to ppGpp. (B) Structure of ppGpp.

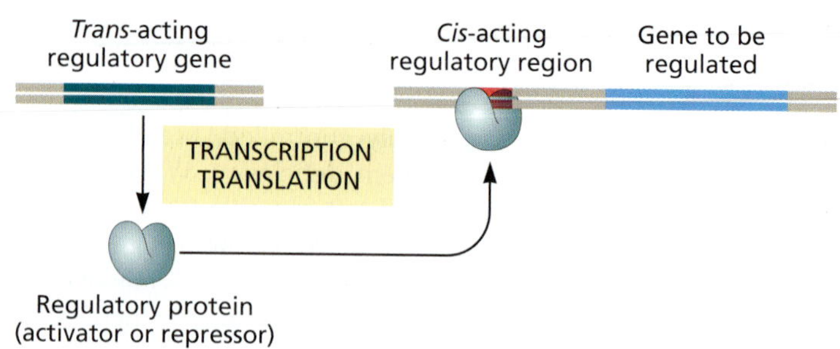

Figure 11.3 General principle of transcriptional regulation: *cis–trans* interactions. Each gene possesses one or more regulatory regions in *cis*, in the vicinity of the gene to be regulated. These are bound by regulatory protein factors, either activators or repressors, which are diffusible molecules that are encoded somewhere else in the genome; that is, they act in *trans*. We now know that the regulatory molecules can in some cases be functional RNA molecules.

11.3 Specific regulation of transcription

Regulation of specific genes occurs through cis–trans *interactions with transcription factors*

It has long been recognized that gene regulation occurs through ***cis–trans* interactions** between *cis*-regulatory DNA elements, in the vicinity of the gene, and *trans*-acting factors, encoded elsewhere in the genome (Figure 11.3). For years, it was believed that all *trans*-acting factors are proteins; we now know that noncoding RNAs can also perform this function.

Transcription factors are activators and repressors whose own activity is regulated in a number of ways

The genome of *Escherichia coli* has been predicted to contain more than 300 genes that encode proteins that bind to promoter regions to regulate transcription. Roughly 50% of these transcription factors or TFs have been studied experimentally. As we see at the end of the chapter, around 60 of these TFs each regulate only one set of genes with common functions; such a set of genes is called an **operon**. Regulation of a single operon is accomplished by highly sequence-specific binding of the TF to that operon's unique promoter. Other TFs may regulate several operons. Such groups of operons share identical or closely similar promoters. A small number of TFs, ~10, control very large sets of genes and have come to be known as **global regulators**. The existence of such classes of TFs reflects a hierarchical organization of gene regulation in bacteria.

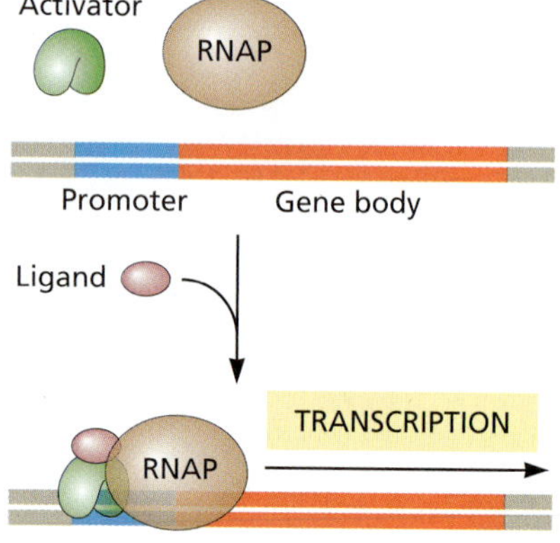

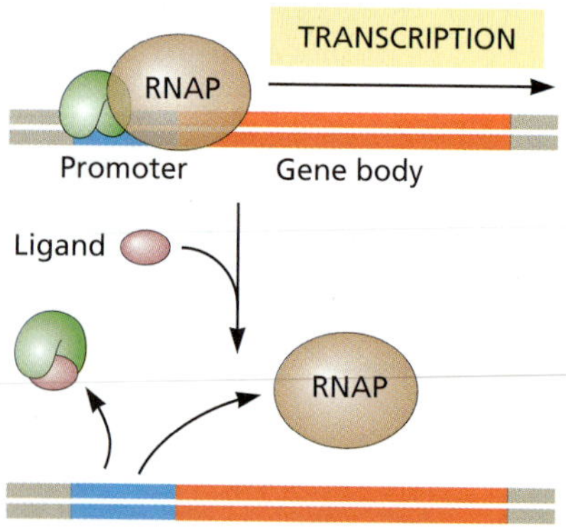

Figure 11.4 Activity of activators can be up- or down-regulated by ligand binding. (A) Up-regulation: the activator is active only if bound to a ligand. (B) Down-regulation: the activator dissociates from RNAP upon binding the ligand, and hence transcription is inhibited.

Numerous studies, carried out over many years, have provided a clear-cut picture of how protein factors can interact with promoter regions in a regulated way. These proteins may function either as **activators**, which stimulate transcription, or as **repressors**, which inhibit transcription. Classical examples of the regulation of three operons in *E. coli* are described below. Before going into the details of these systems, we should point out that both activators and repressors are themselves regulated by the binding of small ligands that transmit environmental signals to the respective gene systems. Interestingly, the action of regulator proteins can be either stimulated or inhibited by the binding of specific ligands (Figure 11.4 and Figure 11.5).

In addition to being controlled by ligand binding, TF activity can be regulated by post-translational modifications. A well-understood example is NarL, one of the global regulators, which can bind its target DNA sequences only when it is phosphorylated. The phosphorylation is carried out by two cognate sensor kinases, cell membrane proteins whose kinase activity is regulated by extracellular nitrite or nitrate ions.

Finally, the activity of some TFs in the cell is controlled by their intracellular concentration; this, in turn, is regulated at the level of their own transcription and/or translation and by their proteolytic degradation.

Several transcription factors can act synergistically or in opposition to activate or repress transcription

Transcription of some genes is regulated by the action of single activators or single repressors on their cognate promoters. For most genes, several TFs act together to affect transcription activity either additively, or synergistically, or in opposition.

There are at least three major, distinct mechanisms of transcriptional activation by single activators (Figure 11.6). These differ by where in the promoter region the activator binds and how it activates transcription. In class I activation, the activator binds to a target that is located upstream of the promoter –35 element and recruits the RNAP by direct interactions with the polymerase C-terminal domain, CTD, of the α subunit. The linker that joins the CTD and the N-terminal domain, NTD, of the α subunit is flexible, so binding of the activator can occur at different distances from the –35 element. Class II activation involves RNAP recruitment via domain 4 of the σ factor. The spatial constraints of σ factor binding provide little flexibility in the positioning of the activators. In the third mechanism, the activator alters the conformation of the promoter to enable simultaneous interactions of RNAP with the –10 and –35 elements. Usually, the DNA is twisted to reorient the two elements for optimal binding of the σ elements. A fourth, somewhat different mechanism depends on the modulation of repressor molecules by activators in a way that prevents the repressor from binding to the DNA.

The three mechanisms of transcriptional repression by a single repressor are steric hindrance, DNA loop formation, and modulation of an activator protein (Figure 11.7). These three mechanisms all act at the promoter region itself or on DNA sequences in close proximity during transcription initiation. A fourth mechanism of repression by single repressors acts during elongation, when the DNA-bound repressor creates a roadblock to the advancing RNA polymerase.

Sometimes a single promoter can serve as an initiation site for a whole operon, a series of linked genes, which often code for proteins that are involved in a single metabolic pathway. In many cases, the activity of the set of genes comprising an operon is controlled at a single locus, called an **operator**. We now describe three such examples that are important in bacterial physiology and that have also played significant roles in developing our understanding of gene regulation.

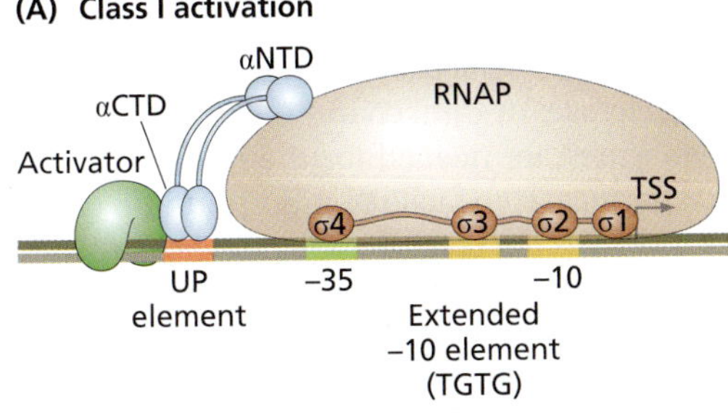

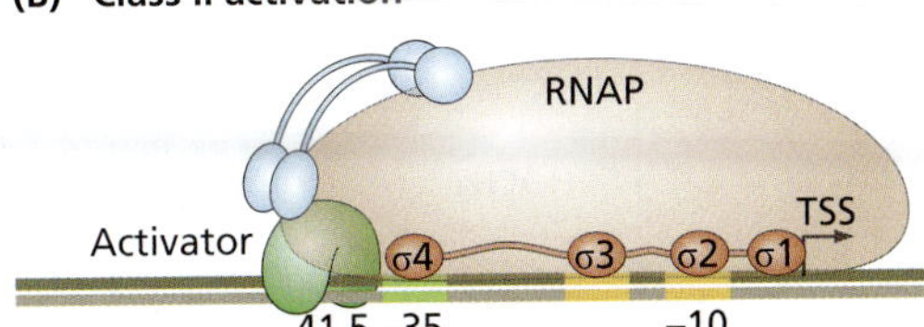

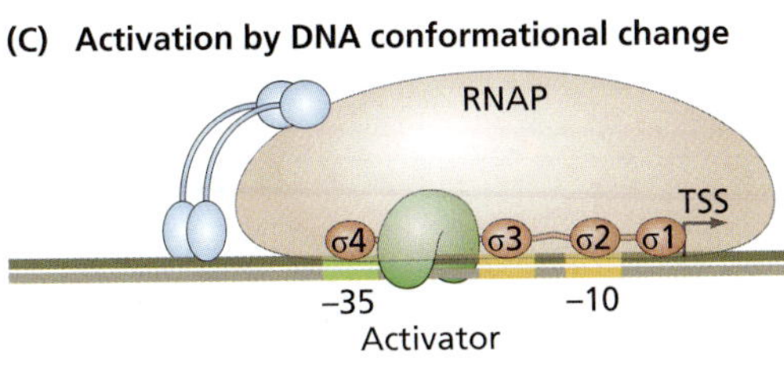

Figure 11.6 Transcriptional activation by a single activator. (A) Class I activation: the best-studied example is the action of CAP at the *lac* promoter. (B) Class II activation: the best-studied example is the activation of a phage λ promoter by CI protein. (C) Transcriptional activation by conformational changes in promoter DNA caused by activator binding. (Adapted from Browning DF & Busby SJW [2004] *Nat Rev Microbiol* 2:57–65. With permission from Macmillan Publishers, Ltd.)

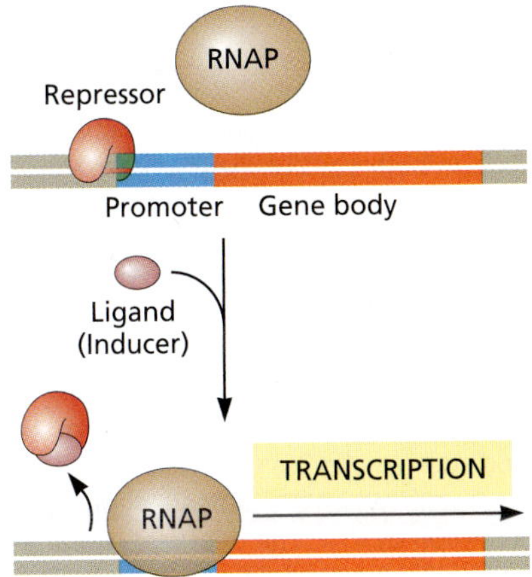

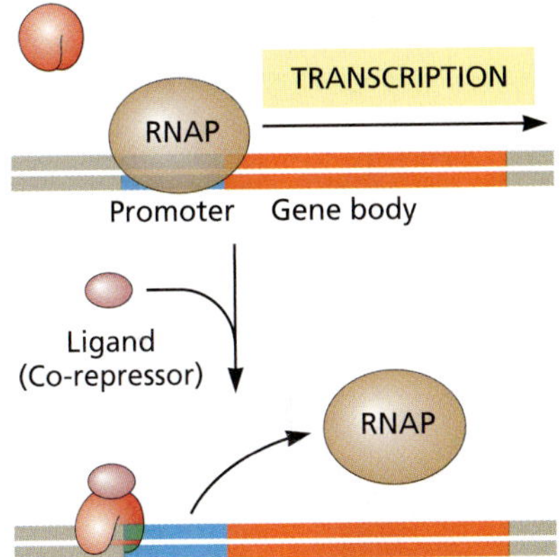

Figure 11.5 Activity of repressors can be up- or down-regulated by ligand binding. (A) Up-regulation: the repressor is active when not bound to a ligand, so ligand binding leads to dissociation of repressor and induction of transcription. The ligand in this case is called an inducer. (B) Down-regulation: binding of a ligand to the repressor is necessary for repressor binding and inhibition of transcription. The ligand in this case is called a co-repressor.

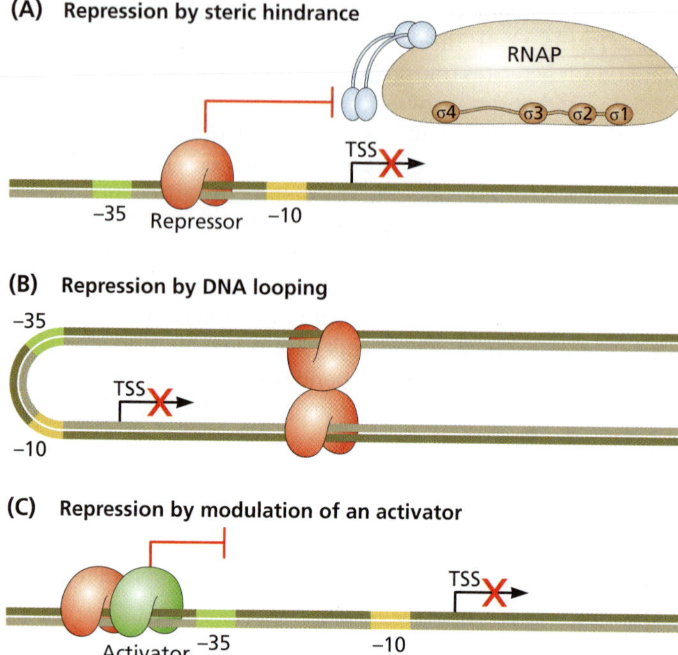

Figure 11.7 Transcriptional repression by a single repressor. (A) Repression by steric hindrance. The repressor-binding site overlaps core promoter elements and blocks promoter recognition by RNAP. The best-studied example is the action of the Lac repressor. (B) Repression by DNA looping. Repressors bind to distal sites on both sides of the promoter and oligomerize, making the intervening DNA loop out and thus preventing initiation. Examples are the Gal repressor at the *gal* promoter and the *ara* operon. (C) Repression by modulation of an activator protein: that is, the repressor functions as an anti-activator. The best-studied examples are CytR-repressed promoters, which are dependent on activation by CAP. The CytR repressor and the CAP activator interact directly. (Adapted from Browning DF & Busby SJW [2004] *Nat Rev Microbiol* 2:57–65. With permission from Macmillan Publishers, Ltd.)

11.4 Transcriptional regulation of operons important to bacterial physiology

The lac *operon is controlled by a dissociable repressor and an activator*

Escherichia coli is an omnivorous microbe that is able to subsist on a number of food sources, including many sugars. Glucose is the sugar of preference but other monosaccharides, such as galactose, or disaccharides, such as lactose (**Figure 11.8**), can be utilized. Specific enzymes are needed for the metabolism of each sugar: in the case of lactose, both a permease, to facilitate transport into the cell, and the enzyme β-galactosidase, which cleaves the disaccharide into galactose plus glucose, are necessary (see Figure 11.8). The genes for these proteins are positioned in tandem on the *E. coli* chromosome, followed by the gene for a third enzyme, a transacetylase (**Figure 11.9**). Although not essential for lactose metabolism, the transacetylase serves

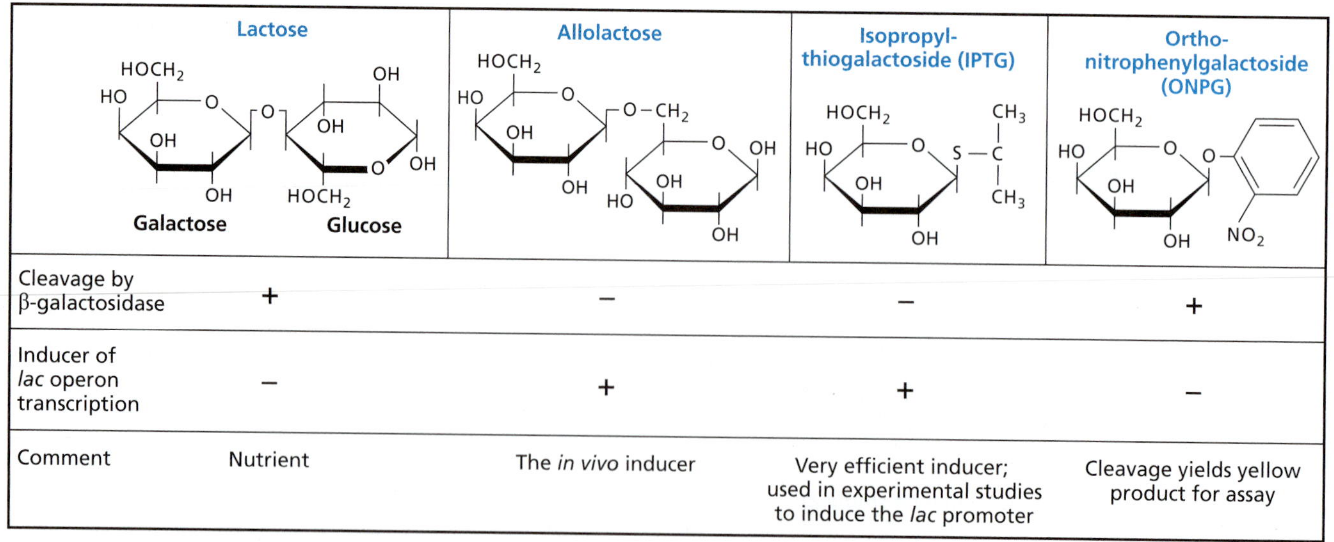

	Lactose	Allolactose	Isopropyl-thiogalactoside (IPTG)	Ortho-nitrophenylgalactoside (ONPG)
	Galactose / Glucose			
Cleavage by β-galactosidase	+	–	–	+
Inducer of *lac* operon transcription	–	+	+	–
Comment	Nutrient	The *in vivo* inducer	Very efficient inducer; used in experimental studies to induce the *lac* promoter	Cleavage yields yellow product for assay

Figure 11.8 Lactose and analogs used in studies of the *lac* operon. Allolactose is the natural, *in vivo* inducer, but IPTG is a much more effective inducer that is used in many experimental studies. For experiments in which detection of β-galactosidase activity is important, ONPG is used as a substrate whose cleavage produces a colored product that can be observed even in cell clones.

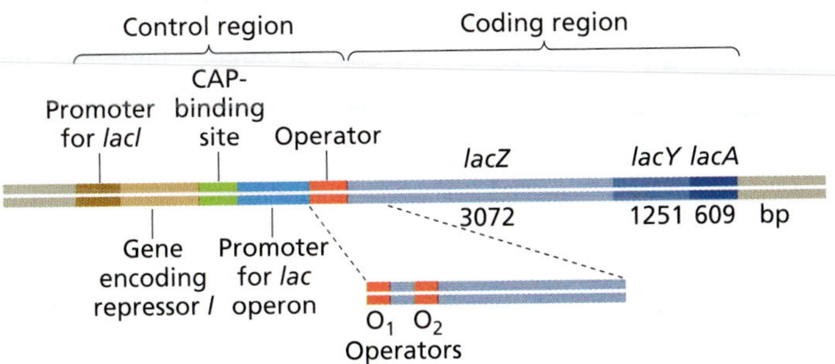

Figure 11.9 Organization of the *lac* operon. The three structural genes that encode the proteins involved in the overall process of lactose metabolism are *lacZ,* which codes for β-galactosidase, the enzyme that hydrolyzes the glycoside bond; *lacY,* which codes for the permease that transports the sugar from the medium through the cell membrane; and *lacA,* which codes for the transacetylase that acetylates β-galactosides that cannot be hydrolyzed, a reaction that eliminates toxic compounds from the cell. The control region contains the promoter; the operator, which binds the Lac repressor; and the gene, with its promoter, that encodes the repressor molecule. The control region also contains a binding site for the activator CAP.

to detoxify aberrant galactosides that are produced as by-products of the pathway. The galactosidase gene is designated *lacZ,* the permease *lacY,* and the transacetylase *lacA;* they lie in the genome in that order. These three genes comprise the ***lac* operon**. The *lacZ* gene is also a useful tool for assaying the success or failure of gene cloning experiments. Often, this gene is inserted along with the desired gene, and cell clones are then tested with a galactoside that produces a distinctive color when cleaved by the β-galactosidase product of the inserted *lacZ* gene.

When an *E. coli* bacterium encounters a medium containing glucose, it will use this nutrient in preference to any other. But if glucose is wholly lacking and lactose is available, the bacterium will turn on the *lacZ, lacY,* and *lacA* genes of the *lac* operon. Just how this switch occurred fascinated Jacques Monod and co-workers in the 1940s and 1950s. Their work led to the discovery of an important mode of gene regulation in bacteria; see Box 11.1 for the history of that discovery. They showed, specifically, that

Box 11.1 Discovery of the Lac repressor: The value of unexpected results In 1940 Jacques Monod, a young graduate student at the Sorbonne, was beginning his research on the ability of bacteria to use different sugars for nutrients. When he fed *E. coli* a mixture of glucose and lactose, he observed a peculiar phenomenon. The bacteria consumed the glucose first and only then turned to lactose. No explanation was available at the time, and Monod's further work was delayed by the Second World War. He served first in the French army, and after the fall of France, he became a leader in the underground resistance in Paris. Fearing for his life if found by the enemy, Monod surreptitiously left the Sorbonne and began to work in the laboratory of André Lwoff at the Institut Pasteur. This was a fortunate choice, as Lwoff was an expert in the field of bacterial metabolism.

With the aid of visiting scientist Melvin Cohen, Monod soon discovered that when the bacteria switched to lactose, they synthesized a new enzyme, β-galactosidase, which catalyzed the cleavage of lactose into glucose and galactose. This enzyme could cleave a number of galactosides (see Figure 11.8). Even more interesting was the fact that there were other galactosides that could not be cleaved but that stimulated, or induced, the production of β-galactosidase.

But how did this induction work? Did the inducer act directly on the gene? A clue came from the discovery by Georges Cohen, also in the Monod group, of a second enzyme that is required for lactose metabolism. This is a permease, which facilitates the entry of lactose through the cell wall into the bacterium. The β-galactosidase and permease are encoded by two separate genes named *lacZ* and *lacY,* respectively. Mutants that affected the possibility of induction invariably influenced both of these genes, never just one or the other. This meant that induction must be governed by a third gene, which was given the symbol *lacI* for inducer. This set of genes, plus one other, is now referred to as the *lac* operon.

How did the presumed protein product of the *lacI* gene work? The most obvious idea was that it was some kind of activator that allowed reading of the gene. Recall that, in 1954, understanding of transcription or even mRNA was still in the future. By now Monod was collaborating with François Jacob, and in 1957, visiting scientist Arthur Pardee joined the team. Together they designed an experiment, often called the PaJaMo experiment, which gave a surprising answer to the question. The PaJaMo experiment was also intended to shed light on the phenomenon of constitutive expression; there existed *E. coli* mutants that strongly expressed the *lac* genes in the absence of inducer or even lactose.

The PaJaMo experiment was both elegant and had far-reaching implications. The experiment depends on observations that go back to 1945 when Joshua Lederberg, at the University of Wisconsin, discovered that bacteria had sexes. One type, called males, is able to transfer part or all of their DNA to another type, called females, in a process known as **conjugation**. The PaJaMo experiment would not have been practicable without the discovery by microbiologist William Hayes of **high-frequency recombinants** or **Hfr**. These Hfr male bacteria conjugate 10,000 times more frequently than the wild type. This allowed experiments in which mixing a number of Hfr males with an excess of females provided almost complete synchronization of the onset of conjugation. The process could be stopped at any point in the DNA transfer by agitation with a blender, leaving a part of the male's DNA within the female. It was found that genes were transferred in a sequential fashion, following their order on the chromosome.

In the PaJaMo experiment (Figure 1A), wild-type Hfr males ($lacI^+$, $lacZ^+$) were conjugated with females ($lacI^-$, $lacZ^-$) in the

(Continued)

Box 11.1 (Continued)

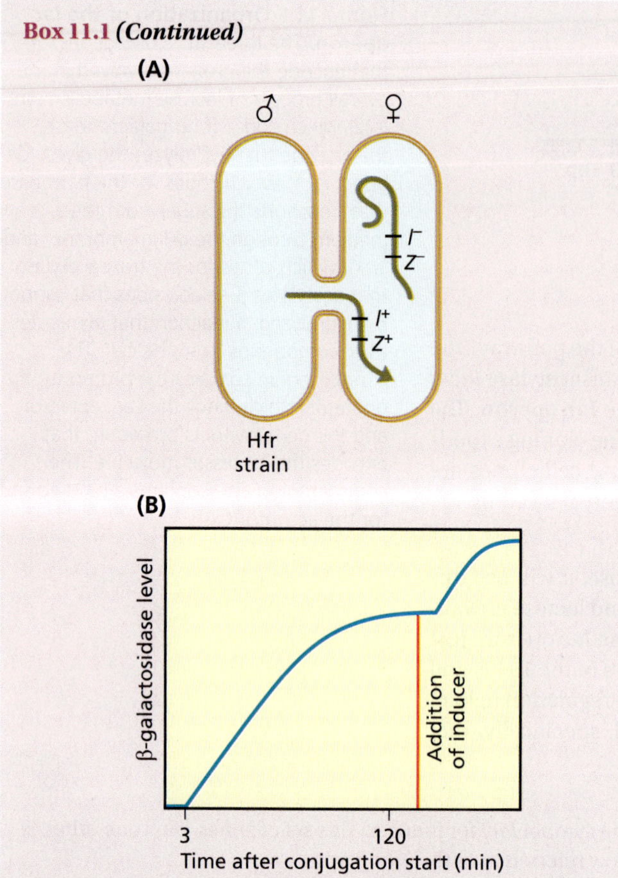

Figure 1 The PaJaMo experiment. (A) The principle: an Hfr high-frequency-recombination male strain of bacteria that was wild type with respect to the *lacZ* and *lacI* genes was allowed to conjugate with an excess of ($lacZ^-$, $lacI^-$) females. During conjugation, part of the male chromosome containing the wild-type genes was transferred to the female cell, allowing the production of β-galactosidase. (B) Time course of β-galactosidase generation. Aliquots of the conjugating bacteria were agitated in a Waring blender at different times after the start of conjugation, and β-galactosidase activity was assayed. After a brief lag period of ~3 min, the activity rises until the repressor has accumulated enough to shut down expression of the *lacZ* gene. After this time point, the enzyme becomes inducible, because the addition of inducer will cause release of this repressor from the DNA.

absence of any inducer. The males were not synthesizing *lac* gene products, and the females could not. The bacteria were assayed for β-galactosidase at different times after the start of conjugation, with the results shown in Figure 1B. After three minutes, expression of the *lacZ* gene began and continued for a while. But then it slowed, and had completely stopped after 120 minutes. This was wholly unexpected. Furthermore, after this time, the system became inducible.

The idea that *lacI* coded for an activator did not fit the data; in fact, the simplest explanation was that the *lacI* product is a repressor. As the male DNA entered the female, expression of gene *lacZ* could begin constitutively, because only a minute amount of repressor was available at first. But as the repressor built up, it shut down the *lac* genes, causing the halt in β-galactosidase synthesis. The fact that synthesis could now resume if an inducer was added could be explained most simply by a role for the inducer in dissociating the repressor from the DNA. In a moment, the whole picture of gene regulation in bacteria had taken on new dimensions.

The PaJaMo experiment had repercussions far beyond the elucidation of gene control in bacteria. The fact that expression of the *lac* genes as proteins was shut off rather quickly when repressor accumulated indicated to some, particularly Jacob, that there must be a short-lived intermediate between DNA and protein. If the message was contained in long-lived ribosomes, as many believed in 1957, why was it not still around even after its production was shut off? This led Jacob to suggest an RNA message, an idea not widely considered for several years.

Another major fallout from this experiment was the concept of allosteric control of protein function (see Box 3.8). If the inducer acted by binding to the repressor, permitting it to dissociate from the DNA, this strongly implied that binding of one kind of molecule to a protein could influence that protein's affinity for another molecule, in this case DNA. Today, allosteric regulation is recognized as a major mode of protein function, regulating a myriad of processes. A monumental paper by Monod, Jean-Pierre Changeux, and Jacob announced this concept. A second paper by Monod, Jeffries Wyman, and Changeux presented a coherent mathematical analysis of the phenomenon. Wyman, an American scientist working in Rome, was primarily concerned with the application of these ideas to the binding of oxygen by hemoglobin.

Finally, the PaJaMo experiment also provides a lesson for all researchers. It is not the experiment that confirms a preconceived idea that is most important and fruitful. Rather, it is the surprise, the unexpected result that can open whole new vistas and establish new paradigms. Never write off the failed experiment. The Nobel Prize in Physiology and Medicine was awarded to Jacques Monod, François Jacob, and André Lwoff in 1965.

the *lac* genes, in the absence of lactose, were essentially turned off by a repressor. The repressor protein binds to the DNA at operator sites near the *lac* promoter and can be displaced only when an **inducer**, such as allolactose, binds to the repressor. Allolactose is a minor product of the weak galactosidase activity that is always present (see Figure 11.8).

We now have the much more complete picture shown in Figure 11.10. All three genes, *lacZ, lacY,* and *lacA*, are transcribed together, with the polymerase starting from the promoter site just upstream from the operator sites. The two operator sites bind two dimers of the repressor protein, which effectively blocks polymerase binding. The gene for the repressor protein itself, *lacI*, is upstream from the *lac* control region and has its own promoter; the gene is designated *lacI* because the protein was at first thought to

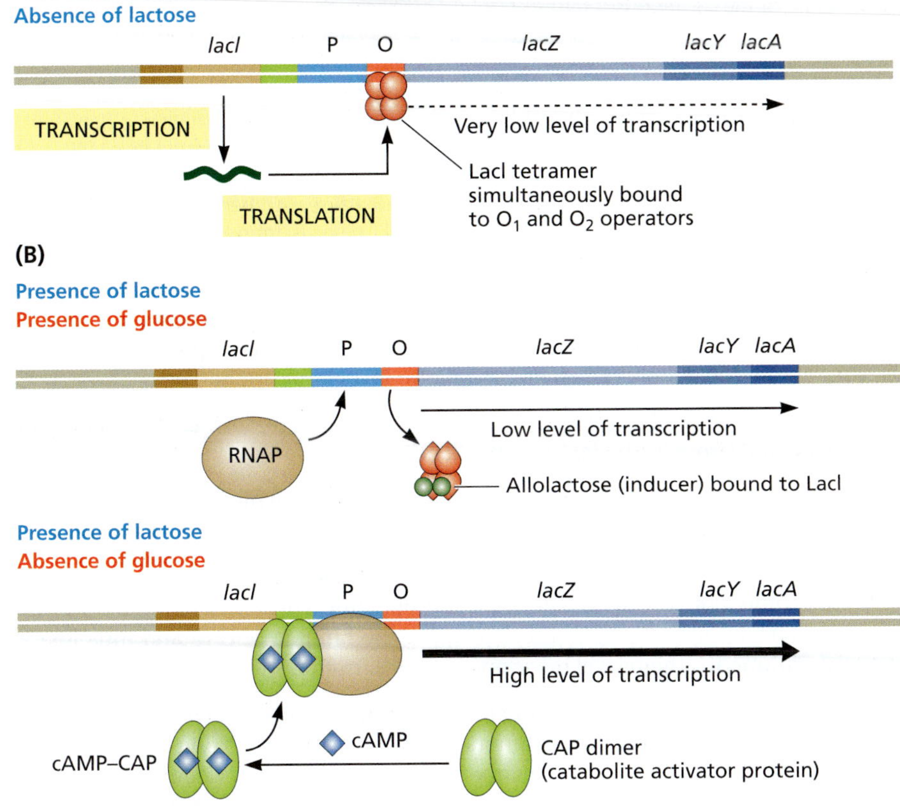

Figure 11.10 Transcriptional activity of the lactose operon is regulated by two sugars. When glucose is present in the medium, independent of the presence or absence of lactose, the *lac* operon is repressed, as glucose is the preferred carbon source. The presence of glucose leads to catabolic repression, which affects the activity of all operons that encode enzymes that catabolize alternative sugars. For the *lac* operon to be active, two conditions must be met: lactose should be present but glucose should not. (A) In the absence of lactose, the *lac* operon is repressed: the LacI repressor is bound to a site that partially overlaps the *lac* promoter, preventing the binding of RNAP. To be specific, LacI binds as a tetramer to two operator sites, O_1 and O_2, the latter of which is within the coding region of *lacZ*, and the intervening DNA between O_1 and O_2 loops out. (B) Behavior of the operon in the presence of lactose. In the presence of glucose, in the upper part, the genes are only slightly expressed. A secondary lactose metabolite, 1,6-allolactose, binds to the repressor, changing its conformation and reducing its affinity for the operator. Allolactose acts as an inducer. For full activation to take place, there should be no glucose in the medium The mechanism of activation in the absence of glucose, in the lower part, involves an activator—catabolite activator protein or CAP, also known as cyclic AMP regulatory protein or CRP—which binds to a regulatory site upstream of the promoter only if cAMP is bound to it. The contact between CAP and RNAP is thought to be through the α subunit of the polymerase. The absence of glucose is sensed by the cell through the intracellular concentration of cAMP: when glucose is high, cAMP levels are low, but when glucose levels are low, there is a high concentration of cAMP. cAMP then binds to CAP to change its conformation and facilitate its binding to DNA.

be an inducer. Between the *lacI* gene and the promoter for the *lac* operon lies another control element, the **catabolite activator protein** or **CAP**-binding site, which is involved in an additional, broader level of control. CAP is also known as the catabolite regulatory protein, CRP. Monod's initial observation was that lactose metabolism would begin only when glucose in the medium had mostly been consumed. When glucose levels in a cell are low, the production of ATP is limited and cyclic adenosine monophosphate, cAMP, accumulates. Cyclic AMP will bind to the CAP protein, favoring a conformational change that facilitates binding of the protein to the CAP-binding site. This, in turn, recruits RNA polymerase to the promoter by direct interaction with the CAP dimer bound to the DNA.

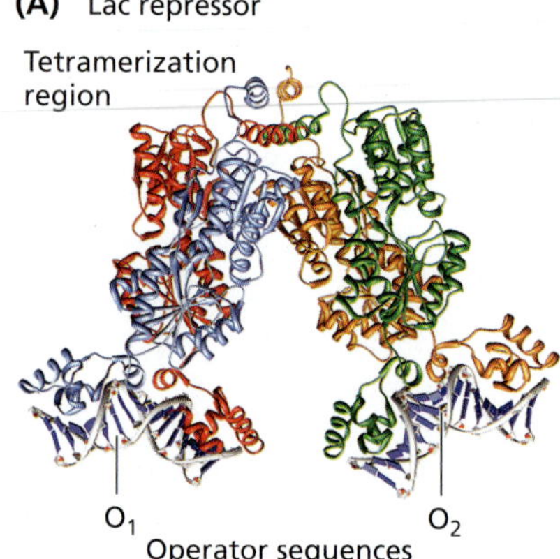

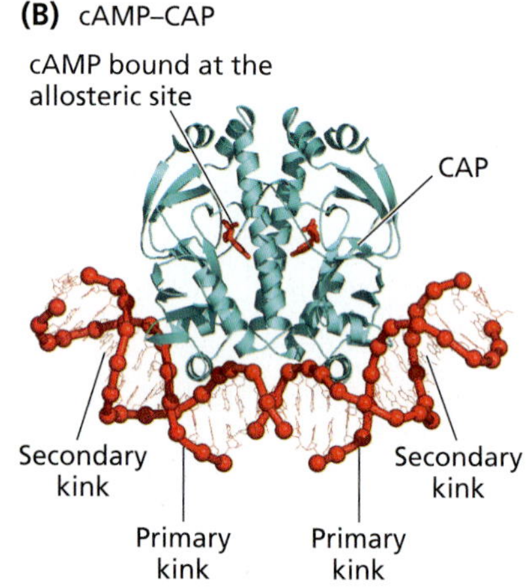

Figure 11.11 Structure of the Lac repressor and the cAMP–CAP dimer bound to DNA. (A) Structure of the Lac repressor tetramer bound to O_1 and O_2. Each dimer, one shown in red/blue and the other in green/orange, binds to one operator site through helix–turn–helix (HTH) motifs interacting with the major groove. The dimers are coupled at the tetramerization region; thus the tetramer binds O_1 and O_2 simultaneously. (B) Structure of the cAMP–CAP dimer bound to DNA. The crystallization DNA fragment contained a single-phosphate gap between positions 9 and 10 of each DNA half-site. The conformational change conferred by cAMP binding brings the two HTH motifs of each monomer together at an appropriate distance for interaction with neighboring major grooves. CAP binding causes 90° bending in DNA. (A, courtesy of Daniel Parente, University of Kansas Medical Center. B, from Lawson CL, Swigon D, Murakami KS et al. [2004] *Curr Opin Struct Biol* 14:10–20. With permission from Elsevier.)

The effects of various combinations of available nutrients on the expression of the *lac* operon can now be readily explained (see Figure 11.10). To summarize:

- In the absence of lactose, levels of inducer will be very low, and the repressor will bind strongly; the *lac* operon will be silent.
- In the presence of both lactose and glucose, RNAP can bind, albeit weakly, and the *lac* genes will be transcribed at a low level.
- When lactose is present but glucose is absent, the inducer will be present and the repressor will dissociate from the operator. Furthermore, cAMP will be present in high concentrations, favoring the binding of CAP and thus RNAP recruitment. A high level of *lac* operon transcription will result.

The structures of the Lac repressor and CAP are shown in **Figure 11.11**. The repressor functions as a tetramer of identical dimers, each dimer binding to one of the operator sites. CAP also binds to the DNA as a dimer. Both repressor and CAP bind to their specific sites via helix–turn–helix or HTH motifs and produce bending in DNA (see Chapter 6). The repressor dimer, in the absence of inducer, can also bind more weakly and nonspecifically to DNA, allowing it to search along the DNA for its specific site. In this condition, the protein does not produce DNA bending. The presence of bound inducer produces a conformational change in which the HTH motifs become disorganized, and binding to DNA is weakened.

Both the Lac repressor and the CAP protein provide excellent examples of allosteric regulation (see Box 3.8). In fact, it was the discovery of the repressor and its control by an inducer that suggested the very idea of allostery to Jacob and Monod. Note that the binding of two small molecules, inducer to repressor and cAMP to CAP, produces opposite effects: the inducer weakens the DNA binding of the repressor, whereas cAMP strengthens CAP's affinity for DNA. Likewise, the allostery of interacting proteins within a heterodimer can be either positive or negative (see Box 3.8). Within the *lac* operon, interaction between repressor dimers to produce the tetramer strengthens their binding at each of the pair of operator sites, and interaction between CAP and the polymerase enhances the DNA binding of both.

Finally, using the example of CAP homodimers and their binding sites across the genome, we introduce the relatively recent concept of **sequence logos** (**Box 11.2**). Sequence logos are a quantitative and visually effective way of presenting sequence conservation information for a family of closely related TF-binding sites or for a family of closely related TFs. This is another example of the power of bioinformatics tools.

Control of the trp *operon involves both repression and attenuation*

Synthesis of the amino acid tryptophan in bacteria involves enzymes that are encoded by a set of five contiguous genes, coding for the tryptophan biosynthesis enzymes themselves plus a leader sequence. As in the *lac* operon, these genes are all transcribed from a single promoter (**Figure 11.12**). Tryptophan is not used very much in proteins, so to synthesize a lot of tryptophan or even to transcribe much messenger when the amino acid is abundant would be wasteful of cellular resources and energy. Therefore, the entire operon is regulated, mostly by allosteric feedback control of the kind mentioned in Box 3.8. As Figure 11.12 shows, there is a repressor protein that

Box 11.2 Sequence logos Sequence logos are graphical representations of the patterns of sequence similarities within multiple sequence alignments. They provide a richer and more precise description of sequence conservation than consensus sequences because they are quantitative. Each logo consists of a stack of letters for each position in the respective sequence: four letters in DNA sequences and 20 letters in protein sequences. The overall height of each stack is a measure of the overall sequence conservation at that position, measured in bits; the height of each letter within each of the stacks reflects the relative frequency or probability of that letter occurring at that position. Figure 1 presents sequence logos for the partners involved in transcriptional activation of the *lac* operon in the presence of lactose and absence of glucose in the medium (see Figure 11.10).

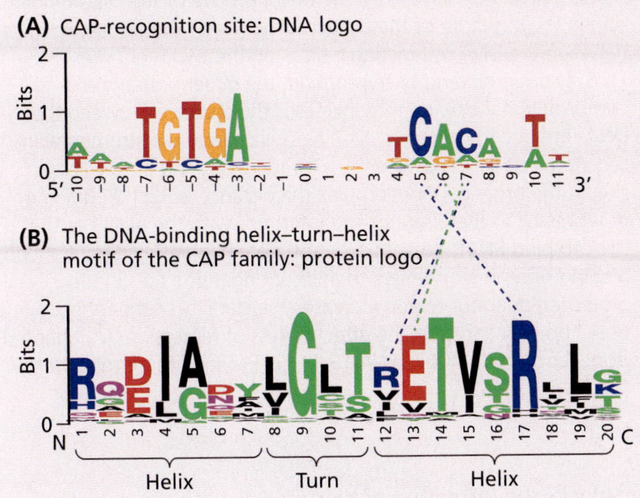

Figure 1 Sequence logos for two DNA binding sites of CAP dimer and for HTH motif of the CAP family. (A) The two DNA-recognition helices of the CAP homodimer insert themselves into consecutive turns of the major groove of the DNA helix. The DNA logo of their binding site is approximately palindromic, which provides two very similar recognition sites, one for each subunit of the dimer. As shown by the heights of the symbols, however, the binding site is not perfectly symmetrical, possibly because of the inherent asymmetry of the operon promoter region. The distance between the two parts of the recognition sequence is 11 bp, or approximately one full turn of the DNA helix: this allows the two monomers of the dimer to bind to the same site of the double helix. The data used to construct this logo consisted of the sequences of 59 binding sites determined by DNase footprinting (see Box 6.4). (B) Helix–turn–helix motif from the CAP family of homodimeric DNA-binding proteins. CAP binds at more than 100 sites within the *E. coli* genome. Residues 1–7 form the first helix, residues 8–11 form the turn, and residues 12–20 form the DNA-recognition helix. The glycine at position 9 is strictly conserved; it is located inside the turn, where packing constraints prevent the insertion of an amino acid with a side chain. Positions 4, 8, 10, 15, and 19 are partially or completely buried in the three-dimensional structure of the protein and therefore tend to be populated by hydrophobic amino acids, shown with black letters. Positions 12, 13, and 17 interact directly with bases in the major groove of the DNA, as shown by connecting lines between the two logos, and thus are critical to the sequence-specific binding of the protein. The data that were used to construct this logo consisted of 100 CAP protein sequences. (Courtesy of Computational Genomics Research Group, University of California, Berkeley.)

can bind to an operator site, blocking the binding of RNAP. In analogy with the Lac repressor, this is coded by a separate gene. The Trp repressor, however, operates in a way that is opposite to the Lac repressor. Binding of the effector molecule, in this case tryptophan itself, promotes binding of the repressor to the operator element, blocking transcription initiation. Thus, tryptophan accumulation nearly shuts down transcription of the genes necessary for its synthesis. Transcription at high tryptophan levels is only about 0.1% of that when no tryptophan is present.

However, there is evidence that not all of this effect is due to binding of the repressor. There exist mutants of the *trp* operon that have wholly effective repressor and operator, yet they transcribe at a rate tenfold higher than wild type under the same tryptophan-poor conditions. This led Charles Yanofsky to discover a whole other level of regulation, overlaid upon repression; this is called **attenuation**, and it operates in an entirely different manner.

The mechanism of attenuation, as it occurs in the *trp* operon, is illustrated in Figure 11.13. This method of regulation acts on the termination of transcription. It depends upon the fact that in bacteria even the earliest stages of transcription are accompanied by attachment of ribosomes to the mRNA: transcription and translation are coupled. Also important is the precise sequence structure near the 5′-end of the operon. Following the start codon is a leader sequence, followed by three sequences in which the transcribed RNA has the potential to form a hairpin. These sequences have the potential to form hairpins in two different places: between segments 2 and 3 or between segments 3 and 4.

Even in the repressed state, there is some initiation of transcription on the *trp* operon. When tryptophan levels are high, the polymerase will transcribe right through the leader sequence, with a ribosome following along the new RNA to effect translation.

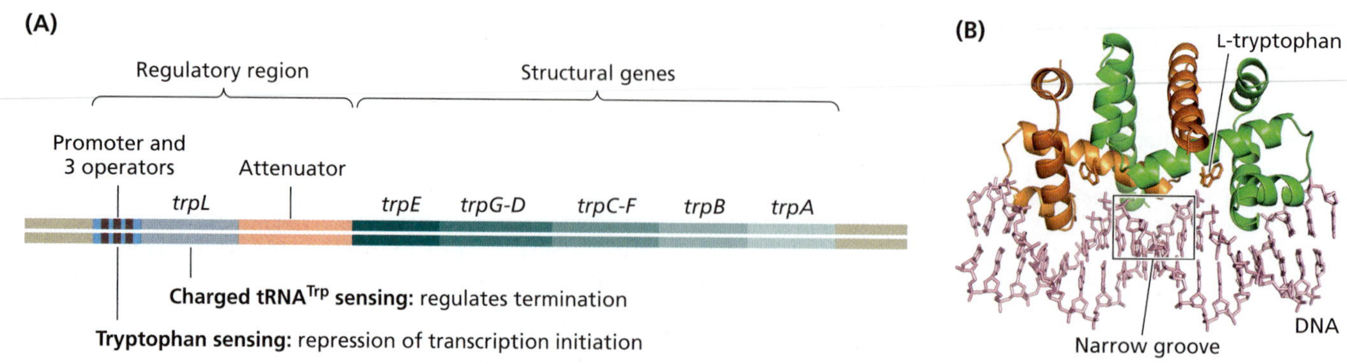

Figure 11.12 Organization and regulation of the *trp* operon in *E. coli*. The genes required for tryptophan biosynthesis are organized and regulated as a single transcriptional unit or operon. (A) Seven genetic elements, *trpEGDCFBA*, are under the control of a regulatory region located at the beginning of the operon. Two pairs of genes are fused: *trpG* with *trpD*, and *trpC* with *trpF*; separate polypeptide segments of the products of these fused genes catalyze different reactions in the tryptophan biosynthesis pathway. The regulatory region senses and responds to two different signals through two different mechanisms. The first signal is the concentration of tryptophan in the cell: the response to tryptophan is exerted through regulation of binding of the Trp repressor to the operator elements. The second signal is the proportion of $tRNA^{Trp}$ that is attached to tryptophan, or charged, which depends on the concentration of free tryptophan in the cell; the response to this signal is exerted through the attenuation mechanism (see Figure 11.13). (B) Trp repressor bound to one of the operator sites. Trp repressor is a homodimer; one subunit is colored green and the other orange. The recognition helices of the DNA-binding HTH motifs insert themselves into the major groove of the DNA binding site. In the structure of the protein dimer itself, the distance between the DNA-interacting helices on the two monomers is 26.5 Å; upon binding of tryptophan, the dimer undergoes a conformational change that increases the distance between the DNA-interacting helices to 32.7 Å. This change is instrumental in securing a good fit between the Trp repressor and a *trp* operator site; in the protein–DNA complex the distance is 32.1 Å. Binding of repressor to the operator site interferes with the ability of the DNA to bind RNAP, preventing transcription initiation. Simultaneous binding of repressor dimers to adjacent operator sites in the promoter–operator region increases the stability of the repressor–DNA complex and thus the effectiveness of repression. (A, adapted from Yanofsky C [2007] *RNA* 13:1141–1154. With permission from The RNA Society.)

This ribosome physically blocks the formation of hairpin 2-3, with the consequence that hairpin 3-4 can form as soon as the polymerase has passed (see Figure 11.13). Hairpin 3-4 is followed immediately by a stretch of seven adenine nucleotides, A, in the DNA, which transcribes into seven uracil nucleotides, U, in the RNA. But this is the exact condition, a hairpin followed by several U residues, that we encountered in Chapter 9 as the intrinsic termination signal. Thus, under these conditions, the polymerase encounters a termination signal and is released, and transcription of the operon is aborted.

Now consider what happens when tryptophan is in short supply and needs to be made. Attenuation must somehow be overridden. Within the leader sequence is a pair of adjacent Trp codons. If tryptophan is scarce, it will be difficult for the ribosome to find enough charged $tRNA^{Trp}$ to allow it to translate the transcript beyond these Trp codons. Accordingly, it stalls within the leader sequence while the polymerase proceeds on its way. The stalled ribosome allows the 2-3 hairpin to form, which disrupts the formation of the 3-4 hairpin (see Figure 11.13). Consequently, the termination signal is not formed, and the polymerase can proceed to transcribe the rest of the *trp* operon.

The same protein can serve as an activator or a repressor: the ara *operon*

Finally, we describe one more common bacterial operon that has some distinctive features. Bacteria can utilize the sugar arabinose, metabolizing it into xylulose 5-phosphate, which can enter the pentose phosphate pathway. The *ara* operon codes for three enzymes involved in this pathway, AraB, AraA, and AraD, plus the regulatory protein AraC. The *araB*, *araA*, and *araD* genes have a separate promoter; *araC* is transcribed in the opposite direction (**Figure 11.14A**). There are four AraC binding sites: $araO_1$ and $araO_2$ are located upstream from the *araC* promoter, and $araI_1$ and $araI_2$ are located between the two promoters.

AraC can bind arabinose, and its choice of DNA binding sites depends upon whether or not arabinose is present. In the absence of arabinose, AraC binds as a dimer between

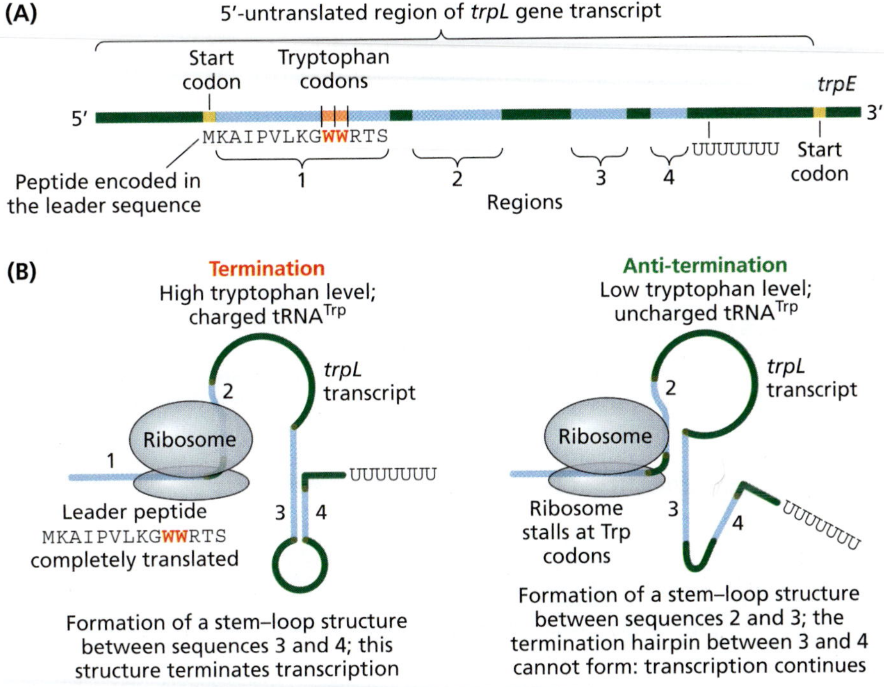

Figure 11.13 Attenuation as a regulating mechanism in the *trp* operon. Functioning of the attenuation mechanism depends on the two-dimensional structure of the leader transcript and on translation, or protein synthesis, in that region, which is coupled to transcription. (A) The first 141 nucleotides of the *trp* operon transcript can fold into three alternative RNA hairpin structures involving regions 1, 2, 3, and 4, bracketed below the transcript. (B) Left, formation of a stem–loop structure between regions 3 and 4 leads to termination of transcription through the typical intrinsic termination process. Right, the competing formation of a stem–loop structure between regions 2 and 3 causes anti-termination; that is, it allows transcription to continue because of the absence of the terminator 3–4 hairpin. The formation of terminator versus anti-terminator structures is controlled through the coupled translation process. Note that the 5′-untranslated region, 5′-UTR, of the *trp* operon transcript contains a small nucleotide sequence that can be translated into a short leader peptide that contains two adjacent tryptophan residues, WW. This peptide sequence is shown alongside the corresponding region of the operon. As soon as the 5′-UTR is transcribed, a ribosome is attached to the ribosome-binding site immediately preceding the start codon and initiates translation. Whenever the level of charged $tRNA^{Trp}$ is adequate for rapid translation of the two Trp codons, translation of the leader peptide is completed, the ribosome dissociates, and hairpin structures 1–2 and 3–4 form: transcription is terminated. When there is not enough tryptophan to charge all $tRNA^{Trp}$, the ribosome stalls at one of the two Trp codons, allowing for formation of the 2–3 hairpin; this, in turn, prevents the formation of the terminator 3–4 structure and the operon is actively transcribed. The regulation of the entire operon is thus subject to a two-level feedback response to the presence of tryptophan, the final product of the synthetic pathway encoded in the operon. (Adapted, courtesy of Michael King, Indiana University.)

$araO_2$ and *$araI_1$*. As shown in **Figure 11.14B**, this forces a loop in the DNA, which has the effect of blocking transcription from the *araBAD* promoter, repressing the operon. That the loop is critical was demonstrated in an elegant way by Lobell and Schleif, who inserted an extra five base pairs between *$araO_1$* and *$araO_2$*. This produced an extra half-twist in the DNA, rotating the *$araO_1$*–*$araO_2$* binding sites by 180° with respect to each other and thus hindering the interaction between AraC and these two binding sites. The effect was to relieve repression of *araBAD*. Adding or subtracting a full 10-base-pair turn had no effect, serving as an elegant control.

In the presence of arabinose, AraC no longer causes loop formation or acts as a repressor. Instead, as shown in Figure 11.14B, it binds as a dimer to the *$araI_1$* and *$araI_2$* sites and promotes adjacent binding of CAP protein. These effects stimulate transcription

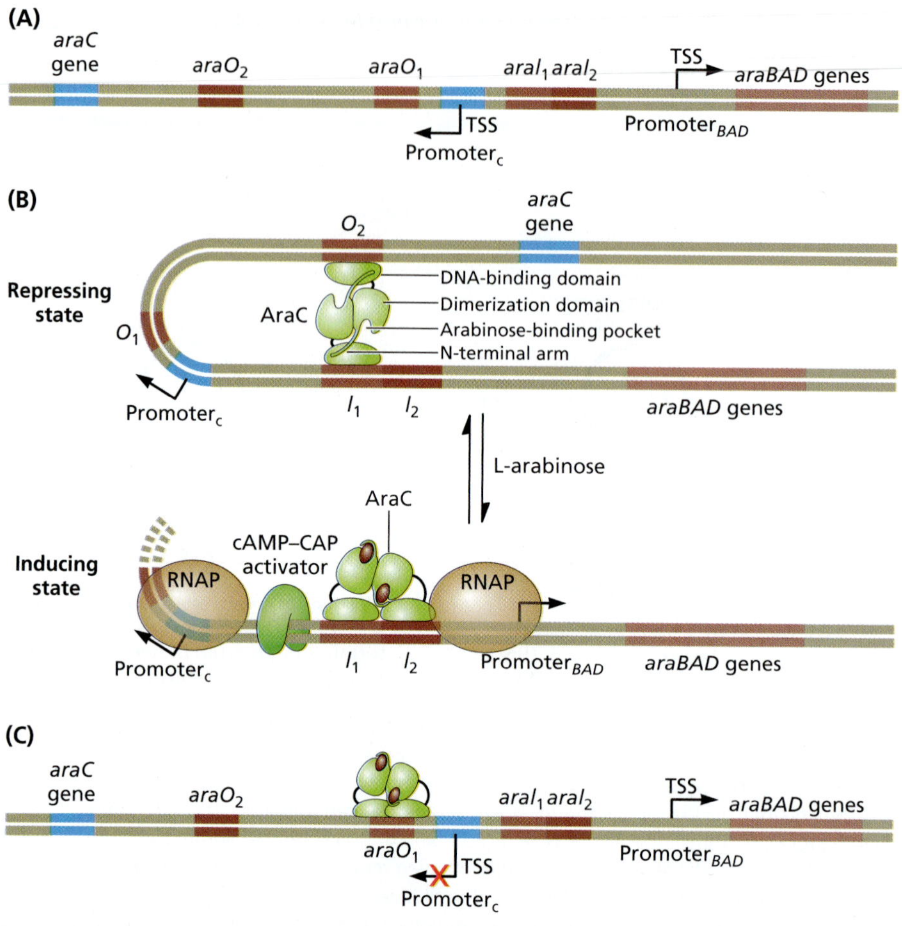

Figure 11.14 Transcriptional control of the *ara* operon in *E. coli*. (A) Map of the arabinose control region. Four binding sites for the transcriptional factor AraC (shown in brown) are located just upstream of the promoter for the arabinose structural genes, *araB*, *araA*, and *araD*. The promoter for the *araC* gene drives transcription from the opposite DNA strand. (B) Negative control of transcription, in the absence of arabinose, and positive control of transcription, in the presence of arabinose, through the action of AraC, which acts as a dimer. Each monomer has a C-terminal DNA-binding domain connected through a flexible linker to the N-terminal dimerization domain, which also contains the arabinose-binding pocket. In the absence of arabinose, the two monomers interact through their N-terminal tails, so that the two C-terminal domains bind to the *araO*$_2$ and *araI*$_1$ regulatory elements, thus forcing the intervening DNA to loop out. The promoter is inaccessible to polymerase binding in this configuration. When arabinose becomes available, it binds to AraC, changing the conformation of the dimer such that it now binds to the *araI*$_1$ and *araI*$_2$ sequences; this opens up the *araBAD* promoter to bind the polymerase. For full transcriptional activation, glucose should be absent from the medium. This condition is sensed by the levels of cAMP in the cell, which increase when glucose is absent, and thus the cAMP-bound positive regulator CAP binds to the promoter region of the *araC* gene to up-regulate its transcription. The response of the cell to lack of glucose is exactly the same as in the case of *lac* operon regulation. (C) Autoregulation of the *araC* gene. This occurs through utilization of the *araO*$_1$ element, which does not participate in regulation of the *araBAD* genes; it instead allows AraC to regulate its own synthesis. As levels of AraC increase, it binds to *araO*$_1$, thus preventing polymerase binding to the *araC* promoter and inhibiting the transcription of its own gene from the bottom DNA strand. (B, adapted from Weldon JE, Rodgers ME, Larkin C & Schleif RF [2007] *Proteins* 66:646–654. With permission from John Wiley & Sons, Inc.)

of the *araBAD* genes. Thus, AraC can act as either a repressor or an activator, depending on the concentration of arabinose.

An overabundance of AraC is avoided by a simple autoregulatory mechanism. Note that *araO*$_1$ is just downstream, and in the *araC* gene orientation, from the *araC* promoter. High concentrations of AraC will tend to bind at this site, repressing its own gene via a roadblock mechanism (**Figure 11.14C**). Note the contrast between *trp* and *ara* operon regulation. In the *trp* operon, the end product of a metabolic pathway regulates transcription of the operon; the *ara* operon is regulated by the first substrate of the metabolic pathway. There are many different modes of transcriptional control.

11.5 Other modes of gene regulation in bacteria

DNA supercoiling is involved in both global and local regulation of transcription

As we see in Chapter 8, the bacterial chromosome is organized in supercoiled loops. The degree of superhelical stress in individual **supercoiled domains** is finely regulated by topoisomerases (see Chapter 4) and by structural proteins that twist, bend, or loop the DNA upon binding. In addition, enzymes that translocate along the DNA, such as RNA polymerase and helicases, create temporary waves of positive supercoiling in front and negative supercoiling in their wake.

The level of **global supercoiling** is different under different environmental conditions. Gyrase, the enzyme that pumps negative superhelical stress into DNA, utilizes ATP to do so. Thus, the overall cellular energetics play a role in determining the global level of supercoiling. It is well known that the ratio [ATP + $^{1}/_{2}$ADP]/[ATP + ADP + AMP], known as **energy charge**, drops when *E. coli* cells are shifted from aerobic to anaerobic conditions. This shift is accompanied by a decrease in the overall level of DNA supercoiling. Under optimal growth conditions, the energy charge is high, ~0.85, and the superhelical density σ equals –0.05. Under anaerobic conditions, σ drops to ~–0.038. As is not difficult to imagine, any control on transcription that is exerted by global levels of supercoiling must be rather crude. However, the factors that control local levels of supercoiling serve to fine-tune the nonspecific global effects.

How can local supercoiling affect transcription? The first mechanism involves changes in the twist of the double-helical DNA. A change in twist changes the base-pair spacing in the affected region and thus the physical distance between two sites. The level of supercoiling that gives optimal expression depends on the number of base pairs in the spacer region between the –35 and –10 regions of the promoter (**Figure 11.15**). The optimal orientation between the two regions, which maximizes the ability of σ^{70} to locate and bind to a promoter, depends on the actual physical length of the –35 to –10 spacer. Promoters with 17-bp spacers have been optimized during evolution for the normal physiological levels of supercoiling. Longer spacers will have to be overtwisted, and shorter spacers will have to be undertwisted, in order to create the optimal orientation of the recognition elements for σ^{70} binding.

DNA supercoiling can also regulate transcription through changes in writhe: short regions can loop out, and long regions can form plectonemes. Both these structures can bring sequence elements into close proximity so that they can interact, either

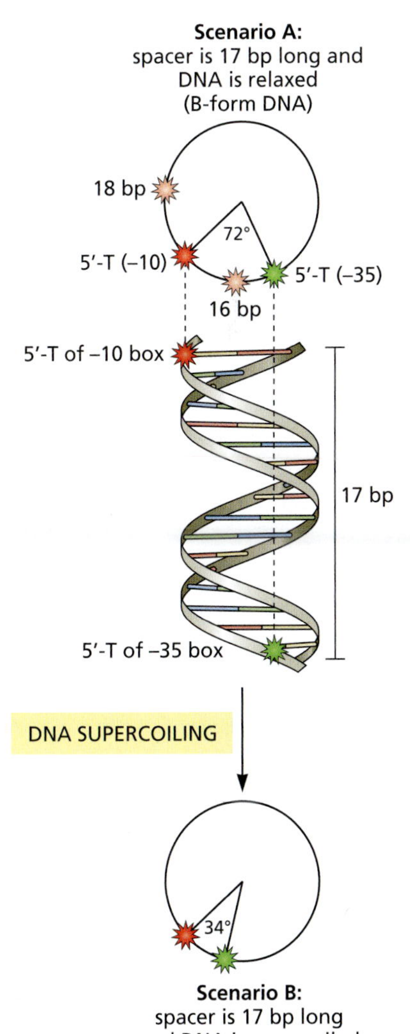

Figure 11.15 Change in the mutual disposition of the –35 and –10 promoter boxes due to changes in helical twist. The boxes are represented by their 5′-T residues, shown as green and red starbursts. The projections of the σ^{70} promoter, represented as circles, are perpendicular to the helical axis. In scenario A, the spacer is 17 bp long and the DNA is relaxed; that is, the average rotation angle is 34.5° per base pair, as in B-DNA. In this case, the angle between the two 5′-Ts in the projection is 72°. For reference, the positions of the –10 box Ts if the spacer were 16 or 18 bp long are also shown, as pink starbursts. In scenario B, the DNA is now negatively supercoiled or untwisted so that the superhelical density σ = –0.065, and the angle between the 5′-Ts of the two promoter boxes decreases to 34°, bringing them much closer to each other in the projection plane. In other words, the two boxes are now much closer to being on the same side of the helix. This scenario presumes that $^{2}/_{3}$ of the change in linking number is absorbed by a change in twist and $^{1}/_{3}$ by a change in writhe. As can be seen from comparison of the two projections, the two boxes in the supercoiled DNA molecule that are spaced 17 bp apart will be located similarly with respect to each other as would the two boxes spaced by 16 bp in the relaxed molecule.

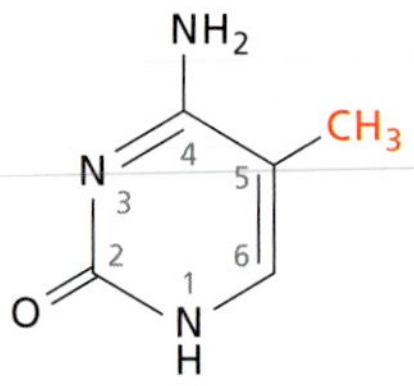

directly or through the proteins bound to them. An example of a loop structure that can inhibit RNAP binding is presented by the regulation of the *ara* operon in the absence of arabinose (see Figure 11.14).

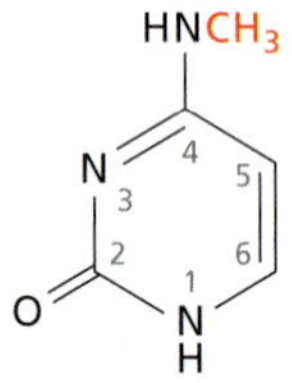

Finally, transcription can be regulated through the formation, under negative stress, of **alternative DNA structures** such as locally denatured regions, cruciforms, Z-DNA, or triplex DNA. If these form downstream of the promoter, transcription elongation will be inhibited. Alternative structures in regulatory regions can also serve as recognition elements for binding of TFs or the polymerase itself.

DNA methylation can provide specific regulation

DNA methylation in bacteria is widespread and can occur on both cytosines and adenines. The three recognized base modifications are 5-methylcytosine, ^{m5}C; N^6-methyladenine, ^{m6}A; and N^4-methylcytosine, ^{m4}C (**Figure 11.16**). The first two modifications occur in bacteria, protists, and fungi, whereas the third is bacteria-specific. These marks can be propagated from generation to generation because of the capacity of the respective enzymes to recognize and act upon hemimethylated DNA. This is the state of the DNA immediately following replication: the parental strand keeps its methyl groups, whereas for a while the newly synthesized DNA strand remains unmethylated. Methyltransferases then add a new methyl group to the new DNA strand (see Figure 12.25B). The specific enzymes that introduce these modifications are DNA cytosine methyltransferase or Dcm for ^{m5}C, DNA adenine methyltransferase or Dam for ^{m6}A, and cell-cycle-regulated methyltransferase or CcrM for ^{m6}A; different bacterial taxa may contain any one of these. ^{m4}C is modified by members of restriction/modification enzyme systems. Thus, DNA methylation is a true epigenetic mark. For more discussion on epigenetics, see Chapter 12. DNA methylation is used by numerous cellular processes, including mismatch repair (see Chapter 22).

As far as transcription regulation is concerned, the methylation of adenine in GATC sequences in *cis*-regulatory elements can either increase or decrease transcription by affecting regulatory protein binding. An interesting example of how the expression of pathogenic genes in uropathogenic *E. coli* is regulated through DNA methylation is presented in **Box 11.3**.

5-methylcytosine
^{m5}C

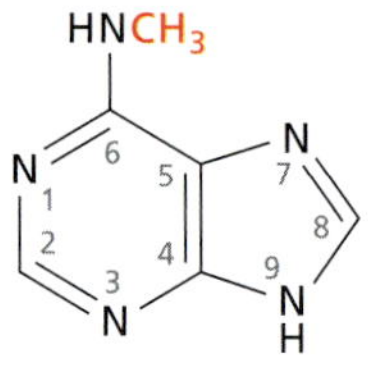

N^6-methyladenine
^{m6}A

N^4-methylcytosine
^{m4}C

Figure 11.16 DNA methylation in bacteria. Bacterial DNA can be methylated at cytosine and adenine nucleotides, as shown in the structures.

11.6 Coordination of gene expression in bacteria

Regulation of gene expression in bacteria is not restricted to the individual behavior of single genes or even operons. Rather, there are multiple levels of coordination. We have already discussed the local control of individual operons. The DNA-binding proteins that regulate specific operons are usually highly specific to the regulatory regions of the respective operon and are present in relatively small amounts, sometimes only a few molecules per cell. A higher level of control is seen when sets of operons form **regulons**, whose expression is regionally controlled. The operons constituting a regulon participate in a common function, such as utilization of nitrogen or carbon, and share common regulators, either activators or repressors, that recognize DNA sequences common to all member operons and respond to nutrient or environmental conditions. The regulon regulators are more abundant than those of individual operons and bind to multiple targets. Regulation of the heat-shock regulon is an example (see Figure 9.11). This regulation involves the use of alternative σ factors in succession, with $\sigma^{32/H}$ being the factor recognizing the promoters of heat-shock operons.

Multiple regulons may also be controlled in coordination; sets of such regulons have been termed **stimulons** or modulons. Operons in a modulon may be under individual controls as well as under the control of common, pleiotropic regulatory proteins. For example, the CAP modulon contains all regulons and operons, including the *lac* operon and *ara* regulon, that are regulated by cAMP-bound CAP; each operon has other regulators as well. Finally, there are **global controls of overall expression patterns** (**Figure 11.17**). The global level of DNA supercoiling represents one such global control.

Box 11.3 Regulation of bacterial pathogenicity by phase variation Phase variation is defined as the reversible generation of variant bacteria that differ in the presence or absence of specific surface antigens. We illustrate the regulation of phase variation with the example of synthesis of pyelonephritis-associated pili or Pap in uropathogenic *E. coli.* Synthesis of pili is either turned ON or OFF, producing two bacterial populations: one with pili that are capable of adhering to the mucosa in the urinary tract, which are pathogenic, and the other without pili, which are nonpathogenic. The switch in Pap expression is controlled at the transcriptional level by a mechanism involving DNA methylation by Dam methyltransferase (see Figure 11.16) and the leucine-responsive regulatory protein Lrp (**Figure 1**).

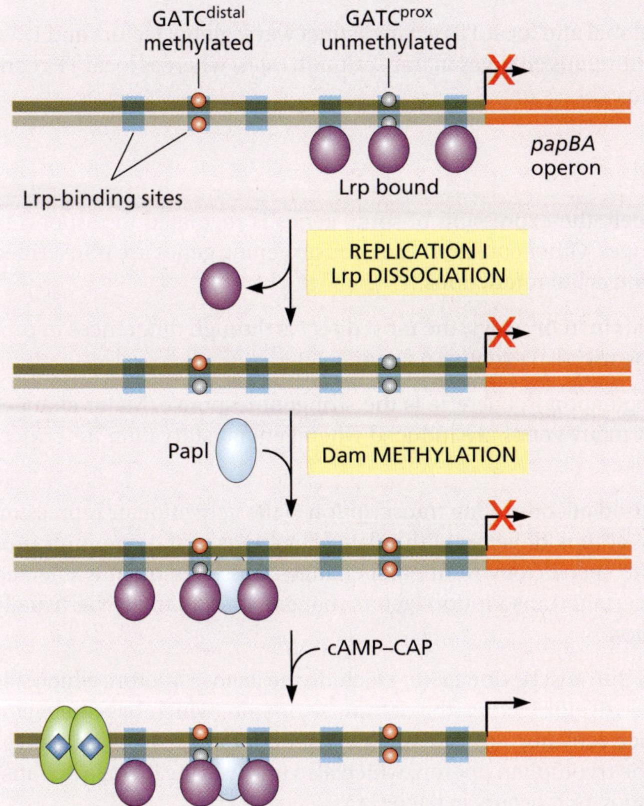

Figure 1 Switching the *pap* operon from OFF to ON phase. The upstream region of the operon contains six binding sites for leucine-responsive regulatory protein Lrp; sites 2 and 5 contain methylatable adenines in a GATC context. When the *papBA* operon is OFF, Lrp is bound to sites 1–3, closest to the transcription start site or TSS, to prevent methylation of GATC^{prox} following passage of the replication fork (see Chapter 19). Lrp binding to sites1–3 reduces the affinity of Lrp for sites 4–6 and RNA polymerase binding. Every round of replication offers a window of opportunity to switch the transcriptional status from OFF to ON, the necessary condition being the presence of PapI. The GATC^{distal} sequence at Lrp binding site 5 is not protected from methylation because Lrp is not bound, and thus it is methylated in the OFF state. The OFF state is perpetuated by the high affinity of Lrp for unmethylated GATC and its inability to bind to its methylated distal counterpart. The switch to the ON state is initiated by the activity of the ancillary protein PapI, which tends to move Lrp to the distal GATC motif; this allows the methylation of Lrp binding site 2 and protects the distal GATC motif from methylation. Thus the ON state is defined by passive demethylation, during replication, of GATC^{distal} and methylation of GATC^{prox}. Activation is dependent on the presence of cAMP–CAP, as already illustrated with the *lac* operon activation example. One of the *pap* products, PapB, activates *papI* transcription, thus providing a positive-feedback loop that propagates the ON state. Note that switching in both directions requires DNA replication because DNA methylation on both strands and passive DNA demethylation occur in two consecutive hemi-steps.

Networks of transcription factors form the basis of coordinated gene expression

The different levels of transcriptional regulation in bacteria must rely on highly interactive networks of transcription factors, TFs. Often two or more TFs control the transcription of a single gene; on the other hand, sets of target genes in regulons are co-regulated by the same set of TFs. It is as if the organism were conducting a sophisticated computer search using a Boolean algorithm with keywords: transcribe if *x* AND *y* but NOT *z*, and so on. By use of bioinformatics tools, it is now possible to unravel transcription regulatory networks, which depict the complex interplay of individual TFs in the overall regulation of transcriptional activity of the bacterial genome.

In *E. coli*, these networks are dominated by about 10 global TFs (see Figure 11.17). Many of these global regulators represent nucleoid-associated factors, such as IHF and H-NS, that control the organization of the bacterial chromosome (see Chapter 8). CAP, the activator of the *lac* and *ara* operons described above, is also a global TF: it regulates numerous operons that form the CAP-regulated modulon. Local TFs usually act in concert with global TFs. Most of the local TFs tend to be encoded in DNA neighboring the regulated gene or operon. This proximity is explained by the fact that small amounts of local factors must find their way to the regulated gene without being diluted in the large cellular volume. Often, this search is along the genome, not through space. As a rule, global TFs do not regulate each other but regulate more

(A)

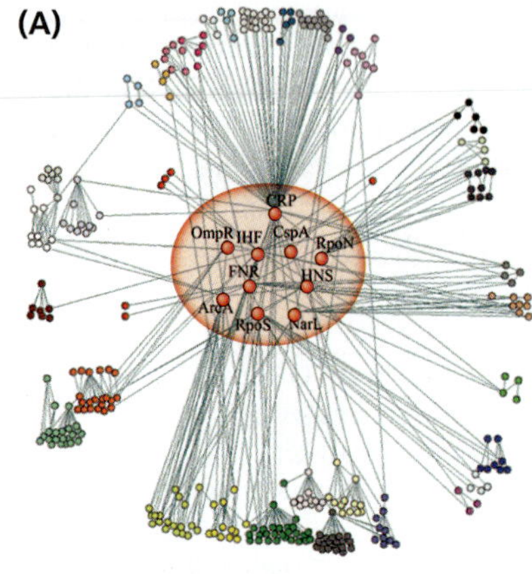

(B)

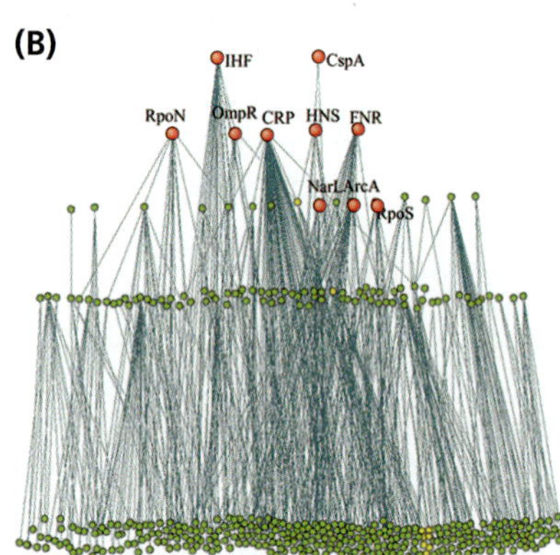

Figure 11.17 Transcriptional regulatory network of *E. coli*. (A) Operons can be organized into modules; different modules are shown in different colors. The 10 global regulators shown inside the oval form the core part of the network. The peripheral modules are connected mainly through the global regulators. (B) A different representation of the hierarchical regulation structure, in which all of the regulatory links are downward. Nodes in the graph are operons. Links show transcriptional regulatory relationships. Global regulators are labeled at the top. (From Ma H-W, Buer J & Zeng A-P [2004] *BMC Bioinformatics* 5:199. With permission from BioMed Central.)

local TFs. In addition, global and local TFs act in distinct ways; global factors and DNA supercoiling induce continuous changes in transcription rates, whereas local TFs control ON/OFF switches in expression.

Key concepts

- Within a bacterial cell, the expression of some genes is regulated to respond to environmental changes. Other constitutive or housekeeping genes are transcribed uniformly to maintain cellular functions.
- Regulation may occur in many ways: the most direct is through differences in promoter strength or the use of alternative σ factors.
- A general type of regulation in bacteria is the stringent response. Major changes in the expression of many genes are induced when general starvation for critical nutrients occurs.
- A more precise method of controlling transcription is the activation or repression of specific operons, groups of genes with related functions and a common promoter, in response to specific environmental changes. This is frequently effected by the presence of certain transcription factors that can bind to the DNA, usually in the promoter region.
- Examples of control by transcription factors include the lactose operon, which has both repressor and activator factors; the arabinose operon, which has one factor that can act as either activator or repressor through a mechanism that involves DNA looping; and the tryptophan operon, which also involves the linkage of translation and transcription that occurs in bacteria.
- Transcription factors are usually allosteric proteins, in which one or another conformational state with specific DNA-binding capabilities is favored by the absence or presence of a small effector molecule.
- Bacterial transcription can also be regulated by the control of DNA supercoiling, on either a global or local scale.
- DNA methylation can also affect the transcription of particular genes.
- Regulation of transcription in bacteria is hierarchical; there are several distinguishable levels of control. The most specific control occurs at the operon level. However, sets of operons with related functions, called regulons, may have common control elements, and several functionally related regulons may be grouped into a few modulons.

Further reading

Books

Dame RT & Dorman CJ (eds) (2010) Bacterial Chromatin. Springer Verlag.

Wagner R (2000) Transcription Regulation in Prokaryotes. Oxford University Press.

Reviews

Balleza E, López-Bojorquez LN, Martinez-Antonio A et al. (2009) Regulation by transcription factors in bacteria: Beyond description. *FEMS Microbiol Rev* 33:133–151.

Browning DF & Busby SJ (2004) The regulation of bacterial transcription initiation. *Nat Rev Microbiol* 2:57–65.

Crooks GE, Hon G, Chandonia J-M & Brenner SE (2004) WebLogo: A sequence logo generator. *Genome Res* 14:1188–1190.

Dillon SC & Dorman CJ (2010) Bacterial nucleoid-associated proteins, nucleoid structure and gene expression. *Nat Rev Microbiol* 8:185–195.

Dorman CJ & Deighan P (2003) Regulation of gene expression by histone-like proteins in bacteria. *Curr Opin Genet Dev* 13:179–184.

Gao R & Stock AM (2010) Molecular strategies for phosphorylation-mediated regulation of response regulator activity. *Curr Opin Microbiol* 13:160–167.

Gruber TM & Gross CA (2003) Multiple sigma subunits and the partitioning of bacterial transcription space. *Annu Rev Microbiol* 57:441–466.

Hatfield GW & Benham CJ (2002) DNA topology-mediated control of global gene expression in *Escherichia coli*. *Annu Rev Genet* 36:175–203.

Henkin TM & Yanofsky C (2002) Regulation by transcription attenuation in bacteria: How RNA provides instructions for transcription termination/antitermination decisions. *BioEssays* 24:700–707.

Jacob F (1966) Genetics of the bacterial cell. In Les Prix Nobel: The Nobel Prizes, 1965 (Grandin K ed). Nobel Foundation.

Magnusson LU, Farewell A & Nyström T (2005) ppGpp: A global regulator in *Escherichia coli*. *Trends Microbiol* 13:236–242.

Pruss GJ & Drlica K (1989) DNA supercoiling and prokaryotic transcription. *Cell* 56:521–523.

Roberts JW (2009) Promoter-specific control of *E. coli* RNA polymerase by ppGpp and a general transcription factor. *Genes Dev* 23:143–146.

Schleif R (2003) AraC protein: A love–hate relationship. *BioEssays* 25:274–282.

Schleif R (2010) AraC protein, regulation of the L-arabinose operon in *Escherichia coli*, and the light switch mechanism of AraC action. *FEMS Microbiol Rev* 34:779–796.

Srivatsan A & Wang JD (2008) Control of bacterial transcription, translation and replication by (p)ppGpp. *Curr Opin Microbiol* 11:100–105.

Stock AM, Robinson VL & Goudreau PN (2000) Two-component signal transduction. *Annu Rev Biochem* 69:183–215.

Travers A & Muskhelishvili G (2005) Bacterial chromatin. *Curr Opin Genet Dev* 15:507–514.

Travers A & Muskhelishvili G (2005) DNA supercoiling—A global transcriptional regulator for enterobacterial growth? *Nat Rev Microbiol* 3:157–169.

van Hijum SAFT, Medema MH & Kuipers OP (2009) Mechanisms and evolution of control logic in prokaryotic transcriptional regulation. *Microbiol Mol Biol Rev* 73:481–509.

van Holde K & Zlatanova J (1994) Unusual DNA structures, chromatin and transcription. *BioEssays* 16:59–68.

Wang J-Y & Syvanen M (1992) DNA twist as a transcriptional sensor for environmental changes. *Mol Microbiol* 6:1861–1866.

Wilson CJ, Zhan H, Swint-Kruse L & Matthews KS (2007) The lactose repressor system: Paradigms for regulation, allosteric behavior and protein folding. *Cell Mol Life Sci* 64:3–16.

Wion D & Casadesús J (2006) N^6-methyl-adenine: An epigenetic signal for DNA–protein interactions. *Nat Rev Microbiol* 4:183–192.

Yanofsky C (2007) RNA-based regulation of genes of tryptophan synthesis and degradation, in bacteria. *RNA* 13:1141–1154.

Experimental papers

Jacob F & Monod J (1961) Genetic regulatory mechanisms in the synthesis of proteins. *J Mol Biol* 3:318–356.

Lewis M, Chang G, Horton NC et al. (1996) Crystal structure of the lactose operon repressor and its complexes with DNA and inducer. *Science* 271:1247–1254.

Lobell RB & Schleif RF (1990) DNA looping and unlooping by AraC protein. *Science* 250:528–532.

Monod J, Changeux J-P & Jacob F (1963) Allosteric proteins and cellular control systems. *J Mol Biol* 6:306–329.

Monod J, Wyman J & Changeux J-P (1965) On the nature of allosteric transitions: A plausible model. *J Mol Biol* 12:88–118.

Soisson SM, MacDougall-Shackleton B, Schleif R & Wolberger C (1997) Structural basis for ligand-regulated oligomerization of AraC. *Science* 276:421–425.

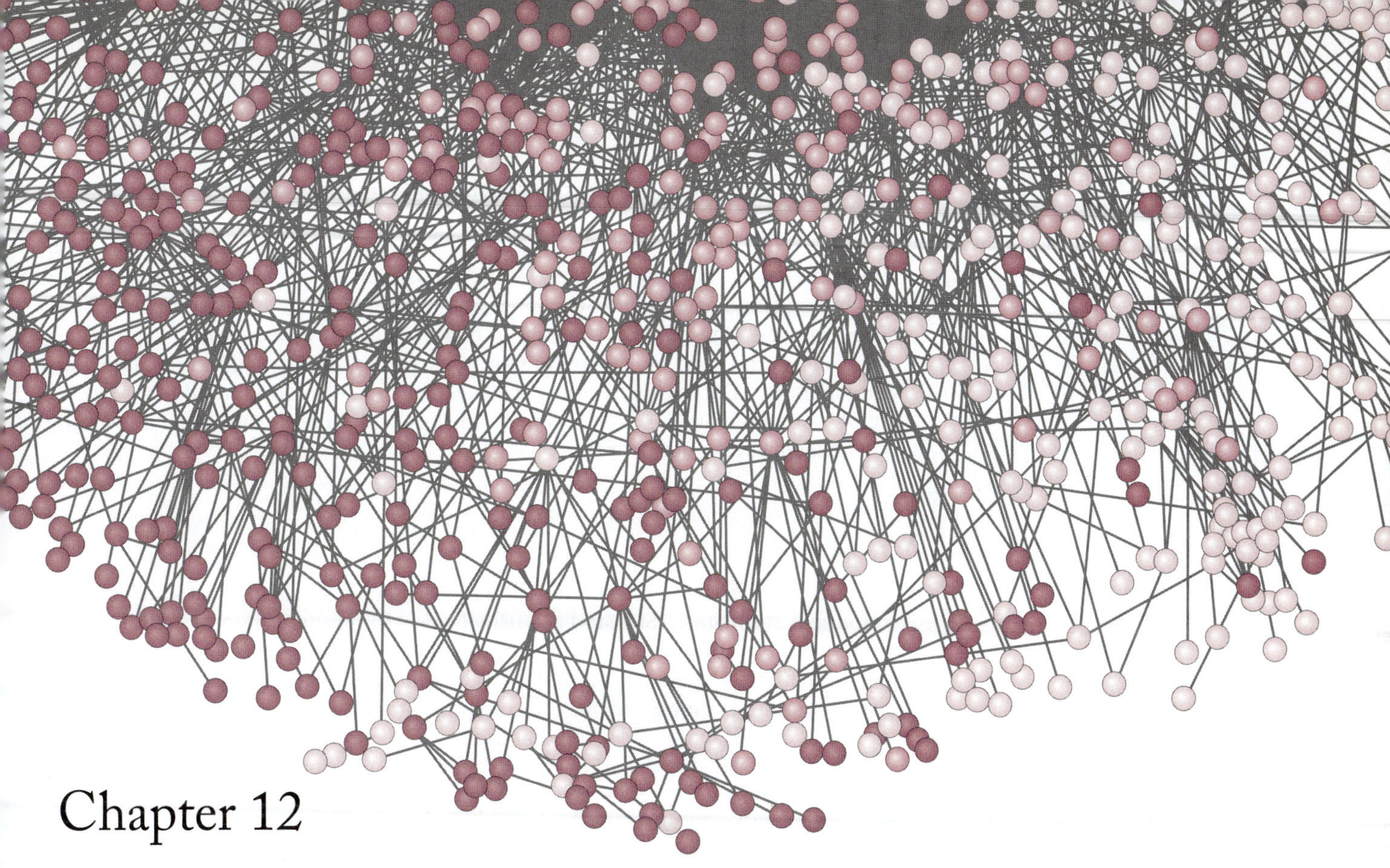

Chapter 12

Regulation of Transcription in Eukaryotes

12.1 Introduction

Eukaryotic organisms must regulate their transcriptional output not only in response to environmental signals, as do bacteria, but also to adapt to programmed developmental or differentiation cues. The diversity of these responses has demanded an enormous complexity in regulatory mechanisms. Nevertheless, the basic principle of transcriptional regulation in bacteria that we presented in Chapter 11 is also valid for eukaryotes: *cis* elements in the DNA interact with *trans* factors, which can be either proteins or noncoding RNA molecules. However, two other levels of regulation were added in the evolution from bacteria to eukaryotes: regulated changes in chromatin structure or in DNA methylation provide a major part of the overall landscape of eukaryotic transcriptional regulation.

Unfortunately, the topic of eukaryotic transcription is made particularly difficult by the associated nomenclature, which has evolved without rules or consistency. Often, very similar proteins in different organisms have been given entirely different names and acronyms at their discoverers' will. Systematization is needed but has only just begun.

In this chapter, we provide the background to and an overview of this complex field, but the sequencing of the human genome has opened whole new vistas. Analysis of the genome sequence combined with studies of cellular RNA populations, transcription-factor binding, chromatin structure, and DNA methylation, provided by the ENCODE project, have begun to unravel the broader functional aspects of the genome. These analyses have provided so much new insight, and promise of more, that we felt it appropriate to incorporate this material separately in Chapter 13.

12.2 Regulation of transcription initiation: Regulatory regions and transcription factors

The demands for exquisite gene regulation in eukaryotes have led to the appearance of complex arrays of regulatory elements, both close/proximal to and sometimes far/distal from the core promoter region. In considering any aspect of transcription in eukaryotes, it must be kept in mind that transcription occurs within the context of chromatin, so the accessibility of any of these regulatory elements is modulated by the local chromatin structure. We discuss the effects of chromatin structure on transcription in detail later in this chapter.

Core and proximal promoters are needed for basal and regulated transcription

The promoter regions of eukaryotic genes are located immediately upstream of transcriptional start sites, TSSs, and are usually several hundred base pairs in length. Frequently, the promoter region is found to be nucleosome-free (see Chapter 13 and this chapter). The entire promoter can be subdivided into **core promoter** region and **proximal promoter** region. The core promoter is the docking site for the basal transcriptional machinery. Recall that this machinery includes the polymerase and any associated factors needed to establish the pre-initiation complex. The core promoter contains sequence elements that are instrumental in the formation of this complex (see Chapter 10). The first characterized eukaryotic promoter contained a TATA box, but as we have seen, there are a variety of core promoters that bind to different protein factors: see Figure 10.7 for Pol II promoters and Figure 10.16 for Pol III promoters. Recent evidence shows that only ~20% of eukaryotic promoters carry TATA boxes.

Proximal promoter regions, which generally lie just upstream of the core promoter, serve to regulate the activity of the core promoter. These promoter regions can also be involved in the regulation of distant genes, as part of complex regulatory networks (see Chapter 13). Proximal promoters are highly gene-specific as they bind a variety of factors that regulate the expression of specific genes. For example, in the human *Hsp70* gene (**Figure 12.1**) both the core promoter and the proximal promoter region are complex, each with numerous sequence elements and protein-binding sites. Regulatory elements often act synergistically; their synergism results from physical interactions between transcription factors or TFs that are bound to them. In yeast, the proximal promoter elements are replaced by a single **upstream activating sequence** or **UAS**; a single UAS can control multiple regulatory elements.

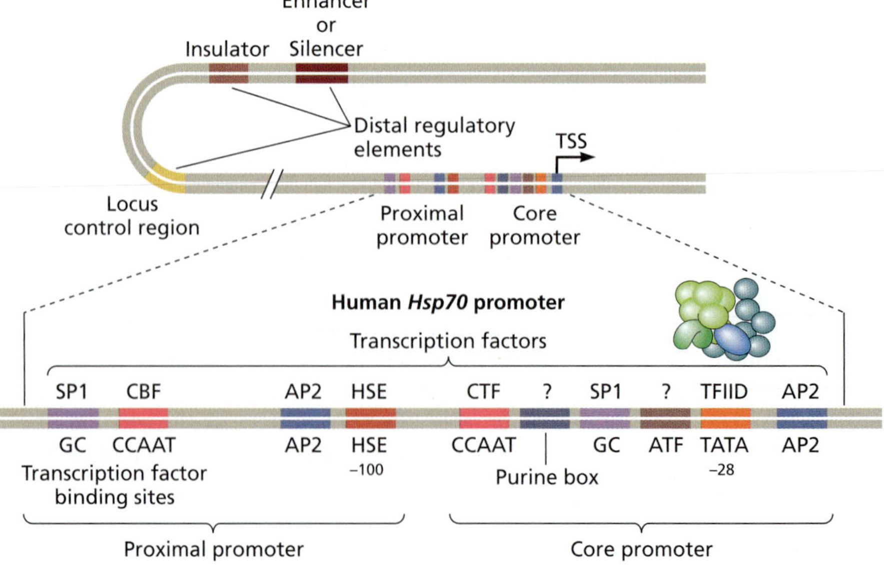

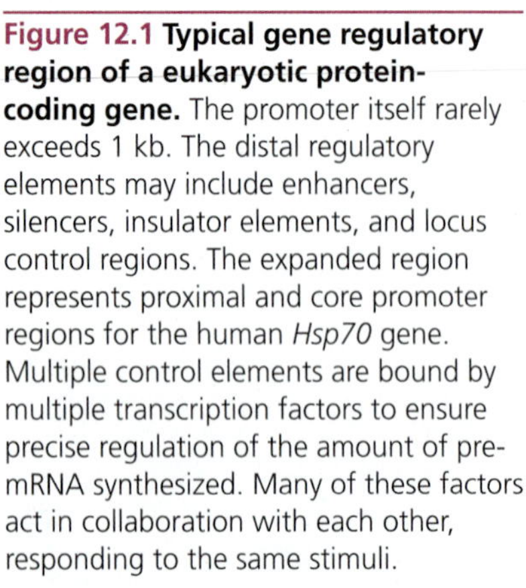
Figure 12.1 Typical gene regulatory region of a eukaryotic protein-coding gene. The promoter itself rarely exceeds 1 kb. The distal regulatory elements may include enhancers, silencers, insulator elements, and locus control regions. The expanded region represents proximal and core promoter regions for the human *Hsp70* gene. Multiple control elements are bound by multiple transcription factors to ensure precise regulation of the amount of pre-mRNA synthesized. Many of these factors act in collaboration with each other, responding to the same stimuli.

Enhancers, silencers, insulators, and locus control regions are all distal regulatory elements

The regulation of eukaryotic gene expression has still more dimensions to it: there are numerous distal regulatory elements outside the promoter that act in response to signaling pathways to either enhance or inhibit the transcription of a particular gene.

Enhancers and silencers, although they have opposite effects on gene activity, seem to share some basic features: (1) they are capable of acting on a gene from long distances, up to several kilobase pairs away; (2) they can be located on either side of the gene or even within an intron; and (3) they act in an orientation-independent manner, that is, these elements would work even if they were flipped over with respect to the sequence they regulate. Each enhancer may be capable of binding a number of transcription factors.

The mechanisms that allow enhancers and silencers to act from a distance are still not entirely clear. Enhancers are better understood, and two models prevail. In the **looping model of enhancer or silencer action**, transcription factors that bind to specific sequences in the enhancer or silencer interact with protein factors that are bound to the promoter, with the intervening DNA looping out. The alternative **scanning or tracking mechanisms of enhancer action** posit that activators originally bound to enhancers move continuously along DNA until they encounter other proteins that are already bound to the promoter. If, as expected, the tracking mechanism involves actually moving along the backbone of the double-helical DNA, then it may affect the superhelical tension in the DNA template and the integrity of nucleosomes. It is now known that many enhancers of developmentally regulated genes can be more than a megabase away from the target gene, sometimes with one or two active genes in between, making the scanning mechanism highly unlikely in such cases. The ENCODE project has revealed some fascinating, previously unrecognized features of enhancers, which are discussed in Chapter 13.

Insulators are regulatory regions that insulate sections of the genome from the spread of either activating or repressing influences from nearby genomic regions. Two major types of insulators have been described: those that possess **enhancer-blocking activity**, that is, they block communications between enhancers and promoters, and those that physically prevent the spread of repressive heterochromatin structures in what is known as **barrier activity**. Figure 12.2 illustrates the enhancer-blocking activity of an insulator that controls mutually exclusive gene transcription at a well-studied imprinted gene locus.

The expression patterns of the two alleles of an **imprinted gene** depend on the origin of the allele: whether it is inherited from the mother or the father. The two alleles are characterized by distinct DNA methylation patterns at specific DNA sequences, and

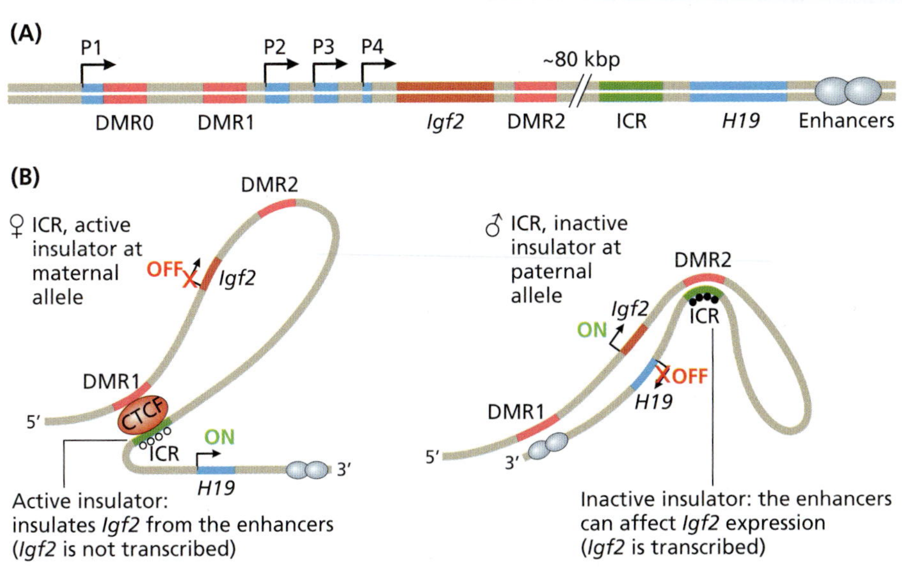

Figure 12.2 Action of enhancer-blocking insulator in control of mutually exclusive gene expression at imprinted gene loci. Differential loop formation in the maternal and paternal alleles of the *Igf2–H19* imprinted locus is responsible for the differing transcriptional status of the *Igf2* and *H19* genes in the two alleles. (A) Map of the region shows *Igf2* and *H19* genes as well as sequences involved in regulation of the locus: differentially methylated DNA regions DMR0, DMR1, and DMR2 and the imprinting control region ICR. Arrows above the blue boxes designate the start sites of alternative promoters. (B) Schematic demonstrating differential gene expression in the maternal and paternal alleles. In the maternal allele, loop formation is mediated by binding of the transcription factor CTCF to the unmethylated ICR and to another differentially methylated region, DMR1, upstream of the *Igf2* gene. This conformation creates an active insulator region that precludes utilization by the *Igf2* promoter of the two adjacent enhancers that are located downstream of the *H19* gene: hence the *Igf2* gene is inactive. In the paternal allele, the ICR is methylated and cannot bind CTCF. As a result, a different spatial conformation is formed that allows the enhancers to affect the expression of *Igf2,* and the enhancer-blocking activity of ICR is lost. The alternative spatial organization of the region in the maternal and paternal alleles regulates the expression of the *Igf2* gene; what regulates the expression of *H19* is unclear.

this distinct pattern of methylation determines whether certain proteins can bind to the sequence in question. Thus, the DNA methylation pattern is the inherited imprint that distinguishes the two alleles. During gametogenesis, specific parental patterns of DNA methylation, either maternal or paternal, are established at imprinted loci. The binding of proteins, or the lack thereof, to the differentially methylated sequences determines the alternative formation of DNA loops; some loops may separate a promoter from an enhancer, thus inhibiting transcription.

Insulators that have barrier activity are even more frequent. A good example is the insulator in the locus control region presented in **Figure 12.3**. In this case, one of the DNase I hypersensitive sites, sites in chromatin that show higher than average sensitivity to DNase I cleavage of DNA, in the region serves as a barrier to the spread of highly compacted heterochromatin from the region containing olfactory genes into the region containing globin genes. This may happen because the hypersensitive site with its bound transcription factors represents a gap in the sequence of nucleosomes (see Chapter 13).

Finally, we consider the **locus control regions** or **LCRs**, probably the most complex of all distal control elements. These regions control the developmentally regulated expression of individual genes within a gene cluster. One of the best-studied examples is the LCR of the mouse β-globin locus. The locus contains four globin genes that are expressed at different stages of erythroid cell differentiation: ε^{y} and $\beta h1$ are expressed in primitive red blood cells during early embryonic development, while β^{maj} and β^{min} are expressed in definitive red blood cells, later during embryogenesis and in the adult organism. The globin genes are embedded in the midst of numerous olfactory receptor genes that are transcriptionally inactive in erythroid cells and possess highly compacted chromatin structure. Two DNase I hypersensitive sites, HS5 and 3′HS1, flank the entire β-globin locus (see Figure 12.3). Figure 12.3 also shows the changes that occur in the organization of this locus during erythroid cell development. These changes are initiated through the action of transcription factors such as EKLF and GATA-1 and are mediated by sequence-specific binding of the transcription factor CTCF to four sites. Interestingly, this figure also shows two of these sites are upstream of the globin genes, in intergenic regions between olfactory receptor genes, while the other two fall within the DNase I hypersensitive sites that flank the locus.

The regulation of the β-globin locus also introduces a new concept in molecular biology, the **chromatin hub**, a dynamic spatial organization or clustering of genes and chromatin regions that affect the transcriptional activity of the resident genes. In erythroid progenitors, which express globin at basal levels, a poised chromatin hub is formed through interactions between CTCF-binding sites and DNase I hypersensitive sites (see Figure 12.3). During erythroid differentiation, the pairs of specific globin genes that become highly activated and the LCR stably interact with the poised chromatin hub to form a functional active chromatin hub or ACH; high β-globin gene expression is LCR-dependent. Clustering of binding sites for transcription factors in the ACH causes local accumulation of cognate proteins and associated positive chromatin modifiers, which are required to drive efficient transcription of the globin genes. During hub reorganization, inactive globin and olfactory receptor genes loop out.

Thus, in summary, an important feature of chromatin hubs is the presence of chromatin loops, formed through protein–protein interactions between proteins bound at different sites. Gene activation is accompanied by dynamic reorganization of poised hubs, leading to the formation of active chromatin hubs.

Some eukaryotic transcription factors are activators, others are repressors, and still others can be either, depending on context

Transcription factors belong to two general categories, activators and repressors of transcription. **Activators** are modular proteins that have distinct **DNA-binding domains** or **DBDs** and **transcriptional activation domains** or **ADs**. Activation domains are the actual functional parts of the TFs. Contrary to some earlier beliefs, ADs do not interact directly with DNA but with protein components of the transcriptional machinery. Other domains, such as those involved in signal responses or

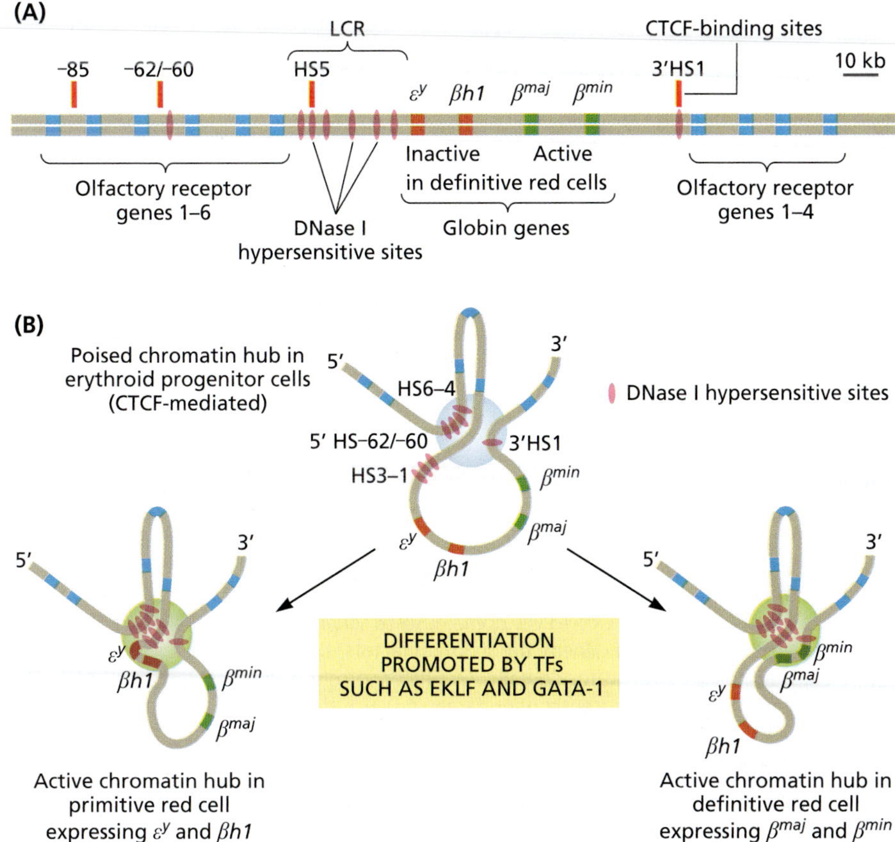

Figure 12.3 Locus control region in the mouse β-globin locus. (A) The locus contains four globin genes that are expressed at different stages of erythroid cell differentiation: ε^y and *βh1*, shown as orange blocks, are expressed in primitive red blood cells during early embryonic development, and β^{maj} and β^{min}, shown as green blocks, are expressed in definitive red blood cells, later during embryogenesis and in the adult organism. Developmental regulation of these genes occurs through the action of the LCR. Two DNase I hypersensitive sites, HS5 and 3′HS1, flank the entire β-globin locus. The domain is embedded in the midst of numerous olfactory receptor genes that are transcriptionally inactive and possess highly compacted chromatin structure. The transcription factor CTCF binds to four sites, shown by red bars; two of these sites are upstream of the globin genes in intergenic regions between olfactory receptor genes. Upon binding of CTCF, the CTCF-binding sites contact each other in both progenitor and erythroid cells to form the chromatin hub. (B) In erythroid progenitors, which express globin at basal levels, a hub is formed through interactions between the upstream 5′ HS–60/–62, the downstream 3′HS1, and hypersensitive sites HS4–HS6 within the 5′ side of the LCR. During erythroid differentiation, which occurs through the action of transcription factors such as EKLF and GATA-1, the β-globin genes that become highly activated and the rest of the LCR stably interact with this substructure to form a functional active chromatin hub or ACH, shown as a green shaded circle; high β-globin gene expression is LCR-dependent. Clustering of binding sites for transcription factors in the ACH causes local accumulation of cognate proteins and associated positive chromatin modifiers, which are required to drive efficient transcription of the globin genes. Inactive globin and olfactory receptor genes, shown in blue, loop out. (Adapted, courtesy of Wouter de Laat, Hubrecht Institute.)

protein-protein interactions, may also be present. In some cases, these domains are part of the same polypeptide chain, but in other instances they are located on separate subunits of multiprotein complexes; Figure 12.4 shows some well-studied examples. The multisubunit organization is beneficial in terms of combinatorial control. However, not all activators are modular in structure: in some proteins, residues that participate in one function are interdigitated with residues performing another function. Representative activators belonging to this class are MyoD, the key developmental regulator of expression of muscle-specific genes, and the glucocorticoid receptor (see Figure 12.4C).

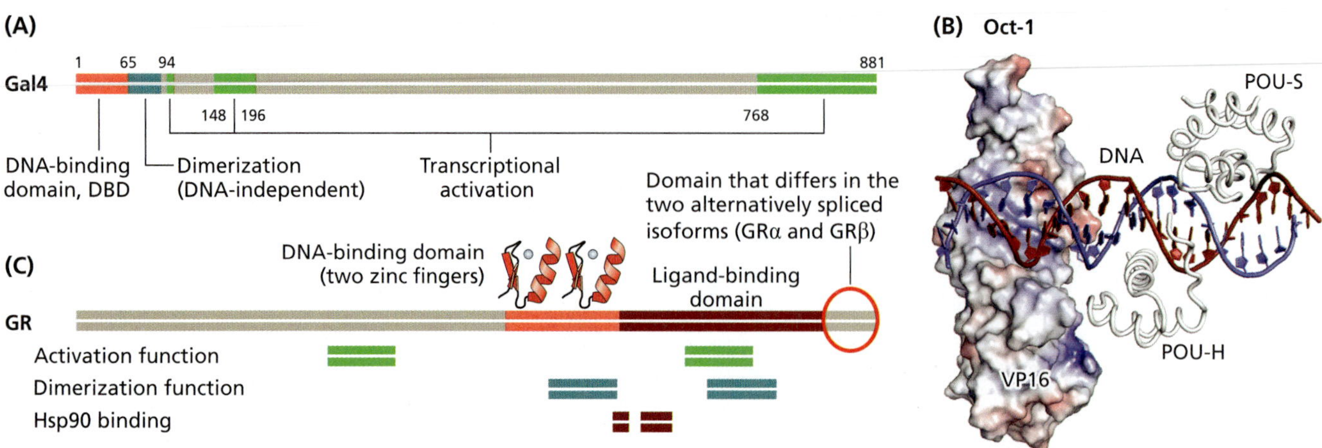

Figure 12.4 Versatility of eukaryotic transcription factors. (A) Gal4 is a positive protein regulator of the expression of a whole battery of galactose-induced genes; these genes encode enzymes that convert galactose to glucose. Gal4 recognizes a 17-base-pair sequence in the upstream activating sequence or UAS of these genes. The DNA-binding domain or DBD is often used as a tool for identification of genome sequences that contain activation domains. For that purpose, the sequence encoding the tested presumptive activation domain is fused, through recombinant DNA technology, to the DNA sequence encoding the DBD of Gal4. If transcription of a DNA template containing the binding sequence for the DBD is stimulated by the resulting fusion protein, *trans*-activating activity of the tested activation domain is indicated. Note that there are three activating regions in Gal4; the region between 94 and 100 is nonfunctional in yeast *in vivo*. (B) Oct-1 is an example of a composite transcription factor, which harbors individual domains on different subunits of a multisubunit complex. In mammals, Oct-1 is a ubiquitous DNA-binding protein that contains a homeodomain and a separate POU domain. The POU domain contains two separate DNA-binding domains, POU-specific, or POU-S, and POU-homeo, or POU-H, joined by a flexible linker. The two domains bind to opposite faces of the DNA. On its own, Oct-1 binds weakly to 8-bp sites or octamers, but this binding is significantly enhanced by interaction with protein partners such as OCA-B or VP16, a herpes simplex virus factor involved in determining the outcome of viral infection. In the model shown, VP16 is bound to a TAATGARAT element that also interacts with the Oct-1 POU domain; VP16 serves as a bridge between Oct-1 and general transcription factors, thus providing Oct-1 with an activator domain. (C) Glucocorticoid receptor, GR, is an example of a transcription factor in which the individual domains are not discernible as such but overlap to different degrees: the same portion of the polypeptide chain can participate in multiple functions. The N-terminal region of GR encompasses a hormone-independent activation domain that is important in recruiting positive or negative regulatory factors. The DNA-binding domain contains two zinc fingers, which bind to the GR elements in the DNA and also play roles in nuclear transport and in dimerization functions. The ligand-binding domain binds the steroid hormone; in addition, it binds Hsp90, which is instrumental in opening the hormone-binding cleft. Hsp90 dissociates once the steroid binds: this occurs in the cytoplasm. The binding of the hormone induces a conformational change in the hormone-dependent activation domain, allowing interaction with co-activators.

We discuss in detail the structure of protein domains that interact with DNA, known as DBDs, in Chapter 6. Much less is known about the structure and function of ADs. These domains are functionally defined as regions of proteins that stimulate transcription when attached to a DBD which is heterologous, that is, from a different protein. They interact either directly or indirectly, through co-activators, with components of the general transcriptional machinery. According to some experts, the activator is misnamed, as it does not directly activate transcription through binding to DNA but rather merely recruits the transcriptional machinery to the gene promoter. This view forms the basis of the **recruitment model for transcriptional activation**, which seems to be supported by much experimental evidence (**Figure 12.5**). As very little is known about the exact mechanisms involved, the classification of activation domains is mainly based on their amino acid composition: acidic, glutamine-rich, and so on.

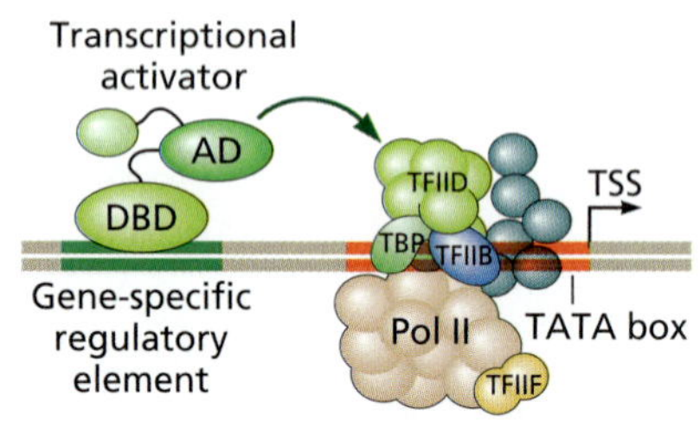

Figure 12.5 Model of transcription activation by recruitment of the basal transcriptional machinery to the promoter. The example shows a TATA-box promoter in a protein-coding gene. The typical activator contains a DNA-binding domain, DBD, and an activation domain, AD, in addition to other domains of various functions. The activation domain acts by recruiting core promoter-binding factors such as TFIID, and hence the polymerase Pol II, to the promoter sequence.

The majority of activators contain more than one AD. In the examples presented in Figure 12.4, Gal4 has three identifiable ADs, whereas the glucocorticoid receptor has two. These domains function additively or synergistically to increase the activation efficiency of each individual domain. Multiple domains increase the probability that any individual domain will interact with its target in the basal transcriptional machinery. In addition, different domains may interact with different targets. In terms of structure, activation domains are mainly α-helical; in some cases, the α-helices are clearly amphipathic (see Chapter 6 and Figure 6.12).

It must also be noted that the same TF can serve as an activator or a repressor, depending on the specific context. Such situations are reminiscent of the regulation of the bacterial *ara* operon, where the protein AraC acts as a repressor in the absence of arabinose or as an activator in the presence of arabinose (see Chapter 11).

Finally, it is clear that combinatorial interactions among TFs are critical for directing tissue-specific gene expression patterns. Transcription factors often interact with each other physically, forming homo- and heterodimers or larger complexes. As pointed out in Chapter 6, such interactions can function in cooperation, enhancing interactions with targets. It has been estimated that ~75% of all metazoan transcription factors heterodimerize with other factors, allowing enormous complexity and subtlety in gene regulation. Contemporary high-throughput methods allow for sensitive and precise detection of such interactions, and this information can be used for the construction of combinatorial transcription regulation atlases (see Chapter 13). Complete TF interaction networks for both humans and mice already exist. From such atlases, it becomes apparent that, in general, highly connected TFs are broadly expressed across tissues, and these TFs are highly conserved between humans and mice. On the other hand, TFs that have few interactions tend to be expressed in tissue-specific patterns.

Regulation can use alternative components of the basal transcriptional machinery

It was long believed that gene-specific transcriptional regulation required gene-specific activators and/or repressors but that the remainder of the factors that participated in formation of the pre-initiation complex—components of the basal transcriptional machinery—were invariant and common to all genes (see Figure 10.9). Clues that this is probably not the case came from the discoveries of different core-promoter architectures that were found to recruit unique combinations of TATA-binding protein, TBP, and various TBP-associated factors or TAFs (see Figure 10.8), even in the same cell. Thus, variant basal complexes could regulate different sets of genes. The picture became even more complicated when cell-type-specific basal factors were described, including cell-specific TAFs and specific TBP-related factors. Moreover, convincing evidence emerged that pointed to major dynamic changes in the basal transcriptional machinery during processes involved in cell differentiation. We illustrate this new aspect of transcriptional regulation with the example of changes that occur during muscle differentiation (**Figure 12.6**).

Mutations in gene regulatory regions and in transcriptional machinery components lead to human diseases

The complexity of *cis–trans* regulation of transcription in eukaryotes creates special problems if one or more of the DNA sequences and/or their protein partners undergo mutational changes. Often such mutations are involved in various diseases; some prominent examples are presented in **Table 12.1**. Our knowledge about such disease connections increases by the day and is expected to lead to better diagnostic tools and treatment options in the future.

12.3 Regulation of transcriptional elongation

The polymerase may stall close to the promoter

Not only formation of the initiation complex but also the process of transcription elongation itself is under complex regulation. It has long been recognized that some genes, like the *HS* heat-shock genes in *Drosophila*, exhibit RNA polymerase **stalling.** That is, RNA polymerase that has just begun transcribing the body of the gene may become stalled at a position very close to the promoter region. The existence of this promoter-proximal stalling was once considered an idiosyncratic feature of *HS* genes, but genomewide studies have revealed that promoter-proximal stalling may be widespread, affecting ~10–15% of all *Drosophila* genes, for example. Parallel studies in human cells indicate a similar widespread occurrence of promoter-proximal stalling. This phenomenon is now considered an important aspect of the

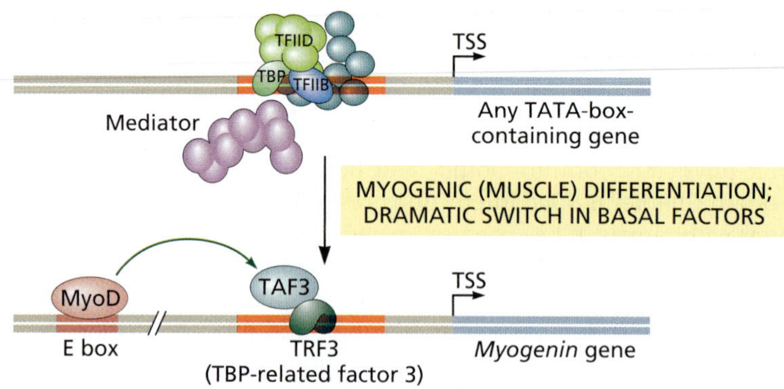

Figure 12.6 Global changes in gene expression during cell differentiation are carried out by significant changes in the general transcriptional machinery. This figure illustrates the example of the myogenin gene promoter. (Top) In myoblasts, the precursor cells to mature multinuclear muscle cells or myotubes, one sees the usual players involved in formation of the pre-initiation complex on TATA-box-containing promoters: TBP, the TAFs that bind to it as a part of TFIID, and Mediator. The process of differentiation is accompanied by drastic changes in the availability of these factors: there is a near-wholesale elimination of the prototypic TFIID, including TBP, and disappearance of Mediator. (Bottom) Active transcription of muscle-specific genes such as myogenin, which encodes the key regulator of myogenic differentiation, is directed by a much pared-down complex consisting of only TAF3, which is a subunit of TFIID, and TBP-related factor 3 or TRF3. TRF3 is nearly identical to TBP in its C-terminal region, which includes the TATA-box-binding and TFIIA- and TFIIB-binding domains, but it differs considerably in the N-terminal region. The myogenin gene is regulated by a classical cell-type-specific transcriptional activator, MyoD, that directly interacts with TAF3.

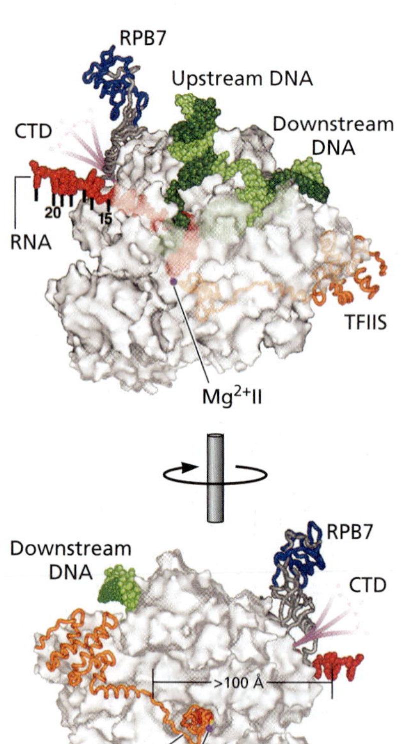

Figure 12.7 Location of TFIIS bound to elongating yeast RNAP II. Portions of RNAP II are rendered semitransparent to reveal RNA, shown in red, DNA, shown in green, and TFIIS, shown in orange, in the internal channels of the enzyme. The approximate length of exiting RNA and the distance between the exiting RNA and TFIIS are indicated. The C-terminal domain or CTD emerging from the enzyme is shown in magenta, the RPB7 subunit of the polymerase is shown in blue and gray, and the weakly bound Mg^{2+} ion required for transcript cleavage is represented by a sphere. (From Palangat M, Renner DB, Price DH & Landick R [2005] *Proc Natl Acad Sci USA* 102:15036–15041. With permission from National Academy of Sciences.)

regulation of inducible genes, as it allows for a very fast induction of transcription when an appropriate signal appears; the stalled transcriptional machinery is ready to go at the signal. The stalling itself requires negative transcription factors, which need to be displaced for the stalled polymerase to resume transcription. This kind of stalling must be distinguished from the transient, unprogrammed pausing that occurs frequently during transcription elongation of any gene (see Chapter 9 and this chapter).

Transcription elongation rate can be regulated by elongation factors

The rate of transcription elongation is not uniform along the gene body. The polymerases often enter pauses of various durations (see Figure 9.4); the exit from some of these pauses requires the action of specific transcription elongation factors. In bacteria, these elongation factors are GreA and GreB; in eukaryotes, the best-studied factor is **TFIIS**. TFIIS acts to overcome transcriptional pausing caused by backtracking of the polymerase enzyme. Backtracking results in the extrusion of the 3′-end of the nascent transcript through a channel in the enzyme, thus misaligning the 3′-OH group and the catalytic site (see Figure 9.4). TFIIS acts by engaging transcribing Pol II and stimulating a cryptic, nascent RNA cleavage activity that is intrinsic to the polymerase. Structural studies have indicated that TFIIS extends from the polymerase surface through a pore to the internal active site, spanning a distance of 100 Å (**Figure 12.7**). Two essential and invariant acidic amino acid residues in TFIIS complement the polymerase active site, acting to properly position a metal ion and a water molecule for the RNA cleavage to occur. TFIIS also induces extensive structural changes in Pol II that realign nucleic acids with the active center.

12.4 Transcription regulation and chromatin structure

What happens to nucleosomes during transcription?

The question of what happens to nucleosomes during transcription has plagued researchers for almost 30 years, but we still do not know exactly how the transcription process deals with the presence of nucleosomes. One thing is clear: the nucleosomes

Table 12.1 Transcriptional regulatory elements and transcriptional machinery components involved in some human diseases. (Adapted from Maston GA, Evans SK & Green MR [2006] *Annu Rev Genomics Hum Genet* 7:29–59.)

Regulatory element	Disease	Affected gene[a]	Mutations (bound factor)
core promoter	β-thalassemia	β-globin	TATA box, CACCC box (EKLF), DCE
proximal promoter	hemophilia	factor IX	CCAAT (C/EBP)
	hereditary persistence of fetal hemoglobin	Aγ-globin	~175 bp upstream of TSS (Oct-1, GATA-1)
	δ-thalassemia	δ-globin	77 bp upstream of TSS (GATA-1)
enhancer	X-linked deafness	*POU3F4*	microdeletions 900 kb
silencer	asthma and allergies	TGF-β	509 bp upstream of TSS (YY1)
insulator	Beckwith–Wiedemann syndrome	*H19–Igf*	CCCTC-binding factor (CTCF)
LCR	α-thalassemia	α-globin	~62 kb deletion upstream of gene cluster
	β-thalassemia	β-globin	~30 kb deletion removing 5′HS2–5

Transcriptional component	Disease	Mutated factor
general transcription factors	xeroderma pigmentosum, Cockayne syndrome, trichothiodystrophy	TFIIH
activators	congenital heart disease	Nkx2-5
	Down syndrome with acute megakaryoblastic leukemia	GATA-1
	prostate cancer	ATBF1
	X-linked deafness	POU3F4
repressors	X-linked autoimmunity–allergic deregulation syndrome	FOXP3
co-activators	Parkinson's disease	DJ1
	type II diabetes mellitus	PGC-1
chromatin remodeling factors	cancer	BRG1/BRM
	retinal degeneration	ataxin-7
	Rett syndrome	MeCP2

[a]POU3F4 = POU domain, class 3, transcription factor 4; TGF-β = transforming growth factor β; Igf = insulin-like growth factor.

must go, at least partially/temporarily, because when the double-helical DNA is wrapped around the histone core, it cannot open to provide a template DNA strand for the polymerase to copy. Different mechanisms for overcoming this problem may operate at promoter regions and within gene bodies, but in the majority of cases, promoter regions are devoid of nucleosomes altogether, forming **nucleosome-free regions** or **NFRs**. These are defined primarily by specific nucleotide sequences that are resistant to bending and thus cannot wrap easily around the nucleosome histone core. Such regions will remain nucleosome-free regardless of the transcriptional state of the linked gene. In cases where inactive promoter regions contain nucleosomes, some kind of remodeling will be needed for activation.

Figure 12.8 illustrates the chromatin structure at the heat-shock protein 70 locus, *Hsp70Ab*, in *Drosophila* and the changes that occur in this structure upon heat-shock treatment. The heat-shock genes provide a useful system in which to study the regulation of transcription of inducible eukaryotic genes. Their expression is induced rapidly, within seconds, and robustly, with ~500-fold increase in mRNA levels, under a variety of stress conditions including heat shock. This rapid response is favored by the presence of a stalled polymerase in the NFR (see Figure 12.8A). The genes undergo numerous highly synchronous changes in chromatin structure upon induction: these changes are recognized cytologically as the appearance of decondensed puffs in polytene chromosomes (see also Figure 2.7 and Figure 2.8).

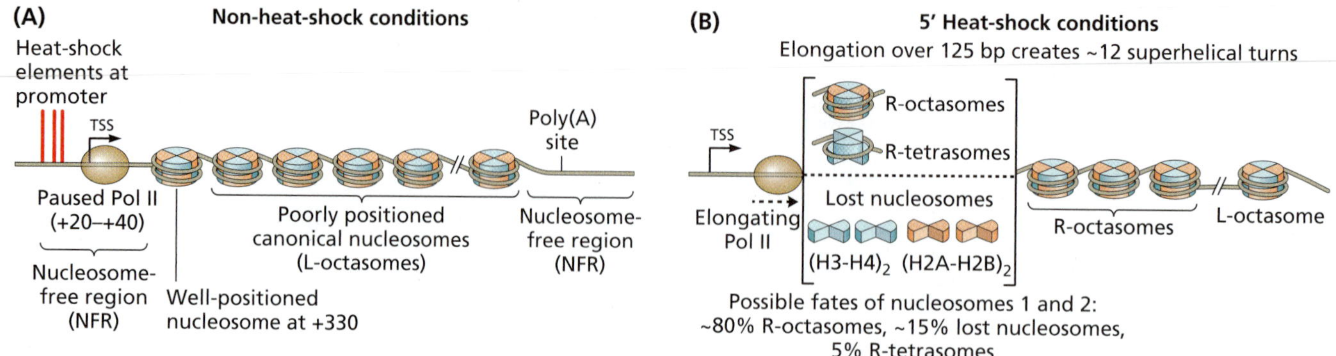

Figure 12.8 Transcription elongation affects nucleosome structure at a distance. This distance effect occurs through the positive supercoiling stress or torque created by the polymerase advancing along the DNA template. (A) Chromatin structure of the *Hsp70Ab* locus in *Drosophila* under normal, non-heat-shock conditions. The heat-shock elements in the promoter provide binding sites for the heat-shock transcription factor HSF, whose binding induces the genes. A number of other proteins including TBP, GAGA, Spt5, PARP-1, and the negative elongation factor NELF are present on the gene. Importantly, the gene harbors a paused RNAP II between positions +20 and +40. The chromatin structure over the *Hsp70Ab* gene is characterized by two nucleosome-free regions or NFRs and a well-positioned first nucleosome; note that nucleosomes within the gene body gradually show less and less defined positioning as one proceeds downstream from the transcription start site, TSS. (B) Changes in chromatin structure of the locus upon heat shock. As a first step, the paused polymerase is released into productive elongation through the action of the positive transcription factor P-TEFb, a kinase that phosphorylates Ser2 on the C-terminal domain of Pol II and NELF. Phosphorylated NELF leaves the complex, and other proteins join and travel with the polymerase along the gene. During active transcription, new unphosphorylated Pol II molecules enter the promoter; these molecules also pause at the promoter-proximal site, but the duration of the stall is dramatically decreased in comparison with that of the inactive gene. Importantly, the positive superhelical stress created as a result of polymerase movement along the DNA template disrupts downstream nucleosomes. In ~5 s, Pol II advances 125 bp, which creates ~12 positive superhelical turns; this is enough to convert six L-nucleosomes into R-octasomes (see Figure 9.15) at a rate far exceeding that of Pol II transcription. (Adapted from Zlatanova J & Victor J-M [2009] *HFSP J* 3:373–378. With permission from Taylor and Francis Group.)

Some classical remodeling, performed by chromatin remodelers, may also occur in the transcribed regions, but most proposed models of how elongation proceeds through nucleosomal regions presume that physical invasion of the nucleosome by the progressing polymerase removes the nucleosomal barrier. Whether the histone octamer is removed as a whole in a single step or is sequentially dismantled and reassembled during passage of the polymerases is not known.

We are now aware that nucleosome disassembly may be driven from a distance, the driving force being the creation of positive superhelical stress in front of the translocating polymerase. Transcription elongation requires relative rotation of the enzyme around the DNA. In the **twin supercoil domain model**, the polymerase is assumed to be fixed, anchored to some nuclear structure such as a transcription factory or the nuclear matrix, so that the DNA must instead rotate in a screwlike fashion inside it. This also suppresses the entangling of the RNA transcript around the DNA that would occur if the polymerase and the transcript it carries were to rotate around the DNA, following the helix backbone. Solving the entangling problem, however, leads to another topological problem: transcription on a fixed, topologically constrained template creates positive supercoiling in the downstream portion of the template and negative supercoiling in the upstream portion. Transcription-coupled supercoiling has been observed *in vitro* and *in vivo*, including in genes transcribed by Pol II. One model of the effect of transcription-induced supercoiling on the integrity of the nucleosome is presented in terms of the *Hsp70* gene in Figure 12.8B. The basic proposition of this model is that the supercoiling induced by the polymerase advance results in a reversal of chirality in downstream nucleosomes. Recall from Chapter 8 that nucleosomes exist as a family of particles of different histone composition and/or of different chirality; that is, the DNA may be wrapped in either a left-handed or a right-handed sense around the histone core. Even if most of the nucleosomes do not immediately fall apart but just change the chirality of the DNA superhelix around their histone octamers, these right-handed particles are highly unstable and can be easily disrupted by independent forces, possibly including the invasion of the polymerase.

Other changes in the nucleosome particle that can occur under positive superhelical tension have also been described: for example, nucleosomes may split into two halves, presumably giving rise to the well-known elevated DNase I sensitivity of transcriptionally active regions.

A number of other models have been proposed to explain the fate of nucleosomes during transcriptional elongation. Most agree that, at a minimum, H2A-H2B dimers must be displaced and then rebind. In one model, the DNA released from the nucleosome by H2A-H2B removal forms a loop that can be occupied by the polymerase as it moves around the nucleosome. As yet, no one model is supported by conclusive evidence.

In any event, the disruption of nucleosomes during transcription may require specific protein factors in addition to the polymerase. Among the proteins that have been shown *in vitro* to aid elongation is the **FACT** or **facilitates chromatin transcription complex**, which interacts with DNA and with histones H2A and H2B. The interaction with H2A-H2B facilitates their release from nucleosome particles in front of the polymerase and their subsequent rebinding after the polymerase has passed. In fact, FACT appears to act as a chaperone for the H2A-H2B dimer.

12.5 Regulation of transcription by histone modifications and variants

Modification of histones provides epigenetic control of transcription

It is now clear that the mechanisms controlling gene transcription are not encoded directly in the nucleotide sequence of the DNA. Rather, they reside outside that sequence, in the form of proteins and their modifications, as well as in postreplicational modifications of the DNA molecule itself. As an everyday analogy of this situation, suppose there were a vast document, which is mostly unimportant to a reader but contains important, readable passages buried within it. Some of these need to be read, others do not. To save the searcher the work of hunting through the vast amount of irrelevant material, one could go in and mark these passages, maybe by underlining the important ones, after the whole document had been prepared. These would be like *epi* marks, from the Greek for on, upon, over: not there in the original but put in later. All those molecular mechanisms that convey regulatory information beyond and above the informational content in the DNA sequence are now grouped under the term **epigenetic regulation**. However, there is a problem with this designation.

The term **epigenetics** was originally introduced by geneticists in the 1930s to describe heritable cellular events that could not be explained by genetic principles. The term was subsequently used to define heritable controls in gene expression that are not based on changes or mutations in DNA. Nowadays, the term has acquired a broader meaning: it is used to designate all postsynthetic modifications, to proteins, DNA, and noncoding RNAs, that affect gene expression and other processes, without taking into account the inheritance of the respective state. This broader definition deviates from the original meaning of the terms *epi* and *genetics*, which focus on inheritance. Under this new, broader definition, all post-translational modifications or PTMs of histones are described as epigenetic. However, since these modifications are very dynamic and are not inherited through mitosis, we prefer not to use the term epigenetic in reference to them. The term is, however, appropriate with respect to DNA methylation, because DNA methylation patterns that regulate transcription are inherited; they pass from mother cell to daughter cells during mitosis.

In terms of gene regulation, several classes of postsynthetic modifications form part of the regulatory mechanisms. We discuss the post-translational modification of histones and DNA methylation separately. In addition, we discuss other players and mechanisms that contribute to regulation, such as chromatin remodelers and long noncoding RNAs or lncRNAs.

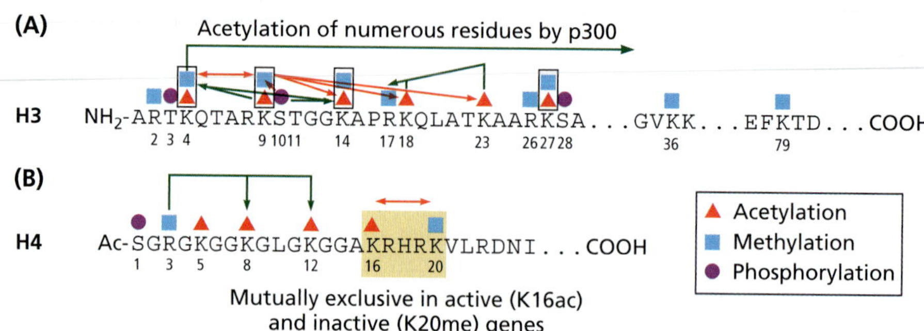

Figure 12.9 Cross-talk in *cis* among PTMs occurring on histones H3 and H4. (A) Histone H3; (B) histone H4. Inhibition of modification is represented by a red arrow; the direction of the arrow shows the directionality of inhibition. For example, H3K9 methylation inhibits acetylation of H3K14, H3K18, and H3K23 and methylation of H3K4. Promotion of modification is marked by a green arrow. Thus, for example, H3K4 methylation promotes acetylation of numerous other residues by the acetyltransferase p300. For methylation to occur on an acetylated residue, the acetyl group has to be removed, shown by boxed red and blue symbols.

Gene expression is often regulated by histone post-translational modifications

As mentioned in Chapter 8, histones can undergo multiple chemical modifications that affect their properties. Modifications that are especially important in transcriptional regulation are acetylation, methylation, phosphorylation, ubiquitylation, and poly(ADP)ribosylation. For the most part, these modifications occur on the N- or C-terminal tails of the histones, which project from the nucleosome and are thus accessible for interactions with other proteins. It turns out that most of these modifications can communicate with each other in terms of stimulating or inhibiting further modifications of other residues on the same or a different histone molecule. Several mechanisms could be responsible for communication among modifications: the initial histone modification may trigger increased or decreased activity of a histone-modifying enzyme, or alternatively, different histone-modifying enzymes may be present in a single protein complex, thus coordinating the simultaneous occurrence of several modification or demodification reactions needed to achieve the desired transcriptional outcome.

When a modification of one amino acid residue affects the modification of another amino acid residue in the same histone molecule, this is known as **cross-talk in *cis*** (**Figure 12.9**). A modified amino acid residue in one histone molecule may also affect the modification pattern of a different histone molecule, residing either in the same nucleosome particle or in a different nucleosome within the chromatin fiber. Such interactions are known as **cross-talk in *trans*** (**Figure 12.10**). These kinds of interactions ensure the coordination of modifications that together lead to a unique transcriptional outcome.

Readout of histone post-translational modification marks involves specialized protein molecules

Three different classes of proteins interact with nucleosomal histones to regulate transcription. First, there are **writers**, enzymes that catalyze various modification reactions. Then, there are **readers**, proteins that recognize specific histone marks. Finally, there are **erasers**, enzymes that remove the PTM marks from histones.

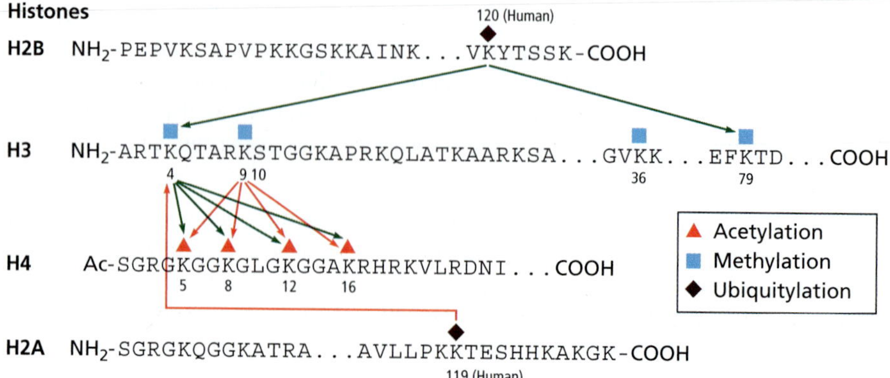

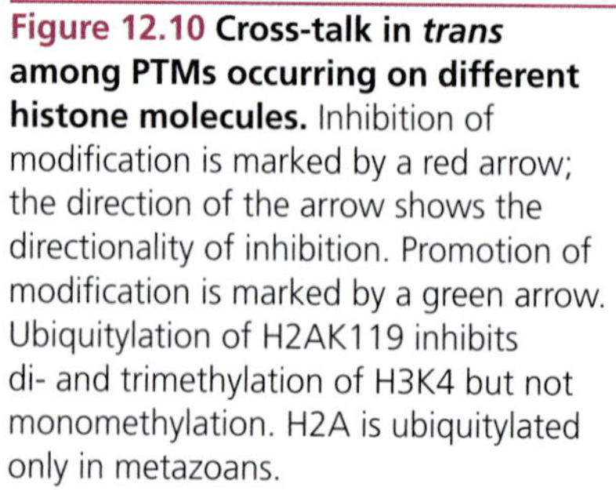
Figure 12.10 Cross-talk in *trans* among PTMs occurring on different histone molecules. Inhibition of modification is marked by a red arrow; the direction of the arrow shows the directionality of inhibition. Promotion of modification is marked by a green arrow. Ubiquitylation of H2AK119 inhibits di- and trimethylation of H3K4 but not monomethylation. H2A is ubiquitylated only in metazoans.

Table 12.2 A list of major histone modification enzymes. The selection of enzymes presented in the table is based on well-known enzymes that are present, in slightly different forms, in both human and yeast. The new nomenclature considered existing close relationships in primary sequence and domain structure/organization: the comparable domain structures may not extend across the entire protein but are clearly recognizable as evolutionarily related. The second consideration, if the domain structure is not recognizable, is sequence homology in the catalytic domains and in substrate specificity. Related enzymes from a single species have been given the same name with a capital letter as a distinguishing suffix, for example, A or B. Related enzymes from different species have been given an identical name but with a different prefix to denote the species of origin: for example, h for human, d for *Drosophila*, Sc for *Saccharomyces cerevisiae*, or Sp for *Schizosaccharomyces pombe*. As an example, not shown in the table, the human demethylase LSD1/BHC110 becomes hKDM1, the *Drosophila* equivalent Su(var)3-3 becomes dKDM1, and the fission yeast equivalent SpLsd1/Swm1/Saf110 becomes SpKDM1. (Adapted from Allis CD, Berger SL, Cote J et al. [2007] *Cell* 131:633–636.)

New name	Human	*S. cerevisiae*	Substrate specificity	Function
K-Methyltransferases, KMTs; Formerly Lysine Methyltransferases				
KMT2		Set1	H3K4	transcription activation
KMT2A	MLL		H3K4	transcription activation
KMT3		Set2	H3K36	transcription activation
KMT3A	SET2		H3K36	transcription activation
KMT4	DOT1L	Dot1	H3K79	transcription activation
KMT6	EZH2		H3K27	polycomb silencing
K-Acetyltransferases, KATs; Formerly Acetyltransferases				
KAT1	HAT1	Hat1	H4K5/12	histone deposition, DNA repair
KAT2		Gcn5	H3K9/14/18/23/36; H2B	transcription activation, DNA repair
KAT2A	hGCN5		H3K9/14/18; H2B	transcription activation
KAT4	TAF1	Taf1	H3>H4	transcription activation
KAT5	TIP60	Esa1	H4K5/8/12/16; H2A	transcription activation, DNA repair
KAT8	HMOF/MYST1	Sas2	H4K16	chromatin boundaries, dosage compensation, DNA repair
KAT9	ELP3	Elp3	H3	

There are many writer enzymes, often with very well-defined specificities. We cannot describe them all, but some information about representative enzymes is provided in Table 12.2. Note that different enzymes have very distinct specificities and modify specific residues on specific histones. The example shown for acetylation is typical in that most modifications are on histone termini that lie on or near the nucleosome surface (Figure 12.11).

A remaining uncertainty in the field is this: what determines which sites on which proteins on which nucleosomes receive which modification marks? To some extent, the question has been answered by the observation that certain modifications favor or disfavor certain other modifications. But it still seems that there is a chicken-and-egg problem here.

In order for histone modifications to have an effect on transcription, they must affect nucleosome structure, chromatin fiber structure, or the interaction of chromatin with other factors. Each of these consequences is probably modulated by the effect of modifications on the interactions of chromatin with other proteins or protein complexes. The proteins that recognize specific histone marks are called readers (Figure 12.12). In general, protein modules that recognize acetylated lysines belong to **bromodomain**

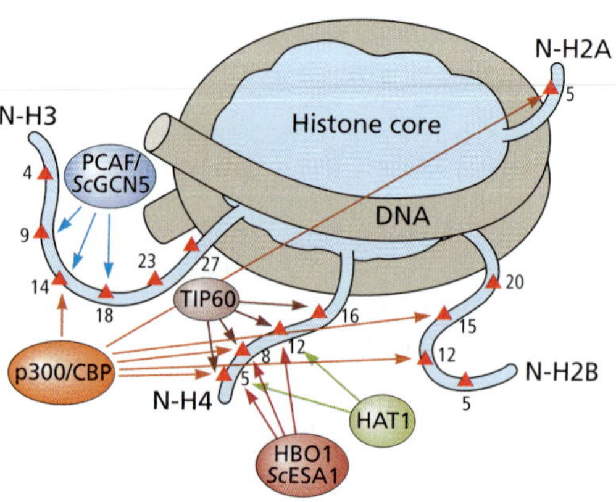

Figure 12.11 Specificity of action of human acetyltransferases. The schematic shows the nucleosome with four protruding histone tails, in which lysine residues that are modifiable by acetylation are depicted as red triangles. As denoted by arrows of different colors, different acetyltransferases modify lysines within a specific sequence context. Some enzymes, such as p300/CBP, have rather broad specificities, modifying numerous residues on several tails. Others modify several residues on a specific histone tail; for example, HBO1 acetylates only the tail of H4, whereas PCAF is specific for the tail of H3. In some cases, the human enzyme has a close ortholog in yeast, denoted by *Sc* in front of the name. Only one residue-specific histone deacetylase has been identified, SIRT2 or *ScSir2* in yeast: it deacetylates H4K16, not shown here.

families; the bromodomain modules are the only known protein recognition motifs that selectively target ε-*N*-acetylation of lysine residues. They were first recognized as domains in a protein encoded by the *Drosophila brm* gene, hence the name bromodomains. The human proteome contains more than 40 proteins with more than 60 diverse bromodomains. These domains are characterized by low sequence identity, but all share a conserved fold comprising a bundle of four α-helices (see Figure 12.12A).

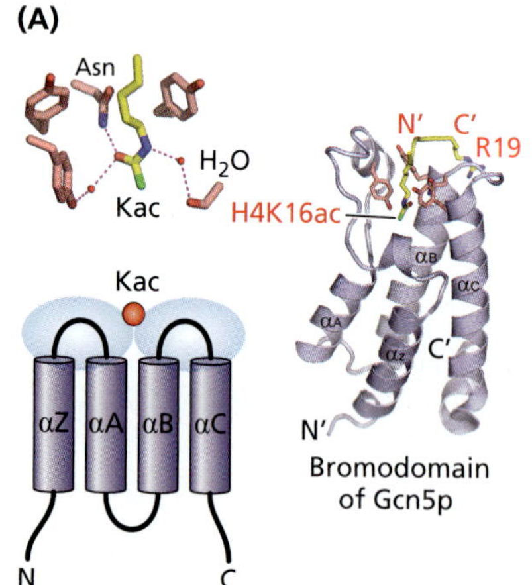

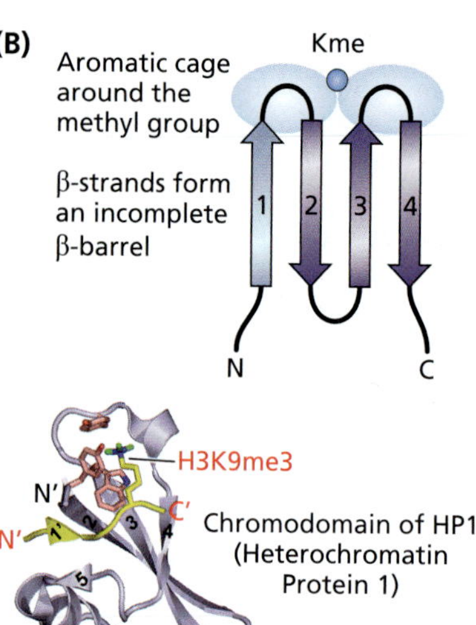

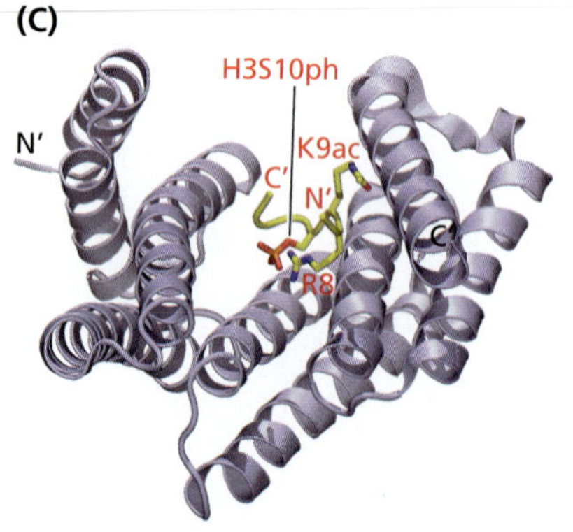

Figure 12.12 Protein domains known as readers recognize and bind post-translationally modified histones. (A) Bromodomain modules bind to acetylated lysines. Bromodomains are present in several histone acetyltransferases, nucleosome remodelers, and TAFs. The overall topology of the fold consists of four α-helices; the loops connecting the helices participate in formation of the acetyllysine or Kac reader pocket. The pocket is essentially hydrophobic and neutral with significant hydrogen (H)-bonding capacity to form direct or water-mediated H-bonds. The particular structure shown is that of the bromodomain of Gcn5p, a histone acetyltransferase or HAT. The existence of HAT activity in an acetyl mark reader suggests a potential mechanism for the spreading of acetylation marks in chromatin. (B) Chromodomain modules bind to methylated residues. Recall from Chapter 8 that lysine and arginine residues can accommodate one, two, or three methyl groups, yielding considerable diversity in the physicochemical properties of the modified proteins. The need to recognize these diverse methylation states led to the existence of numerous variants of a basic binding module, the chromodomain. (Top) The basic fold of chromodomains consists of incomplete β-barrels, with the connecting loops forming aromatic cages around the modified residue. (Bottom) In the structure given, that of the chromodomain of heterochromatin protein 1 or HP1, the domain has three core strands, strands 2–4, and one orphaned β-strand, strand 5. Upon complex formation, the histone peptide completes the β-barrel by introducing an extra β-strand, strand 1′, sandwiched between strands 2 and 5 and shown in yellow. (C) Protein 14-3-3 reads phosphoserine in the H3Ser10 context. Phosphorylation adds a bulky, negatively charged phosphate moiety to the -OH group of amino acid side chains, substantially expanding the ion-pairing and H-bonding capacities of the modified residues. The mammalian 14-3-3 protein family that recognizes the phosphorylation mark plays roles in signal transduction, chromosome condensation, and apoptosis. The phosphate-bearing histone peptide from the N-terminal tail of H3, H3S10ph, is buried in the V-shaped 14-3-3, an all-α-helical protein; the phosphate group forms multiple contacts and its charge is neutralized by the basic side chains of two arginine residues in 14-3-3. (Adapted from Taverna SD, Li H, Ruthenburg AJ et al. [2007] *Nat Struct Mol Biol* 14:1025–1040. With permission from Macmillan Publishers, Ltd.)

Protein modules that recognize methylated residues form families of **chromodomain** or chromatin organization modifier motifs. Recall from Chapter 8 that lysine and arginine residues can accommodate one, two, or three methyl groups, yielding considerable diversity in the physicochemical properties of the modified proteins. The need to recognize these diverse methylation states led to the existence of numerous variants of the chromodomain. Proteins that read the methylation marks belong to a large superfamily, the Royal superfamily, which is subdivided into proteins that read higher and lower methylation states. The basic fold of chromodomains consists of incomplete β-barrels, with the connecting loops forming aromatic cages around the modified residue (see Figure 12.12B).

Finally, phosphoserine marks have their own designated readers. Phosphorylation adds a bulky, negatively charged phosphate moiety to the -OH group of serine side chains, substantially expanding the ion-pairing and hydrogen-bonding capacities of the modified residues. Here, we present only the structure of the protein 14-3-3 reader domain (see Figure 12.12C); we return to phosphoserine readers in Chapter 22, where we describe the action of BRCA1 and phosphorylated histone variant H2A.X in the repair of double-strand DNA breaks.

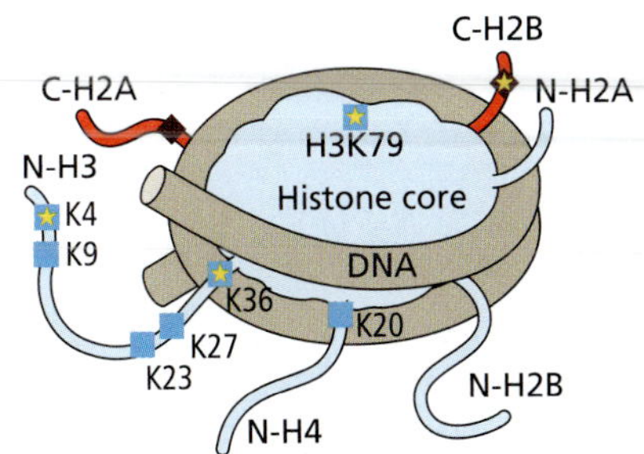

Figure 12.13 Histone marks that are commonly present in transcriptionally active or inactive genes. Histone acetylation at lysine residues is a characteristic feature of active genes. These post-translational modifications may act by changing the stability of the nucleosome particle and/or the interactions between nucleosomes in the context of chromatin fiber structure or by interacting with reader proteins that contain modification-recognition modules.

Post-translational histone marks distinguish transcriptionally active and inactive chromatin regions

The major markings used to distinguish active from inactive genes include acetylation, methylation, and ubiquitylation of histones. Histone phosphorylation is also observed during the transcriptional activation of some specific genes, but its general role in transcription regulation remains controversial because high levels of phosphorylation, especially in histones H1 and H3, are observed during chromosome compaction in mitosis, and compaction is considered incompatible with active transcription. The roles of two other PTMs, poly(ADP)ribosylation and ubiquitylation, are discussed separately because of their peculiarities.

It has been known for about half a century that transcriptionally active chromatin is enriched in acetylated histones, yet even today we do not fully understand the structural consequences of this or other histone modifications. Nevertheless, much more information has been gathered by studying individual gene systems and, more recently, by genomewide localization studies of individual histone marks (see Chapter 13). As a result, a clearer picture has emerged about the presence or absence of at least some modifications in active and inactive genes (**Figure 12.13**).

It has also become obvious that practically all modifications are distributed in gradients along eukaryotic genes (**Figure 12.14**). The case for the different methylation levels at residue K4 of histone H3 is particularly striking: trimethylation is prevalent at the beginning of the transcribed DNA region and falls off quickly and dramatically toward the end of the region. At the same time, H3K4me2 and H3K4me1 show different gradual changes along the gene. We still do not understand why these gradients exist: are they needed for regulation of transcription, or do they occur as a consequence of the elongation process? Nevertheless, the very fact that they are created, by different enzymes, is amazing. Understanding the complexity of the regulatory system that controls the activity of these enzymes remains a challenge for the future.

Some genes are specifically silenced by post-translational modification in some cell lines

It is clear that histone PTMs also play crucial roles in processes that lead to gene silencing. **Gene silencing** is defined as a permanent, irreversible loss of the ability of a gene to be transcribed. Cytologically, silenced genes are organized as dense **constitutive heterochromatin**. Constitutive heterochromatin is distinguished from a less dense form known as **facultative** heterochromatin. Genes packed as facultative heterochromatin are not expressed but can revert to active expression. These types of heterochromatin differ significantly from each other and from transcriptionally active euchromatin in the nature and extent of their histone modifications; they are also characterized by

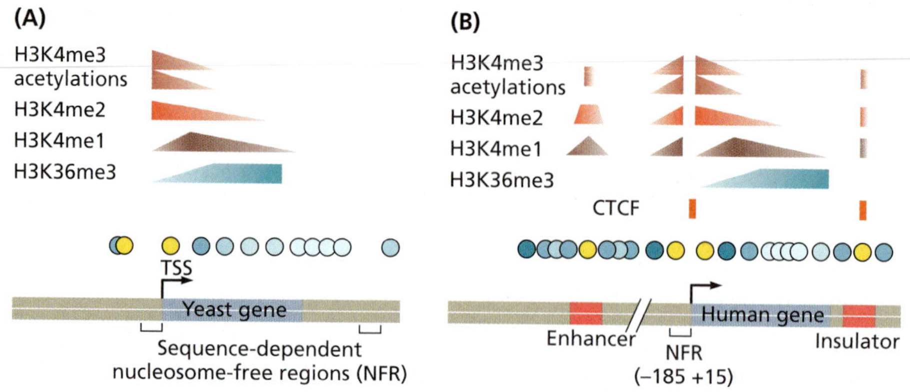

Figure 12.14 Partial chromatin maps of typical transcriptionally active genes in yeast and humans. In both yeast and humans, the TSS is embedded in a nucleosome-free region or NFR. The existence of such regions is determined by the underlying DNA sequence; for example, it may contain poly(dA/dT) tracts. In addition, the NFR is flanked on both sides by nucleosomes containing the histone H2A replacement variant H2A.Z, depicted by yellow circles. The stringency of nucleosome positioning is depicted with increasing intensity of blue color. Note that nucleosomes close to the beginning of the gene are well-positioned but nucleosomes further down the gene lose their strict positioning. It is believed that the positioning signals are determined by the presence of the H2A.Z-containing nucleosome at the beginning of the gene; this nucleosome may serve as a boundary element for the positioning of both downstream and upstream nucleosomes. (A) Typical yeast gene, showing the distribution of some histone marks associated with transcription. Note that the different histone modifications appear as characteristic gradients along the gene. Thus, for example, H3K4 trimethylation is maximal at the start of the gene and diminishes to zero toward the middle of the gene. By contrast, H3K4 monomethylation is relatively weak at the beginning of the gene, then picks up and decreases again. (B) Typical human gene, with an enhancer and an insulator linked to it. Note again the gradients in the distribution of histone marks that exist along the gene. Note also the existence of clear differences in the modification status of enhancers and insulators. Both enhancers and insulators contain H2A.Z nucleosomes and are marked by H3K4 monomethylation but not by trimethylation. In addition, the enhancers are enriched in the acetyltransferase p300, which is absent from promoters and not shown here. Thus the chromatin signature of enhancers and promoters is different enough to allow prediction of the location and function of novel enhancers. Another notable feature of both the NFR and of insulators is significant enrichment in transcription factors, such as CCCTC-binding factor, CTCF. Both enhancers and insulators contain DNase I hypersensitive sites, which are generally attributed to more relaxed chromatin structure, eventually the absence of nucleosomes, and binding of TFs. (Adapted from Rando OJ & Chang HY [2009] *Annu Rev Biochem* 78:245–271. With permission from Annual Reviews.)

the presence or absence of specific *trans*-acting factors and other chromatin components. Finally, they differ in the presence of bound RNA components and in their DNA methylation status. We have summarized these characteristic features in a concise format in **Table 12.3**.

Next we discuss a few well-understood examples of gene silencing. Note the involvement, in each specific case, of a particular histone modification. We must state, though, that this knowledge is still at the descriptive stage; we do not understand the exact molecular mechanisms that underlie these phenomena.

Polycomb protein complexes silence genes through H3K27 trimethylation and H2AK119 ubiquitylation

Polycomb complexes bring about silencing of hundreds of genes that encode crucial developmental regulators in a wide variety of plants and animals. Classic examples are the **homeobox** genes, which were first discovered in *Drosophila* and then shown to regulate organ development and body form in many eukaryotic organisms. These genes are active in early development but are silenced when they are no longer needed later in development. Two critical repressive complexes, **polycomb repressive complexes** (**PRC1** and **PRC2**), are recruited to genes to be silenced. Both complexes

Table 12.3 Molecular characteristics of euchromatin, constitutive heterochromatin, and facultative heterochromatin. (Adapted from Trojer P & Reinberg D [2007] *Mol Cell* 28:1–13.)

		Histone modifications	Chromatin components and *trans*-acting factors	DNA methylation	RNA component
		Euchromatin			
	hyperacetylation	H3K4me2/3 H3K36me3	ATP-dependent chromatin remodelers; H3.3, H2A.Z, H2ABbd	–	–
		Constitutive Heterochromatin			
	hypoacetylation	H3K9me3 H4K20me3		+	+
		Facultative Heterochromatin			
local gene silencing	hypoacetylation	H3K9me2 H4K20me1 H2AK119ub1	PRC1, PRC2, and other PcG proteins; HP1γ, MBT proteins[a]	?	?
long-range silencing, such as of *Hox* gene clusters	hypoacetylation	H3K27me2/3 H4K20me3 H2AK119ub1	PRC1, PRC2, and other PcG proteins	+	+
autosomal imprinted genomic loci	hypoacetylation	H3K9me2/3 H3K27me3 H4K20me3	PRC2; macroH2A; CTCF	+[b]	+
inactive X chromosome, Xi	hypoacetylation	H3K9me2 H3K27me3 H4K20me1 H2AK119ub1	PRC1, PRC2, and other PcG proteins; macroH2A; CULLIN3/SPOP[c]	+	+

[a]MBT, malignant brain tumor-domain proteins are low-methylation-state-specific readers. The MBT-domain-containing protein specifically interacts with histone H1.4 methylated at lysine 26, compacting chromatin.

[b]Associated with the inactive allele.

[c]CULLIN3/SPOP is a ubiquitin U3 ligase that ubiquitylates macroH2A on the inactive X chromosome; this modification of macroH2A is important for association with Xi.

contain subunits that have specific histone-modifying activities (**Figure 12.15**). PRC2 recognizes specific sequence elements in the DNA, whereas the recruitment of PRC1 usually occurs through interactions with H3K27me3, a modification introduced by PRC2. It must be noted that PRC1 and PRC2 act through a variety of mechanisms, only some of which involve their histone-modifying activities.

Heterochromatin formation at telomeres in yeast silences genes through H4K16 deacetylation

Another well-understood example of silencing mechanisms that involve histone PTMs is the formation of heterochromatic structures at yeast telomeres (**Figure 12.16**). Three proteins are involved. The initial step depends on the action of an unusual enzyme, **Sir2** or **silent information regulator 2**, the founding member of a family of NAD-dependent histone deacetylases. During the reaction, nicotinamide adenine dinucleotide, NAD, is hydrolyzed to *O*-acetyl-ADP-ribose, which promotes the binding of multiple Sir3 molecules to the Sir2–Sir4 complex (see Figure 12.16). This facilitates heterochromatin spreading. As a curious note, Sir2 is considered a longevity factor in both yeast and humans, and as such it attracts much attention.

HP1-mediated gene repression in the majority of eukaryotic organisms involves H3K9 methylation

Organisms other than budding yeast have a different mechanism for gene silencing. The process involves the action of **HP1** or **heterochromatin protein 1**, first discovered

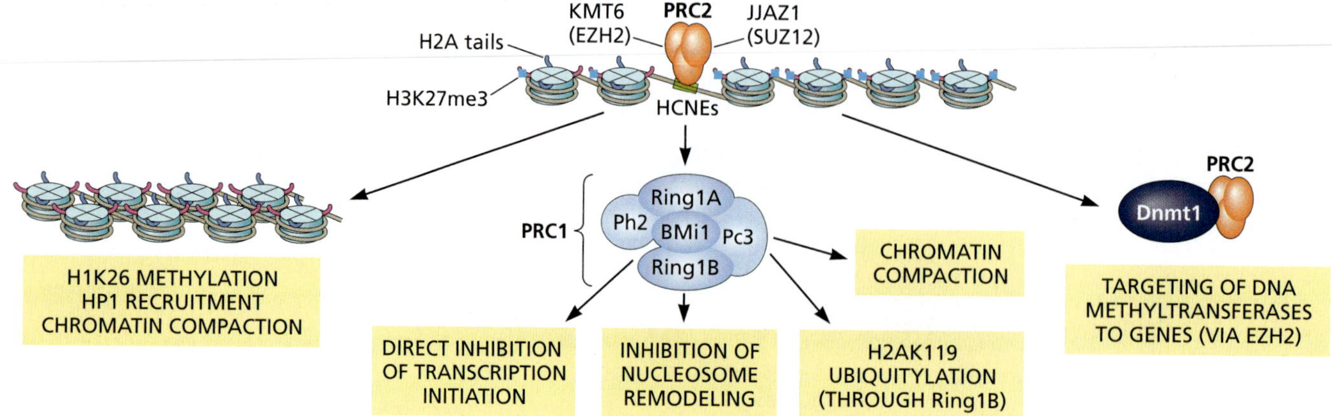

Figure 12.15 Epigenetic gene silencing by polycomb protein complexes. Binding of polycomb repressive complex 2, PRC2, to polycomb-group or PcG target genes induces H3K27 trimethylation through the activity of the KMT6 enzyme. KMT6 was previously known as enhancer of zeste homolog 2 or EZH2. H3K27me3 is recognized by a second polycomb complex, PRC1, through the chromodomain of Pc3. The recruitment of PRC1 might act through different mechanisms, all leading to gene silencing. Here, HCNEs stands for highly conserved noncoding DNA sequence elements in genes silenced by polycomb complexes. One such mechanism involves the ubiquitylation of H2AK119; the downstream consequences of this modification are not yet understood. Protein subunits of both PRC2, subunits EZH2 and JJAZ1/SUZ12, and of PRC1, subunit BMi1, are highly overexpressed in certain human cancers; high levels of EZH2 are associated with poor prognosis and indicative of the metastatic stage of the disease.

in *Drosophila*. The domain structure of HP1 is characterized by the presence of two closely related domains—a chromodomain, CD, and a chromo-shadow domain, CSD (**Figure 12.17**)—that perform different functions. CD specifically recognizes methylated lysines in H3. CSD, on the other hand, interacts with other proteins, including the methylation enzyme SUV3-9 in *Drosophila* and its orthologs in mouse and human. This domain also interacts with histones H3, H4, and H1 to strengthen the interaction

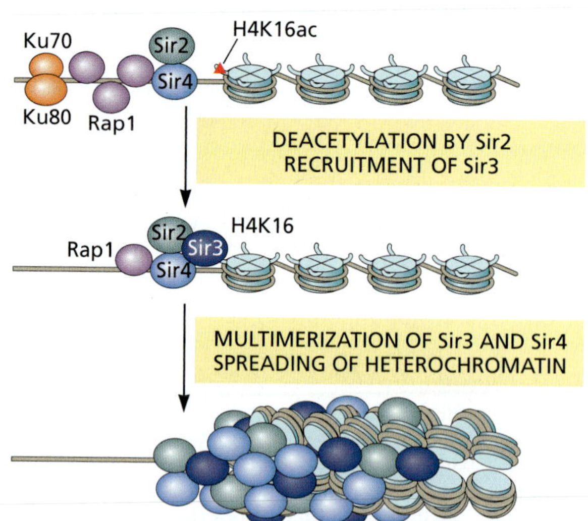

Figure 12.16 Stepwise model for assembly of heterochromatin at telomeres in budding yeast. The initial step involves recruitment of Sir2–Sir4 complexes to DNA through interactions between Sir4 and the telomere-bound proteins Ku70/Ku80 and Rap1, repressor activator protein, a sequence-specific protein that can function either as a repressor or as an activator of transcription, depending on the binding-site context. A series of 16–20 Rap1-binding sites within telomeres ensures Rap1 binding and thus recruitment of Sir2–Sir4. In the second step, the acetyl group at H4K16 on the adjacent nucleosome is removed by Sir2, the founding member of an interesting family of nicotinamide adenine dinucleotide- or NAD-dependent histone deacetylases. Sir3 is recruited via interactions with Rap1, Sir4, and the deacetylated histone tails. Multimerization of Sir3 and Sir4 leads to spreading of the Sir complex along nucleosomes to create compact chromatin structures at telomeres. Slight overexpression of Sir2 in yeast extends the life span by about 30%; the human homolog SIRT1 is also implicated in longevity.

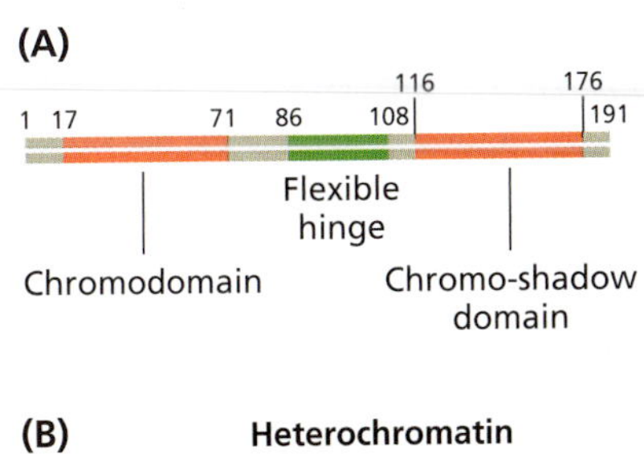

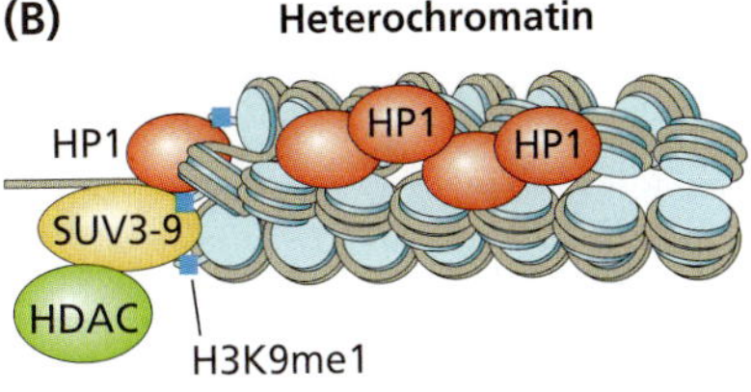

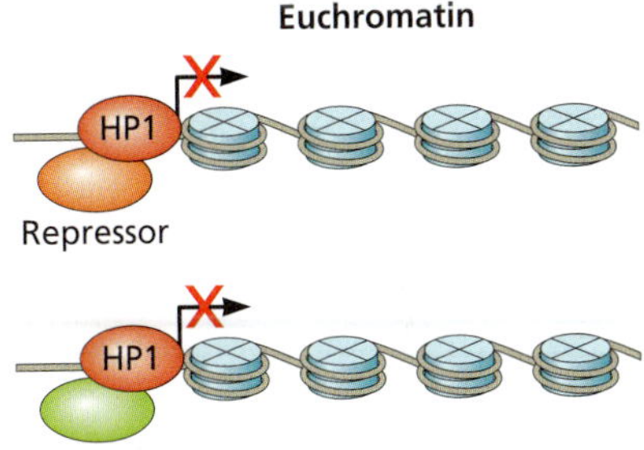

Figure 12.17 HP1-mediated repression of transcription in organisms from *S. pombe* to humans. (A) Domain structure of heterochromatin protein 1, HP1, with the evolutionarily conserved chromodomain, CD, and chromo-shadow domain, CSD. The CD recognizes and binds to H3K9me/me2/me3. The flexible hinge shown in green interacts with RNA, DNA, and chromatin and contains a number of regulatory phosphorylation sites. The CSD is the site of dimerization and protein binding. (B) Various mechanisms of transcriptional repression in heterochromatin and euchromatin. Heterochromatization involves the formation of highly compacted chromatin structures through numerous HP1 molecules. In addition, heterochromatin contains deacetylated and methylated histones; the involved writer enzymes are depicted. In euchromatin, HP1 repression can occur through the formation of similar highly compacted chromatin structures, not shown here, or by short-range action over very short stretches of chromatin fibers, sometimes even at the level of a single nucleosome. Repression in euchromatin can also occur via repressive interactions with components of the basal transcriptional machinery, such as TAFII130 or TFIID, at promoter regions. HP1 can also activate transcription, usually via interactions with transcriptional activators.

with chromatin; with DNA methyltransferases Dnmt1 and Dnmt3a; with MeCP2; with histone deacetylases or HDAC; and with other closely related isoforms of HP1. CSD is also the interaction interface with transcriptional activators and components of the basal transcriptional machinery, such as TFIID. HP1 participates in the formation of heterochromatin by recognizing mono-, di-, or trimethylated H3K9. Unexpectedly, it can also repress or activate genes in euchromatin through the mechanisms depicted in Figure 12.17. Thus the protein is stuck with a name that does not accurately reflect its seemingly opposing functions in gene regulation. It is not alone: many other proteins have been described that can act as either activators or repressors, depending on the specific context.

Poly(ADP)ribosylation of proteins is involved in transcriptional regulation

Poly(ADP)ribosylation, or **PARylation**, is a very distinct, bulky post-translational modification of proteins that significantly affects the transcriptional status of eukaryotic genes. The modification consists of adding one or more adenosine diphosphate-ribose or ADP-ribose units from donor NAD^+ molecules onto target proteins (**Figure 12.18A**). This reaction is catalyzed by the enzyme **poly(ADP-ribose) polymerase 1**, known as **PARP-1**, and its homologs. The enzyme has a modular structure comprising three domains (**Figure 12.18B**) and can modify itself in a reaction known as automodification. Modification of other acceptor proteins is known as heteromodification. The enzyme is inactive for **ADP-ribosylation** reactions until an appropriate stimulus appears. Automodification then activates the enzyme for further PARylation reactions. A unique aspect of PARylation is that, in addition to the standard covalent modification, many proteins, including histones, can undergo noncovalent modification (see Figure 12.18B).

PARP-1 is extremely abundant, with ~1–2 million molecules per cell, second only to histones. It participates in numerous biological processes, such as DNA damage detection and repair, DNA methylation and gene imprinting, insulator activity, and chromosome organization. These apparently disparate roles of PARP-1 can be divided into two major categories: emergency responses and housekeeping. The emergency function occurs after DNA damage and involves the numerous PARP-1 molecules normally present in the nucleus in inactive, non-PARylated form; these non-PARylated PARP-1 molecules rapidly become auto-poly(ADP)ribosylated upon DNA injury. The housekeeping role is played under normal unstressed conditions and involves only the few PARP molecules that are PARylated under such conditions. The main housekeeping role is transcriptional regulation, through a wide variety of mechanisms (**Figure 12.19**).

The first role for PARylation in the regulation of transcription to be recognized was the modulation of chromatin structure. Chromatin structure is modulated either by direct binding of PARP-1 to chromatin or by PARylation of a number of chromatin proteins, some of which are released when modified; more details are given in the legend

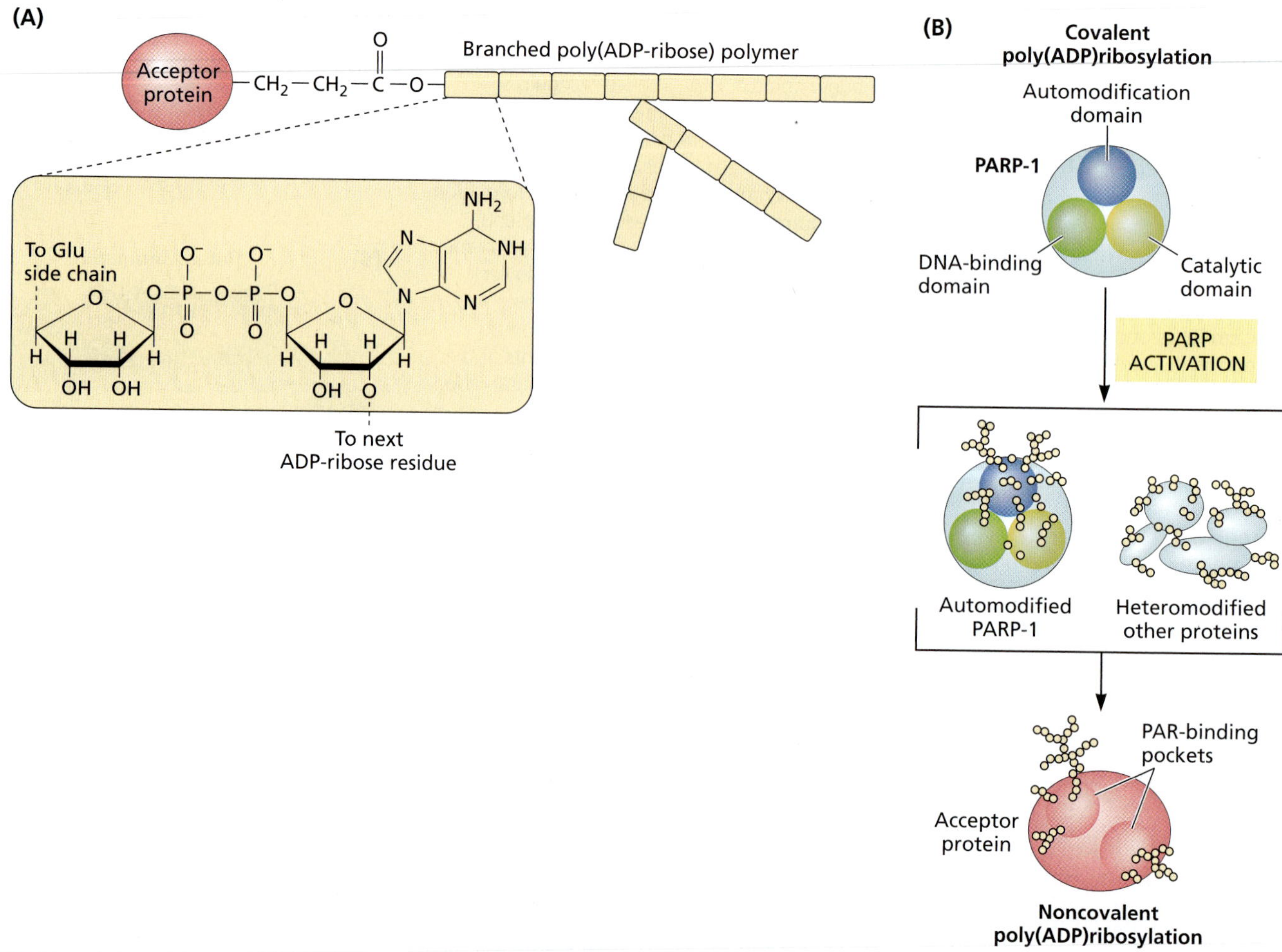

Figure 12.18 ADP-ribose polymerization reaction. (A) Structure of poly(ADP-ribose) or PAR polymer attached to an acceptor protein. The polymer chain is attached to glutamate residues within the acceptor protein. The polymers are usually long, up to 200 residues, and highly branched. Mono(ADP-ribose) adducts also exist. (B) Covalent and noncovalent poly(ADP)ribosylation or PARylation. Covalent PARylation begins with activating the catalytic activity of poly(ADP-ribose) polymerase 1 or PARP-1, usually in response to nicks in the DNA. The enzyme has a modular structure comprising multiple independently folded domains: the three most prominent domains are an N-terminal DNA-binding domain or DBD, a central automodification domain, and a C-terminal catalytic domain. The enzyme builds numerous long and branched poly(ADP-ribose) polymers onto its automodification domain; it also modifies numerous other proteins in a process known as heteromodification. A number of additional acceptor proteins can be modified by noncovalent attachment of PAR polymers to specific binding pockets. The interacting polymers could be still hosted on the automodified PARP-1 or on heteromodified proteins; alternatively, noncovalent modification could occur through protein binding of free polymers that are transiently available in the cell because of the activity of enzymes that cleave the polymer, PAR glycosylases. PARP-1 is the founding member of a family of at least 17 members that share homology with its catalytic domain.

to Figure 12.19. In addition, PARP-1 can serve in traditional activator or co-regulator roles. A fourth mechanism is involved in insulator function, through PARylation of the insulator CTCF protein (see Figure 12.2). In all these scenarios, it remains unclear how poly(ADP)ribosylation is targeted to specific genes and how such a bulky and poorly defined modification can have so many specific effects.

Histone variants H2A.Z, H3.3, and H2A.Bbd are present in active chromatin

Several of the nonallelic replacement variants described in Chapter 8 are thought to be involved in the regulation of transcription, or at least transcribability. The role of **histone H2A.Z** in this context (Figure 12.20) is particularly intriguing for several reasons: (1) H2A.Z is significantly enriched in promoter regions genomewide; (2) there are two H2A.Z-containing nucleosomes flanking the NFR at the transcription start site in at least two-thirds of all genes; (3) H2A.Z levels anticorrelate with transcription in

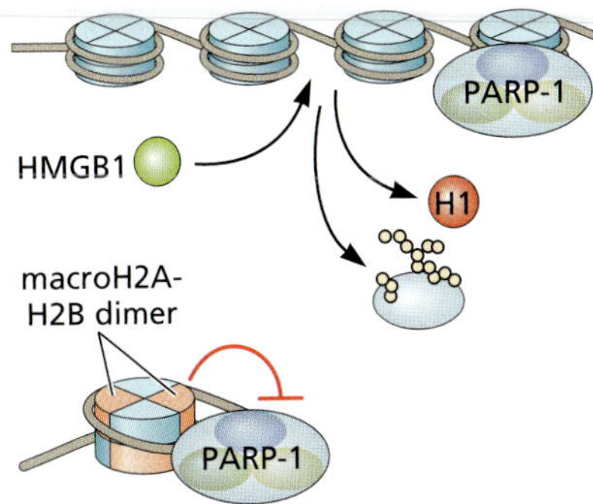

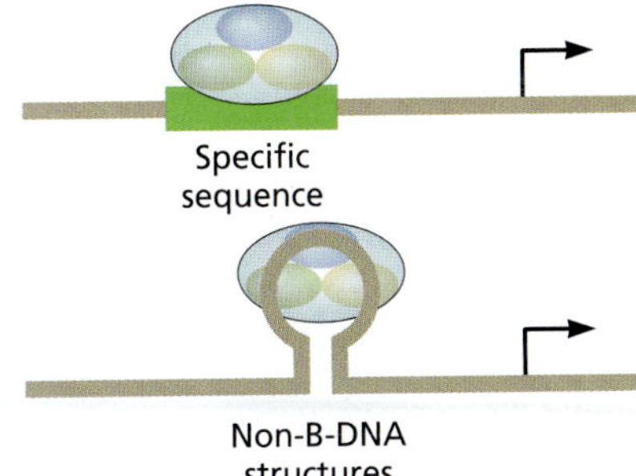

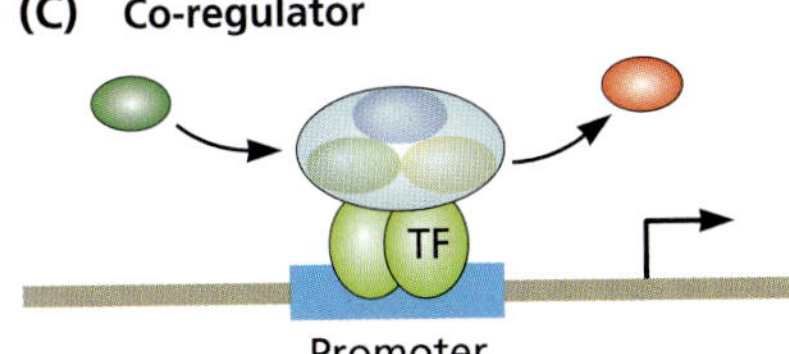

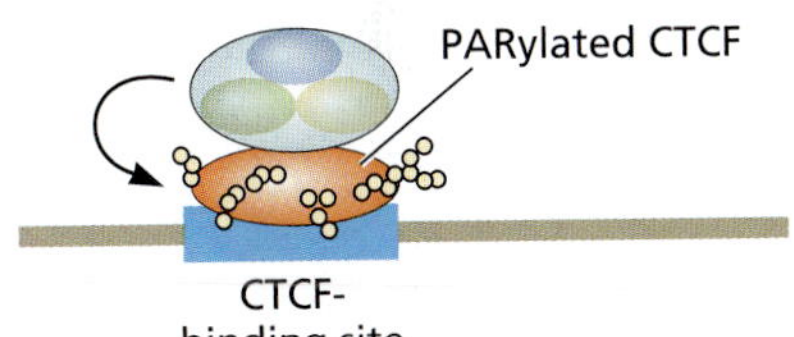

Figure 12.19 Poly(ADP-ribose) polymerase regulates transcription through multiple mechanisms. (A) (Top) PARP-1 modulates chromatin structure and composition by directly competing with H1, or by removing H1 by PARylation from its binding site on the nucleosome, or by stimulating the binding of HMGB1, a chromatin non-histone protein that stimulates transcription, or by releasing other proteins after modifying them. PARylated proteins acquire a highly negative charge that is incompatible with chromatin or DNA binding. (Bottom) PARP-1 also interacts with nucleosomes that contain the histone variant macroH2A; the interaction leads to inhibition of the enzymatic activity of PARP-1, with consequences for chromatin structure and function in the specific regions containing macroH2A. (B) PARP-1 can serve a typical activator function by binding to specific sequences in enhancers or to non-B-DNA structures such as hairpins, cruciforms, cross-overs, or double-strand breaks. (C) PARP-1 can be a co-activator or co-repressor, depending on the context. As a co-regulator, it may function as an exchange factor, promoting the release of some factors and the recruitment of other factors, in a promoter-specific way. (D) PARP-1 is involved in the function of insulator elements by PARylating the insulator factor CTCF.

yeast, whereas in humans they actually correlate with transcription; and (4) H2A.Z-containing nucleosomes have very high turnover rates, making them among the most popular nucleosomes known. The mechanisms of H2A.Z action remain to be elucidated.

The other replacement histone variant that has been associated with transcriptional activity is **histone H3.3**. Recall from Chapter 8 that replacement variants can be synthesized and incorporated into chromatin throughout the cell cycle, in contrast to the canonical histones H2A and H3.3, which are produced only in S phase. H3.3 marks actively transcribed regions, where nucleosomes are constantly disassembled and reassembled (see Figure 12.20). Genomewide, the two nucleosomes flanking the NFR in promoters contain H3.3 as well as H2A.Z. The exact properties and functions of these hybrid nucleosomes remain to be established. In addition, H3.3-containing nucleosomes are enriched over *cis*-regulatory boundary elements in the *Drosophila* genome, suggesting that chromatin structure in these regions is in constant flux, probably as part of a mechanism to keep *cis* elements exposed to factor binding. Thus, it remains unclear whether the presence of H2A.Z and H3.3 variants is just a reflection of high nucleosome turnover or whether H2A.Z and H3.3 endow certain structural characteristics upon the nucleosomes or chromatin fibers that directly affect their behavior. Recent *in vitro* results suggest that the answer may actually be different for the two variants.

There is another very interesting histone variant, **H2A.Bbd** or **Barr body-deficient histone H2A**, that is largely excluded from the inactive X chromosome or Barr body of mammals. Moreover, its genome deposition pattern overlaps with regions of histone H4 acetylation, suggesting that it is associated with transcriptionally active euchromatic regions. The polypeptide is relatively short, lacking the C-terminal tail and a portion of the docking domain of canonical H2A; the docking domain is the portion of H2A that interacts with H3, thus stabilizing the nucleosome. H2A.Bbd is considered the most specialized of all of the histone variants known to date because of its very low sequence identity with H2A, only 48%. The protein is very rapidly evolving.

Nucleosomes that contain H2A.Bbd show high turnover rates, in agreement with the view that the presence of H2A.Bbd destabilizes nucleosomes. Indeed, histone octamers containing H2A.Bbd organize only 120–130 bp of DNA in a nucleosome, leaving ~10 bp at each end free from interactions with histones. Such nucleosomes are organized in more relaxed fiber structures.

All of these properties would seem to underlie a function in transcriptionally active chromatin. *In vitro* studies using reconstituted arrays of H2A.Bbd-containing nucleosomes do show more efficient transcription than arrays of canonical nucleosomes. It is still unclear, however, whether the variant facilitates transcription *in vivo*. It may have other roles, such as an involvement in mammalian spermatogenesis, as part of the mechanism of replacement of histones by protamines. This replacement also requires histone acetylation, suggesting coordinated action of histone variants and histone PTMs.

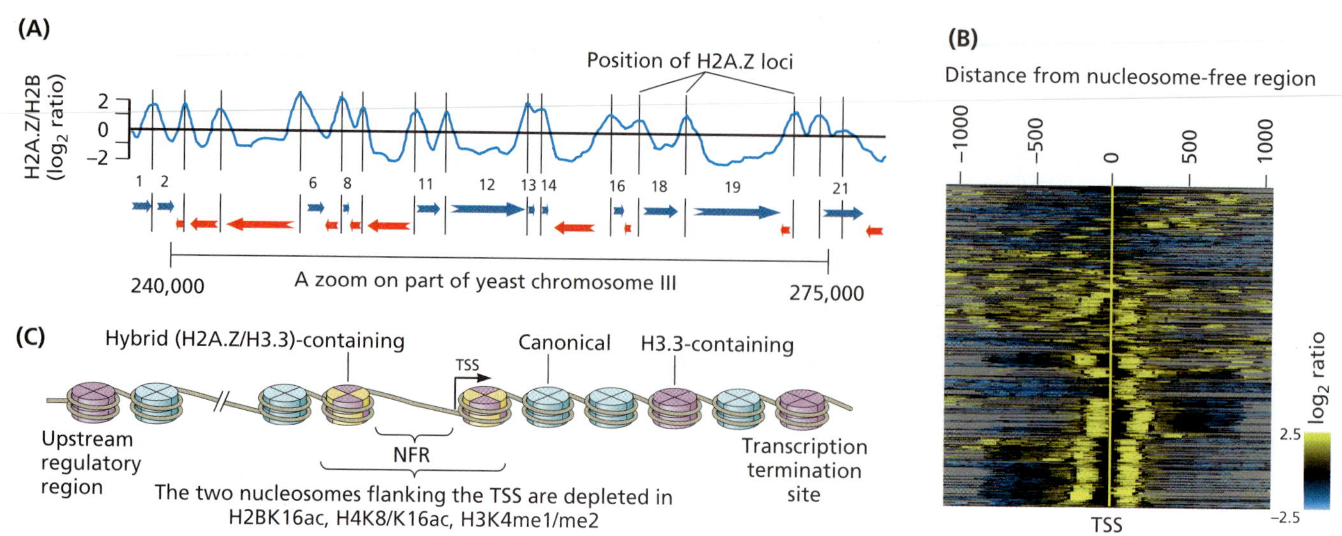

Figure 12.20 Genomic localization of histone variants H3.3 and H2A.Z. (A) Genomewide location of H2A.Z on ~40 kbp of yeast chromosome III. Genes occupying the region are represented by blue and red arrows that denote the direction of transcription. The centers of the H2A.Z loci are marked by vertical lines. Note the significant enrichment in promoter regions: there is a H2A.Z locus upstream of every open reading frame or ORF. Intergenic regions that do not contain promoters, such as regions between convergent genes, are not enriched in H2A.Z, and intergenic regions between divergently transcribed genes contain two separable H2A.Z loci. (B) High-resolution mapping of H2A.Z-containing nucleosomes in 2000-bp regions surrounding the nucleosome-free region or NFR in each promoter. Genes present in the database are aligned by their NFRs, with each row representing a single promoter region. Yellow indicates H2A.Z enrichment; blue denotes H2A.Z depletion. Note that the two nucleosomes surrounding the TSS contain H2A.Z. (C) Schematic of genomic localization of the two variants, H2A.Z and H3.3, believed to be associated with actively transcribed regions. Note that H3.3 is enriched not only in the gene body but also in the upstream regulatory regions. Repressed genes also contain H3.3 in these regulatory regions and in the promoter region: thus H3.3 enrichment may be a characteristic feature of genomic regions that are in flux, with nucleosomes constantly assembling and disassembling, and may not be a specific marker of active transcription. (A, adapted from Guillemette B, Bataille AT, Gévry N et al. [2005] *PLoS Biol* 3:e384. With permission from Luc Gaudreau, Université de Sherbrooke. B, from Raisner RM, Hartley PD, Meneghini MD et al. [2005] *Cell* 123:233–248. With permission from Elsevier.)

MacroH2A is a histone variant prevalent in inactive chromatin

MacroH2A is another recently discovered histone H2A variant that is present only in vertebrates. It occurs abundantly on inactive X chromosomes, and, in general, its presence in chromatin leads to transcriptional silencing. On average, one out of every 30 nucleosomes contains macroH2A. As illustrated in Figure 8.11, macroH2A contains a long C-terminal non-histone region termed the macrodomain. The macrodomain recruits PARP-1 to chromatin and inhibits its enzymatic activity, with many ensuing consequences for transcription. In addition, macroH2A-containing nucleosomes are more stable than their canonical counterparts, which may also be a factor in their inhibition of transcription. Additional mechanisms of gene silencing may include interference with TF binding and chromatin remodeling.

Problems caused by chromatin structure can be fixed by remodeling

Given that chromatin structure and its modification can have major effects on transcription, it is not surprising that mechanisms exist to remodel chromatin. **Chromatin remodeling** is an active process in which remodeling complexes use the energy of ATP hydrolysis to introduce changes in the structure of the nucleosomal particle and/or the location of the particle with respect to the underlying DNA sequence. Although the mutual disposition or spacing of nucleosomes along DNA has the potential to affect folding of the chromatin fiber, most efforts have been directed toward understanding remodeler action at the level of individual nucleosomes, especially in promoter regions. Thus, chromatin remodelers are also termed nucleosome remodelers.

Chromatin remodelers form a large superfamily containing at least three or four subfamilies; each subfamily contains multiple complexes in each species. In humans,

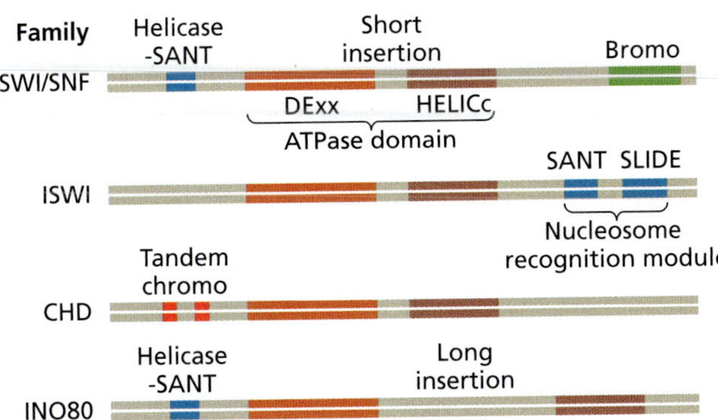

Figure 12.21 Defining ATPase subunits of remodeler families. The conserved ATPase domain is split into two parts characterized by specific amino acid sequences: DExx and HELICc. Each family has distinct unique domain(s) residing on one or both sides of the ATPase domain, which define the unique functions of the respective families. The unique domains recognize differently modified histones: bromodomains recognize acetylated histones, whereas chromodomains bind to methylated histones. The SANT-SLIDE domain in ISWI binds unmodified histone tails and DNA. The helicase-SANT domain interacts with actin and actin-related proteins, which are subunits of some remodeling complexes of undefined function.

four different subfamilies have been recognized. Although the subfamilies differ considerably, each is characterized by the shared ATPase subunit (Figure 12.21):

- **SWI/SNF** or **switch/sucrose nonfermenting**: the name originates from the yeast genes identified by mutations that affect the ability of yeast to switch mating type or to use sucrose as a carbon source. Compositions of human and prototypic yeast SWI/SNF complexes are presented in Figure 12.22; note that the two species share some subunits, especially the subunits that carry out ATP hydrolysis, but also contain distinctive subunits.
- **ISWI** or **imitation switch** subfamily: contains four different complexes, RSF, hACF/WCFR, hCHRAC, and WICH.
- CHD or chromo-helicase-DNA-binding protein subfamily: at least nine members of the subfamily, CHD1–9, have been recognized in humans. CHD3 is a

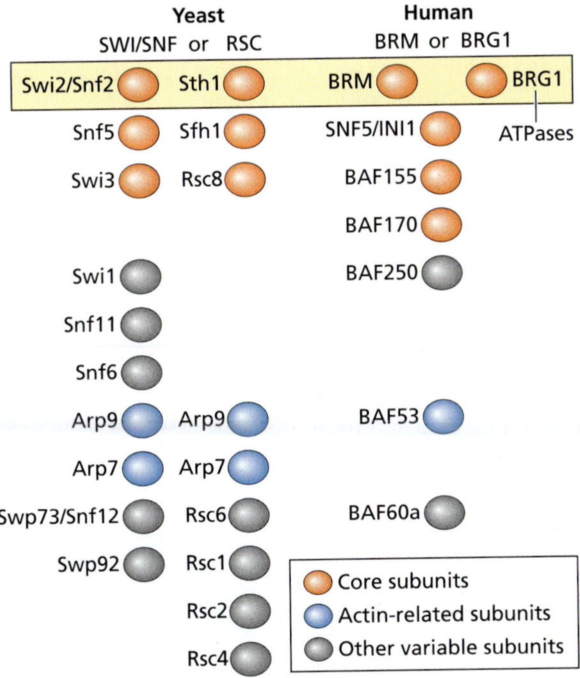

Figure 12.22 Yeast and human SWI/SNF complexes. The columns show the subunit composition of each complex; the rows indicate the homologs and/or orthologs of each individual subunit. Note that each complex possesses core subunits, one of which is the subunit with ATPase activity, as well as variable subunits. The core subunits are necessary for full activity *in vitro*. The role of the variable subunits is not well understood; they may participate in directing the specificity of the complex through protein–protein interactions. Curiously, some subunits carry sequence similarity to the contractile muscle protein actin; actin is also involved in the formation of the cytoskeleton and of nuclear skeletal structures. The amino acid sequences of the ATPase subunits BRM and BRG1 are 75% identical, and both are widely expressed. Nonetheless, the presence of these subunits in a complex is mutually exclusive.

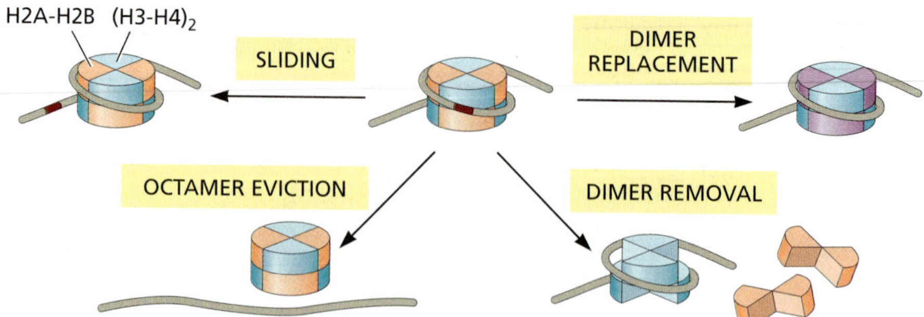

Figure 12.23 Models of nucleosome remodeling. Remodelers can act in a variety of ways, depending on the biological context. They can slide the histone octamer with respect to the underlying DNA sequence, so as to allow accessibility of factor-binding sites. They can totally evict the histone octamer, an event that frequently occurs in promoters of genes undergoing transcriptional activation. They can remove one or both H2A-H2B dimers during transcriptional elongation. This activity is not unique to remodelers and can sometimes be performed by unrelated protein complexes that contain histone chaperones: a well-studied example is FACT, facilitates chromatin transcription complex. Finally, remodelers can replace existing H2A-H2B dimers with dimers containing variant histones, such as H2A.Z.

component of the Mi2/NuRD complex that contains both histone deacetylase and nucleosome-dependent ATPase subunits. Current models suggest that the Mi2/NuRD complex functions primarily in repression of transcription.

- INO80/SRCAP or inositol-requiring/SNF-2-related CREB-binding activator protein.

Remodelers can act in a variety of ways (**Figure 12.23**); in fact, each subfamily is characterized by the specific outcome of the remodeling activity. **Figure 12.24** illustrates the proposed mode of action of the best understood remodeler, ISWI, which causes sliding of the histone octamer with respect to the underlying DNA sequence. Other remodelers change the internal nucleosome structure, destabilizing the particle, which may lead to histone replacement or dissociation. Such destabilization may be crucial in allowing access to nucleosomal DNA by the machineries that utilize DNA as their substrate.

Finally, there is a well-documented link between mutations in genes coding for chromatin remodeling factors and disease, especially cancer. Thus, mutations in the hSNF5/INI1 subunit of the human SWI/SNF complex are characteristically present in aggressive pediatric tumors as well as in some acute leukemias. Strong support for considering *hSNF5* a tumor-suppressor gene comes from the observation that heterozygous knockout mice develop tumors. Mutations in BRG1, the ATPase subunit of human SWI/SNF, have been identified in several cancer cell lines; *BRG1*$^{+/-}$ mice are also predisposed to tumors, again confirming the role of BRG1 as a tumor suppressor.

Endogenous metabolites can exert rheostat control of transcription

It has been recently recognized that small-molecule metabolites can regulate transcription as well as translation (see Chapter 17), either directly or by affecting chromatin structure or remodeling. Because the level of activity will vary in a continuous manner with metabolite concentration, this is sometimes called **rheostat control**.

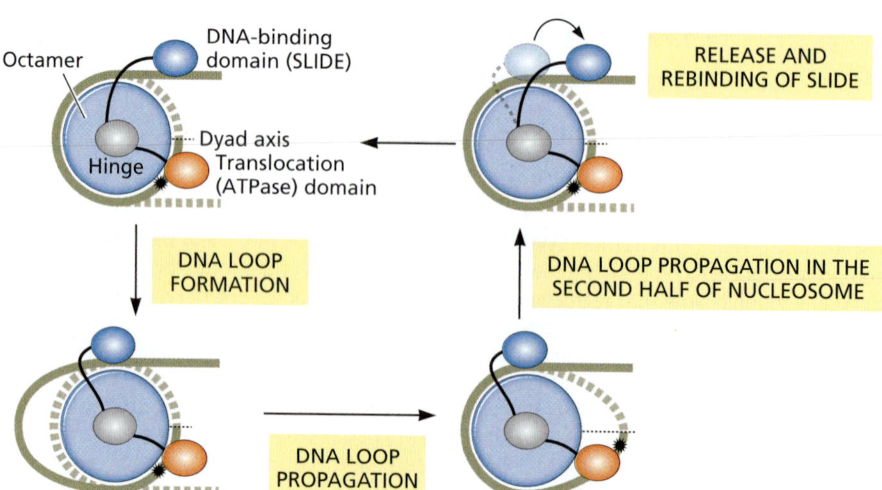

Figure 12.24 Loop or bulge model for the sliding remodeling action of ISWI. The second gyre of DNA is presented as a broken line to reinforce the perspective of DNA wrapping. The black starburst provides a reference point on the DNA to facilitate visualization of the translocation of DNA along the octamer surface. Concerted action of the DNA-binding SLIDE domain, bound to the linker region, and the translocation or ATPase domain, bound close to the dyad axis, generates a small bulge that propagates on the nucleosomal surface. The loop is initially created by the DNA-binding domain, which pushes DNA into the nucleosome. Generation of the loop is accompanied by conformational changes in the SLIDE domain. The directional propagation of the loop is then conducted by the ATPase domain, which remains anchored at its position on the nucleosome: it draws the DNA from the linker and pumps it toward the dyad and then further in the second half of the particle. Histone–DNA contacts are being broken at the leading edge of the loop and re-formed at its lagging edge. (Adapted from Clapier CR & Cairns BR [2009] *Annu Rev Biochem* 78:273–304. With permission from Annual Reviews.)

Direct regulation occurs when metabolites affect the activity of activators or repressors. Thus, for example, estrogen receptors activate gene transcription only when steroid hormones are bound to them. Similarly, interactions of the C-terminal binding protein or CtBP with activators or repressors are controlled by NAD^+/NADH. Chromatin-mediated control by metabolites depends on the presence or concentration of metabolites that are used to add or remove marks on histones and other chromatin components. Thus, for example, both histone and DNA methylation require *S*-adenosylmethionine as a donor of methyl groups; acetyl-CoA is needed for the acetylation of lysine residues on histones; and the Sir2 deacetylase needed for heterochromatin silencing in yeast is dependent on the presence of NAD. NAD is, of course, the precursor needed to build poly(ADP-ribose) polymers onto proteins. Such metabolite-induced alterations in protein function and chromatin structure may allow fine-tuning of gene expression in response to changes in the levels of cellular metabolites and, in a broader sense, to the environment.

12.6 DNA methylation

DNA methylation by **DNA methyltransferase** (**DNMT**) enzymes is the major epigenetic modification of DNA. In eukaryotes, it is achieved through the transfer of methyl groups from *S*-adenosylmethionine to cytosine, converting it into 5-methylcytosine, ^{m5}C (**Figure 12.25**). The modification preferentially occurs on CpG dinucleotides. As cytosine methylation does not affect base pairing, the deposition of this postreplicational mark can regulate processes without affecting the genetic information in DNA. In addition, mechanisms exist that propagate this modification throughout DNA replication (**Table 12.4** and Figure 12.25B), thereby passing the modification pattern from mother to daughter cells during mitosis. In this sense, and in contrast to histone PTMs, this mark can be considered truly epigenetic: it carries regulatory information that is not encoded in the DNA sequence and is inherited from cell generation to cell generation.

DNA methylation patterns are not directly inherited during sexual reproduction in many organisms; they are erased during early development and later established *de novo* (**Figure 12.26**). The mechanisms that instruct the somatic cells in the new organism to establish patterns are not clear. It is clear, though, that the DNA methylation patterns are established in totally defined ways; thus there must be mechanisms, albeit indirect and unknown, to ensure the preservation of these patterns in the new

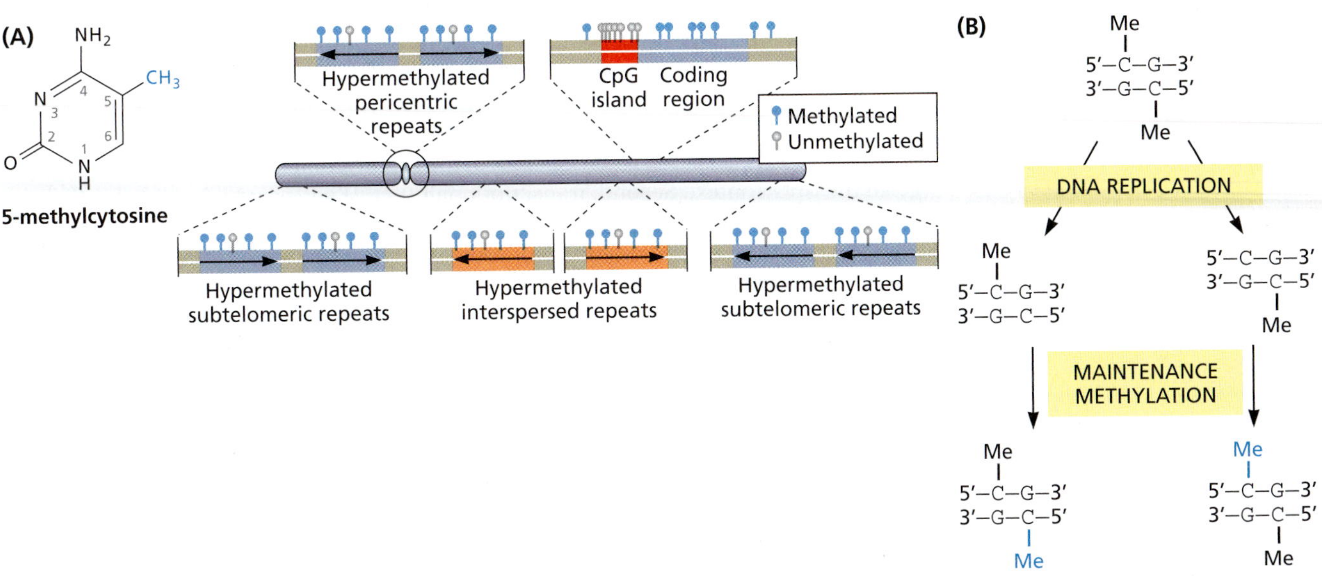

Figure 12.25 DNA methylation. (A) Schematic of a chromosome with representative regions that are rich in CpG dinucleotides, shown as lollipops, whose methylated or unmethylated pattern is required for normal cellular functions. The structure of methylated cytosine is shown to the left. (B) Maintenance of DNA methylation of CpG sites following DNA replication. The newly synthesized DNA strand is unmethylated for a while; then methyl groups are added to it by use of the opposite strand as a template. (A, adapted, courtesy of Paola Caiafa, University La Sapienza.)

Table 12.4 DNA methyltransferases in mammals.

Mammalian enzyme	Function	Phenotype of homozygous deletion mutants
Dnmt1	maintenance, functions after replication on hemimethylated DNA to restore existing methylation patterns; associates with histone deacetylase	*Dnmt1*$^{-/-}$ [a] die *in utero*; DNA in these embryos is hypomethylated and imprinted genes are biallelically expressed; embryonic stem (ES) cells from *Dnmt1*$^{-/-}$ mice are viable and capable of *de novo* methylation
Dnmt3a	*de novo* methylase	*Dnmt3a*$^{-/-}$ mice die at 4 weeks; *Dnmt3a*$^{-/-}$ embryos and ES cells exhibit demethylation of centromeric satellite repeats; ES cells from *Dnmt3a*$^{-/-}$ mice are viable and capable of *de novo* methylation
Dnmt3b[b]	*de novo* methylase	*Dnmt3b*$^{-/-}$ mice die *in utero*; ES cells from *Dnmt3b*$^{-/-}$ mice are viable and capable of *de novo* methylation; *Dnmt3b*$^{-/-}$ embryos and ES cells do not exhibit demethylation of centromeric satellite repeats

[a]The designation $^{-/-}$ refers to lack of both alleles or homozygosity of the respective gene.

[b]Heterozygous mutations affecting the catalytic domain of DNMT3b have been identified in humans with ICF syndrome, which stands for immunodeficiency, centromere instability, and facial anomalies. It is characterized by variable reduction in serum immunoglobulin levels, which causes severe susceptibility to infectious diseases in childhood. Facial abnormalities include hypertelorism, increased distance between the eyes; low-set ears; Mongolian eye folds; and macroglossia, unusual enlargement of the tongue.

organism. Although in vertebrates ^{m5}C has occasionally been found on cytosines followed by a C, an A, or a T, the best substrate for the addition of methyl groups is cytosine in the CpG or C-phosphate-G dinucleotide.

DNA methylation patterns in genomic DNA may participate in regulation of transcription

The first complete map of the human DNA methylome, published in 2009, examined the methylation state of 27 million CpG locations on the 23 pairs of human chromosomes. It also looked for C methylation in the context of the other dinucleotides. The effort relied on the use of bisulfate treatment and high-throughput sequencing; treatment of methylated DNA with sodium bisulfate converts the nonmethylated cytosines to uracil while leaving the methylated cytosines untouched. Because of limitations in the method, the sequencing effort amounted to sequencing the equivalent of the entire genome 57 times. This effort was worthwhile because it produced some rather surprising results. First, there were dramatic differences between pluripotent embryonic cells and differentiated cell types such as fibroblasts in terms of which dinucleotides are methylated: 99.98% of all methylation occurs at CpG dinucleotides in fibroblasts, whereas ~25% of the methylation in stem cells does not occur in the CpG context. Second, transcriptionally active genes contain undermethylated CpGs. Importantly, if terminally differentiated cells are engineered to revert to pluripotent stem cells, they regain the unusual modifications at sites different from CpGs. The question that still needs to be addressed is whether the observed differences are consequences of differential gene activity or are actively involved in the regulation of transcription.

Differentiated cells are characterized by an uneven distribution of methylated cytosines along chromosomes. Of particular interest are **CpG islands**, which despite being unusually rich in this dinucleotide somehow evade being heavily methylated (**Figure 12.27**). Furthermore, the presence of methylated DNA in promoters is directly linked to silencing of the downstream genes. Numerous mechanisms suggesting how promoter DNA methylation may affect gene expression have been proposed, including interplay between DNA methylation and histone modifications or chromatin remodeling. DNA methylation could also directly affect chromatin structure by compacting the chromatin fiber. Indeed, it has been convincingly demonstrated that chromatin regions that contain methylated cytosines are more compact than their unmodified counterparts. However, DNA methylation alone is not sufficient: the compaction also seems to require the presence of bound linker histones.

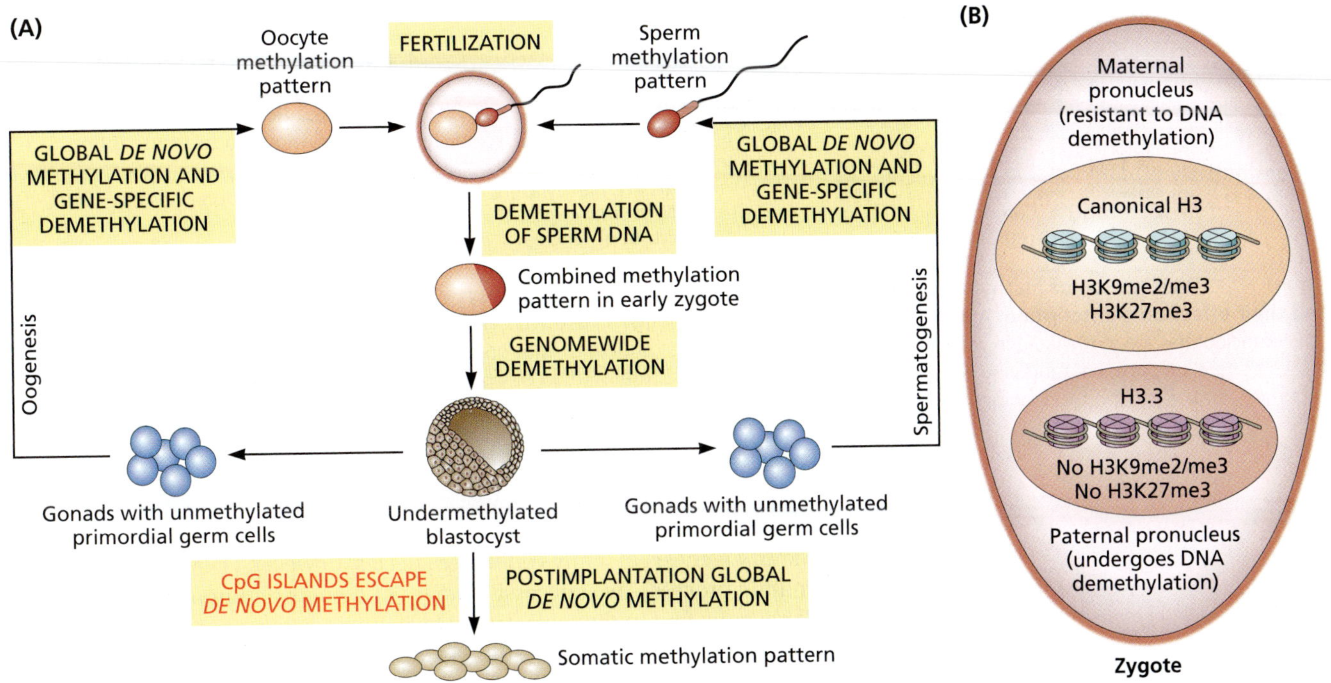

Figure 12.26 Dynamics of DNA methylation patterns during mouse development. DNA methylation patterns are heritably propagated in somatic lineage cells, but during the development of an organism, methylation is a very dynamic process. (A) The sperm genome is rapidly demethylated following fertilization; this creates the combined methylation pattern in the early zygote. The global loss of DNA methylation in the sperm-derived pronucleus occurs independently of DNA replication and thus involves active DNA demethylation. During the first two to three cleavage divisions, the level of methylated cytosines decreases further and stays low through the blastula state; the loss of methylation in the maternal pronucleus is a passive, replication-dependent event. Postimplantation, the embryo genome undergoes *de novo* methylation; the CpG islands remain mostly unmethylated. The primordial germ cells also remain unmethylated; during gametogenesis, specific parental (maternal or paternal) patterns of DNA methylation are established at imprinted loci. (B) Correlation between DNA demethylation and chromatin structure in the early zygote. The paternal and maternal pronuclei behave very differently in terms of DNA demethylation: the former undergoes the process, while the latter is resistant to DNA demethylation. They also differ in several chromatin-structure characteristics. Maternal chromatin contains inactive chromatin marks and canonical H3, whereas paternal chromatin is devoid of these inactive chromatin marks and contains the replacement histone variant H3.3. As embryonic transcription of many genes starts only after the first cell division, the differences in chromatin structure between maternal and paternal pronuclei are probably not transcription-related but serve to allow access of the DNA demethylation machinery to DNA.

Carcinogenesis alters the pattern of CpG methylation

Importantly, carcinogenesis alters the pattern of CpG methylation in two opposing ways: the genome as a whole becomes hypomethylated whereas CpG islands, which regulate the expression of housekeeping genes, including a variety of tumor-suppressor genes, cell-cycle-related genes, DNA mismatch-repair genes, hormone-receptor genes, and others, become hypermethylated. Hypermethylated CpG islands lead to silencing of the linked genes, which has catastrophic consequences: the cells become malignantly transformed. DNA hypermethylation is now the best-characterized epigenetic change in tumors, and it is found in virtually every type of human neoplasm. Silencing of tumor-suppressor genes by promoter hypermethylation is at least as common as silencing caused by genetic mutations. Numerous genes have been shown to undergo promoter hypermethylation in cancer, but a unique profile of hypermethylated CpG islands defines each neoplasia. Thus, hypermethylation in a limited number of gene markers can be used for cancer diagnostics; attempts are also being made to use these patterns as a predictor of response to treatment. Finally, drugs are being developed that should lead to the demethylation of hypermethylated promoters and thus to the reactivation of silenced tumor-suppressor genes. Unfortunately, these drugs suffer from high cytotoxicity levels because they are not selective for the cancer-causing genes but affect numerous other genes, leading to undesirable side effects.

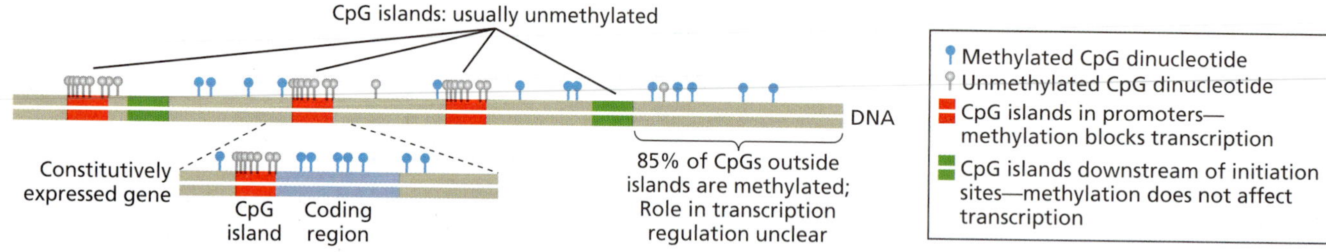

Figure 12.27 Distribution of methylated CpGs in genomic DNA and their effect on transcription. Approximately 70–80% of all CpG dinucleotides are methylated in vertebrate genomes. mCpG, shown as blue lollipops, are randomly distributed throughout the genome but are excluded from regions that have unusually high CpG density, known as CpG islands. Most of the CpG islands are associated with gene promoters and are maintained unmethylated, shown as light gray lollipops, in all types of somatic cells. Aberrant methylation of CpG islands occurs in cancer cells and leads to the silencing of tumor suppressors and other essential genes.

DNA methylation changes during embryonic development

As we have said, DNA methylation patterns are stably inherited during mitosis. During early embryogenesis in many vertebrate species, however, DNA methylation is a highly dynamic process. Two successive waves of demethylation occur in the fertilized egg or zygote: first in the male pronucleus, immediately after fertilization, and then during the transition to the blastocyst stage of development (see Figure 12.26). It is noteworthy that the paternal and maternal pronuclei which undergo differential changes in DNA methylation are also characterized by clear-cut differences in their histone methylation patterns, especially at K9 and K27 of histone H3. They also differ in their content of the active histone variant H3.3, which is present only in the paternal pronucleus. DNA methylation levels are restored by *de novo* methylation following implantation. Gametogenesis is also characterized by dynamic and highly selective changes in methylation patterns. Global demethylation and remethylation is, however, not obligatory in all vertebrates, with zebrafish being a notable exception. A more limited drop in methylation occurs in *Xenopus laevis*. How and why all these changes occur is one of the unresolved problems in development. Another unresolved question concerns a purely structural issue: DNA methylation and demethylation must occur in the context of chromatin. Thus there must be challenges in terms of which CpG would be accessible in the context of the nucleosome particle and/or chromatin fiber structure and compaction. These issues have not been addressed yet.

DNA methylation is governed by complex enzymatic machinery

There are two classes of enzymes that convert cytosine to methylcytosine in the context of CpGs. One enzyme, **DNA methyltransferase 1** or **Dnmt1**, recognizes and methylates the unmodified C in hemimethylated sites generated during DNA replication, thus preserving the methylation pattern of the genome on both strands. In higher eukaryotes, this process occurs within a minute or two after replication. Two other enzymes, **Dnmt3a** and **Dnmt3b**, are responsible for introducing new methyl groups onto DNA at sites in which neither strand was previously methylated: these are the enzymes involved in global *de novo* methylation during development (see Figure 12.26). For more information on the important roles of all three Dnmts *in vivo*, see Table 12.4. In terms of structure, all Dnmts contain a conserved catalytic domain and additional domains, which are mainly responsible for protein–protein interactions (**Figure 12.28**).

The enzymes that put the methylation mark on CpGs are well understood, but persistent efforts to identify the enzyme(s) that actively demethylates DNA have created what is probably the most contentious area of research in the past decade. Several enzymes were described that reportedly performed this function, only to be quickly rejected. The long-sought-after, mysterious pathway may have finally been found: in 2010, it was reported that the active removal of methyl groups from cytosine actually occurs in steps, in a rather indirect, convoluted way (**Figure 12.29**). The first step involves deamination of cytosine, with the production of thymidine; in a second step,

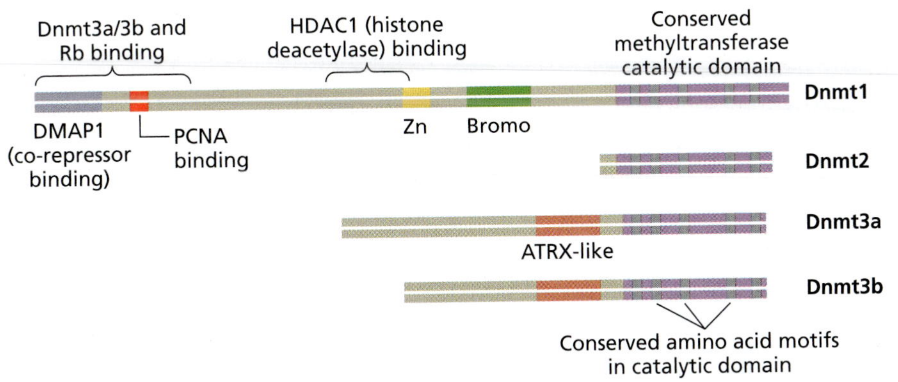

Figure 12.28 Domain structure of mammalian DNA methyltransferases. All known DNA methyltransferases contain a conserved catalytic domain with conserved amino acid motifs. Each DNA methyltransferase interacts with a number of regulatory proteins; some of them modulate enzymatic activity, while others participate in transcriptional repression mechanisms. When DNA methyltransferase 1 represses transcription, it interacts with histone deacetylases such as HDAC1 either directly or through other proteins, such as co-repressor DMAP1 or retinoblastoma protein Rb.

the resulting mismatched T-G base pair is recognized by the base excision repair pathway (see Chapter 22), and the mismatched T is replaced by C. In this way, the cell avoids the challenge of having to break a very strong chemical bond, C–CH_3.

There are proteins that read the DNA methylation mark

As is customary in molecular biology, if there is a specific mark on any molecule that is added following its synthesis, there are players and mechanisms that recognize the mark. These molecules perform effector functions: they translate the signal provided by the mark into biochemical outcomes. DNA methylation is no exception. Methylated regions of the genome are recognized by a variety of proteins, dubbed **methyl-CpG-binding proteins**. All of these share a common methyl-CpG-binding domain or MBD. The MBD proteins are listed in **Table 12.5**, which provides information on structures of the respective proteins, characteristic features of the DNA binding site, specific effects on transcription, and, finally, *in vivo* expression and localization of each protein. **Box 12.1** describes one of these proteins, MeCP2, whose mutations contribute to the neurodevelopmental disorder known as Rett syndrome. Undoubtedly, more MBD proteins will be discovered in the future.

12.7 Long noncoding RNAs in transcriptional regulation

Noncoding RNAs play surprising roles in regulating transcription

Finally, we describe a newly recognized, unexpected mode of transcriptional regulation. As already discussed in Chapter 10, one of the biggest surprises that has emerged from studying transcription at the genomewide level is the pervasive transcriptional activity throughout the genome. As a recent review puts it, "transcription is achieved using a breathtaking number of transcription events." Approximately 180,000 mouse cDNAs have been identified, whereas a mere ~20,200 protein-coding genes are known. Less than 3% of the human genome encodes proteins, leaving many of the remaining transcribed RNA sequences with unknown functions. Analysis of the mouse and human transcriptomes thus reveals a huge number of transcripts that do not encode proteins, hence their name **noncoding RNAs** or **ncRNAs**. A major question still lingers in the transcription field: are these ncRNAs mere transcriptional noise, resulting from initiation at spurious promoters that have arisen serendipitously in the genome, or

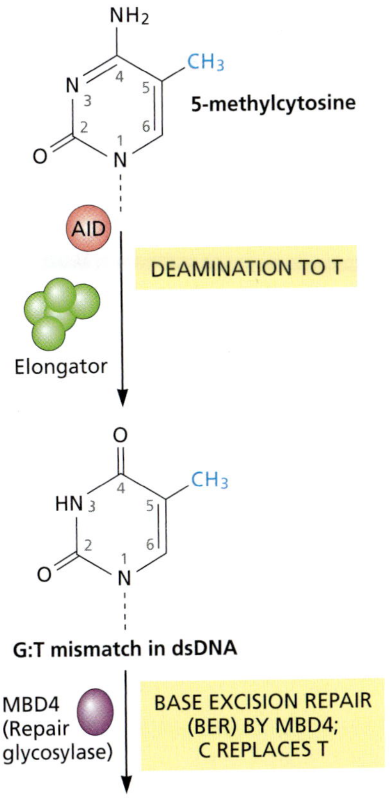

5-methylcytosine
AID
Elongator
DEAMINATION TO T
G:T mismatch in dsDNA
MBD4 (Repair glycosylase)
BASE EXCISION REPAIR (BER) BY MBD4; C REPLACES T

Figure 12.29 Active DNA demethylation in mammals. Active demethylation occurs in a number of biological contexts, including development and gene activation. DNA demethylation occurs in several steps: first, activation-induced cytidine deaminase or AID deaminates methylcytosine, with the production of thymidine. Additional proteins such as the Elongator complex, which is implicated in transcription elongation, may be required. Then the resulting mismatched T is thought to be replaced by cytosine via the base excision repair or BER pathway, which involves the action of MBD4, a thymine-specific glycosylase. This mechanism is active in primordial germ cells to effect the global genome demethylation observed there. AID is also implicated in the rapid removal of methyl groups from methylated promoters of genes that undergo transcriptional activation.

Table 12.5 Methyl-CpG-binding proteins.

Protein	Features of protein structure and DNA binding	Effect on transcription; other properties	*In vivo* expression and localization
MBD1	contains MBD domain and CxxCxxC motifs; has several splice variants	represses transcription from a methylated promoter *in vitro* and *in vivo*	expressed in somatic tissues but not embryonic stem (ES) cells
MBD2a	contains MBD domain and $(Gly\text{-}Arg)_{11}$ $[(GR)_{11}]$ domain	represses transcription; component of Mi2/NuRD deacetylase complex	co-localizes with heavily methylated satellite DNA in mouse cells
MBD2b	truncated version of MBD2a that lacks the $(GR)_{11}$ domain; translation starts at a second methionine codon in MBD2a	represses transcription; component of Mi2/NuRD deacetylase complex	expressed in somatic tissues but not ES cells
MBD3	contains MBD domain and a C-terminal stretch of 12 Glu residues; mammalian MBD3 does not bind methylated DNA *in vivo* or *in vitro*	component of Mi2/NuRD and SMRT/HDAC5-7 deacetylase complexes	expressed in somatic tissues and ES cells
MBD4	contains MBD domain and a repair domain, T-G mismatch glycosylase	thymine glycosylase that binds to deamination product at methylated CpG sites	co-localizes with heavily methylated satellite DNA in mouse cells; expressed in somatic tissues and ES cells
MeCP2	contains MBD domain and a transcriptional repression domain; binds to a single symmetrically methylated CpG	represses transcription from a methylated promoter *in vitro* and *in vivo*; participates in several co-repressor complexes, Sin3a/HDAC1-2, NCoR/Ski, and Rest/CoRest; activates transcription from both methylated and unmethylated promoters	co-localizes with heavily methylated satellite DNA in mouse cells; expressed in somatic tissues and ES cells

do some of them have bona fide cellular functions? If yes, what are these mysterious functions whose existence we did not suspect until now? This is the dark matter of molecular biology. There is another possibility: the functional importance of ncRNA transcription may lie in the process of transcription itself rather than in the products. Such views presume that the changes occurring in chromatin structure during transcription are what matters, since transcribed regions are seemingly more open and this openness facilitates further rounds of transcription. Thus, it is possible that transcription of ncRNAs helps to maintain a transcriptionally competent state for the transcription of functional RNA- and protein-coding genes. In the few cases where we have information, the function of these molecules seems remarkably diverse (**Figure 12.30**).

The sizes and genomic locations of noncoding transcripts are remarkably diverse

Noncoding RNAs may be relatively short or relatively long, with the arbitrary dividing line set at ~200 nucleotides. There are several well-defined classes of short ncRNAs; we describe their biogenesis in Chapter 14 and the roles of others in Chapter 17. Here we discuss the properties and transcription regulatory functions of some long **noncoding RNAs** or **lncRNAs**, all recently described.

Many ncRNAs tend to be transcribed away from the 5′- or 3′-ends of protein-coding genes; for the lncRNAs, there is significant concentration near promoters, initial exons, and initial introns. At least some of these RNA molecules are both capped and polyadenylated, as are regular mRNA molecules (see Chapter 15). Some also contain short, stable stem–loop structures that are deemed important in protein binding. The proteins bound could be either basal or gene-specific transcription factors, elongation factors, and/or chromatin remodelers. Figure 12.30 presents several well-understood lncRNAs

Box 12.1 Methyl-CpG-binding protein 2 and Rett syndrome
Rett syndrome or RTT, an X-linked neurodevelopmental disorder affecting ~1 in 15,000 live female births, is the main cause of mental retardation in girls. It is characterized by an initial period of apparently normal development followed by regression, with manifestation of many autistic features and loss of acquired language and motor skills. Next, stereotypic hand movements and gait abnormalities become evident; these are usually accompanied by numerous other problems. A major breakthrough in understanding the etiology of the disease has come with the discovery that ~80% of those with Rett syndrome possess numerous mutations in *MECP2*, the gene encoding a prominent member of the methyl-CpG-binding proteins in humans (see Table 12.5). The mutations lead to production of a defective protein. The spectrum of mutations is unusually broad, in terms of both their genetic character and their distribution throughout the protein molecule (**Figure 1**). Despite the large number of different mutations—218 mutations in more than 2100 patients reported by 2003—there are eight mutational hot-spots, mainly affecting arginine.

Significant advances in understanding MeCP2 action came from the development of several mouse models of the disease that showed the symptoms of the human condition, including hand wringing. Some of the mouse models were *MECP2* nulls, others contained point mutations or truncations, while still others overexpressed the protein in neurons. Importantly, transgene expression of *MECP2* in mutant mice rescued the Rett syndrome phenotype, pointing to the causal relationship between MeCP2 dysfunction and the disease.

Initially, it was thought that MeCP2 acts as a repressor of brain-specific genes by binding to methylated CpGs in their promoters, but recent genomewide studies have shown that MeCP2 is both an activator and a repressor of transcription (**Figure 2**). Expression studies revealed MeCP2-mediated activation of 2184 genes and repression of 377 genes. Additional studies indicated that ~60% of the MeCP2-binding sites are located outside genes and only ~6% in CpG islands; only 6% of MeCP2-bound promoters contain methylated CpGs. Thus, in less than 5 years, the prevalent repressor model was substituted by a set of very different models (see Figure 2). These new findings changed the views that Rett syndrome treatment should focus on repressing the brain genes that are derepressed in patients because of their MeCP2 dysfunction. Rather, the focus should be on preserving the totality of MeCP2 activities. The challenges are enormous but not insurmountable.

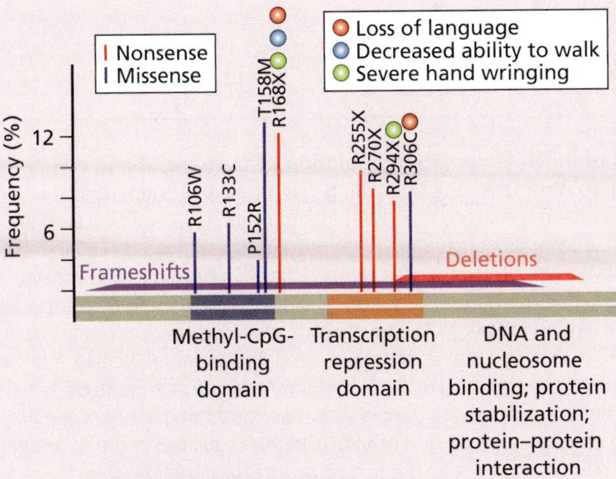

Figure 1 Overall structure of MeCP2. MeCP2 is ~60% unstructured, with nine polypeptide segments predicted to acquire secondary structure upon forming complexes with binding partners, both DNA and proteins. The best-characterized domains are depicted in the schematic, which also shows mutational hot-spots in Rett syndrome patients. Note that different mutations lead to different clinical outcomes.

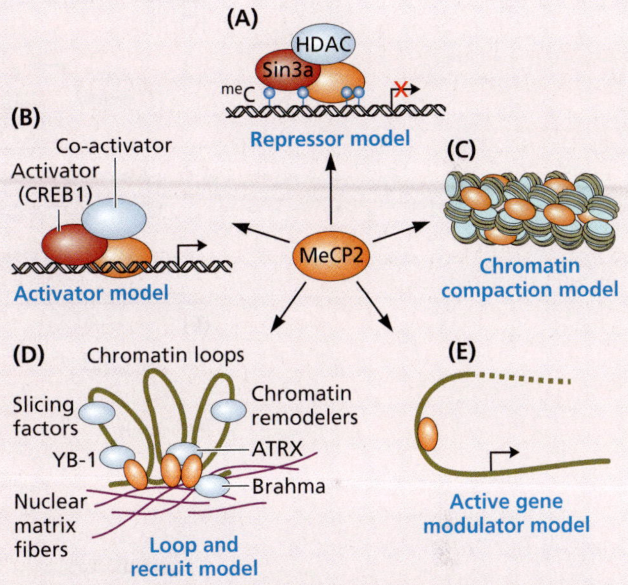

Figure 2 Diversity of models for the molecular mechanisms of action of MeCP2 *in vivo*. (A) Repressor model: MeCP2, shown as an orange oval, binds to methylated cytosines, shown as blue lollipops, in CpG dinucleotides in promoter regions and acts by recruiting co-repressor complexes, such as Sin3a, and histone deacetylases, HDAC. (B) Activator model: MeCP2 binds to promoter regions, activating transcription through interactions with activators, such as CREB1. (C) Chromatin compaction model: MeCP2 associates with itself and DNA to form dense chromatin structures, consistent with its localization to nuclear heterochromatin. (D) Loop and recruit model: MeCP2 binds to components of the nuclear matrix and forms chromatin loops; MeCP2 recruits RNA splicing and chromatin remodeling factors such as ATRX. (E) Active gene modulator model: genomewide localization studies showed that MeCP2 binds predominantly, ~60%, to intergenic sites and acts at a distance. Note that each of these models describes a different function of MeCP2: thus the models are not mutually exclusive. The challenge is to determine which of the diverse MeCP2 roles are essential for development of the postnatal brain, which is obviously affected in Rett syndrome patients.

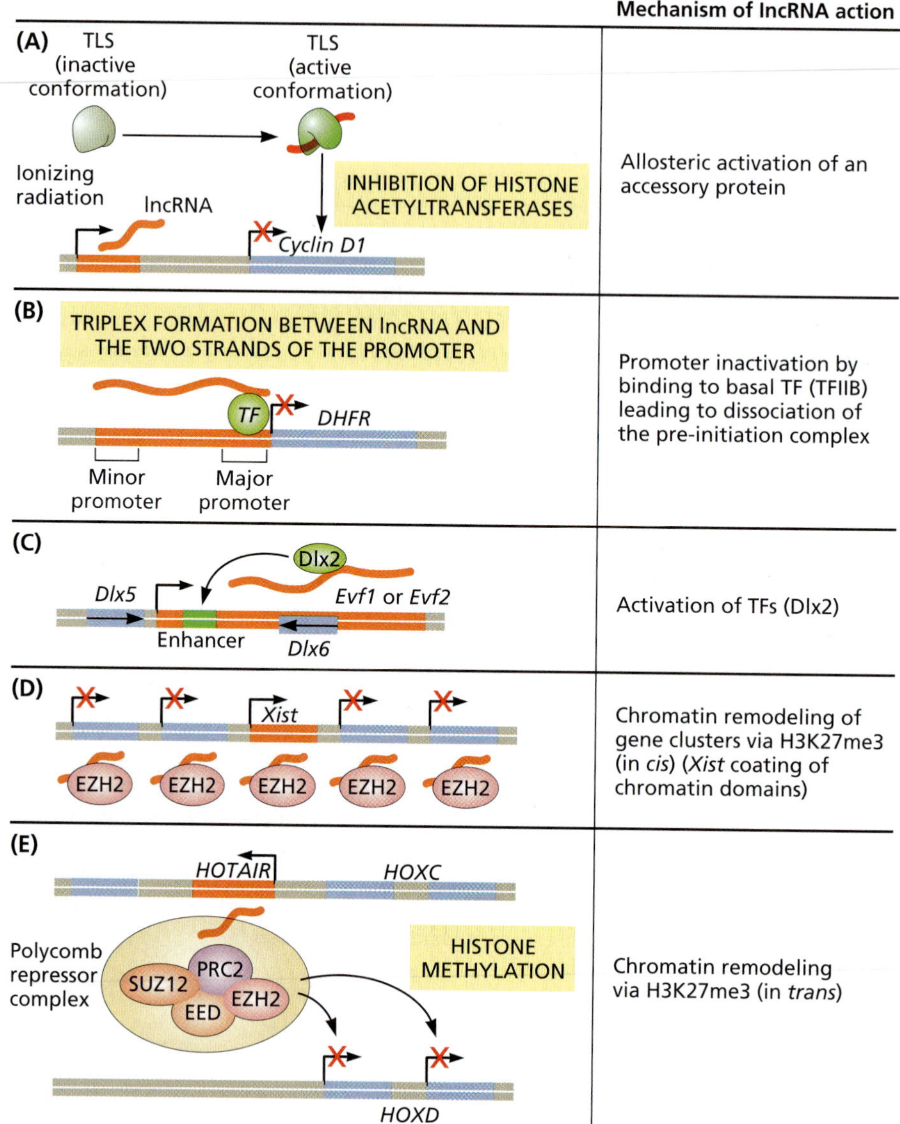

Figure 12.30 Mechanisms of lncRNA-mediated regulation of transcription in eukaryotes. (A) In response to stress, the long noncoding RNA or lncRNA transcribed upstream of the *Cyclin D1* gene interacts with the RNA-binding protein TLS, translocated in liposarcoma, and induces a conformational change in the protein. The activated TLS inhibits histone acetyltransferase activities and thus inhibits *Cyclin D1* transcription. (B) The promoter of the *DHFR* or dihydrofolate reductase gene is inhibited by binding of the lncRNA to the basal transcription factor TFIIB, preventing initiation. The mechanism also involves the formation of a stable triplex complex, purine-purine-pyrimidine, between the single-stranded lncRNA and the double-stranded *DHFR* promoter. (C) The lncRNA *Evf1–Evf2* interacts with the homeodomain protein Dlx2 to activate the enhancer serving *Dlx5* and *Dlx6*, two genes that are involved in neuronal differentiation and migration. Expression of *Evf2* is highly regulated in the developing mouse brain. (D) Epigenetic silencing of gene clusters in *cis* by lncRNAs: the example shown is *Xist*, the ncRNA crucial for X-chromosome inactivation in mammals. *Xist* coats the X-chromosome that is to be stably inactivated. It is believed that *Xist* coating establishes a specialized nuclear compartment devoid of Pol II; it also interacts with EZH2, the subunit of PRC2 that introduces repressive methylation marks on H3K27. (E) Epigenetic repression of genes in *trans*. The *HOTAIR* lncRNA is transcribed within the human *HOXC* gene cluster; it then directly targets epigenetic modifiers, such as PRC2, to the *HOXD* gene cluster. Other histone-modifying complexes, such as G9a methyltransferase, may also be targeted.

that function as regulators of transcription. The mechanisms involved are incredibly varied and show no common feature. Figure 12.31 illustrates another fascinating example of how one lncRNA mediates the apoptotic response of cells to stress, by repressing the transcription of pro-survival genes. The diversity of mechanisms is amazing. One thing is certain: we have seen just the tip of the iceberg as far as lncRNAs and their functions are concerned. More insights were gleaned from the ENCODE project, which systematically analyzed the entire human genome and transcriptome (see Chapter 13).

12.8 Methods for measuring the activity of transcriptional regulatory elements

Almost all our definitions of regulatory elements in the eukaryotic genome depend on methods that use recombinant DNA technology to incorporate the DNA element of interest into some sort of construct and then to measure the transcriptional activity of the construct after its introduction into living cells (see Chapters 5 and 13). Although it took years to come up with such methods, at present we have at our disposal a whole battery of various strategies to study all kinds of regulator elements (**Box 12.2**). Nevertheless, a big question still remains to be answered: do the elements identified in these reporter assays really function as expected *in vivo*?

Box 12.2 Measuring the activity of transcriptional regulatory elements *in vivo* Assays that measure the activity of transcriptional regulatory elements are based on the use of reporter gene constructs, whose transcriptional activity can be easily monitored (**Figure 1**). The most frequently used reporter genes are chloramphenicol acetyltransferase or CAT, β-galactosidase, or luciferase genes. CAT detoxifies the antibiotic chloramphenicol by acetylation, preventing its binding to the ribosome (see Chapter 16). Cells that express CAT can grow on medium containing the antibiotic. β-Galactosidase catalyzes the hydrolysis of β-galactosides into monosaccharides. X-gal, a colorless modified galactose sugar, is metabolized by the enzyme to an insoluble product, 5-bromo-4-chloroindole, which is bright blue and thus functions as an indicator of enzymatic activity. Finally, luciferase, an enzyme derived from the firefly, is responsible for oxidation of luciferin pigment, a reaction that is accompanied by the production of light or bioluminescence.

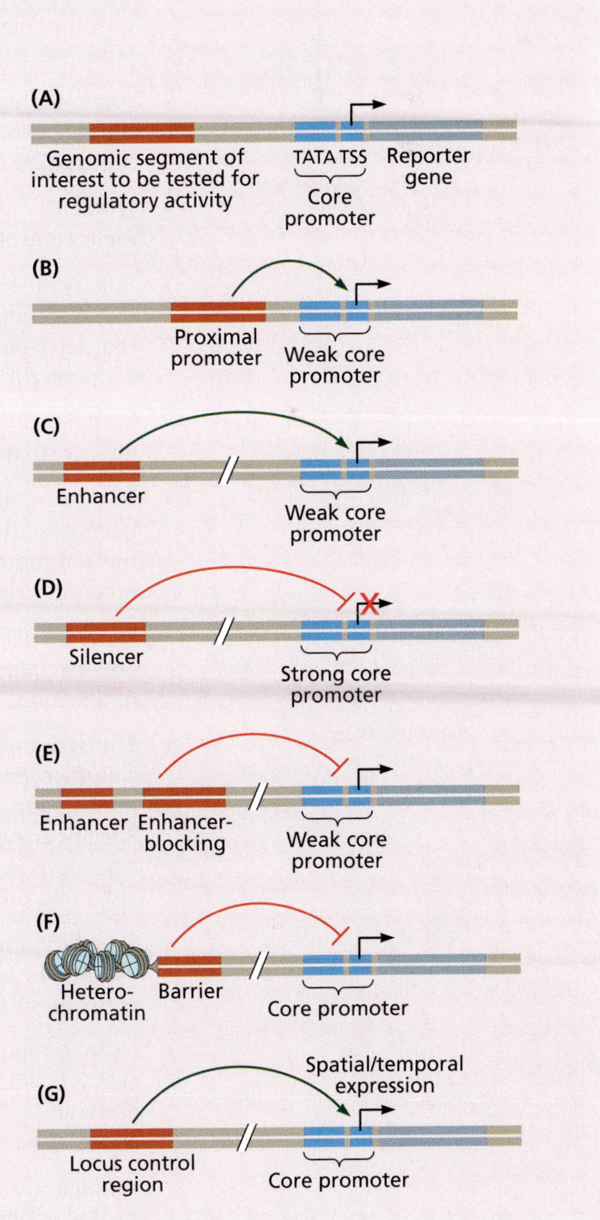

Figure 1 Functional *in vivo* assays to measure the activity of transcriptional regulatory elements. (A) Regions to be tested for regulatory activity are cloned into a plasmid bearing the reporter gene; the constructs are transfected, either transiently or stably, into cultured cells; and the activity of the reporter gene is monitored. If the segment is tested for core promoter activity, then it is placed immediately upstream of the reporter gene lacking its endogenous promoter, not shown here. (B) Testing for proximal promoter elements: an increase in transcription is expected. (C–D) Testing for enhancers or silencers would require the use of appropriate strength promoters. (E–F) Testing for two different types of insulator activities. A potential enhancer-blocking element should be active when it is cloned between an existing enhancer and the gene. A barrier element should block the spreading of heterochromatin structures from neighboring regions. This assay requires stable integration into the genome. A barrier element should insulate the gene construct from position effects, that is, the gene should always be active wherever in the genome the integration occurred. (G) Definitive identification of a locus control region, LCR, also requires stable integration. The LCR should confer regulated expression of the linked gene, independently of where the construct integrates. (Adapted from Maston GA, Evans SK & Green MR [2006] *Annu Rev Genomics Hum Genet* 7:29–59. With permission from Annual Reviews.)

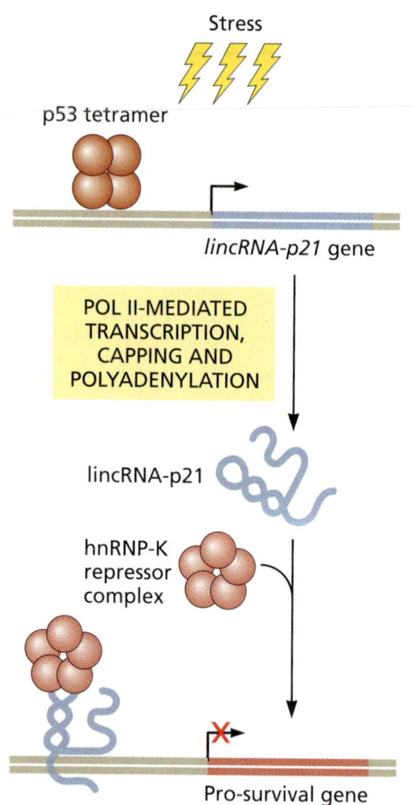

Figure 12.31 Participation of lincRNA in p53-mediated transcriptional repression. *p53* is an important tumor-suppressor gene mutated in around 50% of human cancers. The p53 protein binds to DNA as a tetramer; it becomes stabilized in response to DNA damage and triggers a complex transcriptional response, activating or repressing the expression of numerous genes. The transcriptional response leads either to cell-cycle arrest or to apoptosis, programmed cell death. The p53-mediated gene repression that is part of apoptosis occurs through a mechanism involving a long intergenic noncoding RNA, lincRNA-p21, so named for its physical proximity to the p53 target gene *p21*, although *p21* is not involved in the lincRNA pathway. The sequence of events is as follows: p53 induces the transcription of lincRNA-p21, which then interacts with a repressive RNA–protein complex, hnRNP-K; the interaction is required for proper localization of hnRNP-K on responsive genes. The repressor complex contains the linker histone variant H1.2 and blocks histone acetylation by the histone acetyltransferase p300. The changes in chromatin organization are the probable cause for transcriptional repression of the linked pro-survival genes, leading to cell death.

Key concepts

- Because of the demands of development and of responding to environmental factors, the regulation of transcription is much more complex in eukaryotes than in viruses or bacteria.
- Promoter regions in eukaryotes are usually split into core and proximal elements; the core contains the polymerase docking site, while the proximal promoter region has regulatory functions.
- Distal DNA regulatory elements include enhancers, silencers, insulators, and locus control regions. Each of these elements acts through its own mechanisms.
- Eukaryotic transcription factors are molecular complexes that bind to DNA regulatory elements. Typically, they contain both DNA-binding and activation domains that act to bind to specific DNA sequences in specific genes and then activate transcription through recruitment of the basal transcriptional machinery.
- Mutations in gene regulatory regions or in the associated protein machinery can give rise to disease states, including cancer.
- Regulation also occurs at the level of transcriptional elongation. Sometimes polymerases are poised near the promoter until a signal releases them to continue transcription. Some elongation factors help the polymerase to overcome temporary transcriptional pausing as it travels along the gene body.
- There are also protein factors that help polymerases to pass through nucleosomes. The mechanisms for transcription of DNA within nucleosomes are still not well understood.
- Transcription in eukaryotes is markedly influenced by post-translational modification or marking of histones on nucleosomes. These marks include acetylation, methylation, phosphorylation, ubiquitylation, and poly(ADP)ribosylation events. Specific enzymes exist for each of these modifications: the enzymes are highly specific for individual amino acid residues on individual histone molecules. There is also cross-talk between such modifications.
- Specific readers of these markings exist that, upon recognizing the marks, initiate the modification of gene expression. Certain modifications, like lysine acetylation, correlate strongly with gene activity. Others are associated with gene silencing.
- Histone replacement variants also play a role in gene regulation. For example, H2A.Z is often found in nucleosomes that flank nucleosome-free regions, which occur around transcription start sites. H3.3 and H2A.Bbd are often associated with active transcription.
- Chromatin structure must sometimes be remodeled to permit transcription. This can be accomplished by a battery of ATP-dependent chromatin remodelers. These

may slide nucleosomes on DNA, modify their structure, or partly dissociate the nucleosomal particle.

- Genes may also be regulated by methylation of cytosine residues, especially at CpG sites, to form 5-methylcytosine. This is considered a true epigenetic modification because it is carried through mitosis.
- Major changes in DNA methylation occur during development, upon differentiation of pluripotent stem cells, and in carcinogenesis.
- There are enzymes for methylating DNA sites and for reading or removing those DNA methylation marks.
- In addition to all the other mechanisms described above, it is now becoming clear that long noncoding RNAs can carry out a remarkable number of regulatory functions.

Further reading

Books

Allis CD, Jenuwein T & Reinberg D (eds) (2007) Epigenetics. Cold Spring Harbor Laboratory Press.

Carey M & Smale ST (2000) Transcription Regulation in Eukaryotes: Concepts, Strategies, and Techniques. Cold Spring Harbor Laboratory Press.

Chapman KE & Higgins SJ (eds) (2001) Essays in Biochemistry, Vol. 37: Regulation of Gene Expression. Portland Press.

Latchman DS (2010) Gene Control. Garland Science.

Zlatanova J & Leuba SH (eds) (2004) Chromatin Structure and Dynamics: State-of-the-Art. Elsevier.

Reviews

Ausió J (2006) Histone variants—the structure behind the function. *Brief Funct Genomic Proteomic* 5:228–243.

Barsotti AM & Prives C (2010) Noncoding RNAs: the missing "linc" in p53-mediated repression. *Cell* 142:358–360.

Bulger M & Groudine M (2010) Enhancers: The abundance and function of regulatory sequences beyond promoters. *Dev Biol* 339:250–257.

Caiafa P, Guastafierro T & Zampieri M (2009) Epigenetics: Poly(ADP-ribosyl)ation of PARP-1 regulates genomic methylation patterns. *FASEB J* 23:672–678.

Clapier CR & Cairns BR (2009) The biology of chromatin remodeling complexes. *Annu Rev Biochem* 78:273–304.

D'Alessio JA, Wright KJ & Tjian R (2009) Shifting players and paradigms in cell-specific transcription. *Mol Cell* 36:924–931.

Davis PK & Brachmann RK (2003) Chromatin remodeling and cancer. *Cancer Biol Ther* 2:23–30.

Elsaesser SJ, Goldberg AD & Allis CD (2010) New functions for an old variant: No substitute for histone H3.3. *Curr Opin Genet Dev* 20:110–117.

Fischle W, Wang Y & Allis CD (2003) Histone and chromatin cross-talk. *Curr Opin Cell Biol* 15:172–183.

Hiragami K & Festenstein R (2005) Heterochromatin protein 1: A pervasive controlling influence. *Cell Mol Life Sci* 62:2711–2726.

Kraus WL (2008) Transcriptional control by PARP-1: Chromatin modulation, enhancer-binding, coregulation, and insulation. *Curr Opin Cell Biol* 20:294–302.

Kraus WL & Lis JT (2003) PARP goes transcription. *Cell* 113:677–683.

Ladurner AG (2006) Rheostat control of gene expression by metabolites. *Mol Cell* 24:1–11.

Maston GA, Evans SK & Green MR (2006) Transcriptional regulatory elements in the human genome. *Annu Rev Genomics Hum Genet* 7:29–59.

Mercer TR, Dinger ME & Mattick JS (2009) Long non-coding RNAs: Insights into functions. *Nat Rev Genet* 10:155–159.

Moazed D (2001) Common themes in mechanisms of gene silencing. *Mol Cell* 8:489–498.

Narlikar L & Ovcharenko I (2009) Identifying regulatory elements in eukaryotic genomes. *Brief Funct Genomic Proteomic* 8:215–230.

Ooi SKT & Bestor TH (2008) The colorful history of active DNA demethylation. *Cell* 133:1145–1148.

Ponting CP, Oliver PL & Reik W (2009) Evolution and functions of long noncoding RNAs. *Cell* 136:629–641.

Ptashne M (2005) Regulation of transcription: From lambda to eukaryotes. *Trends Biochem Sci* 30:275–279.

Rando OJ & Chang HY (2009) Genome-wide views of chromatin structure. *Annu Rev Biochem* 78:245–271.

Sanz LA, Kota SK & Feil R (2010) Genome-wide DNA demethylation in mammals. *Genome Biol* 11:110.

Schübeler D (2009) Epigenomics: Methylation matters. *Nature* 462:296–297.

Shahbazian MD & Grunstein M (2007) Functions of site-specific histone acetylation and deacetylation. *Annu Rev Biochem* 76:75–100.

Simon JA & Kingston RE (2009) Mechanisms of polycomb gene silencing: Knowns and unknowns. *Nat Rev Mol Cell Biol* 10:697–708.

Sparmann A & van Lohuizen M (2006) Polycomb silencers control cell fate, development and cancer. *Nat Rev Cancer* 6:846–856.

Suganuma T & Workman JL (2008) Crosstalk among histone modifications. *Cell* 135:604–607.

Taverna SD, Li H, Ruthenburg AJ et al. (2007) How chromatin-binding modules interpret histone modifications: Lessons from professional pocket pickers. *Nat Struct Mol Biol* 14:1025–1040.

Trojer P & Reinberg D (2007) Facultative heterochromatin: Is there a distinctive molecular signature? *Mol Cell* 28:1–13.

van Holde KE, Lohr DE & Robert C (1992) What happens to nucleosomes during transcription? *J Biol Chem* 267:2837–2840.

Zlatanova J & Thakar A (2008) H2A.Z: View from the top. *Structure* 16:166–179.

Zlatanova J & Victor J-M (2009) How are nucleosomes disrupted during transcription elongation? *HFSP J* 3:373–378.

Experimental papers

Heintzman ND, Hon GC, Hawkins RD et al. (2009) Histone modifications at human enhancers reflect global cell-type-specific gene expression. *Nature* 459:108–112.

Heintzman ND, Stuart RK, Hon G et al. (2007) Distinct and predictive chromatin signatures of transcriptional promoters and enhancers in the human genome. *Nat Genet* 39:311–318.

Ravasi T, Suzuki H, Cannistraci CV et al. (2010) An atlas of combinatorial transcriptional regulation in mouse and man. *Cell* 140:744–752.

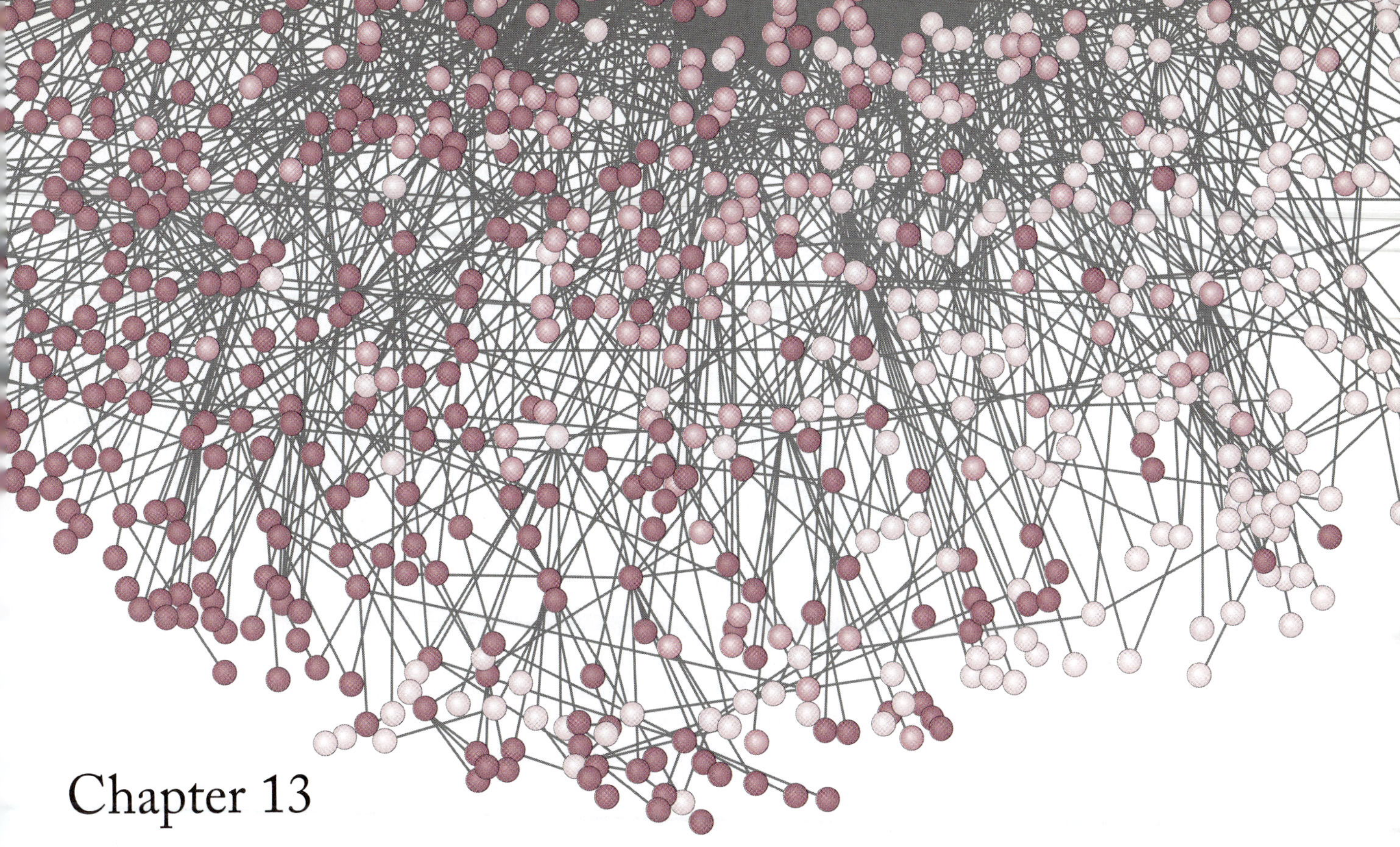

Chapter 13

Transcription Regulation in the Human Genome

13.1 Introduction

The year 2012 saw a revolution in molecular biology when the full results of the ENCODE or Encyclopedia of DNA Elements project were released. This project originated shortly after the completion of the Human Genome Project. ENCODE aimed at nothing less than to mine the entire human genome sequence for an encyclopedia of elements characteristic of various genomic functions, known as functional elements. These elements have immense diversity and include, among many others, transcription start sites or TSSs, promoters, enhancers, nucleosome locations, and methylation sites. The first output from ENCODE, published in 2007, was an analysis of only a selected 1% of the genome. The development of powerful new methods provided remarkable progress, allowing 100% of the genome to be analyzed in the next few years. In total, 147 different human cell types were analyzed. This required the collaboration of hundreds of scientists around the world, each centering on special aspects of the larger problem.

In Chapter 12, we outlined the knowledge about transcription regulation that had been obtained through the use of traditional reductionist methodologies. In these approaches, when genetic elements of functional interest were identified, they were systematically altered, for example, truncated or mutated at predefined positions, and then introduced into living cells, where their function was investigated by traditional methods. As we have seen, these approaches have yielded a wealth of information on the functions of specific genomic regions of limited lengths and the factors that interact with them.

Rapid full-genome sequencing allows deep analysis

Researchers gradually recognized that functional elements were characterized by common biochemical or biophysical features, called signatures. This realization led to the development of genomewide methods, both experimental and computational, that look for these specific biochemical signatures in order to scan whole genomes for the occurrence and distribution of individual functional elements. The biochemical signature strategy became the cornerstone of the ENCODE project's efforts to identify all functional elements in the human genome that are specified by the genomic sequence. In this chapter, we present those major results of the second, production stage of the ENCODE project that are relevant to transcription. Some of the findings were expected on the basis of knowledge accumulated from studies of gene-specific systems, whereas others revealed a complex picture of transcription regulation and provided a wealth of new insights. Excitingly, the mine of information has not been exhausted; in fact, its full potential has hardly been touched. We are concerned here only with what has been learned to date about transcription and its regulation.

The material and methods covered in this chapter will become more and more important in the future development of molecular biology, but this content is necessarily more advanced. We are at the frontier right at the moment, looking into unknown territory. ENCODE results are mentioned in several other chapters, but those who wish to more fully understand the cutting edge of molecular biology should focus on this chapter and the references herein.

13.2 Basic concepts of ENCODE

ENCODE depends on high-throughput, massively processive sequencing and sophisticated computer algorithms for analysis

Locating thousands or millions of functional elements of a particular type requires that the genome be fragmented in some way. Then, fragments containing the sequence elements must be separated and sequenced so as to locate them in the known sequence of the whole genome. The techniques for isolation are varied and too detailed to recount here; we give an example in Box 13.1. Then there is the problem of sequencing

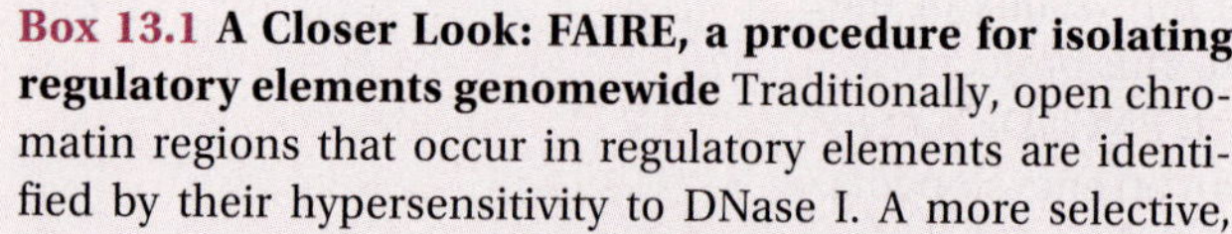

Box 13.1 A Closer Look: FAIRE, a procedure for isolating regulatory elements genomewide Traditionally, open chromatin regions that occur in regulatory elements are identified by their hypersensitivity to DNase I. A more selective, simple, and highly efficient technique has recently been introduced for the isolation and analysis of active regulatory elements from eukaryotic genomes. The technique, termed **FAIRE** or **formaldehyde-assisted isolation of regulatory**

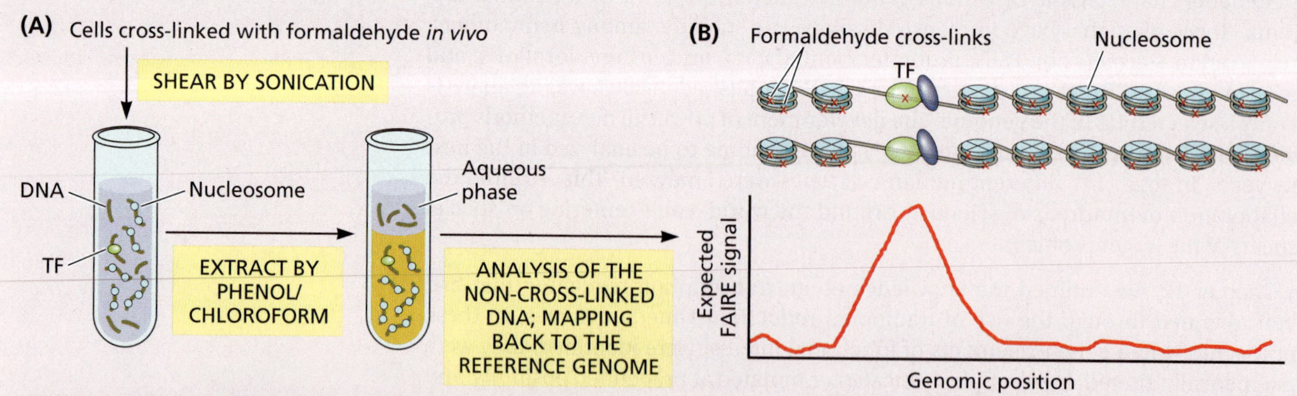

Figure 1 FAIRE-seq depends on formaldehyde cross-linking efficiency. (A) Overview of FAIRE, or formaldehyde-assisted isolation of regulatory elements: DNA recovered in the aqueous phase following phenol/chloroform extraction of the cross-linked and sheared sample is subjected to high-throughput sequencing or analyzed in microarrays. TF, transcription factor. (B) Cross-linking captures only a portion of potential protein–DNA interactions. Given that histone–DNA interactions constitute the majority of cross-linkable interactions in the genome, these interactions are captured preferentially; interactions between other DNA-binding proteins and DNA will be only occasionally captured by cross-linking. The two rows represent two cells in a population. (Adapted from Giresi PG & Lieb JD [2009] *Methods* 48:233–239. With permission from Elsevier.)

elements, consists of protein-DNA covalent cross-linking with formaldehyde, fragmentation of DNA by shearing, and phenol/chloroform extraction of the non-cross-linked DNA fragments (Figure 1A). The DNA that is organized within nucleosomes has greater formaldehyde cross-linking efficiency than the open chromatin regions that contain DNA-binding proteins. Thus, the non-cross-linked DNA, which is recovered in the aqueous phase, originates from nucleosome-depleted regions (Figure 1B). These regions are also coincident with the location of DNase I hypersensitive sites as well as TSSs, active promoters, enhancers, and insulators (Figure 2).

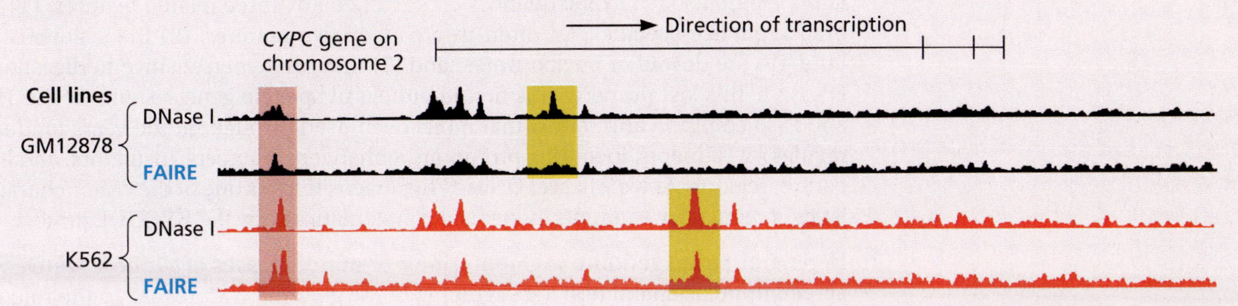

Figure 2 Coincidence of open chromatin regions as identified by DNase I hypersensitivity and FAIRE-seq in two cell lines. The *CYPC* gene is the only gene in this ~90 kb subsection of human chromosome 2. Yellow shading shows open chromatin regions identified by both methods in just one of the cell lines; note that open regions are also present in the body of the gene. Reddish shading shows open regions identified by both methods in both cell types. (Adapted from Song L, Zhang Z, Grasfeder LL et al. [2011] *Genome Res* 21:1757–1767. With permission from Linyung Song.)

thousands or millions of fragments. This would have been wholly impossible by the classical techniques described in Chapter 4. However, recent decades have produced a host of high-throughput methods that can provide rapid, automated sequencing. Most of these are also massively parallel, meaning that they can analyze thousands of samples simultaneously. Again, to center on any one of the currently used methods would be pointless, for improvements and new methods are appearing all the time. Just to give an idea of current capabilities, there are techniques that can simultaneously read 50,000 samples, each of length up to several thousand base pairs, in less than an hour, with better than 99.9% accuracy. The cost is less than one dollar per one million bases.

The ENCODE project integrates diverse data relevant to transcription in the human genome

The ENCODE project has systematically mapped features in the genome that relate to transcription: transcribed regions, transcription factor or TF binding sites, chromatin structure and histone modifications, and DNA methylation. In addition, the project addressed issues concerned with both evolutionary conservation of regulatory elements and sequence variations in these elements. The results are of unprecedented magnitude and are impossible to cover in a textbook. Still, we present here at least some of the data and the conclusions of these studies to give the reader an appreciation of the direction in which molecular biology is heading, and to expand on the more familiar facts presented in Chapter 12.

The project analyzed numerous aspects of functional genome organization, often providing overlapping data sets centering around a specific genomic element. For example, the analyses of promoters and TSSs provide information on many related features of the chromatin organization in these elements—nucleosome positioning, histone modification patterns, and so on—while at the same time investigating the connectivity of these elements with other recognized regulatory elements. We follow this pattern of presenting the most significant findings of the ENCODE project, focusing on specific elements and whatever properties were interrogated. We begin with a general overview of the genome organization in terms of distinguishable elements that are relevant to transcription. In addition to the extreme sequencing demands mentioned above, interpretation of the enormous mass of raw data, especially when interrelationships between functional elements are addressed, has necessitated the construction of a host of new computer algorithms.

13.3 Regulatory DNA sequence elements

Seven classes of regulatory DNA sequence elements make up the transcriptional landscape

The project discovered numerous candidate regulatory elements that are physically associated with one another and have potential for regulating gene expression. Regulatory elements are DNA sequences characterized by three related features: (1) they bind sequence-specific TFs, often in a cooperative manner; (2) the segments that bind TFs are devoid of nucleosomes; and (3) they are hypersensitive to digestion by DNase I. This last property was first identified in specific gene systems in the 1980s and is so common and robust that it has been used throughout the years to identify regulatory elements, including promoters, enhancers, silencers, insulators, and locus control regions. As we will see, DNase I hypersensitivity is one of the major characteristics of regulatory elements interrogated systematically by the ENCODE project.

In general, seven **genome segmentations**, or **major classes of genome states**, were agreed upon as characterizing segments of the genome with regard to the role they play in transcription: transcribed genes, T; their transcription start sites, TSS, and promoters, PF; two classes of predicted enhancers, strong enhancers E and weak enhancers WE; CTCF-binding sites, CTCF; and transcriptionally repressed regions, R (**Figure 13.1A**). The established notion that active promoters and transcribed genes go together with TSS was confirmed. In addition, three active distal states were recognized. Two of these were labeled as predicted enhancers and predicted weak

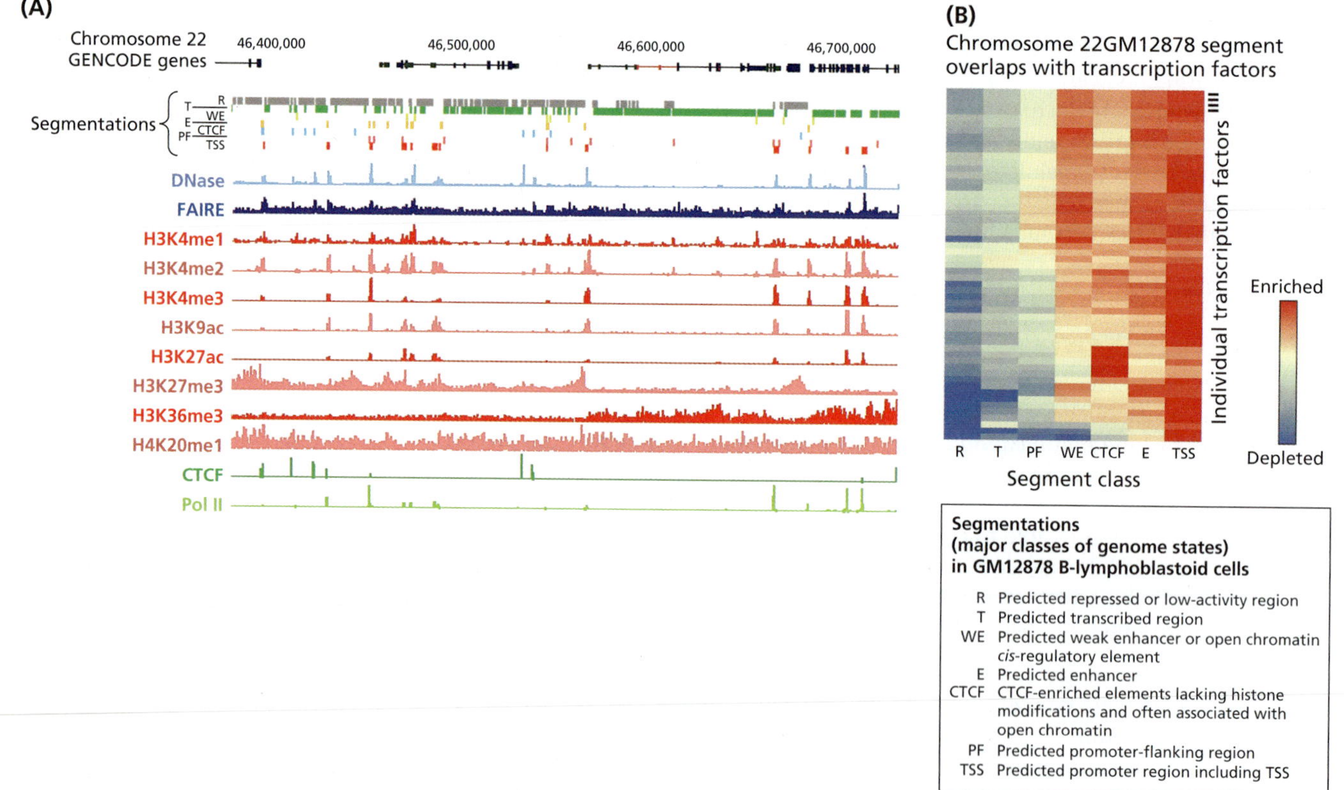

Figure 13.1 Major classes of genome transcriptional states. (A) Seven segmentation classes of genome states and their characteristic features of human chromosome 22. FAIRE, or formaldehyde-assisted isolation of regulatory elements (see Box 13.1), isolates nucleosome-depleted genomic regions by exploiting the difference in cross-linking efficiency between nucleosomes and sequence-specific regulatory factors. The CTCF-enriched element is composed of CTCF-binding sites that lack histone modifications. They are often associated with open chromatin but may also have other functions, for example, as insulators. (B) Association of selected transcription factors or TFs with combined segmentation states. Each horizontal line in the heat map corresponds to an individual TF. In concordance with conventional views, both transcription start sites, TSS, and enhancers, E and WE, are highly enriched in TFs. CTCF-enriched elements are also significantly enriched in TFs. Abbreviations for the segmentation classes are as in part A. (From The ENCODE Project Consortium [2012] *Nature* 489:57–74. With permission from Macmillan Publishers, Ltd.)

enhancers because they occur in regions of open, DNase I-sensitive chromatin with high levels of the histone modification H3K4me1. A high content of this specific histone modification has been seen in the majority of enhancer elements analyzed on a gene-to-gene basis. The third active state has high CTCF binding. CTCF is a multifunctional protein (see Chapter 10) that can serve as a transcription factor, an insulator, and a master organizer of the genome. In all of these roles, CTCF is intimately involved in transcription regulation. Finally, the repressed state includes sequences belonging to actively repressed or inactive, permanently repressed quiescent chromatin. The enhancer and transcribed gene states are highly cell-specific and undoubtedly reflect the long-recognized fact that cell types differ in the specific portion of the genome they express.

Figure 13.1A illustrates the organization of a selected region of human chromosome 22, with its characteristic DNase sensitivity patterns, histone post-translational modifications, and patterns of CTCF and Pol II occupancy. This figure also shows nucleosome-depleted segments, as determined by the formaldehyde-assisted isolation of regulatory elements or FAIRE technique (see Box 13.1). **Figure 13.1B** shows the linkage between transcription factor occupancy and the seven segmentation classes. The mechanistic explanation of how these chromatin patterns are involved in gene activity remains to be elucidated.

13.4 Specific findings concerning chromatin structure from ENCODE

Millions of DNase I hypersensitive sites mark regions of accessible chromatin

DNase I hypersensitivity is a common feature of all regulatory elements, reflecting their open chromatin structure. The long-held view that DNase I hypersensitivity is linked to the binding of TFs has been confirmed beyond doubt by the genome-wide analysis of DNase I hypersensitivity included in the ENCODE project. This analysis, performed by DNase-seq on 125 cell types, identified 2.89 million unique, non-overlapping DNase I hypersensitive sites or DHSs, the overwhelming majority of which lie far from TSSs. **Figure 13.2** provides an illustrative example of the coincidence between DHSs and transcription factor occupancy for a selected region of human chromosome 19. This figure also shows the distribution of DHSs over the recognized gene annotations in the GENCODE gene set, and it points to the high

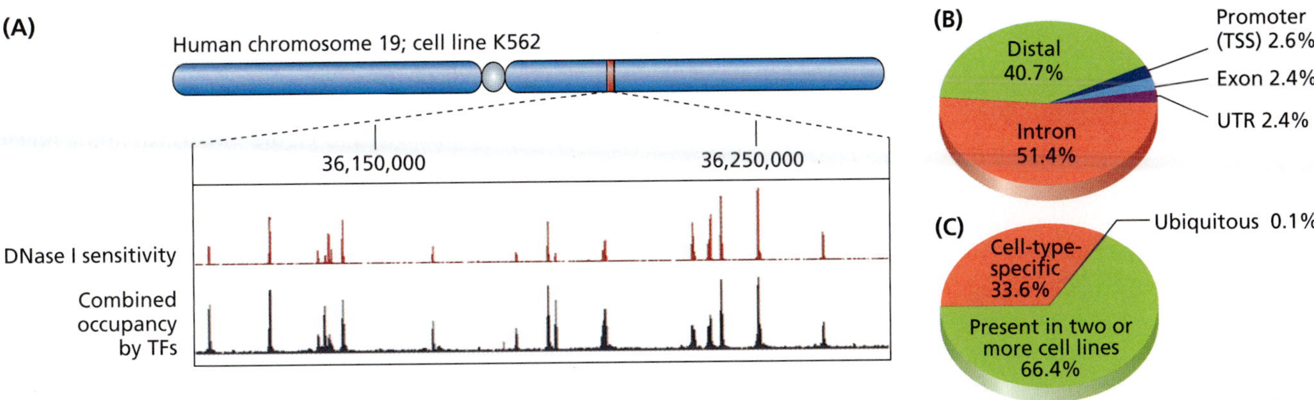

Figure 13.2 General features of accessible chromatin landscape as determined by DNase I hypersensitivity or DNase-seq. (A) Chromatin accessibility is driven by TF binding. Data for a 175-kb region of chromosome 19 obtained from cell line K562 show the coincidence of regions hypersensitive to DNase I and those bound by TFs; the lower graph is the cumulative sum of 45 TFs tested by ChIP-seq. (B) Distribution of 2,890,742 DNase I hypersensitive sites or DHSs with respect to GENCODE gene annotations. Promoter DHSs are defined as those located within 1 kb upstream of a TSS. Note that promoter DHSs are a small portion of all DHSs, with practically the same percentage present in exons and untranslated regions, UTRs. The majority sites are about evenly distributed between intronic and intergenic, or distal, regions. (C) Cell specificity of DNase I hypersensitive sites. The majority of sites are shared among two or more cell types, but a large proportion are cell-type-specific, and only a very small minority are present in all cell types. (Adapted from Thurman RE, Rynes E, Humbert R et al. [2012] *Nature* 489:75–82. With permission from Macmillan Publishers, Ltd.)

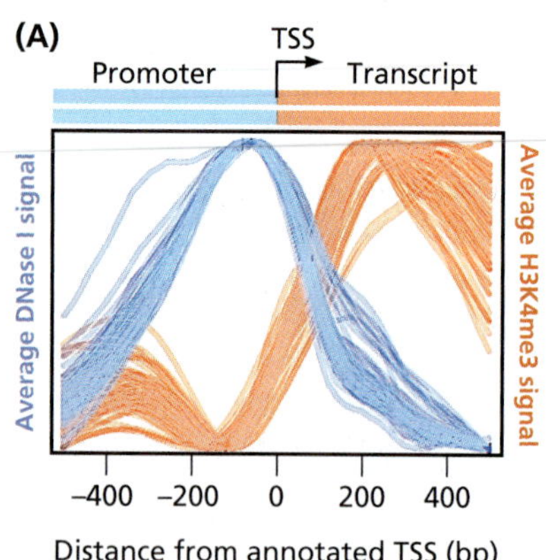

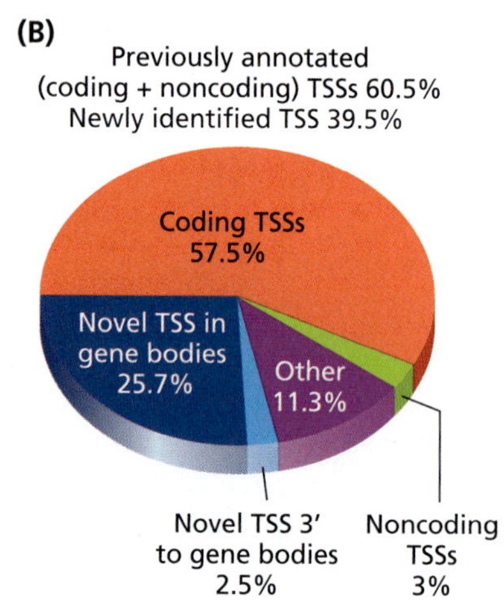

Figure 13.3 Invariant directional signature of two features of promoter chromatin structure. The features shown are DNase I sensitivity and presence of H3K4me3. (A) The same biological samples from 56 cell types were analyzed by ChIP-seq to show the presence of H3K4me3 and for DNase I sensitivity. Averaged H3K4me3 signal, shown in orange, is plotted against averaged DNase sensitivity signal, shown in blue, across 10,000 randomly selected TSSs oriented 5′ → 3′. Each curve is for a different cell type. Note that the patterning of these two chromatin features is highly directional, or asymmetric, and has a precise relationship to the TSS. The pattern is consistent with a rigidly positioned nucleosome immediately downstream from the promoter DHS. (B) A pattern-matching search across the genome for the pattern shown in part A was used to identify novel promoters. The overall distribution of TSSs in the genome is presented in the pie chart. It may be highly significant that ~60% of the promoters that were newly identified in annotated gene bodies or 3′ to them are oriented antisense to the annotated direction of transcription. (A, from Thurman RE, Rynes E, Humbert R et al. [2012] *Nature* 489:75–82. With permission from Macmillan Publishers, Ltd.)

degree of cell specificity of DHSs. It is significant that only 0.1% of all DHSs are ubiquitously present in all human cell types, underlining the involvement of DHSs in cell differentiation.

DNase I signatures at promoters are asymmetric and stereotypic

The recognized involvement of promoters in gene regulation (see Chapter 12) was the impetus behind the considerable effort that has gone into the genomewide characterization of these elements. The focus within the ENCODE project was on the hypothesis that RNA output can be predicted from patterns of chromatin structure and/or TF binding in and around promoters. In general, and consistent with earlier-held views, two distinct types of promoters were recognized: broad, CG-rich TATA-less promoters and narrow, TATA-box-containing promoters. See Figure 10.6 for a description of dispersed and focused promoters.

The discovery of new promoters was driven by biochemical knowledge obtained by DNase I sensitivity assays, mapping of 5′-ends of transcripts, and the recognition that promoters are highly enriched in H3K4me3. Well-annotated promoters were systematically studied for these properties: DNase I cleavage profiles were plotted against ChIP-seq data for H3K4me3 for 56 cell types, revealing that these features have highly stereotypical, asymmetric patterns (**Figure 13.3A**). The promoter regions immediately preceding TSSs were found to be highly sensitive to DNase I cleavage, whereas the region that codes for the transcript contains a wave of high H3K4me3 content, which dwindles as the distance from the TSS increases. This genomewide pattern was then used in computational approaches to scan the entire genome for the possible presence of such patterns. In total, 113,622 distinct putative promoters were identified, 39.5% of which were new. The distribution of all promoters, both previously annotated and newly recognized, in the genome is presented in **Figure 13.3B**. We find a novel feature: a significant portion of new promoters is found within gene bodies and 3′-regions of already annotated genes. These are oriented either sense or antisense to the annotated direction of transcription.

The location and orientation of the newly discovered promoters is highly reminiscent of the situation already observed in the mouse genome, where pervasive transcription was first recognized (see Chapter 10 and Figure 10.18 and Figure 10.19). Although the functions of these numerous transcripts interleaved with the mRNA portions of genes remain to be elucidated, it is clear that pervasive transcription is a common property, at least of highly complex metazoans.

Another important feature recognized by the ENCODE project is the highly stereotypic structural motif in the immediate upstream vicinity of TSSs. There is a robust 50 bp footprint region in the center of the region of high DNase sensitivity that typifies promoters (**Figure 13.4**). Genomic DNase I footprinting performed on 41 cell types identified 8.4 million distinct footprints. *De novo* motif discovery methods recovered ~90% of known TF-binding motifs, together with hundreds of novel motifs; many of these display high cell selectivity, suggesting their involvement as regulators of differentiation.

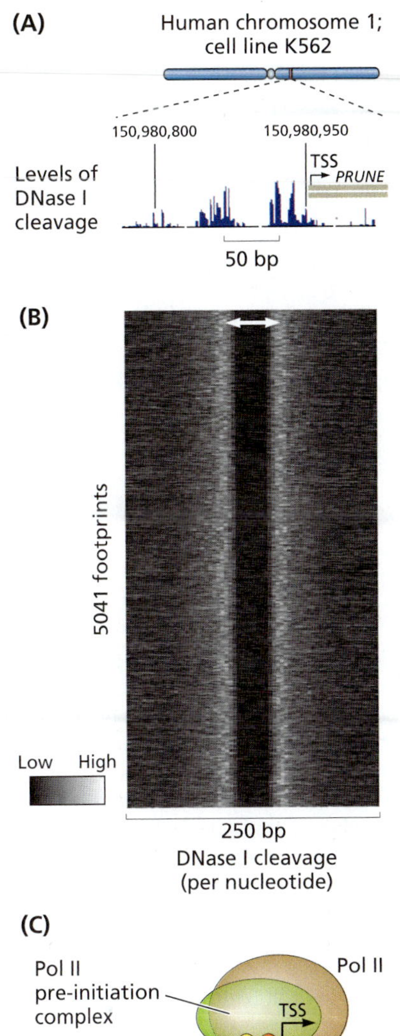

Figure 13.4 Sites of transcription initiation exhibit a highly stereotypic chromatin structural motif of ~80 bp. The motif consists of a ~50 bp central DNase I footprint flanked symmetrically by ~15 bp regions of uniformly elevated levels of DNase I cleavage. (A) Promoter region of the gene *PRUNE* on chromosome 1 is presented as an example. Note the tight spatial coordination of the motif with the TSS of the gene. (B) Heat map of the per-nucleotide DNase I cleavage patterns over 5041 instances of this stereotypical signature. (C) Schematic presentation of the molecular structure underlying the signature. This interpretation was derived from the finding that there are two distinct peaks of evolutionary conservation within the central footprint, not shown here, compatible with binding sites for paired canonical sequence-specific TFs. The TSS is localized precisely within the footprint. (Adapted from Neph S, Vierstra J, Stergachis AB et al. [2012] *Nature* 489:83–90. With permission from Macmillan Publishers, Ltd.)

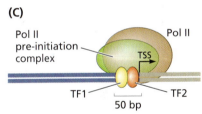

Nucleosome positioning at promoters and around TF-binding sites is highly heterogeneous

To understand nucleosome positioning at promoter sites genomewide, the ENCODE researchers generated data by digestion of chromatin DNA with micrococcal nuclease, MNase, followed by DNA sequencing for two cell lines. In addition to analyzing regions around TSSs in these cell lines, they also related the nucleosome positioning signals to the binding sites of 119 DNA-binding proteins across a large number of cell lines.

Traditionally, data quantifying relationships among genomic signals are presented in the aggregation plots. In these plots, the signal of interest is averaged for each position within a predefined window around the center of an anchor site; the anchor sites are all aligned at the location of a feature that they share in common. In the specific example we consider, TSSs or the location of TF-binding sites serve as the aligned features of the anchor sites, while the nucleosome positions are the signal whose shape, magnitude, and asymmetry is being analyzed. This method, despite its popularity and effectiveness, suffers from a major disadvantage. The averaging over all anchor sites of a particular type, such as all TSSs, produces a misleading aggregate that could obscure underlying heterogeneities that may be of major biological significance. To avoid this drawback, a novel methodology, termed clustered aggregation tool or CAGT, was introduced (**Figure 13.5**).

The application of CAGT to the analysis of nucleosome positioning around TSSs revealed 17 clusters of distinct patterns. **Figure 13.6** shows the 11 clusters that each contained >2% of the TSSs. Broadly, the clusters fall into two categories in terms of their regularity of nucleosome positioning with respect to the TSS: either upstream or downstream. Surprisingly, no cluster had equally strong positioning on both sides of the TSSs, suggesting that the aggregate plot (see top-left plot in Figure 13.6A) is an averaging artifact. Attempts to relate these diverse patterns to the level of transcriptional activity of each cluster showed no consistent positioning features. A general trend can be discerned between high expression levels and the particularly pronounced nucleosome positioning peaks, either upstream or immediately downstream of the TSSs. Much remains to be learned about how or whether this structural heterogeneity plays a role in transcriptional regulation.

The chromatin environment at regulatory elements and in gene bodies is also heterogeneous and asymmetric

Since gene regulation is governed by interplay of nucleosome positioning, histone modifications, and TF binding, the ENCODE project analyzed the relationships among these features genomewide. Again, data analyses were performed by use of the CAGT methodology (see Figure 13.5). Some sample data are presented in **Figure 13.7**. As can be seen, the DNase hypersensitivity signal is predominantly symmetric around protein-binding sites. This is not surprising in view of the fact that DNase I cleaves preferably at open chromatin right next to and on both sides of the bound TF. By contrast, the distribution of nucleosome positioning is strikingly asymmetric, with 90% of TFs exhibiting pronounced asymmetry in nucleosome positioning around >90% of their binding sites. This kind of asymmetry is similar to the asymmetry in

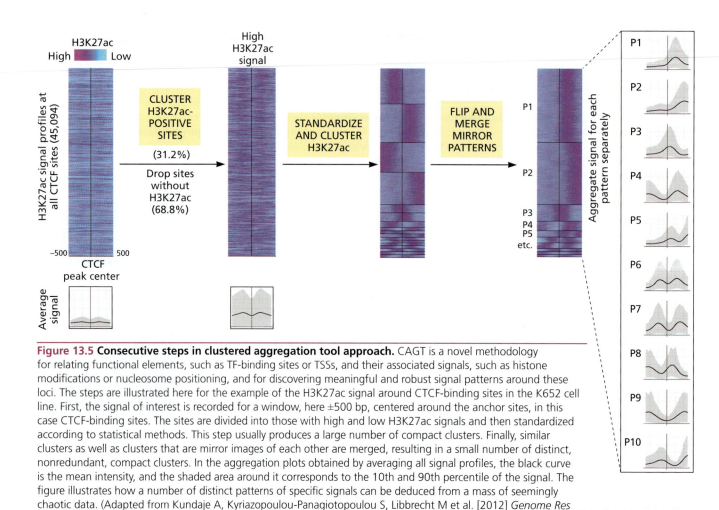

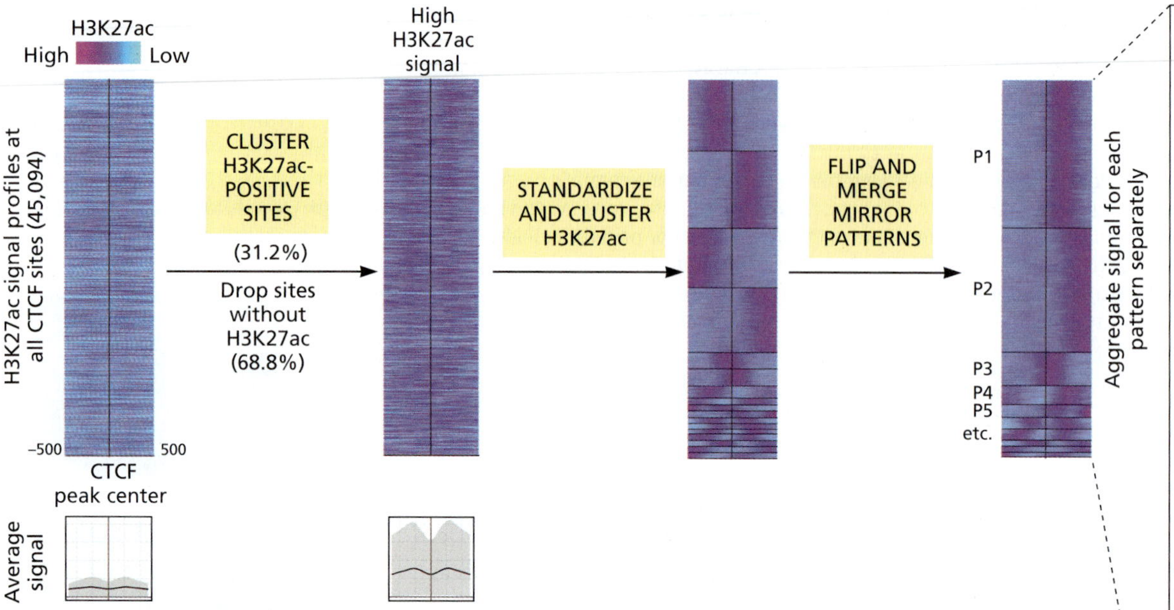

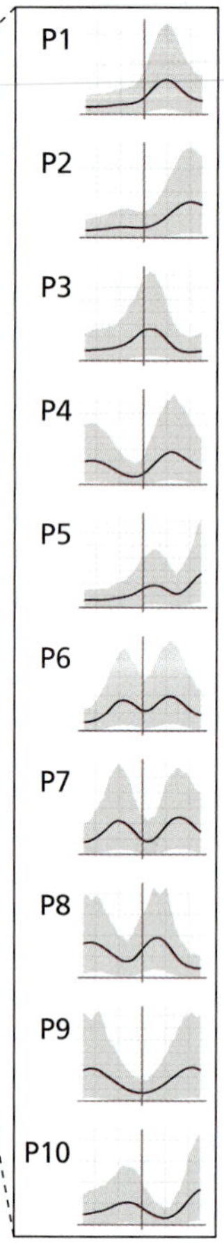

Figure 13.5 Consecutive steps in clustered aggregation tool approach. CAGT is a novel methodology for relating functional elements, such as TF-binding sites or TSSs, and their associated signals, such as histone modifications or nucleosome positioning, and for discovering meaningful and robust signal patterns around these loci. The steps are illustrated here for the example of the H3K27ac signal around CTCF-binding sites in the K652 cell line. First, the signal of interest is recorded for a window, here ±500 bp, centered around the anchor sites, in this case CTCF-binding sites. The sites are divided into those with high and low H3K27ac signals and then standardized according to statistical methods. This step usually produces a large number of compact clusters. Finally, similar clusters as well as clusters that are mirror images of each other are merged, resulting in a small number of distinct, nonredundant, compact clusters. In the aggregation plots obtained by averaging all signal profiles, the black curve is the mean intensity, and the shaded area around it corresponds to the 10th and 90th percentile of the signal. The figure illustrates how a number of distinct patterns of specific signals can be deduced from a mass of seemingly chaotic data. (Adapted from Kundaje A, Kyriazopoulou-Panagiotopoulou S, Libbrecht M et al. [2012] *Genome Res* 22:1735–1747. With permission from Anshul Kundaje.)

nucleosome positioning around TSSs described above and probably indicates that the asymmetry around TSSs is determined by TF binding at these sites. The only notable exception to the norm of asymmetric nucleosome positioning around TF-binding sites occurs around binding sites for the CTCF–cohesin complex; why this is so has yet to be determined. All histone marks at promoter regions or in gene bodies show a highly asymmetric distribution.

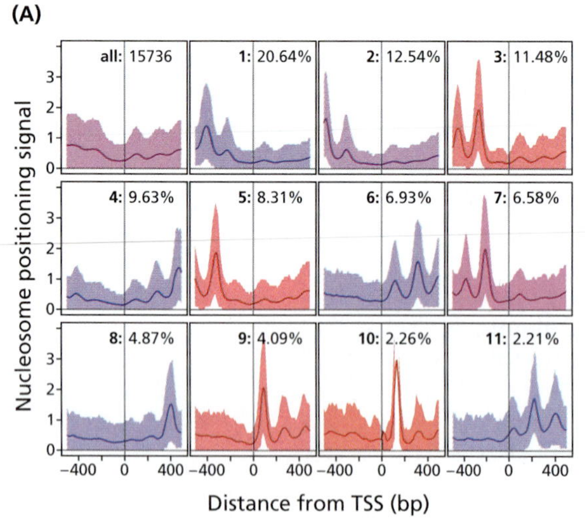

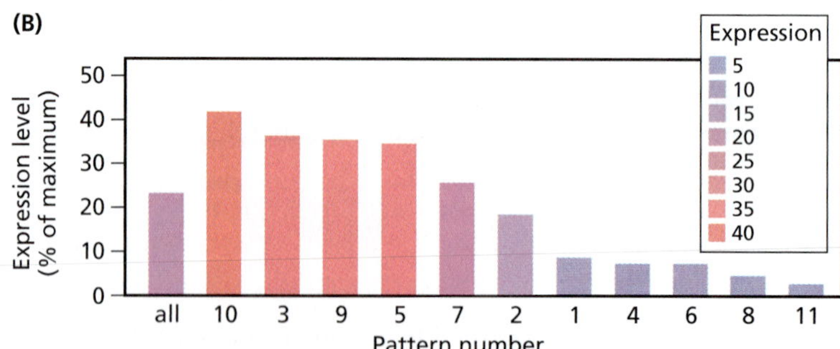

Figure 13.6 Nucleosome positioning patterns around TSSs in the K562 cell line. (A) The top-left plot, labeled all, is a traditional aggregation plot of the nucleosome positioning signal in a 1001-bp window centered on each of the 15,736 TSSs annotated by GENCODE. The rest of the plots show the patterns of nucleosome positioning uncovered by CAGT, ordered by the percentage of TSSs that follow each pattern. All TSSs are reoriented so that the direction of transcription is from left to right. (B) Each of these patterns and the corresponding expression level of the TSSs clustered in these patterns. The colors in parts A and B match. (Adapted from Kundaje A, Kyriazopoulou-Panagiotopoulou S, Libbrecht M et al. [2012] *Genome Res* 22:1735–1747. With permission from Anshul Kundaje.)

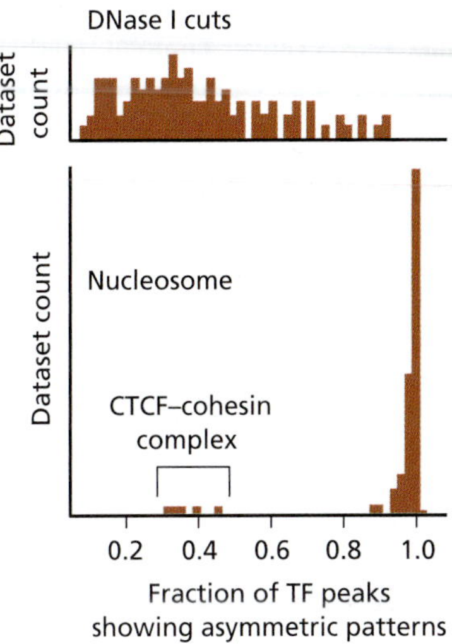

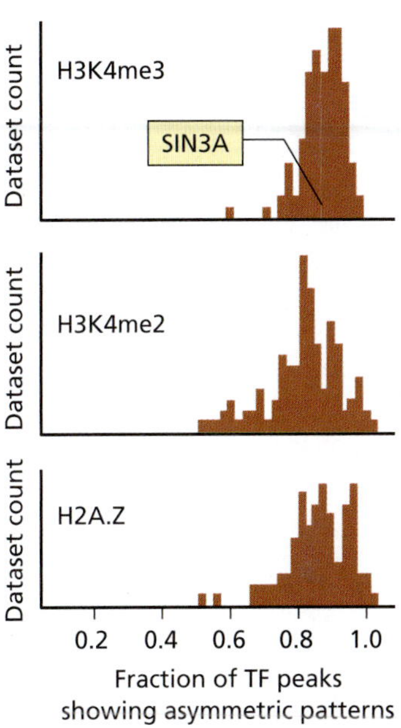

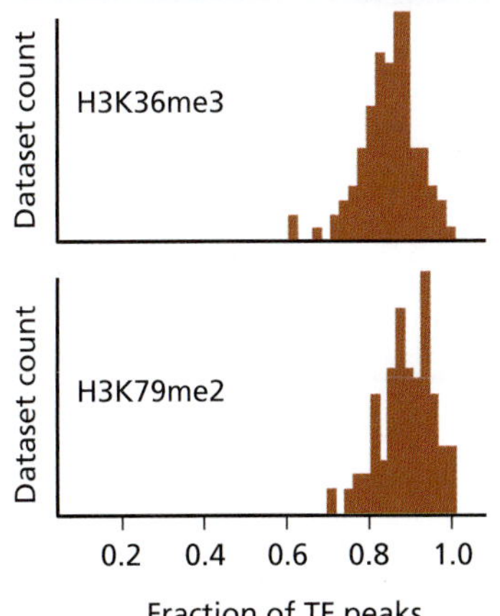

Figure 13.7 Shape asymmetry for DNase I hypersensitivity, nucleosome positioning, and specific histone marks around TF-binding sites. For each combination of TF and chromatin mark, the fraction of high-signal binding sites in asymmetric CAGT clusters (see Figure 13.5) was calculated. Results are averaged, over all available data sets for the same TF and the respective chromatin mark in all cell lines examined, and then plotted as a histogram of the asymmetry fractions over all factors for each mark. Dataset count is shown on the *y*-axis. Note that DNase I patterns are symmetric overall; by contrast, nucleosome positioning is highly asymmetric, with the fraction of asymmetric sites approaching 1.0. The arrow from the boxed SIN3A TF points to the asymmetry of that factor. (Adapted from Kundaje A, Kyriazopoulou-Panagiotopoulou S, Libbrecht M et al. [2012] *Genome Res* 22:1735–1747. With permission from Anshul Kundaje.)

Thus, overall, heterogeneity and asymmetry of chromatin marks around TF-binding sites is the norm, likely reflecting functional asymmetries in regulatory elements. A substantial fraction of TF-binding sites and sites for polymerase binding mark switch points where chromatin structure is different on the two sides of the respective binding site. The rules of regulatory genomics are only now beginning to be unraveled.

13.5 ENCODE insights into gene regulation

Distal control elements are connected to promoters in a complex network

The ENCODE project performed experiments to compare the dynamics of appearance of DNase sensitivity of known cell-selective enhancers with that of the promoters of their target genes, when these genes are activated. The researchers found that in many cases both elements became hypersensitive at the same time following gene activation, an observation that prompted them to analyze the possible connection between promoters and enhancers genomewide. They studied the patterning of 1,454,901 distal DHSs, where distal means a DHS separated from a TSS by at least one other DHS. They used 79 diverse cell types and correlated the DNase I signal at each distal position with the signal at all promoters within ±500 kb. A total of 578,905 DHSs were highly correlated with at least one promoter. These data provided an extensive map of candidate enhancers controlling specific genes. Experimental validation of this connectivity was provided by identifying chromatin interactions by the chromosome conformation capture carbon copy or 5C technique (**Box 13.2**). The connectivity derived from both DNase I hypersensitivity studies and 5C is presented in **Figure 13.8A** which uses the example of the phenylalanine hydroxylase gene, *PAH*. **Figure 13.8B** shows the genomewide pattern of connectivity among these control elements. Most promoters were connected to more than one distal DHS, and vice versa, most distal DHSs interacted with more than one promoter. The number of distal DHSs connected with a particular promoter provided, for the first time, a quantitative measure of the overall regulatory complexity of that particular gene.

Visualization of the regulatory networks, most of which seem to involve looping interactions among the interacting partners, in the human genome is still a challenge. **Figure 13.9** presents one attempt to show the complex networks governing the behavior of just one specific interrogated region. Although it is almost impossible to extract individual interactions from such plots, the visualization makes it very clear that the interactions among promoters, in this case TSSs, and distal elements are many and extremely complex. Importantly, these networks at a given genome region are cell-type-specific and robust, allowing for the prediction of functional behaviors. The difficulties in interpreting such patterns point to the need for improved methods of data presentation for complex systems.

Transcription factor binding defines the structure and function of regulatory regions

We are already aware of the role TFs play in regulation of eukaryotic transcription (see Chapter 12). The ENCODE analysis of the entire human genome confirmed this role and revealed novel features of TFs and their cooperation in defining the transcriptional status of a gene, and more broadly in shaping cellular identity.

Box 13.2 A Closer Look: 5C methodology, a massively parallel solution for mapping interactions between genomic elements genomewide The **5C technology**, chromosome conformation capture carbon copy, was developed as an expansion of the 3C method, chromosome conformation capture, introduced in 2002 to detect physical interactions between genomic sequences. 3C uses formaldehyde cross-linking to trap covalently interacting segments in the genome, followed by restriction endonuclease treatment and ligation of the interaction segments. The ligated products are then quantified individually by polymerase chain reaction, PCR (see Figure 10.22). 3C is particularly suited for relatively small-scale analysis of interactions between a set of candidate sequences. PCR detection is, however, not suitable for large-scale mapping of previously unidentified chromatin interactions. To perform such mapping, the 5C method was introduced, in which highly multiplexed ligation-mediated amplification, LMA, is used to first copy and then amplify parts of the complex 3C libraries produced as the end product of the 3C methodology. LMA is widely used to detect and amplify specific target sequences by use of primer pairs that anneal next to each other on the same DNA strand; only primers annealed next to each other can be ligated. Inclusion of universal tails that contain strong promoter sequences at the end of these primers allows subsequent amplification. LMA-based approaches can be performed at high levels of multiplexing, using thousands of primers in a single reaction. The amplified libraries are analyzed by microarrays or DNA sequencing (**Figure 1**).

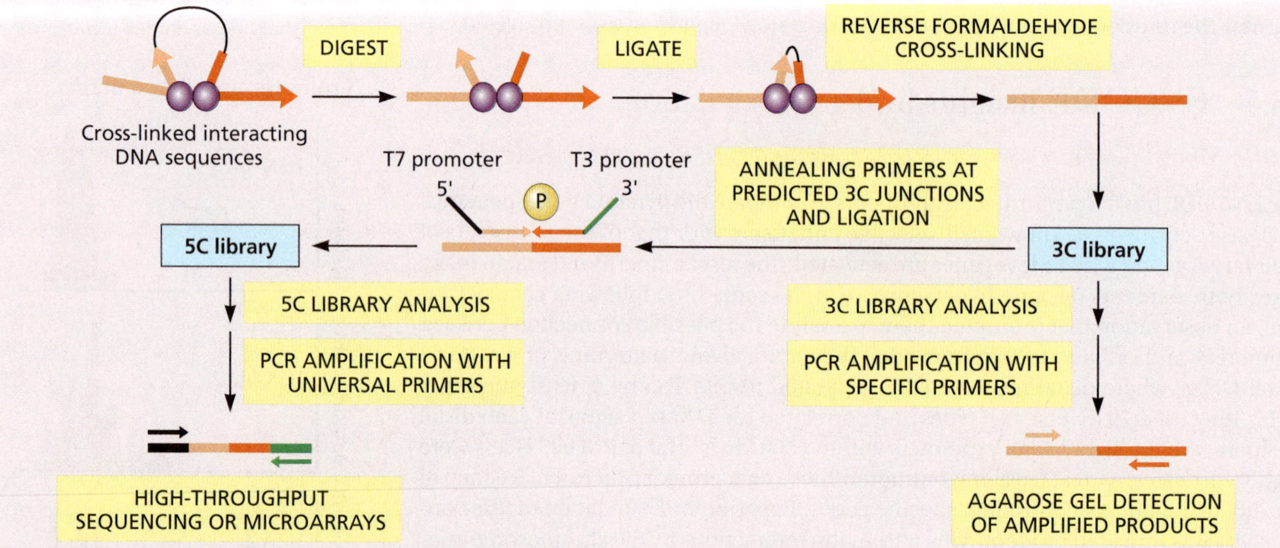

Figure 1 5C detects ligation products in 3C libraries. A 3C library is generated by conventional methods and then converted into a 5C library by annealing and ligating 5C oligonucleotides in a multiplex setting. The new 5C libraries are then analyzed by sequencing or in microarrays. 5C libraries are produced by annealing 5C primers at predicted 3C junctions in a multiplex setting, followed by specific ligation of annealed primers with an NAD-dependent DNA ligase. The universal tails of 5C primers are illustrated as black and green lines and contain T7 and T3 promoter sequences, respectively. These promoters are used to amplify libraries in a single PCR step. (Adapted from Dostie J, Richmond TA, Arnaout RA et al. [2006] *Genome Res* 16:1299–1309. With permission from Josee Dostie.)

TF binding to regulatory elements in the genome protects the underlying DNA sequence from DNase I cleavage, creating the footprints seen in the high-resolution electrophoretic gels used to analyze DNase I cleavage patterns (see Box 6.4). Footprinting was originally used to study known gene-specific *cis*-regulatory sequences and led, among other things, to the discovery of the first human sequence-specific TF, SP1.

It is now possible to perform genomewide mapping of DNase I footprints. Efficient footprinting in large genomes requires focused analysis of a small fraction, 1–3%, of the genome that is characterized by a substantial concentration of DNase I cleavage sites. Analyses of sequences that are enriched for DNase I cleavage across 41 cell types identified an average of ~1.1 million footprints, 6–40 bp long, per cell type. The majority of DHSs (99.8%) contained at least one footprint, indicating that DHSs do not simply represent open or nucleosome-free chromatin regions but are constitutively populated with DNase I footprints. DNase I footprints are distributed throughout the genome in the quantitative pattern depicted in **Figure 13.10**. Importantly, footprints are deficient in DNA methylation: the CpG dinucleotides contained within DNase I footprints are significantly less methylated than CpGs in non-footprinted regions of the same DHS (**Box 13.3**).

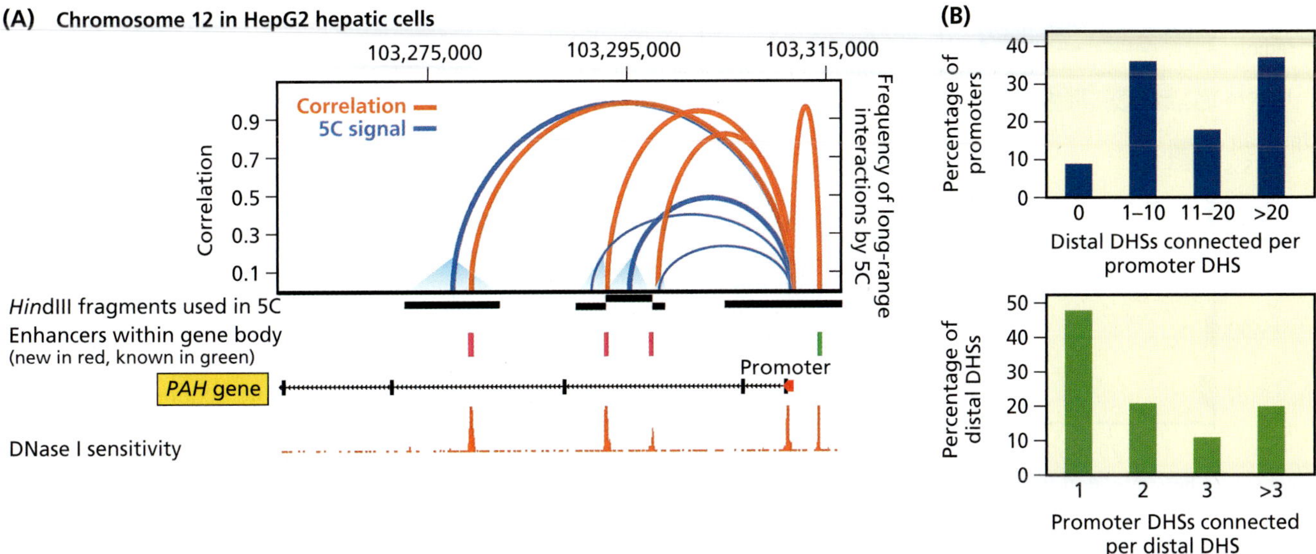

Figure 13.8 Promoter-to-distal DHS connectivity for the *PAH* gene. (A) Cross-cell-type correlation, shown as orange arcs as measured on the left *y*-axis, of distal DHSs and *PAH* promoter closely parallels chromatin interactions measured by 5C, shown as blue arcs as measured on the right *y*-axis. Black bars indicate *Hin*dIII fragments used in 5C assays. Known and new enhancers are shown as green and red vertical bars, respectively. (B) Proportions of connected elements. (Top) Each promoter is connected to the number of DHSs. Note that more than half of the promoters are connected to more than 11 DHSs. (Bottom) Each distal DHS is connected to 1, 2, 3, or >3 promoters. (Adapted from Thurman RE, Rynes E, Humbert R et al. [2012] *Nature* 489:75–82. With permission from Macmillan Publishers, Ltd.)

In addition to interrogating DNase I footprints, the ENCODE project mapped the binding sites of 119 different DNA-binding proteins genomewide in 72 cell types by use of ChIP-seq methods. These included canonical or sequence-specific TFs, histone-modification enzymes, chromatin remodelers, a number of components of RNA polymerases II and III, and their associated basal TFs. Overall, ENCODE identified 636,336 binding regions across all studied cell types, covering 8.1% of the genome. In addition, the project looked for known and new DNA-binding motifs. Both high- and low-affinity binding sites were recognized; moreover, it was possible to identify TF target regions that were indirectly bound by TFs through interactions with other factors.

Transcription factors interact in a huge network

One of the major findings of the ENCODE project concerns the elements responsible for cell-selective transcription regulation and the complex combinatorial patterns

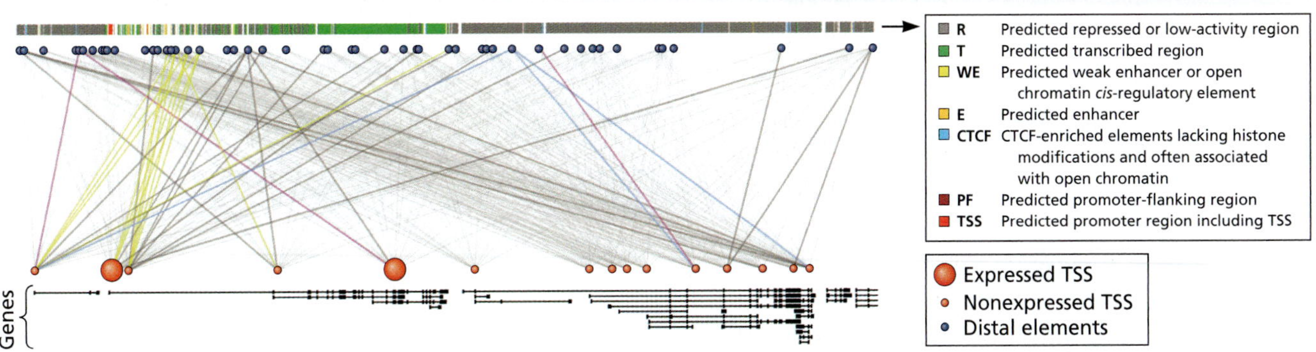

Figure 13.9 Network of looping interactions as inferred from 5C. Chromosome conformation capture carbon copy methodology is known as 5C. The genomic region shown, ENr132, is one of the regions interrogated by the ENCODE pilot project, which analyzed 1% of the human genome; the cell line is K562. Distal elements and the TSSs are positioned according to genomic coordinates; GENCODE gene annotation is indicated at the bottom. Thin gray lines show all interactions that were interrogated. Colored lines show significant looping interactions between TSS and distal elements. Most importantly, the interactions are not exclusively one-to-one, suggesting that multiple genes and distal elements can assemble in large clusters. (Adapted from Sanyal A, Lajoie BR, Jain G & Dekker J [2012] *Nature* 489:109–113. With permission from Macmillan Publishers, Ltd.)

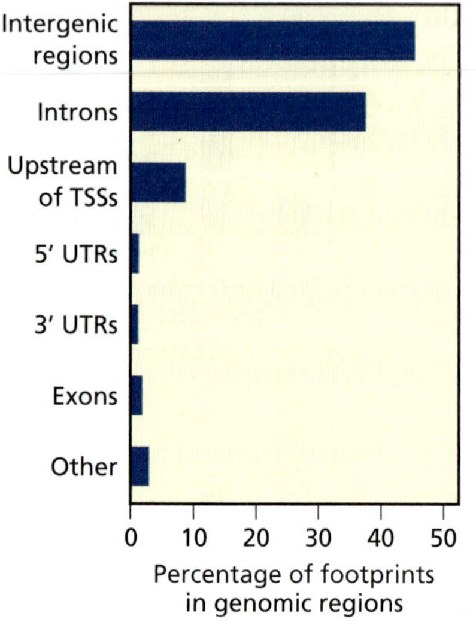

Figure 13.10 Distribution of DNase I footprints among different elements of the genome. There is an enrichment of footprints in recognized control regions located in intergenic portions of the genome and upstream of TSSs, commensurate with the DNase I cleavage densities observed in these regions. A significant portion of DNase I footprints occurs in introns. Of interest, and unexpectedly, 2% of footprints were localized within exons. The functional significance of the footprints in introns and exons is unclear.

of TFs that are bound to these elements. As stated by John Stamatoyannopoulos, "Although ENCODE was conceived as a genomic annotation project fundamentally focused on the linear organization of sequence elements, it is now becoming clear that connectivity between linear elements is an intrinsic part of this annotation – from splicing, to long-range chromatin interactions, to transcription factor networks."

It is possible to obtain a systems-level perspective of TF interactions by constructing networks; we have seen relatively simple networks in previous chapters. For the purposes of visualization, the networks consist of nodes and edges. The nodes of the network are TFs, and the edges designate cross-regulatory relationships between one TF and another. The connectivity network of the 119 human TFs studied by ENCODE

Box 13.3 A Closer Look: How are DNA methylation patterns studied on a genomic scale? The recognized involvement of methylation of cytosines in the context of CpG dinucleotides has led to efforts to analyze this modification in an efficient and cost-effective way. As a result, numerous methods have been designed, including the most frequently used treatment of samples with sodium bisulfite at high temperature and low pH. Under these conditions, unmethylated cytosines are specifically deaminated and converted into uracil, while methylated cytosines are left unchanged. When subjected to subsequent PCR amplification, uracil is replaced by thymine, giving rise to methylation-specific single-nucleotide alteration; this is detectable by conventional sequencing and alignment against a reference sequence (Figure 1A).

Recently, a number of technical advances, such as the use of microarrays and high-throughput sequencing, have increased the scale at which methylation can be assessed. In all large-scale methods, it is important to first enrich for methylated DNA. This can be done by affinity purification of methylated DNA by using antibodies against ^{5me}CpG or against methyl-binding proteins. Although affinity purification can identify ^{5me}CpG-containing DNA fragments, it cannot determine the methylation status of individual CpGs along the fragments. Thus, a major advance came when the bisulfite conversion protocol was adapted to high-throughput sequencing; the protocol was first successfully applied to the assessment of the *Arabidopsis* methylome at single-CpG resolution. However, even that protocol would be prohibitively costly for a similar

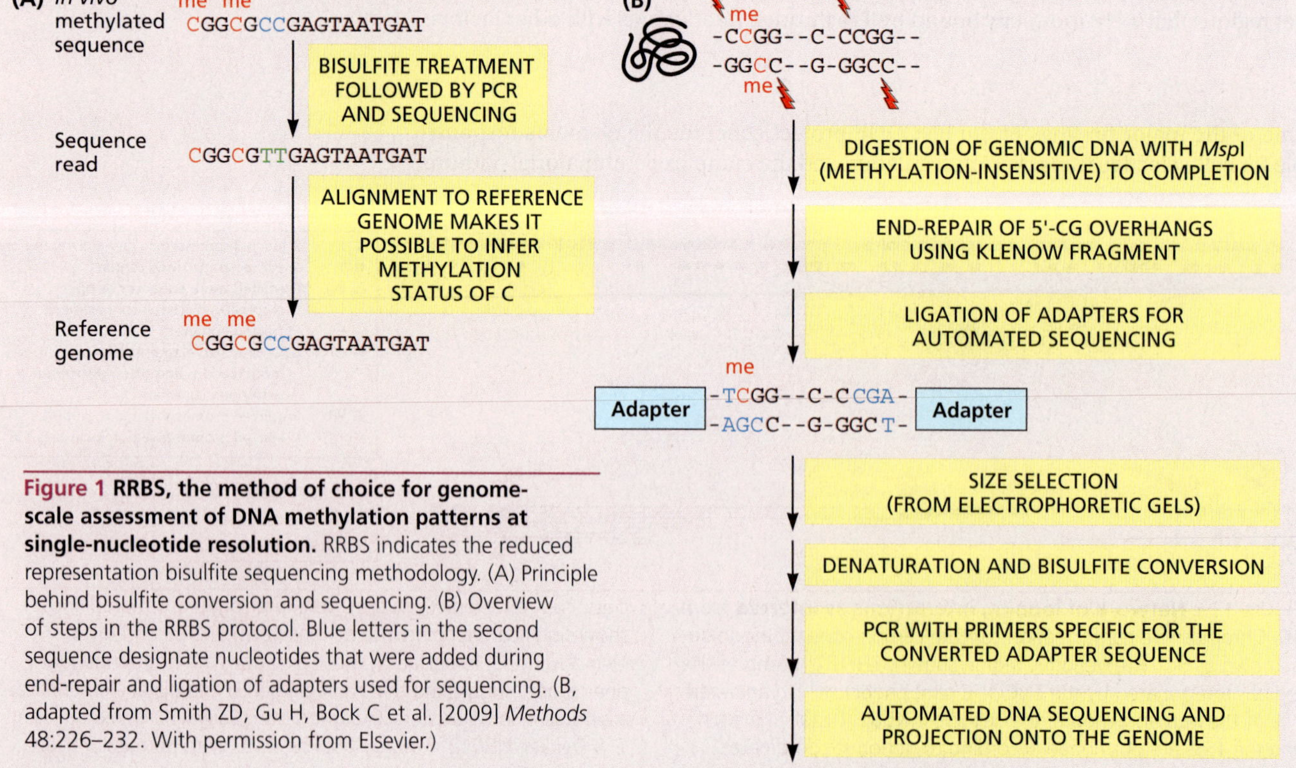

Figure 1 RRBS, the method of choice for genome-scale assessment of DNA methylation patterns at single-nucleotide resolution. RRBS indicates the reduced representation bisulfite sequencing methodology. (A) Principle behind bisulfite conversion and sequencing. (B) Overview of steps in the RRBS protocol. Blue letters in the second sequence designate nucleotides that were added during end-repair and ligation of adapters used for sequencing. (B, adapted from Smith ZD, Gu H, Bock C et al. [2009] *Methods* 48:226–232. With permission from Elsevier.)

analysis of larger genomes, such as that of humans, which is ~30-fold larger than the *Arabidopsis* genome.

Another modification of the method, **reduced representation bisulfite sequencing** or **RRBS**, enriches the libraries to be analyzed by digesting genomic DNA with restriction nucleases that are specific for CpG-containing motifs. Digestion with methylation-insensitive restriction enzymes makes use of the fundamental CpG asymmetry present within vertebrate genomes (see Figure 12.26 and Figure 12.27). The vast majority of CpG dinucleotides occur infrequently; thus, if genomic DNA is fragmented by sonication, it may often be the case that individual fragments contain no CpGs, so sequencing these fragments will be in vain. Restriction-based enrichment ensures that every sequence read will contain information regarding at least one CpG, purely because CpG is part of the restriction enzyme's target sequence. *MspI*, for example, will cleave CCGG following the first C, so libraries generated through *MspI* digestion are expected to provide information about the methylation state of nearly 90% of the CpG islands within the mouse genome. The individual steps of RRBS are presented in **Figure 1B**.

via ChIP-seq can be roughly divided among three levels (**Figure 13.11**). The top-level TFs are the most powerful factors, each regulating many other factors. The factors at the bottom are more regulated than regulating. Connectivities for specific factors in specific cell types can be found at www.regulatorynetworks.org.

The main conclusion from these studies is that human TFs associate in many fashions; different combinations of factors bind near different targets and the binding of one factor often affects the choice of binding partner by others. Often, transcription factors exhibit different co-association patterns in gene-proximal and gene-distal regions.

Another powerful approach to constructing TF networks makes use of genomewide maps of *in vivo* DNase I footprints. By performing systematic analysis of TF footprints in the proximal regulatory regions of each TF gene, researchers developed a paradigm for comprehensive, unbiased mapping of TF networks. Iterating this paradigm across diverse cell types provides a powerful system for the analysis of TF network dynamics in a complex organism. A schematic representation of the footprint approach is given in **Figure 13.12**, and the cell-type specificity revealed by the regulatory networks is illustrated in **Figure 13.13** and **Figure 13.14**.

TF-binding sites and TF structure co-evolve

As discussed above, the application of genomic DNase I footprinting across regulatory genomic sequences defines the recognition landscape of hundreds of DNA-binding proteins. Mining footprint sequences for recognition motifs has nearly doubled the

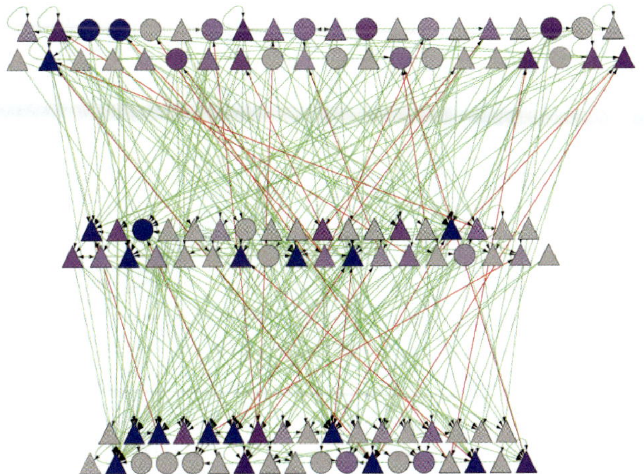

Figure 13.11 Arrangement of the 119 human transcription factors studied by ENCODE in a hierarchical network. Top-level factors are of the executive type, regulating many other factors; bottom-level factors are of the foreman type, more regulated than regulating. Nodes are transcription factors, where sequence-specific TFs are represented by triangles and non-sequence-specific factors by circles; lines depict connectivity. (From Gerstein MB, Kundaje A, Hariharan M et al. [2012] *Nature* 489:91–100. With permission from Macmillan Publishers, Ltd.)

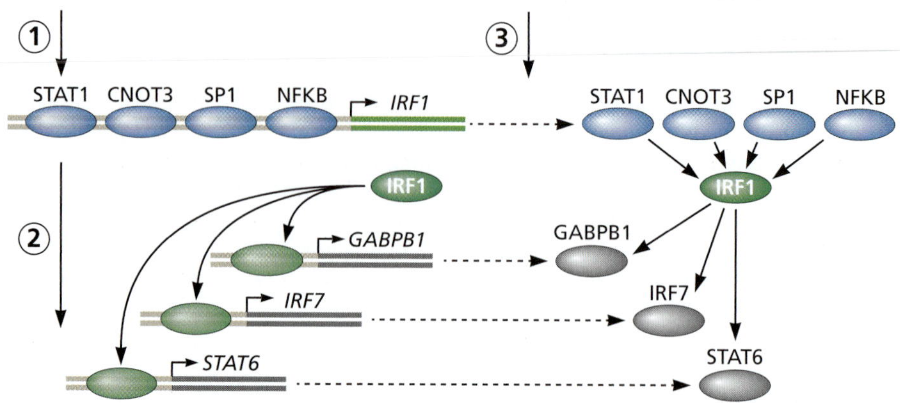

Figure 13.12 Construction of comprehensive transcriptional regulatory networks by use of DNase I footprints. (Step 1) Collect genomic DNase I footprints from 41 diverse cell types; infer the identities of TFs occupying footprints by using well-annotated databases of TF-binding motifs. (Step 2) Use DNase I footprints to determine occupied binding elements within promoter-proximal regions of each TF gene, within ±5 kb of TSS; identify other TFs targeted by the protein product of the interrogated TF gene. (Step 3) Generate network for this specific TF; repeat in 41 cell types using all 475 TFs with recognition motifs. Each TF node is a gene that has regulatory inputs, defined in terms of TF footprints within its proximal regulatory regions, and regulatory outputs, or footprints of that TF in the regulatory regions of other TF genes. The inputs and outputs comprise the regulatory network interactions or edges. The example given is the *IRF1* gene in Th1 cells. (Adapted from Neph S, Stergachis AB, Reynolds A et al. [2012] *Cell* 150:1274–1286. With permission from Elsevier.)

human ***cis*-regulatory lexicon**, identifying numerous new *cis*-regulatory elements with characteristic structural and functional features. Of importance, the *cis*-regulatory compartment that comprises all *cis*-acting regulatory sequences has co-evolved with TFs, so that the structures of the DNA elements closely fit the structures of the TFs that binds to them. Figure 13.15 illustrates this point with the specific example of USF1, upstream stimulatory factor 1. It also points out that all recognized sequences

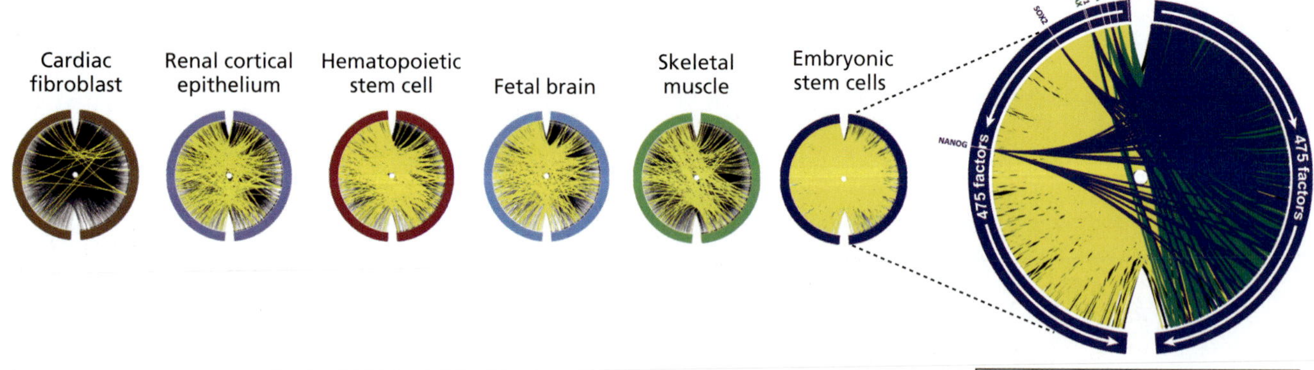

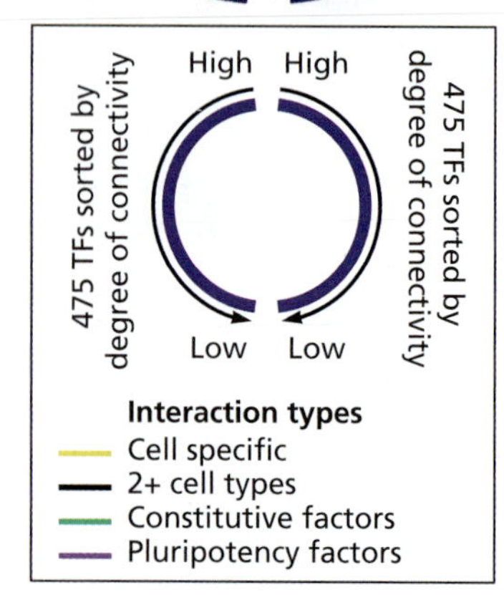

Figure 13.13 Cell-specific versus shared regulatory interactions in TF networks in six cell types. Some of the cell types presented in this figure and in Figure 13.14 are the same. Each half of each circular plot is divided into 475 points, one for each TF; the points are not visible on this scale. Lines connecting the left and right half-circles represent regulatory interactions between each factor and the factor(s) with which it interacts. The order of TFs along each half-circle is sorted in descending order according to the degree of connectivity—that is, the number of connections to other TFs—in the embryonic stem cell (ESC) network. The close-up view of the ESC network on the right highlights the interactions of four pluripotent factors shown in purple, KLF4, NANOG, POU5F1, and SOX2, and four constitutive factors shown in green, SP1, CTCF, NFYA, and MAX. (Adapted from Neph S, Stergachis AB, Reynolds A et al. [2012] *Cell* 150:1274–1286. With permission from Elsevier.)

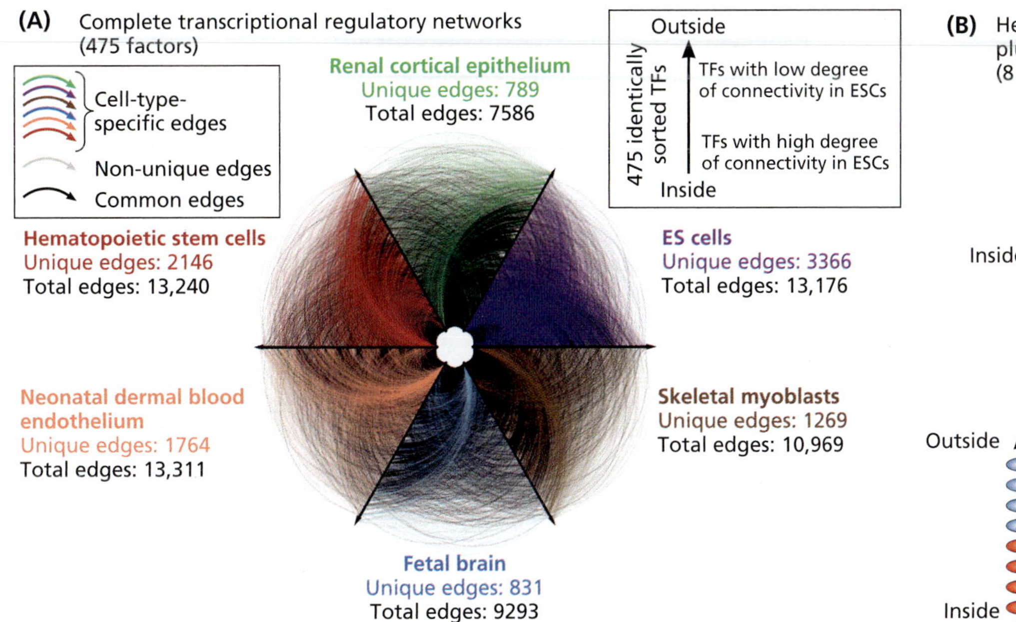

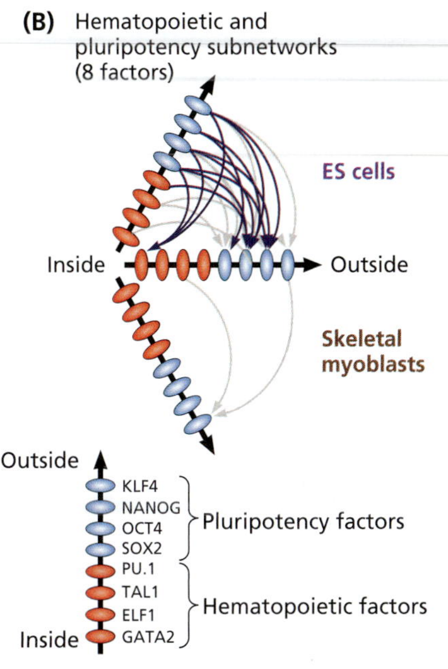

Figure 13.14 Transcriptional regulatory networks show marked cell-type specificity. (A) Cross-regulatory interactions between all 475 TFs in regulatory networks of six diverse cell types. The factors are arranged in the same order along each axis; regulatory interactions are directed clockwise. Embryonic stem (ES) cells, ESCs. (B) To help explain the complete network in part A, we have presented cross-regulatory interactions between only four pluripotency factors, which are present in stem cells and determine their ability to differentiate into numerous cell types later in development, and four hematopoietic factors, which regulate genes involved in the differentiation of blood cell types, for two of the cell types. TFs are arranged in the same order on each axis, as presented at the bottom. Regulatory interactions from regulator to regulated are shown by clockwise arrows. (Adapted from Neph S, Stergachis AB, Reynolds A et al. [2012] *Cell* 150:1274–1286. With permission from Elsevier.)

that bind the factor *in vivo* share a common nucleotide-level DNase I signature. The DNase cleavage pattern closely parallels the topology of the protein-DNA interface, including a marked depression in DNase I cleavage at nucleotides directly involved in protein-DNA contacts and increased cleavage at exposed nucleotides. Thus, the high-resolution aggregated DNase I cleavage signature reflects fundamental features of the protein-DNA interaction interface. Interestingly, conservation of DNA residues in vertebrates closely correlates with the DNase I cleavage pattern, implying that regulatory DNA sequences have evolved to fit the morphology of the TF-DNA binding interface.

DNA methylation patterns show a complex relationship with transcription

As we have already seen in Chapter 12, DNA methylation is a major participant in controlling gene expression, more particularly in gene silencing. Recognizing the importance of CpG methylation, ENCODE researchers quantitatively measured methylation for several million CpG dinucleotides, focusing on those that fall within DHSs in 19 cell types. The aim was to gain some understanding of the connection between DNase I hypersensitivity and DNA methylation patterns. Two broad classes of sites were

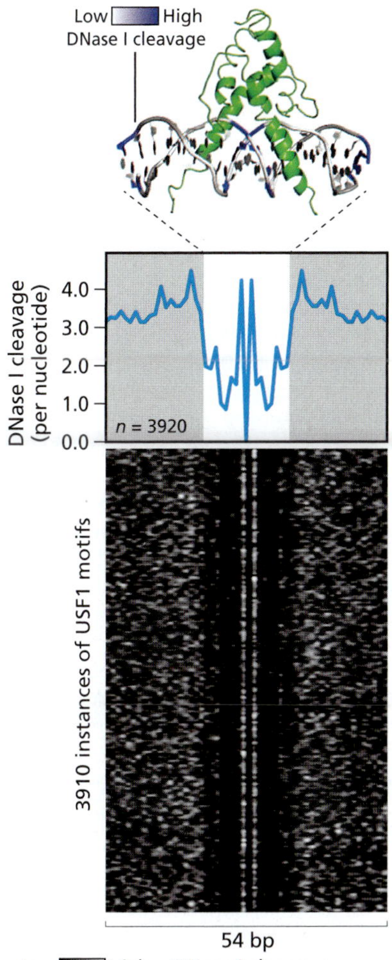

Figure 13.15 DNase I footprint structure parallels the structure of the cognate TF. The co-crystal structure of a TF, upstream stimulatory factor or USF1, bound to its specific binding site is positioned above the average nucleotide-level DNase I cleavage patterns. Nucleotides that are sensitive to DNase I cleavage are colored blue in the co-crystal structure. The heat map at the bottom shows the DNase I signature at each individual USF1 binding motif; these individual signatures have been averaged to produce the blue line profile shown. (Adapted from Neph S, Vierstra J, Stergachis AB et al. [2012] *Nature* 489:83–90. With permission from Macmillan Publishers, Ltd.)

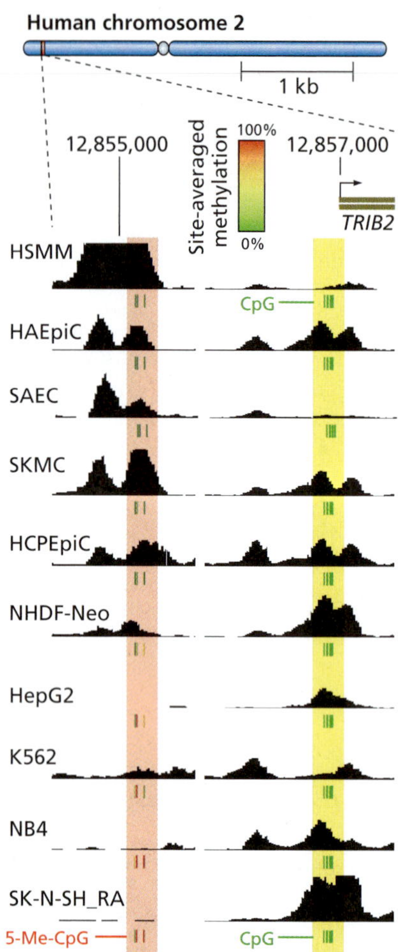

Figure 13.16 DNA methylation patterns as determined by RRBS and DNase I accessibility. RRBS indicates the reduced representation bisulfite sequencing methodology. The example presented is a region of chromosome 2 containing the *TRIB2* gene, which was interrogated in 10 cell types as listed on the left. The *x*-axis represents the region, and the *y*-axis is the level of DNase I accessibility. CpGs are represented by small vertical bars, which are colored according to the scale shown. Reddish shading highlights a region in which accessibility decreases quantitatively as methylation increases. Yellow shading highlights a region in which there is little change in CpG methylation across the cell types and low correlation between accessibility and methylation. (Adapted from Thurman RE, Rynes E, Humbert R et al. [2012] *Nature* 489:75–82. With permission from Macmillan Publishers, Ltd.)

observed as seen in **Figure 13.16**: those with a strong inverse cross-cell-type correlation between DNA methylation and chromatin accessibility, indicated by reddish shading, and those with variable chromatin accessibility but constitutive hypomethylation, indicated by yellow shading. Further analysis provided an overview of the distribution of various methylation classes across cell types (**Figure 13.17A**). Importantly, for the class of sites that were differentially methylated in various cell types and showed an associated DNase I sensitivity, increased methylation was almost uniformly negatively correlated with chromatin accessibility (**Figure 13.17B**). The researchers argued that this association resulted from passive deposition of methyl groups after the regulatory DNA was vacated by TFs.

Another link was revealed between DNA methylation and occupancy of CTCF. We have already pointed out in Chapter 12 that the occupancy of CTCF-binding sites is strongly cell-selective. Comparison of genomewide data obtained by ChIP-seq and bisulfite sequencing data across 19 diverse cell types indicated that ~40% of the cell-specific sites show differential DNA methylation. In particular, occupancy of CTCF-binding sites quantitatively increases as local CpG methylation decreases. This result is in concordance with earlier *in vitro* observations that methylation hinders CTCF binding at certain sequence elements.

13.6 ENCODE overview

What have we learned from ENCODE, and where is it leading?

Aside from the many specific revelations detailed above, what are the main lessons obtained to date from ENCODE? It seems to us that they are twofold. First, it is now clear that human cells utilize, in some fashion, much more of the genomic information encoded in their DNA than was hitherto expected. Perhaps it will turn out to be 100%, when all cell types have been interrogated. At any rate, junk DNA is no longer a useful term. To be sure, we do not know the functional importance of even a fraction of the transcripts that have been identified, and unless new screening methods are devised, we will not know for a very long time. But if there is any practical lesson from the history of molecular biology, it is that powerful new methods often appear quickly when needed.

A second surprise is the finding that the vast majority of the newly detected functional entities and interactions among them are cell-type-specific. Perhaps this revelation should not have been surprising, given the remarkable number of very different varieties of cells present in the adult human. A complicating aspect of this is that the function of a given gene or regulatory element may vary, depending on the cellular milieu in which it exists. This will make the unscrambling of interconnected regulatory pathways even more complex but additionally rewarding. Perhaps the major impact of this line of research will be in the areas of developmental biology and evolution. There are already hints that the fundamental reason for the difference in size between mammalian and invertebrate genomes lies in this difference in cell-type diversity. Now it can be analyzed and quantified. It may be that a new age in biology has been born.

Certain methods are essential to ENCODE project studies

For a monumental project such as ENCODE to be meaningful and successful, it is important to develop and standardize experimental and computational methods to be used by the involved research community. We have already covered the more specific methods at the appropriate places throughout the chapter. We will now list and briefly introduce the more common methods, following the format used in the master publication of the second, production phase of the ENCODE project. To fully describe all these techniques would in itself require a very large book.

- **RNA-seq:** isolation of different RNA subpopulations, often combining different purification techniques for different RNA fractions, followed by high-throughput sequencing.
- **CAGE:** capture of the 5′-caps on mRNAs, followed by high-throughput sequencing from a small tag adjacent to the 5′-methylated caps (see Box 10.4).

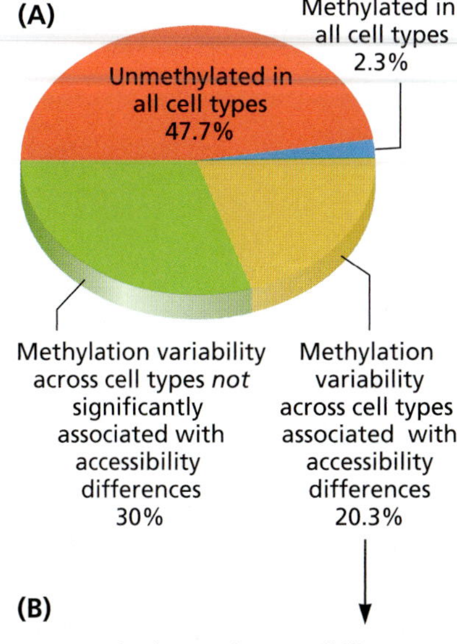

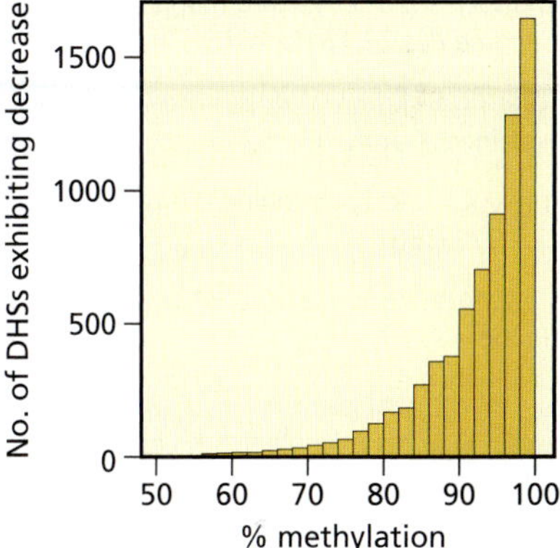

Figure 13.17 Global characterization of the effect of methylation on chromatin accessibility. The effect of methylation on chromatin accessibility was surveyed at 34,376 DHSs for which RRBS data are available. (A) Cross-cell-type analysis revealed that the majority of sites are unmethylated in all cell types, whereas a small portion of sites are methylated in all cell types. Sites that exhibit methylation variability across cell types form two distinct classes: those in which methylation variability across cell types is associated with chromatin accessibility differences and those in which it is not. (B) In the population of sites that exhibit variability in cross-cell-type methylation and associated differences in accessibility, shown as the yellow section in part A, there is a strong inverse correlation between the level of methylation and accessibility. (Adapted from Thurman RE, Rynes E, Humbert R et al. [2012] *Nature* 489:75–82. With permission from Macmillan Publishers, Ltd.)

- **RNA-PET:** simultaneous capture of the 5′-caps and poly(A) tails on mRNAs, indicative of a full-length RNA, followed by high-throughput sequencing from each end.
- **ChIP-seq:** cross-linking of DNA to bound proteins in chromatin *in vivo*, fragmentation of the DNA, and selection of fragments bound to a specific protein by use of antibodies specific to that protein. The enriched sample is subjected to high-throughput sequencing. In addition to identifying bound proteins, the technique can be used to identify features such as specific histone post-translational modifications or DNA methylation. In the latter case, the antibodies directly recognize methylated CpGs or specific methyl DNA-binding proteins (see Box 6.6).
- **DNase-seq:** cleavage of chromatin by DNase I occurs preferentially at sites that are exposed and is thus used as a probe for open chromatin. Openness is defined by the absence of nucleosomes and the binding of TFs. The cut segments are then subjected to high-throughput sequencing to determine the location of hypersensitive sites on the known genomic sequence.
- **FAIRE-seq, formaldehyde-assisted isolation of regulatory elements:** isolation of nucleosome-depleted genomic regions by exploiting the difference in cross-linking efficiency between nucleosomal histones and DNA, which is highly efficient, and sequence-specific regulatory factors and DNA, which is characterized by low efficiency. Cross-linking is followed by phenol extraction and sequencing of DNA fragments in the aqueous phase (see Box 13.1).
- **5C, chromosome conformation capture carbon copy:** a modification of the 3C method introduced in Chapter 10 (see Figure 10.22), designed to convert physical chromatin interactions into specific ligation products that can be quantified by PCR-based methods. The method has been expanded for large-scale parallel detection of chromatin interactions (see Box 13.2).
- **RRBS, reduced representation bisulfite sequencing:** conversion of unmethylated cytosines in a DNA sequence to uracil by treatment with bisulfite, followed by sequencing to determine the methylation status of individual cytosines in a quantitative manner. In RRBS, the assay is preceded by treatment of the genomic DNA with restriction endonucleases that cut around CpG dinucleotides and enrichment for CpGs; thus the genome is reduced to a relatively small CpG-rich portion to be further analyzed by bisulfite sequencing (see Box 13.3).

Key concepts

- The ENCODE project has made use of the total human genome sequence and sophisticated data analysis algorithms to probe transcriptional functional elements and their regulation in depth.
- The project defines seven major classes of elements: transcribed genes, their transcription start sites and promoters, two classes of predicted enhancers (strong and weak enhancers), CTCF-binding sites, and transcriptionally repressed regions.

- DNase hypersensitive sites or DHSs indicate more open chromatin structure and correlate strongly with transcribed genes, transcription start sites, promoters, and strong enhancers.
- Nucleosomes are often precisely positioned about TF-binding sites near TSSs.
- Footprints in the DHS regions correspond to the binding of specific TFs. The topographies of these DHS regions match the TF structures, indicating co-evolution.
- TFs and other elements are widely and multiply interconnected, in a cell-type-dependent manner.
- DNA methylation is correlated in complex ways with the seven classes of regulatory DNA sequence elements identified by ENCODE and with TF occupancy.
- Although most of the genome is transcribed, the pattern of transcribed genes and their control is highly dependent on cell type.

Further reading

Reviews

Chanock S (2012) Toward mapping the biology of the genome. *Genome Res* 22:1612–1615.

Ecker JR, Bickmore WA, Barroso I et al. (2012) Genomics: ENCODE explained. *Nature* 489:52–55.

Frazer KA (2012) Decoding the human genome. *Genome Res* 22:1599–1601.

Stamatoyannopoulos JA (2012) What does our genome encode? *Genome Res* 22:1602–1611.

Experimental papers

Arvey A, Agius P, Noble WS & Leslie C (2012) Sequence and chromatin determinants of cell-type-specific transcription factor binding. *Genome Res* 22:1723–1734.

Ball MP, Li JB, Gao Y et al. (2009) Targeted and genome-scale strategies reveal gene-body methylation signatures in human cells. *Nat Biotechnol* 27:361–368.

Cheng C, Alexander R, Min R et al. (2012) Understanding transcriptional regulation by integrative analysis of transcription factor binding data. *Genome Res* 22:1658–1667.

Derrien T, Johnson R, Bussotti G et al. (2012) The GENCODE v7 catalog of human long noncoding RNAs: Analysis of their gene structure, evolution, and expression. *Genome Res* 22:1775–1789.

Djebali S, Davis CA, Merkel A et al. (2012) Landscape of transcription in human cells. *Nature* 489:101–108.

Dong X, Greven MC, Kundaje A et al. (2012) Modeling gene expression using chromatin features in various cellular contexts. *Genome Biol* 13:R53.

Dostie J, Richmond TA, Arnaout RA et al. (2006) Chromosome Conformation Capture Carbon Copy (5C): A massively parallel solution for mapping interactions between genomic elements. *Genome Res* 16:1299–1309.

The ENCODE Project Consortium (2012) An integrated encyclopedia of DNA elements in the human genome. *Nature* 489:57–74.

Gerstein MB, Kundaje A, Hariharan M et al. (2012) Architecture of the human regulatory network derived from ENCODE data. *Nature* 489:91–100.

Harrow J, Frankish A, Gonzalez JM et al. (2012) GENCODE: The reference human genome annotation for The ENCODE Project. *Genome Res* 22:1760–1774.

Kundaje A, Kyriazopoulou-Panagiotopoulou S, Libbrecht M et al. (2012) Ubiquitous heterogeneity and asymmetry of the chromatin environment at regulatory elements. *Genome Res* 22:1735–1747.

Natarajan A, Yardimci GG, Sheffield NC et al. (2012) Predicting cell-type-specific gene expression from regions of open chromatin. *Genome Res* 22:1711–1722.

Neph S, Stergachis AB, Reynolds A et al. (2012) Circuitry and dynamics of human transcription factor regulatory networks. *Cell* 150:1274–1286.

Neph S, Vierstra J, Stergachis AB et al. (2012) An expansive human regulatory lexicon encoded in transcription factor footprints. *Nature* 489:83–90.

Pei B, Sisu C, Frankish A et al. (2012) The GENCODE pseudogene resource. *Genome Biol* 13:R51.

Sanyal A, Lajoie BR, Jain G & Dekker J (2012) The long-range interaction landscape of gene promoters. *Nature* 489:109–113.

Schaub MA, Boyle AP, Kundaje A et al. (2012) Linking disease associations with regulatory information in the human genome. *Genome Res* 22:1748–1759.

Thurman RE, Rynes E, Humbert R et al. (2012) The accessible chromatin landscape of the human genome. *Nature* 489:75–82.

Vernot B, Stergachis AB, Maurano MT et al. (2012) Personal and population genomics of human regulatory variation. *Genome Res* 22:1689–1697.

Wang H, Maurano MT, Qu H et al. (2012) Widespread plasticity in CTCF occupancy linked to DNA methylation. *Genome Res* 22:1680–1688.

Wang J, Zhuang J, Iyer S et al. (2012) Sequence features and chromatin structure around the genomic regions bound by 119 human transcription factors. *Genome Res* 22:1798–1812.

Whitfield TW, Wang J, Collins PJ et al. (2012) Functional analysis of transcription factor binding sites in human promoters. *Genome Biol* 13:R50.

Yip KY, Cheng C, Bhardwaj N et al. (2012) Classification of human genomic regions based on experimentally determined binding sites of more than 100 transcription-related factors. *Genome Biol* 13:R48.

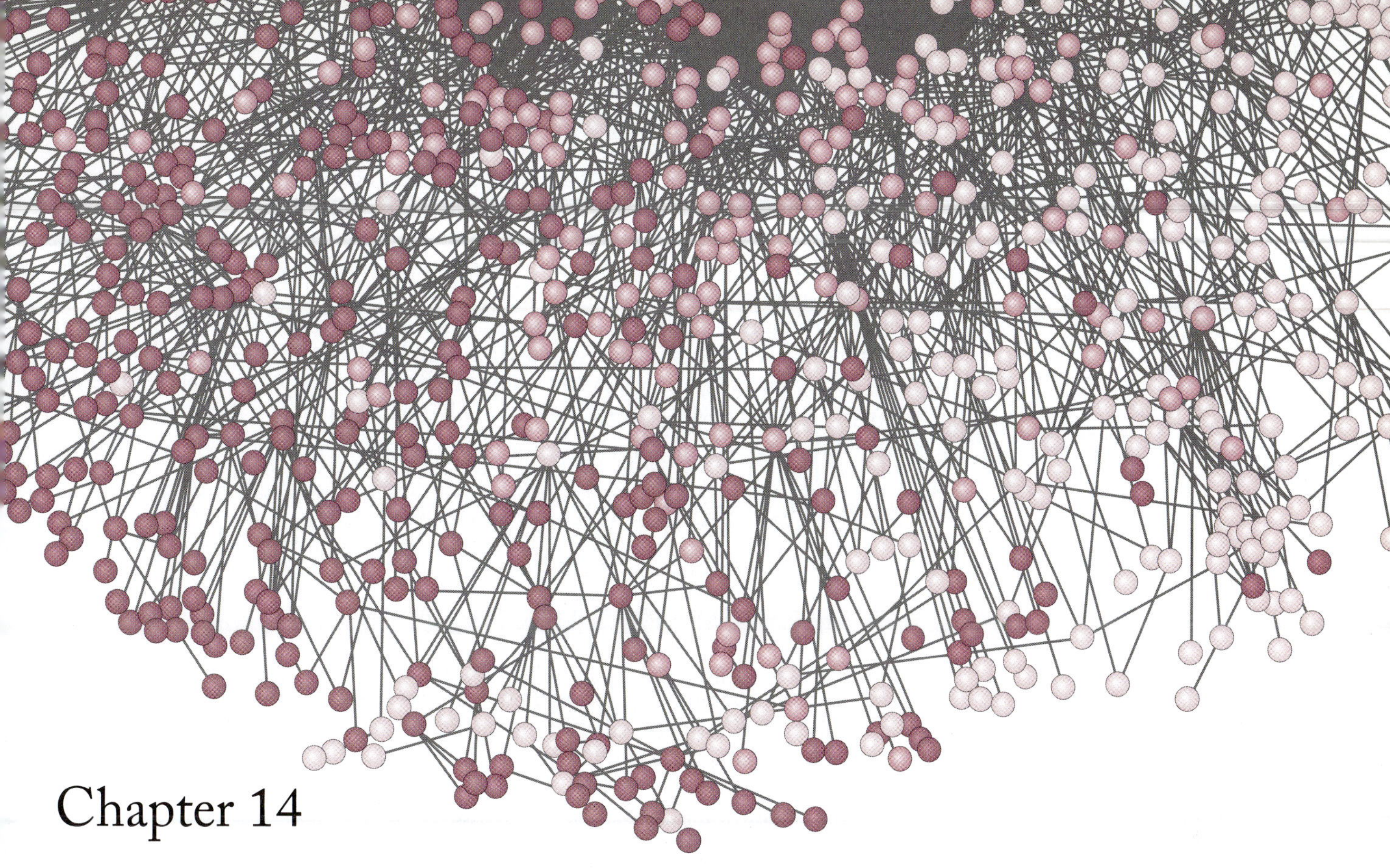

Chapter 14

RNA Processing

14.1 Introduction

Most RNA molecules undergo post-transcriptional processing

In earlier chapters, we have seen that large portions of the genome can be accurately copied into RNA. In most cases, however, these primary transcripts are still not appropriate to the cell's needs and must be modified. The transformation of primary transcripts into mature functional RNA molecules is collectively called **RNA processing**. Additional processes occur to degrade RNA molecules that have reached the end of their physiologically relevant lifetimes or to degrade RNA molecules that are defective or misfolded. Virtually every cellular RNA undergoes one or, more often, several forms of processing.

There are four general categories of processing

The first category involves removal of nucleotides from the primary transcript; this may involve substantial portions of the transcript, as, for example, in the removal of introns from eukaryotic messenger RNA, mRNA. The second category adds nucleotides to the 3′- and 5′-termini of RNA chains in a template-independent manner. Third, RNAs can be edited by removal or insertion of nucleotides within the sequence originally prescribed by the DNA. Finally, bases can be modified covalently by a number of different enzyme-catalyzed reactions.

Eukaryotic RNAs exhibit much more processing than bacterial RNAs

This chapter covers the best-characterized processing reactions. The main focus will be on eukaryotic mRNA because bacterial RNAs exhibit few post-transcriptional modifications. The modifications that do occur in bacteria mainly involve cutting up long tandem transcripts to produce functional transfer RNA, tRNA, and ribosomal RNA, rRNA, molecules.

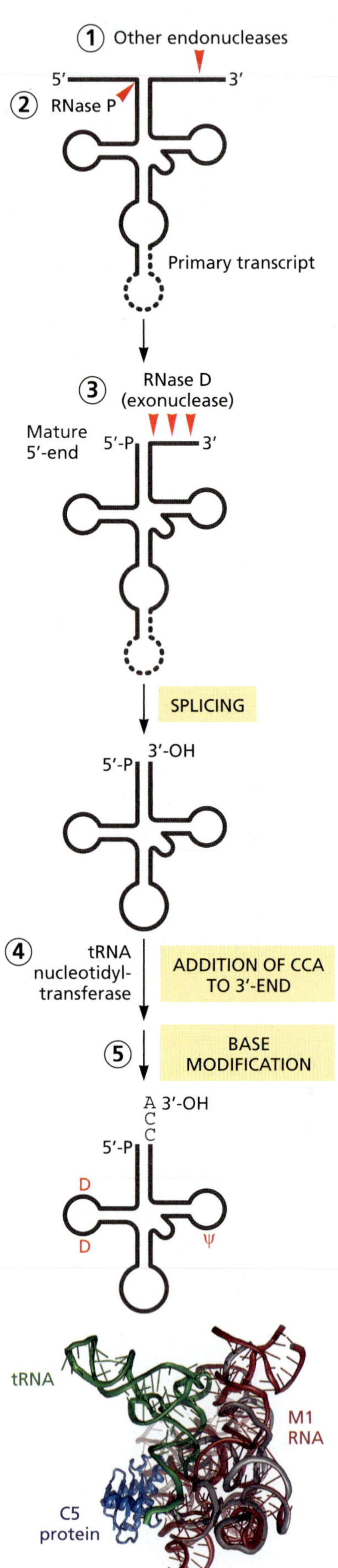

Bacterial protein-coding mRNAs are produced in the cytoplasm and can immediately attach to ribosomes and begin to be translated, without undergoing any processing. Eukaryotic mRNA, on the other hand, is produced in the nucleus and must be transported to the cytoplasm for translation. Furthermore, the directly transcribed message in eukaryotes almost invariably contains introns, which must be spliced out to produce the mature, translatable forms of mRNA. Two other modifications are common: capping of the 5′-end and addition of a 3′-polyadenylate or poly(A) tail. In describing eukaryotic mRNA processing, we first consider the modifications at the 5′- and 3′-ends, and then we will describe the complex reactions that carry out and regulate splicing.

14.2 Processing of tRNAs and rRNAs

tRNA processing is similar in all organisms

The processing of tRNAs always involves nuclease cleavage of pre-tRNAs from primary transcripts. In bacteria such as *Escherichia coli*, some tRNA genes are embedded in the gene cluster that encodes rRNAs, but the majority of tRNA genes are clustered together in groups of one to seven, surrounded by lengthy flanking sequences. The processing of the primary transcripts originating from these tRNA gene clusters is a multistep process that involves several endo- and exonucleases (**Figure 14.1**). After the polynucleotide chain cleavage reactions that trim the chain to its final length, **tRNA nucleotidyltransferase** comes into play to add the universal triplet CCA to the 3′-end of the trimmed chain. The 2′- or 3′-OH groups of the terminal adenosine, A, serve as attachment sites for the amino acid during protein synthesis (see Chapters 15 and 16). The final step in tRNA processing involves chemical reactions that modify some of the bases; these modifications include methylation, thiolation, reduction of uracil to dihydrouracil, and pseudouridylation.

The enzyme that creates the mature 5′-end of all tRNAs, **RNase P**, was the first ribozyme ever to be described (**Box 14.1**). Sidney Altman made the discovery and was honored with a portion of the 1989 Nobel Prize in Chemistry for his work. RNase P is a most unusual enzyme: it consists of a 377-nucleotide-long, highly structured RNA molecule and a small protein (see Figure 14.1). The big surprise came when it was proven, albeit under nonphysiological conditions, that the protein is not needed for the reaction: catalysis is carried out by the RNA component.

All three mature ribosomal RNA molecules are cleaved from a single long precursor RNA

The three rRNA molecules 23S, 16S, and 5S, which together with numerous ribosomal proteins form the structure of the bacterial ribosome, are initially transcribed as a very long pre-rRNA molecule (**Figure 14.2**). The ribosomal structure contains only one

Figure 14.1 Prokaryotic tRNA processing from primary transcripts that contain several tandem tRNA precursors. The steps are catalyzed as follows: (Step 1) Endonucleases cleave the pre-tRNAs from primary transcripts. (Step 2) RNase P cleaves the transcript on the 5′-end of each tRNA sequence, releasing monomeric tRNA precursors with mature 5′-ends. (Step 3) RNase D, an exonuclease, reduces the length of the 3′-end. At the same time a loop, shown as a dashed line, is removed from the tRNA precursor and the cut ends are respliced. (Step 4) tRNA nucleotidyltransferase adds the universal CCA triplet to the 3′-end of the tRNA; this is an example of the addition of nucleotides without template DNA. (Step 5) The final step in the process, base modification, involves various specific enzymes that catalyze specific modification reactions. (Bottom) Crystal structure of *E. coli* RNase P bound to tRNA. The structure of the RNA component of the enzyme, M1 RNA, reveals a number of coaxially stacked helical domains; this structure is highly conserved in archaea, bacteria, and eukaryotes. The tRNA substrate is shown in green, and the protein component of the enzyme, C5, is shown in blue. *In vivo,* both RNA and protein components are required for activity. Under certain *in vitro* conditions, such as high Mg^{2+} concentrations, the protein is not needed for catalysis and the RNA acts as a true ribozyme. (Crystal structure from Wikimedia.)

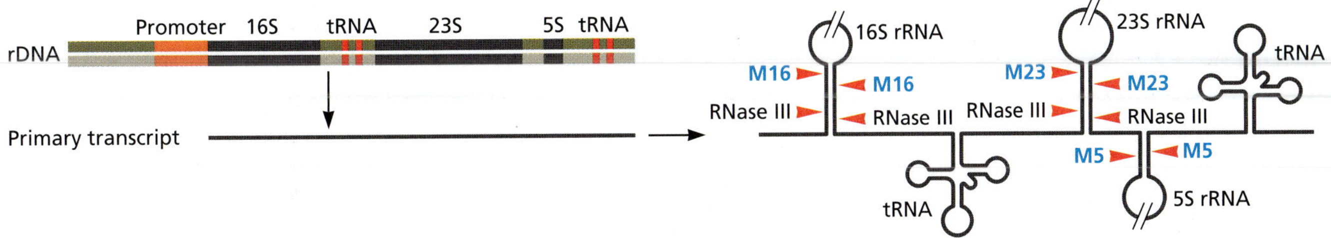

Figure 14.2 Processing of ribosomal RNA precursors in *E. coli*. The primary transcript contains a copy of each of the three ribosomal RNAs and may also contain several interspersed tRNA precursors. The 5′- and 3′-ends of mature rRNA are found in base-paired regions; initially, these base-paired regions are processed by RNase III, followed by specific endonucleases: M16 endonuclease for 16S rRNA, endonuclease M23 for 23S rRNA, and M5 endonuclease for 5S rRNA. Processing is coupled to ribosome assembly.

copy of each of the three molecules, so having them in a single transcript ensures that they are generated in the right stoichiometry. The long primary transcript is processed to give rise to the individual rRNA molecules. The first enzyme involved is RNase III, which introduces double-strand breaks at the bases of the stem-loop structures that contain the 16S, 23S, and 5S sequences. Further processing steps involve endonucleases that are specific for each of the three rRNA sequences and take place only after the first steps of ribosomal assembly, which occur co-transcriptionally (see Chapter 15). Processing of eukaryotic rRNAs follows a very similar path from a long precursor to yield the 28S, 18S, and 5.8S RNA molecules present in the eukaryotic ribosome. It relies extensively on the action of the exosome complex.

Box 14.1 Ribozymes Much RNA processing *in vivo* is catalyzed by RNA molecules, called ribozymes, that have many of the properties of protein enzymes. The existence of such molecules was suggested as early as 1968, in independent papers by Francis Crick and Leslie Orgel. Both were concerned with a conundrum about the origin of life: which came first, the proteins that catalyze nucleic acid transactions or the nucleic acids that code for those proteins? The suggestion that RNA molecules, which can have three-dimensional structures as complex as those of proteins, might carry out both catalysis and information transfer seemed a way out of the dilemma. The idea was not considered seriously for many years, principally because nucleic acids lack the varied side chains generally involved in enzyme function. Furthermore, there was no experimental evidence to support the hypothesis.

A major breakthrough came in 1982, when Thomas Cech and collaborators discovered that the intron in *Tetrahymena* ribosomal RNA could both excise itself and resplice the rRNA remnants without the aid of any protein co-factor. The intron was, then, a ribozyme. This breakthrough, although at first greeted with some skepticism, soon led to the discovery of a variety of ribozymes and to the RNA world model for early evolution.

Many other ribozymes are also involved in self-cleaving reactions; thus they might not be considered true analogs of enzymes, which remain intact after the catalytic event, ready to repeat the process. There are, however, RNA molecules that meet this more stringent criterion. The ribozyme shown in Figure 1 is a truncated version of the *Tetrahymena* rRNA intron, which will very effectively catalyze, in an acid-base reaction, the transfer reaction shown. A model for the transfer process itself is shown in Figure 2. The presence of metal ions, especially Mg^{2+}, appears to stabilize the pentacoordinated phosphorus transition state.

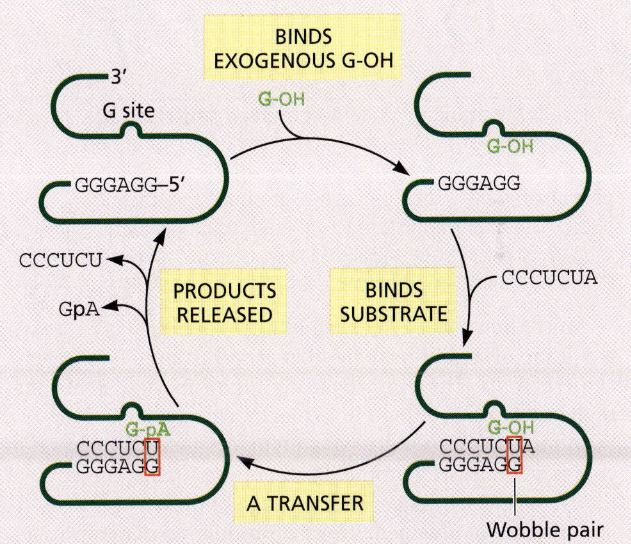

Figure 1 Enzymatic activity of the free ribozyme corresponding to *Tetrahymena* group I rRNA intron. The ribozyme's G-binding site or G site can noncovalently bind a free G-OH as co-factor. At the 5′-end of the ribozyme, the GGAGGG motif can bind the oligonucleotide CCCUCUA through Watson–Crick base pairing, a coupling that includes a wobble G-U pair. The overhanging A is transferred to the G-OH, forming GpA, which is released along with the truncated substrate. The overall reaction, in the direction shown, is CCCUCUA + G → CCCUCU + GA. (Adapted from Hougland JL, Piccirilli JA, Forconi M et al. [2006]. In The RNA World, 3rd ed. [Gesteland R, Cech TR & Atkins JA eds], pp 133–199. With permission from Cold Spring Harbor Laboratory Press.)

(Continued)

Box 14.1 *(Continued)*

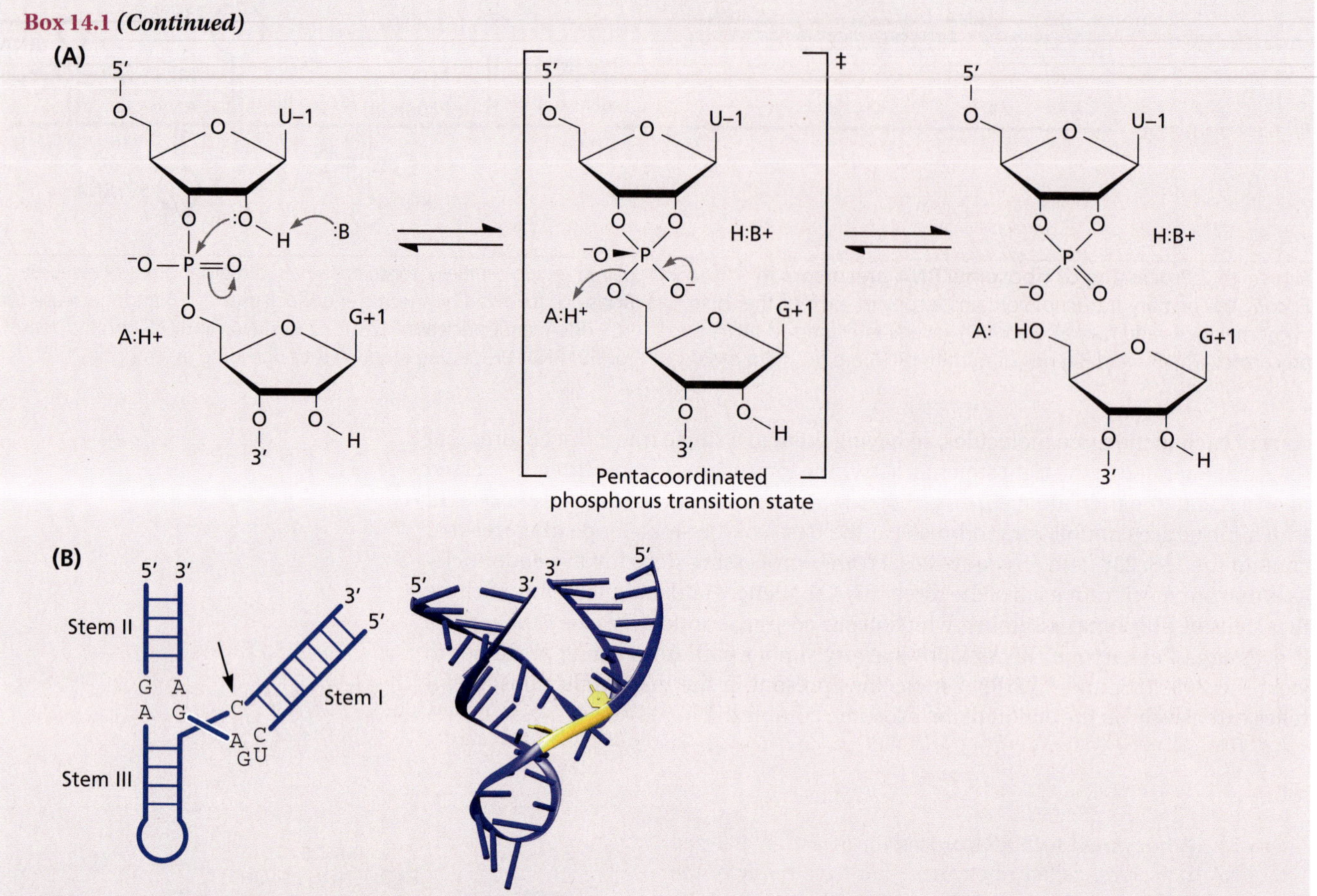

Figure 2 Mechanism of RNA-catalyzed self-cleavage. (A) Chemistry of the reaction. The 2′-hydroxyl of the ribose carries out nucleophilic attack on the adjacent scissile or cleavable phosphate. The reaction proceeds through a pentacoordinated phosphorus transition state to generate products with 2′,3′-cyclic phosphate and 5′-hydroxyl termini. (B) Structure of the hammerhead ribozyme. (Left) Secondary structure: the nucleotides important for catalytic activity are indicated, and the cleavage site is indicated by an arrow. (Right) Tertiary structure based on crystallographic data. The nucleotides flanking the scissile bond are shown in gold. (Adapted from Doudna JA & Cech TR [2002] *Nature* 418:222–228. With permission from Macmillan Publishers, Ltd.)

Most known ribozymes fall into two general classes. First is a group of small ribozymes that carry out simple self-cleavage reactions. These are typically found in viral RNAs that replicate by the rolling-circle mechanism (see Chapter 19). Such replication produces a tandem, covalently linked set of copies of the RNA genome. Each copy contains one copy of the ribozyme sequence. Self-cleavage of these copies results in a number of RNA genes for the production of new viruses. A well-studied example is the hammerhead ribozyme found in some plant viruses (see Figure 2B). The X-ray structure shows a Y-shaped molecule, with both the cleavage site and the residues essential for cleavage near the junction of the Y. The cleavage reaction yields a 3′-OH group and a 5′-cyclic phosphate.

These products are fundamentally different from the products of the second major class of ribozymes, those involved in the splicing out of introns, principally group I and group II introns. For these ribozymes, the cleavage step must produce a 3′-OH on the 5′-exon and a 5′-phosphate on the 3′-exon to allow resplicing of the exons. An example is the *Tetrahymena* rRNA intron described above.

Sidney Altman and Thomas R. Cech were awarded the 1989 Nobel Prize in Chemistry "for their discovery of catalytic properties of RNA."

14.3 Processing of eukaryotic mRNA: End modifications

Processing of the eukaryotic precursors to mRNA usually involves three distinct steps: 5′-end capping; splicing, or removal of introns and splicing together of exons; and polyadenylation, or addition of poly(A) tails to the 3′-end of the transcript (**Figure 14.3**). We first discuss the modifications that occur at the 5′- and 3′-ends before considering the complexities of splicing. Not all mRNAs experience all types of processing; thus, for example, the histone genes do not contain introns, so no splicing occurs to produce the mature mRNA. Most histone mRNAs also do

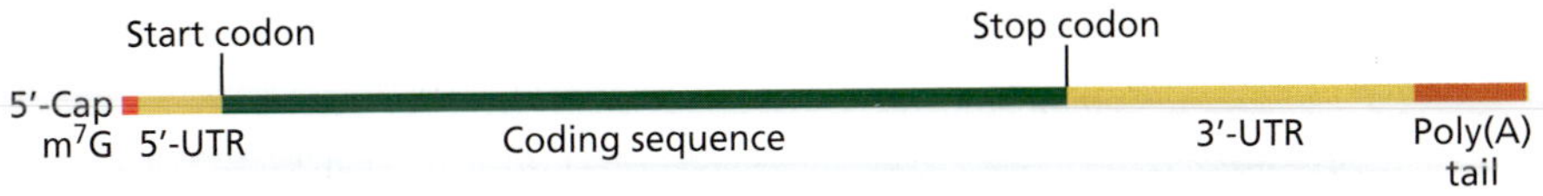

Figure 14.3 Structure of a typical eukaryotic mRNA following processing. The coding region and two flanking untranslated regions or UTRs are included. The coding region contains both exons and introns in the primary transcripts, but only exons are present in the mature mRNA, which is ready to be translated on the ribosome. Note the cap structure at the 5′-end and the poly(A) tail at the 3′-end of the molecule. Not all mRNAs are polyadenylated; metazoan histone mRNAs are a notable example of mRNAs that lack a poly(A) tail.

not undergo polyadenylation; the creation of their mature 3′-end involves other specialized pathways.

Eukaryotic mRNA capping is co-transcriptional

Capping is the addition of a methylated guanosine monophosphate cap at the 5′-end of the mRNA precursor via an unusual 5′–5′ triphosphate linkage (**Figure 14.4**). Capping occurs in the nucleus, very early during transcription, when the length of the nascent RNA chain is less than 30 nucleotides. The cap performs multiple roles: it protects mRNA from exonucleolytic degradation from the 5′-end, it creates the appropriate substrate for splicing, and it serves as the site for initiation factor attachment during translation initiation. The enzymatic activities that perform the capping reaction are attached to the C-terminal domain of Pol II, as are many of the other proteins that couple transcription to processing. Once the cap structure is in place, a dimeric Cap-binding

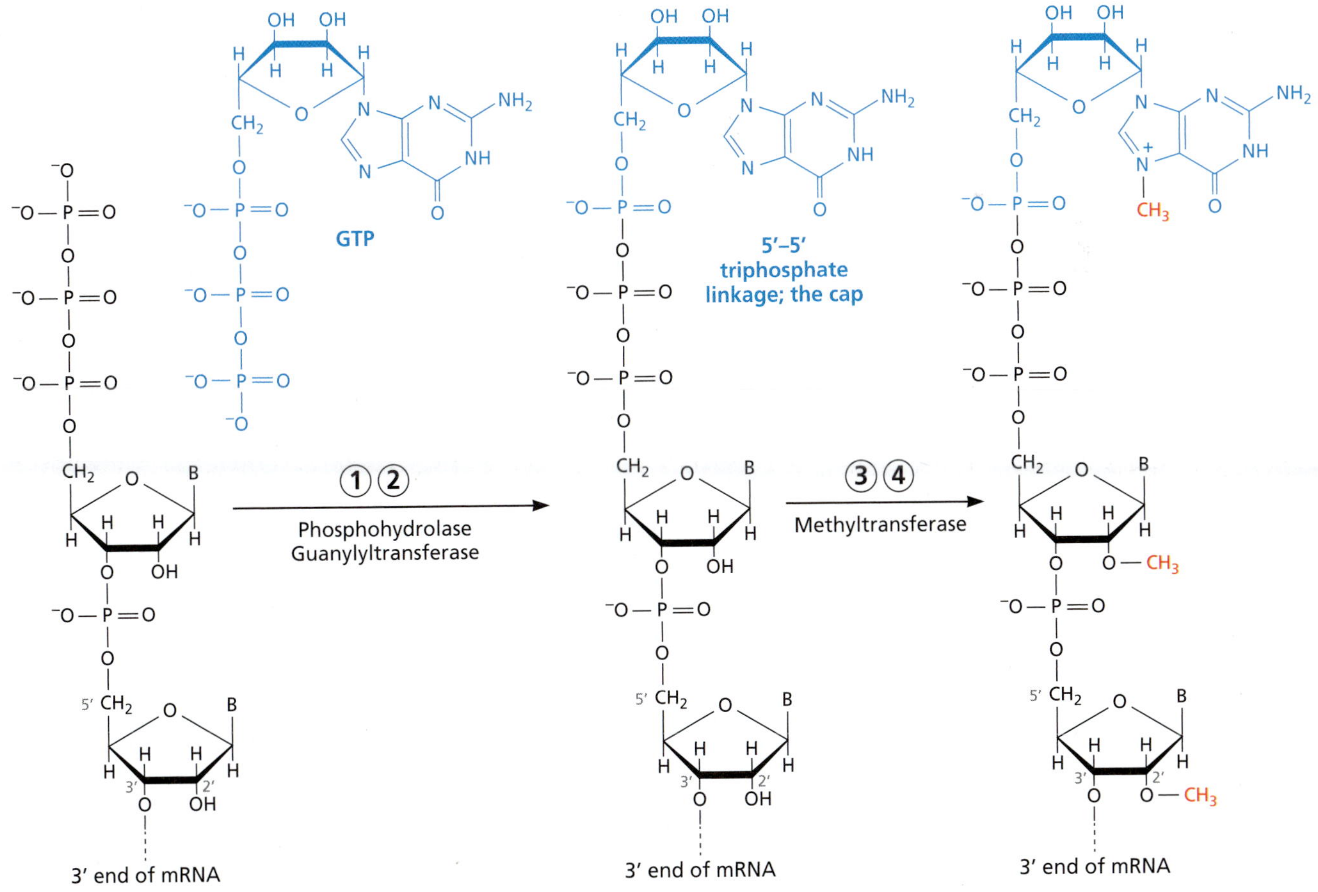

Figure 14.4 Formation of a cap at the 5′-end of a eukaryotic mRNA precursor. The reaction proceeds in several steps: (Step 1) A phosphohydrolase catalyzes removal of the phosphate group at the 5′-end of the precursor. (Step 2) The 5′-end then receives a GMP group from GTP with release of pyrophosphate, in a reaction catalyzed by guanylyltransferase. (Step 3) The base of the guanylate group is methylated at N-7. (Step 4) The 2′-hydroxyl groups of the terminal and penultimate ribose groups of the precursor may also be methylated.

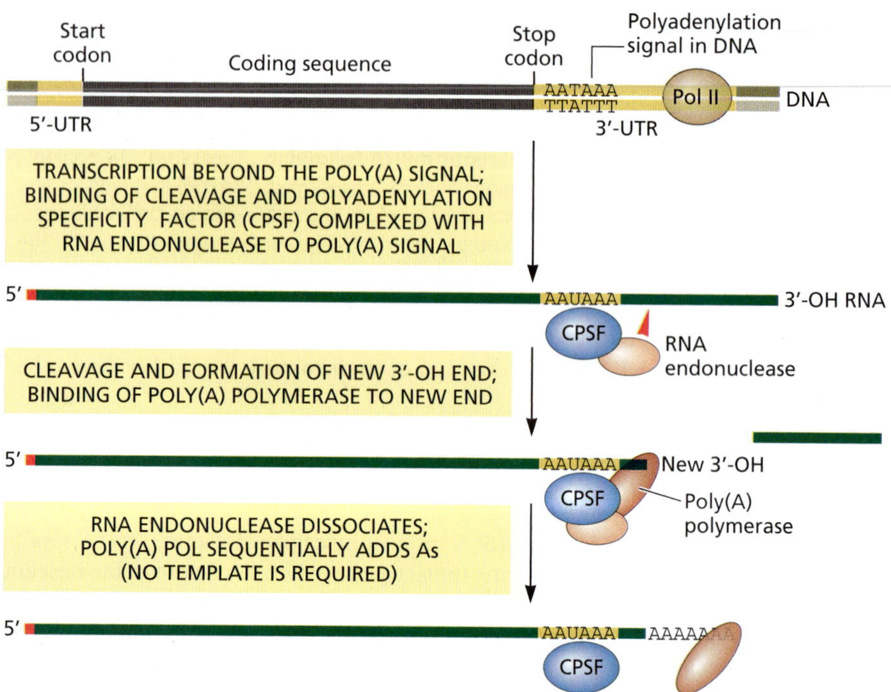

Figure 14.5 Polyadenylation of mRNA precursors.

complex, CBC, attaches to it, and stays bound until the mRNA is exported to the cytoplasm. CBC serves to recruit the transcription-export complex TREX to the mRNA early in the process.

Polyadenylation at the 3′-end serves a number of functions

Addition of a string of consecutive adenylate or A residues to the 3′-end of mRNA, known as polyadenylation, is an important step in eukaryotic mRNA processing. **Figure 14.5** depicts the process and the major proteins involved. The discovery that the actual site of poly(A) addition is situated 5′ to the transcription termination site was unexpected and led to the discovery of a poly(A) signal in the DNA that is located in what will be transcribed into the 3′-untranslated region or UTR of the mRNA. The actual poly(A) signal in the primary transcript, AAUAAA, acts in conjunction with a poly(U) sequence downstream to form the active poly(A) signal. The poly(A) signal in the message interacts with a protein factor dubbed CPSF, cleavage and polyadenylation specificity factor, which is complexed with an RNA endonuclease. This endonuclease cleaves the RNA to form the 3′-OH group needed for the polymerization activity of the special poly(A) polymerase PAP, which adds the A residues in a template-independent process.

Poly(A) tails are highly heterogeneous in length, ranging from a few adenylate residues to more than 200–300. The length varies with biological species, developmental stage, and type of RNA. An important role of the poly(A) tail is believed to be stabilization of the message against exonucleolytic degradation: mRNAs that need to be around for a long time contain long poly(A) tails. A classic example is the globin mRNA in mammalian red blood cells. These cells lose their nuclei when they enter the peripheral blood circulation and therefore have to depend for their function entirely on long-lived mRNA in the cytoplasm. Thus, their globin mRNAs live as long as the cells themselves, around 3 months. This is certainly a record in longevity; bear in mind that mRNAs generally have short half-lives, measured in minutes. Such a short half-life would seem to allow more flexible regulation of gene expression: genes that need to be expressed only for brief periods of time produce unstable messages and vice versa.

The poly(A) tail shortens progressively during the lifetime of the mRNA; the shortening is executed by exonucleases and starts even before the mRNA leaves the nucleus. Where such a shortening of an mRNA is undesirable, there is a cytoplasmic mechanism

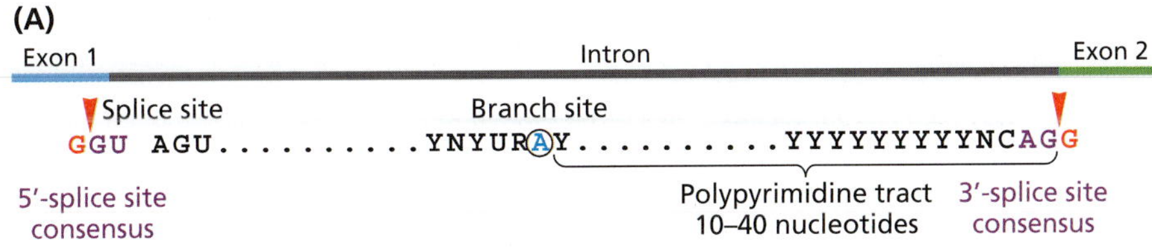

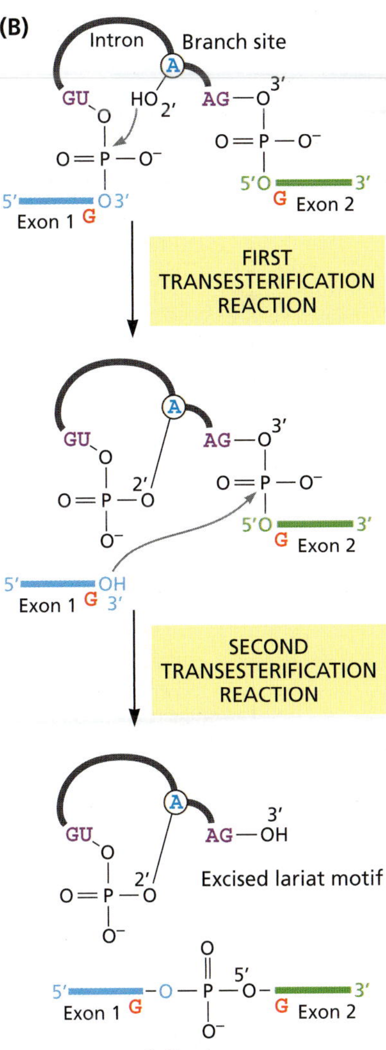

Figure 14.6 Splice sites and chemistry of the two-step splicing reaction. (A) Typical nucleotide sequences at the 5′- and 3′-splice sites and in the intron regions that are essential for the splicing reaction to occur. The intron is flanked by G residues in the exons; the intron itself contains conserved dinucleotides, GU at its 5′-end and AG at its 3′-end. An A residue situated in the intron serves as a branch site, see part B; it is usually close to the 3′-splice site, separated from it by a 10–40-nucleotide-long polypyrimidine tract. (B) The splicing reaction involves two consecutive transesterification steps. In the first step, the 2′-OH group of the A residue in the branch site attacks the 5′-splice site as a result, both 2′- and 3′-OH groups of the branch-site A are engaged in phosphodiester bonds, thus creating a lariat structure that includes the intron and the downstream exon 2. During the second reaction, the newly created 3′-OH at the terminal G of exon 1 attacks the 3′-splice site; as a result, the two exons are linked together and the intron is released as a shortened lariat, which is then destroyed.

that restores the length of the poly(A) tail. This is achieved through the action of a cytoplasmic polyadenylation element that is present in the 3′-UTR of some mRNAs. A full discussion of cytoplasmic modification is presented when we consider regulation of translation (see Chapter 17).

How does the poly(A) tail stabilize the message? In the nucleus, stabilization is achieved through the binding of a poly(A)-binding or PAB protein to the tail that protects it from exonucleolytic degradation. PAB also helps the polymerase to synthesize long tails. Once a short poly(A) tail has been synthesized, PAB binds to it and forms a quaternary complex with CPSF, PAP, and the substrate RNA. This complex transiently stabilizes PAP binding, thus supporting rapid processive catalysis. Another class of PABs is cytoplasmic and in some cases required for poly(A) removal.

Additional functions for the poly(A) tail have been recognized. In bacteria and during eukaryotic nuclear surveillance, the addition of a poly(A) tail initiates mRNA decay. The tail binds to a protein, Nab2, which positions the 3′-end of the export-competent mRNA in close proximity to the nuclear pore channel. Thus the tail plays a crucial role in mRNA export. In addition, it enhances translation initiation through an incompletely understood mechanism.

14.4 Processing of eukaryotic mRNA: Splicing

The splicing process is complex and requires great precision

Most eukaryotic genes contain noncoding sequences, known as introns, that need to be removed from the primary transcript. The flanking exons must then be spliced together. This must be done with great precision so as not to cause frameshifts. To meet these strict requirements, a complex and highly regulated nuclear machine, named the **spliceosome**, has evolved. For the spliceosome to work properly, it has to recognize precisely the boundaries between exons and introns. To facilitate this recognition, conserved sequences have evolved that mark the boundaries on both the exon and the intron sides (Figure 14.6). Consensus sequences have been derived for the boundary sequences. However, these sequences are very short and do not bind the spliceosome very tightly, which can allow alternative splicing to occur. Spliceosome binding is enhanced by the existence of the **branch site**, which consists of an A residue embedded in a somewhat conserved internal intron sequence. The branch site is located relatively close to the 3′-splice site, separated from it by a polypyrimidine tract of 10–40 nucleotides that serves as a binding site for some of the spliceosome proteins. The splicing reaction occurs through two consecutive transesterification reactions, as detailed in Figure 14.6 and Figure 14.7.

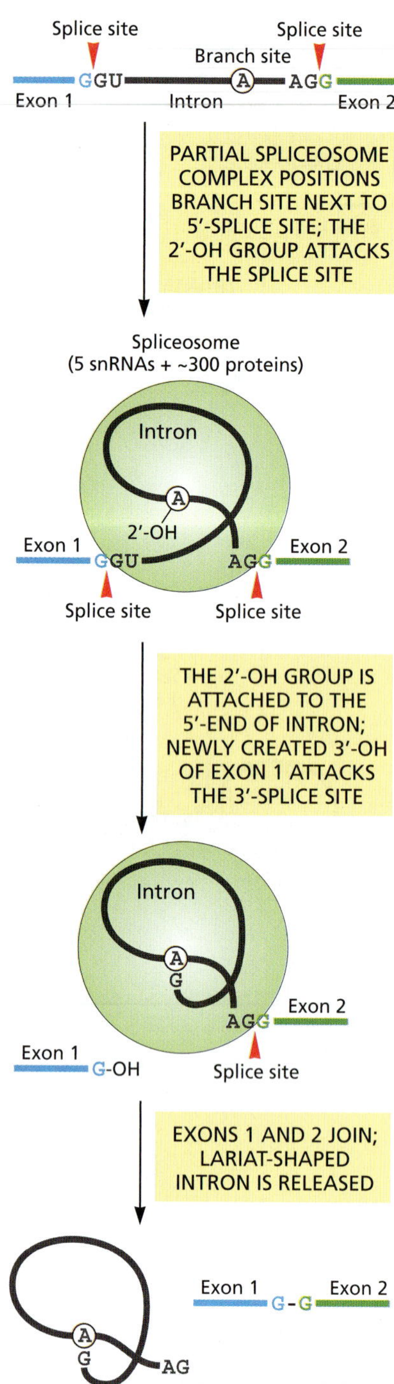

Figure 14.7 Overview of the splicing reaction. Intron removal in mRNA precursors is performed by a specialized ribonucleoprotein or RNP machine, termed the spliceosome. The schematic also indicates the main chemical reactions, depicted in Figure 14.6B, catalyzed by the spliceosome. (Adapted from Sharp PA [1987] *Science* 235:766–771. With permission from American Association for the Advancement of Science.)

Splicing is carried out by spliceosomes

The spliceosome is a huge macromolecular machine that contains five small nuclear RNAs or snRNAs, known as U1, U2, U4, U5, and U6, complexed with 41 different proteins that constitute ~45% of the mass of the particle. The U snRNAs have been named for uracil, the base prominent in all five members of the family. There is extensive intrachain base-pairing in all U snRNAs (**Figure 14.8A**), and many of the bases are post-transcriptionally modified. U snRNAs are very abundant, present at around 100,000 copies per nucleus. This number is not surprising considering the level of splicing that almost every precursor mRNA undergoes and the large average number of introns in each pre-mRNA. A curious, and still unexplained, fact is that four of the five U snRNAs are transcribed by Pol II; the exception is U6, which is synthesized by Pol III. Overall, more than 300 proteins have been recognized that bind to the spliceosome; some of those proteins act as splicing factors, whereas the function of most of these proteins is yet to be identified.

Most studies on splicing have been performed in *in vitro* systems, usually using simple synthetic mRNA containing a single intron and two flanking exons as substrate. These investigations led to the view that the spliceosome complex is assembled, during the process of transcription, in a stepwise manner (**Figure 14.9**). The binding of U2 small nuclear ribonucleoprotein, snRNP, to the branch site is assisted by the protein dimer U2AF, which recognizes and binds to the conserved polypyrimidine tract.

Extensive *in vitro* studies also helped to identify the roles of the individual spliceosome components and led to acquisition of electron microscopy (EM) images of individual A, B, and C complexes (**Figure 14.8B**). Nevertheless, experiments in which native spliceosome particles were isolated in a form bound to nascent mRNA during processing produced micrographs that present a somewhat different picture of the structure. **Supraspliceosome** particles that contain four closely packed monomeric spliceosomes have been recognized by EM (**Figure 14.8C**). Further cryo-EM imaging and reconstitutions led to a relatively high-resolution structure of these native monomeric spliceosomes. These studies allowed the creation of a model that possibly explains how four splicing events can take place on a long pre-mRNA simultaneously.

Splicing can produce alternative mRNAs

Alternative splicing is a process that uses some splice sites while neglecting others; it can also use cryptic sites that are located internally in exons or introns. Thus, alternative splicing can be viewed as the suppression of optimal splice sites and/or the use of those that are suboptimal or cryptic. The outcome of such selective splice-site usage is the creation of alternative mRNA forms that deviate in sequence and/or structure from the mature mRNA that is produced through conventional splicing. In principle, this should also lead to the generation of alternative variants of protein sequence. As we shall see, it sometimes does. The alternative forms may differ from the conventionally spliced forms in a variety of different ways.

Alternative splicing was discovered in individual gene systems some years ago, but only recently, with the advent of genomewide methods, have we come to appreciate its pervasiveness. Indeed, more than 90% of human pre-mRNAs undergo alternative splicing events, leading to the existence of families of mature mRNA molecules and, eventually, of closely related protein isoforms. Most studies have addressed the heterogeneity of the mRNA molecules that result from alternative splicing of a single mRNA precursor. Technically, this is achieved by creating cDNA libraries from the mRNA populations in a single cell and then cloning and sequencing the cDNA clones. In few cases, however, have the actual protein isoforms been identified and their structure or function determined. The characterization of protein isoforms remains an enormous challenge because of the lack of appropriate high-throughput methods.

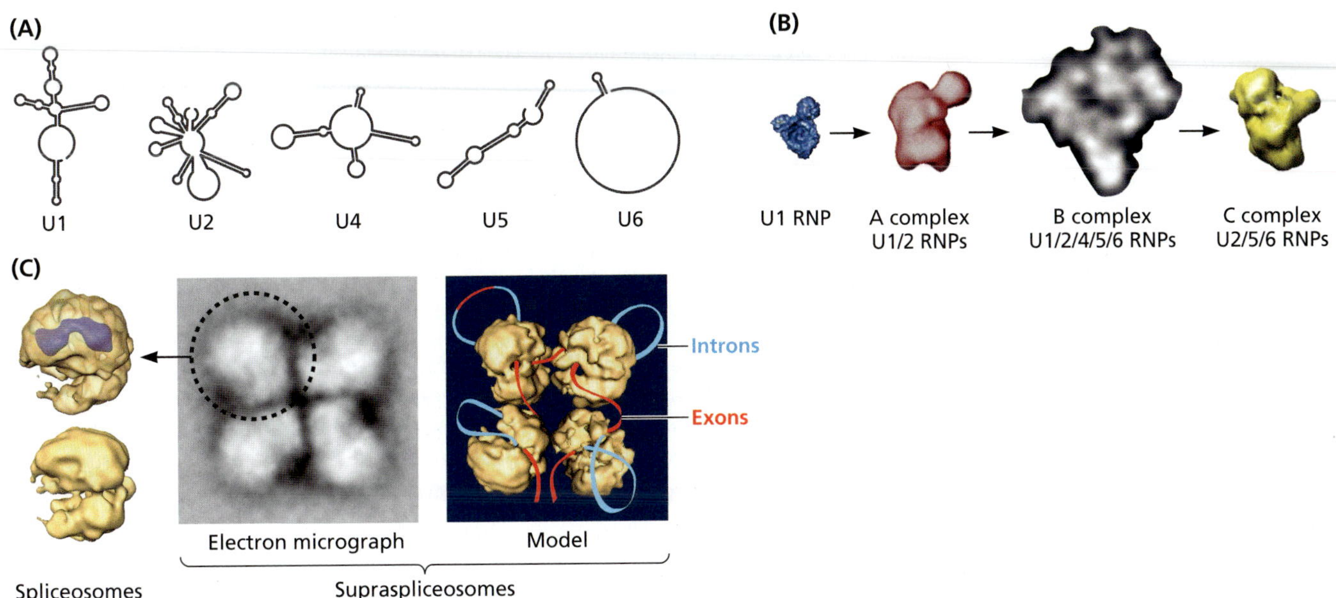

Figure 14.8 The spliceosome: A structural view. (A) Secondary structures of the five snRNAs involved in formation of the spliceosome. (B) Reconstructions based on EM images of the structures of individual RNP complexes participating in the stepwise assembly and splicing cycle, as observed *in vitro*. (C) Structure of the native spliceosome, isolated from a nuclear fraction that contained 85% of all nascent Pol II transcripts. The particles associated with the transcripts were large tetrameric structures, as visualized by EM in the middle: these were termed supraspliceosomes. The structure of the monomeric spliceosome, as reconstructed from cryo-EM images, is presented on the left. Two distinct subunits are distinguishable, interconnected with a tunnel running between them; the tunnel is large enough to allow the pre-mRNA to pass through. The purple region represents the position of the five spliceosomal snRNAs. On the right is a model of the supraspliceosome. The supraspliceosome presents a platform onto which the exons can be aligned and splice junctions can be checked before splicing occurs. When a pre-mRNA is not yet processed, it is folded and protected within the cavities of the spliceosome. When splicing occurs, the RNA is proposed to unfold and loop out. The existence of the supraspliceosome allows for simultaneous splicing of four exons in the pre-mRNA. Exons are shown in red; introns are shown in blue. An alternative exon is depicted in red in the upper-left corner of the model. (A, adapted, courtesy of Wellcome Trust Sanger Institute. B, from Jurica MS [2008] *Curr Opin Struct Biol* 18:315–320. With permission from Elsevier. C, from Sperling J, Azubel M & Sperling R [2008] *Structure* 16:1605–1615. With permission from Elsevier.)

The computational tools used to date suggest that very few of the potential isoforms are functional, since alternative splicing often leads to deleterious changes in protein structure. Thus, the accepted view that alternative splicing creates a large diversity of proteins—of altered or eventually even of different functions, known as neofunctionalization—is based on studies of a limited number of genes and may require a reevaluation. Despite this uncertainty, it is clear that alternative splicing itself is a frequent occurrence that is regulated according to the physiological needs of the cell. According to the most recent data from the ENCODE project in 2012, each human gene produces, on average, 6.3 differently spliced transcripts. The fact that alternative splicing can explain the presence of multiple forms of a protein found in higher organisms may indicate an important role for this phenomenon in evolution (Box 14.2).

Box 14.2 Alternative splicing and evolution Recent analysis of the whole genomes of a wide variety of species has produced a surprising and perplexing result: the number of protein-coding genes does not correlate well with complexity of organismal structure and behavior. As pointed out in Chapter 7, humans do not have a great many more protein-coding genes than lower invertebrates such as worms. In fact, we have far fewer genes than most scientists expected. It may well be that utilization of alternative splicing has played a major role in developing the complex structures of higher organisms. To take an example, humans utilize distinct tropomyosins in a variety of tissues, from striated muscle to brain. Yet only one gene need be carried, because of alternative splicing. In general, mice and humans seem to have a higher degree of alternative splicing than do fruit flies and nematodes. A note of caution is in place here: this assertion is based on genomewide computational methods, which are still being optimized. In particular, these estimates are based on far fewer data in flies and nematodes than in humans.

We have also frequently noted the modular structures of many proteins: transcription factors, for example, where one kind of DNA-binding domain may be coupled to a wide variety of activator or repressor domains. One way in which such diversity could be generated from a limited vocabulary of domain modules may be through *trans*-splicing. Finally, we note that although such speculations seem reasonable, they are without experimental test and likely to remain so for some time.

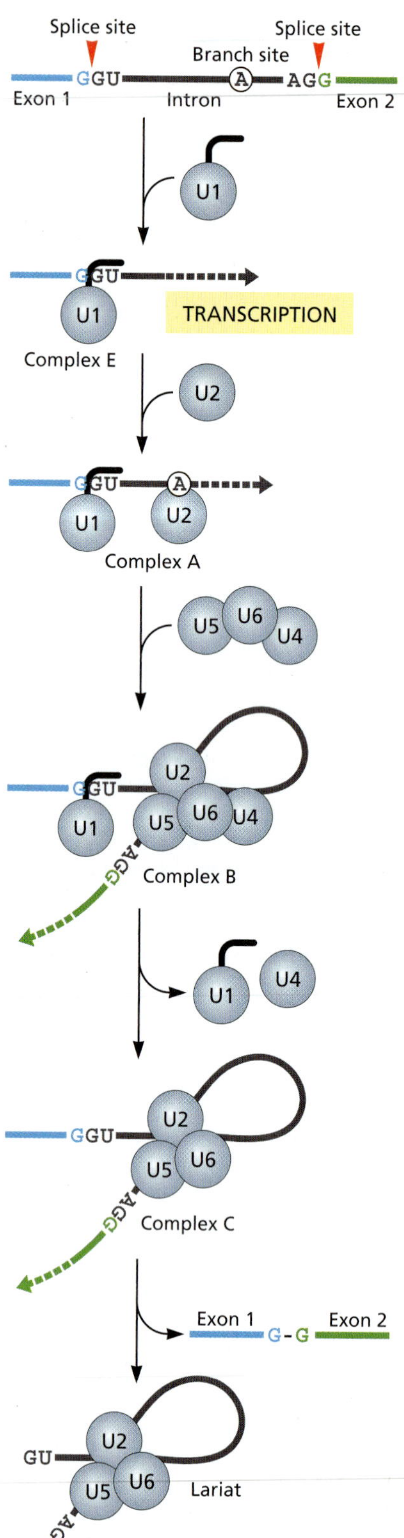

Figure 14.9 Stepwise spliceosome assembly during transcription. As soon as the 5′-splice site exits the transcription complex, U1 snRNP binds to it through base-pairing between GGU and U1 RNA; the resulting complex is known as complex E, for early. In the next step, U2 snRNP binds to the branch site in an ATP-dependent process to form complex A. The complex also contains the protein factor U2AF, not shown here, which helps to recruit the U2 RNP. When the 3′-splice site emerges from the transcription complex, U4, U5, and U6 bind as a triple RNP complex to form complex B. Following release of U1 and U4 snRNPs and conformational transitions in the U2–U5–U6 complex, the catalytically active complex C is formed. Both of the chemical reactions that are involved in splicing occur on complex C. Finally, the U2–U5–U6 complex dissociates from the lariat structure and the individual RNP complexes can be reused; the lariat structure is destroyed.

Alternative splicing can occur through a variety of mechanisms; the major types are outlined in **Figure 14.10**. In addition to these four mechanisms, alternative mRNAs can arise through the use of alternative promoters or alternative polyadenylation sites (**Figure 14.11**). Examples of specific mechanisms as they occur in specific systems are presented in **Box 14.3**.

There is a well-established connection between alternative splicing and certain types of cancer (**Table 14.1**). **Box 14.4** is dedicated to a brief description of this connection and provides some prominent examples.

Tandem chimerism links exons from separate genes

The mechanisms of alternative splicing described and illustrated thus far concern the classical case in which one pre-mRNA, the product of transcription of one gene, gives rise to alternative mRNA products. Recently, researchers have come to realize that alternative splicing may involve exons from two or more neighboring genes, or from annotated genes and 5′-exons from previously unannotated genes that lie very far away. The 5′-ends of 399 genes were mapped from the region of the human

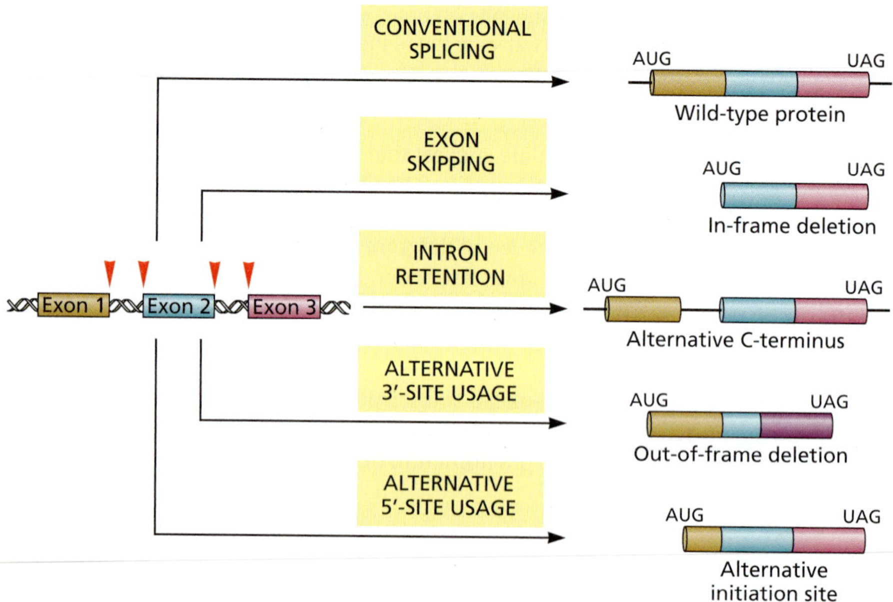

Figure 14.10 Alternative splicing can occur through a variety of mechanisms that lead to different protein products. Exon skipping, also known as the exon cassette mode, is the most common known alternative splicing mechanism in which one or more exons are excluded from the final gene transcript, leading to shortened mRNA variants. In intron retention, also known as the intron retaining mode, an intron is retained in the final transcript. Intron retention can lead, depending on the length of the intron, to a frameshift and the creation of an altered amino acid sequence downstream of the intron. In humans, ~15% of a set of 20,687 known genes have been reported to retain at least one intron. Finally, the utilization of alternative 3′- and/or 5′-splice sites can lead to the creation of a great variety of proteins. In this splicing mechanism, two or more alternative 5′-splice sites compete for joining to two or more alternative 3′-splice sites.

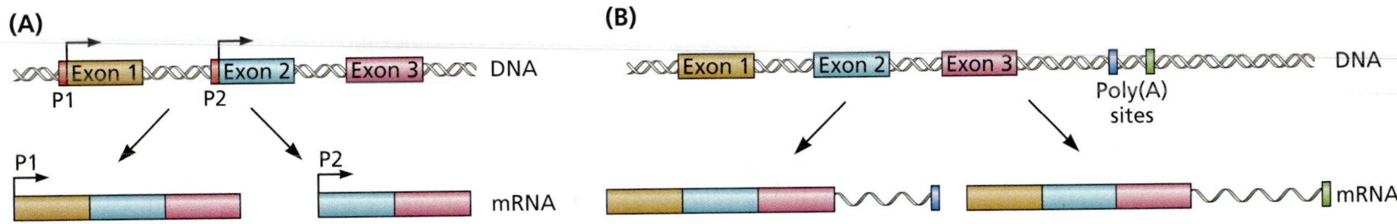

Figure 14.11 Alternative selection of promoters and cleavage–polyadenylation sites. (A) The production of isoforms of human p53 is an example of the use of three alternative promoters: the resultant protein isoforms are either full-length or truncated at their N-termini. (B) Some genes, such as that encoding tropomyosin, have two alternative polyadenylation sites, which can be used to produce mRNAs that differ in their 3′-termini.

Box 14.3 A Closer Look: Examples of mechanisms and outcomes of alternative splicing The high incidence of alternative splicing produces numerous alternative mRNA molecules and possibly some functional protein isoforms. Here we present some well-studied examples that illustrate the complexity of the processes in multicellular eukaryotes. Figure 1 depicts a case where the utilization of alternative 5′- and 3′-splice sites bordering constitutive and alternative exons creates more than 500 different mRNA molecules from a single primary transcript.

Figure 2 illustrates the diversity and tissue specificity of protein isoforms produced via alternative splicing mechanisms.

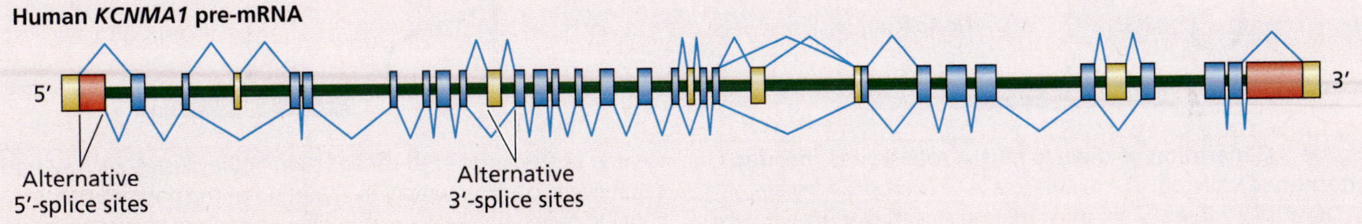

Figure 1 Alternative splicing through the use of alternative 5′- and 3′-splice sites. The human *KCNMA1* pre-mRNA provides the example shown. Blue boxes represent constitutive exons; other colors represent alternative exons. Possible splicing patterns are indicated as blue connecting lines. The alternative splicing mechanism makes use of multiple alternative 5′- and 3′-splice sites as marked. Alternative splicing creates >500 mRNA isoforms from a single pre-mRNA molecule. (Adapted from Nilsen TW & Graveley BR [2010] *Nature* 463:457–463. With permission from Macmillan Publishers, Ltd.)

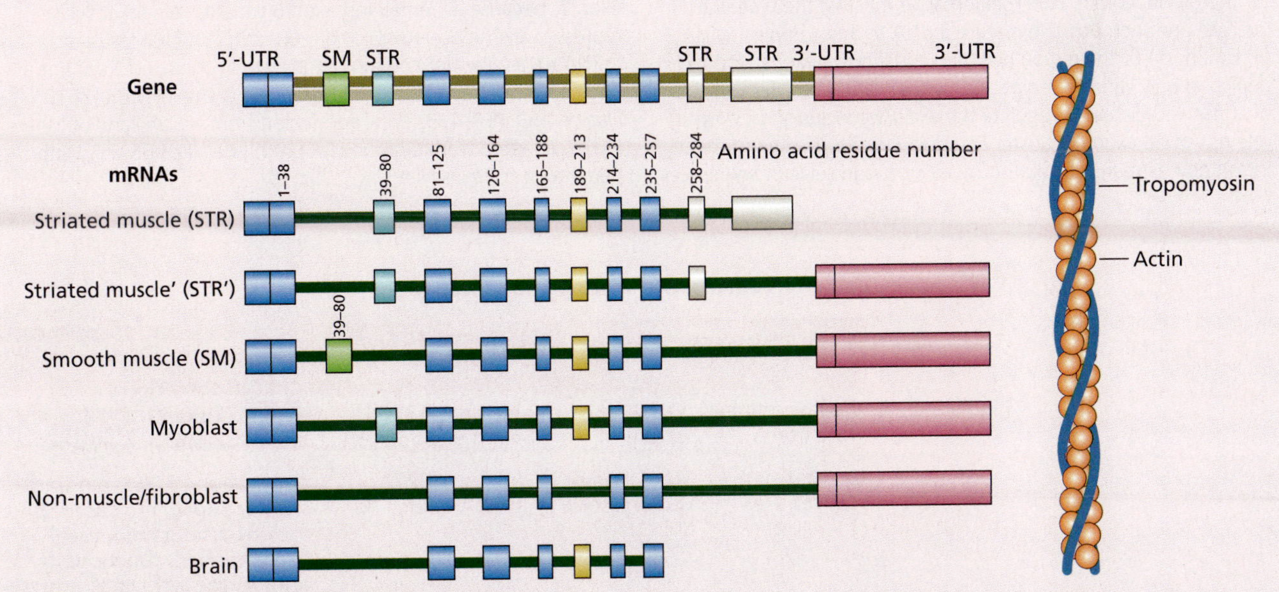

Figure 2 Alternative splicing of the tropomyosin gene. Tropomyosin is an actin-binding protein that regulates the binding of myosin to actin; this interaction is important in muscle contraction and other actomyosin functions. Distinct tropomyosin variants are expressed in different tissues, with some exons selectively present or absent in some tissue-specific isoforms. In addition to the common form of alternative splicing, exon skipping, the production of some isoforms involves the utilization of alternative polyadenylation sites (top and bottom variants). (Adapted from Breitbart RE, Andreadis A & Nadal-Ginard B [1987] *Annu Rev Biochem* 56:467–495. With permission from Annual Reviews.)

(Continued)

Box 14.3 *(Continued)* Finally, **Figure 3** describes the extreme case of producing close to 40,000 mRNA variants from a single gene. The physiological relevance of such extensive alternative splicing remains to be established.

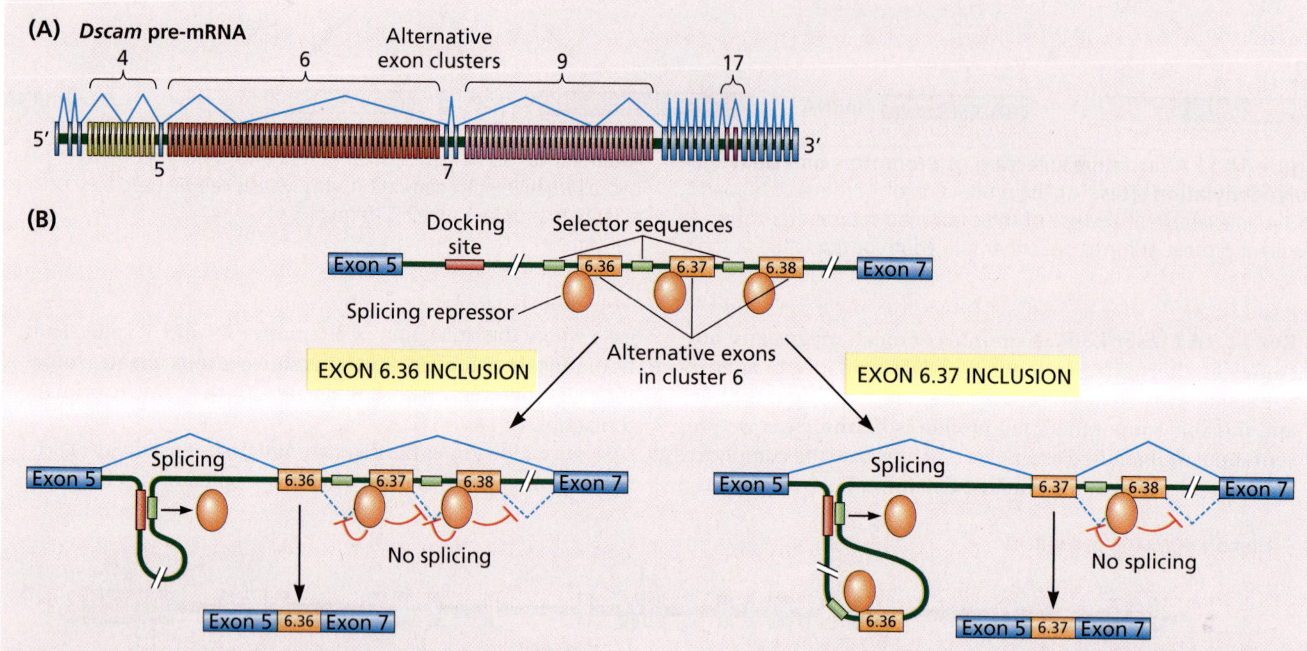

Figure 3 Generation of diverse mRNA repertoires through alternative splicing. The extreme case of *Drosophila Dscam*, or Down syndrome cell adhesion molecule, is shown. Blue boxes represent constitutive exons, while other colors show alternative exons. The possible splicing patterns are indicated as blue connecting lines. (A) *Dscam* pre-mRNA contains clusters 4, 6, 9, and 17, which have 12, 48, 33, and 2 variable exons, respectively, for a total of 95 alternative exons; note that in each cluster only one exon is utilized in the mature mRNA product. Combination of the 95 alternative exons with the 20 constitutive exons allows for the production of 38,016 alternative mRNAs; the blue lines indicate one such possible mature mRNA as an example. (B) Expansion of the region between constitutive exons 5 and 7, depicting only variable regions 6.36, 6.37, and 6.38. Two classes of conserved elements regulate the exclusive inclusion of only one out of 48 exons in cluster 6: a docking site, located in the intron downstream of constitutive exon 5, and selector sequences, located upstream of each variant exon in the cluster. Each selector sequence is complementary to a portion of the docking site, capable of juxtaposing one, and only one, alternative exon to the upstream constitutive exon 5. A key player in this model is a splicing repressor that is bound to each alternative exon, which inhibits splicing. Base pairing between the docking site and a given selector sequence inactivates, by an unknown mechanism, the splicing repressor on the downstream exon and consequently activates the splicing of that downstream exon to exon 5. The exon that is joined to exon 5 can only be spliced to constitutive exon 7, because all remaining exon 6 variants remain actively repressed by the bound splicing repressor. The final result is an mRNA molecule that contains exon 5, only one exon 6 variant, and exon 7. (A, adapted from Nilsen TW & Graveley BR [2010] *Nature* 463:457–463. With permission from Macmillan Publishers, Ltd. B, adapted from Graveley BR [2005] *Cell* 123:65–73. With permission from Elsevier.)

Table 14.1 Alternative splicing and cancer.

Disease	Gene	Mutated sequence	Consequences of mutation on alternative splicing
hepatocellular carcinoma	*CDH17*, LI cadherin	intron 6 A35G exon 6 codon 651	exon 7 skipping exon 7 skipping
prostate cancer	*KLF6*, tumor suppressor	intron 1 G27A, IVSDA allele	generation of ISE[a], binding site for SRp40; increased splice variant production; novel splice variants functionally antagonize wtKLF6's[a] growth-suppression properties
breast and ovarian cancer	*BRCA1*, tumor suppressor	exon 18 G5199T or E1694X multiple other mutations	disruption of ESE[a] leading to exon 18 skipping effects on splicing enhancers and silencers

[a]ISE, intronic splicing enhancer; ESE, exonic splicing enhancer; wt, wild type.

Box 14.4 The connection between alternative splicing and cancer Protein splice variants can display functions very different from those of the conventionally spliced molecules. It is important that the right variants are expressed at the right time and in the right place. It has been estimated that at least 15% of mutations that cause genetically inherited diseases affect pre-mRNA splicing. Traditionally, single-base changes in exons have been considered in terms of their effect on the quality of the encoded protein. We now know that some point mutations can either disrupt or create splice sites, whereas others affect functional exonic enhancers or silencers. Such mutations affect alternative splicing and change the spectrum of protein isoforms produced. Mutations that affect splice sites and/or splicing regulatory sequences have been described in numerous cancers (see Table 14.1), but only in a few cases has a cause–effect relationship been established.

The human oncosuppressor* p53 *gene encodes multiple isoforms of p53

p53 is a major oncoprotein involved in numerous biochemical pathways (see Chapter 3 and Figure 3.13). Importantly, mutations in *p53* have been found in more than 50% of human tumors. The *p53* mRNA undergoes complex alternative splicing (Figure 1). Nine different proteins are expected to be produced when the identified alternative mRNAs are translated. Normally, these are expressed in a tissue-dependent manner. Of relevance here, human breast tumors are characterized by a set of isoforms that differ from those expressed in normal breast tissue; moreover, individual tumors possess distinct isoforms. One p53 isoform lacks 36 internal amino acid residues. This isoform possesses a number of unusual properties: it recognizes and binds to only a subset of the usual p53 target genes, and moreover, it associates with target genes only during the S phase of the cell cycle, in contrast to wild-type p53, which binds to promoter DNA only in other phases of the cell cycle. The internally deleted isoform regulates the transcription of genes, such as *p21* and *14-3-3*, that are involved in the control of cell proliferation and cancer.

Mutations in AT-binding transcription factor 1 lead to prostate cancer

The tumor-suppressor gene that encodes AT-binding transcription factor 1, ATBF1, is often mutated in sporadic prostate cancer. A wide range of mutations have been identified, among them frameshift mutations that disrupt protein structure altogether; nonsense mutations that remove some zinc fingers and homeodomains, thus affecting the ability of the encoded protein to bind DNA; and mutations that disrupt normal splicing of the pre-mRNA. These latter mutations often involve deletions in the polypyrimidine tracts that are located in introns, close to

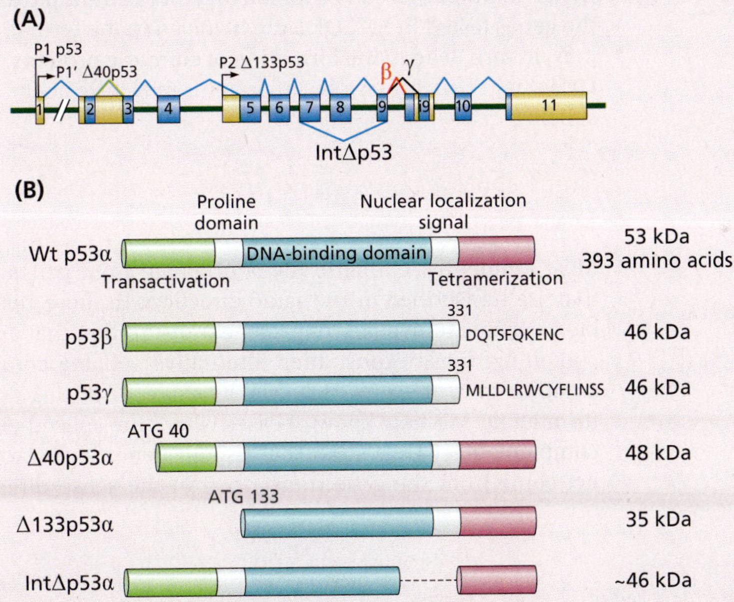

Figure 1 The human *p53* gene encodes multiple isoforms of p53. (A) Gene structure, with boxes indicating exons: noncoding sequences are in yellow, coding sequences are in blue. Total RNA extract from normal human colon was used for specific amplification of capped mRNA: the amplified RNA was subjected to reverse transcription and polymerase chain reaction or PCR, cloned, and sequenced. The results revealed the existence of three promoters, P1, P1′, and P2, with P2 localized in exon 4. When transcription is initiated at these alternative promoters, the mRNAs that are produced encode proteins lacking 40 or 133 amino acids from the N-terminus, respectively, designated Δ40p53 and Δ133p53. Three additional splice variants involve intron 9. (B) The wild-type protein, also called p53α, is 393 amino acids long; two variants lack the tetramerization domain and replace it with 10 amino acid residues, in p53β, or a distinct sequence of 15 amino acids, in p53γ. The same alternative splicing at intron 9 occurs in the N-terminal-shortened variants Δ40p53 and Δ133p53 to produce Δ40p53β, Δ40p53γ, Δ133p53β, and Δ133p53γ, not shown here. Thus nine different proteins are expected to be produced when the identified alternative mRNAs are translated. The p53 variants are expressed in normal human tissue in a tissue-dependent manner. Finally, another isoform, IntΔp53α, which is also derived by alternative splicing, lacks 66 internal amino acid residues, residues 237–322, which comprise a considerable portion of the DNA-binding domain and the linker regions between the DNA-binding and tetramerization domains. (Adapted from Prives C & Manfredi JJ [2005] *Mol Cell* 19:719–721. With permission from Elsevier.)

(Continued)

Box 14.4 *(Continued)* the splicing border (Figure 2). In some cases, ATBF1 also appears to undergo transcriptional down-regulation. At the molecular and cellular levels, ATBF1 expression is associated with a reduced rate of cell proliferation, up-regulation of the *p21* tumor-suppressor gene, and down-regulation of the AFP oncoprotein.

Whatever the molecular change may be that leads to the presence of alternatively spliced forms in tumors, in many cases overexpression of the tumor-associated isoforms is sufficient to cause malignant transformation of cells in culture. This establishes the causal relationship between these isoforms and carcinogenesis.

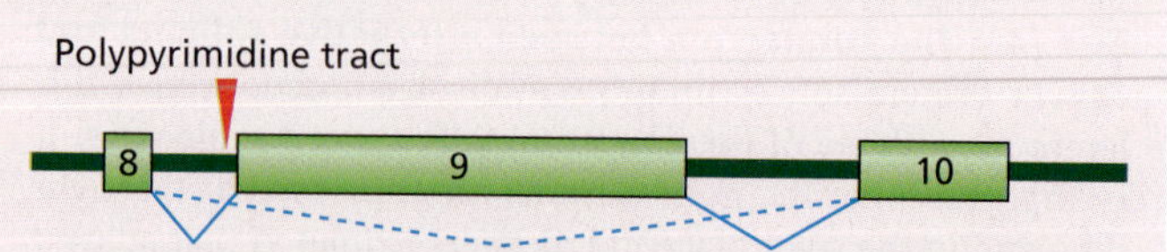

Figure 2 Aberrant splicing of *ATBF1* pre-mRNA. This kind of splicing was observed in prostate cancer cell lines that harbor deletions in the polypyrimidine tract in intron 8. Solid and dashed lines indicate normal and abnormal splicing, respectively. Exons are shown as green boxes. (Adapted from Sun X, Frierson HF, Chen C et al. [2005] *Nat Genet* 37:407–412. With permission from Macmillan Publishers, Ltd.)

genome selected in the initial phase of the ENCODE project. This analysis revealed that many genes use alternative 5′-ends that lie tens or hundreds of kilobase pairs away from the annotated 5′-ends. Often, other genes are located between the annotated genes and these distal, previously unannotated 5′-ends. As a consequence of long-range transcription, multiple exons from separate protein-coding genes can be spliced together, creating intergenic splicing products. The creation of these kinds of intergenic products has been termed **tandem chimerism**. Two examples, one involving two neighboring genes and the other affecting widely separated exons, are presented in Figure 14.12. Thus, exons that have been considered until recently as discrete modules of a specific gene should now be viewed as more general functional modules that can be joined together in multiple RNA molecules. About 65% of the genes tested in ENCODE are involved in the formation of chimeric RNAs. These may, in turn, lead to the formation of chimeric proteins. Indeed, much of the modular domain structure we see in proteins may have had its evolutionary origin in such splicing.

Trans-*splicing combines exons residing in the two complementary DNA strands*

Most frequently, chimeric RNAs originate from genes on the same chromosome that are transcribed in the same direction. In these cases, the underlying molecular mechanism involves long-range transcription that creates a very long transcript containing these exons. Then alternative splicing comes into play to create new combinations of exons. However, alternative splicing does not rely on this mechanism alone. We now know of cases where the spliced exons originate from the two complementary DNA strands of a single gene. Some exons in the final mRNA product come from the gene transcribed in the sense direction, whereas other exons

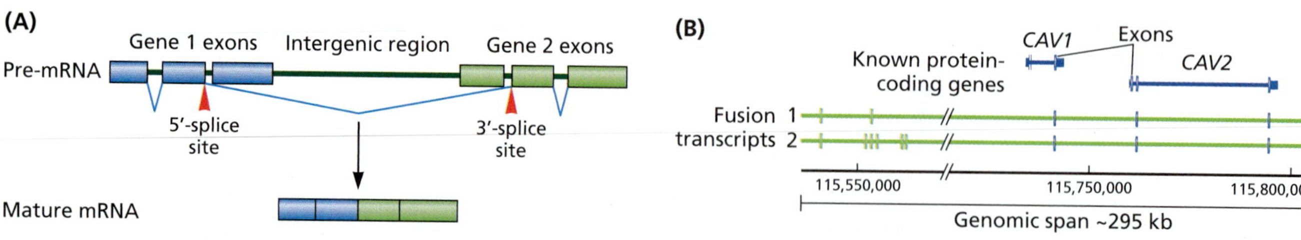

Figure 14.12 Tandem chimerism, long-range splicing events based on long-range transcription. (A) In this case, the transcribed region spans two consecutive genes and the intergenic region between them. Splicing of the pre-mRNA involves a 5′-splice site in the upstream gene and a 3′-splice site in the downstream gene; thus, the intergenic region is removed from the mature mRNA. (B) Two different fusion transcripts combine selected exons from the caveolin *CAV1* and *CAV2* genes with novel, unannotated 5′-exons that lie far from the two known genes. The creation of such fusion transcripts is very common in both human and *Drosophila* genomes. In the pilot phase of the ENCODE project, 65% of the genes tested are involved in the formation of such chimeric RNAs. Exons are shown as vertical bars and introns as horizontal lines; slanted lines indicate a gap of ~200 kbp. (A, adapted from Akiva P, Toporik A, Edelheit S et al. [2006] *Genome Res* 16:30–36. With permission from Cold Spring Harbor Laboratory Press. B, adapted from Kapranov P, Willingham AT & Gingeras TR [2007] *Nat Rev Genet* 8:413–423. With permission from Macmillan Publishers, Ltd.)

originate from antisense transcripts of the same gene. This phenomenon is known as ***trans*-splicing**. The repertoire of known *trans*-splicing events constantly increases: separate and remotely located genes on the same chromosome or even on a different chromosome may contribute some of their exons to a new product. The latter occurrences exclude the mechanism of initial transcription of very long primary transcripts, followed by alternative splicing, as a plausible explanation for the formation of these *trans*-spliced mRNAs. The molecular mechanism of *trans*-splicing remains to be elucidated.

14.5 Regulation of splicing and alternative splicing

Splicing is subject to fine-tuned regulation, which occurs through a number of different mechanisms. Since the mechanisms that regulate splicing are also those that participate in alternative splicing decisions, we will treat them together.

Splice sites differ in strength

Because the utilization of different splice sites depends on relatively weak protein-RNA and protein-protein interactions, recognition of specific sites can be easily and subtly modified by control of these interactions. When the sequence of a splice site deviates somewhat from the consensus shown in Figure 14.6A, the site is used with reduced efficiency because these weaker sites have less affinity for their spliceosomal protein partners. The selective utilization of these weaker sites constitutes the main mechanism through which alternative splicing occurs and is regulated.

Exon–intron architecture affects splice-site usage

Unexpectedly, the length of the exons and introns has emerged as a factor in determining splice-site usage. The recognition of splice sites is most efficient when introns and exons are small. Genomewide, this is an important regulatory feature, especially in view of the broad distribution of intron lengths in the human genome. In the **intron-definition** model, assembly of the spliceosome occurs at the 5′- and 3′-splice sites of a given intron. The contrasting **exon-definition** model posits that the initial recognition of splice sites occurs across an exon, at splice sites flanking the exon. The two models differ in the possible orientations of interactions between the U1 and U2 small nuclear ribonucleoproteins, or snRNPs, across the intron or across the exon. In the latter case, U1 bound to the 5′-splice site of the downstream intron interacts with U2 bound to the 3′-splice site of the upstream intron. The exon-definition model was recognized when scientists moved from *in vitro* investigations of simple artificial splicing substrates, exon–intron–exon, to multi-intron substrates. **Figure 14.13** illustrates the notion behind intron and exon definition. Kinetic experiments demonstrated that splice-site preference across introns is lost when intron

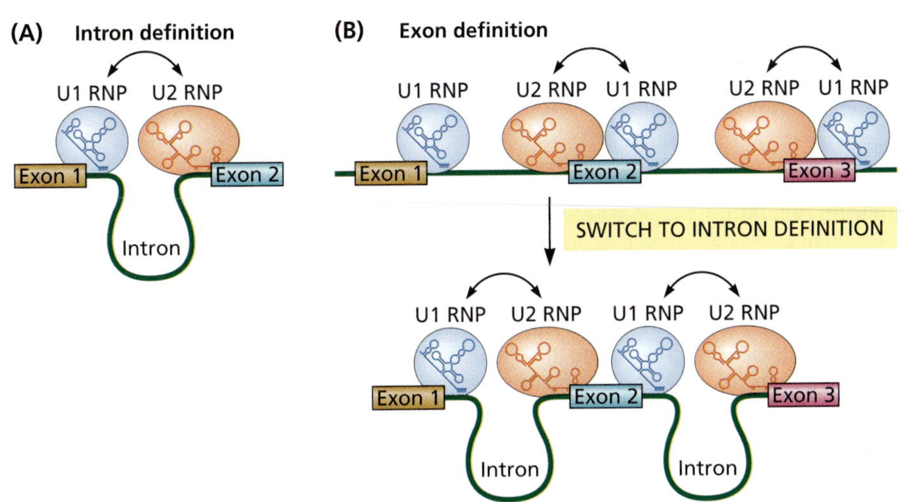

Figure 14.13 Intron and exon definition in spliceosome assembly. (A) Intron definition: in a pre-mRNA molecule containing one intron and two exons, interaction between U1 RNP and U2 RNP, as well as auxiliary factors, occurs across the intron. (B) Exon definition: in a pre-mRNA containing multiple exons and introns, the initial interactions between factors bound at the splice sites occur across the exon. The two transesterification reactions of splicing occur across an intron, that is, in an intron-defined complex. Hence, once an exon-defined complex is formed, it needs to switch to an intron-defined complex for the reactions to take place. At least one factor that is involved in the switch has been identified: the polypyrimidine-tract-binding protein PTB. (Adapted from Schellenberg MJ, Ritchie DB & MacMillan AM [2008] *Trends Biochem Sci* 33:243–246. With permission from Elsevier.)

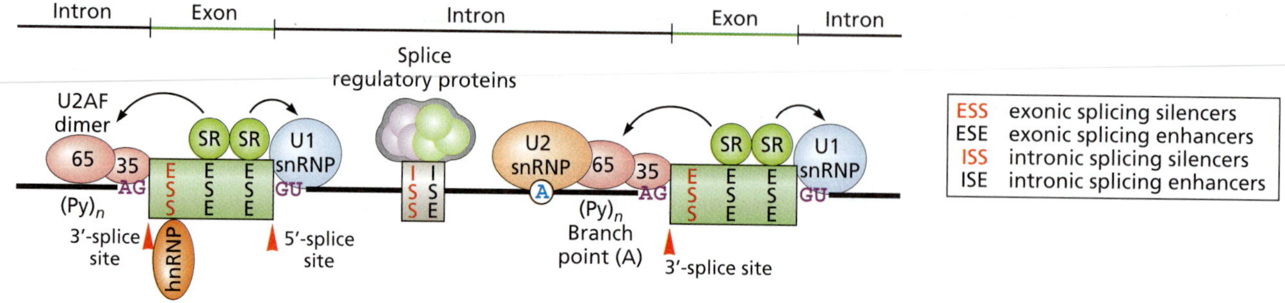

Figure 14.14 Overview of *cis–trans* regulation of splicing. The removal of introns is guided by specific protein factors that interact with specific nucleotide sequences in *cis–trans* interactions. Such sequences are present in both exons, shown in green, and introns, shown in black. Splice regulatory proteins include SR or Ser-Arg proteins, which act by binding exonic splicing enhancers, ESE, and heterogeneous nuclear ribonucleoproteins or hnRNPs, which bind to the exonic, ESS, or intronic, ISS, splicing silencers to repress splicing. (Adapted from Schwerk C & Schulze-Osthoff K [2005] *Mol Cell* 19:1–13. With permission from Elsevier.)

sites are larger than 200–250 bp; beyond this length, splice sites are recognized across exons. Intron definition is much more efficient than exon definition, and so intron size profoundly influences the probability of exon inclusion or skipping in the final mRNA generated during alternative splicing of exons with weak splice sites. Finally, both experimental and computational approaches showed that the length of the upstream intron is more important in alternative splicing than the length of the downstream intron. Thus, exon–intron architecture defines the very mechanism of splice-site recognition and affects the frequency of alternative splicing events.

Cis–trans *interactions may stimulate or inhibit splicing*

In addition to the sequences of the splice sites, other sequences can affect the efficiency of splicing by forming *cis–trans* interactions. These sequences, found in both introns and exons, can exert either a stimulatory or an inhibitory effect on splicing. Depending on their location and effect, they are termed **exonic splicing enhancers, ESE; exonic splicing silencers, ESS; intronic splicing enhancers, ISE;** or **intronic splicing silencers, ISS** (Figure 14.14). The enhancers and silencers are relatively short conserved sequences of ~10 bp, found in isolation or in clusters, that affect the use of weak splice sites. These regulatory elements are operationally identified if mutations in their sequences lead to enhancement of splicing, for silencers, or to inhibition of splicing, for enhancers. As in all cases of *cis–trans* regulation, the *cis* elements are recognized by and interact with proteins encoded somewhere else in the genome, known as *trans*-factors.

There are two general categories of regulatory proteins that affect splicing. The first class comprises Ser-Arg or **SR proteins**, which are usually activators of splicing, although under certain circumstances they can also serve as repressors. Members of this protein class share similar organization: they possess one or two RNA-binding domains at their N-terminus and a variable-length domain at the C-terminus that contains repeats of RS or Arg-Ser dipeptides. RS domains serve the activation function and are extensively phosphorylated. Activation of splicing occurs by enhancing the recruitment of spliceosome components to the 5′- and 3′-splice sites (Figure 14.15A).

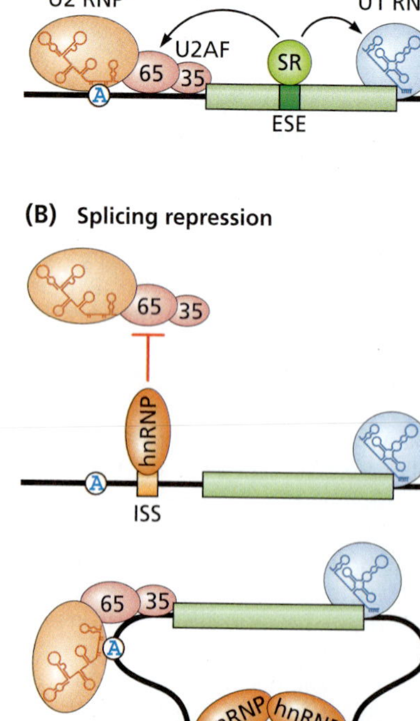

Figure 14.15 Proposed models for the most common mechanisms of traditional splicing regulation. (A) SR or Ser-Arg proteins bind to exon enhancers, ESEs, to stimulate binding of U1 snRNP to the 5′-splice site and binding of the U2 auxiliary factor U2AF, a heterodimer of 65 and 35 kDa proteins, to the polypyrimidine tract and the conserved AG at the 3′-splice site. U2AF in turn guides the U2 snRNP to the branch-point A nucleotide. (B) Two alternative, non-mutually-exclusive models for inhibition of splicing by hnRNPs, which bind to silencer elements occurring in both introns and exons. hnRNP binding can interfere with the binding of U2AF to the 3′-splice site, as shown in the top drawing. Alternatively, hnRNPs can bind to ISSs in the introns flanking an exon and then interact with each other, which leads to looping out of the intervening exon. This exon is excluded from the final mature mRNA. (Adapted from Graveley BR [2009] *Nat Struct Mol Biol* 16:13–15. With permission from Macmillan Publishers, Ltd.)

The effect of post-translational modifications in SR proteins on their activity is illustrated with the example of the SRp38 protein, which participates in inhibition of splicing during stress (Box 14.5). In addition, SR proteins participate in numerous other aspects of mRNA metabolism, such as nonsense-mediated mRNA decay, nuclear export, and translation.

The second class of *trans*-regulators acts through binding to silencers; usually they inhibit splicing, although they can sometimes act as activators too. These proteins are members of a large, structurally diverse group of RNA-binding proteins, usually complexed with small RNA molecules, hence their name **heterogeneous nuclear ribonucleoproteins** or **hnRNPs**. hnRNPs may act through a number of mechanisms, including blocking the recruitment of spliceosome snRNPs, looping out exons, or multimerization along exons; the first two mechanisms are illustrated in Figure 14.15B.

Cis–trans interactions also constitute a major part of alternative splicing regulation, since the concentrations of splicing activators or repressors, as well as those of spliceosomal components, can be regulated in physiologically meaningful ways and can simultaneously modify splicing at many loci. Thus, these concentrations may differ between different terminally differentiated cell types, may change during differentiation and development programs, or may fluctuate during the cell cycle. As a result of these changes, exon inclusion or exclusion may be differentially regulated to reflect the needs of the cell.

Box 14.5 Cellular stress, RNA splicing, and the role of post-translational modifications in SR proteins The cell is constantly exposed to factors that challenge the integrity of the genome, known as genotoxic stress, or that cause protein denaturation, lipid peroxidation, or disturbance in the cellular redox state. The best-characterized stress response, which is triggered by a variety of stresses, involves the transcriptional activation of a set of heat-shock genes encoding molecular chaperones (see Chapter 3). Another target of several stressors is the inhibition of pre-mRNA splicing, which may occur by affecting any of the large number of weak protein–protein and protein–RNA interactions involved in the regulated assembly of the spliceosome. Although the inhibition of splicing observed upon heat shock is almost immediate and very robust, it is still highly selective. First, splicing inhibition cannot affect the actual expression of the heat-shock genes, since the majority of these genes, including the abundant HSP70, do not contain introns. Second, the primary transcripts from those chaperone genes that do contain introns, such as HSP90a, HSP90b, and HSP27, are properly spliced in heat-shocked human cells, as these chaperones are needed to counteract the protein denaturation that results from exposure to high temperatures.

One component of splicing inhibition relies on the inactivation of an activity termed HSLF, heat-shock labile splicing factor, which functions in assembly of the U4–U5–U6 tri-snRNP (see Figure 14.9). A separate component in the inhibition is governed by the phosphorylation status of an unusual SR protein family member, SRp38. A distinguishing feature of SRp38 is its ability to function as a potent splicing repressor when dephosphorylated in M phase or in response to heat shock. The molecular mechanism underlying this repressive action is presented in Figure 1.

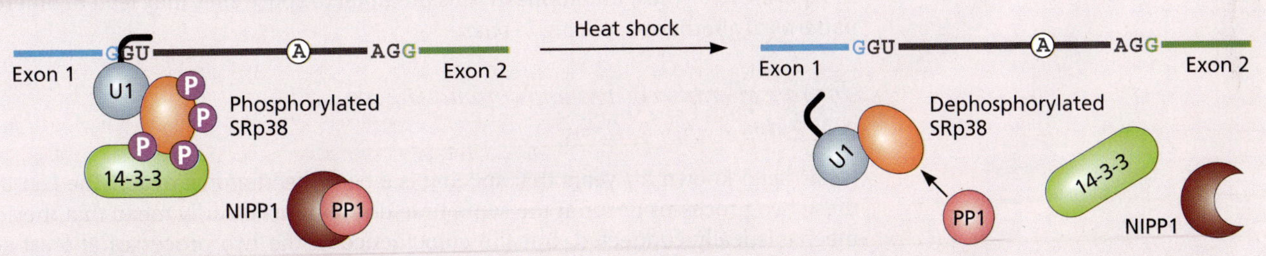

Figure 1 Phosphorylation state of SRp38 affects the interaction of U1 RNP with pre-mRNA. In normal unstressed cells, phosphorylated SRp38 stabilizes the interaction of U1 RNP with the 5′-splice site. The phosphorylated state is protected by binding of 14-3-3 protein to the phosphorylated residues; for structural information on 14-3-3, see Figure 12.12C. PP1, protein phosphatase 1, which is responsible for dephosphorylation of numerous protein targets, is inactivated by binding to NIPP1, nuclear inhibitor of PP1. Heat shock promotes the dissociation of 14-3-3 from SRp38 and of NIPP1 from PP1, resulting in dephosphorylation of SRp38; stable interaction of dephosphorylated SRp38 with U1 RNP prevents its association with the 5′-splice site, thus inhibiting splicing. (Adapted from Biamonti G & Caceres JF [2009] *Trends Biochem Sci* 34:146–153. With permission from Elsevier.)

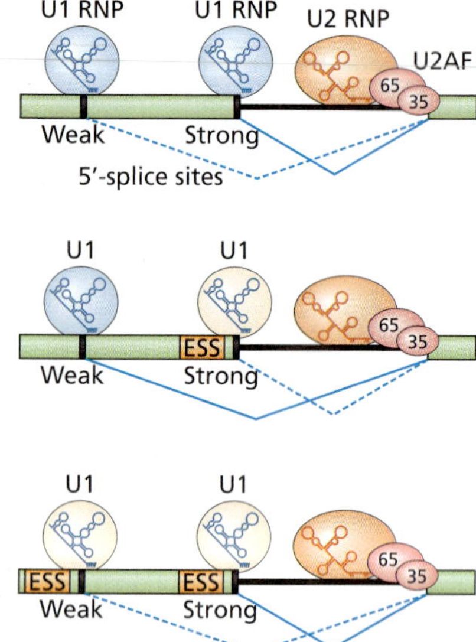

Figure 14.16 Regulation of alternative splicing without regulators. Experiments performed in T. Nilsen's laboratory uncovered a new type of regulation of alternative splicing that occurs without the involvement of auxiliary splicing factors of the SR or hnRNP classes. The authors designed a synthetic splicing substrate containing two competing 5′-splice sites, one weak and one strong, and a single 3′-splice site. *In vitro*, the strong downstream 5′-splice site was predominantly used in the absence of a splicing silencer, despite the fact that U1 snRNP binds to both 5′-splice sites. A splicing silencer, ESS, was created by inserting randomized sequences and selecting for predominant use of the weak site. Importantly, the newly created ESS does not prevent U1 snRNP from binding to the strong 5′-splice site, though the configuration in which U1 snRNP interacts is altered, depicted here as a change in color for the U1 snRNP particles. When the ESS is present near both 5′-splice sites, it alters the association of U1 snRNP with both 5′-splice sites in the same way. As a result, efficient splicing to the strong downstream 5′-splice site is restored. These observations suggested that the newly created silencers do not impair the ability of U1 snRNP to recognize and bind to the 5′-splice sites but rather alter the efficiency with which the U1 snRNP–5′-splice site complex engages in splicing; the efficiency is altered because of subtle changes in the way U1 snRNP interacts with the splice site. Thus, the sequences flanking the 5′-splice site themselves regulate alternative splicing events without the need for auxiliary protein factors. (Adapted from Graveley BR [2009] *Nat Struct Mol Biol* 16:13–15. With permission from Macmillan Publishers, Ltd.)

RNA secondary structure can regulate alternative splicing

We often depict the structure of mRNA or its precursors as a straight or a wavy line. This is, of course, an oversimplification, as we now know that significant portions of any pre-mRNA and its mature counterpart are double-stranded helices, forming the stems of hairpin structures of various lengths. The kinetic stability of such secondary structures will determine their half-lives. Structures that persist may modulate splice-site recognition and usage. For example, local secondary structures may interfere with splicing if they conceal splice sites or enhancer-binding sites from their binding partners. The opposite effect will occur if local RNA structures mask splicing repressors. Thus, the existence and stability of secondary RNA structures adds another level of complexity to the already extremely complex regulation of splicing.

Sometimes alternative splicing regulation needs no auxiliary regulators

A recently discovered regulatory mechanism does not involve auxiliary splicing factors of the SR or hnRNP classes; sequences close to splice sites can affect splicing by changing the configuration of the U1 snRNP that interacts with the 5′-splice site. The *in vitro* experiments that led to the discovery of this kind of regulation are illustrated in **Figure 14.16**. Thus, mutations in sites proximal to splice sites may lead to different patterns of alternative splicing *in vivo*.

The rate of transcription and chromatin structure may help regulate splicing

It has been known for years that splicing is a co-transcriptional event. The fact that these two processes occur at the same time does not necessarily mean that they are mechanistically connected, but the coincidence of the two processes at least sets the stage for mechanistic coupling. Indeed, such a coupling has been indicated in many experimental studies, and it has numerous possible connections to chromatin structure.

Two models exist to account for the coupling of transcription and splicing. In one of these models, the C-terminal domain or CTD of the advancing polymerase serves as a landing pad to recruit various splicing factors (**Figure 14.17**). Hence, this model came to be known as the recruitment model. In fact, recruitment of splicing factors may also occur through direct interactions with chromatin components. The second

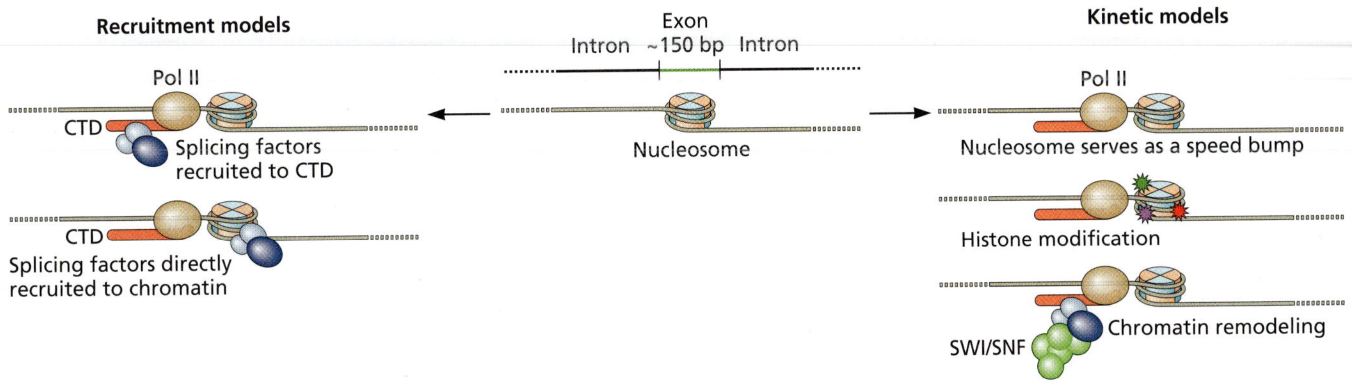

Figure 14.17 Multiple connections between splicing, transcription, and chromatin structure. The center schematic illustrates increased nucleosome occupancy at exons. Note that because of the relatively short sequences in exons, most frequently a single nucleosome organizes an entire exon. (Left) The recruitment models state that Pol II, via its C-terminal domain or CTD, helps to bring the splicing machinery to the mRNA; alternatively, recruitment may occur through direct binding to chromatin components. (Right) Possible scenarios of kinetic models. The mere presence of a nucleosome can slow down the progression of the polymerase, serving as a speed bump; histone modifications can affect the rate of polymerase movement; and finally, remodeling of chromatin structure by remodelers such as the SWI/SNF complex can also affect the rate of transcription.

model of coupling is known as the kinetic model. It states that the rate of movement of the polymerase along the gene is important for recognition of weak splice sites: the more slowly the polymerase progresses, or if it pauses, as we know it often does, the greater the opportunity for weak splice sites to be recognized and used by the spliceosome.

Since the rate of transcription is dependent on the chromatin structure of the underlying template, the coupling between transcription and splicing may also involve chromatin organization. In the past decade, numerous scattered studies have supported this idea. Both bona fide ATP-dependent chromatin remodelers and histone modification enzymes have been shown to affect splicing regulation. Recent advances in bioinformatics have allowed comparisons between different genomewide sets of experimental data obtained by independent methods. One data set identified sequences that are organized as nucleosomes. The other data set was obtained by chromatin immunoprecipitation using antibodies against various histone modifications, followed by sequencing of the immunoprecipitated DNA.

Importantly, it was established that there is an enrichment, ~1.5-fold, of nucleosome occupancy in exons as compared to introns. On the surface, this level of enrichment may seem small, and thus probably insignificant, but simple calculations indicate that it could be important. For example, a stretch of DNA 4800 bp in length would accommodate 30 nucleosomes of repeat length ~160 bp, which is the shortest nucleosome repeat length described *in vivo*. If the number of nucleosomes on the same stretch of DNA is reduced 1.5-fold, to 20, then the nucleosome repeat length would jump to 240. It may be just a coincidence that this nucleosome repeat length is the largest found in nature. In any case, the difference between the two cases would be in linker length of 80 bp, which would have profound effects on both the structure of the chromatin fiber and the accessibility of the resident DNA to protein binding. It may be important that introns are not densely populated by nucleosomes; according to current ideas about the participation of introns in gene regulation (see Chapter 13), a looser chromatin structure may help this function. It is also intriguing that the average exon length in metazoans is ~140–150 bp, exactly the length of DNA organized in a nucleosome. So the majority of exons contain a single nucleosome, which may carry specific histone modifications. Indeed, there are differences in the histone modification patterns of exonic and intronic nucleosomes, even after nucleosome occupancy is accounted for.

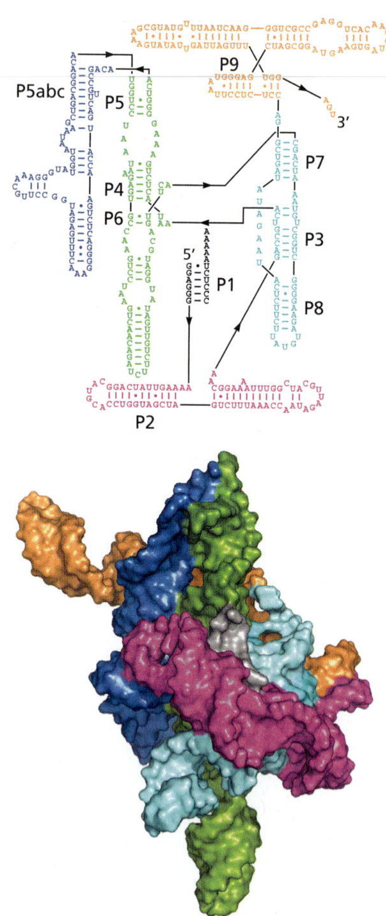

Figure 14.18 Representative secondary and tertiary structures of a group I intron. This intron is from the ciliate *Tetrahymena thermophila*. The color scheme in the secondary structure corresponds to that used in the three-dimensional structure. The intron is highly structured, containing nine double-stranded helices or paired regions, numbered P1–P9. The structure is highly evolutionarily conserved: the crystallographically resolved structures from *Tetrahymena*, the purple bacterium *Azoarcus*, and the *Staphylococcus aureus* phage Twort are almost superimposable. This conservation of the structure is impressive when one bears in mind the very poor sequence conservation, apart from a few crucial nucleotides located at the active or G site. The intron misfolds substantially *in vitro* and needs chaperone proteins to help proper folding *in vivo*. (Adapted from Jarmoskaite I & Russell R [2011] *Wiley Interdiscip Rev: RNA* 2:135–152. With permission from John Wiley & Sons, Inc.)

14.6 Self-splicing: Introns and ribozymes

A fraction of introns is excised by self-splicing RNA

Most pre-mRNAs use spliceosomes for splicing, but some employ self-splicing mechanisms catalyzed by the very intron that needs to be excised. RNase P and self-splicing introns were the first **ribozymes** to be discovered in the early 1980s (see Box 14.1). Ribozymes are so termed because they possess enzymatic activity in the RNA molecule itself. They are often bound to proteins that serve to stabilize the complex folded RNA structures that are needed for catalysis. The initial small group of known ribozymes, which were considered exceptional in structure and biochemical activities, quickly grew to include other molecules, such as hammerhead, hairpin, and hepatitis delta virus ribozymes. Moreover, the enzymatic activity that creates the peptide bond during protein synthesis also turned out to be a ribozyme (see Chapter 16).

Self-splicing introns occur in many organisms, but most are not essential for viability, unlike RNase P. They may be viewed as selfish genetic elements that found a way to propagate themselves by persisting in the genome as normal components of intron-containing genes but splicing themselves out at the RNA level, so that they do not bring about the destruction of their hosts. This view may be changing, however, as most of the DNA sequences that were previously considered junk may actually participate in gene regulation (see Chapter 13).

There are two classes of self-splicing introns

There are two major classes of self-splicing introns: group I and group II. The two groups differ mainly in their co-factor requirements: group I introns use a guanosine molecule as a co-factor, whereas group II generally use an internal adenosine, similar to the A in the branch site that is involved in spliceosome-catalyzed splicing. The second distinguishing feature is their quite different structures.

Group I introns are very abundant, with more than 2000 members described, mainly in bacteria and lower eukaryotes; they are rare in animals, and to date none have been found in archaea. They catalyze their own excision from mRNA, tRNA, and rRNA precursors. Their structure is formed by a specific arrangement of nine double-helical elements, called paired regions or P, capped by loops and connected by junctions; a tenth helix is formed during the reaction itself (Figure 14.18 and Figure 14.19). The catalytic G-binding site is located in helix P7. Some group I introns require a protein for activity; well-studied examples come from *Neurospora crassa* and *Saccharomyces cerevisiae*. The proteins do not, however, directly participate in catalysis; they are involved in stabilizing the catalytic core by reinforcing long-range interactions between individual P elements.

Group II introns are phylogenetically unrelated to group I introns; they are mainly found in the mRNA, tRNA, and rRNA of organelles in fungi, protists, and plants and in the mRNA of bacteria. Group II introns possess an unusually diverse repertoire of chemical activities, including catalysis of 2′–5′ phosphodiester bond formation and the ability to reinsert themselves into DNA with the help of intron-encoded proteins. The latter process is known as retrotransposition and is discussed in detail in Chapter 21. Structurally, group II introns possess six helical domains (Figure 14.20). The most conserved domain, DV, consists of a 30–34-nucleotide stem-loop structure and contains a highly conserved catalytic triad of nucleotides, located exactly five base pairs away from the two-nucleotide bulge in the helix. This structural arrangement is strikingly similar to that of U6 spliceosomal RNA. This similarity, together with other similarities that occur during the self-splicing process, such as the use of an internal A, rather than an external co-factor as in the case of group I introns, and the formation of a lariat structure, led to the hypothesis that group II introns and the spliceosome machinery may be evolutionarily connected. More in-depth analysis is needed to prove or disprove this hypothesis. Finally, we note that group II introns are not, in fact, true catalysts, as the entire intron is degraded once self-splicing occurs.

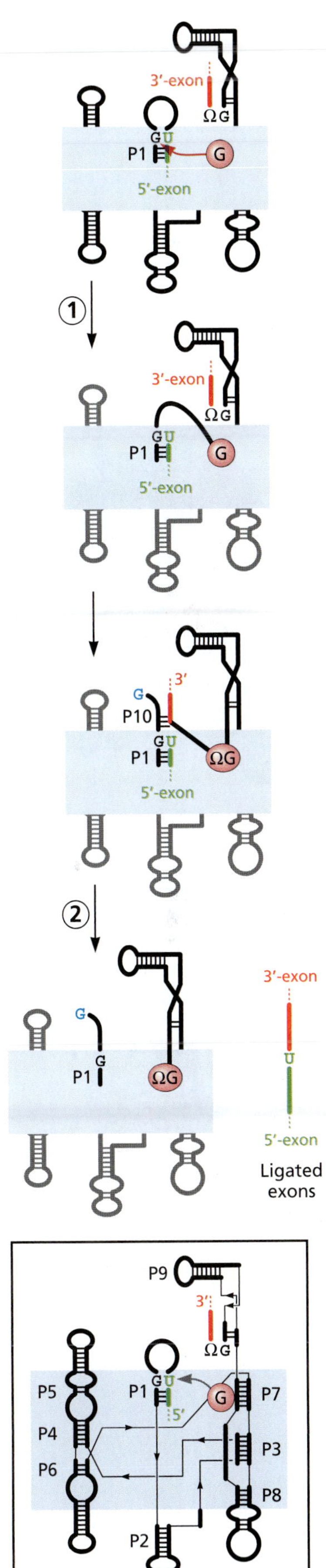

Figure 14.19 Mechanism of group I intron self-splicing. The example shown is the *Azoarcus* intron. The shaded box delimits the catalytic core of a group I intron. This is shown in more detail in the box at the bottom, which depicts the conserved secondary structure, with the nine helices forming the catalytic core and the branching peripheral elements. Introns are shown in black, with the 5′-exon in green and the 3′-exon in red. Only the peripheral elements that branch out of the catalytic core are shown for simplicity. The 5′-splice site contains a conserved GU pair, where G is in the intron and U is in the exon; a conserved guanine, termed ΩG, is at the 3′-terminal position of the intron. The splicing reaction occurs in two catalytic steps: (Step 1) The 3′-OH group of a guanosine co-factor binds to the intron at the G-binding site, in P7, and attacks the 5′-splice site. After the reaction occurs, the G is covalently linked to the 5′-end of the intron. (Step 2) A conformational change occurs: another helix, P10, is formed, which involves base pairing between the intron and the 3′-exon. Recognition of the 3′-splice site is achieved, in part, by ΩG, which displaces the guanosine from the G-binding site. Then the 3′-OH group of ΩG attacks the 3′-splice site. The two exons are ligated and the intron is released. (Adapted from Vicens Q & Cech TR [2006] *Trends Biochem Sci* 31:41–51. With permission from Elsevier.)

14.7 Overview: The history of an mRNA molecule

Proceeding from the primary transcript to a functioning mRNA requires a number of steps

We have described many ways in which mRNA molecules are processed; it is useful here to stand back and look at the overall process. The history of a typical pre-mRNA from synthesis to export to the cytoplasm is presented in **Figure 14.21**. It is clear that the cell has evolved a set of highly complex mRNA processing events that are intimately connected to transcription. Practically all individual steps are subject to regulation, the result being the production of the right type of mRNA in the right amount, at the right place, at the right time.

After mRNA molecules have been properly processed, they must be made available to the translational machinery in the cell cytoplasm for the synthesis of proteins. In bacteria, this presents no problems, as both transcription and translation occur in the cytoplasm. In eukaryotes, however, mRNA that has been transcribed and processed in the nucleus must be transported to the cytoplasm through the membranes of the nuclear envelope. There exists a whole molecular machine to accomplish and regulate this transport.

Finally, all mRNA molecules must eventually be degraded. There are two quite different reasons for this. First, even completed and processed RNA molecules may contain errors in sequence, processing, or packaging that would impede proper cell function. These must be removed. Second, it is not desirable for the cell to maintain mRNA molecules active in protein synthesis continually. Proper control of cellular function requires the shutting off of protein synthesis that is no longer needed. Recall that the clue that led Jacob and Monod to postulate the existence of mRNA was the evidence of a short-lived intermediate in gene expression. Many small RNA molecules that play regulatory roles are also short-lived. The only long-lived RNAs in most cells are those involved in the mechanisms of protein synthesis, such as rRNA and tRNA. Therefore, there must be a selective mechanism for the degradation of even competent mRNAs in the cytoplasm.

mRNA is exported from the nucleus to the cytoplasm through nuclear pore complexes

Both export and import of macromolecules or their complexes to and from the nucleus occur through the nuclear pore complexes or NPCs (**Box 14.6**). Despite unique features characteristic of specific trafficking pathways, all these processes share common features and mechanisms. All cargo passes through the channel of the nuclear pore with the help of soluble carrier proteins in a three-step process: (1) generation of a cargo-carrier complex in the donor compartment of the cell,

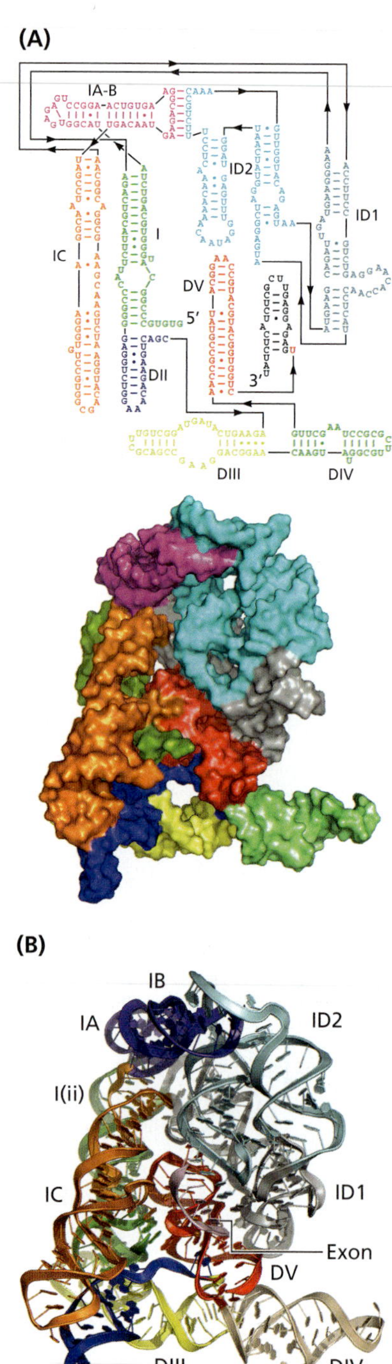

Figure 14.20 Secondary and tertiary structures of a group II intron. The example shown is from the halophilic and alkaliphilic eubacterium *Oceanobacillus iheyensis*. This bacterium was isolated from seabed mud at a depth of 1 km off the coast of Japan. A crucial step in getting the intron in its native, catalytically active form was to isolate it immediately after it had gone through both steps of splicing, thus ensuring that it is properly folded. This procedure is a departure from the routine purification of RNAs for structural analysis, which involves denaturing polyacrylamide gel electrophoresis followed by renaturation steps. (A) The color scheme in the secondary structure corresponds to that used in the three-dimensional (3D) structure. The conserved DV domain involved in catalysis is depicted in red. (B) The same structure is presented as a ribbon diagram to show the RNA helices within DV, shown in red, forming the catalytic core and the bound exon, shown in purple, with the rest of the domains encapsulating this active site. Note that different colors are used to indicate domain IA-B in the two 3D structures. (A, adapted from Jarmoskaite I & Russell R [2011] *Wiley Interdiscip Rev: RNA* 2:135–152. With permission from John Wiley & Sons, Inc. B, from Toor N, Keating KS & Pyle AM [2009] *Curr Opin Struct Biol* 19:260–266. With permission from Elsevier.)

(2) passage of the complex through the NPC, and (3) release of the cargo in the target compartment, followed by recycling of the carrier back to the donor compartment. The passage itself is a Brownian or random-walk process facilitated by the FG nucleoporins that line the NPC channel; see Box 16.4 for discussion of the concept of ratcheted Brownian motion. The majority of nuclear trafficking pathways that transport proteins or small RNA molecules employ members of the β-karyopherin superfamily of proteins as carriers. mRNA, however, uses the Mex67–Mtr2 heterodimer (**Figure 14.22**). We shall refer to the RNA–carrier complex as messenger ribonucleoprotein or mRNP.

There are also mechanisms that allow recognition of the donor and target cell compartments, thereby ensuring that appropriate assembly of the cargo–carrier complex or release of cargo occurs. When β-karyopherins are used as carriers, compartment recognition is achieved through the nucleotide state of the Ran GTPase, which is bound to GTP in the nucleus and is maintained in a GDP-bound state in the cytoplasm. In mRNA export, it is believed that the transition from one compartment to the other is associated with extensive remodeling of the mRNP complex by two distinct DEAD-box helicases, nuclear Sub2 and cytoplasmic Dbp5, which are thus the compartment-recognition molecules. Nuclear Sub2 is needed for remodeling of the cargo–carrier complex so that one of the adaptor proteins used to recruit Mex67–Mtr2 to the mRNP is released from the export-competent mRNA. The cytoplasmic DEAD-box helicase Dbp5 remodels the mRNP on the cytoplasmic side of the NPC so that the Mex67–Mtr2 carrier is released for recycling (see Figure 14.22).

RNA sequence can be edited by enzymatic modification even after transcription

In some situations the sequences of pre-mRNAs are edited, actually changed by insertion, deletion, or chemical modification of residues. The insertion and deletion of residues appear to be restricted to mitochondrial RNA of certain protozoa, such as trypanosomes. In these cases, insertion or deletion of short oligomers of U can occur at specific locations in the sequence. The sequences are dictated by guide RNAs, oligomers complementary to the target RNA but with a mismatched bulge. The guide RNA maintains connection while the mRNA is cleaved and oligo(U) is inserted or deleted. A ligase then reseals the modified message.

A quite different kind of editing is observed in some higher organisms, including mammals. In some cases adenosine can be converted to inosine, or cytidine to uridine (**Figure 14.23**). The enzymes that catalyze these reactions contain RNA-binding domains that recognize specific sequences a few nucleotides away from the site of modification. Although not common, the amino acid sequence changes effected by such editing may have significant effects. For example, there exists evidence that amyotrophic lateral sclerosis or ALS, also known as Lou Gehrig's disease, may involve

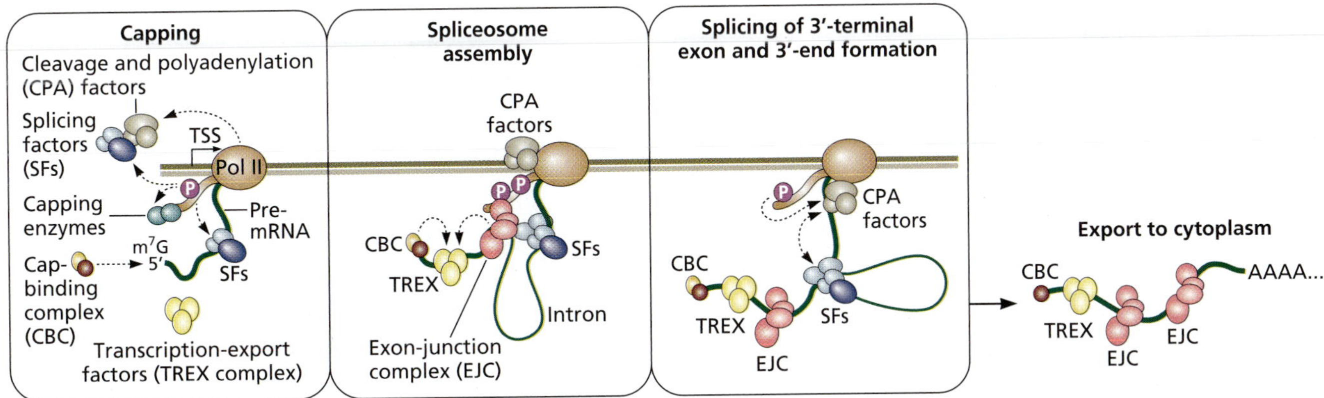

Figure 14.21 Overview of co-transcriptional RNA processing. Pre-mRNA is represented by a green line: thicker at the exon portions and thinner at the introns. The three adjacent boxes represent the composition of protein complexes bound to either the polymerase, mainly through its CTD, or the nascent transcript. The complexes perform specific RNA processing functions during specific stages of polymerase movement along the gene. Dashed arrows denote interactions that stabilize the complexes or perform the respective enzymatic function. (Left box) 5′-End capping. Capping occurs as soon as the 5′-end of the RNA transcript emerges from the RNA polymerase Pol II; the capping enzymes are recruited via Ser5 phosphorylation of the CTD. Once the cap structure is on, the cap-binding complex, CBC, binds to it and recruits the transcription-export complex, TREX. Splicing factors, SFs, and some of the CPA or cleavage and polyadenylation components also join the complex at this stage. (Middle box) Spliceosome assembly. Assembly at the first intron is enhanced by protein factors that bind to both the CTD and the nascent RNA, thus bringing the first and second exons into close proximity. The exon-junction complex or EJC is recruited by the splicing machinery and is deposited just upstream of the exon–exon junction. The TREX complex is now stably associated with nascent RNA through interactions with the CBC, SFs, and/or the EJC. (Right box) Splicing of the 3′-terminal exon and formation of the 3′-end of mRNA. These two processes occur when transcription approaches the end of the gene, after the final intron and 3′-end exons have been transcribed. Recruitment of the CPA machinery occurs on the CPA signal. The schematic on the far right shows the proteins bound to the processed mRNA when exported to the cytoplasm. Note that many of the proteins still remain bound and might affect subsequent processes. (Adapted from Pawlicki JM & Steitz JA [2010] *Trends Cell Biol* 20:52–61. With permission from Elsevier.)

Box 14.6 Nuclear pore complexes Nuclear pore complexes, NPCs, are huge protein assemblies, probably the largest in eukaryotic cells. Their diameter has been measured at ~125 nm and their mass at ~125 MDa in metazoa and ~60 MDa in yeast. Their core structure, the spoke-ring complex, possesses eight-fold symmetry and is sandwiched between a cytoplasmic and a nuclear ring (Figure 1 and Figure 2). The eight spokes surround the channel through which molecules pass back and forth between the cytoplasm and the nucleus.

Biochemically, the NPC is made of ~30 different proteins, termed **nucleoporins**, grouped into three major classes. The

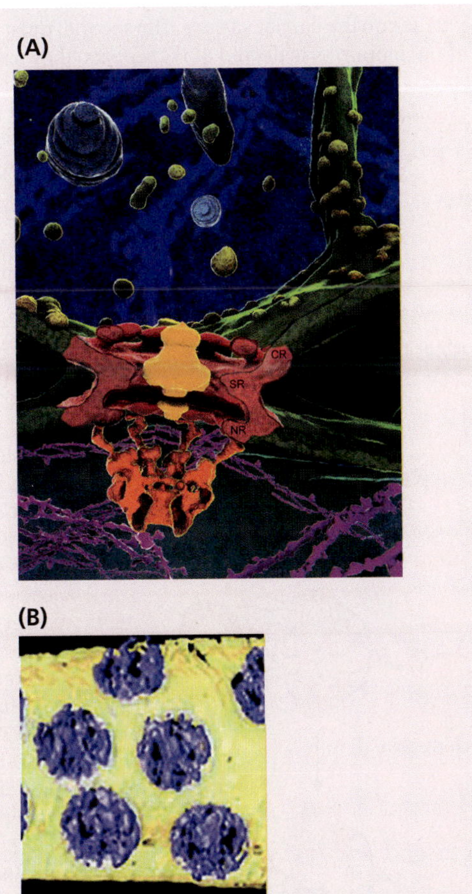

Figure 1 Nuclear pore complexes, a structural overview. (A) Schematic view of the nuclear pore complex, NPC, embedded in the nuclear envelope, shown in green. SR, spoke–ring complex; CR, cytoplasmic ring; NR, nuclear ring. At the nuclear face, a basketlike structure shown in orange emanates from the central framework shown in red. Lamin filaments, shown in purple, are attached to the NPC and to the inner nuclear membrane. The nuclear envelope is continuous with the endoplasmic reticulum and is decorated with ribosomes, shown in yellow, on the outer membrane. (B) To analyze NPC structure in a native state, transport-active intact nuclei of *Dictyostelium* were visualized by cryo-electron tomography and nuclear subvolumes were computationally analyzed *in silico*. In this representation, the NPCs are in blue and the surrounding nuclear membrane is in yellow. The number of NPCs was ~45/mm^2. (A, from Elad N, Maimon T, Frenkiel-Krispin D et al. [2009] *Curr Opin Struct Biol* 19:226–232. With permission from Elsevier. B, from Beck M, Förster F, Ecke M et al. [2004] *Science* 306:1387–1390. With permission from American Association for the Advancement of Science.)

(Continued)

Box 14.6 (Continued)

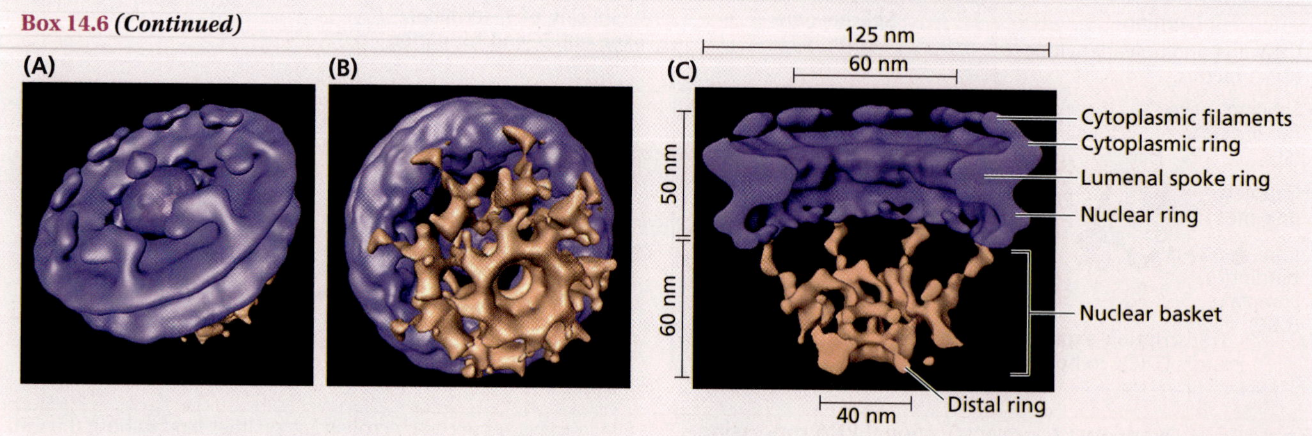

Figure 2 Nuclear pore complexes, a structural overview. (A) Cytoplasmic face, with cytoplasmic filaments arranged around the central channel; the filaments are kinked and point toward the central plug/transporter, representing cargo in transit. (B) On the nuclear face, the distal ring of the basket is connected to the nuclear ring by nuclear filaments. (C) Cutaway view of the NPC, with the plug removed to expose the actual pore structure. (From Beck M, Förster F, Ecke M et al. [2004] *Science* 306:1387–1390. With permission from American Association for the Advancement of Science.)

first class consists of FG nucleoporins, so named because of the Phe-Gly-rich tandem repeats that they contain. These repeats are unstructured and fill up the transport channel, forming a dense brushlike structure or, according to other models, a cross-linked hydrogel, which serves as a barrier to free movement of macromolecules larger than 40 kDa. Thus, larger molecules must exploit carrier molecules and active, energy-consuming processes to pass through the NPC. FG nucleoporins interact directly with the cargo carriers through hydrophobic patches on the carrier surface. The second, most prevalent class of nucleoporins is devoid of FG repeats; they are the structural constituents of the pores. Finally, the third class of nucleoporins, Nups, consists of integral membrane proteins that anchor the pore to the nuclear membrane.

Most nucleoporins are symmetrically located on both sides of the structure, but some are bound to either the cytoplasmic or the nuclear side. These are thought to be involved in ensuring the directionality of transport (see Figure 14.22). Some of these participate in mRNA transport specifically; others may serve in the transport of small nuclear RNA or rRNA, while still others may be shared by all RNA molecules.

a defect in editing of the mRNA encoding a protein involved in calcium conductance in neural membranes.

Finally, tRNA molecules undergo extensive post-transcriptional modification involving a variety of base modifications at numerous sites. These are discussed in detail in Chapter 15.

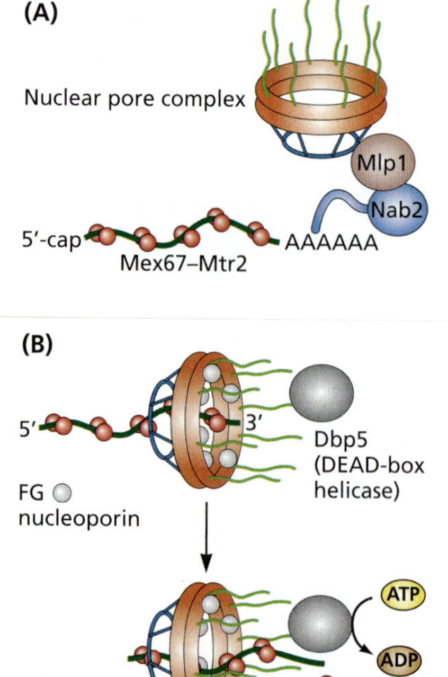

Figure 14.22 Nuclear export of mRNA in yeast. (A) Export-competent mRNA, with some export proteins bound to it. Additional proteins accompany mRNP to the cytoplasm but are not directly involved in the transport, such as the TREX complex that coordinates many of the steps during transcription and processing, exon-junction complexes, cap-binding complex, etc. (see Figure 14.21). The carrier complex Mex67–Mtr2 binds mRNA only weakly and is recruited to the message by proteins serving as adaptors, Yra1 and Nab2. Yra1 is dissociated from mRNA following recruitment of the Mex67–Mtr2 carrier and before export. The dissociation requires the action of a nuclear DEAD-box helicase, Sub2, not shown here, which somehow remodels the mRNP, probably by changing the conformation of mRNA. Nab2 positions the 3′-end of the export-competent mRNA at the Mlp1 component of the nuclear basket to help mRNA thread into the entrance of the NPC channel. (B) Brownian ratchet model for mRNA transport through the NPC. The Mex67–Mtr2 carriers bound to export-competent mRNP interact with the FG nucleoporins that line the NPC channel to facilitate movement of the mRNP back and forth by thermal or Brownian motion. When one of the Mex67–Mtr2 carriers reaches the cytoplasmic face of the NPC, it is removed by the cytoplasmic DEAD-box helicase Dbp5, whose ATPase activity is stimulated by other factors bound to the same face of the pore. Removal of the carrier functions as a molecular ratchet, as it does not allow the portion of the mRNP to which it was bound to go back into the channel. Hydrolysis of ATP to ADP makes the ratchet work in one direction. These steps are repeated several times until the entire mRNP enters the cytoplasm. The released carrier molecules are recycled back into the nucleus to participate in the export of another mRNP complex. (Adapted from Stewart M [2007] *Mol Cell* 25:327–330. With permission from Elsevier.)

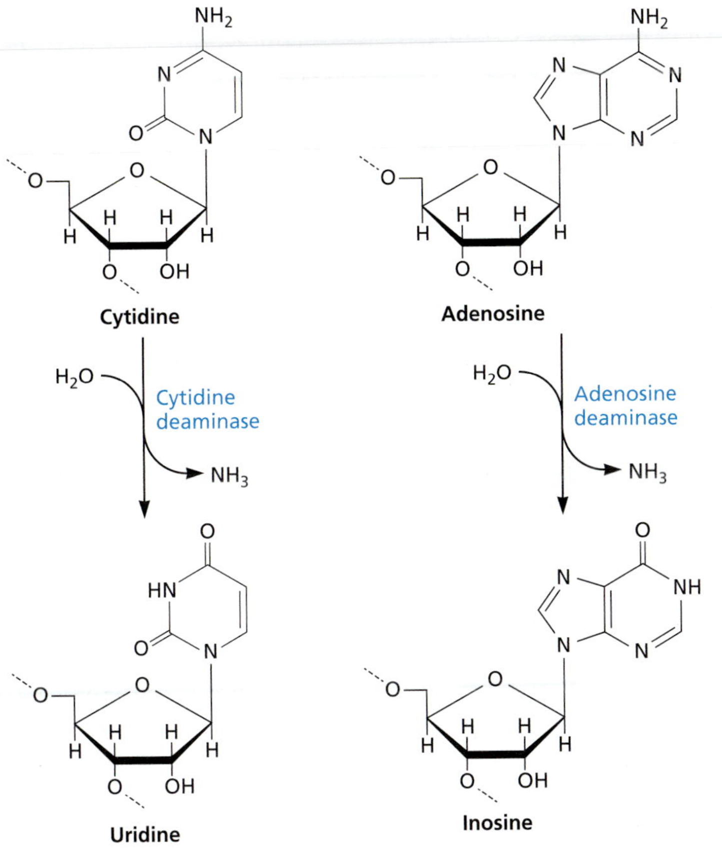

Figure 14.23 Editing by deamination in mammals.

14.8 RNA quality control and degradation

Bacteria, archaea, and eukaryotes all have mechanisms for RNA quality control

The importance of fully functional RNA molecules has led to the evolution of numerous quality-control mechanisms in all three domains of life. Quality control acts at many steps before, during, and after translation. If, for example, mRNA is incompletely or incorrectly spliced or polyadenylated, it will be degraded in the nucleus instead of being exported to the cytoplasm. This applies to aberrant rRNA and tRNA molecules also. RNA molecules are also degraded in the cytoplasm at the end of their useful life. We first introduce the structure of the main protein complexes used for RNA degradation in both circumstances (**Figure 14.24** and **Figure 14.25**). In archaea

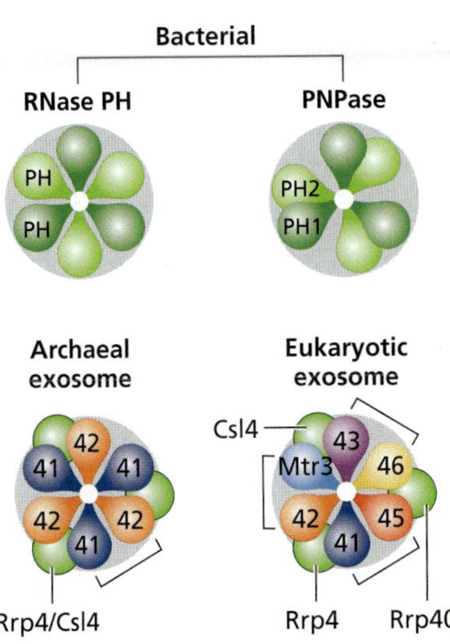

RNA_n + inorganic phosphate (P_i) $\longrightarrow$ RNA_{n-1} + nucleoside 5′-diphosphate

Figure 14.24 Conserved architecture of RNA degradation complexes among bacteria, archaea, and eukaryotes. Bacteria contain two distinct complexes. Lighter and darker shades of green in the bacterial RNase PH complexes indicate the inverse orientation of neighboring subunits; the combined shapes in PNPase indicate PH domains from the same polypeptide chain. All schematics show the structures from the bottom; note the ring shape of the complexes with a hole in the middle. The RNA-binding caps of the archaeal and eukaryotic exosomes are depicted in green behind the subunits forming the rings. The archaeal structure is a homotrimer of heterodimer Rrp41–Rrp42; the three cap proteins that stabilize the ring structure can be either Rrp4, Csl4, or a combination of these. The eukaryotic exosome is a complex of three distinct heterodimers, Rrp41–Rrp45, Rrp46–Rrp43, and Mtr3–Rrp42; the dimers are held together by the cap proteins Rrp40, Csl4, and Rrp4. In addition to stabilizing the complex, the cap proteins contain RNA-binding domains that interact with the RNA to be degraded. The archaeal and eukaryotic ring subunits are related by sequence and structure to the bacterial RNase PH enzyme. The PH domains in these enzymes function in a phosphorolytic reaction, as shown below the schematics. (Adapted from Lykke-Andersen S, Brodersen DE & Jensen TH [2009] *J Cell Sci* 122:1487–1494. With permission from The Company of Biologists.)

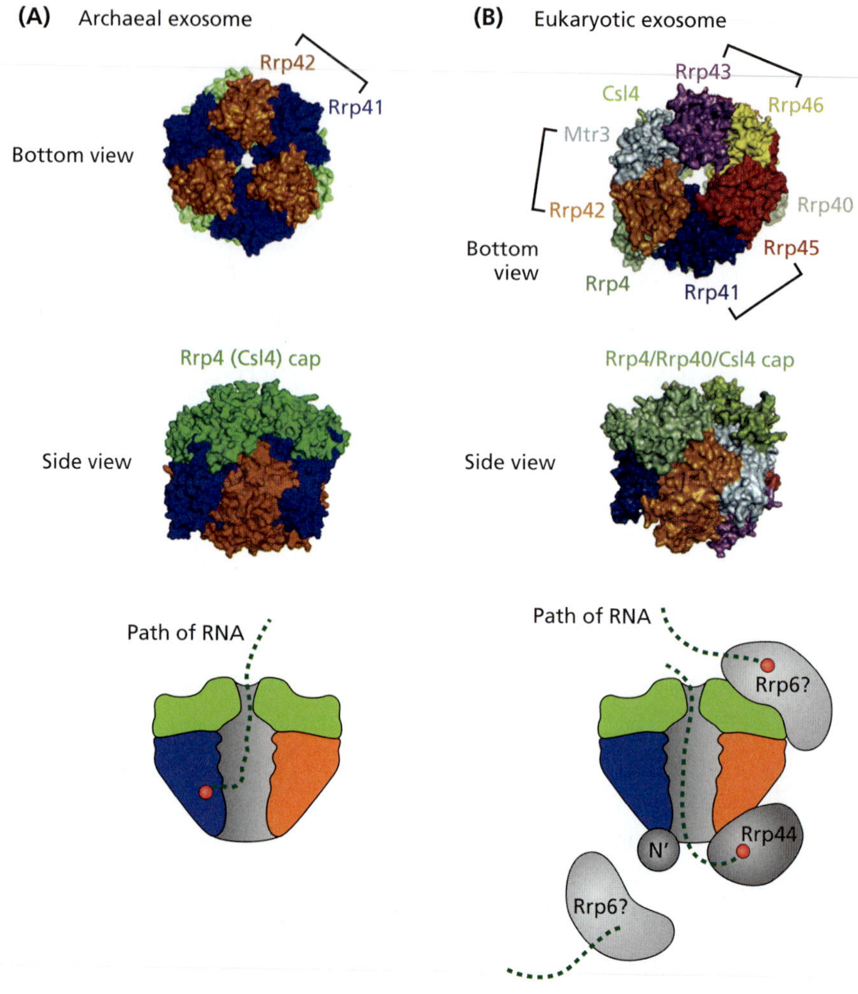

Figure 14.25 Conserved structures of RNA-degrading complexes from archaea and eukaryotes. The individual proteins are color-coded. (A) Structure of the *Archaeoglobus fulgidus* exosome containing Rrp41, Rrp42, and Rrp4. (B) Human exosome structure. The schematics below the structures represent sliced side views, showing the central cavity and the path of the RNA to the active site, marked by a red dot. The eukaryotic exosome does not possess enzymatic activities in the main ring structure; rather, the active center is located on Rrp44, and the N-terminal head domain of Rrp44 interacts with Rrp41. RNA may access the active site through the central channel or through a path that is independent of the exosome core. The position of the other subunit with enzymatic activity, Rrp6, is unclear. (Adapted from Schmid M & Jensen TH [2008] *Trends Biochem Sci* 33:501–510. With permission from Elsevier.)

and eukaryotes, the complexes that accomplish this are known as **exosome complexes**. Exosome complexes are related, both in sequence and structure, to the bacterial RNase PH, an enzyme that uses inorganic phosphate to mediate cleavage of RNA. All RNA degradation complexes of this class contain a ring of six protein subunits, with a hole in the middle through which the RNA to be degraded passes. The archaeal and eukaryotic exosome rings are not stable on their own and require additional cap proteins to stabilize their structures. The exosome complexes degrade RNA exonucleolytically, in a $3' \rightarrow 5'$ direction.

RNA degradation in bacteria occurs as a succession of steps (**Figure 14.26**). Four distinct enzymatic processes are involved, each catalyzed by one or more enzymes: endonucleolytic cleavage, oligoadenylation, exonucleolytic cleavage, and helicase action. The addition of an oligo(A) tail and the helicase activity are required for the removal of stable secondary structures that would otherwise preclude normal exonucleolytic degradation.

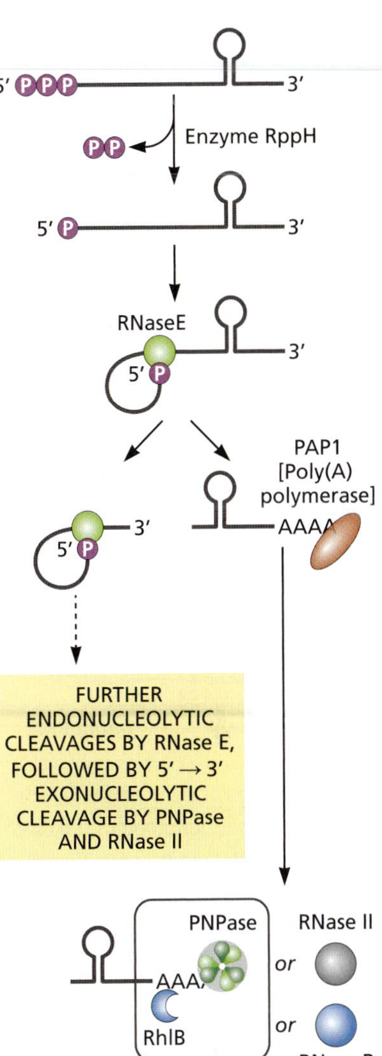

Figure 14.26 RNA degradation pathways in bacteria. RNA turnover is initiated by removal of a pyrophosphate, PP, from the 5′-end of the RNA and recognition of the resulting 5′-monophosphate by the endonuclease RNase E, which cleaves the mRNA into two pieces. The 5′-fragment is further degraded by the combined action of RNase E and the exonucleolytic activities of PNPase and RNase II. When the 3′-end happens to be protected by a stem–loop structure and proteins associated with it, it is first oligoadenylated by the bacterial poly(A) polymerase PAP1. The oligo(A) tail recruits the bacterial degradosome, PNPase plus the helicase RhlB, or is degraded by one of two other hydrolytic 3′ → 5′ exonucleases, RNase II or RNase R. Even with the wide range of phosphorolytic and hydrolytic enzymes available, degradation in bacteria seems to be predominantly hydrolytic. (Adapted from Lykke-Andersen S, Brodersen DE & Jensen TH [2009] *J Cell Sci* 122:1487–1494. With permission from The Company of Biologists.)

Archaea and eukaryotes utilize specific pathways to deal with different RNA defects

During evolution, the pathways involved in degrading RNA have become more complex, highly regulated, and divergent, with distinct pathways dedicated to the degradation of RNA carrying particular molecular defects. Degradation can occur either in the nucleus or in the cytoplasm. Some degradation occurs co-transcriptionally or immediately after transcription is terminated, as soon as processing and/or RNA packaging defects are detected. Nuclear degradation can occur in all three major RNA classes: rRNA, tRNA, and mRNA. A detailed description of nuclear degradation of faulty RNA molecules is presented in **Figure 14.27**.

In the cytoplasm, degradation occurs either at the end of the useful lifetime of an RNA molecule or to destroy faulty RNA molecules that have escaped nuclear surveillance and should not be translated. The three major processes, nonsense-mediated decay or NMD, no-go decay or NGD, and **non-stop decay** or **NSD**, are all intimately involved with translation and are discussed in Chapter 17.

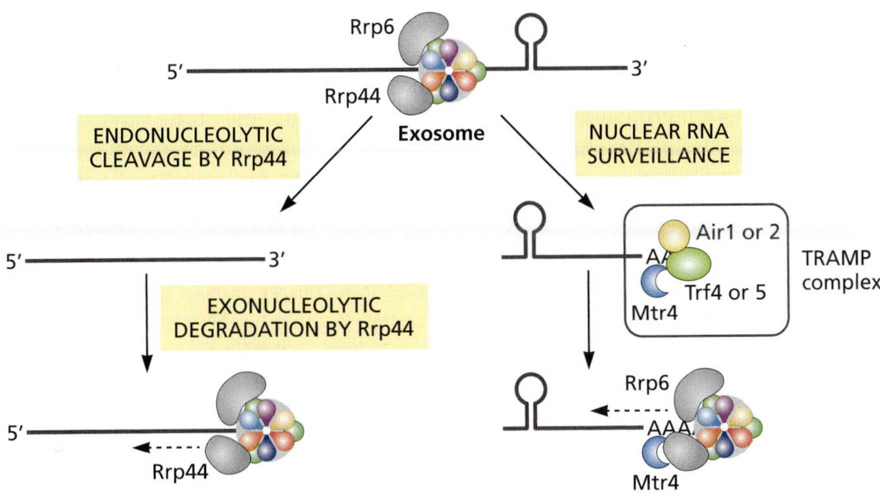

Figure 14.27 Nuclear RNA degradation pathways in eukaryotes. In eukaryotes, RNA degradation begins with endonucleolytic cleavage by exosome component Rrp44 and 3′-oligoadenylation by the nuclear TRAMP complex, which consists of a helicase, Mtr4; one of two poly(A) polymerases, Trf4 or Trf5; and one of two RNA-binding proteins, Air1 or Air2. The Mtr1 helicase can also associate directly with the exosome. TRAMP is also involved in 3′-end processing of stable RNAs that are not destined for degradation. Thus, it is likely that TRAMP and the exosome survey the entire RNA population but degrade only transcripts that lack protective secondary structures or certain RNA-binding proteins. (Adapted from Lykke-Andersen S, Brodersen DE & Jensen TH [2009] *J Cell Sci* 122:1487–1494. With permission from The Company of Biologists.)

Table 14.2 Three major types of small silencing RNAs.

	Type	Organisms	Length (nt)	Function
miRNA	microRNA	viruses, protists, algae, plants, animals	20–25	mRNA degradation; inhibition of translation
siRNA	small interfering RNA	all eukaryotes, mainly plants	21–24	post-transcriptional silencing of transcripts and transposons; in some cases, silencing of transcription
piRNA	Piwi-interacting RNA	metazoa	21–30	transposon regulation and unknown functions; maintain germline stem cells and promote their division

14.9 Biogenesis and functions of small silencing RNAs

All ssRNAs are produced by processing from larger precursors

Small silencing RNAs, ssRNAs, are a heterogeneous group of RNAs that perform distinct functions in silencing gene expression at the level of transcription, post-transcriptional regulation, or translation. The defining features of ssRNAs are short length, usually not exceeding 30 nucleotides, and association with members of the **Argonaute** family of proteins, which serve the effector function in the silencing pathways. Argonaute proteins are ribonucleases characterized by two domains: Piwi, a ribonuclease domain, and PAZ, an ssRNA-binding domain. The function of the ssRNA is to guide the effector Argonaute proteins to their nucleic acid targets. Table 14.2 lists some characteristics of the three major ssRNA classes. It should be noted that the mechanism of action of these small RNAs cannot be strictly defined as postsynthetic RNA processing as it does not represent a covalent change in RNA. However, their biosynthesis, in every case, involves processing of larger RNA precursors, which warrants the discussion of their biogenesis here.

As far as the regulation of target mRNA is concerned, all ssRNAs use a similar mechanism of interaction with their targets. As Figure 14.28 illustrates, the ssRNA interacts with its target by base pairing, which can be either very extensive or partial. Actually, the extent of base pairing determines the mechanical outcome of the silencing reaction: extensive base pairing leads to destruction of mRNA, whereas partial base pairing results in inhibition of translation. In either case, the biological outcome is the silencing of gene expression.

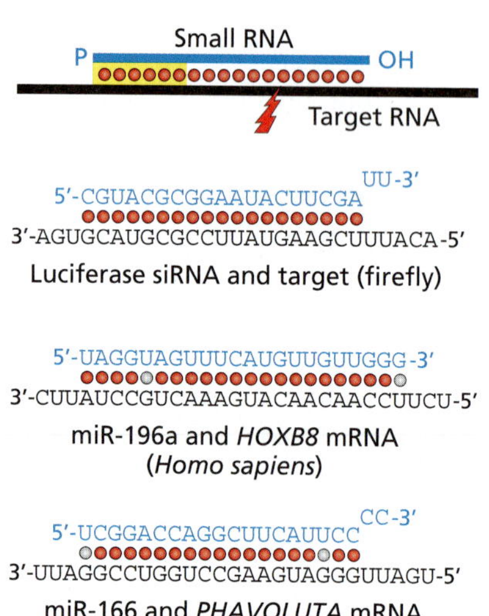

(B) Inhibition of translation

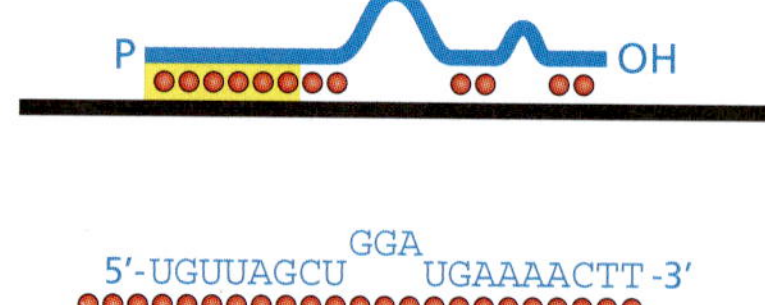

CXCR4 siRNA and target

UC C A C A A G

5′-UUCCCUGAG UGUGA-3′

3′-UCCAGGGACUCAACCAACACUACC-5′

Lin-4 miRNA and *Lin-14* mRNA (*C. elegans*)

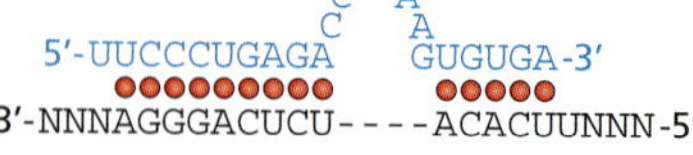

Lin-4 miRNA and *Lin-28* mRNA (*C. elegans*)

Figure 14.28 Two modes of small RNA binding to their target mRNAs determine distinct mechanisms of action. (A) Extensive base pairing of a small RNA to the 3′-UTR of a target mRNA guides catalytically active Argonaute proteins to specific mRNA molecules; the Argonaute proteins then cut a single phosphodiester bond in the mRNA, triggering its destruction. This mode of binding is customary for plants and some mammalian miRNA. The three examples given show the extensive base pairing that occurs between three ssRNAs and their respective targets. (B) Partial base pairing between a small RNA and the 3′-UTR of a target mRNA tethers an Argonaute protein to its mRNA target; the miRNA–Argonaute complex prevents translation. The first miRNA to be discovered, *Lin-4* miRNA in *Caenorhabditis elegans*, acts in this way on two closely related mRNAs, transcribed from the *Lin-14* and *Lin-28* genes. Note that the complexes are slightly different in the miRNA nucleotides that base-pair with the corresponding mRNAs. During larval development, *Lin-4* coordinates the down-regulation of LIN-14 and LIN-28 protein concentrations, which in turn regulates the expression of stage-specific developmental events. In both parts of the figure, miRNA contains a short seed sequence at its 5′-end, marked with yellow highlighting, that contributes most of the energy for target binding; that is, it is the specificity determinant for target selection. The small size of the seed region allows a single miRNA to regulate many different genes. (Adapted from Zamore PD & Haley B [2005] *Science* 309:1519–1524. With permission from American Association for the Advancement of Science.)

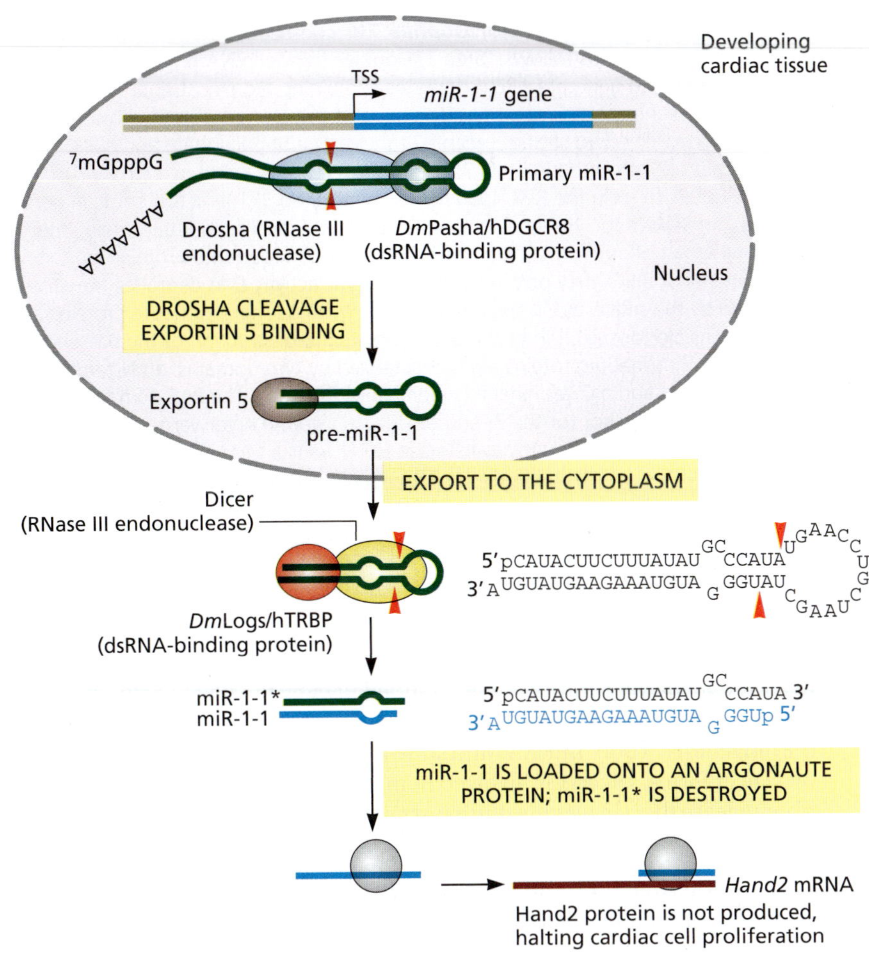

Figure 14.29 Processing of a microRNA involved in gene regulation. This microRNA example is from developing cardiac tissue in *Drosophila*. The consecutive stages of processing are as follows: (Step 1) Transcription factors SRF or serum response factor and MyoD stimulate Pol II transcription of the *miR-1-1* gene. (Step 2) The primary transcript is processed by the RNase III endonuclease Drosha with its dsRNA-binding partner Pasha, or partner DGCR8 in humans. The product, pre-miR-1-1, contains a 2-nt single-strand 3′-overhang, which is recognized by exportin 5 for transport into the cytoplasm. (Step 3) In the cytoplasm, a second RNase III endonuclease, Dicer, with its own dsRNA-binding partner, makes a second pair of cuts to liberate the miRNA–miRNA* duplex. The mature 21-nt-long miRNA is loaded, by a specialized loader, onto an Argonaute family member protein; the miRNA* chain is destroyed. (Step 4) The Argonaute protein, an effector of small-RNA-directed silencing, is now guided to the 3′-untranslated region of the mRNA, in this case *Hand2* mRNA, by the miRNA, where the protein represses translation of the Hand2 protein. This halts cardiac cell proliferation. (Adapted from Zamore PD & Haley B [2005] *Science* 309:1519–1524. With permission from American Association for the Advancement of Science.)

There are three major types of small silencing RNAs:

1. **MicroRNAs** or **miRNAs**. The first miRNA to be discovered was the *lin-4* gene product in *Caenorhabditis elegans*, which is involved in regulating the expression of two important developmental genes, *lin-14* and *lin-28*. This discovery was quickly followed by descriptions of a whole range of miRNA molecules in a wide variety of organisms. To date, ~7000 miRNA genes have been identified in animals and their viruses; at least another ~1600 exist in plants. We illustrate miRNA biogenesis using the example of silencing of the *Hand2* gene in developing cardiac tissue in *Drosophila* (Figure 14.29). The process is very similar for all miRNAs, in terms of both the succession of steps and the proteins involved. In general, the process is characterized by the action of the cytoplasmic RNase III endonuclease **Dicer**.

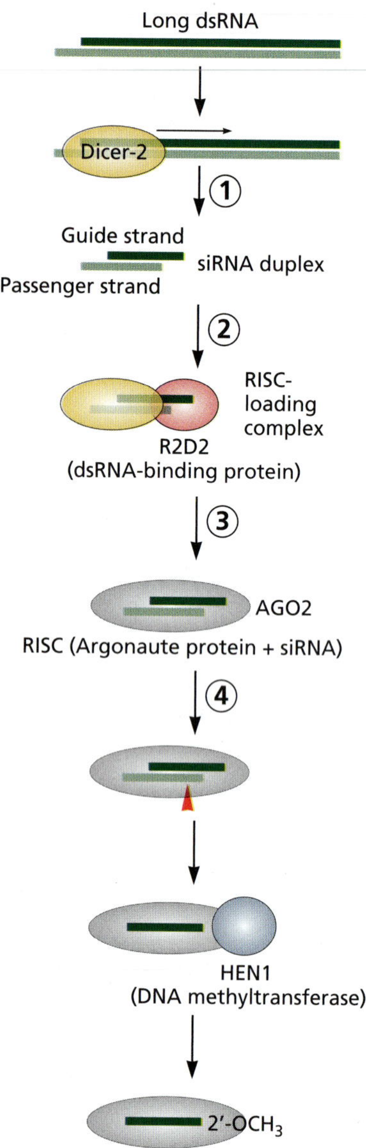

Figure 14.30 Processing of siRNAs in *Drosophila*. The consecutive stages of processing for small interfering RNAs are as follows: (Step 1) Double-stranded RNA precursors, dsRNA, are processed by Dicer-2 to generate siRNA duplexes containing a guide strand, which is of physiological relevance as it directs the silencing, and a passenger strand, which is subsequently degraded. The two strands, each ~21 nt, carry a 5′-phosphate and a 3′-OH group, and each has a 2-nt overhang at its 3′-end. (Step 2) Dicer-2 partners with a dsRNA-binding protein, R2D2, to form the RISC-loading complex. (Step 3) The active entity in gene silencing is RISC, an RNA-induced silencing complex of AGO2, an Argonaute protein, and siRNA. siRNA guides Argonaute proteins to their RNA targets. Argonaute proteins are the catalytic components of RISC: they possess endonucleolytic activity that degrades the mRNA that is recognized by the siRNA by complementary base pairing. The Argonaute proteins are also partially responsible for selection of the guide strand and destruction of the passenger strand of the siRNA. Argonaute proteins are characterized by two domains: an N-terminal PAZ domain of ~20 kDa and a C-terminal Piwi domain of ~40 kDa. The PAZ domain interacts with RNA, serving as the anchor for the 3′-end of siRNAs. (Step 4) Afterward, the passenger strand is destroyed, and the DNA methyltransferase HEN1 adds a methyl group to the 2′-OH group of the siRNA, which stabilizes the RNA. Finally, the siRNA carrying the catalytic Argonaute protein interacts with the target mRNA to cleave it; this step is not shown. (Adapted from Ghildiyal M & Zamore PD [2009] *Nat Rev Genet* 10:94–108. With permission from Macmillan Publishers, Ltd.)

2. **Small interfering RNAs** or **siRNAs.** These are also called short interfering RNAs or silencing RNAs. A characteristic distinguishing feature of siRNAs is that they are derived from double-stranded RNA by the RNase activity of Dicer (**Figure 14.30**). Mammals and *C. elegans* have a single Dicer that participates in the biogenesis of both miRNAs and siRNAs; *Drosophila*, on the other hand, has two Dicer forms. Dicer-1 is involved in making miRNA, whereas Dicer-2 is the ribonuclease that cleaves the double-stranded RNA in the early stages of siRNA biogenesis.

3. **Piwi-interacting RNAs** or **piRNAs.** These were discovered in 2001 in the *Drosophila* germline, where they repress transposons, thus stabilizing the germline genome. piRNA sequences are very diverse, with more than 1.5 million distinct piRNAs identified in *Drosophila*. These are clustered to a few hundred genomic regions. It remains to be established how many of these identified piRNAs have physiological significance. Their source seems to be extremely long ssRNA transcripts, 100,000–200,000 nucleotides, usually antisense. piRNAs are distinct from other small interfering RNAs in that they bind Piwi proteins, a clade of the Argonaute protein superfamily, and do not require Dicer for their biogenesis. The process of piRNA biogenesis is described in **Figure 14.31**.

Key concepts

- Many kinds of RNA molecules must undergo one or more forms of post-transcriptional processing before they can play their appropriate roles in the cell.
- In bacteria, mRNA can be used directly for translation without processing, but functional tRNAs and rRNAs are generated by cleavage and trimming of tandem transcripts. tRNA molecules also undergo a variety of base modifications.
- The pre-mRNAs synthesized in eukaryotic nuclei undergo a series of modifications before they are exported to the cytoplasm. These include 5′-capping, removal of introns and splicing of exons, 3′-polyadenylation, and in some cases editing: insertion or deletion of nucleotide residues and/or chemical modification of bases.
- Capping is the addition of GMP at the 5′-end of a message, in a 5′–5′ orientation. The cap recruits proteins that protect the message from exonucleases, aids in transport of the processed message to the cytoplasm, and serves as the site of ribosomal attachment during initiation of translation.
- Polyadenylation involves the addition of a poly(A) tail to the 3′-end of the message. The tail, with recruited proteins, protects this end from exonucleolytic attack.

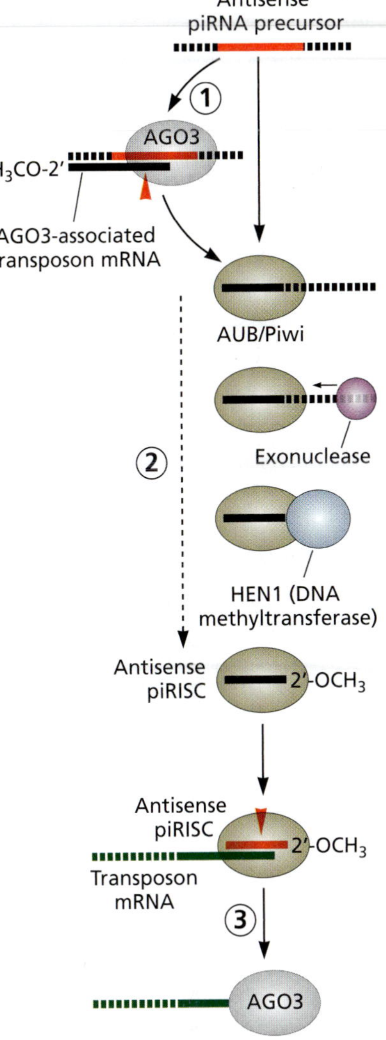

Figure 14.31 Current model for piRNA biogenesis. piRNAs are thought to derive from ssRNA precursors and do not go through a dicing step. The current model for their biogenesis is derived from the sequences of piRNAs bound to three Argonaute proteins belonging to the Piwi Argonaute clade: Piwi, Aubergine or AUB, and AGO3 in *Drosophila*. piRNA bound to Piwi and AUB is typically antisense to transposon mRNA, while piRNA bound to AGO3 corresponds to a portion of the transposon mRNA itself; that is, sense. The first 10 nucleotides of antisense piRNAs are frequently complementary to the sense piRNAs found in AGO3. This unexpected complementarity is part of a mechanism for piRNA amplification that is activated only after transcription of transposon mRNA. The consecutive stages of processing are as follows: (Step 1) The piRNA precursor binds AGO3-associated transposon mRNA and cleaves it across from position 10 of the antisense piRNA guide. The sense piRNA can, in turn, guide cleavage of the antisense piRNA precursor transcript. (Step 2) Several substeps lead to creation of the antisense piRISC, which can further interact with transposon mRNA. (Step 3) The 5′-end of the cleaved transposon mRNA product loads onto AGO3. (Adapted from Ghildiyal M & Zamore PD [2009] *Nat Rev Genet* 10:94–108. With permission from Macmillan Publishers Ltd.)

- Removal of most introns and resplicing of exons require complex nuclear machines called spliceosomes. These recognize 3′- and 5′-splice sites, plus an internal intron site, catalyze cleavage, and then religate adjacent exons. The spliceosome is a huge complex of a number of RNA molecules and many proteins.
- In a few cases, introns are self-splicing; they excise themselves from the RNA while ligating exons. In such cases, the RNA in the intron has the catalytic power, even though proteins may help to maintain the structure. Such RNA molecules that act like protein enzymes are called ribozymes.
- In many cases, splicing can take alternate routes, adding or excluding exons or using alternative or cryptic splice sites. Such alternative splicing may occur within a given gene or may even involve exons from distant genes. Alternative splicing has the effect that one gene can often produce different protein products in different tissues or at different stages of development.
- Regulation of alternative splicing involves a number of factors, including splice-site strength, protein enhancers or silencers, RNA secondary structure, and chromatin structure.
- Processed eukaryotic mRNAs are exported to the cytoplasm through nuclear pores, via a ratcheted Brownian motion mechanism.
- After mRNA molecules have been processed in all of these ways, they can still be degraded if they are imperfect or if the cell no longer requires the protein product. Such degradation can occur in either the nucleus or cytoplasm.
- Small silencing RNAs such as microRNA, small interfering RNA, and Piwi-interacting RNA, are generated from much larger gene products.

Further reading

Books

Gesteland RF, Cech TR & Atkins JF (eds) (2006) The RNA World, 3rd ed. Cold Spring Harbor Laboratory Press.

Lilley DMJ & Eckstein F (eds) (2007) Ribozymes and RNA Catalysis. RSC Publishing.

Reviews

Bentley DL (2005) Rules of engagement: Co-transcriptional recruitment of pre-mRNA processing factors. *Curr Opin Cell Biol* 17:251–256.

Black DL (2003) Mechanisms of alternative pre-messenger RNA splicing. *Annu Rev Biochem* 72:291–336.

Chen M & Manley JL (2009) Mechanisms of alternative splicing regulation: Insights from molecular and genomics approaches. *Nat Rev Mol Cell Biol* 10:741–754.

Doma MK & Parker R (2007) RNA quality control in eukaryotes. *Cell* 131:660–668.

Doudna JA & Cech TR (2002) The chemical repertoire of natural ribozymes. *Nature* 418:222–228.

Fedor MJ (2008) Alternative splicing minireview series: Combinatorial control facilitates splicing regulation of gene expression and enhances genome diversity. *J Biol Chem* 283:1209–1210.

Ghildiyal M & Zamore PD (2009) Small silencing RNAs: An expanding universe. *Nat Rev Genet* 10:94–108.

Gingeras TR (2009) Implications of chimaeric non-co-linear transcripts. *Nature* 461:206–211.

Graveley BR (2009) Alternative splicing: Regulation without regulators. *Nat Struct Mol Biol* 16:13–15.

Hamma T & Ferré-D'Amaré AR (2010) The box H/ACA ribonucleoprotein complex: Interplay of RNA and protein structures in post-transcriptional RNA modification. *J Biol Chem* 285:805–809.

Kapranov P, Willingham AT & Gingeras TR (2007) Genome-wide transcription and the implications for genomic organization. *Nat Rev Genet* 8:413–423.

Köhler A & Hurt E (2007) Exporting RNA from the nucleus to the cytoplasm. *Nat Rev Mol Cell Biol* 8:761–773.

Kornblihtt AR, Schor IE, Allo M & Blencowe BJ (2009) When chromatin meets splicing. *Nat Struct Mol Biol* 16:902–903.

Licatalosi DD & Darnell RB (2010) RNA processing and its regulation: Global insights into biological networks. *Nat Rev Genet* 11:75–87.

Lykke-Andersen S, Brodersen DE & Jensen TH (2009) Origins and activities of the eukaryotic exosome. *J Cell Sci* 122:1487–1494.

Nilsen TW & Graveley BR (2010) Expansion of the eukaryotic proteome by alternative splicing. *Nature* 463:457–463.

Schellenberg MJ, Ritchie DB & MacMillan AM (2008) Pre-mRNA splicing: A complex picture in higher definition. *Trends Biochem Sci* 33:243–246.

Schmid M & Jensen TH (2008) The exosome: A multipurpose RNA-decay machine. *Trends Biochem Sci* 33:501–510.

Schwartz S & Ast G (2010) Chromatin density and splicing destiny: On the cross-talk between chromatin structure and splicing. *EMBO J* 29:1629–1636.

Sperling J, Azubel M & Sperling R (2008) Structure and function of the pre-mRNA splicing machine. *Structure* 16:1605–1615.

Srebrow A & Kornblihtt AR (2006) The connection between splicing and cancer. *J Cell Sci* 119:2635–2641.

Stewart M (2010) Nuclear export of mRNA. *Trends Biochem Sci* 35:609–617.

Toor N, Keating KS & Pyle AM (2009) Structural insights into RNA splicing. *Curr Opin Struct Biol* 19:260–266.

Vicens Q & Cech TR (2006) Atomic level architecture of group I introns revealed. *Trends Biochem Sci* 31:41–51.

Zamore PD & Haley B (2005) Ribo-gnome: The big world of small RNAs. *Science* 309:1519–1524.

Experimental papers

Alt FW, Bothwell ALM, Knapp M et al. (1980) Synthesis of secreted and membrane-bound immunoglobulin μ heavy chains is directed by mRNAs that differ at their 3′ ends. *Cell* 20:293–301.

Cech TR, Zaug AJ & Grabowski PJ (1981) *In vitro* splicing of the ribosomal RNA precursor of *Tetrahymena*: Involvement of a guanosine nucleotide in the excision of the intervening sequence. *Cell* 27:487–496.

Early P, Rogers J, Davis M et al. (1980) Two mRNAs can be produced from a single immunoglobulin μ gene by alternative RNA processing pathways. *Cell* 20:313–319.

Fire A, Xu S, Montgomery MK et al. (1998) Potent and specific genetic interference by double-stranded RNA in *Caenorhabditis elegans*. *Nature* 391:806–811.

Fong N & Bentley DL (2001) Capping, splicing, and 3′ processing are independently stimulated by RNA polymerase II: Different functions for different segments of the CTD. *Genes Dev* 15:1783–1795.

Guerrier-Takada C, Gardiner K, Marsh T et al. (1983) The RNA moiety of ribonuclease P is the catalytic subunit of the enzyme. *Cell* 35:849–857.

Kruger K, Grabowski PJ, Zaug AJ et al. (1982) Self-splicing RNA: Autoexcision and autocyclization of the ribosomal RNA intervening sequence of *Tetrahymena*. *Cell* 31:147–157.

Moss EG, Lee RC & Ambros V (1997) The cold shock domain protein LIN-28 controls developmental timing in *C. elegans* and is regulated by the *lin-4* RNA. *Cell* 88:637–646.

Olson S, Blanchette M, Park J et al. (2007) A regulator of *Dscam* mutually exclusive splicing fidelity. *Nat Struct Mol Biol* 14:1134–1140.

Sperling R, Sperling J, Levine AD et al. (1985) Abundant nuclear ribonucleoprotein form of CAD RNA. *Mol Cell Biol* 5:569–575.

Tress ML, Martelli PL, Frankish A et al. (2007) The implications of alternative splicing in the ENCODE protein complement. *Proc Natl Acad Sci USA* 104:5495–5500.

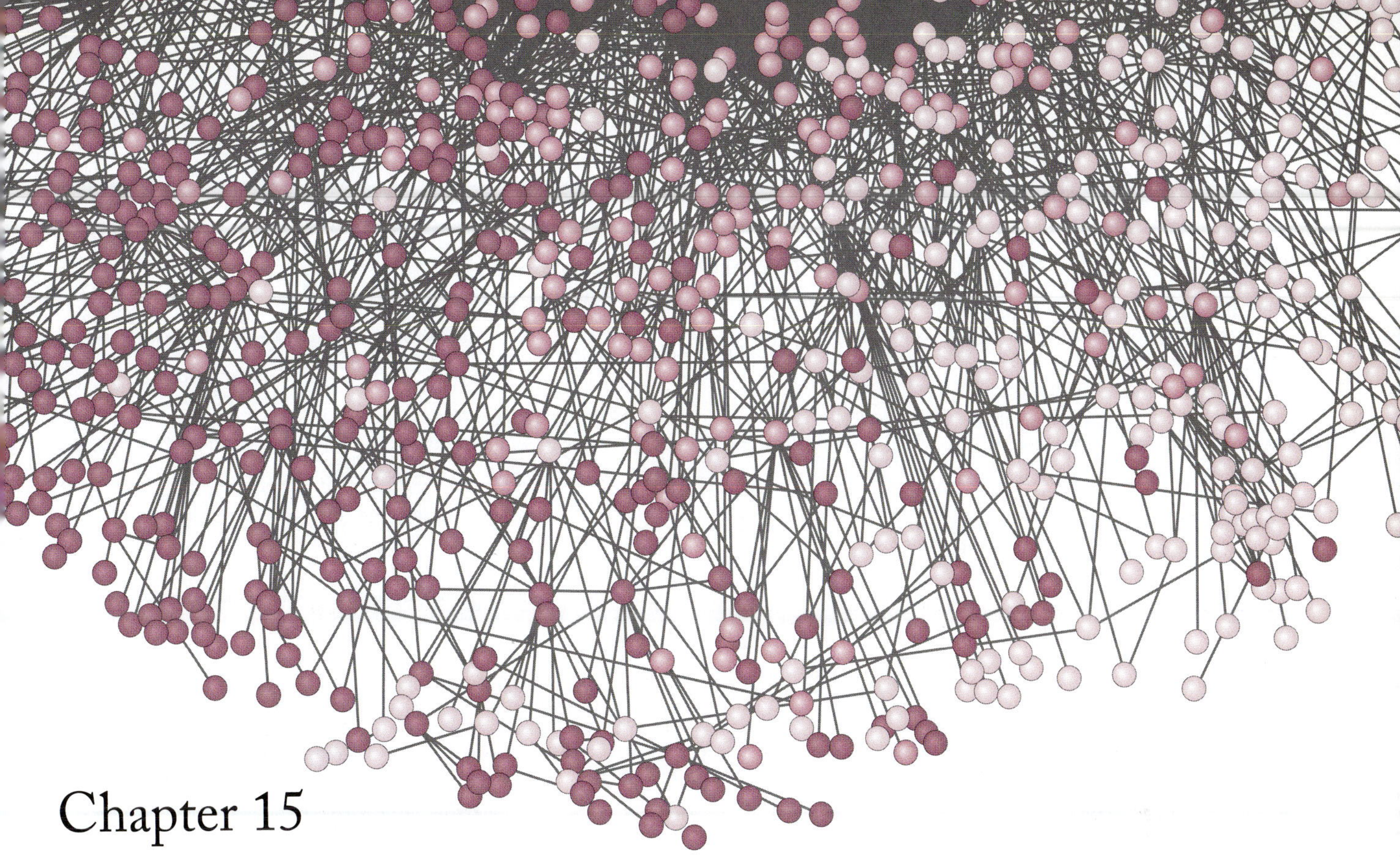

Chapter 15

Translation: The Players

15.1 Introduction

To this point, this book has mainly been concerned with transcription. This process, although certainly complex in details, is at base fundamentally simple. It involves making a polynucleotide strand complementary to another, using base-pairing rules like those that dictate DNA helices. However, to take a further step and use an RNA sequence to specify a polypeptide sequence poses a fundamentally more difficult problem.

In transcription, both the sending and receiving of genetic information use a single language, that of nucleotides, but the transfer of information from RNA to proteins is much more complicated. It requires the coordinated use of the nucleic acid language during codon–anticodon recognition and also recognition between nucleic acids and proteins during the initial state of creating charged transfer RNAs, tRNAs. A preliminary view of translation is presented in Chapter 7. In this chapter we focus on the details of translation, describing the molecular players in the process, and then in Chapter 16 we cover the phases of the translation process itself. In Chapter 17, we reveal what is known about the complex regulation of translation and also delineate the major co- and post-translational processing and modifications that determine final protein structure and function.

15.2 A brief overview of translation

Three participants are needed for translation to occur

As described in Chapter 7, translation requires three kinds of participants: messenger RNA or mRNA, which provides the message to be translated; a set of tRNAs, each charged with an appropriate amino acid; and a ribosome, as a platform and catalyst for the process.

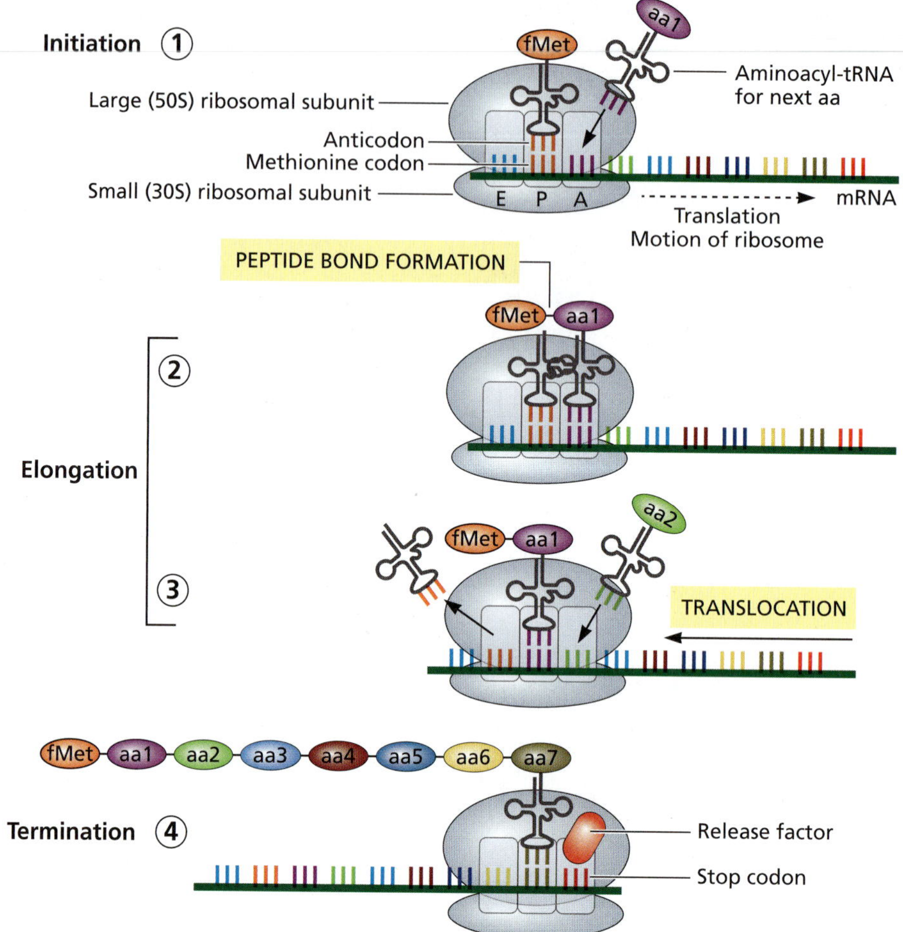

Figure 15.1 Overview of translation. Schematic of the ribosome with its three sites: exit or E site interacts with uncharged, deacylated tRNA; peptidyl or P site interacts with tRNA that carries the nascent polypeptide chain; and acceptor or A site interacts with tRNA that carries the incoming amino acid (aa). (Step 1) A special initiator tRNA, carrying the amino acid formylmethionine, fMet, in bacteria or methionine in eukaryotes, binds to the P site of the ribosome that accommodates the start codon. A new aminoacyl-tRNA or aa-tRNA then joins the complex, entering into the A site. The incoming amino acid must be cognate: that is, the anticodon on the carrier tRNA must correspond to the codon in the message, to ensure truthful transmission of the information encoded in the mRNA into the sequence of amino acids in the polypeptide chain. (Step 2) Peptide bond formation. (Step 3) Translocation of the ribosome, with its bound tRNAs, with respect to the mRNA, so that the next codon is now in the A site, ready to accept the next amino acid specified by the mRNA codon. (Step 4) When a stop codon enters the A site, translation is terminated with the help of special release factors.

Let us first begin with a rather simplified schematic of translation, to put into perspective the roles of these major players. The schematic in Figure 15.1 depicts, in an elementary way, the major stages in the translation process: initiation, elongation, and termination. As we see in Chapter 16, peptide bond formation and translocation are part of a three-step microcycle that adds each amino acid to the nascent polypeptide chain during elongation. We present initiation, elongation, and termination steps separately here because of the importance of each step in understanding the overall process.

All of these steps occur on the ribosome, a complex RNA–protein machine that acts as both stage and director for translation. The ribosome carries three sites called P or peptidyl, A or acceptor, and E or exit, which can each accommodate a tRNA molecule (see Figure 15.1). Translation initiates with the binding of a special initiator tRNA or tRNAi to the P site. The initiator tRNA must carry the amino acid formylmethionine, in bacteria, or methionine, in eukaryotes. In either case, this amino acid is encoded in the messenger RNA by a methionine codon, which must be aligned on the ribosome at the P site. Thus, during initiation, the anticodon on the initiator tRNA must interact with one of the initiation methionine codons on the mRNA. The codons that are capable of this interaction are AUG, GUG, and UUG in bacteria and just AUG in eukaryotes. At this point, a new amino acid is brought to the A site on the ribosome by its own tRNA. The transfer of sequence information is ensured by the fact that the accepted incoming tRNA with its bound amino acid carries an anticodon that corresponds to the mRNA codon in the A site. At this point we have two adjacent amino acids, specified by neighboring codons in the mRNA, that are positioned appropriately for the catalytic step of peptide bond formation. During this step, the amino acid carried by the initiator tRNA, during initiation, or the nascent polypeptide, during subsequent elongation, is transferred from the P site to attach to the tRNA in the A site. It is counterintuitive that the entire growing polypeptide chain, which during elongation may become hundreds or even thousands of amino acid residues long, would be transferred onto the lone amino acid in the A site, rather than the new amino acid jumping onto the stationary peptide chain stationed at the P site, but this is the way it occurs; the chemistry of the reaction and the conformational transitions in the ribosome are detailed in Chapter 16.

The next important step involves translocation movements of the ribosome, with its bound tRNAs, with respect to the mRNA, so that now a new codon is placed into the A site. During translocation, the tRNA, now carrying a chain that has grown by one amino acid, moves from the A into the P site, and the deacylated or discharged tRNA moves from the P site into the third ribosomal site, the E site, from where it leaves the ribosome. When these movements are all complete, the polypeptide chain has grown by one amino acid and the ribosome is ready for a new cycle of addition, with the A site ready to accept the new **aminoacyl-tRNA** or aa-tRNA specified by the new mRNA codon in this site.

The final step in translation is termination of synthesis, which occurs when a stop codon enters the A site. This step does not require the participation of tRNA of any kind. Instead, it occurs through the action of specific protein release factors, which can be accommodated in the A site across from a stop codon. Termination of polypeptide chain synthesis is followed by dissociation of all bound components and splitting of the ribosome into its subunits, so that re-initiation begins on the small subunit.

With this very simplified synopsis of the process, we turn now to detailed examination of the players: tRNAs, mRNAs, and ribosomes. In Chapter 16 we show how they work together to synthesize a protein.

15.3 Transfer RNA

Only a couple of years after Crick proposed his adaptor hypothesis (**Box 15.1**), experimenters discovered both the adaptor molecules and the enzymes involved in attaching the amino acids to the adaptors. The adaptors turned out to be contained in the soluble RNA fraction as named by their discoverer Mahlon Hoagland,

Box 15.1 The RNA Tie Club and the adaptor hypothesis The RNA Tie Club, founded in 1954 by James Watson and George Gamow, was a scientific gentleman's club of select members who discussed issues relating to the genetic code and how it was read. The club consisted of 20 members, each representing one amino acid, with an additional four honorary members representing the four nucleotides. The Club met twice a year in an informal setting, but in between meetings, members wrote notes to each other in which they discussed their ideas.

After George Gamow's idea of the existence of a code in which several nucleotides would code for one amino acid was accepted, the question remained how such a code would work. Abstract attempts were made to fit the amino acid sequence of a polypeptide chain directly onto DNA as the physical form by which the code is read. In a note written to the RNA Tie Club in 1955, *On Degenerate Templates and the Adaptor Hypothesis*, Francis Crick argued that this was not possible and proposed his bold ideas of how the transfer of information from nucleic acids to proteins might work.

In his note, Crick argued that the pure physical dimensions of the DNA, with base-pair spacing of ~3.4 Å in fiber direction, are not compatible with the ~3.7 Å distance from one amino acid to the next in the polypeptide chain. In addition, if the physical-chemical nature of the amino acid side chains is considered, there are no obvious complementary features on the nucleic acid. Crick suggested the existence of a small adaptor molecule that would chemically bind the amino acid, through the action of special enzymes, and combine it with the nucleic acid template. In Crick's words: "In its simplest form there would be 20 different kinds of adaptor molecules, one for each amino acid, and 20 different enzymes to join the amino acid to their adaptors." The actual name of the hypothesis, the adaptor hypothesis, was proposed by Sydney Brenner, with whom Crick discussed the idea.

Nowadays, we know that the mysterious adaptor molecules are tRNAs and the special enzymes are the aminoacyl-tRNA synthetases.

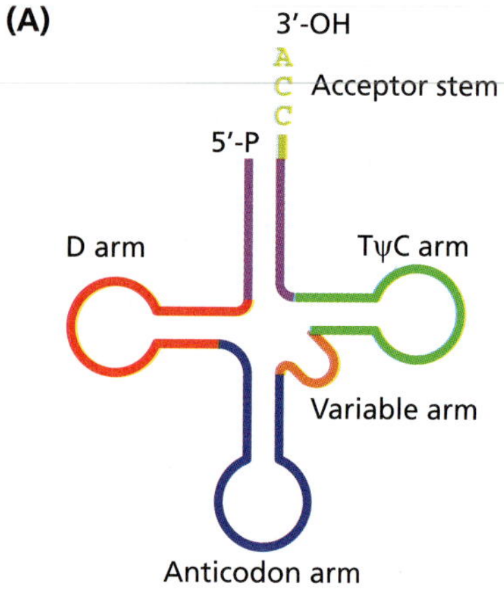

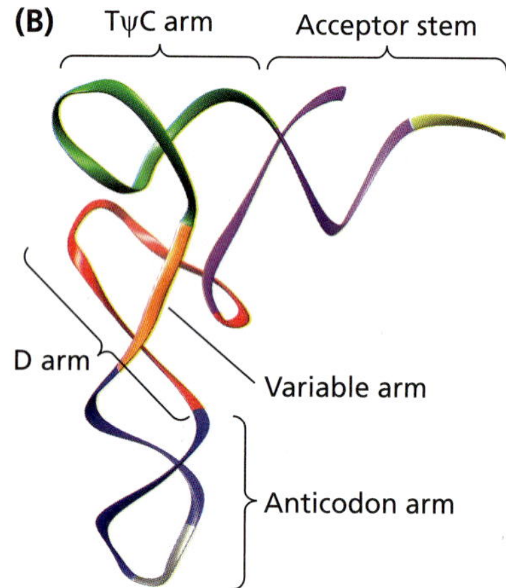

Figure 15.2 Generalized cloverleaf secondary structure and three-dimensional (3D) folding of tRNA. Distinct regions in (A) the cloverleaf model and (B) the 3D structure are color-coded in the same scheme. Some arms derive their names from specifically modified residues that occur frequently in these parts of the polynucleotide: D, dihydrouridine; T, ribothymidine; ψ, pseudouridylate. The amino acid attaches through its COOH group to the 3′- or the 2′-OH groups of the ribose of the terminal adenine, A, residue in the universal CCA triplet at the acceptor stem's terminus. The anticodon arm contains the triplet anticodon, which specifies the amino acid that is to be incorporated into the growing peptide chain by interacting with the codon of the mRNA.

working in Paul Zamecnik's laboratory in 1958; they were later termed transfer RNAs or tRNAs. Hoagland also recognized the existence, in crude cellular extract, of **aminoacyl-tRNA synthetases** that specifically attach an amino acid to its respective cognate tRNA.

tRNA molecules fold into four-arm cloverleaf structures

Each cell, whether bacterial, archaeal, or eukaryotic, contains a set of tRNA molecules, each of which mediates incorporation of one of the 20 canonical amino acids present in proteins into the polypeptide chain during translation. We know that there are 61 meaningful codons that specify the 20 amino acids. Recall from Chapter 7 that because of the redundancy of the genetic code, some amino acids are encoded by more than one codon. Does this mean that there are 61 different types of tRNA in each cell? The answer is no, because some tRNAs can recognize more than one codon; the molecular basis for this is defined by the wobble hypothesis proposed by Crick in 1966. In *Escherichia coli*, for example, there are 40 tRNA types to accommodate the 20 amino acids and the 61 meaningful or sense codons. tRNA molecules containing different anticodons that specify the same amino acid are termed **isoacceptor tRNA** molecules. The specificity of a tRNA in terms of the amino acid it carries is designated by a superscript, for example, $tRNA^{Thr}$. The three remaining codons usually serve as stop codons, but in a few situations they can act as codons for unusual amino acids (see Chapter 7).

tRNAs are relatively short RNA molecules, usually 73–74 nucleotides long, whose sequence allows the formation of four different arms by intramolecular base pairing. The **cloverleaf structure** representations of two-dimensional (2D) tRNA structures are useful for visualizing the general patterns of hydrogen bonding and for denoting the individual arms (**Figure 15.2** and **Figure 15.3**). These figures also show that the three-dimensional (3D) folding of tRNA is more complex than the 2D patterns might suggest. Two of the arms are named to reflect their function: the anticodon and acceptor arms. The other two, the TψC arm and the D arm, are given their respective names because

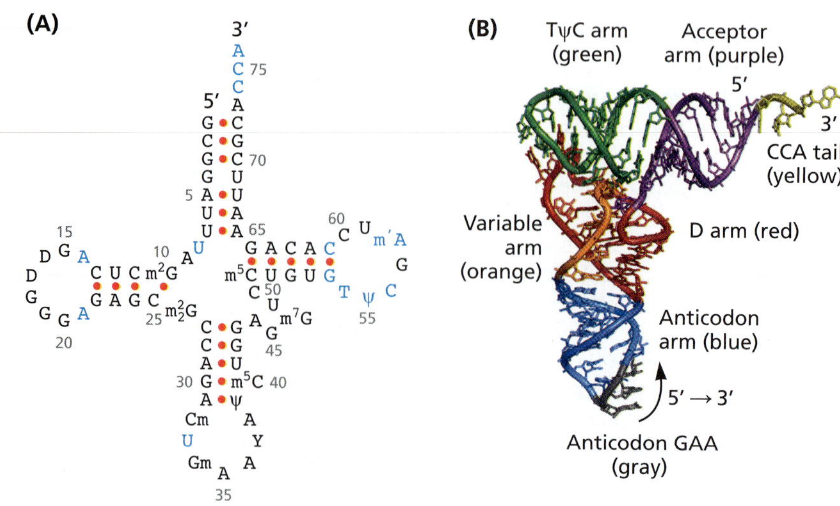

Figure 15.3 Primary, secondary, and tertiary structure of $tRNA^{Phe}$ from yeast. (A) Nucleotide sequence and cloverleaf secondary structure, with two-dimensional base-pairing depicted by red dots. The bases in blue are invariantly present in all tRNAs. (B) The crystal structure reveals the three-dimensional organization of the helices, with the individual domains represented in different colors. The structure is stabilized by non-Watson–Crick base-pairing between the TψC loop and the D loop; stacking interactions align the TψC arm with the acceptor arm and the D arm with the anticodon arm. Note the exposed bases that form the anticodon; this organization helps the interaction with the codon on the mRNA. (Adapted from Wikimedia.)

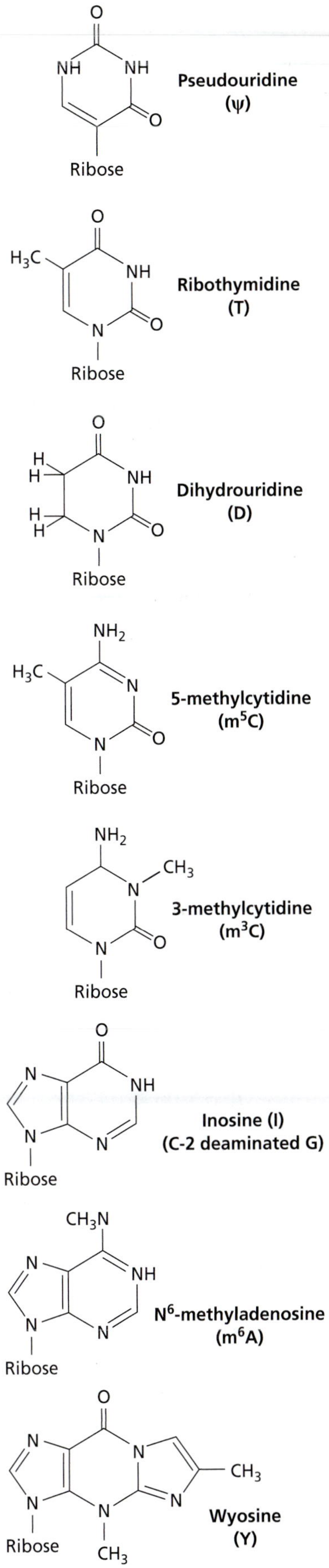

Figure 15.4 Some typical modified nucleosides in tRNA.

they contain a substantial fraction of invariant positions and modified bases in which tRNAs are especially rich. The chemical structures of some typical modified nucleosides in tRNA are given in **Figure 15.4**.

The anticodon arm carries the anticodon in the loop of the arm; the three bases that form the anticodon are exposed to the solution (see Figure 15.3), which facilitates their interaction with the mRNA codon during protein synthesis. It is important to understand that when the anticodon pairs with the codon, a very short stretch of double-stranded A-form RNA helix is formed, following the rules of complementarity and anti-parallelism that govern double-helical nucleic acid structure. This means that when writing or depicting codon–anticodon interactions, care has to be exercised to reflect these rules. For example, in **Figure 15.5**, the anticodon in the tRNA is given in the usual 5′ → 3′ direction, but the codon in the mRNA is written in the 3′ → 5′ direction, UCG; the correct reading of the codon, 5′ → 3′, in this case is, of course, GCU.

As mentioned above, there are fewer tRNAs than sense codons, meaning that several anticodons, and thus several tRNAs, must be capable of recognizing the same codon during translation. It turns out, as proposed by Crick and later proven experimentally, that the 5′-base of the anticodon can wobble, making alternative, non-Watson–Crick hydrogen bonds with several different bases at the 3′-position of the codon. The general wobble rules are given in Figure 15.5, which also illustrates the wobbling of the modified inosine in the 5′-position of the anticodon of yeast alanine tRNA.

The acceptor arm at the top or 3′-end of the cloverleaf structure contains the universal triplet CCA as a single-strand extension to the arm. Recall from Chapter 14 that this triplet is added post-transcriptionally to all tRNAs with the help of a specific enzyme, tRNA nucleotidyltransferase, and that no template sequence is needed for this addition. The 2′- or 3′-OH group of the terminal adenosine, A, serves as an acceptor for the amino acid during the aminoacylation reaction.

The other arms play a role in the 3D folding and stabilization of the molecule that give rise to its characteristic L-shape (see Figure 15.2); both non-Watson–Crick base pairing and stacking interactions participate.

tRNAs are aminoacylated by a set of specific enzymes, aminoacyl-tRNA synthetases

In order to serve as an adaptor molecule between mRNA and an amino acid, tRNA has to carry the **cognate amino acid** at its acceptor arm and interact with the respective codon on the mRNA via its anticodon arm. How is the appropriate amino acid attached? How is the fidelity of the genetic code ensured? In other words, how is the amino acid selected so that its carrier tRNA has the anticodon that specifies this, and no other, amino acid? The answer is, by the specificity of action of a group of enzymes called **aminoacyl-tRNA synthetases**.

As pointed out by Christian de Duve, the scientist who shared the 1974 Nobel prize with Albert Claude and George E. Palade "for their discoveries concerning the structural and functional organization of the cell," the accuracy of translation of the genetic code depends on the precision of two successive independent matchings: first, that of amino acids with tRNAs, and second, that of charged tRNAs with ribosome-linked messenger RNA. The second recognition event is rather straightforward, as it involves only complementary interactions between nucleotide sequences, the codon and anticodon triplets. The matching between amino acids and tRNA is quite different: it is indirect and bilingual. The aminoacyl-tRNA synthetase enzymes that do the matching use recognition between structural features of the enzyme itself and structural features of the tRNA that they charge. De Duve suggested the existence of a second genetic code imprinted into the structure

(A)

5′-base at tRNA anticodon	3′-base at mRNA codon
A	U
C	G
G	C or U
U	A or G
I	A or C or U

(B) Anticodon arm of tRNA^{Ala}

5′---m^{2}GCUCCCUUIGC^{m}IψGGGAGA--- 3′

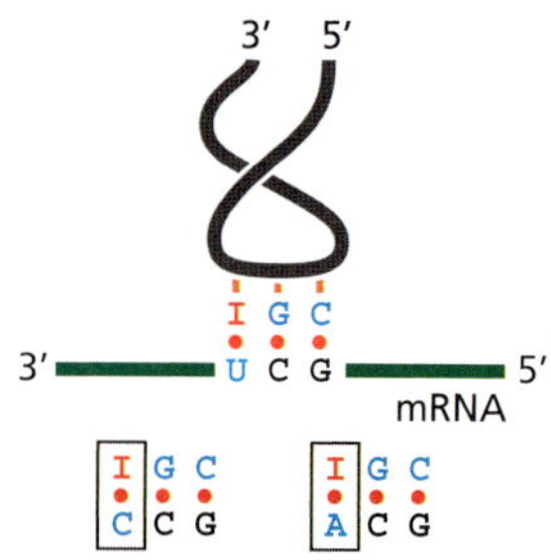

I G C / C C G   I G C / A C G

Inosine at the 5′-position of the anticodon can pair with U, C, or A at the 3′-position of the codon on mRNA

Figure 15.5 Wobble rules illustrated by the example of yeast tRNA^{Ala}. (A) Wobble rules as proposed by Crick in 1966. These rules have been extended on the basis of experimental data to include differently modified bases in the anticodon region. (B) The partial sequence presented is that of the anticodon arm, in blue letters. The wobble base hypoxanthine, which is the base portion of the modified nucleoside inosine (see Figure 15.4) in the anticodon, shown in red, can pair with U, C, or A in the codon, according to the rules in part A.

of aminoacyl-tRNA that matches the amino acid with the cognate tRNA; he even introduced a term for these features, paracodons. He found the term convenient for defining a code, different from the codon–anticodon interactions, that involves protein–tRNA recognition.

Aminoacylation of tRNA is a two-step process

Before delving into the recognition problem, we first consider the chemistry of the tRNA aminoacylation reaction. This reaction takes place in two steps (**Figure 15.6**): (1) activation of the amino acid and (2) its subsequent addition to the CCA end of the tRNA. Both steps take place on the aminoacyl-tRNA synthetase and can occur on either the 2′- or 3′-hydroxyl of the A residue. The choice between amino acid attachment to a 3′- or a 2′-OH is determined by the specific structural class to which a given aminoacyl-tRNA synthetase belongs. Class I synthetases contain a characteristic Rossmann fold catalytic domain, act as monomers, and couple the aminoacyl group to the 2′-OH of the tRNA. Class II synthetases share an anti-parallel β-sheet fold flanked by α-helices, are mostly dimeric or multimeric, and prefer the 3′-OH group of the tRNA. If the amino acid is attached initially to the 2′-OH group, it is then shifted to the 3′-position in an additional step, as only 3′-attached amino acids can serve as substrates in translation. Despite the considerable structural differences between the two enzymatic classes, tRNA binding involves an α-helical structure that is conserved between them.

Aminoacyl-tRNA synthetases possess a catalytically active site, where ATP binds to activate the amino acid, and two separate sites for binding of amino acid and tRNA (**Figure 15.7**). The crystal structures of complexes comprising synthetases and their cognate tRNA are available; an example is presented in **Figure 15.8**. The majority of the contacts are with the acceptor and the anticodon stem, but other interactions are also important for the recognition of specific tRNAs through the **identity elements in tRNAs** (**Figure 15.9**). Identity elements are defined by their ability to present the correct tRNA to the correct enzyme in an *in vitro* reaction. A major step forward was the identification, by Paul Schimmel's laboratory in 1988, of a single base pair, G3U70, in the **acceptor stem** of *E. coli* tRNA^{Ala} as the identity element critical for binding to a cognate aminoacyl-tRNA synthetase. Moreover, this specificity could be conferred onto two other tRNAs by introduction of this base pair into their acceptor stems. As shown by the tRNA^{Ala} example in Figure 15.9, other base pairs in the acceptor stem also participate in the identification process. There was a major surprise concerning identity elements: although the anticodons are indeed used as such elements in many cases, some tRNAs can happily recognize their cognate enzyme without the participation of their unique anticodon triplets.

Adenosine–P–P–P + $^-$O(O=)C–CH(R)–NH_3^+
Amino acid
(1) → P–P
Adenosine–P~O–C(=O)–CH(R)–NH_3^+
Aminoacyl adenylate (activated amino acid)
tRNA (CCA, OH 2′, OH 3′)
(2) → AMP
3′-aminoacyl-tRNA
⇅
2′-aminoacyl-tRNA

Figure 15.6 Aminoacylation of tRNA: The chemistry. (Step 1) The nucleophilic carboxylate group of the amino acid attacks the γ-phosphorus of adenosine triphosphate, ATP. The pyrophosphate that is released during this step is immediately hydrolyzed by an abundant enzyme, pyrophosphatase, to two molecules of inorganic phosphate. This prevents the amino acid activation reaction from going into reverse. (Step 2) The 3′-OH or 2′-OH group of the terminal A in the universal CCA tail of tRNA attacks the carboxylate of the activated amino acid, releasing adenosine monophosphate, AMP, and creating an ester linkage between the amino acid and the ribose. This is the form in which the amino acids are brought to the ribosome for incorporation into the nascent polypeptide chain. Both steps occur on the aminoacyl-tRNA synthetase.

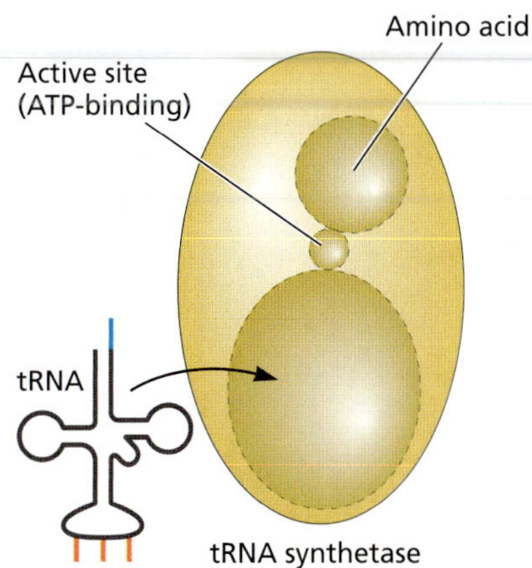

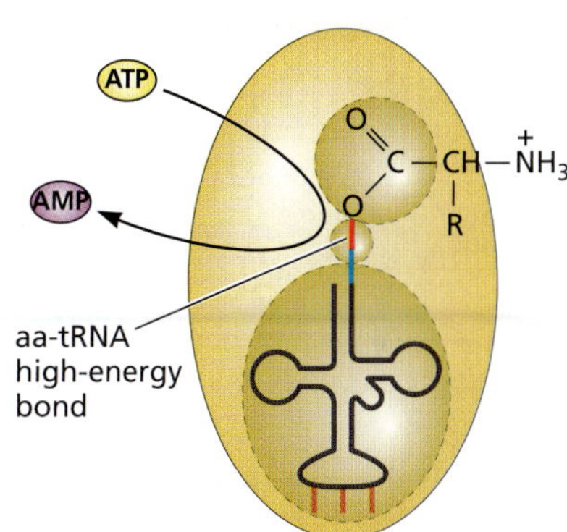

Figure 15.7 Schematic of the overall structure of tRNA synthetase. The catalytic ATP-binding site and binding sites for tRNA and amino acid are depicted in the upper part. ATP and the amino acid bind first, followed by tRNA; the blue line depicts the single-stranded universal CCA end. The lower part illustrates the structure at the end of catalysis, before release of the aminoacyl-tRNA. Often, a single aminoacyl-tRNA synthetase recognizes several isoacceptor tRNAs.

Fortified with the concept of identity elements, we now return to the structures and dynamic networks in tRNA synthetase complexes (see Figure 15.8). It is known that the charging reaction is stimulated by specific interactions between the enzyme and the tRNA identity elements, but it is not exactly clear how this occurs, especially when the sites involved are far apart. Theoretical analysis has solved the coupling problem: allosteric signals pass between the anticodon and the distant catalytic site, with many possible paths of transferring the signal. The amino acid residues and nucleotides involved in these communication pathways are evolutionarily conserved.

Quality control or proofreading occurs during the aminoacylation reaction

Molecular recognition allows the possibility that an amino acid structurally similar to the cognate one can be erroneously bound by the wrong tRNA. Why can mischarging occur? The aminoacylation enzymes have to deal with two separate recognition issues: they need to correctly recognize both the cognate amino acid and its tRNA partner. The choice of tRNA poses no major problem, because the aminoacyl-tRNA synthetases make extensive contacts with the tRNA over a large surface area, between 2500 and 5700 Å^2, allowing for numerous specific interactions. The real challenge for the enzyme is to recognize the cognate amino acid. Amino acids are small and very similar in structure and chemical reactivity, differing only in their side chains. The problematic amino acids are those whose side chains differ only slightly, for example, by just one methylene or OH group. These similar amino acids are usually described as **near-cognate amino acids**, to distinguish them from the noncognate amino acids that are significantly different from the cognate ones. The spontaneous rate of misincorporation of near-cognate amino acids is expected to be rather high. The error rate *in vivo* is, however, at least several orders of magnitude lower than that expected. What is the secret?

Biochemical and structural studies led to the discovery, in some aminoacyl-tRNA synthetases, of a second catalytic center that can perform hydrolytic release of a noncognate amino acid even after it has been linked to the tRNA. This center is thus responsible for the **proofreading** activity, or **editing** activity, of the enzymes. A historic account of the evolution of ideas of how proofreading occurs is given in Box 15.2, which focuses on only one, albeit major, aspect of proofreading: steric selection of the cognate amino acid through the **double-sieve mechanism**.

In totality, quality control during the aminoacylation of tRNA involves a rather complex set of processes: some acting after activation of the amino acid but before its attachment to the tRNA, known as pre-transfer mechanisms, and others acting after

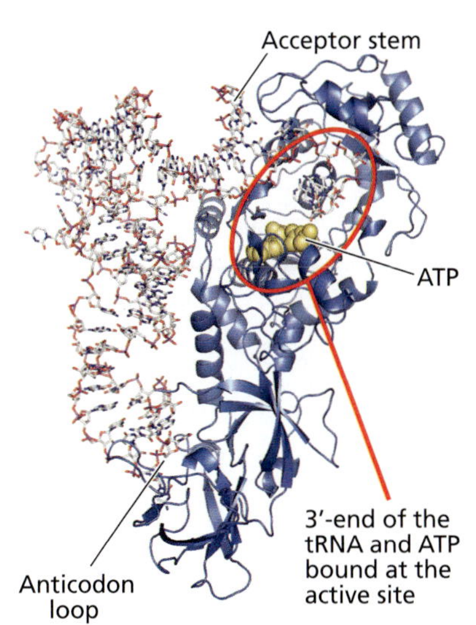

Figure 15.8 *E. coli* tRNA^{Gln} bound to glutaminyl-tRNA synthetase. The enzyme is in the ribbon representation, and the tRNA is shown as a stick model. Note that the enzyme interacts with both the tRNA acceptor stem and the anticodon. Molecular dynamics simulations indicate that, upon binding, the tRNA induces conformational changes throughout the protein–tRNA interface and within the catalytic site. The charging reaction is stimulated by interactions between the enzyme and the tRNA identity elements, which, for the most part, are located far away from the tRNA acceptor stem. Thus there must be an allosteric signal passed between the anticodon and the distant catalytic site. The amino acid residues and tRNA nucleotides involved in these allosteric interactions are evolutionarily conserved.

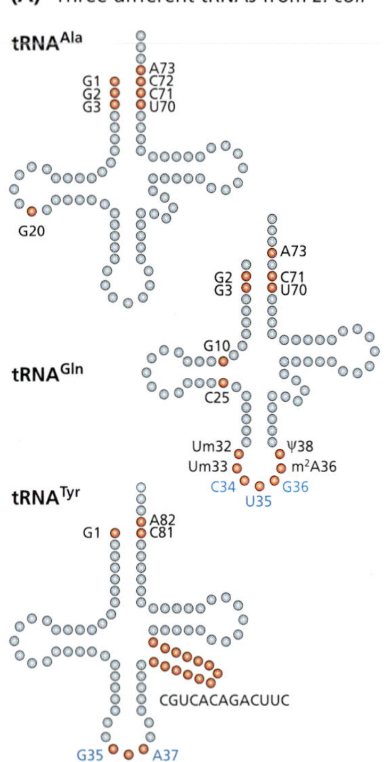

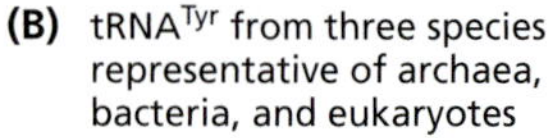

(B) tRNATyr from three species representative of archaea, bacteria, and eukaryotes

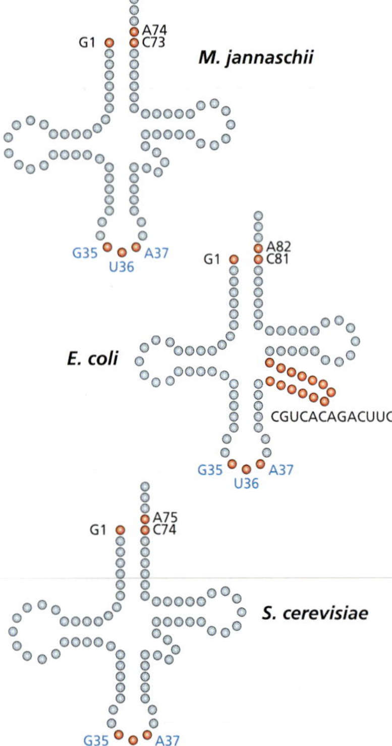

Figure 15.9 Diversity of the major identity elements in tRNAs that identify the tRNA to the cognate tRNA-synthetase. (A) Three different tRNAs—tRNA^{Ala}, tRNA^{Gln}, and tRNA^{Tyr}—in *E. coli* illustrate the diverse locations and chemical nature of the respective identity elements. The identity elements are depicted as red dots. Note that the anticodon nucleotides, shown as blue letters, are not always involved in the recognition of some cognate tRNAs, for example, tRNA^{Ala}. Generally, nucleotides in the acceptor stem confer a major portion of the specificity of recognition. (B) Identity elements in tRNA^{Tyr} molecules from *Methanocaldococcus jannaschii*, an archaeon; *E. coli*, a bacterium; and *Saccharomyces cerevisiae*, a eukaryote. Note that evolutionarily conserved elements are present in all species.

formation of the aminoacyl-tRNA adduct, known as post-transfer mechanisms. **Figure 15.10** also provides a background schematic illustrating the sequence of events that occur during the normal process of aminoacylation and its connection to the downstream events in protein synthesis. To complicate things even more, the post-transfer mechanisms can act in *cis* or in *trans*, depending on whether the editing involves the same enzyme molecule that performed the original aminoacylation reaction, acting in *cis*, or whether another aminoacyl-tRNA synthetase molecule is used to edit the mischarged tRNA following its release from the original enzyme, acting in *trans* (**Figure 15.11**). The process involving a second enzyme is known as resampling and is used in other steps of protein synthesis (see Chapter 16). Finally, the *trans* pathway of post-transfer editing may involve freestanding editing factors, that is, proteins other than the aminoacyl-tRNA synthetases (see Figure 15.11).

Interestingly, removal of D-stereoisomers of amino acids from tRNA is performed by a special class of such factors, called D-aminoacyl-tRNA deacylases. Although the intrinsic editing sites in aminoacyl-tRNA synthetases are highly conserved among living things, the *trans*-acting factors are much more diverse and may have arisen more than once during evolution. Thus, although D-aminoacyl-tRNA deacylases are universally required for cell viability, three unrelated classes have been found: one in bacteria and some eukaryotes, another in archaea and plants, and a third in cyanobacteria.

The existence and proper functioning of proofreading during charging of tRNAs with their cognate amino acids is of extreme importance to the fidelity of translation. It is one of only two major steps that exert control over the fidelity of translation; the second step occurs during codon–anticodon recognition at the elongation step of translation (see Chapter 16). If for some reason this step is not working properly and mischarging occurs, the consequences for the organism could be drastic. **Box 15.3** describes the connection between the proofreading activity of aminoacyl-tRNA synthetases and human diseases.

Insertion of noncanonical amino acids into polypeptide chains is guided by stop codons

An important noncanonical amino acid is the formylmethionine that is used for initiating new polypeptide chains in bacteria, mitochondria, and chloroplasts (see Chapter 16). The insertion of this amino acid does not require an additional aminoacyl-tRNA synthetase because addition of the formyl group occurs after tRNAi^{Met} is charged with methionine by the regular methionyl-tRNA synthetase.

For a long time it was believed that, except for *N*-formylmethionine, only the 20 canonical amino acids could be inserted into proteins. It is now known that at least two additional amino acids, selenocysteine and pyrrolysine (**Figure 15.12**), can be incorporated internally into polypeptide chains. Selenocysteine is a cysteine analog in which the sulfur is replaced with its immediate downstairs neighbor in the periodic table, selenium. Selenocysteine is found in several essential enzymes, and the activities of these enzymes depend on its presence. Selenoproteins have been described in all three domains of life, with 25 representatives in humans. Pyrrolysine-containing proteins are less broadly distributed, being restricted to just a handful of archaea and bacteria.

In both cases, insertion of these two extra amino acids into the polypeptide chain is guided by stop codons in a specific mRNA sequence context and is preceded by a complex sequence of biochemical reactions that attach the respective amino acid to their cognate tRNAs. These designated tRNAs contain anticodons that would recognize the context-dependent stop codons during translation (see Figure 15.12). In addition, once formed, these specific aminoacylated tRNAs need specific elongation factors to bring them to the A site of the ribosome during translation elongation.

Experimental technologies are now so advanced that it is possible to introduce even **unnatural amino acids** into proteins (Box 15.4). This capability is just starting to affect a number of areas of protein research, bioengineering, and other areas. Such modification is expected to become a major source of designed proteins that have highly useful properties.

Box 15.2 The double-sieve mechanism of proofreading in aminoacyl-tRNA synthetases: An ever-evolving story The idea that there is a problem in molecular recognition that may affect the fidelity of protein synthesis was floating around as early as 1958, several years before molecular biologists understood the process of translation. Both Linus Pauling and Francis Crick predicted that there must be a mechanism for correcting errors, that is, proofreading, in the selection of structurally similar amino acids for incorporation into proteins. Pauling noted that for similar amino acids, such as isoleucine and valine that differ only by a single methylene group, the error in molecular recognition should be much higher than proteins allow: thus a proofreading mechanism should exist.

Later experimental data illustrated the problem in real numbers. In proteins such as ovalbumin and globin, the misincorporation of valine instead of isoleucine occurred in only 1 out of 3000 instances; that is, the error rate was $\sim 3 \times 10^{-4}$. This rate could not be explained by the rate of activation of the two amino acids or by the relative affinity of the activated amino acid for the Ile-tRNA synthetase: activated valine had only a 150-fold weaker affinity than activated isoleucine. The mystery was solved by solid experimental data from Michael Chamberlin, Robert Baldwin, and Paul Berg (Figure 1).

In 1977, Alan Fersht proposed the double-sieve mechanism, arguing that "the 'strong' force in specificity is steric repulsion: whereas a smaller substrate can always rattle around in a larger cavity, it is energetically very difficult to cram a larger substrate into a cavity built for a smaller one." The double-sieve hypothesis for sorting amino acids (Figure 2) states that there are two sieves that function in succession; the first one is a coarse sieve that allows smaller objects to pass through the holes, whereas larger objects are rejected. The second, finer

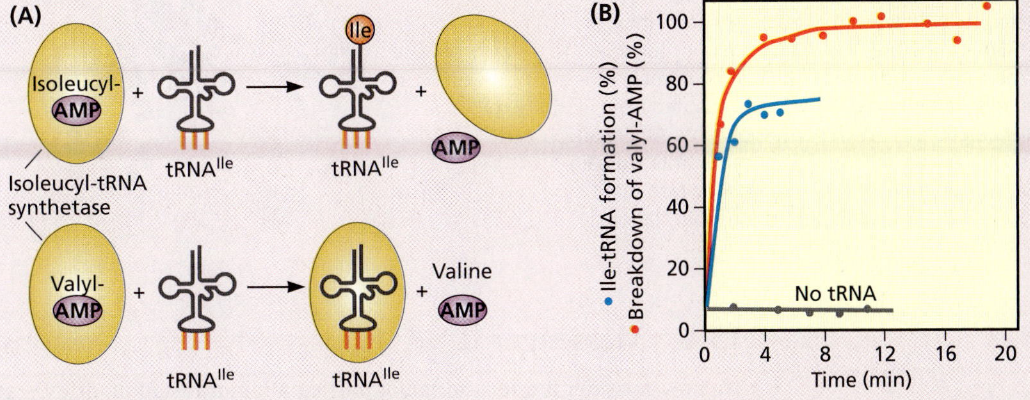

Figure 1 Biochemical experiments of Baldwin and Berg indicating the existence of intrinsic proofreading activity in isoleucyl-tRNA synthetase. (A) Schematic of the experiment. Purified complexes of Ile-tRNA synthetase carrying either activated isoleucine or activated valine, with a radioactive label on the AMP moiety, were incubated with $tRNA^{Ile}$ in the presence of alkaline phosphatase. The phosphatase was added to follow the release of AMP from the complexes by the production of labeled inorganic phosphate. The products of the reaction that were released into the medium were either charged Ile-tRNA or valine. (B) Experimental data showing the time course of the two reactions. Note that there was no release of radioactivity if $tRNA^{Ile}$ was not added; equally, *E. coli* $tRNA^{Val}$ and yeast $tRNA^{Ile}$, which do not accept isoleucine in the reaction catalyzed by *E. coli* Ile-tRNA synthetase, did not lead to hydrolysis of Val-AMP. Thus, the reaction is strictly dependent on the cognate tRNA species. (B, data from Baldwin AN & Berg P [1966] *J Biol Chem* 241:839–845.)

(Continued)

Box 15.2 *(Continued)*

sieve performs the same function in reverse but for only those objects that passed through the first sieve: smaller objects pass through while the object of desired size is retained. Thus the job of discriminating amino acids according to their size is accomplished.

Forty years after the double-sieve hypothesis was put forward, high-resolution crystallographic data have clearly demonstrated how it works, at least for the coarse sieve. We now know that the catalytic site for aminoacylation acts as the coarse sieve, activating at a significant rate only amino acids that are smaller than or equal in size to the cognate one. The editing hydrolytic site may be the fine sieve, destroying the products of amino acids that are smaller than the cognate one. Structural data from 2010 suggest, however, that the functional positioning of substrates, rather than steric exclusion, is actually key to the mechanism of discrimination at the editing site. A strategically positioned catalytic water molecule is excluded to avoid hydrolysis of the cognate substrate via an RNA-mediated substrate-assisted catalysis mechanism at the editing site. The mechanistic proof of the critical role of RNA in proofreading activity is a completely unique solution to the problem of the cognate–noncognate selection mechanism. Thus, the double-sieve steric model evolved into a different kind of double sieve, with a coarse steric sieve at the catalytic site and a functional sieve at the editing site.

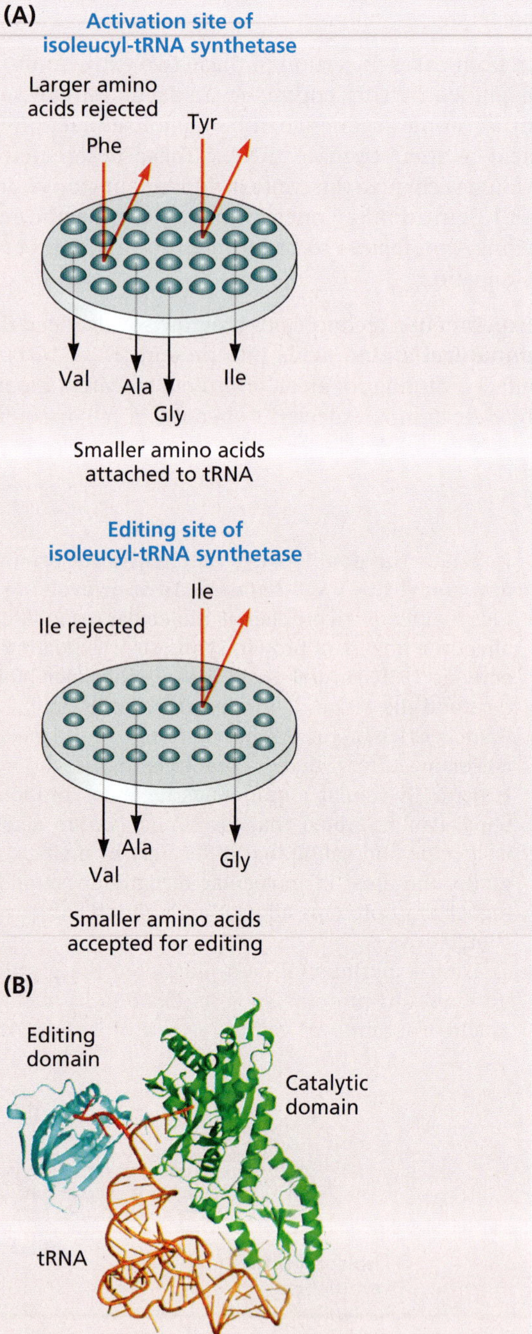

Figure 2 Double-sieve mechanism for steric selection of cognate amino acids by aminoacyl-tRNA synthetase. The structure of a tRNA synthetase, with catalytic and editing domains, is also shown. (A) In the first step, cognate and small near-cognate amino acids will be activated and then attached to the tRNA; the larger amino acids will not be able to fit sterically within the active center and will thus be rejected. In the second step, a reverse-sieving occurs in the editing center. Isoleucine, the cognate amino acid, is too large to enter the editing site and will be rejected from editing; whereas the smaller near-cognate amino acids will enter the editing site and be subjected to hydrolysis. (B) Ribbon representation of archaeal Thr-tRNA synthetase; the red CCA end of tRNA is flipped from the catalytic domain to the editing domain when serine is being mischarged onto $tRNA^{Thr}$. (A, adapted from Fersht AR [1998] *Science* 280:541. With permission from American Association for the Advancement of Science. B, from Hussain T, Kamarthapu V, Kruparani SP et al. [2010] *Proc Natl Acad Sci USA* 107:22117–22121. With permission from National Academy of Sciences.)

15.4 Messenger RNA

We now consider the second major participant in translation, mRNA. mRNAs in both bacteria and eukaryotes are of variable length and stability. Bacterial mRNAs are ready for use as soon as they are transcribed, but eukaryotic mRNAs have undergone extensive processing by the time they are accepted for translation in the cytoplasm (see Chapter 14). The length of an mRNA after splicing is, of course, determined mainly by the length of the protein product encoded. Its stability is carefully controlled so that the encoded protein is made available only when it is needed and so that protein synthesis can respond to external and internal signals (see Chapters 14 and 17).

Given what we know about other RNA structures, it should not be surprising that mRNAs may contain hairpin structures formed by self-complementary stretches of the polynucleotide chain (**Figure 15.13**). Because the channel within the ribosome is

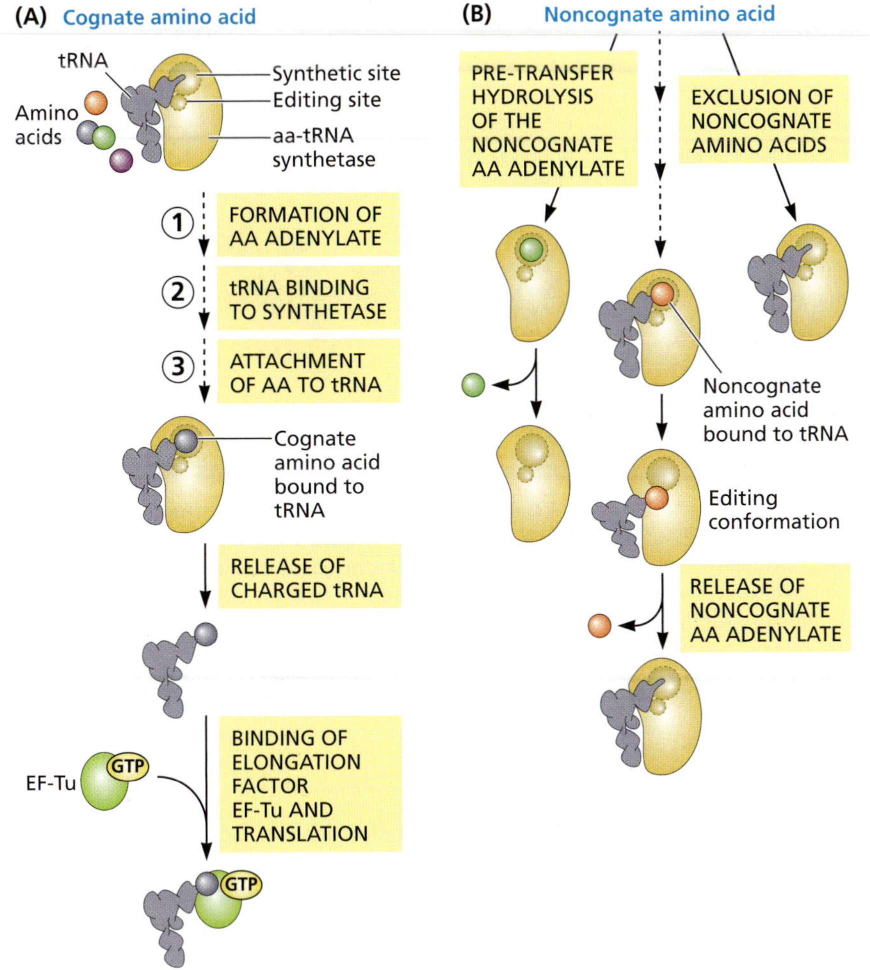

Figure 15.10 Quality control during the formation of an aminoacyl-tRNA occurs in two stages. (A) Desirable outcome of the tRNA charging reaction that results in the formation of a tRNA molecule carrying its cognate amino acid. (B) Quality control that prevents tRNA charging with a noncognate amino acid acts at either of two different stages, pre- and post-transfer, that is, before and after attachment of the activated amino acid to the tRNA. First, the synthetic site excludes amino acids that are larger than the cognate one or that cannot establish proper specific contacts with the enzyme. Another pre-transfer mechanism may act by simply hydrolyzing the labile mischarged aminoacyl (aa) adenylate. The hydrolysis can occur at the editing site of aminoacyl-tRNA synthetase or spontaneously in solution after the noncognate aa-AMP is expelled from the enzyme. Second, if a noncognate amino acid becomes bound to the tRNA, then there is a movement of the amino acid from the synthetic site into the editing site, where the RNA–amino acid ester linkage is hydrolyzed. This movement is possible because of the flexibility of the CCA end of the acceptor stem of the tRNA. (Adapted from Cochella L & Green R [2005] *Curr Biol* 15:R536–R540. With permission from Elsevier.)

too narrow to allow the entrance of highly folded RNAs, tertiary structures must be dissolved upon or before entry. This is probably spontaneous, for such structures, products of chance complementarity, are likely to be small.

The Shine–Dalgarno sequence in bacterial mRNAs aligns the message on the ribosome

Many bacterial mRNAs share a common element at their 5′-ends that helps them to bind to the ribosome during initiation and positions the initiation codon in close

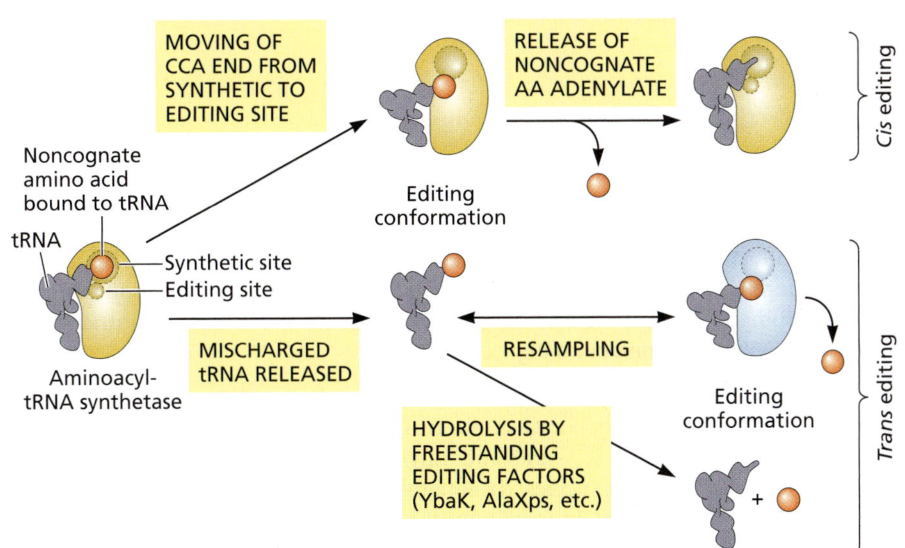

Figure 15.11 Discrimination between *cis* and *trans* post-transfer editing following mischarging. The *cis* pathway involves the action of an editing site on the same aminoacyl-tRNA synthetase molecule that catalyzed the original aminoacylation reaction. In *trans* editing, there are two possible scenarios, both of which occur following release of the mischarged aa-tRNA from the original synthetase: (1) the mischarged tRNA binds to another synthetase molecule, directly using its editing site, or (2) freestanding editing factors perform the hydrolysis reaction.

Box 15.3 A Closer Look: Proofreading activity of aminoacyl-tRNA synthetases, translational fidelity, and human disease The incorporation of wrong amino acids into the polypeptide chain can give rise to misfolded proteins; these elicit complex cellular responses, including enhanced transcription of protein chaperone genes, inhibited translation, and ultimately cell death. Misfolding occurs, among other reasons, because the incorporation of erroneous amino acids introduces ambiguity in the usage of the genetic code; that is, the same codon specifies the incorporation of more than one amino acid, at random positions through the chain, resulting in the production of what are called statistical polypeptides. Currently there is no direct evidence that defects in the editing mechanisms working at the level of aminoacyl-tRNA synthetases cause human disease. Nevertheless, clear-cut examples from human cells in culture or from mice raise the possibility that editing defects may also lead to disease in humans.

One interesting study in mutant mice demonstrates that low levels of mischarged tRNA can lead to intracellular accumulation of misfolded proteins in terminally differentiated Purkinje neuronal cells of the brain. The experiments were performed on homozygous *sticky* mutant mice, which are characterized by the rough, unkempt, sticky appearance of their fur. With aging, follicular dystrophy and patchy hair loss appear and mild tremors develop, which progress to overt ataxia, or gross lack of coordination of muscle movement. Histological analysis of the brains of these mice revealed extensive Purkinje cell loss. The brains of the mutant mice were positive for apoptotic markers. Further analysis demonstrated that the *sticky* mutation was in fact a missense point mutation in the editing domain of $tRNA^{Ala}$. As **Figure 1** shows, this mutation affected base 2201 in exon 16, resulting in an alanine to glutamic acid substitution at amino acid 734, which is evolutionarily conserved. The compromised proofreading activity of the enzyme during aminoacylation of tRNAs affects the overall translation fidelity, leading to neurodegeneration in the neuron cells.

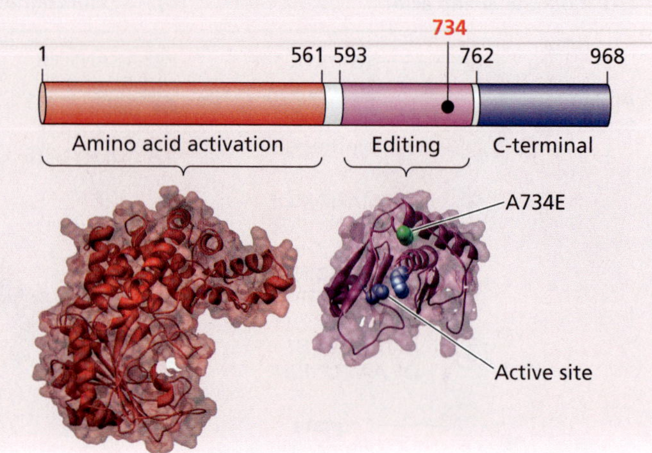

Figure 1 Catalytic domains of $tRNA^{Ala}$ synthetase and their three-dimensional structures. The amino acid activation domain, shown in red, is modeled on the *Aquifex aeolicus* alanyl-tRNA synthetase enzyme, and the editing domain, shown in purple, is modeled on the *Pyrococcus horikoshii* freestanding homolog of the alanyl-tRNA editing domain. Blue spheres show the location of critical active-site residues. Ala734 is represented by a green sphere. (Adapted from Lee JW, Beebe K, Nangle LA et al. [2006] *Nature* 443:50–55. With permission from Macmillan Publishers, Ltd.)

proximity to the P site of the ribosome (see Chapter 16). This is known as the **Shine–Dalgarno sequence** or **SD sequence** and is located several nucleotides upstream of the start codon. It acts by base-pairing with a conserved region near the 3′-end of 16S rRNA in the small ribosomal subunit (**Figure 15.14**). Note, however, that there are bacterial mRNAs that lack the SD sequence in the 5′-UTR or even lack the 5′-UTR altogether; these are dubbed leaderless mRNAs. Most archaeal mRNAs also contain Shine–Dalgarno sequences, but a considerable proportion of these mRNAs use their first start codon to initiate translation.

Eukaryotic mRNAs do not have Shine–Dalgarno sequences but more complex 5′- and 3′-untranslated regions

The situation in eukaryotes is quite different. We see in Chapter 14 that eukaryotic mRNA undergoes several important modification steps during its transition from a primary transcript in the nucleus to a mature functional mRNA in the cytoplasm. These modifications include 5′-end capping, 3′-end polyadenylation, and frequently splicing. Here we present a much more detailed view of a typical eukaryotic mRNA, also describing functional regions and elements in both the 5′- and 3′-untranslated regions or UTRs of the molecule (**Figure 15.15**).

The length of the 5′-UTR is more or less constant throughout eukaryotes, ranging from ~100 to ~200 nucleotides. By contrast, the length of the 3′-UTR is much more variable, from ~200 nucleotides in plants and fungi up to 800 nucleotides in vertebrates, including humans. Within a species, the lengths of both 5′- and 3′-UTRs vary considerably for different messages. Interestingly, the DNA regions corresponding to UTRs may contain introns.

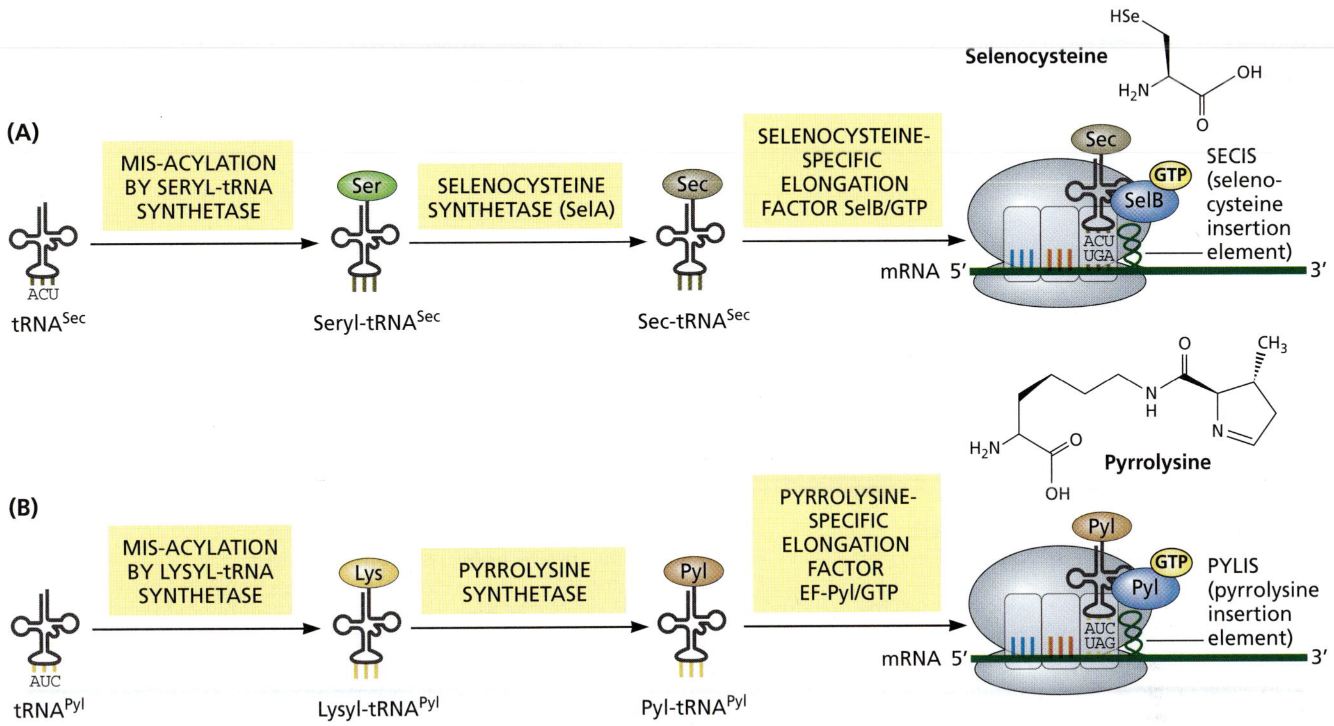

Figure 15.12 Co-translational insertion of unusual amino acids into the nascent polypeptide chain. (A) Selenocysteine, Sec; (B) pyrrolysine, Pyl. In both cases, the incorporation is guided by stop codons, which are followed by specific sequences in the mRNA that form secondary structures. Similar mechanisms have been implicated in the insertion of these two amino acids. First, a special cognate tRNA carrying the anticodon that recognizes the stop codon is mis-acylated by a normal aminoacyl-tRNA synthetase; this is a mis-acylation, for the enzyme should not normally recognize this anticodon. Then the noncognate amino acid bound to the cognate tRNA is converted, with the help of specific enzymes, into the cognate selenocysteine or pyrrolysine, respectively. The binding of these amino acids to the A site of the ribosome requires specific elongation factors or EFs, because the canonical EF, EF-Tu, would not recognize these structures. The binding of the specific EFs is aided by nearby insertion elements. Bacteria differ from archaea and eukarya in the exact location of the selenocysteine insertion element or SECIS; in bacteria, SECIS is located adjacent to the UGA, whereas in archaea and eukaryotes, it is in the 3′-untranslated region.

Box 15.4 Introducing unnatural amino acid residues into proteins: Expanding the genetic code If we had a way of inserting unnatural amino acids, of our choosing and design, at selected points in proteins, we would have a powerful tool for changing the properties of a protein at will. This option is now available, and it opens many vistas.

The story goes back to the early 1960s, when students in Seymour Benzer's lab were looking for a peculiar kind of mutant that had been expected: one in which an amino acid coding codon is mutated into a stop codon, aborting formation of the polypeptide chain. One student, named Harris Bernstein, left for the day, saying "If you find it, name it after me." His colleagues found one such mutant in which a UGG codon is converted to UAG; they named it the amber mutant because Bernstein is German for amber. Other mutants, named ochre and opal, soon followed.

A little later, Mario Capecchi and Gary Gussin discovered suppressor tRNAs, which could recognize an internal codon mutated to a stop signal but put an amino acid in this position. For example, the UAG codon could be recognized by a mutant *tRNA^{Tyr}, which would insert a tyrosine and allow chain elongation to continue.

Flash forward almost 40 years, to the realization by Peter Schultz and co-workers that this scheme might be tricked into inserting wholly unnatural amino acids into an amber site. With modern technology, it is no problem to make site-specific mutations to an amber codon. Similarly, suppressor tRNAs could be convinced to accept unusual amino acids if the appropriate aminoacyl-tRNA synthetase were available. These could be generated by a selective evolution and selection, using coupled antibiotic resistance. At the present time, over 30 different unnatural amino acids have been incorporated into a wide variety of proteins.

These have been used for a number of projects and purposes such as the specific placement of fluorescence resonance energy transfer or FRET donors and acceptors to follow protein dynamics, adding precisely placed ^{19}F as an NMR probe, modifying protein bioluminescence, and modifying the catalytic activity of enzymes. An example of the latter was published by Ryan Mehl's laboratory. They modified the enzyme nitroreductase in the vicinity of its active site. This native enzyme is used to activate two prodrugs in cancer therapy, called CB1954 and LH7. Figure 1A shows the battery of natural and unnatural amino acids inserted, and Figure 1B shows the corresponding activities for LH7. Importantly, there was >30-fold improvement of enzyme activity over that of the native active site.

(Continued)

Box 15.4 *(Continued)*

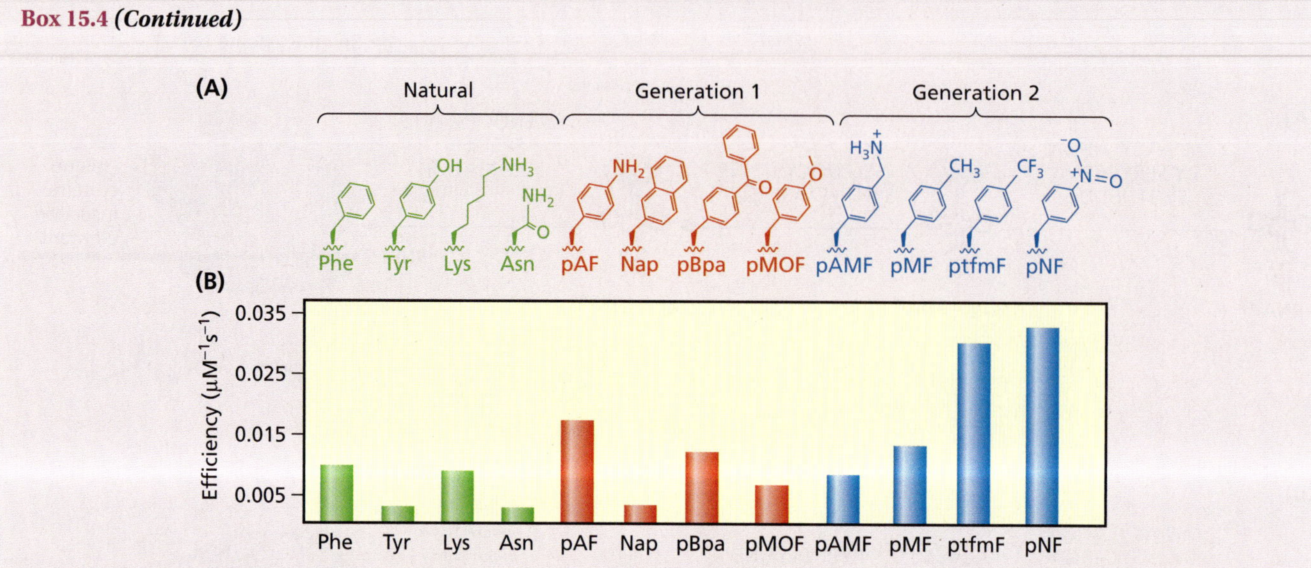

Figure 1 Unnatural amino acids improve the activity of an enzyme beyond that naturally available. (A) Nitroreductase modifications: natural and unnatural amino acids incorporated at site 124 of nitroreductase. Generation 1: pAF, *p*-aminophenylalanine, Nap, naphthylalanine, pBpa, *p*-benzoylphenylalanine, pMOF, *p*-methyoxyphenylalanine. Generation 2: pAMF, *p*-aminomethylphenylalanine, pMF, *p*-methylphenylalanine, ptfmF, *p*-trifluoromethylphenylalanine, pNF, *p*-nitrophenylalanine. (B) Improvement of prodrug LH7 activator nitroreductase catalytic efficiency upon incorporation of unnatural amino acids. Incorporation of unnatural amino acids has led to a significant increase in activity over that of the native enzyme. (Adapted from Jackson JC, Duffy SP, Hess KR & Mehl RA [2006] *J Am Chem Soc* 128:11124–11127. With permission from American Chemical Society.)

As described in Chapter 14, alternative UTRs can be formed through the use of alternative transcription start sites, polyadenylation sites, or splice donor and/or acceptor sites. The production of messages that encode the same protein but contain variable UTRs contributes to translational regulation and thus helps to determine overall gene expression patterns.

Overall translation efficiency depends on a number of factors

It is well established that gene expression is controlled at many different levels. This becomes obvious when the abundance of mRNAs and protein levels in a cell are compared. The abundance of less than 25% of proteins, including many secreted proteins, correlates with the abundance of their encoding mRNA. For other proteins, there can be huge differences in the abundance of the mRNA and its encoded protein. These differences are mainly attributed to differences in the frequency of translation initiation for different mRNAs; they are also influenced by mRNA and protein stability. Structural features at the 5′-UTR control mRNA translation efficiency, mainly through the presence of stable secondary structures, and sometimes proteins bound to them, that inhibit translation initiation. Additional control is exerted by elements known as **internal ribosome entry sites** or **IRES**. Under conditions in which normal cap-dependent translation initiation is impaired—for example, during stress, apoptosis, or mitosis—some cellular mRNAs can initiate translation at IRES. Use of IRES for initiation does not require the cap structure or any of the factors needed to remove secondary structures, such as the cap-binding subunit eIF4E of the eIF4F complex (see Box 16.1 and Figure 16.6). Some viral mRNAs also initiate translation in an IRES-dependent manner. Finally, 5′-UTRs sometimes contain short upstream reading frames; whether translation products of these sequences actually exist is unclear.

3′-Untranslated regions contain the conventional signal for polyadenylation (see Chapter 14). In addition, they may contain a **cytoplasmic polyadenylation element**

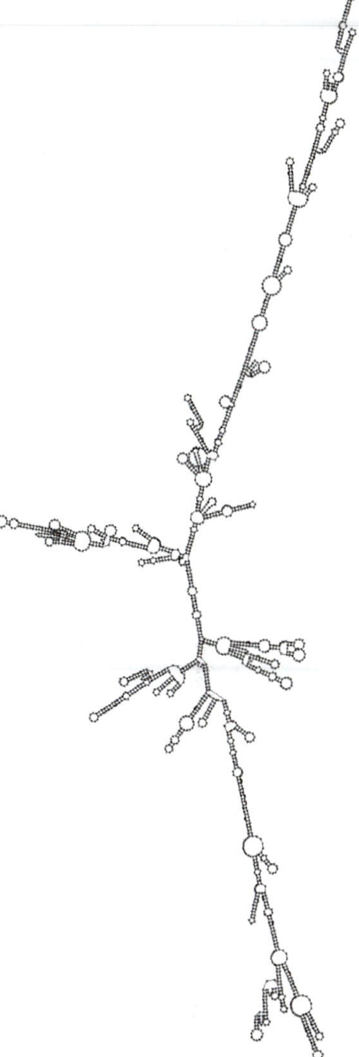

Figure 15.13 Model of the secondary structure of mRNA. The polynucleotide chain folds back on itself to form numerous short double helices, which are separated by loops of different size. The secondary structure is stabilized by binding of numerous proteins. The mRNA presented encodes the human epididymal protease inhibitor EPPIN, which plays a critical role in sperm function and male fertility. (From Ding X, Zhang J, Fei J et al. [2010] *Hum Reprod* 25:1657–1665. With permission from Oxford University Press.)

or **CPE**, which is responsible for the lengthening of poly(A) tails on mRNAs that have already lost a portion of the tail and need to be further stabilized, a process that takes place in the cytoplasm. CPE action requires the binding of a specific protein factor, CPE-binding protein. 3′-Untranslated regions also contain sequences that determine the subcellular localization of the mRNA.

Finally, an interesting feature of some 3′-UTRs is that the cell transcribes small antisense RNAs that are complementary to portions of the UTR; these are known as microRNAs. These antisense RNAs are believed to participate in translational control, either by directly inhibiting translation or by decreasing the stability of the mRNA. These issues are addressed in more detail in Chapter 17.

15.5 Ribosomes

Ribosomes, initially called microsomes, were first observed as early as 1941 by dark-field microscopy (**Box 15.5**), and later in the mid-1950s by electron microscopy. Cell biologist George Palade received a share of the 1974 Nobel Prize in Physiology or Medicine for his work on ribosomes.

The ribosome is a two-subunit structure comprising rRNAs and numerous ribosomal proteins

In all three domains of life, ribosomes consist of two subunits, small and large, which have different functionalities. The small subunit serves as a landing pad for the mRNA during the initiation of new polypeptide chains and also contains the decoding center, known as DC. This is the site where anticodons on the tRNAs interact with the respective codons on mRNA during the selection of the appropriate amino acid to be added to the growing chain. The large subunit contains the peptidyltransferase center or PTC, the region where two interacting entities—the peptidyl-tRNA, at the end of the growing peptide chain, and the incoming aminoacyl-tRNA—come in close proximity and proper orientation for transfer of the new amino acid residue with formation of a new peptide bond. The large subunit also contains the exit tunnel, through which the nascent polypeptide chain leaves the ribosome and enters the surrounding cytoplasm; more about the tunnel structure and function can be found in Chapter 16.

The ribosome subunits are ribonucleoprotein or RNP particles containing one or more ribosomal RNAs, rRNAs, and many different proteins (**Figure 15.16**). The small subunit in bacteria and archaea sediments as a 30S particle, whereas in eukaryotes

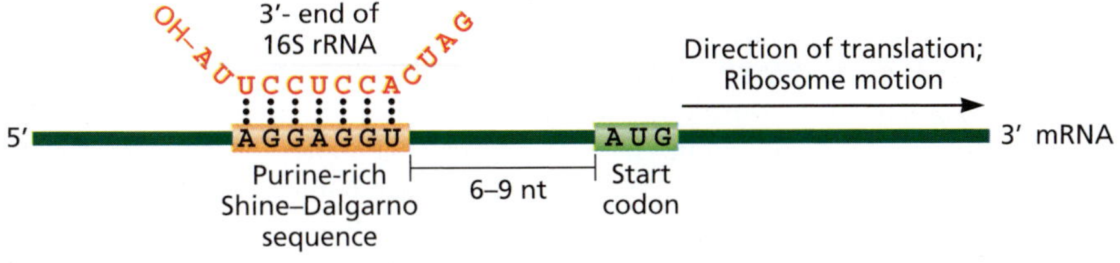

Figure 15.14 Ribosome-binding sites at the 5′-end of prokaryotic mRNA. The ribosome-binding site, also known as the Shine–Dalgarno sequence, is located 6–9 nucleotides upstream of the start codon and helps to position the codon at the ribosome's P site, that is, to establish the correct reading frame. It does so by base-pairing with the 3′-end of 16S rRNA in the small subunit of the ribosome.

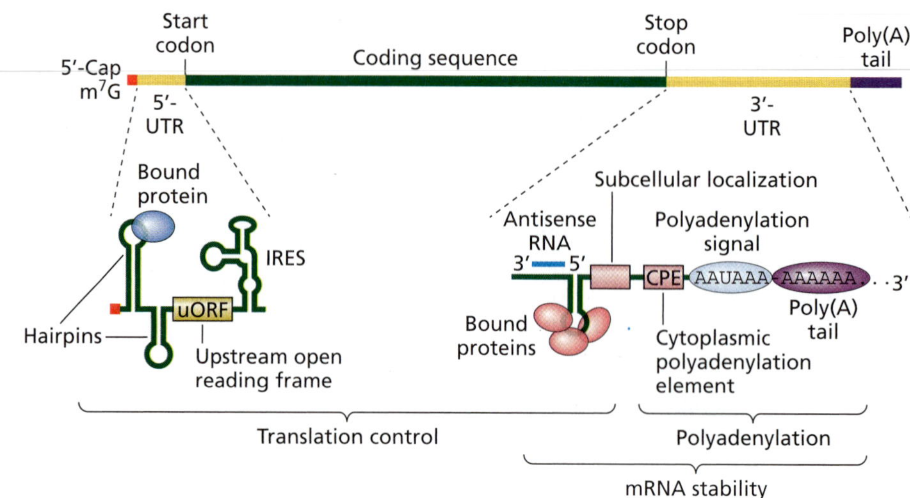

Figure 15.15 Structure of a typical mature eukaryotic mRNA. The coding region and two flanking untranslated regions are included. Note the existence of a cap structure at the 5′-end and a poly(A) tail at the 3′-end. The expanded regions below the mRNA depict structures that can be found in the untranslated regions and their functions. IRES, internal ribosome entry sites, are usually found in noncapped mRNAs. (Adapted from Mignone F, Gissi C, Liuni S & Pesole G [2002] *Genome Biol* 3:reviews0004. With permission from BioMed Central.)

the corresponding small subunit has a sedimentation coefficient of 40S. The large subunit in bacteria and archaea is a 50S particle, whereas that in eukaryotes sediments at 60S. The eukaryotic ribosomal subunits contain both more RNA and more proteins than their bacterial or archaeal counterparts (see Figure 15.16). Two eukaryotic organelles, mitochondria and chloroplasts, carry their own ribosomes, which resemble those of bacteria. This is often cited as an argument for the symbiotic origin of organelles.

Functional ribosomes require both subunits, with specific complements of RNA and protein molecules

In all types of cells, one small and one large subunit combine to form the intact ribosome, which is a 70S particle in bacteria and archaea and an 80S particle in eukaryotic

Box 15.5 The long history of the ribosome Ribosomes were first observed in 1941, when Albert Claude at the Rockefeller Institute began dark-field microscopic studies of the cytoplasm of various cell types. This technique can reveal objects below the microscope's resolution limit, visible as light-scattering dots. In all of the cell types, Claude found the cytoplasm to be rich in tiny particles that he called microsomes. Over the next decade, methods were developed to separate and isolate these particles by preparative sedimentation, and analysis showed them to consist of both protein and RNA. Consequently, they gained the name ribosomes.

A major breakthrough came in 1956 when Howard Schachman, using the analytical ultracentrifuge (see Box 3.1), demonstrated that yeast ribosomes sedimented as homogeneous particles with a sedimentation coefficient of 80S. Remarkably, they could be reversibly dissociated into two subunits, of 60S and 40S. A couple of years later, Alfred Tissières and James Watson showed comparable results for *E. coli* ribosomes, except that the ribosomes and subunits had sedimentation coefficients of 70S, 50S, and 30S. Many researchers at that time suspected that ribosomes might have a structure like that of RNA viruses, with only one or a few coat proteins.

Extensive studies on the number and nature of ribosomal proteins, carried out in many labs, soon showed this proposed structure to be completely incorrect; a powerful tool in these analyses was two-dimensional gel electrophoresis (**Figure 1**). Furthermore, other research in the 1960s clearly demonstrated the ribosomes' role in protein synthesis. Greatly facilitating this work was the discovery, by Peter Traub and Masayasu Nomura, in 1968, that the 30S subunit could be reconstituted from RNA plus proteins. The final phase in this work took place in the 1970s, when new techniques allowed the sequencing of both ribosomal RNAs and ribosomal proteins.

Thus, by about 1980, a great deal was known about the composition of ribosomes, yet virtually nothing was understood about their internal structure or mode of function. The obvious next step would be to determine the fine structure by X-ray diffraction, but most scientists believed this to be beyond feasibility. No such enormous, complex structure had ever been solved by this method. Consequently, a great deal of effort was spent in the 1980s to infer some ideas about the internal structure of ribosomes by indirect methods: studies of protein–protein or protein–RNA cross-linking, immunoelectron microscopy, and studies of neutron scattering, to name a few. None of these methods were able to provide the structural detail needed to reveal the mechanisms of ribosome function.

During this whole decade, starting in 1980, Ada Yonath, working in Heinz Wittmann's laboratory in the Max Planck Institute for Molecular Genetics in Berlin, learned how to prepare crystals of ribosome subunits that were suitable for X-ray diffraction. This was an excellent venue for such work, both because of its facilities and especially because of its direction by Wittmann, another pioneer in ribosome research. Only a few researchers followed her lead, and it was almost 20 years later, in 1998, when Thomas Steitz and his group at Yale published the first low-resolution, 9 Å structure of the 50S ribosomal subunit. Within two years, three groups, including those of Steitz, Yonath, and Venkatraman Ramakrishnan at Cambridge, had produced subunit structures with resolutions in the range of 2–4 Å. These permitted further analysis of the 5.5 Å structure of the whole 70S ribosome determined by Harry Noller of the University of California Davis in 1999. The story culminated with the publication of a study by Ramakrishnan of the 70S ribosome complete with mRNA and tRNAs. For their pioneering "studies of the structure and function of the ribosome," Thomas A. Steitz, Ada E. Yonath, and Venkatraman Ramakrishnan were awarded the Nobel Prize in Chemistry in 2009.

The availability of high-resolution structures has provided a leap forward in our understanding of ribosome function. For example, studies of ribosomes with bound antibiotics have clarified how many antibiotics work (see Box 16.2). Dynamic studies of internal motions during ribosome function, using techniques such as single-pair fluorescence resonance energy transfer, FRET, and cryo-electron microscopy, can now be interpreted at the submolecular level. In many ways, the ribosome has become one of the best-understood molecular machines.

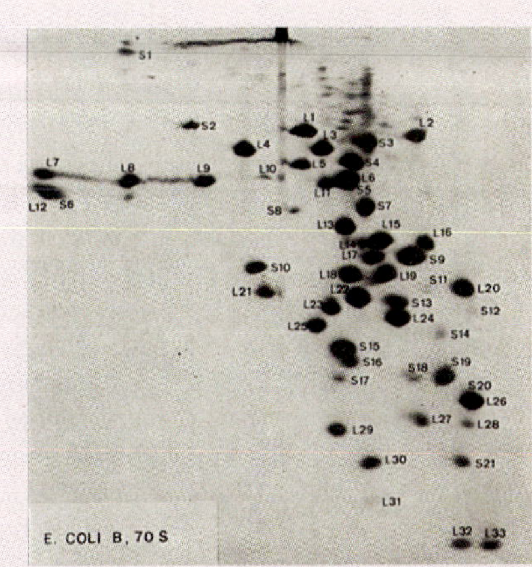

Figure 1 Two-dimensional gel electrophoretogram of 70S ribosomal proteins of *E. coli*. First dimension of electrophoresis, 4% acrylamide, pH 8.6; second dimension, 18% acrylamide, pH 4.6. (From Kaltschmidt E & Wittmann HG [1970] *Proc Natl Acad Sci USA* 67:1276–1282. With permission from National Academy of Sciences. Image kindly provided by H-J Rheinberger.)

cells. Here we focus on the bacterial ribosome because of the wealth of knowledge about its structure and function, obtained by a variety of biochemical methods, microscopic techniques, and crystallography. The general characteristics are much less well understood for eukaryotic ribosomes.

Each subunit contains a large RNA molecule. In bacteria, there is one 16S RNA molecule in the small subunit and one 23S RNA molecule in the large subunit. These rRNA molecules contain post-transcriptionally modified bases. In *E. coli*, the mature 16S contains 11 modified residues, 10 methylations, and one pseudouridine; the 23S rRNA contains 25 modifications comprising 14 methylations, nine pseudouridines, one methylated pseudouridine, and one unknown modification. These large RNA molecules may be represented as complex two-dimensional structures containing numerous intramolecular stem-loops, as seen in the computational model presented in **Figure 15.17**. Experimental validation of such models comes from high-resolution crystal structures, which provide three-dimensional structures of the RNA, from which we can further refine and improve the accuracy of the two-dimensional predictions. The large subunits also carry smaller rRNAs: a 5S RNA in bacteria and both 5S and 5.8S RNAs in eukaryotes. Each ribosomal subunit contains a number of different proteins (see Figure 15.16). These can be completely separated by 2D gel electrophoresis, and all have been sequenced. They constitute about 40% of the particle mass.

The approximate three-dimensional structures of ribosomes were initially derived from EM imaging, by both traditional transmission electron microscopy and cryo-EM, with ever-increasing resolution of structural details (**Figure 15.18**). After it became

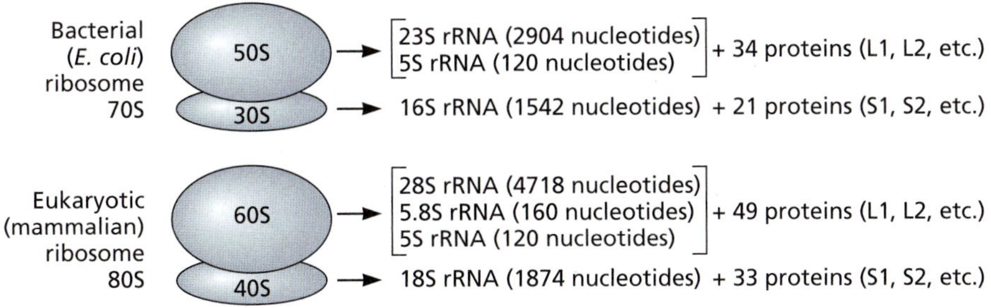

Figure 15.16 The ribosome: An assembly of one large and one small RNP subunit. Each subunit contains rRNA molecules and a set of bound ribosomal proteins. The two subunits and their RNAs are usually referred to by their sedimentation coefficients or S values. The sequence of the eukaryotic 5.8S rRNA corresponds to the 3′-end of the bacterial 23S rRNA.

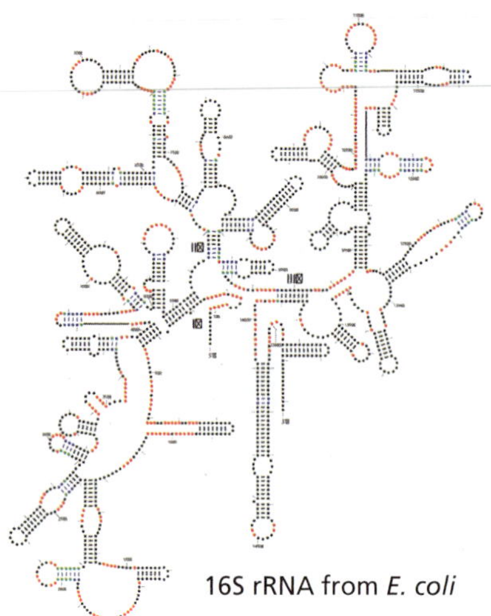

Figure 15.17 Computational model of 2D structure of the 16S rRNA from *E. coli*. The overall aim of such models is to predict the folding of a linear sequence of nucleotides into a thermodynamically stable, biologically active 3D structure. The challenge is to distinguish the correct folding pattern from a large number of possible conformations. The 1500 nucleotides of 16S rRNA can form about 15,000 possible helices, with less than 100 actually occurring in the final structure; for the 23S rRNA, the number of possible helices goes up to ~50,000, with only 150 in the final structure. The possible number of combinatorial arrangements of all possible helices is extremely large: ~4.3×10^{393} for 16S rRNA and 6.3×10^{740} for 23S rRNA. These numbers are far greater than the number of particles in the universe. Tertiary interactions that are layered onto the secondary structures are even harder to predict theoretically. The structural model presented is that of Noller–Woese–Gutell from the early 1980s. Nucleotides are presented as dots of different colors, indicating positions that are unchanged, in black, or changed, in red, blue, and green, between the original and current models. Crystallographic data of the 30S and 50S ribosomal subunits have been used to improve the accuracy of this and other computational models. (From Gutell RR, Lee JC & Cannone JJ [2002] *Curr Opin Struct Biol* 12:301–310. With permission from Elsevier.)

possible to crystallize the individual subunits or the intact ribosome, with or without bound ligands, numerous crystal structures, again at increasing resolution, have been published (**Figure 15.19** and **Figure 15.20**; see Box 15.5). The 30S particle has approximately half the molecular mass of the 50S subunit, and its structurally distinguishable regions have been named the head, shoulder, and platform or body (see Figure 15.18). When viewed from the side that interfaces with the large subunit in the intact particle, the head has a narrow neck region, which is slightly bent over the shoulder, forming a cleft for binding of the mRNA. The decoding site is located at the bottom of the cleft, in the platform.

The 50S subunit has a characteristic shape; when viewed from the interface, the subunit shows a prominent central protuberance, which includes the 5S rRNA and its associated protein partners, and two flexible arms or stalks on the sides. The L1 stalk, formed by protein L1, plays a role in the translocation step that follows the addition of each new amino acid to the chain (see Figure 16.13). The other flexible stalk is labeled L7/L12; L7 is an acetylated form of L12. This stalk is instrumental in bringing the incoming aminoacyl-tRNA to the ribosome.

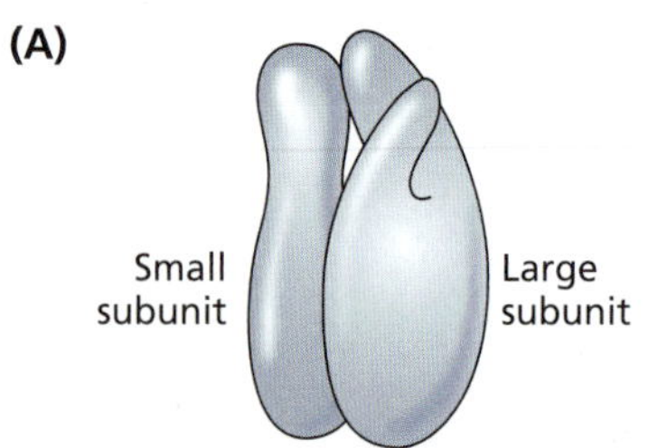

The small subunit can accept mRNA but must join with the large subunit for peptide synthesis to occur

The first complex that forms at the beginning of synthesis of each polypeptide involves the 30S subunit, the mRNA, and a special initiator tRNA. At this point, the 50S subunit joins the complex to form the intact ribosome, which functions during the future stages of translation, discussed in detail in Chapter 16. Association of the two subunits occurs through a complex and dynamic network of interactions along their interface. The interface is relatively free of protein; the flexibility of the rRNA chains allows for constant rearrangements of contacts or intersubunit RNA bridges to ensure proper communication between subunits and coordination of their movements.

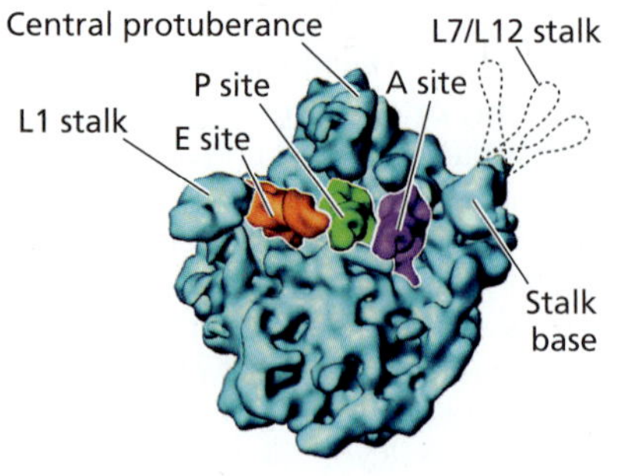

Figure 15.18 Structural models of the bacterial ribosome derived from EM imaging. (A) Model based on low-resolution electron microscopic or EM images shows the 70S ribosome, consisting of a large and a small subunit. (B) Models derived from high-resolution cryo-EM images. Interface views of (top) 30S and (bottom) 50S subunit are shown: in each case, we are looking directly at the surface that will interact with the other subunit. Positions of the three tRNA-binding sites, A or aminoacyl site, P or peptidyl site, and E or exit site, are indicated; both subunits contribute to formation of the sites. Other portions of the structure are also marked. The L7/L12 stalk is normally invisible because of its flexibility. The position and even number of arms of the L7/L12 stalk that make contact with the small subunit are difficult to resolve and are only suggested here by the dotted lines. (B, from Frank J [2003] *Genome Biol* 4:237. With permission from BioMed Central.)

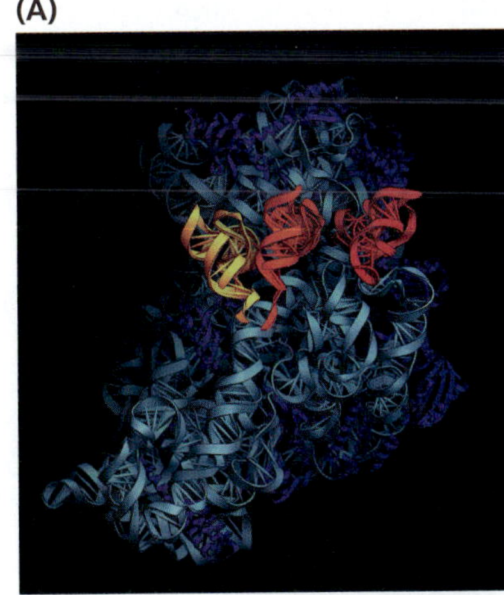

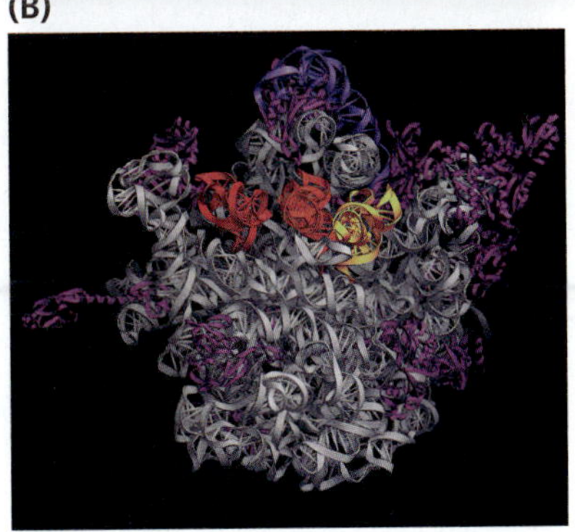

Figure 15.19 Crystallographically derived structure of *Thermus thermophilus* ribosome at 5.5 Å resolution. Interface views of (A) 30S and (B) 50S subunit, showing the position of the three tRNAs. Different molecular components are colored: 16S rRNA, cyan; 23S rRNA, gray; 5S rRNA, lighter blue at the top of the 50S subunit; 30S proteins, darker blue; 50S proteins, magenta. The three tRNA-binding sites are shown as A, gold; P, orange; and E, darker orange. (From Yusupov MM, Yusupova GZ, Baucom A et al. [2001] *Science* 292:883–896. With permission from American Association for the Advancement of Science.)

As mentioned above, the complete ribosome contains three sites for binding of tRNAs, seen both in the cryo-EM structures and in crystals (see Figure 15.18, Figure 15.19, and Figure 15.20). These are called the **A site** or **aminoacyl site**, the **P site** or **peptidyl site**, and the **E site** or **exit site**. Each site extends across the interface to involve both subunits. They are therefore recognizable in the isolated subunits, where they are sometimes referred to as half-sites. We have briefly mentioned the role of these sites in the overview presented in Figure 15.1, and we further elaborate on their functions in Chapter 16.

Ribosome assembly has been studied both in vivo *and* in vitro

Formation of the ribosomal particle involves a complex series of highly coordinated steps. For a ribosome to form, the cell needs to first synthesize numerous proteins and precursors to rRNA. The primary RNA transcript has to mature through processing and modification steps (see Chapter 14), and then the components need to be assembled into a functional particle. The process has been well studied in bacteria, although some major questions linger even there. One of these concerns the highly coordinated, controlled synthesis of ribosomal components. It is well known that growth of the bacterial cell is highly dependent on its protein synthetic capacity. The fraction of existing ribosomes that engage in translation at any given time is constant, at around 80%, and the rate of the synthetic process itself does not vary significantly with cellular growth rate. Thus, the only way to increase protein synthesis in response to growth demands is by increasing the total number of ribosomes in the cell. The magnitude of this increase can be amazing. A cell that is not growing, because of nutritional limitations or other unfavorable environmental conditions, may contain as few as 2000 ribosomes. At first sight this number may seem high, but to put it in perspective, a rapidly growing cell can easily possess 70,000–100,000 ribosomes. Note that for a bacterial cell to increase its number of ribosomes rapidly, it must increase, in a coordinated fashion, the production of three complex RNA molecules and 55 different proteins.

While the regulatory issues involved in ribosome biogenesis *in vivo* will be the subject of major research efforts for some time, *in vitro* studies using purified components have been remarkably successful and have allowed the construction of **assembly maps for ribosome biogenesis** (Figure 15.21). The main conclusion of these studies is that assembly occurs in steps and that proteins bind to the rRNAs at different times and in a fixed order of succession. Depending on whether a given protein binds to the naked RNA molecule or requires a partial RNA–protein complex as a landing platform, we now distinguish three categories of proteins: primary, secondary, and tertiary (see Figure 15.21A). As Figure 15.21B indicates, there are defined intermediates in the process. Importantly, there is an obligatory directionality in the assembly reaction, with the 5′-end of the rRNA binding proteins first and the 3′-end last. Such directionality implies that the assembly reaction occurs while the rRNA is being transcribed; this conjecture has been supported by *in vivo* experiments.

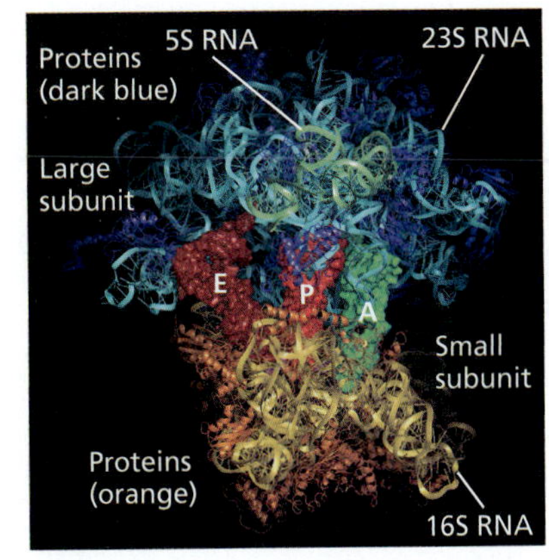

Figure 15.20 Intact 70S ribosome complexed with mRNA and tRNA at 2.8 Å resolution. Nestled between the two subunits are the A-site tRNA, shown in green; P-site tRNA, shown in red; and E-site tRNA, shown in reddish brown. The mRNA at these three sites is just barely visible as a magenta strand. (From Ramakrishnan V [2008] *Biochem Soc Trans* 36:567–574. With permission from Biochemical Society.)

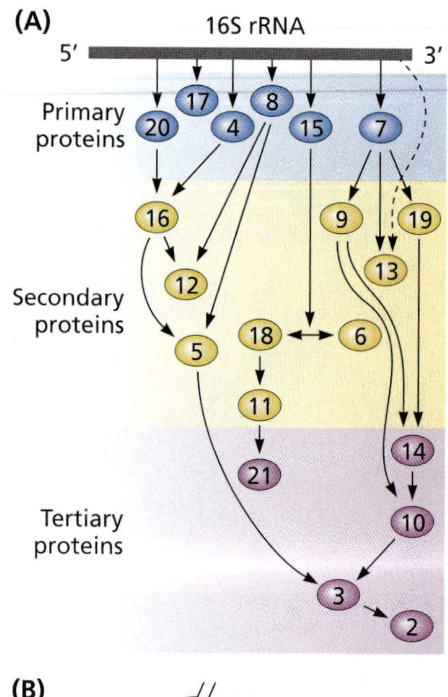

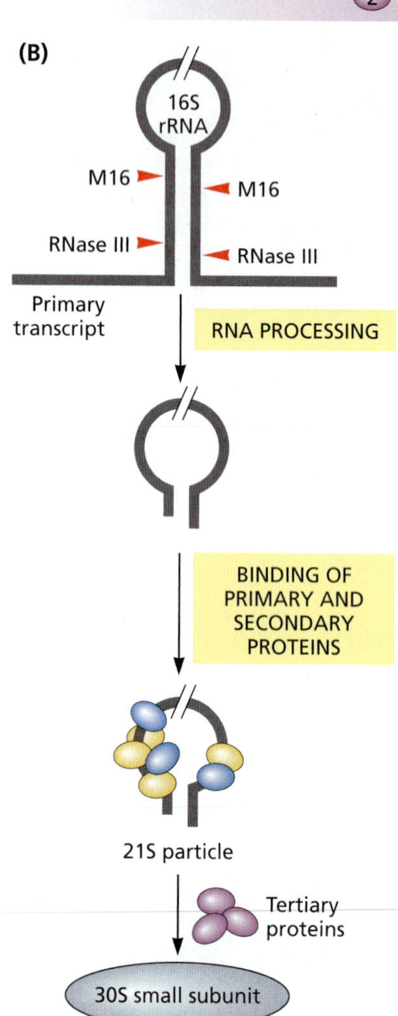

Figure 15.21 Assembly map of the 30S ribosomal subunit derived from *in vitro* experiments. (A) 30S subunits can be reconstituted *in vitro* from purified 16S rRNA and a mixture of ribosomal proteins from crude cellular extracts or from individually purified or recombinant proteins. The numbers are a shortened version of the ribosomal protein name; the arrows show interactions between the proteins. Protein binding to rRNA is cooperative and hierarchical, with early binding events forming binding sites for the proteins that bind later. Three categories of proteins can be distinguished: primary proteins, which bind directly to RNA; secondary proteins, whose binding requires that primary proteins are already bound to rRNA; and tertiary proteins, which need at least one primary and one secondary protein for proper association with the complex. The *in vitro* assembly reaction proceeds with 5′→ 3′ polarity, suggesting that protein binding occurs co-transcriptionally; *in vivo* experiments showed that this is indeed the case. (B) The upper portion of the schematic reminds us that *in vivo* ribosome assembly is preceded by co-transcriptional processing of the primary transcript. The *in vitro* assembly process occurs through a couple of well-defined intermediates. The first such intermediate is the 21S particle, which consists of RNA and primary and secondary proteins. The 21S particle then binds tertiary proteins, forming the 30S subunit. (A, adapted from Kaczanowska M & Rydén-Aulin M [2007] *Microbiol Mol Biol Rev* 71:477–494. With permission from American Society for Microbiology.)

Technically sophisticated studies have directly approached the *in vitro* assembly process. Figure 15.22 gives an account of one such approach. The single-particle EM image analysis in this work provided results that, in combination with data from biochemical studies, allowed the construction of assembly pathways. Not totally unexpectedly, there are a couple of major roads to assembly, but minor pathways also exist in parallel or branch from the major pathways. The multiplicity of pathways and the similar thermodynamic properties or free-energy levels of the intermediates sometimes lead to dead-end products. Production of such products is probably minimized *in vivo* by co-factors or chaperones.

In contrast to bacteria, where ribosome biogenesis occurs spontaneously, eukaryotes use a very complex process to form mature ribosome subunits, which requires numerous auxiliary factors. In addition, all three RNA polymerases (see Chapter 10) are involved: Pol I transcribes the precursor to 28S, 5.8S, and 18S rRNAs, Pol III produces 5S rRNA, and Pol II transcribes the genes for ribosomal proteins and other auxiliary protein factors required for the assembly.

The structures of rRNA and ribosomal proteins have been highly conserved from yeast to humans; there is also a high degree of evolutionary conservation of the non-ribosome factors involved in ribosome biogenesis. More than 150 such auxiliary factors have been identified, mainly through proteomic approaches that are based on mass spectrometry. Protein factors include endo- and exonucleases, pseudouridine synthases, and methyltranferases, all of which process and modify the pre-rRNAs; RNA helicases and RNA chaperones, which mediate RNP folding and remodeling; and GTPases and ATPases, which facilitate association and dissociation of protein factors. A multitude of small nucleolar ribonucleoproteins or snoRNPs also participate in the process of ribosome maturation by guiding the site-specific conversion of uridine to pseudouridine in rRNA. This is accomplished by direct base-pairing of snoRNAs with specific rRNA sequences, leading to exposure of the single uridine to be modified by pseudouridine synthase. Other snoRNAs guide the formation of methylated bases.

Our knowledge of eukaryotic ribosome biogenesis is far from complete. What we know is summarized in Figure 15.23. The schematic emphasizes the participation of at least five different classes of molecules, both RNAs and proteins, in the formation of a large precursor complex, pre-90S. A sequence-specific cleavage in the precursor rRNA leads to the formation of separate precursor particles to the 60S and 40S subunits, which then mature and are transported through the pores in the nuclear membrane independently.

Despite their necessary complexity, ribosomes have remained remarkably conserved in both structure and function over evolutionary history. Eukaryotic ribosomes are somewhat more complicated than those of bacteria and archaea, but their basic

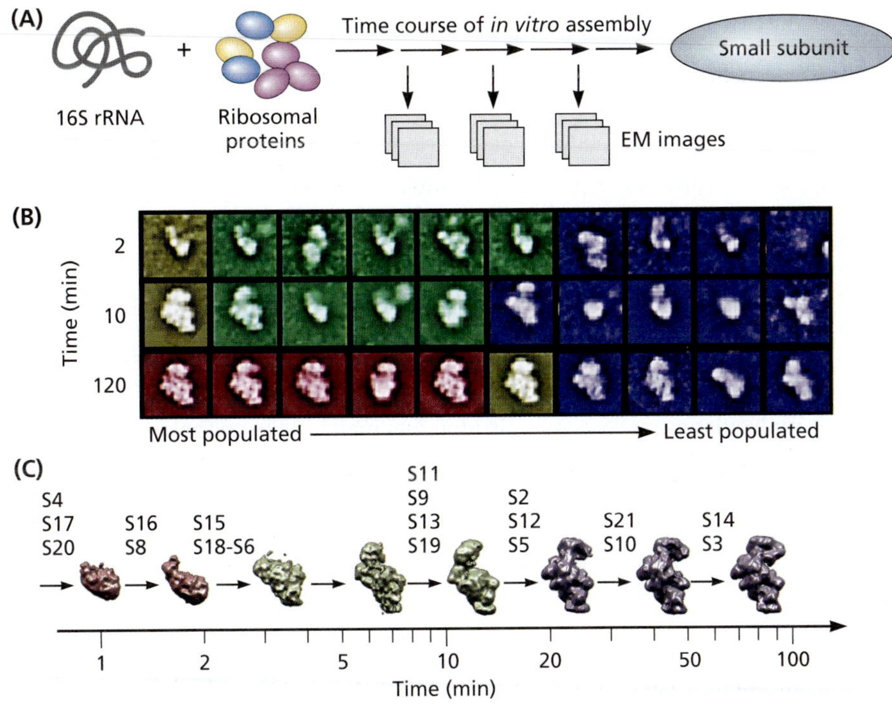

Figure 15.22 Visualizing ribosome assembly by time-resolved single-particle EM. Single-particle EM image analysis can resolve heterogeneous populations of molecules and classify them into homogenous subpopulations that can be seen in both two and three dimensions. (A) Schematic of the experimental approach. Synchronous 30S subunit assembly was initiated *in vitro* from purified components, and the assembly intermediates were visualized at different time points during the folding and assembly process. More than 1 million snapshots of assembling 30S subunits were analyzed, which led to the identification of 14 assembly intermediates; the population flux of these intermediates was monitored over time. The EM results were integrated with mass spectrometry data to construct a ribosome assembly mechanism that incorporates binding dependencies, rate constants, and structural characterization of the intermediate structures. These kinds of integrated approaches provide insights that individual methods cannot, and they define the path of future research. (B) Sample snapshots of some intermediate structures: each row of the array corresponds to a single time point, with the various classes of particles color-coded. Particle averages are ranked from left to right according to the percentage of particles that populate a given class. (C) Combination of kinetic, thermodynamic, and single-particle EM data allowed the construction of assembly pathways. One such major pathway is depicted, but other productive parallel pathways exist. Note that the pathway shown corresponds approximately to that deduced from other experiments. Some of the particle intermediates seem to be dead-end products that cannot lead to a functional 30S subunit. The co-transcriptional nature of 30S assembly and the presence of co-factors *in vivo* likely minimize the formation of such unproductive intermediates. (Adapted from Mulder AM, Yoshioka C, Beck AH et al. [2010] *Science* 330:673–677. With permission from American Association for the Advancement of Science.)

structure and mechanism of function are essentially the same as those of their bacterial ancestors. Their biogenesis is more complex, but this may arise in part from the fact that they must be exported from the nucleus. Protein synthesis originated in the cytoplasm and remains there in all organisms today.

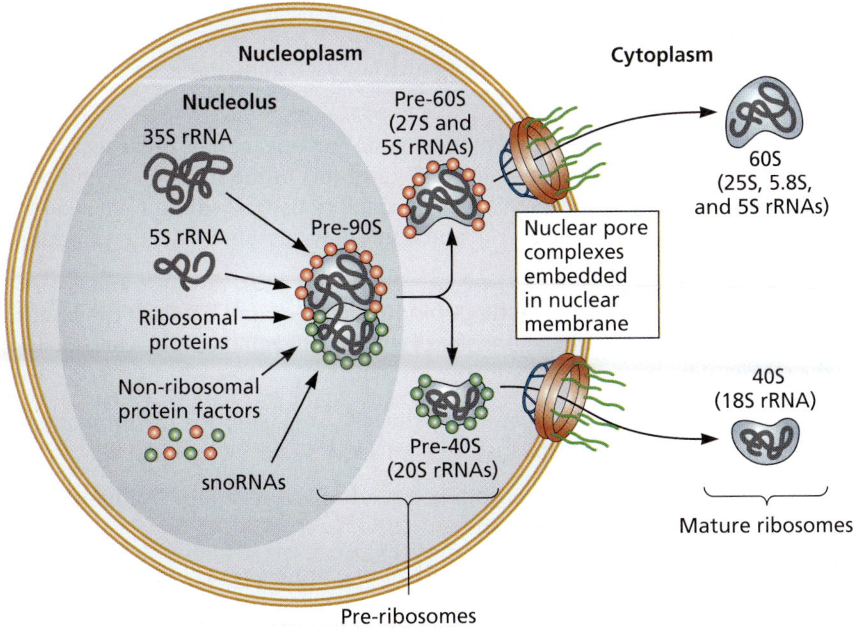

Figure 15.23 Schematic for initial assembly, maturation, and export of the 40S and 60S ribosomal subunits in yeast. After the pre-90S complex forms in the nucleolus, the constituent primary transcript 35S rRNA is cleaved in the spacer regions between the sequences of 18S and 5.8S rRNA, which leads to formation of the pre-ribosomal subunits pre-60S and pre-40S. The subsequent maturation of the two subunit precursors and their transport to the cytoplasm occur independently for the two subunits. The numerous auxiliary factors needed for various steps in the reaction leave the particles following export from the nucleus. Note that the mature large subunit contains 25S rRNA, in contrast to the 28S rRNA characteristic of mammals. (Adapted from Tschochner H & Hurt E [2003] *Trends Cell Biol* 13:255–263. With permission from Elsevier.)

Key concepts

- Translation, in which messenger RNA is read to yield protein chains, is more complex than transcription because it requires both RNA–RNA and RNA–amino acid recognition.
- Translation involves three phases: initiation, elongation, and termination. Elongation is a repetitive cyclic process, with one cycle for the addition of each amino acid to the nascent polypeptide chain.
- There are three major participants in translation: the mRNA that carries the message, the tRNA or adaptor molecules that carry the amino acids to be incorporated in the chain, and the ribosome, which provides an active platform on which the process occurs.
- tRNA molecules carry both an anticodon that matches the codon on the message and an amino acid to be incorporated in the nascent peptide chain.
- The charging of tRNAs with amino acids is catalyzed by a set of enzymes called aminoacyl-tRNA synthetases. The amino acid is first activated by adenylation and then attached to the acceptor end of the corresponding tRNA.
- Fidelity of amino acid–tRNA matching is ensured by various modes of proofreading.
- There are a few special cases in which amino acids outside of the normal code table are incorporated. Most important is the *N*-formylmethionine used with a start codon in bacteria. Other examples are selenocysteine and pyrrolysine, whose incorporation into the chain uses stop codons, and the bioengineering of unnatural amino acids into proteins.
- Functional ribosomes in all three domains of life are composed of two subunits. These are designated by their sedimentation coefficients: 30S and 50S subunits give rise to 70S ribosomes in bacteria and archaea, while 40S and 60S subunits give 80S ribosomes in eukaryotes.
- All ribosomes are ribonucleoprotein particles, each containing several RNAs and many proteins.
- Initiation of translation is different in bacteria and eukaryotes. In most cases, bacteria use a special sequence near the 5′-terminus of the mRNA, which complements a sequence in the 16S rRNA of the small ribosomal subunit. In eukaryotes, the ribosome attaches near the cap and hunts downstream for the first methionine codon.
- Each ribosome contains three tRNA-binding sites, shared between the large and small subunit. Unidirectional transfer of tRNAs between these sites occurs during the elongation phase of translation.
- Termination of translation involves both specific stop codons and termination factors that bind to the ribosome instead of tRNAs.
- *In vivo*, biogenesis requires coordinated synthesis of several RNA molecules and the whole set of ribosomal proteins.
- Assembly of bacterial ribosomes from RNAs and proteins is a complicated process, yet it can be carried out *in vitro*. Assembly of eukaryotic ribosomes is even more complex and requires RNA processing and auxiliary protein factors.

Further reading

Books

Nierhaus KH & Wilson DN (2004) Protein Synthesis and Ribosome Structure: Translating the Genome. Wiley–VCH.

Rodnina M, Wintermeyer W & Green R (eds) (2011) Ribosome Structure, Function, and Dynamics. Springer Verlag.

Reviews

Antonellis A & Green ED (2008) The role of aminoacyl-tRNA synthetases in genetic diseases. *Annu Rev Genomics Hum Genet* 9:87–107.

de Duve C (1988) The second genetic code. *Nature* 333:117–118.

Hernández G (2009) On the origin of the cap-dependent initiation of translation in eukaryotes. *Trends Biochem Sci* 34:166–175.

Hernández G, Altmann M & Lasko P (2010) Origins and evolution of the mechanisms regulating translation initiation in eukaryotes. *Trends Biochem Sci* 35:63–73.

Ibba M & Söll D (2004) Aminoacyl-tRNAs: Setting the limits of the genetic code. *Genes Dev* 18:731–738.

Kaczanowska M & Rydén-Aulin M (2007) Ribosome biogenesis and the translation process in *Escherichia coli*. *Microbiol Mol Biol Rev* 71:477–494.

Ling J, Reynolds N & Ibba M (2009) Aminoacyl-tRNA synthesis and translational quality control. *Annu Rev Microbiol* 63:61–78.

Mignone F, Gissi C, Liuni S & Pesole G (2002) Untranslated regions of mRNAs. *Genome Biol* 3:reviews0004.

Park SG, Schimmel P & Kim S (2008) Aminoacyl tRNA synthetases and their connections to disease. *Proc Natl Acad Sci USA* 105:11043–11049.

Ramakrishnan V (2008) What we have learned from ribosome structures. *Biochem Soc Trans* 36:567–574.

Reynolds NM, Lazazzera BA & Ibba M (2010) Cellular mechanisms that control mistranslation. *Nat Rev Microbiol* 8:849–856.

Schimmel P (2008) Development of tRNA synthetases and connection to genetic code and disease. *Protein Sci* 17:1643–1652.

Experimental papers

Baldwin AN & Berg P (1966) Transfer ribonucleic acid-induced hydrolysis of valyladenylate bound to isoleucyl ribonucleic acid synthetase. *J Biol Chem* 241:839–845.

Ban N, Nissen P, Hansen J et al. (2000) The complete atomic structure of the large ribosomal subunit at 2.4 Å resolution. *Science* 289:905–920.

Cate JH, Yusupov MM, Yusupova GZ et al. (1999) X-ray crystal structures of 70S ribosome functional complexes. *Science* 285:2095–2104.

Lee JW, Beebe K, Nangle LA et al. (2006) Editing-defective tRNA synthetase causes protein misfolding and neurodegeneration. *Nature* 443:50–55.

Mulder AM, Yoshioka C, Beck AH et al. (2010) Visualizing ribosome biogenesis: Parallel assembly pathways for the 30S subunit. *Science* 330:673–677.

Nureki O, Vassylyev DG, Tateno M et al. (1998) Enzyme structure with two catalytic sites for double-sieve selection of substrate. *Science* 280:578–582.

Voss NR, Gerstein M, Steitz TA & Moore PB (2006) The geometry of the ribosomal polypeptide exit tunnel. *J Mol Biol* 360:893–906.

Wimberly BT, Brodersen DE, Clemons WM Jr et al. (2000) Structure of the 30S ribosomal subunit. *Nature* 407:327–339.

Yusupov MM, Yusupova GZ, Baucom A et al. (2001) Crystal structure of the ribosome at 5.5 Å resolution. *Science* 292:883–896.

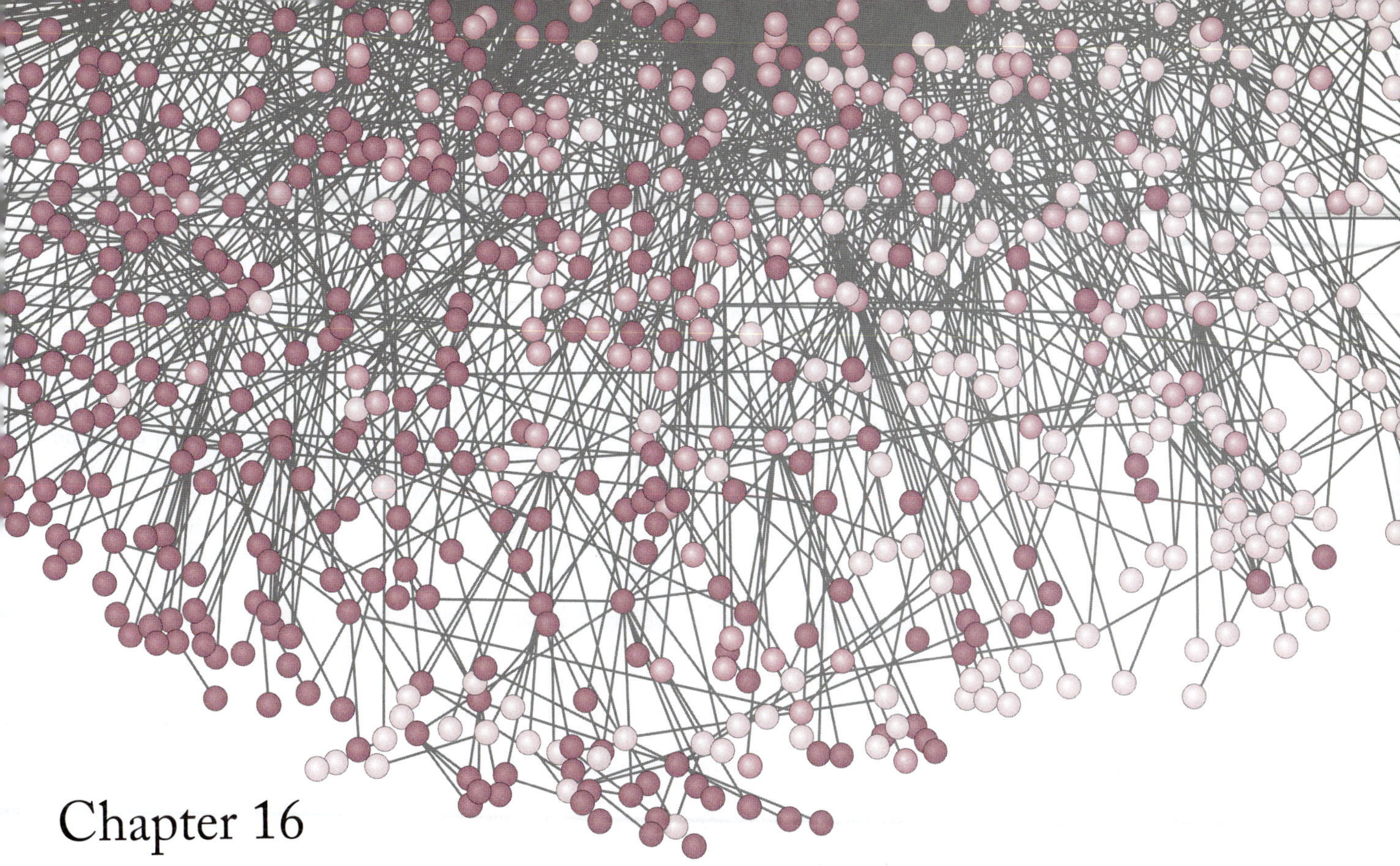

Chapter 16

Translation: The Process

16.1 Introduction

In Chapter 15, we describe the three major players in the process of translation. In this chapter, we cover the process itself from both a structural and a mechanistic point of view. We focus on bacterial translation, which is better understood and less complex than translation in eukaryotes. Much of this understanding stems from the abundant information gathered in recent years by a powerful combination of methods, including X-ray crystallography, cryo-EM, bulk kinetic studies, single-pair FRET, and computational modeling. Together, these studies have provided the first clear mechanistic picture of the process.

16.2 An overview of translation: How fast and how accurate?

To meet cellular needs, a process such as translation must be both fast enough to maintain the product at necessary levels, especially in cases when the proteins turn over quickly, and accurate enough to deliver a usable product. It might seem intuitive that speed and accuracy are competitive, but an overall view of the transfer of information in the cell does not support such a conclusion. For example, DNA replication proceeds at ~1000 nucleotides (nt) per second, and even on first pass maintains an error frequency, r, of only 1 in 100,000: $r = 1 \times 10^{-5}$. With proofreading, the error frequency drops to 1×10^{-10} (see Chapter 19). Translation is much slower: in bacteria, around 15 amino acid residues are added to a nascent polypeptide chain per second; in eukaryotes, the rate is probably about half this or less. Nevertheless, the error frequency, even with proofreading, is $\sim 1 \times 10^{-4}$. Understanding the velocity of translation in bacteria is complicated by the fact that it is coupled to transcription: typically, bacteria transcribe ~45 nt of mRNA per second; this is equivalent to ~15 codons per second. Thus, the mRNA elongation rate matches the rate of translation. So we do not know whether the rate of translation evolved to match that of transcription, or vice versa, in these organisms. Of course, in eukaryotes, no such coupling is possible because transcription and translation occur in different compartments of the cell.

We can gain an idea of the effects of an error frequency of 1×10^{-4} by a simple calculation. If the chance of any one residue being incorrect is r, then the chance of the first one being right is $1 - r$, and the chance that the first two are both right is $(1 - r)(1 - r)$. For a protein of N residues, the chance P of it being wholly correct is $(1 - r)^N$. A typical protein chain is ~300 residues in length; this gives $P = 0.97$, or 97%, for $r = 1 \times 10^{-4}$. If r were as large as 1×10^{-2}, only 5% of all 300-residue products would turn out completely as encoded by the mRNA. This is about the value that would be expected if translation proceeded without proofreading. Thus, proofreading is essential. Errors in the translational machinery can have significant medical consequences (**Box 16.1**). We note, however, that a higher error frequency in proteins than in RNA or DNA is tolerable. One incorrect protein molecule in the cell does not have nearly the same impact as does a single DNA or even RNA error, either of which will ultimately result in the synthesis of many incorrect protein molecules.

So how does the cell make sufficient proteins? The cell compensates for the slow rate of protein synthesis by using a huge number of ribosomes as protein factories. Rapidly growing bacterial cells contain ~100,000 ribosomes. A typical mammalian cell may easily have more than a million ribosomes that allow it to synthesize the billion or so protein molecules present in the cell. Not only do eukaryotic cells contain very many protein molecules, but these are in flux, being constantly degraded and synthesized. So the synthetic machinery is always active.

Large mRNAs can be translated by many ribosomes simultaneously; the presence of many active ribosomes on a single message creates **polyribosomes**, seen in the electron micrograph presented in **Figure 16.1**. As we have discussed, translation and transcription are coupled in bacteria: once a newly synthesized portion of mRNA extrudes from the exit channel of the RNA polymerase, ribosomes attach to it and begin to translate it. This means that, unlike translation in eukaryotes, bacterial translation is not complicated by the frequent formation of secondary structures in the RNA. Such structures never have a chance to form. Each mRNA is usually translated multiple times; in *Escherichia coli*, the average is around 30 times. How often

Box 16.1 Initiation and elongation translation factors are possible oncoproteins The first evidence implicating translation factors in oncogenesis was the observation that overexpression of eIF4E, the mRNA cap-binding factor (see Figure 16.6), transforms mouse NIH 3T3 cells grown in culture. These cells are immortalized, that is, they have the ability to divide indefinitely in culture but cannot form solid tumors when grafted into immunocompromised or nude mice. Upon overexpression of eIF4E, they acquire the ability to grow as solid tumors in such animals; that is, they become malignantly transformed. Abnormally high levels of eIF4E are observed in a wide variety of primary human cancers; the gene encoding the factor is frequently amplified in some tumors. Usually, elevated eIF4E levels are associated with poor prognosis and an increased chance of recurrence after initial therapy. Tumor growth in mice is significantly decreased by administering antisense oligonucleotides against eIF4E, without any noticeable side effect. This antisense DNA sequence is now in clinical trials as a cancer treatment.

Elongation factors eEF1A1 and eEF1A2 are also viewed as oncoproteins. The path of discovery of their transforming properties was long and winding, and there are still unresolved issues. Much of the earlier work lacked important control experiments; evidently, at the time it was not appreciated that a wild-type translation factor could be an oncoprotein.

eEF1A1 and eEF1A2 are isoforms of eEF1A that share more than 95% identity at the DNA and protein levels. They are believed to have the same enzymatic function in protein translation of recruiting a charged tRNA to the A site of the ribosome; that is, they are functional equivalents of bacterial EF-Tu. eEF1A1 is ubiquitously expressed, whereas eEF1A2 expression is restricted to heart, brain, and skeletal muscle. The reason for this differential expression is unclear. Surprisingly, of these two isoforms, only eEF1A2 is an experimentally proven oncoprotein, although eEF1A1 is also overexpressed in tumors. Overexpression of wild-type eEF1A2 transforms cells in culture, endowing them with tumor-formation properties when grafted into nude mice. Other elongation factors may also play roles in oncogenesis.

Whereas overexpression of eEF1A1 or eEF1A2 seems to correlate with accelerated cell growth in cancer, their underexpression accompanies the slowed growth of senescent cells. When normal mammalian cells stop dividing in culture and become senescent, they continue to have normal metabolic functions, albeit at slowed rates. Senescent human fibroblasts in culture do exhibit a decrease in eEF1A1 mRNA levels and in catalytic activity. In intact mice, the rate of protein elongation decreases significantly with age, by up to 80%. Whether there is a causal relationship between eEF1A1 or eEF1A2 levels and/or activity and aging remains to be established experimentally.

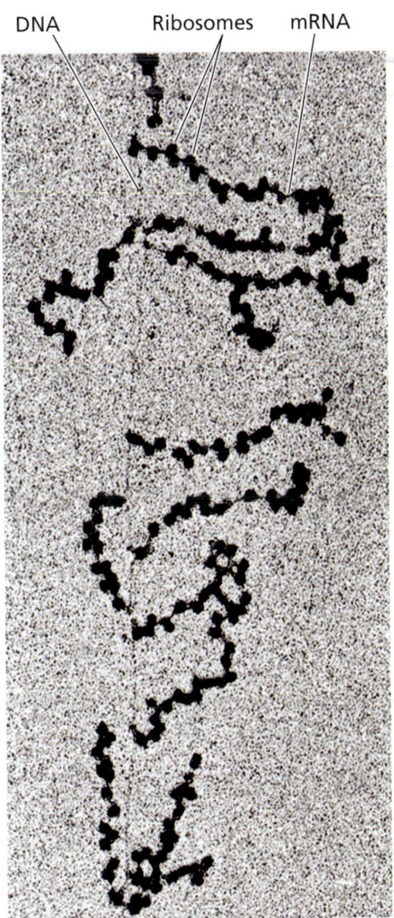

Figure 16.1 Electron micrograph illustrating polyribosomes and showing that transcription and translation in bacteria are coupled. The length of the nascent mRNA transcripts increases from bottom to top along the DNA strand, indicating that this is the direction of transcription. Ribosomes bind to the 5′-end of the mRNA molecules during their synthesis, as soon as a short portion of the message appears from the exit channel of the RNA polymerase. The mRNA is seen to be covered with translating ribosomes, forming polyribosomes. The long polyribosomes at the end of the gene are released from the DNA when transcription terminates. (Courtesy of Oscar L. Miller, University of Virginia.)

each message is utilized varies considerably and is strictly controlled by the cell, depending on the need for a particular protein at a given moment. In eukaryotes, this control occurs in part at the level of mRNA stability, which at the molecular level is determined by the 3′-poly(A) tail and other elements in the 3′-untranslated region or UTR of the message.

Finally, mRNAs can be classified as **monocistronic** when they encode a single polypeptide or **polycistronic** when a single RNA chain encodes several polypeptides. Polycistronic messages are common in bacteria, which have a mechanism for the utilization of every initiation codon in the mRNA; each initiation codon is preceded by its own Shine–Dalgarno sequence. Recent studies have revealed the very rare presence of polycistronic mRNAs in eukaryotes too.

In general, translation in archaea closely resembles that in bacteria. The process in eukaryotes is somewhat different and generally more complex. In part, this may be explained by the uncoupling of transcription and translation in eukaryotes, and also by the requirement to protect mRNA molecules from the activities of nuclease enzymes during and after their journey from the nucleus to the cytoplasm.

16.3 Advanced methodology for the analysis of translation

Before describing in detail the individual steps in the process of translation, we briefly discuss some special methods that have been responsible for the recent rapid advances in understanding translation: single-particle cryoelectron microscopy or cryo-EM, single-pair fluorescence resonance energy transfer or spFRET, and X-ray diffraction. Each of these methods has advantages, drawbacks, and complementarity in providing both still snapshots of the various complexes formed at different steps in translation and dynamic real-time information on how these steps occur. The principles of these methods have already been introduced in previous chapters; here we focus on their application to studying translation. It must be understood, however, that much of our knowledge of ribosome function stems from a long background of kinetic studies, using methods such as stopped-flow or quenched-flow analysis. The newer work has allowed visualization of these processes on a molecular scale.

Cryo-EM allows visualization of discrete kinetic states of ribosomes

In cryo-EM (see Chapter 1), the samples are not attached to surfaces, as they are in most other imaging techniques, but instead are suspended in a layer of vitrified water, a state of water obtained by very rapid cooling with liquid nitrogen. The individual molecules or complexes are oriented randomly in this layer, and ice crystals do not form, so no distortions of the structures are expected. The original images obtained from the microscope are of low contrast, but highly sophisticated image-reconstruction methods allow the subsequent production of a three-dimensional density map of an average particle, based on thousands of projections of individual particles. Thus, the method as a whole can be considered an ensemble averaging, although the original data come from single molecules or complexes. If the sample is structurally heterogeneous, which is often the case, then computational classification methods can divide the data into homogeneous subsets. Not only can the structures of the subsets be deduced but also their relative abundance. A similar classification of EM images is used to describe the pathways of 30S ribosome subunit assembly in Figure 15.22.

The application of cryo-EM to ribosome structure and dynamics using single-particle reconstruction has been challenging, mainly because of the difficulties in synchronizing the molecular population. Synchronization can be achieved by trapping the majority of the complexes in the same state of the process by using ribosome-targeting antibiotics (**Box 16.2**) or nonhydrolyzable or slowly hydrolyzable analogs of GTP. There are five major steps in translation that involve GTP hydrolysis (**Table 16.1**). Replacing available GTP with GTP analogs that have the same ability to bind to the appropriate locations in the ribosomal complex but do not undergo hydrolysis has been very useful, not only in structural studies but also in biochemical experiments seeking to determine the kinetic parameters of the process.

Some disadvantages of cryo-EM include difficulties in seeing peripheral structures and components that are highly mobile. However, the main drawback is the relatively low resolution of the images, which precludes chain tracing of individual protein or nucleic acid molecules. Here, X-ray structures come to the rescue. Whenever such atomic structures are available, it is possible to fit them into cryo-EM-derived molecular models. Fitting tools are becoming more and more sophisticated, thus greatly facilitating the interpretation of the data.

X-ray crystallography provides the highest resolution

The power of crystallography in providing high-resolution atomic structures that form the basis for understanding the function of molecules and complexes has been emphasized throughout this book. As seen in Chapter 15, we now have ribosome structures of amazing, sometimes overwhelming detail. Several important states of the ribosome and its two subunits, alone or with bound ligands, have been resolved crystallographically, providing a wealth of information to guide biochemical and genetic studies. In addition, we now have structures, to atomic resolution, of practically all ribosome-targeting antibiotics complexed with ribosomes. These structures help to elucidate the molecular mechanisms of antibiotic action; they are also powerful tools for understanding ribosome structure and function and for drug design.

There are drawbacks to X-ray crystallography too. Because of the constraints imposed by packaging the molecules or complexes into a crystal lattice, the structure in the crystal is not always physiologically relevant. Molecular interactions in the lattice may

Box 16.2 Ribosome function and antibiotics Today, there is a huge and ever-growing number of both natural and semi-synthetic compounds that are known to inhibit bacterial growth. There are several distinct classes of antibiotics each distinguished by their mechanism of action. Some inhibit cell wall synthesis, others affect DNA replication or transcription (see Box 9.4 and Box 19.6), and still others target protein synthesis. In the latter case, there are antibiotics that inhibit almost every step of the elongation cycle (**Figure 1**). The availability of such a broad range of translation-inhibiting antibiotics has been instrumental in treating numerous infectious diseases in clinical practice. Antibiotics that target the translational apparatus are in fact the most widely used anti-infection drugs.

In addition, antibiotics provide a powerful tool for investigating the fundamental mechanisms of protein synthesis. The clinical and research applications of antibiotics led to the solution of numerous structures of antibiotics bound to the 30S and 50S subunits. Today, practically all major classes of ribosome-targeting antibiotics have been visualized in complexes with ribosomes. The structures revealed that antibiotics target highly conserved, functional centers of the bacterial ribosome, including the path of tRNA and mRNA movement through the small subunit, the peptidyl transferase center (PTC), and the exit tunnel in the large subunit.

There is a perpetual need for new antibiotics because of the development of antibiotic resistance. At least three classes of antibiotic-resistant pathogens are emerging that constitute a major threat to public health and health care costs: (1) methicillin-resistant *Staphylococcus aureus* or MRSA; (2) multi-drug- or pan-drug-resistant Gram-negative bacteria including *Acinetobacter baumannii, Escherichia coli, Klebsiella pneumoniae,* and *Pseudomonas aeruginosa*; and (3) multidrug-resistant *Mycobacterium tuberculosis* strains. Most other drugs will be just as effective in the future as they are today, but the inevitable rise of resistance will erode the utility of today's antibiotics. Prospects for finding new antibiotics for Gram-negative bacterial pathogens are especially poor because of problems of membrane-blocked entrance of drug into the cell and the activity of efflux pumps that expel much of the remainder.

There are obstacles to developing new antibiotics: their production and marketing is not as profitable for pharmaceutical companies as for drugs that treat chronic diseases. Thus, although resistance is definitely on the rise, antibiotic discovery and development are on the decline. Between 1962 and 2000, no major classes of antibiotics were introduced.

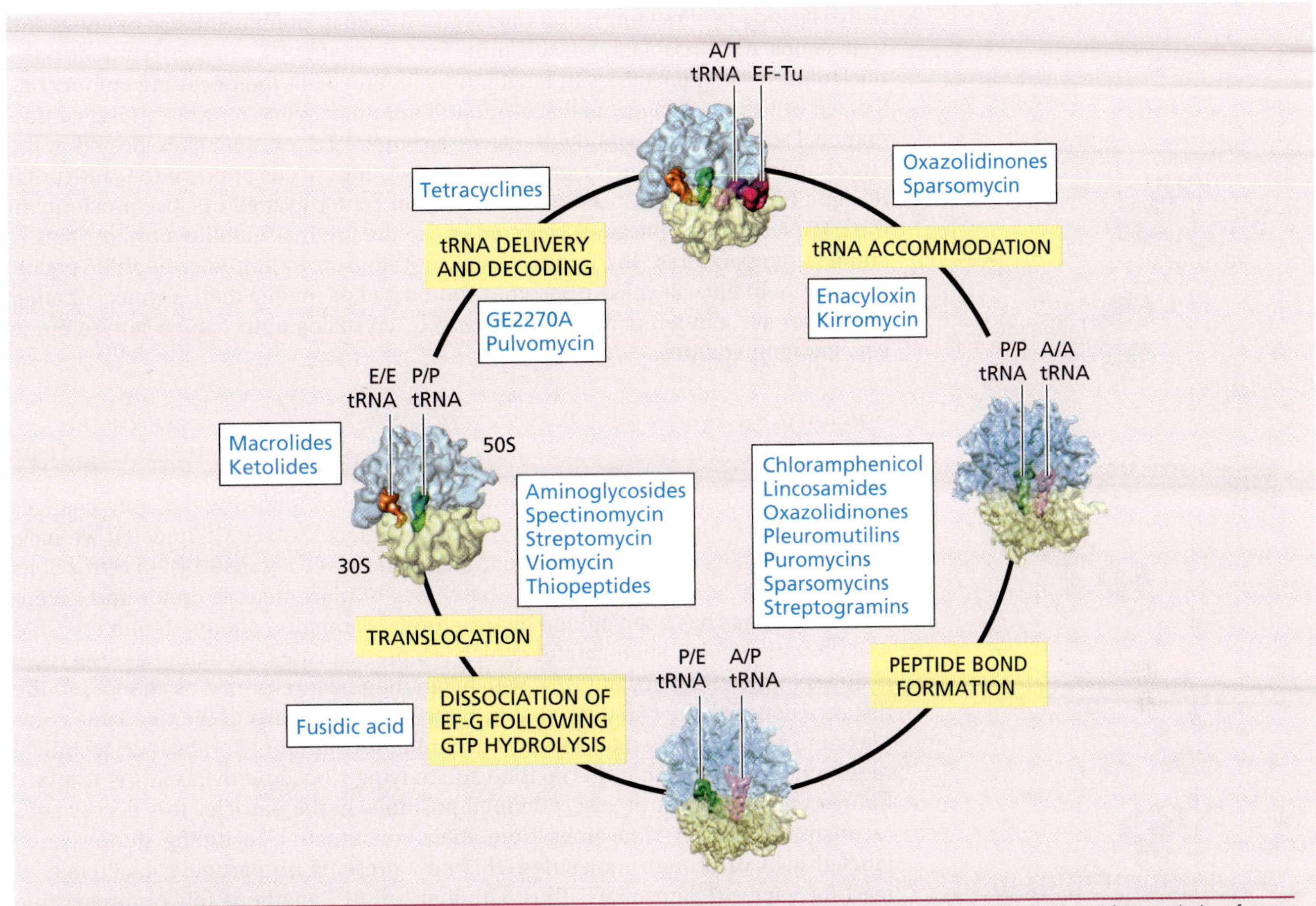

Figure 1 Inhibition of translation by antibiotics. A generalized schematic of the translational elongation cycle. Beside each major step in the cycle are listed antibiotics that specifically interfere with that step. (Adapted from Blanchard SC, Cooperman BS & Wilson DN [2010] *Chem Biol* 17:633–645. With permission from Elsevier. Structures, from Frank J & Gonzalez RL Jr [2010] *Annu Rev Biochem* 79:381–412. With permission from Annual Reviews.)

decrease the range of observable conformations. The reverse is also true: interactions may be seen in the crystal that are not likely to be present in the molecules in their native environment, that is, in the cytoplasm. For huge complex structures such as ribosomes, it is often necessary to use modified or truncated ligands instead of the

Table 16.1 Steps in bacterial translation that require the hydrolysis of GTP carried on protein factors. In all cases, the protein factor involved possesses catalytic activity, which is stimulated by the GTPase center of the large subunit. Protein factors that are bound to the ribosome complex in a GTP form are released following GTP hydrolysis, which induces a conformational change in the factor that is instrumental in its release. aa-tRNA, aminoacyl-tRNA.

Step in translation	Factor involved	Role
initiation	IF2	formation of 30S initiation complex; brings initiator tRNA to P site; promotes subunit joining to form 70S initiation complex
elongation, delivery of incoming aa-tRNA to A site in 70S ribosome	EF-Tu	hydrolysis occurs following correct base-pairing between codon and anticodon on incoming aa-tRNA; EF-Tu release is followed by accommodation
elongation, translocation	EF-G	EF-G–GTP binding stabilizes hybrid conformation of the ribosome, P/E and A/P states; hydrolysis returns the ribosome into classical P/P and E/E conformation, ready for the next round of elongation
termination, peptide release	RF3	accelerates dissociation of RF1 or RF2 from the ribosome following peptide release; GTP hydrolysis releases the factor from the ribosome
termination, ribosome recycling	RRF–EF-G	GTP hydrolysis promotes disassembly of the ribosome, yielding free 50S subunit and 30S–mRNA–deacylated tRNA complex

native ligands to allow formation of crystals. A case in point is the use of anticodon stem–loops, ASL, in isolation to substitute for full-length tRNAs. These synthetic polynucleotides, ~30 nucleotides in length, interact only with the **decoding center**, **DC**, located in the small ribosomal subunit and are not capable of transmitting conformational signals to the peptidyl transferase center, PTC, or other sites located in the large subunit. In cells, such signal transduction does occur upon codon–anticodon recognition, from the DC to the GTP activation center located in close proximity to the PTC. A final complication can come from the internal mobility of ribosomes at ambient temperatures. The frequent choice of ribosomes from thermophilic organisms can help here, as these ribosomes should be less mobile than is typical of other ribosomes at room temperature. Alternatively, crystallography can be carried out at very low temperatures.

Single-pair fluorescence resonance energy transfer allows dynamic studies at the single-particle level

We have already pointed out the power of spFRET in following dynamic changes of the distance between dye labels on macromolecules (see Chapter 1); we have also illustrated its use for the study of transcription (see Figure 9.14). In terms of its application in studying translation, the choice of placement of donor and acceptor fluorophores has been guided by structural information obtained from cryo-EM and crystallography. Such careful choice is a prerequisite for obtaining observable changes in the efficiency of FRET, which can then be interpreted as changes in the distance between the two dyes, that is, between the portions of the ribosome complex and/or its ligands that undergo dynamic fluctuations. An absolute technical requirement for applying the method to studying ribosome dynamics is to place fluorescent markers at precisely defined positions in the particle. This necessitates reconstituting intact ribosomes from their constituents, including fluorescently labeled RNA or protein molecules. This now presents no particular challenge, at least for bacterial ribosomes. Thus, although spFRET methods rely on preexisting structural information, they are able to provide dynamic information that complements the information that can be obtained from a series of static snapshots captured throughout the process of interest.

spFRET is limited in that it can measure one, and only one, distance determined by the exact position of the fluorescent probes. Now methods are being developed to simultaneously follow two distances; this can be achieved by monitoring FRET signals between one donor and two distinct acceptor fluorophores that emit at discernibly different wavelengths; this is known as three-color FRET. Despite this progress, numerous labeling schemes must be employed to follow the complex choreography

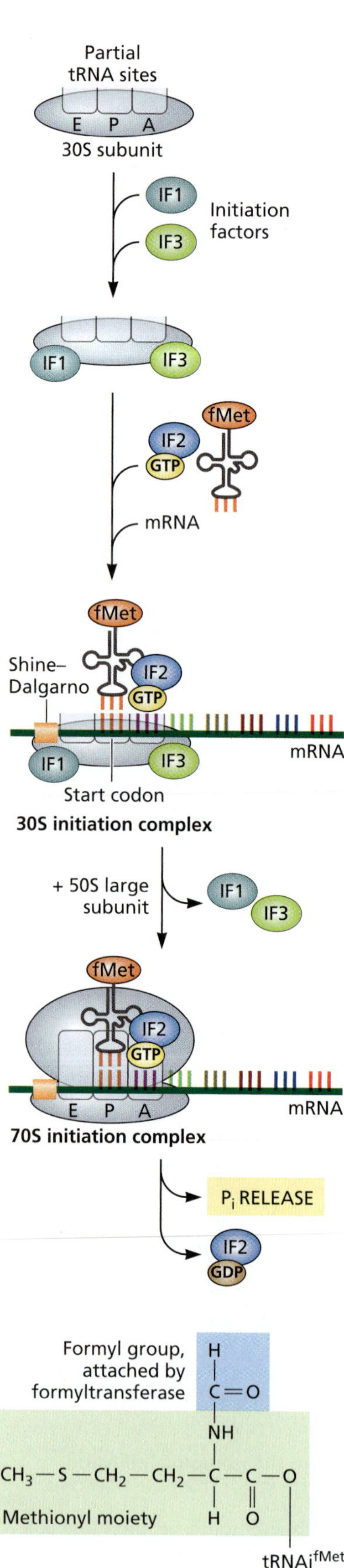

Figure 16.2 Formation of bacterial 70S initiation complex. Initiation requires the ribosome to position the initiator fMet-tRNAi^{fMet} over the start codon of mRNA, which is located in the P half-site of the ribosome. This precise positioning involves the action of three initiation factors, IF1, IF2, and IF3, the roles of which remain largely unclear. The process probably begins with binding of initiation factor IF3 to the small subunit. The intact ribosome will have split into its small and large subunits at the end of the previous round of translation, following translation termination. IF3 binding occurs following the dissociation of the 50S and 30S subunits, which stimulates the release of deacylated tRNA and mRNA from the small subunit at the end of translation. IF3 prevents the large subunit from re-associating, that is, it acts as an anti-association factor; in addition, it ejects any tRNA other than fMet-tRNAi^{fMet} from the P site. The role of IF1 is less clear. At the next step, IF2–GTP specifically recognizes the charged initiator tRNA; it binds to the 30S subunit and facilitates binding of fMet-tRNAi^{fMet}. The 30S complex interacts with mRNA by recognizing the Shine–Dalgarno sequence. The complex containing the small subunit, fMet-tRNAi^{fMet}, and mRNA is known as the 30S initiation complex; the three initiation factors are still bound. In the next step, the 50S subunit joins the complex to form the 70S initiation complex. Exactly how and when the bound initiation factors are released is unclear. IF3 and IF1 must leave before the 50S subunit joins; the most recent data indicate that IF2 promotes subunit joining and most probably remains bound to the 70S complex until GTP bound to IF2 is hydrolyzed. The structure of fMet is shown at the bottom.

of the ribosome in action. In addition, the time resolution of the measurements is sometimes below the rate at which a process occurs. Thus very fast processes, such as the actual peptidyl transfer, might simply be impossible to monitor. Finally, to define intraribosome motions unambiguously by this approach, one dye, preferably the donor, should be in a position in the structure that does not move; the other fluorophore should undergo motion.

16.4 Initiation of translation

The overall process of translation is usually divided into three major phases: initiation, elongation, and termination. Some researchers define a separate fourth step, dissociation of the intact 70S ribosome into its constituent subunits, allowing them to be reused for synthesis of another polypeptide chain. Here, we describe these steps in the order of occurrence during translation. As stated above, the discussion focuses first on the bacterial process, and then some major differences that occur in eukaryotic translation will be pointed out, so far as we have information on them.

Initiation of translation begins on a free small ribosomal subunit

Initiation requires first the positioning of the **initiator tRNA**, charged with formylmethionine in bacteria, fMet-tRNAi^{fMet}, or just methionine in eukaryotes, Met-RNAi^{Met}, at the P half-site of the small ribosomal subunit. The mRNA must also be bound, with its codon in register with the anticodon. In bacteria, the mRNA alignment depends on base-pairing between the 3′-end of 16S rRNA and the Shine–Dalgarno sequence located just upstream of the mRNA start codon. Three initiation factors, IF1, IF2, and IF3, must also bind (**Figure 16.2**). The initiator tRNA, tRNAi, is special because it recognizes and binds to the AUG start codon but does not interact with the same codon when it is internally located in the message. AUG is the only codon that specifies methionine. Its role in the message, either serving as a start codon or coding for an internal methionine, is discriminated by using two different methionyl-tRNAs that differ in primary sequence and, as a consequence, in the structure of one of the arms. The initiator tRNA is the only tRNA that is accepted by the 30S subunit at initiation, binding to the P half-site present in the free 30S subunit; all other acylated, charged tRNAs can bind only to the fully assembled ribosome. Thus, each cell contains at least two methionyl-tRNAs, each dedicated to a specific role in supplying methionine for incorporation into the polypeptide chain: one for initiation and the other for elongation of the chain by one methionine residue.

In bacteria, the methionine attached to the initiator tRNA is formylated to form *N*-formylmethionine. The formyl group is added by a dedicated enzyme, formyltransferase or transformylase, after charging of the tRNA. In almost all cases, the formyl group is removed during chain elongation. For many proteins, the methionine itself is also removed later, so the first amino acid at the N-terminus of the polypeptide chain is rarely a methionine. In archaea and eukaryotes, the methionine is not formylated, but the discrimination between AUG start and internal codons is still performed by the use of two structurally different methionine tRNAs.

Initiation is a multistep process (see Figure 16.2). Although we now know the probable sequence of events during initiation, the process as a whole remains mechanistically poorly defined, mainly because of the paucity of structural data. The first step in bacteria is the binding of IF3 to the 30S subunit. This actually occurs at the end of the previous round of protein synthesis, after the two subunits split from each other. The binding of other initiation factors, mRNA, and the charged tRNAi leads to the formation of the 30S initiation complex. The roles of the various initiation factors are only partially understood.

Cryo-EM provides details of initiation complexes

Further insights have been provided by the resolution of a partial 30S initiation complex by cryo-EM (**Figure 16.3**), yielding valuable information about how IF2 interacts

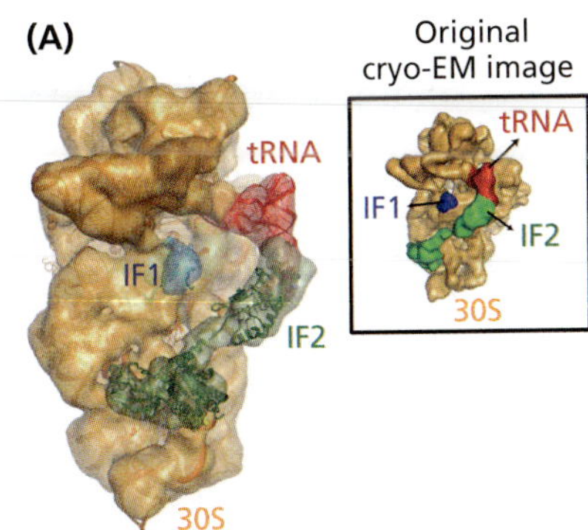

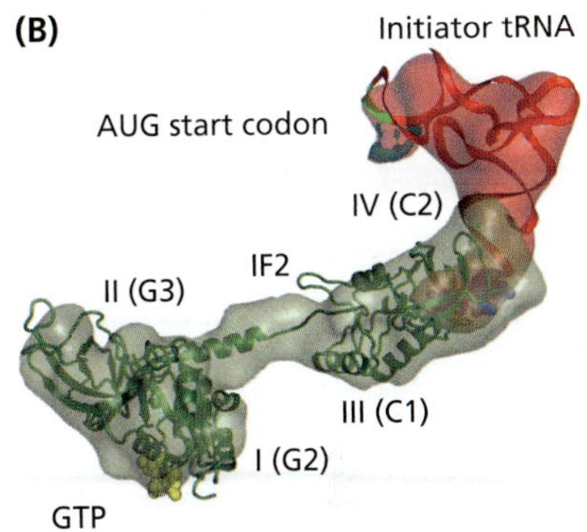

Figure 16.3 Structure of a partial 30S translation initiation complex. The cryo-EM map and the derived molecular model disclose the positions and interactions of IF1, IF2, and fMet-tRNAi^{fMet} on the 30S subunit; IF3 is missing from this structure. To facilitate interpretation, the crystal structures of 30S–IF1, tRNA, and an IF2 homology model, aIF5B, have been fitted to the cryo-EM map. (A) View from the subunit interface. The inset shows the original cryo-EM image reconstruction. (B) View of the IF2–fMet-tRNAi^{fMet} subcomplex, slightly rotated to the left compared to the entire structure on top. The tRNA decoding stem is bent toward the mRNA; the 3′-CCA end of the tRNA is kinked toward IF2 domain IV (C2). The anticodon is shown in green; the residual molecular density next to it corresponds to the AUG start codon, shown in darker green. The stability of the complex is ensured by two interactions: (1) the tRNA decoding stem being buried in the 30S P half-site and (2) the C-domain of IF2 interacting strongly with the acceptor end of fMet-tRNAi^{fMet}. The structure rationalizes the rapid activation of GTP hydrolysis upon 30S initiation complex–50S joining: the GTP-binding domain of IF2 would directly face the GTPase-activated center of the 50S subunit, and the largest part of IF2 is complementary in shape to the 50S surface, favoring subunit association. (From Simonetti A, Marzi S, Myasnikov AG et al. [2008] *Nature* 455:416–421. With permission from Macmillan Publishers, Ltd.)

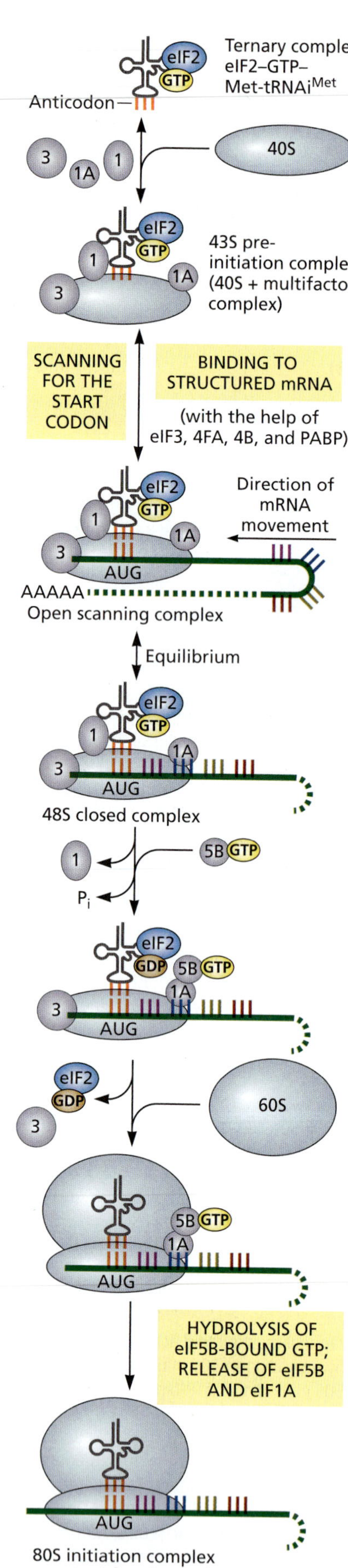

with the face of the 30S subunit that will later interact with the 50S subunit. IF2 stretches across this interface, contacting not only the anticodon stem but also the acceptor end of fMet-tRNAi^{fMet}. Further cryo-EM work has shown that, after subunit joining, the G domain of IF2 interacts with the GTPase-stimulating center of the large ribosomal subunit; this interaction promotes the GTPase activity of IF2. It is worth noting here that in all steps of translation that are driven by GTP hydrolysis, the appropriate catalytic GTPase center lies within the respective factors involved. In all cases, the ribosome, with its GTPase center, serves to stimulate these activities (see Table 16.1). Within the structure of a 70S initiation complex that contains fMet-tRNAi^{fMet} and IF2 bound to a nonhydrolyzable GTP analog, IF2 is still bound to the GTPase center but has lost contact with fMet-tRNAi^{fMet}, which is now properly positioned, in the P site, for initiation.

Start site selection in eukaryotes is complex

Initiation in eukaryotes is much more complex and involves at least 12 initiation factors, composed of more than 20 polypeptides (**Figure 16.4**). The need for so many eukaryotic initiation factors, eIFs, is explained by important mechanistic differences between **start site selection** in bacteria and that in eukaryotes. As we have already seen, the initial positioning of the AUG initiation codon in the mRNA close to the P half-site in the 30S subunit is accomplished by the Shine–Dalgarno sequence in the 5′-end of the mRNA, which base-pairs with a portion of the 16S rRNA. Such a sequence exists in front of almost every initiation codon contained in a polycistronic message. No such complementary sequences are found in the interaction of eukaryotic mRNAs with ribosomal RNA. Instead, the pre-initiation complex performs a unidimensional search for a start codon within the 5′-UTR of the message. Usually, but not always, the 5′-most AUG is used to initiate a polypeptide chain. The scanning process responds to a set of poorly understood molecular signals. The importance of such signals is obvious, as initiation at an incorrect codon can produce a product that is useless for the cell; even worse, such products may be toxic.

Met-tRNAi^{Met} is delivered to the P half-site of the 40S small eukaryotic ribosomal subunit in the form of a ternary complex with eIF2 and GTP. The ternary complex and the eIFs participating in this step bind to 40S to form the **43S pre-initiation**

Figure 16.4 Initiation of translation in eukaryotes. In the first step of this multistep process, the ternary complex eIF2–GTP–Met-tRNAi^{Met} binds to the small 40S ribosomal subunit. This involves the active participation of three initiation factors, which bind cooperatively to the small subunit. eIF1A is the homolog of bacterial IF1 and eIF3 is a scaffolding factor for the assembly, made up of 13 subunits in mammals. eIF1 is a functional homolog of bacterial IF3: although the two proteins share no sequence homology, they bind at the same location on the small subunit and can function in heterologous initiation assays. This first complex has a sedimentation coefficient of 43S. If the mRNA contains no stable secondary structures, the 43S complex can bind directly to the message without the need for additional factors. Additional factors are required, however, to bind and scan structured mRNA; for simplicity, these factors are named but not depicted in the figure. In either case, this open complex scans along the mRNA from the 5′-end, until Met-tRNAi^{Met} base-pairs with the AUG codon in the partial P site of the small subunit. The open complex is in dynamic equilibrium with the closed complex, which is not capable of scanning but investigates the codon in the P site; when an AUG codon is identified, the equilibrium shifts toward the closed state and tRNA binds to AUG. Upon recognition of the start codon, a conformational change occurs in the ribosomal complex, and eIF1 dissociates. eIF1's departure from the complex, or at least its moving away from its rRNA binding site while still interacting with other factors, is instrumental in phosphate (P_i) release, making hydrolysis irreversible and thus committing the pre-initiation complex to initiation at the codon currently in the P site. In the next step, eIF5B–GTP binds to the complex, making important contacts with the C-terminal tail of eIF1A. The binding of eIF5B–GTP to the 48S complex catalyzes the association with the large subunit to form the 80S initiation complex; at this stage, the bound initiation factors are displaced by the large subunit. In the final step, GTP hydrolysis, catalyzed by the large ribosomal subunit, leads to eIF5B dissociation, leaving the complex ready for elongation. (Adapted from Jackson RJ, Hellen CUT & Pestova TV [2010] *Nat Rev Mol Cell Biol* 11:113–127. With permission from Macmillan Publishers, Ltd.)

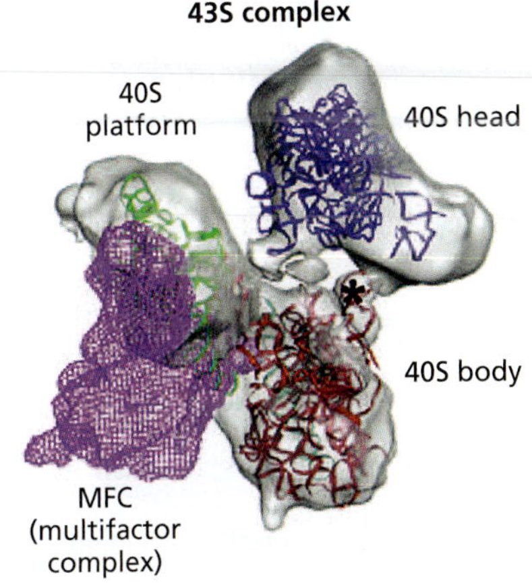

Figure 16.5 Multifactor complex of eukaryotic initiation factors. The MFC may help to reconfigure 40S ribosomal subunit domains to increase the accessibility of the mRNA-binding channel, thus promoting initiation. (Top) The cryo-EM reconstruction of the MFC-bound 40S subunit has been fitted with the atomic models for the head, platform, and body. MFC is shown in purple. (Bottom) The cryo-EM image displays the 40S subunit in a similar orientation. (From Gilbert RJC, Gordiyenko Y, von der Haar T et al. [2007] *Proc Natl Acad Sci USA* 104:5788–5793. With permission from National Academy of Sciences.)

complex, the cryo-EM structure of which is presented in Figure 16.5. It has been suggested that binding of the multifactor complex, MFC, of initiation factors to the 40S subunit creates a structure in which the head of the small subunit can occupy a broad spectrum of positions, indicating that it is highly mobile. This mobility may reconfigure the head–platform–body relationship in the small subunit, leading to remodeling of the mRNA-binding channel and increased accessibility for the mRNA.

The next step in eukaryotic initiation involves binding of additional factors to create the **48S open scanning pre-initiation complex**. This complex forms only on mRNA molecules containing stable secondary structures in their 5′-UTR. If the message does not contain such structures, the 43S complex can bind directly to it, without the need for additional factors. According to some models, the mRNA at this stage is circularized through interactions of poly(A)-binding protein, PABP, with the 3′-poly(A) tail of the message and the eIF4F 5′-cap binding complex. The function of the 48S complex is to remove mRNA secondary structures that would otherwise inhibit the scanning process (Figure 16.6). Although either eukaryotic or bacterial mRNAs could possibly form secondary structures, such structures are much more likely to be found in eukaryotic mRNAs, because they have more time to form during transcription and upon or after delivery to the cytoplasm. Recall that the 5′-ends of bacterial mRNAs are attached to ribosomes immediately upon synthesis in the coupling between transcription and translation (see Figure 16.1). As an observant reader will see upon comparing the schematics in Figure 16.4 and Figure 16.6, there is uncertainty about exactly when the eIF4F complex functions: before or after the formation of the 43S pre-initiation complex. Different researchers have proposed different models. As far as the position of eIF4G is concerned, cryo-EM reconstruction of a partial eukaryotic initiation complex (Figure 16.7) indicates that the location of eIF4G is exactly at the edge of the 40S subunit, where the 5′-end of the mRNA resides.

As Figure 16.4 shows, the 48S open scanning complex is in a dynamic equilibrium with the **48S closed complex**, which investigates the identity of the codon in the P half-site. Upon identification of the start codon, the phosphate product of GTP hydrolysis is released, making further steps in the process irreversible; this commits the pre-initiation complex to initiation using the start codon in the P site. An 80S initiation complex that is competent for elongation is subsequently formed by addition of the 60S subunit.

Much information concerning initiation, particularly with respect to the search for the start codon and the use of alternative AUG codons, has been obtained by the technique of **toeprinting** (Box 16.3).

16.5 Translational elongation

Elongation is the most complex of the three stages of translation. Addition of each new amino acid to the nascent peptide occurs through a three-step microcycle: (1) positioning of the correct, cognate aminoacyl-tRNA in the A site of the ribosome, or decoding; (2) formation of the peptide bond, or peptidyl transfer; and (3) **translocation**, or shifting of the mRNA with respect to the ribosome by precisely one codon. The two tRNAs in the P and A sites simultaneously translocate to the E and P sites, respectively. A schematic of the whole cycle is shown in Figure 16.8. Current knowledge of each individual step described in this section is mainly derived from the three methods described earlier in this chapter.

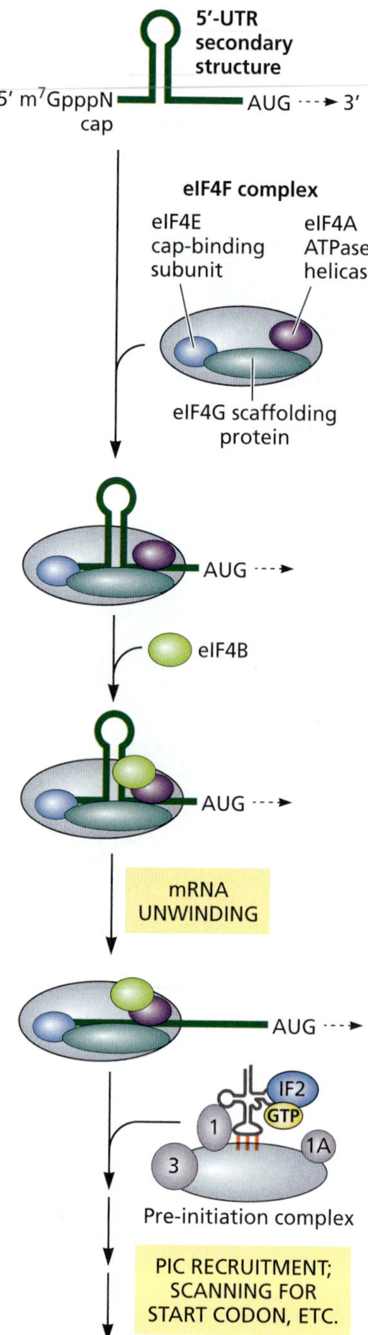

Figure 16.6 Mechanisms of action of eIF4F in initiating translation of mRNAs that contain stable secondary structures. The eIF4F complex binds to the 5'-cap structure of the mRNA through its cap-binding subunit eIF4E. The helicase eIF4A is thought to unwind the secondary structures in the 5'-untranslated region, UTR, of the message to create a landing pad for the ribosome. An RNA-binding protein, eIF4B, assists the action of the helicase. At this point, the 43S pre-initiation complex, PIC, can assemble. (Adapted from Sonenberg N [2008] *Biochem Cell Biol* 86:178–183. With permission from Canadian Science Publishing.)

Decoding means matching the codon to the anticodon-carrying aminoacyl-tRNA

At the beginning of the microcycle, the ribosome contains a peptidyl-tRNA, a tRNA carrying the growing polypeptide chain, in the P site; or it may be an initiator tRNA if the ribosome is just at the end of the initiation phase. The A site is empty, ready to accept the next aminoacyl-tRNA (aa-tRNA) in accordance with the codon–anticodon interactions. The incoming amino acid is delivered to the A site as a ternary complex of aminoacyl-tRNA with **elongation factor Tu**, **EF-Tu**, bound to GTP. Selection of the appropriate amino acid by codon–anticodon pairing occurs in the decoding center of the small subunit. This step is probably rate-limiting in determining the rate of elongation, for many factor–aminoacyl-tRNA complexes are competing for the same site, but only a few of them are cognate.

The observed accuracy of selecting the cognate aa-tRNA cannot be accounted for by just the energetics of hydrogen-bond formation between the three bases of the codon and the three bases of the anticodon. In fact, the codon–anticodon bonding energy is low compared to the total energy of interaction between the EF-Tu–tRNA complex and the ribosome. Early studies from the 1960s showed that the stability of trinucleotide codon–anticodon interactions in solution is weak and suggested that the ribosome itself must stabilize these interactions. It was not clear, however, whether the ribosome merely adds to the stability of the codon–anticodon interactions or plays some role in further augmenting the specificity of the cognate interactions.

Genetic studies with *E. coli* mutants that have antibiotic-specific phenotypes provided evidence that the ribosome controls the accuracy of decoding through multiple interactions with the incoming aa-tRNA. Recent structural studies further clarified the contribution of the ribosome to the selectivity of decoding. Indeed, three universally conserved bases of the 30S subunit of the ribosome were found to interact with the minor groove of the first two base pairs of the codon–anticodon helix, and thus the correct interactions are reinforced by the 30S subunit's ability to recognize the geometry of the base pairs. As stated by V. Ramakrishnan, "such close monitoring of base-pairing geometry by the ribosome does not occur at the wobble position, consistent with the degeneracy of the genetic code." These extra interactions between the codon–anticodon helix and the ribosome induce a domain closure in the 30S subunit, leading to acceleration of the forward steps of decoding. This has been considered an example of the induced-fit model of enzyme activity. Conversely, if the ribosome fails to find a proper codon–anticodon pairing, the release of incorrectly inserted aa-tRNAs is favored, thereby increasing the fidelity of translation.

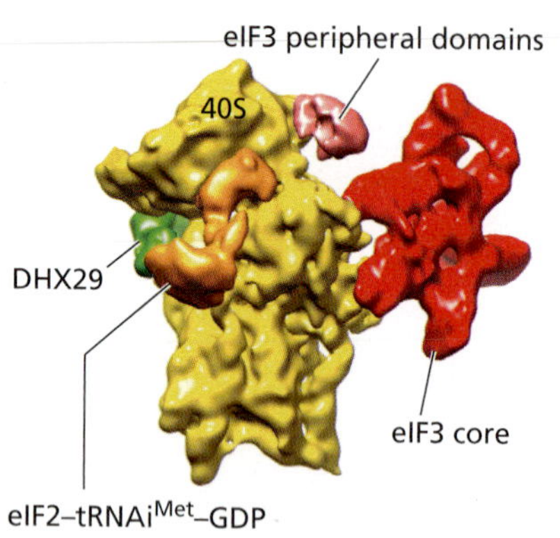

Figure 16.7 Architecture of 43S eukaryotic pre-initiation complex resolved by cryo-EM at 11.6 Å resolution. The 43S complex was assembled *in vitro* from a mixture of native and recombinant proteins or complexes and purified by gel filtration. The cryo-EM images were reconstructed into the structure shown, with the location of the triple complex eIF2–tRNAi^{Met}–GDP, eIF3, and DHX29, a protein required for scanning on structured mRNA, clearly identifiable. No density corresponding to eIF1 was found, although the protein was present in the assembled complex. It was suggested that this factor dissociated during deposition of the complex to the surface. The multiple ribosomal contacts established by eIFs and DHX29 did not cause any major conformational change in the 40S ribosomal subunit. (From Hashem Y, des Georges A, Dhote V et al. [2013] *Cell* 153:1108–1119. With permission from Elsevier.)

Box 16.3 Toeprinting: Primer extension inhibition by ribosome–mRNA complexes Much information concerning the motion of ribosomes on mRNA, and the roles of co-factors and other molecules, has been obtained from a technique called toeprinting. The somewhat facetious name was given in the same spirit as fingerprinting and footprinting, which we have already encountered. The principle is very simple (**Figure 1**). *In vitro*, ribosomes are allowed to associate with a known-sequence mRNA and begin tracking along, and in some cases translating, this mRNA. After the ribosomes halt, because of either addition of an inhibitor or lack of a critical component, the 3′-portion of the mRNA is reverse-transcribed into a DNA strand by use of reverse transcriptase (see Chapter 20). The primer used to bind near the 3′-end of the mRNA is either radioactively or fluorescently labeled, thereby labeling the DNA product. The reverse transcriptase will proceed only until it bumps into the ribosome, about 15 nucleotides downstream from the P site.

A control reverse transcription is run, using the same primer but naked mRNA. The single-stranded DNA fragments are isolated and electrophoresed side-by-side, along with a series of size markers. This technique allows a measure of the distance of the ribosome from the 5′-end of the mRNA.

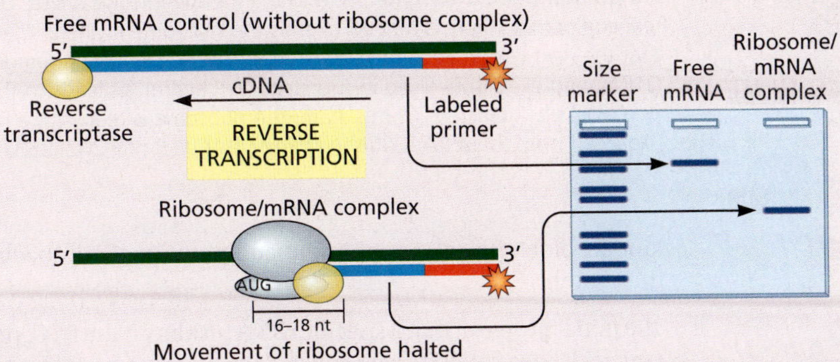

Figure 1 Principle of toeprinting. The ribosome has moved onto the mRNA, but its movement has been blocked at the point indicated. The presence of the ribosome complex prevents the reverse transcriptase from moving further than this point, resulting in a shorter reverse transcript, shown as a blue line, than that produced from the naked mRNA control. The distance the ribosome has moved along the mRNA is measured by comparing lengths of the reverse transcripts on electrophoretic gels; the set of size markers run on the side helps in determining these lengths.

Accommodation denotes a relaxation of distorted tRNA to allow peptide bond formation

Within the EF-Tu–GTP complex, the conformation of the tRNA is distorted; see **Figure 16.9** for an overview and **Figure 16.10** for details. However, it relaxes to its thermodynamically stable conformation upon dissociation of the EF-Tu factor from the ribosome, which occurs following hydrolysis of the bound GTP. This process, termed **accommodation**, places the chemical groups that participate in formation of the peptide bond into an appropriate position for the reaction to occur. Relaxation of the distorted tRNA form can therefore contribute to the driving force for translation, but note once again the role of energy stored in GTP, which is released upon hydrolysis.

Peptide bond formation is accelerated by the ribosome

Peptide bond formation or peptidyl transfer between the aa-tRNA in the A site and the peptidyl-tRNA in the P site occurs in the **peptidyl transferase center, PTC,** of the 50S subunit. In this reaction, the α-NH_2 group of the incoming aa-tRNA nucleophilically attacks the carbonyl carbon of the ester bond linking the peptide moiety to peptidyl-tRNA (**Figure 16.11**). If peptidyl transfer is not catalyzed, it proceeds at a rate of ~10^{-4} M^{-1} s^{-1} at room temperature. The ribosome speeds up the reaction by 10^6–10^7-fold. The question is, how is this accomplished? Is the ribosome directly involved in chemical catalysis, employing residues in domain 5 of the 23S rRNA as general acids and bases, according to a long-held view, or does it just stabilize a transition state? Alternatively, can it act by simply bringing the two substrates into close proximity and correct orientation? Progress in structural studies as well as

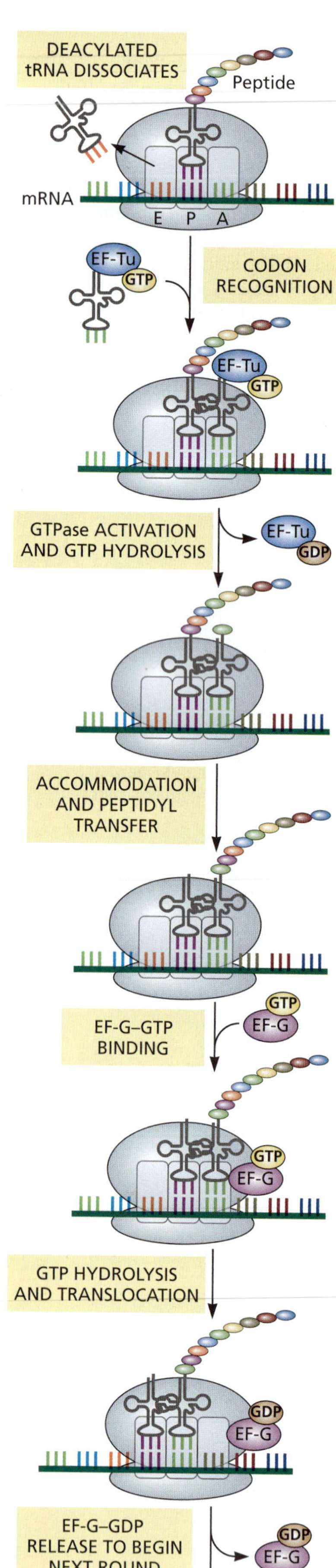

Figure 16.8 Translation elongation in bacteria. During initiation, the initiator tRNA carrying fMet is bound to the P site. Elongation is a cyclic process in which the addition of each incoming amino acid to the growing polypeptide chain occurs in three major steps: positioning of the correct aminoacyl-tRNA (aa-tRNA) into the A site of the ribosome, formation of the peptide bond, and translocation or shifting of the mRNA with respect to the ribosome over the length of a codon; note that during translocation the two tRNAs in the P and A sites also translocate. Other steps that occur alongside these three major steps are necessary to create favorable conditions for the major steps. An incoming acylated or charged tRNA is delivered to the A site by elongation factor EF-Tu bound to GTP. Correct base-pairing or recognition between the codon and the anticodon activates the GTPase center of the ribosome and GTP hydrolysis occurs, leading to a conformational change in EF-Tu and its release from the ribosome. This release in turn leads to conformational changes in the rRNA and the tRNA, which optimally orient the two tRNAs for the peptidyl transfer reaction to occur. During the peptidyl transfer reaction, the nascent polypeptide is moved from the tRNA in the P site to the aa-tRNA in the A site. Once the peptidyl transfer reaction occurs, there is a need to shift the ribosome in the 3′-direction on the mRNA, so that the next codon becomes available for decoding, that is, binding of a cognate aa-tRNA. GTP-bound EF-G is involved in this step: GTP hydrolysis leads to the movements of the two tRNAs as depicted. At this state, the ribosome is ready for a new round of elongation. (Adapted from Steitz TA [2008] *Nat Rev Mol Cell Biol* 9:242–253. With permission from Macmillan Publishers, Ltd.)

kinetic, biochemical, genetic, and computational approaches have finally provided the answer.

The initial proposal that a specific rRNA residue in the PTC, residue A2451, acts in general acid-base catalysis was rejected when it was found that this residue was dispensable for the peptidyl transfer reaction. High-resolution crystallographic data further showed that the atom supposedly involved in the catalysis, atom N3 of residue A2451, was not within the required distance of the nucleophile. Finally, kinetic data also indicated that the peptidyl transfer reaction is not strongly pH-dependent, providing solid evidence that there is no acid-base catalysis.

Kinetic experiments aimed at uncovering the catalytic mechanism of peptide bond formation require a reconstituted *in vitro* translation system that can truthfully recapitulate what happens *in vivo*. In such a system, the reaction rate can be probed under different conditions of pH, temperature, ion availability, and so on. But there is a problem with such *in vitro* experiments: how does one ensure that the rate of product formation measured reflects the chemistry step and not the many complex steps of substrate binding, accommodation, and other conformational rearrangements? The first-order rate constant for accommodation of aa-tRNA into the A site is ~10 s^{-1}, and formation of the peptide bond occurs very rapidly after that. Obviously, accommodation is the rate-limiting step in product formation. Thus, it is possible to use kinetic approaches only if accommodation and the actual formation of the peptide bond can be uncoupled.

A couple of ingenious approaches achieved just that. In the first approach, the reactive nucleophilic α-NH_2 group was replaced by a much less reactive OH group. Care was taken to make sure that this replacement did not significantly affect the reaction pathway. The rate of reaction was now on the order of 10^{-3} s^{-1}, far below that of accommodation. A second approach was to use substrate analogs that can bind to the A site rapidly and do not require accommodation. The antibiotic puromycin could be used as such an analog in a system that contained the 70S ribosome, mRNA, and initiator fMet-tRNAi^{fMet} or peptidyl-tRNA in the P site.

The current view as to how the ribosome accelerates the reaction is presented in **Figure 16.12A**. An extensive network of hydrogen bonds between nucleotides in the PTC and the two reactants positions and orients them for the reaction to occur. This network presumably accounts for most of the reaction rate enhancement provided by the ribosome. Some large-subunit ribosomal proteins, L16 and L17, also aid the RNA components in facilitating the reaction. It is worth noting that there is also a postulated rotatory motion of the A-tRNA 3′-end along a path confined by the PTC walls that helps in the positional catalysis of peptide bond formation. The actual

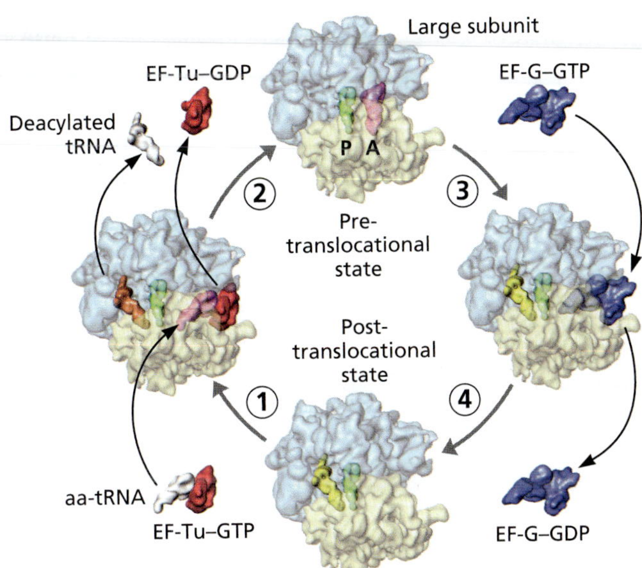

Figure 16.9 Elongation cycle: Structural overview. The schematic is based on cryo-EM imaging and three-dimensional image reconstruction used to determine ribosomal structures and positions of EF-Tu and EF-G; the structures of these two elongation factors themselves are derived from crystallographic studies. The ribosome is viewed from the top, with the small subunit, shown in pale yellow, below the large subunit, shown in pale blue. We consider the whole cycle as involving a series of steps between conformational states. We begin with the complex at the bottom of the figure in the post-translocational state. It is ready to accept a new aminoacyl-tRNA, shown in gray, which is presented to the ribosome as a ternary complex that also includes the elongation factor EF-Tu and a GTP, shown in red. This step, which we call step 1, produces the structure on the left: the ribosome with a bound ternary complex, shown in magenta and red. It has been suggested that when the ternary complex binds to the ribosome, the E-site tRNA moves further away from the P site to the E2 site, shown in brown. The snapshot shown here represents part of the decoding step when the aminoacyl-tRNA whose anticodon matches the next codon in the mRNA is selected to enter the A site, accompanied by GTP hydrolysis and conformational changes. If the codon and anticodon match, EF-Tu with the GDP formed by the hydrolysis of GTP leaves the ribosome, accompanied by a relaxation in the formerly strained conformation of the A-site RNA, termed accommodation in (step 2); the E-site tRNA also leaves the ribosome. The ribosome is now in the pre-translocational state, shown at the top, with tRNAs in the A and P sites shown in magenta and green, respectively. After peptidyl transfer, the nascent peptide, not shown, is covalently attached to the A-site tRNA. On the right, following step 3, the elongation factor EF-G in complex with GTP, shown in dark blue, has bound to the ribosome to facilitate the translocation of the peptidyl-tRNA to the P site, shown in green, and the deacylated tRNA to the E site, shown in yellow. The translocation is induced by GTP hydrolysis accompanied by large transient conformational changes in the EF-G and the ribosome. In step 4, the release of EF-G after GTP hydrolysis leaves the ribosome in the post-translocational state, shown at the bottom, to close the elongation cycle. (From Cheng RH & Hammar L [eds] [2004] Conformational Proteomics of Macromolecular Architecture. With permission from World Scientific Press.)

chemical catalysis involves the 2′-OH group of the peptidyl-tRNA, which is well positioned to extract protons from the NH_2 and donate them to the leaving group (**Figure 16.12B**).

The formation of hybrid states is an essential part of translocation

Peptide bond formation is followed by a complex series of steps that translocate the mRNA by precisely one codon in the 3′ → 5′ direction (**Figure 16.13**). Simultaneously, the tRNA from the A site has to move to the P site, and the now deacylated tRNA has to move from the P site to the E site. Translocation is initiated by spontaneous reversible movements of the tRNAs with respect to the 50S subunit, through the formation of **hybrid states**, in which a tRNA is bound simultaneously to one site in the small subunit and a different site in the large subunit. For example, the peptidyl-tRNA is bound simultaneously to the A half-site in the 30S subunit and the P half-site in the 50S subunit: see states 9, 10, and 11 in Figure 16.13. The coordinated, reversible shifting of tRNAs into and out of hybrid states can be thought of as a consequence of spontaneous partial rotation or sliding of the small and large subunits relative to one another (**Box 16.4**). Only upon EF-G–GTP binding and GTP hydrolysis is the reaction made irreversible and moved in a

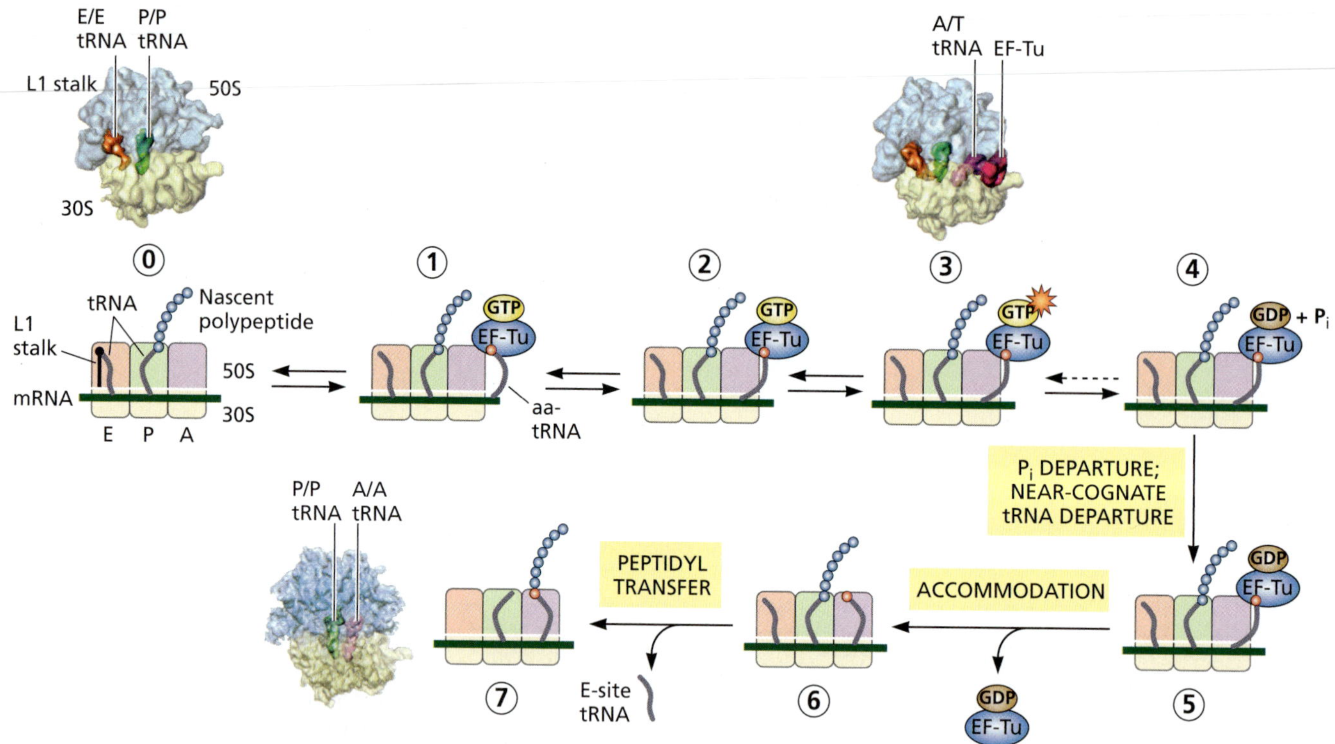

Figure 16.10 Discernible distinct states and reversible versus irreversible steps in decoding and peptidyl transfer. Schematic is based on biochemical kinetic studies, single-pair FRET experiments, and cryo-EM maps. (State 0) Post-translocational state in which the A site is empty; the P site contains a peptidyl-tRNA bound in a classical P/P configuration, denoting the 30S P and 50S P sites, respectively; and the E site is occupied by a deacylated tRNA bound in the classical E/E configuration in direct contact with the open L1 stalk, which is a highly mobile domain within the 50S subunit. (State 1) Elongation factor Tu in a ternary complex with aa-tRNA and GTP bound to the ribosome via the L7/L12 stalk; the rest is as in state 0. During the transition to state 2, the aa-tRNA anticodon probes for the mRNA codon in the decoding center, DC, close to the A site of the small subunit. (State 2) The aa-tRNA is engaged with the codon at the DC. In the absence of ligands, the DC is conformationally dynamic, but the formation of a cognate codon–anticodon complex in the DC stabilizes a specific local conformation of the otherwise disordered DC. In addition, there is a global conformational change in the 30S subunit, termed domain closure. During the transition to state 3, the cognate and a fraction of near-cognate ternary complexes stay bound for sufficient time to induce GTP activation of EF-Tu. The noncognate and a fraction of near-cognate ternary complexes are rejected, as their binding to the ribosome is unstable. This is the first step of selection of the cognate aa-tRNA. (State 3) EF-Tu is activated for GTP hydrolysis. Recognition of a cognate aa-tRNA at the DC leads to formation of the A/T configuration of the ternary complex, in which the aa-tRNA anticodon engages the codon at the DC while the aa-tRNA acceptor stem remains bound to EF-Tu. This binding results in a large conformational change of the tRNA body compared to its known X-ray structure: there is a kink and twist in the anticodon stem to allow the acceptor stem to remain bound to EF-Tu. Thus, it is highly likely that the aa-tRNA itself participates in transmitting a codon–anticodon recognition signal from the DC to the 50S subunit GTPase-associated center, inducing GTP hydrolysis. (State 4) EF-Tu is bound in the GDP-P_i state. Departure of the inorganic phosphate released upon GTP hydrolysis makes the process irreversible. (State 5) EF-Tu is bound in the GDP state. The conformational change in EF-Tu–GDP that occurs upon GTP hydrolysis and P_i release induces departure of the elongation factor. This is the second step of selection of the cognate aa-tRNA, since near-cognate aa-tRNA also leaves the ribosome at this stage. (State 6) The aa-tRNA is now accommodated in the classical A/A configuration of the A site. This accommodation occurs because EF-Tu–GDP dissociation releases the stress on the tRNA, bringing it back to its thermodynamically stable structure. This conformational change repositions the acceptor stem, with its bound amino acid, to promote peptide bond formation. Now the reactive carbonyl group at the C-terminal end of the peptidyl-tRNA is optimally positioned with respect to the α-NH_2 group of the incoming aa-tRNA for the nucleophilic attack that transfers the nascent polypeptide from the P-site-bound peptidyl-tRNA to the A-site-bound aa-tRNA. (State 7) Following departure of the E-site deacylated tRNA, which according to recent evidence may occur earlier than indicated here, and peptidyl transfer, a pre-translocational complex forms, with the nascent polypeptide now covalently linked to the A-site tRNA; the P-site tRNA is deacylated. (Adapted from Frank J & Gonzalez RL Jr [2010] *Annu Rev Biochem* 79:381–412. With permission from Annual Reviews.)

forward direction. Thus, the translocation steps are another example of **Brownian ratcheting** (discussed in Box 9.1 and Box 16.4). Exactly how GTP hydrolysis leads to movements that restore the canonical P/P and E/E conformations is still unclear in structural terms.

The positioning of the two irreversible steps, peptidyl transfer and translocation, also contributes to the fidelity of translation. Even after the initial acceptance of a

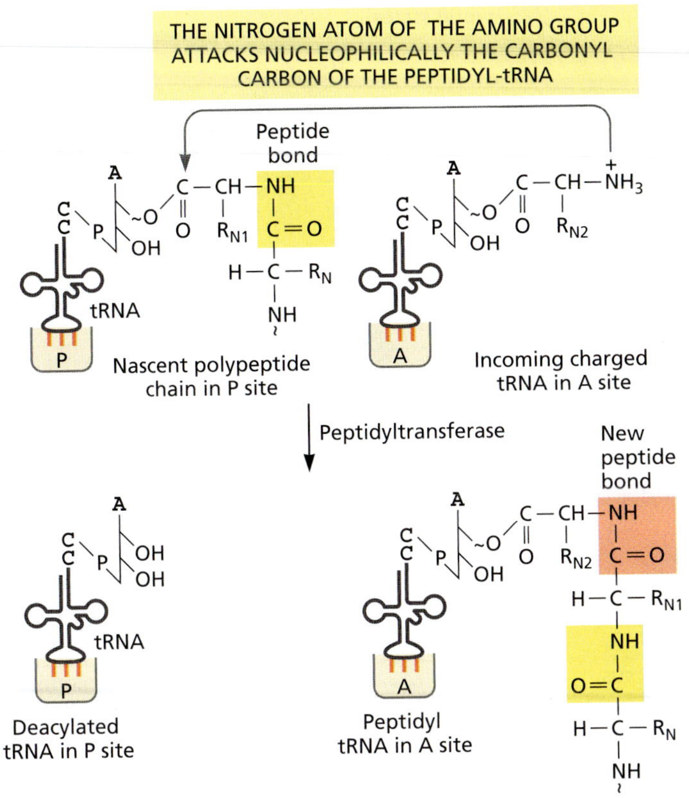

Figure 16.11 Formation of a peptide bond. Decoding, or matching of the codon with a cognate anticodon, occurs at the decoding center, close to the A site in the small subunit, whereas actual peptide bond formation occurs on the 23S rRNA of the large subunit. For simplicity, only the small subunit A and P half-sites are depicted here.

charged tRNA into the A site, the essential step between states 2 and 3 in Figure 16.10, activation for GTP hydrolysis, is favored only by correct matching of the incoming tRNA to the appropriate codon. The fact that the same substate may be sampled more than once may also contribute to fidelity; this came to be known as **kinetic proofreading**. Further proofreading occurs in the subsequent steps leading up to peptidyl transfer.

Structural information on bacterial elongation factors provides insights into mechanisms

We have already introduced the two elongation factors involved in the elongation microcycle: EF-Tu brings the incoming aa-tRNA into the A site of the ribosome, and **EF-G** is instrumental in translocation. The high-resolution structural information

(A)

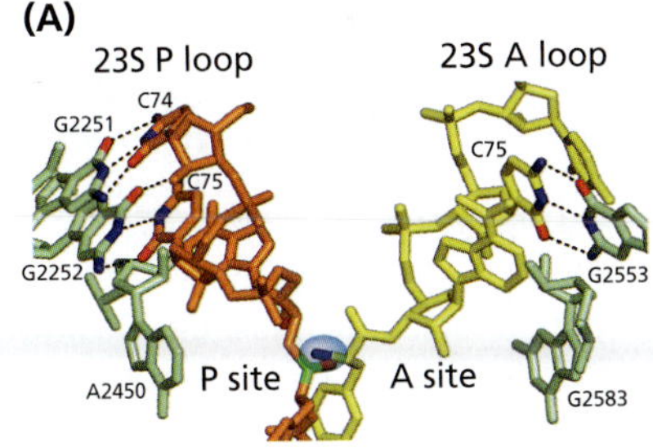

(B)

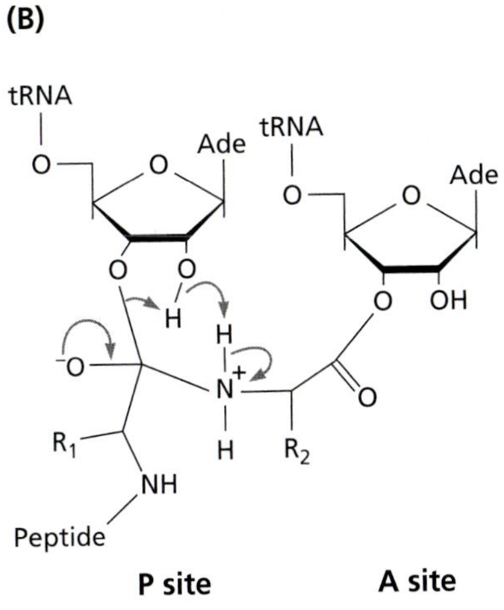

Figure 16.12 Structure of peptidyl transferase center and chemistry of peptide bond formation. The PTC is located in the 50S ribosome subunit. (A) Acceptor arms of the A- and P-site tRNA substrates are located in a cleft on the 50S interface side. The universally conserved CCA ends are oriented and held in place by interactions with residues of 23S rRNA near the active site. In the P site, residues C74 and C75 of the tRNA are base-paired with residues G2251 and G2252 of the P loop, a secondary structure in 23S rRNA. The CCA end of the aa-tRNA in the A site is also involved in base-pairing with bases from the A loop secondary structure of 23S rRNA: C75 base-pairs with G2553. The 3′-terminal A76 residues of both tRNAs interact with bases in 23S rRNA. No metal ion is seen in the immediate vicinity of the site of reaction. Importantly, the α-NH_2 group of the A-site substrate, highlighted by a blue oval, is positioned for nucleophilic attack on the carbonyl carbon of the ester bond that links the peptide moiety of the P-site substrate, shown in bright green, to the ultimate A in CCA. (B) Possible mechanism of peptidyl transfer in which the intermediate of the reaction breaks down into products. The concerted proton shuttle involves the 2′-OH of A76 of peptidyl-tRNA: it simultaneously accepts a proton from the α-NH_2 and donates one to the 3′-O leaving group. Confirming the role of the 2′-OH is the observation that substitution of the 2′-OH by H or F reduces the reaction rate by at least 10^6-fold. (A, from Beringer M & Rodnina MV [2007] *Mol Cell* 26:311–321. With permission from Elsevier. B, adapted from Schmeing TM & Ramakrishnan V [2009] *Nature* 461:1234–1242. With permission from Macmillan Publishers, Ltd.)

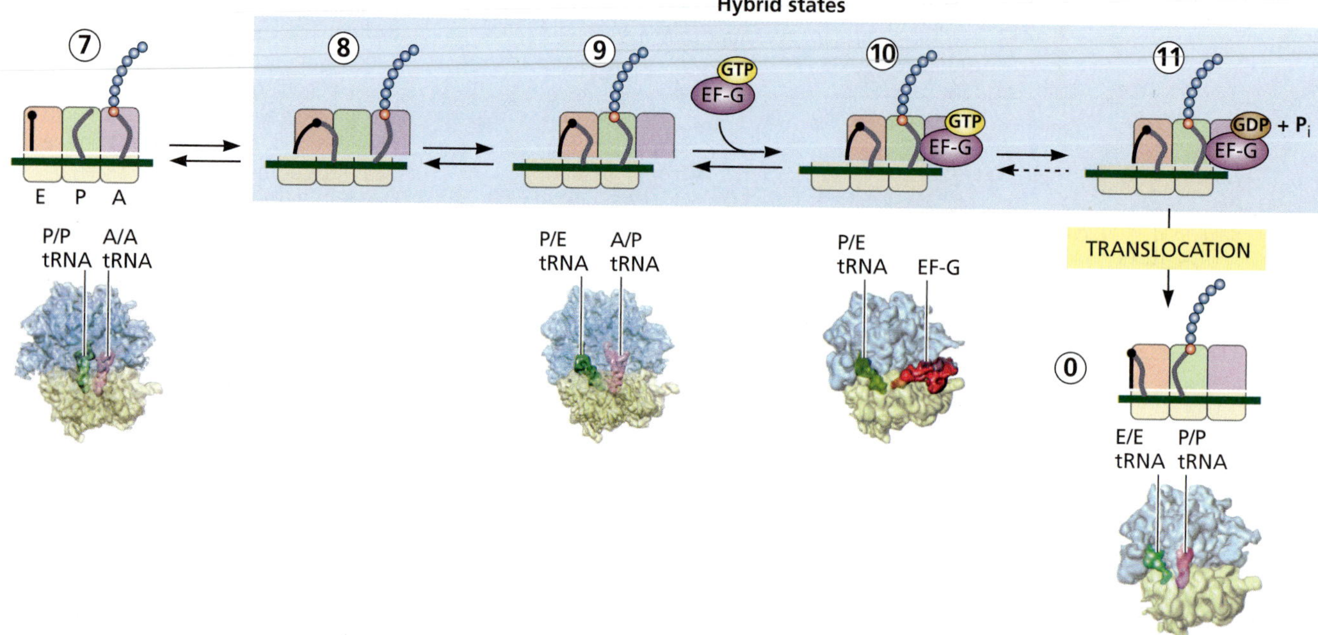

Figure 16.13 Discernible distinct states and reversible versus irreversible steps in translocation. Schematic is based on biochemical kinetic studies, single-pair FRET experiments, and cryo-EM maps. The numbering of states continues the numbering of earlier translation states in Figure 16.10. (State 7) Pre-translocational complex. The nascent polypeptide is covalently linked to the A-site tRNA, the P-site tRNA is deacylated, and the E site is unoccupied. The tRNAs are in their classical A/A and P/P positions, and the L1 stalk is in an open conformation. (State 8) The pre-translocational complex is in an intermediate state of ratcheting: the ribosome is in a semirotated state and the tRNAs are in an intermediate state with classical A/A and hybrid P/E positions. The L1 stalk is in a closed position, forming a direct contact with the hybrid P/E tRNA. (State 9) The ribosome is in a rotated state, tRNAs are in hybrid A/P and P/E configurations, and the L1 stalk is in a closed conformation where it directly interacts with the hybrid P/E tRNA. State 9 can be reached directly from state 7, bypassing the intermediate state 8. (State 10) Elongation factor EF-G–GTP is bound, stabilizing the hybrid complex. (State 11) Following GTP hydrolysis, EF-G is bound in the GDP-P_i state. The transition to the 0 state occurs upon release of the GDP-bound EF-G and is irreversible. The ribosome returns to the nonrotated position, the newly formed peptidyl-tRNA and the newly deacylated tRNA move into the classical P/P and E/E configurations, and the L1 stalk moves back into the open position. (State 0) Post-translocational complex, ready to begin a new cycle. (Adapted from Frank J & Gonzalez RL Jr [2010] *Annu Rev Biochem* 79:381–412. With permission from Annual Reviews.)

Box 16.4 A Closer Look: Detailed studies of the elongation cycle: The Brownian ratchet mechanism The picture we have presented in the text provides the broad outlines of the ribosomal elongation cycle. So far as it goes, it is essentially correct. It has long been recognized, however, that the cycle must be more complex and must involve movement of portions of the ribosome itself. Although X-ray diffraction studies have provided vital information about the structure of the ribosome and its interaction with other players, the two additional methods described in the text, cryo-EM and spFRET, have contributed significantly to deepening our understanding of the dynamics of the process.

The kind of insight gained from such studies is shown in Figure 16.10 and Figure 16.13. Rather than the 4 steps illustrated in Figure 16.9, the new experiments identify up to 11 distinct steps in the cycle. The distinct states connected by these steps include not only the classical occupancies of the A, P, and E sites but also hybrid states, in which the tRNAs are bound to one site on the 30S subunit and a different site on the 50S. There are also recognized shifts in ribosomal structure, including rotation of the body of the 30S subunit with respect to the 50S, and lateral motions of segments of each subunit.

A rather surprising result from this research is the realization that many of the intermediate substates are in fact in dynamic equilibrium; the transitions indicated by two-way arrows in Figure 16.10 and Figure 16.13 are reversible. This means that the ribosome–mRNA–tRNA complex can shuttle back and forth between such substates. Nevertheless, the overall process is irreversible because of certain key steps indicated by one-way arrows. These occur upon GTP hydrolysis, which provides the thermodynamic driving force for the overall cycle. Once such a step is passed, the process cannot be reversed. The whole system constitutes what is referred to as a Brownian ratchet. Thermal energy is sufficient to shake the system from one substate to another, but once the ratchet falls—that is, an irreversible step occurs—the system is prevented from going backward. The overall direction must be forward, and the second law of thermodynamics is obeyed (see Box 9.1).

Just how broad and uniform the distribution of substates can be is suggested by **Figure 1**, which shows the distribution of rotation angles of the 30S subunit around a pivot point of 16S rRNA in post-translational states. If samples were held at 4°C just before freezing for cryo-EM, the particles seem to be in a rather stable ground conformation, with a major peak in the vicinity of the zero angle as shown in Figure 1B. At 18°C, a distribution with two strong maxima is observed, indicating two much-preferred rotational conformations. Samples taken at higher temperatures show broader distributions. In fact, the distribution at 37°C, the physiological temperature, is almost flat, showing no strong preference for any particular angle: the ribosome seems to be showing almost free internal rotation at this temperature.

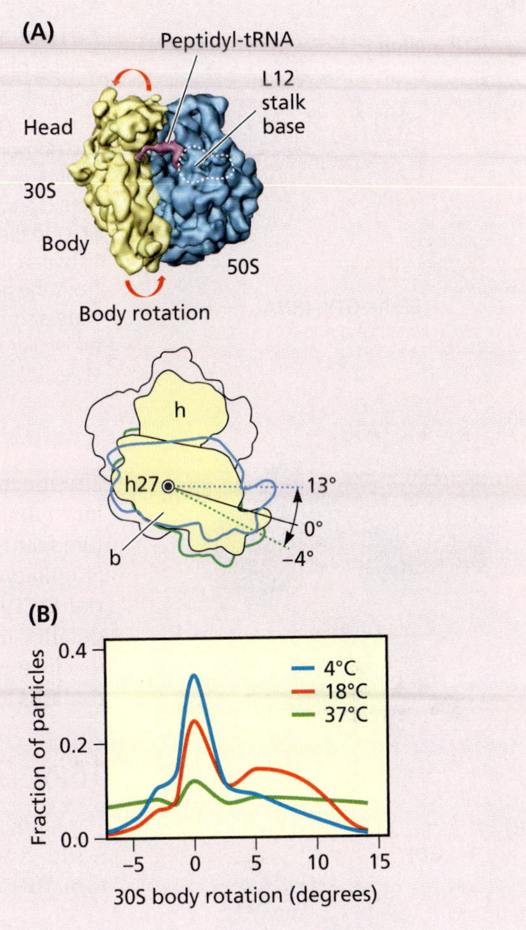

Figure 1 Ribosome dynamics. The dynamics shown were deduced from cryo-EM imaging of ~2 million individual post-translocational complexes at three different temperatures. (A) (Top) Three-dimensional reconstruction of post-translocational complexes from unsorted cryo-EM images taken at 18°C. (Bottom) The rotation of the body is schematically presented: the body rotates around a pivot point at helix 27 of 16S rRNA and is independent of the head movements. (B) Temperature dependence of 30S body rotation, with the fraction of particles plotted against the rotation angle. The observed strong dependence of ribosome dynamics on temperature is expected if the ribosome is a Brownian machine. (Adapted from Fischer N, Konevega AL, Wintermeyer W et al. [2010] *Nature* 466:329–333. With permission from Macmillan Publishers, Ltd.)

that shows exactly how these elongation factors bind to the ribosome is shown in **Figure 16.14**. Two points are worth special note. First, the tRNA in the EF-Tu complex is bent, occupying a hybrid A/T state; the molecule springs back into its normal conformation upon GTP hydrolysis and EF-Tu release. Second, domains III and IV of EF-G mimic the shape of the tRNA in its complex with EF-Tu. This similarity is the structural basis of EF-G action in the translocation step.

There is an exit tunnel for the peptide chain in the ribosome

During elongation, the growing polypeptide chain is extruded from the ribosome through an exit tunnel in the 50S ribosomal subunit. The description of translation would be incomplete without providing some information on the exit tunnel and its presumed role in protein folding. The position of the tunnel in the 50S subunit and its overall shape are presented in **Figure 16.15**. The channel is much too narrow, and possesses especially constricted portions, to allow protein folding beyond α-helix formation. As Thomas Steitz puts it: "The extremely porous nature of the ribosomal RNA structure leads one to compare it to that of a sponge; however, the possible robustness of the structure may be more appropriately compared to that of the Eiffel Tower. The many crisscrossing RNA helical rods make specific interactions with each other and with proteins to stabilize and reinforce a particular structure, thereby making it extremely unlikely that the tunnel can expand and contract by 10–20 Å." Space-filling models from X-ray diffraction studies show that the ribosome has a densely packed structure. Furthermore, direct structural studies with polypeptides occupying the tunnel show no expansion. We discuss how proteins are folded once the chain leaves the tunnel in Chapter 3.

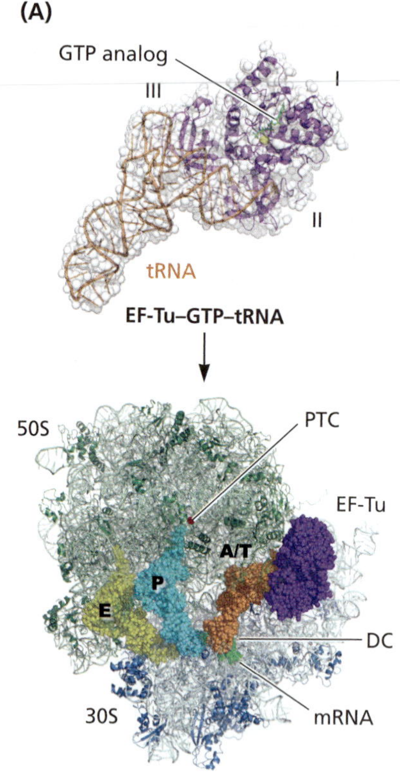

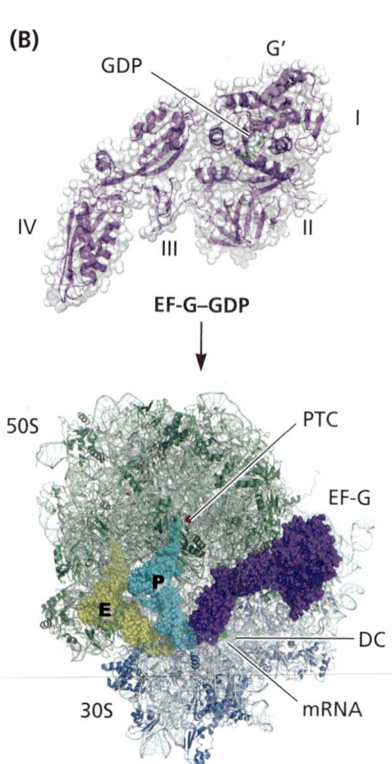

Figure 16.14 Structures of EF-Tu and EF-G and their location in the ribosome structure. (A) (Top) *Thermus aquaticus* EF-Tu bound to Phe-tRNA^{Phe} in the presence of a nonhydrolyzable analog of GTP. Such analogs or other small molecules are usually used to trap the majority of the molecules at the same frozen state, so that no further reactions can occur. (Bottom) Crystal structure of the ribosome complexed with EF-Tu and aa-tRNA. As long as EF-Tu is bound to the ribosomes and the tRNA, it bends the tRNA into the A/T state. EF-Tu release, upon GTP hydrolysis, allows the tRNA to spring into the A site, placing the bound amino acid in the peptidyl transferase center, PTC. (B) (Top) *Thermus aquaticus* EF-G bound to GDP. (Bottom) EF-G in the post-translocational state of the ribosome. Note the surprising similarity in shape between the ternary EF-Tu–GTP–aa-tRNA complex and EF-G: domain IV of EF-G protrudes from the body of the molecule in a manner similar to that of the tRNA anticodon stem–loop. Note also the similarity in how the two entities bind to the ribosome. This similarity forms the molecular basis of EF-G action in the translocation step. (From Nakamura Y & Ito K [2011] *Wiley Interdiscip Rev RNA* 2:647–668. With permission from John Wiley & Sons, Inc.)

Translation elongation in eukaryotes involves even more factors

The mechanism of the elongation step of translation in eukaryotes is rather similar to that in bacteria, though not as thoroughly studied. As in initiation, more protein factors are involved and some of these are present as isoforms. A good example is eukaryotic elongation factor 1A or eEF1A, the functional equivalent of bacterial EF-Tu, which is present as two very similar isoforms: eEF1A1 and eEF1A2. As detailed in Box 16.1, one of these isoforms has proven oncogenic properties, whereas its close cousin seems to be innocent. The reason(s) for this mysterious difference remains to be elucidated.

16.6 Termination of translation

The elongation microcycles continue until a stop codon in the mRNA moves into the A site. A series of steps follows that involve termination factors termed **release factors**, **RFs**; a **ribosome releasing (or recycling) factor**, **RRF**; and the already familiar

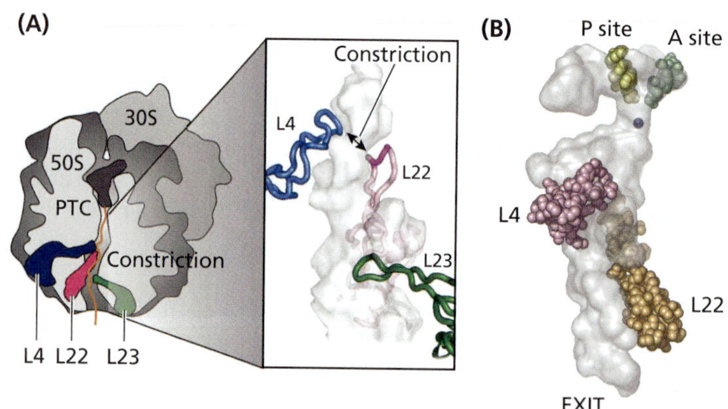

Figure 16.15 Exit tunnel in the 50S subunit. The tunnel is 80–100 Å long, with a diameter of ~10 Å at its narrowest point and 20 Å at its widest point. The tunnel resembles a tube, which is able to accommodate a peptide stretch of ~30 amino acids in extended conformation and up to 60 amino acids in an α-helical conformation. It is believed that protein folding beyond the level of an α-helix is extremely unlikely. (A) Path of the nascent chain through the ribosomal exit tunnel. (Left) ribosome in gray, sliced along the tunnel; the nascent polypeptide chain, shown in orange, extends from the peptidyl transferase center, PTC, to the exit site. The three ribosomal proteins interacting with the nascent chain are color-coded. (Right) View of the outside surface of the tunnel, showing the loops of ribosomal proteins L4 and L22, which form the narrowest constriction along the tunnel. Protein L23, shown in green, is at the exit of the tunnel. (B) Diagram showing the positions of landmarks; note that the structure is slightly rotated with respect to that in part A. The tRNA-binding cleft is at the top, and the exit is at the bottom. The dark blue dot represents the active-site α-amino group. (A, from Kramer G, Boehringer D, Ban N & Bukau B [2009] *Nat Struct Mol Biol* 16:589–597. With permission from Macmillan Publishers, Ltd. B, from Voss NR, Gerstein M, Steitz TA & Moore PB [2006] *J Mol Biol* 360:893–906. With permission from Elsevier.)

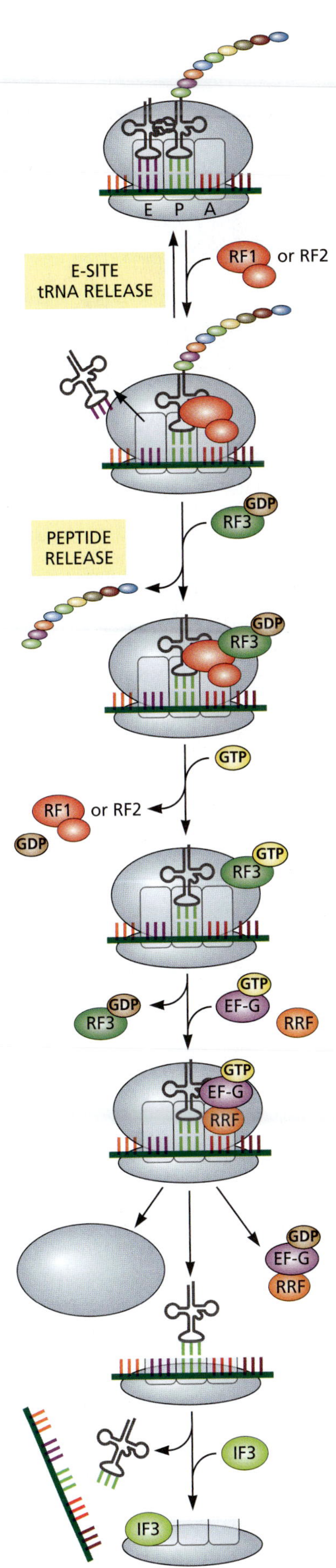

Figure 16.16 Termination of translation in bacteria. A stop codon, indicated by red nucleotides, recruits either release factor 1 or 2, RF1 or RF2, to mediate hydrolysis and release of the peptide from the tRNA in the P site. This functions as a signal to recruit RF3–GDP. Exchange of GDP for GTP on RF3, followed by GTP hydrolysis, is thought to release RF1 or RF2. The residual complex between the ribosome, mRNA, and a deacylated tRNA in the P site is disassembled by binding of ribosome releasing factor RRF and elongation factor EF-G. GTP hydrolysis causes 50S subunit dissociation. Initiation factor 3, IF3, is required for tRNA dissociation from the P site and mRNA release, which occur at the same rate, presumably simultaneously. (Adapted from Steitz TA [2008] *Nat Rev Mol Cell Biol* 9:242–253. With permission from Macmillan Publishers, Ltd.)

EF-G-GTP and initiation factor IF3 (**Figure 16.16**). Successive binding of these factors leads to release of the polypeptide chain and dissociation of the 70S ribosome into its subunits.

Special protein release factors recognize stop codons in the mRNA. In bacteria, two release factors, RF1 and RF2, which are class I factors, recognize the three stop codons (**Figure 16.17**). Only a single release factor, eRF1, exists in eukaryotes; it recognizes all three stop codons and its amino acid sequence is not related to RF1

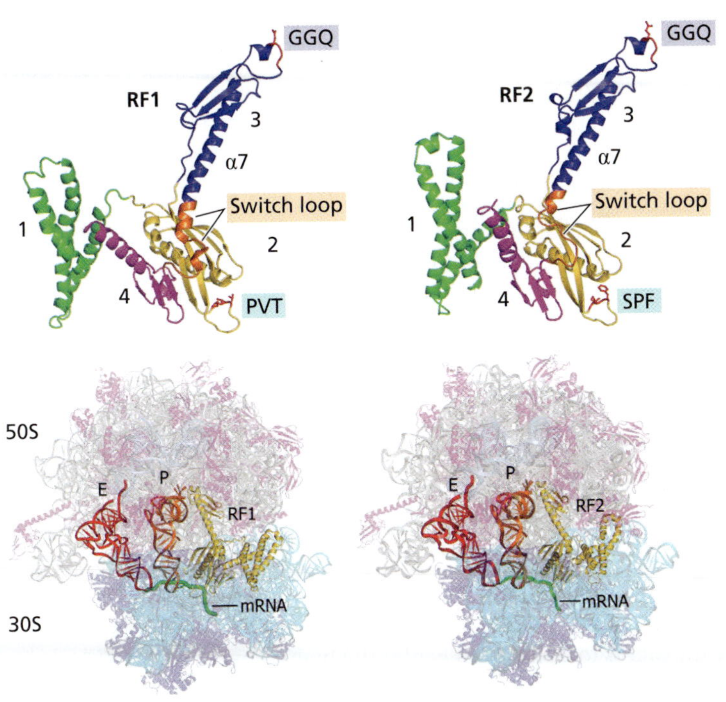

Figure 16.17 Structures of RF1 and RF2 termination complexes as resolved by X-ray crystallography. (Top) Structures of RF1 and RF2 from *Thermus thermophilus* in their ribosome-bound or open conformations. The free unbound factors have a much more compact, closed conformation that opens up upon ribosome binding. Functionally important portions of the molecules are boxed. Determinants for codon specificity of the individual factors are located in domain 2 of the factors; these determinants contain conserved tripeptides, P(A/V)T in RF1 and SPF in RF2. These recognize the stop codons UAG for RF1, UGA for RF2, and UAA for both factors. It should be noted that the specificity of these recognitions has not been confirmed by recent crystallographic studies and hence is suspect. The conserved GGQ motif in domain 3 is present in bacteria, archaea, and eukaryotes and has been implicated in peptidyl-tRNA hydrolysis, through the main-chain amide of the Q residue. Finally, the switch loop between domains 3 and 4 forms a rigid connector that places domain 3 and its GGQ motif in contact with the peptidyl-tRNA ester linkage in the peptidyl transferase center of the 50S subunit. (Bottom) In the structures of the ribosome bound to the respective RFs, the factors are rotated ~180° from the views above. (From Korostelev A, Asahara H, Lancaster L et al. [2008] *Proc Natl Acad Sci USA* 105:19684–19689. With permission from National Academy of Sciences.)

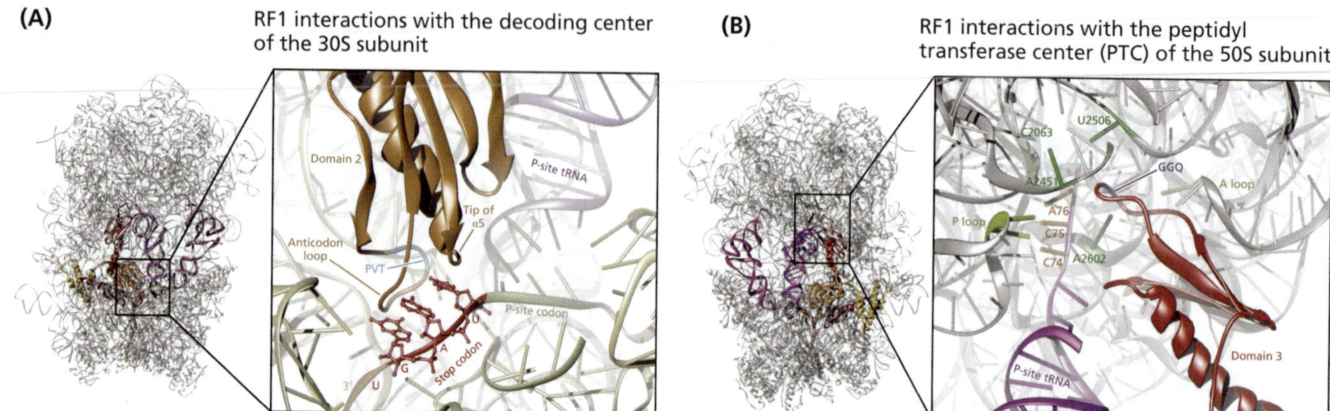

Figure 16.18 Interaction of RF1 with the ribosome. (A) Interaction of RF1 with the decoding center of the 30S subunit. The stop codon UAG is surrounded by a loop containing the tripeptide PVT, shown in light blue; this tripeptide has been implicated by genetic experiments as conferring specificity for recognition of UAG. In RF2, the tripeptide anticodon motif is SPF; it specifically recognizes the alternative stop codon UGA, while both factors recognize the third stop codon UAA. The tripeptide motif and its surrounding amino acid residues are situated close to the stop codon and are particularly close to the second and third nucleotides of this codon. In addition, the tip of helix α5 approaches the first codon base and may be involved in the recognition of uridine, the base common to all three stop codons. Stop codon recognition by RFs occurs without proofreading but is extremely accurate, more so than decoding of tRNA during elongation. (B) Interaction of RF1 with the peptidyl transferase center, PTC, of the 50S subunit and loops A and P from 23S rRNA, shown in lime green. The highly conserved GGQ loop implicated in peptide release is surrounded by conserved bases of the PTC, shown in green; the tip of this loop faces residue A76 of the P-site tRNA. It has been proposed that binding of RF1 and RF2 induces a conformation of the PTC that is crucial for specifically selecting water for nucleophilic attack of the ester bond on peptidyl-tRNA. In other words, hydrolysis is only sterically helped by the GGQ loop and not performed by it. Mutation of the universally conserved glutamine Q, the only long side chain that can span the distance to the ribose of A76, does not affect RF activity. Mutation analysis has also indicated that A2602 plays a role in peptide hydrolysis but not in peptide bond formation. The GGQ loop is disordered in the isolated crystal structure but is ordered in the ribosome. (From Petry S, Brodersen DE, Murphy FV IV et al. [2005] *Cell* 123:1255–1266. With permission from Elsevier.)

and RF2 of bacteria. Both genetic and structural studies have identified putative **tripeptide anticodon motifs**, P(A/V)T in RF1 and SPF in RF2, that participate in recognition of the stop codons. Downstream bases in the mRNA affect the termination efficiency of class I RFs. The overall structures of RF1 and RF2, and their positions in the ribosome structure, are presented in Figure 16.17. **Figure 16.18** provides a close-up view of the interactions of RF1 with both the decoding center of the 30S ribosomal subunit and the peptidyl transferase center of the 50S subunit.

For the polypeptide chain to be released, the bond between the last amino acid and its cognate tRNA must be severed. The reaction is like that of peptidyl transfer, except that the transfer is to a water molecule. The highly conserved GGQ loop of class I RFs is involved in the reaction by providing sterically important interactions that aid in the hydrolysis.

RF3 aids in removing RF1 and RF2

The class II release factor RF3 binds to the ribosome in a GDP-bound state (see Figure 16.16). The binding presumably induces an exchange of GDP for GTP. The GTP-bound form induces conformational changes in the ribosome likely to destabilize the binding of RF1 or RF2, thus leading to their dissociation. GTP hydrolysis then releases RF3 from the ribosome. The cryo-EM image shown in **Figure 16.19** reveals the effect of RF3 binding on the conformation of bound RF1.

Ribosomes are recycled after termination

After the ribosome terminates a polypeptide chain, the deacylated tRNA and the mRNA stop codon end up in the P site (see Figure 16.16). Mechanisms exist to remove them from the ribosome and to dissociate it into its constituent subunits, so that a new polypeptide chain can be synthesized by the recycled subunits (see

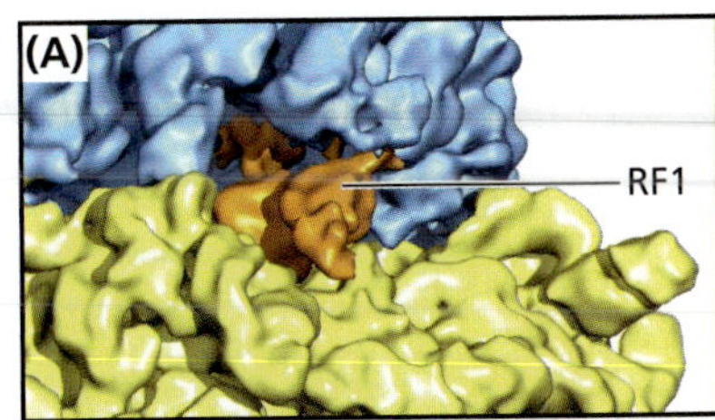

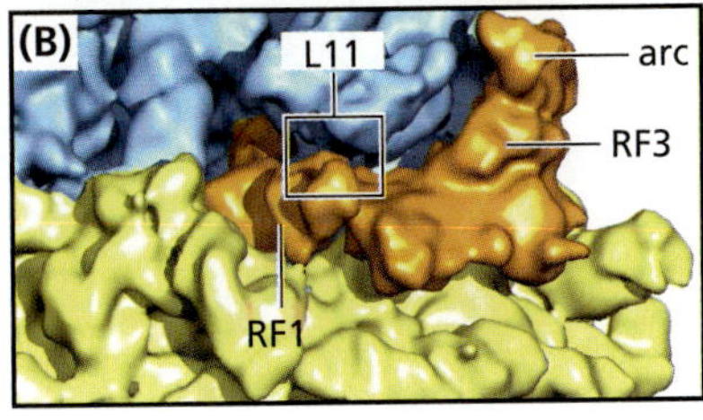

Figure 16.19 Termination complexes of ribosomes bound to release factors. The model is based on high-resolution cryo-EM imaging and single-particle-based reconstruction. The 50S subunit is in blue and 30S is in yellow. (A) Ribosome with RF1 alone. (B) Ribosome with RF1 and RF3. L11 is ribosomal protein 11, and arc represents one of the domains of RF3. Comparison of the two complexes reveals a large conformational change in RF1 in response to RF3 binding, resulting in formation of a bridge from RF3 via RF1 to L11. This, and accompanying changes in the ribosome conformation, is proposed to promote RF1 or RF2 dissociation. (From Pallesen J, Hashem Y, Korkmaz G et al. [2013] *eLife* 2:e00411. With permission from eLife Sciences Publications, Ltd.)

Figure 16.16). In bacteria, this process occurs with the participation of three factors: ribosome releasing factor and two other factors that we have encountered: EF-G–GTP, which ratchets the ribosome during translocation, and IF3, which acts as an anti-association factor during initiation.

The structure of RRF is presented in **Figure 16.20**. The shape of free RRF is an almost perfect mimic of the shape of tRNA structure, which led to the proposal that this structural mimicry is important for binding to the ribosome. Further structural and biochemical studies have shown, however, that this is not the case. RRF domain 1 spans the A and P sites in the 50S subunit rather than mimicking the anticodon stem-loop that binds to the small subunit. The other domain is flexible; this flexibility has been implicated in ribosome splitting. We have received the lesson that mere structural similarities can be deceiving. Finally, and not too surprisingly, IF3 is required for dissociation of the tRNA from the P site and concomitant release of the message. The participation of IF3 in ribosome recycling thus couples the last step of protein synthesis to the first, making the entire process of protein synthesis more efficient.

The details of ribosome recycling in eukaryotes and some of the factors involved are presented in **Figure 16.21**. Disassembly of the 80S ribosome in eukaryotes requires the three initiation factors eIF1, eIF1A, and eIF3 and a number of special factors. eIF3 is a complex of more than 13 subunits in mammals and 5 subunits in yeast; it possesses an anti-association activity similar to that of bacterial IF3. The post-termination ribosome is first split into two parts: the 60S subunit and the 40S subunit, which is bound to tRNA and mRNA. The release of P-site deacylated tRNA is promoted by eIF1 and the subsequent release of the mRNA by eIF3j, a loosely associated subunit of eIF3.

Our views of translation continue to evolve

The past few decades of research have markedly changed our views as to the role of the ribosome in protein biosynthesis. It is neither an inert platform for the process, as was once thought, nor a catalytic ribozyme, a more recent view. Rather, it seems to be a dynamic stage, which can both place the participants of translation in their proper positions for reaction and aid in their necessary motions to take them through the cycle. There is surely more to be learned, especially with respect to the more complicated process in eukaryotes.

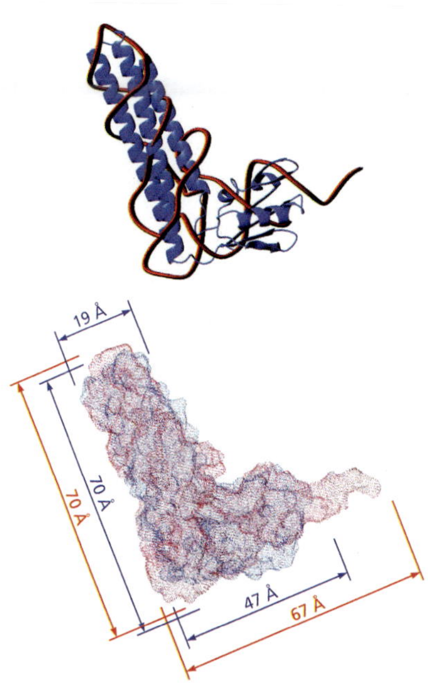

Figure 16.20 Crystal structure of ribosome releasing factor from *Thermotoga maritima* at 2.55 Å resolution. The function of the RRF, in conjunction with EF-G, is to recycle ribosomes after each round of protein synthesis. RRF has a rather unusual L-shaped structure in which one domain is formed by a long three-helix bundle, whereas the other domain is a three-layer β/α/β sandwich. The molecule can be superimposed almost perfectly upon that of a tRNA, with the exception of the amino-acid-binding 3′-end of the tRNA. The tRNA mimicry is almost perfect, more so than, for example, domains III and IV of EF-G, which imitate the shape of the tRNA bound to EF-Tu (see Figure 16.14). (Top) Ribbon representation of the superposition of RRF, in blue, and yeast $tRNA^{Phe}$, in red. (Bottom) Surface representation of the same superposition, using the same color scheme. (From Selmer M, Al-Karadaghi S, Hirokawa G et al. [1999] *Science* 286:2349–2352. With permission from American Association for the Advancement of Science.)

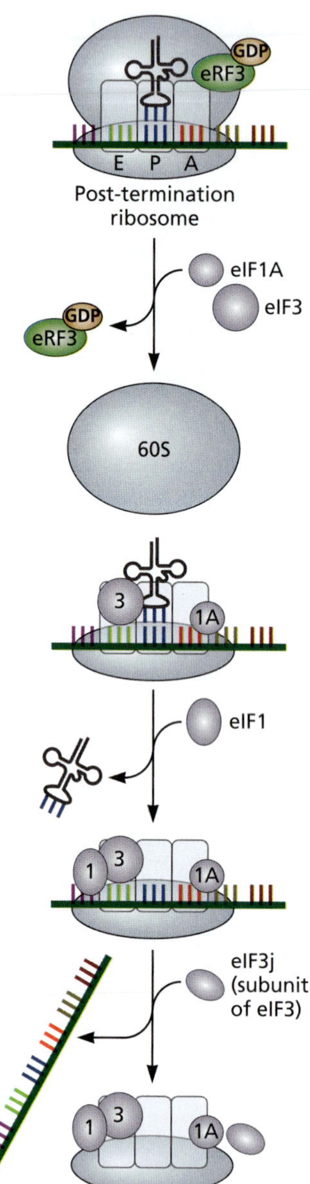

Figure 16.21 Model for post-termination 80S ribosome disassembly in eukaryotes. In eukaryotes, only one release factor, eRF1, recognizes all three stop codons; otherwise, termination is similar to that in bacteria. Disassembly of a post-termination 80S ribosome is accomplished by initiation factors eIF1, eIF1A, and eIF3. (Adapted from Pisarev AV, Hellen CUT & Pestova TV [2007] *Cell* 131:286–299. With permission from Elsevier.)

Key concepts

- We now know a great deal about the process of translation from extensive kinetic studies over many years, from high-resolution X-ray diffraction studies of the ribosome and its complexes, and from newer techniques such as single-particle reconstruction from cryo-EM imaging and single-pair FRET.
- The whole process of translation can be divided into three major phases: initiation, elongation, and termination. Each is a complex, multistep process, involving three sites on the ribosome designated A for acceptor, P for peptidyl, and E for exit. Both ribosomal subunits contribute to each site.
- Initiation, in bacteria, begins when the 30S subunit of the ribosome attaches the mRNA in such a way that the P site aligns with the start codon for the initiator formylmethionine tRNA, fMet-tRNAi^{fMet}. In many cases, this is accomplished by aligning a particular common sequence in the bacterial message, the Shine–Dalgarno sequence, with a complementary ribosomal RNA sequence. GTP-bound initiation factor IF2 brings the charged fMet-tRNAi^{fMet} to the P site of the ribosome; GTP is later hydrolyzed to release the factor from the complex. Two more initiation factors are needed for formation of the 30S pre-initiation complex. Initiation ends with addition of the 50S ribosomal subunit.
- In eukaryotes, there is no Shine–Dalgarno sequence. The 40S subunit binds to the message downstream from the cap site and tracks until it finds the first AUG codon, which serves for a start. Again, the participation of many protein factors and the hydrolysis of GTP are required, culminating in the addition of the charged initiator methionine-tRNA, Met-tRNAi^{Met}, and the 60S subunit to assemble the 80S initiation complex.
- Elongation is also a multistep process. Two major transitions are involved. The first is peptidyl transfer, in which the growing polypeptide chain is transferred from the tRNA in the P site to the tRNA charged with the new amino acid to be added to the chain; the new aa-tRNA is brought to the A site by GTP-bound elongation factor EF-Tu. The second transition is translocation, in which the two tRNAs move from the P and A sites to the E and P sites, respectively; the mRNA moves on the ribosome by one codon length.
- Many of the intermediate steps in elongation are reversible, but the key steps of peptidyl transfer and translocation are made irreversible by GTP hydrolysis. Thus, the ribosome behaves like a Brownian ratchet.
- During elongation, there are opportunities for proofreading the incoming aa-tRNA, so as to increase the fidelity of protein synthesis.
- The critical catalytic step of peptidyl transfer appears to be catalyzed by RNA rather than by ribosomal proteins. The actual catalysis is performed by the 2′-OH group of A76 of the peptidyl-tRNA, which is well positioned by the ribosome to extract protons from the NH_2 of the incoming A-site aa-tRNA and donate them to the leaving group. The role of the ribosome is to create the appropriate environment for the reaction to occur.
- The growing polypeptide chain is extruded through a tunnel in the large subunit. This tunnel is wide enough to probably accommodate an α-helix but not most proteins in their fully folded conformations.
- Termination of transcription occurs when a stop codon is encountered. This triggers the sequential recruitment of release factors, which results in the hydrolytic release of the polypeptide chain and eventually dissociation of the ribosome into its subunits, ready for another round.

Further reading

Books

Nierhaus KH & Wilson DN (2004) Protein Synthesis and Ribosome Structure: Translating the Genome. Wiley–VCH.

Rodnina M, Wintermeyer W & Green R (eds) (2011) Ribosomes Structure, Function, and Dynamics. Springer-Verlag.

Reviews

Aitken CE, Petrov A & Puglisi JD (2010) Single ribosome dynamics and the mechanism of translation. *Annu Rev Biophys* 39:491–513.

Bashan A & Yonath A (2008) Correlating ribosome function with high-resolution structures. *Trends Microbiol* 16:326–335.

Beringer M & Rodnina MV (2007) The ribosomal peptidyl transferase. *Mol Cell* 26:311–321.

Blanchard SC (2009) Single-molecule observations of ribosome function. *Curr Opin Struct Biol* 19:103–109.

Blanchard SC, Cooperman BS & Wilson DN (2010) Probing translation with small-molecule inhibitors. *Chem Biol* 17:633–645.

Fischbach MA & Walsh CT (2009) Antibiotics for emerging pathogens. *Science* 325:1089–1093.

Frank J & Gonzalez RL Jr (2010) Structure and dynamics of a processive Brownian motor: The translating ribosome. *Annu Rev Biochem* 79:381–412.

Jackson RJ (2007) The missing link in the eukaryotic ribosome cycle. *Mol Cell* 28:356–358.

Kramer G, Boehringer D, Ban N & Bukau B (2009) The ribosome as a platform for co-translational processing, folding and targeting of newly synthesized proteins. *Nat Struct Mol Biol* 16:589–597.

Marshall RA, Aitken CE, Dorywalska M & Puglisi JD (2008) Translation at the single-molecule level. *Annu Rev Biochem* 77:177–203.

Mitchell SF & Lorsch JR (2008) Should I stay or should I go? Eukaryotic translation initiation factors 1 and 1A control start codon recognition. *J Biol Chem* 283:27345–27349.

Petry S, Weixlbaumer A & Ramakrishnan V (2008) The termination of translation. *Curr Opin Struct Biol* 18:70–77.

Ramakrishnan V (2010) Unraveling the structure of the ribosome (Nobel Lecture). *Angew Chem Int Ed Engl* 49:4355–4380.

Rodnina MV, Beringer M & Wintermeyer W (2007) How ribosomes make peptide bonds. *Trends Biochem Sci* 32:20–26.

Rodnina MV & Wintermeyer W (2001) Fidelity of aminoacyl-tRNA selection on the ribosome: Kinetic and structural mechanisms. *Annu Rev Biochem* 70:415–435.

Rodnina MV & Wintermeyer W (2009) Recent mechanistic insights into eukaryotic ribosomes. *Curr Opin Cell Biol* 21:435–443.

Rodnina MV & Wintermeyer W (2010) The ribosome goes Nobel. *Trends Biochem Sci* 35:1–5.

Schmeing TM & Ramakrishnan V (2009) What recent ribosome structures have revealed about the mechanism of translation. *Nature* 461:1234–1242.

Steitz TA (2008) A structural understanding of the dynamic ribosome machine. *Nat Rev Mol Cell Biol* 9:242–253.

Wilson DN (2009) The A–Z of bacterial translation inhibitors. *Crit Rev Biochem Mol Biol* 44:393–433.

Yonath A (2005) Antibiotics targeting ribosomes: Resistance, selectivity, synergism, and cellular regulation. *Annu Rev Biochem* 74:649–679.

Zaher HS & Green R (2009) Fidelity at the molecular level: Lessons from protein synthesis. *Cell* 136:746–762.

Experimental papers

Ban N, Nissen P, Hansen J et al. (2000) The complete atomic structure of the large ribosomal subunit at 2.4 Å resolution. *Science* 289:905–920.

Fischer N, Konevega AL, Wintermeyer W et al. (2010) Ribosome dynamics and tRNA movement by time-resolved electron cryomicroscopy. *Nature* 466:329–333.

Gromadski KB & Rodnina MV (2004) Kinetic determinants of high-fidelity tRNA discrimination on the ribosome. *Mol Cell* 13:191–200.

Korostelev A, Asahara H, Lancaster L et al. (2008) Crystal structure of a translation termination complex formed with release factor RF2. *Proc Natl Acad Sci USA* 105:19684–19689.

Petry S, Brodersen DE, Murphy FV IV et al. (2005) Crystal structures of the ribosome in complex with release factors RF1 and RF2 bound to a cognate stop codon. *Cell* 123:1255–1266.

Schmeing TM, Voorhees RM, Kelley AC et al. (2009) The crystal structure of the ribosome bound to EF-Tu and aminoacyl-tRNA. *Science* 326:688–694.

Simonetti A, Marzi S, Myasnikov AG et al. (2008) Structure of the 30S translation initiation complex. *Nature* 455:416–420.

Voss NR, Gerstein M, Steitz TA & Moore PB (2006) The geometry of the ribosomal polypeptide exit tunnel. *J Mol Biol* 360:893–906.

Wimberly BT, Brodersen DE, Clemons WM Jr et al. (2000) Structure of the 30S ribosomal subunit. *Nature* 407:327–339.

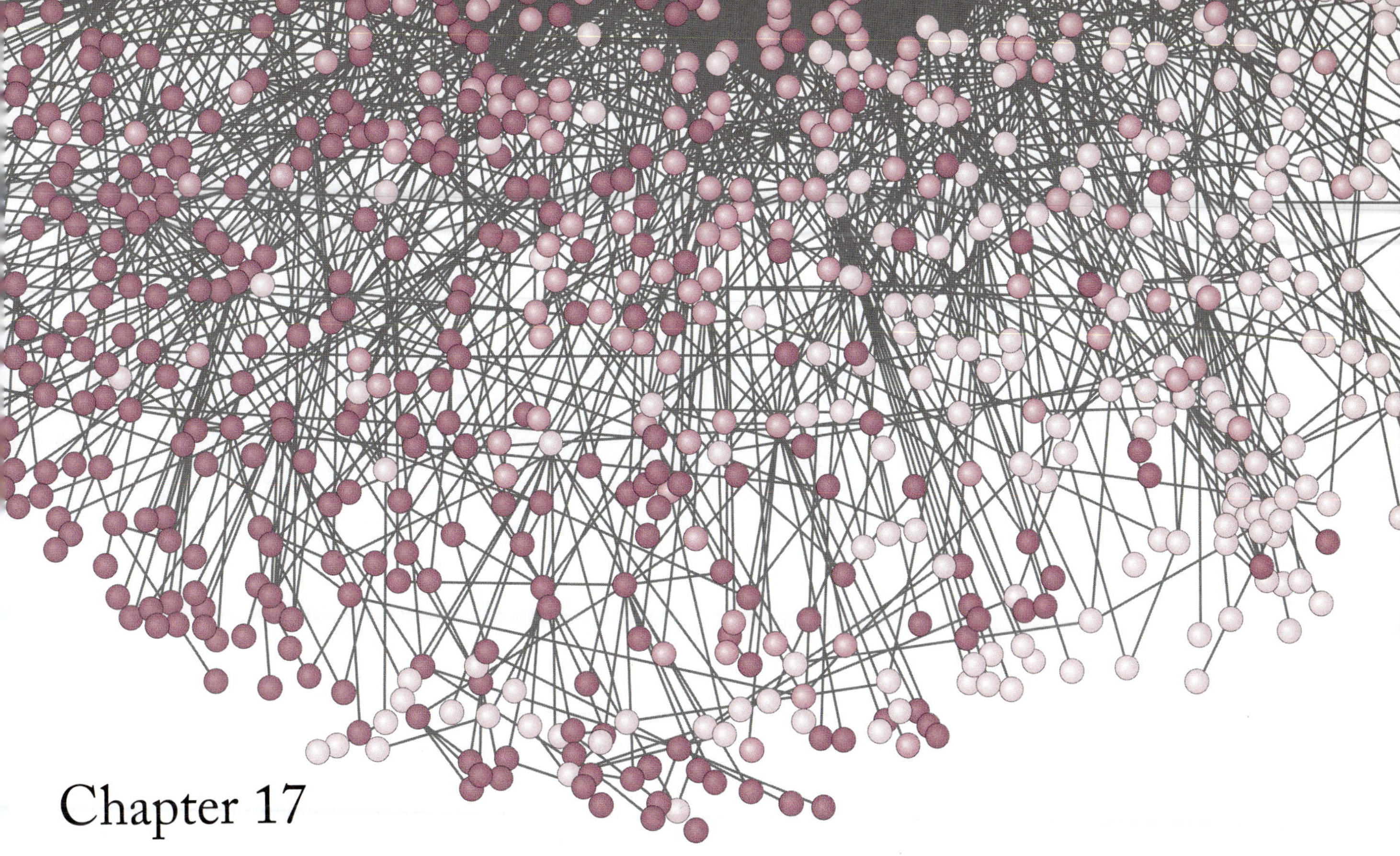

Chapter 17

Regulation of Translation

17.1 Introduction

Cells must continually change the profile of the proteins that they contain, in terms of both the types of proteins and their relative abundances. These changes are required in response to changes in environmental conditions and, in eukaryotes, to integrate signals that control development and differentiation. A major portion of regulation occurs at the level of transcription, but in this chapter we demonstrate that regulation at the level of translation is no less prevalent and important. Translational regulation can occur at several levels, from control of the availability of ribosomes to adjustments to the rate of translational initiation and elongation. The abundance, accessibility, and stability of mRNAs are also factors. Although translation elongation is very similar in bacteria and eukaryotes, initiation is rather different in these domains of life, especially in terms of the way the mRNA is presented to the appropriate site of the ribosome. These mechanistic differences have led to different regulatory pathways. Thus we make a clear distinction, when necessary, between bacteria and eukaryotes. We begin by describing a very general mode of regulation: control of the abundance of ribosomes.

17.2 Regulation of translation by controlling ribosome number

Ribosome numbers in bacteria are responsive to the environment

Bacteria live in environments that can change dramatically. To survive and adapt, they have developed a wide range of regulatory mechanisms that control practically every cellular process, including transcription, translation, and the stability of mRNA molecules and of proteins. The most general, nonspecific way to affect protein synthesis in bacteria is strict and precise control over the number of ribosomes in each cell. Depending on the environmental conditions, which dictate appropriate rates of growth, bacteria can have as few as 2000 ribosomes or as many as 100,000 ribosomes per cell.

This number will depend upon both the rate of synthesis of ribosomal components and the rate of degradation of ribosomes. Controls of rRNA and **ribosomal protein** or **r-protein** synthesis have been thoroughly studied, whereas ribosome degradation has been little explored. Recently, it has been clearly shown that ribosomes are very stable and do not degrade under exponential growth conditions. On the other hand, once growth slows down, for example, because of exhaustion of nutrient supply, degradation of rRNA and hence ribosomes commences. In other recent studies, Murray Deutscher's laboratory has presented evidence that whereas the 70S ribosome is relatively resistant to endoribonucleases, the 30S and 50S subunits are very vulnerable. This suggests a simple yet elegant mechanism. When growth, and hence translation, is slowed, 30S and 50S particles accumulate because of dissociation of 70S particles at the termination of the elongation cycle (see Chapter 16). The subunits are then susceptible to degradation. The degradation of large numbers of ribosomes can then provide an excellent source of nutrients under stress conditions.

Synthesis of ribosomal components in bacteria is coordinated

Ribosome synthesis requires coordinated synthesis of rRNAs and r-proteins. Control of these processes relies on the transcription of rRNA genes and the translation of mRNAs that encode r-proteins. In *Escherichia coli*, seven rRNA operons are highly transcribed under conditions of rapid growth: *rrnA*, *rrnB*, *rrnC*, *rrnD*, *rrnE*, *rrnG*, and *rrnH*. Indeed, transcription from rRNA operons is estimated to constitute more than 50% of total RNA synthesis under these conditions. The operons are almost identical in sequence and some contain tRNA genes in addition to genes for ribosomal RNAs. We have already presented the structure and control of these operons in previous chapters. Here we consider the control over activity of the operons that encode the ribosomal proteins.

In *E. coli*, there are 19 ribosomal-protein or r-protein operons. In most cases these are clusters of several r-protein genes, coding for both large-subunit L proteins and small-subunit S proteins. Often, other proteins involved in translation, such as EF-Tu and EF-G, are also encoded by part of these operons. The genes encoding the various subunits of the RNA polymerase holoenzyme, α, β, β', and σ^{70}, are also in these clusters. The main mechanism for coordinating the rate of r-protein synthesis within each operon is autogenous feedback regulation, presented in **Figure 17.1**. One of the proteins synthesized as a result of translation of the polycistronic mRNA can bind both to the corresponding rRNA molecule in the ribosome and to a recognition sequence in the 5′-regulatory sequence of the operon mRNA. Alternatively, this regulatory sequence may be located between genes in the operon. Binding inhibits translation of the entire polycistronic message through formation of mRNA structures that occlude the translation start site. To date, 10 repressor proteins that regulate 10 of the 19 r-protein operons have been identified.

Although the feedback mechanism neatly accounts for the coordinated synthesis of a number of r-proteins, the picture is not yet complete. First, not all operons encoding r-proteins can be regulated in this way: operons that encode r-proteins that do not bind directly to rRNA cannot use feedback mechanisms. Recall from Chapter 15 that only some of the r-proteins bind directly to the rRNA; others need already bound proteins to be able to join the nascent ribosomal particle. These secondary and tertiary

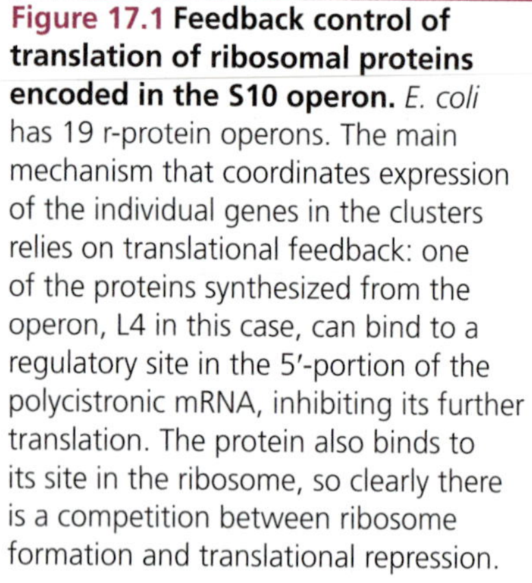

Figure 17.1 Feedback control of translation of ribosomal proteins encoded in the S10 operon. *E. coli* has 19 r-protein operons. The main mechanism that coordinates expression of the individual genes in the clusters relies on translational feedback: one of the proteins synthesized from the operon, L4 in this case, can bind to a regulatory site in the 5′-portion of the polycistronic mRNA, inhibiting its further translation. The protein also binds to its site in the ribosome, so clearly there is a competition between ribosome formation and translational repression.

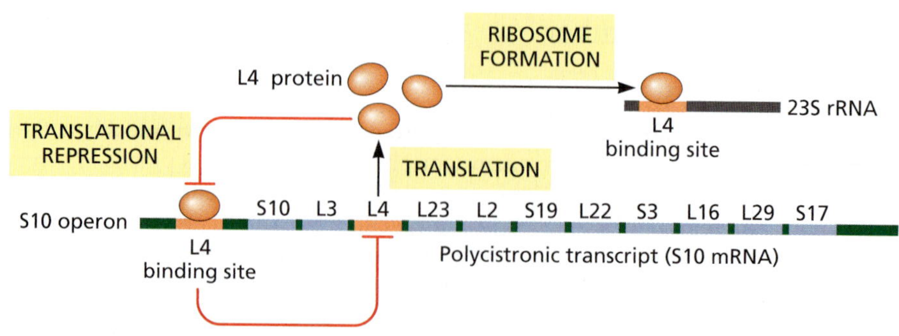

r-proteins are regulated either at the transcriptional level or through protein degradation. Second, it is still unclear how the synthesis of groups of r-proteins that are encoded by different operons is co-regulated or how the synthesis of rRNAs is coordinated with that of the ribosomal proteins.

Regulation of the synthesis of ribosomal components in eukaryotes involves chromatin structure

As discussed in Chapter 15, the biogenesis of eukaryotic ribosomes is very complex, even before actual assembly of the ribosomal subunits. This is because synthesis of ribosomal components requires the participation of all three eukaryotic polymerases and is thus subject to different kinds of regulation.

In eukaryotes, the repetitive genes encoding rRNAs exist in two distinct epigenetic states; that is, they are packaged in distinct chromatin structure. These different chromatin states correspond to two functionally distinguishable sets of rRNA genes, those that are transcriptionally active and those that are transcriptionally repressed. Surprisingly, a fraction of the rRNA genes is silent even in proliferating cells, which have high demands for protein synthesis. Because of the repetitive nature of rRNA genes, two strategies for regulating rRNA are conceptually possible: (1) changes to the rate of transcription from each active gene or (2) changes to the number of repeats that are actively transcribed. Most short-term regulation affects the rate of transcription through mechanisms of the first type: pre-initiation complex formation, promoter escape, etc. On the other hand, redistribution between active and inactive rDNA copies occurs during development and differentiation, that is, during long-term regulation.

The proportion of eukaryotic rRNA genes that are actively transcribed can be identified by a psoralen cross-linking assay (**Box 17.1**). Active, psoralen-accessible genes, which are associated with nascent pre-rRNA, are free of regularly spaced

Box 17.1 Psoralen cross-linking and the structure of rDNA chromatin Psoralens are natural compounds, tricyclic furocoumarins, that are present in plants. They are usually extracted from a Chinese herb, *Psoralea corylifolia*, although they are also found in other plants, such as figs, parsnips, limes, parsley, and celery. These compounds become intercalated between DNA base pairs (**Figure 1A**); upon UV irradiation, they form covalent adducts with thymine, both intrastrand and interstrand. These psoralen adducts cause replication arrest and are used in the treatment of psoriasis and vitiligo.

Psoralen has been used extensively to study chromatin structure. In general, the structure of euchromatic, transcriptionally active genes is more open, and thus more accessible to psoralen cross-linking, than that of the more compact heterochromatic, inactive genes, which are less cross-linkable. Because cross-linked DNA fragments migrate more slowly on agarose gels than non-cross-linked DNA, active genes can be distinguished from inactive ones by their electrophoretic behavior. Psoralen cross-linking, when performed on cells or isolated nuclei from a number of eukaryotic organisms, has shown that there are two classes of rDNA. Active gene copies, which are associated with nascent pre-rRNA, lack regularly spaced nucleosomes. By contrast, inactive copies of genes exhibit regular nucleosome spacing and are not associated with either Pol I or transcription factors of Pol I. The distinction between these two classes of ribosomal genes was also directly visualized by a combination of immunostaining and immunofluorescence *in situ* hybridization or immuno-FISH analysis, which is able to detect the localization of rDNA and Pol I-specific upstream binding factor, UBF, simultaneously (see Chapter 10). The psoralen cross-linking experiment was performed as outlined in **Figure 1B**, which also shows its main results.

Thus, EM studies and psoralen cross-linking experiments seemed to indicate that the actively transcribed rRNA genes are devoid of nucleosomes. Alas, the situation is not that straightforward. Using the power of yeast genetics, Nick Proudfoot's laboratory created a yeast strain that had a reduced number of rDNA copies, all of them actively transcribed; the strain was appropriately dubbed AA, for all active. Chromatin immunoprecipitation analysis showed the presence, albeit in nonstoichiometric amounts, of both histones H3 and H2B in the rDNA chromatin in these strains. Furthermore, the micrococcal nuclease or MNase ladders in this strain were very similar to those in the wild-type isogenic strains. Recall from Chapter 8 that MNase preferentially cleaves linker DNA between nucleosomes to create a ladder of fragments in multiples of ~200 bp on electrophoretic gels. When mononucleosomes were isolated from both strains and probed for the presence of rDNA by PCR or Southern blotting, they both showed rDNA sequences organized as nucleosomes. Interestingly, and in accordance with earlier results, the spacer regions between the 18S, 5.8S, and 28S genes in the locus were organized in well-spaced nucleosomal particles. Finally, chromatin immunoprecipitation using antibodies

(Continued)

Box 17.1 (*Continued*)

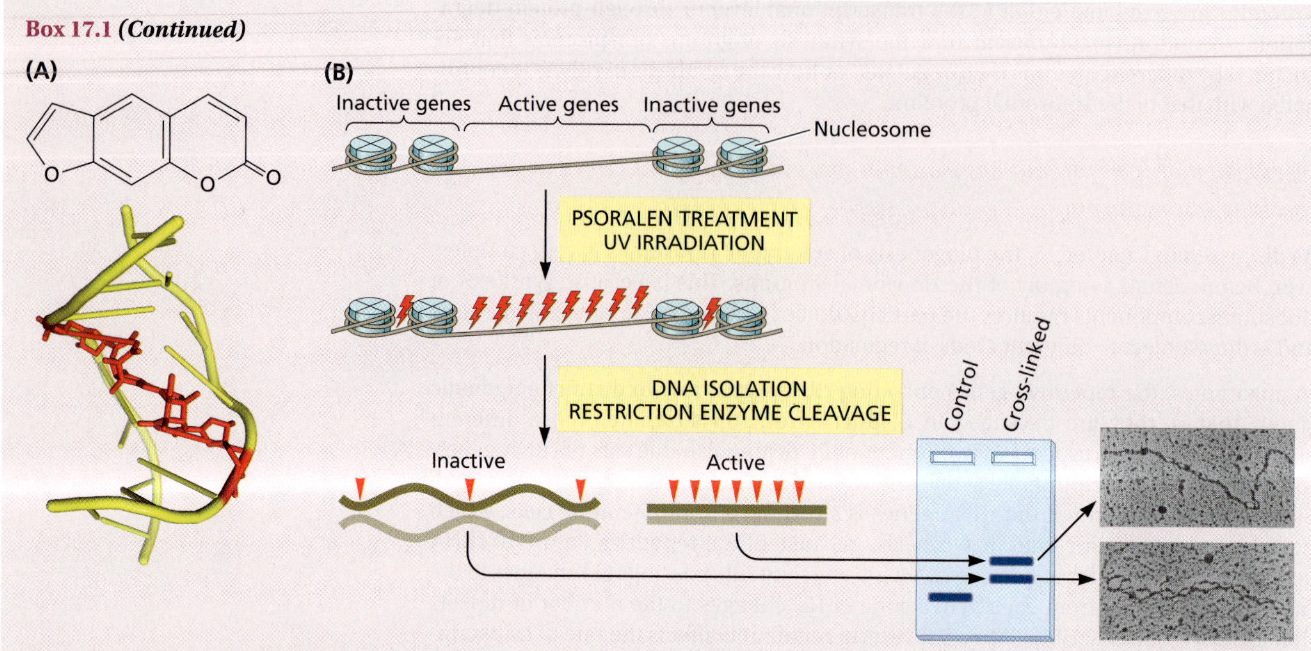

Figure 1 Psoralen cross-linking. (A) Psoralen structure and double-stranded DNA interstrand cross-linking. (B) Psoralen photo-cross-linking of ribosomal RNA genes reveals two classes of genes, inactive and active, with different chromatin organization. Intact yeast cells, or isolated nuclei, were irradiated with UV-A light in the presence of psoralen. DNA was purified and digested with appropriate restriction enzymes and then separated by gel electrophoresis. The DNA fragments migrate according to the amount of psoralen incorporated: the slow-migrating DNA band is derived from active rRNA genes, whereas the fast-migrating DNA band is derived from inactive rRNA genes. Non-cross-linked DNA was run in parallel as a control. DNA from the two bands was eluted and spread for EM under denaturing conditions. The presence of interstrand cross-links does not allow separation of the two DNA strands at the position of the cross-links. Thus, microscopy shows that the active, slow-migrating fraction comprises heavily cross-linked double-stranded DNA, indicating accessibility of the DNA at virtually all points. On the other hand, the inactive fraction shows only periodic cross-linking and regularly spaced single-stranded bubbles, characteristic of nucleosomal chromatin. (B, EM images, from Dammann R, Lucchini R, Koller T & Sogo JM [1993] *Nucleic Acids Res* 21:2331–2338. With permission from Oxford University Press.)

against certain chromatin remodelers showed that these remodelers were present in rDNA chromatin. Thus, it was concluded that active rDNA exists in a dynamic chromatin structure of unphased nucleosomes and that chromatin remodelers help in transcribing the existing nucleosomes. The lesson from all of these studies is clear: do not draw conclusions from data obtained by a single method. Use as many different techniques as possible, and try to reconcile results that are seemingly at odds. They may not be; they may just reflect limitations in methodology.

nucleosomes, although they do contain some nucleosomes (see Box 17.1). Their inactive counterparts contain regularly spaced nucleosomes and are devoid of Pol I or its associated transcription factors. Active and inactive genes also differ in their DNA methylation patterns, which can be assessed through the use of methylation-sensitive and methylation-insensitive restriction enzyme pairs (**Box 17.2**). In addition, they differ in the post-translational modifications of their resident histones. The inactive rDNA genes are predominantly methylated in the promoter and enhancer regions. The effect of this methylation is chromatin-specific: methylation does not affect the transcription of naked rDNA templates. In terms of post-translational histone modifications, active genes are associated with acetylated histones H3 and H4, as well as with histone H3 that is trimethylated on lysine 4. Inactive rDNA genes are associated with methylated lysines 9, 20, and 27 on H3. These histone modifications conform to the more general patterns observed for other active and inactive genes (see Chapter 12). Finally, we now know that there are chromatin remodeling complexes that either promote rDNA transcription, for example, Cockayne syndrome protein B, or inhibit rDNA transcription, for example, NoRC.

How are the syntheses of different eukaryotic r-proteins coordinated with each other and with the syntheses of rRNAs? The synthesis of r-proteins appears to be regulated

Box 17.2 Methylation of CpG dinucleotides in promoter regions of rRNA genes: The transcription connection It has been known for many years that DNA methylation in the promoter regions of many eukaryotic genes correlates with the transcriptional status of these genes (see Figure 12.27). In some gene-specific cases, this correlation has been extended to a causal relationship: methylation leads to transcriptional repression. Does this general relationship also hold true for sets of active and inactive rRNA genes? The answer is yes, as illustrated in Figure 1. In addition to these experiments, the correlation between promoter methylation and transcriptional inactivity was strengthened by *in vivo* experiments, in which cells were treated with an inhibitor of cytosine methylation, 5-aza-2′-deoxycytidine or aza-dC: the treatment stimulated transcription, establishing a mechanistic link between DNA methylation and repression of rRNA gene copies. In addition, when naked rDNA was assembled into chromatin, methylated templates were not transcribed.

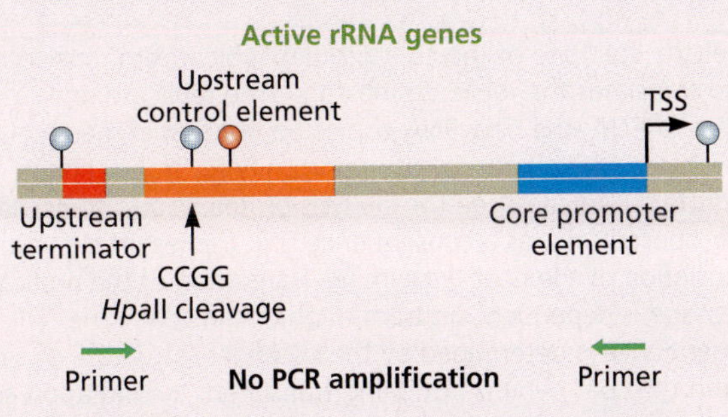

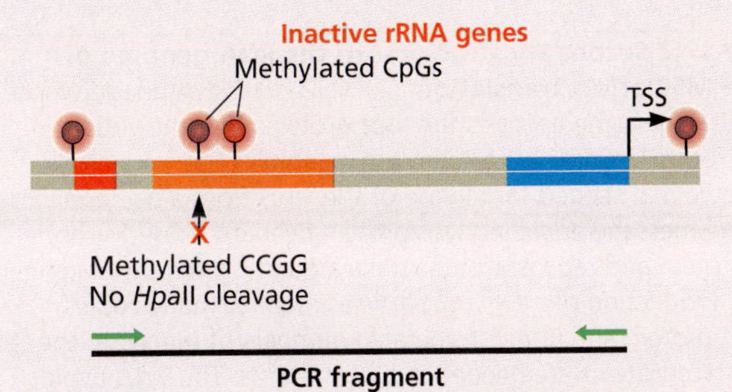

Figure 1 Active and silent rRNA gene copies have different DNA methylation patterns. Organization of the mouse rDNA promoter, with its major control elements, is depicted. The lollipops indicate the position of CpG dinucleotides at −167, −143, −133, and +8. The red lollipop marks the position of a critical CpG dinucleotide, whose methylation is sufficient to impair binding of the transcription factor UBF to nucleosomal rDNA, thus impairing transcription. The methylation status of a DNA fragment containing the CpG dinucleotide, which is the preferred substrate for adding methyl groups to the cytosine, can be assessed by cleavage by methylation-sensitive restriction enzymes: *Hpa*II would cleave the sequence CCGG only if the second C is unmethylated. Methylation-insensitive restriction enzymes are used as controls. If cleavage occurs, as would be the case for active rDNA copies, then no PCR product is generated because the primer pair used for PCR amplification is split between fragments. If the CCGG sequence is methylated, as it is in inactive gene copies, then *Hpa*II cannot cleave, and a PCR product can be generated subsequently.

differently in yeast and mammals. The yeast r-protein genes behave as a precisely coordinated transcriptional cluster in which the transcription factor Rap1 binds to one or two sites in the promoter regions of all the genes.

By contrast, transcriptional regulation does not seem to play a significant role in the control of mammalian r-protein synthesis; rather, regulation is at the translational level. Sequence comparisons of r-protein genes revealed a rather distinctive structure around the transcription start site, TSS, which is always positioned within a pyrimidine stretch, with cytosine, C, being the first nucleotide base in the message. Recall that the vast majority of mRNAs start with a purine, most often an adenine. For example, the S6 gene has the following pyrimidine-rich sequence around its TSS: 5′-TGGC**C**CTCTTTTCC-3′, with the first C in the message set in boldface type. Genes whose corresponding mRNAs begin with such a 5′-terminal oligopyrimidine or TOP sequence are translationally regulated. The TOP sequence represents the major *cis*-acting element that interacts with the La *trans*-acting factor. La proteins are RNA-binding proteins that interact with an extensive variety of cellular RNAs. They promote the metabolism of Pol III transcripts by binding to their common UUU-3′-ends; the mechanism of their involvement in the translation of r-protein mRNA is less well understood.

In addition to the TOP sequence, the mammalian r-protein mRNAs are characterized by relatively short untranslated regions, a ~40-nt-long 5′-UTR and ~35-nt-long 3′-UTR. The 3′-UTR may play a role in regulation by collaborating with the TOP sequence in a poorly understood manner.

Finally, it has been observed that in yeast some r-protein synthesis is inhibited by blocking of splicing of transcripts. This is an autoregulatory mechanism, in which the protein itself can bind to the pre-mRNA and block splicing.

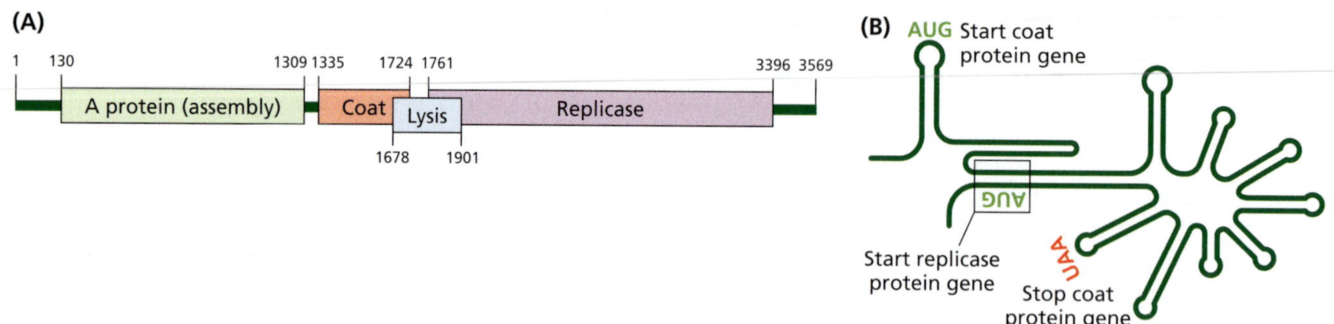

Figure 17.2 Secondary structures in the RNA genome of phage MS2 affect translation. (A) MS2 is an RNA phage whose small RNA genome encodes the four proteins that the virus needs for its life cycle. The A protein assembles the virus, while lysis protein is needed for release of the virus from a bacterial cell. The message encoding lysis protein overlaps those for both coat protein and replicase but is translated in a different reading frame. Production of new viral particles requires many copies of coat protein and also considerable numbers of replicase; the other two proteins are needed in tiny amounts. The virus avoids translating the entire polycistronic message as a whole, which would produce the four proteins in equal amounts. The ribosome-binding site at the 5′-end of the RNA is normally blocked by folding. This leads to initial utilization of an attachment sequence near the start site of the coat protein, which from time to time also translates the replicase message. The replicase replicates the viral RNA; the new RNA copies do not fold immediately, which provides a window of opportunity for translation of the A protein. The message for the lysis protein is also translated only sporadically, as a consequence of a frameshift slip during translation of the coat protein. (B) Translation of the replicase message is dependent on translation of the coat gene. This dependence is determined by the secondary structure of the RNA: when the coat gene is not being translated, the initiation codon for the replicase is buried in a base-paired region, shown in a box. When the coat gene is translated, this base pairing is disrupted and the replicase message can now be translated, although not at every coat-protein reading.

17.3 Regulation of translation initiation

Regulation of translation initiation is ubiquitous and remarkably varied

Although control of ribosome numbers is an effective mechanism in responding to massive or slow changes in the environment, it is not a quick or selective response. Controlling translation initiation is the fastest and most economical way to change the protein profile in a cell. Furthermore, such control can be highly selective, regulating the production of only certain proteins. Thus, it is found at every major level of evolution and takes a wide variety of forms. Bacteria and their viruses, or phages, make proficient use of secondary structures in the mRNA to control initiation. Figure 17.2 depicts a most sophisticated, classical example of how a simple RNA phage, MS2, controls the production of very different amounts of proteins, all encoded by the same polycistronic RNA molecule. Bacteria also make use of **riboswitches** and of small noncoding regulatory RNAs, such as microRNAs and short interfering RNAs. Such mechanisms will be presented in detail later in this chapter. However, the major emphasis here will concern eukaryotes, as most of the knowledge on translation initiation relates to these organisms.

Regulation may depend on protein factors binding to the 5′- or 3′-ends of mRNA

As detailed in Chapter 16, translation initiation in eukaryotic cells is a complex, multistep process with alternative pathways, which allows regulation at many points. Here, we first describe the regulation of cap-dependent initiation, before we turn to the more infrequent initiation at **internal ribosome entry site** or **IRES** elements, which are mainly used for protein synthesis under stress conditions. Finally, we consider some other, completely different modes of regulation of initiation.

Cap-dependent regulation is the major pathway for controlling initiation

Cap-dependent initiation requires binding of **eukaryotic initiation factor 4F** or **eIF4F**, which is a complex of three eIF proteins E, G, and A, to the 5′-cap structure in the mRNA (Figure 17.3). This binding can be eliminated by members of a protein

(A) Active translation

43S pre-initiation complex

eIF4F complex

eIF4E m^{7}GpppN

AUG

eIF4G eIF4A

PABP

Coding region

AAAAAAAAAAAAAAA

(B) Repression

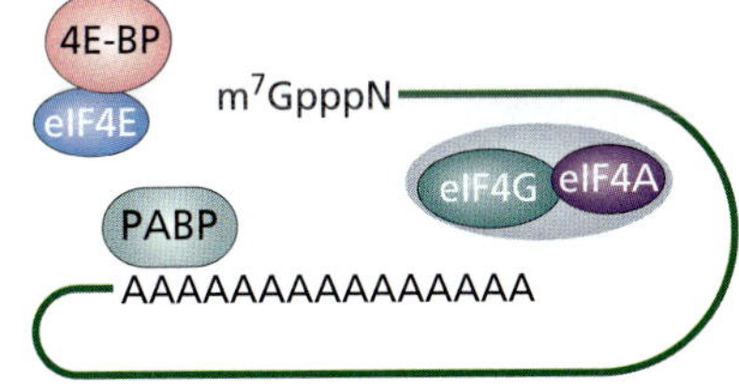

(C) Repression

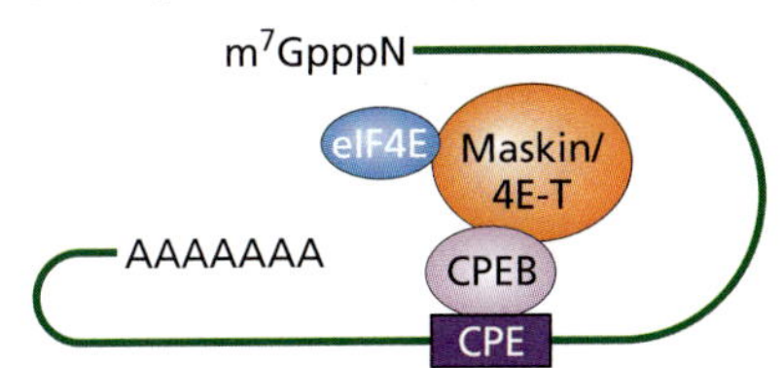

Figure 17.3 Mechanisms of cap-dependent translational repression by soluble or tethered eIF4E-binding proteins. (A) Cap-dependent initiation in eukaryotes requires interaction of the 5′-cap with the initiation factor complex eIF4F. In addition to the cap-binding protein eIF4E, the complex contains the RNA helicase eIF4A and the scaffolding protein eIF4G. The latter interacts with the cap-binding protein and PABP, the poly(A)-binding protein, circularizing the template for efficient translation. (B) A widespread mechanism of translational repression involves competition between eIF4G and members of a family of related 4E-binding proteins or 4E-BP for binding to eIF4E. Phosphorylation of 4E-BP weakens its interaction with eIF4E. (C) Other mechanisms are more specific to the particular mRNA whose translation needs to be repressed. For example, in *Xenopus* oocytes, mRNAs that contain cytoplasmic polyadenylation elements, CPE, in their 3′-untranslated region are repressed by displacement of eIF4G by the protein Maskin, which in turn interacts with the cytoplasmic polyadenylation element-binding protein or CPEB. In early oogenesis, Maskin is absent and repression of translation of maternal mRNAs occurs through the use of a different protein, 4E-T or 4E transporter. (Adapted from Sonenberg N & Hinnebusch AG [2009] *Cell* 136:731–745. With permission from Elsevier.)

family of eIF4E-binding proteins, which bind strongly to eIF4E, thus disassembling the eIF4F complex and repressing translation initiation. This mechanism is rather general, as there are many members of the eIF4E-binder family. In certain specific cases, this mechanism is superimposed on a mechanism that uses *cis*-acting elements in the 3′-UTR of the message and *trans*-acting factors that help in recruiting the eIF4E-binding proteins.

Historically, the first mechanism found to be involved in the regulation of eukaryotic initiation makes use of a common initiation factor, **eIF2α**, and its phosphorylation during the stress response. Repression of translation via eIF2α phosphorylation is the dominant pathway in the stress response because a variety of stress signals activate four different kinases, all of which phosphorylate eIF2α on Ser51 (**Figure 17.4**). Interestingly, and importantly, the mechanism does not repress translation of all mRNAs; the messages that code for proteins that must be present during the actual stress response are translated through an alternative cap-independent mechanism.

Initiation may utilize internal ribosome entry sites

Recall from Chapter 15 that certain mRNAs contain IRES in their 5′-UTR (see Figure 15.15). The nucleotide sequences of these sites form stable secondary structures that can be used to initiate translation through the binding of *trans*-acting factors and other proteins (**Figure 17.5**).

IRES may coexist with 5′-caps or may be the only elements that initiate translation. An example of an mRNA that contains both types of initiation elements, a cap and two IRES, is the human c-*myc* mRNA (**Figure 17.6**). This mRNA makes alternative use of the two IRES to initiate the translation of different open reading frames or ORFs that exist on a single mRNA. Actually, this arrangement is not exceptional; it is reminiscent of the polycistronic mRNAs that were once considered to be present only in bacteria. Small ORFs have been identified in the 5′-UTRs of a considerable number of RNA messages. Their functionality is incompletely defined. It is not even clear whether some are ever translated. The c-*myc* example, in which the two IRES elements control the translation of two alternative ORFs, at least proves that these additional ORFs are translated into proteins. Whether these proteins are functional or not remains to be seen, although simple logic would probably suggest that if such a sophisticated mechanism of translational regulation has evolved, more than one functional protein from a single message must be produced. There is, however, no assurance that evolution has followed any form of logic.

An even more interesting situation is encountered in the regulated translation of the two major regulators of apoptosis: X-chromosome-linked inhibitor of apoptosis, XIAP, and apoptotic protease-activating factor, Apaf1. The physiological, cellular, and molecular biology of **apoptosis**, also known as programmed cell death, is described in Box 18.2. Under specific stress conditions, both the negative regulator, XIAP, and the positive regulator, Apaf1, of apoptosis undergo a transition from cap-dependent

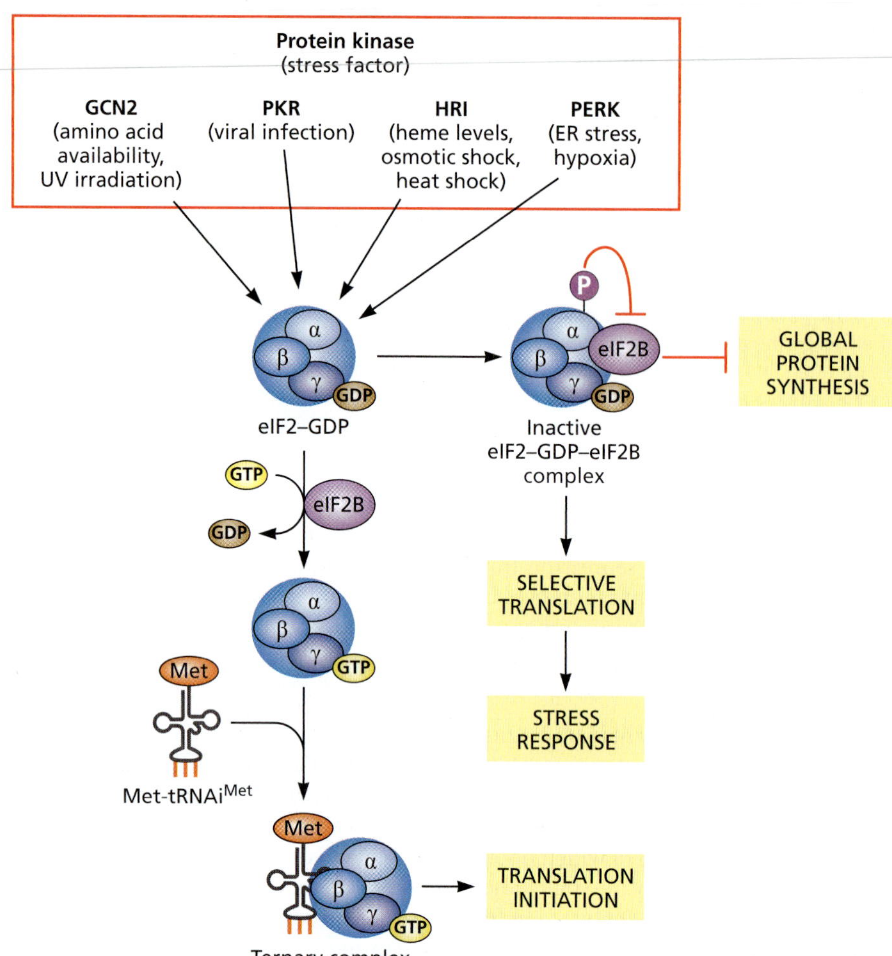

Figure 17.4 Phosphorylation of eIF2α integrates the translational responses of eukaryotic cells to various stress factors. Four distinct protein kinases—GCN2, general control non-derepressible 2; PKR, protein kinase RNA; HRI, heme-regulated inhibitor kinase; and PERK, protein kinase RNA-like endoplasmic reticulum (ER) kinase—sense different types of stress signals and phosphorylate eIF2α. eIF2α is one subunit of the three-subunit eIF2 complex. Together with GTP and the methionine-charged initiator tRNA, eIF2 forms the ternary complex that delivers the initiator tRNA to the ribosome during initiation. Phosphorylation of eIF2α inhibits GDP–GTP exchange by reducing the dissociation rate of nucleotide exchange factor eIF2B. As a result, global initiation is inhibited. However, selective translation of a subset of mRNAs continues; these messages code for proteins that mediate the stress response. (Adapted from Holcik M & Sonenberg N [2005] *Nat Rev Mol Cell Biol* 6:318–327. With permission from Macmillan Publishers, Ltd.)

to IRES-dependent initiation. One target of caspases, which are the proteases that transduce or execute death signals in apoptosis, is the familiar eIF4G, the scaffolding protein in the eIF4F complex. Its destruction shuts down all cap-dependent initiation. IRES-dependent initiation is, however, not affected by the caspase-mediated cleavage of eIF4G. Thus, during apoptosis, both XIAP and Apaf1 switch to IRES-dependent initiation and can thus be synthesized. In both cases, complex interaction networks lead to the appropriate outcome, cell survival or cell death, depending on the exact demands in each specific situation.

5′–3′-UTR interactions provide a novel mechanism that regulates initiation in eukaryotes

Recently, a novel and rather unusual mechanism has been described that uses the formation of dsRNA regions between complementary sequences in the 5′- and 3′-UTRs, thus circularizing the mRNA (**Figure 17.7**). These double-stranded regions bind to

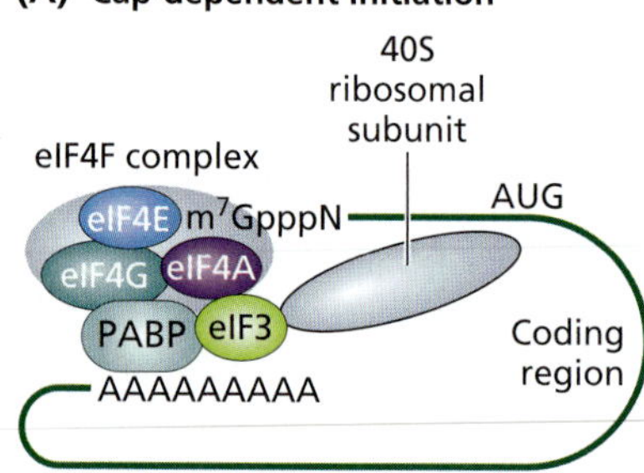

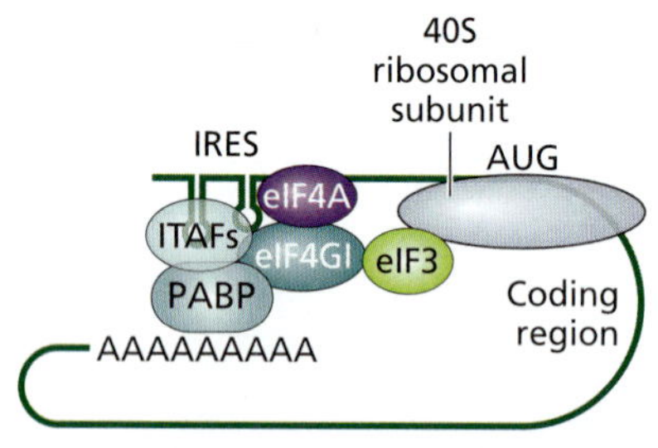

Figure 17.5 Comparison of cap- and IRES-dependent translation initiation in eukaryotes. (A) Cap-dependent initiation uses the cap structure at the 5′-end of mRNA to recruit the eIF4F complex, which consists of cap-binding protein eIF4E, RNA helicase eIF4A, and scaffolding platform eIF4G, to initiate the assembly of the pre-initiation complex. Poly(A)-binding protein PABP circularizes the mRNA through its interaction with eIF4G. (B) IRES-dependent translation initiation uses 5′-UTR sequences that form stable secondary stem–loop structures. IRES bind *trans*-acting factors, ITAFs, and proteolytic fragments of eIF4GI to initiate translation. (Adapted from Holcik M & Sonenberg N [2005] *Nat Rev Mol Cell Biol* 6:318–327. With permission from Macmillan Publishers, Ltd.)

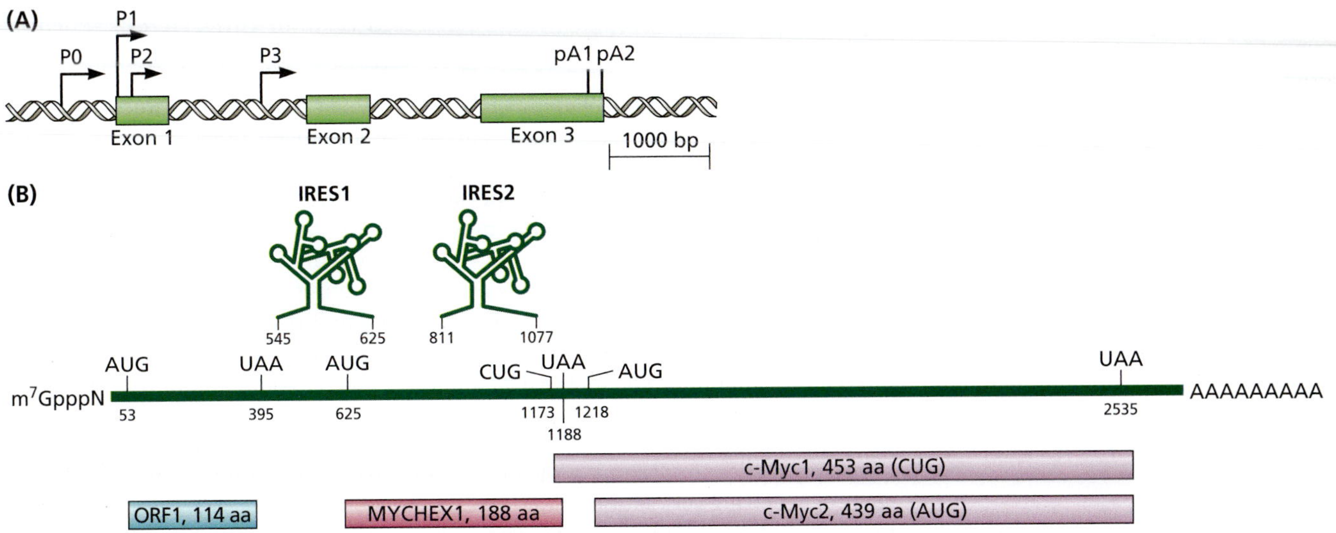

Figure 17.6 Structure of human c-*myc* genomic locus, mRNA transcribed from promoter 0, and protein products of its translation. The c-*myc* proto-oncogene encodes a transcription factor that regulates the expression of genes controlling cell proliferation, differentiation, and apoptosis. The gene is transcribed from four alternative promoters, P0–P3, with P1 and P2 accounting for 90% of c-*myc* transcripts in normal cells. In Burkitt lymphoma, transcription from P0 is greatly enhanced, reaching 100% of transcripts in some lymphoma cell lines. (A) Schematic of the locus, with locations of the three exons, four promoters, and two alternative polyadenylation sites pA1 and pA2. (B) Three open reading frames, ORFs, in the c-*myc* P0 mRNA. c-Myc1 and c-Myc2 are initiated from two closely situated initiation codons; curiously, the resulting proteins seem to have distinct roles in control of cell proliferation. Two ORFs and two IRES elements are present upstream of the c-*myc* sequences. IRES2 promotes cap-independent translation of the two c-Myc proteins. IRES1 allows for cap-independent initiation of the second ORF, MYCHEX1. Thus, c-*myc* P0 mRNA is a eukaryotic polycistronic mRNA that translates two different ORFs under the control of two independent IRES elements. (Adapted from Nanbru C, Prats A-C, Droogmans L et al. [2001] *Oncogene* 20:4270–4280. With permission from Macmillan Publishers, Ltd.)

specific proteins that stimulate translation. The specific example analyzed in Michael Kastan's laboratory focuses on regulation of the tumor-suppressor protein p53. The *p53* gene is involved in DNA damage responses and is mutated in more than 50% of sporadic human tumors; inheritance of germline *p53* mutations leads to an increased susceptibility to cancer. The increase in p53 protein levels following DNA damage and other cellular stresses has been largely attributed to stabilization of the protein, which is very unstable in nonstressed cells. In addition to being a tumor suppressor, p53 is an extremely important eukaryotic protein that plays a remarkable number of roles in the cell.

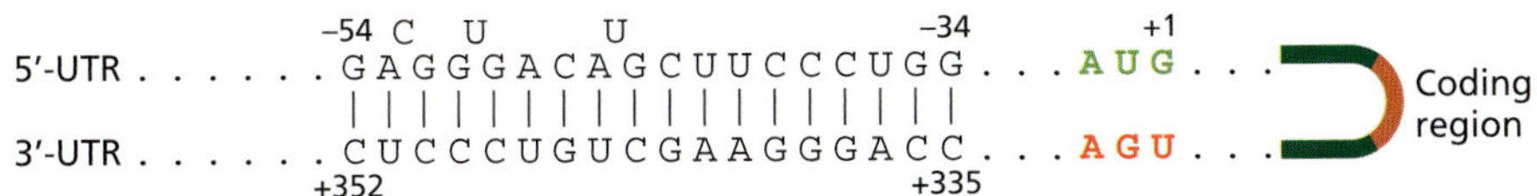

Figure 17.7 Regulation of tumor-suppressor protein p53 translation by 5′–3′-UTR interactions. In addition to the stabilization of p53 that occurs following DNA damage, there is an increased translation of *p53* mRNA that results from the binding of a specific protein, RPL26, to the 5′-UTR of the message, enhancing the association of *p53* mRNA with polysomes and thus its translation. RPL26 actually interacts with a double-stranded RNA region containing complementary sequences of the 5′- and 3′-UTRs. The initial clue to the existence of such a structure came from mathematical modeling aiming at predicting the minimum free-energy secondary structure of the message. Experimental validation came from mutating as few as three bases in either of the two complementary UTR sequences, which abrogated the ability of RPL26 to stimulate translation. Thus, this is a novel mechanism of regulating translation, which has the potential to modulate protein levels, by use of oligonucleotides that disrupt the double-stranded helix. (Adapted from Chen J & Kastan MB [2010] *Genes Dev* 24:2146–2156. With permission from Cold Spring Harbor Laboratory Press.)

Riboswitches are RNA sequence elements that regulate initiation in response to stimuli

Riboswitches are sequence elements in the 5′-UTRs of mRNAs that directly monitor physiologically relevant signals and subsequently regulate transcription or translation according to the needs of the cell (**Table 17.1**). They were first discovered in bacteria, where they are of major importance. Riboswitches that sense and react to elevated temperatures, known as thermosensors, are the simplest switches. Because the secondary structure of RNA is very sensitive to temperature, a simple stem–loop structure can perform the role of a sensor. At low temperature, the sensor adopts a conformation that masks ribosome-binding sites and prevents ribosome binding. At elevated temperature, local melting of secondary structure elements occurs, exposing the ribosome-binding site. The two best documented instances of thermosensors in action are found in heat-shock genes and virulence genes that are activated during pathogenic invasion of animal hosts. We have illustrated the translational activation of virulence genes in *Listeria*, a bacterial pathogen that causes food-borne infections, some of which are very serious (**Figure 17.8**). The fatality rate of *Listeria* infections may reach 25% of infected individuals; compare that to the mortality rate of *Salmonella* infections, which is usually less than 1%.

A more general class of riboswitches, which respond to metabolites, was described in 2002. Evgeny Nudler's laboratory at New York University, in collaboration with scientists in Russia, discovered that the leader region of nascent RNA chains in *Bacillus* can form structures that bind small metabolite molecules that are synthesized by proteins coded by the operons being transcribed. In other words, there is feedback regulation of these genes by the final products of their activity. The specific pathway they studied involves mechanisms akin to attenuation. In other words, it works through termination hairpin formation (see Chapter 11, specifically Figure 11.13). Ronald Breaker's laboratory at Yale, working on the same operon, this time in *E. coli*, reported that mRNAs that encode enzymes involved in thiamine or vitamin B1 biosynthesis can directly

Table 17.1 Major classes of riboswitches that participate in control of gene expression at the level of translation. Other riboswitches such as the tRNA-sensing or T-box riboswitches control gene expression at the level of transcription termination in a process akin to attenuation (see Chapter 11).

Class	Group	Natural ligand	Size (nt)	Occurrence
thermosensors			variable	phages, bacteria, eukaryotes
metabolites	coenzymes	thiamine pyrophosphate, TPP	100	bacteria, archaea, fungi, plants
		flavin mononucleotide, FMN	120	bacteria
		S-adenosyl-methionine, SAM	60–105	bacteria
		adenosylcobalamin, AdoCbl	200	bacteria
	amino acids	lysine	175	some bacteria
		glycine	110	bacteria
	bases	guanine, hypoxanthine	70	Gram-positive bacteria
		adenine	70	bacteria
magnesium		Mg^{2+}	70	Gram-negative bacteria

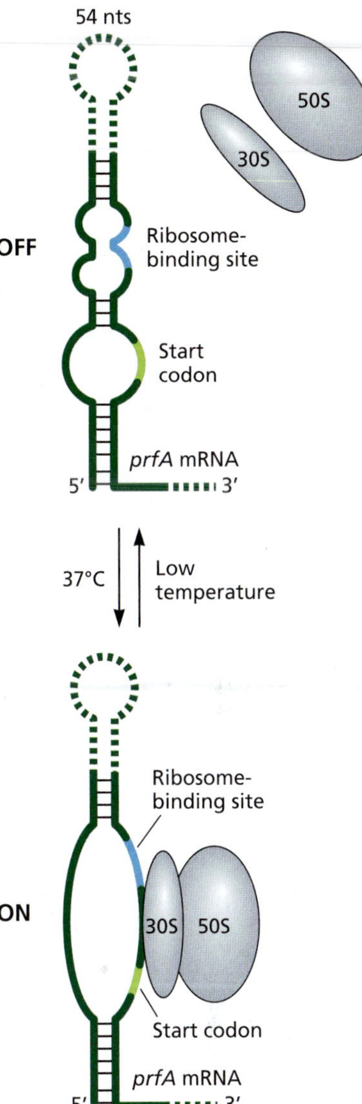

Figure 17.8 Action of a temperature-sensing riboswitch or thermosensor. Temperature sensing is exemplified here by the activation of virulence genes in the bacterial pathogen *Listeria monocytogenes*. Thermosensors are the simplest riboswitches described to date. They control multiple adaptation reactions to ambient temperature. A thermosensor can be as primitive as a stem–loop structure that bears the ribosome-binding site and the initiation or start codon. At low temperatures, that is, the off state, the conformation masks the ribosome-binding site, preventing ribosome binding. When the temperature is raised, that is, the on state, the structure melts locally, exposing the ribosome-binding site for ribosome binding and hence translation. The example presented is the *prfA* mRNA from the bacterium *L. monocytogenes*, which is known to be responsible for a rare but potentially lethal food-borne infection known as listeriosis. (Adapted from Serganov A & Patel DJ [2007] *Nat Rev Genet* 8:776–790. With permission from Macmillan Publishers, Ltd.)

bind thiamine, without the need for protein factors. The complex formed between the mRNA and the effector molecule sequestered the ribosome-binding site or Shine–Dalgarno sequence; thus the effect was clearly at the level of translation initiation.

Riboswitches that react to small metabolites constitute the largest category of riboswitches, sensing coenzymes, nucleotide bases, amino acids, and other small molecules or ions. Another, smaller category senses the concentration of certain ions, such as Mg^{2+}. In all cases, the signal alters secondary structure elements in the riboswitch, and these alterations allosterically change downstream RNA structures to execute the task at hand. Thus, riboswitches consist of two parts: a sensing region that is involved in recognition of the specific effector, and an expression platform that executes the command coming from the sensing region. The structural reorganization of the sensing region that occurs upon sensor binding results in the formation of specific structures in the expression platform.

The excitement in the field that was initiated by data from biochemical and phylogenetic analyses soon led to structural investigations by both NMR and crystallography. These structural studies were largely concerned with the very high specificity of recognition between the riboswitches and their small-molecule ligands. Structures for practically every major class of riboswitches complexed with their ligands are now available. Several conclusions have already been drawn: (1) The overall shape and relative orientations of the structural elements are defined by helical junctions (**Figure 17.9**). (2) Junctions and adjacent nucleotides directly contribute to regulation of gene expression by sequestering translational regulatory elements. (3) Riboswitches often involve helical bundles, formed by parallel RNA helices that are packed together by long-range interactions. It is believed that the highly specific recognition of even small ligands results from the many contacts formed by almost complete enveloping of the ligand by the RNA. We illustrate some of these principles using the example of the thiamine pyrophosphate-sensing riboswitch, the only riboswitch so far found in all three domains of life (see Figure 17.9).

It is important to note that, at the RNA level, riboswitches perform the kind of allosteric functions that we often associate with regulatory proteins. The capabilities of specific RNA structures are far broader than was initially imagined.

MicroRNAs can bind to mRNA, thereby regulating translation

In Chapter 14, we introduced **microRNAs** or **miRNAs** mainly from the viewpoint of their biogenesis. We also presented the two different ways in which miRNAs interact with their target sequences in the 3′-UTR of the mRNA that they regulate. Extensive base-pairing leads to destruction of the targeted mRNA molecules, whereas partial base-pairing leads to inhibition of translation (see Figure 14.28). In this chapter, we further discuss the types of miRNA target sites that exist in mRNAs and then the actual mechanism of miRNA action.

The abundance of miRNAs in metazoa has led to massive sequencing efforts. The availability of the resulting miRNA sequence database, in conjunction with similar databases of mRNA sequences, led to the recognition that there are several different types

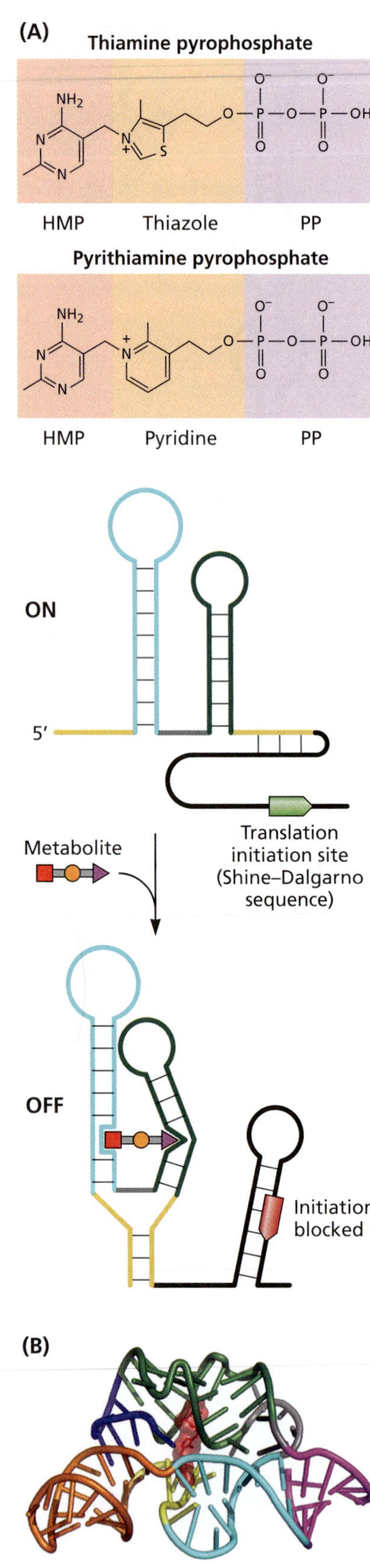

Figure 17.9 Action of a metabolite-sensing riboswitch. Metabolite sensing is exemplified by the thiamine pyrophosphate-specific riboswitch and its potential as a drug target. Thiamine pyrophosphate, TPP, is a coenzyme derived from vitamin B1. HMP is 4-amino-5-hydroxymethyl-2-methylpyrimidine. The translation of enzymes involved in TPP synthesis is under control of riboswitches in their mRNAs. (A) These riboswitches adopt alternative conformations, depending on whether TPP is bound or not. The structural elements of the metabolite-sensing or aptamer domain of the riboswitch are differently colored. The expression platform, shown in black, carries the signals, whose function is regulated by the TPP. In this case, these are nucleotide sequences in the mRNA that determine the point of initiation of translation. In the absence of bound TPP, that is, the on state, the sensing domain is folded to expose the initiation site, shown in green, and the message is translated. When TPP is bound, that is, the off state, the sensing domain folds into a hairpin that occludes the initiation site. Thus, a high concentration of TPP will inhibit more synthesis. The riboswitch acts as an on/off switch in a feedback mechanism. (B) Structural model of the TPP riboswitch sensing domain from *E. coli thiM* mRNA bound to its ligand. The elements are colored as in part A, with TPP shown in red. The sensing domain consists of two large helical domains that run in parallel and a short helix, P1, connected by a junction, shown in gray. TPP binds the riboswitch in an extended conformation and positions itself between and perpendicular to the helical domains, such that opposite ends of TPP are each bound into a specific RNA pocket. Thus, the TPP riboswitch can be considered as a molecular ruler that measures the length of the ligand; TPP analogs lacking one or both phosphates cannot reach far enough and cannot control translation effectively. The fact that the central part of TPP does not contribute to the specificity of recognition is being exploited in drug design. The compound pyrithiamine pyrophosphate, which differs from TPP in the central part only, can interact with the riboswitch, substituting for TPP, and thus can down-regulate the expression of thiamine-related bacterial genes, starving the microbe of TPP. The drug does not harm humans because no such riboswitch has been detected in the human genome. (A, adapted, courtesy of GR Kantharaj, Bangalore University. B, from Serganov A, Polonskaia A, Phan AT et al. [2006] *Nature* 441:1167–1171. With permission from Macmillan Publishers, Ltd.)

of miRNA target sites in mRNAs (Figure 17.10). The majority of sites belong to the canonical category, in which seven or eight nucleotides in the 3′-UTR interact by base-pairing with the seed region in the miRNA. A seed region is a short sequence at the 5′-end of miRNA that contributes most of the energy for target binding; that is, it determines the specificity of interaction between the miRNA and its target sequence in the mRNA (see Chapter 14). Although 6-mer sites also seem to exist, these are rejected by bioinformatics algorithms because of the large chance that these short sequences are conserved in the targets by chance. Supplementary sites contain, in addition to seed-sequence interactions, several adjacent base pairs. Finally, compensatory sites are longer and compensate for single-nucleotide bulges or mismatches in the seed region.

Once the miRNA is processed into its final mature form (see Figure 14.29), it interacts with a member of the Argonaute or Ago family of proteins. The number of these proteins varies with the organism: *Drosophila* has two Ago proteins, while humans have four. Individual representatives of the family may have slightly different functions and do not necessarily complement each other in some functions. Ago proteins are constituents of the **miRNA-induced silencing complex**, **miRISC**, which also contains protein GW182 and the **poly(A)-binding protein**, **PABP**. The domain structures of these three miRISC constituent proteins in humans are presented in Figure 17.11.

The important role of seed pairing with the target mRNA can be understood by looking at interactions of the miRNA seed sequence with the Ago protein (Figure 17.12). When Ago binds miRNA, it preorganizes the seed sequence to favor efficient base-pairing with the target region of the mRNA. The Ago-bound seed region adopts a single-stranded A-form helix, which strongly favors interaction with the single-stranded mRNA. Figure 17.13 depicts the conformational rearrangements that must occur in Ago to allow interactions beyond the seed sequence.

How does miRISC lead to silencing of gene expression? The topic is still unresolved, but it is believed that miRISC may act to silence gene expression via two different pathways that influence cap-dependent translation initiation. It may directly disrupt the PABP–eIF4G interactions, which leads to reduced recruitment of the pre-initiation complex to the mRNA. In a different pathway, the complex may favor deadenylation of the poly(A) tail. Deadenylation is the first step toward mRNA decay, via either

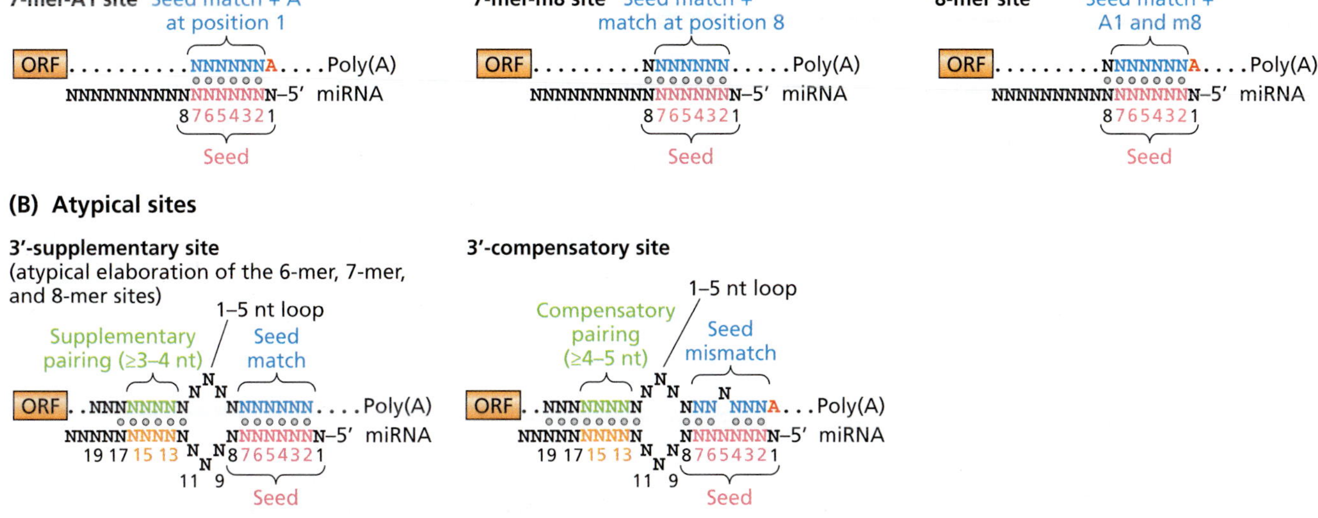

Figure 17.10 Types of miRNA target sites in mRNAs. Gray dots indicate contiguous Watson–Crick base pairing. (A) Canonical 7–8 nt seed-matched sites. (B) Atypical sites. In 3′-supplementary sites, the 3′ extra pairing optimally centers on nucleotides 13–16 of miRNA. The stability of this extra pairing is determined more by the pairing geometry than by predicted thermostability: at least three or four contiguous Watson–Crick pairs uninterrupted by bulges, mismatches, or wobbles should be present. Despite the expectation that supplementary sites may be frequent, since the additional base pairing outside the seed region would enhance both binding affinity and specificity, there is as yet no convincing evidence that these sites are widely used in *Drosophila* or mammals. Finally, pairing to the 3′-portion of the miRNA that is used to compensate for a single nucleotide bulge or mismatch in the seed region gives rise to what is known as 3′-compensatory sites. In these sites, the length of the additional paired region may be significant, four or more base pairs. (Adapted from Bartel DP [2009] *Cell* 136:215–233. With permission from Elsevier.)

5′ → 3′ or 3′ → 5′ decay pathways. Mechanistic details of this model of miRISC action are presented in **Figure 17.14**.

Finally, we must note that deregulation of miRNA biogenesis and/or function has profound effects on the proper functioning of cells and may lead to disease states. Two convincing examples have been recently reported that establish a connection between miRNAs and diabetes (**Box 17.3**). Bearing in mind the sheer number and abundance of miRNAs in the cell, it seems likely that many more such connections will be found.

17.4 mRNA stability and decay in eukaryotes

One factor that determines protein levels in the cell is the stability of their respective mRNAs. These mRNAs are constantly being synthesized and degraded, and messages that encode different proteins may have very different half-lives, ranging from minutes to hours and, in exceptional cases, even months. In Chapter 14, we introduced the basis of RNA degradation in both bacteria and eukaryotes. We discussed the conserved architecture of an important RNA degradation complex, the eukaryotic exosome and its equivalents in bacteria. Here, we focus on the sophisticated systems that eukaryotic cells use to degrade mRNAs at the end of their useful life; we also describe the pathways that degrade faulty messages. Although, strictly speaking, mRNA degradation cannot be viewed as regulation of the process of translation per se, it determines, together with direct translational regulation, the ultimate level of proteins in the cell. Hence, it is appropriate to discuss it here. The stability of mRNA is by no means a minor method of control; it has been estimated to influence up to 50% of gene expression changes responding to cellular signals.

The two major pathways of decay for nonfaulty mRNA molecules start with mRNA deadenylation

The two major pathways of degrading mRNA, 5′ → 3′ and 3′ → 5′, are summarized in **Figure 17.15**. Importantly, both pathways start with **deadenylation**, shortening of the

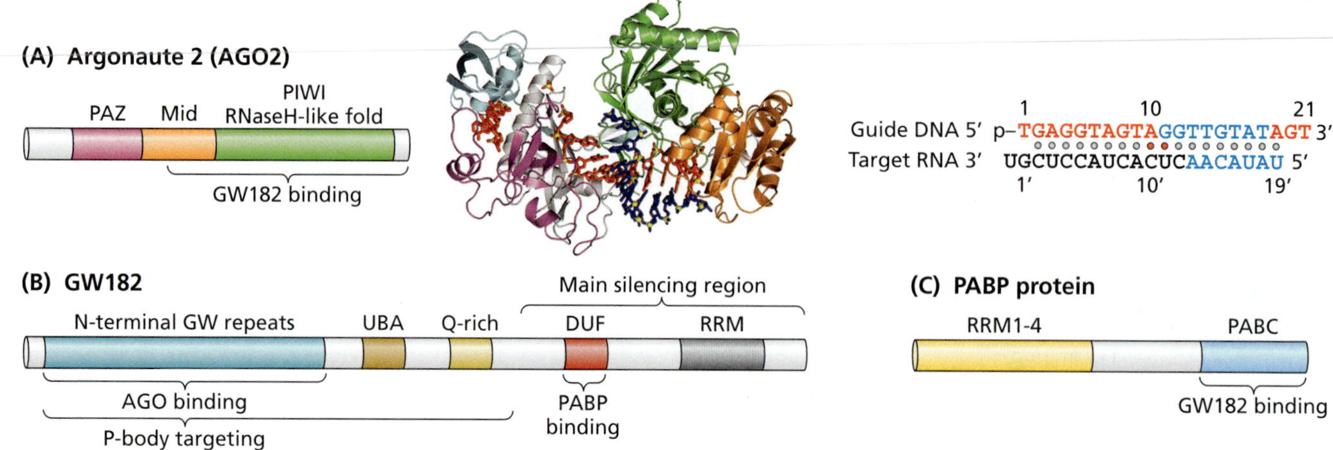

Figure 17.11 Domain structures of the three major protein components of miRISC in humans. (A) Argonaute 2, or AGO2, contains three evolutionarily conserved domains, PAZ or Piwi–Argonaute–Zwilli, Mid, and PIWI, which interact with the 3′- and 5′-ends of the miRNA. The PIWI domain is competent for endonucleolytic cleavages of RNA targets that show perfect base-pairing with the miRNA seed domain. The crystal structure is that of a bacterial *Thermus thermophilus* paralog of an Argonaute protein, complexed with a 21-nt DNA strand in red and with a 19-nt target RNA in blue. The structure identifies the DNA in a binding channel between the PAZ- and PIWI-containing lobes of the AGO scaffold. The two mismatches, shown with red dots, between the guide DNA and the target sequences were engineered to prevent the cleavage reaction, allowing crystallization of the complex. Disordered segments of the bases on both strands that could not be traced in the structure are shown in blue letters in the sequence. (B) GW182 has rather unusual GW, glycine-tryptophan, repeats in its N-terminal region. This region also contains ubiquitin-associated UBA domains and glutamine-rich or Q-rich domains and is responsible for targeting the protein to P bodies (see text). The C-terminal part contains a DUF motif and RNA recognition motifs, RRM, and is the effector domain that mediates translational repression and deadenylation of mRNA. The crystal structure of the DUF domain bound to the C-terminal domain of PABP has been resolved; mutations at the interface of the two proteins impair mRNA deadenylation in mammalian cell extracts, suggesting a contribution of the GW182–PABP interaction to miRNA gene silencing. (C) Poly(A)-binding protein, PABP, contains four RNA recognition motifs and a conserved C-terminal domain, which is implicated in miRNA silencing through its interactions with the GW182 effector domain. (Adapted from Fabian MR, Sonenberg N & Filipowicz W [2010] *Annu Rev Biochem* 79:351–379, and Wang Y, Juranek S, Li H et al. [2008] *Nature* 456:921–927. With permission from Annual Reviews and Macmillan Publishers, Ltd.)

poly(A) tail. Deadenylation is thus the major rate-limiting step. It is also important to note that deadenylation can be reversed; there are pathways by which shortened poly(A) tails can be extended. This implies a more direct role of the poly(A) tail in translational regulation.

Following deadenylation, the pathways diverge: the mRNA could be decapped and then degraded through the action of the abundant 5′ → 3′ exonuclease Xrn1; alternatively, it could be subjected to 3′ → 5′ exonucleolysis by the exosome (see Figure 14.25 and Figure 14.26 for a detailed description of exosome structure and function). Many mRNAs experience both 5′ → 3′ and 3′ → 5′ degradation following the initial deadenylation step. Both pathways involve decapping of the mRNA; this occurs at different stages of the process and is executed by different enzymes.

The 5′ → 3′ pathway is initiated by the activities of the decapping enzyme Dcp2

In the 5′ → 3′ pathway, the decapping enzyme **Dcp2** is extensively regulated, both positively and negatively, by numerous factors (see Figure 17.15). The domain structure of Dcp2 and the crystal structure of its complex with its positive co-factor Dcp1 are shown in **Figure 17.16**. This structure makes it clear why and how Dcp1 serves as a positive regulator: it reorients a portion of Dcp2 toward the active site, promoting a transition from an open to a more closed and active conformation that allows decapping of mRNA.

Other positive regulators act as *trans*-acting factors binding to *cis*-acting elements in the 3′-terminal portion of the mRNA. Two common activation mechanisms work through the AU-rich elements, ARE, in the 3′-untranslated regions of certain mRNAs (**Figure 17.17A**) and through oligo(U) tracts at the 3′-end of mRNA (**Figure 17.17B**).

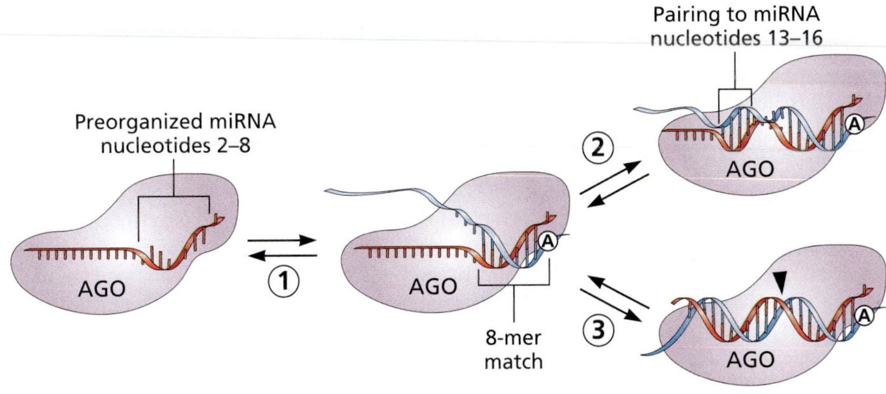

Figure 17.12 Putative structural models for recognition of target mRNA by the Argonaute–miRNA complex. MicroRNA is bound to AGO such that the bases of nucleotides 2–8 from the 5′-end of miRNA, the seed sequence, are preorganized to favor efficient base pairing with the target mRNA. AGO binding leads to the formation of an A-form single-stranded helix, which enhances both the affinity and specificity of the seed sequence, shown in red, for the matched mRNA segment, shown in blue. The length of the helix is optimal to provide sufficient interaction without excessive helical distortion. Nucleotide 1 is twisted away from the helix and is permanently unavailable for pairing; nucleotides 9–11 face away from the incoming mRNA and do not participate in nucleation of the interaction between the two RNA partners. (Step 1) Recognition of the seed site by target mRNA; note that the interaction is limited to the seed region, where a short double-helical structure forms. (Step 2) In the supplementary 3′-pairing, the mRNA can base-pair with nucleotides 13–16, forming a short helical segment; note that the miRNA and mRNA are base-paired but are not wrapped around each other into a double helix. (Step 3) Conformation of the extensively base-paired complex, which is a rare occurrence. When additional base-pairing occurs outside the seed region, nucleation occurs at the seed match and then spreads to other regions of the miRNA. In this two-state model for target RNA recognition, nucleation followed by propagation, substantial conformational accommodation must occur in AGO to allow extensive Watson–Crick pairing. This complex is suitable for cleavage of the mRNA. AGO locks down the extensively paired duplex, placing the active site, shown by a black arrowhead, in a position that allows cleavage of the mRNA. (Adapted from Bartel DP [2009] *Cell* 136:215–233. With permission from Elsevier.)

Both pathways involve the action of the LSm1-7 protein complex, a heteroheptameric ring. The history of how the LSm and Sm proteins were discovered, their structure, and the connection to numerous diseases, including systemic lupus erythematosus and prostate and breast cancers, are presented in **Box 17.4**. The same complex, LSm1-7, participates in an unusual pathway that degrades mRNAs at the end of S phase. Synthesis of most histones is tightly coupled to DNA replication, and thus their messages have to be destroyed at the end of S phase or whenever DNA replication is inhibited by drugs or other factors. The pathway of histone mRNA degradation is a variation of the U-tract mechanism depicted in Figure 17.17B.

There is still another mechanism that involves stem-loop structures toward the very 5′-end of the 5′-UTR. These elements enhance the recruitment of Dcp2 and thus stimulate decapping, presumably without additional protein involvement (**Figure 17.17C**).

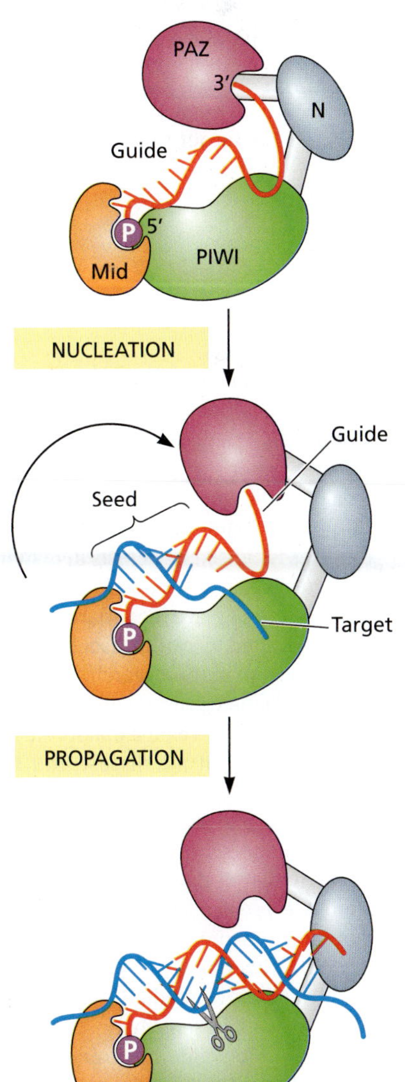

Figure 17.13 Conformational rearrangements in Argonaute that allow interactions beyond the seed sequence. Before target recognition, miRNA is presumably bound to the AGO protein along its entire length in order to prevent its nucleolytic degradation. Both termini of the miRNA, labeled here as the guide, are anchored in their respective binding pockets within AGO: the 5′-end is in a cleft between the PIWI and Mid domains, whereas the 3′-terminus is located in the PAZ domain, toward its N-terminus. After nucleation, the protein has to release its firm grip on the RNA to allow further spreading of the interaction. AGO undergoes a pronounced conformational change to a more open conformation. This is achieved by rotating the N-terminal and PAZ domains away from the lobe that contains the Mid and PIWI domains (curved arrow in middle panel). (Adapted from Jinek M & Doudna JA [2009] *Nature* 457:405–412. With permission from Macmillan Publishers, Ltd.)

(A) Active translation

(B) Deadenylation and inhibition of initiation

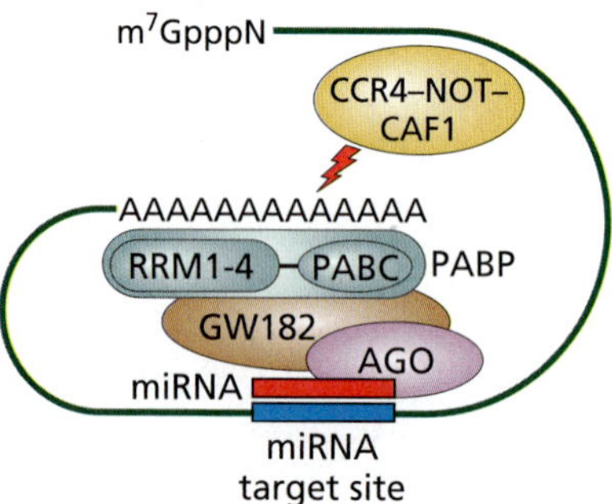

Figure 17.14 Interactions of protein components of miRISC in cap-dependent translational initiation that lead to miRNA-mediated gene silencing. (A) Cap-dependent initiation requires interaction of the cap-binding protein eIF4E, which is part of the eIF4F complex, with the 5′ mRNA cap. eIF4G plays the role of a protein scaffold; it binds to both eIF4A and PABP, thus promoting circularization of the template. Circularization enhances the recruitment of the 43S pre-initiation complex to the 5′-UTR of mRNA. (B) miRNA specifically base-pairs with a sequence in the target mRNA, bringing along the effector molecules of miRISC. miRNA interacts with AGO, which binds GW182; the DUF domain (see Figure 17.11B) of GW182 in turn interacts with the C-terminus of PABP, which sequesters the poly(A) tail to the vicinity of miRISC. The poly(A) tail is now properly located for deadenylation, mediated by the CCR4–NOT–CAF1 deadenylase complex, where CCR4 is carbon catabolite repression 4 protein; NOT1 is negative on TATA-less protein, and CAF1 is CCR4-associated factor. Both AGO and GW182 are required for miRNA-mediated deadenylation. Experimental tethering of GW182 to mRNA alleviates the requirement for AGO, indicating that AGO acts as a scaffold to recruit GW182, the actual effector protein for gene silencing. GW182 recruits the CCR4–NOT deadenylase complex. Another deadenylase often associated with the complex is CAF1. Deadenylation also requires direct interaction of GW182 with PABP. Following deadenylation, the mRNA cap is removed by the decapping DCP1–DCP2 complex, which leads to 5′ → 3′ exonucleolytic cleavage of the message by Xrn1. In addition, miRISC inhibits initiation due to disruption of the PABP–eIF4G interactions and reduced 43S pre-initiation complex recruitment. A possible block of initiation through antagonizing 60S subunit joining has also been suggested. (Adapted from Jinek M, Fabian MR, Coyle SM et al. [2010] *Nat Struct Mol Biol* 17:238–240. With permission from Macmillan Publishers, Ltd.)

The 3′ → 5′ pathway uses the exosome, followed by a different decapping enzyme, DcpS

Complete deadenylation makes the 3′-end of the mRNA accessible to the exosome, which then can degrade until close to the 5′-end; decapping is then carried out by DcpS, a scavenger enzyme, which can act only on relatively short RNAs that have been produced by the exosome. DcpS activity affects several other cellular processes such as cap-proximal pre-mRNA splicing. DcpS has been implicated in the etiology of spinal muscular atrophy, through negative regulation of expression of the *SMN2* gene (**Box 17.5**).

Box 17.3 MicroRNAs and diabetes Diabetes is a widespread chronic metabolic disease that affects millions of people worldwide. The underlying causes of the disease are defects in the production, secretion, and signaling pathways of insulin, a small protein hormone. Type 1 diabetes is due to self-destruction of the insulin-producing β cells in the pancreas; type 2 diabetes is caused by defects in insulin production and action. Diabetes is a very serious condition, especially because of the numerous and grave secondary complications that include, but are not limited to, cardiovascular disease and kidney failure.

Despite years of research, the molecular mechanisms underlying diabetes remain ill-understood. Recently, numerous reports have unraveled a major role for a variety of microRNAs in the etiology of the disease and its complications. Various miRs, as microRNAs are now designated for short, are critical regulators of diverse metabolic processes, such as insulin synthesis and secretion, glucose metabolism, and lipid metabolism. The list of diabetes-related miRs is growing by the day. Here, we present important examples from Markus Stoffel's laboratory at the Institute of Molecular Systems Biology in Zurich, Switzerland.

The two miRs studied were miR-103 and miR-107, which regulate insulin sensitivity. These two microRNAs differ by only one nucleotide at position 22 and are up-regulated in mice with either genetic or diet-induced obesity. Northern

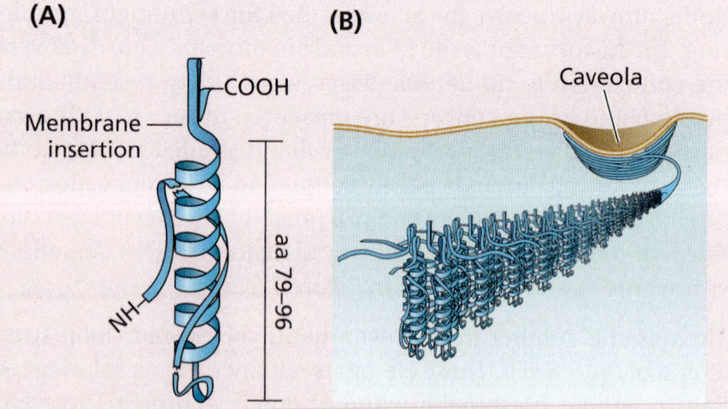

Figure 1 Structure of caveolin-1 in monomeric and heptameric forms. (A) Monomeric form: a model of the α-helix between amino acids 79 and 96 is shown. The region between amino acids 1 and 80 is shown wrapped around the helix; there are truncated portions at both ends of this region whose location is uncertain. (B) Heptameric form embedded in the plasma membrane: a model shows how the heptameric structures formed through lateral interactions between the α-helices of the monomers assemble further to make the 10-nm-thick filament of the caveolae coat. (Adapted from Fernandez I, Ying Y, Albanesi J & Anderson RGW [2002] *Proc Natl Acad Sci USA* 99:11193–11198. With permission from National Academy of Sciences.)

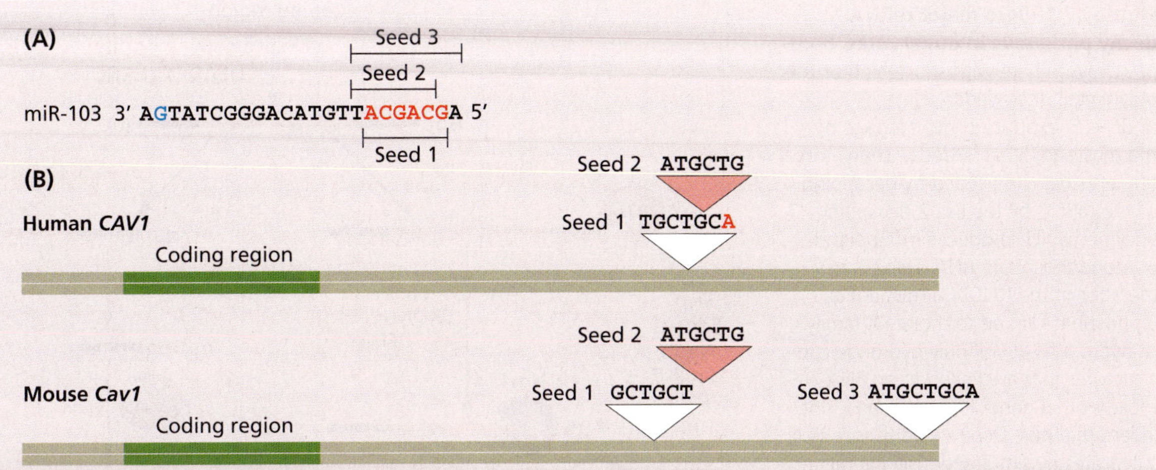

Figure 2 Caveolin-1 expression is regulated by miR-103 and miR-107. (A) Sequence of miR-103, with identified seed sequences. The sequence of miR-107 differs only in the second nucleotide, shown in blue, which is C instead of G. (B) Schematic of human and mouse caveolin mRNAs, with coding sequences and the location and sequence of seed targets in the 3′-UTRs. Human caveolin mRNA has two target sites, whereas mouse mRNA has three; the sequence presented by a red triangle is strictly conserved between the two species. Interaction of these seed targets with the seed regions in miR-103 and miR-107 leads to down-regulation of caveolin expression and destabilization of the insulin receptor. (Adapted from Trajkovski M, Hausser J, Soutschek J et al. [2011] *Nature* 474:649–653. With permission from Macmillan Publishers, Ltd.)

blot analysis indicated 2-3-fold up-regulation of miR-103 and miR-107 in the livers of both types of obese mice. These miRs were also elevated in liver biopsies of human patients, notably in those who suffered from conditions associated with diabetes. Injection of recombinant adenovirus that expressed miR-107 into wild-type mice produced high glucose levels and decreased insulin sensitivity. Conversely, miR-107 silencing enhanced insulin sensitivity.

Further experiments identified the direct target gene as *CAV1*, which encodes **caveolin-1**, a critical regulator of the insulin receptor. Caveolin-1 is the key protein component of caveolae, 50–100 nm invaginations in the plasma membrane first visualized by George Palade and Eichi Yamada in the 1980s. Caveolae are specialized lipid rafts containing specific lipids that perform a number of signaling functions, among them insulin signaling. The structures of the caveolin-1 monomer and heptamer are presented in Figure 1.

Figure 2 depicts the seed sequences in miR-103 and miR-107 that participate in molecular recognition of the target sequences in caveolin-1 mRNA. Figure 3 illustrates how caveolin-1 interacts with the insulin receptor. Stoffel and coauthors suggest that the miR-103/miR-107-mediated down-regulation of caveolin-1 expression affects insulin signaling through this interaction: a reduced level of caveolin diminishes the number of insulin receptors in caveolae-enriched plasma membrane microdomains, thus reducing downstream insulin signaling.

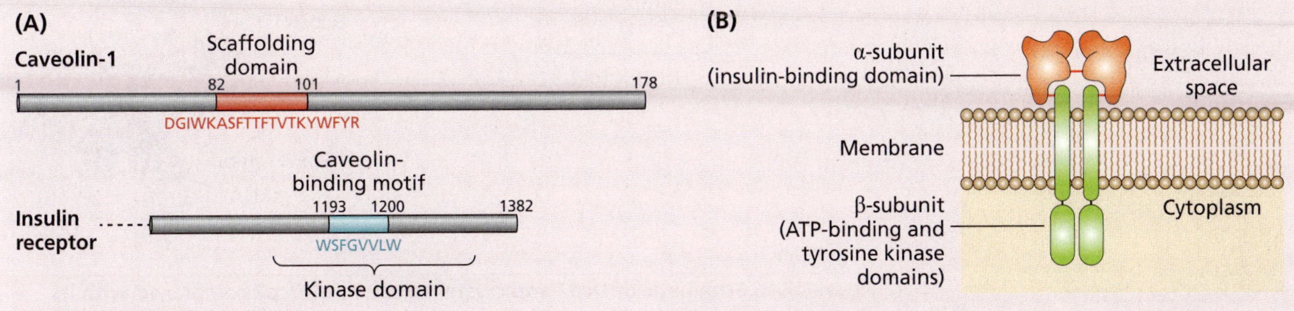

Figure 3 Schematic of the proposed caveolin-1–insulin receptor interaction. (A) Residues 82–101 of caveolin-1 represent the caveolin scaffolding domain, which mediates protein–protein interactions between caveolin-1 and numerous signaling molecules, including the insulin receptor. The insulin receptor, not shown to scale, contains a characteristic caveolin-binding amino acid motif, which is found within most caveolin-interacting proteins. The insulin receptor precursor 1–1382 undergoes post-translational modification to remove the signal sequence, 1–27, and is then cleaved into the α-subunit, 28–758, and the β-subunit, 763–1382. The kinase domain occupies positions 1023–1298. (B) Overall structure of the insulin receptor, which consists of two α- and two β-subunits connected by disulfide bridges, shown as red bars. The α-subunits occupy the extracellular space, where they interact with the ligand, insulin. The β-subunits contain the transmembrane helix and the kinase domain. Upon insulin binding, the protein undergoes autophosphorylation and then phosphorylates numerous signaling molecules. (A, adapted from Cohen AW, Combs TP, Scherer PE & Lisanti MP [2003] *Am J Physiol Endocrinol Metab* 285:E1151–E1160. With permission from American Physiological Society. B, adapted, courtesy of A. Malcolm Campbell, Davidson College.)

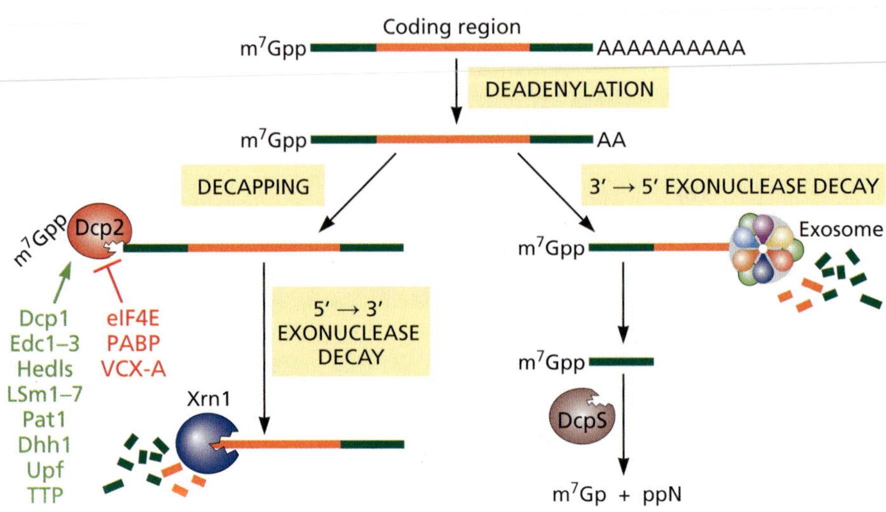

Figure 17.15 Two major mRNA decay pathways in eukaryotes. Both pathways initiate with shortening of the poly(A) tail, followed by either 5′ → 3′ or 3′ → 5′ exonucleolytic cleavage of the message. 5′ → 3′ decay, shown on the left side, is preceded by decapping of the mRNA by the Dcp2 decapping enzyme, which produces m^7Gpp and 5′-monophosphate RNA. Dcp2 is a member of the NUDIX or nucleotide diphosphate linked to moiety X family of hydrolases; it will only hydrolyze cap structures that are linked to an RNA moiety of, in general, more than 25 nt in length. Thus, Dcp2 recognizes mRNA substrates by simultaneously interacting with their cap and the RNA body. Dcp2 is regulated by numerous proteins, either positively, shown in green, or negatively, shown in red. Once mRNA is decapped, it is further degraded by the 5′ → 3′ exonuclease Xrn1. 3′ → 5′ decay, shown on the right side, involves the exonucleolytic action of the exosome. When the length of the RNA body becomes less than 10 nt, decapping occurs through the action of the DcpS or scavenger decapping enzyme. (Adapted from Li Y & Kiledjian M [2010] *Wiley Interdiscip Rev RNA* 1:253–265. With permission from John Wiley & Sons, Inc.)

There are additional pathways for mRNA degradation

For completeness, we must note that there exist additional, less frequently used pathways for mRNA degradation. In a few cases of specific messages, poly(A) degradation can be bypassed and decapping can proceed directly. A potentially very efficient mode of degradation involves endonucleolytic cleavage of an mRNA by specialized endoribonucleases. Degradation then proceeds from both of the new ends, in both 5′ → 3′ and 3′ → 5′ directions. Such enzymes must, obviously, be under stringent control.

Unused mRNA is sequestered in P bodies and stress granules

In eukaryotic cells, translationally repressed or nontranslating mRNAs often accumulate. These can be sequestered in special cytoplasmic aggregates, of which there are

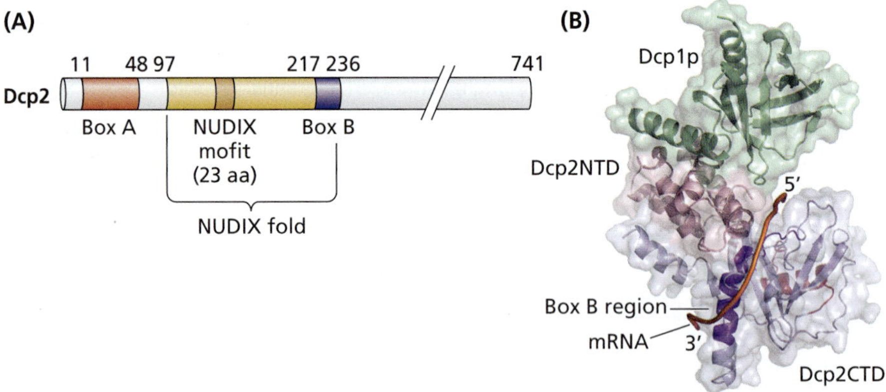

Figure 17.16 Schematic of Dcp2 and crystal structure of Dcp2 complexed with its positive co-factor Dcp1. The structure shown is from *Schizosaccharomyces pombe*. (A) Dcp2 contains three conserved regions: the NUDIX motif contains the catalytic site; box A is important for the catalytic fidelity of decapping, to generate m^7GDP exclusively and not m^7GMP, and for interaction with Dcp1; and box B is a part of the NUDIX fold and is important for RNA binding. (B) Surface view of the Dcp1–Dcp2 complex shows the putative RNA-binding channel with a modeled 12-mer RNA, indicating the proposed path of the RNA body; box B is shown in purple. The model explains the preference for longer RNA substrates: the RNA needs to be at least 12 residues long in order to bind to both the active site and box B. Dcp1 is believed to stimulate Dcp2 by orienting the N-terminus of Dcp2 (Dcp2NTD) toward the active site, promoting a transition from an open conformation to a more active closed conformation. (A, adapted from Wang Z, Jiao X, Carr-Schmid A & Kiledjian M [2002] *Proc Natl Acad Sci USA* 99:12663–12668. With permission from National Academy of Sciences. B, from She M, Decker CJ, Svergun D et al. [2008] *Mol Cell* 29:337–349. With permission from Elsevier.)

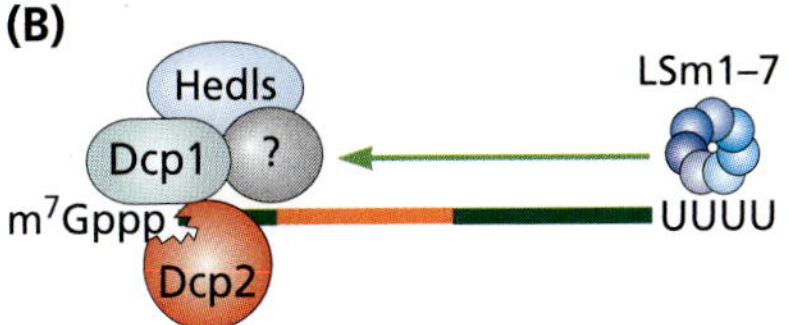

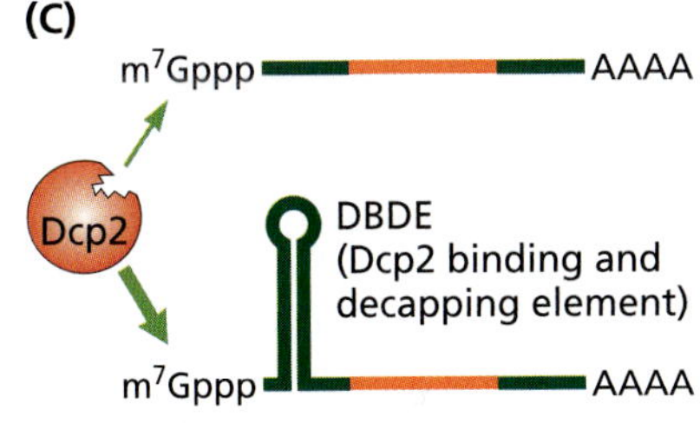

Figure 17.17 *Cis*-acting RNA elements that promote activation of Dcp2. (A) Activation of decapping by AU-rich elements, ARE, found in the 3′-untranslated region of certain mRNAs, mainly those encoding transcription factors and proteins that promote cell proliferation. AREs act through binding of a set of proteins, TTP being one of them, that directly or indirectly recruit Dcp2. LSm1–7 (see Box 17.4) may be involved in the recruitment. (B) Activation of decapping by a short U-tract at the 3′-end of mRNA. U-tracts can be generated following miRNA-directed cleavage of mRNA in plants and mice; these U-containing fragments correspond to the 5′ cleavage product. Stimulation by U-tracts is mediated by LSm1–7 that binds the U-tract. An important example of the involvement of U-tract/LSm1 interactions in mRNA decay is the histone mRNA decay that occurs at the end of the S phase, upon inhibition of DNA synthesis. (C) mRNA molecules that contain a stem–loop structure within the first 10 nucleotides recruit Dcp2 more efficiently to stimulate decapping. (Adapted from Li Y & Kiledjian M [2010] *Wiley Interdiscip Rev RNA* 1:253–265. With permission from John Wiley & Sons, Inc.)

two types: **P bodies**, or **processing bodies**, and **stress granules**. P bodies contain the major components of the RNA degradation machinery, especially those of the 5′ → 3′ pathway, and are believed to be the actual sites of mRNA decay. P bodies also contain proteins involved in specific mRNA degradation pathways, such as **nonsense-mediated decay** or NMD, ARE-mediated mRNA decay or AMD (see Figure 17.17A), and miRNA-directed gene silencing. Stress granules contain, among other proteins, translation initiation factors. An image of a cell that has been subjected to oxidative stress and contains both P bodies and stress granules is presented in **Figure 17.18**. The figure also lists the major protein components of the two particles.

The presence of P bodies and stress granules and the accumulation of mRNAs within them, as well as the presence of free mRNAs in the cytoplasm, reflect a dynamic flow

Box 17.4 Sm proteins and the like: The connection to systemic lupus erythematosus The first **Sm protein** was discovered in a patient with systemic lupus erythematosus, SLE, who was diagnosed in 1959 and died of the disease in 1969 at the age of 22. SLE is a serious autoimmune disease in which the immune system produces antibodies against the body's own macromolecules, mainly double-stranded DNA and histones but also a variety of other proteins. The patient was found to be producing an antibody against a then-unknown protein, which was named Sm after her. As it turned out, ~30% of SLE patients produce antibodies to that same protein or its very close relatives, LSms or like-Sm proteins.

Sm proteins and the like constitute a highly conserved family of proteins, with paralogs found in bacteria and archaea, that

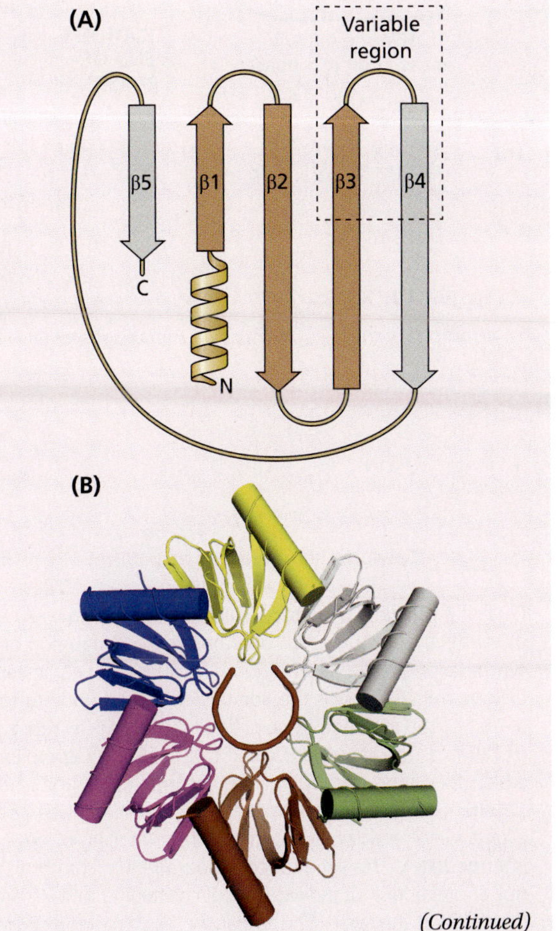

Figure 1 Structure of LSm1–7 heteroheptameric complex. (A) Secondary structure of a single protein subunit. It is a small five-strand anti-parallel β-sheet, with the β-strands forming a barrel-like structure; at the N-terminus there is a short α-helix of 2–4 turns. Two sequence motifs have been identified through sequence comparisons among members of the LSm family. The first motif SM1, shown in light brown, is 32 amino acids long and corresponds to β1, β2, and β3 strands; motif SM2, shown in light gray, is 14 amino acids long and corresponds to β4 and β5 strands. These two motifs are separated by a nonconserved region of variable length in different representatives of the protein family. (B) Model of the bacterial Hfq hexamer torus, with each protein subunit presented in a different color; the brown wire arc represents a short RNA oligonucleotide bound close to the central hole. Hfq belongs to the large family of Sm and LSm proteins. The eukaryotic LSm1–7 rings contain seven different subunits. (A, adapted, courtesy of Robert Plaag, Wikimedia.)

(Continued)

Box 17.4 *(Continued)*

function in numerous aspects of RNA metabolism. Sm forms stable complexes with the U-rich small nuclear RNAs that are part of the spliceosome. LSms bind to a broad spectrum of RNA molecules and influence the fate of these molecules. Two distinct heteroheptameric ring complexes exist that differ in only one of the seven proteins forming the ring structure (**Figure 1**). The first complex, LSm2–8, localizes to the nucleus and functions in processing pre-mRNA, pre-tRNA, and pre-rRNA. The second complex, LSm1–7, is located in the cytoplasm and plays a key role in mRNA decay. Its role is especially prominent in the pathways that involve *cis*-acting activating elements (see Figure 17.17).

The LSm1–7 complex also plays a role in the decay of histone mRNAs; these mRNAs are the only messages in eukaryotes that are not polyadenylated (**Figure 2**). The levels of mRNAs encoding all five classes of histones are tightly coupled to DNA replication, which occurs in S phase. The half-life of the messages is altered rapidly and in coordination in response to changes in the rate of DNA replication, so as to ensure balanced availability of DNA and histones to form chromatin. Regulation of the half-life of histone mRNA is mediated by a unique conserved sequence at the 3′-end of all histone messages; this sequence forms a short stem–loop structure that binds a specialized protein, stem–loop-binding protein or SLBP.

Interestingly, a connection has been found between LSm1 and human cancers, such as prostate and breast cancer. The expression of LSm1 was found to be down-regulated in prostate cancer cells. The breast cancer connection involves ~15–20% of breast cancers that have amplification of a chromosomal region, 8p11–12, that includes the gene for the LSm1 protein. Two major results have confirmed the oncogenic role of LSm1. First, overexpression of the gene in normal mammary gland epithelial cells led to the malignant transformation of these cells; in other words, it conferred an ability to proliferate in a growth-factor-independent way and to form colonies on soft agar plates. Second, inhibition of LSm1 production in a breast cancer cell line led to a dramatic decrease in colony growth on soft agar. Further microarray analysis indicated numerous genes with altered expression as a result of elevated levels of LSm1, including genes with recognized roles in cell-cycle regulation and cell proliferation.

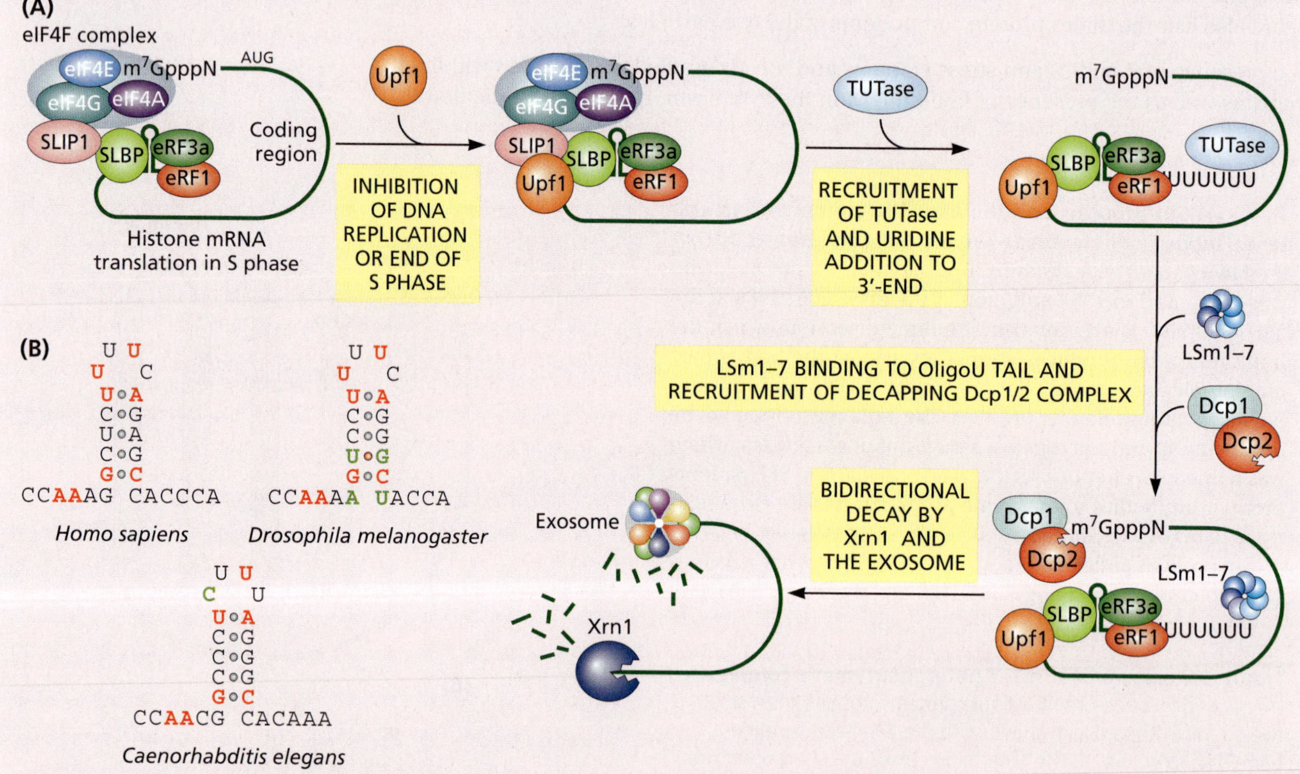

Figure 2 Model for histone mRNA degradation. (A) Sequence of events that occurs upon pharmacological inhibition of DNA synthesis or at the end of a normal S phase. An early step in histone mRNA decay is the addition of several uridines to the 3′-end of the message by specific terminal uridylyltransferases or TUTases. The U-tract is bound by the LSm1–7 complex, which stimulates decapping. The decapped message is then simultaneously degraded from the 5′- and the 3′-ends. These events were revealed by cloning and sequencing of decay intermediates. This simultaneous degradation from both ends is not unprecedented: individual ARE-containing mRNAs are also degraded in this way. (B) Comparison of stem–loop sequences and structures in three model organisms: humans or *Homo sapiens*, the fruit fly *Drosophila melanogaster*, and the nematode worm *Caenorhabditis elegans*. Nucleotide positions that are crucial for SLBP binding in mammals are shown in red. Deviations from the consensus stem–loop are shown in green. Watson–Crick base pairing is indicated by gray dots; the red dot in the *Drosophila* sequence marks the absence of such base pairing. (A, adapted from Mullen TE & Marzluff WF [2008] *Genes Dev* 22:50–65. With permission from Cold Spring Harbor Laboratory Press. B, adapted from Marzluff WF, Wagner EJ & Duronio RJ [2008] *Nat Rev Genet* 9:843–854. With permission from Macmillan Publishers, Ltd.)

Box 17.5 A Closer Look: Decapping, X-linked mental retardation, and spinal muscular atrophy mRNA decapping is an important regulator of gene expression, now linked to at least two neurological disorders, X-linked mental retardation and spinal muscular atrophy. In both disorders, the defects lie in genes or proteins that affect the function of the two decapping enzymes, Dcp2 and DcpS.

X-linked mental retardation is a complex disorder usually characterized by three features: generally impaired intellectual functioning with IQ < 70, limitations in social and interpersonal skills and the ability to live independently, and onset before the age of 18 years. The disease affects 2–3% of the population in developed countries. The genetic cause of mental retardation has been recognized as a deletion of the *VCX-A* gene, which encodes the VCX-A or variable charge X-linked protein. This protein is highly basic, and the second exon contains 30-bp repeats of unknown function whose number varies between 1 and 14 even among normal individuals. The protein is ubiquitously expressed in primate tissues, including the brain, and binds to a subset of mRNA molecules involved in neuronal development. Interestingly, no obvious orthologs have been recognized in lower mammals or other organisms.

The function of VCX-A was derived from ingenious experiments in which the protein was expressed in rat primary hippocampal neurons, cells that do not express the protein because it occurs naturally only in primates. Note that VCX-A is only one member of the VCX/Y protein family, which is encoded by four distinct genes on the X chromosome and two identical genes on the Y chromosome. In these experiments, neurite projections were promoted in the rat neurons; moreover, knock-down of the genes in a human neuroblastoma cell line inhibited neuronal cell differentiation. At the molecular level, VCX-A specifically binds the 5′-end of capped mRNA to prevent Dcp2-mediated decapping; recall from Figure 17.15 that VCX-A is one of the three known negative regulators of Dcp2. Interestingly, this binding increases upon association with Dcp2. The cap-binding property of VCX-A also inhibits translation initiation, most likely by competing with the cap-binding initiation factor eIF4E for cap access. The resulting nontranslating mRNAs are sequestered in stress granules (see Figure 17.18).

Spinal muscular atrophy or **SMA** is a disorder that affects the control of muscle movement. There are numerous types of the disease that affect mostly young children, but adults are not spared either. SMA can be inherited, depending on the type, in an autosomal recessive pattern, an autosomal dominant manner, or an X-linked pattern. The disease occurs in 1 in ~6000 individuals. The clinical manifestation of SMA is atrophy of the muscles used for walking, sitting up, and controlling head movement. In severe cases, the muscles involved in breathing and swallowing can be affected, usually leading to death. At the cellular level, SMA is characterized by loss of motor neurons, which are specialized nerve cells in the spinal cord and the brain stem.

At the molecular level, in the autosomal recessive pattern of inheritance cases, SMA is caused by deletion or mutation of both copies of the *SMN1* gene, which codes for the essential SMN1 or survival motor neuron protein. In the human genome, there is a second copy of the gene, *SMN2*, which is transcriptionally active, but most of the time it is spliced incorrectly as a result of a single nucleotide mutation. This mutation leads to the production of mRNAs that lack exon 7, so that ~90% of the translated proteins are truncated and nonfunctional. The remaining 10% of the SMN2 pre-mRNAs are spliced correctly to produce wild-type full-length protein molecules. The presence of these proteins partly compensates for the missing SMN1 protein and thus modulates the severity of the disease.

Scientists are looking for ways to increase the expression of the human *SMN2* gene to levels that would help overcome the *SMN1* genetic defects. In these efforts, pharmaceutical companies have joined forces with university researchers to search for small-molecule compounds that increase *SMN2* expression. A cell-based reporter assay was utilized to study a hybrid cell line, between a mouse spinal cord cell and a mouse neuroblastoma, that could be propagated indefinitely in culture. The hybrid cells were transformed with a construct that contained a fragment of the human *SMN2* promoter linked functionally to the bacterial β-lactamase gene, whose expression can be easily and conveniently monitored. The resulting cell line was screened for increased β-lactamase activity after stimulation with nonselective histone deacetylase inhibitors, which are known to increase the transcription of 2% of all genes, including *SMN2*. Once it was established that the construct and the cell line behaved as expected, >550,000 compounds were tested for up-regulation of the gene promoter. A quinazoline compound was identified that doubled SMN2 levels. The surprise came when protein microarray scanning with a radiolabeled quinazoline probe detected DcpS, the scavenger decapping enzyme (see Figure 17.15), as a binder of the compound. Quinazoline binding holds the enzyme in an open, catalytically incompetent conformation, thus inhibiting decapping.

In discovering the link between SMA and DcpS, researchers had pursued a complex chain of connections: the disease, the *SMN1* and *SMN2* genes, the idea that overexpressing the *SMN2* gene may help to relieve the symptoms, the identification of a compound that caused *SMN2* overexpression in isolated cells, and the identification of its molecular target as DcpS. The path is surely convoluted but it led to the identification of DcpS as a novel therapeutic target for treatment of SMA. This is, of course, just the first step in a long process toward creating a drug for treatment of humans.

of mRNAs that occurs between the cytoplasm and these bodies; each of these cellular aggregates contains mRNAs in a different state and complexed with different proteins. The mRNA cycle through P bodies and stress granules is depicted in **Figure 17.19**. Accumulation of transiently unused mRNAs into P bodies, and then into stress granules, serves as a mechanism to store and protect some mRNAs for future use. Thus, some of the mRNA molecules stored in P bodies are degraded, whereas others are rescued for future use. It should be noted that mRNA molecules in a particular state do not necessarily form microscopically observable bodies. mRNAs can exist in a nontranslating form in the cytoplasm without being aggregated into morphologically distinct structures.

Stress granule (red dots) major components
poly(A)-mRNA
Translation initiation (40S, eIF4E, eIF4G, eIF3, eIF2)
Translation control (CPEB, PABP, DHH1)
mRNA decay (DHH1, Staufen)
snRNP assembly, RNA processing, Ubiquitin ligase
Scaffold (FAST)

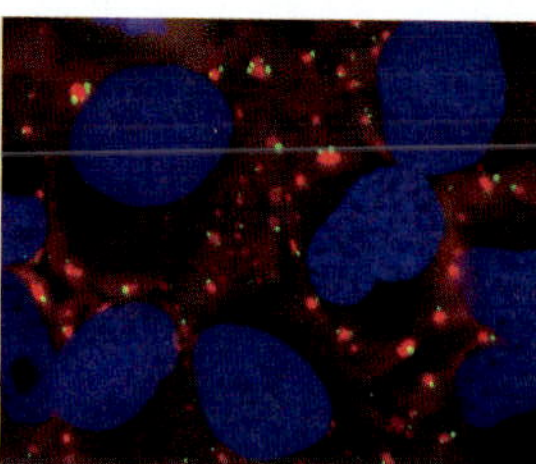

P body (green dots) major components
mRNA
5'→ 3' exonuclease (XRN1)
Deadenylation (CCR4, CAF1, NOT1–4)
Decapping (LSM1–7, DCP1–2, PAT1, DHH1)
Translation control (eIF4E, eIF4E-T, PAT1, DHH1, CPEB)
Nonsense-mediated decay (SMG5,7, UPF1)
miRNA pathway (miRNA, AGO1–4, GW182)
Scaffold (FAST)

Figure 17.18 Stress granules and processing bodies. Immunofluorescence image of human HeLa cells that were subjected to oxidative stress. The cells were fixed and stained with a polyclonal anti-eIF3 antibody, red, and with an antiserum that recognizes Dcp1, green; nuclear DNA was stained with Hoechst dye, blue. The yellow spots denote transient docking, or partial co-localization, of some stress granules and P bodies. Stress granules contain aggregates of translationally stalled poly(A)-mRNA; pre-initiation complexes, including the small 40S ribosomal subunit; and various RNA-binding proteins involved in regulation of translation and decay of specific mRNAs. Mitotic cells fail to form stress granules or P bodies when under stress. (Image, from Sivan G, Kedersha N & Elroy-Stein O [2007] *Mol Cell Biol* 27:6639–6646. With permission from American Society for Microbiology.)

Cells have several mechanisms that destroy faulty mRNA molecules

The eukaryotic cell attempts to control rigorously the quality of all three types of RNA molecules that play a role in the process of translation: mRNAs, tRNAs, and rRNAs. It has evolved specific mechanisms to deal with every possible faulty situation. As described in Chapter 14, surveillance of mRNA processing is carried out in the nucleus. Nevertheless, incorrect RNA molecules may be generated, exported to the cytoplasm, and enter the translation process. Here, we outline the three most frequent mechanisms for destroying mRNA molecules that have specific defects.

mRNA molecules that contain premature stop codons are degraded through nonsense-mediated decay or NMD

Premature stop codons have long been thought to produce truncated proteins; currently, we believe that messages containing premature stop codons are destined to be degraded at the very first round of translation, without causing the accumulation of truncated and potentially toxic proteins. Up to a third of alternative splicing events in humans create a premature stop codon and elicit NMD; thus many researchers are of

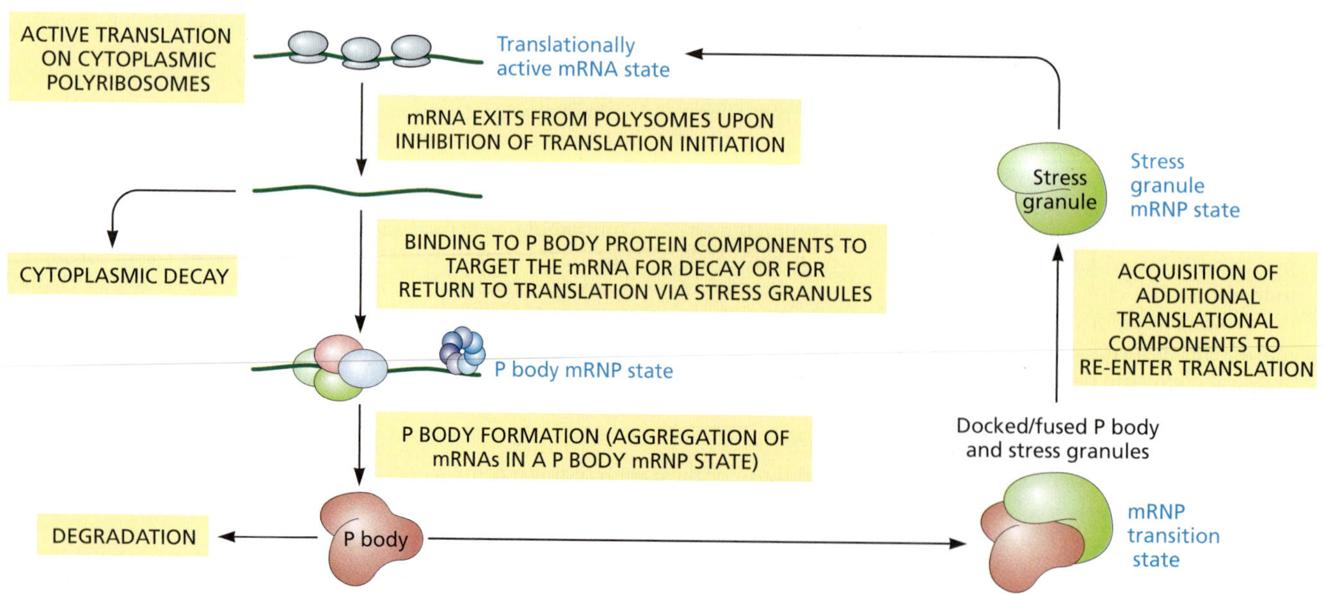

Figure 17.19 Flow of mRNAs through three distinct cytoplasmic states. The three states depicted are a translationally active state in polysomes, the P body mRNP state, and the stress granule mRNP state. Transitions from one state to the other are accompanied by remodeling of the RNP complexes: dissociation of some protein components and association of others. (Adapted from Buchan JR & Parker R [2009] *Mol Cell* 36:932–941. With permission from Elsevier.)

(A) First (pioneering) round of translation on normal mRNA

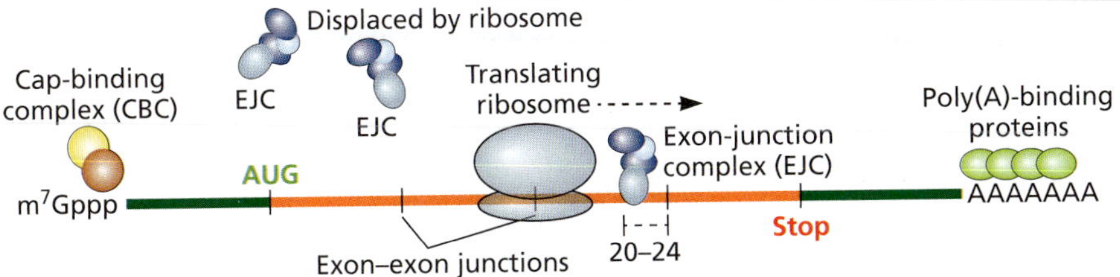

(B) Pioneering round of translation on mRNA containing a premature stop codon

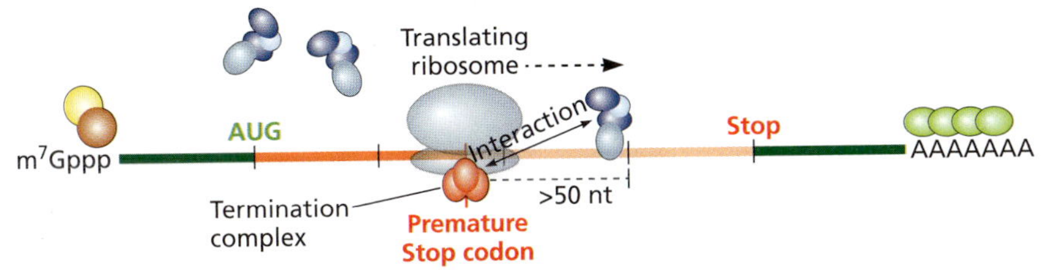

Figure 17.20 Molecular mechanisms of mammalian nonsense-mediated decay or NMD. NMD occurs during the first, pioneering round of translation. A premature termination codon is recognized by its spatial relationship to an exon–exon junction. (A) During the translation of normal mRNA, the ribosome proceeds along the polynucleotide chain, displacing the exon-junction complexes, EJC, bound ~20–24 nt upstream of exon–exon junctions. When the ribosome reaches a normal stop codon, a termination complex forms and translation is terminated. (B) If the mRNA contains a premature stop codon, the translating ribosome recognizes it as such and forms a termination complex at this position. The next exon–exon junction and the EJC normally bound ~20 nt upstream of it are now >50 nt downstream of the termination complex, allowing direct interaction between the premature termination complex and the EJC, which in turn initiates the decay reaction. In addition, the process involves the conserved Upf proteins and their interaction with the termination complex. Depending on the organism and cell type, NMD can target aberrant mRNAs for decapping and 5′ → 3′ degradation by the exonuclease Xrn1p, endonucleolytic cleavage, or accelerated deadenylation and 3′ → 5′ degradation by the exosome. (Adapted from McGlincy NJ & Smith CWJ [2008] *Trends Biochem Sci* 33:385–393. With permission from Elsevier.)

the opinion that the appearance of premature stop codons does not represent random noise in the splicing process but instead functions as a means of post-transcriptional gene regulation. A strong argument supporting this view is the fact that knockout mice lacking a major component of the NMD pathway die in the uterus. The process involves the **exon-junction complexes** or **EJC** that become bound to exon–exon junctions during splicing (Figure 17.20).

No-go decay or NGD functions when the ribosome stalls during elongation

The existence of this mechanism for quality control was recognized in 2006 in the laboratory of Roy Parker. The mechanism functions when the ribosome stalls at certain positions in the message. The mRNA is then cleaved near the stall site and degraded; the ribosomal subunits are released and can be recycled. NGD bears mechanistic resemblance to the normal termination process. Indeed, the two evolutionarily conserved proteins Dom34 and Hbs1 that initiate **no-go decay** are related to eRF1 and eRF3 that function as release factors during normal termination (see Chapter 16). However, NGD can also occur independently of Dom34 and Hbs1, by a mechanism that remains unclear. A working model of NGD is presented in Figure 17.21.

Non-stop decay or NSD functions when mRNA does not contain a stop codon

The cell has also evolved a special mechanism to deal with messages that do not contain a stop codon. If these messages were to be translated, they would give rise to longer proteins that would contain additional sequences at their C-termini, now translated from the 3′-UTRs. The mechanism of NSD is conserved from yeast to man

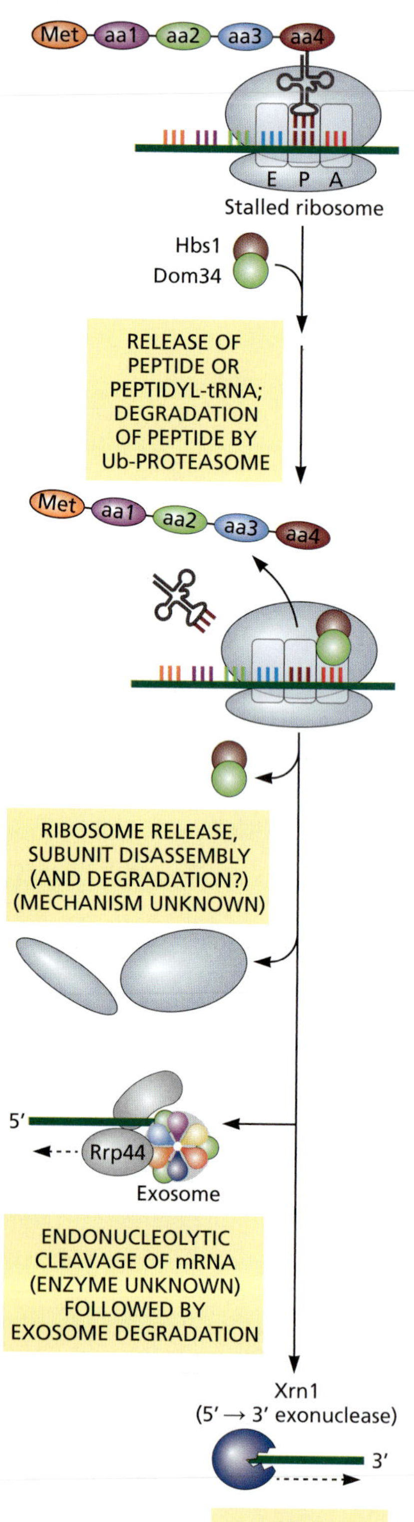

Figure 17.21 Working model for the mechanism of mammalian no-go decay or NGD. NGD functions during prolonged translation elongation pausing, which leaves the acceptor or A site on the ribosome, the site of entrance of the charged tRNA, unoccupied. This allows the Hbs1–Dom34 complex to bind to this site, which leads to release of the peptide or of the peptidyl-tRNA. The process is akin to normal translation termination; in fact, Hbs1 and Dom34 are related to eRF1 and eRF3, the normal eukaryotic release factors involved in translation termination. It is not yet known what distinguishes normal pauses in translation that serve a biological role from abnormal pauses that require NGD. (Adapted from Harigaya Y & Parker R [2010] *Wiley Interdiscip Rev RNA* 1:132–141. With permission from John Wiley & Sons, Inc.)

and requires the exosome, the Ski7 adaptor protein, and the SKI or Superkiller complex (**Figure 17.22**).

17.5 Mechanisms of translation

The level of protein production in cells is regulated in many ways, involving both transcriptional and translational control. In this chapter, we have concentrated on the latter, and even here we find a number of very different mechanisms. These include regulating the number of ribosomes available for synthesis, control of initiation of translation, or regulation by riboswitches or microRNAs. Finally, the possibility of translation depends on the stability of individual mRNAs. The multitude of transcriptional and

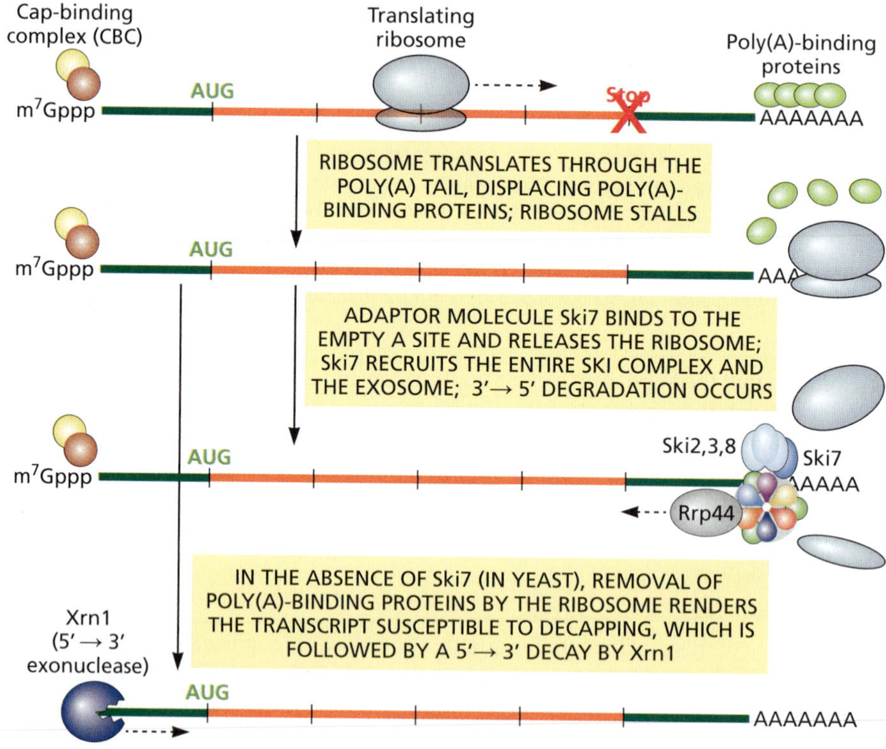

Figure 17.22 Molecular mechanisms of non-stop decay or NSD. mRNA transcripts that lack a stop codon can be generated in a number of ways, including mutations that remove the stop codon, mRNA breakage, or premature polyadenylation. Experimental evidence points to two distinct pathways of NSD, depending on whether a specific adaptor molecule, Ski7, is present. The C-terminus of Ski7 is structurally similar to domains in the translation elongation factor 1A and the release factor eRF3, as is protein Dom34 that participates in the no-go pathway; thus it can bind to an empty A site in the ribosome, thereby releasing the ribosome. Ski7 then recruits the exosome and the associated SKI or Superkiller complex which comprises Ski2, Ski3, and Ski8, and activates the exosome. The rest of the pathway involves the 3′ → 5′ exonucleolytic action of the exosome. In the absence of Ski7, an alternative pathway functions in *Saccharomyces cerevisiae*. (Adapted from Garneau NL, Wilusz J & Wilusz CJ [2007] *Nat Rev Mol Cell Biol* 8:113–126. With permission from Macmillan Publishers, Ltd.)

translational regulators is amazing in its total complexity and subtlety. Perhaps we should not be amazed, as this is what it takes to allow any complex and adaptable organism to function in life: to survive, replicate, and evolve.

Key concepts

- The most general, nonspecific translational regulation involves controlling the number of ribosomes in the cell.
- Ribosome numbers in bacteria can be reduced by degradation of the 30S and 50S subunits, which are released at the end of the translation cycle and not immediately reused if the need for translation diminishes.
- Production of new ribosomes depends upon concerted synthesis of rRNA and r-proteins. Synthesis of at least some bacterial ribosomal proteins is regulated by feedback control of r-protein operons.
- In eukaryotes, only a fraction of the rRNA genes are normally transcribed, as regulated by chromatin structure and DNA methylation.
- The most rapid and specific way to regulate translation is through control of initiation. In bacteria and phage, this often depends on dealing with secondary structure of the mRNA.
- In eukaryotes, control over initiation can be exerted by protein binding to the cap sequence or to internal ribosome entry sites or to the 3′-UTR.
- Riboswitches are secondary structural elements in certain mRNAs whose conformation can respond directly to temperature, metabolites, or specific ions, to regulate ribosome accessibility to the message.
- MicroRNAs, by recruiting the Argonaute complex to sequences in the 3′-UTRs of mRNA molecules, may either inhibit initiation of translation or degrade the message.
- Translation is also sensitive to mRNA stability. Decay of most mRNAs begins with deadenylation and then may involve degradation from either the 5′- or the 3′-end.
- In eukaryotes, some mRNAs that have stalled in translation can be stored in stress granules; those destined for degradation, for whatever reason, are often found in processing particles known as P bodies.
- Faulty mRNAs are destroyed by a number of mechanisms, depending on the error: a premature stop codon leads to nonsense-mediated decay, stalling of the ribosome leads to no-go decay, and a missing stop codon leads to non-stop decay.

Further reading

Books

Hershey JWB (ed) (2009) Progress in Molecular Biology and Translational Science, Volume 90: Translational Control in Health and Disease. Elsevier.

Hershey JWB, Sonenberg N & Matthews MB (eds) (2012) Protein Synthesis and Translational Control. Cold Spring Harbor Laboratory Press.

Nierhaus KH & Wilson DN (2004) Protein synthesis and ribosome structure. Translating the genome. Wiley–VCH.

Reviews

Balagopal V & Parker R (2009) Polysomes, P bodies and stress granules: States and fates of eukaryotic mRNAs. *Curr Opin Cell Biol* 21:403–408.

Bao Q & Shi Y (2007) Apoptosome: A platform for the activation of initiator caspases. *Cell Death Differ* 14:56–65.

Bartel DP (2009) MicroRNAs: Target recognition and regulatory functions. *Cell* 136:215–233.

Buchan JR & Parker R (2009) Eukaryotic stress granules: The ins and outs of translation. *Mol Cell* 36:932–941.

Caldarola S, De Stefano MC, Amaldi F & Loreni F (2009) Synthesis and function of ribosomal proteins: Fading models and new perspectives. *FEBS J* 276:3199–3210.

Fabian MR, Sonenberg N & Filipowicz W (2010) Regulation of mRNA translation and stability by microRNAs. *Annu Rev Biochem* 79:351–379.

Filipowicz W, Bhattacharyya SN & Sonenberg N (2008) Mechanisms of post-transcriptional regulation by microRNAs: Are the answers in sight? *Nat Rev Genet* 9:102–114.

Garneau NL, Wilusz J & Wilusz CJ (2007) The highways and byways of mRNA decay. *Nat Rev Mol Cell Biol* 8:113–126.

Gottesman S (2005) Micros for microbes: Non-coding regulatory RNAs in bacteria. *Trends Genet* 21:399–404.

Grundy FJ & Henkin TM (2006) From ribosome to riboswitch: Control of gene expression in bacteria by RNA structural rearrangements. *Crit Rev Biochem Mol Biol* 41:329–338.

Harigaya Y & Parker R (2010) No-go decay: A quality control mechanism for RNA in translation. *Wiley Interdiscip Rev RNA* 1:132–141.

Holcik M & Sonenberg N (2005) Translational control in stress and apoptosis. *Nat Rev Mol Cell Biol* 6:318–327.

Jinek M & Doudna JA (2009) A three-dimensional view of the molecular machinery of RNA interference. *Nature* 457:405–412.

Kaczanowska M & Rydén-Aulin M (2007) Ribosome biogenesis and the translation process in *Escherichia coli*. *Microbiol Mol Biol Rev* 71:477–494.

Li Y & Kiledjian M (2010) Regulation of mRNA decapping. *Wiley Interdiscip Rev RNA* 1:253–265.

McGlincy NJ & Smith CWJ (2008) Alternative splicing resulting in nonsense-mediated mRNA decay: What is the meaning of nonsense? *Trends Biochem Sci* 33:385–393.

McStay B & Grummt I (2008) The epigenetics of rRNA genes: From molecular to chromosome biology. *Annu Rev Cell Dev Biol* 24:131–157.

Pop C & Salvesen GS (2009) Human caspases: Activation, specificity, and regulation. *J Biol Chem* 284:21777–21781.

Riedl SJ & Salvesen GS (2007) The apoptosome: Signalling platform of cell death. *Nat Rev Mol Cell Biol* 8:405–413.

Serganov A & Patel DJ (2007) Ribozymes, riboswitches and beyond: Regulation of gene expression without proteins. *Nat Rev Genet* 8:776–790.

Serganov A & Patel DJ (2008) Towards deciphering the principles underlying an mRNA recognition code. *Curr Opin Struct Biol* 18:120–129.

Sonenberg N & Hinnebusch AG (2007) New modes of translational control in development, behavior, and disease. *Mol Cell* 28:721–729.

Sonenberg N & Hinnebusch AG (2009) Regulation of translation initiation in eukaryotes: Mechanisms and biological targets. *Cell* 136:731–745.

Song M-G, Li Y & Kiledjian M (2010) Multiple mRNA decapping enzymes in mammalian cells. *Mol Cell* 40:423–432.

Tang X, Tang G & Ozcan S (2008) Role of microRNAs in diabetes. *Biochim Biophys Acta* 1779:697–701.

Experimental papers

Alnemri ES, Livingston DJ, Nicholson DW et al. (1996) Human ICE/CED-3 protease nomenclature. *Cell* 87:171.

Chen J & Kastan MB (2010) 5′–3′-UTR interactions regulate p53 mRNA translation and provide a target for modulating p53 induction after DNA damage. *Genes Dev* 24:2146–2156.

Herrero AB & Moreno S (2011) Lsm1 promotes genomic stability by controlling histone mRNA decay. *EMBO J* 30:2008–2018.

Johansson J, Mandin P, Renzoni A et al. (2002) An RNA thermosensor controls expression of virulence genes in *Listeria monocytogenes*. *Cell* 110:551–561.

Mironov AS, Gusarov I, Rafikov R et al. (2002) Sensing small molecules by nascent RNA: A mechanism to control transcription in bacteria. *Cell* 111:747–756.

Morita MT, Tanaka Y, Kodama TS et al. (1999) Translational induction of heat shock transcription factor σ^{32}: Evidence for a built-in RNA thermosensor. *Genes Dev* 13:655–665.

Mullen TE & Marzluff WF (2008) Degradation of histone mRNA requires oligouridylation followed by decapping and simultaneous degradation of the mRNA both 5′ to 3′ and 3′ to 5′. *Genes Dev* 22:50–65.

Nanbru C, Prats A-C, Droogmans L et al. (2001) Translation of the human *c-myc* P0 tricistronic mRNA involves two independent internal ribosome entry sites. *Oncogene* 20:4270–4280.

Piir K, Paier A, Liiv A et al. (2011) Ribosome degradation in growing bacteria. *EMBO Rep* 12:458–462.

Singh J, Salcius M, Liu SW et al. (2008) DcpS as a therapeutic target for spinal muscular atrophy. *ACS Chem Biol* 3:711–722.

Trajkovski M, Hausser J, Soutschek J et al. (2011) MicroRNAs 103 and 107 regulate insulin sensitivity. *Nature* 474:649–653.

Wang Y, Juranek S, Li H et al. (2008) Structure of an argonaute silencing complex with a seed-containing guide DNA and target RNA duplex. *Nature* 456:921–926.

Winkler W, Nahvi A & Breaker RR (2002) Thiamine derivatives bind messenger RNAs directly to regulate bacterial gene expression. *Nature* 419:952–956.

Zundel MA, Basturea GN & Deutscher MP (2009) Initiation of ribosome degradation during starvation in *Escherichia coli*. *RNA* 15:977–983.

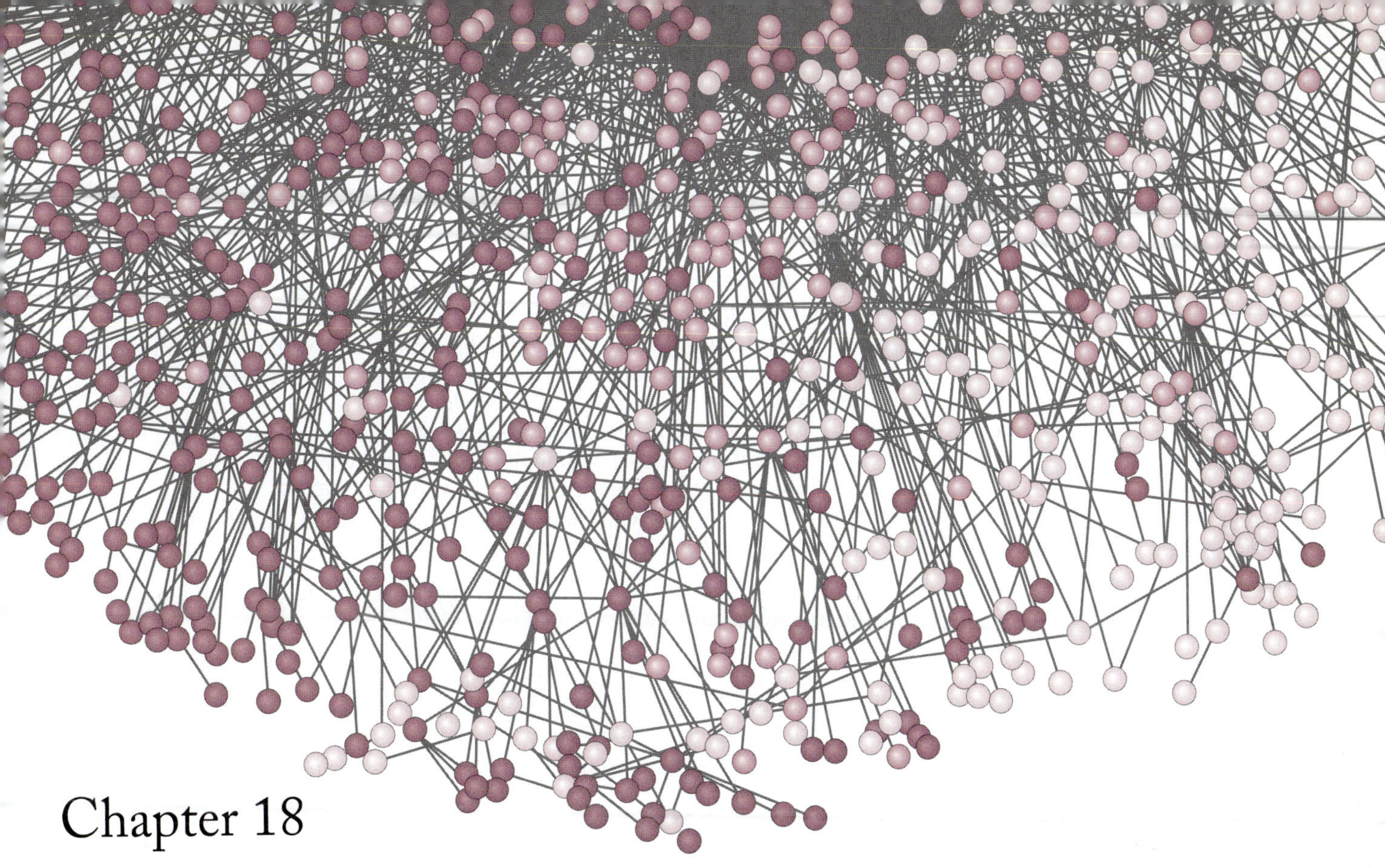

Chapter 18

Protein Processing and Modification

18.1 Introduction

Throughout this book, we trace the remarkable path by which a DNA sequence in a cell dictates the production of a corresponding protein sequence. However, the needs of the cell are so many and varied that the story does not end at this point. Many protein structures are modulated after the gene sequence has been read by transcription and translation. This further modulation has the general title of post-translational processing, and it takes a number of forms. Some proteins are proteolytically cleaved into pieces, and sometimes these pieces can be even spliced together in new arrangements. In some cells or tissues, proteins must be destroyed, either selectively or wholesale. Furthermore, a variety of modifications of side chains can influence the properties and interactions of proteins, often in reversible ways. Overall, the consequences of protein processing are enormously varied: processing can dictate the cellular destination of a new protein molecule, modify its functional properties, or prescribe its degradation.

Many of the modifications we describe in this chapter take place in specific, membrane-bound compartments such as the endoplasmic reticulum and the Golgi apparatus. The proteins that are processed must get into and out of these organelles. In fact, the passage of proteins through membranes in a directed way often determines where the protein will end up in the cell, or even outside the cell. Thus, we must consider that the first stage in protein post-translational processing frequently involves transport into or through membranes. To explain how this transport occurs, we preface this chapter with a brief description of biological membranes and transport.

18.2 Structure of biological membranes

Biological membranes are protein-rich lipid bilayers

All biomembranes have a common structure: a bilayer of lipid molecules with hydrophobic tails buried in the interior of the bilayer and hydrophilic head groups on the two surfaces (**Figure 18.1A**). Most of the lipid molecules within biomembranes are derivatives of fatty acids, with hydrocarbon tails 14–26 carbons in length and with varying degrees of unsaturation (**Figure 18.1B**). One important lipid that does not conform entirely to this general description is **cholesterol**, which is a sterol (**Figure 18.1C**). But even cholesterol has a more hydrophobic and a more hydrophilic end and aligns itself perpendicular to the membrane plane; it is also more rigid than the fatty acids and conveys that rigidity to membranes in which it is enriched. All biomembranes also contain attached or embedded proteins.

The head groups, which are linked by ester or phosphodiester bonds to the carbonyls of the fatty acids, include a variety of hydrophilic moieties, such as sugars, alcohols, and amino acids, and are sometimes charged. Note that phosphodiester bonds also contribute negative charges to the membrane surface. All of these hydrophilic groups will invite water molecules into the outer regions of the membrane, yielding an overall cross-section composition like that shown in **Figure 18.2**. While lateral motion within the membrane is easy, motion across the membrane by either head groups or any hydrophilic molecule is very difficult. Because of the physical flexibility and variety of their constituents, biomembranes are not crystallinelike under physiological conditions but form a fluid structure with a great deal of lateral motion. Aggregates of more regularly packed molecules, often including cholesterol, form rafts that can diffuse as stable entities within the bilayer.

Numerous proteins are associated with biomembranes

Many proteins are associated with biomembranes. A few examples are shown in **Figure 18.3**, and a partial list of their functions is given in **Table 18.1**. Membrane proteins can be divided into two major structural classes: peripheral membrane proteins and integral membrane proteins. As the name implies, peripheral proteins are tethered to one of the two surfaces of the bilayer (see Figure 18.3A). The majority of such proteins lie within the aqueous solution adjoining the bilayer. Peripheral proteins usually attach through a moiety that can bond either electrostatically or through van der Waals interactions with lipid head groups; alternatively, they may become membrane-associated through interactions with some integral protein. Sometimes a hydrophobic tail embedded in the membrane core acts as a tether.

Integral membrane or transmembrane proteins have some or much of their structure embedded in the hydrophobic core of the bilayer (see Figure 18.3B). Sometimes, a single α-helix traverses the membrane. More often, the polypeptide chain passes back and forth through the membrane several times. The membrane-spanning domains of integral membrane proteins mostly fall into one of two major classes: α-helix bundles and β-barrels. In either case, the portion of the protein in direct contact with

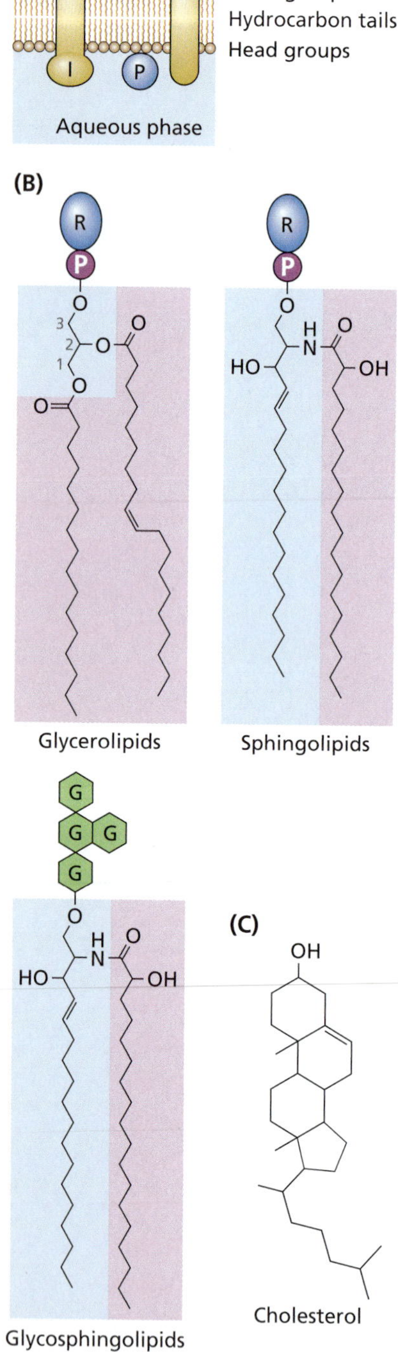

Figure 18.1 Biomembranes. (A) The common bilayer structure in all biomembranes. P denotes peripheral membrane proteins; I denotes integral membrane proteins. (B) Structures of typical membrane lipids. Glycerolipids are based on glycerol, shown by blue shading, with two C16–C18 fatty acid chains, shown by purple shading. A *cis*-double bond is usually present in the fatty acid linked to the second OH group of glycerol; this bond kinks the acyl chain, which leads to decreased packing density of the lipid. A phosphate attached at the third OH group can carry a head group, marked R, which can be neutral or carry positive or negative charge. Sphingolipids are based on sphingosine, in animals, shown by blue shading, with a saturated C16–C26 fatty acid, shown by purple shading, linked to the nitrogen. Glycosphingolipids have the same basic structure, with glucose or galactose constituting the head group, to which extra monosaccharides can be added. (C) Structure of cholesterol, a derivative of sterol. Cholesterol is present in mammals; other sterol derivatives are found in fungi and plants. (B–C, adapted from Holthuis JCM & Levine TP [2005] *Nat Rev Mol Cell Biol* 6:209–220. With permission from Macmillan Publishers, Ltd.)

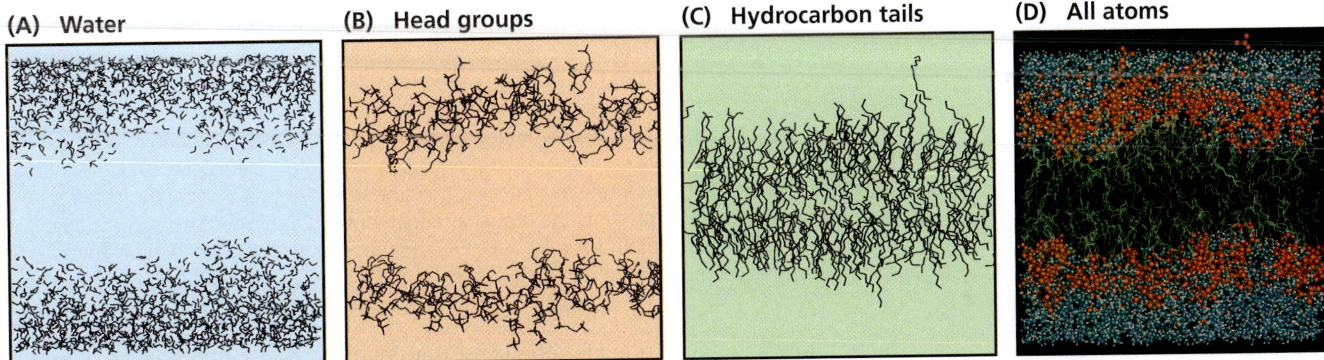

Figure 18.2 Atomic model of hydrated phosphatidylcholine bilayer. The model does not represent experimental data but is a computer simulation based on parameters obtained by several techniques, including NMR, X-ray diffraction, and neutron diffraction. It incorporates 100 molecules of phosphatidylcholine and 1050 molecules of bulk-phase water on each side, which are allowed to move on a picosecond time scale. Production of a model covering 100 ps of simulated time took several weeks of computer computation. Shown are (A) water, showing bulk concentration outside the membrane and penetration into the head-group region, down to the carbonyl groups; (B) polar regions of the phospholipid; (C) hydrocarbon tails of the phospholipids; and (D) the entire system, consisting of A–C depicted together with the following color code: polar atoms are shown as orange spheres, water oxygens are shown in blue, water hydrogens are shown in white, and hydrocarbon chains are shown as green stick models. A–C are stick representations. (From Chiu SW, Clark M, Balaji V et al. [1995] *Biophys J* 69:1230–1245. With permission from Elsevier.)

the membrane core is hydrophobic. Many integral membrane proteins function as pores, carriers, or pumps that transmit substances, ranging from ions to whole protein molecules, from one compartment to another. Sometimes only a signal rather than a substance is transmitted: a ligand binds to the extramembrane domain on one side, which transmits, via conformational change, an allosteric signal to the opposite domain, invoking a response there. Often, peripheral membrane proteins interact with the extramembrane domains of integral proteins to form functional complexes.

18.3 Protein translocation through biological membranes

In this section, we are especially concerned with two questions:

1. How are newly synthesized proteins transported through membranes so that they reach their appropriate cell compartment?
2. How are integral membrane proteins inserted into membranes and properly folded?

Protein translocation can occur during or after translation

In Chapter 16, we followed protein synthesis up to the point where the new polypeptide chain emerges from the ribosome tunnel. What happens next? At this juncture, major decisions have to be made concerning the protein's fate and final destination. In bacteria and archaea, many proteins are destined to remain in the cytoplasm in which they were synthesized, but others will be secreted through the plasma membrane or

Table 18.1 Types and functions of integral membrane proteins.

Type	Function	Mode of operation	Examples
pores	mainly ion transport	driven by ion concentration gradient; may open or close	voltage-gated Na^+/K^+ channels in nerve cells; Ca^{2+} channel in muscle cells
carriers	small molecules, substrates	driven by concentration gradient, sometimes by substrate, sometimes by co-transporter or anti-transporter	glucose transport; ATP/ADP exchange in mitochondria
pumps	ions, small or large molecules	driven by free-energy release by ATP hydrolysis, often against the substrate concentration gradient	light-driven H^+ transport by rhodopsin; SecY/SecA protein transport in bacteria

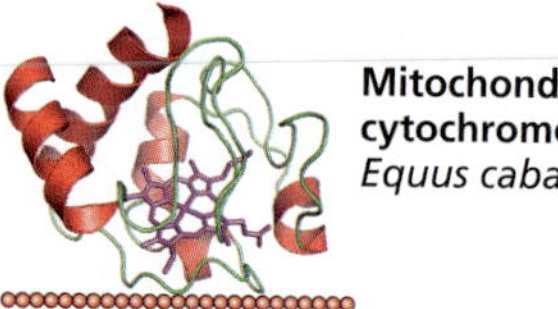

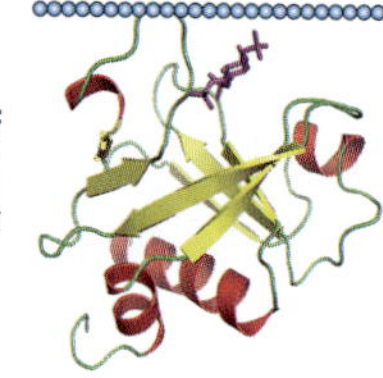

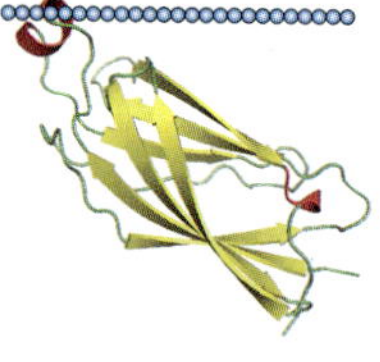

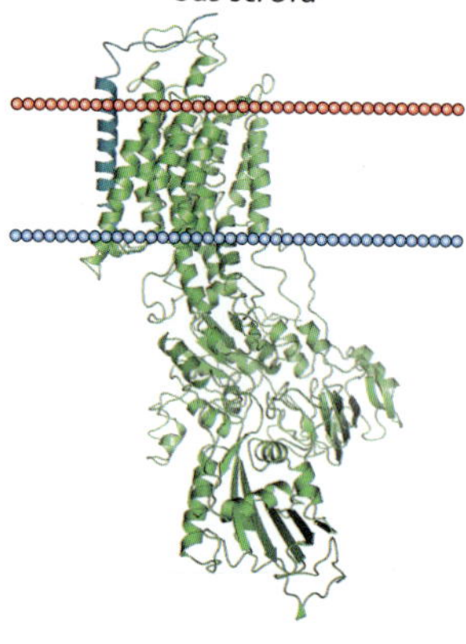

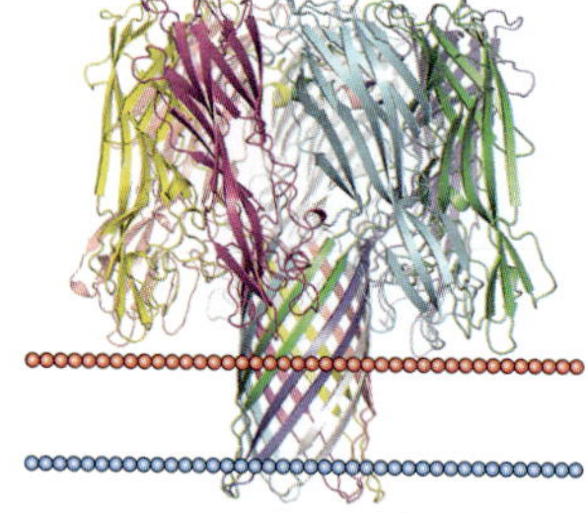

Figure 18.3 Examples from the two classes of membrane proteins. The identity of each protein and its biological source is indicated. (A) Peripheral membrane proteins can be exclusively or predominantly α-helices (top), mixed (middle), or β-strands (bottom). (B) Integral membrane proteins, or transmembrane proteins, fall into two major classes according to the secondary structures of their transmembrane portions: α-helix bundles (top and middle) or β-barrel (bottom). There are several thousand entries in the Protein Data Base, PDB, describing the structures of membrane proteins, but their exact orientations in biological membranes are unknown. This figure presents the spatial arrangement of the protein structures in lipid bilayers as determined computationally. Red dots denote an outer membrane; blue dots denote an inner membrane.

incorporated therein. In eukaryotic cells, the possibilities are much more complex (Figure 18.4). Some proteins will be cytoplasmic, some are destined for various organelles or their membranes, and still others are secreted. Certain cells in the pancreas, for example, are largely devoted to the export of enzymes into the digestive system. Most of these fates involve the transport, or **translocation**, of new polypeptides, some of them quite hydrophilic, across hydrophobic membranes. How is this accomplished?

The mechanisms of protein translocation across membranes can generally be classified as post-translational or co-translational, depending on whether the translocation occurs after synthesis is complete and the polypeptide is released from the ribosome or during polypeptide chain elongation in the ribosome. In either case, translocation involves specialized channels in membranes, known as **translocons** (Figure 18.5). Translocons are evolutionarily conserved heterotrimeric protein complexes that undergo significant motions during translocation. Thus, the central pore through which the polypeptide passes is usually closed by a plug, a portion of one of the transmembrane domains; for translocation to occur, the plug has to move away from the pore. Translocons serve the same function in post- and co-translational translocation; they just interact with several different complexes that bring the protein to the pore.

Membrane translocation in bacteria and archaea primarily functions for secretion

Because bacteria and archaea lack organelles, most of their proteins remain in the cytoplasm in which they were synthesized. Nevertheless, many bacteria secrete selected proteins, either into the surrounding medium or into the periplasmic space between the plasma membrane and the cell wall (see Figure 18.4). How are such proteins selected? In fact, each is marked by a **signal sequence** at its N-terminus. This is usually 15–30 amino acid residues in length, with a positive charge at the N-terminus followed by a string of about a dozen hydrophobic residues (Figure 18.6). Similar signal peptides are used by eukaryotic cells. It should be noted that not all secretory proteins possess signal peptides, and proteins that lack these sequences do not use the classical translocation routes. Fibroblast growth factors and interleukins are eukaryotic representatives of such secretory proteins, and there are numerous bacterial proteins that use signal-peptide-independent secretion systems.

Post-translational translocation pathways are used by a larger fraction of proteins in simpler organisms, such as bacteria. Post-translational translocation is typical of soluble secretory proteins, whose signal sequences are only moderately hydrophobic and thus escape recognition by the signal recognition particle. The marked, unfolded protein is taken up by a cytoplasmic complex called SecA, which then pairs with the translocon in the plasma membrane (Figure 18.7). SecA uses ATP hydrolysis to move a molecular finger, thereby pushing the polypeptide chain through the pore. When the channel is not occupied by a transiting polypeptide, it is both more constricted and closed by a molecular plug so as to prevent the leakage of ions and small molecules from the cell (see Figure 18.5).

Membrane translocation in eukaryotes serves a multitude of functions

Much of the membrane traffic that accompanies protein processing and modification in eukaryotic cells is from the cytoplasm into the lumen of the endoplasmic reticulum

Figure 18.4 Schematic of the conserved principle of protein exports through membranes. (A) In bacteria and archaea, proteins are transported from the cytosol into the periplasm or the extracellular environment. (B) In eukaryotes, proteins are transported from the cytoplasm, where synthesis is initiated, to the endoplasmic reticulum (ER) lumen, then to the Golgi complex, and finally, depending on the nature and site of action of the particular protein, to the cellular membrane, extracellular space, or other cellular compartments such as the lysosome. Transport across membranes can occur post-translationally or co-translationally, and in both cases it involves conserved transmembrane channels: SecYEG in bacteria, SecYEβ in archaea, and Sec61 in eukaryotes. (Adapted from Cross BCS, Sinning I, Luirink J & High S [2009] *Nat Rev Mol Cell Biol* 10:255–264. With permission from Macmillan Publishers, Ltd.)

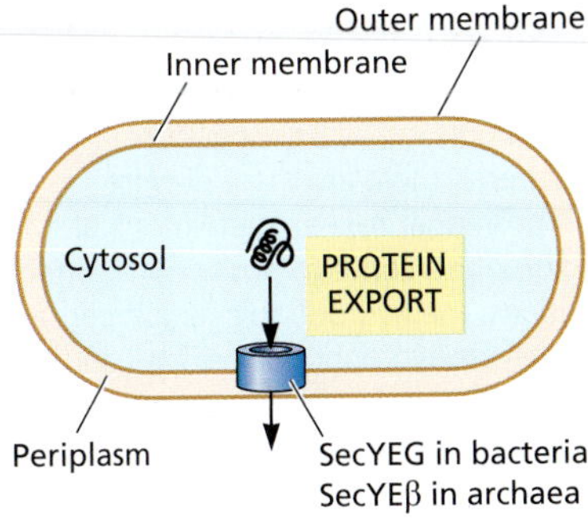

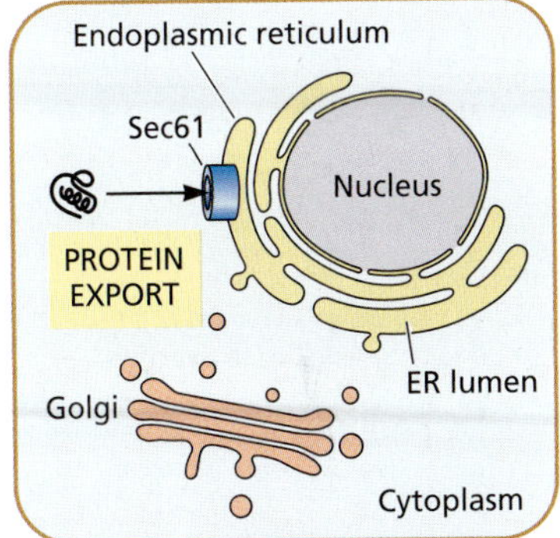

or ER (see Figure 18.4). This organelle provides a transfer point for passage to the Golgi apparatus, where final selection of destinations occurs. Golgi is a highly dynamic organelle, consisting of numerous, usually stacked cisternae with recognizable morphology, with distinguishable *cis*- and *trans*-faces. A historical account of the discovery of the Golgi apparatus and a summary of our present-day knowledge of its organization and function are presented in **Box 18.1**.

Most eukaryotic translocation is co-translational; it is coupled directly to translation. Indeed, the outer surfaces of the rough ER are studded with ribosomes that are engaged in protein synthesis. The process begins when a ribosome synthesizes enough of the polypeptide chain to allow the signal sequence to emerge from the ribosome tunnel (**Figure 18.8**). This is a signal for the ribosome to associate with a **signal recognition particle** or **SRP**. The SRP contains several proteins and a short piece of RNA, termed S RNA, about 300 nt long and having a sedimentation coefficient of 7S (**Figure 18.9**).

Translation halts at this point, while the ribosome–SRP complex finds a membrane marker called the SRP receptor protein. The whole complex then binds to the translocon Sec61 (see Figure 18.5). The ribosome tunnel is positioned over the membrane pore, SRP and receptor dissociate, and translation resumes, now pushing the polypeptide through the pore. When translation is complete, the signal peptide is cleaved and degraded by a special signal peptidase.

Some eukaryotic proteins are translocated by a post-translational mechanism (**Figure 18.10**). The marked polypeptide, escorted by chaperones, encounters a pore containing Sec61, as well as a Sec62/Sec63 complex. Passage through the pore is believed to be driven by repeated binding of an Hsp70-type ATPase called BiP. Segments of the polypeptide chain pass through the pore by Brownian motion and are prevented from retreating when BiP clamps down following ATP hydrolysis (see Figure 18.10). Thus, the mechanism here is a Brownian ratchet. It is interesting to note that the three mechanisms of translocation are very different (**Table 18.2**), but all depend ultimately on the hydrolysis of nucleoside triphosphates. Why evolution has produced three quite different mechanisms to achieve the same end is unclear.

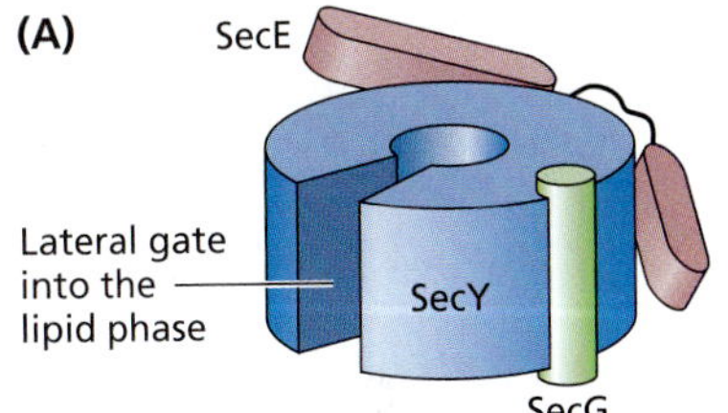

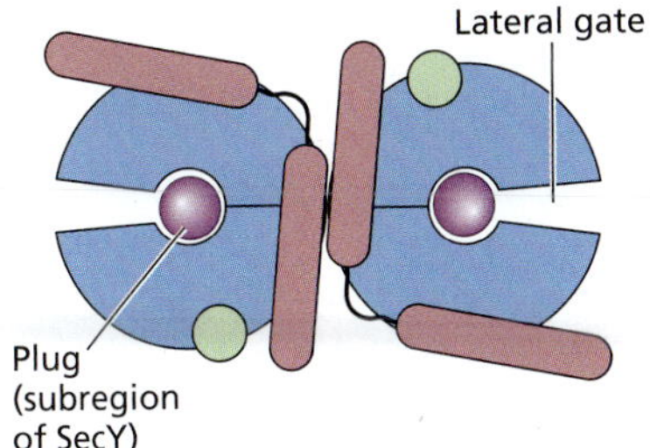

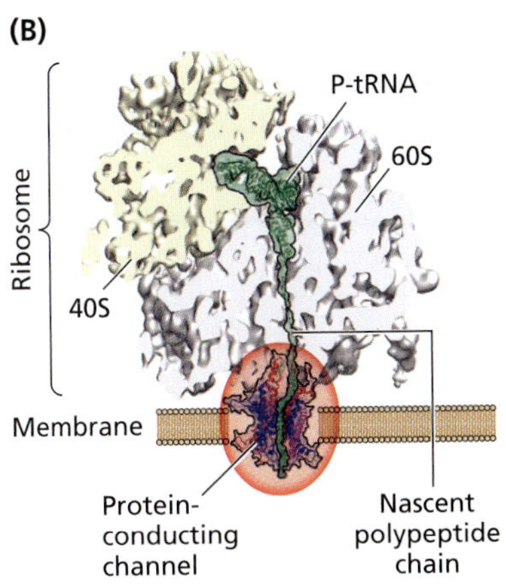

Figure 18.5 The translocon: Evolutionarily conserved protein translocation machinery. Translocons are universal conduits through membranes for both co-translational and post-translational protein delivery. (A) In bacteria, the translocon is a heterotrimeric complex consisting of SecY, SecE, and SecG, located within the inner plasma membrane. When there is no protein in the translocon, its aqueous conduit or pore is plugged by a portion of the second transmembrane domain of SecY; the plug moves away during translocation. The arrangement of the transmembrane helices in SecY forms a lateral gate into the lipid phase of the membrane. This gate is used for insertion of integral membrane proteins into the membrane. (B) In eukaryotes, the complex that forms translocons in the endoplasmic reticulum membrane consists of Sec61α, Sec61β, and Sec61γ, which correspond to SecY, SecG, and SecE, respectively. The schematic represents a translating and translocating ribosome–Sec61 complex from yeast, as deduced from cryo-EM analysis. The translocon, circled in red, is at the end of the exit tunnel of the ribosome. As in bacteria, a central pore is visible in the idle translocon complex. (A, adapted from Cross BCS, Sinning I, Luirink J & High S [2009] *Nat Rev Mol Cell Biol* 10:255–264. With permission from Macmillan Publishers, Ltd. B, from Becker T, Bhushan S, Jarasch A et al. [2009] *Science* 326:1369–1373. With permission from American Association for the Advancement of Science.)

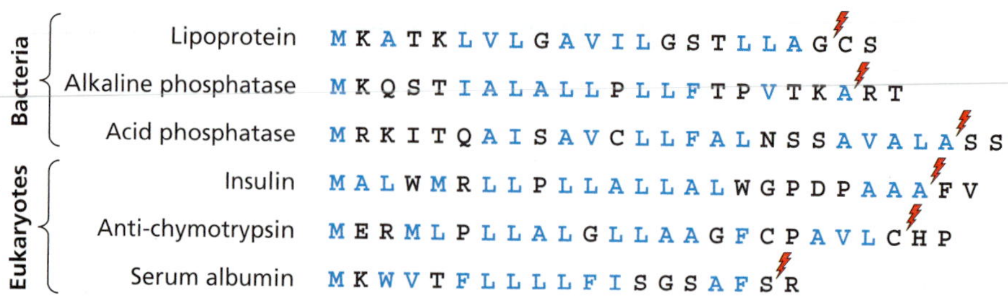

Figure 18.6 N-terminal signal sequences in bacteria and eukaryotes. The signal peptides are typically 15–30 amino acids long. There is no simple consensus sequence; in general, there are three distinct compositional zones: the N-terminal region often contains positively charged residues; the hydrophobic region contains at least six hydrophobic residues, shown in blue; and the C-terminal region consists of polar uncharged residues. Red lightning bolts denote cleavage sites.

The situation is further confused by the fact that co-translational translocation is dispensable in yeast but post-translational is not.

Integral membrane proteins have special mechanisms for membrane insertion

Proteins that are destined to be integrated into a membrane, rather than to pass through it, appear to utilize the same basic mechanisms that we have described above for translocation. Proteins that span membranes do so with stretches of hydrophobic residues on the order of 20 amino acids; in an α-helix, this corresponds to ~3 nm, the approximate thickness of the hydrocarbon-rich center of the membrane (see Figure 18.2). Apparently, such stretches do not pass head-first through the membrane during translocation but instead are passed sideways, through a lateral gate in the translocon (see Figure 18.5). At times, this must involve some molecular calisthenics, because both flipped and unflipped orientations are found in multipass integral membrane proteins. Mechanisms for flipping segments have been proposed, but definitive information is not yet available.

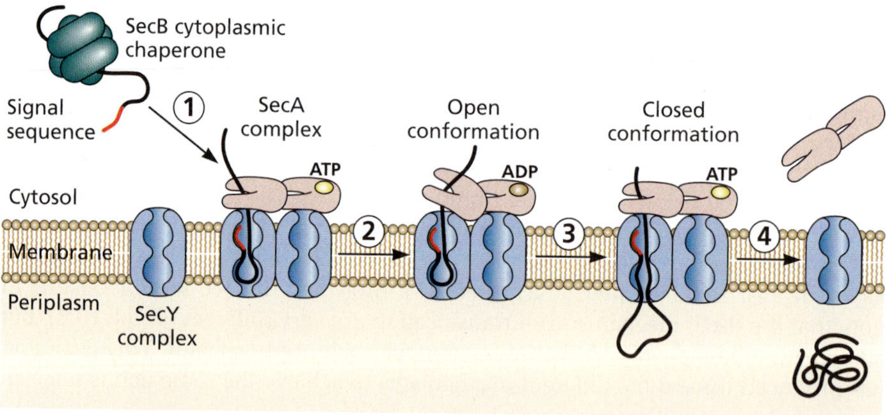

Figure 18.7 Model of post-translational translocation in bacteria. In this pathway, the SecY translocon complex partners with the cytosolic ATPase SecA. SecA possesses two nucleotide-binding folds that bind ATP between them and that move relative to each other during the ATP hydrolysis cycle. Other parts of the SecA molecule also move with respect to each other, creating two alternating conformations, open and closed. The sequence of steps is as follows: (Step 1) The translocation substrate is brought to the channel in a form that is bound by the cytosolic homotetrameric chaperone SecB; this form is frequently referred to as the translocation-competent conformation, which is thought to be loosely folded to allow easy threading through the channel. Upon binding to the channel, the chaperones are released. SecA accepts the polypeptide, probably binding both the signal sequence and the segment following it. SecA binds to Sec Y, inserting the polypeptide chain into the membrane. (Step 2) SecA changes into the open conformation, releasing its grip on the adjacent peptide sequence. The transfer requires ATP hydrolysis. (Step 3) With rebinding of ATP, the closed complex is re-formed, grips a further segment of polypeptide, and pushes it through the membrane. (Step 4) The cycle is repeated until the entire polypeptide is translocated, at which point SecA moves away from the channel. (Adapted from Rapoport TA [2007] *Nature* 450:663–669. With permission from Macmillan Publishers, Ltd.)

Box 18.1 A Closer Look: The Golgi complex, still an enigma

The Golgi complex or apparatus, or simply Golgi, was discovered in 1897 by physician and scientist Camillo Golgi during investigations of the nervous system. He originated a method for staining nervous tissue by impregnation with silver. This staining method allowed Golgi to view the paths of brain neurons for the first time. Golgi was awarded the 1906 Nobel Prize in Physiology or Medicine, jointly with Santiago Ramón y Cajal, "in recognition of their work on the structure of the nervous system."

A fascinating account of the past 30 years of Golgi research with an eye to the future was presented in an essay by James Rothman of Yale University, on the occasion of his American Society of Cell Biology 2010 award. Despite all these years of concerted research effort in a number of laboratories worldwide, Golgi is still an enigmatic organelle, with many knowns but an even greater number of unknowns. In response to some trivial explanations of why Golgi exists, Rothman wrote "it seemed to me (in the late 1960s), and still does, that there must be a deeper explanation for this organelle, a core concept we were still missing to explain." Even in 2000, after some considerable success in identifying Golgi components and their function, mainly coming from biochemistry and reconstituted systems, Rothman admitted "it began to seem that nothing about the Golgi could be resolved with clarity" and that "somehow the forest may have been missed for the trees." These rather gloomy views from a major Golgi expert are based on the lack of appropriate technical advances that can be applied toward understanding Golgi function at the molecular level.

What is needed are high-resolution techniques that would allow direct visualization of the dynamic behavior of individual protein components when they move back and forth through Golgi and also in the surrounding tubular structures, including the endoplasmic reticulum, and the endosomes. Electron microscopy has the resolution needed but provides only static images. Observing proteins tagged with GFP, green fluorescent protein, under the fluorescence microscope provides the necessary time resolution, but the spatial resolution is inadequate for these kinds of studies: ~50 nm resolution is needed, whereas only ~250 nm resolution is available. It is hoped that recent advances in superresolution microscopy may bridge this gap. The problem of inadequate resolution is compounded by the fact that the Golgi cisternae are very closely stacked, with center-to-center distances of ~100 nm. Using the naturally unstacked Golgi cisternae of yeast can partially solve that problem, but only partially, as the functions of mammalian Golgi, for example, have evidently evolved significantly to satisfy the more complex demands of a multicellular organism. Fortunately, Rothman and other researchers in the field now believe that the development of appropriate methods is in sight.

What is Golgi?

Golgi is a dynamic organelle whose functions include receiving and sorting newly synthesized proteins and lipids for transportation to their final destination, whether inside or outside the cell. Some post-translational processing, such as glycosylation and proteolytic cleavage, also occurs in the Golgi. Morphologically, the organelle is organized as several stacked cisternae and has two recognizable faces: the *cis*-face that accepts proteins arriving from the endoplasmic reticulum and the *trans*-face on the opposite side, where cargo leaves the Golgi cisternae for different destinations inside or outside the cell. In addition to anterograde or forward, *cis*-to-*trans* movement, reverse retrograde flow, which could be *trans*-to-*cis* or *trans*-to-ER, occasionally occurs. Some of the distinct components that participate in anterograde or retrograde flows have been identified, but exactly how and why these distinct directions of movement occur is unclear. A schematic of the structure and contemporary microscopic and tomographic images are presented in **Figure 1**.

Several classes of proteins that participate in different Golgi functions have been characterized. Among them are the

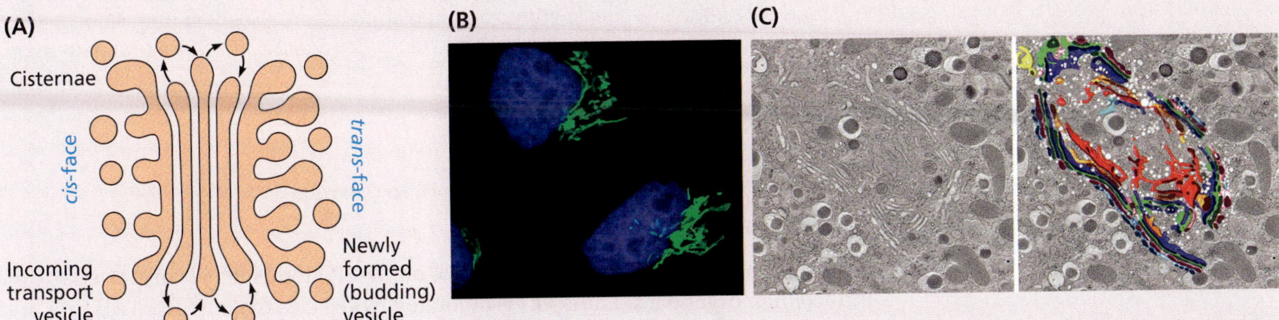

Figure 1 The Golgi. (A) Schematic of stacked cisternae of the Golgi apparatus, with the *cis*- and *trans*-faces marked. (B) Fluorescence light microscopic image of fixed human HeLa cells, showing the location of the Golgi-resident glycosylation enzyme GalNAc-T2 labeled with green fluorescent protein. Nuclei were stained with 4′,6-diamidino-2-phenylindole, DAPI, which appears blue. This image emphasizes the close association of the Golgi with the nucleus. (C) Tomographic image slice from a section of an insulin-secreting β pancreatic mouse cell. In the copy on the right, the cisternae are artificially stained with different colors to show the organization of the structure, going from *cis* to *trans*: light blue, pink, green, dark blue, gold, and red. Small white dots represent Golgi-associated vesicles. (A, adapted from Alberts B, Bray D, Hopkin K et al. [2014] Essential Cell Biology, 4th ed. With permission from Garland Science. B, from Grabenbauer M, Geerts WJC, Fernandez-Rodriguez J et al. [2005] *Nat Methods* 2:857–862. With permission from Macmillan Publishers, Ltd. C, from Emr S, Glick BS, Linstedt AD et al. [2009] *J Cell Biol* 187:449–453. With permission from The Rockefeller University Press.)

(Continued)

Box 18.1 *(Continued)*

well-studied SNARE proteins, which have a well-understood role in membrane fusion (see Figure 18.11). Interestingly, SNARE proteins have been identified as targets for botultum and tetanus toxins, highly specific proteases that block synaptic vesicle fusion. Botulin is produced by the bacterium *Clostridium botulinum* and is the most powerful neurotoxin known. Another major class of proteins, the golgins, regulates the structure and function of the *trans*-Golgi network. Their structure and the pathways they participate in are depicted in **Figure 2**.

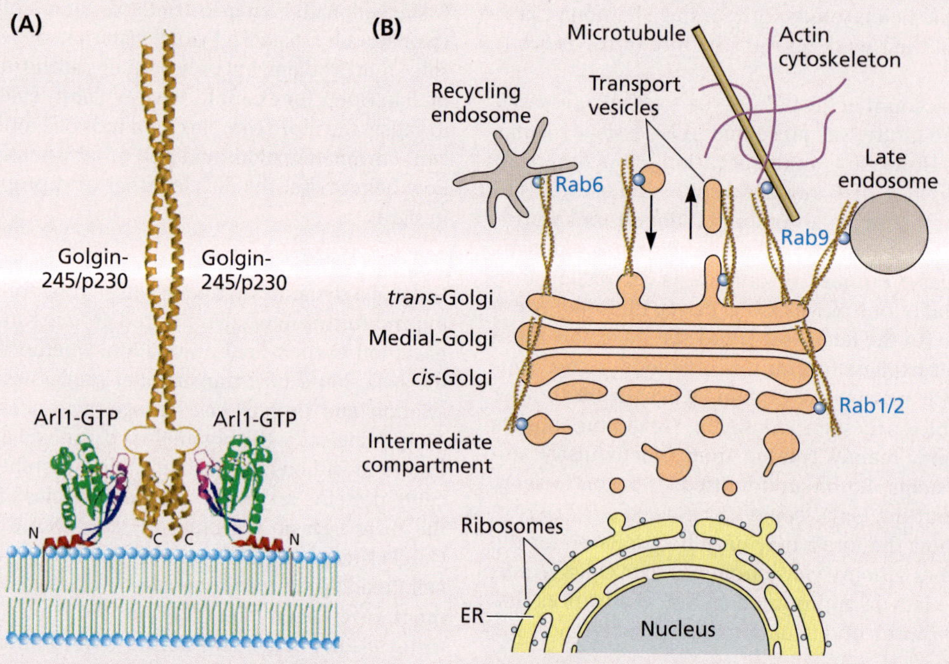

Figure 2 Golgins. A model for the binding of golgins to membranes and their proposed functions in regulating the structure and function of the *trans*-Golgi network, TGN, is illustrated. (A) Both *cis*- and *trans*-sides of the Golgi stacks are associated with networks of tubular structures, representing the cargo entry and exit points. Golgins are coiled-coil peripheral membrane proteins associated with the cytoplasmic side of the Golgi membranes. Their extensive coiled-coil regions form long filaments, up to ~200 nm, that provide interaction platforms for the numerous proteins that participate in various pathways. Golgins are homodimers recruited to Golgi membranes through the interactions of their GRIP domains with a membrane-associated small GTPase known as Arl1, ADP-ribosylation factor-like protein 1. (B) Once bound to membranes, golgins interact with numerous partners, including cytoskeletal proteins such as tubulin, actin, or proteins that serve as linkers between them and members of the Rab family of small G proteins. Rab proteins connect golgins to other TGN membranes, such as those of the endosomes or the *cis*-Golgi network; they also tether incoming cargo carrier vesicles before membrane fusion. Finally, golgins are involved with the biogenesis of cargo transporters for outgoing transport. (Adapted from Goud B & Gleeson PA [2010] *Trends Cell Biol* 20:329–336. With permission from Elsevier.)

Vesicles transport proteins between compartments in eukaryotic cells

Once a protein has crossed the membrane of the ER, it is usually transported to other intracellular organelles. Transport is mediated by vesicles that bud from a donor organelle, travel some distance along cytoskeletal tracks, and then dock and fuse with the membrane of the target organelle. The molecular machineries involved in vesicular trafficking and fusion are highly conserved among all eukaryotes.

The key players during vesicle fusion are called **SNARE proteins**, a large protein superfamily of relatively small and mostly membrane-bound proteins. Classically, SNAREs have been classified as vesicle or v-SNAREs, which are incorporated into the membranes of transport vesicles during budding, and target or t-SNAREs, which are located in the membranes of target compartments. More recent nomenclature takes into account some structural features of the SNARE proteins and instead divides them into R-SNAREs and Q-SNAREs. R-SNAREs contribute an arginine or R residue to a particular portion of the SNARE complex whereas Q-SNAREs provide a glutamine or

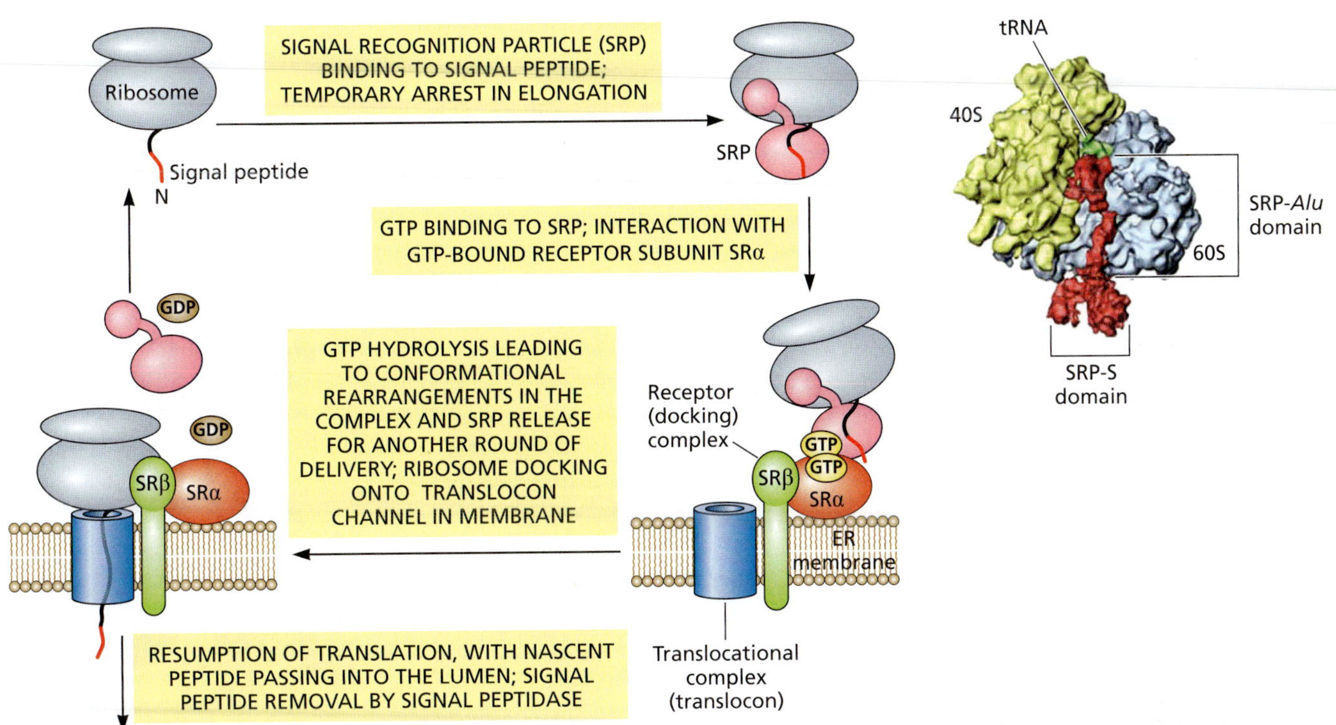

Figure 18.8 Signal recognition particle-dependent protein targeting cycle. Translocation of eukaryotic proteins into the lumen of the endoplasmic reticulum, ER, begins as soon as the signal peptide of the nascent polypeptide chain emerges from the ribosome tunnel in the 60S subunit and is recognized by the signal recognition particle, SRP. Recognition of the signal is based on its hydrophobicity and secondary structure, rather than on sequence specificity; the central region of the signal sequence forms an α-helix in the exit tunnel of the ribosome. The SRP portion that interacts with the signal sequence mainly comprises a hydrophobic cleft enriched in Met residues, known as the M domain, in the SRP54 protein. The cryo-EM structure of the ribosome–nascent chain complex bound to SRP is shown on the right. Translation is temporarily arrested because of competition between SRP and elongation factors for binding to the ribosome. Binding of the SRP to the signal sequence causes structural rearrangements in the SRP core, allowing GTP binding. Following signal sequence recognition, the SRP–ribosome–nascent chain complex is recruited to the membrane via its interaction with SRα, a subunit of the membrane receptor complex. For this to occur, both SRP and SRα need to be in their GTP-bound forms. The GTPase activities of the two proteins are synchronized by the formation of a symmetrical dimer, which creates a composite active site, with the GTPases reciprocally activating each other. For the ribosome to switch from SRP binding to translocon binding, GTP hydrolysis must occur; the conformational rearrangements that occur upon GTP hydrolysis lead to release of SRP and docking of the ribosome on the translocon. Once the nascent peptide is in the translocon, translation is resumed; the nascent peptide then passes into the lumen, where the signal peptide is removed. (Left, adapted from Batey RT, Rambo RP, Lucast L et al. [2000] *Science* 287:1232–1239. With permission from American Association for the Advancement of Science. Right, from Halic M, Becker T, Pool MR et al. [2004] *Nature* 427:808–814. With permission from Macmillan Publishers, Ltd.)

Q residue. The organization of SNARE proteins in membranes and their mechanistic role in vesicle membrane fusion are presented in **Figure 18.11**.

Parenthetically, SNARE proteins are known targets for bacterial neurotoxins, such as botulin and tetanus toxin (see Box 18.1). This reflects their well-established role in mediating the docking of synaptic vesicles with the presynaptic membrane as part of the transmission of signals between nerve cells.

18.4 Proteolytic protein processing: Cutting, splicing, and degradation

Some proteins are synthesized as precursor molecules that need to undergo post-translational processing, sometimes involving proteolytic cleavage, to attain their functional mature form. Others are subject to protein splicing, a reaction very much analogous to the post-transcriptional splicing that occurs on RNA primary transcripts (see Chapter 14). Finally, there are situations where proteins need to be destroyed, either selectively or wholesale. We describe such modes of processing separately.

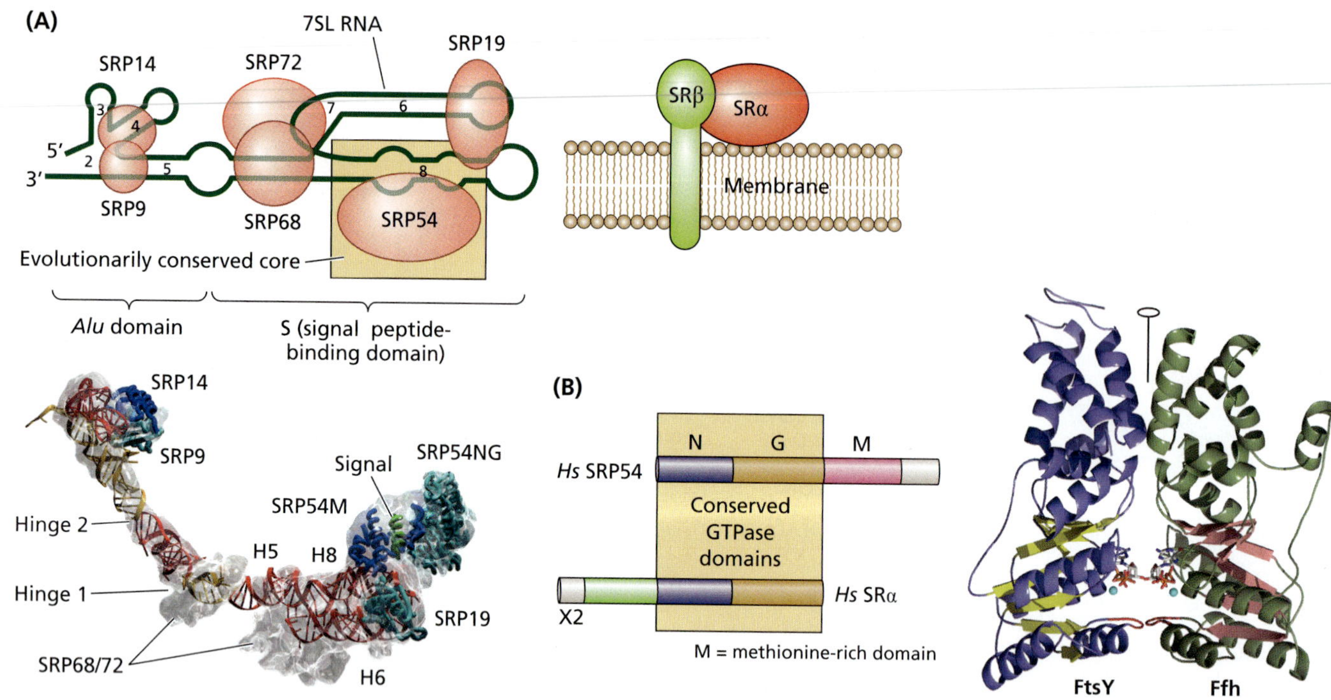

Figure 18.9 Composition and structure of human SRP and its membrane receptor. (A) (Top left) In mammals, SRP consists of one 7S RNA molecule, 300 nucleotides long, and six proteins. The complex is divided into two domains on the basis of nuclease cleavage: the *Alu* and S domains. The regions near the 5′- and 3′-ends of the mammalian SRP RNA are similar to the dominant *Alu* family of middle-repetitive sequences of the human genome. These sequences bind the SRP9–14 heterodimer and are involved in elongation arrest. The S domain binds to signal sequences and to receptors in the membrane, thereby promoting translocation. The SRP core, shown by a box, is universally conserved and consists of two proteins SRP54 and SRα, as well as helix 8 of the RNA. The molecular model of SRP is presented below the schematic. (Top right) The SR receptor is a dimer of two GTPases: SRα is a peripheral membrane protein, and SRβ is an integral membrane protein. (B) Domain structure of SRP54 and SRα. Both proteins are GTPases and share conserved GTP domains NG, an α-helical N domain packed against a G domain similar to those of other GTPases. The structure on the right is that of bacterial heterodimeric Ffh–FtsY NG domain complex; Ffh and FtsY are the bacterial counterparts of SRP54 and SRα, respectively. The α-helices in FtsY are shown in blue, and the β strands in yellow; the corresponding structures in Ffh are shown in green and brown. The two GTPase active sites are in direct apposition to form an active site chamber at the center of the G domains, with the bound nucleotides shown in stick representation. The structure is highly symmetric, and all secondary structure elements adopt the same orientation in both proteins. Conformational changes function allosterically to communicate to the other domains in the SRP, its receptor, and the translocon. Also, observations that the two GTPases behave as reciprocal GTPase-activating proteins can be understood by the formation of a shared catalytic chamber. (Adapted from Grudnik P, Bange G & Sinning I [2009] *Biol Chem* 390:775–782. With permission from Walter de Gruyter. A, bottom, from Halic M, Becker T, Pool MR et al. [2004] *Nature* 427:808–814. With permission from Macmillan Publishers, Ltd. B, right, from Egea PF, Shan S, Napetschnig J et al. [2004] *Nature* 427:215–221. With permission from Macmillan Publishers, Ltd.)

Proteolytic cleavage is sometimes used to produce mature proteins from precursors

Many proteins are shortened following synthesis to remove portions of the molecules that are not needed for the function of the mature protein. Thus, signal peptides at the N-termini of some proteins, which help them to traverse membranes, are removed early in the translocation process. In some cases, this is the only proteolytic cleavage needed to produce the final form of the protein. Other protein precursors undergo a series of very specific cleavages by specific enzymes to reach their functional form. The classic example of insulin processing is presented in **Figure 18.12**. The polypeptide chain initially synthesized is 110 amino acid residues long; the final form, which comprises two separate polypeptide chains connected by disulfide S-S bridges, contains only 51 residues. The first precursor, preproinsulin, is translocated through the ER, and then the leader sequence is cleaved. This allows the protein to fold and S-S bridges to form, with the production of proinsulin. Two additional specific endopeptidases then act in succession to remove the middle section, known as the C-peptide, and a couple of basic residues from the new ends of the chains. The existence of proinsulin, which is not active as a hormone, is believed to provide a physiological

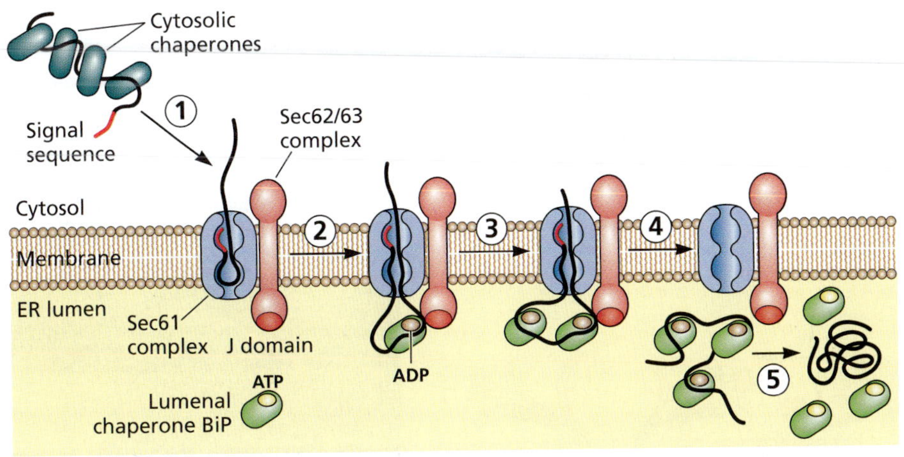

Figure 18.10 Post-translational translocation in eukaryotes involves a ratcheting mechanism. In this pathway, the Sec61 translocon complex partners with the membrane-bound complex Sec62/63 and, on the lumenal side, with the chaperone BiP, a member of the Hsp70 family of ATPases. The sequence of steps is as follows: (Step 1) The substrate is brought to the channel bound by chaperones in a translocation-competent conformation. Upon binding to the channel, the chaperones are released. (Step 2) The polypeptide is translocated by a ratcheting mechanism: it slides back and forth in the channel by Brownian motion until its binding to the ER lumenal chaperone BiP prevents reverse movement back into the cytoplasm; that is, BiP binding serves as a ratchet. Thus, the overall motion is unidirectional. ATP-bound BiP is in an open peptide-binding conformation; it interacts with a specific J domain of Sec63. This interaction causes rapid ATP hydrolysis and closure of the peptide-binding pocket around the translocation substrate. BiP interacts with practically any polypeptide that emerges from the channel; that is, it has very low binding specificity. (Step 3) As the length of the translocation substrate in the lumen increases, additional BiP molecules bind to it. (Steps 4–5) The polypeptide is now completely in the ER lumen; exchange of ADP for ATP opens the peptide-binding pocket to release the peptide and BiP. The signal peptide is removed. (Adapted from Rapoport TA [2007] *Nature* 450:663–669. With permission from Macmillan Publishers, Ltd.)

advantage: the inactive form can be stored in high concentrations and then quickly converted to the active form when needed.

Many enzymes require cleavage in order to attain their mature, functional forms. Such enzymes include a number of digestive-tract proteases, such as elastase, carboxypeptidase, and chymotrypsin, which need to be kept as inactive precursors before being secreted into the digestive system. Other examples are found among the caspases, discussed in this chapter, and also include the enzymes of the blood-clotting cascade (see Box 21.4). In this cascade, each member of the chain of enzymes must be activated by enzymatic attack by the previous member, ultimately yielding active thrombin, which cleaves fibrinogen to produce the clotting fibrin. Clearly, protein modification by chain cleavage plays a major role in biology.

Some proteases can catalyze protein splicing

Proteases are usually regarded as rather mundane enzymes because they catalyze a relatively simple reaction, the hydrolysis of peptide bonds. Many proteases participate

Table 18.2 Mechanics and energetics of peptide translocation.

Process	Mechanism	Free-energy source
post-translational translocation of bacterial proteins	a molecular finger pushes polypeptide through a pore	ATP hydrolysis to drive finger motion
post-translational translocation of eukaryotic proteins	ratcheting by successive BiP binding	ATP hydrolysis to clamp BiP onto the polypeptide
co-translational translocation of eukaryotic proteins	process of translation in ribosome drives the polypeptide through a pore	GTP hydrolysis to promote ratcheting in the ribosome

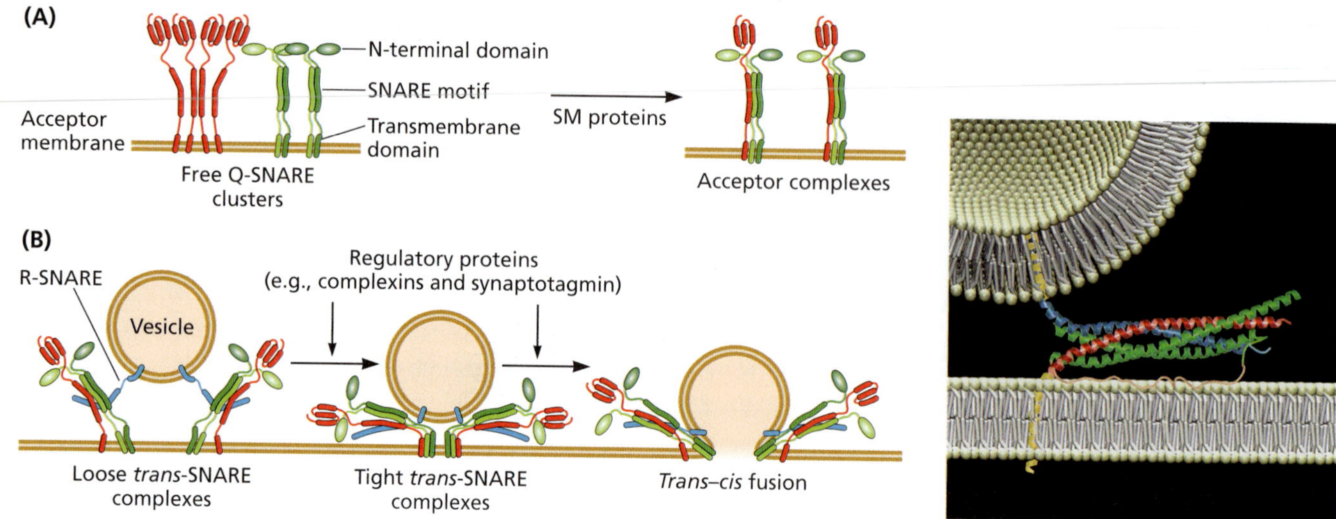

Figure 18.11 Model for vesicle docking and fusion. As an example, three kinds of Q-SNAREs are depicted on an acceptor membrane, in two shades of green and in red, and an R-SNARE, in blue, on the vesicle. (A) The SNAREs on the acceptor membrane are clustered within nanodomains, the stability of which depends on cholesterol; they assemble into acceptor complexes with the help of a group of SM proteins, which act as clasps that bind to both Q- and R-SNARE components, not depicted here. (B) The acceptor complexes then interact with the vesicular R-SNAREs to nucleate the formation of a four-helical *trans*-complex: the initial loose complex, in which only the N-terminal portions of the SNARE motifs interact, is zipped up into a tight complex, in which the SNARE motifs interact along their lengths, as shown in the structure to the right. As a result of zipping, mechanical force is exerted on the membranes, which might overcome the energy barrier for fusion. During fusion, the strained *trans*-complex relaxes into a *cis*-configuration. (Adapted from Jahn R & Scheller RH [2006] *Nat Rev Mol Cell Biol* 7:631–643. With permission from Macmillan Publishers, Ltd. B, right, courtesy of Dirk Fasshauer, University of Lausanne.)

in protein degradation at the end of a protein's useful life; this role may require little regulation. On the other hand, there are proteolytic processes that are highly regulated: these participate in cell signaling, in the turnover of regulatory proteins that must function over only short periods of time, and in protein maturation and activation. Mechanistically, proteolytic enzymes should be capable of performing the hydrolysis reaction in reverse: they should be able to catalyze peptide bond ligation. However, this reaction is thermodynamically disfavored in the aqueous medium of the cell; it can proceed only if free energy is available from a coupled, spontaneous reaction. In fact, such a coupled reaction is involved in **protein splicing**, where the energy is provided by simultaneous proteolysis of a nearby peptide bond, all within the protease active site.

In protein splicing, different portions of a polypeptide chain are cut and then spliced together to form a new molecule. Protein splicing is one of several **transpeptidation** reactions, which are defined as the transfer of an amino acid or a peptide from one peptide chain to another. Transpeptidation is best characterized in bacteria, where it occurs as part of the process that attaches proteins to peptidoglycans in the cell wall.

Experiments since 2004 in animal and plant cells have revealed that protein splicing also occurs in eukaryotes. The reason that this modification remained undetected in eukaryotes for so long is that it does not leave a mark on the protein, and can be detected only by comparing the protein sequence with that of its gene or precursor protein. The first convincing example of eukaryotic protein splicing came from studies of the mammalian immune system. In an attempt to identify cancer-specific antigens on the cell surface that are recognized by T lymphocytes, researchers identified peptides bound to major histocompatibility complex, MHC, proteins that did not correspond to contiguous segments of any protein expressed in the cell. In this case, the protease that performed the transpeptidation reaction was the proteasome itself. Importantly, the proteasome-mediated protein splicing reaction is not limited to tumor cells. A model explaining how protein splicing may occur is presented in **Figure 18.13**.

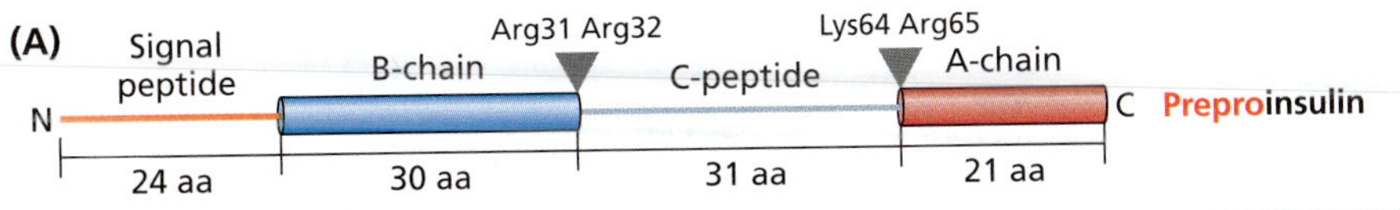

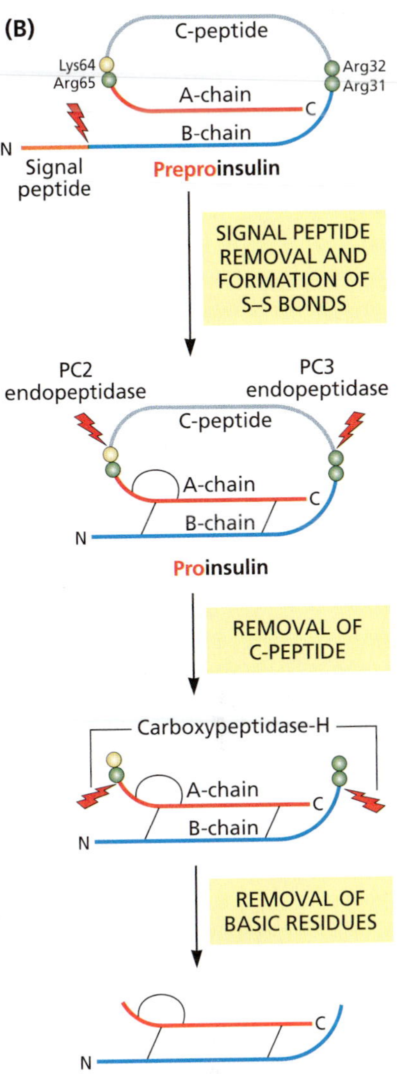

Figure 18.12 Post-translational processing from a preproinsulin precursor to the mature functional insulin. (A) Structure of preproinsulin and (B) processing pathway that creates mature insulin. Folding of preproinsulin probably requires prior removal of the signal peptide. Only then can the correct disulfide bonds be formed between SH groups, indicated by black lines.

Another example of protein splicing is the **circular permutation** of the jack bean lectin known as concanavalin A or ConA (**Figure 18.14**). Lectins are sugar-binding proteins with high specificity for the sugar moieties they bind; ConA binds mannose/glucose. They are widespread in plants, with a well-established role in the processing of seed storage proteins. In the case of ConA, the protease that both cleaves some peptide bonds and creates new ones belongs to a class of proteases called legumains. These proteases specifically cleave peptide bonds at asparagine residues; they are also known as asparaginyl endopeptidases. The final product of this series of steps is a protein that represents a circular permutation of the original amino acid sequence.

Today, we know of more than 100 proteins that adopt a circular form. Examples from mammals, plants, and bacteria are presented in **Figure 18.15**. In most cases, the physiological significance of circularization remains to be elucidated. One thing is certain: "tying up the loose proteins ends," in the words of D. J. Craik, creates proteins of extremely high stability and activity. Circular proteins can withstand boiling, extremes of pH, and the action of proteolytic enzymes with their structure and function intact. As Craik puts it, they are "a tough crowd."

Controlled proteolysis is also used to destroy proteins no longer needed

There are many instances in the life of a cell when a particular protein is no longer needed; indeed, its continued presence may be deleterious. Examples are found in proteins involved in regulation of the cell cycle. Likewise, there are situations in eukaryotic development when particular cells, or even whole tissues, are no longer appropriate. A classic example is found in human development, where the embryonic webbing tissue between digits must be fully destroyed before birth. These two kinds of situations require different mechanisms for protein hydrolysis.

An important example of a mechanism for highly regulated and specific protein destruction is the ubiquitin-initiated proteasome system, which uses covalent tagging of unneeded proteins to mark them for destruction. This is described in the following section. Much less specific proteolysis is a part of apoptosis, or programmed cell death, which can destroy the proteins of entire cells or tissues. Here a family of proteases, termed **caspases**, is activated by appropriate signals. Apoptosis is described in **Box 18.2**.

18.5 Post-translational chemical modifications of side chains

Modification of side chains can affect protein structure and function

A much more subtle way of modifying protein structure and/or function is through chemical modification of side chains on a previously synthesized protein molecule. These modifications are important regulators of protein activity. The numerous different side-chain modifications are all reversible. Each uses its own enzymatic machinery to modify specific amino acids, usually in a sequence-specific context. We covered poly(ADP)ribosylation and methylation, in the context of histone proteins and their role in controlling gene expression, in Chapter 12. Other modifications were mentioned in Chapter 3 (see Table 3.3). Here, we discuss in a more systematic manner the major post-translational modifications: phosphorylation, acetylation, glycosylation,

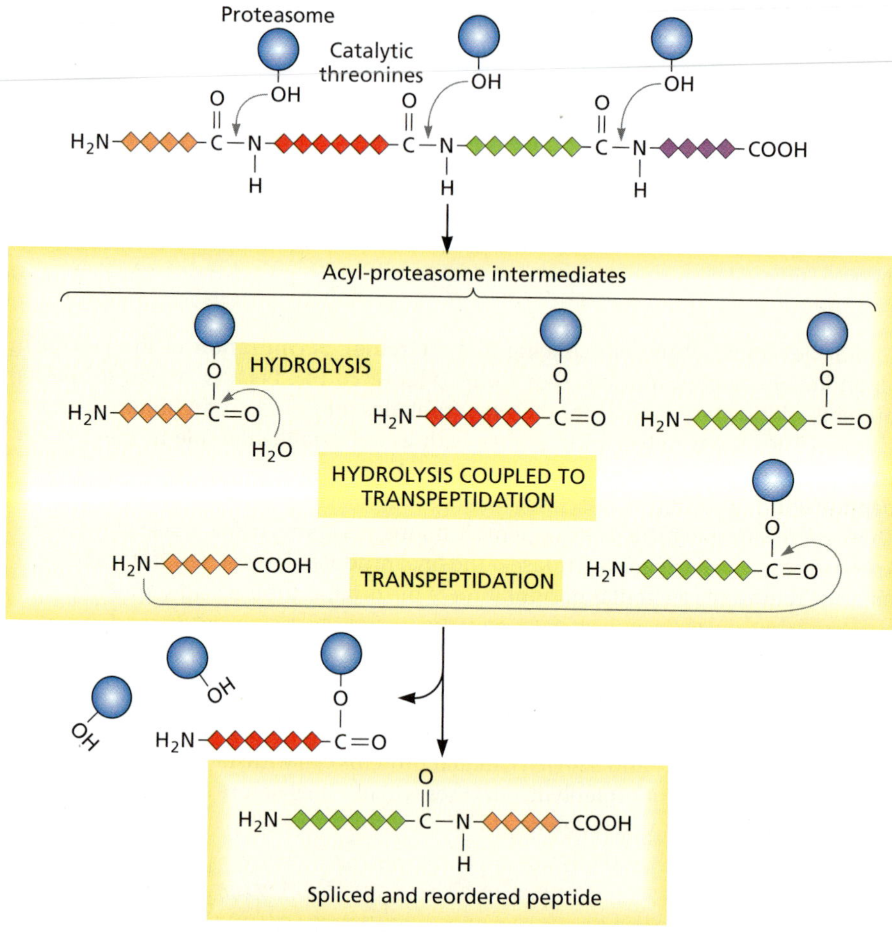

Figure 18.13 Model for protein splicing reaction inside the proteasome. Blue spheres represent the catalytically active β-subunits of the 20S core proteasome (see Figure 3.22) with the hydroxyl group from the N-terminal threonine side chain. Diamonds represent amino acid residues in the sequence of the polypeptide chain that undergoes splicing, with sections color-coded for clarity. The splicing reaction occurs through transpeptidation, involving the production of transient acyl-enzyme intermediates that are rapidly hydrolyzed during proteolysis. The N-terminal amino groups of the peptides in the intermediates compete with water molecules to attack the ester bond in the C-termini of other intermediates, resulting in a transpeptidation reaction that produces the spliced peptide. This mechanism also allows the reordering of the peptide fragments before religation. In the reaction shown, a four-amino-acid segment shown in orange and a six-residue segment shown in green are excised and ligated in reverse order. The religation reaction of the isolated green and orange peptide segments of the final peptide cannot occur unless the C-terminal peptide tail, shown in purple, is present. These observations indicate that the proteasome can catalyze peptide bond formation only when the process is coupled to peptide bond hydrolysis, which provides the energy needed for the formation of the new peptide bond. The model also shows that the splice site need not be highly conserved: once an intermediate is formed at the protease active site, ligation to any free N-terminus can readily occur. (Adapted from Warren EH, Vigneron NJ, Gavin MA et al. [2006] *Science* 313:1444–1447. With permission from American Association for the Advancement of Science.)

ubiquitylation, and sumoylation. Most of these modifications do not act in isolation: there is extensive cross-talk, either positive or negative, among them. Methods are being developed that make it possible to determine which modifications coexist on a single molecule. The data obtained by such methods will be invaluable in deepening our understanding of the interplay among modifications and their biological roles. It may well be that the relative paucity of genes in vertebrate genomes is compensated for, in part, by the extensive and highly selective use of post-translational modifications and their interactions.

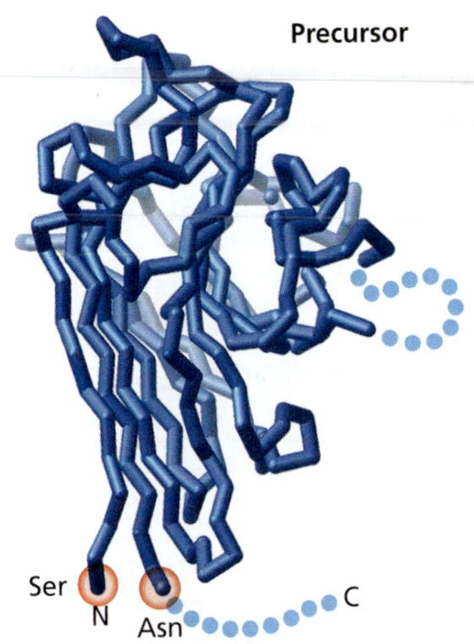

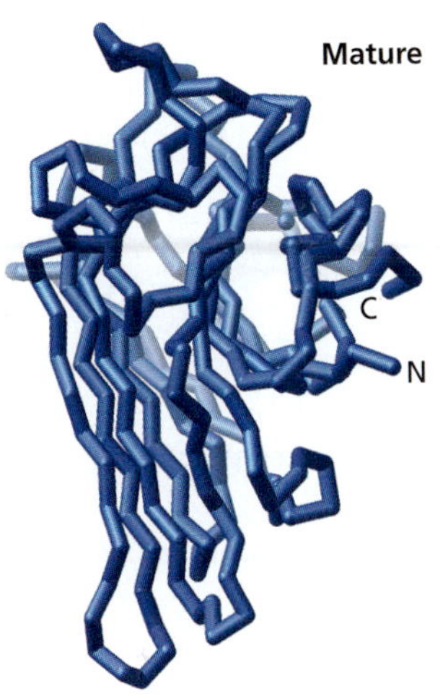

Figure 18.14 An example of protease-catalyzed protein splicing. The example shown is post-translational circular permutation of concanavalin A or ConA, a lectin protein, 237 amino acid residues long, that is found in the seeds of jack beans, *Canavalia ensiformis*. The structure of ConA is composed mainly of β-sheets, with only minor helical regions. The precursor form of the protein, synthesized by ribosomes, possesses an extra tail and an extra loop, shown by blue dots. The mature form is produced by clipping off the loop to produce the new ends of the chain and by sealing up the two original ends of the chain into a new loop. The enzyme responsible for the cleavages belongs to the legumain family of proteases that cleave peptide bonds at asparagine residues; these enzymes are alternatively known as asparaginyl endopeptidases. Note that the splicing leads to circular permutation of the protein sequence. (From Goodsell DS [2010] Molecule of the Month [10.2210/rcsb_pdb/mom_2010_4].)

Phosphorylation plays a major role in signaling

Phosphorylation was one of the first post-translational modifications to be discovered; for a historical account, see **Box 18.3**. In this reversible modification, a phosphate group is added to a serine or threonine residue, with a small percentage of modifications also occurring on tyrosines and histidines. Addition of the phosphate group is carried out by **kinases**, and its removal is catalyzed by **phosphatases**. Phosphorylation is used by the cell to regulate a wide range of processes, by regulating the activity of the enzymes that participate in these processes. Before the discovery of the regulatory role of phosphorylation, it was believed that enzymes are simply regulated by their turnover, that is, by the rate at which they are synthesized and degraded. However, because these processes are relatively slow, they could not account for the quick regulation that occurs, for example, in response to fast environmental changes. Phosphorylation attracted considerable attention when Edmond Fischer and Edwin Krebs discovered the first cascade of hormonal regulation, in which each successive enzyme in the cascade was a kinase that was activated by phosphorylation by a preceding kinase (see Box 18.3). We know today that such phosphorylation cascades form the molecular basis of practically all signaling pathways, and there are dozens of them. A cascade can produce, from a very few starting signals, a sudden large increase in the final product.

The power of contemporary methods, both experimental and computational, allows an unprecedented look at protein modifications in general and phosphorylation in particular. Kinase–substrate relationships have been studied by protein chip analysis, large-scale genetic screens using kinase overexpression strategies, and large-scale genetic interaction analyses. The latter analyses record negative genetic interactions, in which two mutations in combination cause a stronger growth defect than expected from two single unrelated mutations, and positive genetic interactions, in which the double mutant is usually healthier than the single mutant. When performed in yeast, such analysis found an unexpectedly large number of positive genetic interactions between kinases, phosphatases, and their substrates. **Figure 18.16** illustrates the kind of interaction maps obtained from genetic analysis.

The significance of phosphorylation becomes especially notable when one looks at results from systems analysis of whole proteomes. A breathtaking number of phosphorylated proteins and sites are identified in humans on the basis of quantitative mass spectrometry (**Figure 18.17**). The majority of proteins contain multiple phosphorylation sites, which can be both singly and multiply phosphorylated: single, 68%; double, 23%; triple, 5%; quadruple, 2%; and more than quadruple, 1%. The distribution of phosphorylation sites, in terms of amino acid modified, is 86% Ser, 12% Thr, and 2% Tyr. When the state of phosphorylation of individual sites within the same protein was followed over time after stimulation of HeLa cells with epidermal growth factor, EGF, most proteins that had an EGF-regulated site had at least one additional phosphopeptide that exhibited different dynamics. This observation confirms earlier studies with some signaling proteins, indicating that phosphorylation typically serves different functions on different sites of the protein. It is also clear that proteins with different locations in the cell can be differentially targeted by phosphorylation. Thus, nuclear proteins and proteins of the cytoskeleton are preferentially modified, whereas those in the mitochondria, for example, are undermodified. The challenge remains to

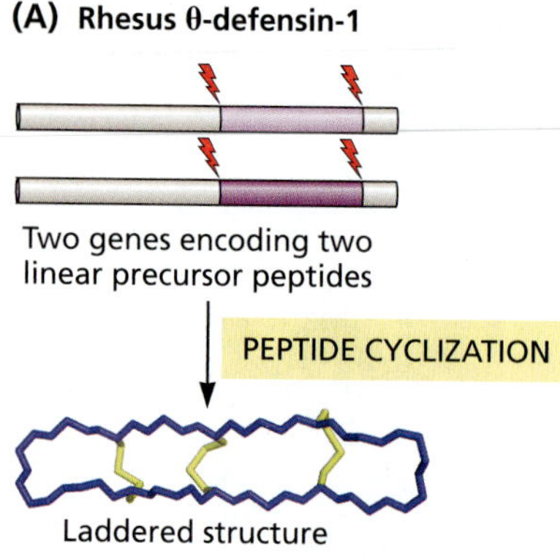

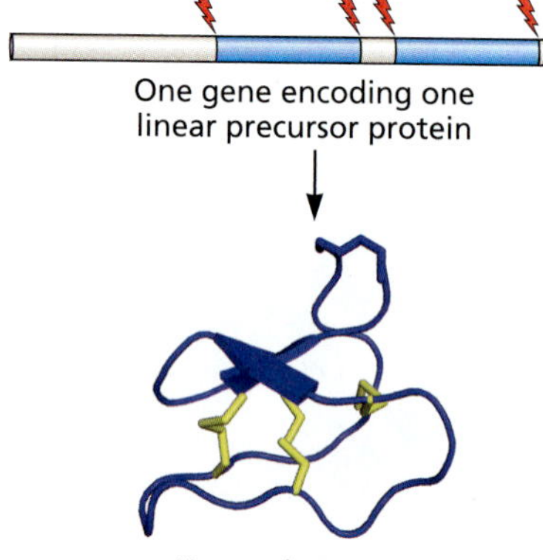

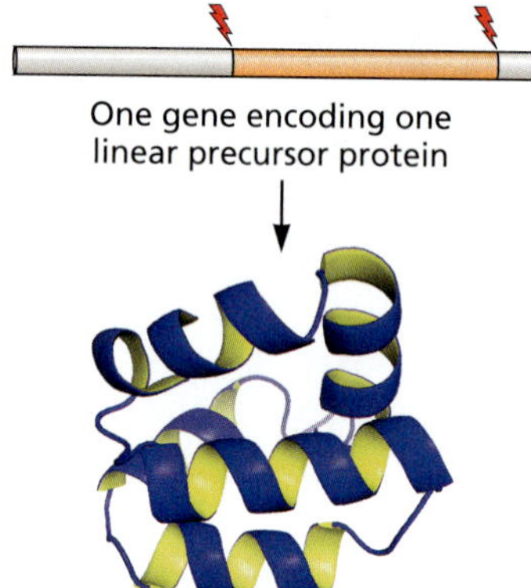

Figure 18.15 Circular proteins in different organisms. These proteins are synthesized as linear precursors by the normal translation of mRNAs; specific excision reactions, denoted by red lightnin bolts, produce fragments that are then spliced head-to-tail. (A) In the case of the defensin from rhesus monkey, two genes each code for half of the 18-amino-acid mature peptide, which is produced by a double head-to-tail ligation reaction. The circular peptide product has the appearance of a ladder that is stabilized by three S–S bridges. (B) The plant cyclotide kalata B1 is the active ingredient of medicinal tea used by tribeswomen in the Congo to facilitate childbirth. The small circular protein has a knotted arrangement of disulfide bonds that contributes to its exceptional stability: the protein withstands boiling, extremes of pH, and proteolytic enzymes. (C) Bacterial bacteriocins are compounds that have antimicrobial activities against a broad range of pathogenic microbes, mainly of the Gram-positive type. They act by disrupting cell membranes. Bacteriocins, with ~35–70 amino acid residues, are exceptionally stable; some, such as AS-48, are circular helix bundles. (Adapted from Craik DJ [2006] *Science* 311:1563–1564. With permission from American Association for the Advancement of Science.)

understand the biological relevance of such data, which are usually used as leads in the experimental validation of individual systems or processes.

Acetylation mainly modifies interactions

Another common post-translational modification of proteins is acetylation, which occurs on lysine residues. Both phosphorylation and acetylation change the charge of a protein molecule, and both may affect protein–protein or protein–DNA interactions. In Chapter 12, we discuss some of the effects that histone acetylation has on gene activity. These effects are either direct acetylation-mediated changes in chromatin structure or effects that are brought about by specific proteins, known as readers of the specific modification (see Figures 12.9–12.14); the enzymes that put the respective modification in place are usually referred to as writers.

As for phosphorylation, proteomewide studies have contributed enormously to our understanding of acetylation. A recent example of such an analysis based on quantitative mass spectrometric data is presented in **Figure 18.18**. The grouping of acetylated proteins according to their function shows the involvement of acetylation in practically every major cellular function, with the number of acetylated proteins being especially large in the cell-cycle and RNA-splicing categories. Even more revealing is the complexity of the acetylated protein–protein networks. The specific example shown in Figure 18.18 depicts the network of proteins involved in DNA repair. It is amazing to realize that, out of the 54 proteins identified as acetylated in this functional category, only five had been previously recognized as undergoing modification. The percentage of proteins identified as acetylated by routine methods as a proportion of those identified by the proteomic analysis is even smaller in other categories such as RNA-splicing, cell-cycle, and ribosomal proteins.

Several classes of glycosylated proteins contain added sugar moieties

The formation of conjugates between polysaccharides and proteins or peptides, known as glycosylation, is a widespread modification that has multiple roles in the cell. In general, such conjugates are classified into three different categories: proteoglycans, peptidoglycans, and glycoproteins.

Proteoglycans are complexes of proteins in which a specific class of unbranched polysaccharide chains is formed by repeating disaccharide units. They are typically present in the connective tissue of multicellular animals. Well-known examples are the complexes containing hyaluronic acid found in the viscous solutions that lubricate joints.

Peptidoglycans are compounds in which the polysaccharides are linked to small peptides, usually forming large cross-linked complexes. They are found in bacterial cell walls. In 1884, bacteriologist Hans Christian Gram introduced a staining method that could distinguish the two major classes of bacteria, based on interaction of the stain with peptidoglycans that have different organization in the two bacterial classes. In

Box 18.2 Apoptosis: Physiological, cellular, and molecular perspectives Apoptosis, or programmed cell death, has been recognized since the middle of the nineteenth century, but it was not until the mid-1960s that apoptotic research became active, when John Kerr and two collaborators at the University of Aberdeen published a seminal paper, in which the term apoptosis was introduced. The term is of Greek origin, from *apo*, from/off/without, and *ptosis*, falling, and translates as dropping off: of petals from flowers, leaves from trees, and so on.

Apoptosis is crucial to the normal development of multicellular organisms, in which the balance between cell proliferation and apoptosis is responsible for shaping tissues and organs, as well as for maintaining homoeostasis in adult tissues. Thus, for example, the embryonic separation of toes and fingers from each other relies on apoptosis of cells located in between. Apoptosis also participates in control of the immune system: T lymphocytes, before being released from the thymus into the bloodstream, are checked to ensure their ability to be effective against foreign antigens and to not attack the organism's own molecules. Ineffective or self-reactive T cells are destroyed by apoptosis.

Inappropriate levels of apoptosis characterize several diseases. Cancer, for example, is viewed as a disease in which too little apoptosis occurs, allowing cancerous cells to proliferate out of control. Moreover, mutations in apoptotic pathways often lead to the formation of cells that are resistant to treatment with radiation or chemicals that would normally trigger apoptosis. Autoimmune diseases, such as rheumatoid arthritis, are also characterized by excessive proliferation of synovial cells that are resistant to apoptotic stimuli. On the other hand, too much apoptosis seems to occur in some neurodegenerative diseases, such as Alzheimer's or Parkinson's diseases, accounting for the progressive loss of neurons.

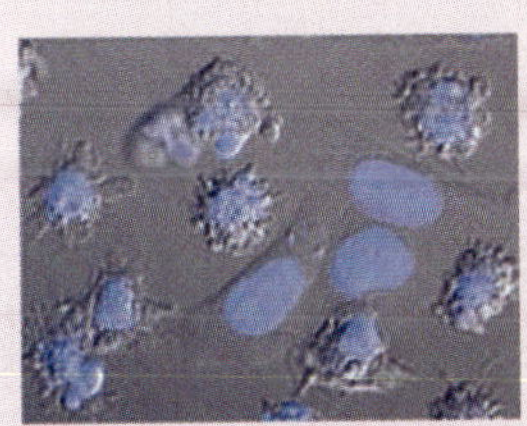

Figure 1 Apoptotic cell at the final stages of apoptosis. HeLa cells grown in culture were subjected to stress. The image is that of a mixture of viable and apoptotic cells. The cells undergoing apoptosis retract and exhibit extensive plasma membrane blebbing. Their nuclei are condensed and/or fragmented into several pieces. Nuclei were stained with Hoechst dye, blue color. (From Taylor RC, Cullen SP & Martin SJ [2008] *Nat Rev Mol Cell Biol* 9:231–241. With permission from Macmillan Publishers, Ltd.)

Apoptotic cells undergo a series of typical morphological changes. (1) The cell shrinks and becomes spherical due to breakdown of the proteinaceous cytoskeleton. The cytoplasm appears dense, and the organelles are tightly packed. (2) Chromatin is condensed into compact patches against

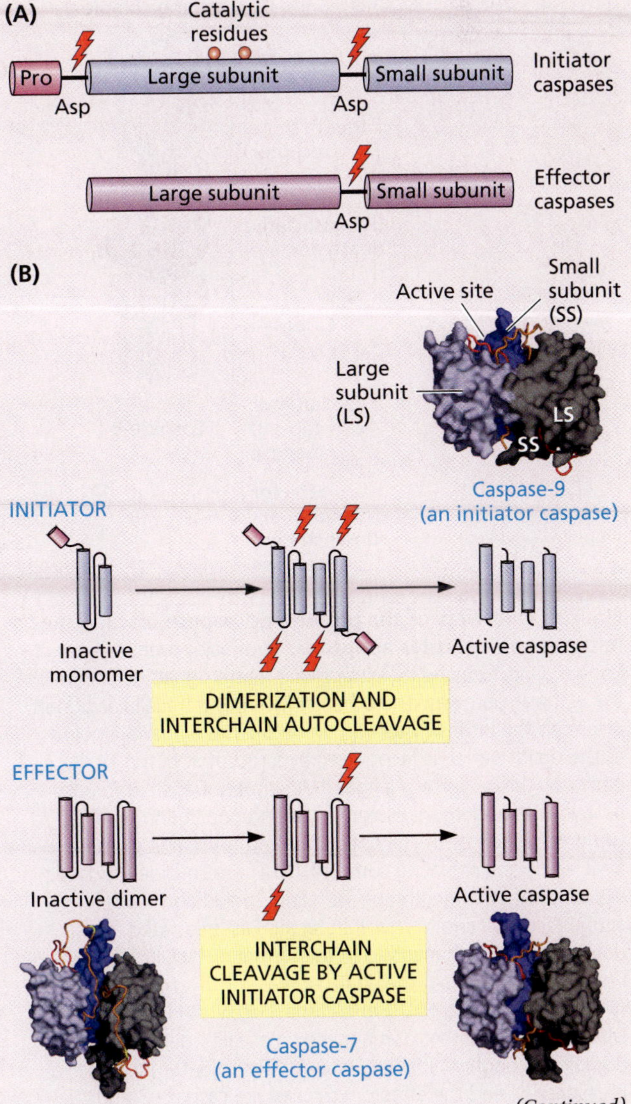

Figure 2 Architecture and activation of the two classes of caspases. All caspases are initially in the form of inactive, single-chain proteins that undergo activation by limited proteolytic cleavage. (A) Domain structure of the inactive or zymogen caspase forms. Initiator caspases are composed of a pro domain and a catalytic domain, which contains two covalently linked subunits. The inactive forms of the effector caspases lack a pro domain. The final mature forms of all caspases contain two separate subunits and are devoid of pro domains.
(B) Processing of zymogens to create mature forms occurs through site-specific proteolytic cleavages following Asp residues: the initiator caspases cleave themselves in a process known as autoproteolysis, whereas the effector caspases are cleaved by the upstream caspases in the pathways. The inactive form of initiator caspases is monomeric; activation occurs through dimerization followed by cleavage. By contrast, the inactive form of effector caspases is an uncleaved dimer, which undergoes cleavage for activation. Following activation, additional proteolysis leads to the mature caspase forms that are more stable and can be subjected to regulation. Surface renderings of the available crystal structures of caspase-9, an initiator caspase, and caspase-7, an effector caspase in both inactive and active forms, are shown above or below the respective schematics. The subunits of each monomer are presented in different shades of the same color. (Adapted from Tait SWG & Green DR [2010] *Nat Rev Mol Cell Biol* 11:621–632. With permission from Macmillan Publishers, Ltd. B, structures, from Riedl SJ & Salvesen GS [2007] *Nat Rev Mol Cell Biol* 8:405–413. With permission from Macmillan Publishers, Ltd.)

(Continued)

Box 18.2 *(Continued)*

the nuclear envelope; this process is known as pyknosis, from Greek *pyknono*, to thicken up, to close or to condense, and is attributed to fragmentation of chromatin DNA into nucleosome-sized fragments. (3) The nucleus breaks into discrete bodies, and the cell membrane begins to form irregular buds or blebs, which eventually form separate membrane-bound particles. In the final stage, these apoptotic bodies are phagocytozed by macrophages (Figure 1).

Apoptosis is a two-step proteolytic process, in which two proteolytic enzymes or caspases are activated in succession. The name caspase is derived from Cys-dependent Asp-specific protease: their catalytic activity is determined by a conserved cysteine residue, or actually a catalytic duo of a cysteine and a histidine, and they have an unusual stringent specificity for cleaving after Asp residues. Thus, they do not degrade proteins totally; rather they produce one or two cuts in a substrate.

Caspases exist as inactive molecules or zymogens, which are themselves activated by limited proteolytic cleavages (Figure 2). The initial cleavage occurs in response to the death signal. Activation of this first caspase, known as an initiator **caspase** or apical caspase, leads to cleavage of the effector caspase or executioner caspase, which then cleaves a vast number of various cellular proteins, leading to cell destruction.

The two major pathways that trigger and execute apoptosis—the **extrinsic and intrinsic pathways of apoptosis**—are depicted in Figure 3. The choice between them is determined by the nature of the death signal. A signal that comes from outside the cell activates the extrinsic pathway. Stress conditions inside the cell lead to the intrinsic pathway. The death signals

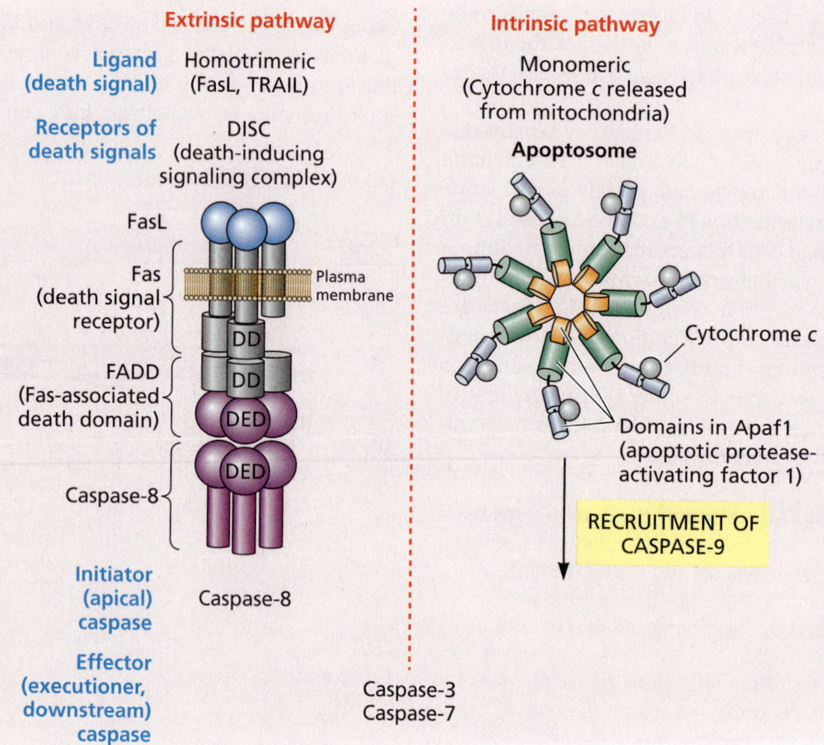

Figure 3 Overview of the proteolytic caspase cascade that initiates and executes apoptosis. Two major pathways are known. One, induced by extracellular death signals, is known as the extrinsic pathway. The other, induced by intracellular signals, is termed the intrinsic pathway. In both cases a ligand, known as the death signal, is recognized by a receptor. In the extrinsic pathway, the receptor is a ligand-dependent transmembrane protein that undergoes oligomerization upon ligand binding. Oligomerization of the receptor, Fas, and formation of the death-inducing signaling complex, DISC, are considered classic examples of transmembrane signaling. In addition to the actual receptor, another protein factor, Fas-associated death domain or FADD, is involved in the formation of DISC. Both Fas and FADD contain the death domain or DD, through which they interact. FADD and caspase-8, which is part of DISC, contain another domain known as DED, death effector domain, through which they interact. The initiator protease is caspase-8, and the effector caspases are caspase-3 and caspase-7. In the intrinsic pathway, which occurs upon cellular stress such as DNA damage, the death signal is the release of cytochrome *c* from mitochondria. The signal is sensed by a soluble cytoplasmic receptor, the apoptosome, which contains a heptamer of Apaf1, apoptotic protease-activating factor 1, and either ATP or dATP. Hydrolysis of bound dATP to dADP leads to structural rearrangements in Apaf1, which are important for further formation of an active apoptosome. dATP nucleotide exchange leads to Apaf1 oligomerization and active apoptosome formation. The active wheel-like structure with bound cytochrome *c* can now bind and activate caspase-9. The initiator protease is caspase-9, and the effector caspases are caspase-3 and caspase-7. Thus the end points of the extrinsic and intrinsic pathways converge. (Left, adapted from Bao Q & Shi Y [2007] *Cell Death Differ* 14:56–65. With permission from Macmillan Publishers, Ltd. Right, adapted from Riedl SJ & Salvesen GS [2007] *Nat Rev Mol Cell Biol* 8:405–413. With permission from Macmillan Publishers, Ltd.)

themselves are different, both in the way they are sensed by the cell and in the first enzymes in the caspase cascade. The two pathways converge at the very last or executional stage, cell degradation.

The fact that proteolysis is irreversible necessitates the regulation of caspase activation. Indeed, once caspases are activated, apoptosis will occur. The cell uses three strategies to regulate caspases: caspase inhibitors, caspase degradation, and decoy inhibitors. In caspase inhibition, a protein or peptide structural mimic of the physiological substrate blocks the substrate-binding cleft. Some viruses use this strategy to defeat the host's natural defense mechanisms. A very selective inhibitory mechanism in humans acts through the negative regulator of apoptosis, XIAP (**Figure 4**). This mechanism of inhibition relies on increased protein levels, which result from an IRES-dependent translation initiation mechanism. An IRES-dependent initiation is also involved in the control of apoptosis by the positive regulator Apaf1. Caspases can be degraded by the proteasome via ubiquitylation, and XIAP's RING domain may participate in this reaction. Finally, decoy proteins carry structural resemblance to caspase pro domains, competing with caspases for activation platforms.

In summary, apoptosis is a very complex, multistep, highly regulated process. It provides examples of proteins, both positive and negative regulators of the process, whose translation under specific circumstances is key to the biological outcome of their action. Translation of these proteins is achieved through the alternative use of cap-dependent and IRES-dependent initiation mechanisms. Much remains to be done to fully understand how this transition from one type of initiation to a different type is regulated.

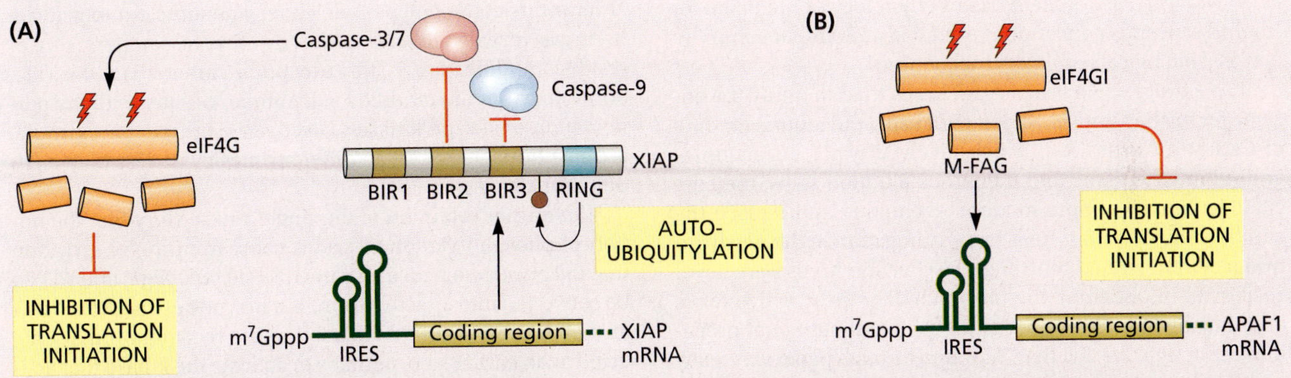

Figure 4 Translational regulation of the two major apoptotic regulators, XIAP and APAF1. XIAP, X-chromosome-linked inhibitor of apoptosis, is a multidomain protein that efficiently and selectively inhibits caspase-9 through its BIR3 or baculoviral IAP repeat domain, and caspase-3 and caspase-7 through its BIR2 domain. The RING or really interesting new gene domain of XIAP is an E3 ubiquitin ligase and automodifies a neighboring region of the protein, in addition to ubiquitylating protein targets to direct them to the ubiquitin–proteasome system. The complex network of interactions leads to either cell survival or cell death, depending on the exact stress condition. (A) Under certain conditions of cellular stress, cap-dependent initiation of protein synthesis is inhibited due to caspase cleavage of the initiation factor eIF4G. Translation of the caspase inhibitor XIAP is not affected because under these conditions its translation initiation is mediated by an IRES at the 5′-UTR of the mRNA. XIAP levels increase despite the reduction in overall protein synthesis, resulting in negative regulation of the caspases and increased survival of cells. Once the stress is removed, the cells can restore their normal metabolic activities. (B) Under other stress conditions, cell death is regulated through a transition from cap-dependent to IRES-dependent initiation, this time by a positive regulator of apoptosis, APAF1. APAF1 forms the skeleton of the soluble receptor platform, the apoptosome, which activates the initiator caspase-9. Caspase-mediated cleavage of eIF4GI gives rise to a specific middle fragment, M-FAG, which selectively enhances IRES-mediated translation of APAF1. Increased APAF1 levels stimulate apoptosis. (Adapted from Holcik M & Sonenberg N [2005] *Nat Rev Mol Cell Biol* 6:318–327. With permission from Macmillan Publishers, Ltd.)

Gram-negative bacteria, the peptidoglycan layer is very thin and is located between the inner and outer plasma membranes. By contrast, Gram-positive bacteria do not possess an outer membrane, and the peptidoglycan cell wall is much thicker; this thick layer retains the Gram stain in these bacteria. Gram staining is still widely used in medical microbiology.

The biological importance of peptidoglycans for bacteria is illustrated by the fact that inhibition of peptidoglycan synthesis leads to total inhibition of growth. The antibiotic **penicillin** binds to, and irreversibly inhibits, one of the enzymes that participates in the biosynthesis of peptidoglycans. This prevents further bacterial growth and proliferation. Sir Alexander Fleming was awarded the 1945 Nobel Prize in Physiology or Medicine for his 1928 discovery of penicillin; he shared the prize with Sir Howard Florey and Sir Ernst Chain, whose work led to the purification of the antibiotic for use in clinical practice.

Box 18.3 Discovery of protein phosphorylation and phosphorylation cascades: The work of Edmond Fischer and Edwin Krebs The discovery of protein phosphorylation, its role as a regulator of enzymatic activity, and the existence of phosphorylation cascades has had a large impact on numerous, seemingly unrelated fields in the life sciences. Cascades are activated in response to environmental stimuli, transforming even small signals into major biochemical responses. Protein phosphorylation was discovered by Edmond Fischer and Edwin Krebs, both from the University of Washington in Seattle, as part of their effort to understand the regulation of a rather inconspicuous enzyme, glycogen phosphorylase. They were interested in this specific enzyme because of its involvement in carbohydrate metabolism and its regulation by hormones (**Figure 1**). At the onset, in the mid-1950s, Fischer and Krebs had no idea that they were dealing with a very fundamental and widespread phenomenon. Their work was recognized with the 1992 Nobel Prize in Physiology or Medicine for "their discoveries concerning reversible protein phosphorylation as a biological regulatory mechanism."

It all started when Fischer and Krebs tried to purify the enzyme for further studies. Rather than using the routine method of Carl and Gertie Cori, which involved water extraction of ground muscle tissue and numerous filtration steps, they decided to "go try a more modern" technique and replace the cumbersome filtration steps by centrifugation in the newly introduced refrigerated centrifuge. Fischer and Krebs were never successful in obtaining the phosphorylase in an active form, and they had no choice but to go back to the original procedure, although they felt that "filtration through paper was really pathetic." In following the Cori procedure, they now checked every single step, analyzing every fraction from beginning to end. To their amazement, it turned out that the original extract contained the enzyme in an inactive form, which was somehow activated through the procedure. Fischer and Krebs burnt filter paper prewashed with dilute acid, to remove possible contaminants that might have been the activation agents, in a muffle furnace and sprinkled ashes from the paper into the original extract. Eventually, they figured out that the agent which activated the inactive phosphorylase was the calcium that contaminated the paper. ATP and Mg^{2+} were also needed for the activation.

The requirement for ATP suggested some sort of phosphorylation reaction, but what was phosphorylated? After acquiring some γ-^{32}P-labeled ATP from Arthur Kornberg, as no radioactive labels were commercially available at that time, Fischer and Krebs realized that the glycogen phosphorylase enzyme itself was actually activated through phosphorylation. Phosphorylation of a single seryl residue close to the N-terminus of the polypeptide chain was sufficient to activate it. As was revealed in later studies by other researchers, single-residue modification is the exception rather than the rule. Some enzymes are modified at a number of sites, with the participation of several kinases. Glycogen synthase, for example, is phosphorylated at seven different sites by no less than eight different kinases.

The further evolution of the single kinase story into the notion of phosphorylation cascades came from the observation that the purified inactive enzyme needed only Mg^{2+} and ATP to be converted into its active form. But in crude extracts, Ca^{2+} was also absolutely required. This led to the recognition that Ca^{2+} acted at an earlier step, perhaps to activate the kinase that activates the phosphorylase. Further research led to a full description of the phosphorylation cascade illustrated in Figure 1B.

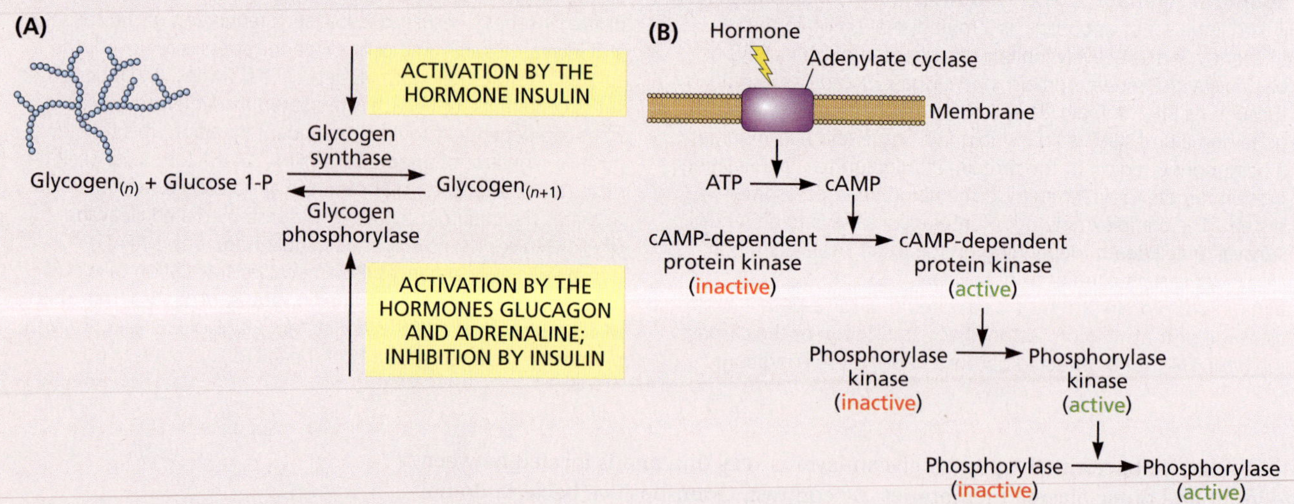

Figure 1 Glycogen phosphorylase and its regulation through phosphorylation. (A) Glycogen phosphorylase is instantly activated under stress conditions, when an organism needs energy for fight or flight. When a stress signal is sent to the brain, it causes the adrenal glands to secrete adrenaline; adrenaline is quickly carried by the blood to stimulate the activity of glucagon, the peptide hormone whose action leads to hydrolysis of stored glycogen to glucose 1-phosphate. Metabolism of glucose 1-phosphate produces energy in the form of ATP. Insulin, on the other hand, inhibits the activity of glycogen phosphorylase and promotes the storage of glucose into glycogen, by stimulating glycogen synthase. (B) Hormonal control of glycogenolysis through a phosphorylation cascade. The adrenal hormone adrenaline triggers the activation of adenylate cyclase, the enzyme that synthesizes cyclic adenosine monophosphate, cAMP, from ATP. cAMP acts as a second messenger that is involved in many of the signaling pathways activated in response to external stimuli. In the specific pathway shown, cAMP activates cAMP-dependent protein kinase, which in turn activates phosphorylase kinase, which in turn activates glycogen phosphorylase. Such phosphorylation cascades form the basis of numerous signaling pathways recognized to date.

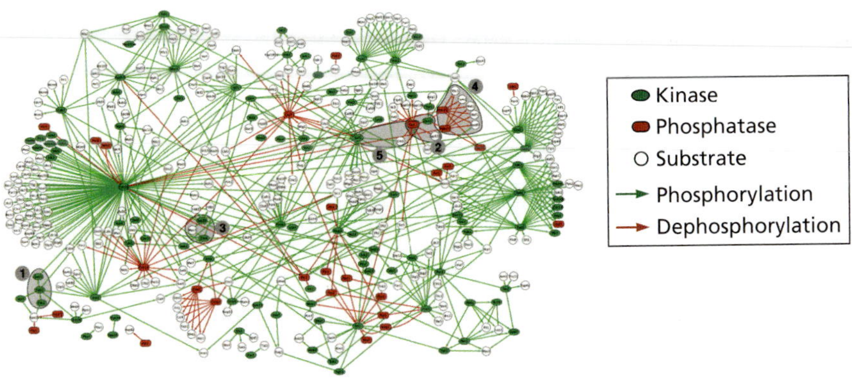

Figure 18.16 Experimentally recognized phosphorylation and dephosphorylation events in yeast. The network contains 795 well-characterized relationships involving kinases, phosphatases, and their substrates and was manually curated by Nevan Krogan and collaborators as a tool for validating the results from their high-throughput genetic-interaction phosphorylation profile. This profile identified 100,000 pairwise genetic interactions including virtually all kinases, phosphatases, and their substrates. Not surprisingly, two kinases, or two phosphatases, can share the same substrate, and kinase/phosphate pairs can have a common substrate. The areas in gray depict such cases, which were targeted for further investigations. (From Fiedler D, Braberg H, Mehta M et al. [2009] *Cell* 136:952–963. With permission from Elsevier.)

Glycoproteins are the most diverse class of protein–sugar conjugates. They contain a diverse set of sugar chains, usually branched; these can vary in length from 1 to more than 30 residues. Sometimes, the carbohydrate moiety can account for more than 80% of the total mass of the complex. The structural diversity of the chains is attributed to several factors, including the identity of the sugars (hexoses, hexosamines, pentoses, and sometimes sialic acid); the type of glycosidic linkages and the sugar atoms they involve; and the branching pattern. The composition and structure of the oligosaccharide can vary even among individual molecules of the same protein; whether this microheterogeneity has any functional roles is still unclear.

Depending on the amino acid residue to which the carbohydrate chain is attached, we discriminate between O-linked glycosylation, in which an oligosaccharide is

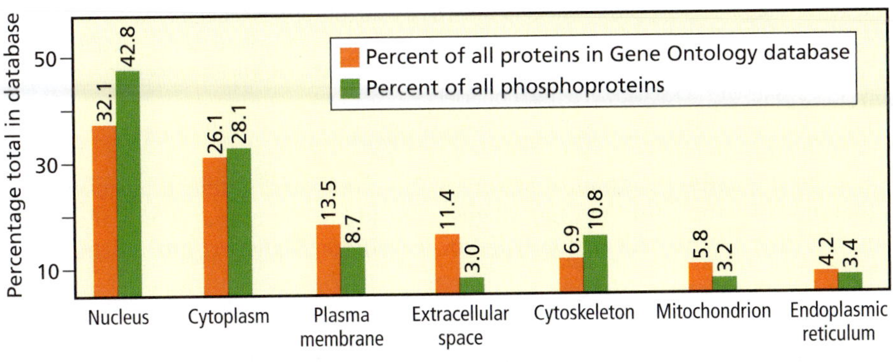

Figure 18.17 Analysis of the *in vivo* human phosphoproteome. This analysis is based on the Gene Ontology or GO database. Using quantitative mass spectrometry, Matthias Mann and collaborators detected 6600 phosphorylation sites on 2244 proteins in HeLa cells. The plot represents the distribution of phosphoproteins, in comparison with the distribution of all proteins in the database, according to their cellular localization. Phosphorylation preferentially targets nuclear proteins, whereas proteins annotated by GO as extracellular, mitochondrial, and plasma membrane are significantly underrepresented. The phosphopeptide screen used was still not comprehensive since lower-abundance proteins might not have been detected. (Adapted from Olsen JV, Blagoev B, Gnad F et al. [2006] *Cell* 127:635–648. With permission from Elsevier.)

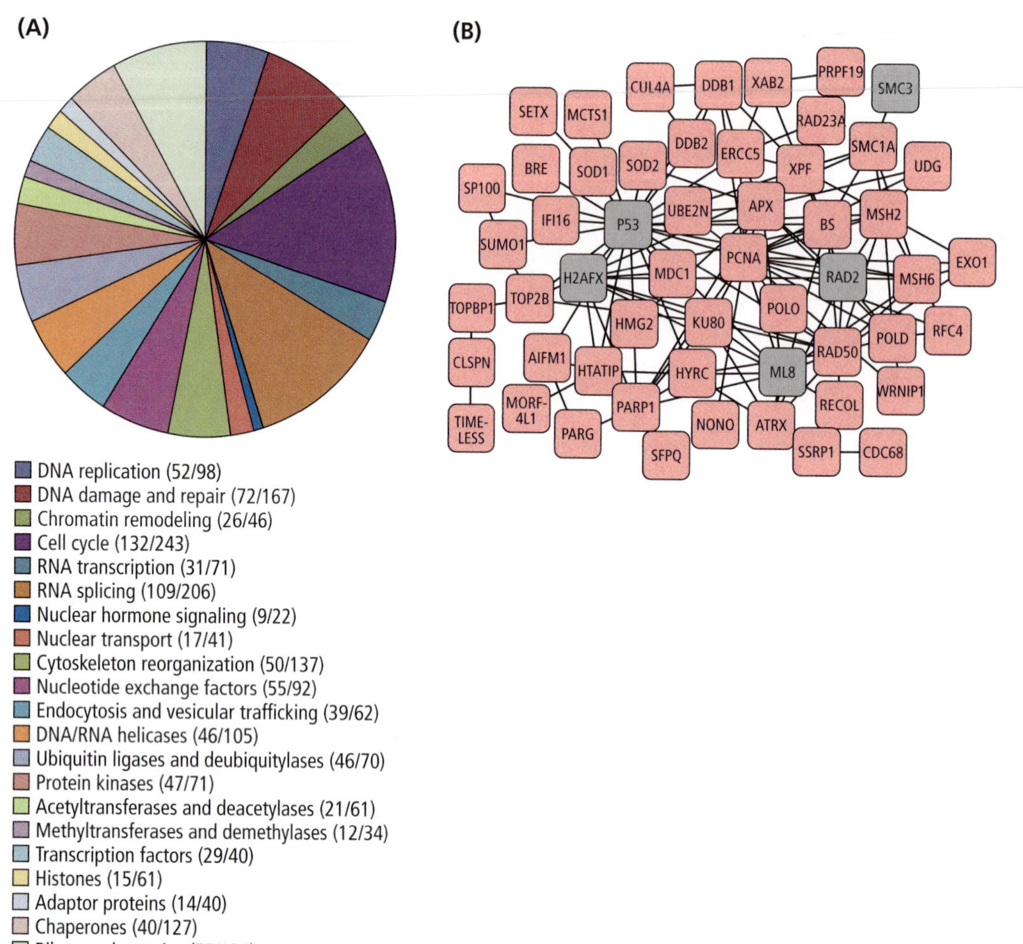

Figure 18.18 Quantitation of acetylation on a proteomic scale. This quantitation is based on quantitative mass spectrometric data. Acetylated peptides from trypsin-digested whole-cell lysates of a human acute myeloid leukemia cell line were enriched for by an antibody against acetyllysine and then subjected to identification by high-resolution mass spectrometry. More than 3600 acetylation sites on 1750 proteins were identified. (A) Number of acetylated proteins and sites grouped into major cellular functional categories. A large number of acetylated proteins, and an even larger number of acetylation sites, are involved in all major nuclear processes, suggesting a role for acetylation in all of these processes. Shown in parentheses are the number of acetylated proteins/number of acetylation sites. (B) The acetylated proteins are connected in dense protein–protein interaction networks; the sample network pictured is that of proteins participating in DNA damage repair. Gray nodes are the proteins previously reported to be acetylated. The figure underlines the paucity of direct experimental data and the power of proteomic approaches in uncovering new important facts; the proteome data serve as a lead for further experimental validation. (Adapted from Choudhary C, Kumar C, Gnad F et al. [2009] *Science* 325:834–840. With permission from American Association for the Advancement of Science.)

added to the hydroxyl group of serine or threonine, and N-linked glycosylation, which involves the addition of an oligosaccharide to the amino group of asparagine (**Figure 18.19**). The major types of vertebrate N-glycans are illustrated in **Figure 18.20**.

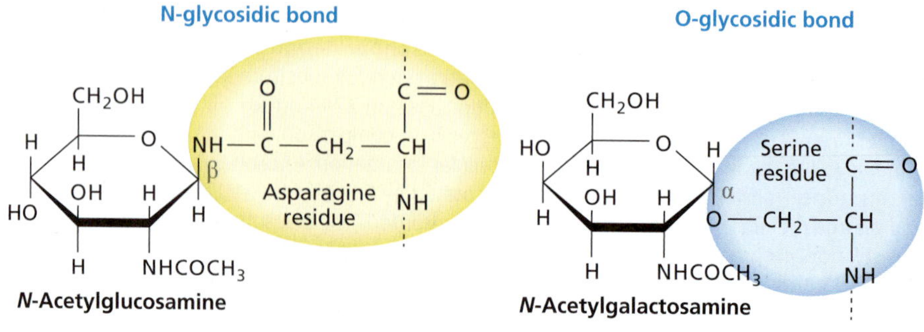

Figure 18.19 N- and O-glycosidic linkages in proteoglycans. *N*-Acetylglucosamine–asparagine is the major N-glycosidic bond; *N*-acetylgalactosamine–serine or –threonine is the major O-glycosidic bond. The N-bond has a β configuration; the O-bond has an α configuration.

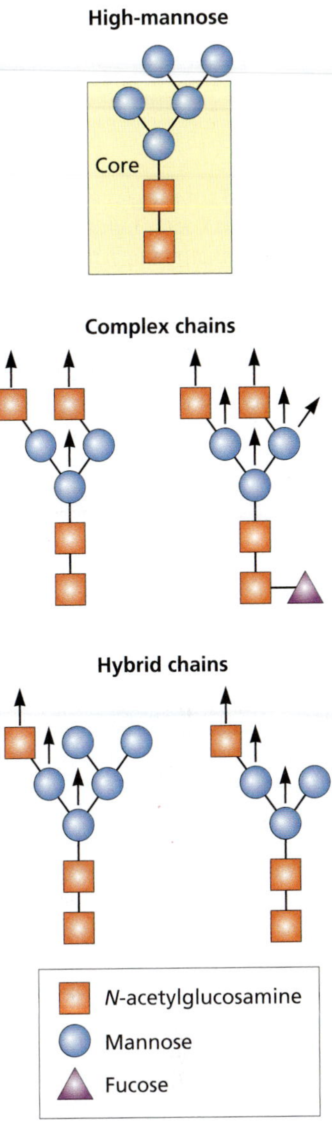

Figure 18.20 Major types of vertebrate N-glycans. The pentasaccharide core common to all N-linked structures is highlighted in the yellow box. The existence of this common core reflects the initial biosynthetic pathway that is shared by all N-linked glycans. Most N-linked oligosaccharides can be divided into three subclasses: high-mannose, hybrid, and complex. High-mannose glycans are synthesized at early stages in the biosynthetic pathways. Complex-chain synthesis involves the removal of some mannose residues from high-mannose chains and the addition of other sugar moieties in a process known as oligosaccharide processing. In hybrid chains, there is extensive branching, with one branch being a high-mannose chain and the other branch of the complex type. Black arrows denote sugar moieties at which the glycan chain can be branched by the addition of other sugars. This involves the formation of more than the normal two glycosidic bonds with C-2, C-3, C-4, or C-6. These extensions and the use of several different sugars produce a large variety of glycan molecules.

Mechanisms of glycosylation depend on the type of modification

In most cases, the initial protein glycosylation reactions occur in the rough ER; the chain is then further modified in the ER and Golgi complex. The exact mechanism depends on the modification type. N-linked glycosylation occurs simultaneously with translation (Figure 18.21). This process involves the assembly of a core oligosaccharide, which is attached to a lipid carrier, dolichol phosphate, on the cytoplasmic side of the ER membrane. This core molecule is then translocated or flipped across the ER membrane in a reaction that is catalyzed by specialized enzymes, **flippases**. Phospholipids diffuse rapidly and freely in the plane of membrane bilayers, but their polar head groups cannot pass easily through the hydrophobic environment of the central portion of the membrane. The cell has evolved flippases to help in such passages. Flipping also occurs rather frequently when phospholipid molecules that are incorporated into the cytoplasmic face of the ER membrane have to be transferred to the exoplasmic face. Once the sugar component that will join the polypeptide is in the lumen of the ER, the complexity of the oligosaccharide chain evolves and is finally added to asparagine residues of a nascent polypeptide. In some instances, further modification of the oligosaccharide occurs in the Golgi complex. O-linked glycosylation occurs through stepwise addition of monosaccharides in either the ER or Golgi complex. Most O-linked oligosaccharides are short, containing only four sugar residues.

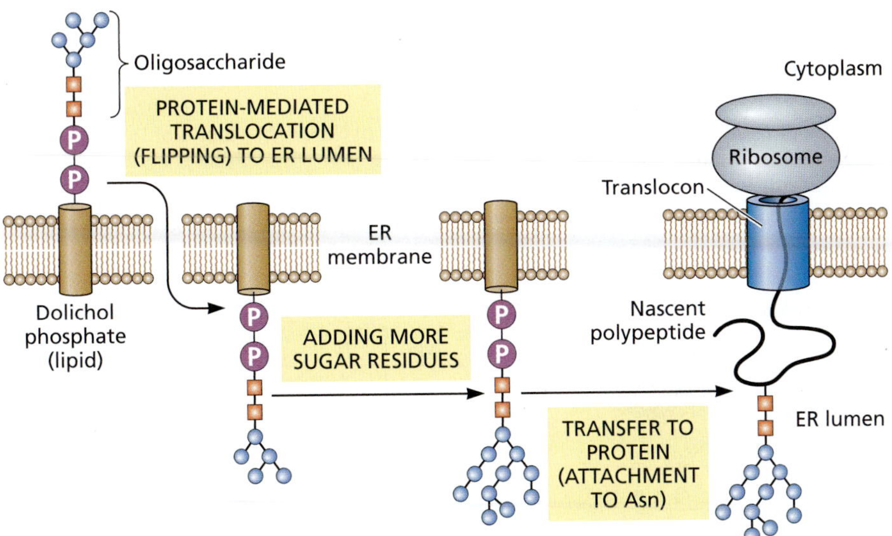

Figure 18.21 Mechanism of N-linked protein glycosylation. In N-linked glycosylation, the addition of an oligosaccharide to the polypeptide occurs by covalent attachment of the oligosaccharide chain to an asparagine side chain (see Figure 18.19). The core oligosaccharide is assembled on a lipid carrier, dolichol phosphate, on the outer ER membrane. This lipid–oligosaccharide complex is flipped across the lipid bilayer by proteins known as flippases. Next, the oligosaccharide is extended by specific enzymes. In the final step, the oligosaccharide is transferred to the nascent polypeptide. (Adapted, courtesy of Graham Thomas, Pennsylvania State University.)

Glycosylated proteins play numerous roles, especially in cell–cell recognition processes and signaling pathways. A classic example of how the diversity of oligosaccharides linked to proteins or lipids on the surface of red blood cells creates individuals of different blood types is presented in **Box 18.4**. It is imperative that the blood group

Box 18.4 Protein glycosylation, blood groups, and blood transfusion Today, blood transfusion is a frequently used and safe lifesaving practice. But it was not like this before the turn of the nineteenth century, when many blood recipients died during or shortly after the procedure. In 1901, scientist Karl Landsteiner discovered human blood groups and the reason that led, in some cases, to clumping or agglutination of the red blood cells of the recipient. He realized that blood clumping was an immunological reaction that occurs when the recipient of a blood transfusion has antibodies against the donor blood cells. For "his discovery of human blood groups," Landsteiner was awarded the 1930 Nobel Prize in Physiology or Medicine.

What is the molecular basis for the existence of blood groups? Here, we explain the situation with the basic ABO blood groups, consisting of groups A, B, AB, and O, although a total of 30 blood group systems are now recognized by the International Society of Blood Transfer. Individual people can carry, on the surface of their red blood cells, one of three different kinds of oligosaccharides—antigens H, A, and B (**Figure 1**). There is a core oligosaccharide, **antigen H**, that characterizes the red blood cells of group O; this group represents the majority of the population, between 30% and 40% in different countries. The oligosaccharides in group A, **antigen A**, and in group B, **antigen B**, are derivatives of the core structure with the addition of some other sugars. The addition reactions are catalyzed by specific enzymes, known as A and B, respectively, which are encoded in the genome and present only in certain individuals.

Why would different individuals possess one or the other form of these enzymes? The answer lies in the organization or expression of their genes. The ABO blood group enzymes are encoded in a single gene, which exists as several alleles. The original gene encoded the A enzyme, which transfers *N*-acetyl-galactosamine to the core oligosaccharide. The other alleles are slightly different and produce enzymes with slightly altered specificity. Thus, the B allele can only transfer galactosamine to the core oligosaccharide. The difference between A and B is in four nucleotides only; in fact, structural information has pinpointed a single amino acid substitution that is responsible

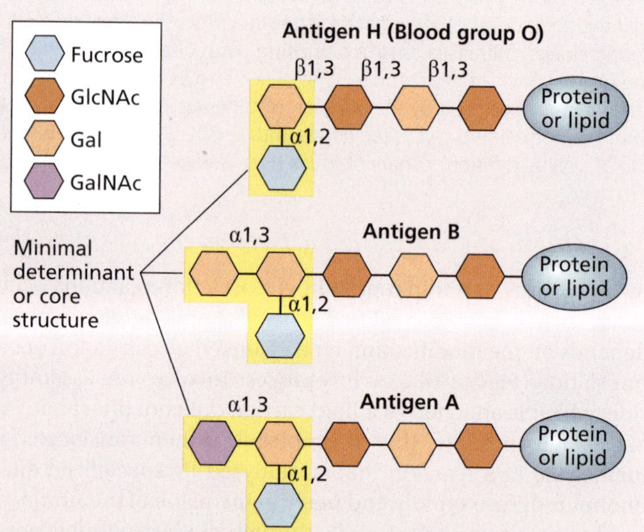

Figure 1 Structures of antigens in ABO blood groups. Most primates exhibit three different kinds of O- or N-linked oligosaccharides on the surface of their red blood cells. The structures shown are only examples of possible structures; the actual structures can vary widely in both the sugar components and in the protein or lipid carriers. The glycans themselves are very heterogeneous; they may contain branches, repeated motifs, or other moieties. The core structure of these varied oligosaccharides is called the H antigen, and this is present on red blood cells of the O group. These oligosaccharides are nonantigenic, so they do not elicit an immune response. The core oligosaccharide can be further modified in a number of ways. Addition of a galactose or Gal residue, in a reaction catalyzed by the B enzyme, gives rise to the B antigen. Addition of *N*-acetylgalactose or GalNAc by the A enzyme forms the A antigen. (Adapted, courtesy of National Institutes of Health.)

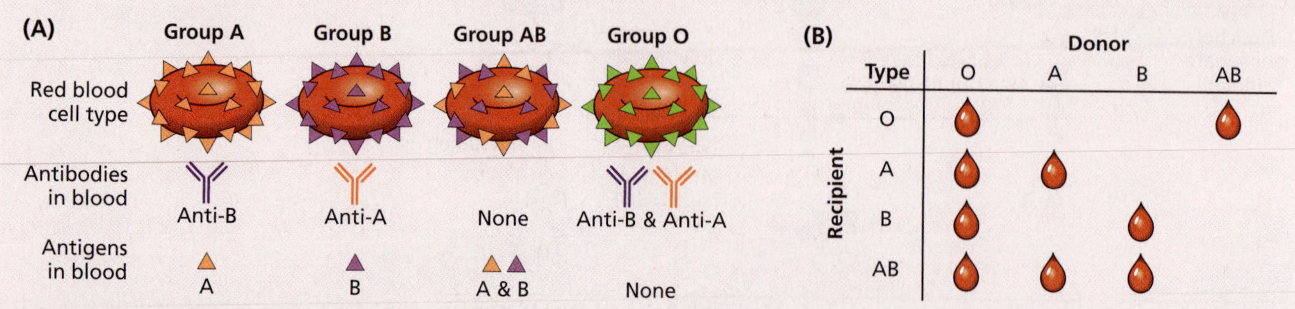

Figure 2 General characteristic features of the four major ABO blood types. (A) ABO blood group antigens present on red blood cells, and antibodies present in serum, of individuals belonging to the different groups. In healthy people, antibodies are not produced against their own oligosaccharide, but antibodies against the other antigen types are made. (B) Basic rules of safe blood transfusion involving a donor and a recipient who belong to different groups. The red drops signify that transfusion is allowed. If, for example, an A-type individual receives blood from a B-type donor, the antibodies in the A-individual blood will react with the transfused red blood cells, leading to aggregation. Thus, transfer from B to A is prohibited.

for the change in specificity. A single-nucleotide deletion near the N-terminus of the coding region creates a nonfunctional version of the enzyme in group O individuals, shifting the reading frame for the rest of the sequence. Thus, people who are homozygous for these nonfunctional alleles, of which there are several, have only the core oligosaccharide, that is, the basic H antigen, on the surfaces of their red blood cells.

Although in general blood type has no effect on growth and development, there are some correlations with disease. A well-known example is the elevated susceptibility of group O individuals to infection with *Vibrio cholerae*, the bacterial agent causing cholera.

All of this knowledge has led to the creation of strict rules for blood transfusion. Figure 2 depicts these rules and shows the characteristic features for each blood type. All types will contain some H antigen, because not 100% of the molecules are further transformed into A or B antigens, but since the H type is not immunogenic, no antibodies against it are produced. Why O-type individuals would have anti-A and anti-B antibodies is less clear. Although the expression of this gene is mono-allelic—otherwise there would not be distinct groups—it is possible that some leakage expression occurs, giving rise to very small amounts of alternative enzymes, which then elicit an immune response.

of both individuals is known before blood is transfused from a donor individual to a recipient in order to avoid deadly agglutination, or clumping, of red blood cells of the recipient.

Ubiquitylation adds single or multiple ubiquitin molecules to proteins through an enzymatic cascade

Ubiquitylation is the covalent conjugation to proteins of a small, 76-amino-acid long protein known as ubiquitin or Ub. A historical account of the discovery of ubiquitin and ubiquitylation is given in Box 18.5. The enzymatic pathway that leads to ubiquitin conjugation to proteins is shown in Figure 18.22.

Box 18.5 Discovery of the ubiquitin–proteasome system
Discovery of the ubiquitin–proteasome system is a particularly instructive example of how ideas in science evolve and paradigms change, following the development of ingenious methodologies that yield results difficult to explain within the framework of existing ideas.

The fact that proteins turn over had not been recognized widely until Rudolf Schoenheimer performed experiments with amino acids labeled with heavy isotopes, such as ^{15}N. His results indicated that proteins are not static but are continuously synthesized and degraded. It took a while for these ideas to be accepted because, until then, there was a strict distinction between food proteins, which are degraded and function to provide fuel, and bodily proteins that, once synthesized, were thought not to turn over and to be subject to only minor wear and tear. Schoenheimer fed ^{15}N-labeled tyrosine to rats and found that (1) ~50% of the label became incorporated into body proteins and (2) only a fraction of the label remained in tyrosine, while the bulk was distributed into other amino acids, mainly in the form of α-amino groups. Thus, not only the proteins but also individual amino acids are in a state of constant flux.

The first intracellular system that was identified as the place where proteins are degraded was the **lysosome**, discovered by Christian de Duve in the mid-1950s. The lysosome is a cytoplasmic organelle surrounded by a membrane; it contains various proteolytic enzymes with a pH optimum in the acidic range. The protein to be degraded is first transported into the organelle; in this way, the proteolytic enzymes are segregated and do not attack the normal protein components of the cytoplasm. The lysosome concept has evolved considerably over the years, and today we talk about the lysosomal/vacuolar system, a discontinuous and heterogeneous intracellular digestive system that also includes structures that do not contain proteolytic enzymes, the early endosome being one example (Figure 1).

First, it was realized that lysosomes undergo a complex maturation process. Second, it was found that the system performs a variety of degradation processes: it digests endogenous proteins and cellular organelles, in processes known as **micro**- and **macroautophagy**, respectively. It also degrades exogenous proteins that are brought to the lysosome via receptor-mediated endocytosis and pinocytosis, or nonspecific engulfment of extracellular fluid, or exogenous particles that are internalized through phagocytosis. The degradation of these exogenous proteins or particles is termed **heterophagy**. Significant progress in understanding autophagy and heterophagy made it clear that neither of these processes selectively degraded specific proteins. Hence, lysosomal degradation cannot explain the very different half-lives of different proteins.

The lack of such specificity and the difficulty in explaining some of the results obtained with inhibitors of lysosomal proteases hinted that a second proteolytic system could exist. Another important fact that could not be explained by lysosomal degradation was the early observation that protein degradation in mammalian cells requires energy. This requirement is paradoxical, because proteolysis itself is an exergonic process. This observation meant that the free energy must be required for another, unknown process in the proteolytic pathway. These ideas were floating around in the mid-1970s, but experimental evidence for an alternative, energy-requiring and protein-specific process was lacking.

In the meantime, as seen in the timeline depicted in Figure 2, ubiquitin was discovered, and an isopeptide linkage between the C-terminal glycine of ubiquitin and the

(Continued)

Box 18.5 *(Continued)*

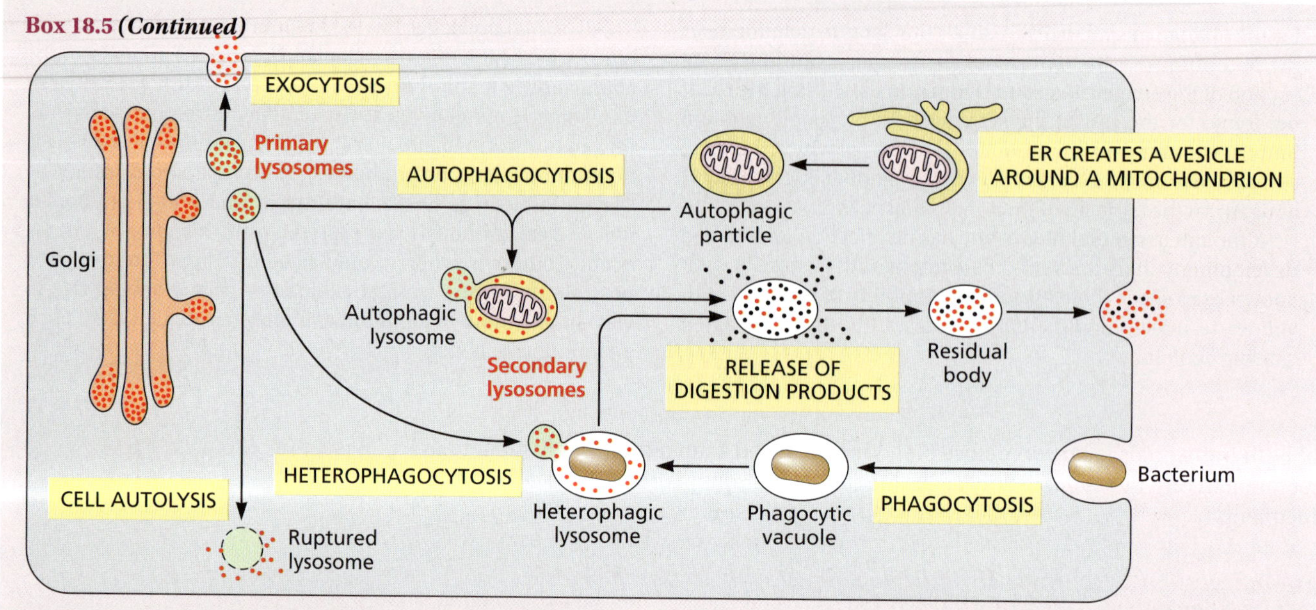

Figure 1 Primary and secondary lysosomes and pathways that lead to various digestion processes in the cell. The primary lysosomes that bud from the Golgi complex may participate in several different pathways: (1) they may transport enzymes to the outside of the cell in a process known as exocytosis; (2) they can participate in the destruction of old and defective organelles, known as autophagocytosis; (3) they can digest foreign bodies, such as bacteria and viruses, and use the products of digestion for food, in a process called heterophagocytosis; or (4) they can cause destruction of the cell itself, known as autolysis.

ε-NH_2 of internal lysine residue K119 in histone H2A was described. This was the first time such bonding was recognized. The existence of isopeptide bonding allows the formation of bifurcated or branched polypeptides and such bonds were later identified in ubiquitin–target protein conjugates. Ubiquitin was initially thought to be present in all living cells,

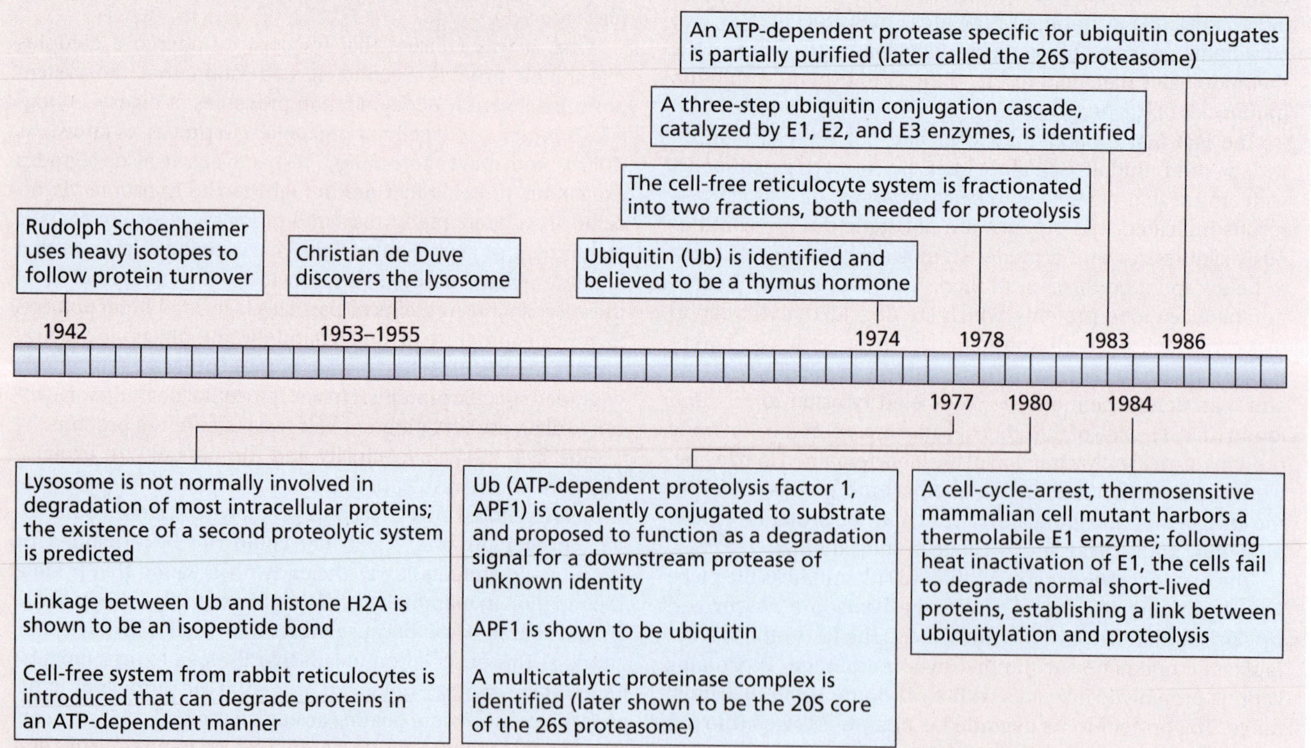

Figure 2 Timeline presenting the most important discoveries in intracellular protein degradation. The timeline is limited to the discovery of the two main, basic facts: covalent ubiquitylation as a degradation signal, and the proteasome machinery. The years that follow focus on specific examples of high physiological importance, such as degradation of cyclin, certain oncoproteins, p53 tumor suppressor, and others. (Adapted from Ciechanover A [2005] *Nat Rev Mol Cell Biol* 6:79–87. With permission from Macmillan Publishers, Ltd.)

hence the name, but published results with *Escherichia coli* turned out to be an artifact: ubiquitin was contaminating the bacterial cell preparations because it was present in the growth medium. So ubiquitin is, strictly speaking, eukaryotic and not ubiquitous.

The description of a cell-free system—from rabbit reticulocytes, which are naturally devoid of lysosomes—capable of degrading proteins in an ATP-dependent manner *in vitro* was an important step forward in proteolysis research. Further fractionation and biochemical characterization of the reticulocyte system showed that there are two separate ATP-requiring activities, both of which are needed for specific proteolysis. The first activity contained the three enzymes of the ubiquitin-conjugating cascade (see Figure 18.22B); the second was the actual protease, the huge 26S proteasome complex (see Chapter 3, specifically Figure 3.22). Thus, all pieces of the puzzle were put together in a beautifully coherent picture of intracellular protein degradation. The significant contributions of Aaron Ciechanover, Avram Hershko, and Irwin Rose toward understanding the ubiquitin-proteasome system were recognized with the 2004 Nobel Prize in Chemistry "for the discovery of ubiquitin-mediated protein degradation."

Proteins can be modified by the addition of a single ubiquitin or several monoubiquitin moieties or by building ubiquitin chains of different structures. In monoubiquitylation, ubiquitin is attached via an **isopeptide bond** between the carboxyl group of its C-terminal glycine and an ε-amino group of a lysine residue on the acceptor protein. Conjugation requires the sequential activity of three enzymes: **E1** or **ubiquitin-activating enzymes**, **E2** or **ubiquitin-conjugating enzymes**, and **E3** or **ubiquitin-protein ligase enzymes**, as depicted in **Figure 18.22** and presented in more detail for the well-studied yeast system in **Figure 18.23**; this figure also provides information on the unusual way that ubiquitin is encoded in the yeast genome, as fusion genes. As Figure 18.23 shows, in yeast there is only one E1 enzyme but 13 E2s and ~200 E3s; to date, at least 1000 ubiquitin ligases have been recognized in mammalian genomes.

Specificity of ubiquitin targeting is determined by a special class of enzymes

Substrate specificity of the reaction is determined by the E3 enzymes, which fall into two large groups: RING-finger domain E3s and HECT domain E3s (see Figure 18.22B). RING stands for really interesting new gene, and HECT stands for homologous to the E6 carboxy-terminus (**Box 18.6**). The grouping is according to the structure of the domain that interacts with E2 to execute Ub transfer from the E2 active-site cysteine to the acceptor protein. In addition to these domains, both types of E3s have substrate-binding domains. There is also a difference in the mechanism of the transfer reaction. RING-finger E3s bind to the substrate and directly transfer the Ub to the acceptor protein, whereas HECT E3s first become ubiquitylated themselves, and only then transfer the ubiquitin moiety to the acceptor proteins (see Figure 18.22B).

RING-finger E3s often act as heterodimers. We have chosen to illustrate this point using the example of an important complex **Ring1b–Bmi1**, two subunits of the repressive polycomb repressor complex 1 or PRC1 (**Figure 18.24**). PRC1 participates in the regulation of developmentally important groups of genes (see Figure 12.15), in X-chromosome inactivation, in carcinogenesis, and in stem-cell renewal. The biological importance of Ring1b is demonstrated by early embryonic lethality in Ring1b knockout mice, whereas the lack of its close relative Ring1a, also a part of PRC1, causes only minor skeletal aberrations. Bmi1, which stands for B-cell-specific Moloney murine leukemia virus integration site 1, is an oncoprotein whose levels are tightly regulated: even a twofold increase induces lymphoma formation. Its RING-finger domain is critical for its tumorigenic functions. PRC1 places a monoubiquitylation mark on Lys119 of histone H2A, as part of its repressive function (see Chapter 12, especially Figure 12.15). Ring1b carries the E3 enzymatic function, which is greatly stimulated by Bmi1. Both Ring1b and Bmi1 contain RING-finger domains, which extensively interact with each other, although only the RING-finger domain of Ring1b has a catalytic function. It is important to note that the structure of the Ring1b–Bmi1 complex is very similar to, almost superimposable with, that of other members of E3 RING-finger heterotypic dimers, including that of the protein product of the breast cancer susceptibility gene 1 complexed with another RING-finger-containing protein, BARD1 (see Box 18.6).

The same enzymatic system that adds monoubiquitin to the acceptor protein builds **ubiquitin chains**. It does this by catalyzing the formation of an isopeptide bond

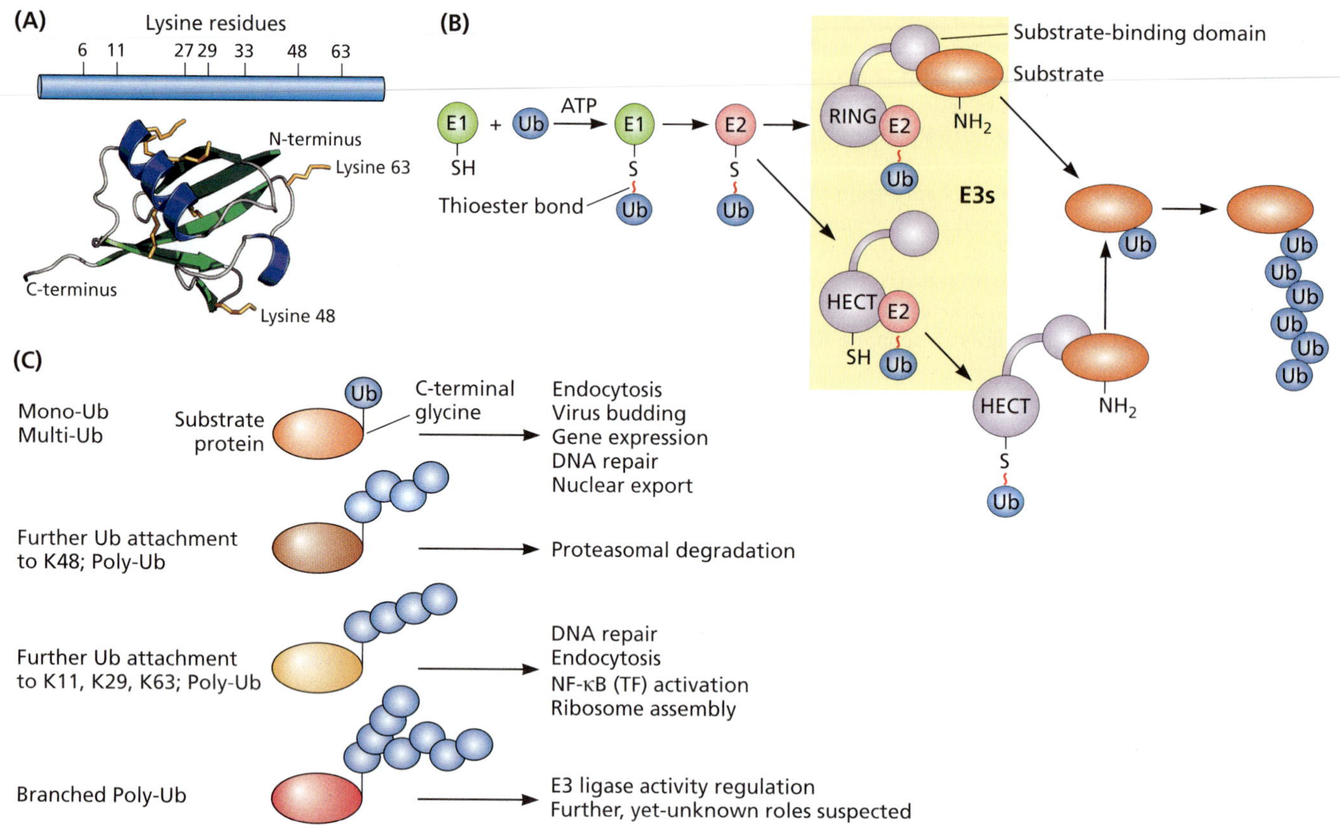

Figure 18.22 Protein ubiquitylation and various functional roles of different Ub modifications. (A) Ubiquitin has seven lysine residues, which are marked on the schematic and shown in orange in the structural model. (B) Ubiquitylation occurs through a series of three enzymatic reactions, with the participation of three classes of enzymes: E1 or ubiquitin-activating, E2 or ubiquitin-conjugating, and E3 or ubiquitin–protein ligase enzymes. There are two types of E3s, RING-finger type and HECT-domain type; they differ in the way they effect Ub transfer from the active-site Cys on the E2 enzyme to the substrate lysine. RING-finger E3s directly transfer Ub to the acceptor lysine; strictly speaking, they are not enzymes that perform chemical catalysis, but rather they promote the formation of Ub chains by properly juxtapositioning the E2 and the substrate. By contrast, HECT E3s first load Ub onto themselves, forming a Ub-thioester intermediate with a Cys located at the C-terminus of the HECT domain, and then transfer the Ub to the substrate. Thus, HECT E3s have intrinsic catalytic activity that is absent from RING-finger E3s. Examples of the two types of E3 are presented in Box 18.6. (C) Depending on how many ubiquitin moieties are covalently linked to each other, one can distinguish mono- or polyubiquitylation; alternatively, some proteins are modified by single ubiquitins attached at numerous protein sites, known as multi-Ub. The first connection between ubiquitin and acceptor proteins is via an isopeptide bond between the C-terminal glycine of Ub and a lysine residue on the acceptor substrate protein. Depending on which of the seven lysine residues of Ub participates in the further addition of ubiquitin moieties to the first one bound, different poly-Ub chains can form; these possess different biological activities. Note that branched structures can also form. (A, bottom, from Wikimedia. B, adapted from Wenzel DM, Stoll KE & Klevit RE [2011] *Biochem J* 433:31–42. With permission from Biochemical Society. C, adapted from Woelk T, Sigismund S, Penengo L & Polo S [2007] *Cell Div* 2:11. With permission from BioMed Central.)

between the C-terminal Gly76 of the conjugated ubiquitin and an ε-amino group of a lysine residue in another ubiquitin molecule. Ubiquitin possesses seven lysine residues, allowing for seven possible homotypic linkage types. In homotypic linkages, the same lysine residue, for example, Lys63 or Lys48, serves as a chain elongation point on each successive ubiquitin moiety (see Figure 18.22A). In addition, multiple heterotypic chains are possible when different Lys residues on individual ubiquitins participate in formation of the chain. Obviously, the seven lysines provide the opportunity for extensive chain branching. As a consequence of this diversity, the ubiquitin chains themselves can be very diverse, in terms of length, linkage, and branching patterns.

Using different linkages can yield very different ubiquitylated structures. Here, we use the example of the two most common homotypic or homogeneous linkages, those that use Lys48 or Lys63 for chain elongation. Structural studies have revealed that Lys48-linked chains are compact, with their ubiquitin moieties packed against

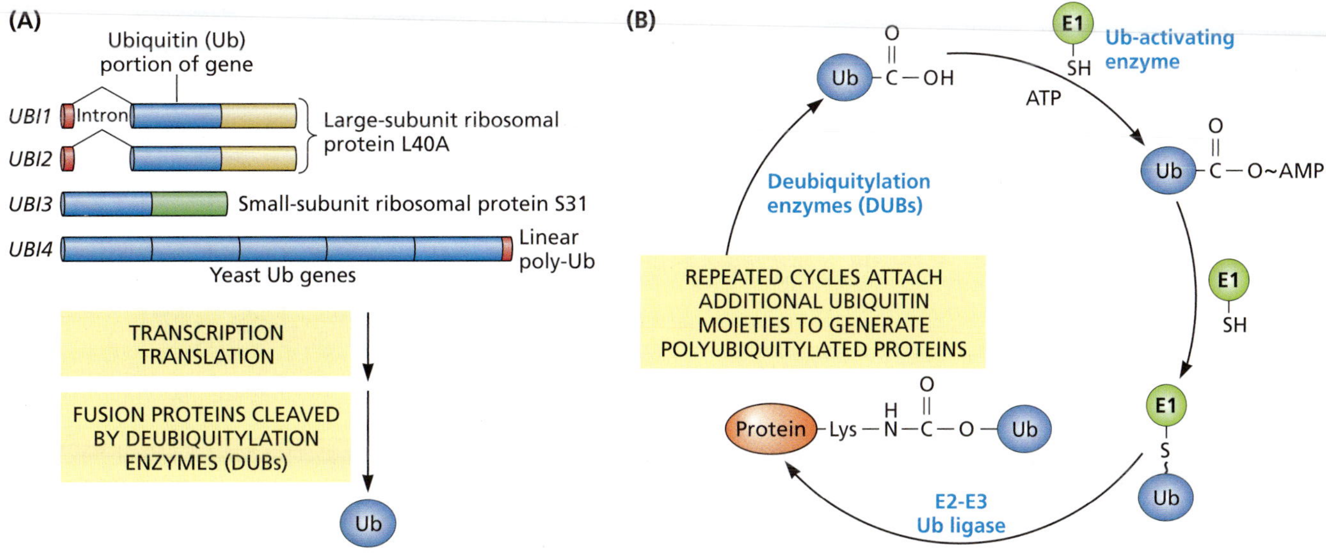

Figure 18.23 Ubiquitin system of *Saccharomyces cerevisiae*. The structure of the ubiquitin system is highly conserved among eukaryotic organisms. (A) In yeast, however, ubiquitin is encoded in a rather unusual way, by fusion genes *UBI1–UBI4*. In three of these genes, the portion that encodes full-length 76-amino-acid ubiquitin is fused to one of two ribosomal protein genes, *L40A* or *S31*. The fourth gene, *UBI4*, encodes a head-to-tail poly-Ub precursor protein, which is cleaved by deubiquitylation enzymes to produce monomeric ubiquitin. This gene is specifically induced under stress conditions; indeed, deletion of *UBI4* results in hypersensitivity to a variety of stresses. (B) Covalent addition of ubiquitin to proteins takes place through a complex mechanism involving three basic enzyme groups, E1, E2, and E3. Yeast has 1 E1, 13 E2s, and ~200 E3s; mammalian genomes contain at least a thousand Ub ligases. In the first step, Ub is activated by E1 or Ub-activating enzyme to produce AMP-Ub, which is then linked to a Cys residue of the activating enzyme E1 by formation of a thioester bond. Ub is then transferred to a Cys residue of one of several E2 or Ub-conjugating enzymes. The final transfer to a lysine residue on a substrate protein requires the action of E3 or Ub-ligating enzyme, which confers substrate specificity to E2. (A, adapted from Özkaynak E, Finley D, Solomon MJ & Varshavsky A [1987] *EMBO J* 6:1429–1439. With permission from John Wiley & Sons, Inc. B, adapted from Mathews CK, van Holde KE, Appling DR & Anthony-Cahill SJ [2013] Biochemistry, 4th ed. With permission from Pearson Prentice Hall.)

each other (**Figure 18.25**). On the other hand, Lys63-linked chains are extended, as are the linear polyubiquitins, the newest addition to the polyubiquitin family, in which ubiquitins are linked by typical peptide bonds between the C-terminus of one ubiquitin and the NH_2 group at the N-terminus of another. The different chain conformations are recognized by different ubiquitin-binding domains or UBDs present in a multitude of cellular proteins, currently estimated to be >150. UBDs are modular elements in the effector proteins of ubiquitylation signaling. The preferences of UBDs for specific linkages are based on multiple interactions: UBDs can synergistically bind multiple ubiquitin molecules or can recognize the linkage

Box 18.6 A Closer Look: E3 ligases and human disease As pointed out in the text, E3 ubiquitin ligases catalyze the final step of the attachment of ubiquitin to acceptor proteins. E3 enzymes are implicated in numerous cellular processes. Thus, it comes as no surprise that genetic alteration and/or abnormal expression patterns of members of either family of these ligases, HECT-domain and RING-finger domain, are often linked to human pathologies.

The founding member of the HECT E3 family is E6-AP or E6-associated protein, also known as UBE3A, which was the first mammalian E3 to be identified and characterized. This 100 kDa protein interacts with the E6 protein of high-risk types of human papillomavirus, HPV, that cause cervical cancer. HPV-induced cervical cancer afflicts about 500,000 women each year worldwide, with a mortality rate approaching 50%. Some 99.7% of cervical cancers are HPV-induced.

The virally encoded E6 protein binds to E6-AP and targets the tumor suppressor protein p53 for polyubiquitylation and proteasomal degradation. Cervical cancer cells harbor wild-type p53, in contrast to most other tumors, which contain mutated copies of the p53 gene (see Figure 17.7); thus, in cervical cancer, the defect is in the extremely low p53 levels that result from increased proteasomal degradation. Interestingly, the E6-AP–E6 complex does not form stably with E6 from low-risk HPV types; hence, p53 is not degraded when the infection involves these low-risk viral types, and cancer does not arise.

The gene for E6-AP has also been linked to **Angelman syndrome**, a human imprinted genetic disorder characterized by mental retardation, seizures, jerky movements, and a number of other neurological characteristics. The disorder was first recognized in 1964–1965 by pediatrician Harry Angelman in school-age children. As in all imprinting, only one of the two

(Continued)

Box 18.6 *(Continued)*

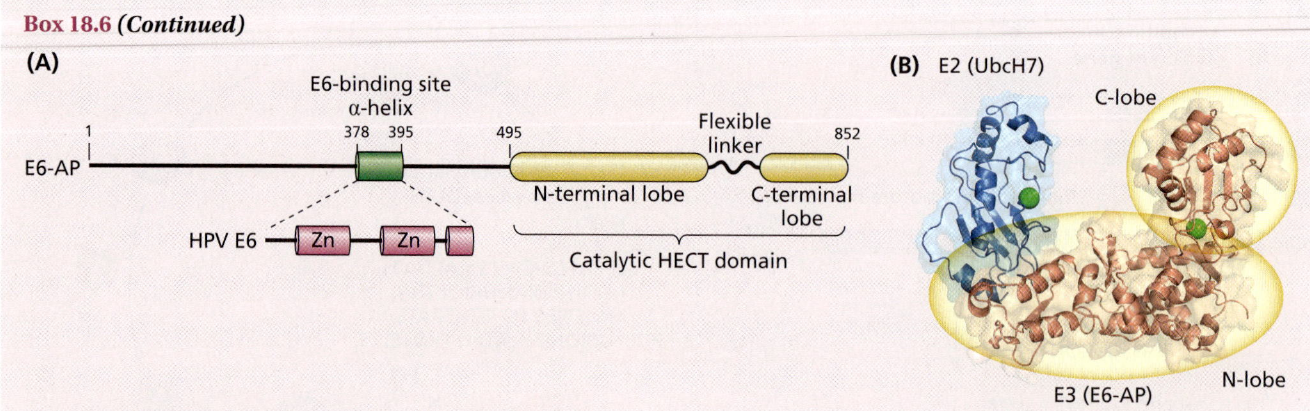

Figure 1 Model of prototype HECT-domain E3 ligase E6-AP. (A) Schematic of E6-AP domain structure, showing the site of binding of papilloma viral protein E6, which is depicted below. (B) Crystal structure of the complex between the E6-AP HECT domain, shown in pink, and the E2 enzyme with which it interacts, shown in blue. The HECT domain of E6-AP contains two lobes, with a broad catalytic cleft at the junction between them; the cleft is the site of Angelman syndrome mutations. The transfer of Ub that occurs between active-site cysteines in E2 and E3 requires that the two cysteines, depicted as green spheres, be in close proximity. In the structure shown, however, the two thiol groups are separated by a significant distance. Thus, a large conformational change in the complex has been suggested to occur during Ub transfer: Ub-loaded E2 first engages the HECT domain N-lobe, and the two lobes then rotate with respect to each other. (A, adapted from Beaudenon S & Huibregtse JM [2008] *BMC Biochem* 9(Suppl 1):S4. With permission from BioMed Central.)

alleles of the gene is expressed, depending on the origin of the allele, maternal or paternal. Recall from Chapter 12 that the molecular basis of such imprinting is differential methylation of the two alleles. In two specific regions of the brain, the hippocampus and the cerebellum, the paternal gene for E6-AP is silenced, and the maternal allele is transcriptionally active. If the maternal allele is lost, generally by large chromosomal deletions, or mutated, then the individual develops the syndrome. Although the natural targets of E6-AP in normal, non-virus-infected cells are practically unknown, it is suggested that lack of ubiquitylation of one or more E6-AP target proteins in the brain is responsible for the phenotype of the syndrome.

The overall structure of E6-AP, with the catalytic HECT domain and the short portion in the middle that binds HPV, is presented in Figure 1A. Figure 1B shows the three-dimensional

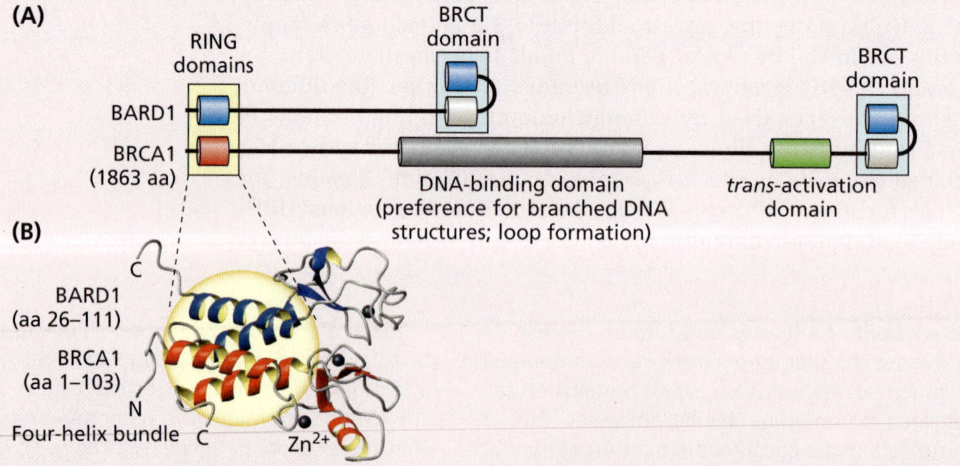

Figure 2 Solution structure of heterodimeric complex between BRCA1 and BARD1. BARD1 is one of five protein partners that interact with the BRCA1 RING-finger domain to stimulate its activity. (A) Schematic of the domain structure of the two proteins. In addition to the RING-finger domain, BRCA1 contains a large DNA-binding domain and a *trans*-activation domain that plays a role in transcriptional activation of a number of genes. The BRCT or BRCA1 C-terminal domain does not have an enzymatic function; it serves as a protein interaction interface. Interestingly, both partners in the complex contain RING fingers and BRCT domains. (B) Structure of the complex between the isolated RING-finger domains. The four-helix bundle is encircled, and the bound Zn^{2+} ions are marked. This structure is highly conserved among other representatives of the RING-finger proteins. One notable example is the complex between the oncoprotein Bmi1 and its partner Ring1b, the protein responsible for monoubiquitylation of histone H2A; this complex is shown in Figure 18.24. (A, adapted from Baer R [2001] *Nat Struct Biol* 8:822–824. With permission from Macmillan Publishers, Ltd. B, from Brzovic PS, Rajagopal P, Hoyt DW et al. [2001] *Nat Struct Biol* 8:833–837. With permission from Macmillan Publishers, Ltd.)

crystallographic model of the complex of HECT domain and its E2 interactive partner UbcH7.

RING stands for really interesting new gene. To demonstrate the structure of the RING-finger E3s, we use the example of the protein encoded by the **breast cancer susceptibility gene *BRCA1***. *BRCA1* encodes a tumor suppressor protein that participates in numerous fundamental cellular processes involved in maintenance of genomic integrity and transcriptional regulation. One of the recognized biochemical activities of BRCA1 is to function as an E3 ligase of the RING-finger type for a number of proteins, including histones H2A and H2B. Interestingly, BRCA1, together with BARD1 or BRCA1-associated RING domain 1, is part of the Pol II holoenzyme and travels with the polymerase during transcription elongation. When Pol II stalls at sites of DNA damage, BRCA1-BARD1 polyubiquitylates stalled RNA polymerases to mark them for degradation, so that repair of the damage can occur. In addition, BRCA1 serves to recruit repair factors.

When the *BRCA1* gene is mutated, breast and ovarian cancers can develop; of the clinically relevant *BRCA1* mutations, ~20% occur within the N-terminal 100 residues in the RING-finger domain. BRCA1 forms stable complexes with a number of proteins, including BARD1 (Figure 2).

itself. The structures of many UBDs bound to ubiquitin have already been solved and show that ubiquitin has a high degree of structural adaptability when binding to its different partners. Interestingly, when ubiquitin structure, in its various complexes with UBDs, is compared to the dynamic spectrum of structures adopted by free ubiquitin in solution, it becomes clear that the structural heterogeneity of the UBD-bound states is represented in the ensemble of free ubiquitin structures sampled over time.

The structure of protein–ubiquitin conjugates determines the biological role of the modification

The complexity of biological outcomes that result from ubiquitylation depends to a large degree on the structure of the conjugated ubiquitin-protein complexes. Consequently, the labeling of proteins with ubiquitin occurs in a highly controlled manner. Defects in ubiquitylation have severe consequences for any organism, and in humans these lead to numerous pathologies. The involvement of two representative examples of the E3 ligases, one belonging to the HECT-domain family and the other to the RING-finger family, in human diseases is described in Box 18.6.

Polyubiquitin marks proteins for degradation by the proteasome

Of the many roles of ubiquitylation (see Figure 18.22C), the best understood is the participation of ubiquitin in proteolytic degradation carried out by the proteasome. Proper folding of a newly synthesized polypeptide chain is an absolute requirement for its biological activity (see Chapter 3). Proteins that do not fold properly are subjected to refolding by specialized machines, chaperones, or are degraded by the proteasome. Proteasomes also degrade proteins that, for whatever reason, can no longer function normally; a stalled Pol II is just one example (see Box 18.6). Finally, proteasomal degradation is the usual way for the cell to degrade proteins, such as transcription

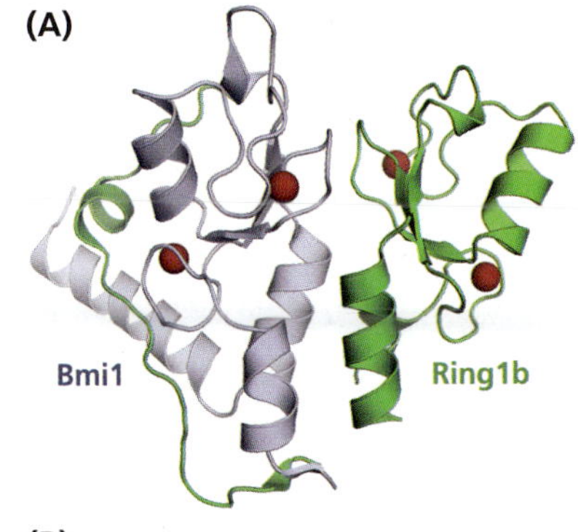

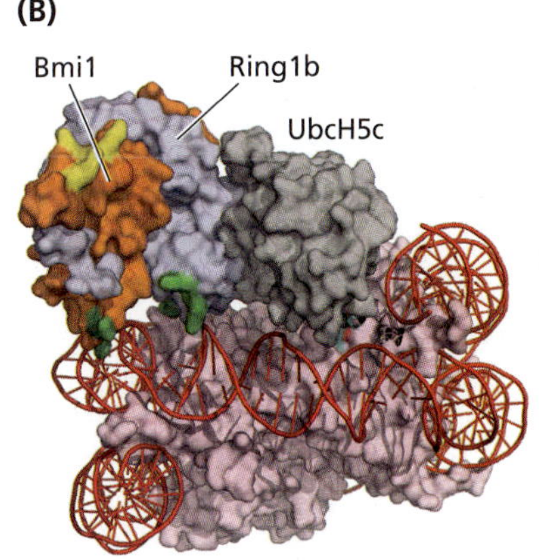

Figure 18.24 Structure of mini Ring1b–Bmi1 heterodimeric complex. (A) Crystal structure of mini Ring1b–Bmi1 heterodimeric complex. The structure contains amino acids 5–115 of Ring1b and 1–102 of Bmi1. These are the portions of the polypeptide chains that fold into RING fingers and coordinate Zn atoms, shown as dark red spheres. The structure of the two RING domains is almost superimposable, and their interaction is extensive, involving a mixture of hydrophobic and polar interactions. The N-terminus of Ring1b wraps around Bmi1. The overall structure of the complex is also extremely similar to other members of the E3 RING-finger heterotypic dimers, including that of the BRCA1–BARD1 complex (see Box 18.6). (B) Model for interaction of Ring1b–Bmi1 complexed with the E2 ubiquitin ligase UbcH5c carrying ubiquitin with the nucleosome. The model explains how the heterodimer facilitates efficient ubiquitylation of lysine 119 of histone H2A by positioning the substrate nucleosome and the enzyme E2 in an optimal configuration. Histones are shown in pink; DNA is in red; Ring1b is in light blue; Bmi1 is in orange; and UbcH5c is in gray; basic surface residues that affect DNA are shaded green, while those that do not affect DNA binding are in yellow. (B, from Bentley ML, Corn JE, Dong KC et al. [2011] *EMBO J* 30:3285–3297. With permission from John Wiley & Sons, Inc.)

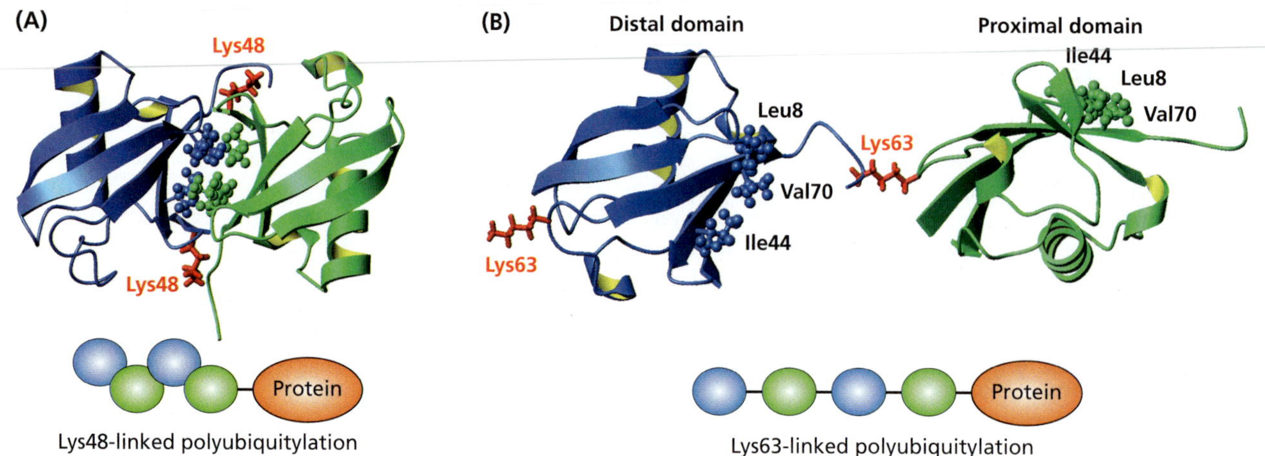

Figure 18.25 Type of ubiquitin linkage and length of Ub chain determine the biological outcome of ubiquitylation. This view is based on the structural differences between chains, which are especially pronounced in those assembled through a single linkage type, that is, homogeneous chains. The structures shown are those of diubiquitin chains linked by either Lys63 or Lys48 and solved by NMR. Chains connected by Lys48 are quite compact, with the ubiquitin moieties packed against each other, whereas Lys63-linked chains are extended. (A) In the compact Lys48-linked structure, hydrophobic residues Leu8, Ile44, and Val70 from one subunit interact with these residues from another ubiquitin molecule. (B) These contacts do not exist in the open Lys63-linked extended structure. The respective tetraubiquitins bound to target proteins are illustrated schematically below the structures. (Adapted, courtesy of David Fushman, University of Maryland.)

factors or cyclins, that need to be retained for only brief periods of time. The Nobel Prize-winning discovery of the ubiquitin–proteasome system is described in Box 18.5.

The polyubiquitin chains that serve as a signal for proteasomal degradation are usually K48-linked, although other linkages may also be involved. How are proteins that are destined for proteasomal degradation recognized by the ubiquitylation machinery, so that they are modified in this specific way? Alexander Varshavsky's laboratory has recognized two interconnected pathways in yeast that allow this recognition: the Arg/N-end rule pathway and the Ub-fusion degradation pathway.

The **N-end rule** states that it is the nature of the N-terminal amino acid residue in a protein that channels the protein for degradation by the proteasome. Certain amino acids, known as primary amino acids, can be directly recognized by the RING-finger E3 ligase Ubr1. This enzyme, in cooperation with the Rad6 E2, builds the Lys48 chain. Other N-terminal amino acid residues, Asp and Glu, can be targeted by Ubr1 only after an arginine residue is added to them by a rather peculiar enzymatic reaction (**Figure 18.26**). These amino acids are termed secondary amino acids. Finally, if the protein contains Asn or Gln, tertiary residues in this nomenclature, as an N-terminal residue, these residues first undergo deamidation reactions to give Asp or Glu, respectively, which then undergo arginylation.

The second pathway, the **ubiquitin-fusion degradation pathway** or **UFD**, is rather limited in scope: it acts on Ub-fusion proteins in which the N-terminal ubiquitin cannot be removed by deubiquitylation. The N-terminal Ub is recognized as a degradation signal by specific ligases (see Figure 18.26B).

In summary, ubiquitylation is a complex, multiform, multifunctional modification, with proven participation in a variety of cellular processes. The ubiquitin–proteasome system has evolved as a highly controlled tool for the specific degradation of cellular proteins.

Sumoylation adds single or multiple SUMO molecules to proteins

Sumoylation of proteins is the most recent addition to the list of known reversible post-translational modifications of proteins. Its discovery in 1995 was followed by intensive research into the structure of the small protein **SUMO**, **small ubiquitin-like modifier**; the enzymes involved in adding and removing the SUMO entity; its target proteins, of which there are already hundreds described; and its role(s).

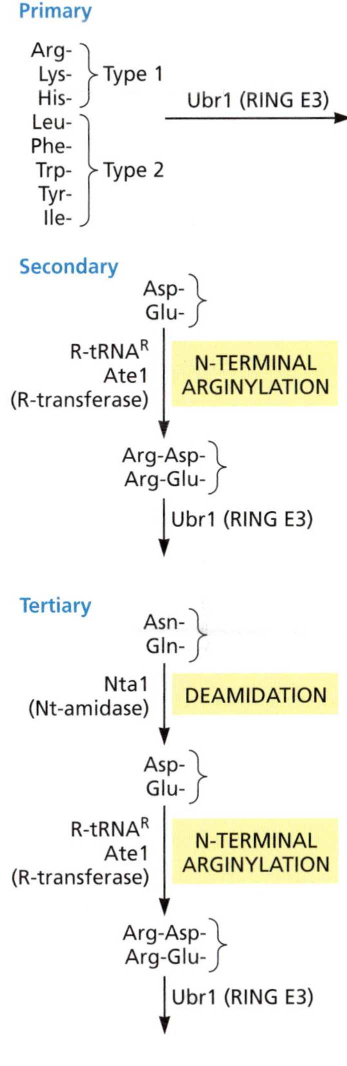

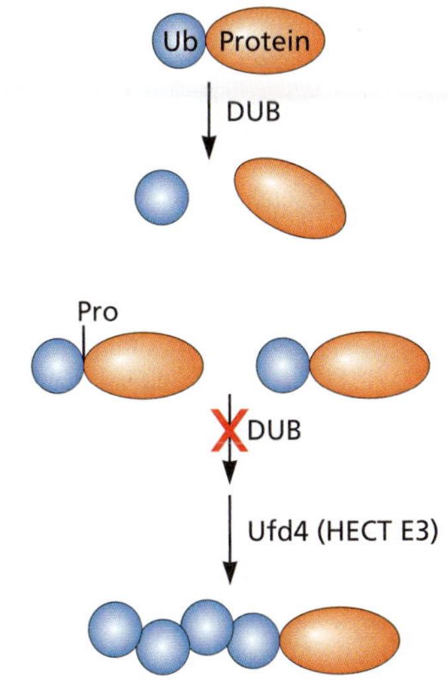

Figure 18.26 Two major pathways in yeast that lead to polyubiquitylation and then proteasomal degradation of protein substrates. (A) Arg/N-end rule pathway. N-terminal amino acids that channel the protein for degradation by the proteasome are called destabilizing residues. These amino acids are classified as primary if they can be directly recognized by an E3, in this case Ubr1, a RING-finger E3. Type 1 and Type 2 amino acids bind to two different substrate-binding sites of Ubr1, which recognize basic and bulky hydrophobic N-terminal residues, respectively; Ubr1 also contains binding sites that recognize internal, that is, non-N-terminal, degradation signals called degrons. Ubr1 and other E3 ligases that recognize specific N-degrons are called N-recognins. Proteins ending in Asp, Glu, Asn, and Gln can be targeted by Ubr1 only after their N-terminus undergoes arginylation by a specific enzyme, Ate1 or Arg-tRNA protein transferase. These residues are called secondary or tertiary, depending on the number of steps that precede their targeting and polyubiquitylation by Ubr1. Secondary amino acids undergo one step, arginylation, whereas tertiary amino acids are first deamidated and then arginylated. (B) Ub-fusion degradation or UFD pathway. Substrates of this pathway include a subset of Ub-fusion proteins in which the N-terminal Ub moiety cannot be cleaved off *in vivo* by deubiquitylation or DUB enzymes. The inability of DUBs to cleave Ub is attributed to the presence of a Pro residue at the Ub–protein junction or to alterations in Ub structure. Retention of the N-terminal Ub results in recognition of this moiety as a primary degradation signal by specific Ub ligases: Ufd4, a HECT E3 enzyme, in combination with Ubc4 or Ubc5 E2 enzymes. (A, adapted from Varshavsky A [2011] *Protein Sci* 20:1298–1345. With permission from The Protein Society.)

SUMO proteins are small, ~12 kDa proteins that resemble ubiquitin in overall tertiary structure (**Figure 18.27**). The similarity with ubiquitin at the level of primary sequence is, however, rather limited, with ~20% identity. All SUMO proteins possess an unstructured stretch of ~10–25 amino acid residues at their N-terminus; this is a unique feature that is believed to function in the formation of SUMO chains, although in the majority of cases only a single SUMO entity is added to target proteins. Some organisms such as yeast, *Caenorhabditis elegans*, and *Drosophila melanogaster* possess only a single SUMO gene, whereas plants and vertebrates have several. Three of the human SUMO proteins, SUMO1-SUMO3, are ubiquitously expressed; the fourth shows some tissue specificity. In most organisms, sumoylation is an essential process; disruption of SUMO1 in mice is embryonically lethal.

The biogenesis of sumoylated proteins is strikingly similar to ubiquitylation: it involves a three-enzyme cascade with E1, E2, and E3, and activation of SUMO before the first conjugation step is required (**Figure 18.28**). In contrast to ubiquitin conjugation, sumoylation involves just a single E2 enzyme. The process starts with proteolytic cleavage of a few amino acids from the SUMO C-terminus—between 2 and 11, with different numbers for the different SUMO proteins—to reveal the Gly-Gly dipeptide used in the conjugation reaction. On the target protein side, a lysine residue is modified. There is a peculiar feature of this modification: in most target proteins only a small percentage, ~5%, of molecules is modified at steady state. Despite the small pool of sumoylated molecules of a given target, the downstream effects of the modifications are dramatic. It seems that rapid cycles of sumoylation and desumoylation allow the entire molecular target population to experience modification over a short period of time.

The research to date has suggested several mechanisms of SUMO action, all of them leading to alterations in the molecular interactions of the modified protein with its protein partners (see Figure 18.28). One well-understood example is the sumoylation which nicely illustrates the need for reversible modification of the target (**Figure 18.29**). This example also illustrates how a small population of modified molecules can function in such an effective way. There is only a narrow time window between release of the modified enzyme from DNA and its demodification by a SUMO isopeptidase; hence the majority of TDG molecules are not modified. In the framework of the entire reaction, the low steady-state level of modification is actually a plus, as the demodified enzyme can start another cycle of repair. The dynamics of modification and demodification also explain the repressive effects of sumoylation on transcription. In this case, the targets are transcription factors that can recruit inhibitory complexes to the gene to be repressed. Once these complexes

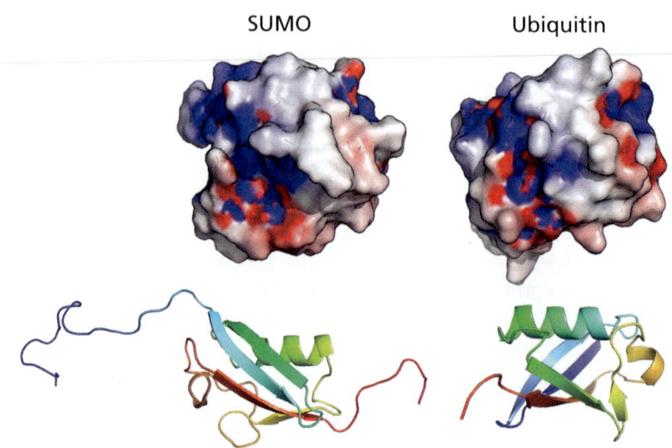

Figure 18.27 Structural similarity between small ubiquitin-related modifier, SUMO, and ubiquitin. (Top) In addition to overall shape similarity, the two proteins share similar distributions of charges on the surface. Gray indicates nonpolar residues; blue indicates basic residues; and red indicates acidic residues. Most protein partners bind through the hydrophobic patch colored in gray. (Bottom) Corresponding ribbon diagrams. (Top, from Winget JM & Mayor T [2010] *Mol Cell* 38:627–635. With permission from Elsevier.)

are recruited, SUMO can be removed without affecting the stability of the repressive complex.

The role of sumoylation is not restricted to the nucleus. Several sumoylation-dependent processes have been described that occur in other cellular compartments. For example, sumoylation is required to maintain the correct balance between mitochondrial fusion and fission. At the ER membrane, it inactivates protein-tyrosine phosphatase 1B, an enzyme that negatively regulates cell proliferation by dephosphorylating key receptor tyrosine kinases. Sumoylation is also involved in regulating the activity of potassium channels and cell surface receptors. As stated by Geiss-Friedlander and Melchior, "every conceivable consequence has been described for target sumoylation." The molecular mechanisms of most of these reactions are in need of further study.

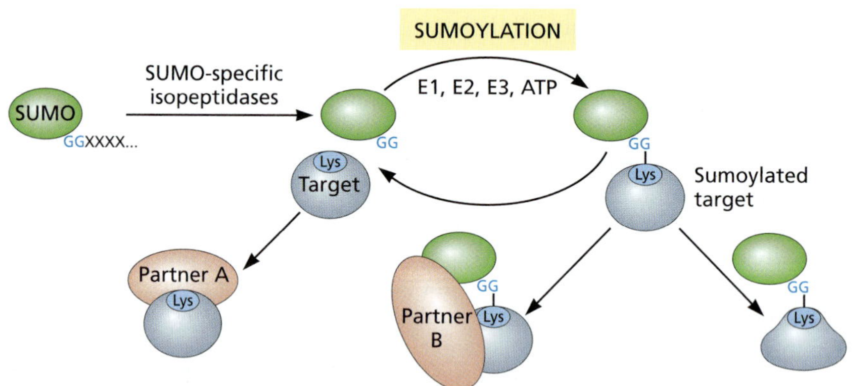

Figure 18.28 Sumoylation. Sumoylation is a highly dynamic reversible modification that exerts its effect by altering the molecular interactions of the sumoylated protein target. The basic modification pathway is very similar to that of ubiquitylation, involving three enzymatic steps in succession and using ATP to activate SUMO. Despite this amazing similarity, there is no overlap in the enzymes used in the conjugation cascade. The first conjugation reaction is preceded by proteolytic processing, which removes between 2 and 11 amino acid residues from the C-terminus of individual members of the SUMO family. This proteolytic cleavage reveals the C-terminal Gly-Gly dipeptide that serves as the attachment site for a lysine residue in the target protein. Sumoylation can affect its target in one of three ways: First, only the unsumoylated target can interact with a protein partner, here called partner A. Second, the sumoylated target can form a landing platform for a protein partner, here called partner B. In this case, the interacting partner contains a specialized motif such as SIM/SBM, SUMO interaction/binding motif, which binds SUMO noncovalently. Third, SUMO binding can alter the conformation, and thus the properties, of the target protein. (Adapted from Geiss-Friedlander R & Melchior F [2007] *Nat Rev Mol Cell Biol* 8:947–956. With permission from Macmillan Publishers, Ltd.)

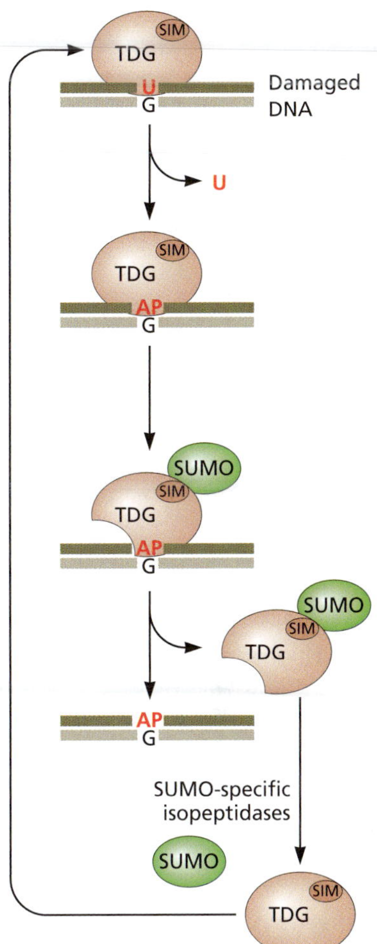

Figure 18.29 Model illustrating the requirement for sumoylation of thymine DNA glycosylase for its action. The role of glycosylases in base excision repair is described in Chapter 22. Here we use thymine DNA glycosylase, TDG, as an example of how reversible sumoylation/desumoylation is needed for each enzymatic cycle in which it participates. TDG binds to G-U or G-T DNA mismatches and removes the mutated base, with the production of an apyrimidinic or AP site. The need for sumoylation comes from the fact that TDG has a high affinity for the AP site and needs to change its conformation to be released from the DNA; the release is a necessary condition for repair to occur. The conformational change in TDG is indicated by a change in shape of the symbol and is affected by noncovalent SUMO binding through the SIM motif present in TDG. Upon release, TDG is rapidly desumoylated, and is now ready to bind to another mismatch in the DNA. (Adapted from Geiss-Friedlander R & Melchior F [2007] *Nat Rev Mol Cell Biol* 8:947–956. With permission from Macmillan Publishers, Ltd.)

18.6 The genomic origin of proteins

It should be apparent from this chapter that the proteins that actually function in a cell are, in many instances, much different from the polypeptides that are directly specified by their genes. Does this mean that the central dogma of molecular biology is invalid? By no means. In every case in which a protein is post-translationally modified, the modification occurs at specific sites in the sequence, and these are gene-dictated. Furthermore, all of these modifications occur through catalysis by specific enzymes, which seek these sites and promote the given modification. These enzymes are themselves gene-dictated. It seems that all of the information, for both synthesis and processing of proteins, still has its origin in the genome.

Key concepts

- Even after translation, many proteins require further processing to be directed to appropriate organelles or to fulfill their physiological function. Such processing may involve covalent cleavage and/or splicing of the chain, or modification of residue side chains.
- Much processing and modification occurs in the endoplasmic reticulum or the Golgi apparatus; this requires directed transport across membranes.
- Biological membranes are constructed of phospholipid bilayers, with a hydrophobic interior and hydrophilic surface. They contain both peripheral and integral proteins.
- Transport of proteins across membranes, known as translocation, utilizes multiprotein complexes called translocons. A number of different mechanisms are employed in bacteria and eukaryotes. In particular, proteins can be translocated either directly from the ribosome, in co-translational translocation, or from the cytoplasm, in post-translational translocation.
- One aspect of protein processing involves directed cleavage of polypeptide chains by specific proteases. Sometimes this is accompanied by splicing of rearranged segments. End-to-end splicing yields circular proteins.
- The vast majority of protein processing involves specific modifications of amino acid side chains. These reactions are catalyzed by specific enzymes and are often reversible.
- Phosphorylation and dephosphorylation are widely employed in enzyme regulation and in cell signaling. In some cases, there are cascades of successive enzyme activations. Phosphorylation occurs primarily at serine and threonine residues.
- Acetylation of lysine residues is mostly used to modulate interactions between proteins or between proteins and nucleic acids.
- Glycosylation involves the addition of saccharide or polysaccharide moieties to serine or threonine, in O-linked glycosylation, or to asparagine, in N-linked

glycosylation. Such modification serves a wide variety of functions, including recognition, as in blood group factors.

- Ubiquitylation is the covalent addition of a small protein, ubiquitin, to lysine side chains of proteins. Monomers or linear or branched polymers of ubiquitin can be added. Ubiquitylation often serves as a marker for protein degradation by the proteasome. Proteins themselves carry information that determines the addition of the ubiquitin degradation signal.
- Sumoylation resembles ubiquitylation in many respects but has more limited function.
- All protein processing and modification is believed to be ultimately determined in the genome, by specification of the sites of modification and the specificity of the enzymes involved.

Further reading

Books

Pollard TD, Earnshaw WC & Lippincott-Schwartz J (2007) Cell Biology, 2nd ed. Saunders.

Walsh CT (2005) Posttranslational Modification of Proteins: Expanding Nature's Inventory. Roberts and Company Publishers.

Reviews

Bernassola F, Karin M, Ciechanover A & Melino G (2008) The HECT family of E3 ubiquitin ligases: Multiple players in cancer development. *Cancer Cell* 14:10–21.

Brown FC & Pfeffer SR (2010) An update on transport vesicle tethering. *Mol Membr Biol* 27:457–461.

Ciechanover A (2005) Proteolysis: From the lysosome to ubiquitin and the proteasome. *Nat Rev Mol Cell Biol* 6:79–87.

Craik DJ (2006) Seamless proteins tie up their loose ends. *Science* 311:1563–1564.

Cross BCS, Sinning I, Luirink J & High S (2009) Delivering proteins for export from the cytosol. *Nat Rev Mol Cell Biol* 10:255–264.

Deshaies RJ & Joazeiro CAP (2009) RING domain E3 ubiquitin ligases. *Annu Rev Biochem* 78:399–434.

Dikic I, Wakatsuki S & Walters KJ (2009) Ubiquitin-binding domains: From structures to functions. *Nat Rev Mol Cell Biol* 10:659–671.

Fischer EH (2010) Phosphorylase and the origin of reversible protein phosphorylation. *Biol Chem* 391:131–137.

Geiss-Friedlander R & Melchior F (2007) Concepts in sumoylation: A decade on. *Nat Rev Mol Cell Biol* 8:947–956.

Grudnik P, Bange G & Sinning I (2009) Protein targeting by the signal recognition particle. *Biol Chem* 390:775–782.

Hunter T (2007) The age of crosstalk: Phosphorylation, ubiquitination, and beyond. *Mol Cell* 28:730–738.

Rapoport TA (2007) Protein translocation across the eukaryotic endoplasmic reticulum and bacterial plasma membranes. *Nature* 450:663–669.

Rothman JE (2010) The future of Golgi research. *Mol Biol Cell* 21:3776–3780.

Südhof TC & Rothman JE (2009) Membrane fusion: Grappling with SNARE and SM proteins. *Science* 323:474–477.

Varshavsky A (2011) The N-end rule pathway and regulation by proteolysis. *Protein Sci* 20:1298–1345.

Wickner W (2010) Membrane fusion: Five lipids, four SNAREs, three chaperones, two nucleotides, and a Rab, all dancing in a ring on yeast vacuoles. *Annu Rev Cell Dev Biol* 26:115–136.

Winget JM & Mayor T (2010) The diversity of ubiquitin recognition: Hot spots and varied specificity. *Mol Cell* 38:627–635.

Experimental papers

Becker T, Bhushan S, Jarasch A et al. (2009) Structure of monomeric yeast and mammalian Sec61 complexes interacting with the translating ribosome. *Science* 326:1369–1373.

Choudhary C, Kumar C, Gnad F et al. (2009) Lysine acetylation targets protein complexes and co-regulates major cellular functions. *Science* 325:834–840.

Fiedler D, Braberg H, Mehta M et al. (2009) Functional organization of the *S. cerevisiae* phosphorylation network. *Cell* 136:952–963.

Focia PJ, Shepotinovskaya IV, Seidler JA & Freymann DM (2004) Heterodimeric GTPase core of the SRP targeting complex. *Science* 303:373–377.

Halic M, Becker T, Pool MR et al. (2004) Structure of the signal recognition particle interacting with the elongation-arrested ribosome. *Nature* 427:808–814.

Kerr JFR, Wyllie AH & Currie AR (1972) Apoptosis: A basic biological phenomenon with wide-ranging implications in tissue kinetics. *Br J Cancer* 26:239–257.

Lomize MA, Lomize AL, Pogozheva ID & Mosberg HI (2006) OPM: Orientations of proteins in membranes database. *Bioinformatics* 22:623–625.

Olsen JV, Blagoev B, Gnad F et al. (2006) Global, *in vivo*, and site-specific phosphorylation dynamics in signaling networks. *Cell* 127:635–648.

Warren EH, Vigneron NJ, Gavin MA et al. (2006) An antigen produced by splicing of noncontiguous peptides in the reverse order. *Science* 313:1444–1447.

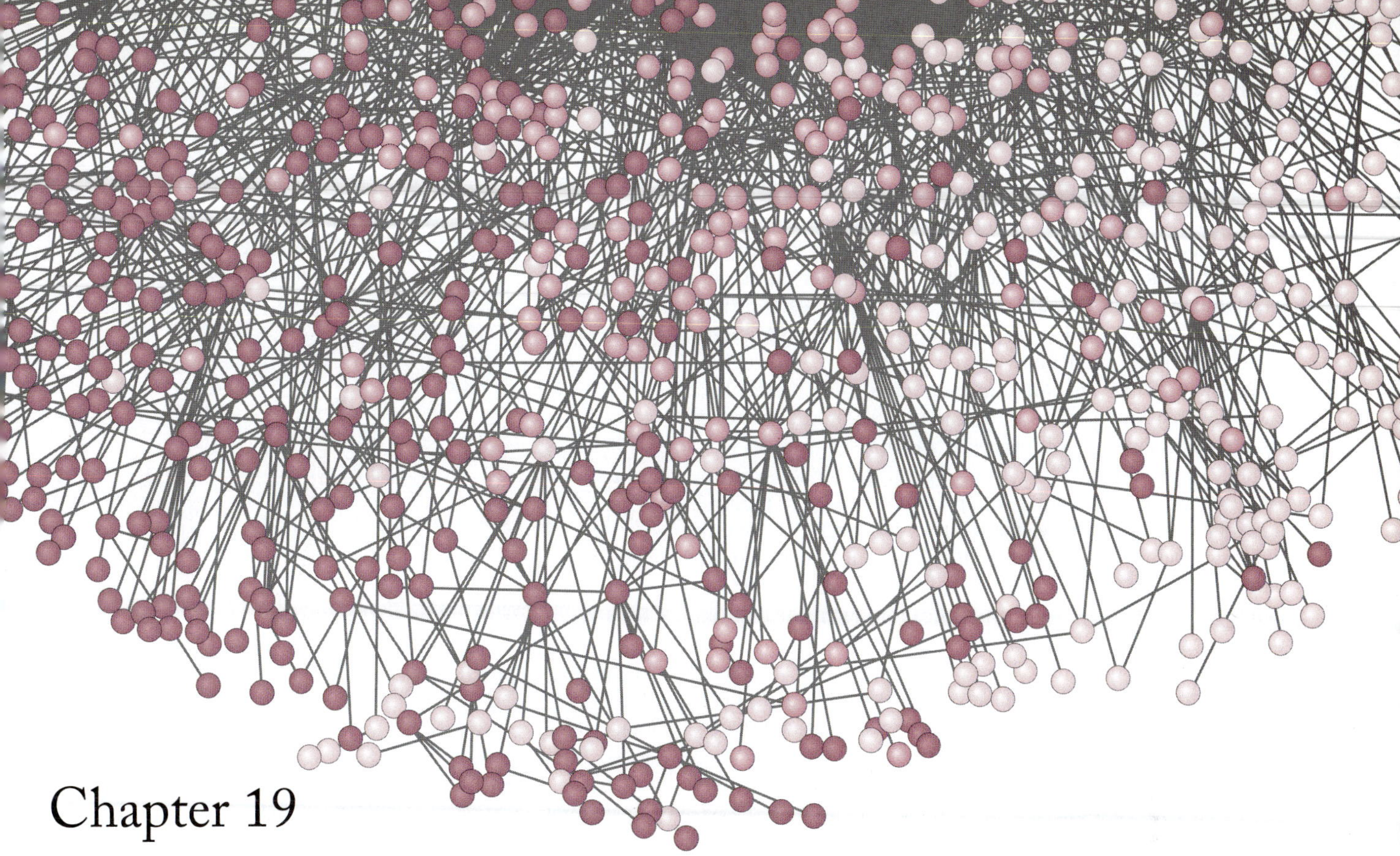

Chapter 19

DNA Replication in Bacteria

19.1 Introduction

Throughout this book, we have demonstrated how information passed from DNA directs the synthesis of specific RNA and protein molecules that are essential for the life of the cell and the organism. When organisms reproduce, they must pass their genetic information accurately to their descendants. Within each organism, dividing cells need to replicate their DNA in a precise manner, so that the two daughter cells inherit exactly the same genetic information from the mother cell. Organisms, and viruses, have evolved complex and highly regulated replication machineries to do just that. The remaining chapters are devoted to the questions of how DNA is replicated, sometimes rearranged, and often repaired. In this chapter, we describe how DNA replication occurs in bacterial cells and viruses, whereas the more complex processes involved in eukaryotic replication, in both mitosis and meiosis, form the focus of Chapter 20.

19.2 Features of DNA replication shared by all organisms

Replication on both strands creates a replication fork

As we know from Chapter 4, the double-helical model of DNA structure, put forth by Watson and Crick in 1953, immediately suggested **semiconservative replication** as one possible mode of DNA replication. The Meselson–Stahl experiments demonstrated that DNA does indeed replicate in this way (**Box 19.1**). Because both strands of the parental DNA helix are replicated, the helix must be unwound and the two strands copied separately. Thus, replication occurs at a Y-shaped structure, termed the **replication fork** (**Figure 19.1A**); the fork moves steadily through a parental DNA helix, producing two daughter helices behind it, which form the two arms of the Y.

Box 19.1 The Meselson–Stahl experiment Occasionally, in the course of science, an experiment is performed that is so elegant in its clarity of design and interpretation that it comes to be regarded as a classic. Such a case was the 1958 experiment by Matthew Meselson and Franklin Stahl, which established the mode of DNA replication. One way in which DNA might replicate is called semiconservative replication (Figure 1) because the products are two double-stranded DNA molecules, each of which contains one of the parent DNA strands and one strand made of newly synthesized DNA. This seems like a reasonable way to copy DNA but is not the only possibility, and, in 1958, at least two other modes were considered. First, replication might be conservative: that is, the whole parental DNA duplex might somehow be copied into a new duplex, composed entirely of new DNA. Second, the DNA might be copied in a patchwork fashion, by dispersive replication, so that all four resulting strands would be mixtures of old and new.

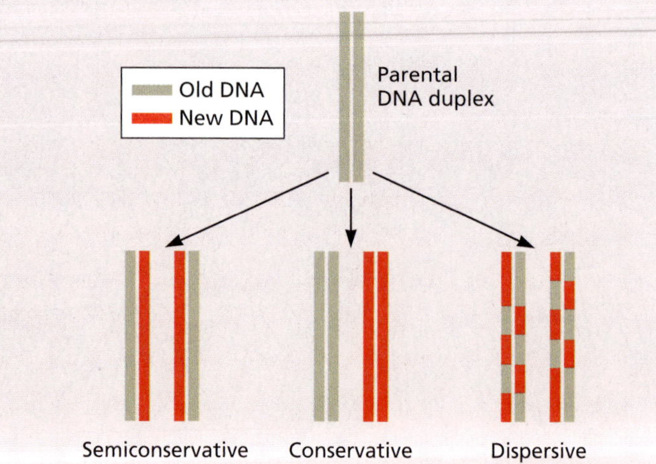

Figure 1 Three modes of DNA replication that were considered as possibilities.

To resolve the question of how DNA is replicated, Meselson and Stahl made clever use of an ultracentrifuge technique that had recently been invented by Jerome Vinograd and coworkers. In this method, DNA is centrifuged in a salt solution that forms a density gradient within the ultracentrifuge cell. If a dense salt, like cesium chloride, is used, there will be a point in the solution column where the density of DNA exactly matches the local salt density. The DNA will migrate to that point and form a band there, and its location can be determined by ultraviolet light absorption. This method can separate DNA molecules of only very small density difference, differences as slight as those produced by different isotopic compositions of the DNAs.

Meselson and Stahl grew the bacterium *E. coli* on a nitrogen source that contained only the heavy isotope ^{15}N. The DNA from these bacteria banded at a precisely defined point in the gradient, whereas bacteria grown on the common isotope ^{14}N banded at a different position (Figure 2). Next, the researchers took bacteria grown in ^{15}N for 14 generations and switched them, for one, two, three, or four generations, to ^{14}N. The result was DNA that all banded at an intermediate density after one generation; after two generations, half of the material had the intermediate density and the other half was light. The DNA distribution in the gradient changed with each generation as depicted in Figure 2. These results are consistent with semiconservative replication and not consistent with conservative replication, which would predict two bands of high and low density after the first generation.

Finally, dispersive replication was ruled out by repeating the centrifugation experiment at high pH, where the DNA strands separate. In this case, only two bands, corresponding to ^{15}N and ^{14}N DNA, were found in any generation. Strands were copied whole. Rarely has science given such unequivocal answers.

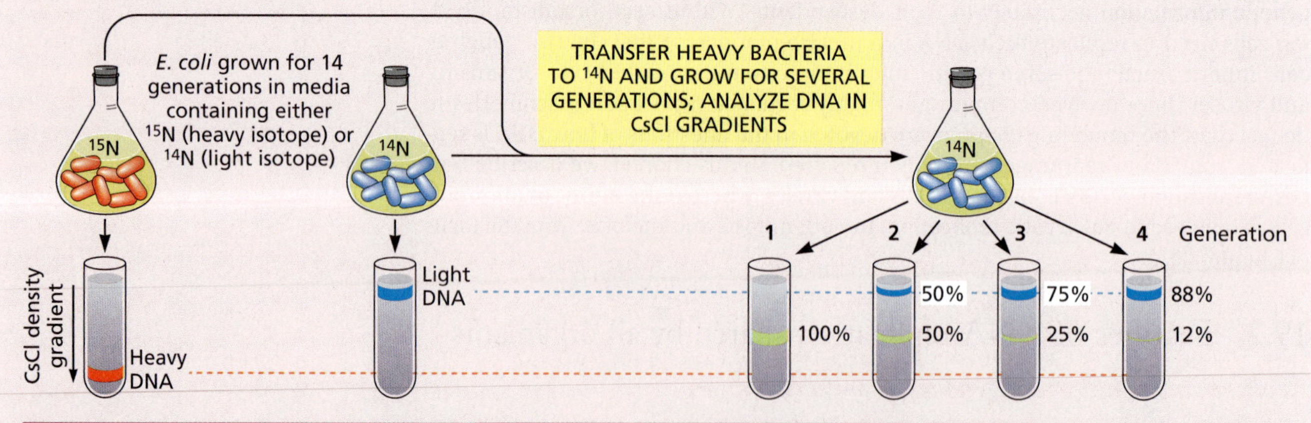

Figure 2 The Meselson–Stahl experiment. The experiment was actually performed in an analytical ultracentrifuge, but the principal is just as schematized here. (Adapted, courtesy of Mariana Ruiz Villarreal, Wikimedia.)

The chemistry of the reaction is such that each new nucleoside monophosphate is added to the free 3′-OH group of the ribose of the preceding nucleoside monophosphate (Figure 19.1B). Thus, replication of each strand is always in the 5′ → 3′ direction. This, and the anti-parallel directions of the two strands in the double helix, raises a serious problem for replication; continuous DNA synthesis is possible on only one of the parental strands. DNA synthesis on the other strand must be by a discontinuous mechanism (Figure 19.2).

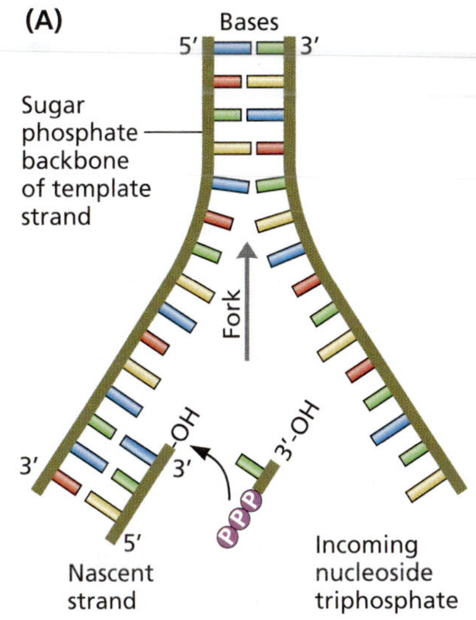

Figure 19.1 DNA replication. (A) DNA polymerase at the replication fork uses deoxynucleoside triphosphates (dNTPs) as substrates for the polymerization reaction; the nucleotides to be added to the nascent DNA chain are selected according to the base-pairing rules that govern DNA structure. The anti-parallel directions of the two strands in the double helix and the requirement for a free OH group allow continuous synthesis on only one of the parental strands; the other strand is synthesized by a discontinuous mechanism. (B) Chemistry of the DNA polymerization reaction. Each new nucleoside monophosphate is added to the free 3′-OH group of the ribose of the preceding nucleoside monophosphate; a phosphodiester bond is created by nucleophilic attack of the 3′-OH group of the nascent DNA strand on the α-phosphate of the dNTP. To prevent reversal of the reaction, the released pyrophosphate is quickly converted to inorganic phosphate by the activity of pyrophosphatase, a very abundant enzyme. (B, adapted from Mathews CK, van Holde KE & Ahern KG [1999] Biochemistry, 3rd ed. With permission from Pearson Prentice Hall.)

Mechanistically, synthesis of new DNA chains requires a template, a polymerase, and a primer

There are three basic requirements for DNA synthesis: (1) a template strand, which provides the nucleotide sequence information needed for the synthesis of a complementary strand; (2) a **polymerase**, which catalyzes the addition of residues to form the new strand; and (3) a **primer**, which is the source of a free 3′-OH group needed to start the action of the DNA polymerase. The need for **priming** of DNA synthesis stems from the fact that DNA polymerase is not capable in itself of starting a new chain; it can only extend an already existing DNA or RNA chain that carries the 3′-OH group to which the incoming deoxyribonucleoside triphosphate can be attached (see Figure 19.1B). By contrast, during transcription, the RNA polymerase enzyme can start a new chain by just matching a nucleoside triphosphate to a complementary base in the template at the point of transcription initiation. Furthermore, as transcription only copies one strand, the synthesis is continuous.

The free 3′-OH groups that are necessary for replication to begin are usually provided by RNA primers. The primers are synthesized by a primase, a DNA-dependent RNA polymerase. In bacteria the primase is encoded by the *dnaG* gene. In eukaryotes, the primers are two-part oligonucleotides composed of RNA followed by DNA (see Chapter 20). In some cases, a priming protein is used. A **nick**, or discontinuity in the phosphodiester bond in one of the DNA strands, can also serve as a point of addition of new units to the chain. Finally, nascent mRNA molecules can serve a priming function in situations in which the replication machinery collides with a co-directional RNA polymerase.

DNA replication requires the simultaneous action of two DNA polymerases

In all organisms, two DNA polymerase complexes are active at the replication fork at any time (see Figure 19.2). One of them moves in a continuous fashion, following the direction of the replication fork movement, to produce the **leading strand**, a complementary copy of the leading-strand template. The leading strand is equivalent to the RNA formed during transcription.

The other polymerase synthesizes the **lagging strand** as a series of short **Okazaki fragments** using the lagging-strand DNA template; see Box 19.2 for a more detailed account of Okazaki's discoveries. Note that both polymerases proceed in the 5′ → 3′ direction, as all polymerases do. Okazaki fragment initiation requires multiple primers (see Figure 19.2). These fragments are later connected to each other, or matured, into the continuous lagging strand. This maturation requires that the primers are removed. In bacteria, this is accomplished by polymerase Pol I, which both removes the primer RNA and inserts DNA in its place. After this, the fragments are ligated together by DNA ligase. The two copies of the elongating polymerase, one for each template strand, are anchored to their templates by auxiliary proteins that prevent them from falling off.

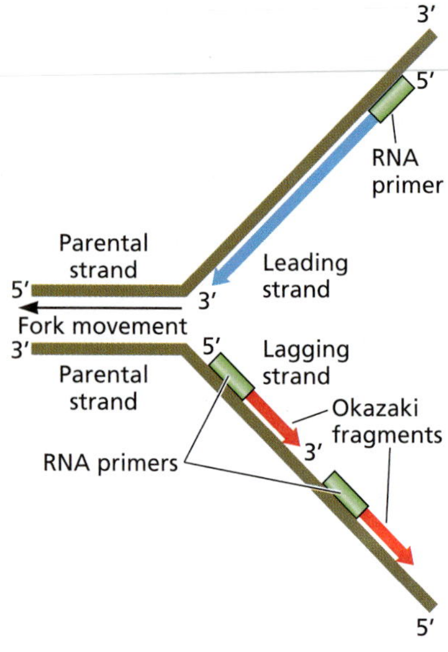

Figure 19.2 Replication fork and priming of DNA synthesis. The addition of each new nucleotide to a free 3′-OH group and the anti-parallel directions of the two strands in the double helix allow continuous synthesis, shown in blue, on only one of the parental strands. The strand that is synthesized continuously is known as the leading strand, and its synthesis moves in the same direction as the replication fork. The other strand, the lagging strand, is synthesized in the opposite direction by a discontinuous mechanism: short Okazaki fragments are first synthesized and then ligated together to form an uninterrupted polynucleotide chain. Leading-strand synthesis requires only one primer, which is synthesized during replication initiation. Lagging-strand synthesis requires multiple primers, each synthesized when a new Okazaki fragment is initiated at the fork. The RNA primers in bacteria are ~10 nucleotides long and are synthesized at the fork by primase, a DNA-dependent RNA polymerase.

Other protein factors are obligatory at the replication fork

The machinery that advances the replication fork along the parental DNA duplex consists of a number of proteins in addition to the DNA polymerase complex. Each fulfills a specific function.

Unwinding of the DNA helix is needed to expose the bases, which are otherwise hidden in the interior of the helix, so that the polymerases can copy the sequence. Unwinding is performed by a **DNA helicase**, an enzyme that moves quickly along one of the template strands to force open the parental helix using the energy of ATP hydrolysis for its translocation. In bacteria, the DnaB helicase moves along the lagging-strand template in a 5′ → 3′ direction. It is not exactly clear whether the helicase acts by a passive or an active mechanism. In the proposed passive mechanism, the helicase makes use of temporary spontaneous breathing motions of the double helix to translocate along a single strand of DNA. In this model, the movement of the helicase into a spontaneously melted region locks this region in a single-stranded conformation; that is, it effectively unwinds the helix. In the proposed active model, the helicase actively unwinds the double helix, using energy in the process; this is additional energy over that needed for translocation of the helicase. This translocation is a unidirectional motion, and thus it would seem essential that an energy source is somehow involved.

Another protein that is important in replication is DNA topoisomerase. This enzyme relieves the positive superhelical stress that accumulates in front of the moving DNA helicase. Recall from Chapter 4 that unwinding of one turn of the double helix results in the production of one compensatory positive superhelical turn. The replication fork moves very rapidly, at about 1000 nucleotides per second. Without topoisomerase

Box 19.2 Okazaki fragments and DNA replication The Meselson–Stahl experiment proved conclusively that DNA replication in *E. coli* is semiconservative: each strand served as a template for the synthesis of a new daughter strand. But it left a puzzle: it was clear how one strand, the leading strand, might be continuously copied by a polymerase moving in the 5′ → 3′ direction right behind the replication fork. But what about the other, lagging strand? For parallel, continuous replication of the lagging strand, a polymerase would have to move in the 3′ → 5′ direction. Even 10 years later, no 3′ → 5′ DNA polymerase had been found in any organism.

A young molecular biologist, Reiji Okazaki, provided the answer. Okazaki had worked on RNA polymerases in Arthur Kornberg's laboratory in the early 1960s. Upon returning to his own laboratory in Japan, Okazaki began experiments that suggested, as early as 1966, that some DNA was replicated discontinuously, in small fragments. By 1968, Okazaki, his wife Tsuneko, and their collaborators were able to present a paper in the *Proceedings of the National Academy of Sciences* that convinced many scientists that lagging-strand synthesis might be discontinuous, involving the repeated initiation and ligation of small pieces of the DNA strand. Indeed, by the Cold Spring Harbor conference in the summer of 1968, a whole section was devoted to intermediates in DNA replication. Headed by an elegant paper by Okazaki, this section included a number of papers supporting the Okazaki model.

The critical experiments that Okazaki's group performed are described in Figure 1. The system utilized *E. coli* infected with T4 bacteriophage, which shuts off bacterial replication and turns on phage DNA replication. Such cells were pulse-labeled with [^{3}H]thymidine for various periods; the DNA was then extracted and analyzed by sucrose-gradient sedimentation under alkaline conditions, in which the DNA was dissociated into single strands. Short label times put almost all of the radioactivity into small DNA pieces; longer labeling led to the accumulation of longer DNAs, in accord with the model (see Figure 1A). In order to demonstrate the ligation process directly, the experiment was repeated with a T4 mutant whose T4 ligase is temperature-sensitive (see Figure 1B). Now, a larger fraction of the DNA remains as short pieces, even after long labeling periods. The same principles were found to apply to *E. coli* DNA synthesis and to DNA synthesis in other bacteria. Further experiments strongly supported the idea that all DNA synthesis was in the 5′ → 3′ direction.

Thus, in a very short time, Okazaki and his group resolved a problem that had confused molecular biologists for a decade.

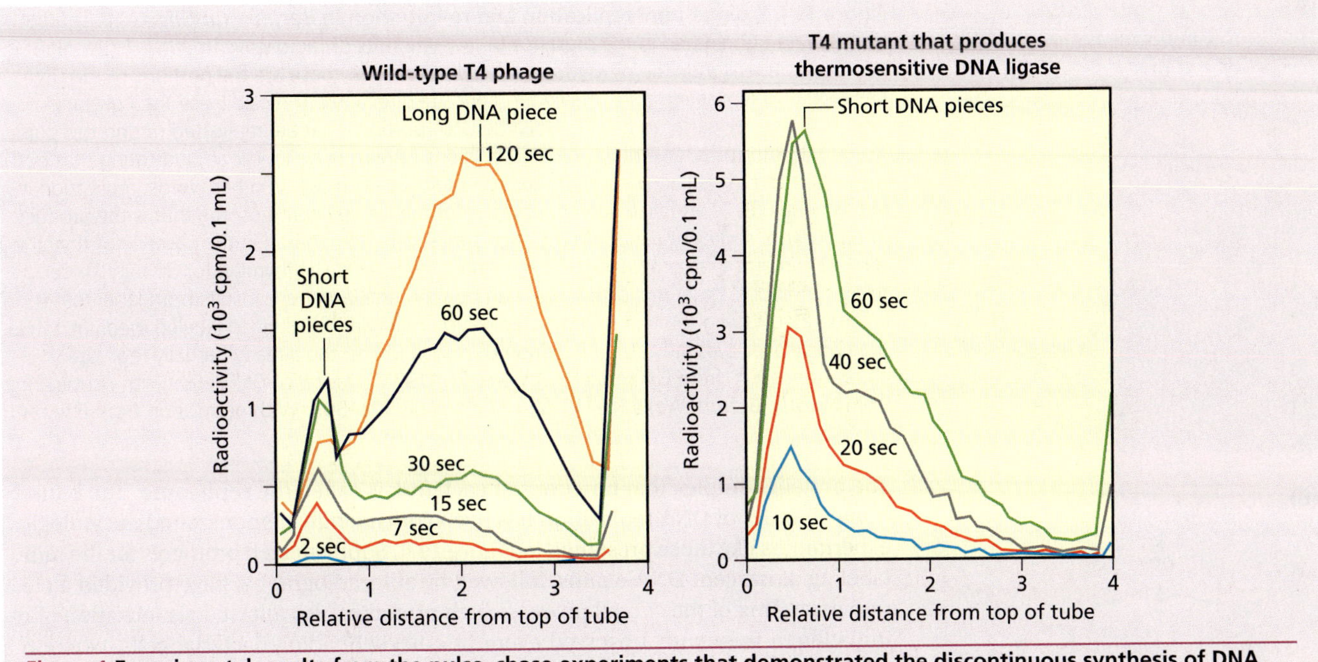

Figure 1 Experimental results from the pulse–chase experiments that demonstrated the discontinuous synthesis of DNA. (Adapted from Weaver RF [2008] Molecular Biology, 4th ed. With permission from McGraw-Hill.)

activity, the level of positive supercoiling would quickly rise to levels that would prohibit further unwinding and movement of the replication fork.

Finally, single-strand DNA-binding proteins, or SSB proteins, cover the lagging-strand template while it is temporarily single-stranded, protecting it from degradation. In addition, SSB binding prevents the formation of unwanted secondary structures; it also holds the DNA in an open conformation with the bases exposed for copying.

DNA polymerase remains bound to the replication fork while synthesizing long stretches of DNA; thus, it acts as a highly processive enzyme. This processivity accounts for the small number of enzymes needed and allows the rapid rate of replication. Processivity is due to **sliding clamp** subunits, which form a ring surrounding the DNA; these are described in more detail below and in Chapter 20. Sliding clamps are evolutionarily conserved and provide a striking example of how protein conformation is sometimes more conserved in evolution than is protein sequence. In this instance, three different proteins provide the same-shaped clamp for DNA.

The interactions among all of these proteins during replication elongation are not only extremely complex but also very dynamic. Accordingly, the use of simpler replication systems, which recapitulate the main features of bacterial and eukaryotic systems, has been instrumental in defining the minimal requirements for rapid and faithful replication of dsDNA molecules. The most widely used model systems are those derived from bacteriophages. The T7 system is probably the simplest system available; it requires only four proteins, three virus-encoded and one host factor, to replicate DNA. The T4 bacteriophage system has provided important information about the sliding clamp and clamp loader. It is important to note, however, that replication in some viruses proceeds by mechanisms that are quite different from those in other organisms.

19.3 DNA replication in bacteria

Bacterial chromosome replication is bidirectional, from a single origin of replication

In most bacteria, the chromosome is a single circular DNA molecule, containing a single **origin of replication**. Replication initiated at this specific sequence is bidirectional: two replication forks form and move in opposite directions (**Figure 19.3**).

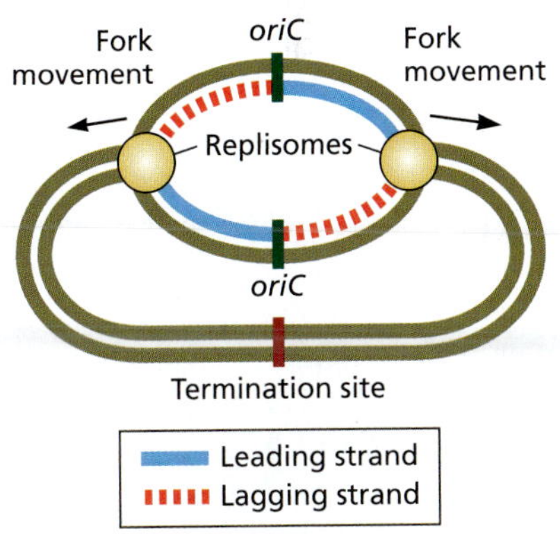

Figure 19.3 Bidirectional DNA replication of bacterial chromosomes. Schematic of the circular chromosome and the process of replication in *E. coli* are shown. Replication is initiated at a specific sequence, *oriC*, with two replication forks moving in opposite directions. The spheres represent replisomes, the protein complexes at the forks that contain the enzyme activities needed for replication to occur. The similarity of the replicating structure to the Greek letter θ has led to this process being called the θ mode of replication.

(A)

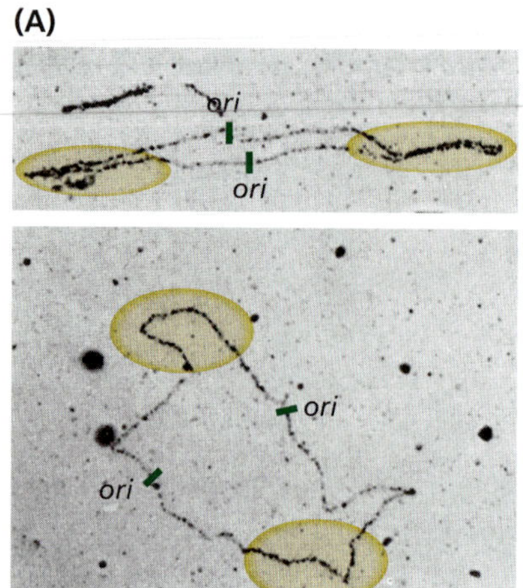

(B)

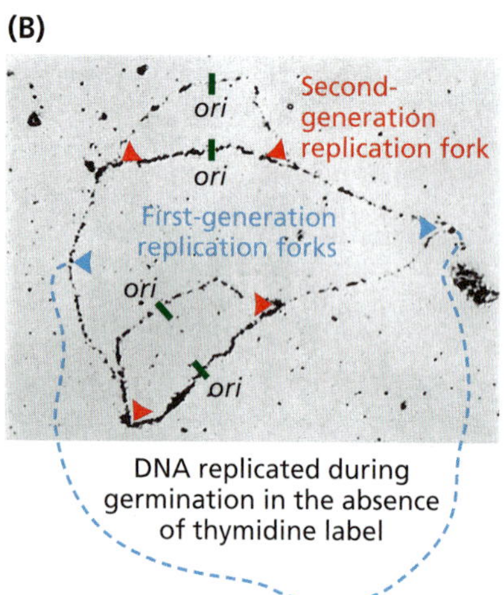

Figure 19.4 Bidirectional replication and re-initiation in *Bacillus subtilis*. (A) *B. subtilis* spores were germinated in the presence of low amounts of radioactive [*methyl*-^{3}H]thymine in order to label lightly the newly formed replication eyes; these are the portions of the DNA between two replication forks moving in opposite directions. The cells were then given a high-radioactivity pulse, to heavily label the DNA portions that are replicated during this pulse. Note that both forks have highly labeled portions, shown by yellow ovals, indicating that both replication forks were active during the high-radioactivity pulse. In other words, replication is bidirectional. (B) To visualize only that portion of the circular chromosome that is the product of replication, spores of *B. subtilis thy⁻ trp⁻* strain were germinated in the absence of thymine for 150 min and then grown in medium containing [*methyl*-^{3}H]thymine for 30 min. Under these conditions, three replication eyes can be seen: the larger eye results from label that is incorporated during the first round of replication following the shift to labeled medium, while the two smaller eyes indicate re-initiation of replication on the already partially replicated chromosome. (A, from Gyurasits EB & Wake RG [1973] *J Mol Biol* 73:55–63. With permission from Elsevier. B, from Wake RG [1972] *J Mol Biol* 68:501–509. With permission from Elsevier.)

The protein complex that functions at each fork is called the **replisome**. The bidirectional manner of DNA replication has been convincingly demonstrated in cytological experiments like those presented in **Figure 19.4**. Sophisticated protocols for the radiolabeling of nascent DNA chains, followed by autoradiography, have provided a clear understanding of the overall process. Such protocols also allowed visualization of re-initiation, a frequently observed event in quickly dividing bacterial cells, wherein a second round of replication begins even before the first round is completed. Thus, the two daughter cells inherit chromosomes that are already partially replicated, speeding up the next round of cell division.

As in all replication processes, many biochemical activities are in play here. We consider first the most important players. We begin with a description of the elongation process, rather than initiation. This is because initiation can be best understood in terms of assembling elongation-competent replisomes.

DNA polymerase III catalyzes replication in bacteria

DNA polymerase III, a multisubunit complex, is the major replicative enzyme in *Escherichia coli* and other bacteria (**Figure 19.5**; **Table 19.1**). The catalytic core consists of three subunits: (1) subunit α provides the polymerization activity (**Figure 19.6**); (2) subunit ε is a 3′ → 5′ exonuclease, which performs co-replicational proofreading; and (3) subunit θ binds to ε and stimulates its activity. The *E. coli* replisome uses twin Pol III core enzymes to copy leading and lagging strands simultaneously.

Sliding clamp β, or processivity factor, is essential for processivity

The structure of the **β clamp** is presented in **Figure 19.7**. Sliding clamp β, or processivity factor, is highly conserved in structure: the ring shape that is necessary to confer processivity to the DNA Pol III core is ubiquitous. In some bacteria, such as *E. coli*, it is a dimer of a three-domain protein, whereas in some phages and in eukaryotes it is a trimer of a two-domain protein (**Figure 19.8A**). Sliding clamps act by embracing the newly formed duplex, comprising the newly synthesized DNA chain and its single-stranded template, and then moving along the template strand together with

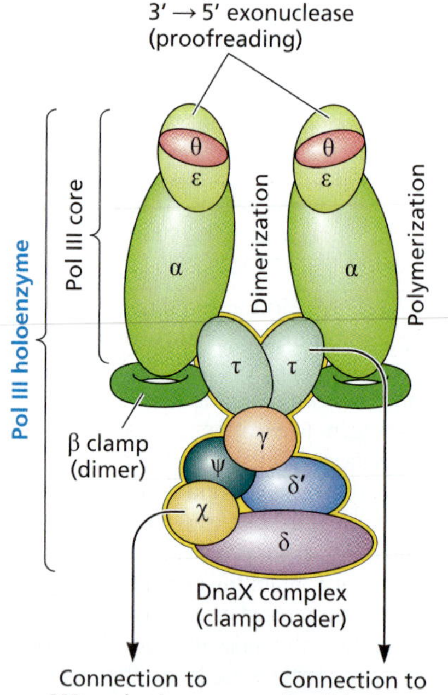

Figure 19.5 Composition of the asymmetric DNA Pol III holoenzyme complex. The holoenzyme, 900 kDa, contains (1) two copies of the catalytic core, which consists of α subunit with polymerase activity, ε subunit with 3′ → 5′ proofreading exonuclease activity, and θ subunit that stimulates exonuclease; (2) two copies of the clamp, each a homodimer of β subunits; and (3) one copy of the five-subunit DnaX complex or clamp loader, which assists in assembly of the replisome and in the binding and dissociation cycle of the lagging-strand polymerase. The schematic does not reflect the exact interactions among the subunits nor their actual shapes or sizes. The χ subunit is not essential for clamp loading; it links the clamp loader to SSB and primase. The ψ subunit is not essential for clamp loading either; it serves as a connector to χ and stabilizes the clamp loader. (Adapted from Mathews CK, van Holde KE, Appling DR & Anthony-Cahill SJ [2012] Biochemistry, 4th ed. With permission from Pearson Prentice Hall.)

Table 19.1 Properties of the two replicative DNA polymerases in *E. coli*.

Polymerase	Gene	Mol mass (kDa)	Family[a]	No. of molecules/cell	Max speed (nt/s)	Processivity	Biochemical activity	Biological function
Pol I	*polA*	103	A	400	16–20	100–200	polymerase; 3′ → 5′ exonuclease; 5′ → 3′ exonuclease	Okazaki fragment maturation with primer degradation
Pol III	*polC*	130	C	10	250–1000	500,000	polymerase; 3′ → 5′ exonuclease	replicative chain elongation

[a]The polymerases are classified in several different protein families on the basis of similarities in primary structures.

the associated polymerase (see Figure 19.6). The structure of the quadruple complex of polymerase, sliding clamp, single-stranded DNA template, and double-stranded helix of template and nascent DNA strands provides a clear picture of how the clamp ensures polymerase processivity (**Figure 19.8B**).

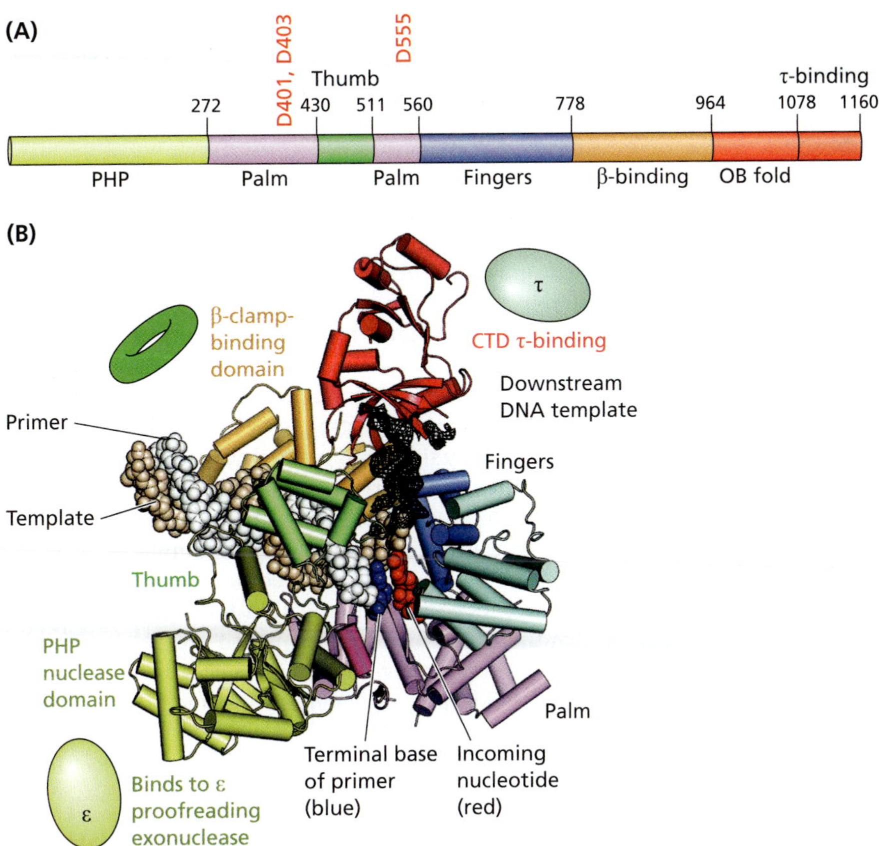

Figure 19.6 Structure of replicating triple complex of *Thermus aquaticus* Pol IIIα at 4.6 Å resolution. The triple complex comprises Pol IIIα, a DNA template strand, and an RNA primer. (A) Molecular organization of Poll IIIα: the numbers define domain borders in the *E. coli* enzyme. Domains are colored as follows: PHP nuclease, greenish-yellow; palm, pink; thumb, green; fingers, blue; β-binding domain, orange; and C-terminal domain (CTD), red. The three active-site acidic residues are indicated in red. (B) Cartoon representation of Pol IIIα with the standard right-hand palm, fingers, and thumb domains. Domains are colored and labeled as in part A. The downstream DNA template strand is represented in black mesh. The 3′-primer terminal base is depicted as blue spheres and the incoming nucleotide is depicted as red spheres. The β-binding domain binds the duplex DNA–RNA and the clamp β; the OB-fold domain in the C-terminus contributes to binding the single-stranded DNA template, and the most C-terminal domain binds τ. The PHP domain provides a binding site for the Mg^{2+}-dependent ε proofreading exonuclease. In some bacteria, this domain contains a second Zn^{2+}-dependent proofreading exonuclease that could function on a variety of substrates, for example, 3′-phosphate-terminated nucleotides. These two complementary exonucleases could work in concert. (B, from Wing RA, Bailey S & Steitz TA [2008] *J Mol Biol* 382:859–869. With permission from Elsevier.)

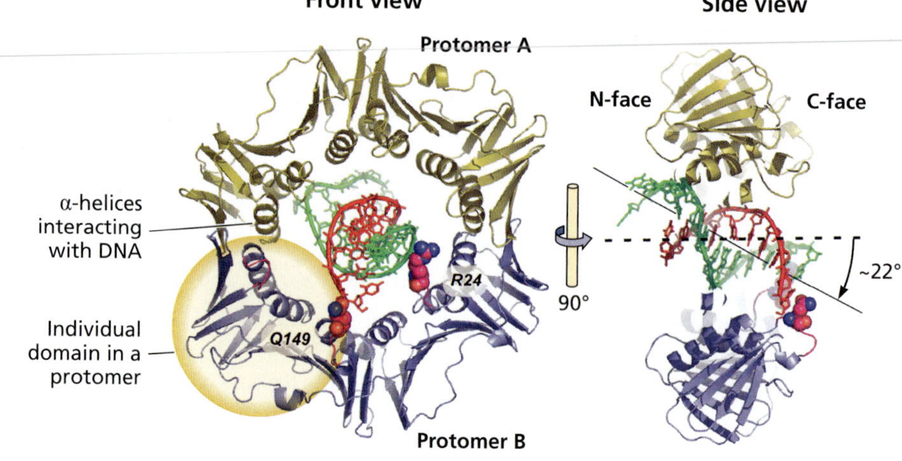

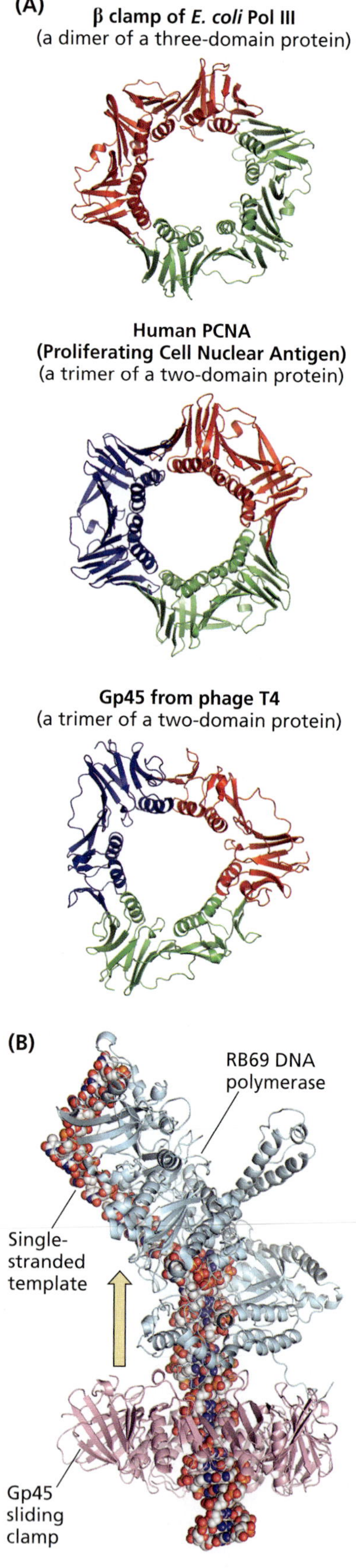

Figure 19.7 Structure of processivity factor, sliding clamp β. The structure shown is the *E. coli* factor complexed with primed DNA. The clamp is a ring-shaped dimer containing two identical protomers, A and B, each containing three individual globular domains and together forming a six-domain ring. The protomers are arranged head-to-tail, which results in structurally distinct surfaces on the two faces of the clamp. The C-terminal face, from which the C-termini project, is implicated in many of the interactions of β clamp with other proteins. The clamp loader and the polymerase compete for binding to the C-terminal face of the β-ring. The DNA is sharply tilted within the central channel; the tilt allows DNA to make contacts with R24 and Q149 on the C-terminal face. (From Georgescu RE, Kim S-S, Yurieva O et al. [2008] *Cell* 132:43–54. With permission from Elsevier.)

The clamp loader organizes the replisome

How are all of these elements structurally organized to allow synchronous replication of two oppositely oriented strands by a single fork? The key organizer is the **clamp loader**, also known as the DnaX complex in bacteria. The clamp loader is a five-subunit structure that is responsible for loading the clamp onto the DNA at the primer–template junctions. For quite a while, there was uncertainty about exactly which subunits in what stoichiometry combine to form the loader; see **Figure 19.9** for more details. In an interesting twist, it was discovered that two of the five subunits are encoded by the same gene, *dnaX*; subunit τ is a full-length gene product, whereas subunit γ is a shortened version missing two of the C-terminal domains. The extra domains IV and V of τ endow the subunit with the ability to simultaneously bind the helicase and the polymerization subunit α of the Pol III core. These protein–protein interactions are absolutely required for the formation of a functional twin-polymerase complex at the replication fork. Researchers in the field have finally agreed that the *E. coli* loader is composed of subunits τ, γ, δ, and δ′, in the stoichiometry $\tau_2\gamma\delta\delta'$. This composition seems logical because two τ subunits are needed to keep the twin Pol III core polymerases moving on both leading- and lagging-strand templates simultaneously. It should be noted that this composition and stoichiometry were derived from *in vitro* experiments; the *in vivo* situation still remains to be elucidated.

The present understanding of how the clamp loader acts to load the β clamp onto DNA in an energy-consuming process is presented in **Figure 19.10** and in more structural detail in **Box 19.3**. Note that the loader and the polymerase compete for the same face of the clamp. In order for the polymerase to bind the β clamp, the clamp loader needs

Figure 19.8 Sliding clamps are evolutionarily conserved. (A) All sliding clamps are ring-shaped structures comprising six domains that surround the double-stranded portion of the newly synthesized double helix. (B) Model of bacteriophage RB69 DNA polymerase, in light blue, bound to DNA and to the C-face of the gp45 sliding clamp. The 3′-end of the nascent DNA strand is positioned in the active center, with the single-stranded region of the template strand extending leftward. The movement of the polymerase with the bound sliding clamp is indicated by the arrow.

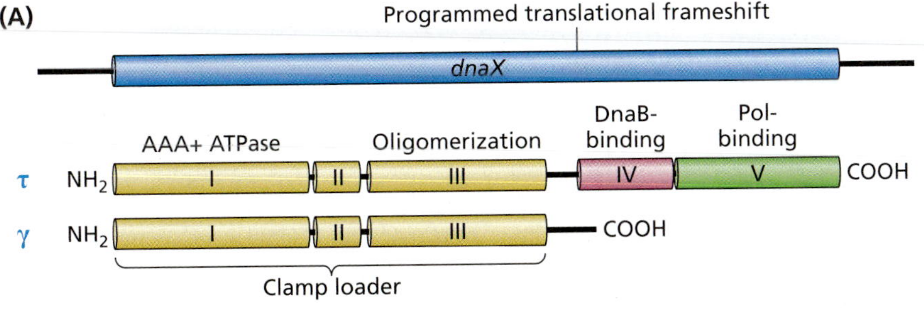

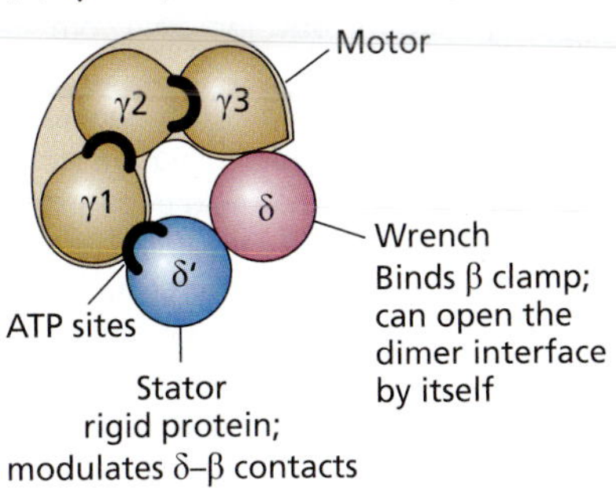

Figure 19.9 *E. coli* clamp loader. (A) Domain structure of the τ and γ subunits, both encoded by the *dnaX* gene. τ is the full-length protein product of 71 kDa, whereas γ is a truncated product of 47 kDa, resulting from a programmed translational frameshift that produces a stop codon. The two polypeptide chains share the first three clamp loader domains; these are ATPases whose activity is needed for the clamp-loading function. Domains IV and V in τ bind the helicase DnaB and the α polymerization subunit of the Pol III core. Thus, only τ has the capacity to bind helicase and core polymerase and thus to serve as a central organizer for the replisome. (B) Generalized structure of the minimal clamp loader or γ complex from *E. coli*. At one time, it was believed that the clamp loader complex contained just three γ subunits, hence the name γ complex. Indeed, the minimal $\gamma_3\delta\delta'$ complex shown here, reconstituted from recombinant subunits, is capable of loading β clamps on appropriate DNA structures. Now a more general term, DnaX complex, is used to refer to the clamp loader since each loader is expected to contain at least two τ subunits to be able to bind two molecules of Pol III core for leading- and lagging-strand synthesis. δ and δ′ also contain domains IV and V. The roles of the subunits in DnaX complex are as follows: the three τ/γ subunits bind and hydrolyze ATP and constitute the motor of the complex, the δ subunit is the wrench that cracks open the β clamp at the dimer interface, and the δ′ subunit is the stator because of its rigidity; its domains assume the same orientation in the free protein and its complexes. (B, adapted from Pomerantz RT & O'Donnell M [2007] *Trends Microbiol* 15:156–164. With permission from Elsevier.)

to leave the clamp. Thus, the clamp loader has two essential functions at the replication fork: it places the clamp onto the appropriate position at the primer-template junction of the DNA; through the τ subunits, it also cross-links between the leading- and lagging-strand polymerases and binds to the replicative helicase, thus serving as a central organizer for the entire replisome.

The full complement of proteins in the replisome is organized in a complex and dynamic way

The overall organization of the core proteins, together with the helicase and the primase at the replication fork, is presented in **Figure 19.11**. This complex constitutes the functional replisome. In addition to the key interactions between the β sliding clamps and Pol IIIα, other specific molecular interactions are important for stability of the holoenzyme at the replication fork. These include interactions between Pol IIIα and τ and, importantly, between τ and the DnaB helicase.

The specific requirement for simultaneous synthesis of the two DNA strands imposes a very peculiar structure on the lagging-strand template: the template folds into a

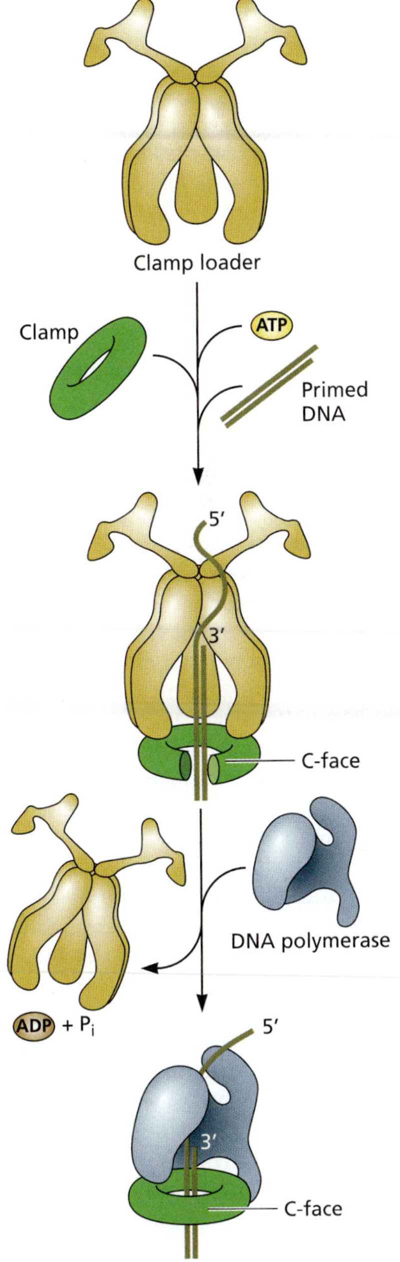

Figure 19.10 Generalized mechanism of clamp loader action in *E. coli*. The multiprotein clamp loader in the presence of ATP binds and opens the ring-shaped sliding clamp. In the ATP-bound state, the clamp loader has a high affinity for primer–template junctions. Binding of DNA stimulates ATP hydrolysis and the clamp loader is ejected, leaving the closed clamp on DNA, properly oriented for use by a replicative DNA polymerase. The loader and the polymerase compete for the same C-terminal face of the clamp. Because of this competition, it is necessary for the clamp loader to leave the clamp so that the replicative DNA polymerase can bind. Various protein partners involved in cell-cycle control, DNA replication, DNA repair, and apoptosis bind to the clamp. (Adapted from Indiani C & O'Donnell M [2006] *Nat Rev Mol Cell Biol* 7:751–761. With permission from Macmillan Publishers, Ltd.)

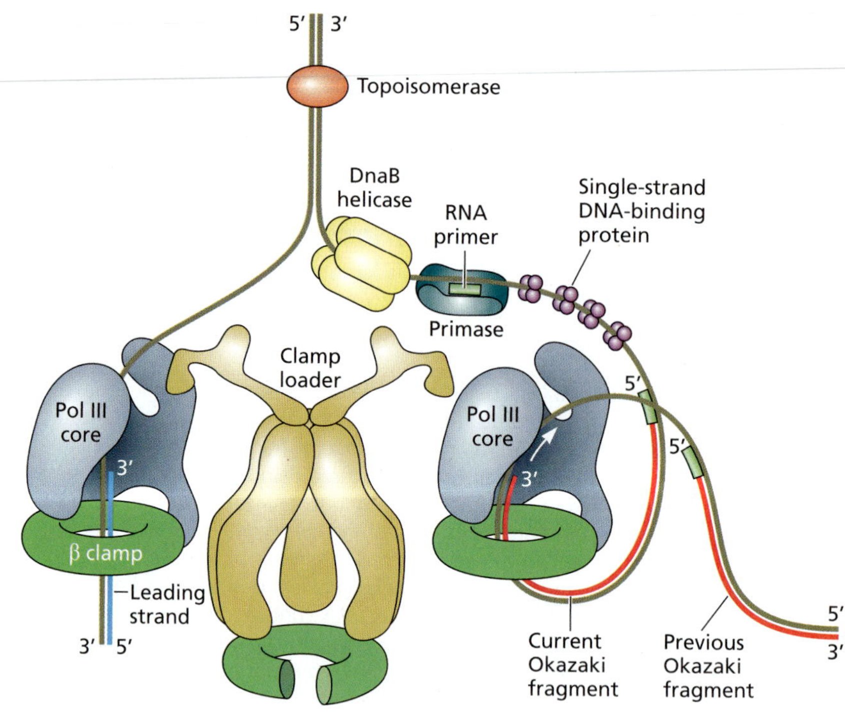

Figure 19.11 Core proteins at the DNA replication fork. Two DNA core polymerases are active at the fork at any one time, allowing synthesis of leading and lagging strands simultaneously. Note the loop or trombone structure formed by the lagging-strand template that allows the two core polymerases to move in the same direction. Both polymerases are anchored to their templates by auxiliary proteins, a sliding clamp and a clamp loader, so that they do not fall off. The other obligatory protein factors at the fork are DNA helicase, which unwinds the parental helix in an ATP-dependent manner; DNA topoisomerase, which relieves the superhelical stress accumulating in front of the moving DNA helicase; primase or DNA-dependent RNA polymerase, which synthesizes the RNA primers; and single-strand DNA-binding or SSB proteins, which cover the single-stranded lagging-strand template to protect it from degradation and hold the DNA in an open conformation with the bases exposed. (Adapted from Pomerantz RT & O'Donnell M [2007] *Trends Microbiol* 15:156–164. With permission from Elsevier.)

Box 19.3 A Closer Look: How the clamp loader works The clamp loader is one of the most remarkable of the many molecular machines we have encountered in this book. It must accept two macromolecular assemblies, the DNA–primer duplex and the sliding clamp. It must then wrap the clamp around the duplex before releasing the complex. It must perform these tasks repeatedly, and rapidly, to maintain the high rate of DNA replication in the cell.

The combination of a model bacteriophage system with high-resolution structural studies has allowed insight into this process. Studies in the laboratories of John Kuriyan and Mike O'Donnell, for example, have investigated the cycle of clamp loading by the bacteriophage T4 loader, which closely resembles that of both bacteria and eukaryotes (Figure 1). The functional portion is a heteropentamer of AAA+ ATPase-type subunits designated A–E, which is held together by a circular collar. ATP/ADP-binding sites lie between adjacent subunits.

X-ray diffraction studies of the clamp loader–clamp complex at different stages in the cycle, and in the presence or absence of ATP analogs, have provided convincing evidence of a mechanism for clamp loader activity (Figure 2). Binding of the clamp loader to ATP that is sequestered between the active domains, or modules, produces a conformational change in the loader that allows it to bind and open the clamp. This structure can then accept the primer–template DNA. A long DNA, as found *in vivo*, will presumably thread out through the gap between domains A and A′ of module A. Entrance of the primer–DNA complex into the core of the clamp loader triggers further ATP hydrolysis and accompanying conformational changes, which allow closing of the

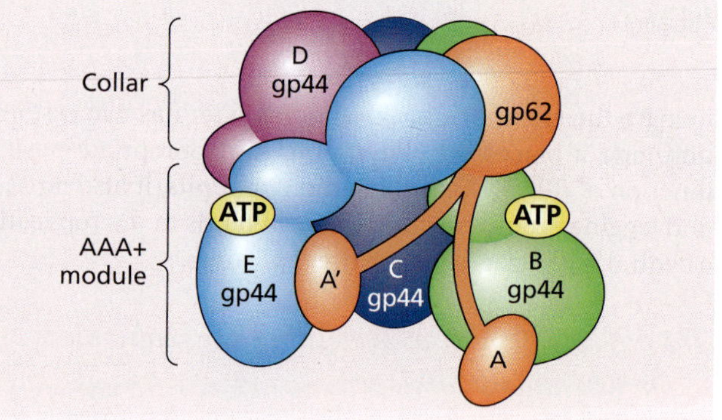

Figure 1 Clamp loader from phage T4. Clamp loaders are highly conserved from phages to bacteria to eukaryotes. They form a subfamily of the AAA+ ATPases superfamily; however, clamp loaders are pentameric, whereas typical AAA+ ATPases are hexameric. It is believed that the lack of the sixth subunit creates a gap in the complex that is essential for specific recognition of primer–template junctions (see Figure 19.10). The five subunits of clamp loaders are in general designated A, B, C, D, and E. In bacteria the subunits are τ, γ, δ, and δ′; in eukaryotes, the subunits form the heteropentameric RFC complex; in phage T4, the clamp loader consists of four copies of gp44 and one copy of gp62. Each subunit consists of three domains. The first two domains form an AAA+ ATPase module: when the five AAA+ ATPase modules are brought together in intact clamp loaders, ATP can be bound at the interface site. The third domain of each subunit is integrated into a circular collar that holds the complex together in the absence of ATP. (Adapted from Kelch BA, Makino DL, O'Donnell M & Kuriyan J [2011] *Science* 334:1675–1680. With permission from American Association for the Advancement of Science.)

clamp and subsequent release of the clamp-primer-DNA complex.

Throughout these steps, ATP binding and hydrolysis appear to function as allosteric triggers for the requisite conformational changes. Binding of the ATP analog was found to be cooperative, which is consistent with an allosteric transition between different conformations depending upon the ATP–ADP ratio. Furthermore, the successive hydrolysis steps provide a driving force to assure that the cycle proceeds only in one direction (see Figure 2).

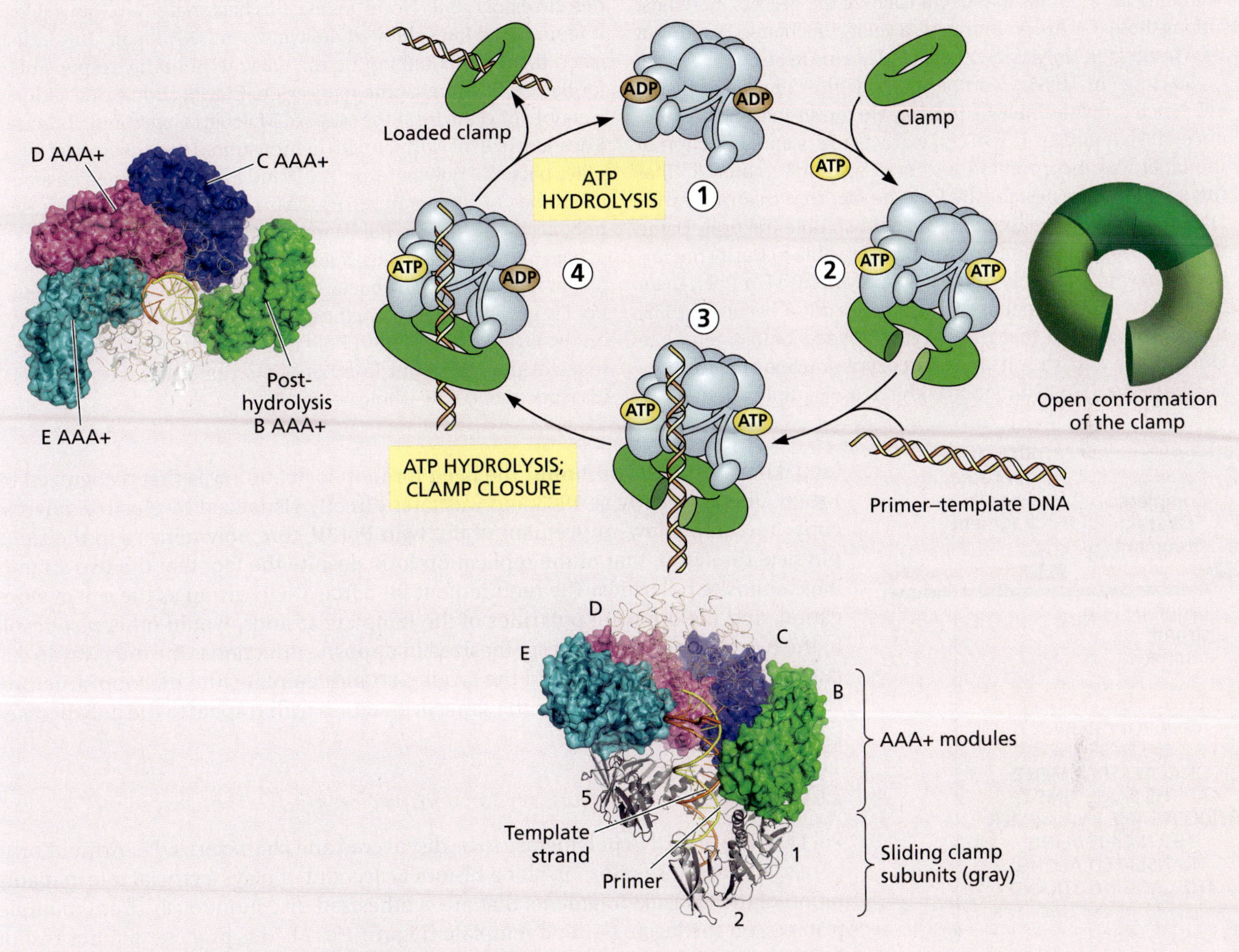

Figure 2 Mechanism of clamp loader action. This mechanism was derived from crystal structures of the complex between ATP-bound T4 clamp loader, an open clamp, and primer–template DNA, a 20-bp double-stranded DNA segment and a 10-nt single-stranded region; an ATP analog was used in lieu of ATP. The clamp loader is presented by only the contour of the entire complex, as shown in Figure 1. The ATP/ADP shown are only the ones seen from the perspective depicted in Figure 1. (State 1) In the absence of bound ATP, the clamp loader AAA+ modules cannot organize into a spiral state. (State 2) ATP binding reconfigures the complex into a spiral shape that can bind and open the clamp. In this state, all AAA+ modules are positioned perfectly to match the clamp-binding sites. (State 3) The primer–template has been threaded through the gaps between clamp subunits I and III and clamp loader subunit gp62 domains A and A′. Shown below is the crystal structure corresponding to state 3. The spiral organization of the AAA+ modules, shown in surface representation, tracks the minor groove of the DNA; gp62 is omitted for clarity. Upon internalization of the DNA, ATP hydrolysis induces changes in clamp subunit interactions. ATP hydrolysis at the B subunit breaks the interface at the AAA+ modules of the B and C subunits, which allows closure of the clamp around the primer–template DNA. (State 4) Further hydrolysis at the C and D subunits dissolves the symmetrical spiral of AAA+ modules, ejecting the clamp loader. Ejection occurs because recognition between the loader and the DNA and the clamp is lost. (Adapted from Kelch BA, Makino DL, O'Donnell M & Kuriyan J [2011] *Science* 334:1675–1680. With permission from American Association for the Advancement of Science.)

Box 19.4 The discovery of bacterial DNA polymerases The final sentence in Watson and Crick's classic 1953 paper on DNA structure hints at the kind of mechanism that might be involved in replication. Indeed, by 1958, Meselson and Stahl had demonstrated that bacterial replication was semiconservative, with each new strand being copied from an old strand that is used as template. But even earlier, others were seeking the mechanism of replication by searching for enzymes that might catalyze the process. Foremost among these was Arthur Kornberg, a young biochemist working at the Washington University School of Medicine in St. Louis.

Starting in 1955, Kornberg used the newly available ^{14}C-labeled thymidine as a marker. He found that in the presence of DNA and an *E. coli* cell extract, a very small fraction of the label was incorporated into acid-insoluble material. That this was DNA was established by the fact that treatment with DNase released the label into the acid-soluble fraction. Using this assay, Kornberg and co-workers were able to purify the relevant enzyme, presently called DNA polymerase I or Pol I. In an impressive series of papers, whose publication began in 1958, they demonstrated that the enzyme needed both a template and primer and that it added deoxyribonucleoside triphosphates in the 5′ → 3′ direction with release of diphosphate. The enzyme was also shown to possess a 3′ → 5′ exonuclease activity. Arthur Kornberg was awarded a share of the 1959 Nobel Prize in Physiology or Medicine for "discovery of the mechanisms in the biological synthesis of DNA."

Ten years later, the whole challenge of understanding DNA replication suddenly became more complicated. John Cairns and Paula de Lucia isolated an *E. coli* mutant that contained a defective Pol I gene. Nevertheless, this bacterial strain was viable; it reproduced but exhibited deficiency in DNA repair. This indicated that Pol I could not be the enzyme primarily responsible for bacterial chromosome replication. Clarification came within a couple of years from the work of Malcolm Gefter and Thomas Kornberg (son of Arthur), who demonstrated the existence of two other bacterial polymerases, Pol II and Pol III. They then created a number of temperature-sensitive mutants for each of these polymerases. Only the Pol III mutants showed restricted growth at high temperatures. Thus, Pol III is the enzyme mostly responsible for replication of the bacterial genome. The major function of Pol I is now known to be in the maturation of Okazaki fragments on the lagging strand. Both Pol I and Pol II are involved in DNA repair. Although Arthur Kornberg's enzyme is not the vital one, his work opened the whole field.

loop, known as the **trombone**. The need for loop formation was first recognized by Bruce Alberts as early as 1983 and was later directly visualized by electron microscopy. The loop allows movement of the twin Pol III core polymerases in the same physical direction, that of the replication fork, despite the fact that the two strands have opposite polarities. The requirement for a free 3′-OH group as the site of elongation, and the opposite polarities of the template strands, would otherwise result in the two new strands being synthesized in opposite directions. The only way to circumvent this problem is to fold the lagging-strand template into the loop structure, so that both strands are now in the same orientation with respect to the polymerases (see Figure 19.11).

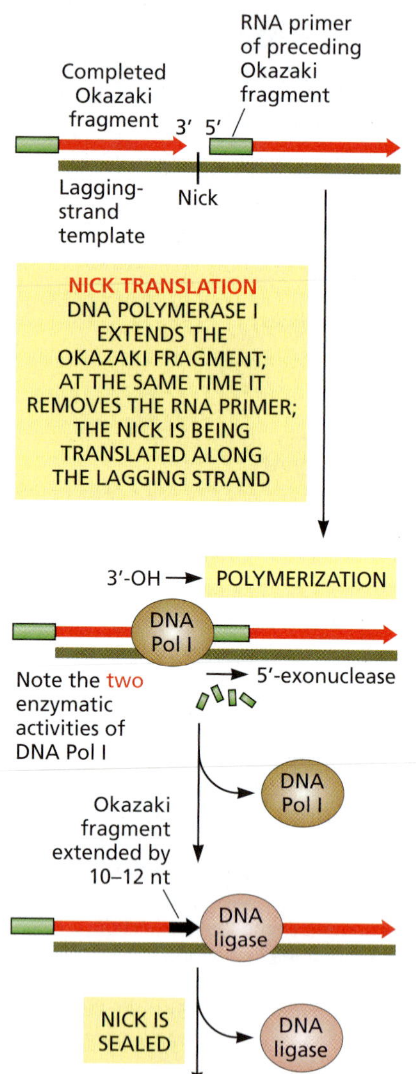

Figure 19.12 Maturation of Okazaki fragments: Combined action of DNA Pol I and DNA ligase.

DNA polymerase I is necessary for maturation of Okazaki fragments

Pol I was the first DNA polymerase to be discovered and characterized in Arthur Kornberg's laboratory; see **Box 19.4** for a historical account. It plays a crucial role in maturation of the Okazaki fragments that are synthesized discontinuously, from multiple primers, on the lagging-strand template (**Figure 19.12**). This process requires two of Pol I's three catalytic activities, all residing in the same polypeptide chain. The protein is a 5′ → 3′ polymerase, which extends the Okazaki fragment synthesized by Pol III, and a **5′ → 3′ exonuclease**, which removes the RNA primer during the nick-translation phase of the maturation process. Pol III does not have such activity.

By treating purified Pol I with the protease subtilisin, it is possible to remove the 5′ → 3′ exonucleolytic activity responsible for primer removal during replication. The large, 605-amino-acid polypeptide fragment that is obtained as a result of subtilisin cleavage is known as the **Klenow fragment** after Hans Klenow, the biochemist who devised the procedure, and has numerous applications in the research laboratory. Some details of these are illustrated in **Box 19.5**.

In addition, Pol I possesses a **3′ → 5′ exonuclease** activity. This serves in proofreading, an activity also present in subunit ε of Pol III. Proofreading is essential during replication, as it ensures the very low error frequency that is necessary during the reproduction of genetic information. During elongation, a wrong base is incorporated once every 10^5 polymerization steps, but the overall error frequency is four orders of

Box 19.5 The Klenow fragment and its use in the laboratory

The Klenow fragment is a large fragment produced when DNA Pol I from *E. coli* is enzymatically cleaved by the protease subtilisin (**Figure 1**). It retains the 5′ → 3′ polymerase activity and the 3′ → 5′ exonuclease activity for proofreading, but it loses its 5′ → 3′ exonuclease activity, which resides in the smaller N-terminal fragment. The Klenow fragment is widely used in DNA manipulations in the laboratory in applications that require DNA synthesis without 5′ → 3′ degradation, some of which are listed below.

- **Synthesis of double-stranded DNA from single-stranded templates.** For example, Sanger dideoxy-mediated sequencing uses single-stranded DNA as a template (see Box 4.10).
- **Filling-in recessed 3′-ends of DNA fragments to create blunt ends on the sticky-end fragments generated by restriction nuclease digestion.** Use of the whole Pol I could degrade the 5′-overhangs before they can be used as a template for the filling-in reactions. The fill-in reaction allows incorporation of radioactive or fluorescent labels at the 3′-ends of the fragments (see Chapter 5).

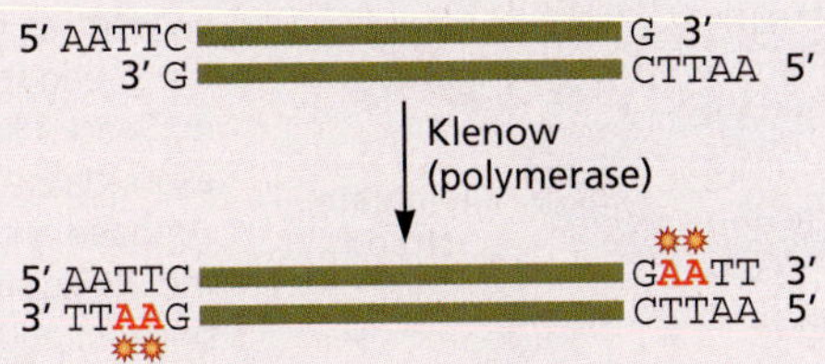

- **Digesting away protruding 3′-overhangs to produce blunt-end DNA fragments.** Removal of nucleotides from the 3′-end will continue, but in the presence of dNTPs, the polymerase activity will lead to the creation of blunt ends. This reaction is usually performed with T4 DNA polymerase, whose exonuclease activity is more efficient.

Klenow
5′ GACGACCT (3′ → 5′ exo) 5′ GACG
3′ CTGC 3′ CTGC

- **Primer extension.** This is a method used to map the 5′-ends of DNA or RNA fragments or to determine the locations of breaks or modified bases in polynucleotide strands. An oligonucleotide primer, usually 5′-end-labeled with ^{32}P or a fluorescent dye, is annealed to a position downstream of the 5′-end. The primer is extended with Klenow fragment or reverse transcriptase. The latter is used mainly for mapping 5′-ends of RNA fragments, as it can utilize both DNA and RNA as a template for synthesizing a DNA strand (see Chapter 20). A prerequisite for primer extension is knowledge of at least a portion of the sequence of interest, so an oligonucleotide primer can be synthesized.

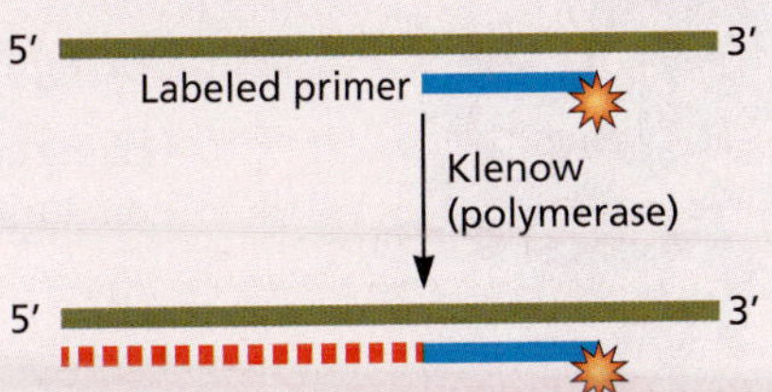

In processes for which the remaining 3′ → 5′ exonuclease activity is undesirable or unnecessary, researchers use a mutated version of Klenow that retains only the polymerase activity. These mutant enzyme forms are termed exo⁻ Klenow.

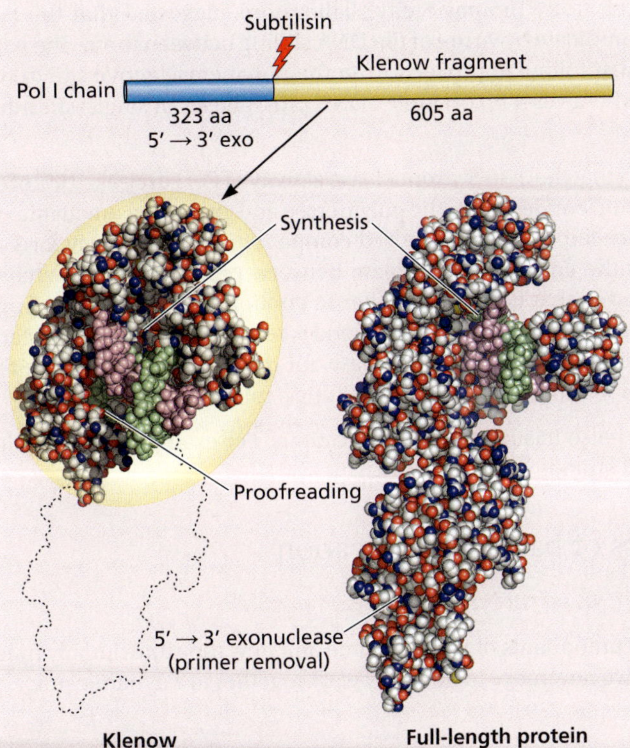

Figure 1 Klenow fragment. Schematic shows the polypeptide chain of Pol I and the two fragments obtained by subtilisin cleavage. The crystal structures, shown in surface representation below the schematic, allow a clear comparison of full-length protein and Klenow fragment. Note the deep crevice in the enzyme that accommodates DNA, shown as light green and pink spheres, which is almost completely surrounded by protein.

magnitude lower, 10^{-9}, because of the combined proofreading activities of both Pol III and Pol I. Some errors are also corrected postreplicatively, in DNA repair processes (see Chapter 22). In perspective, during each round of replication of the human genome, some 3.2×10^9 bp, only about one error, on average, is transmitted to one of the two daughter cells.

The proofreading activity of Pol I is well understood because high-resolution information is available to describe the structure of the crystallized Klenow fragment with suitable DNAs complexed with each of its two active sites (**Figure 19.13**). Nevertheless, understanding the Pol I proofreading process required finding answers to a puzzling question: how can the polymerase active site and the 3′ → 5′ exonuclease

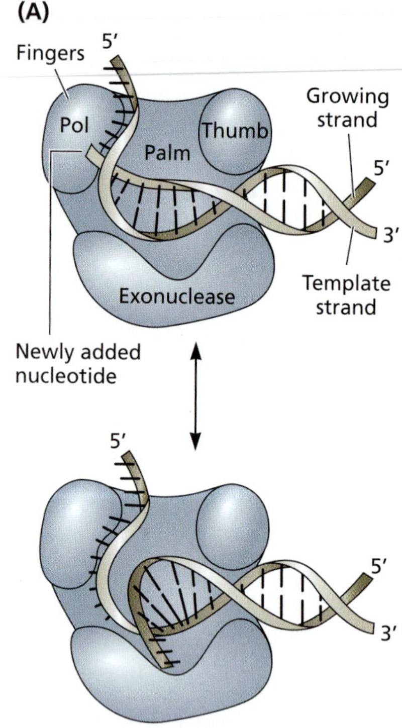

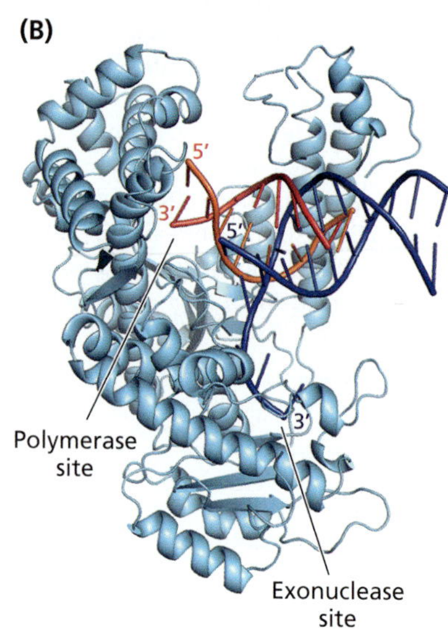

Figure 19.13 Proofreading: Removal of a mismatched nucleotide during DNA synthesis. (A) Schematic model of DNA polymerase I–DNA structure during polymerization (top) and during proofreading (bottom). The model shows the fingers, palm, and thumb domains common to all DNA polymerases. When a mismatched nucleotide is added to the growing 3′-end, the polymerase is retarded, leaving the mismatched nucleotide at the end of the duplex between the template strand and the nascent strand. The slowing down of the polymerase movement allows spontaneous melting of the end of the DNA duplex and releases the 3′-end to contact the exonuclease site, where the excision of the wrong nucleotide occurs. (B) Space-filling model of Klenow fragment with two separate active sites, polymerase and exonuclease. The 3′-end of the primer strand in polymerizing mode, shown in red, is double-stranded and lies near three catalytically important carboxylates in the polymerase active site; the 3′-end of the primer strand in editing mode, at the exonuclease site, shown in blue, is single-stranded. (A, adapted from Baker TA & Bell SP [1998] *Cell* 92:295–305. With permission from Elsevier.)

active site work together to assure that mismatched base pairs incorporated at the polymerase active site are edited out at the 3′ → 5′ exonuclease active site, some 25–30 Å away? Research from Thomas Steitz's laboratory suggested that the two active sites might communicate by virtue of the DNA sliding between them. The path that the 3′-terminus must follow to proceed from the polymerase active site to the exonuclease active site involves 4 bp of duplex DNA plus 4 bases of single-stranded frayed end.

A hint is given by the fact that the 3′ → 5′ exonuclease activity of Pol I excises ~10% of all correctly incorporated nucleotides; thus, the polymerase and exonuclease activities of Pol I are in what Steitz called a "delicately posed competition" for the newly formed 3′-terminus. How does the enzyme discriminate between proper and mismatched bases at the 3′-terminus, so that it knows whether to continue polymerization or go for excision? The fact that melting of 4 bp at the 3′-end is required for this end to reach the exonuclease site means that the structural basis for discrimination could be the increased propensity for melting of a duplex containing a mismatch.

Finally, intact DNA Pol I also has laboratory applications. For example, it is used for internal labeling of DNA strands (**Figure 19.14**).

19.4 The process of bacterial replication

The replisome is a dynamic structure during elongation

Having considered the components of the replisome and how they individually function during elongation, we now move to how they work together in a dynamic process.

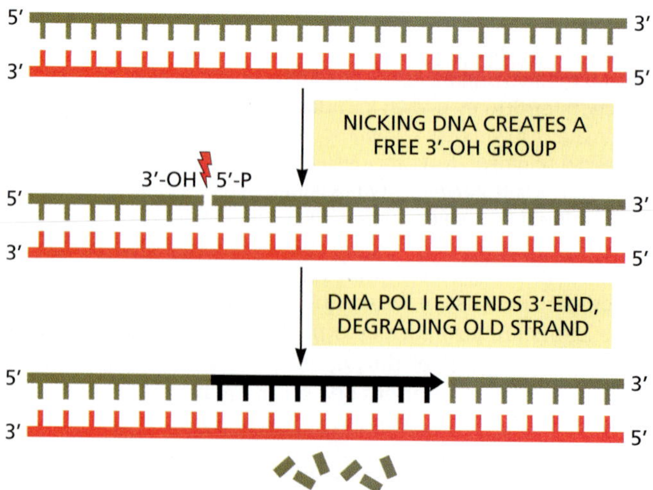

Figure 19.14 Use of DNA Pol I for nick translation in laboratory manipulations. The method is primarily used for internal labeling of DNA strands with labels such as radioactivity or fluorescence.

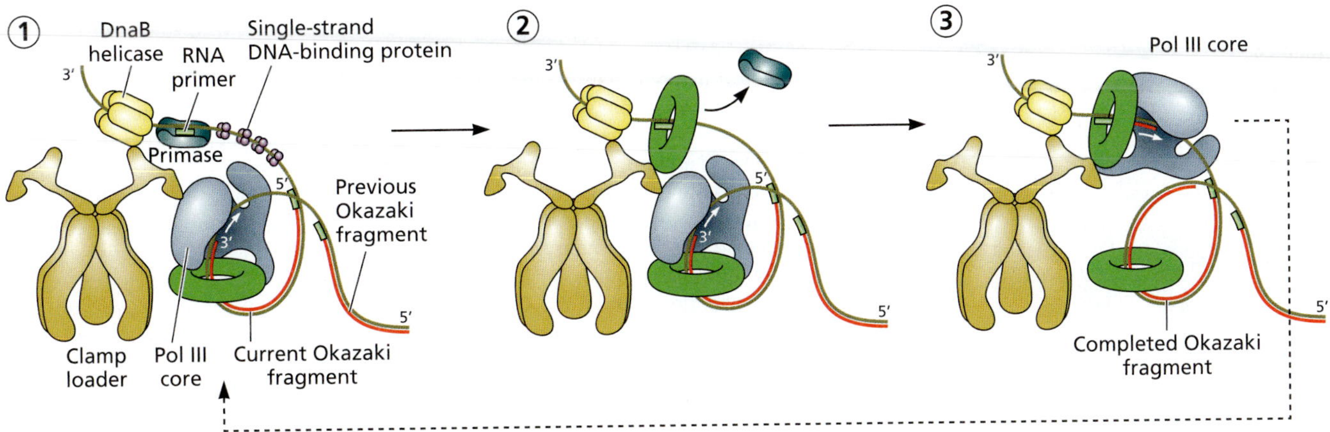

Figure 19.15 Replisome dynamics during fork progression in *E. coli*. The *E. coli* replisome uses twin Pol III core enzymes to copy leading and lagging strands simultaneously. For clarity, this schematic shows the dynamics of the replisome only on the lagging-strand template. The leading-strand polymerase uses only one sliding clamp; the lagging-strand polymerase is recycled to a new clamp for each Okazaki fragment. (Step 1) Synthesis of a current Okazaki fragment; a completed, previously synthesized Okazaki fragment is also depicted. The RNA primase is positioned in close proximity to the replication fork through interactions with the helicase DnaB. It is shown in the process of synthesizing the next primer. (Step 2) Loading of a new sliding clamp at the newly primed site; the clamp loader uses ATP to crack open and assemble the β clamp at this site (mechanism not shown here). Upon clamp assembly, the clamp loader departs, leaving the clamp behind. The primase dissociates from the DNA upon completion of primer synthesis. (Step 3) Cycling of the laggi ng-strand polymerase from the completed Okazaki fragment to the new clamp loaded at the 3′-terminus of the primer. Lagging-strand Pol III dissociates from the clamp when it hits the 5′-end of the previously extended Okazaki fragment. The polymerase is then recycled for synthesis of a new Okazaki fragment. The old β clamp is left behind on the complete Okazaki fragment. (Adapted from Pomerantz RT & O'Donnell M [2007] *Trends Microbiol* 15:156–164. With permission from Elsevier.)

We have noted that the two twin Pol III core polymerases are in constant contact through the τ subunits of the holoenzyme. The situation is actually more complex, as there is constant recycling of the lagging-strand polymerase, which must move from a completed Okazaki fragment to a new clamp assembled at the 3′-end of the primer that initiates the next Okazaki fragment (**Figure 19.15**). The clamp loader plays a central role by maintaining the overall integrity of the replisome, despite its complex internal dynamics.

Another problem arises at the site of primer synthesis. We know that the primase must be associated with the helicase in order to function, but the primers are synthesized in the same direction as synthesis of the lagging strand, the direction opposite to the movement of the helicase. Moreover, primer synthesis is slow in comparison to the movement of the helicase. There are three possible ways in which this problem could be solved (**Figure 19.16**), and it seems that different organisms use one or another of these mechanisms. First, the replisome may pause to wait for the primer to be synthesized. Second, there could be a temporary release of the

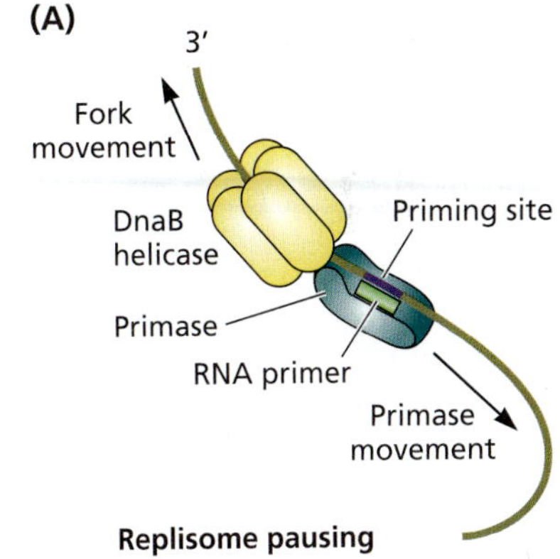

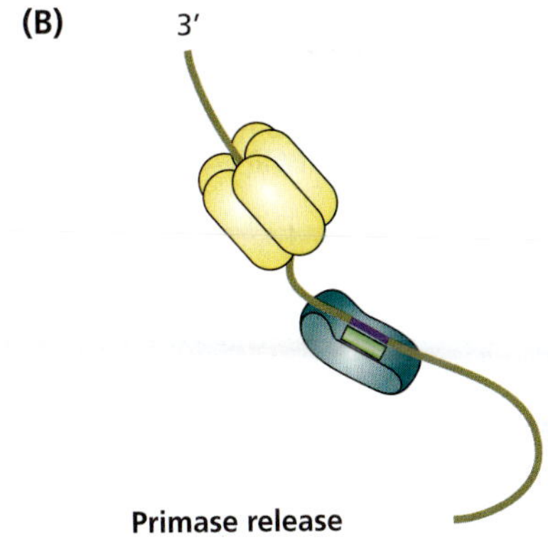

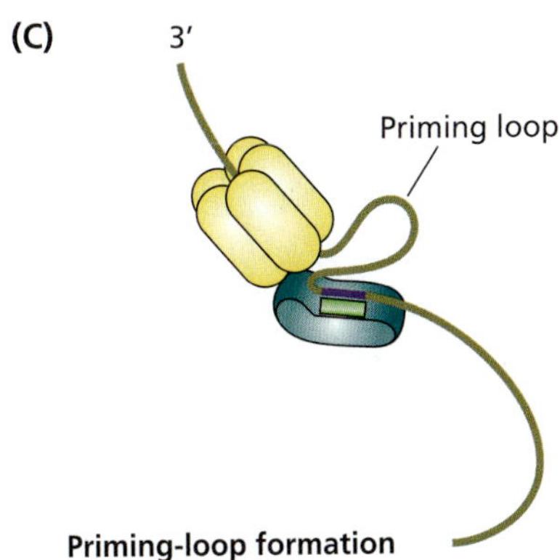

Figure 19.16 Models for DNA priming. All three models for solving the directionality problem have experimental data to support them. (A) The pausing scenario envisions that the replisome pauses for primer synthesis to occur; this has been shown to occur in the T7 phage replisome (see Box 19.7). (B) In the second model, the primase, once clamped onto the lagging-strand template by the helicase, can be temporarily released from its interaction with the helicase to synthesize the primer; this release is known to occur in the *E. coli* replisome. (C) The third model envisions the formation of a temporary loop, which allows the replisome to move forward while the normal primase–helicase interaction persists. The priming loop eventually collapses into the lagging-strand trombone loop, probably when the primer is transferred from the primase to the lagging-strand polymerase. This model has been supported by single-molecule experiments in T7 and T4 replication systems. (Adapted from Dixon NE [2009] *Nature* 462:854–855. With permission from Macmillan Publishers, Ltd.)

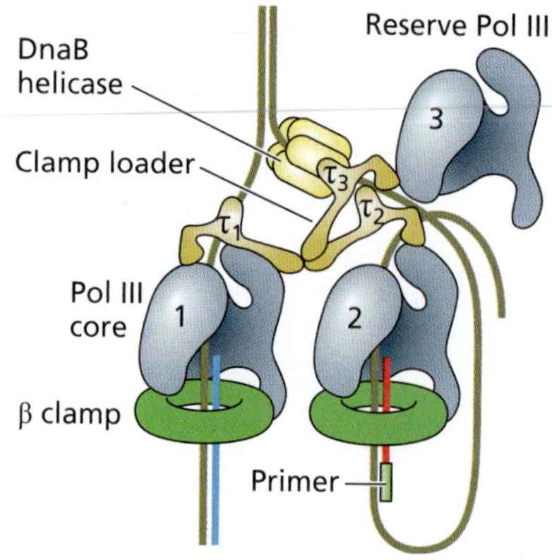

THREE POLYMERASES ARE COUPLED TO EACH OTHER AND TO THE HELICASE VIA INTERACTION WITH THE τ SUBUNITS OF THE CLAMP LOADER COMPLEX

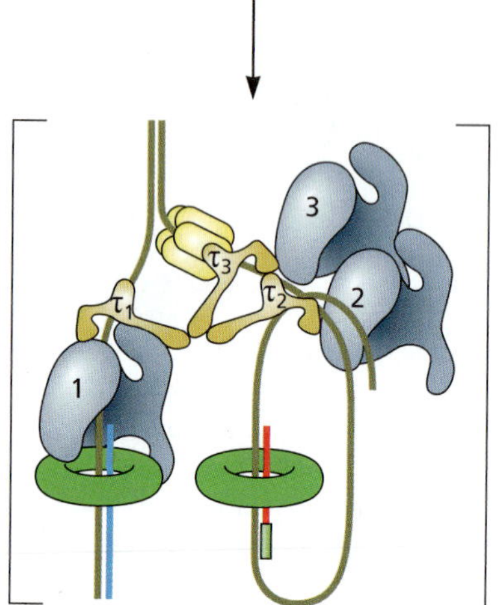

TRANSIENT DISSOCIATION OF ONE OF THE POLYMERASES, LEAVING A β CLAMP AND A 3′-END FREE TO ENGAGE REPAIR FACTORS

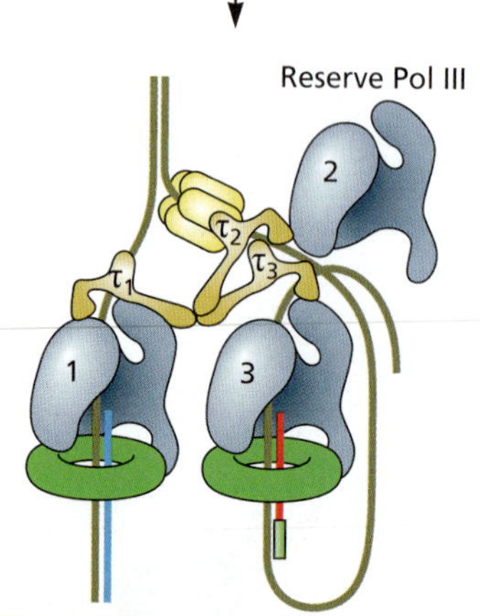

RE-ASSOCIATION OF DNA POLYMERASE WITH THE CLAMP, WITH POLYMERASE EXCHANGE

Figure 19.17 Model of the three-polymerase *E. coli* replisome. *In vitro* studies of T7 and T4 replication reveal an unexpectedly high processivity. A possible explanation is that the polymerase experiences only transient dissociations from the substrate while remaining bound to the fork through interactions with the helicase. Importantly, the helicase can also bind a third reserve or spare polymerase that can quickly attach to the template to take over synthesis of the lagging strand immediately. Researchers looked at the situation in *E. coli*. By use of four reconstituted forms of clamp loaders that are equally active in loading the β clamp—$\gamma_3\delta\delta'$, $\tau_3\delta\delta'$, $\gamma_2\tau\delta\delta'$, and $\gamma\tau_2\delta\delta'$—they demonstrated that Pol III holoenzyme assembles into a particle containing three polymerases. The trimeric $\tau_3\delta\delta'$ polymerase produces slightly shorter Okazaki fragments than the $\gamma\tau_2\delta\delta'$ complex, which suggests that the trimeric enzyme has greater efficiency in utilizing the primer of the lagging strand. It is proposed that two polymerases can function on the lagging strand, with the spare polymerase taking over when the active polymerase becomes temporarily arrested. Thus, all three polymerases are simultaneously engaged in DNA replication. The schematic uses the same symbols for the relevant subunits of the holoenzyme as Figure 19.11; for simplicity, only the τ subunits of the clamp loader are shown. (Adapted from Lovett ST [2007] *Mol Cell* 27:523–526. With permission from Elsevier.)

primase from its interactions with the helicase. Finally, as recently demonstrated by single-molecule experiments, a small priming loop can form, nested in the larger lagging-strand template's trombone loop. The priming loop eventually collapses into the trombone loop.

There is still another level of complexity to the dynamic replisome. It turns out that a second mechanism, in addition to the sliding clamp's movement together with the polymerase, contributes to the high processivity of the replisome complex (see Figure 19.8). The replisome might actually carry three, not two, core polymerases and there could be free exchange, or switching, between the lagging-strand polymerase and a spare polymerase that is bound to the helicase. The spare core polymerase might take over lagging-strand synthesis when the active polymerase becomes temporarily arrested, for whatever reason. This mechanism is especially important when the low concentration of Pol III in the cell is considered; with just ~10 copies per cell, it would be very hard to continue synthesis if the lagging-strand polymerase is lost into solution. A detailed model for a **three-polymerase replisome** is presented in Figure 19.17.

A final interesting twist in the dynamics of the replisome has to do with situations in which the fast-moving replisome catches up with, or collides with, the ~20-times-slower RNA polymerase when the two polymerases are using the same DNA strand as template for DNA replication and transcription. This situation, described as co-directional or rear-end collision, contrasts with the head-on collision that occurs when the two polymerases move in opposite directions on each of the two strands. While head-on collisions lead to replication fork arrest and induce DNA recombination (see Chapter 21), co-directional collisions do not block fork progression. RNA complexes are simply bypassed in an interesting mechanism that involves permanent release of RNA polymerase, temporary dissociation of Pol III from the DNA, and subsequent use of the nascent mRNA molecule as a primer for continued DNA replication (Figure 19.18). It is worth pointing out that the existence of this bypass mechanism could explain why essential genes and most transcription units in bacteria have been observed to be encoded by the leading-strand template. It also suggests natural selection against head-on collisions. Co-directional collisions are also selected for in human cells.

Knowledge of bacterial DNA replication is instrumental in designing strategies to inhibit bacterial growth in clinical practice. Some of these are described in Box 19.6.

19.5 Initiation and termination of bacterial replication

There are well-defined mechanisms for both starting and terminating DNA replication. For example, the bacterial cell possesses a single, well-regulated **replication origin**, known as ***oriC*** in *E. coli*. As we have seen, replication forks proceed bidirectionally from this position, and in the circular bacterial chromosome, they terminate at some point ~180° away, on the other side of the circle (Figure 19.19).

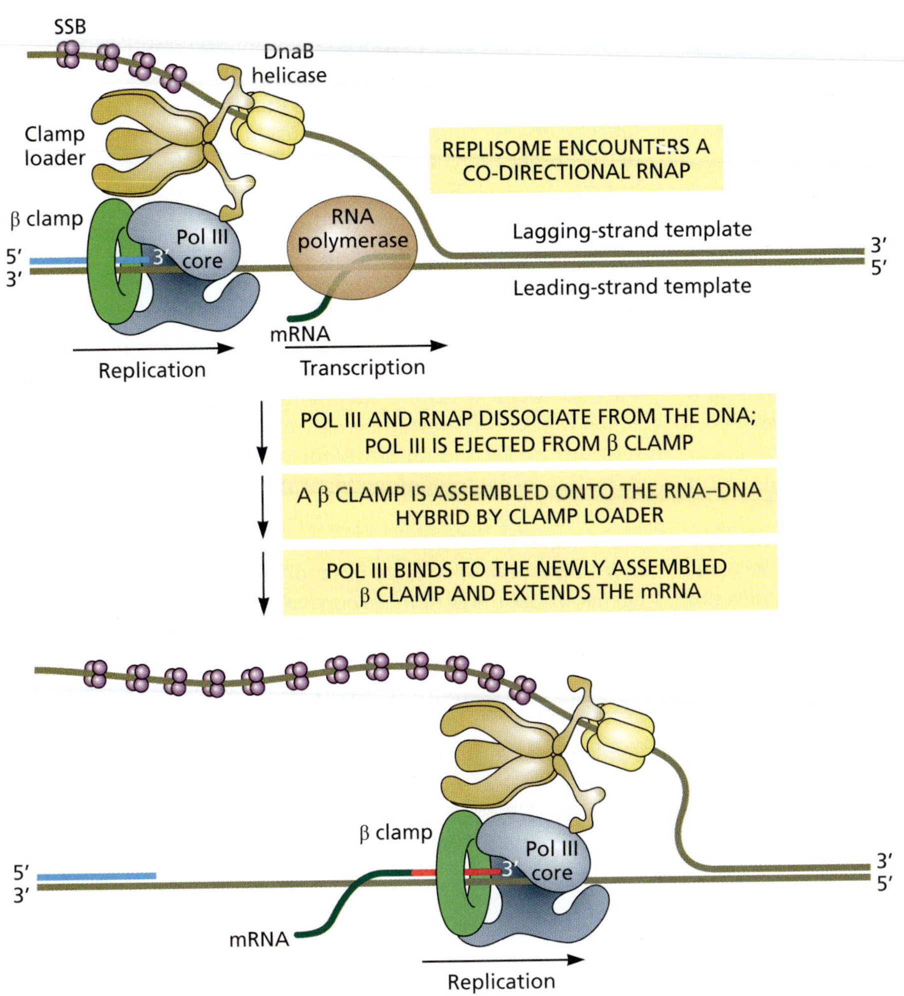

Figure 19.18 Model for bypass of co-directional RNA polymerase by the replisome. Co-directional collision of the replisome with RNA polymerase (RNAP) results in mRNA takeover. RNA polymerase transcribes the leading-strand template in the same direction as movement of the DNA replication fork. Collision of Pol III and the RNA polymerase causes displacement of both polymerases from the DNA template, but the newly formed mRNA transcript remains in position and is then used as a primer for DNA extension by the newly assembled leading-strand polymerase. The model has precedent in normal synthesis of the lagging strand, in which Pol III rapidly hops from a clamp on the completed Okazaki fragment to a newly assembled clamp on an RNA primer–DNA hybrid (see Figure 19.15). (Adapted from Pomerantz RT & O'Donnell M [2008] *Nature* 456:762–766. With permission from Macmillan Publishers, Ltd.)

Initiation involves both specific DNA sequence elements and numerous proteins

In considering the process of initiation, the factors that participate in it, and how it actually occurs, we shall focus on the process as defined in *E. coli*, where it is best understood. The origin of replication is a sequence of ~250 bp, composed of multiple

Box 19.6 Drug inhibition of DNA replication in clinical practice The quest for drugs that can inhibit bacterial cell growth, and thus infection, focuses on the inhibition of processes such as cell-wall biosynthesis, RNA and protein synthesis, and, last but not least, DNA replication. Several different strategies are being used to inhibit DNA replication, including the use of intercalators, such as doxorubicin, or compounds that inhibit some of the enzymes that are involved at various stages of replication. Drugs used to inhibit bacterial replication in the clinic should be carefully tested to not inhibit the human replicative enzymes, to avoid side effects.

Fluoroquinolones, such as ciprofloxacin, are the only class of synthetic antimicrobial agents presently in clinical use that are direct inhibitors of bacterial DNA synthesis. These drugs have broad activity against Gram-positive and Gram-negative aerobic organisms but are not effective against anaerobes. For the time being, these are the drugs of choice for treatment and prophylaxis of anthrax, caused by the toxin produced by *Bacillus anthracis*. Fluoroquinolones inhibit both DNA gyrase and topoisomerase IV, which are topo II enzymes. DNA gyrase introduces negative supercoils into DNA (see Chapter 4); this function is essential for both DNA replication and transcription. The primary function of topo IV is the decatenation of multiply linked daughter chromosomes during the terminal stages of DNA replication. Because these agents inhibit both enzymes, it is believed that there is a much lower likelihood that bacteria will develop resistance against them. The drugs bind to the complex of each of these enzymes with DNA, blocking the progress of the DNA replication enzyme complex.

Gyrase and topo IV are also the targets of aminocoumarin antibiotics, natural products of various *Streptomyces* strains; these antibiotics bind the enzymes with higher affinity than fluoroquinolones. Recent biochemical and X-ray crystallographic evidence shows that the aminocoumarin antibiotic simocyclinone D8 inhibits gyrase by a new mechanism; it interacts with two separate pockets of the enzyme and thereby prevents binding to DNA.

(Continued)

Box 19.6 *(Continued)* Another enzyme pair targeted by drugs is the helicase–primase complex. Such drugs have been developed and are expected to be highly effective against *B. anthracis* and *Staphylococcus aureus*; these drugs are in preclinical trials.

Hydroxyurea, **HU**, inhibits cell growth by targeting ribonucleotide reductase, RNR, the essential enzyme that reduces ribonucleotides to deoxyribonucleotides. By doing so, it depletes cells of DNA precursors, causing arrest of the DNA replication fork. The actual response to HU treatment consists of a sequence of events that initially ensure the survival of the cell but eventually lead to cell death. Figure 1 illustrates these events and highlights the importance of gene regulation and the interplay of multiple cellular systems in the response.

HU also affects eukaryotic cells, as RNR is an essential enzyme in eukaryotes too. In clinical practice, HU is used to treat cancer, sickle cell anemia, and psoriasis, among other diseases. Since HU affects both diseased and normal proliferating cells, it is highly cytotoxic, so the course of treatment must be carefully monitored and controlled. HU is also widely used in the laboratory to synchronize cells in the cell cycle. HU treatment arrests the cell population in early S phase; removal of the drug then allows the cells to move synchronously through the cell cycle.

Finally, targeting other enzymes that are present only in certain bacteria species could be very useful in selective inhibition of DNA replication in only those cells, not affecting the normal human bacterial flora. A recent example of this is targeting of the replication initiator RctB in the family Vibrionaceae, which comprises numerous aquatic bacterial species and includes several human pathogens such as *Vibrio cholerae*, the cause of cholera. All organisms in this family have two chromosomes, and replication of the smaller one depends on *rctB*, a gene that is restricted to the Vibrionaceae. Researchers at Harvard Medical School identified a compound, vibrepin, which blocks the unwinding of the origin of replication present on the second chromosome, apparently by promoting formation of large nonfunctional RctB complexes. Although vibrepin also appears to have targets other than RctB, the findings suggest that RctB is a useful drug target for the creation of Vibrionaceae-specific antimicrobial agents. *Vibrio*-specific agents, unlike the antibiotics currently used in clinical practice, will not engender resistance in the normal human flora or in non-*Vibrio* environmental microorganisms.

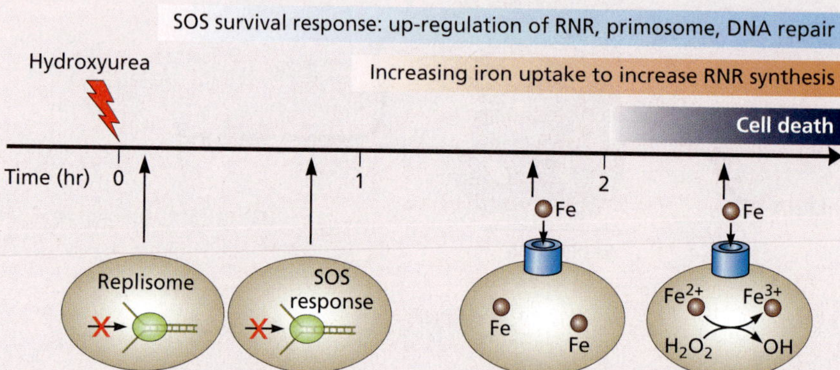

Figure 1 Sequence of cellular responses to hydroxyurea exposure involving regulation of gene expression. Hydroxyurea or HU exposure triggers a protective response consisting of DNA damage repair; up-regulation of ribonucleotide reductase, RNR, to compensate for its inhibition by HU; and up-regulation of the primosome, the system that may help to restart stalled DNA replication forks. After 2 hours of HU exposure, however, cells begin to die rapidly, mainly as a result of hydroxyl radical formation. Cells up-regulate iron-uptake genes—iron is part of the catalytic center of RNR—presumably in an effort to counter the effects of HU; in doing so, they catalyze the formation of hydroxyl radicals via Fenton chemistry. Thus, bacterial cells specifically respond to RNR inhibition in a way that ultimately leads to their own death. (Adapted from Bollenbach T & Kishony R [2009] *Mol Cell* 36:728–729. With permission from Elsevier.)

9-bp repeat elements. These elements are called **DnaA boxes**, as they provide the binding sites for the sequence-specific **initiator protein DnaA**; see Figure 19.20 for a more detailed description of the DnaA-binding sites. Initiation absolutely requires DnaA and two additional proteins: the DNA hexameric **helicase DnaB** and the **helicase loader DnaC**. DnaA and DnaC form multisubunit right-handed protein filaments; their structures are shown in Figure 19.20. Oligomerization of DnaA is instrumental in melting the AT-rich DNA unwinding elements in the DUE region at *oriC* (Figure 19.21 and Figure 19.22) by forming a positive toroidal DNA superhelix on the outside of the DnaA filament. Why DnaC would form a very similar structure is less clear. Both DnaA and DnaC are ATPases of the AAA+ (ATPases Associated with

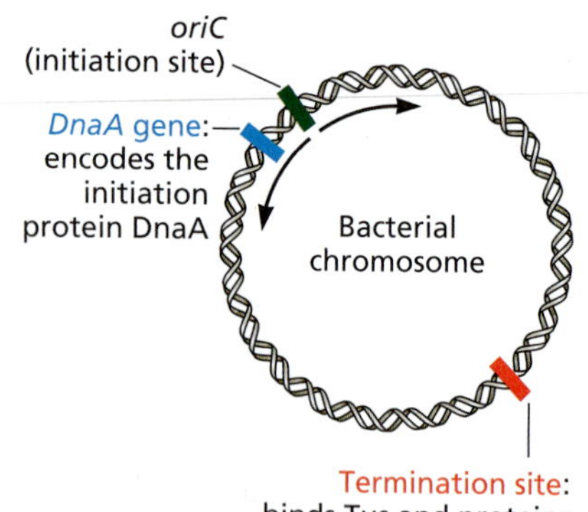

Figure 19.19 Simplified schematic of circular *E. coli* chromosome. Sites where DNA replication is initiated, *oriC*, and terminated are shown.

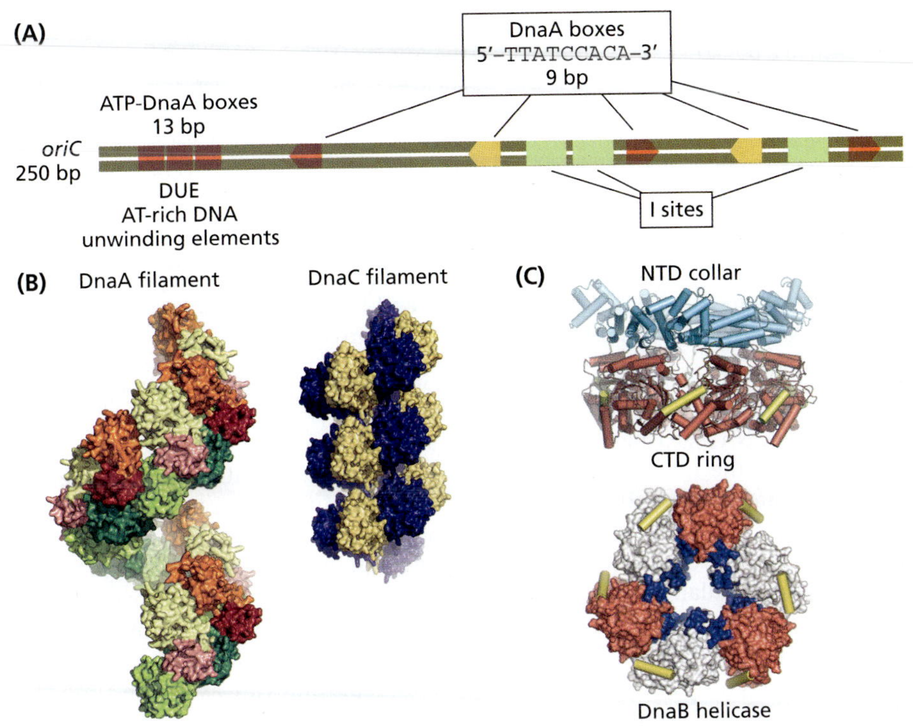

Figure 19.20 *OriC* of *E. coli* chromosome and the proteins that participate in replication initiation. (A) *OriC* has three types of DnaA binding sites: (1) high- and low-affinity DnaA boxes, shown as orange and yellow arrowheads, respectively, with the orientation of each DnaA box shown relative to that of the top DNA strand; (2) I sites that differ slightly from the box consensus and are interspersed among the DnaA boxes; and (3) ATP-DnaA boxes that are similar to I sites and only bind ATP–DnaA. (B) (Left) ATP-bound DnaA forms a spiral filament. Individual domains of the protein are highlighted in three different colors: red, green, and orange; individual subunits are distinguished from each other by darker or lighter shading. Two turns of the superhelix are shown. (Right) Nucleotide-bound ATPase domain of DnaC forms a higher-order, spiral assembly that exhibits a different rise and pitch from that of DnaA. Three turns of the superhelix are shown, with different subunits colored blue or gold to distinguish them from one another. Filaments in both structures are rendered on the same scale to highlight their different sizes. (C) DnaB helicase is a hexameric ring structure. (Top) Side view of the DnaB hexamer in ribbon representation. Each monomer has a C-terminal domain or CTD, shown in dark red, and an N-terminal domain or NTD, shown in light blue, that are connected by a flexible linker, shown in yellow. The hexamer has a double-layered ring structure in which the NTDs, residues 1–152, pack into a rigid triangular collar seated on a more loosely packed ring of CTDs, residues 186–454. (Bottom) Surface representation of CTD rings of the helicase. Alternate subunits are colored white and red, predicted DNA-binding loops are in blue, and linker helices are shown as yellow cylinders. (A, adapted from Mott ML & Berger JM [2007] *Nat Rev Microbiol* 5:343–354. With permission from Macmillan Publishers, Ltd. B, courtesy of James Berger, Johns Hopkins University. C, from Bailey S, Eliason WK & Steitz TA [2007] *Science* 318:459–463. With permission from American Association for the Advancement of Science.)

various cellular Activities) superfamily. ATP hydrolysis by both DnaA and DnaC is a prerequisite for function. The similarities in the structure and enzymatic activities of DnaA and DnaC suggested that the cell may use DnaC as a molecular adaptor that employs ATP-activated DnaA as a docking site for recruitment and correct orientation of the helicase DnaB at replication origins. Two molecules of the helicase

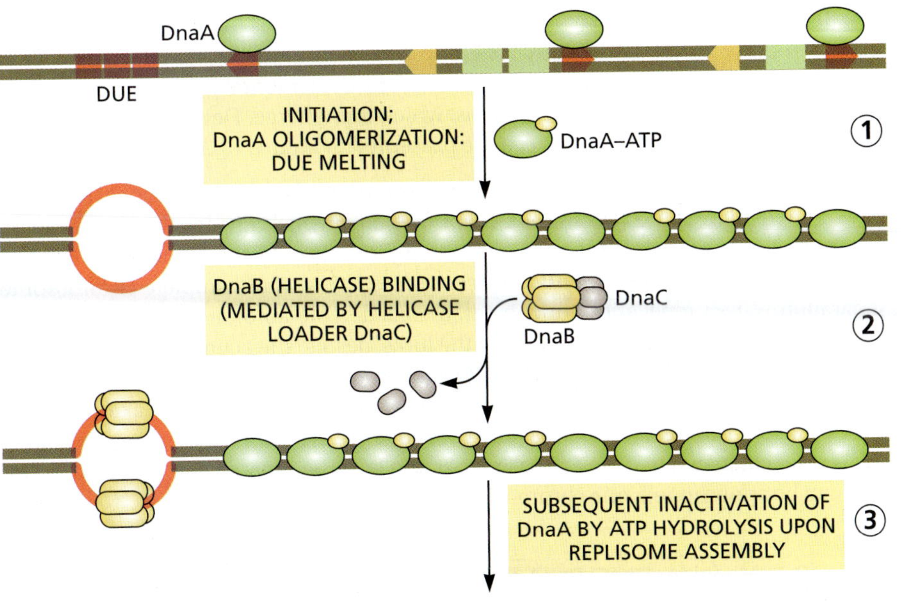

Figure 19.21 Initiation of replication at *oriC* in *E. coli*. DnaA is bound to the three high-affinity DnaA boxes, shown as orange arrowheads, throughout the cell cycle; it interacts with the weaker sites only at the start of initiation, when ATP binds to DnaA. Initiation of replication involves a number of steps: (Step 1) DnaA–ATP molecules bind to the origin, where they oligomerize into the large nucleoprotein complex shown in Figure 19.20 and facilitate the melting of the adjacent DNA unwinding elements or DUE. (Step 2) Helicase DnaB binds to the unwound single strands with the help of the helicase loader, DnaC. (Step 3) DnaA is inactivated by ATP hydrolysis, which is stimulated by additional regulatory mechanisms, upon replisome assembly.

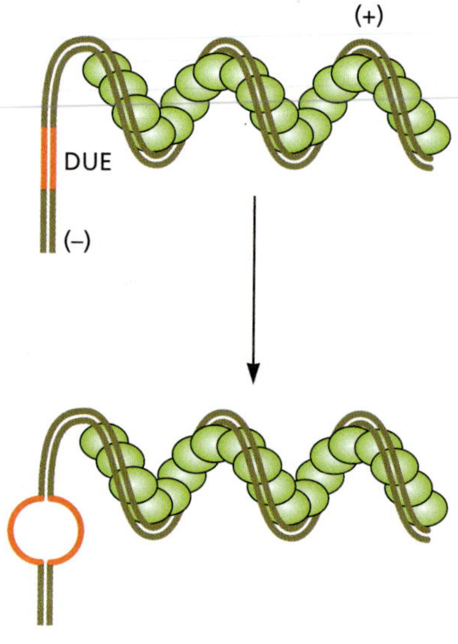

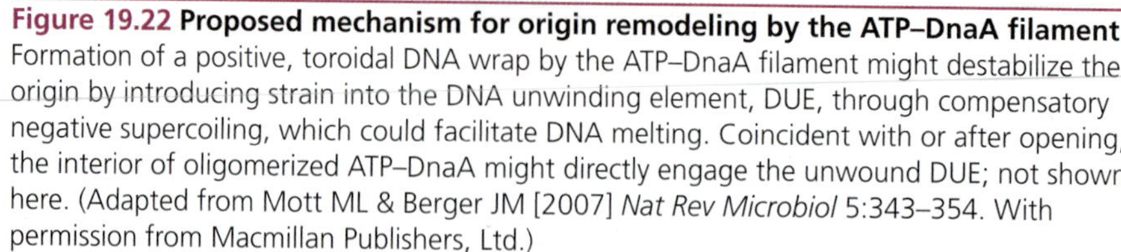
Figure 19.22 Proposed mechanism for origin remodeling by the ATP–DnaA filament. Formation of a positive, toroidal DNA wrap by the ATP–DnaA filament might destabilize the origin by introducing strain into the DNA unwinding element, DUE, through compensatory negative supercoiling, which could facilitate DNA melting. Coincident with or after opening, the interior of oligomerized ATP–DnaA might directly engage the unwound DUE; not shown here. (Adapted from Mott ML & Berger JM [2007] *Nat Rev Microbiol* 5:343–354. With permission from Macmillan Publishers, Ltd.)

are bound, on opposite strands, and oriented in opposite directions. These will provide the docking sites for the two replisome complexes to be assembled at *oriC*. A model for possible cross-talk between DnaA and DnaC filaments, as suggested by James Berger and collaborators, is presented in **Figure 19.23**.

It is becoming increasingly clear that abundant bacterial nucleoid proteins, known primarily for their ability to compact the bacterial chromosome by bending and bridging the DNA (see Chapter 8), also play significant roles in the initiation of replication. Thus, the non-sequence-specific heat-unstable or HU protein has been shown to dramatically enhance DnaA-mediated *oriC* unwinding *in vitro*, probably through its ability to bend and destabilize double-stranded DNA. The participation of two other DNA-bending proteins, IHF or integration host factor and Fis or factor for inversion stimulation, seems to be more DNA-sequence specific, because binding sites for these two proteins are interspersed among DnaA-binding sites at *oriC*. The sequence of events that underlie the dynamic interplay among these proteins and DnaA is presented in **Figure 19.24**. Once two replisomes are assembled into the structures described above, bilateral elongation can begin.

Termination of replication also employs specific DNA sequences and protein factors that bind to them

Despite the importance of termination of replication and its role in chromosome segregation and cell division, the location of the sites where replication terminates is ill-defined, and both the nucleotide sequences and the proteins that interact with them differ from bacterium to bacterium; that is, there is hardly any evolutionary conservation. In fact, researchers today talk about **termination zones** rather than termination sites, and these zones can occupy a significant region of the entire chromosome, at least 5%. Termination, in general, is not merely the coincidental arrival of two replication forks traveling in opposite directions, which then merge. Details of the organization of the termination zone in *E. coli* are presented in **Figure 19.25**.

Interestingly, the DNA in the bacterial chromosome has strand compositional skew: the base composition of the two strands is not the same, which divides the chromosome into two segments bisected by *oriC* and the *dif* loci. The *dif* site is the chromosomal nexus for the recombination and decatenation reactions that complete chromosome separation. When replication forks meet, a considerable amount of supercoiling will have accumulated between the fronts of the forks, despite the action of topoisomerase. Recall from Chapter 9 that movement of RNA polymerase along the double helix leads to accumulation of supercoiling stress, positive in front and negative in its wake; an analogous principle is in operation here. The consequence is that the two daughter helices will end up catenated or linked to each other and must be separated or unlinked before they can be apportioned, one chromosome each, to the daughter cells.

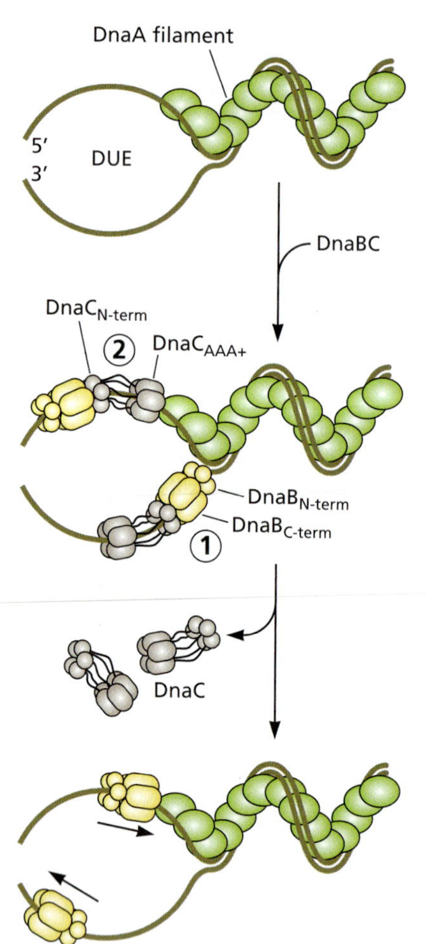

Figure 19.23 Model for DnaC–DnaA cross-talk and deposition of DnaB helicase. The DnaA nucleoprotein forms to one side of a DNA unwinding element, DUE; the question then arises as to how an asymmetric initiation assembly facilitates the symmetric loading of two DnaB helicases, shown in yellow. The model suggests that this is done through the ability of DnaC, shown in gray, to directly engage an available end of the DnaA oligomer. (Top) DnaA has assembled at *oriC* and has melted the DUE. (Middle) (Step 1) Helicase-loading on the bottom DUE strand is facilitated by direct DnaA–DnaB interactions. (Step 2) The helicase destined for the top strand is recruited through a specific interaction between DnaC and ATP-charged DnaA. (Bottom) ATP hydrolysis leads to loss of DnaC; both DnaB helicases are now free to migrate to their proper positions at the fork. (Adapted from Mott ML, Erzberger JP, Coons MM & Berger JM [2008] *Cell* 135:623–634. With permission from Elsevier.)

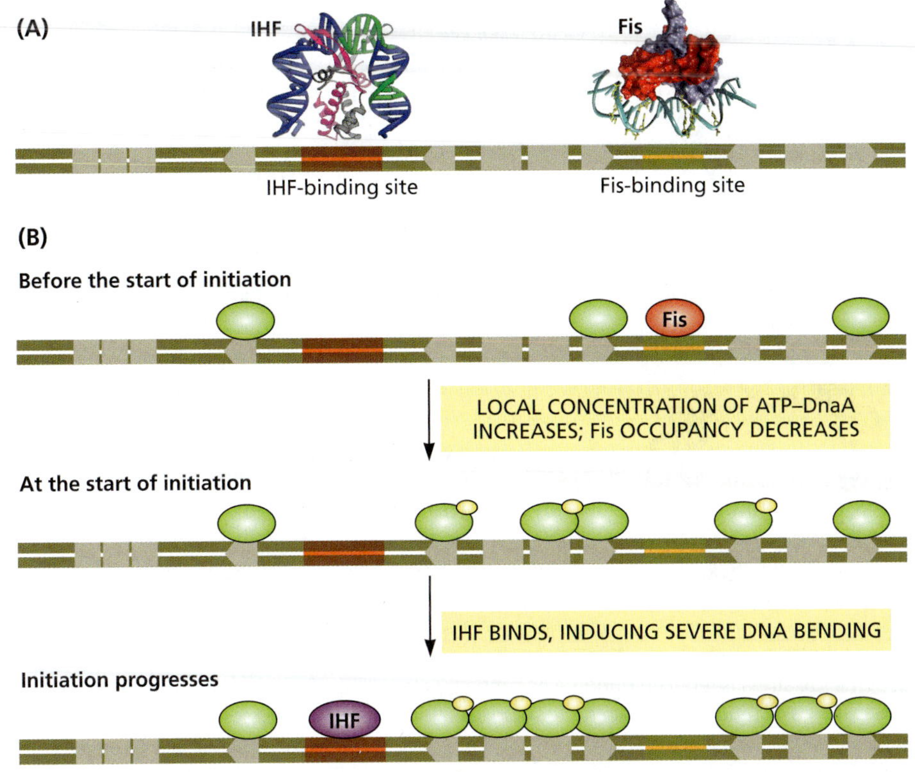

Figure 19.24 Model for dynamic interplay between DNA architectural proteins, IHF and Fis, and DnaA during replication initiation. (A) Map of *oriC* that shows, in addition to DnaA-binding sites, the binding sites for integration host factor, IHF, and for Fis. These proteins induce DNA bending when bound to their respective DNA binding sites, as seen in the structures above the binding sites. All DnaA-binding sites, whether high or low affinity, are shown in gray. (B) Before the start of initiation in the default state, DnaA is bound only to high-affinity DnaA boxes; Fis is also bound. At the start of initiation, the local concentration of ATP-bound DnaA increases and that of Fis decreases. As initiation progresses, IHF binds, causing the DNA to bend to 180°. This bending may assist in the interaction of DnaA with the weaker binding sites, so that the final right-handed nucleoprotein complex can form and unwind the DNA unwinding element, DUE. (Adapted from Mott ML & Berger JM [2007] *Nat Rev Microbiol* 5:343–354. With permission from Macmillan Publishers, Ltd. A, top left, from Lynch TW, Read EK, Mattis AN et al. [2003] *J Mol Biol* 330:493–502. With permission from Elsevier. A, top right, from Nowak-Lovato K, Alexandrov LB, Banisadr A et al. [2013] *PLoS Comput Biol* 9:e1002881. With permission from Public Library of Science.)

The accepted view of termination stems from the identification of **termination sites** or ***Ter* sites**, sequences that arrest replication fork movement only in one direction; that is, these sites are polar. Ten *Ter* sites have been identified in *E. coli*, *TerA–TerJ*, and these are organized into two opposed groups (see Figure 19.25). The two groups each define a **fork trap**: *TerC*, *TerB*, *TerF*, *TerG*, and *TerJ* are oriented to block only clockwise-moving forks, whereas *TerA*, *TerD*, *TerE*, *TerI*, and *TerH* are oriented to block only counterclockwise fork movement. Termination can occur at different sites, depending on which fork, counterclockwise- or clockwise-moving, arrives first at a polar *Ter* site. Most frequently, replication terminates in the region between *TerA* and *TerC*.

Ter sequences bind a protein called **Tus, termination utilization substance**, in *E. coli* (**Figure 19.26**). In *Bacillus subtilis*, the only other bacterium in which termination has been studied in detail, *Ter* binds to RTP, replication terminator protein. The two proteins, Tus and RTP, do not share sequence or structure homology. The bound proteins serve as an asymmetric roadblock to the movement of the helicase.

There are many questions still surrounding the functional significance of the *Ter* sites and bacterial termination in general—Why is the replication trap so large, with the innermost sites spaced ~270 kb apart? If the supposedly redundant *Ter* sites serve as a backup for the innermost sites, as suggested by some, why are they found so far away,

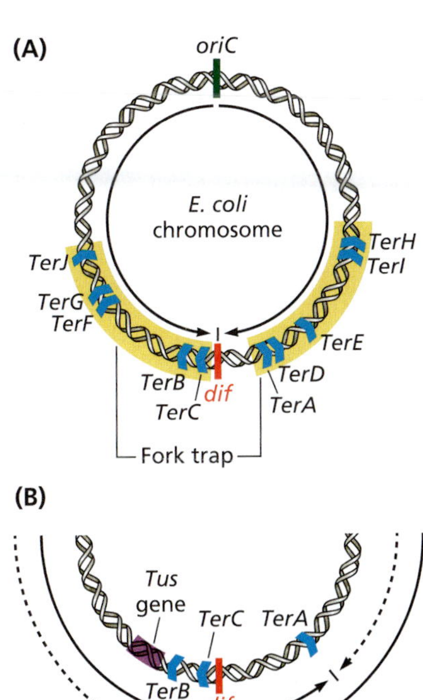

Figure 19.25 Termination of DNA replication in *E. coli*. (A) Ten *Ter* sites have been identified in *E. coli*, known as *TerA–TerJ*. The two opposed groups of polar *Ter* sites form the structure known as a fork trap: *TerC*, *TerB*, *TerF*, *TerG*, and *TerJ* are oriented so as to block clockwise-moving forks, whereas *TerA*, *TerD*, *TerE*, *TerI*, and *TerH* are oriented to block counterclockwise fork movement. (B) Expanded view of the inner region of the fork trap. The continuous arrows indicate the movement of the first fork to be halted, at *TerA* for a counterclockwise-moving fork or at *TerC* for a clockwise-moving fork, depending on which fork arrives first at the respective site. Dashed arrows indicate the movement of the second fork to arrive in each case. Thus, termination would occur at different sites: at *TerA* if the clockwise replication fork were delayed for some reason or at *TerC* if the counterclockwise fork were delayed. Most frequently, replication will terminate in the region between *TerA* and *TerC*. (Adapted from Duggin IG, Wake RG, Bell SD & Hill TM [2008] *Mol Microbiol* 70:1323–1333. With permission from John Wiley & Sons, Inc.)

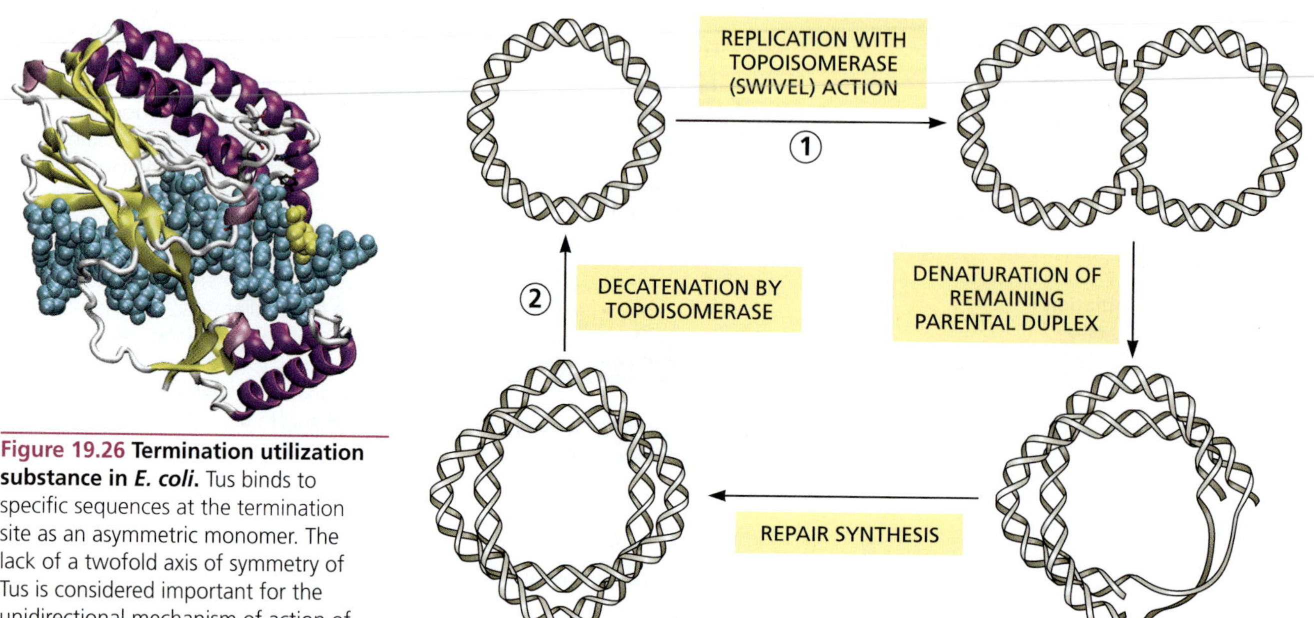

Figure 19.26 Termination utilization substance in *E. coli*. Tus binds to specific sequences at the termination site as an asymmetric monomer. The lack of a twofold axis of symmetry of Tus is considered important for the unidirectional mechanism of action of the *Ter* sites, which create polar arrest in the movement of the replication fork. The Tus–*Ter* complex arrests translocation and unwinding activity of helicase DnaB on the DNA duplex through physical contacts between Tus and the helicase. A central basic cleft makes contact with the major groove of *Ter*, deforming the DNA from B-form geometry. Two interdomain β-strands of Tus are involved in DNA recognition. (From Mulcair MD, Schaeffer PM, Oakley AJ et al. [2006] *Cell* 125:1309–1319. With permission from Elsevier.)

Figure 19.27 Two-stage model for unlinking of replicating DNA. (Step 1) Positive superhelical stress is reduced during elongation by the action of topoisomerases. Nevertheless, because of a number of logistic and topological difficulties during the terminal stages of replication, the final stretch of dsDNA is still overtwisted. At this point, a helicase or a helix-destabilization protein converts the DNA into a catenane of two separate rings, which are still interlinked. The gaps in these molecules are filled in by repair synthesis, after which the action of topoisomerase II unlinks the rings (Step 2). (Adapted from Adams DE, Shekhtman EM, Zechiedrich EL et al. [1992] *Cell* 71:277–288. With permission from Elsevier.)

with some actually closer to *oriC* than to the supposed terminus? Why can the *tus* gene be deleted with no obvious phenotype? Why is there no evolutionary conservation of the proteins that bind to *Ter* sites? Bioinformatics analyses hint that termination most likely occurs at or near the *dif* site. *Ter* sites may also participate in halting replication forks that originate from DNA repair events.

As the two forks approach *Ter* sites, they may not pass one another but must leave a segment of double-stranded parental DNA between them (**Figure 19.27**). It is proposed that this segment melts, allowing replication of the two single-stranded regions and ligation. This produces two catenated or interlinked double-stranded daughter circles, which can be decatenated or separated by type II topoisomerases.

The mechanism that has evolved to replicate DNA is truly remarkable. It manages to copy at a very high rate, while maintaining extraordinary accuracy. This is accomplished despite the fact that the two strands are handled differently at the replication fork so that they can be copied in the same direction. The complex of proteins that accomplishes this is self-assembling and highly processive. The efficiency of the mechanism is attested to by the fact that it appears to have been maintained, in basically the same form, throughout evolutionary history from bacteria to humans.

19.6 Bacteriophage and plasmid replication

The mechanisms described above apply in general to bacteria and eukaryotes, but the specialized genomes and lifestyles of some viruses and plasmids require special mechanisms. Unlike bacteria or eukaryotic cells, bacteriophage must rapidly produce multiple copies of their genomes within the host cell. Sometimes, this has the consequence that the viral replication machinery is much simpler than that found in bacteria or eukaryotes, which has been used to facilitate fundamental studies of replication. A case in point is bacteriophage T7 (**Box 19.7**), which utilizes

Box 19.7 Bacteriophage T7 replication system: A handy tool to crack open a challenging process Phage T7 is a lytic phage that infects *E. coli*. Its relatively large genome of 39,936 bp encodes around 50 proteins, three of which act in conjunction with a cell host factor to perform viral DNA replication. The need for only four proteins for viral replication allowed the reconstitution of an *in vitro* system for examining basic aspects of DNA replication. Despite its relative simplicity, the process of replication in phage T7 mimics more complex bacterial and eukaryotic systems: initiation occurs at a single origin, replication is bidirectional, and lagging-strand DNA synthesis proceeds through the usual Okazaki fragment synthesis and maturation and requires the formation of a replication loop (see Figure 19.11). The protein players are (1) gp5, T7 gene 5 protein, a DNA polymerase tightly associated with its processivity factor, the *E. coli* protein thioredoxin or Trx; (2) gp4, T7 gene 4 protein, a hexameric helicase–primase; and (3) gp2.5, T7 gene 2.5 protein, a single-stranded DNA-binding protein. T7 replication does not need clamps, because the polymerase itself provides half of the clamp and the other half is supplied by thioredoxin (**Figure 1**). Replisome assembly occurs without accessory proteins such as clamp loaders, in contrast to the situation in phage T4, *E. coli*, and eukaryotes. The economical use of protein components in phage T7 is also exemplified by gp4, which provides both helicase and primase activities in a single protein, whereas in other systems these activities are provided by separate proteins. Thus, T7 has evolved an efficient and economical mechanism for the replication of its DNA, one that helps to define the minimal requirements for rapid and faithful replication of a duplex DNA molecule.

During leading-strand synthesis gp5–Trx undergoes multiple conformational changes as it moves along the template and senses incoming dNTPs that will fit correctly to each exposed base in the template strand. Gp5 is a nonprocessive enzyme that can add only a few nucleotides per binding event; Trx binding increases its processivity ~100-fold. Thioredoxin is a peculiar candidate for such a function. This small protein has many biological functions, but most involve oxidation and reduction. The synthesis and maturation of each Okazaki fragment occurs through extension of tetranucleotide primers synthesized by the primase domain of gp4. To be functional *in vivo*, other accessory proteins, such as a 5′ → 3′ exonuclease, gp6, and a DNA ligase, gp1.3, are also needed to complete the process of maturation. gp4 assembles as a hexamer on the single-stranded lagging-strand template and translocates unidirectionally, 5′ → 3′, using the energy of dNTP hydrolysis. Like other helicases, the hexamer possesses NTP-binding sites located at the interface of the subunits. The helicase has a high affinity for the replicating DNA polymerase.

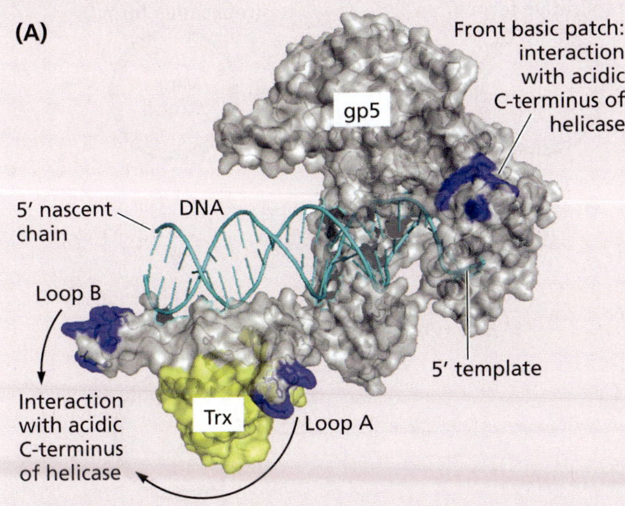

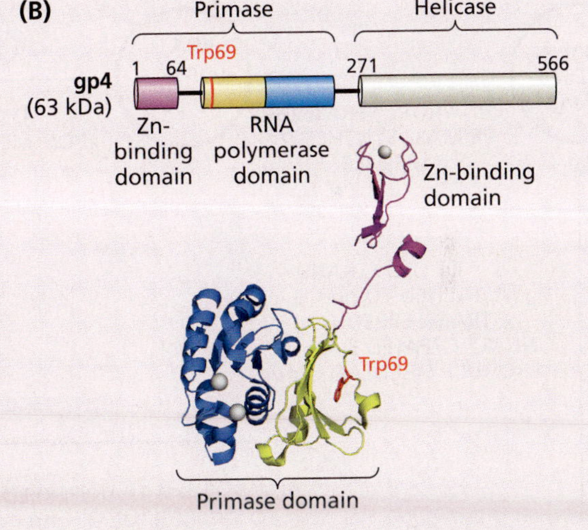

Figure 1 Structures of gp5–Trx complex and helicase–primase gp4. (A) Crystal structure of gp5 polymerase–thioredoxin complex, gp5–Trx, bound to a primer–template. Four basic residues are located within a solution-exposed basic patch, which forms the interaction surface for the acidic C-terminus of gp4 helicase; this interaction is critical for the initiation of leading-strand synthesis. Basic loops A and B are located in the processivity factor Trx-binding domain; they also provide an interaction surface for the acidic C-terminal tails of gp4 and the single-stranded DNA-binding protein gp2.5. This electrostatic interaction between gp5–Trx and gp4 holds gp5–Trx molecules that only transiently dissociate from the template during replication, thus further increasing the processivity from 5 kb to greater than 17 kb. (B) gp4 houses both helicase and primase activity in a single polypeptide chain, with the primase domain in the N-terminal half and the helicase domain in the C-terminal half. The ribbon structure represents the primase portion of gp4. The Cys_4 Zn-binding motif is important for recognition of primase sites, short specific sequences in DNA that serve as templates for primers. The primase domain also contains the catalytic site. Residue Trp69 is critical for delivery of the tetraribonucleotide primer from the primase to the gp5–Trx polymerase complex to initiate DNA synthesis. The helicase plays a pivotal role in T7 replication systems, providing the binding site for both leading- and lagging-strand polymerases. This association allows both strands to be synthesized in the same direction and at identical rates, analogous to the situation in bacterial replication where the two core Pol III complexes are linked to each other mainly by the clamp loader. (A, from Zhang H, Lee S-J, Zhu B et al. [2011] *Proc Natl Acad Sci USA* 108:9372–9377. With permission from National Academy of Sciences. B, adapted from Zhu B, Lee S-J & Richardson CC [2010] *Proc Natl Acad Sci USA* 107:9099–9104. With permission from National Academy of Sciences.)

(Continued)

Box 19.7 *(Continued)*

Figure 2 depicts, in schematic form, the overall structure of the T7 replisome, with its two polymerase gp5-Trx complexes, the gp4 helicase–primase, and the single-stranded DNA-binding protein gp2.5. The helicase unwinds double-stranded DNA to generate the two templates for leading- and lagging-strand synthesis. The primase catalyzes the synthesis of tetraribonucleotide primers, which are necessary for the initiation of each Okazaki fragment, and transfers them to the DNA polymerase. gp2.5 coats the lagging-strand template to prevent the formation of unwanted secondary structures.

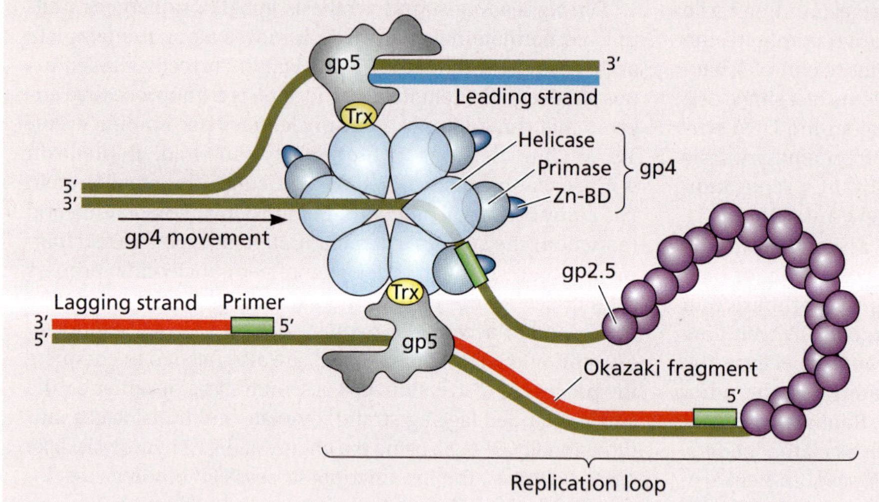

Figure 2 Model of T7 replisome. The model emphasizes its similarities with the *E. coli* replisome; compare with Figure 19.11. The ability to reconstitute the replication system with only four proteins, two of which form the tight complex gp5–Trx, makes the system a handy tool for studies of the basic features of bacterial replication. Zn-BD, Zn-binding domain. (Adapted from Lee S-J & Richardson CC [2011] *Curr Opin Chem Biol* 15:580–586. With permission from Elsevier.)

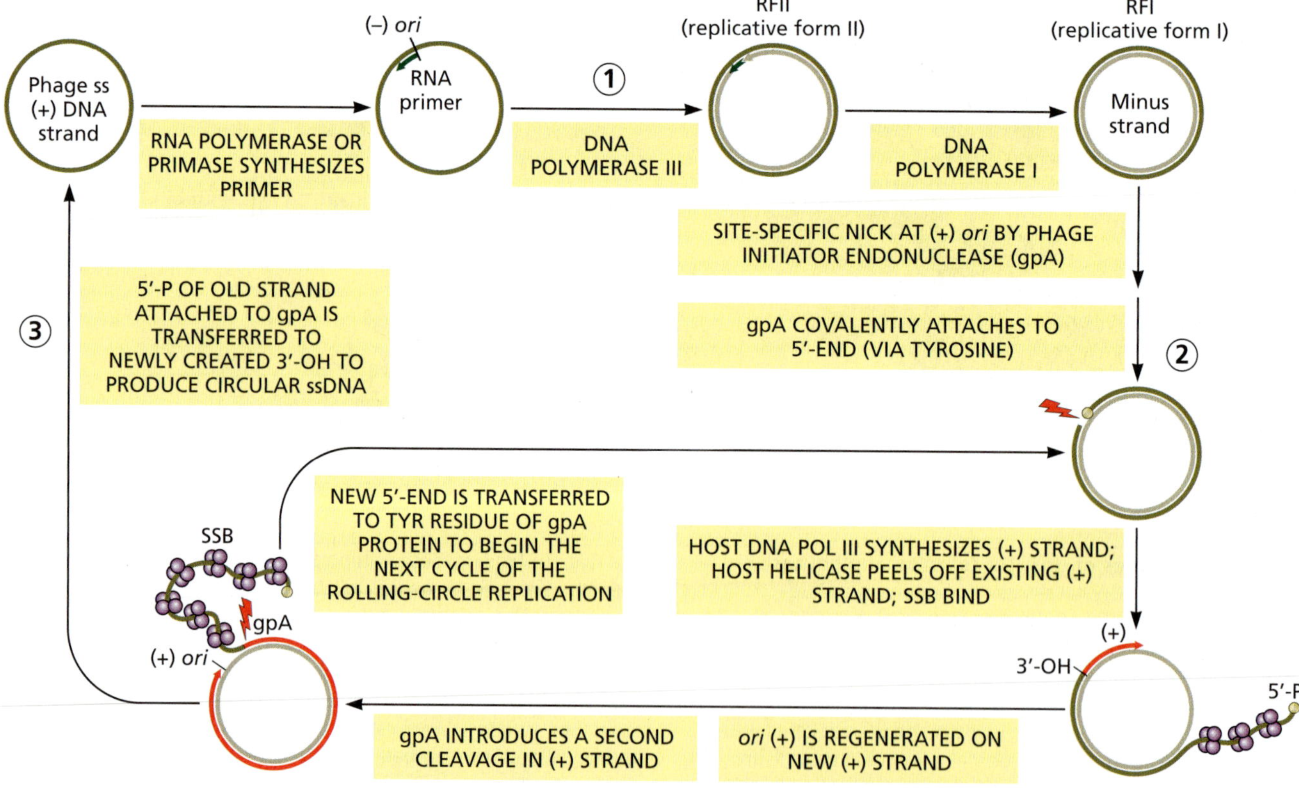

Figure 19.28 Rolling-circle replication of ssDNA phage φX174. The process of replication can be divided into three steps: (Step 1) conversion of the ssDNA phage genome into a double-stranded form, known as replicative form I or RFI; (Step 2) multiplication of RFI by a rolling-circle mechanism; and (Step 3) generation of an ssDNA genome for packaging into new phage particles. RFI is used as a template for transcription, which in turn initiates synthesis of the viral proteins. The DNA genome is synthesized in ~10 s, with more than 20 circles released from a single rolling-circle intermediate. For simplicity, the schematic shows the immediate production of a single-length genome; however, usually this mechanism gives rise to long concatemers, which are then cleaved to fragments of single-genome size. The process relies mainly on the use of host proteins, with the important exception of the phage gpA protein, an initiator endonuclease involved in two processes, as indicated in the schematic. The first cleavage reaction requires superhelical RFI, produced by host gyrase, whereas the second occurs on a relaxed template.

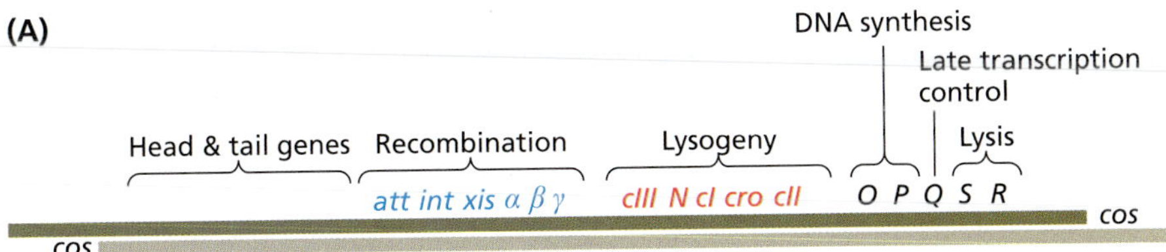

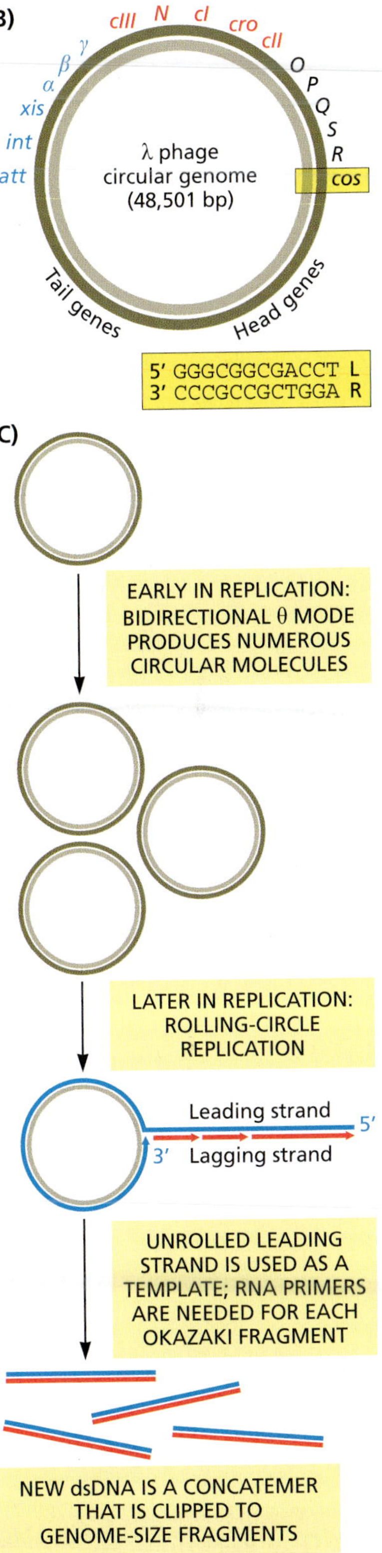

Figure 19.29 λ phage replication. (A) In the virion or mature viral particle, the dsDNA genome is linear. At both ends it contains *cos* sequences, single-stranded 5′-extensions of ~200 nt that contain complementary portions. (B) These complementary sequences, shown in the bright yellow box, are used to base-pair the ends of the genome during the circularization phase of the life cycle, following entry into the host cell. (C) λ phage replication switches from early regular bidirectional replication to rolling-circle replication later in the life cycle.

a replication complex consisting of only a few proteins. This is very useful for *in vitro* studies. We describe two examples of DNA replication mechanisms in phages and plasmids here.

Rolling-circle replication is an alternative mechanism

The genomes of certain small phages are organized as circles of single-stranded DNA or ssDNA. These phages use **rolling-circle replication** to replicate their genomes. Two systems have been extensively studied: the spherical phage φX174 (**Figure 19.28**) and the filamentous phage M13. The process in φX174 involves multiple events that are usually classified into three steps: (1) conversion of the ssDNA genome to a double-stranded form, known as replicative form I or RFI; (2) rolling-circle replication of RFI; and (3) generation of an ssDNA genome for packaging into phage particles. Note that the entire replication process depends heavily on the use of host proteins; the only essential phage-encoded protein is gpA, the initiator endonuclease that introduces a site-specific nick into the double-stranded RFI to provide bacterial Pol III with the free 3′-OH group needed for elongation.

Many small plasmids of Gram-positive bacteria, which do not have ssDNA phages, are multiplied by the rolling-circle mechanism. These plasmids encode an initiator protein that introduces a site-specific nick. The initiator protein shares sequence similarity with gpA and recognizes a similar nucleotide sequence for cutting. These similarities are interpreted as an indication of evolution from a common ancestor.

Phage replication can involve both bidirectional and rolling-circle mechanisms

In certain cases, replication is a combination of the bidirectional replication typical of circular bacterial chromosomes and rolling-circle replication. One well-studied example is the phage λ genome (**Figure 19.29**). The genome in the virion is linear. As we see in the next chapter, the replication of linear genomes encounters what is called the end problem. If an RNA primer is added at each end to begin replication, the removal of these primers leaves each daughter strand incomplete. The phage λ genome evades this problem in the following way: first it undergoes circularization upon entry into the host cells. The now-circular genome is initially replicated bidirectionally in order to quickly produce numerous circular genomes, which can be transcribed and translated to provide essential viral proteins. Later in the cycle, the genome switches to rolling-circle replication, which results in long concatemeric structures that are subsequently cut into genome-size linear fragments, suitable for packaging into new phage particles.

Key concepts

- Replication in bacteria, eukaryotes, and most viruses proceeds in very similar ways; it is semiconservative, copying both complementary strands.
- Because polymerases copy only in the 5′ → 3′ direction, one strand, the leading strand, can be synthesized continuously whereas the other, lagging strand must be synthesized discontinuously.
- Parental strands are separated by a helicase, forming the replication fork.
- Synthesis on both strands is initiated from 3′-OH groups on primers, which are short stretches of RNA in bacteria and viruses. Lagging-strand synthesis is initiated repeatedly, forming Okazaki fragments.
- The major enzyme for bacterial DNA replication is DNA polymerase III. Two copies of this enzyme are present in the protein complex at the replication fork, the replisome. One synthesizes the leading strand, and the other, the lagging strand.
- Maturation of the lagging strand requires the enzyme DNA polymerase I, which has an exonuclease activity to remove the primer and can then fill in the gaps with DNA chains. The fragments are then connected by a ligase.
- Processivity of replication is assured by sliding clamps, which form rings about the DNA behind the polymerases to inhibit their release from the template.
- Sliding clamps are wrapped about the DNA by a multi-ATPase complex called the clamp loader. This also serves as a scaffolding protein complex to hold the replisome together.
- The structure of the replisome, and the necessity to allow for coordinated synthesis of both strands, require the formation of a trombone-loop conformation in the lagging-strand template.
- Fidelity in bacterial replication is assured by the proofreading capabilities of both Pol I and the Pol III complex. Together, these provide an error frequency of about 10^{-9}.
- Initiation of bacterial replication proceeds bidirectionally from, usually, a single initiation region. This contains binding sites for the initiation proteins DnaA and DnaC. These recruit the helicase DnaB to begin unwinding the duplex.
- Termination of replication in bacteria occurs in zones containing *Ter* sites, each capable of binding the protein Tus, which can halt a replication fork traveling in a specific direction.
- After forks have been halted, melting of the remaining short parental duplex region is followed by completion of replication on these single strands. This leaves two interlocked daughter duplexes. These are separated by topoisomerase activity.
- While the above description applies, in the broadest sense, to DNA replication in bacteria, eukaryotes, and many viruses, some viruses employ a quite different strategy called rolling-circle replication or utilize a combination of bidirectional and rolling-circle replication.

Further reading

Books

Cox LS (ed) (2009) Molecular Themes in DNA Replication. RSC Publishing.

Kornberg A & Baker TA (1992) DNA Replication, 2nd ed. University Science Books.

Kušić-Tišma J (ed) (2011) Fundamental Aspects of DNA Replication. InTechOpen.

Reviews

Alberts B (2003) DNA replication and recombination. *Nature* 421:431–435.

Bollenbach T & Kishony R (2009) Hydroxyurea triggers cellular responses that actively cause bacterial cell death. *Mol Cell* 36:728–729.

Dixon NE (2009) DNA replication: Prime-time looping. *Nature* 462:854–855.

Duggin IG, Wake RG, Bell SD & Hill TM (2008) The replication fork trap and termination of chromosome replication. *Mol Microbiol* 70:1323–1333.

Indiani C & O'Donnell M (2006) The replication clamp-loading machine at work in the three domains of life. *Nat Rev Mol Cell Biol* 7:751–761.

Labib K & Hodgson B (2007) Replication fork barriers: Pausing for a break or stalling for time? *EMBO Rep* 8:346–353.

Langston LD, Indiani C & O'Donnell M (2009) Whither the replisome: Emerging perspectives on the dynamic nature of the DNA replication machinery. *Cell Cycle* 8:2686–2691.

Lee S-J & Richardson CC (2011) Choreography of bacteriophage T7 DNA replication. *Curr Opin Chem Biol* 15:580–586.

Lovett ST (2007) Polymerase switching in DNA replication. *Mol Cell* 27:523–526.

McHenry CS (2011) DNA replicases from a bacterial perspective. *Annu Rev Biochem* 80:403–436.

Mott ML & Berger JM (2007) DNA replication initiation: Mechanisms and regulation in bacteria. *Nat Rev Microbiol* 5:343–354.

Pomerantz RT & O'Donnell M (2007) Replisome mechanics: Insights into a twin DNA polymerase machine. *Trends Microbiol* 15:156–164.

Steitz TA (1999) DNA polymerases: Structural diversity and common mechanisms. *J Biol Chem* 274:17395–17398.

Wang T-CV (2005) Discontinuous or semi-discontinuous DNA replication in *Escherichia coli*? *BioEssays* 27:633–636.

Experimental papers

Bailey S, Eliason WK & Steitz TA (2007) Structure of hexameric DnaB helicase and its complex with a domain of DnaG primase. *Science* 318:459–463.

Cooper S & Helmstetter CE (1968) Chromosome replication and the division cycle of *Escherichia coli* B/r. *J Mol Biol* 31:519–540.

Georgescu RE, Kim S-S, Yurieva O et al. (2008) Structure of a sliding clamp on DNA. *Cell* 132:43–54.

Hamdan SM, Johnson DE, Tanner NA et al. (2007) Dynamic DNA helicase-DNA polymerase interactions assure processive replication fork movement. *Mol Cell* 27:539–549.

Kelch BA, Makino DL, O'Donnell M & Kuriyan J (2011) How a DNA polymerase clamp loader opens a sliding clamp. *Science* 334:1675–1680.

Mott ML, Erzberger JP, Coons MM & Berger JM (2008) Structural synergy and molecular crosstalk between bacterial helicase loaders and replication initiators. *Cell* 135:623–634.

Nossal NG, Makhov AM, Chastain PD II et al. (2007) Architecture of the bacteriophage T4 replication complex revealed with nanoscale biopointers. *J Biol Chem* 282:1098–1108.

Okazaki R, Okazaki T, Sakabe K et al. (1989) *In vivo* mechanism of DNA chain growth. *Cold Spring Harbor Symp Quant Biol* 33:129–143.

Okazaki R, Okazaki T, Sakabe K et al. (1968) Mechanism of DNA chain growth. I. Possible discontinuity and unusual secondary structure of newly synthesized chains. *Proc Natl Acad Sci USA* 59:598–605.

Pomerantz RT & O'Donnell M (2008) The replisome uses mRNA as a primer after colliding with RNA polymerase. *Nature* 456:762–766.

Sugimoto K, Okazaki T & Okazaki R (1968) Mechanism of DNA chain growth, II. Accumulation of newly synthesized short chains in *E. coli* infected with ligase-defective T4 phages. *Proc Natl Acad Sci USA* 60:1356–1362.

Wing RA, Bailey S & Steitz TA (2008) Insights into the replisome from the structure of a ternary complex of the DNA polymerase III α-subunit. *J Mol Biol* 382:859–869.

Yang J, Zhuang Z, Roccasecca RM et al. (2004) The dynamic processivity of the T4 DNA polymerase during replication. *Proc Natl Acad Sci USA* 101:8289–8294.

Zhang H, Lee S-J, Zhu B et al. (2011) Helicase-DNA polymerase interaction is critical to initiate leading-strand DNA synthesis. *Proc Natl Acad Sci USA* 108:9372–9377.

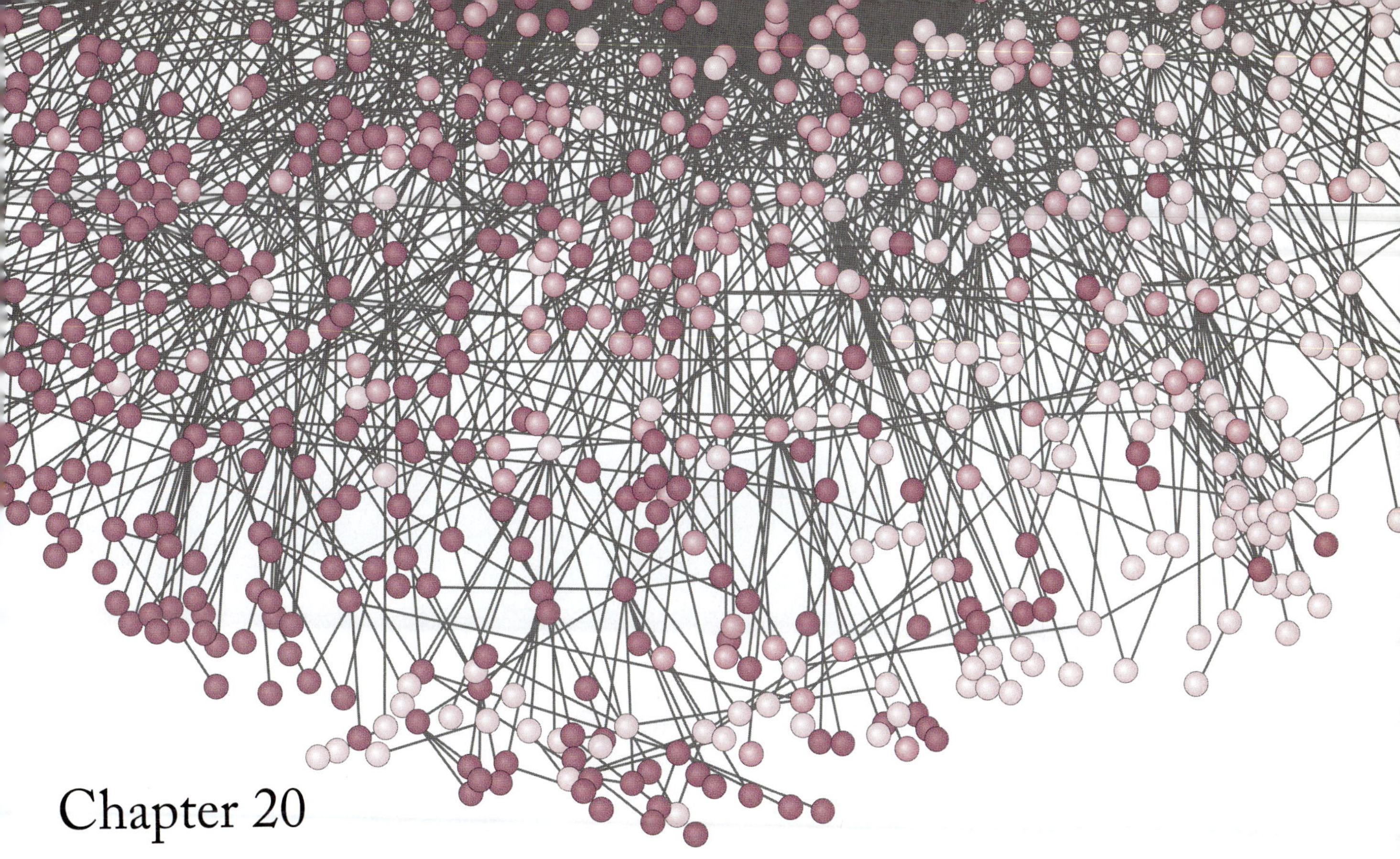

Chapter 20

DNA Replication in Eukaryotes

20.1 Introduction

For several reasons, DNA replication in eukaryotes is inherently more complex than in bacteria. First, the much greater size of the eukaryotic genome necessitates the use of multiple origins of replication in order to complete the process in a reasonable time. The rate of fork movement in higher eukaryotes has been estimated at ~100 base pairs per second, approximately tenfold slower than in bacteria. If bidirectional replication proceeded from only one source in each human chromosome, several days would be needed to copy the whole genome, as compared to the few hours observed.

Second, eukaryotic nuclear DNA is packaged into chromatin (see Chapter 8), whose nucleosomal and higher-order structure must surely complicate both the initiation and elongation stages of replication. Since chromatin structure is a major determinant in the regulation of gene expression, it is essential that this structure also be reproduced upon DNA replication. Finally, chromatin structure carries epigenetic information, which must in some way be transmitted during somatic cell division. Unless replication is carefully monitored, such information can be scrambled and lost. In this chapter, we see how DNA replication in eukaryotes deals with these additional challenges.

20.2 Replication initiation in eukaryotes

Replication initiation in eukaryotes proceeds from multiple origins

The multiplicity of replication origins in eukaryotes ranges from hundreds in yeast to tens of thousands in metazoa, between 30,000 and 50,000 according to recent estimates (**Figure 20.1**). The existence of numerous origins of replication that are active at the same time produces the typical appearance of replication foci in the eukaryotic

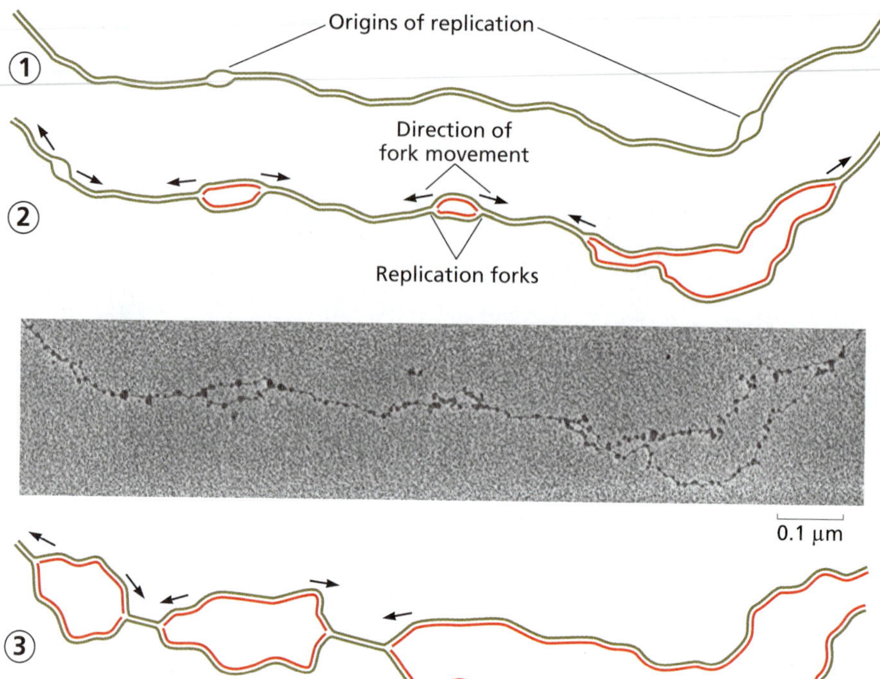

Figure 20.1 Eukaryotic chromosomes are replicated bidirectionally from numerous origins of replication. The same portion of a DNA molecule is depicted at successive time points during replication, marked by Arabic numerals. Green lines represent the two parental DNA strands; red lines represent the newly synthesized DNA strands. Each replication origin gives rise to two replication forks that move away from each other in opposite directions. Corresponding to schematic 2 is an electron micrograph showing DNA replicating in an early *Drosophila melanogaster* embryo. The particles visible along the DNA are nucleosomes. (Adapted from Alberts B, Bray D, Hopkin K et al. [2009] Essential Cell Biology, 4th ed. With permission from Garland Science. Micrograph, courtesy of Victoria Foe, University of Washington.)

nucleus (**Figure 20.2**). There are numerous ways to visualize active origins, such as use of radioactive nucleoside triphosphate or NTP precursors and autoradiography, precursors carrying fluorescent dyes, or derivative precursors such as bromodeoxyuridine, BrdU, that can be visualized by fluorescently labeled antibodies. More recent methods make use of recombinant constructs expressing protein components of the replisome tagged with green fluorescent protein, GFP (see Figure 20.2). At the molecular level, origins can be located by a number of techniques (**Box 20.1**); at the genomewide level, they can be located by chromatin immunoprecipitation, ChIP, against protein components of the initiation complexes.

The large number of eukaryotic origins can be classified in three general categories (**Figure 20.3**): **constitutive**, which are active in all cells under all conditions as set by transcription and/or chromatin structure constraints; inactive or dormant, which are practically always inactive under normal conditions but can be woken or activated in stress conditions or during cell differentiation; and flexible, which are activated randomly in individual cells of the same cell population. These origins show flexibility in another aspect, too: if some origins are mutated or permanently inactivated through another mechanism, nearby origins can be activated or become more efficient. Flexible origins form the largest category and are usually clustered along the DNA. The

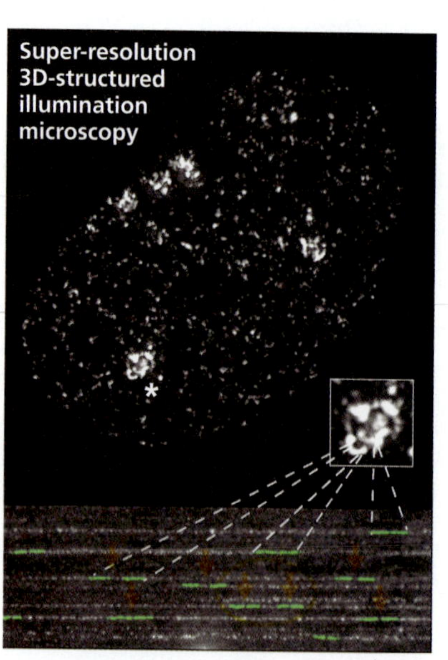

Figure 20.2 Visualization of replication foci. Human HeLa cells expressing GFP-tagged proliferating cell nuclear antigen, also called PCNA or the clamp, were imaged by very-high-resolution microscopy. The image shows the spatial organization of replication foci, appearing as dots, in the nucleus in late S phase. The inset shows a twofold magnification of the area containing a replicating heterochromatic region, marked by an asterisk in the image. Even in these highly condensed regions, there are a large number of individual replicating sites. DNA fibers stretched by combing on glass surfaces are shown below the cell image; DNA combing refers to techniques that straighten DNA molecules on glass slides, so that they run as parallel fibers. Green lines show replicon units pulse-labeled with specific nucleotides; green arrows indicate the origin of replication of each replicon. Some replicons are close together in clusters that initiate replication coordinately; these are circled. Comparison of the number of labeled replicons in combed DNA fibers with the number of replicons in the confocal image indicates that one replication focus corresponds to a spatially organized cluster of replicons, as indicated by the dashed lines. (Adapted from Chagin VO, Stear JH & Cardoso MC [2010] *Cold Spring Harb Perspect Biol* 2:a000737. With permission from Cold Spring Harbor Laboratory Press.)

molecular mechanisms involved in the random activation of a particular flexible origin are still under investigation.

Replication in eukaryotes is tightly coupled to the cell cycle. We present a brief discussion of the cell cycle in Chapter 2, and a more detailed picture, including cycle regulation, is given in Box 20.2.

Box 20.1 Mapping origins of replication Bacteria and most viruses each have a single, unique origin of replication, so locating this origin is relatively simple. An approximate location can be obtained by electron microscopy, which reveals the bubble surrounding the origin. Its location can be determined, and hence the origin approximately located, by repeating the analysis on samples cut by one or more restriction endonucleases. Higher precision as to the origin itself can be obtained by pulse-labeling replicating DNA, followed by autoradiography. Label will be concentrated about the origin.

A relatively simple method for mapping origins of replication, that does not require specialized equipment or techniques, depends on the behavior of bubbles and of branched and linear molecules under different conditions of two-dimensional gel electrophoresis. A schematic of the method as applied to a homogeneous population of replicating circular DNA molecules is shown in Figure 1. The molecules are cleaved by a restriction nuclease at a site well removed from the putative origin of replication. Each molecule so cleaved will contain a replication bubble. Restriction fragments surrounding the putative origin are electrophoresed in a first direction under conditions that separate primarily on the basis of total DNA length. The material on the gel is then re-electrophoresed through a gel that separates largely on the basis of conformation: circular or branched molecules are selectively retarded. The separated DNA is then hybridized against an appropriate labeled probe.

The curve that traces molecules that have been replicated to different extents depends on where the origin is located with respect to the restriction site. If it is close to the center of the fragment, the bubble will expand symmetrically until both ends are reached, as shown by the blue curve in Figure 1. But if the origin is located asymmetrically, as shown by the red curve in Figure 1, a point will be reached at which one fork reaches an end before the other and the bubble will be cut at one end. The molecule will then be converted into a Y-shaped structure, which is retarded within the gels differently than a bubble, resulting in the break seen between the red and green curves in Figure 1. The Y-structure will approach a linear structure that is twice as long as the original DNA before replication is complete.

The methods described above can reveal only approximate locations of origins. To approach base-pair resolution, it is necessary to carry out replication assays on **replicons** that have undergone site-directed mutagenesis.

It is easy to see how these methods can be applied to genomes with only one or a few origins. But how might we approach eukaryotic genomes that may carry tens of thousands of origins?

A technique recently developed in the laboratory of Joyce Hamlin allows the isolation of a library of replication sites from the whole genome. Whole genomic DNA, at the desired stage of replication, is cleaved by restriction endonucleases to yield a mix of fragments of appropriate size. Some of these will contain replication bubbles, and some will not. To isolate the former, the digest is mixed into polymerizing agarose; as the agarose chains grow, some of them penetrate the loops of the bubbles (Figure 2). The gel is electrophoresed, which allows linear fragments and most branched structures to be removed, but the bubbles are stuck. They can be recovered by enzymatic digestion of the agarose matrix. Libraries that are prepared in this way can be used as microarray probes or sequenced by high-throughput methods.

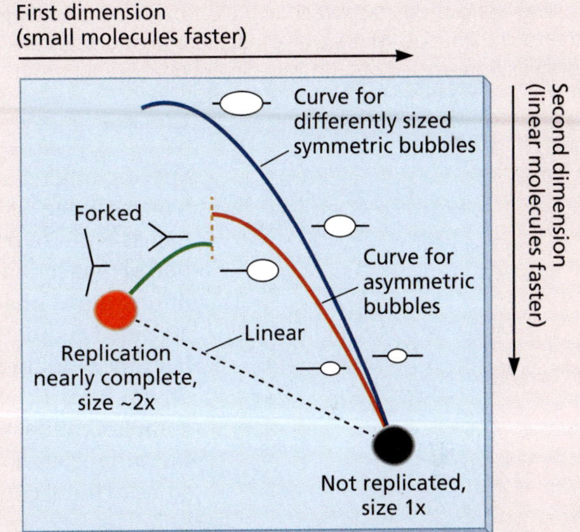

Figure 1 Gel electrophoretic mapping of replication start sites. A mix of replicating fragments, at different stages of replication, is run on the gel and detected by hybridization with a labeled probe. Electrophoresis in the first dimension is in a gel where mobility depends mostly on the size of the DNA fragment. Fragments that have replicated the most will move more slowly and lie to the right. Electrophoresis in the second dimension is under conditions where the movement of bubblelike or forked molecules is much retarded. Expected behavior is depicted for linear molecules of different sizes ranging from 1×, not replicated, to 2×, replication nearly complete, by a black broken line; for bubbles of different sizes that have expanded symmetrically within a DNA fragment by a blue curve; for differently sized bubbles that have expanded asymmetrically within a DNA fragment by a red curve; and for forked molecules by a green curve. As we examine the curve corresponding to a collection of asymmetric bubbles, we see a break point at which asymmetric bubbles are opened by having one fork pass through; this transforms asymmetric bubbles into Y structures. This allows location of the origin on the fragment. (Adapted from Mesner LD & Hamlin JL [2009] *Methods Mol Biol* 521:315–328. With permission from Springer Science and Business Media.)

(Continued)

Box 20.1 *(Continued)*

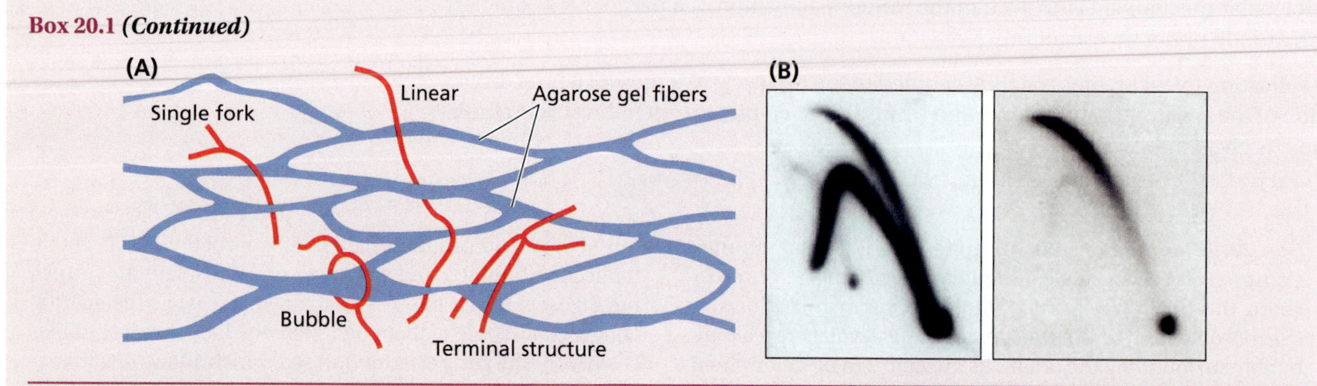

Figure 2 Genomewide bubble-trapping protocol used to isolate replication origins or bubbles from eukaryotic cells. (A) Four different forms of DNA restriction fragments are present in a DNA digest from replicating cells: bubbles, single forks, linear molecules, and X-shaped terminal structures. The circular nature of restriction fragments that contain replication bubbles allows these fragments to be trapped within the matrix of gelling agarose, so that subsequent electrophoresis of the gel cannot pull them out, whereas all other kinds of structures are readily electrophoresed out. (B) Two-dimensional gel patterns of replication intermediates from Chinese hamster ovary cells, CHO, isolated in very early S phase before trapping, shown on the left, and after trapping, shown on the right. The lower curve in the gel to the left corresponds to all of the forked molecules produced by the opening of transcription bubbles at one end by the digestion. As the right image shows, these are almost quantitatively removed, >95%, by the trapping procedure, leaving only the bubbles. (Adapted from Mesner LD & Hamlin JL [2009] *Methods Mol Biol* 521:315–328. With permission from Springer Science and Business Media.)

Specific origins are activated during specific phases—early, middle, or late—of the S phase of the cell cycle, producing the changing pattern of appearance of replication foci as S phase progresses (Figure 20.4). Figure 20.5 illustrates the temporal activation of replication origins during S phase. Complexes that are assembled at the origins during the M phase of the preceding mitosis and during G_1 phase can fire or become activated at different stages of the S phase. The pertinent question here is, what regulates the selection of the origins that fire during those different stages of S phase? We do not have a clear picture yet, but it is obvious that replication timing has multiple complex connections with static properties of genomic regions such as GC content, the subnuclear location of the regions, and their transcriptional activity. We discuss some of these complex interrelationships in Box 20.3.

Eukaryotic origins of replication have diverse DNA and chromatin structure depending on the biological species

There is a vast variety of eukaryotic origins, depending on the species. In budding yeast, *Saccharomyces cerevisiae*, origins correspond to defined DNA sequences termed **autonomous replicating sequences** or **ARSs** (Figure 20.6). Sequences fused

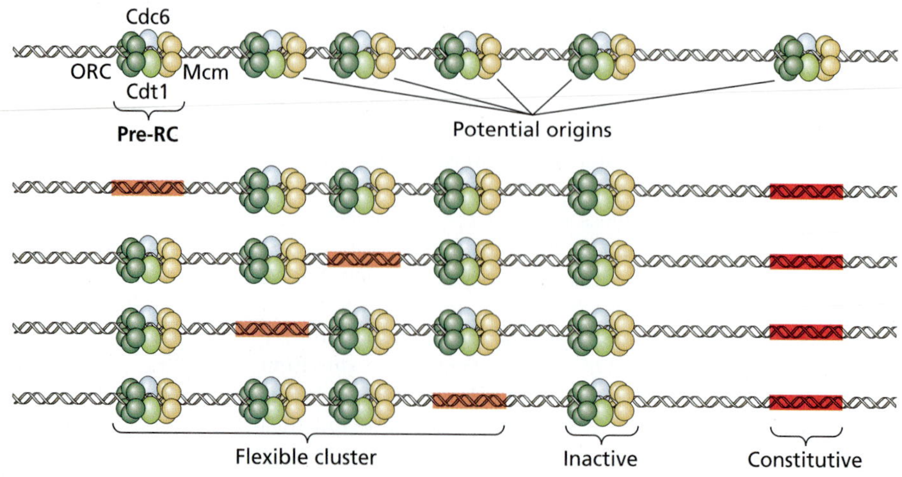

Figure 20.3 Types of DNA replication origins. Potential origins are set during M/G_1 by assembly of the pre-replication complex or pre-RC. Selection of origins that will fire during the following S phase occurs in G_1. (Adapted from Méchali M [2010] *Nat Rev Mol Cell Biol* 11:728–738. With permission from Macmillan Publishers, Ltd.)

Box 20.2 Regulation of the cell cycle Mitotic cell division was discovered by cytogeneticist Walther Flemming in 1879, and within a few years the general features of the cell cycle were recognized. Indeed, this represented one of the triumphs of early light microscopy. It was not until the 1950s, however, that the role of DNA replication in this scenario could be fully appreciated. For a description of the cell cycle itself, see Box 2.2.

Even with this clarification, the mechanisms that regulate the cell cycle remained obscure for decades and in fact were little studied. The first major advances came from the work of Leland Hartwell in the 1970s. Hartwell was studying *Saccharomyces cerevisiae*, the budding yeast, using the appearance of buds and increase in cell size to measure progress through the cell cycle. A particularly interesting subset of temperature-sensitive mutants corresponded to modifications, such as blocks or premature transitions, in the cell cycle itself. These Hartwell termed **cell division cycle mutants**, abbreviated ***cdc***. Over the years, a large collection of *cdc* mutants in various organisms was assembled, but their mode of function remained obscure.

Important clues began to come from the work of Paul Nurse, initially working with Murdoch Mitchison. Nurse used the fission yeast *Schizosaccharomyces pombe*, whose linear growth could be more easily measured than the expansion and budding of *S. cerevisiae*. An important discovery, in 1980, was a *wee* mutant that went into mitosis at a much smaller cell size, though in the same metabolic state, as the wild type. Mutants like this clearly indicated the existence of checkpoints in the normal cycle. Equally important was the discovery that many such mutations affected protein kinases, strongly suggesting that cell-cycle regulation involved phosphorylation and dephosphorylation events.

Just how this might be controlled was revealed by Timothy Hunt in one of the great serendipitous discoveries in molecular biology. Hunt was not even looking at cell-cycle regulation; he was examining the control of translation in marine invertebrates. As a model, he and colleagues studied the accumulation of proteins in sea urchin eggs following fertilization. The first cleavage divisions following fertilization in this organism are synchronous and Hunt was surprised to see a class of proteins that not only increased in amount at a certain point in the cell cycle but were also specifically degraded at a later point. These he called **cyclins**, and he soon found them in other organisms as well. It was soon revealed, in many labs, that cyclins were associated with cdc kinases.

In summation, the work of Hartwell, Nurse, and Hunt has provided a mechanistic basis for understanding cell-cycle regulation. In 2001 they were awarded the Nobel Prize in Physiology or Medicine "for their discoveries of key regulators of the cell cycle."

Now we know that the cell-cycle regulation involves a complex interplay between kinases and phosphatases. Cyclin-dependent kinases, or CDKs, are serine or threonine protein kinases that are involved not only in cell-cycle regulation (**Figure 1**) but also in other processes such as transcription. The activity of CDK is regulated through phosphorylation by other upstream kinases and, significantly, by association with function-specific cyclins. In turn, the CDK–cyclin complexes

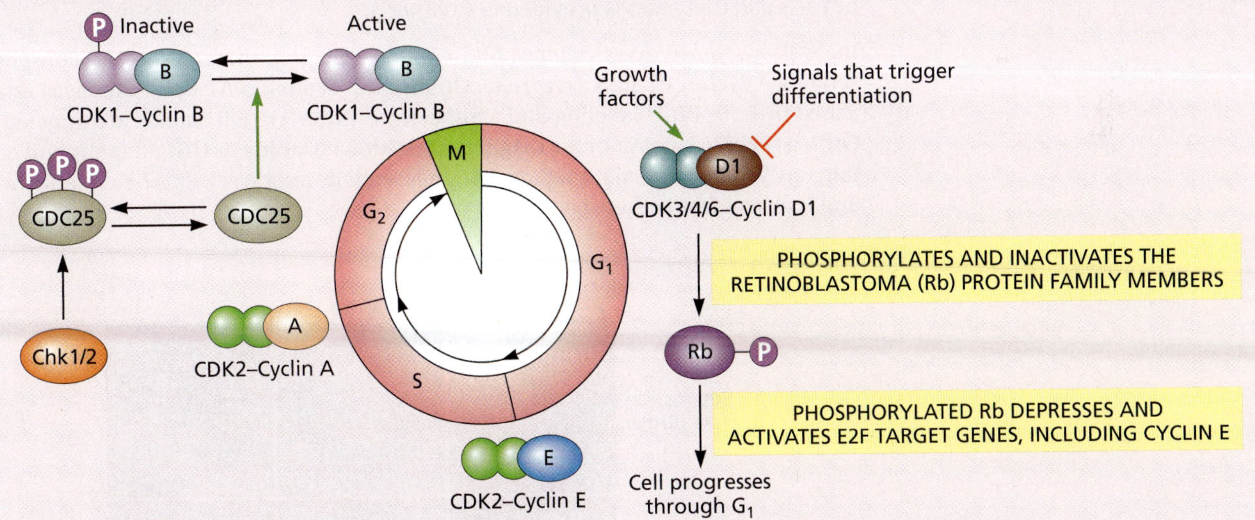

Figure 1 Cell-cycle regulation: An overview. Entry and progression through the four phases of the cell cycle is controlled by distinct CDK–cyclin heterodimeric complexes. D-type cyclins such as cyclin D1, acting in complexes with CDK3, CDK4, or CDK6, regulate events in early G_1 phase; CDK2–cyclin E triggers S phase; the CDK2–cyclin A complex regulates the completion of S phase; finally, CDK1–cyclin B is responsible for transition to mitosis. Some complexes can be inhibited by specific inhibitors, for example, cyclin D-associated kinases are inhibited by a group of proteins belonging to the INK4 or inhibitor of CDK4 family, whereas cyclin E and cyclin A kinases are inhibited by $p21^{waf1}$, $p27^{kip1}$, and $p57^{kip2}$. The decision of whether a cell should proliferate, in response to growth factors, or differentiate, in response to differentiation signals, is made during the G_1 phase of the cell cycle. The decision to initiate mitosis is regulated by the CDK1–cyclin B complex, which needs to be activated from its phosphorylated form by the phosphatase activity of CDC25, cell division cycle 25, which in turn is activated by another phosphatase, not shown here. CDC25 activity is inhibited by phosphorylation by Chk1/2 or mitotic checkpoint kinase 1/2, preventing premature entry into mitosis.

(Continued)

Box 20.2 *(Continued)*

are inhibited through reversible binding of CDK inhibitors and through cyclical degradation of cyclins during the cell cycle. Oscillations in the levels of the four cell-cycle-dependent cyclins are presented in Figure 2. The inhibitors act on CDK–cyclin complexes that spontaneously adopt an active conformation upon heterodimerization. Heterodimers that do not spontaneously remodel into an active conformation upon complex formation are regulated via co-factor or substrate binding.

An example of the latter is the CDK4–cyclin D1 complex, which may be activated by nuclear translocation, substrate binding, or phosphorylation of specific residues. This complex is of special interest because of its connection to numerous human cancers. An unchecked or hyperactivated CDK4–cyclin D1 pathway may be responsible for enhanced cellular proliferation due to the role of this complex in the G_1 checkpoint.

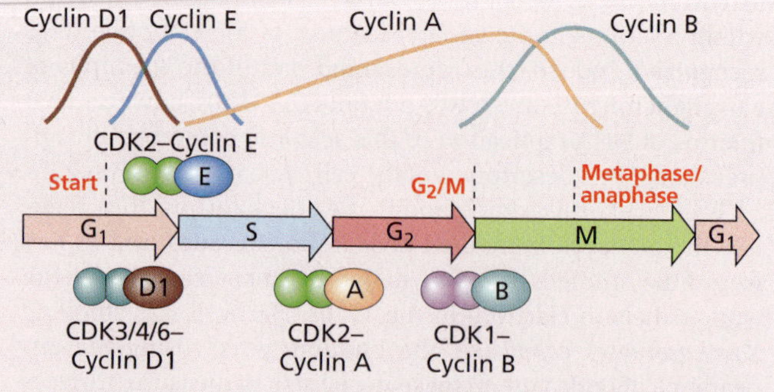

Figure 2 Levels of cyclins during phases of the cell cycle. The cyclin subunits of the CDK–cyclin complexes are degraded by highly specific ubiquitin-mediated proteolysis. (Adapted from Hochegger H, Takeda S & Hunt T [2008] *Nat Rev Mol Cell Biol* 9:910–916. With permission from Macmillan Publishers, Ltd.)

to an ARS gain the ability to replicate in yeast. This can be used to advantage when large eukaryotic genes must be cloned under conditions in which they will be properly expressed, replicated, and processed. The routinely used bacterial cloning vectors cannot accomplish the cloning of large genes, since they only allow the insertion and propagation of relatively small DNA fragments. To clone such large genes, it is necessary to construct yeast artificial chromosomes or YACs that contain, in addition to the usual selective markers, centromeric and telomeric sequences that create a small functioning chromosome in the recipient cell. The replication of these small chromosomes is ensured by introducing ARS into the constructs. A more detailed discussion of YACs and their uses is provided in Chapter 5.

In *Schizosaccharomyces pombe*, ARS elements do not share a specific consensus sequence as they do in *S. cerevisiae*. Origins are located in AT-rich islands, as established by both BrdU labeling and ChIP assays. These AT-rich islands are targeted by Orc4, a specific subunit of the **origin recognition complex** or **ORC**; this subunit contains an AT-hook domain, a region that preferentially interacts with AT-rich sequences, and is absent from other species.

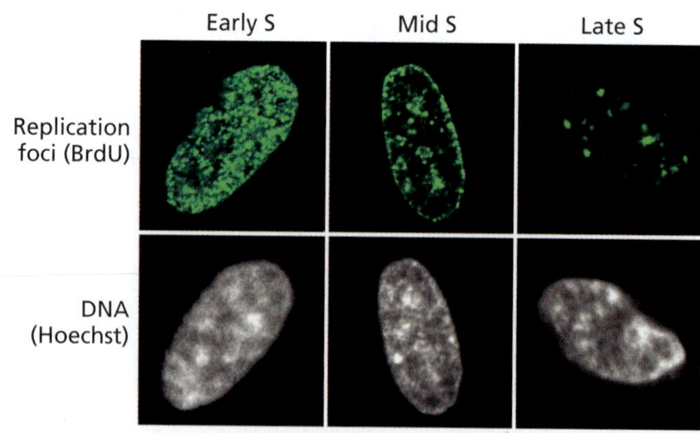

Figure 20.4 Patterns of replication foci as the cell progresses through S phase. Foci were visualized by immunodetection of DNA that has been pulse-labeled for 30 min with bromodeoxyuridine, BrdU. Bulk DNA was stained with Hoechst dye. Early in S phase, hundreds of small foci are distributed throughout the nuclear volume; in mid S phase, foci are preferentially located around the nucleoli and the nuclear periphery; finally, in late S phase only several clusters of foci are seen in heterochromatic regions. (From Méndez J [2009] *Crit Rev Biochem Mol Biol* 44:343–351. With permission from Informa Healthcare.)

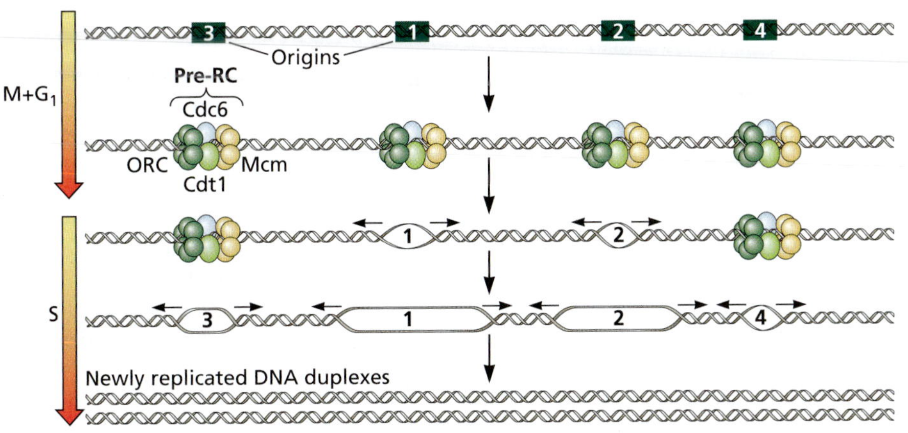

Figure 20.5 Temporal activation of replication origins during S phase. Pre-replication complexes or pre-RC are assembled on origin sequences during M and G_1 phases. Activation or firing of these origins subsequently occurs throughout S phase; origins are classified as early-, mid-, and late-activating depending on whether they become active during the early, mid, or late stages of S phase. The schematic represents a portion of the genome where all potential origins are activated. In reality, only a small fraction of all potential origins is used at each cell cycle. (Adapted from Méchali M [2010] *Nat Rev Mol Cell Biol* 11:728–738. With permission from Macmillan Publishers, Ltd.)

In higher eukaryotes, there seems to be little or no recognized sequence specificity at origins. The use of ChIP to locate replication origins in higher eukaryotes has been less successful than in other species, presumably because of the abundance of some of the proteins involved in pre-initiation complexes or because some of these proteins

Box 20.3 A Closer Look: Is replication timing correlated with transcription? For some time, there has been discussion about a possible connection between replication timing in the course of S phase and transcriptional activity of resident genes, presumably those involving chromatin structure. Many of the published data were, however, obtained on a few genes in scattered, unrelated cell lines, which has made interpretation difficult and generalizations impossible. To add to the confusion, no evidence exists for such an association in either budding or fission yeast, but the situation seems to be different in metazoans. Genomewide methods came to the rescue. Using strategies like that depicted in Figure 1, researchers found a complex picture of regulations and correlations, and it became clear that cellular differentiation is accompanied by coordinated changes in

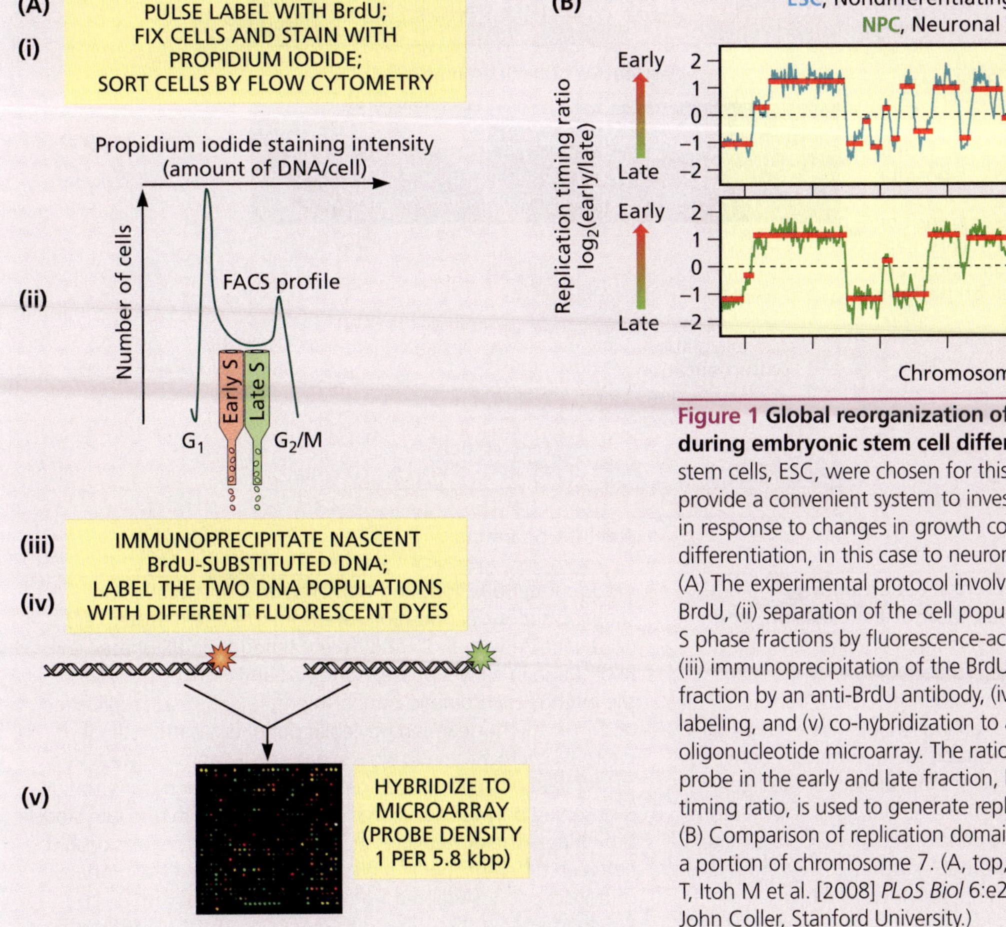

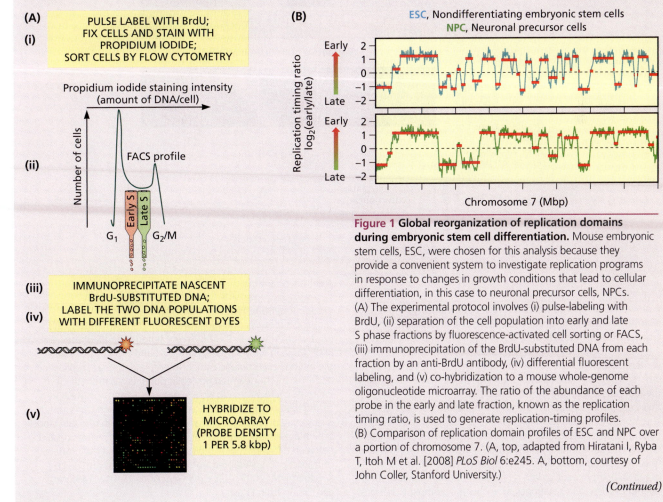

Figure 1 Global reorganization of replication domains during embryonic stem cell differentiation. Mouse embryonic stem cells, ESC, were chosen for this analysis because they provide a convenient system to investigate replication programs in response to changes in growth conditions that lead to cellular differentiation, in this case to neuronal precursor cells, NPCs. (A) The experimental protocol involves (i) pulse-labeling with BrdU, (ii) separation of the cell population into early and late S phase fractions by fluorescence-activated cell sorting or FACS, (iii) immunoprecipitation of the BrdU-substituted DNA from each fraction by an anti-BrdU antibody, (iv) differential fluorescent labeling, and (v) co-hybridization to a mouse whole-genome oligonucleotide microarray. The ratio of the abundance of each probe in the early and late fraction, known as the replication timing ratio, is used to generate replication-timing profiles. (B) Comparison of replication domain profiles of ESC and NPC over a portion of chromosome 7. (A, top, adapted from Hiratani I, Ryba T, Itoh M et al. [2008] *PLoS Biol* 6:e245. A, bottom, courtesy of John Coller, Stanford University.)

(Continued)

Box 20.3 *(Continued)*

replication timing and transcription. Correlated changes also take place in subnuclear repositioning of replication foci (see Figure 20.2) and occur in large, megabase-sized domains that do not reflect any changes of chromatin structure at the local level.

The mammalian genome is partitioned into **isochores**: chore, Greek for combining form; and iso, equal or identical; these are large regions, >300 kbp, of DNA with homogeneous, either high or low, GC content and gene density. This homogeneity of GC content in isochores contrasts with the heterogeneity over the entire genome. The existence of isochores has been demonstrated by sedimentation analysis. Mammalian genomes are a mosaic composed of GC-rich regions and GC-poor regions, with some GC-rich regions having a GC content of ~60%, while GC-poor regions may contain as little as ~30% GC. Today, we recognize five families of isochores that differ in overall GC content; in addition, the high-GC isochores are gene-rich, whereas the low-GC isochores are relatively gene-poor.

In terms of replication timing in S phase, the GC-rich, gene-rich isochores are replicated early in S phase, whereas those that are GC-poor and gene-poor replicate late, as depicted in **Figure 2**. There is, however, no correlation between replication timing and transcription within each of these two categories of isochores. Thus, all genes that are replicated in the first third of S phase have equally high probability of being expressed; even large changes in replication timing within this period are not accompanied by changes in subnuclear position or transcriptional competence. Those isochores in which GC-richness and gene-richness have an inverse relationship (mid–low and low–mid in Figure 2) are frequently subject to regulation of replication timing during differentiation (see Figure 1).

Thus, a complex relationship exists between replication timing, subnuclear positioning of the replication foci, and transcription. We understand this relationship only superficially. As David Gilbert and colleagues put it: "Our understanding of replication timing remains a fragmented set of half-truths that are currently impossible to integrate into absolutes...We are waiting for our various half-truths to intersect and reveal a more complete picture."

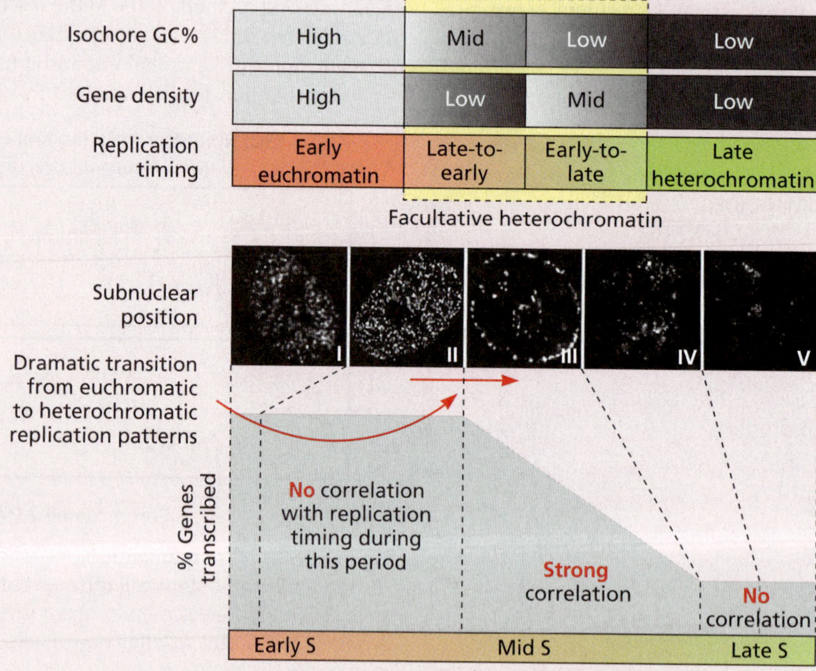

Figure 2 Relationship between replication timing regulation, isochore properties, subnuclear position, and transcription in metazoans. In this figure, isochores in the mammalian genome are categorized according to their GC content and gene density. The four isochore groups behave differently in terms of replication timing, subnuclear location of the replication foci, and the correlation, or lack thereof, between their replication timing and transcriptional activity of their genes. The groups that change replication timing during differentiation belong to the mid-GC–low gene density and low-GC–mid gene density categories: the mid–low group tends to change from late to early replication as cells differentiate, whereas the low–mid group shifts from early to late replication. Changes in replication timing for these two categories of isochores accompany changes in subnuclear positions and transcriptional potential. Images showing subnuclear positions are from Chinese hamster cells pulse-labeled with BrdU at different times during S phase. Replication in early S phase, patterns I and II, takes place within the interior euchromatic compartment; later S phase replication occurs at the nuclear and nucleolar periphery, pattern III, or at internal heterochromatic blocks, patterns IV and V. Note that a strong relationship between replication timing and transcription is observed only for genes that replicate during mid to late stages of S phase. The dotted lines in the figure show the relationship between different spatial replication patterns and the probability of transcription. (Adapted from Hiratani I, Takebayashi S, Lu J & Gilbert DM [2009] *Curr Opin Genet Dev* 19:142–149. With permission from Elsevier.)

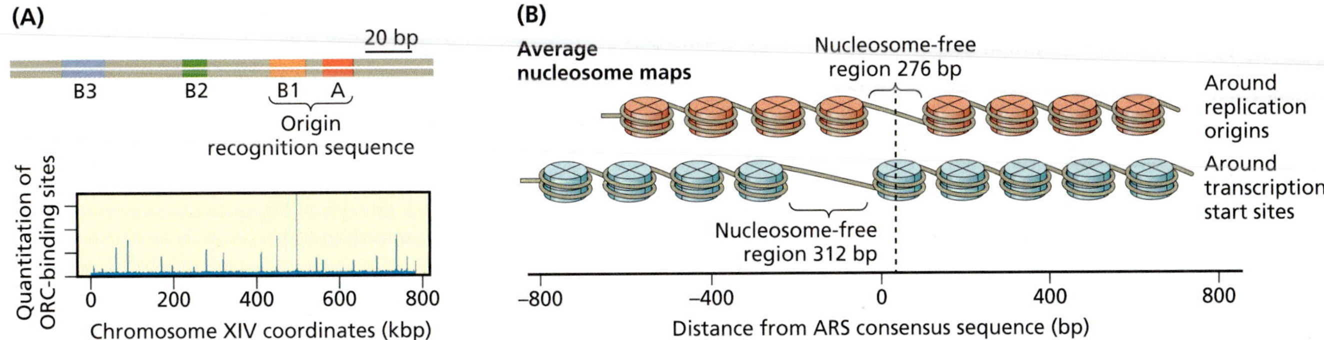

Figure 20.6 DNA and chromatin structure of an origin of replication in *S. cerevisiae.* (A) DNA structure. (Top) Four regions in the 100–300 bp yeast origins, named A, B1, B2, and B3 in order of decreasing effect on replication, are sufficient for autonomous replicating sequence or ARS activity. Mutant ARSs were constructed in which regions in the ARS had been systematically replaced by an unrelated sequence; these were then transformed into yeast and the effects on ARS function were tested *in vivo*. Box A contains an 11-bp ARS consensus sequence, TTTTATATTTT. The origin recognition sequence comprising boxes A and B1 binds the origin recognition complex, ORC; the B2 box is where melting of the double-stranded DNA helix is initiated; and the B3 box is believed to introduce bending in ARS, thus helping in binding of the ABF1 protein factor. (Bottom) High-throughput sequencing combined with chromatin immunoprecipitation to identify ORC and nucleosome localization across the yeast genome shows a precise localization of yeast origins, illustrated here for chromosome XIV. Further studies revealed that replication origins are associated with an asymmetric nucleosome-free region flanked by well-positioned nucleosomes. (B) Comparison of averaged chromatin structure around 222 replication origins and a random set of 222 transcription start sites. The nucleosome-free regions for origins are, on average, narrower than those for promoters and are centered 36 bp to the right of the first nucleotide of box A of the origin. (A [top], adapted from Bielinsky A-K & Gerbi SA [1998] *Science* 279:95–98. With permission from American Association for the Advancement of Science. A [bottom], adapted from Eaton ML, Galani K, Kang S et al. [2010] *Genes Dev* 24:748–753. With permission from Cold Spring Harbor Laboratory Press. B, adapted from Berbenetz NM, Nislow C & Brown GW [2010] *PLoS Genet* 6:e1001092.)

may also possess other functions. There is some evidence for preference for AT-rich sequences, which may imply a preference for certain DNA structures at origins. Also, many origins seem to reside in unmethylated CpG islands.

In any event, it is firmly established that the first prerequisite for a functioning eukaryotic origin is that it binds a multiprotein hexameric complex called the ORC. The ORC must occupy a nucleosome-free region or NFR on the DNA (see Figure 20.6). Given that we now believe that preferences for nucleosome positioning are in some way partially dictated by the underlying DNA sequence, the qualifications for ORC sites may be understood in such terms. The nucleosome-free regions at replication origins bear an interesting similarity to the nucleosome-open regions found at many transcription start sites (see Chapter 12). Nevertheless, a detailed comparison between the NFRs at origins and transcription start sites shows that they have different lengths and slightly different locations of the positioned nucleosomes in the flanking regions.

There is a defined scenario for formation of initiation complexes

A defined sequence of events prepares origins for activation. **Pre-replication complexes** or pre-replicative complexes, **pre-RCs**, are assembled on origin sequences during M and G_1 phases, and these are then activated at various times during S phase (**Figure 20.7**).

First, the origins are recognized by the ORC: while still in M phase, the ORC recruits two proteins, Cdc6 and Cdt1, that are needed for recruiting of the hexameric **Mcm2–7 complexes**. One Mcm2–7 complex is bound on each side of the bound ORC, in opposite orientations. These complexes are the helicases that serve to unwind DNA in the two divergent replication forks. The site is now said to be licensed for initiation. In practice, this means that it can now bind other essential factors (see Figure 20.7) to produce first a pre-RC and subsequently, at the junction between G_1 and S phase, the pre-initiation complex or pre-IC. The more detailed structural views presented in **Figure 20.8** should be helpful in understanding the process at a more mechanistic level. These figures also provide information on the conformational changes experienced by Mcm2–7, as well as the role of the recently

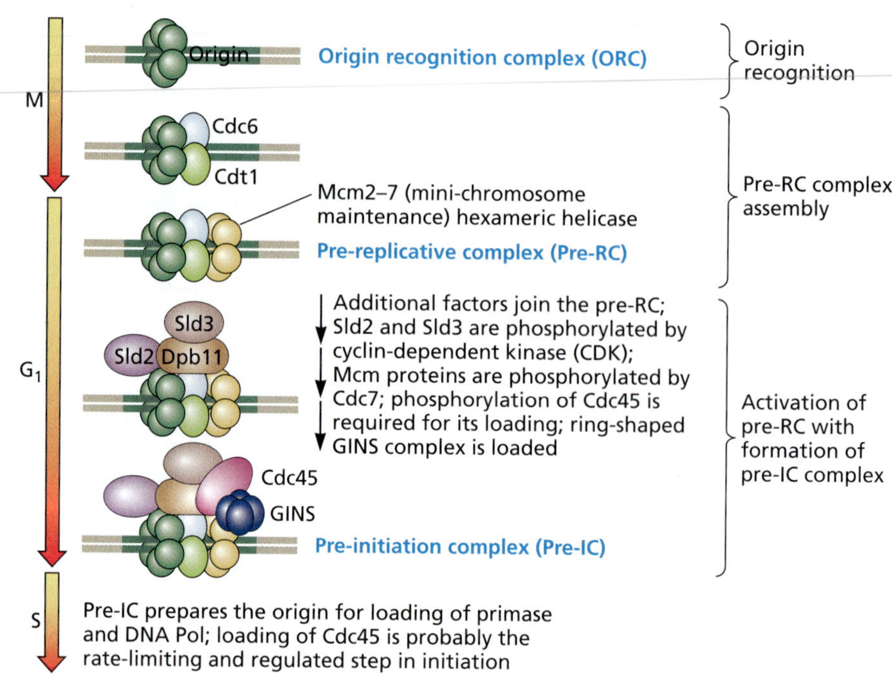

Figure 20.7 Sequence of events leading to initiation of replication in eukaryotes. In eukaryotes, the origins of replication are set by a three-step process: (1) recognition of the origin by the origin recognition complex, ORC; (2) assembly of the pre-replication complex, pre-RC; and (3) activation of the pre-RC, leading to formation of the pre-initiation complex, pre-IC. The pre-IC is ready to accept primase and DNA polymerase. The pre-RC contains two Mcm2–7 helicases, one on each side of the bound ORC; only one Mcm2–7 complex is shown here for simplicity. The schematic also depicts the phases of the cell cycle when the respective events occur. This figure does not attempt to show the structural aspects of the complexes. (Adapted from Boye E & Grallert B [2009] *Cell* 136:812–814. With permission from Elsevier.)

discovered **GINS complex**, in establishing the final clamping of the helicase about the DNA. GINS stands for go, ichi, nii, and san: five, one, two, and three in Japanese, after the four related subunits of the complex, Sld5, Psf1, Psf2, and Psf3. The pre-initiation complex is now ready to accept the polymerase and to begin replication.

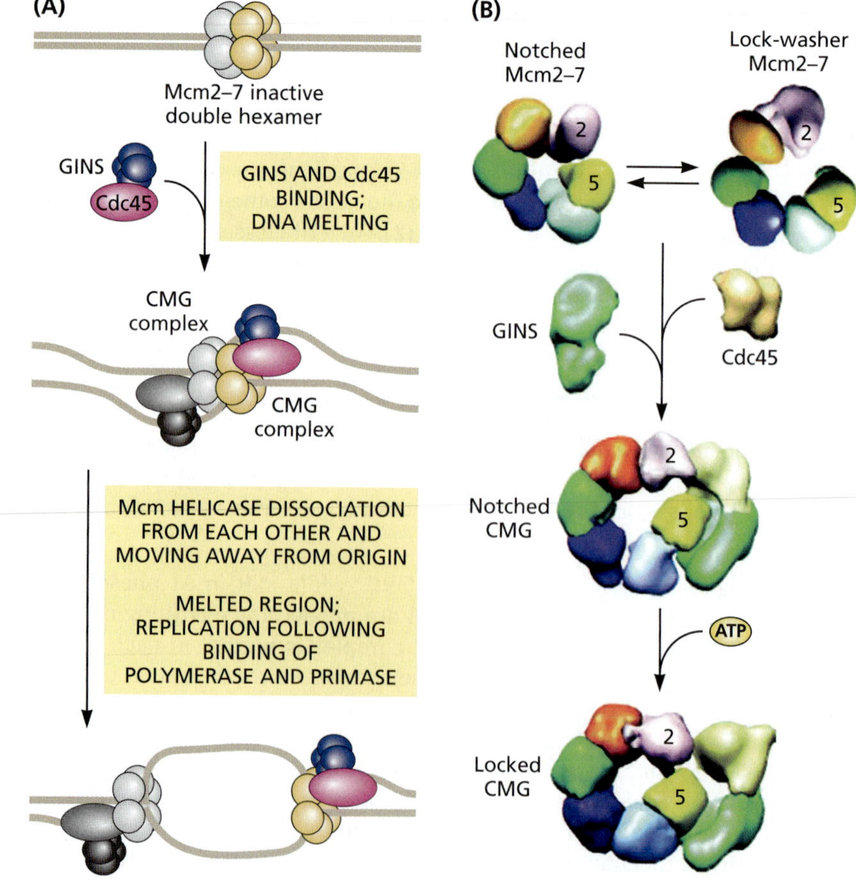

Figure 20.8 Model for activation of replication fork helicase and structure of CMG complex. (A) Helicase activation model. The helicase Mcm2–7 is assembled onto a double-stranded DNA origin of replication as an inactive double hexamer. Several protein factors help binding of GINS and Cdc45 to Mcm2–7, thus assembling the two CMG complexes, each of which comprises Cdc45, the six Mcm components, and the four GINS. DNA is partially melted; the two CMG complexes dissociate from each other and move away from the origin in opposite directions. The origin of replication in the melted region binds primase and DNA polymerase to start replication. (B) Structural basis for Mcm2–7 helicase activation by Cdc45 and GINS as determined by single-particle electron microscopy. Free Mcm2–7 can exist in a dynamic equilibrium between open, lock-washer and notched, planar configurations. In each form, there is a discontinuity between subunits Mcm2 and Mcm5. Binding of Cdc45–GINS stabilizes the notched state, whereas ATP binding promotes ring closure. (Adapted from Costa A, Ilves I, Tamberg N et al. [2011] *Nat Struct Mol Biol* 18:471–477. With permission from Macmillan Publishers, Ltd.)

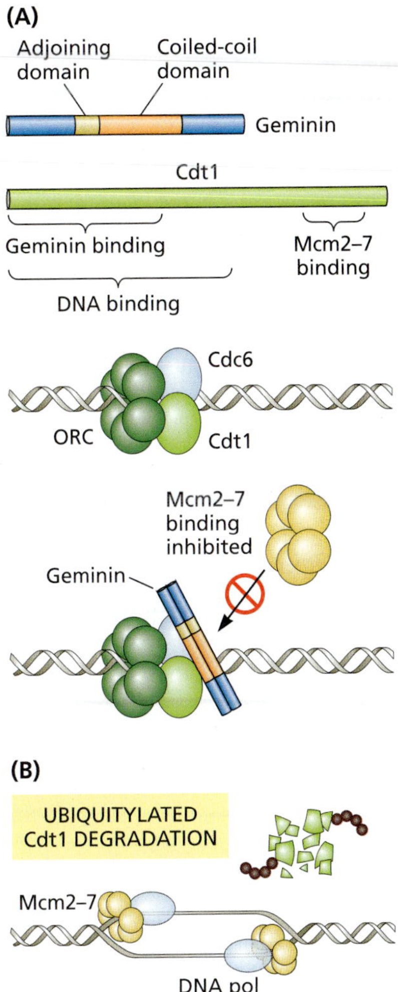

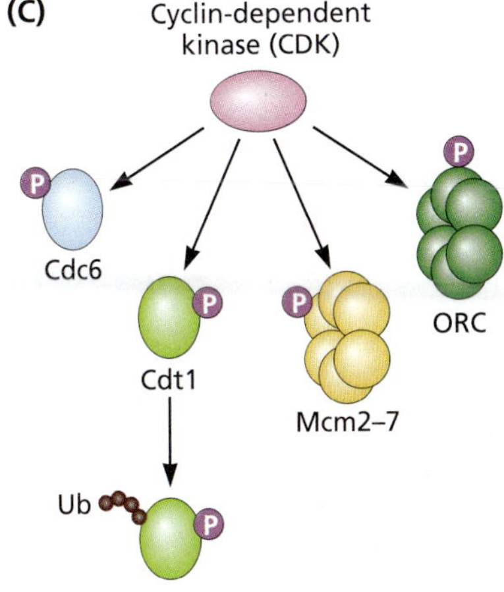

Figure 20.9 Three possible mechanisms for inhibition of re-replication or second-round pre-RC formation. (A) Geminin inhibits Cdt1 function at origins by inhibiting Cdt1 binding, by inhibiting Cdt1 binding to Mcm2-7. The top schematic depicts the domains of geminin and Cdt1 that are important for their interaction. Geminin dimerizes through a coiled-coil domain in its central region, providing a binding interface for a single molecule of Cdt1. Geminin accumulates during S, G_2, and M phases of the cell cycle and is degraded at the metaphase-to-anaphase transition. (B) Replication-dependent origin inactivation: Cdt1 degrades in S phase in a replication-dependent way. (C) Cyclin-dependent kinase or CDK, which has high activity in the G_2 and M phases, phosphorylates and inactivates or destabilizes pre-RC components Cdc6, Cdt1, Mcm2–7, and/or ORC, thus preventing the formation of new pre-RCs or destabilizing old pre-RCs. This regulatory mechanism might be important in preventing a whole new round of replication, starting in G_2. On the flip side, stabilization of Cdc6 in human cells is critical when cells enter the cell cycle from a quiescent state. This is the major pathway in yeast and is important also in metazoans. (Adapted from Machida YJ, Hamlin JL & Dutta A [2005] *Cell* 123:13–24. With permission from Elsevier.)

Re-replication must be prevented

There is one additional aspect of initiation that needs to be mentioned. With many origins of replication simultaneously active, it is essential that none of them fires twice in a single round of replication. Recall that any origin sites already reproduced on the daughter duplexes are potentially capable of accepting ORCs. If one of these were to initiate another round prematurely, the DNA structure would be disastrously tangled. As shown in **Figure 20.9**, there are mechanisms to prevent just this possibility.

The major set of mechanisms that prevent re-replication involves members of the cyclin-dependent kinase or CDK family, which phosphorylate and inactivate, or destabilize, a number of protein factors in the pre-RC: Cdc6, Cdt1, Mcm2–7, and/or ORC (see Figure 20.7). Note that Mcm2–7 can also be phosphorylated by another kinase, Cdc7. CDK activity peaks in G_2 and M phases, thus preventing the formation of new pre-RCs in G_2 or destabilizing old pre-RCs in M. Another pathway involves Cdt1 degradation in S phase; such degradation occurs in a replication-dependent way.

A more recently recognized mechanism for preventing re-replication uses the protein geminin as an inhibitor of Cdt1 function at origins. The mechanism depends on cycling of the amounts of geminin available throughout the cell cycle: its levels are high in S, G_2, and M phase and low in G_1 phase. The degradation of geminin occurs during the metaphase–anaphase transition and is proteasome/ubiquitylation-mediated. Low levels of geminin in G_1 allow binding of Mcm2–7 to its recruiter, Cdt1.

Histone methylation regulates onset of licensing

Recent reports have identified the methylation of histone H4 at lysine 20, H4K20me1, and the enzyme that puts this modification mark in place, PR-Set7, as key positive regulators of the onset of licensing in mammalian cells. Levels of both the enzyme and the modification are cell-cycle-regulated, being high during M and G_1 phases and dropping when S phase begins. Proteolytic degradation of PR-Set7 is needed to prevent re-replication. Even more interesting is the fact that targeting of the enzyme to non-origin sites on chromatin fibers is sufficient to induce the H4K20 modification and assembly of the pre-RC.

20.3 Replication elongation in eukaryotes

Eukaryotic replisomes both resemble and significantly differ from those of bacteria

In general, the basic mechanisms of replication elongation in eukaryotic cells are very similar to those in bacteria. These include replication only in the $5' \rightarrow 3'$ direction and thus the necessity for both continuous, or leading-strand, and discontinuous, or lagging-strand, synthesis. The second shared feature is the requirement for a primer that is synthesized by a specialized enzyme, **Pol** α, which forms a complex with primase.

Table 20.1 Three replicative polymerases in the eukaryotic nucleus. The table presents the three replicative polymerases in yeast, *S. cerevisiae*, with their subunit compositions. Numbers in parentheses are the molecular masses of the respective subunits. The largest subunit in each complex contains the polymerase activity and, for Pol δ and Pol ε, the exonuclease activity. The primase activity resides in the Pri1 subunit of Pol α/primase. (Adapted from TA & Burgers PM [2008] *Trends Cell Biol* 18:521–527.)

	Pol α/primase	Pol δ	Pol ε
biochemical activity	polymerase; primase	polymerase; 3′ → 5′ exonuclease	polymerase; 3′ → 5′ exonuclease; dsDNA binding
process	initiation of replication; initiation of Okazaki fragments; primer synthesis	lagging-strand synthesis, elongation and maturation of Okazaki fragments; proofreading of errors made by Pol α; DNA repair	leading-strand synthesis; replication checkpoint
subunits	four: Pol1 (167), Pol12 (79), Pri1 (48), Pri2 (62)	three: Pol3 (125), Pol31 (55), Pol32 (40)[a]	four: Pol2 (256), Dpb2 (78), Dpb3 (23), Dpb4 (22)
processivity, inherent	moderate	low	high
processivity, with PCNA	moderate	high	high
error frequency	10^{-4}–10^{-5}	10^{-6}–10^{-7}	10^{-6}–10^{-7}

[a]Other organisms have an additional subunit, PolD4 (12) in humans.

Despite the overall similarity in the bacterial and eukaryotic polymerization reactions, there are some major differences in the organization and dynamics of the replisome. Whereas bacterial elongation proceeds through the use of identical twin polymerases, the core Pol III polymerases held together by the clamp loader, leading- and lagging-strand syntheses in eukaryotes use two different polymerases: **Pol ε** for the leading strand and **Pol δ** for the lagging strand.

Thus, there are three replicative multisubunit DNA polymerases in the eukaryotic nucleus (**Table 20.1**). Despite having defined and distinct roles, they all share similarities in structure (**Figure 20.10**). A unifying feature of all three is the presence of a common functional core formed by the C-terminal domain, CTD, of the respective catalytic subunit and one conserved accessory subunit, B. This core serves as a scaffold for the organization of the other subunits, which differ for the three polymerases.

Eukaryotic primase initiates as frequently as its bacterial counterpart, one primer per second, but it has evolved significantly from its bacterial counterpart, synthesizing **two-part primers** that consist of RNA followed by ~20 nt of single-stranded DNA (ssDNA). This can be accounted for by the fact that the primase is part of a larger complex called the **primosome** (**Figure 20.11**). The primosome consists of the RNA-generating primase, itself a heterodimer of a large and a small subunit, and DNA polymerase α. There is a switch from the primase to polymerase activity. This polymerase is low fidelity, with an error frequency of 10^{-4}–10^{-5}, compared to 10^{-6}–10^{-7} for the elongating polymerases, because it does not possess proofreading capability. It is believed that errors introduced by Pol α/primase are later corrected by Pol δ, the polymerase that takes over from Pol α/primase to synthesize the lagging-strand Okazaki fragments.

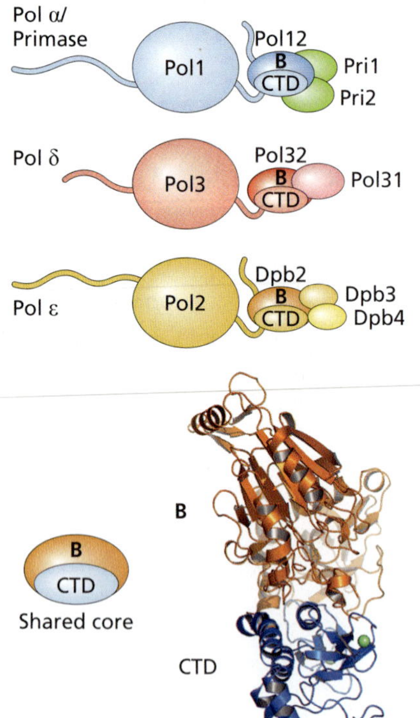

Figure 20.10 Subunit organization of three nuclear replicative yeast DNA polymerases. The three polymerases, Pol α/primase, Pol δ, and Pol ε, share a common functional heterodimeric core, which helps in recruiting the other accessory factors, not shown here. This core consists of the C-terminal domain or CTD of the respective catalytic subunit, Pol1, Pol2, or Pol3, and the accessory regulatory B subunit. Of their different cohorts of accessory subunits, only the B subunit is present in all three polymerase complexes. The crystal structure represents the yeast Pol α CTD–B subunit complex; the green sphere is one of the two zinc atoms in the CTD. Both the CTD and B subunits are highly evolutionarily conserved and are indispensable for growth. The catalytic subunits Pol1, Pol2, and Pol3 are phylogenetically related. (Adapted from Klinge S, Núñez-Ramírez R, Llorca O & Pellegrini L [2009] *EMBO J* 28:1978–1987. With permission from John Wiley & Sons, Inc.)

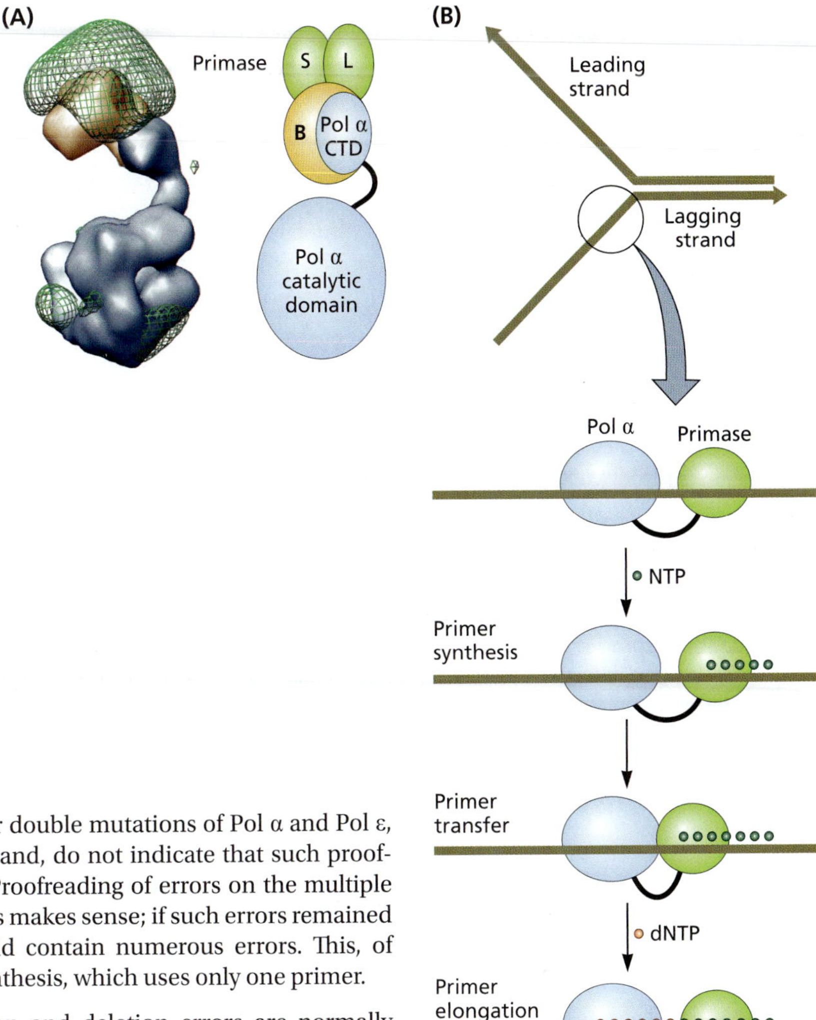

Figure 20.11 Molecular architecture of yeast primosome and steps in RNA–DNA primer synthesis. (A) The overall three-dimensional architecture is an asymmetric, dumbbell-shaped particle, with the two catalytic activities, primase and polymerase, residing in separate lobes, tethered together by a highly flexible linker. The primase, shown as green mesh, is a heterodimer consisting of large and small subunits, L and S. The schematic on the right explains the structure. The B subunit, shown as a yellow sphere, is an accessory subunit present in all three eukaryotic polymerase assemblies and is clearly conserved in eukaryotic organisms. (B) Steps in the process of synthesis of the hybrid RNA–DNA primer. Physical association of the primase and polymerase activities provides the basis for their tight functional coupling, allowing for an intramolecular mechanism of RNA primer transfer between the active sites of the primase and Pol α. After unit-length primer synthesis by the primase, the 3′-terminus of the template-bound primer is internally translocated to the active site of Pol α without being released into solution. (Adapted from Núñez-Ramírez R, Klinge S, Sauguet L et al. [2011] *Nucleic Acids Res* 39:8187–8199. With permission from Oxford University Press.)

Interestingly, genetic studies involving single or double mutations of Pol α and Pol ε, the polymerase that synthesizes the leading strand, do not indicate that such proofreading occurs for the leading-strand primer. Proofreading of errors on the multiple primers synthesized for lagging-strand synthesis makes sense; if such errors remained unrepaired, the resulting DNA duplexes would contain numerous errors. This, of course, is not the case for the leading-strand synthesis, which uses only one primer.

In addition to incorrect base-pairing, insertion and deletion errors are normally expected in a replication process. These can also be introduced by other processes such as DNA repair and recombination. There is an additional type of error that appears to be unique to DNA replication. This is the consequence of slippage, which can easily happen when repeated sequences are copied. Slippage causes a number of genetic diseases (**Box 20.4**).

Box 20.4 DNA replication slippage, expandable repeats, and human disease Cells use a variety of mechanisms, including checkpoint controls, to ensure that the entire genome is replicated accurately, only once per cell cycle. There is no such thing, however, as a perfectly working mechanism, and factors that interfere with DNA replication can jeopardize genomic integrity. Such factors are generally classified into three groups: (1) exogenous factors that damage DNA or deplete nucleotide pools; (2) genetic factors that lead to mutations in genes that perform replication functions; and (3) intrinsic factors such as DNA-binding proteins, transcriptional units, unusual DNA structures, and intrinsically slow zones of replication.

The fact that unusual DNA structures can form at short repeated sequences and interfere with DNA replication, as well as with DNA recombination and repair, explains why expansions of simple DNA repeats lead to more than 30 human hereditary conditions. Figure 1 illustrates, in the context of a generic human gene, the kinds of disease-causing repeats and their localization within genes. Most of these repeats are triplet, but tetra- and pentanucleotide repeats have also been described; a dodecanucleotide repeat has been found in a gene that causes the disease progressive myoclonic epilepsy.

Normal alleles of disease-associated genes contain either very short repetitive runs, termed short-normal alleles, or longer repeats with several stabilizing interruptions, called long-normal alleles; thus, for example, when the $(CGG)_n$ run that causes fragile X syndrome (see Figure 1) is interrupted by AGG inserts, the long repeat does not cause the disease. When the length of an uninterrupted repetitive run exceeds a threshold of ~100–150 bp, further expansion becomes more likely with each cell generation, and disease ensues. There are exceptions to this general rule; for example, the polyalanine-coding repeats that are responsible for numerous diseases do not expand more than 1.5-fold, and their expansion threshold is extremely low, between 30 and 60 bases.

Numerous kinds of unusual DNA structures can form at repeated sequences; these include hairpins, tetrahelical structures stabilized by G quartets (see Chapters 4 and 8), triple

(Continued)

Box 20.4 (Continued)

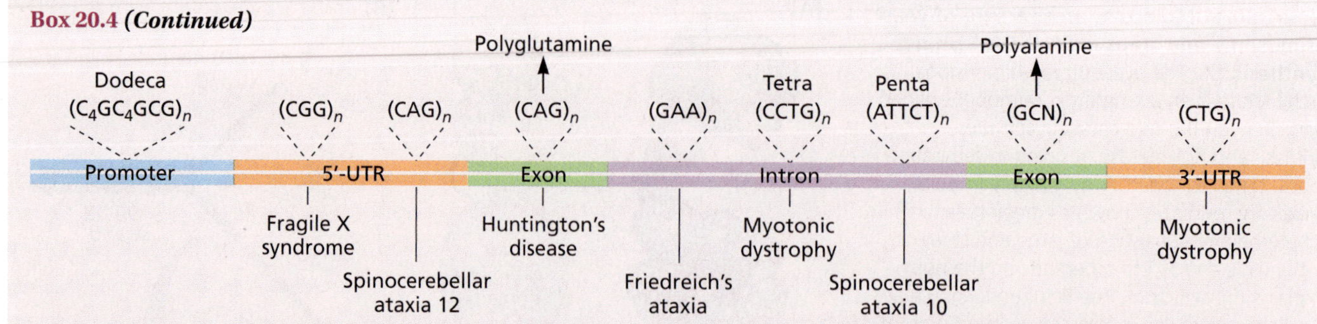

Figure 1 Location of expandable disease-causing repeats within a fictitious human gene. For any given hereditary disorder, only one repeat expands in a particular gene. Expandable repeats can be located in various regions of the genes in which they reside. When they reside in coding regions, they give rise to proteins that contain poly(amino acid) runs; when they are in other regions, they do not directly affect protein structure but exert their effects through disrupting the regulatory mechanisms that reside in these regions. All repeats depicted in the schematic have been associated with at least one disorder; we have listed only the most common diseases. (Adapted from Mirkin SM [2007] *Nature* 447:932–940. With permission from Macmillan Publishers, Ltd.)

helices, and DNA-unwinding elements. The formation of many of these structures is enhanced by negative superhelical stress. It is important to realize that, because of the difference in hairpin-forming potential between the two DNA strands, the template and the nascent strands end up being asymmetrical (**Figure 2**). The location of the unusual structures on either the template or the nascent strand gives rise to different outcomes following two rounds of replication of the repeat-containing DNA segment: either contraction of the repeated length resulting from the presence of a structure on the template strand or expansion resulting from a structure on the leading strand (see Figure 2). Replication stalling at these structures is believed to be the first step that leads to repeat instability.

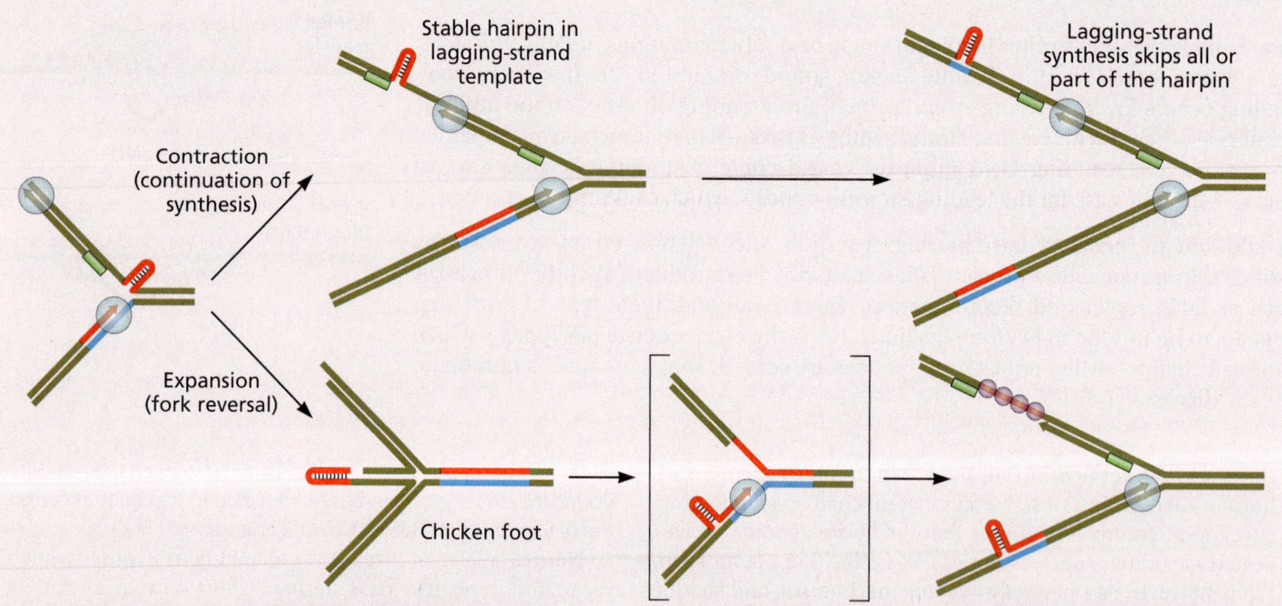

Figure 2 Model for repeat instability based on replication fork stalling and restarting within repetitive runs. The exact outcome of the process depends on whether the repeat sequence, RS, is present in the nascent strand, resulting in expansions, or in the template strand, resulting in contractions. The RSs are shown in red and their complements in blue; the DNA polymerases are represented by circles. Abnormal structures, such as the hairpins in this example, can form on the repeat sequences when they are single-stranded. What happens to them depends on how the replication machinery reacts to such sequences. As shown in the upper pathway, the replication fork may simply move past this segment. This will leave the RS on the lagging strand momentarily single-stranded; it can fold into a hairpin. The next Okazaki fragment synthesis may skip all or part of this hairpin, leading to the formation of one daughter strand that is missing the repeats. In the next round of replication, this change will be fixed in the genome. Alternatively, as shown in the lower pathway, the replication fork may arrest, perhaps because the hairpin starts to form. The fork may then actually reverse, undoing some synthesis and extruding a chicken-foot structure. This leaves the segment of RS from the leading strand in a single-stranded form, so it folds into a hairpin. As leading-strand replication resumes, the additional length of the RS can be copied into the leading strand. This time a second round of replication will produce one daughter duplex with an expanded repeat sequence. (Adapted from Mirkin SM [2007] *Nature* 447:932–940. With permission from Macmillan Publishers, Ltd.)

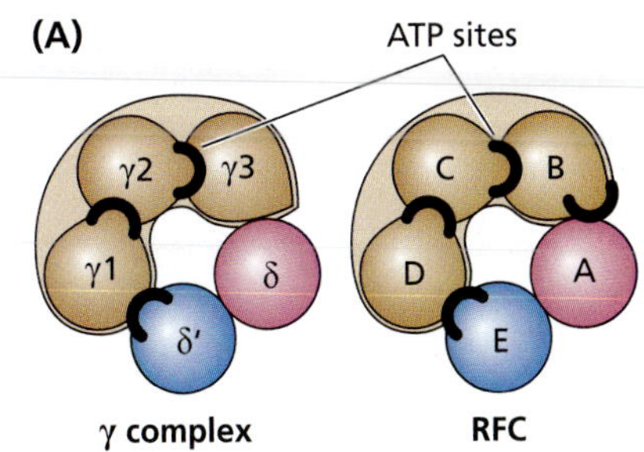

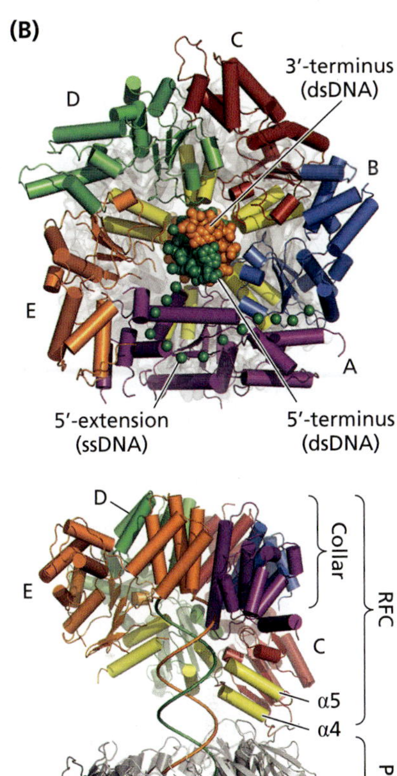

Figure 20.12 Structure of eukaryotic clamp loader and clamp. (A) The γ complex in *Escherichia coli* and replication factor C, RFC, clamp loaders in eukaryotes have similar spiral-shaped subunit architectures; the ATP sites on both complexes are located at subunit interfaces. The DnaX complex has three ATP sites; RFC includes four ATP sites. (B) Model of RFC–PCNA bound to primed DNA: co-crystal of the protein complex with modeled primed DNA. (Top) View from the collar of RFC, with the collar removed for clarity; a potential exit path for the 5′-end of the template strand is indicated by green spheres. (Bottom) Cutaway view of the complex, showing tracking of the DNA duplex by two of the helices, α4 and α5, of each subunit, colored yellow. (A, adapted from Pomerantz RT & O'Donnell M [2007] *Trends Microbiol* 15:156–164. With permission from Elsevier.)

Other components of the bacterial replisome have functional counterparts in eukaryotes

The processivity factor for the eukaryotic polymerases is **proliferating cell nuclear antigen** or **PCNA**, the counterpart of the bacterial β clamp. Recall from Chapter 19 that these two clamps are very similar in structure (see Figure 19.8) and ensure the highly processive nature of replication by encircling the double-stranded DNA portion formed by the template strand and the nascent daughter strand. The clamp loader in eukaryotes is **replication factor C** or **RFC**, whose overall structure and function are again highly similar to those of the pentameric clamp loader DnaX in bacteria. **Figure 20.12** presents these similarities in a schematic form.

Finally, the heterotrimeric **replication protein A** or **RPA** serves as the ssDNA-binding protein, counterpart of the bacterial protein SSB; and the helicase that opens the dsDNA at the replication fork is the heterohexameric Mcm2–7 complex, whose bacterial counterpart is DnaB. Recall that two copies of Mcm2–7 bind in opposite orientation on the two sides of the origin recognition complex at the beginning of S phase (see Figure 20.7) to ensure the bidirectional movement of two forks from a single origin.

Eukaryotic elongation has some special dynamic features

The details of replication elongation in eukaryotes are presented in a schematic form in **Figure 20.13**. The first two steps in replication elongation in eukaryotes, synthesis of the primer and recruitment of PCNA by the clamp loader RFC, are the same for the leading and lagging strand. The processes diverge when different replicative polymerases are recruited for each strand, Pol ε for the leading and Pol δ for the lagging strand. Pol δ also plays a role in Okazaki fragment maturation (**Figure 20.14** and **Figure 20.15**); it participates in the formation of short flaps in both the short- and long-flap pathways. The major difference between eukaryotic and bacterial replication stems from the fact that the former must occur in the context of chromatin. This raises questions as to how this structure can be traversed and what features are transmitted to daughter cells.

20.4 Replication of chromatin

Chromatin structure is dynamic during replication

It is evident from just a glance at chromatin structure (see Chapter 8) that the replication of DNA in eukaryotes faces severe complications. Not only must the parental nucleosomal structure be traversed, it also must be reproduced with fidelity in both daughter duplexes, even maintaining epigenetic information. It is perhaps no wonder that the process has been, until recently, characterized by uncertainty and controversy on fundamental issues. Now a consensus is beginning to emerge (**Figure 20.16**), largely on the basis of the discovery of histone chaperones that play a major role in proper chromatin replication (**Table 20.2**).

First, it should be clear that the bulky DNA replication apparatus cannot pass through an intact nucleosome during elongation. The nucleosome must be displaced or dissociated. Electron microscopy of the well-defined simian virus 40 or SV40 replication system reveals a nucleosome-free region just ahead of the fork, as well as evidence of the perturbation of the next nucleosome. These effects could arise from any or all of the following: (1) unwinding of DNA by the helicase; (2) destabilization of the

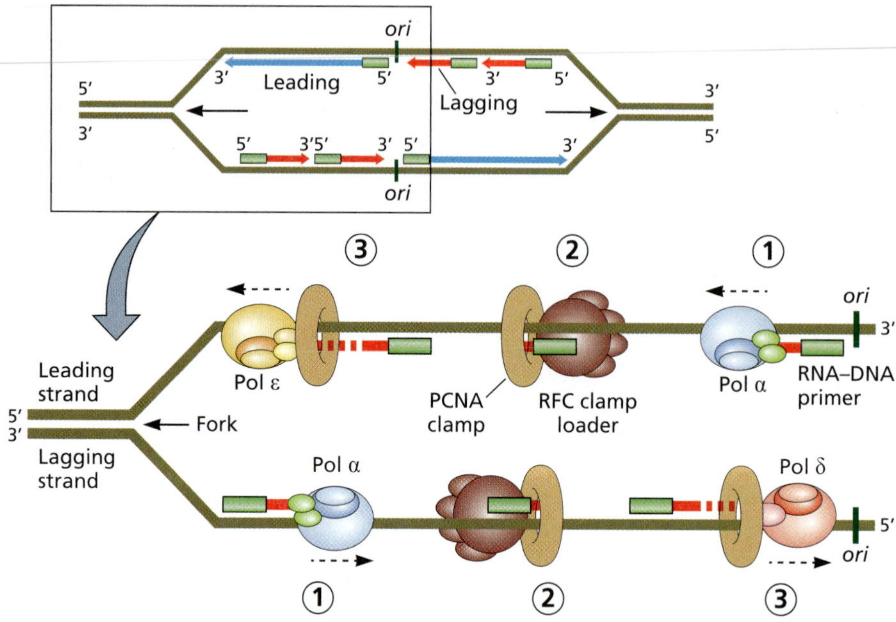

Figure 20.13 Replication elongation in eukaryotes. The schematic shows the steps that lead to synthesis of leading and lagging strands in eukaryotic DNA replication. The bottom schematic represents the processes occurring in the boxed region of a replication eye, shown at the top. The same processes occur in the other half of the replication eye, with reversal of the leading and lagging strands. (Step 1) RNA–DNA primer synthesis by Pol α/primase. (Step 2) Replication factor C, also known as RFC or the clamp loader, displaces polymerase α and recruits proliferating cell nuclear antigen, also called PCNA or the clamp. (Step 3) PCNA recruits the respective polymerase: Pol ε on the leading strand or Pol δ on the lagging strand; elongation by the newly recruited polymerases then proceeds. (Adapted from Henneke G, Koundrioukoff S & Hübscher U [2003] *EMBO Rep* 4:252–256. With permission from John Wiley & Sons, Inc.)

nucleosome by the accumulating positive stress in front of the moving DNA polymerase, in a scenario reminiscent of the situation in front of an elongating RNA polymerase (see Figure 12.8); or (3) acceptance of histones by chaperones.

Histone chaperones may play multiple roles in replication

Chaperones may be envisioned as being actively engaged in removing histones, playing a facilitative or storage role, or both. There are two chaperones that accept H2A-H2B dimers: FACT and NAP1 (see Table 20.2). The former we have already encountered in Chapter 12, where it was shown to facilitate transcription. Because FACT also interacts with the helicase, it could provide a mechanism for the reconstitution of nucleosomes

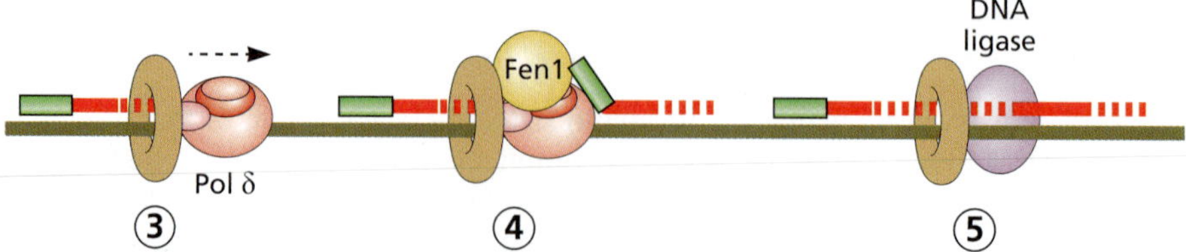

Figure 20.14 Steps in Okazaki fragment maturation in eukaryotes. Step numbering continues from Figure 20.13. (Step 3) Pol δ on the lagging strand continues elongation. (Step 4) Displacement of the RNA portion of the primer strand by Pol δ. When Pol δ encounters the 5′-end of the RNA primer from the previous Okazaki fragment, it continues synthesis for 1–2 nucleotides, displacing a portion of the primer into a single-stranded RNA flap. (Step 5) Cutting or removal of the 5′ displaced flap by flap endonuclease 1, Fen1, and sealing of the nick by DNA ligase. Cleavage of the RNA flaps may occur in one of two pathways: (1) Fen1 cleaves short flaps, as shown here, or (2) long flaps are coated by the ssDNA-binding replication protein A, RPA, and sequentially cleaved by Dna2 nuclease and Fen1; this pathway is shown in Figure 20.15. (Adapted from Henneke G, Koundrioukoff S & Hübscher U [2003] *EMBO Rep* 4:252–256. With permission from John Wiley & Sons, Inc.)

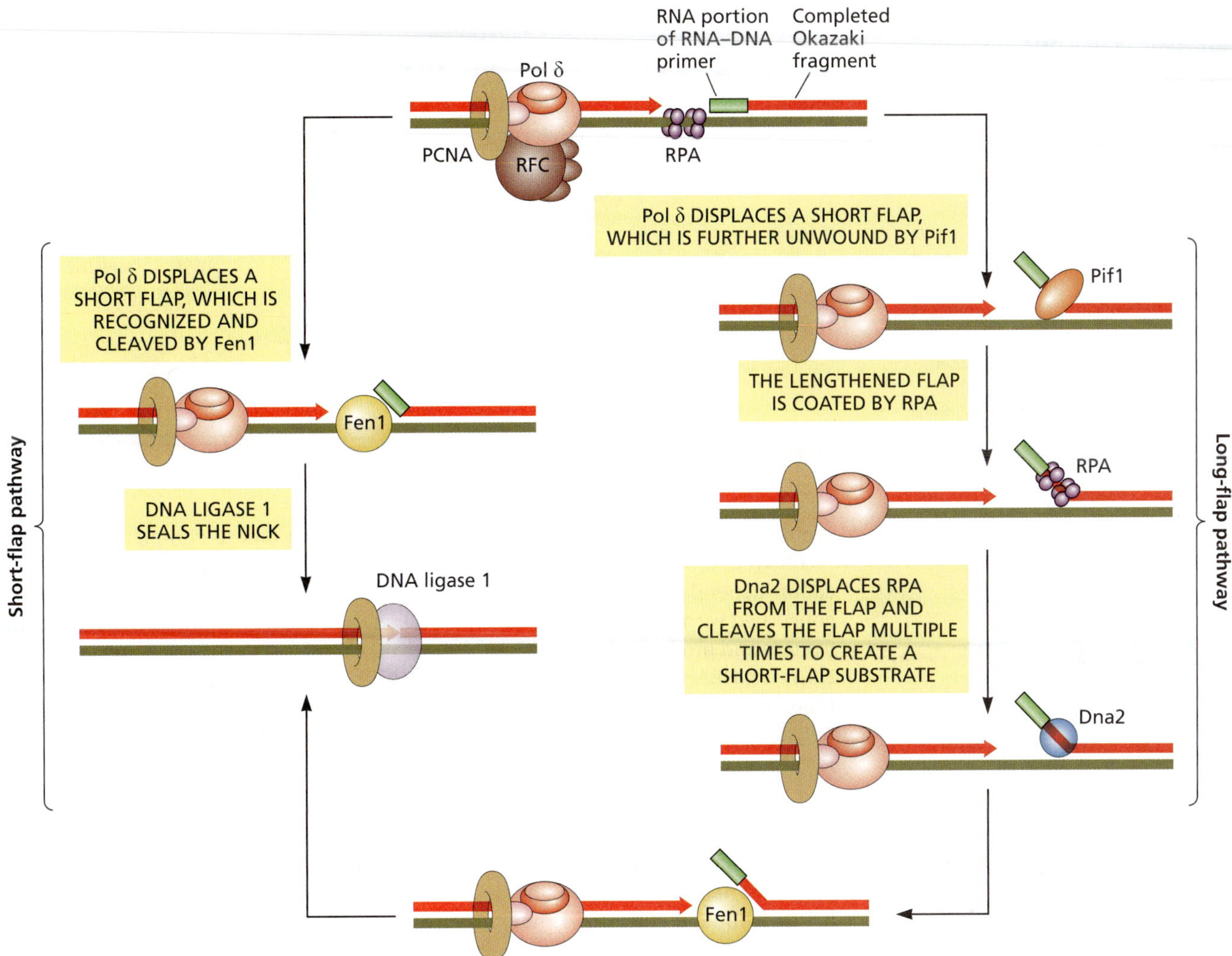

Figure 20.15 Mechanism of short- and long-flap pathways participating in Okazaki fragment maturation in eukaryotes. Pol δ consists of three subunits—Pol3, Pol31, and Pol32—in budding yeast and four subunits in fission yeast and metazoans; the fourth subunit functions to stabilize the holoenzyme. The other factors involved are PCNA or proliferating cell nuclear antigen, the clamp; RFC or replication factor C, the clamp loader; RPA or replication protein A, the ssDNA-binding protein complex; Pif1 or petite integration frequency 1, a 5′ → 3′ DNA helicase promoting rapid flap formation; Dna2, a helicase/nuclease, with the nuclease activity providing the essential function during lagging-strand DNA replication; Fen1 or flap endonuclease 1; and DNA ligase 1. Short- and long-flap pathways are depicted on the left and right sides of the schematic, respectively. (Adapted from Henneke G, Friedrich-Heineken E & Hübscher U [2003] *Trends Biochem Sci* 28:384–390. With permission from Elsevier.)

by transmitting H2A-H2B dimers across the fork and/or by recruiting newly synthesized dimers to the replication machinery.

Another chaperone, ASF1, has an affinity for H3-H4 and certainly can escort newly synthesized H3-H4 dimers to the fork. Its role with parental $(H3\text{-}H4)_2$ tetramers is controversial. During nucleosome dissolution, the linker histone H1 must also be lost, but it is apparently picked up by another chaperone, NASP. At present, we visualize the replication fork with a clear space ahead, formed by the dissociation of a parental nucleosome, and the histones from that nucleosome sequestered by appropriate chaperones (see Figure 20.16). The next question is, how is the complete chromatin structure reconstituted behind the fork on both daughter duplexes?

Both old and newly synthesized histones are required in replication

For a long time, it has been clear that parental histones are recycled randomly onto the leading and lagging strands. However, these histones represent at best only half the requisite amount needed for both daughter duplexes. New histones must be

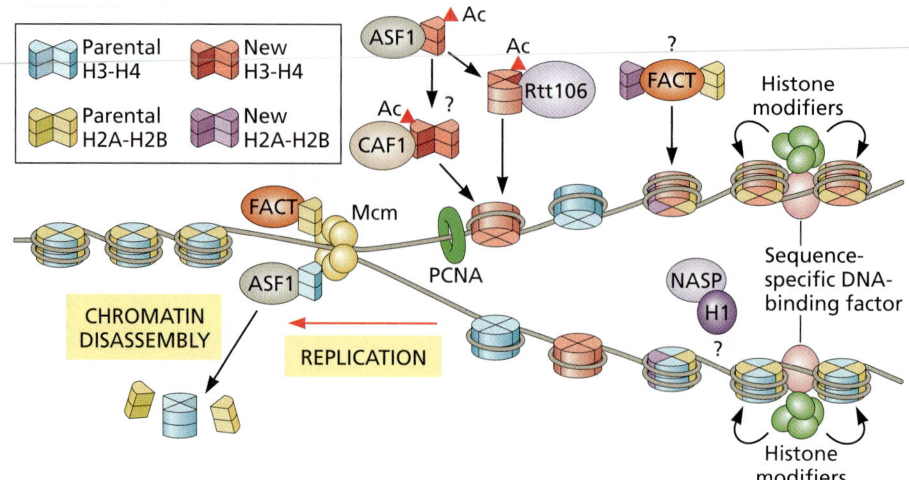

Figure 20.16 Model for replication of chromatin during DNA replication. Much of the model is speculative, as indicated by the question marks. For example, it does not deal with the uncertainty regarding the transfer of intact $(H3\text{-}H4)_2$ tetramers from parental chromatin. The model postulates that nucleosome modification in new chromatin is determined by sequence-specific DNA-binding factors, which then recruit specific histone modifiers, as well as chromatin remodelers, not shown here. After being so marked and modified, both daughter fibers are considered to be old chromatin. (Adapted from Ransom M, Dennehey BK & Tyler JK [2010] *Cell* 140:183–195. With permission from Elsevier.)

synthesized in the cell and delivered to the replication fork to complete chromatin reconstruction. There are two kinds of histone synthesis in most eukaryotes: replication-dependent synthesis and replacement synthesis. It is the former that we are concerned with at this point. Replication-dependent histone synthesis occurs only in S phase and produces the canonical histone variants that make up most nucleosomes: H3.1, H4, H2A, H2B, and the family of linker histone variants H1. The other set of histones are called replacement variants and include H3.3, CENP-A, H2A.Z, and H2A.X. Recall from Chapters 8 and 12 that these are synthesized throughout the cell cycle and may be inserted into preexisting chromatin.

From the structure of the nucleosome (see Chapter 8), it seems evident that histones H3 and H4 must be the first to be assembled on the daughter DNA duplexes, forming the $(H3\text{-}H4)_2$ tetramer, the core of the nucleosome. The tetrasome is a well-defined particle with DNA wrapped about $(H3\text{-}H4)_2$, but it seems that new H3 and H4 are first presented to the DNA as heterodimers bound to the chaperone ASF1. The X-ray structure of this complex (**Figure 20.17**) reveals that the interaction between ASF1 and H3-H4 blocks the histone surface that would otherwise interact with another H3-H4 dimer to form the tetramer. Thus, ASF1 cannot deliver an intact tetramer to start a new nucleosome. It has been suggested that another essential histone chaperone, CAF1, can carry the H3-H4 dimer partner and accept a dimer from ASF1, thereby producing a tetramer that can be transferred to newly replicated DNA (see Figure 20.16).

Table 20.2 Major roles of histone chaperones.

Histone cargo	Chaperone	Recognized functions in DNA replication	Additional interactions with
H3-H4	ASF1	chromatin assembly and disassembly; promotes H3K56 acetylation	CAF1, RFC, Mcm
	CAF1	chromatin assembly; heterochromatin silencing	ASF1, PCNA, Rtt106
	Rtt106	chromatin assembly; heterochromatin silencing	CAF1
H2A-H2B	FACT	chromatin assembly and disassembly	Mcm, RPA, DNA Pol I
	NAP1	chromatin assembly and disassembly	
H1	NASP	chromatin assembly	

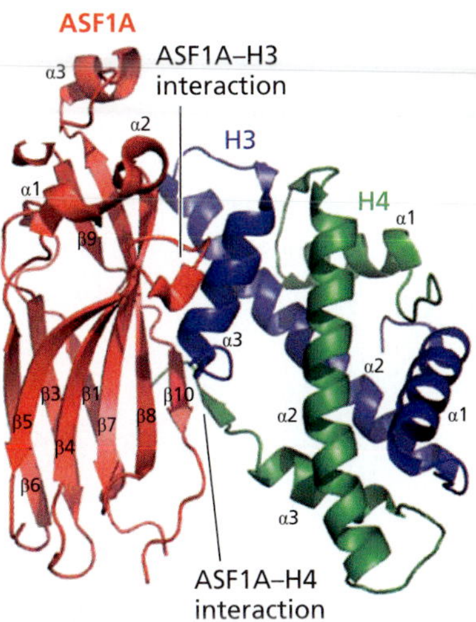

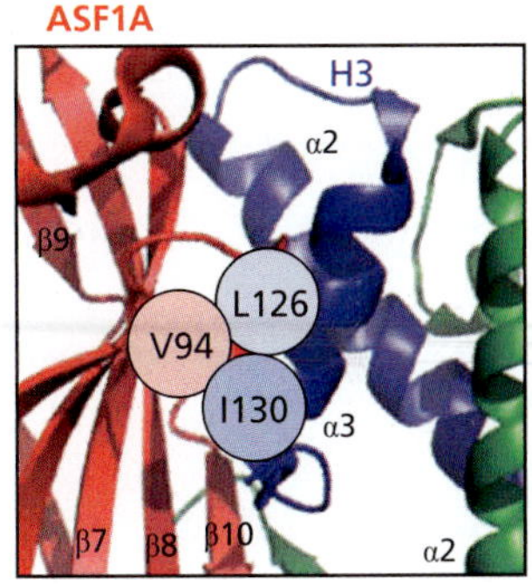

Figure 20.17 Structure of human ASF1A–H3-H4 triple complex. The box at the bottom shows a close-up view of the interaction regions, highlighting the hydrophobic residues that are important in the binding of ASF1A and H3-H4 dimer. Importantly, this part of the H3-H4 dimer surface is the same region that alternatively interacts with a second H3-H4 dimer in the nucleosome. Thus, ASF1 cannot carry a histone tetramer. (Adapted from De Koning L, Corpet A, Haber JE & Almouzni G [2007] *Nat Struct Mol Biol* 14:997–1007. With permission from Macmillan Publishers, Ltd.)

Such a mechanism seems reasonable for the insertion of all-new tetramers, but it does not answer the question of how parental tetramers are transferred to daughter duplexes behind the fork. Are they broken down by ASF1 and then reassembled as described above? Or are they transferred intact, perhaps via CAF1? The first mode might result in some mixed old–new tetramers, and there is evidence against this. Nevertheless, intact transfer might lead to difficulties in conserving epigenetic information. It must be emphasized that the cell has clear ways to tell old H3 and H4 from newly synthesized H3 and H4. Post-translational marks are attached to the new histones by acetylation at lysines 9, 14, and 56 on H3 and lysines 5 and 12 on H4. These modifications are removed as chromatin matures, but they are essential for proper chromatin replication.

Transfer of the linker histone, H1, has not been as thoroughly studied. There is evidence that H1 binding to the chaperone NASP (see Figure 20.16) facilitates loading of the histone onto H1-depleted chromatin.

Epigenetic information in chromatin must also be replicated

In Chapter 12, we saw that much of the information governing gene expression in transcription appears to be coded into modifications of chromatin structure. This information, which is specific to cell and tissue types in metazoans, must be preserved or sometimes specifically modified through somatic cell divisions. Such epigenetic information can be stored in chromatin in a variety of ways, from the placement of nucleosomes along the DNA and higher-order chromatin folding, to the placing of histone variants, to post-translational modifications of histones or methylation of DNA. How is such a variety of often very specific changes preserved through the wholesale process of DNA replication?

The positioning of nucleosomes on DNA is dictated, at least in part, by certain motifs in the DNA sequence itself, yet these do not appear to be strong determinants. It seems likely that the most favorable final positioning can be achieved only with the aid of remodeling factors, which are known to play a role in chromatin reconstitution *in vivo*. The placement of linker histones and non-histone proteins may also participate in nucleosome placement. A particular example is the protein HP1, which is abundant in heterochromatic regions.

Preservation of the multitude of very specific covalent marks, such as acetylation, methylation, phosphorylation, etc., is difficult to explain. Note that most of these marks are found on H3 and H4. When chromatin is being replicated, only the parental H3 and H4 will carry such modifications. These are to be distinguished from the special acetylation marking that is given to newly synthesized H3 and H4 to mark them as new. These marks are not a part of the epigenetic marking pattern and are removed in chromatin maturation. The question is, how are all of the other modifications that exist on the old histones reproduced on the new ones? Possible answers depend on the model assumed for the transmission of old H3-H4. If old tetramers are split into dimers before being transferred intact to daughter duplexes and each pairs with a new dimer, then all nucleosomes will have at least one set of markers. These could presumably recruit enzymes to the proper places to reconstruct the original pattern. There is, however, good evidence against mixed tetramers, so we must consider the other alternative. If H3-H4 tetramers are transferred intact from the parental strands, then half of the nucleosomes on the daughters will be properly marked, but half will be naïve. Marking these naïve tetramers would seem practicable only in the domain sense; a region could be acetylated in a certain way, for example, depending on the preponderance of such acetylation on old H3 and/or H4 in that domain. It is hard to see how very specific, nucleosome-to-nucleosome patterns could be regenerated, but we know little about the patterns at this level.

There is evidence that the processes by which chromatin receives its various epigenetic marks are interrelated and in some sense cooperative. For example, the formation of heterochromatic regions is favored by a very specific H3 methylation, which helps recruit the protein HP1. HP1, in turn, recruits the specific methylase to catalyze this modification. Thus, the formation of heterochromatin domains with their characteristic condensed structure can spread over broad regions.

20.5 The DNA end-replication problem and its resolution

Each chromosome in a eukaryotic nucleus contains a single, linear DNA duplex, which must have two ends. The existence of chromosome ends causes problems in replication, because they should suffer gradual shortening with each cycle of replication. The DNA end-replication problem was first recognized in the early 1970s, independently by James Watson and by Alexey Olovnikov, when it was realized that the requirement of all cellular DNA polymerases for a **primer** meant that DNA replicated by a lagging-strand mechanism would shorten, when the terminal RNA primer is degraded.

During S phase, linear chromosomal DNA is copied by replication forks that move from an interior position on the chromosome toward the ends. Leading-strand synthesis can theoretically copy the parental strand all the way to its last nucleotide. Discontinuous lagging-strand synthesis by polymerase/primase copies the respective parental strand (**Figure 20.18**), primed by RNA primers. The RNA primers are removed from each Okazaki fragment, and the internal gaps are filled in by extension of the discontinuous DNA and subsequent ligation. Removal of the most distal RNA primer, however, leaves a gap at the 5′-terminus, in the telomeric region of the chromosomes. Following subsequent rounds of DNA replication, this growing gap will result in progressively shorter daughter strands. Eventually, this erosion could extend into essential, coding regions of the genome. Additional erosion of telomeric DNA results from post-replicative processing of chromosomes ends. How does the cell deal with this problem?

Telomerase solves the end-replication problem

The key to understanding why chromosome ends are not progressively shortened came with the discovery that telomeres consist of multiple repeats of simple sequences. The subsequent discovery of **telomerase**, the RNA–protein enzyme complex that has the ability to add those sequences so as to elongate chromosome ends, provided the complete solution; see **Box 20.5** for a historical account of this Nobel Prize-winning discovery. Each telomerase complex contains a small RNA molecule characterized by a sequence complementary to that of the telomere repeat sequence (**Figure 20.19**). This portion of the RNA always remains single-stranded, despite the relatively complex secondary and tertiary structures of the rest of the RNA molecule.

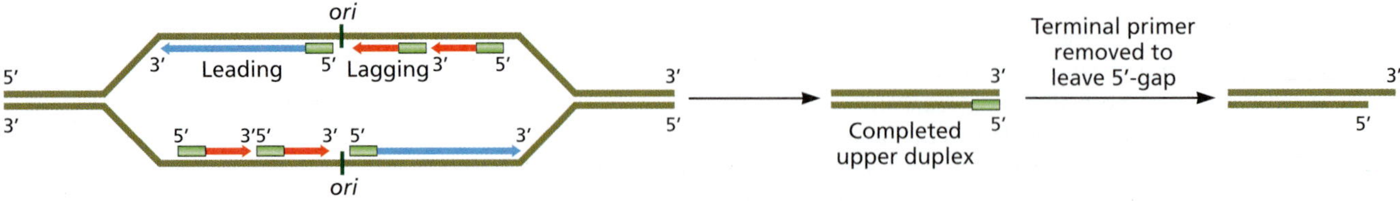

Figure 20.18 The DNA end-replication problem. The requirement of all cellular DNA polymerases for a primer should lead to shortening of DNA that is replicated by a lagging-strand mechanism, once the terminal RNA primer is degraded. This DNA end-replication problem is illustrated to the right for the upper of the two daughter duplexes. Following subsequent rounds of DNA replication, if only the semiconservative DNA replication machinery operates, as shown here, this gap will result in progressively shorter daughter strands.

Box 20.5 Telomeres, aging, and cancer The DNA end-replication problem resisted explanation for over a decade, until the groundbreaking work by Elizabeth Blackburn, in the lab of Joseph Gall. Blackburn and Gall cleverly chose the protozoan *Tetrahymena*, which has two nuclei, for their studies. Each *Tetrahymena* cell has a micronucleus, with five normal chromosomes, and a macronucleus, where the five chromosomes are chopped into hundreds of bits. This means lots of chromosome ends and thus telomeres. By 1978, Blackburn had shown that the telomeric DNA in *Tetrahymena* consisted of multitudes of repeats of the simple sequence TTGGGG. About this time, collaboration was established with Jack Szostak, who was able to demonstrate a similar but more complex situation in yeast.

With her own laboratory, and with a strong collaborator in Carol Greider, then a graduate student, Blackburn was ready to attack the question of how these repeated sequences were added to the DNA ends. Progress was rapid: in 1985 they demonstrated the existence of the enzyme telomerase; in 1987, they showed that it contained RNA; and two years later, they had sequenced the RNA and demonstrated that it could serve as a template to add the repeats successively. A few years later, Greider demonstrated that the enzyme was processive. But it took until 1996 before Joachim Lingner and Thomas Cech were able to purify the enzyme so that its structure could be determined. Blackburn, Greider, and Szostak were awarded the Nobel Prize in Physiology or Medicine in 2009 "for the discovery of how chromosomes are protected by telomeres and the enzyme telomerase."

Telomeres and aging

Most cell types are deficient in telomerase, except briefly in S phase; exceptions are germ cells, stem cells, and cancer cells. This means that most somatic cells are slowly losing telomere length as they repeatedly divide throughout life. When telomeres shrink beyond a certain limit, about 100 repeats, processes are triggered that lead to cell senescence and death. An obvious implication of this is that many of the degenerative processes we associate with aging may have their cause in wilting telomeres. Could a dose of telomerase provide extended life?

Many researchers have become fascinated with this idea. Initial studies were not encouraging. Mice in which the telomerase gene had been knocked out seemed to do very well without it. But then it was found that the mouse lines from which the knockout mice had been derived had unusually long telomeres. When the experiment was repeated on mice with human-length telomeres, degenerative disease and earlier death was marked. Most impressive are experiments in which telomerase has been switched on in aging mice. These mice lived about 40% longer than controls and had improved cognition and fertility. It must be emphasized that we have no evidence that these results extrapolate to humans.

Telomerase and cancer

Cancer is characterized by unlimited cell division, to the point at which certain cancer cell lines are virtually immortal. Apparently, the usual progression to senescence and death does not apply here. Furthermore, the great majority of cancer cells have high levels of telomerase. They express the telomerase gene constitutively, not just in S phase. All of this suggests that if we could find a nontoxic telomerase inhibitor, we might have a useful cancer drug. This perception has not been lost on drug companies, which are engaged in a massive competition for such a find. There are, in fact, promising candidates now undergoing clinical trials.

In addition, the protein part of the telomerase has the thumb-palm-fingers structure typically present in all DNA and RNA polymerases. Its reverse transcriptase activity synthesizes stretches of DNA using the telomerase RNA as a template. **Figure 20.20** depicts the proposed mechanism by which multiple copies of the telomere tandem repeats are synthesized by the telomerase. Multiple repeats are added by a **slippage** mechanism in which one repeat is synthesized and then the enzyme slips along and repositions itself at the new end of the chromosome and repeats the process. This continued re-extension of the telomeric DNA compensates for the chromosome-end

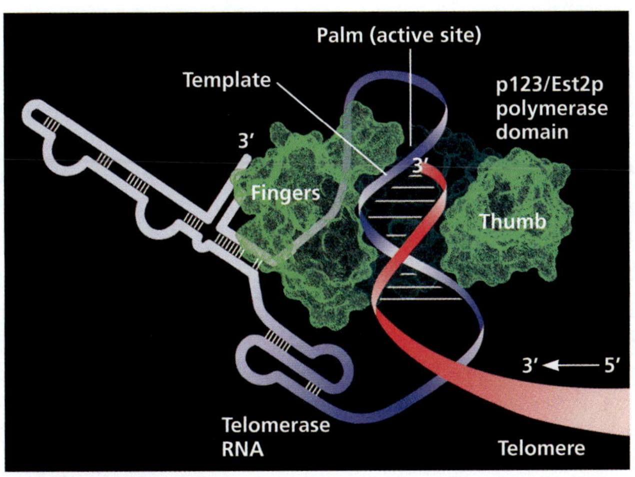

Figure 20.19 Telomerase contains both an RNA component and a protein part. The RNA molecule contains a sequence complementary to that of the telomere repeat sequence. Telomerase RNAs can vary in length from 146 to 1544 nt and they can adopt a characteristic secondary and, probably, tertiary structure; only the portion complementary to the telomere sequence is always single-stranded. It serves as a template against which multiple copies of the telomere tandem repeats are synthesized. (Adapted from Lingner J, Hughes TR, Shevchenko A et al. [1997] *Science* 276:561–567. With permission from American Association for the Advancement of Science.)

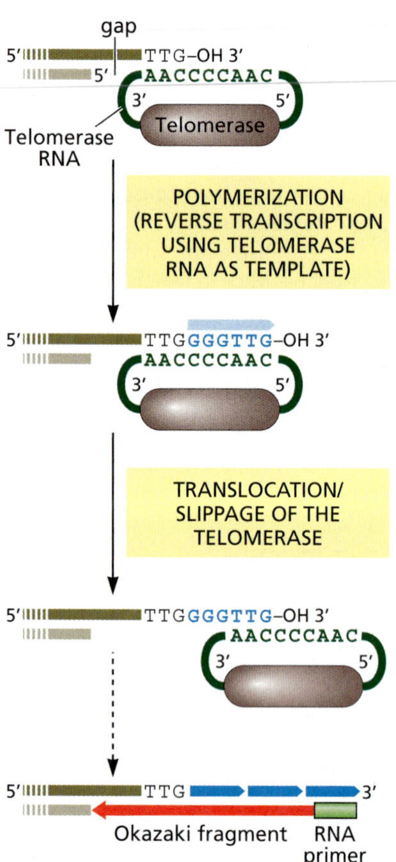

Figure 20.20 Proposed mechanism for synthesis of telomeric DNA. Each repeat can be added by a slippage mechanism, in which one repeat of telomeric ssDNA is synthesized and then the enzyme slips along, repositions itself at the new end of the chromosome, and repeats the process. The extended 3′-end can then act as template for new Okazaki fragment synthesis. Note that the telomere is extended but still has a 3′-overhang when the end RNA primer is removed. (Adapted from Greider CW & Blackburn EH [1989] *Nature* 337:331–337. With permission from Macmillan Publishers, Ltd.)

losses and prevents erosion into coding regions of telomere-proximal genes. If chromosomes are degraded below a critical telomere length of around 100 nucleotides, replicative potential is lost.

Alternative lengthening of telomeres pathway is active in telomerase-deficient cells

Most normal somatic cells do not need infinite replicative potential and hence repress telomerase activity to limit cell replication. By contrast, telomerase is up-regulated in many cancer cells (see Box 20.5), which enables their unlimited proliferation. Many types of cancer cells, however, are telomerase-minus. In such cells **homologous recombination** or **HR** is used to increase telomere length through the alternative lengthening of telomeres or ALT pathway. ALT produces highly heterogeneous telomere lengths, as seen in the fluorescent images of chromosomes in ALT-positive cells (**Figure 20.21A**).

ALT functions in the context of the **shelterin complex** that is present at all telomere ends (see Figure 8.30). The shelterin complex includes TRF1 and TRF2 dimer proteins that bind to the double-stranded portion of the telomere and POT1 protein that binds to the single-stranded G-rich overhang, as well as TIN2 and TPP1 proteins that serve to bridge these proteins. POT1 suppresses normal DNA repair processes that would recognize the single-stranded overhang as DNA damage. Shelterin also protects the ends from degradation.

The mechanism(s) involved in ALT are very poorly understood. It is believed that ALT occurs through the formation of an **intertelomeric D-loop** in which the 3′-overhang of one telomere invades the telomeric duplex of another chromosome (**Figure 20.21B**); lengthening occurs by use of the sequence information in the second chromosome. The resulting intertelomeric D-loop requires resolution, which involves helicases such as **Werner (WRN)** and **Bloom (BLM) helicases**. WRN and BLM interact physically and functionally with the critical telomere-binding and maintenance protein TRF2, a subunit of shelterin. Mutations in WRN and BLM lead to the premature-aging Werner syndrome and to Bloom syndrome, respectively. These diseases are covered in Chapter 22, where the roles played by these two helicases in homologous recombination are discussed in more detail.

It is known for certain that ALT requires dimers of two coiled-coil proteins SMC5 and SMC6, as well as a set of proteins that interact with them (**Figure 20.22**). Two other proteins belonging to the structural maintenance of chromosomes or SMC family of proteins, SMC1 and SMC3, form the backbone of cohesin (see Figure 8.25). Cohesin is the complex that forms in S phase around the two sister chromatids in a replicated chromosome to keep these chromatids together until mitosis. Two other SMC proteins, SMC2 and SMC4, are the major structural component of condensin I, the complex that is instrumental in the formation and maintenance of the compacted structure of mitotic chromosomes.

The SMC5–SMC6 complex contains a protein subunit, MMS21/NSE1, which is an E3 SUMO ligase. It is believed that this subunit sumoylates numerous components of the **promyelocytic leukemia bodies** or **PML**, the cytologically recognizable structures where ALT occurs (see Figure 20.22). PML bodies facilitate the homologous recombination process by bringing together shelterin-decorated telomeres, the SMC5–SMC6 complex, and the numerous proteins that perform the homologous recombination reaction.

Figure 20.21 Alternative lengthening of telomeres pathway, ALT, functions in some cancer cells that are telomerase-minus. In telomerase-minus cancer cells, homologous recombination or HR is used to increase telomere length through the alternative lengthening of telomeres or ALT pathway. (A) Visualization of telomeres in telomerase-positive cells, top, and ALT cells, bottom, by fluorescence *in situ* hybridization, FISH. Fluorescently labeled probes for the telomeric DNA sequence were hybridized to metaphase chromosome spreads. In telomerase-expressing cells, telomere length is very homogenous and telomeres are found at the ends of all chromosomes, but in ALT cells, telomeres are very heterogeneous in length and some chromatids lack telomeres. (B) Current ALT model: the 3′-telomeric tail on one telomere invades the telomeric duplex of another chromosome. Lengthening occurs by use of the information in the second chromosome. (A, from Chung I, Osterwald S, Deeg KI & Rippe K [2012] *Nucleus* 3:263–275. With permission from Taylor & Francis.)

(A)

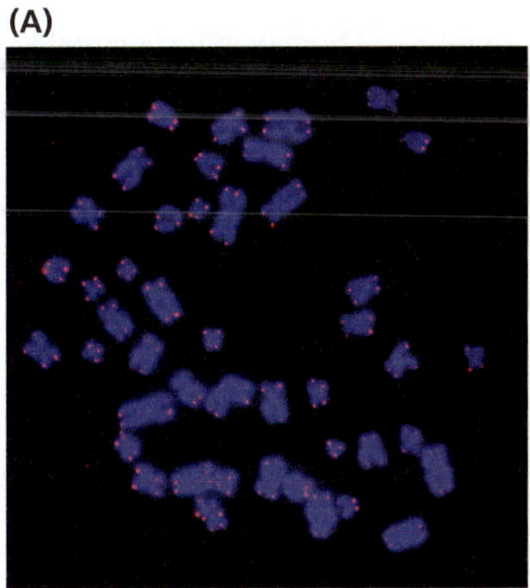

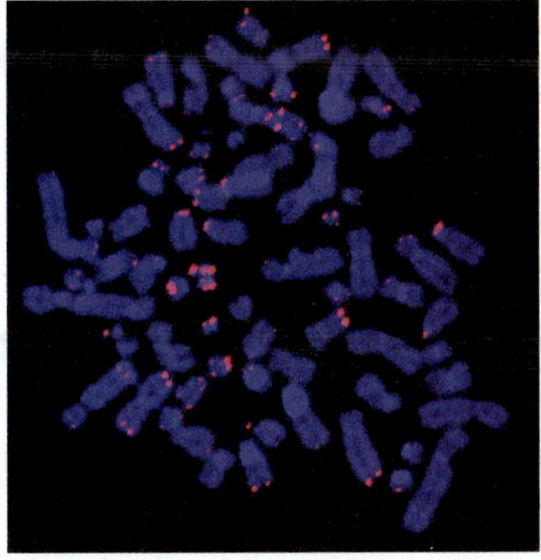

(B)

20.6 Mitochondrial DNA replication

Mitochondria and chloroplasts contain their own DNA. We discuss the more thoroughly studied mitochondrial genome. Although the size of the mitochondrial genome accounts for only 0.0005% of the genome in humans, it is densely packed with ribosomal and tRNA genes, as well as essential genes that code for components of the respiratory chain. Together, these genes make up almost 0.1% of the total number of human genes. Because mitochondria are so universally important and because they are presumably derived from bacterial symbionts, their DNA and its replication are of considerable interest. Like bacteria, mitochondria do not have histones or the chromatin structure of eukaryotic nuclear DNA. Instead, mitochondrial DNA, mtDNA, is condensed with HU-like proteins similar to those of bacteria.

Are circular mitochondrial genomes myth or reality?

For years, the established view on the physical structure of mitochondrial genomes was that they were circular, like those of bacteria. For the purpose of discussing the mechanisms of mitochondrial genome replication, we have chosen to present this conventional view of the mitochondrial genome in human cells, with its resident RNA- and protein-coding genes. The D-loop region is believed to contain promoters and replication origins and to interact with certain proteins that maintain the structure (**Figure 20.23**).

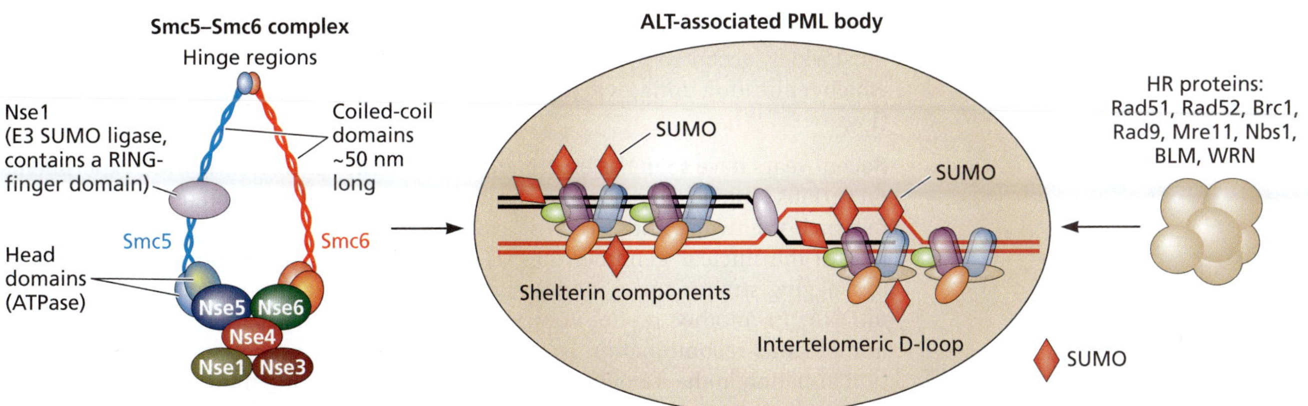

Figure 20.22 Smc5–Smc6 and promyelocytic leukemia or PML bodies in ALT. Smc5–Smc6 plays roles in double-strand break repair, restart of collapsed replication forks, and maintenance of rDNA integrity. The structure presented here is hypothetical and is based on the Smc1–Smc3 cohesin complex. NSE stands for non-SMC element proteins; Nse5 and Nse6 are not found in humans. Loading of Smc5–Smc6 to chromatin is likely coupled to replication. In cells undergoing ALT, telomeres associate with PML bodies known as ALT-associated PML bodies or APBs. PML bodies are dynamic nuclear structures that are involved in numerous cellular processes. They facilitate post-translational modifications and can localize proteins to their sites of action. Many components of PML bodies are sumoylated. Sumoylation affects protein stability, protein–protein interactions, and subcellular localization. PML bodies facilitate homologous recombination, HR, by bringing together telomeres, the Smc5–Smc6 complex, and HR proteins. In ALT cells, the Smc5–Smc6 complex and HR proteins associate with PML bodies in the G_2 phase, when sister chromatids are available for HR. Sumoylation of shelterin components RAP1, TIN2, TRF1, and TRF2 by MMS21 recruits or maintains telomeres at APBs and promotes telomere HR. (Adapted from Murray JM & Carr AM [2008] *Nat Rev Mol Cell Biol* 9:177–182. With permission from Macmillan Publishers, Ltd.)

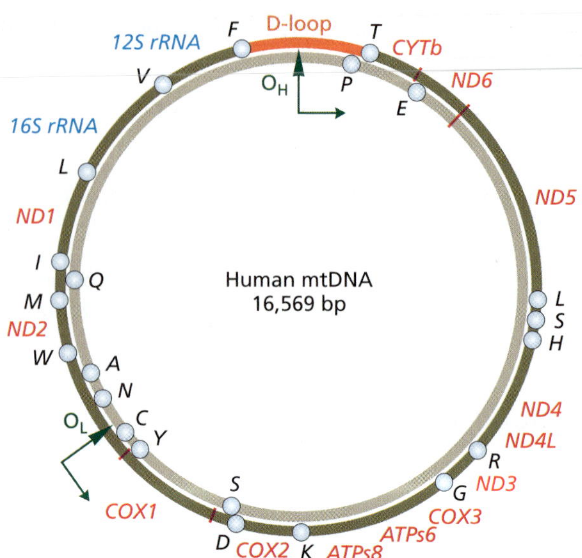

Figure 20.23 Human mitochondrial genome in the conventional circular representation. Mitochondrial DNA encodes 13 essential protein components of the respiratory chain, shown in red lettering. Two rRNA genes, 12S and 16S rRNA, and 22 tRNA genes, denoted by blue dots and the single-letter amino acid code, are interspaced among the protein-coding genes. These provide the necessary RNA components for protein synthesis inside the mitochondrion. The D-loop, a 1.1 kb noncoding region, is involved in the regulation of transcription and replication of the molecule and is the only region not directly involved in the synthesis of respiratory-chain polypeptides. The D-loop contains a third polynucleotide strand believed to be the initial segment generated by replication of the heavy strand. Apparently, it arrests shortly after initiation and is often maintained for some period in that state. O_H and O_L are the origins of heavy- and light-strand mtDNA replication. The origin of L-strand replication is displaced by approximately two-thirds of the genome and is located within a cluster of five tRNA genes. (Adapted, courtesy of Center for the Study of Mitochondrial Pediatric Diseases.)

More recently, though, this notion has been shaken to its core by the advent of new data acquired by new methodology. In the yeast *Candida albicans*, some plants, and human heart cells, the mitochondrial genomes are complex networks of highly branched, variably sized subfragments. Genome-size molecules are very rare and circular molecules are not detected at all in such preparations (**Figure 20.24**).

Models of mitochondrial genome replication are contentious

There are very few examples of processes that have been so controversial as the replication of the mitochondrial genome, and, to this day, the mechanism remains unresolved. The only point of agreement among researchers is that the mitochondrial genome is replicated by **Pol** γ, a homotetramer with polymerase and 3′ → 5′ exonuclease activities and high processivity. In terms of its biochemical activities, Pol γ is similar to the Pol III core in bacteria and to Pol δ, the lagging-strand polymerase in eukaryotes. Early replication models proposed that replication occurred via the formation of a θ structure, similar to that found in the bidirectional replication of λ phage DNA (see Chapter 19). Further research led to the most widely accepted transcription-initiation strand-displacement model or SD, in which replication initiates at two unidirectional origins, each specific for one strand (**Figure 20.25**).

Recent years have seen the emergence of more models resulting from experiments on various biological systems. These include the strand-coupled replication model, in which the two strands are synthesized simultaneously, very much like replication of the nuclear genome; the RITOLS model, RNA incorporation throughout the lagging strand; and a model that proposes initiation from various sites along mtDNA. In another model known as recombination-driven replication or RDR, homologous recombination is considered as a likely process for mtDNA replication initiation in the mitochondria of yeast and plants, and more recently in human heart cells.

Complex networks of linear molecules like those found in the mitochondria of human heart tissue seem to be replicated by both RDR and SD mechanisms, with SD occurring at the branch points, exposing single-stranded regions that persist in mitochondrial and chloroplast genomes. Interestingly, RDR and SD synthesis may account for the abnormal process of telomere lengthening in some cancer cells.

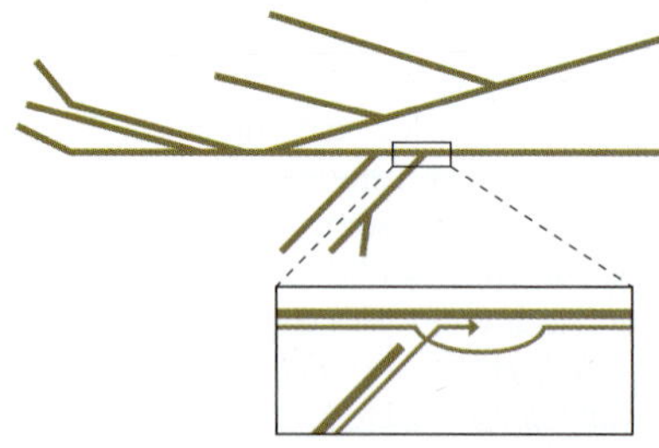

Figure 20.24 Alternative structures of chloroplast and mitochondrial genomes. Bushy, multigenomic chromosomes in plant chloroplasts are shown. Strand-displacement synthesis occurs at the branch points. (Adapted from Bendich AJ [2010] *Mol Cell* 39:831–832. With permission from Elsevier.)

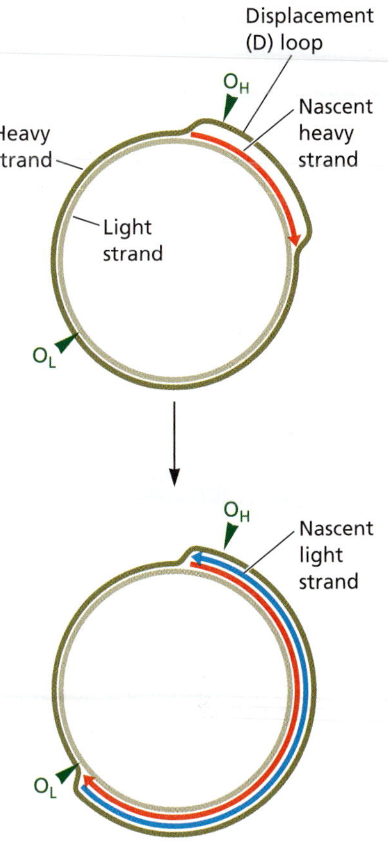

Figure 20.25 Strand-displacement model for circular mitochondrial DNA replication. Replication is continuous, unidirectional from both origins of replication, and begins in the D-loop region. The two DNA strands of the mitochondrial genome are distinguishable by their density during centrifugation and are thus termed heavy and light. As replication commences, the parental heavy strand is displaced by the nascent heavy strand, forming the displacement or D-loop that has been visualized by electron microscopy. When replication of the heavy strand has progressed around a substantial part, nearly two-thirds, of the circular molecule, replication of the light strand begins from its own origin. The initiation requires RNA priming: remnants of inefficient primer excision give rise to the few ribonucleotides frequently found with mature mtDNA. (Adapted from Brown TA, Cecconi C, Tkachuk AN et al. [2005] *Genes Dev* 19:2466–2476. With permission from Cold Spring Harbor Laboratory Press.)

20.7 Replication in viruses that infect eukaryotes

The large number of viruses that infect eukaryotic organisms maintain their genomes in a number of different ways. Some, like SV40, have circular double-stranded DNA genomes, so they can replicate by the θ mechanism in the manner we have described.

RNA viruses, however, have only RNA for a genome. These include many of the common human pathogens, causing illnesses from the common cold to **AIDS** to cancer. These viruses may have single-stranded or double-stranded RNA genomes, and there are a remarkable variety of mechanisms to replicate them. We discuss here an important class of RNA viruses, the **retroviruses**, because they use a DNA intermediate in replication.

Retroviruses use reverse transcriptase to copy RNA into DNA

A very unusual replication mechanism has evolved for replication of the RNA genomes of retroviruses. These viruses have an ssRNA molecule as a genome. The retroviral life cycle involves the integration of a dsDNA copy of this ssRNA viral genome into the genome of the host. The synthesis of the dsDNA copy of the RNA genome requires the action of reverse transcriptase or RT, an enzyme whose discovery was initially met with much skepticism and even derision. The fascinating story of this discovery, a Nobel Prize-winning feat, is presented in **Box 20.6**.

Box 20.6 Retroviruses and reverse transcriptase Retroviruses are an unusual class of RNA viruses that are capable of integrating a DNA copy of their RNA into the host genome, allowing their retention even through cell division. They include the **human immunodeficiency virus type 1, HIV-1,** as well as a number of oncogenic viruses. The history of the discovery of retroviruses goes all the way back to 1911, when Peyton Rous discovered that an extract from chicken muscle tumors was capable of inducing tumors in other chickens. Remarkably, the infective material was an example of the recently discovered viruses.

In the late 1950s, Howard Temin, a graduate student of Renato Dulbecco, began to reveal the true and unusual nature of the Rous sarcoma virus, RSV. Until then, tumor viruses could be assayed only by awkward, nonquantitative *in vivo* techniques. Together with a postdoctoral fellow, Harry Rubin, Temin devised a simple *in vitro* method for following viral transformation in cell culture.

Although it was well known that the infecting virus contained only an RNA genome and no DNA, infection induced morphological changes in the host cell that were determined by viral genetic information. These changes were heritable through cell division, and therefore encoded in DNA. Furthermore, inhibitors of transcription blocked the formation of new viruses, indicating reading from DNA; and inhibition of DNA synthesis early in infection blocked the whole virus life cycle. On the basis of this evidence, in 1964 Temin proposed the provirus hypothesis, which stated that the genetic information from the virus was somehow present in the infected cell in the form of DNA.

This posed a huge problem. The provirus could exist only if there were a mechanism to transcribe RNA into DNA. According to most scientists' reading of the central dogma, the flow of genetic information in the cell occurred only from DNA to RNA to protein. At any rate, Temin's hypothesis was initially met with skepticism, even derision. His ideas finally received the credit due in 1970 when Temin and David Baltimore, working independently, simultaneously demonstrated the existence of an enzyme in retroviruses, termed reverse transcriptase, that could transcribe RNA into a DNA copy. In 1975, Temin, Baltimore, and Dulbecco were awarded the Nobel Prize in Physiology or Medicine "for their discoveries concerning the interaction between tumour viruses and the genetic material of the cell." It seems appropriate that Rous had also received the Nobel Prize: in 1966, 55 years after his discovery.

The genome and structure of a typical retrovirus, HIV, are depicted schematically in Figure 8.2. The viral genome carries three classes of genes: *gag* genes, which code for internal capsid proteins; *pol*, which provides functional enzymes; and *env*, coding for envelope proteins. There are usually two copies of the RNA genome within the capsid, accompanied by the reverse transcriptase protein.

(Continued)

Box 20.6 *(Continued)*

The life cycle of the virus is depicted in Figure 1. Upon infection of a host cell by fusion with its membrane, the reverse transcriptase is activated and replicates one complementary strand of DNA. The primer for this replication is, surprisingly, a tRNA from the host. The transcribed genomic RNA is then degraded by the RNase H function of the reverse transcriptase (see Figure 20.27) which must be activated by a virus-coded protease. This is why protease inhibitors can function as anti-HIV drugs. The single-stranded DNA can then be copied by reverse transcriptase to form a complementary strand, resulting in the formation of a duplex DNA, the provirus.

The provirus has long terminal repeats or LTRs (see Figure 20.26), which promote its insertion into the genome of the host cell. The LTRs can also function as efficient promoters for transcription of the viral genome, which leads to the production of more virus particles, which can then infect other cells. Because the provirus may insert at many sites in the genome and may then be rearranged or mutated, it may influence the activity of host genes, including those involved in cell growth and regulation. This is why some retroviruses can be oncogenic.

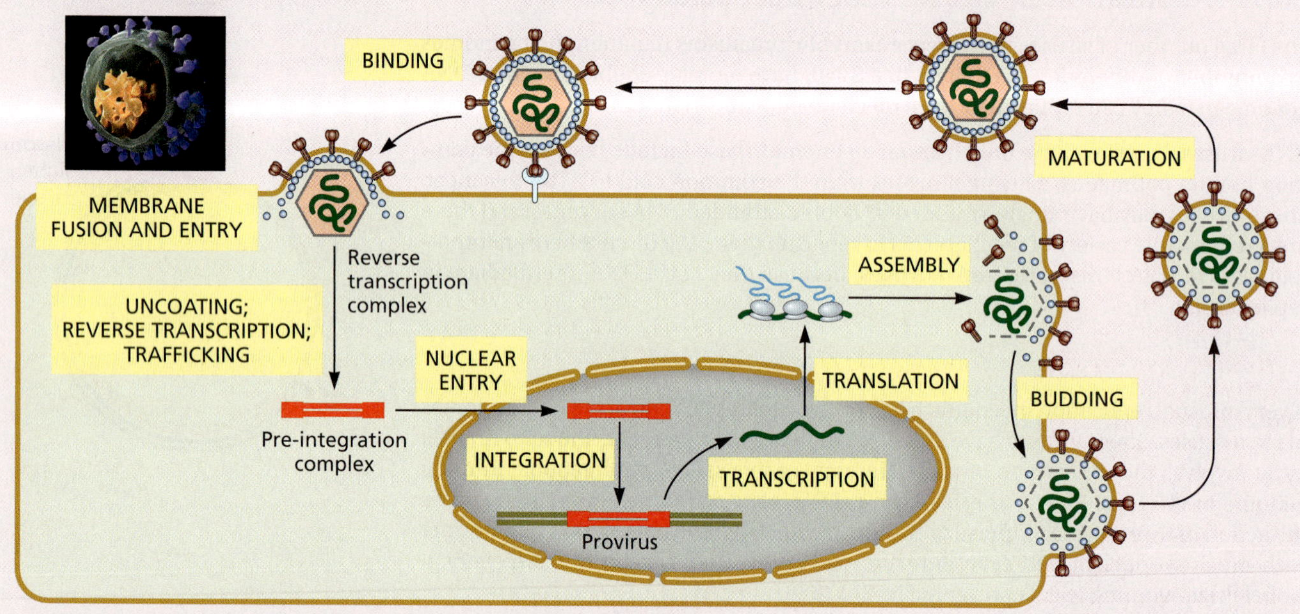

Figure 1 Retroviral life cycle. The stages are depicted as arrows connecting the molecules and complexes. Interactions between viral and host-cell factors occur at every stage of the viral life cycle, although many are still poorly understood. The image shows the three-dimensional structure of a single simian immunodeficiency virus, SIV, obtained by cryo-electron tomography; the architecture of the SIV surface spikes, shown in blue, is similar to that of HIV. (Adapted, courtesy of Kate Bishop, Francis Crick Institute. Inset, top left, from White TA, Bartesaghi A, Borgnia MJ et al. [2010] *PLoS Pathog* 6:e1001249.)

The process of creating the dsDNA copy of the retroviral genome is extremely complex (Figure 20.26). It involves the synthesis of two distinct single strands of DNA, a (–)-strand and a (+)-strand. The (–)-strand is synthesized first by using the viral RNA genome as a template; the (+)-strand is a copy of the (–)-strand and is initiated later in the process, while the synthesis of the (–)-strand is still ongoing. Complicated as it is, synthesis of the dsDNA copy depends on mechanisms and polymerase activities with which we are already familiar: the incoming NTP is attached to the free 3′-OH group of the nascent polynucleotide chain, primers are needed for chain initiation, and the RT enzyme possesses the thumb-palm-fingers structure typical of polymerase enzymes (Figure 20.27). The DNA copy that is ultimately synthesized possesses two **long terminal repeats** or **LTRs** that are not present in the viral genome. These LTRs are required for integration of the dsDNA into the host genome.

There are, however, several features that are unique to RT activities that deserve special mention (see Figure 20.27). First, two different kinds of primers are used for synthesis of the two strands. The (–)-strand uses a host tRNA molecule as a primer, whereas the (+)-strand is initiated on a specific sequence of the viral RNA, the polypurine tract or PPT. This sequence is not susceptible to cleavage by the RNase H activity of the RT, which is responsible for degrading the original RNA strand immediately after it has been copied into the (–)-strand.

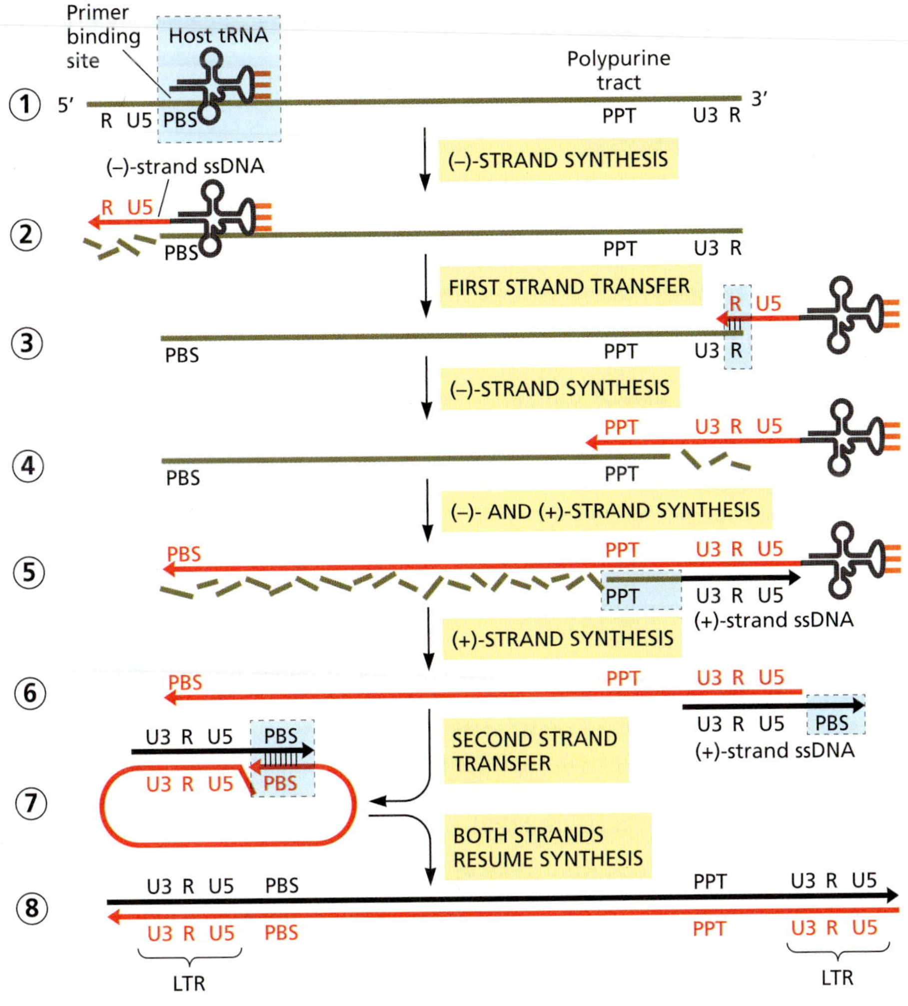

Figure 20.26 Expansion of HIV genomic RNA into DNA to create LTRs. The unusual features of the process are boxed in yellow. The steps are as follows: (Step 1) Formation, by reverse transcription, of a complementary ssDNA strand, the (–)-strand, is initiated on the 3′-end of a primer of host tRNA. (Step 2) Extension of the primer continues to the 5′-end of the RNA genome to generate (–)-strand ssDNA; the 5′-end of the RNA is degraded by RNase H. (Step 3) (–)-Strand DNA is translocated to the 3′-end of genomic RNA to complete (–)-strand DNA synthesis; base-pairing between R at the 3′-end of the RNA genome and R at the 3′-end of the (–)-strand ssDNA mediates the translocation reaction. (Step 4) (–)-Strand DNA synthesis continues; the RNA strand of the hybrid formed between the (–)-strand DNA and the 3′-end of the genomic RNA is susceptible to the action of RNase H, which degrades the RNA strand. (Step 5) (–)-Strand DNA synthesis continues; (+)-strand synthesis is initiated from the PPT or polypurine tract fragment of genomic RNA. PPT is resistant to RNase H action and serves as an efficient primer for the initiation of (+)-strand synthesis. (Step 6) (+)-Strand synthesis copies a portion of the primer tRNA to create a DNA copy of the primer binding site, PBS, at the 3′-end of (+)-strand ssDNA; the primer tRNA is removed by RNase H. (Step 7) (+)-Strand ssDNA must transfer to the 3′-end of (–)-strand DNA to complete viral replication; the transfer is mediated by base-pairing between the complementary copies of PBS on (+)- and (–)-strand DNA. (Step 8) Following the second strand transfer, both strands resume DNA synthesis, with each strand using the other as a template until the dsDNA, with the LTR ends needed for integration into the cellular genome, is fully synthesized. (Adapted from Basu VP, Song M, Gao L et al. [2008] *Virus Res* 134:19–38. With permission from Elsevier.)

Second, there are two specific translocation events, when portions of already synthesized single strands are transferred from one end of the respective template to the other end. These translocation events occur because of the linear nature of the templates and the fact that synthesis is initiated at sites located close to the end of the templates; thus, the newly synthesized stretch of ssDNA must jump to the other end of the template to complete the synthesis. How can this occur? The answer lies in the

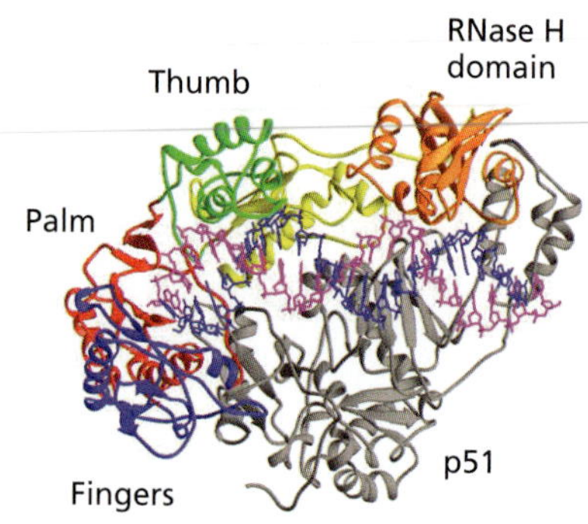

Figure 20.27 Structure of HIV reverse transcriptase complexed with RNA and DNA. Reverse transcriptase, RT, is a heterodimer of two subunits, p66 and p51. The latter is shown in gray. The connection between the two catalytic domains is shown in yellow, and the RNase H domain is shown in orange. The DNA primer strand is colored pink; the RNA template strand is colored blue. The nucleic acids are bent near the thumb of p66. The p66 subunit has an N-terminal polymerase domain and a C-terminal RNase H domain. The polymerase domain catalyzes polymerization of the (–)-strand ssDNA, which is a copy of the RNA genome, and its further conversion to dsDNA. The RNase H domain catalyzes the degradation of the RNA template. The polymerization domain includes the three subdomains—fingers shown in blue, palm shown in red, and thumb shown in green—present in all DNA and RNA polymerases. These clasp the RNA–DNA duplex during the process of RNA-dependent DNA polymerization. The fingers and the thumb form the walls of the nucleic acid binding cleft; the palm domain contains the active site. In the polymerization mode, RT interacts with both RNA and DNA and with incoming dNTPs, with the polymerase active site aligned at the 3′-end of a primer. In the RNA-cleavage mode, it interacts with RNA–DNA, with the active site of the RNase H domain aligned with the RNA strand. p51 is produced by proteolytic cleavage of p66 and corresponds to the polymerase domain of p66, although it does not possess polymerase activity. It stabilizes p66 and binds the tRNA primer. (Courtesy of Stephen Hughes, National Institutes of Health.)

existence of sequences that are repeated in the template and can serve as sites where base-pairing between the ssDNA and its respective template can occur, presumably upon transient circularization of the molecule. For example, the first translocation event involves the R sequences located at the 5′- and 3′-ends of the RNA; copying of the 5′-R sequence into the (–)-strand DNA creates the complementarity needed for interaction between the (–)-strand and the R sequence at the 3′-end of the RNA template. Similar complementarity and base-pairing is utilized during the second translocation event (see step 7 in Figure 20.26).

Key concepts

- Replication in eukaryotes proceeds bidirectionally from many origins. In budding yeast these origins have a defined consensus sequence, but in higher eukaryotes they do not.
- Origins are often clustered in loci, and different ones fire at different times in S phase.
- Origins are made competent or licensed for replication by binding an origin recognition complex, which recruits other proteins to form a pre-replicative complex; this in turn is transformed into a pre-initiation complex, which can accept a polymerase and begin elongation.
- Because many origins are functioning simultaneously, it is essential to have mechanisms that prevent refiring of an origin once it has been copied.
- Elongation in eukaryotes has many similarities to elongation in bacteria but uses three polymerases: one for priming, one for the leading strand, and one for the lagging strand.
- Chromatin structure introduces many complications into eukaryotic replication: nucleosomes must be disassembled ahead of the fork and reassembled on both daughter duplexes.
- Even though old histones are recycled, an equal quantity of new histones must be provided. These are synthesized during S phase and carry acetylation marks at specific lysine residues to mark them as new.
- Histones are transported to the newly replicated DNA by a battery of specific histone chaperones.
- It is also necessary that nucleosome arrangement and epigenetic marking be re-established on the new chromatin. How this occurs is not yet fully understood.

- Termination of replication of linear chromosomes also presents problems; removal of the final RNA primer from the lagging strand leaves a 5′-gap, which would, with each replication cycle, shorten chromosomal ends. This would mean that the genome would be permanently damaged after a number of cell divisions.
- The problem is resolved by the existence of telomeres, which are multiple copies of a simple repeat at the ends of chromosomes. As the telomeres are shortened, they can be restored by an enzyme known as telomerase or by a recombinant pathway termed alternative lengthening of telomeres, ALT.
- Mitochondria and chloroplasts of eukaryotic cells have DNA that is not sequestered by histones, and they require only one polymerase to replicate, in a fashion more like bacterial genomes.
- An unusual form of replication is found in retroviruses. These have an ssRNA genome that is replicated into dsDNA, in a process catalyzed by an enzyme called reverse transcriptase. The dsDNA copy of the RNA genome can then be integrated into the cellular genome, where it can replicate as part of the cellular DNA.

Further reading

Books

Blow JJ (ed) (1996) Eukaryotic DNA Replication. Oxford University Press.

Cox LS (ed) (2009) Molecular Themes in DNA Replication. RSC Publishing.

DePamphilis ML (ed) (2006) DNA Replication and Human Disease. Cold Spring Harbor Laboratory Press.

DePamphilis ML & Bell SD (2010) Genome Duplication: Concepts, Mechanisms, Evolution, and Disease. Garland Science.

Kušić-Tišma J (ed) (2011) Fundamental Aspects of DNA Replication. InTechOpen.

Reviews

Aladjem MI (2007) Replication in context: Dynamic regulation of DNA replication patterns in metazoans. *Nat Rev Genet* 8: 588–600.

Annunziato AT (2005) Split decision: What happens to nucleosomes during DNA replication? *J Biol Chem* 280:12065–12068.

Aparicio T, Ibarra A & Méndez J (2006) Cdc45-MCM-GINS, a new power player for DNA replication. *Cell Div* 1:18.

Balakrishnan L & Bambara RA (2011) Eukaryotic lagging strand DNA replication employs a multi-pathway mechanism that protects genome integrity. *J Biol Chem* 286:6865–6870.

Basu VP, Song M, Gao L et al. (2008) Strand transfer events during HIV-1 reverse transcription. *Virus Res* 134:19–38.

Bendich AJ (2010) The end of the circle for yeast mitochondrial DNA. *Mol Cell* 39:831–832.

Botchan M (2007) Cell biology: A switch for S phase. *Nature* 445:272–274.

Boye E & Grallert B (2009) In DNA replication, the early bird catches the worm. *Cell* 136:812–814.

Burgers PM (2009) Polymerase dynamics at the eukaryotic DNA replication fork. *J Biol Chem* 284:4041–4045.

Burgess RJ & Zhang Z (2010) Histones, histone chaperones and nucleosome assembly. *Protein Cell* 1:607–612.

Chagin VO, Stear JH & Cardoso MC (2010) Organization of DNA replication. *Cold Spring Harb Perspect Biol* 2:a000737.

Duderstadt KE & Berger JM (2008) AAA+ ATPases in the initiation of DNA replication. *Crit Rev Biochem Mol Biol* 43:163–187.

Errico A & Costanzo V (2010) Differences in the DNA replication of unicellular eukaryotes and metazoans: Known unknowns. *EMBO Rep* 11:270–278.

Gilbert DM (2010) Evaluating genome-scale approaches to eukaryotic DNA replication. *Nat Rev Genet* 11:673–684.

Hanawalt PC (2007) Paradigms for the three Rs: DNA replication, recombination, and repair. *Mol Cell* 28:702–707.

Hayashi MT & Masukata H (2011) Regulation of DNA replication by chromatin structures: Accessibility and recruitment. *Chromosoma* 120:39–46.

Henneke G, Koundrioukoff S & Hübscher U (2003) Multiple roles for kinases in DNA replication. *EMBO Rep* 4:252–256.

Hiratani I, Takebayashi S, Lu J & Gilbert DM (2009) Replication timing and transcriptional control: Beyond cause and effect—part II. *Curr Opin Genet Dev* 19:142–149.

Johnson A & O'Donnell M (2005) Cellular DNA replicases: Components and dynamics at the replication fork. *Annu Rev Biochem* 74:283–315.

Kaguni LS (2004) DNA polymerase γ, the mitochondrial replicase. *Annu Rev Biochem* 73:293–320.

Kunkel TA & Burgers PM (2008) Dividing the workload at a eukaryotic replication fork. *Trends Cell Biol* 18:521–527.

Machida YJ, Hamlin JL & Dutta A (2005) Right place, right time, and only once: Replication initiation in metazoans. *Cell* 123:13–24.

Margueron R & Reinberg D (2010) Chromatin structure and the inheritance of epigenetic information. *Nat Rev Genet* 11:285–296.

McMurray CT (2010) Mechanisms of trinucleotide repeat instability during human development. *Nat Rev Genet* 11:786–799.

Méchali M (2010) Eukaryotic DNA replication origins: Many choices for appropriate answers. *Nat Rev Mol Cell Biol* 11:728–738.

Méndez J (2009) Temporal regulation of DNA replication in mammalian cells. *Crit Rev Biochem Mol Biol* 44:343–351.

Mesner LD & Hamlin JL (2009) Isolation of restriction fragments containing origins of replication from complex genomes. *Methods Mol Biol* 521:315–328.

Mirkin EV & Mirkin SM (2007) Replication fork stalling at natural impediments. *Microbiol Mol Biol Rev* 71:13–35.

Mirkin SM (2007) Expandable DNA repeats and human disease. *Nature* 447:932–940.

Murray JM & Carr AM (2008) Smc5/6: A link between DNA repair and unidirectional replication? *Nat Rev Mol Cell Biol* 9: 177–182.

Pomerantz RT & O'Donnell M (2007) Replisome mechanics: Insights into a twin DNA polymerase machine. *Trends Microbiol* 15:156–164.

Ransom M, Dennehey BK & Tyler JK (2010) Chaperoning histones during DNA replication and repair. *Cell* 140:183–195.

Wigley DB (2009) ORC proteins: Marking the start. *Curr Opin Struct Biol* 19:72–78.

Experimental papers

Bowman GD, O'Donnell M & Kuriyan J (2004) Structural analysis of a eukaryotic sliding DNA clamp–clamp loader complex. *Nature* 429:724–730.

Costa A, Ilves I, Tamberg N et al. (2011) The structural basis for MCM2–7 helicase activation by GINS and Cdc45. *Nat Struct Mol Biol* 18:471–477.

Eaton ML, Galani K, Kang S et al. (2010) Conserved nucleosome positioning defines replication origins. *Genes Dev* 24:748–753.

Greider CW & Blackburn EH (1985) Identification of a specific telomere terminal transferase activity in *Tetrahymena* extracts. *Cell* 43:405–413.

Greider CW & Blackburn EH (1989) A telomeric sequence in the RNA of *Tetrahymena* telomerase required for telomere repeat synthesis. *Nature* 337:331–337.

Hiratani I, Ryba T, Itoh M et al. (2008) Global reorganization of replication domains during embryonic stem cell differentiation. *PLoS Biol* 6:e245.

Klinge S, Núñez-Ramírez R, Llorca O & Pellegrini L (2009) 3D architecture of DNA Pol α reveals the functional core of multi-subunit replicative polymerases. *EMBO J* 28:1978–1987.

Takayama Y, Kamimura Y, Okawa M et al. (2003) GINS, a novel multiprotein complex required for chromosomal DNA replication in budding yeast. *Genes Dev* 17:1153–1165.

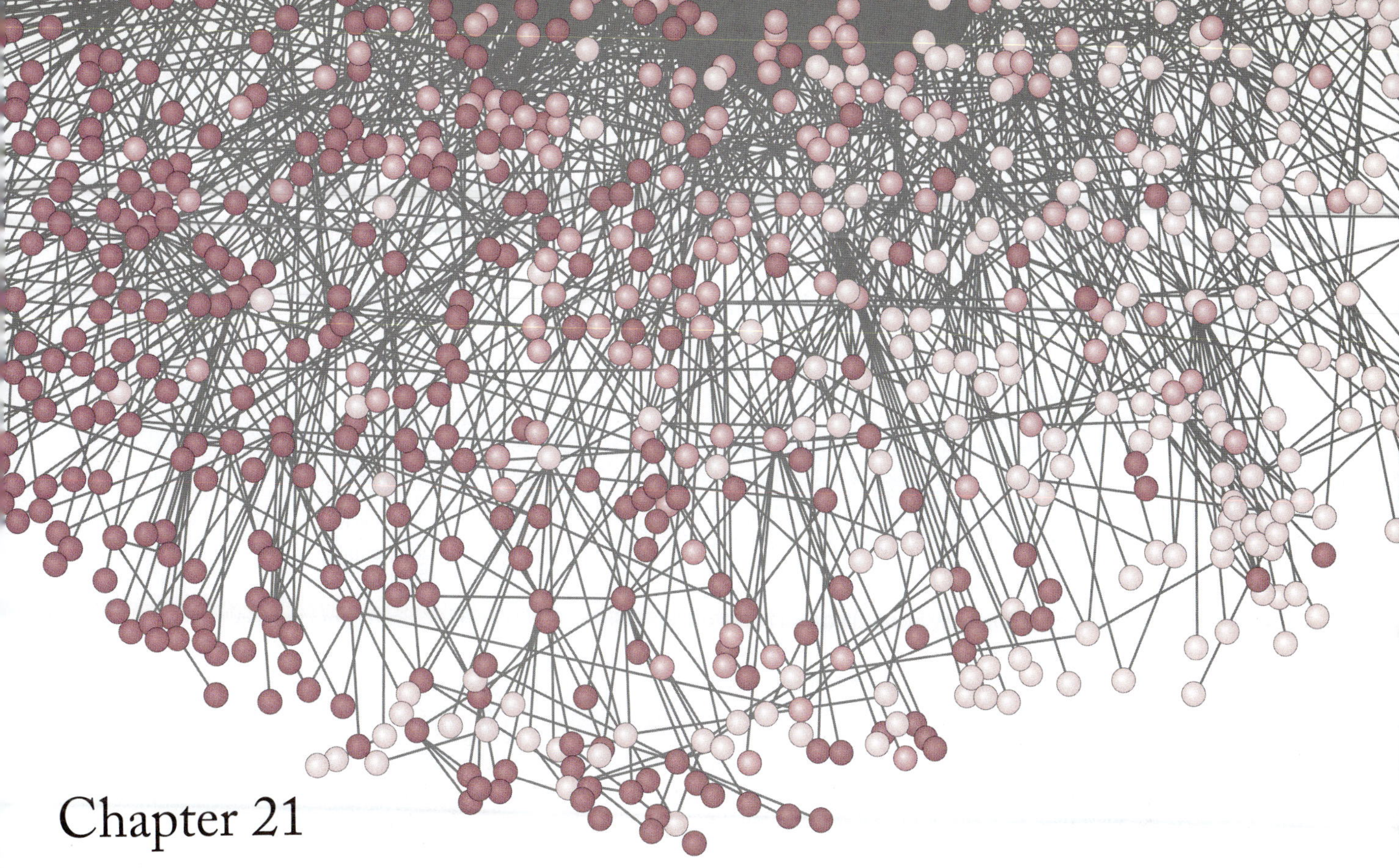

Chapter 21

DNA Recombination

21.1 Introduction

It has been known for some time that the genome is not a static entity with genes arranged along chromosomes in some permanent, immobile fashion. Instead, genes and other genetic elements can change their locations within the genome, and such changes may bring about alterations in gene expression. Of equal or even greater importance is the exchange of segments of DNA between homologous parental chromosomes during meiosis (see Chapter 2). Any change in location must, by necessity, involve cutting of both strands of the DNA, recombining the pieces, and then reestablishing the continuity of the phosphodiester backbones by the action of ligases.

Although the hypotheses regarding recombination date back at least to the early 1900s, the process was not understood for at least another half-century. A molecular mechanism for the breakage and rejoining of chromosomes was experimentally established by the now-classical experiments of Meselson and Weigle in 1961; for an account of this work, see Box 21.1. Exchanges of stretches of nucleotide sequences are broadly defined as **DNA recombination**. The process is widespread; it occurs in all known organisms and takes several distinct forms. We discriminate between homologous recombination or HR, **site-specific recombination**, and **nonhomologous recombination** depending on whether or not regions of homology between the exchanging partner DNA duplexes are required and, to some extent, on the length of these homologous regions. This chapter discusses the biological roles and mechanisms involved in these processes, focusing mainly on eukaryotic cells.

21.2 Homologous recombination

Homologous recombination involves the exchange of DNA sequences within large regions of homology between DNA molecules. If the exchanging regions are not identical, it provides a means of introducing genetic variability into the genome. In other circumstances HR can be a mode of DNA repair.

Box 21.1 The Meselson–Weigle experiment By 1961, the fact of genetic recombination was clearly recognized, but its mechanism remained wholly obscure at the molecular level. One hypothesis suggested that genetic recombination might involve the breaking and rejoining of DNA duplexes. Matthew Meselson and Jean Weigle set out to test this idea, using mutants of the bacteriophage λ. Upon infecting one bacterium, phage particles replicate and infect and kill surrounding bacteria. When a dilute solution of phage particles is spread on a lawn of *E. coli* on agar on a Petri plate, they form characteristic plaques. Wild-type phage give large, rough plaques, but the two mutants employed in these experiments, called mi^- and c^-, formed small and clear plaques, respectively. The wild-type phenotype can then be designated mi^+, c^+.

Meselson and Weigle grew c^-, mi^- double-mutant phage on a medium enriched in heavy isotopes ^{13}C and ^{15}N, and c^+, mi^+ wild-type phage were grown on normal medium containing ^{12}C and ^{14}N. A mixture of heavy phage and light wild-type phage were then allowed to infect bacteria growing on the usual light isotopes (**Figure 1**). Phage were harvested from these bacteria and then sedimented to equilibrium in a salt gradient. This procedure was somewhat similar to that used in the classic Meselson–Stahl experiment (see Box 19.1) except that whole phage were used, not just DNA, and sedimentation was in tubes from which differently banding fractions could be removed.

Some of the resulting phage had high density like the double mutants, and others had low density like wild type, but a substantial number were found at intermediate density, indicating mixed or recombinant DNA (see Figure 1). The phage were still infectious at this point and could be assayed by plaque formation. It was found that samples of intermediate density could contain phage showing either the c^-, mi^+, or c^+, mi^- phenotype. This observation strongly suggested that recombination between these marker mutations, by DNA breaking and rejoining, had occurred when the phage DNA was being duplicated in the bacterium.

A further, unexpected result provided deeper insight as to the mechanisms of recombination. At very high dilutions, each plaque can be initiated by a single bacteriophage infecting a single bacterium. Yet Meselson's lab sometimes observed mottled plaques from the progeny, indicating mixed phenotypes from a single infection. How could this be? The only reasonable explanation was that the recombination occurring during replication in bacteria sometimes resulted in heteroduplex regions in the recombinant DNA. In these regions, the sequence on one region of one strand came from one infecting phage, while the corresponding sequence on the other strand came from another. Wherever there was a mismatch in the two strands, subsequent replication would produce two differing duplex DNAs.

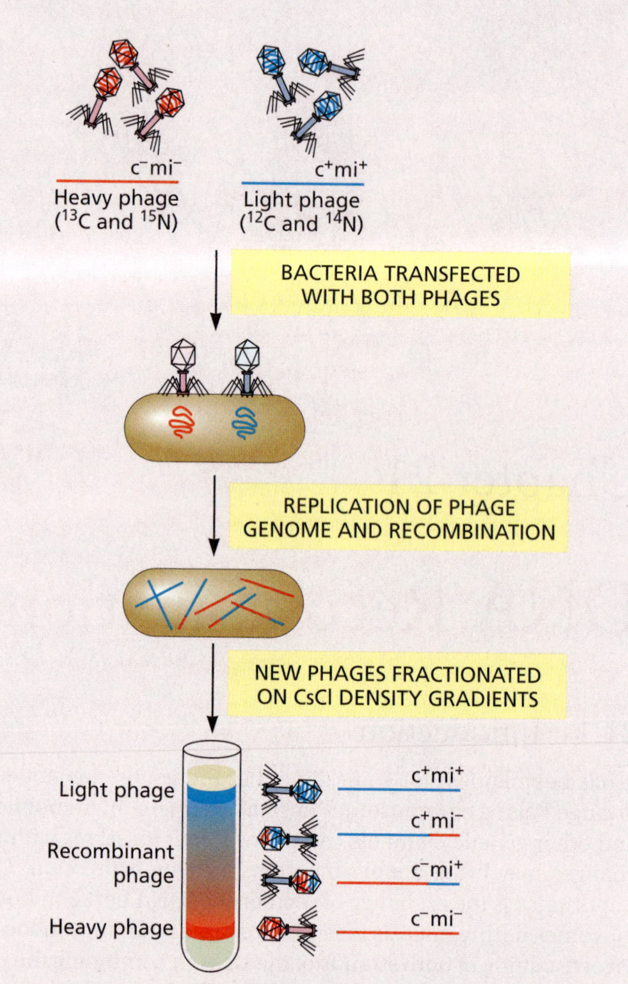

Figure 1 Schematic description of the Meselson–Weigle experiment. (Adapted from Mathews CK, van Holde KE, Appling DR & Anthony-Cahill SJ [2012] Biochemistry, 4th ed. With permission from Pearson Prentice Hall.)

Homologous recombination plays a number of roles in bacteria

Homologous recombination is the basis for major DNA repair processes in bacteria; it occurs only following DNA replication, when an intact copy of a sequence is available to serve as a source of information for the repair process. DNA repair is examined in detail in Chapter 22.

Homologous recombination is also used in **horizontal gene transfer** to exchange genetic material between different strains and species of bacteria and viruses. The ability of cells in natural populations to take up and incorporate homologous DNA from outside the cell has been maintained during evolution; it is beneficial because it provides the genetic variation needed for natural selection. There are three mechanisms for acquiring new DNA: conjugation, transformation, and transduction (see Chapter 2). In **conjugation**, DNA from one bacterial cell is transferred, concomitantly with its replication, to another cell, through a process akin to the sexual process in eukaryotes. Transformation involves the entry of pieces of naked DNA, or DNA–protein complexes, from the environment into the cell. Experiments with bacterial transformation proved that DNA is the carrier of genetic information (see Box 4.1). In transduction, pieces of DNA from one bacterium are transferred to another bacterium via a bacteriophage intermediate. Viruses that infect bacteria can accidently pick up pieces of bacterial DNA and transfer them into a newly infected bacterial cell; the newly introduced piece of DNA can recombine with the host bacterial chromosome.

The recombination mechanisms have enormous practical use in modern molecular biology as they can be exploited to introduce, modify, or delete desired genes in particular species (see Chapter 5).

Homologous recombination has multiple roles in mitotic cells

Exchange of homologous alleles is important in meiosis, but in mitotic cells it is vital to maintain genomic stability in the face of DNA damage. Homologous recombination participates in several different processes that are responsible for the maintenance of genomic stability. First, HR is the main pathway for the repair of double-strand breaks, DSBs, and other lesions such as interstrand cross-links. In cycling cells, DSB repair involves use of the intact sister chromatid as an information donor; hence DSBs are repaired primarily in S and G_2 phases when the sister chromatids become available. The process is relatively error-free. The alternative pathway for DSB repair is the error-prone nonhomologous DNA end-joining mechanism (see Chapter 22).

Second, HR is the mechanism for rescue of a collapsed DNA replication fork. Replication of a damaged DNA template that contains nicks or adducts leads to breakage and collapse of the replication fork. The newly synthesized sister chromatid is the source of information that is necessary to correct the error. The participation of HR in DSB repair and replication fork restart explains why mutations in genes that participate in HR have repeatedly been observed to result in extreme sensitivity to DNA-damaging agents.

Finally, HR lies at the core of the alternative lengthening of telomeres or ALT pathway, which elongates shortened telomeres without the participation of telomerase (see Chapter 20).

Meiotic exchange is essential to eukaryotic evolution

Homologous recombination plays a vital role in evolution. It is the major mechanism for generating genetic diversity among siblings resulting from sexual reproduction. Genetic diversity is created by exchange of the maternal and paternal alleles within gamete precursor cells, producing the continual recombining of inherited traits (see Chapter 2). In addition, HR has a purely mechanistic function: it ensures proper segregation of homologous chromosome pairs at the first meiotic division through the formation of crossovers known as chiasmata.

21.3 Homologous recombination in bacteria

HR is best understood in bacteria, mainly from experiments with *Escherichia coli*. An overview of the process is presented schematically in **Figure 21.1**. HR is initiated by the formation of DSBs in the DNA; these breaks result from either DNA-damaging agents or, during meiotic recombination, from a programmed process. Long single-stranded regions known as 3′-overhangs are then created in each of the resulting DNA fragments in a process known as end resection, which is catalyzed by the protein complex RecBCD. The single-stranded 3′-overhang is covered by a helical protein filament formed by multimerization of individual subunits of the protein RecA. The RecA-covered single strand then invades the intact homologous dsDNA—this second copy is present in the cell only following DNA replication—and performs a search for the region of homology. Once the homologous regions are properly aligned, the process can take one of two alternative pathways. The first pathway involves the formation and resolution of **four-way junctions**, or **4WJ**, also known as **Holliday junctions**, or **HJ**, between the two duplexes with the formation of **crossovers**. Resolution of the

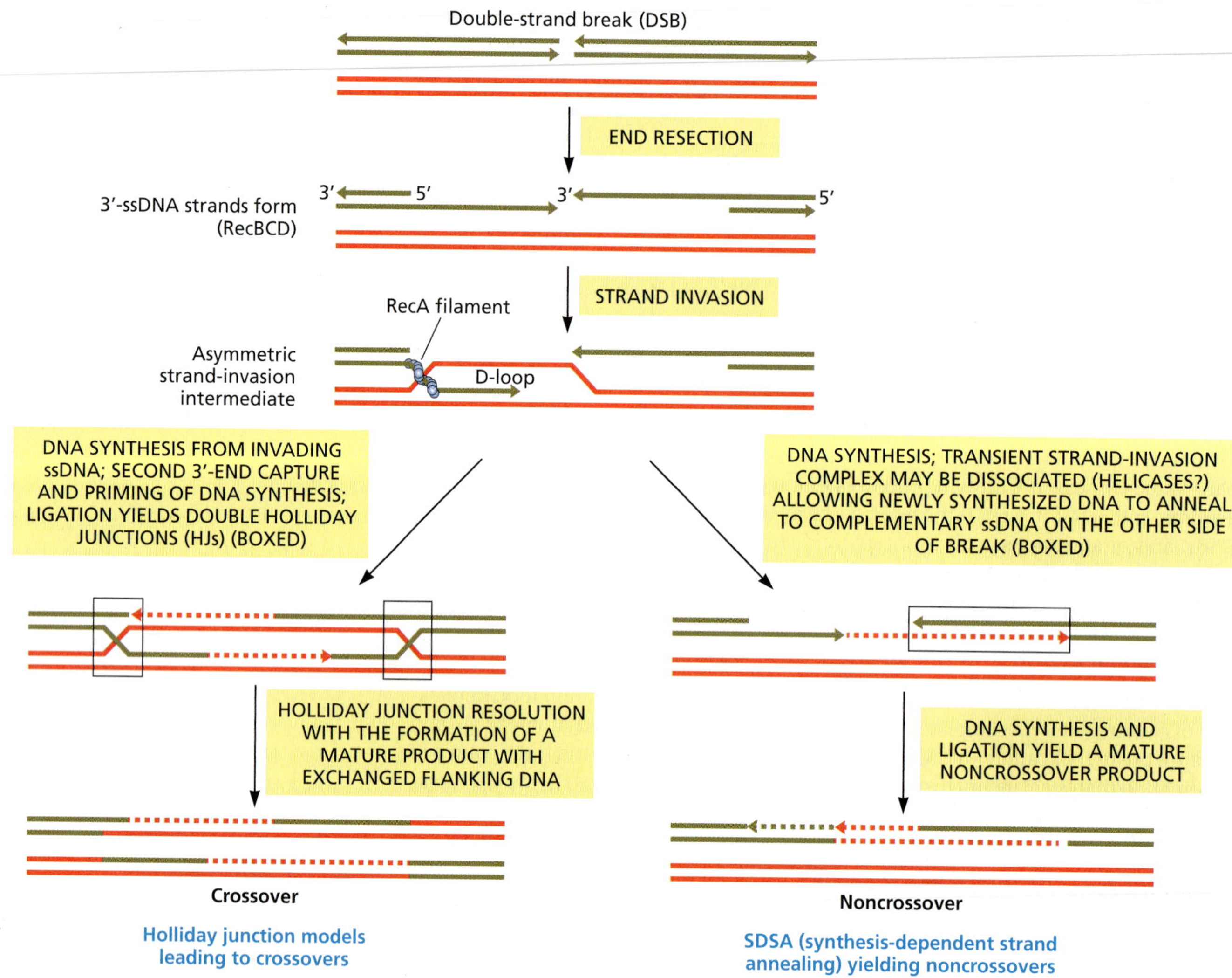

Figure 21.1 Overview of homologous recombination. HR is initiated by double-strand breaks or DSBs in the DNA. Once DSBs are introduced, the protein complex RecBCD resects the 5′-ends of each strand in the resulting double-stranded fragments, forming single-stranded 3′-overhangs. The 3′-ssDNA strand is covered by a helical filament of RecA, which allows the strand to invade into an intact homologous dsDNA, from a sister chromatid or a homologous chromosome. The resulting D-loop or displacement loop can then be processed by two different pathways. The formation of Holliday junctions and their resolution lead to the creation of crossovers. The alternative pathway, known as synthesis-dependent strand annealing or SDSA, yields noncrossover products. Note, however, that even in such noncrossover products there is a short stretch of hybrid DNA. (Adapted from San Filippo J, Sung P & Klein H [2008] *Annu Rev Biochem* 77:229–257. With permission from Annual Reviews.)

junctions by cutting and religating the appropriate strands will yield duplexes that have exchanged segments. The second pathway has come to be known as synthesis-dependent strand annealing or SDSA and leads to noncrossover products.

End resection requires the RecBCD complex

Removal, or resection, of the 5′-ends of each strand of the double-stranded fragments formed by DSBs is mediated by **RecBCD,** a multifunctional enzyme with very rapid helicase activity and single-strand endonuclease and exonuclease activities. RecBCD creates the single-stranded region for RecA binding and strand exchange. A detailed model for the action of RecBCD is presented in Figure 21.2, and details of the molecular structure of the complex and its scaffolding subunit RecC are given in Figure 21.3.

The RecBCD complex comprises three distinct subunits. RecB is a 3′ → 5′ helicase. Helicases are defined according to the direction in which they move; helicases always move on single-stranded stretches of nucleic acids toward the junctions

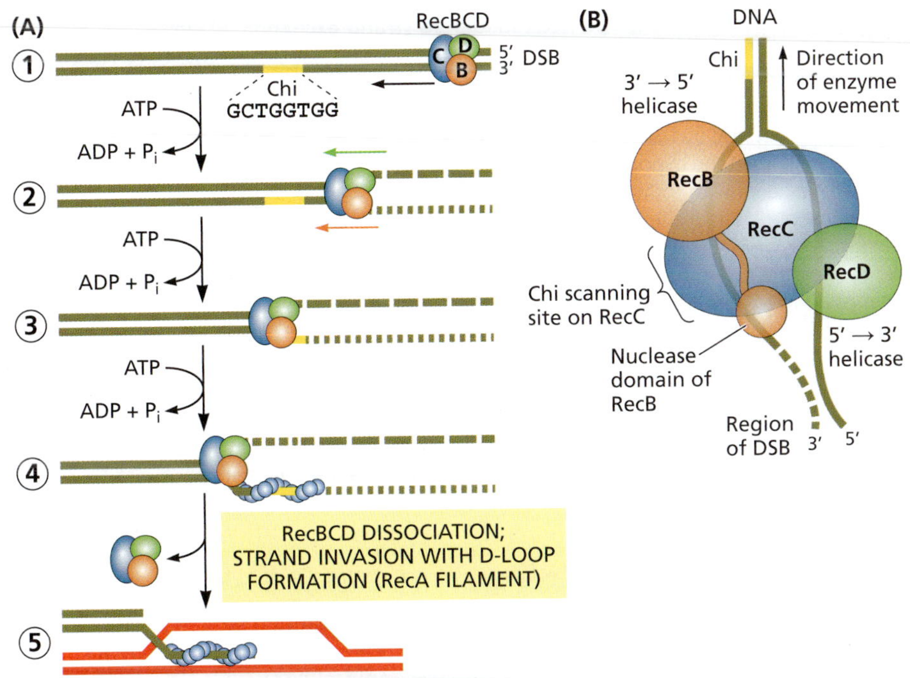

Figure 21.2 Roles of RecBCD, Chi sites, and RecA in initiating homologous recombination in bacteria. (A) A number of steps are involved in the initiation of HR by RecBCD. (Step 1) RecBCD binds to the double-strand break and initiates unwinding. RecB and RecD are helicases that move in the same direction on opposite strands of the duplex. RecB also has nuclease activity. RecC organizes the complex and recognizes the Chi or crossover hotspot instigator sequence to slow the complex movement. (Step 2) ATP-dependent unwinding progresses and is accompanied by nucleolytic cleavage of the DNA strands: the 3′-end is cleaved frequently, and the 5′-tail much less often. (Step 3) The Chi sequence is reached; the RecC subunit binds tightly to the 3′-tail, and RecBCD pauses. The Chi sequence is recognized as ssDNA, that is, after the duplex is unwound. RecC binding to Chi prevents further digestion of this strand. The final cleavage event on the 3′-tail occurs at Chi or within a few bases to the 3′-side of Chi. (Step 4) The 5′-tail is now able to access the nuclease site on RecB more frequently and is degraded more fully. RecA is loaded onto the 3′-tail to form RecA filament. (Step 5) RecBCD dissociates and the RecA filament initiates strand invasion into the intact homologous duplex, shown in red. (B) Schematic of the RecBCD complex, depicting the functions of the individual subunits. (Adapted from Singleton MR, Dillingham MS, Gaudier M et al. [2004] *Nature* 432:187–193. With permission from Macmillan Publishers, Ltd.)

with double-stranded regions, where they act to unwind the double helices. RecB also possesses a nuclease domain that is instrumental in digesting away one of the strands to create the single-stranded 3′-overhang. Initially, the RecB nuclease is very active on the 3′-end, which is cleaved frequently; however, it occasionally interrupts the continuous endonucleolytic chewing of the 3′-terminated strand to attach to the 5′-terminated strand and take an endonucleolytic bite from it. The long tether connecting the helicase and nuclease domains provides sufficient freedom for the nuclease to perform both of these functions. RecD is a 5′ → 3′ helicase that moves along the other strand. Finally, RecC is a protein that serves two functions, both

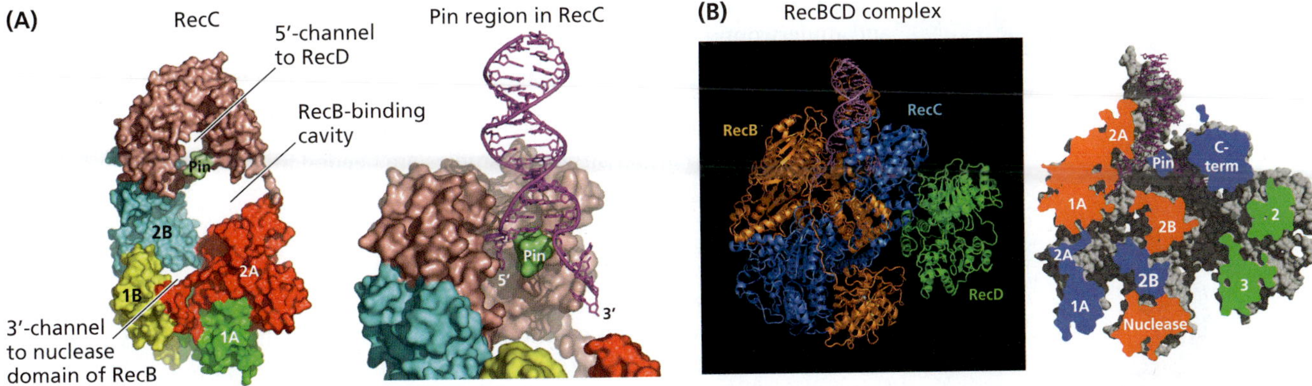

Figure 21.3 Structure of RecC and its complex with RecB and RecD. (A) Space-filling representation of RecC, showing the channels through the protein; this structure has been described as resembling Swiss cheese. The largest channel accommodates one of the domains of RecB. The other two are pathways along which the single-stranded tails of the DNA run to or from helicase subunits RecB and RecD. RecC consists of three domains. Surprisingly, domains 1 and 2 have the architecture of SF1-type helicases, despite the lack of recognizable helicase features in the primary structure. In fact, RecC is *not* a helicase, as it has lost key catalytic amino acid residues. Nevertheless, the helicase-like domains may provide the potential ssDNA-binding site needed for recognition of the Chi sequence. Domain 3 makes intimate contacts with each of the separated DNA strands running on either side of a pin that splits the duplex before the single strands are fed to the RecB and RecD helicases; see enlargement of domain 3 on the right. This splitting function contributes to both the speed and the processivity of the entire complex, transforming the two helicases into the fastest molecular machinery yet discovered: RecBCD unwinds the DNA at ~1000 bp per second. (B) RecBCD complex in ribbon representation. A long linker connects the nuclease domain of RecB with the rest of the protein, which allows the flexibility needed for the domain to cleave both 3′- and 5′-termini. The cutaway view to the right helps to visualize the channels through the complex. The domains are colored as in the ribbon diagram. (Adapted from Singleton MR, Dillingham MS, Gaudier M et al. [2004] *Nature* 432:187–193. With permission from Macmillan Publishers, Ltd.)

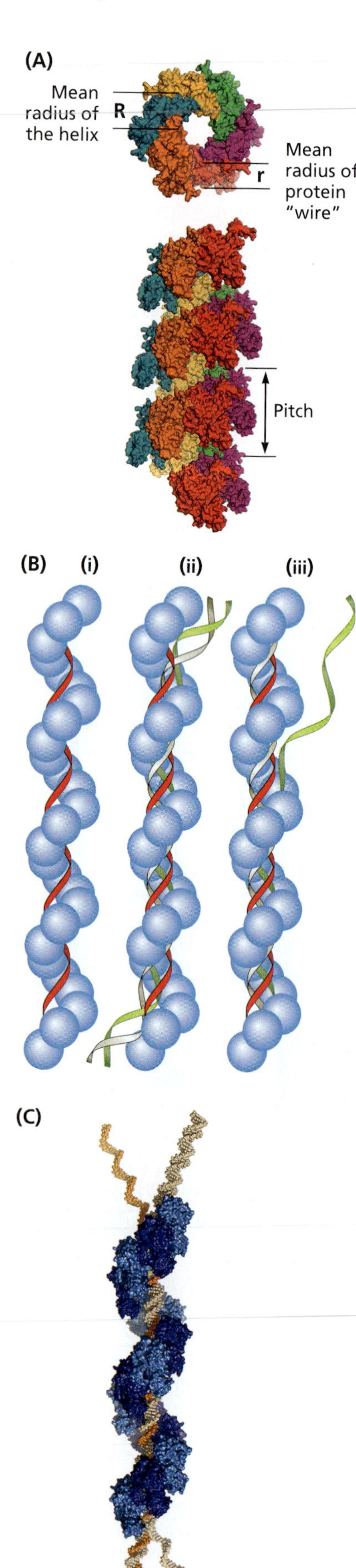

Figure 21.4 RecA filaments and RecA-mediated strand exchange. (A) Top and side views of RecA filament built from 19 individual subunits; the side view shows three complete turns of the RecA helix. Binding of ATP at a subunit interface results in large conformational changes that are coupled to interactions with the substrates. (B) RecA-mediated strand exchange: (i) ssDNA; (ii) triple-stranded DNA, where the original ssDNA has been proposed to transiently wrap in the minor groove of the duplex DNA; and (iii) RecA displacing the green strand, coincident with formation of a new dsDNA depicted in red and yellow. (C) This model depicts the structure of the helical filament RAD51, the human homolog of RecA, enclosing three strands of DNA. Note the conservation of the filament structure from *E. coli* to humans. (B, adapted from Mathews CK, van Holde KE & Ahern KG [1999] Biochemistry, 3rd ed. With permission from Pearson Prentice Hall.)

nonenzymatic: it provides a scaffold for organizing the entire complex—its peculiar Swiss-cheese molecular structure is shown in Figure 21.3—and it recognizes a specific sequence, **crossover hotspot instigator** or **Chi**, and binds to it. This interaction causes the complex to pause at the Chi site, preventing further digestion of the 3′-tail (see Figure 21.2).

Strand invasion and strand exchange both depend on RecA

Invasion of the complementary duplex by the single-stranded stretch created by the action of RecBCD and the formation of a D-loop intermediate are central to the entire process of HR. This phase of the process allows the broken DNA to search for homology within its intact double-stranded homolog, the sequence of which is used as a template for the repair process. Strand invasion is mediated through the action of **RecA**, a globular protein that forms a right-handed helical filament around ssDNA (**Figure 21.4A**). The RecA–ssDNA filament is the active entity that can then invade double-helical DNA in search of homology. The invasion may lead to formation of a transient triple helix, with the original ssDNA being wrapped in the minor groove of the duplex DNA (**Figure 21.4B**). In the next step, RecA displaces the strand in the duplex DNA that is the equivalent to the broken ssDNA, allowing the formation of a new duplex form between the ssDNA strand and its complementary strand. The human homolog of RecA, Rad51, forms a structure very similar to that formed by its bacterial counterpart (**Figure 21.4C**).

It has been recognized for some time that the dsDNA in the RecA filament adopts a conformation that is distinct from the standard B-form double helix; it is stretched and underwound, presumably to facilitate homology recognition. High-resolution structural studies have revealed that the postsynaptic RecA filament, that is, the filament that embraces dsDNA, has a very peculiar structure (**Figure 21.5**). The double helix itself is very different from the B-form DNA, with triplets of stacked base pairs, separated by gaps. Further structural studies are needed in order to fully understand the strand-exchange reaction.

Much concerning homologous recombination is still not understood

Mechanistically, several aspects of this phase of HR remain poorly understood. How can the two homologs find each other in the genome? The problem is especially aggravated in eukaryotic cells, whose large genome sizes require reliable mechanisms to facilitate the homology search. Finding a needle in a haystack may seem like a trivial undertaking in comparison. In eukaryotes at least, it could be that the homologous pairing not only occurs following damage but also is an enduring and general feature of nuclear architecture, which facilitates HR whenever and wherever damage occurs. The specific organization of interphase chromosomes, known as the **Rabl configuration**, may hold the key. In the Rabl configuration (**Box 21.2**), the centromeres cluster at one end of the nucleus with the chromosome arms extending to telomeres that abut the nuclear envelope at the opposite end. Such an organization will situate allelic or homologous loci at the same nuclear altitude or distance from the closest spindle pole. Thus, the homologs will be always closer together than they would be if the chromosomes were totally randomly distributed throughout the nuclear volume.

Figure 21.5 High-resolution structure of pre- and postsynaptic RecA filaments. (A) Structure of presynaptic filament: $RecA_6$–$(dT)_{18}$ complexed with $(ADP\text{-}AlF_4\text{-}Mg)_6$. The $ADP\text{-}AlF_4\text{-}Mg$ molecules are colored gold. The six RecA protomers are numbered from the N-terminal RecA and are presented in different colors. $(dT)_{18}$ is shown in red; the DNA structure in isolation is shown next to the complex. (B) Structure of postsynaptic filament: $RecA_5$–$(ADP\text{-}AlF_4\text{-}Mg)_5$–$(dT)_{15}$–$(dA)_{12}$ complex. The structure was obtained by soaking $RecA_5$–$(ADP\text{-}AlF_4\text{-}Mg)_5$–$(dT)_{15}$ in a solution containing the complementary $(dA)_{12}$ strand. $(dT)_{15}$ is shown in red; $(dA)_{12}$ strand is shown in magenta. The repeating unit is now a triplet of stacked base pairs, with adjacent base-pair triplets separated by a gap as in the presynaptic complex; see the DNA structure in isolation to the right. There are limited contacts between RecA and the complementary strand, which makes heteroduplex formation highly dependent on base pairing, ensuring HR fidelity. (C) Structures of dsDNA and ssDNA are shown for comparison. (A and B, left, from Chen Z, Yang H & Pavletich NP [2008] *Nature* 453:489–494. With permission from Macmillan Publishers, Ltd.)

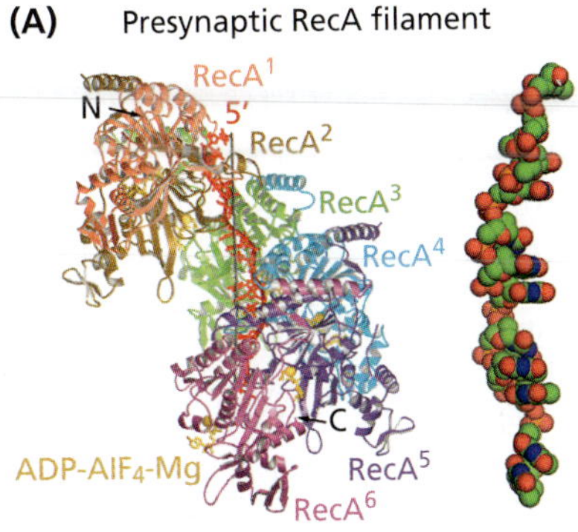

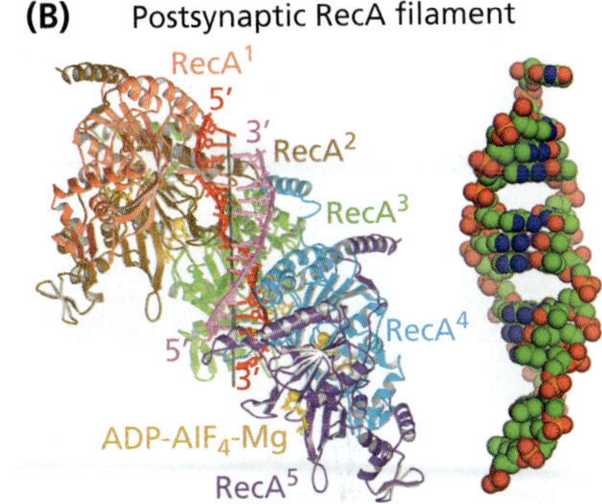

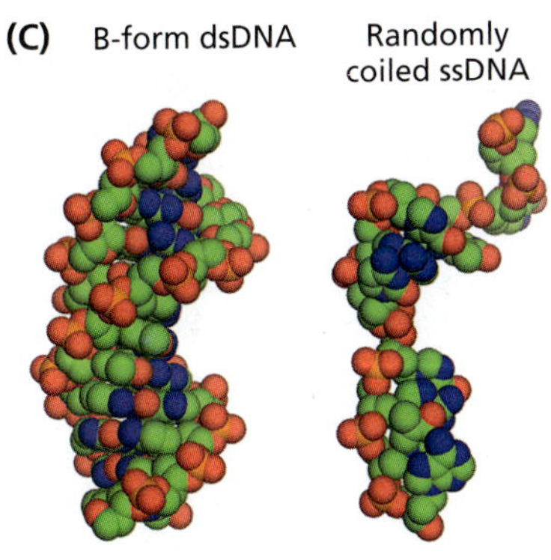

Single-molecule studies from the Kowalczykowski laboratory provide further information. They demonstrate that the efficiency and rate of recombination depend on the three-dimensional (3D) conformation of the DNA target. The data indicate that recombination occurs through a series of transient weak contacts, which are formed and broken until maximum homology is attained. These transient contacts are most easily sampled and exchanged if the target DNA is in a more compacted conformation. In a sense, this is also in accord with the Rabl configuration model, which emphasizes the proximity of homologous regions.

The second major challenge in understanding HR concerns the exact mechanism of strand exchange. RecA is an ATPase, but ATP hydrolysis is *not* required for either D-loop formation or the strand-exchange reaction. Some of the ATP that is hydrolyzed by RecA is coupled to filament assembly and disassembly (**Figure 21.6**). This may not be the major role of ATP hydrolysis, however, as the hydrolysis occurs uniformly throughout the filament and disassembly takes place at the end only. Conceivably, ATP hydrolysis is needed for the unidirectional DNA strand exchange that occurs *in vivo*; recall that most unidirectional processes in the cell require ATP energy. In a three-strand *in vitro* exchange reaction (**Figure 21.7**), RecA promotes exchanges in either direction when ATP is not hydrolyzed. ATP hydrolysis forces the reaction to proceed only in the 5′ → 3′ direction relative to the strand that was originally bound by the RecA filament. It must be noted, however, that the Rad51-mediated reaction in eukaryotes occurs in both directions. ATP hydrolysis may also promote DNA strand exchange through barriers, such as stretches of heterologous sequences embedded in homologous sequences.

Box 21.2 Higher-order genome organization in the interphase nucleus: Does it help in finding a needle in a haystack? For any homologous recombination to occur, whether during DSB repair in interphase or during meiosis, the task of bringing together homologous regions that can recombine is daunting, probably even more so than finding a needle in a haystack. Simple calculations based on the length of ssDNA regions created at the site of DSBs, for which the homologous target duplex DNA needs to be located and engaged, and the size of genomes show that in yeast the task is equivalent to finding a 20-cm-long stretch in a 1-km-long piece of string. With increasing genome size, the numbers are even more staggering. For example, in human cells, the 20-cm stretch must be located within some 2500 km: a daunting task indeed. Additional complexities that arise from the molecular mechanisms involved make the situation even worse; in the case of meiotic recombination, for example, the correct interaction partners, either homologous chromosomes or sister chromatids, must be sought out. It is clear that the gross genome organization in the interphase nucleus or during meiosis is a factor in the search for homologous regions. There are also data to indicate that the 3D conformation of the target DNA plays a major role. Despite these realizations, we still do not fully understand the mechanism of homology searches *in vivo* and the contribution of genome organization to the rate of this process. A prerequisite for such understanding is knowledge of genome organization.

The spatial organization of eukaryotic genomes in the interphase nucleus has been the focus of numerous studies since the late 1880s. In 1885, Carl Rabl, an anatomist who skillfully used the limited capabilities of the light microscope of the time, was the first to recognize that interphase chromosomes do not lose their identity even though they are no longer visible when the cell exits mitosis. His ideas were extended in the years that followed, and the configuration of chromosomes in the interphase nucleus came to be known as the Rabl configuration. The Rabl configuration, as observed in budding yeast, *Drosophila*, and plants, but in general not in mammals, is characterized by a segregation of centromeres at one end of the nucleus, with the chromosome arms extending

(Continued)

Box 21.2 *(Continued)*

to telomeres that abut the nuclear envelope. The position of the centromere cluster is inherited from anaphase, where centromeres point toward poles, with telomeres dragging behind (**Figure 1**).

Highly sophisticated genomewide methods have been applied to describe the topologies and spatial relationships of interphase chromosomes in yeast. A method was developed in William Noble's laboratory at the University of Washington at Seattle that identifies chromosomal interactions by coupling the chromosome capture-on-chip or 4C method (see Figure 10.22) with massively parallel sequencing. The method was applied to generate a map of the haploid yeast genome at kilobase resolution (**Figure 2**). The map represents a coarse-grained image, a snapshot that ignores the dynamic nature of chromosomes. An additional constraint is that the population-averaged data obtained by the technology cannot distinguish high-probability interactions occurring in a small fraction of cells from those that occur at low probability in a majority of cells. Overall, extensive regional and higher-order folding was observed, with interchromosomal contacts anchored by centromeres. These contacts may facilitate the search for homology needed for HR to occur.

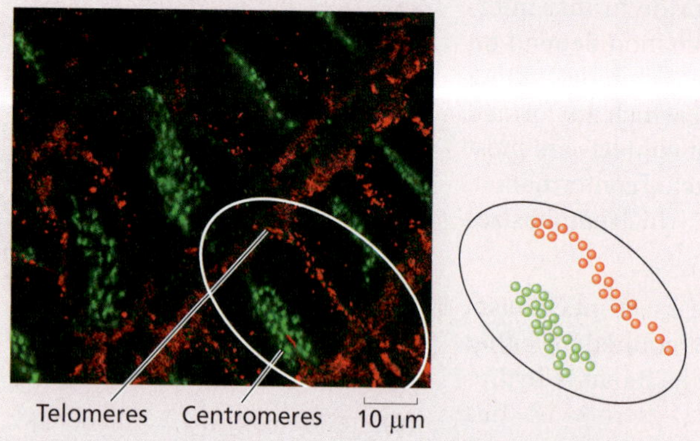

Figure 1 Rabl chromosome organization in wheat root tissue as revealed by fluorescence microscopy. Centromeres, shown in green, and telomeres, shown in red, are labeled by fluorescence *in situ* hybridization, FISH, and are located at opposite sides of the nuclei. The repetitive pattern is due to the presence, in root tissue, of numerous aligned cells and nuclei. A diagrammatic interpretation of the organization is presented on the right. Bar = 10 μm. (Adapted from Schubert I & Shaw P [2011] *Trends Plant Sci* 16:273–281. With permission from Elsevier.)

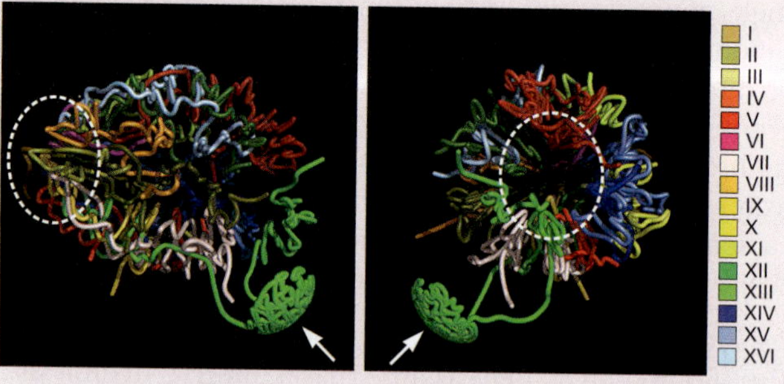

Figure 2 Three-dimensional model of the yeast genome. Two different viewing angles are shown. Each of the 16 yeast chromosomes is depicted in a different color, as indicated to the right. All chromosomes cluster via centromeres at one pole of the nucleus, shown by a dashed oval. Chromosome XII, which carries the rRNA genes, exhibits a striking conformation that implicates the nucleolus, marked by a white arrow, as a strong barrier to interaction between DNA sequences at either end. After exiting the nucleolus, the remainder of chromosome XII interacts with the long arm of chromosome IV. (From Duan Z, Andronescu M, Schutz K et al. [2010] *Nature* 465:363–367. With permission from Macmillan Publishers, Ltd.)

Finally, it was rather astonishing to discover that RecA is actually a multifunctional protein that performs numerous reactions in the bacterial cell (see Figure 21.7). In addition to the reactions illustrated in the figure, RecA participates in replication fork regression, sometimes called chicken-foot structure formation, during replication fork restart. It also stimulates translesion DNA replication by specialized DNA polymerases (see Chapter 22). Curiously, in addition to participating in various kinds of DNA transactions, RecA also has peculiar co-protease functions. Several proteins undergo autocatalytic cleavage reactions, which may lead, depending on the protein,

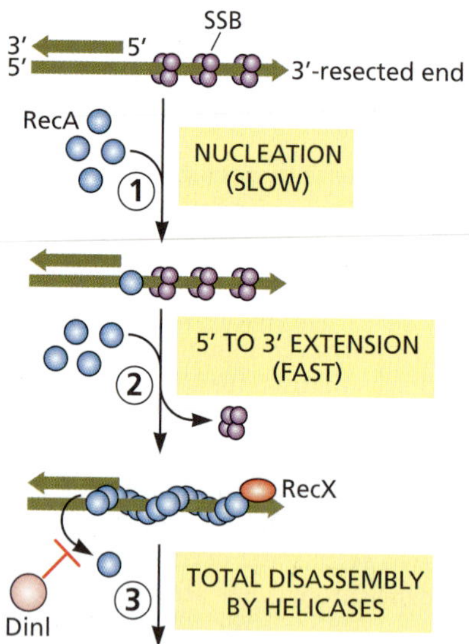

Figure 21.6 Assembly and disassembly of RecA filaments. The assembly process occurs in several steps. (Step 1) Nucleation is slow because the ssDNA overhang is covered by single-stranded DNA-binding proteins, SSBs. *In vivo*, the process is accelerated by the action of RecF, RecO, and RecR proteins, collectively known as RecFOR. (Step 2) The filament quickly extends, primarily on the 3′-end, to cover long stretches of DNA, up to several thousand nucleotides. Growth of the filament is fast, 2–20 subunits per second, and occurs by simple reversible association of ATP-bound RecA protomers. The RecX protein binds to the growing end, where it limits filament growth. RecX is, in turn, regulated by RecF, which antagonizes its activity. (Step 3) Disassembly occurs primarily at the 5′-end, by an active mechanism coupled to ATP hydrolysis. Disassembly is relatively slow, ~1.1 monomers per second; it is somewhat faster when the filament encompasses dsDNA. The protein DinI stabilizes the filament by interfering with such disassembly. When the RecA filament is no longer needed, it is removed from the DNA by helicases such as UvrD (see Chapter 22). Another protein, RdgC, blocks strand exchange if bound to the homologous duplex DNA used for recombination.

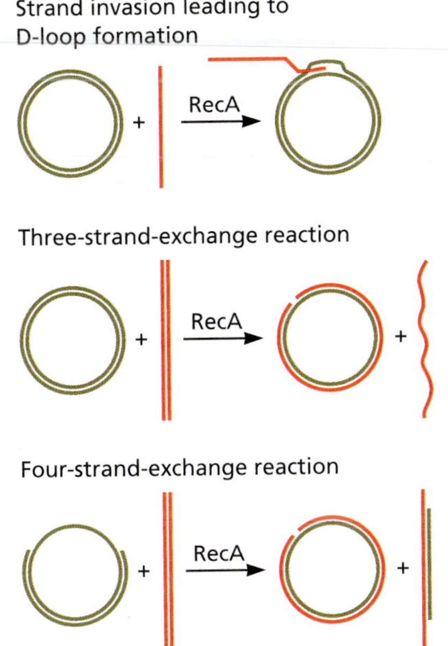

Figure 21.7 RecA performs numerous reactions in the bacterial cell. Three of the reactions involved in DNA transactions are shown here. In addition, RecA participates in replication fork regression, also described as chicken-foot structure formation, during replication restart. It also stimulates translesion DNA replication by specialized DNA polymerases (see Chapter 22). (Adapted from Cox MM [2007] *Nat Rev Mol Cell Biol* 8:127–138. With permission from Macmillan Publishers, Ltd.)

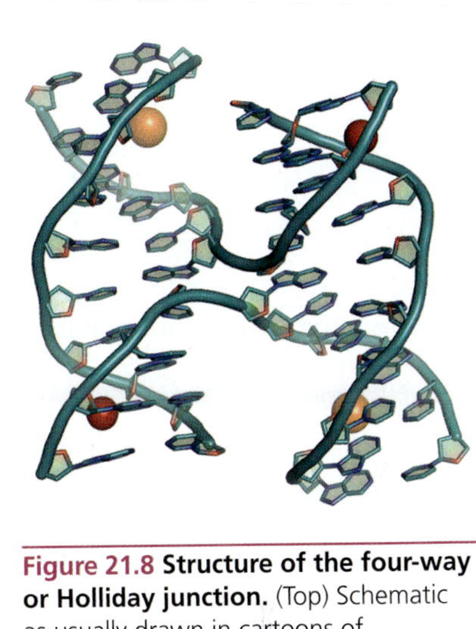

Figure 21.8 Structure of the four-way or Holliday junction. (Top) Schematic as usually drawn in cartoons of recombination. (Bottom) Atomic model based on a crystal structure. Colored spheres represent various ions, such as sodium, calcium, and magnesium.

to activation or inactivation. These reactions are greatly enhanced when the protein is bound within the RecA filament groove.

Holliday junctions are the essential intermediary structures in HR

Holliday junctions, HJ, also known as four-way junctions, 4WJ, are formed as intermediates in the pathway that leads to the formation of crossover products (see Figure 21.1). They were first recognized on the basis of purely genetic studies performed by Robin Holliday in the early 1960s. The structure has more recently been studied by electrophoresis and by crystallography under a variety of ionic conditions. The results revealed that the 4WJ is highly dynamic, switching between closed and open forms, depending on the environmental conditions and on the presence of protein ligands (**Figure 21.8**). **Box 21.3** is especially helpful in visualizing the two major conformations and the transition between them. The open conformation, or something very close to it, seems to be favored in the cellular environment.

Box 21.3 What is a Holliday junction and how is it resolved? The schematic representation that is conventionally used to depict a Holliday junction between recombining DNA duplexes (**Figure 1**) provides no information concerning the actual physical structure involved, nor how it can participate in branch migration or be resolved into recombined duplexes. This representation indicates only that a crossover has occurred, that repair synthesis has been completed, and that the four strands are interlinked. The actual structure that is formed can take several three-dimensional forms while remaining topologically unchanged.

First, imagine rotating one duplex nearly 180° with respect to the other. This produces a physically possible structure, the closed or folded junction, that can exist under certain solution conditions. Here the duplexes run side-by-side, nearly in parallel, with a neat crossing between them. The open or unfolded junction structure is favored, however, especially in the presence of protein RuvA. Details of this structure are shown in Figure 21.8 and Figure 21.10.

In the open conformation, migration of the junction, driven by RuvB, can occur (see Figure 21.9). The junction is resolved by cleavage by the endonuclease RuvC. This can potentially cleave symmetrically along either of two perpendicular axes to yield either recombined or nonrecombined products. In the cell, the Holliday junction recombination pathway always leads to the production of recombined or crossover products (see Figure 21.1).

(Continued)

Box 21.3 (*Continued*)

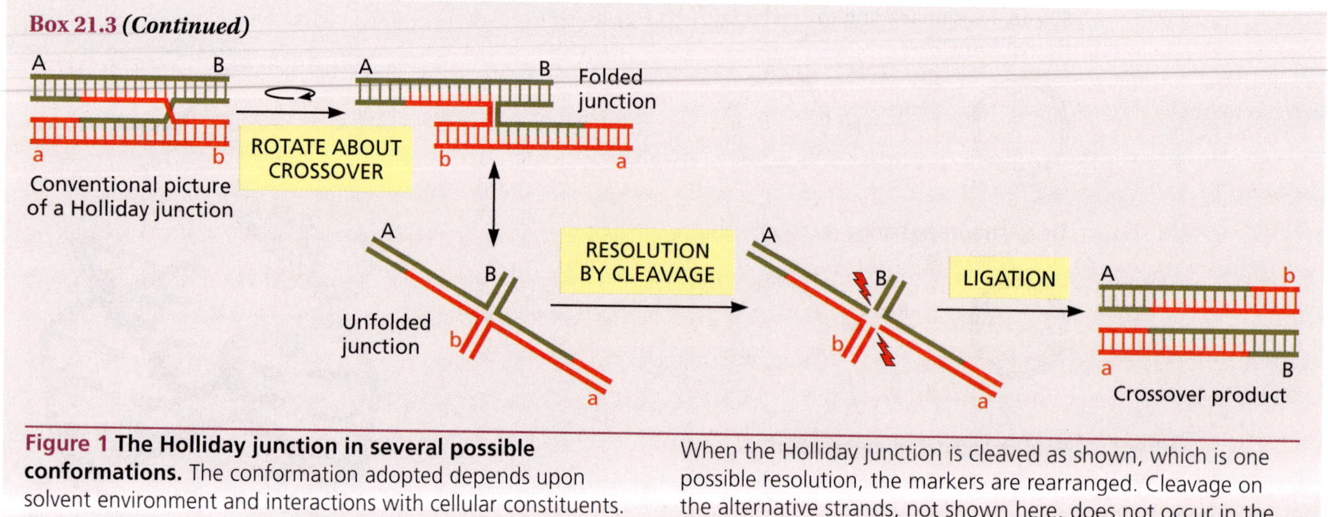

Figure 1 The Holliday junction in several possible conformations. The conformation adopted depends upon solvent environment and interactions with cellular constituents. The symbols A, B and a, b represent pairs of allelic markers. When the Holliday junction is cleaved as shown, which is one possible resolution, the markers are rearranged. Cleavage on the alternative strands, not shown here, does not occur in the cell.

Once formed, the joint in the 4WJ can move along the duplex in a process called **branch migration** (**Figure 21.9**). The extent of branch migration determines the location and length of strand exchanges. Branch migration is an extremely fast, ATP-hydrolysis-dependent process, ~5000 bp/s. It can occur in either direction and is mediated by a complex of two proteins, **RuvA–RuvB complex**. RuvA recognizes and

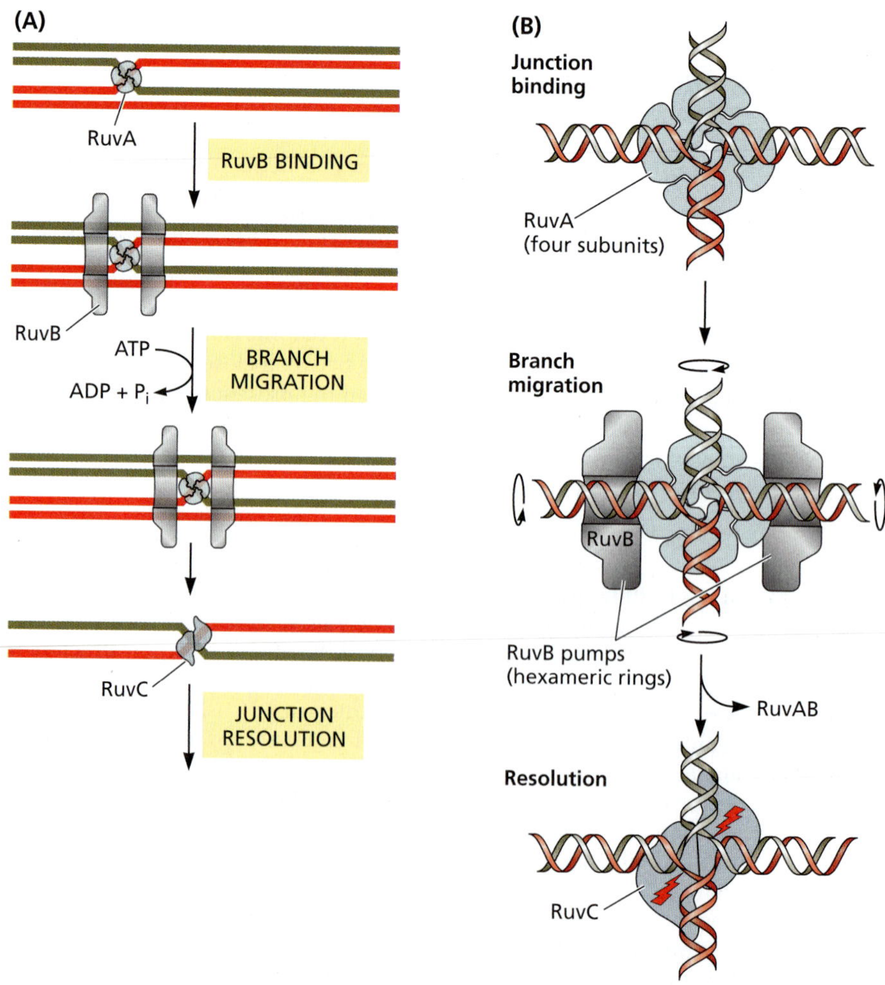

Figure 21.9 Branch migration during the Holliday junction pathway of HR. (A) Branch migration starts with the junction being recognized by RuvA. RuvA binding is followed by binding of the two RuvB pumps or motors. RuvAB promotes branch migration in a reaction requiring hydrolysis of ATP, and RuvC cleaves Holliday junctions. (B) Arrangement of protein and DNA components during the three stages of recombination outlined in part A. The proteins are shown schematically, based on their known structures. The two RuvB hexameric rings are presented in cross section to visualize the DNA passing through their centers. RuvC active sites are marked as zigzags. (B, adapted from Rafferty JB, Sedelnikova SE, Hargreaves D et al. [1996] *Science* 274:415–421. With permission from American Association for the Advancement of Science.)

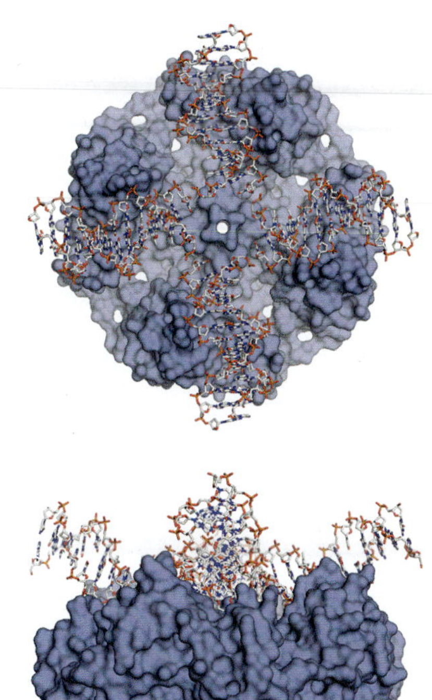

Figure 21.10 Co-crystal structure of *E. coli* RuvA complexed with a Holliday junction. The junction was obtained by annealing of four synthetic oligonucleotides. (Top) Top view, looking down the protein–DNA interface along the fourfold symmetry axis. DNA is shown in stick representation, with O atoms in red, P in yellow, N in blue, and C in white. The overall structure has an open concave architecture. RuvA tetramer is shown in surface representation. (Bottom) Side view of the complex. The fourfold axis lies parallel to the plane of the paper.

binds the open form of the 4WJ; the structure of the DNA–protein complex is shown in Figure 21.10. This binding is followed by RuvB binding on both sides of the junction. RuvB is the molecular motor that rotates two of the four DNA arms of the junction in opposite directions (Figure 21.11). This action results in rotational movement of the other two strands that are thus being drawn into the junction. Rotatory molecular motors, while not common, are not unknown in biology; consider the motor for the bacterial flagellum, for example. The process that ends the branch migration is the resolution of the junction by a resolvase, **RuvC**, which cleaves the DNA at symmetrical positions across the junction (see Box 21.3 and Figure 21.9). There are obviously two possible pairs of opposed cleavage sites, at a 90° angle to one another; the resulting recombination will depend on which pair is cleaved. The usual cleavage is indicated by the zigzag in Figure 21.9.

21.4 Homologous recombination in eukaryotes

Proteins involved in eukaryotic recombination resemble their bacterial counterparts

The eukaryotic recombinase, the counterpart of the bacterial RecA, is **Rad51**. Rad51 belongs to a group of proteins encoded by the *RAD52* epistasis group. Epistatically grouped genes participate in the same biological pathway and are most frequently defined by analysis of double mutants. Mutations in the *RAD52* group of genes lead to hypersensitivity to DNA-damaging agents, more specifically to those causing the formation of DSBs. This presumably reflects the importance of HR in eukaryotic DNA repair. In 1970, the term *rad*, referring to their X-ray irradiation sensitivity, was introduced for these mutants, with the numbers 50 and above reserved for the individual genes that, when mutated, exhibit the irradiation-sensitivity phenotype (Table 21.1).

The eukaryotic recombinase Rad51 possesses all the features characteristic of RecA, including the formation of a right-handed multimeric helix (see Figure 21.4). Rad51, like its bacterial counterpart RecA, needs to bind to the recessed ssDNA ends to form the helical filament for strand invasion. However, these ends are already covered, to protect them from degradation, by replication protein A, RPA, or its bacterial counterpart SSB. Recombination mediator proteins come to the rescue. These proteins share common features: they physically interact with their respective recombinase, Rad51 or RecA, and they preferentially bind to ssDNA over dsDNA. Only a very small amount of mediator is needed to overcome the inhibitory effects of the ssDNA-binding proteins, presumably because the mediators are only necessary for the removal of one or two of these molecules, the rest being removed by the growth of the Rad51 or RecA helical filament. Several such proteins have been

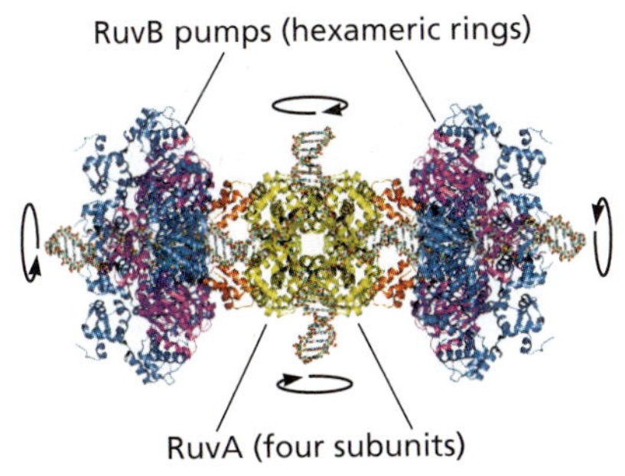

Figure 21.11 Overall architecture of RuvA–RuvB–Holliday junction ternary complex. RuvA forms a core structure, shown in yellow; the hexameric complex RuvB is shown in blue; and the DNA strands forming the Holliday junction are presented as ball-and-stick structures. RuvB is a motor protein that rotates the two arms of the junction in opposite directions as denoted by the arrows: this forces branch migration by driving the rotational movement of the other two strands toward the junction. The DNA is spooled to left and right, with the upper and lower arms being drawn into the center and eventually out through the RuvB pumps. (From Yamada K, Miyata T, Tsuchiya D et al. [2002] *Mol Cell* 10:671–681. With permission from Elsevier.)

Table 21.1 Homologous recombination factors that function with the recombinases Rad51 and/or Dmc1 in yeast and humans.

Saccharomyces cerevisiae	*Homo sapiens*	Biochemical function and characteristic features
MRX complex (Mre11–Rad50–Xrs2)	MRN complex (Mre11–Rad50–NBS1)	DNA binding and nuclease activities; associated with DSB end resection
	BRCA2	Recombination mediator; binds to ssDNA and interacts with RPA, Rad51, Dmc1
Rad52[a]	RAD52	Recombination mediator; binds to ssDNA and interacts with RPA, Rad51
Rad54	RAD54	ATP-dependent dsDNA translocase
Rdh54	RAD54B	induces superhelical stress in dsDNA; stimulates D-loop formation; interacts with Rad51
Rad55–Rad57	RAD51B–RAD51C RAD51D–XRCC2 RAD51C–XRCC3	Rad55–Rad57 and RAD51B–RAD51C are recombination mediators

[a]Recombination-mediator activity has been found in the yeast protein only.

described (see Table 21.1), but we focus on the two best-studied mediators, Rad52 and BRCA2.

Rad52 is a well-understood example of a recombination mediator. Detailed biochemical and structural studies have revealed that Rad52 has two distinct DNA-binding sites (**Figure 21.12**): the first site binds to ssDNA, and the second site binds to either ds- or

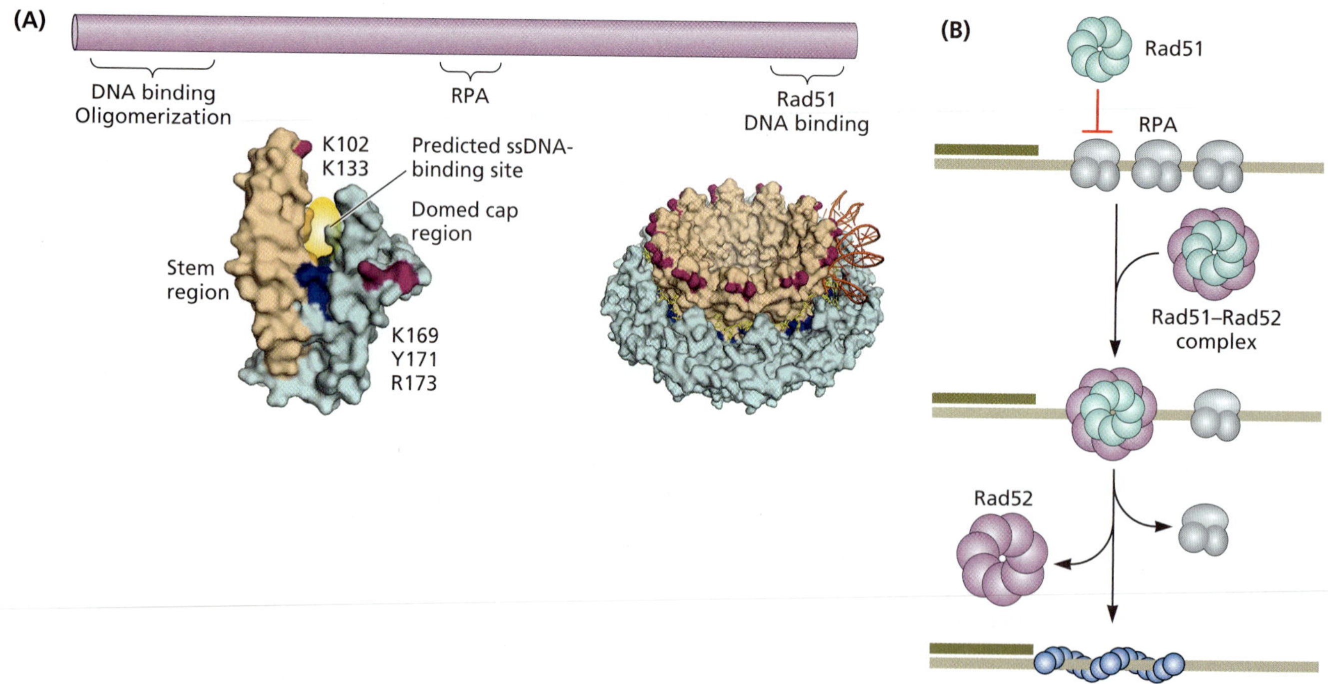

Figure 21.12 Recombination mediator Rad52: Structure and role in the delivery of Rad51 to ssDNA substrate during HR. (A) Top, domain structure of the Rad52 monomer. Bottom, surface views of human Rad52$^{1-212}$ oligomerization domain protomer and of the 11-subunit ring of Rad52. The basic residues, shown in dark blue, clustered at the bottom of the groove formed between the stem and domed cap region constitute the ssDNA-binding site, shown in yellow in the ring structure. The basic residues shown in magenta participate in binding of dsDNA. dsDNA runs along the rim of the stem region, with sites for ssDNA and dsDNA binding closely aligned with each other. (B) To perform its recombination mediator role, Rad52 forms a complex with Rad51 that delivers Rad51 to RPA-coated ssDNA; this seeds assembly of the presynaptic complex. Polymerization of additional Rad51 molecules results in further displacement of RPA from the DNA. (B, adapted from San Filippo J, Sung P & Klein H [2008] *Annu Rev Biochem* 77:229–257. With permission from Annual Reviews.)

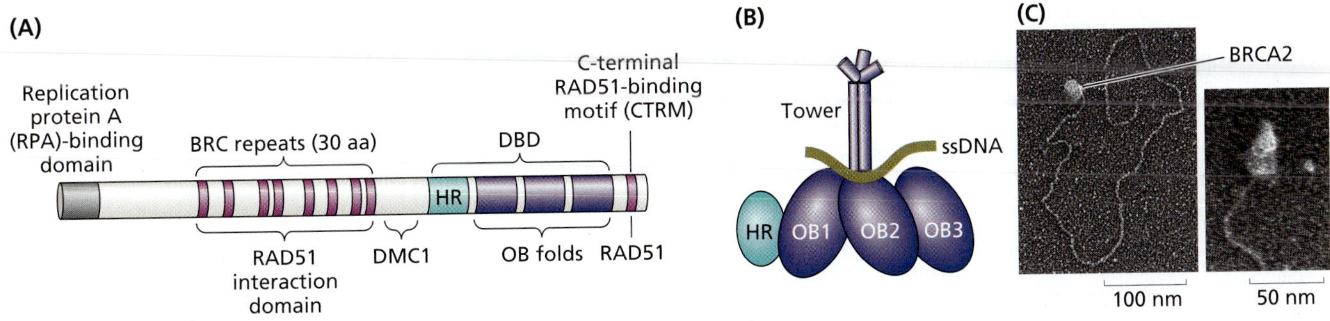

Figure 21.13 Domain structure of human BRCA2. (A) Domains are marked on the top and protein interaction motifs on the bottom. Some single-point mutations within individual BRC motifs are associated with familial early-onset cancer. The DNA-binding domain, DBD, consists of a helix-rich (HR) domain and three oligonucleotide-binding or OB folds. (B) Arrangement of domains in DBD, as derived from crystal structures. OB2 and OB3 are packed in tandem, whereas OB1 is packed with OB2 in the opposite orientation. Two anti-parallel long helices protrude from the core structure to form the tower emerging from OB2; the tower is topped by a three-helix bundle that may bind to duplex DNA. BRCA2 facilitates recruitment of RAD51 to sites of processed DSBs and enhances RAD51-promoted strand invasion. (C) Binding of BRCA2 to tailed duplex DNA, as visualized by EM. The BRCA2 protein was incubated with a long, linear duplex DNA containing a 54-nt ssDNA overhang at one end. BRCA2 localizes to such ssDNA tails. The higher-magnification image reveals the dumbbell-shaped BRCA2 dimers bound to ssDNA tails. (A–B, adapted from Holloman WK [2011] *Nat Struct Mol Biol* 18:748–754. With permission from Macmillan Publishers, Ltd. C, from Thorslund T, McIlwraith MJ, Compton SA et al. [2010] *Nat Struct Mol Biol* 17:1263–1265. With permission from Macmillan Publishers, Ltd.)

ssDNA. In addition, Rad52 binds to Rad51, forming a stable complex that is now capable of displacing RPA from the ssDNA; remember that Rad51 cannot, on its own, bind to RPA-coated ssDNA. Studies in which amino acid residues in either of Rad52's two DNA-binding sites were mutated showed that both sites are necessary for D-loop formation. The present understanding is that Rad52 serves to catalyze the replacement of RPA by Rad51.

The second well-characterized recombination mediator is BRCA2. This protein is best known as a tumor suppressor, mutated forms of which are known to be involved in human diseases such as familial breast and ovarian cancer and the cancer-prone syndrome Fanconi anemia. As is the case for the *rad* genes, mutations in BRCA2 confer a marked hypersensitivity to DNA-damaging agents. Because of its important connection to disease, the protein has been studied extensively. Its domain structure and its interactions with DNA are well characterized (**Figure 21.13**). Like Rad52, the BRCA2 protein interacts with ssDNA tails; it also forms stable complexes with Rad51, a condition necessary for its recombination mediator function (**Figure 21.14**).

HR malfunction is connected with many human diseases

Because HR is known to participate in a wide variety of processes, it comes as no surprise that improper functioning of HR mechanisms leads to numerous human disorders. We illustrate this point by discussing the involvement of HR in hemophilia A. Two defective HR mechanisms lead to the formation of a nonfunctional product of a gene that encodes protein factor 8 in the blood-clotting biochemical cascade. **Box 21.4** further discusses blood clotting and the role of defective HR in hemophilia.

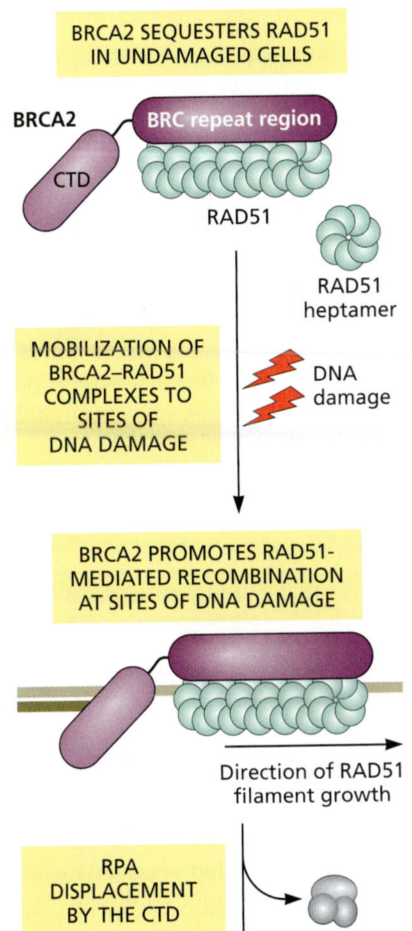

Figure 21.14 BRCA2 mediator activity in RAD51-mediated HR. In undamaged cells, RAD51 is bound or sequestered to BRCA2 in an inactive complex. DNA damage targets the RAD51–BRCA2 complex to the junction between duplex DNA and the recessed ssDNA overhang, which is coated with the heterotrimeric replication protein A, RPA. RPA is displaced, with RAD52 playing an essential role in the process; it is also possible that the BRC repeat region stimulates RAD51-mediated strand exchange, in a reaction separate from RPA displacement. CTD, C-terminal domain. (Adapted from Shivji MKK, Davies OR, Savill JM et al. [2006] *Nucleic Acids Res* 34:4000–4011. With permission from Oxford University Press.)

Box 21.4 Hemophilia A and genetic recombination

Hemophilias are genetic diseases that interfere with blood clotting. Clotting can be initiated along two pathways. One, known as the intrinsic pathway, is started by tissue damage; the other, initiated by damage to blood vessels, is known as the extrinsic pathway. In each case a successive series of proteolytic activation steps leads to the polymerization of fibrin to form a clot (**Figure 1**). A critical point in these pathways is that at which factor X, denoted F10 in current nomenclature, is activated by either F7 or F9. Here the pathways merge, and either one now leads to clot formation.

For F9 to function, it needs the participation of the antihemophilic factor F8. F8 must itself be activated by cleavage by thrombin (**Figure 2**); since active thrombin is a later product in the series, there is strong positive feedback at this point. F8 is the determinant of many varieties of hemophilia A, the classic and most dangerous form. This is because the gene for F8 is carried on the X chromosome: thus, females have two copies but males have only one. A woman heterozygous for aberrant F8 will not experience severe hemophilia but she will be a carrier for the disease, since she may transmit her aberrant gene to male descendants. Queen Victoria of England was such a carrier, and many of the crowned heads in Europe descended from her. Some were hemophiliac; others transmitted the condition to their male heirs. Thus hemophilia A is sometimes called the royal disease.

Hemophilia A can result from any of a large number of point mutations in F8 or proteins that interact with it. However, the most disastrous form results from the presence of sequences in F8 introns that have almost identical sequences located outside the gene. There are at least two large introns that harbor such sequences. One is intron 1 (**Figure 3A**). Looping and recombining splits the F8 gene into two pieces, oriented in opposite directions; these cannot code for any F8 protein. This site, however, accounts for only about 1% of all hemophilia A. The real culprit is in intron 22 (**Figure 3B**). Recombination of this sequence produces the same kind of result but accounts for around 40% of all hemophilia A.

Hemophilia has long been a target of therapy. Unfortunately, concentrates of F8 from blood donors turned out too frequently to be contaminated with HIV. When recombinant DNA technology was developed, it became possible to produce cloned F8; this is used today. At the same time, efforts are underway to introduce somatic gene therapy.

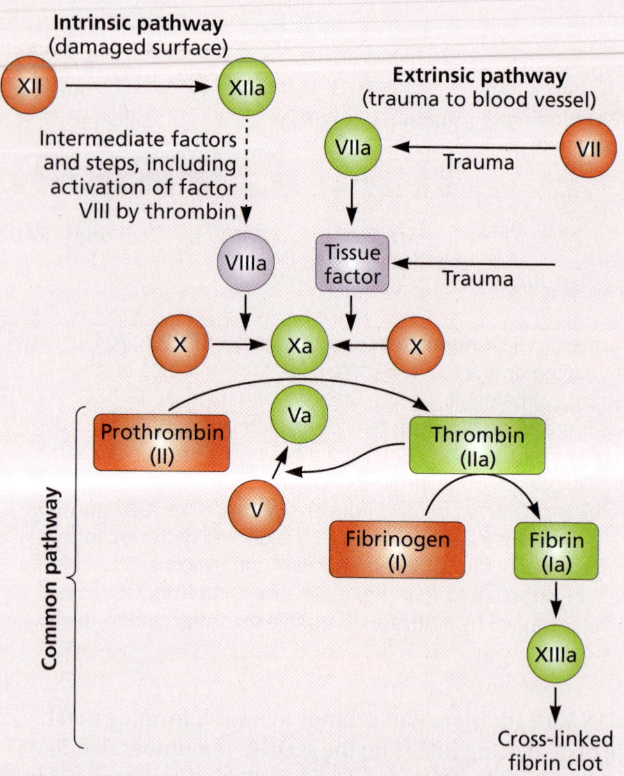

Figure 1 Blood clotting or coagulation cascade. Each factor in the pathway is a serine protease that exists in an inactive form: activation of the protease occurs by proteolysis by an already activated upstream protease. Of the two pathways depicted, the primary pathway for initiation of blood coagulation is the tissue factor or extrinsic pathway. The coagulation factors are usually indicated by Roman numerals, with a lowercase a suffix designating the active form. In our schematic, the inactive proteases are boxed in red and the active ones in green. (Adapted from Mathews CK, van Holde KE, Appling DR & Anthony-Cahill SJ [2012] Biochemistry, 4th ed. With permission from Pearson Prentice Hall.)

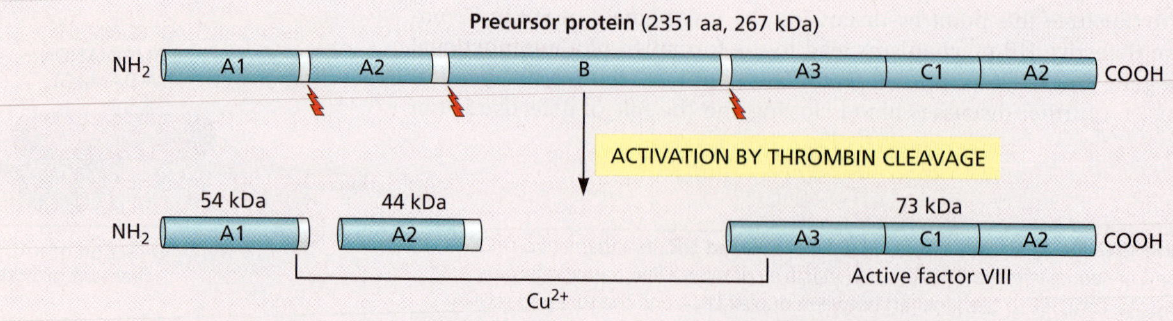

Figure 2 Activation of coagulation factor VIII by stepwise proteolytic cleavage by thrombin. The final three peptides remain associated, stabilized by a copper ligand.

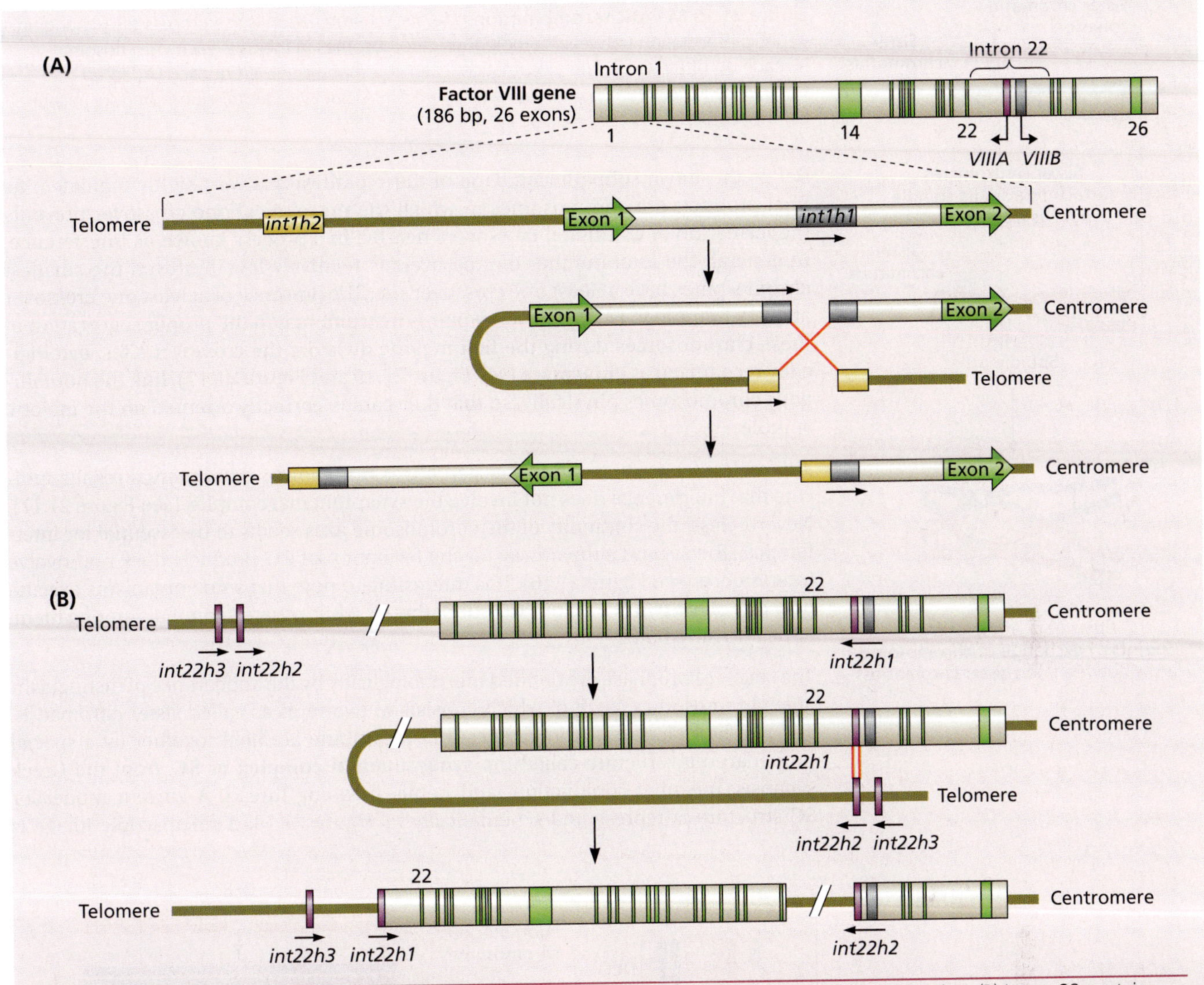

Figure 3 Homologous recombination events lead to pathological inversions of factor VIII gene. The gene contains 26 exons, shown as green boxes, and 25 introns; the two critical introns, 1 and 22, are marked. Intron 22 contains two further genes, *VIIIA* and *VIIIB*; arrows indicate the direction of their transcription. Horizontal arrows in the figure indicate the orientation of the repeated sequences. (A) Intron 1 contains a sequence, *int1h1*, that is repeated outside of the gene, *int1h2*. Homologous recombination between these two repeats explains the origin of pathological gene inversion. (B) Intron 22 contains a sequence, *int22h1*, with similarities to two sequences, *int22h3* and *int22h2*, that are distal to the gene. Intrachromosomal HR leads to the formation of crossover structures, with inversion of exons 1–22 with respect to exons 23–26. For simplicity, only the consequences of recombination with *int22h2* are shown, but *int22h3* can also recombine, with a similar outcome. (Adapted from Graw J, Brackmann H-H, Oldenburg J et al. [2005] *Nat Rev Genet* 6:488–501. With permission from Macmillan Publishers, Ltd.)

Meiotic recombination allows exchange of genetic information between homologous chromosomes in meiosis

Meiotic recombination creates genomic diversity among siblings and ensures proper segregation of homologous chromosome pairs during the first meiotic division. In meiotic recombination, the partners that exchange information are the homologous chromosomes that pair during the first meiotic prophase (**Figure 21.15**). Any pair of homologous sequences in the two homologous chromosomes may be used as substrates for the actual recombination event.

The final outcome of the meiotic recombination depends on which of three possible pathways are followed. Noncrossover or NCO products are formed by the SDSA pathway (see Figure 21.1 and **Figure 21.16**). Crossovers, CO, arise only from the two subpathways that involve HJs and their resolution: different sets of proteins

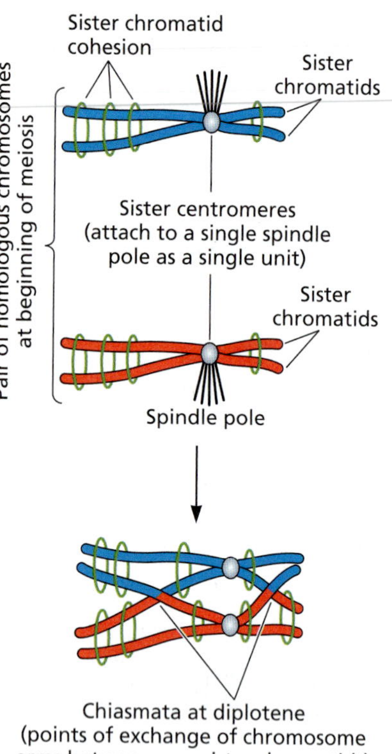

Figure 21.15 Meiotic recombination. Meiotic recombination is the process of exchanging genetic information between homologous chromosomes in meiosis. Each chromosome is made up of a pair of sister chromatids; each chromatid is a DNA duplex. (Adapted from Neale MJ & Keeney S [2006] *Nature* 442:153–158. With permission from Macmillan Publishers, Ltd.)

define these two subpathways. One of these pathways is very highly regulated: its final products are chromosomes in which the presence of one crossover prevents the formation of additional crossovers nearby, in a process known as interference. In general, the total number of crossovers is relatively low, but even the smallest chromosomes have at least one crossover site. The presence of at least one crossover at every homologous chromosome pair is instrumental in the proper segregation of these chromosomes during the first meiotic division: the crossover sites, cytologically recognized as **chiasmata** (see Figure 21.15 and Figure 21.17), link the homologous chromosomes physically, so that they can be correctly oriented on the meiotic spindle.

The molecular mechanism of interference is not known; experimental results indicate that interference does not involve the synaptonemal complex (see Figure 21.17). Nevertheless, the continuity of the chromosome axes seems to be essential for interference. The second subpathway for the formation of CO products does not involve interference (see Figure 21.16). It is interesting to note that some organisms execute only one of the two alternative CO subpathways while others, mammals among them, appear to have both.

The stages of prophase are defined morphologically by the appearance of distinguishable chromosomes, each of which consists of two replicas called sister chromatids. The two homologous chromosomes that recombine are held together by a special proteinaceous structure called the **synaptonemal complex** or **SC**, from the Greek synapsis, meaning conjunction, and nema, meaning thread. A current model for SC structure is represented schematically in Figure 21.18. A microscopic image is

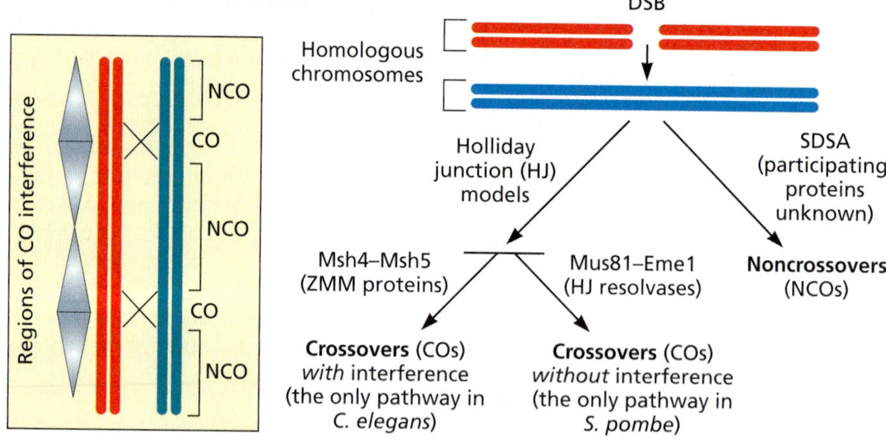

Figure 21.16 Three different outcomes of meiotic HR result from three different pathways. The synthesis-dependent strand-annealing pathway, SDSA, leads to noncrossover products, NCO. Crossovers, CO, are produced by two different mechanisms, each involving different proteins. The number and position of COs are highly regulated. In some organisms, the presence of one CO prevents additional COs from occurring nearby, in a process known as interference. A zone of interference, shown by gray triangles in the boxed inset, is centered on the positions of DSBs that give rise to CO. The level of interference is highest at the CO site and declines with distance. Interference requires the Msh4–Msh5 complex, which forms a sliding clamp that specifically interacts with HJs. Other organisms use a different HJ pathway that is not subject to interference control. The protein complex involved comprises Mus81 and Eme1, evolutionarily conserved proteins that possess HJ resolvase or endonuclease activity. Some organisms, including budding yeast, plants, and mammals, appear to have both pathways. (Adapted from Cromie GA & Smith GR [2007] *Trends Cell Biol* 17:448–455. With permission from Elsevier.)

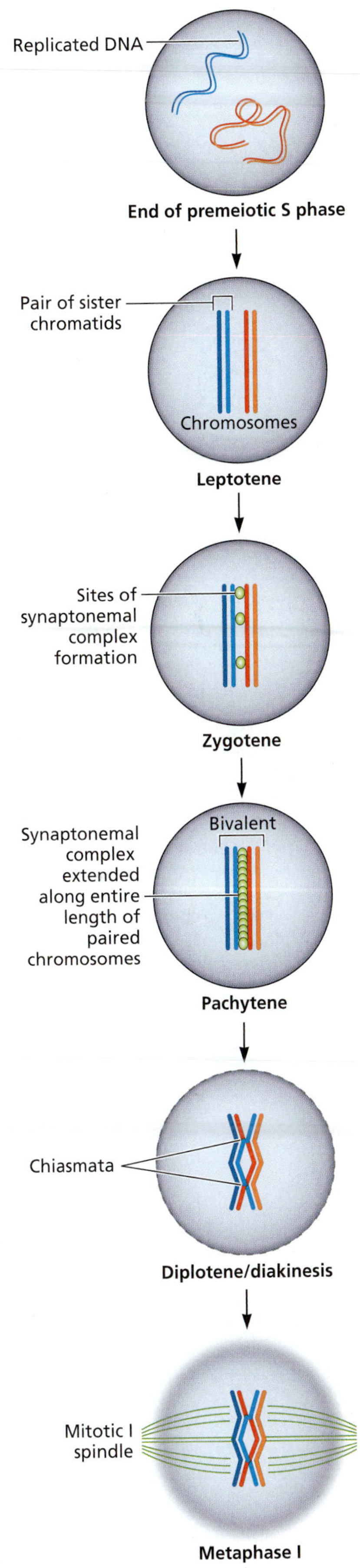

Figure 21.17 Overview of meiotic recombination. The three basic steps that occur during prophase of meiosis I are chromosome pairing, synapsis, and recombination. Only one pair of homologous chromosomes is presented for simplicity, with the sister chromatids shown in different shades. Leptotene: chromosomes begin to condense and homologs are aligned. Zygotene: initiation of synapsis between homologs, that is, building of a synaptonemal complex, at a few sites. Pachytene: synapsis is completed to produce bivalents. Interhomolog recombination occurs during both zygotene and pachytene. Diplotene: chromosomes separate but are still held together by chiasmata, or recombination spots. Breakdown of the nuclear envelope signals the end of prophase and is followed by formation of a mitotic spindle. Sister chromatid cohesion along the chromosome arms is released at anaphase I, and homologous chromosomes separate to move to the two spindle poles. (Adapted from Page SL & Hawley RS [2003] *Science* 301:785–789. With permission from American Association for the Advancement of Science.)

shown in Figure 21.19. The SC structure comprises two kinds of elements: lateral and central. The lateral elements undergo a complex transformation process from their precursor protein structures, the axial elements that hold the two sister chromatids in each homologous chromosome together. The main protein component of the axial elements is cohesin. Recall from Chapter 8 that cohesin embraces the two sister chromatids from DNA replication until mitosis (see inset in Figure 21.18). The evolution from axial elements to lateral elements involves the addition of several more proteins to the protein set that forms the axial structures.

The sequence of events in meiotic recombination is outlined in Figure 21.17. The initial pairing of the homologous chromosomes occurs in only one or a few regions; then pairing extends over the entire chromosome length in a process called **synapsis**. When the two chromosomes are linked by the SC along their entire length, they are called **bivalents**. The exchange between the two homologous non-sister chromatids that occurs in these structures involves breakage, or DSB formation, and reunion, or ligation, between chromatids. Next, chromosomes begin to separate; it is during this process that visible chiasmata are discernible under the microscope.

An important unanswered question concerns the mechanism of recognition and pairing of homologous chromosomes during the onset of meiosis. It is known that homologs take a similar amount of time to pair in different organisms; that is, pairing is independent of the genome size, over several orders of magnitude. This fact rules out simple, linear searching mechanisms, in which a searching sequence slides along another DNA after an initial contact is made. Linear searches are estimated to be very slow: for yeast, such a search may take a few hours, but for wheat, it would require several thousand hours, longer than the life of the plant.

In principle, reducing the search space by limiting the search to particular sequences or chromosome regions could provide the answer. Different species may exploit different sequences to that end. Some species, such as *Drosophila melanogaster* and *Caenorhabditis elegans*, contain specialized chromosome pairing sites where synapsis is initiated. Others, including wheat, may use centromeres as pairing sites. The most frequent sequences that cluster at the onset of meiosis in numerous species are centromeres. Centromeres cluster by attaching to the nuclear envelope at one end of the nucleus to form a structure known as a **meiotic bouquet**. In species where the clustering is tight, the chromosomes resemble a bouquet of flowers with gathered stems, hence the name (Figure 21.20). We must note, however, that the degree of centromeres clustering is highly variable among species, suggesting that mere attachment to the nuclear envelope could be at the heart of the bouquet mechanism for reducing the search space.

The formation of DSBs in meiosis is a programmed event, in contrast to the random occurrence of such breaks during DNA damage, for example, by X-rays. The DSB is introduced by a specialized protein, Spo11, in a rather complex reaction that carries similarities to the action of topoisomerases (Figure 21.21). All meiotic recombination pathways are initiated in the same way.

Another essential protein player in the process is Rad51, the recombinase that is also instrumental in mitotic recombination. In addition, meiotic recombination utilizes a

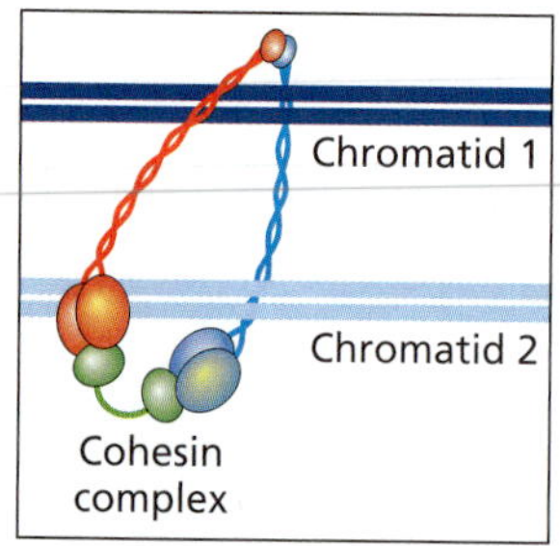

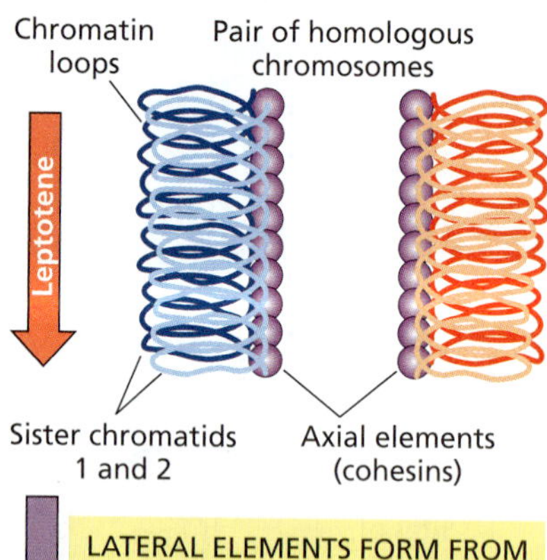

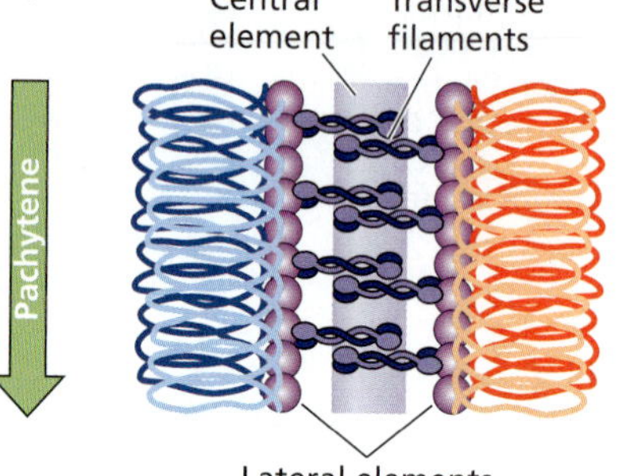

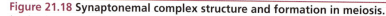

Figure 21.18 Synaptonemal complex structure and formation in meiosis. (Top) Schematic of two homologous chromosomes, each consisting of two sister chromatids held together by a single proteinaceous axis and by cohesin complexes, as seen in the boxed inset. (Bottom) Meiotic chromosomes held together by SC. The two axial elements are converted into lateral elements by additional proteins. The lateral elements are held together by a central element, forming a tripartite proteinaceous structure. Loops of DNA emerge from these lateral elements. The protein composition of the individual elements is as follows: axial elements contain Rec8, STAG3–Rec11, SMC1, and SMC3, also known as cohesins; lateral elements contain axial element proteins plus structural proteins SCP2 and SCP3 and HORMA-domain proteins Hop1–HIM3–Asy1. Transverse filaments contain Zip1–SCP1–C(3)G–SYP1.

meiotic-specific recombinase, Dmc1, denoted as DMC1 in humans, which contributes critical but nonoverlapping functions to Rad51. Interestingly, Dmc1 forms two different structures on ssDNA. Depending on the absence or presence of ATP, stacked rings or right-handed helical filaments, respectively, will be formed (Figure 21.22). Perhaps not surprisingly, only the helical filaments have recombinase activity, as in mitotic recombination.

The number and position of crossovers that are the end product of the two alternative HJ-mediated subpathways are highly regulated through several different mechanisms. The first level of control determines the distribution of DSBs, which is nonrandom in the genome. The clustering of DSBs at particular genomic sites has led to the definition of **recombinational hotspots**, with each hotspot active in a minor fraction of cells, typically from 0.01% to ~10% at most. Interestingly, DSBs in non-hotspot regions are recombinationally silent and are repaired in the same ways as normal DNA damage. Recent genomewide localization studies have employed high-resolution chromatin immunoprecipitation techniques using antibodies to meiosis-specific recombination proteins, such as Spo11 and Dmc1. These techniques were followed by high-throughput sequencing to show the existence and distribution of hotspots along chromosomes. In addition, these studies looked at nucleosome occupancy at such spots, with rather surprising results. It turns out that the chromatin organization at such hotspots is dramatically different in different organisms. In yeast, most DSBs map within the nucleosome-depleted regions of promoters (see Chapter 12), whereas in mice there is a slight enrichment in nucleosome occupancy over hotspot centers. Finally, there is hotspot enrichment, in both yeast and mouse, of histone H3 trimethylated at lysine 4, H3K4me3. Recall from previous chapters that this specific histone modification is typically present in transcriptionally active genes. Nevertheless, a straightforward relationship between transcriptional activity and DSBs does not seem to exist.

Other levels of regulation of the usage of programmed DSBs exist. As we already know, the formation of DSBs is the first obligatory and common step for all HR events, whatever the nature of the final product, COs or NCOs. Whether these DSBs are used to create COs or NCOs, or are just repaired by the conventional DNA repair pathways, is controlled by the choice of the pathway. Practically nothing is known about the factors that determine that choice.

21.5 Nonhomologous recombination

Transposable elements or transposons are mobile DNA sequences that change positions in the genome

Nonhomologous recombination is the major mechanism that moves DNA sequences around the genome. It permits a DNA sequence to be lifted or copied from one site and placed into another that exhibits no homology to the original location. The existence

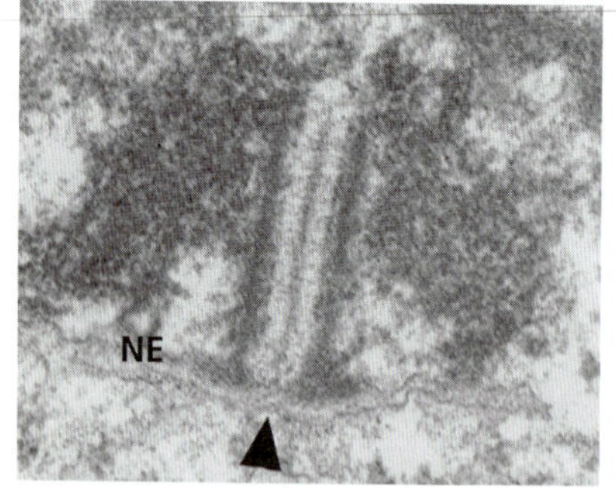

Figure 21.19 Micrograph of pachytene synaptonemal complex. Longitudinal section through a mouse spermatocyte nucleus at pachytene shows the synaptonemal complex and normally condensed chromatin. NE, nuclear envelope. (From Kouznetsova A, Benavente R, Pastink A & Höög C [2011] *PLoS One* 6:e28255.)

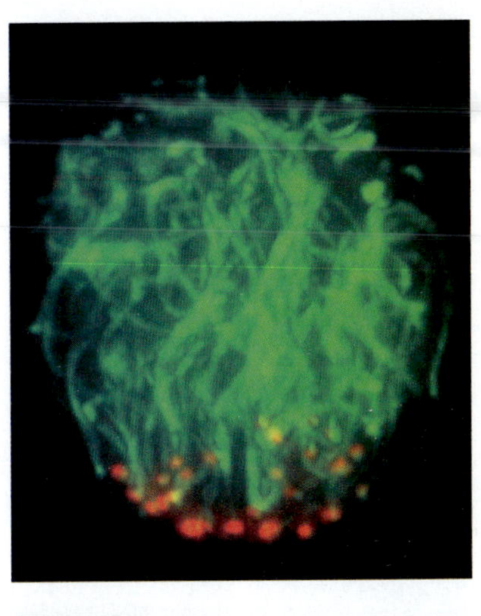

Figure 21.20 Bouquet configuration in early meiotic prophase nuclei of mouse spermatocytes. Centromeres are shown in red, and the chromosome axes are in green. The respective structures were visualized by immunochemistry, using staining with antibodies specific to protein markers present exclusively in either centromeres or chromosome axes. The centromeres aggregate in one main group at the nuclear periphery, while the chromosome axes extend toward the nuclear space. (From Berrios S, Manieu C, López-Fenner J et al. [2014] *Biol Res* 47:16. With permission from BioMed Central.)

of discrete, independent, mobile sequences, known as **transposable elements** or **transposons**, was discovered in maize, *Zea mays*, in 1948 by Barbara McClintock. An account of her visionary work and the reluctance with which it was accepted is presented in Box 21.5. Transposons are abundant, scattered throughout the genomes of many plants and animals, and can constitute a considerable portion of the DNA. Thus, in mammals, transposons occupy ~40–45% of the genome; the percentage is even higher in some plants, ~60% in maize. The largest fraction known is in the frog *Rana esculenta*, 77%. Transposons are also present in bacteria and unicellular eukaryotes but comprise a much smaller portion of the genome: transposons in *E. coli* constitute ~0.3% of the genome, and in *Saccharomyces cerevisiae*, 3–5%. The frequency of transposition varies among the various transposable elements, usually between 10^{-3} and 10^{-4} per element per generation. This is higher than the spontaneous mutation rate of 10^{-5}–10^{-7}.

Many transposons are transcribed but only a few have known functions

Although we have learned quite a lot about transposon structure and the transposition mechanisms, we are still struggling to understand the biological significance of these sequences. Why are they there? What, if anything, do they contribute to the organisms carrying them? Their presence is certainly a heavy burden on the cell that needs to replicate these elements together with the functional portion of the genome. For quite a while, these elements were simply considered molecular parasites or selfish DNA for using the resources of the cells to simply propagate themselves from generation to generation. They were also dubbed **junk DNA** because no obvious function had been revealed. However, analysis of the whole human genome in the ENCODE project (see Chapter 13) has shown that roughly 80% of that genome is transcribed in some cells at some time. Although many of these transcripts are as yet of unknown function, their very existence casts doubt on the concept of junk or selfish DNA. This is an area where we may hope for further clarification in the near future.

From an evolutionary perspective, at least, the presence of transposons could be beneficial, conferring a selective advantage by DNA rearrangements. Obviously, mobile DNA is a slow but potent source of mutations, so it may have provided a method, over millions of years, to shuffle and rearrange the genome, giving rise to

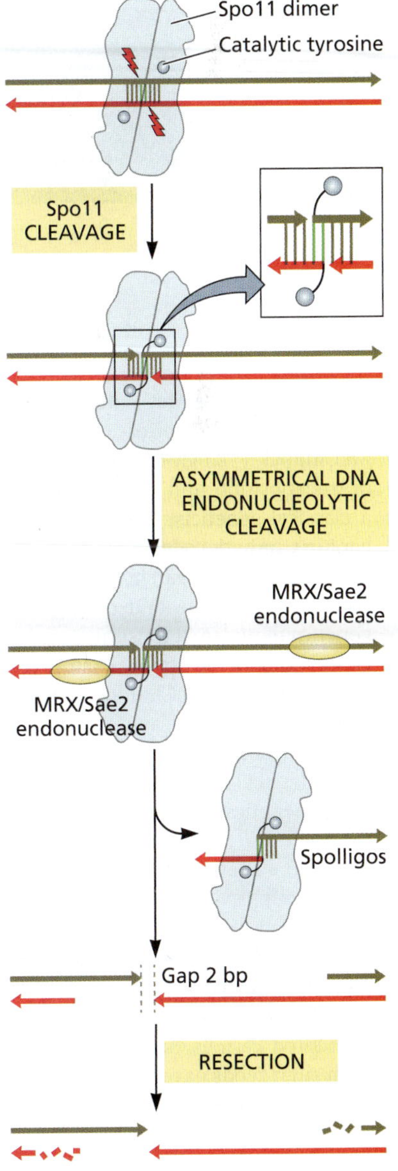

Figure 21.21 Formation of programmed DSBs by action of the Spo11 dimer and its numerous partners. This reaction initiates all meiotic recombination pathways. The two monomers attack the phosphodiester DNA backbone at adjacent base pairs. Cleavage of the backbone results in the formation of a covalent phosphotyrosyl bond between the catalytic tyrosine residue, shown by a gray circle, and the 5′-end of the DNA; this reaction is reminiscent of the mode of action of topoisomerases, although no topoisomerase activity has been detected in Spo11. The reaction also produces 3′-OH groups and nicks in both strands that are offset by 2 bp, as shown in the boxed inset. The covalently trapped Spo11 is removed by endonucleolytic cleavages that occur several nucleotides downstream from the protruding 5′-ends and that are catalyzed by the MRX/Sae2 endonuclease, shown as yellow ovals. The released Spo11 is bound to two oligonucleotides of different lengths, termed by some spolligos. The DNA released is further subjected to the universal resection mechanism that produces the ssDNA stretches needed for binding of Rad51 recombinase. (Adapted from Cole F, Keeney S & Jasin M [2010] *Genes Dev* 24:1201–1207. With permission from Cold Spring Harbor Laboratory Press.)

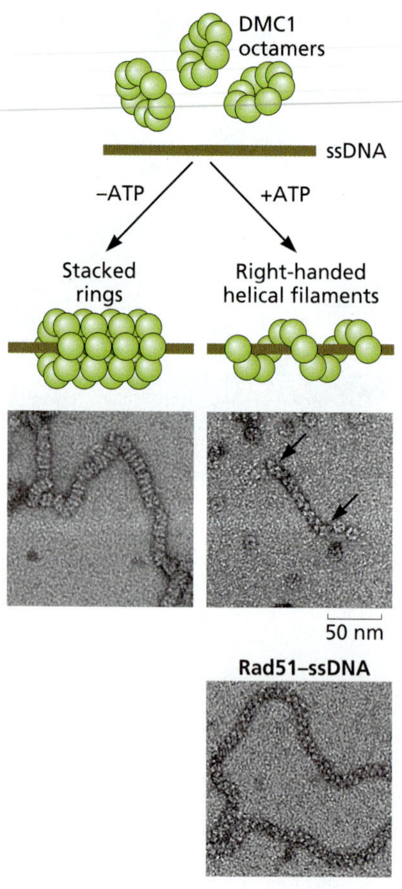

Figure 21.22 Meiotic-specific recombinase Dmc1. This recombinase is denoted DMC1 in humans. In solution, Dmc1 exists as octameric rings that bind to ssDNA in different ways, depending on the absence or presence of ATP, shown to the left and right, respectively. Only the helical filaments have recombinase activity. Some researchers have proposed that stacked rings might be converted into active filaments, probably upon ATP binding. This possibility is unlikely, however, because such a conversion would require dismantling and reorientation of the ring-forming subunits. The EM images below the respective schematics show the appearance of the structures; arrows mark the filament ends. A helical Rad51–ssDNA filament is shown for comparison. (Top, adapted from Sung P & Klein H [2006] *Nat Rev Mol Cell Biol* 7:739–750. With permission from Macmillan Publishers, Ltd. Bottom, from Sehorn MG, Sigurdsson S, Bussen W et al. [2004] *Nature* 429:433–437. With permission from Macmillan Publishers, Ltd.)

the diversity that drives evolution. Some of this diversity, for example, might arise from displacement of genes with respect to their regulatory sequences. Our genome is currently filled with many old, inactive mobile elements, left as a legacy of our gradual evolution. Thus, transposons may form the DNA pool needed for natural selection to occur.

In addition, newer research has provided indications that, at least in some organisms, such as the single-cell pond-dwelling organism *Oxytricha*, junk DNA may not be junk after all but may actually perform functions that are central to early development. *Oxytricha* undergoes massive genome reorganization during development, and transposons seem to play a role in this regrouping and in regulating gene activity. This observation gives credence to the idea that the primary role of all the extra DNA in higher eukaryotes might be involved in their complex development.

In general, transposons create rearrangements of the genome. In doing this, they may cause deletions or inversions during the transposition event itself. In addition, they serve as a substrate for HR systems since transposition creates multiple copies on the same or different chromosomes.

Box 21.5 Barbara McClintock and jumping genes One of the most remarkable stories in the history of science is Barbara McClintock's single-handed, long-unappreciated discovery of transposition and transposons. McClintock was trained as a classical geneticist during the first quarter of the twentieth century, only decades after Mendel's work was rediscovered. She began her research career at Cornell University in 1926, concentrating then, and through the rest of her life, on the genetics of maize. It must be noted that this was a quarter of a century before the physical nature of genes was realized. Genes, during much of McClintock's career, were points on chromosome maps that were associated with phylogenetic traits.

During the next two decades, McClintock brilliantly applied classical methods to elucidate a wide variety of aspects of maize genetics, ranging from centromeres to telomeres, from replication to recombination. She gained major status in her field and was elected to the U.S. National Academy of Sciences in 1944.

After her move to the Cold Spring Harbor Laboratory of the Carnegie Institute of Washington in 1942, McClintock's most remarkable discovery was made. Working with a gene that dictated color in maize kernels, she mapped a mutant that gave uncolored kernels. Surprisingly, some kernels in the progeny from such seeds gave a variegated pattern, with small colored spots upon a white background.

It is easiest for us to understand this result, and its implication in modern terms, by looking at later research and the recognition of genes as stretches of DNA that code for specific proteins. The gene designated *C* in **Figure 1** is critical to color production; it codes for an enzyme that is required to synthesize the pigment anthocyanin. When the gene *Dissociator* or *Ds* is transposed into *C*, no enzyme can be produced, and white kernels result. On the other hand, a further displacement of *Ds* in an individual somatic cell in the developing kernel can reactivate *C* in a clone of cells derived from it. This yields a colored spot in the kernel.

Ds cannot transpose itself; it requires the activity of a transposase, coded in another gene called *Activator* or *Ac*. *Ac* is capable of autotransposition and of mobilizing other elements, very much like the situation with the well-characterized human LINEs and SINEs (see Figure 21.24).

The genetic definition of the *Ac–Ds* system and its transposition properties were all deduced by Barbara McClintock using classical genetic methods in the period between 1942 and 1948. The work was reported in a 1950 paper in *Proceedings of the National Academy of Sciences* and at the following Cold Spring Harbor Symposium on Quantitative Biology. Unfortunately, the idea of jumping genes was pure heresy to most geneticists at the time. Despite her stature in the field and the solidity of her evidence, McClintock's idea was received with skepticism and even scorn. Even by 1953 she "had already concluded that no amount of published evidence would be effective."

Yet science has a way of righting wrongs. In the 1960s and 1970s an increasing body of evidence, involving many different organisms, vindicated her position. In 1983, Barbara McClintock was the sole awardee of the Nobel Prize in Physiology or Medicine "'for her discovery of mobile genetic elements."

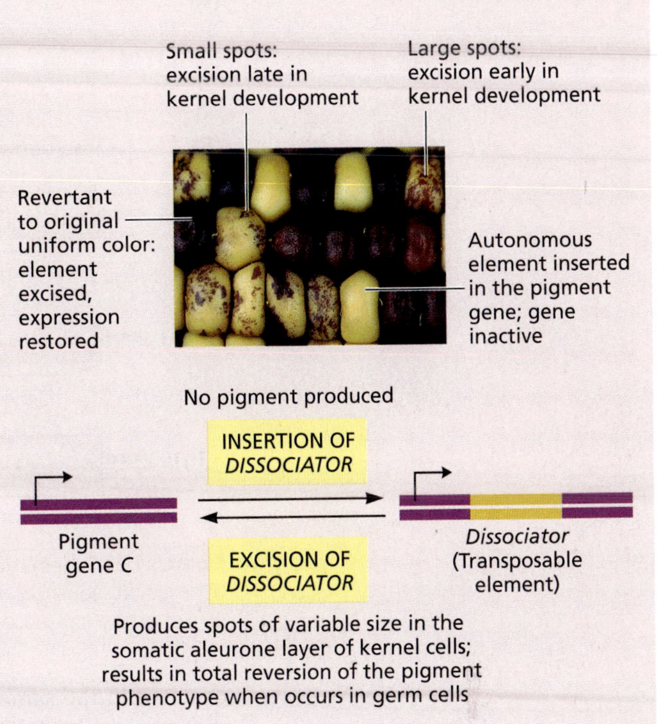

Figure 1 Transposable elements affect maize kernel color. McClintock noticed that kernels on a maize ear may show unstable variegated phenotypes, with some kernels showing no color, others exhibiting dark uniform color, and still others displaying a spotted appearance with dark spots on a white background. She explained this phenomenon by the behavior of some genetic elements that jump around the genome, or jumping genes. She also realized that the product of another gene, *Ac* or *Activator*, is needed for jumping to occur. Today we know that this gene encodes the transposase. (Top, courtesy of Robert Martienssen, Cold Spring Harbor Laboratory. Bottom, adapted from Feschotte C, Jiang N & Wessler SR [2002] *Nat Rev Genet* 3:329–341. With permission from Macmillan Publishers, Ltd.)

There are several types of transposons

There are several criteria according to which transposons are classified. The simplest system is based on their general organization and the types of genes they carry. In bacteria, we recognize simple **insertion sequences, IS**, and **composite transposons** (Table 21.2 and Figure 21.23). All bacterial elements contain inverted repeats at the ends, which are required for the transposition to occur. In addition, they all carry a gene encoding a **transposase**, the enzyme that performs the cleavage and ligation necessary for transposition. Thus, they carry within themselves a critical element for their behavior. The composite transposons can also carry other genes, which usually confer antibiotic resistance. These extra genes are located in the central region, which is flanked by an IS element on each end. The IS elements can be

Table 21.2 Characteristic features of some insertion sequences, IS, or composite transposons in *Escherichia coli*.

Transposon	Size (bp)	Target (bp)	Inverted repeat (bp)	Resistance conferred
IS1	768	9	23	
IS2	1327	5	41	
IS4	1428	11–13	18	
IS10R	1329	9	22	
Tn5	~5700	9		kanamycin
Tn10	~9300	9		tetracycline
Tn2571	~23,000	9		chloramphenicol, streptomycin, sulfonamides, mercury

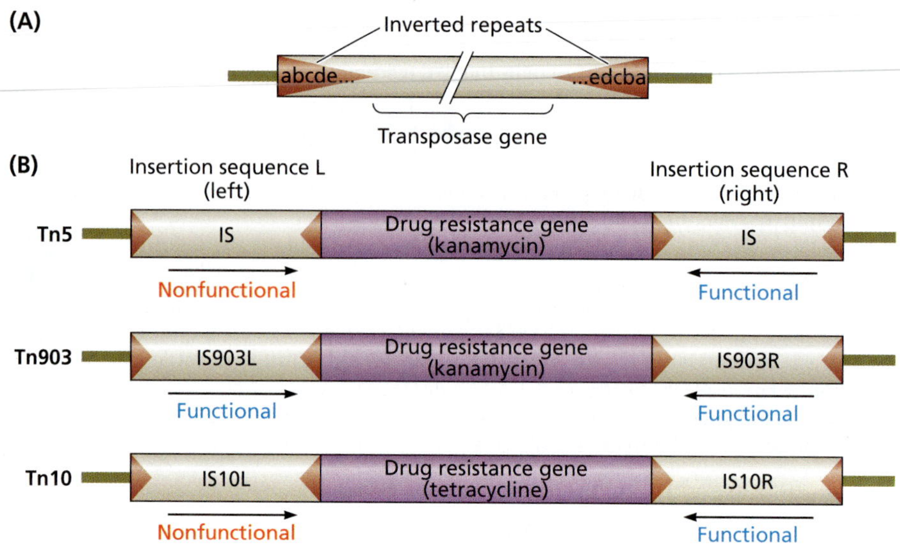

Figure 21.23 Schematics of two types of transposable elements in bacteria. (A) Insertion sequences have inverted repeats at their ends and a transposase gene at their center. (B) Composite transposons: three well-studied examples, Tn5, Tn903, and Tn10, are shown.

oriented in the same direction or inverted; some IS elements may be functional, that is, capable of mediating transposition, whereas others are nonfunctional.

The number and types of transposons in eukaryotic cells are staggering, and the most useful classification is based on their mechanism of transposition (**Figure 21.24** and **Table 21.3**). Some elements use a cut-and-paste route: in other words, a DNA element in the genome is excised from its original position and then inserted into a different position in the genome. Others copy the elements without excising the mother copy and then insert the daughter copy into the new position. These transposons are known as DNA transposons or class II transposons because they involve only DNA transactions. No currently active DNA transposons have been identified in mammals.

Table 21.3 Human transposons.

Element	No. of copies (×1000)	Total length (Mb)	% of genome	Activity
LTR retrotransposons	443	227	8.3	
LINEs	868	558	20.4	
LINE-1	516	462	16.9	active
LINE-2	315	88	3.2	
LINE-3	37	8	0.3	
SINEs	1558	360	13.3	
Alu	1090	290	10.6	active with LINE-1
MIR and MIR3	468	69	2.5	
SVA	2.76	4	0.15	active with LINE-1
DNA transposons	294	78	2.8	

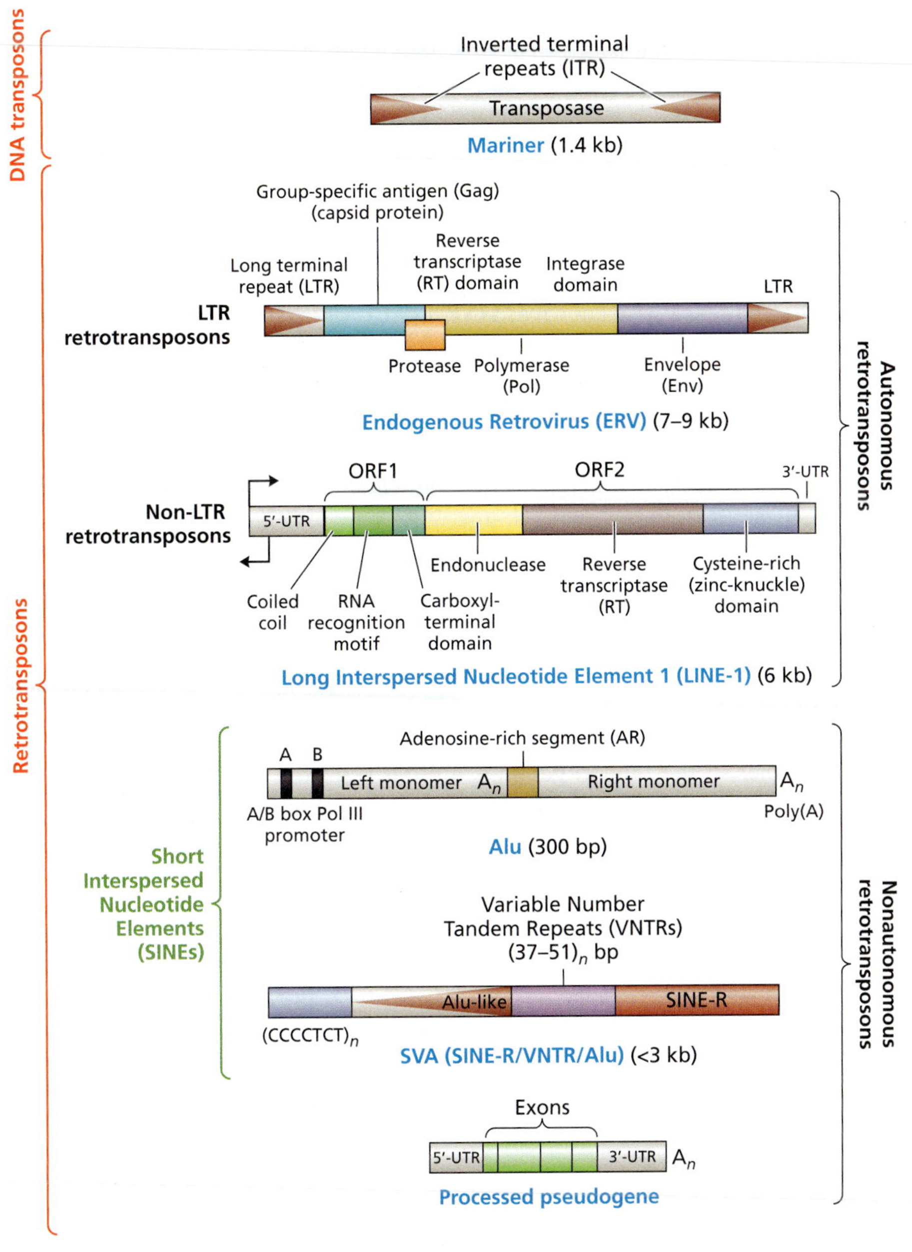

Figure 21.24 Classes of transposons and specific examples in the human genome. Examples are classified according to their mechanisms of transposition. In addition to the specific information for each class of transposons, provided by the figure itself, several features of the long interspersed nucleotide element 1 or LINE-1 are worth mentioning. LINE-1 possesses two open reading frames or ORFs, both of which are required for transposition. ORF1 encompasses a coiled-coil domain, an RNA recognition motif, and a basic carboxyl-terminal domain: the coiled-coil domain participates in ORF1p protein trimer formation, while the other two domains bind to nucleic acids. The endonuclease and reverse transcription activities of ORF2 are also essential for transposition, but the function of the C-terminus is unclear. ORF1p and ORF2p preferentially associate with their respective mRNAs to form ribonucleoprotein or RNP particles that participate in transposition. The LINE-1 5′-untranslated region or UTR contains two promoters. There is an internal Pol II promoter that directs sense transcription of the element. In addition, it contains a potent antisense promoter whose transcript contains a portion of the 5′-UTR and genomic sequences flanking the 5′-end. The products of the sense and antisense transcription presumably form dsRNAs that regulate LINE-1 transposition by RNA-interference-based mechanisms. (Adapted from Beck CR, Garcia-Perez JL, Badge RM & Moran JV [2011] *Annu Rev Genomics Hum Genet* 12:187–215. With permission from Annual Reviews.)

The second major type of eukaryotic transposons, known as **retrotransposons** or class I transposons, are those that pass through an RNA stage, also multiplying by a copy-and-paste mechanism but now involving an RNA intermediate. In the first stage of the process, the DNA element is transcribed into RNA; in the second stage, the RNA is reverse-transcribed into DNA, which is then integrated into the genome at a new position. Reverse transcription is catalyzed by a reverse transcriptase, which is encoded by the transposon itself. Retrotransposons behave very similarly to retroviruses, such as HIV.

Retrotransposons fall into two major categories, **autonomous retrotransposons** and **nonautonomous retrotransposons**, depending on whether they contain all the information needed for transposition or require other functional transposable elements to help them transpose. Each of these two categories is subdivided according to the presence of specific sequence elements. Autonomous retrotransposons may or may not carry long terminal repeats, hence the LTR and non-LTR distinction; both encode reverse transcriptase. **LTR retrotransposons** include endogenous retroviruses, relics of past infection of the germline that have lost their ability to re-infect due to a nonfunctional envelope gene.

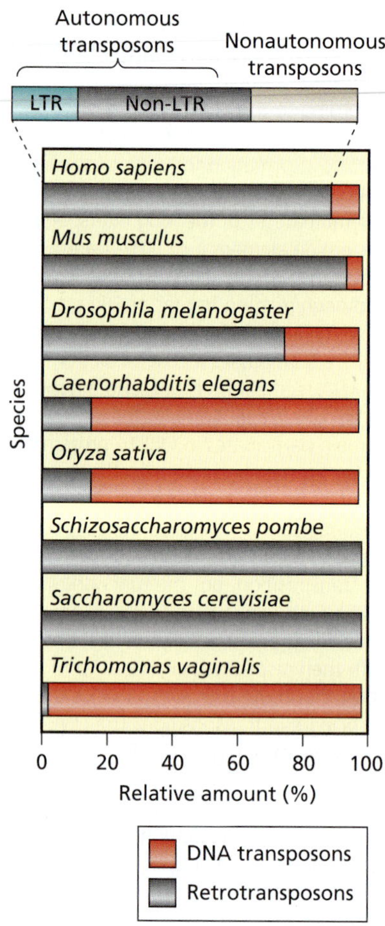

Figure 21.25 Relative amount of DNA transposons and retrotransposons in some eukaryotic genomes. The expansion of the human retrotransposon portion at the top depicts relative amounts of the three major retrotransposon types. (Adapted from Feschotte C & Pritham EJ [2007] *Annu Rev Genet* 41:331–368. With permission from Annual Reviews.)

Most of the members of the non-LTR retrotransposon group, whose number is estimated at half a million, ~17% of the entire human genome, have lost their ability to act in transposition. A well-understood non-LTR transposon is the **long interspersed nucleotide element 1**, also known as **LINE-1** or **L1**, which is the only currently active transposon that can mobilize or transpose itself and all other elements. LINEs are transcribed by Pol II.

The nonautonomous category contains **short interspersed nucleotide elements** or **SINEs**, inactive transposons that rely exclusively on LINE-1 for transposition. These elements fall into two categories: Alu elements and SVA elements. Alu elements account for ~10% of the human genome; they contain two monomeric sequences derived from the signal recognition particle 7SL RNA and an internal A/B box Pol III promoter. The monomeric sequences bind SRP9–14 proteins, like they do in the signal recognition particle itself. SVA elements, short for SINE-R/VNTR/Alu, have a composite structure and are probably transcribed by Pol II.

The nonautonomous category also includes **processed pseudogenes**. Processed pseudogenes arise through occasional usage of L1-encoded proteins to mobilize mature mRNAs to new genomic locations. Interestingly, most pseudogenes, ~10,000 copies in the human genome, are derived from genes highly expressed in the germline, such as housekeeping and ribosomal protein genes. Because pseudogenes lack functional promoters, most of them are not transcriptionally active, although some pseudogenes are expressed, probably by co-opting nearby promoters.

Representatives of the nonautonomous category do not code for reverse transcriptase; they are transcribed by Pol III.

It is curious to note that the relative amounts of DNA transposons and retrotransposons vary enormously among species (**Figure 21.25**). In humans, retrotransposons prevail, and the autonomous category is dominant over the nonautonomous one.

DNA class II transposons can use either of two mechanisms to transpose themselves

There are two mechanisms by which DNA transposons can move: nonreplicative and replicative (**Figure 21.26**). In the nonreplicative mechanism, which is a cut-and-paste-type mechanism, the transposase makes a staggered cut at the target site, which can be sequence-specific or random; this cut produces sticky ends. Then the transposase cuts out the DNA element from a distant genome location and ligates it into the protruding DNA ends of the target site. The resulting gaps are filled by DNA polymerase, and the integrity of the DNA backbone is restored by the action of a DNA ligase. This results in target-site duplication; actually, the existence of short direct repeats serves to identify insertion sites in specific genome regions or genomewide.

The second mechanism, replicative transposition, involves a replication step, in which the sequence at the donor site is copied in S phase and inserted at a new target site that has not yet been replicated. When the target site with the insert is replicated at a later stage, the number of transposons actually doubles. Thus, with time, replicative transposition leads to ever-increasing transposon numbers. This mechanism involves two types of enzymes: the usual transposase, which acts at the ends of the original transposon, and a resolvase, which acts on the duplicated copy.

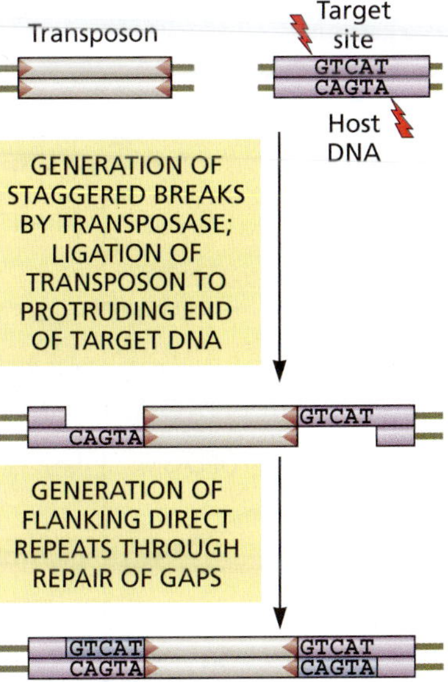

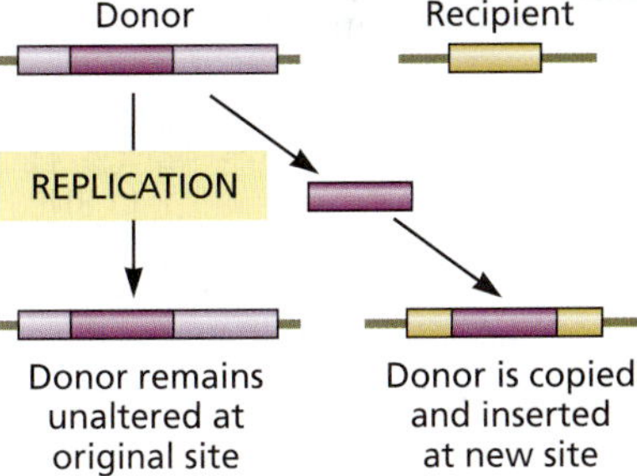

Figure 21.26 Mechanisms of DNA transposition. (A) Nonreplicative or cut-and-paste mechanism; (B) replicative or copy-and-paste mechanism.

Transposase is an enzyme that performs both the excision and reinsertion steps in transposition. The process can be described in three steps: (1) Two copies of the enzyme bind to the transposon DNA at the inverted repeats located at the ends of the transposon; they form a dimer following binding. (2) The two ends are brought together, closing the transposon into a big loop, and the transposase cuts the transposon DNA at both ends. (3) The enzyme finds a new location on the DNA and reinserts the transposon. To do so, it must cut the DNA at the new site. A diagram and structure of a sample bacterial transposase bound to the ends of the excised element are presented in **Figure 21.27**. The enzyme acts as a dimer, with each monomer bound to one of the inverted repeats at the ends of the insertion sequence. The dimerization is responsible for bringing the ends to be cut in close proximity, looping out the intervening IS sequences; the DNA is then cut at both ends, releasing the element for insertion into the target site.

Retrotransposons, or class I transposons, require an RNA intermediate

As might be expected, the mechanism for transposing retrotransposons is a more complicated process, as it involves both forward RNA transcription and reverse transcription of the RNA into a dsDNA copy, which can then be inserted into the genome. In addition, the process requires the participation of proteins encoded by the element, and these have to be translated from the mRNA after it has been exported to the cytoplasm. Transcription and reverse transcription subsequently occur in the nucleus. The mere spatial separation of these processes adds a level of complexity to an already complex process. We illustrate the process for the best-studied LINE-1 element in humans (**Figure 21.28**).

The integration process itself occurs by target-primed reverse transcription or TPRT (see Figure 21.28). Integration of the element into the genome is initiated by sequence-specific nicking of the target site by the *ORF2*-encoded endonuclease. The newly generated 3′-OH group serves as a primer; thus, polymerization occurs directly onto the host DNA. The L1 mRNA serves as a template. For this to occur, the message needs to be in contact with the target site; this steady contact is ensured by base pairing between the poly(A) tail of the message and the single-stranded T-rich portion of the target that was created by ORF2p.

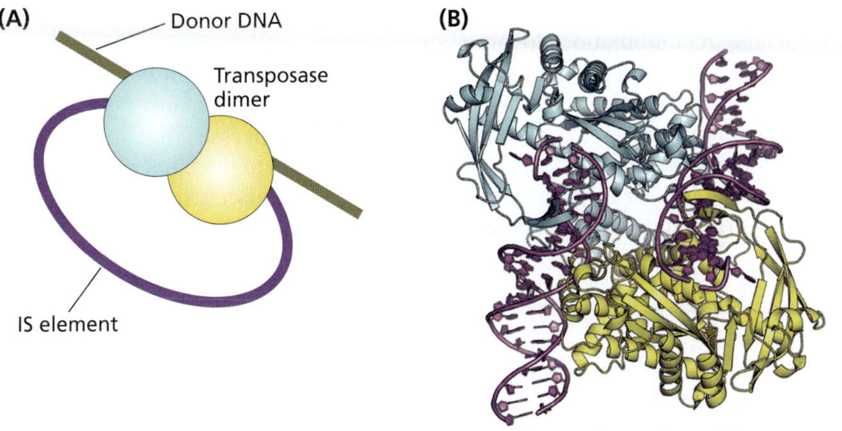

Figure 21.27 Cut-and-paste mechanism of transposase action. (A) The transposase cuts out the transposon DNA and moves it to a different place. (B) The structure shows two copies of the enzyme holding the two severed ends of the DNA; the actual loop of DNA is quite large, ~5700 bp long. The enzyme shown is a bacterial transposase that moves Tn5.

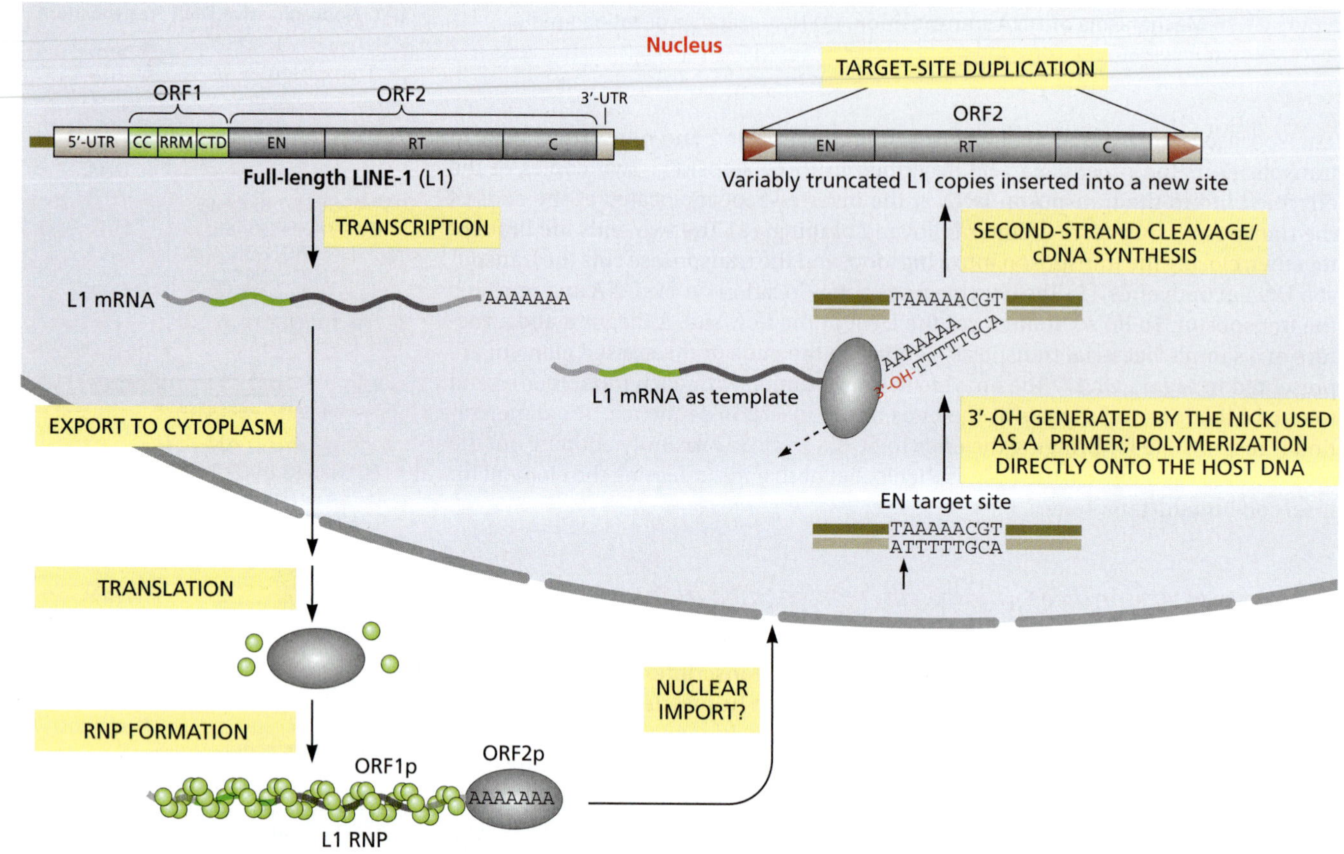

Figure 21.28 Target-primed reverse transcription, TPRT, integrates L1 into the genome. The process begins with transcription of L1 elements and export of the mRNA into the cytoplasm, where the two proteins encoded by ORF1 and ORF2 are translated. These proteins then interact with the mRNA to produce a ribonucleoprotein or RNP particle. Once this particle is imported into the nucleus by an unknown mechanism, nicking of the host chromosome at a specific sequence by the ORF2-encoded endonuclease, EN, initiates the integration of the element into the host genome by TPRT. The process uses the newly created 3′-OH group as a primer for the RT activity of ORF2; the mRNA is held in place by base-pairing interactions between the target site and the poly(A) tail at the 3′-end of the mRNA. The rest of the steps are poorly elucidated. The final result of integration is often a 5′-truncated L1 copy flanked by target-site duplications. The two L1-encoded proteins, ORF1p and ORF2p, are presumably hijacked by nonautonomous elements like Alu, SVA, and occasionally mature mRNAs to mediate their integration *in trans*. (Adapted from Beck CR, Garcia-Perez JL, Badge RM & Moran JV [2011] *Annu Rev Genomics Hum Genet* 12:187–215. With permission from Annual Reviews.)

21.6 Site-specific recombination

Site-specific recombination involves only limited sequence homology between recombining partners; thus, it occupies an intermediate position between homologous and nonhomologous recombination. The process does not require RecA or Rad51 recombinases. Two relatively well-understood processes that depend on site-specific recombination are the integration of phage λ into the bacterial genome and the rearrangement of immunoglobulin genes that occurs during differentiation of antibody-producing B-cells in the immune system.

Bacteriophage λ integrates into the bacterial genome by site-specific recombination

Site-specific recombination was first observed in phage λ as a mechanism for its integration into a specific site of the host chromosome, known as lysogeny. Phage λ is the best-studied representative of **temperate phages** (Figure 21.29). These viruses can exist in one of two modes: they either replicate in the host bacterium immediately after infection and cause cell lysis, which is called lytic mode, or they can integrate their genome into the bacterial chromosome, remaining dormant for many generations while their DNA is replicated as part of the bacterial chromosome, which is

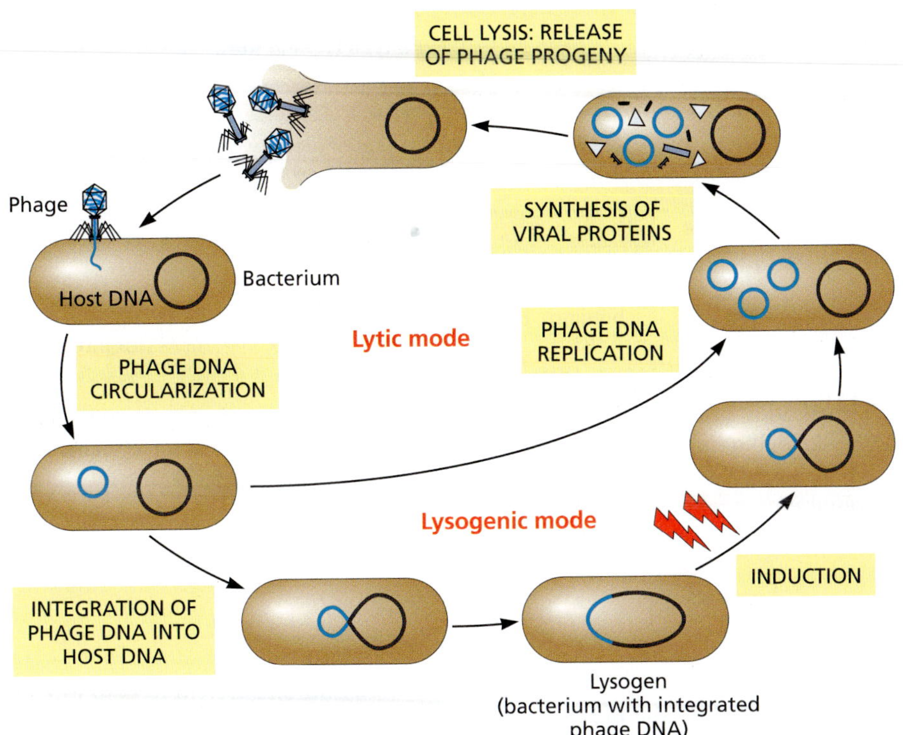

Figure 21.29 Site-specific recombination. This type of recombination during the lysogenic cycle of temperate phages is exemplified by phage λ.

called lysogenic mode. The integrated phage is known as a **prophage**, and the bacterium containing the integrated viral genome is known as the lysogen. Prophages can be induced to enter the lytic cycle by DNA-damaging agents, for example. Excision of the circular phage chromosome during induction is a reversal of the integration process; however, excision requires an additional protein known as Xis or excisionase.

The mechanism of integration (Figure 21.30) makes use of two ***att*** **sites** or attachment sites that share a short 15-bp region of homology: *attB*, in the bacterial genome, and *attP*, in the phage genome. Two proteins, the phage protein **integrase** and the bacterial protein integration host factor or IHF, are absolutely required. Additional bacterially encoded proteins bind to *attP* (Figure 21.31). The two homologous regions are recognized by the integrase, which creates a 4WJ or Holliday junction as an intermediate in

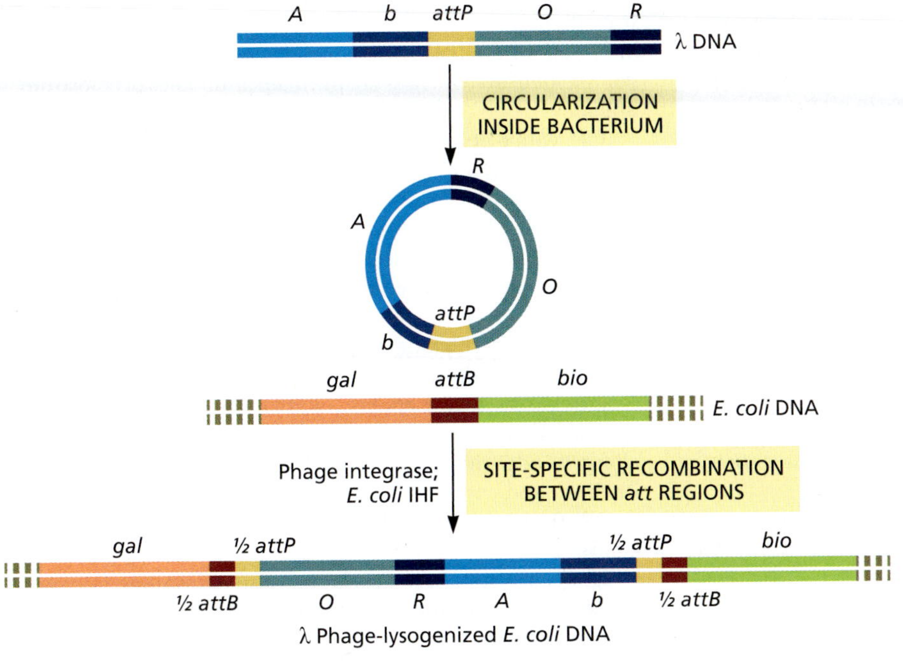

Figure 21.30 Site-specific recombination leads to establishment of lysogeny in bacteriophage λ. The λ phage linear chromosome circularizes upon entry into the bacterial cell by using the *cos* sites at the end of the linear chromosome. Recombination takes place between the *attP* site in the phage genome and the *attB* site in the bacterial genome; these sites share a limited 15-bp region of homology. *attB* is located between genes involved in galactose utilization and biotin synthesis. Two proteins, phage integrase and bacterial integration host factor or IHF, are needed for the reaction to occur. Detailed maps of where these proteins bind in both *attP* and *attB* are shown in Figure 21.31. (Adapted from Mathews CK, van Holde KE, Appling DR & Anthony-Cahill SJ [2012] Biochemistry, 4th ed. With permission from Pearson Prentice Hall.)

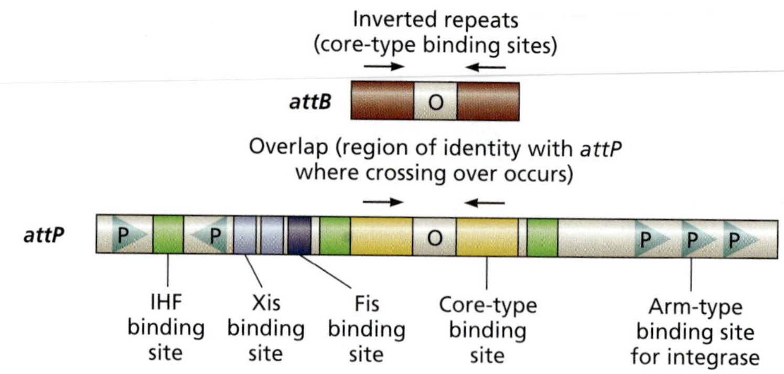

Figure 21.31 Binding of the site-specific recombinase λ integrase to *attP* and *attB* attachment sites. The integrase recognizes both *attB* and *attP*; these sites are quite different apart from the short stretch of identity, O, where crossing over occurs. *attB* is a simple site encompassing two short inverted repeats, called core-type binding sites, flanking the overlap region. *attP* is more complex, containing several adjacent binding sites called arm-type sites; it also contains binding sites for other factors involved in integration and excision. (Adapted from Groth AC & Calos MP [2004] *J Mol Biol* 335:667–678. With permission from Elsevier.)

the integration process. IHF participates in the integration reaction by bending DNA at the site of binding by 180° (see Chapter 8). An interesting observation concerns the state of phage DNA: it must be supercoiled for recombination to occur. The mechanistic requirement for supercoiling is not well understood, nor are the roles of the additional bacterial proteins that bind to *attP*.

Immunoglobulin gene rearrangements also occur through site-specific recombination

Vertebrates have developed a highly sophisticated immune system to fight foreign substances invading the organism, including viruses and bacteria. As part of the immune response, specialized B-cells produce **immunoglobulins** or **antibodies**; these are protein molecules highly specific to the invading molecule. A brief background review on how the immune system works is given in **Box 21.6**. The current

Box 21.6 A Closer Look: Immunoglobulins and polyclonal and monoclonal antibodies The production of antibodies, or immunoglobulin molecules, constitutes the major portion of the immune response of an animal to foreign substances. Sometimes, a molecule that normally occurs within the body is recognized as foreign or, in the terminology of immunology, as non-self; then antibodies are produced against it, leading to autoimmune diseases such as lupus erythematosus.

Immunoglobulins, abbreviated Ig, are produced by specialized cells of the immune system, called B-cells, with one B-cell and its progeny producing only one type of antibody specific to only one antigenic determinant. An antigenic determinant or **epitope** is the entity that is recognized by the immune system as non-self and induces antibody production. There are two types of epitopes in proteins, those that consist of an uninterrupted portion of the primary polypeptide sequence, termed sequential, and 3D epitopes formed by the 3D proximity of amino acid residues that are not sequential in the chain.

Immunoglobulins are made of two heavy chains and two light chains, all held together by disulfide bonds (**Figure 1**). Each chain contains a constant domain, C, and a **variable domain, V**. Constant domains are the same in all antibody molecules of a given class, whereas variable domains confer specificity to a given antigen. Carbohydrates attach to the heavy chain and help to determine the destinations of antibodies in tissues; they also stimulate secondary responses such as phagocytosis.

Several different kinds of heavy chains exist, yielding the following classes of antibodies with different localization and functionalities:

IgG, with γ heavy chains, can readily pass the blood vessel walls and the placenta to protect the fetus. One IgG variant is attached to B-cell surfaces. IgGs trigger secondary immune responses, known as the complement system, that destroy foreign cells. IgGs have the highest serum concentration, ~1 g/dL, and the longest half-life, 21 days.

IgA, with α heavy chains, is found in body secretions such as saliva, sweat, and tears and along the gastrointestinal and respiratory tracts, where the antibodies are arranged along the surface of cells to interact with antigens, preventing the antigens from directly attaching to cells. The invading substance is then swept out of the body together with IgA. They can trigger the complement system. IgAs are the main antibodies of colostrum and milk.

IgD, with δ heavy chains, is found on the surface of B-cells, where the antibodies serve as antigen receptors. IgDs participate in class switching, in which a B-cell changes the class of antibodies it produces. During this process, the constant region of the heavy chain is changed, but the variable region stays the same. Since the variable region does not change, class switching does not affect antigen specificity. What changes is the interaction with different effector molecules that use different pathways to destroy the antigen.

IgE, with ε heavy chains, is associated with allergic responses, known as immediate hypersensitivity. They bind to allergens and trigger histamine release from mast cells in epithelium and connective tissue. IgEs have the lowest serum concentrations, around 5 μg/dL, and the shortest half-life, 2 days.

IgM is involved in the early response to invading microorganisms. They are the largest antibodies, pentamers, whose

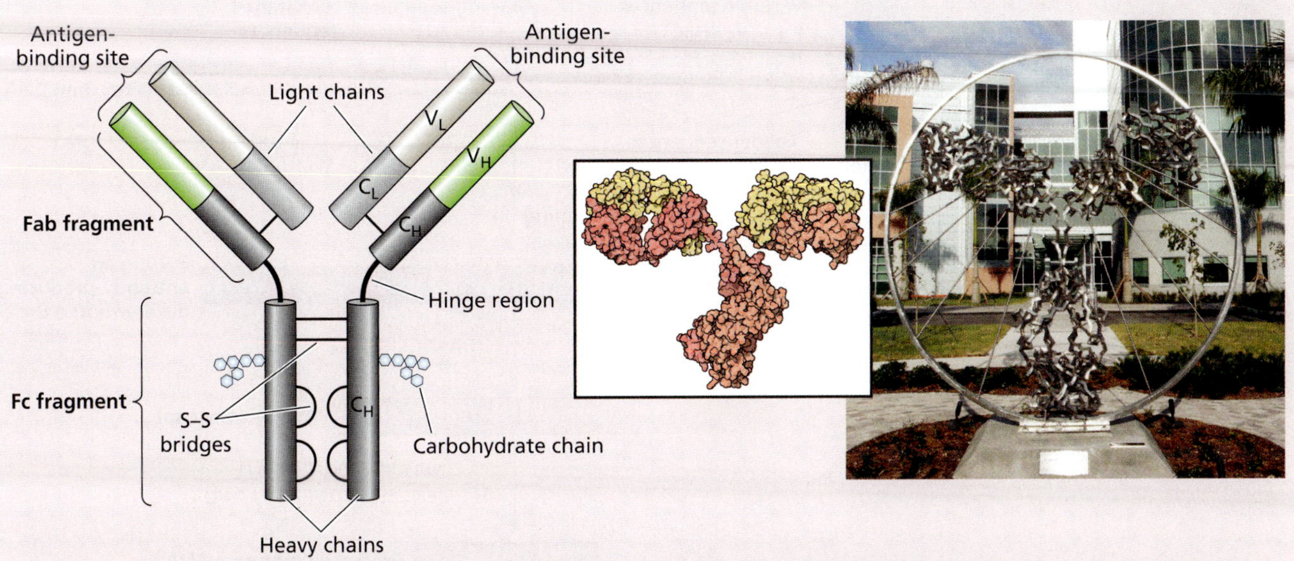

Figure 1 Immunoglobulin structure. Proteolytic cleavage at the hinge regions in the laboratory produces monovalent Fab fragments, which are widely used as research reagents. The Fc fragment in the intact Ig molecule functions as an effector to signal macrophages to attack. The boxed inset shows a space-filling model of an antibody molecule, with the two heavy chains in red and orange and the two light chains in yellow. The picture at the right is an artistic rendering of the molecule by quantum physicist-turned-sculptor Julian Voss-Andreae. The stainless-steel sculpture named Angel-of-the-West was erected in 2008 in front of the Florida campus of the Scripps Research Institute and symbolizes the protective or angel function of immunoglobulins. (Middle, courtesy of David Goodsell, The Scripps Research Institute. Right, courtesy of Julian Voss-Andreae, Wikimedia.)

monomers are held together by disulfide bridges and a joining chain. The large size restricts IgMs to the bloodstream. IgMs trigger the complement system.

Two types of antibodies can be produced; both are widely used in research and the clinic. **Polyclonal antibodies** or PAbs represent a mixed population of antibodies that recognize numerous epitopes. They are produced by an organism in the course of a normal immune response. By contrast, **monoclonal antibodies** or MAbs are produced by a single B-cell or its identical progeny. They are specific only for a given antigenic determinant and can be produced by a specific protocol in the laboratory (Figure 2). The 1984 Nobel Prize in Physiology or Medicine was awarded to Niels Jerne, Georges Köhler, and César Milstein "for theories concerning the specificity in development and control of the immune system and the discovery of the principle for production of monoclonal antibodies."

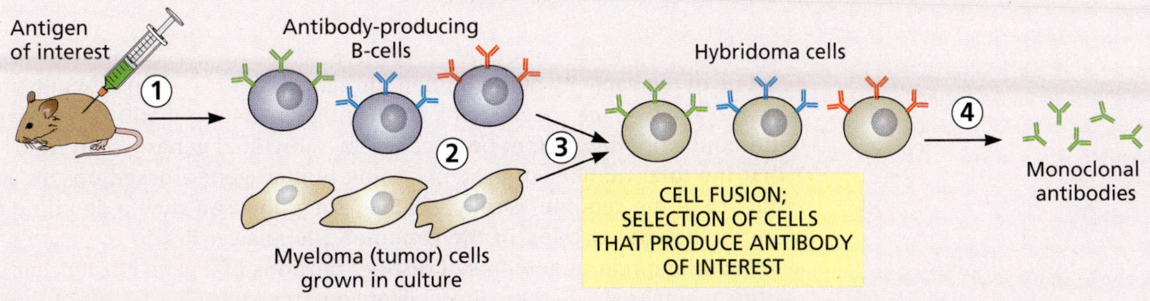

Figure 2 Production of monoclonal antibodies by hybridoma technology. The procedure involves several steps. (Step 1) A mouse is immunized with the antigen of interest; the mouse responds by producing B-cells in the spleen that secrete antibodies against the antigen. (Step 2) The spleen is removed and the B-cells are isolated. These are highly heterogeneous cell populations that contain individual B-cells, each producing an antibody specific to a single epitope on the antigen. These cells are, however, short-lived in culture. (Step 3) Isolated B-cells are fused to myeloma cells, cancerous B-cells that can multiply indefinitely in culture and are preselected to not produce antibodies. The resulting hybrid cells, called hybridomas, grow at the rate of myeloma cells but also produce large amounts of the desired antibody. Unfused plasma cells and myeloma cells die out because of the use of selective growing medium. (Step 4) Hybridomas that produce antibodies specific to a given epitope are selected and grown in bulk. Hybridomas can be frozen and shipped to other research and clinical laboratories, where they can be further propagated, serving as an unlimited source for the monoclonal antibody of interest.

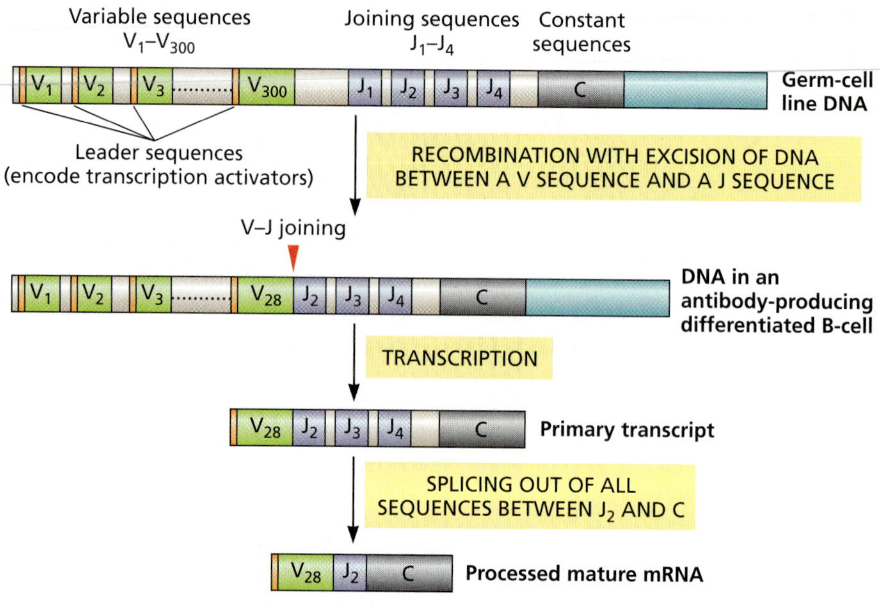

Figure 21.32 Gene rearrangements for production of κ light chains. Gene organization in a germ-cell line is shown at the top. Germ cells are not differentiated, so they do not produce antibodies. In antibodies, each light chain is encoded by noncontiguous sequences on the same chromosome: V, variable; J, joining; and C, constant. In humans there are ~300 different V sequences, each encoding the first 95 amino acid residues of the variable region. Each V sequence is preceded by its leader sequence, shown in orange, which contains transcription activator sequences that are not expressed in the germline; the V sequences cluster on the chromosome. Each of the four J sequences encodes the last 12 amino acid residues of the variable region; they form a separate cluster. Finally, there is one C region. During differentiation of one antibody-producing clone of B-cells, gene rearrangement occurs: the final mature mRNA and the polypeptide will contain one V, one J, and one C region. The DNA sequences that are excised in the recombination process—in the example shown, the sequences between V_{28} and J_2—are permanently lost from all progeny of this cell line. Note, however, that the V sequences upstream of the junction, in this case V_1–V_{28}, and the downstream J sequences, in this case J_2–J_4, remain in the DNA. The other steps that lead to production of a functional antibody molecule occur at the level of transcription and RNA splicing. Transcription uses the leader sequence of only the V sequence that had been joined to the J sequence, thus producing an mRNA precursor that contains only one V sequence. Removal of the extra J sequences occurs during primary transcript splicing. (Adapted from Mathews CK, van Holde KE, Appling DR & Anthony-Cahill SJ [2012] Biochemistry, 4th ed. With permission from Pearson Prentice Hall.)

estimate of the capacity of the human immune system is ~10 million distinct antibody molecules, each specific for a given antigenic determinant. This huge number of immunoglobulins cannot be encoded by individual genes; recall from Chapter 7 that the total number of protein-coding human genes, based on the complete sequence of the genome, is only ~20,500. Then how can such a daunting number of antibodies be encoded in the genome? The answer lies in the use of unique, albeit random, rearrangements of distinct portions of a gene cluster that exists in the precursors to the B-cells. This cluster contains multiple variants of the variable portion of the immunoglobulin gene. The rearrangements occur during differentiation such that each individual differentiated B-cell carries a final rearranged gene that encodes the synthesis of only one specific antibody molecule. The immune response then involves fast and robust proliferation of this specific B-cell type, by clonal expansion, to produce large amounts of antibodies to neutralize that specific antigen.

Immunoglobulins consist of two heavy and two light chains connected by S–S bridges (see Box 21.6). Each chain comprises two different domains, variable and constant, connected by a short segment called the joining sequence. We illustrate the process of gene rearrangement on one class of light chains, κ or kappa; however, similar

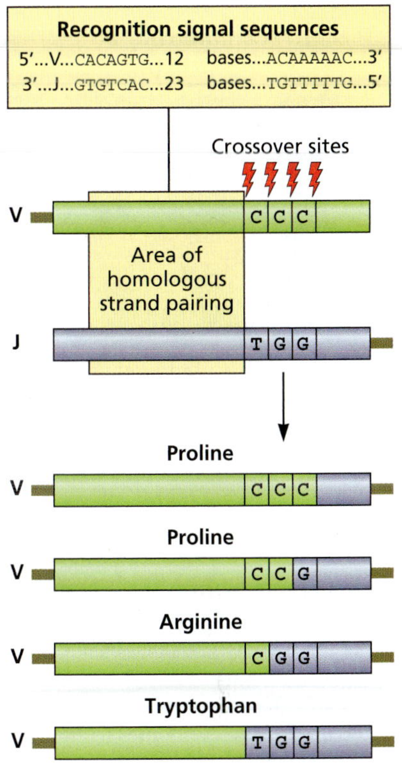

Figure 21.33 Generation of additional diversity by recombination of V and J sequences. Site-specific recombination between recognition signal sequences found at the 3′-side of each V and the 5′-side of each J produces additional diversity. The identity of the final product depends on the exact point of cutting and splicing, that is, location of the crossover sites, within the terminal trinucleotide sequences of the V and J that are being joined. The mechanism enlarges the antibody pool by a factor of 2.5, the average number of different amino acids encoded by four random triplets. (Adapted from Mathews CK, van Holde KE, Appling DR & Anthony-Cahill SJ [2012] Biochemistry, 4th ed. With permission from Pearson Prentice Hall.)

rearrangements are involved in the formation of the genes encoding the other classes of chains. The sequence of events involved in the production of mature mRNA for a specific κ chain is outlined in Figure 21.32. Note that in addition to rearrangements at the gene level, such as V–J joining, the primary transcript needs to undergo splicing of additional sequences, between the J sequence that happened to be joined to V and the C sequence. The final mature mRNA contains only one V, one J, and one C sequence.

Curiously, nature has come up with an additional mechanism that enlarges the antibody pool even further. This mechanism depends on the exact way that V and J sequences recombine. The process is akin to λ phage integration and depends on the existence of limited sequence homology on the 3′-side of each V and the 5′-side of each J (Figure 21.33); these sequences are used for site-specific recombination. The process begins by catalyzing DSBs between the homologous sequences by two proteins, RAG1 and RAG2. During the repair of these DSBs, additional diversity is created.

We have seen how organisms use genetic recombination to produce an enormous number of antibodies to deal with foreign invasion. But sometimes the invaders use the same tactics to frustrate the immunological defense. An elegant example is the trypanosome *Trypanosoma brucei*, the parasitic agent of sleeping sickness (Box 21.7).

Box 21.7 A Closer Look: Parasite antigenic variation and human sleeping sickness African trypanosomiasis, also known as sleeping sickness, is a disease, fatal if untreated, caused by the parasite *Trypanosoma brucei* and its close relatives. The disease develops in two stages. The first, hemolymphatic stage is characterized by fever, swelling of the lymph nodes, headaches, joint pains, and itching. More extensive damage, including anemia and endocrine, cardiac, and kidney dysfunction, occurs if the disease is left untreated. The second, neurological stage occurs when the parasite passes through the blood–brain barrier into the brain and causes symptoms of confusion, lack of coordination, and severe disruption of the sleep cycle; hence the popular name of the disease. Progressive mental deterioration leads to coma and death.

The life cycle of the parasite alternates between two hosts: the African tsetse fly and mammals (Figure 1). In both hosts, the parasite colonizes through rapid proliferation of dividing

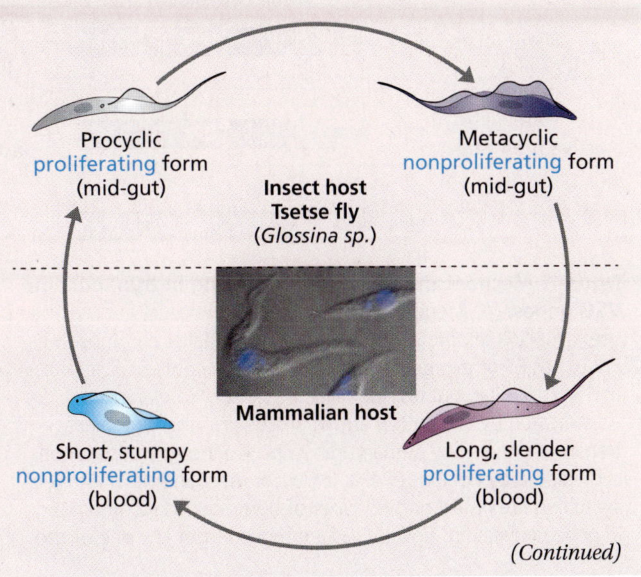

Figure 1 *Trypanosoma brucei* and its life cycle. *T. brucei* is the unicellular parasite that causes sleeping sickness or African trypanosomiasis. The life cycle of trypanosomes alternates between the insect host, tsetse fly, and a mammalian host. The long, slender bloodstream forms present in the mammalian bloodstream undergo continuous variation of the variable surface glycoprotein, VSG. This transformation helps the parasite to survive continuous attacks by the strongly reacting immune system. (Inset) Phase contrast image of *T. brucei*, stained with DAPI, which produces a blue color, to visualize the DNA. The small blue extranuclear spot is mitochondrial DNA in the kinetoplast. (Adapted from Pays E, Vanhollebeke B, Vanhamme L et al. [2006] *Nat Rev Microbiol* 4:477–486. With permission from Macmillan Publishers, Ltd. Inset, from Field MC & Carrington M [2004] *Traffic* 5:905–913. With permission from John Wiley & Sons, Inc.)

(Continued)

Box 21.7 *(Continued)*

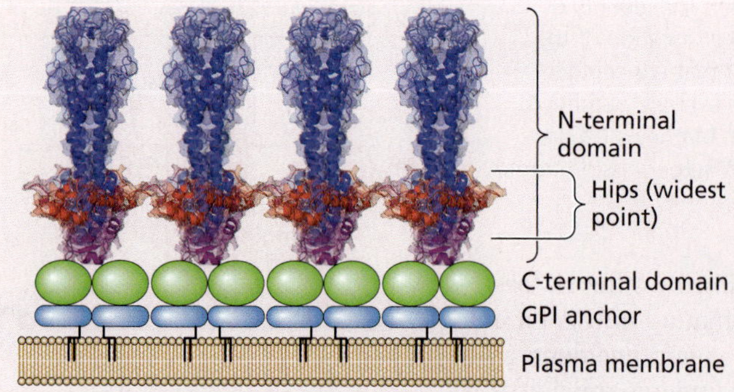

Figure 2 Model of the surface of the bloodstream-form trypanosome. The surface of the trypanosome is essentially a very dense monolayer of VSG molecules, which exist primarily as stable homodimers. The VSG molecules are anchored to the plasma membrane by complex sugar molecules with a fatty acid chain, glycosylphosphatidylinositol or GPI. The density of the VSG is an accurate representation of the *in vivo* situation in the bloodstream form, where the plasma membrane contains ~1×10^7 copies of the same VSG molecule. Significantly, despite substantial primary structural variation among the products of the >1000 gene copies, all VSGs appear to share a common tertiary structure. The arrangement of the VSGs is such that it completely blocks access of antibody molecules to the parasite; thus, the only antigens eliciting an immune response in the host are the VSG molecules themselves, in particular their N-terminal domains. (Courtesy of Mark Carrington, University of Cambridge.)

trypanosome forms. These forms transform into nondividing or quiescent cells that are preprogrammed for cellular differentiation once they transfer from one host to the other.

The surface of the trypanosome is covered by a dense coat of ~1×10^7 molecules of variable surface glycoprotein or VSG (**Figure 2**). The antibodies to a specific VSG kill 99% of the carrier parasites; ~1% survive due to periodic switching to a new VSG form. This switching makes the infection chronic, with the newly coated parasite immune to the antibody response to the old coat. Switching is possible because the *T. brucei* genome encodes >1000 VSG genes that are located in the subtelomeric regions in the 11 large, megabase-sized chromosomes. *T. brucei* also has 1–10 intermediate-size chromosomes and ~100 minichromosomes, which harbor further VSG genes at the telomeres. The parasite expresses these genes monoallelically, one gene at a time, which allows it to keep the other genes as a switching reserve.

Two general strategies have evolved to perform the switching reaction (**Figure 3**). The first is a purely transcriptional mechanism that is not associated with DNA rearrangements: a single gene is expressed, and then its expression is silenced and another gene is activated. The second mechanism is recombination-based; hence its relevance to this chapter.

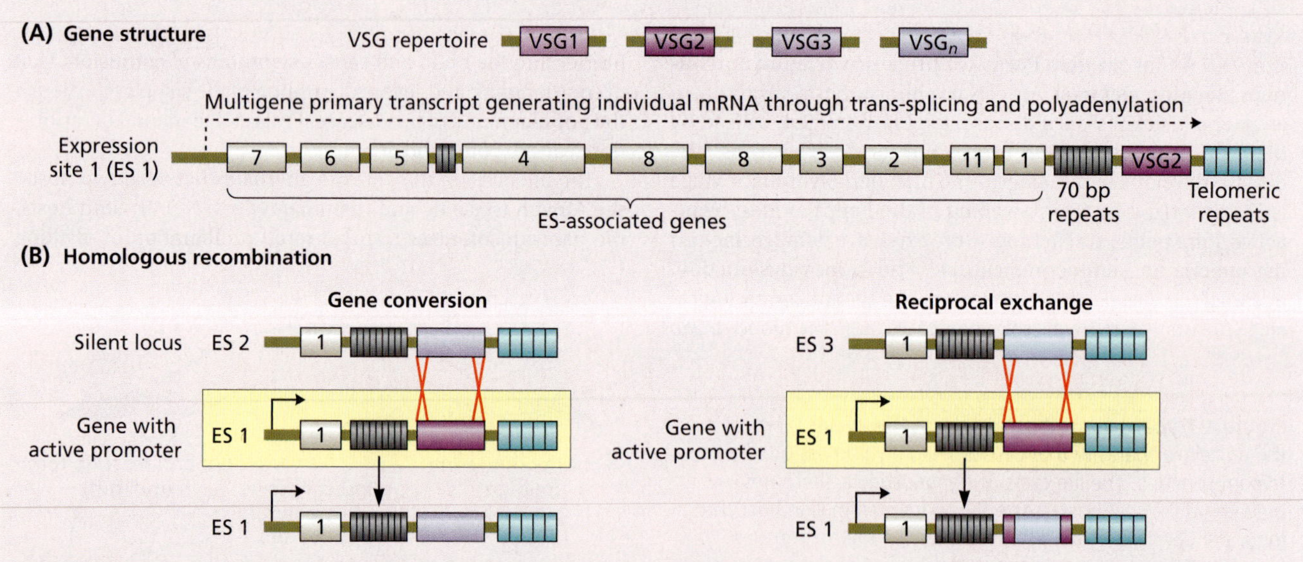

Figure 3 Mechanisms of antigen switching in trypanosome VSG genes. (A) Monoallelic gene expression occurs at only one site, usually a telomeric site on one of the smaller chromosomes, known as ES or expression site. The genome contains a set of 15–20 similar, but not identical, potential expression sites. As indicated by the dotted arrow, all genes in the region are transcribed in a long primary transcript. Curiously, transcription is mediated by Pol I, not Pol II. (B) Major mechanisms of antigen switching are based on two types of homologous recombination. In gene conversion, novel VSGs are moved from any one of the silent loci into the transcriptionally active locus. In reciprocal recombination, new functional genes can be constructed by combining segments of two or more VSG genes or segments of pseudogenes. Transcriptional switching occurs by a process known as *in situ* activation, in which expression of the active ES is turned off and expression of a previously silent ES is turned on. As part of the transcriptional control, active ES sites are depleted of nucleosomes. (Adapted from Pays E, Vanhollebeke B, Vanhamme L et al. [2006] *Nat Rev Microbiol* 4:477–486. With permission from Macmillan Publishers, Ltd.)

Key concepts

- The genome is not static but subject to many kinds of rearrangements. These are grouped under the general title of DNA recombination. Depending upon the homology or lack thereof between the moving DNA and its target site, we distinguish between homologous, site-specific, and nonhomologous recombination.
- Homologous recombination or HR involves exchange of DNA segments between regions of largely, but not necessarily exactly, homologous sequences.
- HR is involved in DNA repair and lengthening of telomeres, but perhaps its most important function is in meiotic cells, where it facilitates exchange of alleles and the proper alignment of homologous chromosomes.
- HR is generally initiated from double-strand breaks or DSBs, followed by resection of the 5′-ends. A single-strand invasion of the intact DNA duplex then occurs, in which the single-strand-binding protein RecA plays a major role. A search for homologous regions then ensues, and the invading single strand forms a duplex with its complementary strand of the duplex DNA, displacing its homologous original strand in the process.
- A fundamental intermediate structure in recombination is the Holliday junction, which allows branch migration and, ultimately, strand exchange in recombination.
- Although some nonhomologous recombination is site-specific, acting through the mutual recognition of small complementary sites, in most cases it occurs through the agency of mobile genetic elements called transposons. These require a special enzyme called transposase to move the transposon to new sites in the genome.
- In bacteria, two types of transposons are found: insertion sequences carry only the transposase gene, whereas composite transposons may also carry other genes, often conveying antibiotic resistance.
- In eukaryotes there are two very different classes of transposons. DNA or class II transposons cut and paste DNA segments, or copies thereof, into new locations. Class I transposons, also called retrotransposons, act through a transcribed RNA that is then reverse-transcribed into dsDNA and inserted into the target site.
- Although transposons constitute a large fraction of the eukaryotic genome, up to almost 80% in some organisms, their function is obscure. Many are not actively transposed and have no other presently known function.
- Site-specific recombination, on the other hand, not only plays the roles mentioned above but also provides great diversity in protein products from a limited genome. Examples are the immune system in vertebrates and the defense against that system used by some parasites.

Further reading

Books

Aguilera A & Rothstein R (eds) (2010) Molecular Genetics of Recombination. Springer.

Leach DRF (1996) Genetic Recombination. Blackwell Scientific.

Smith PJ & Jones CJ (eds) (2000) DNA Recombination and Repair. Oxford University Press.

Tsubouchi H (ed) (2011) DNA Recombination: Methods and Protocols. Springer.

Reviews

Babushok DV & Kazazian HH Jr (2007) Progress in understanding the biology of the human mutagen LINE-1. *Hum Mutat* 28:527–539.

Barzel A & Kupiec M (2008) Finding a match: How do homologous sequences get together for recombination? *Nat Rev Genet* 9:27–37.

Beck CR, Garcia-Perez JL, Badge RM & Moran JV (2011) LINE-1 elements in structural variation and disease. *Annu Rev Genomics Hum Genet* 12:187–215.

Biémont C & Vieira C (2006) Genetics: Junk DNA as an evolutionary force. *Nature* 443:521–524.

Cox MM (2007) Motoring along with the bacterial RecA protein. *Nat Rev Mol Cell Biol* 8:127–138.

Cromie GA & Smith GR (2007) Branching out: Meiotic recombination and its regulation. *Trends Cell Biol* 17:448–455.

Feschotte C, Jiang N & Wessler SR (2002) Plant transposable elements: Where genetics meets genomics. *Nat Rev Genet* 3:329–341.

Field MC & Carrington M (2004) Intracellular membrane transport systems in *Trypanosoma brucei*. *Traffic* 5:905–913.

Goodier JL & Kazazian HH Jr (2008) Retrotransposons revisited: The restraint and rehabilitation of parasites. *Cell* 135:23–35.

Groth AC & Calos MP (2004) Phage integrases: Biology and applications. *J Mol Biol* 335:667–678.

Harper L, Golubovskaya I & Cande WZ (2004) A bouquet of chromosomes. *J Cell Sci* 117:4025–4032.

Holloman WK (2011) Unraveling the mechanism of BRCA2 in homologous recombination. *Nat Struct Mol Biol* 18:748–754.

Kowalczykowski SC (2008) Structural biology: Snapshots of DNA repair. *Nature* 453:463–466.

Lichten M & de Massy B (2011) The impressionistic landscape of meiotic recombination. *Cell* 147:267–270.

Neale MJ & Keeney S (2006) Clarifying the mechanics of DNA strand exchange in meiotic recombination. *Nature* 442:153–158.

Page SL & Hawley RS (2003) Chromosome choreography: The meiotic ballet. *Science* 301:785–789.

Pays E, Vanhollebeke B, Vanhamme L et al. (2006) The trypanolytic factor of human serum. *Nat Rev Microbiol* 4:477–486.

San Filippo J, Sung P & Klein H (2008) Mechanism of eukaryotic homologous recombination. *Annu Rev Biochem* 77:229–257.

Schubert I & Shaw P (2011) Organization and dynamics of plant interphase chromosomes. *Trends Plant Sci* 16:273–281.

Stockdale C, Swiderski MR, Barry JD & McCulloch R (2008) Antigenic variation in *Trypanosoma brucei*: Joining the DOTs. *PLoS Biol* 6:e185.

Sung P & Klein H (2006) Mechanism of homologous recombination: Mediators and helicases take on regulatory functions. *Nat Rev Mol Cell Biol* 7:739–750.

West SC (2003) Molecular views of recombination proteins and their control. *Nat Rev Mol Cell Biol* 4:435–445.

Experimental papers

Ariyoshi M, Nishino T, Iwasaki H et al. (2000) Crystal structure of the Holliday junction DNA in complex with a single RuvA tetramer. *Proc Natl Acad Sci USA* 97:8257–8262.

Biswas T, Aihara H, Radman-Livaja M et al. (2005) A structural basis for allosteric control of DNA recombination by λ integrase. *Nature* 435:1059–1066.

Chen Z, Yang H & Pavletich NP (2008) Mechanism of homologous recombination from the RecA–ssDNA/dsDNA structures. *Nature* 453:489–494.

Duan Z, Andronescu M, Schutz K et al. (2010) A three-dimensional model of the yeast genome. *Nature* 465:363–367.

Forget AL & Kowalczykowski SC (2012) Single-molecule imaging of DNA pairing by RecA reveals a three-dimensional homology search. *Nature* 482:423–427.

Kagawa W, Kagawa A, Saito K et al. (2008) Identification of a second DNA binding site in the human Rad52 protein. *J Biol Chem* 283:24264–24273.

Sehorn MG, Sigurdsson S, Bussen W et al. (2004) Human meiotic recombinase Dmc1 promotes ATP-dependent homologous DNA strand exchange. *Nature* 429:433–437.

Singleton MR, Dillingham MS, Gaudier M et al. (2004) Crystal structure of RecBCD enzyme reveals a machine for processing DNA breaks. *Nature* 432:187–193.

Story RM, Weber IT & Steitz TA (1992) The structure of the *E. coli* recA protein monomer and polymer. *Nature* 355:318–325.

Yamada K, Miyata T, Tsuchiya D et al. (2002) Crystal structure of the RuvA–RuvB complex: A structural basis for the Holliday junction migrating motor machinery. *Mol Cell* 10:671–681.

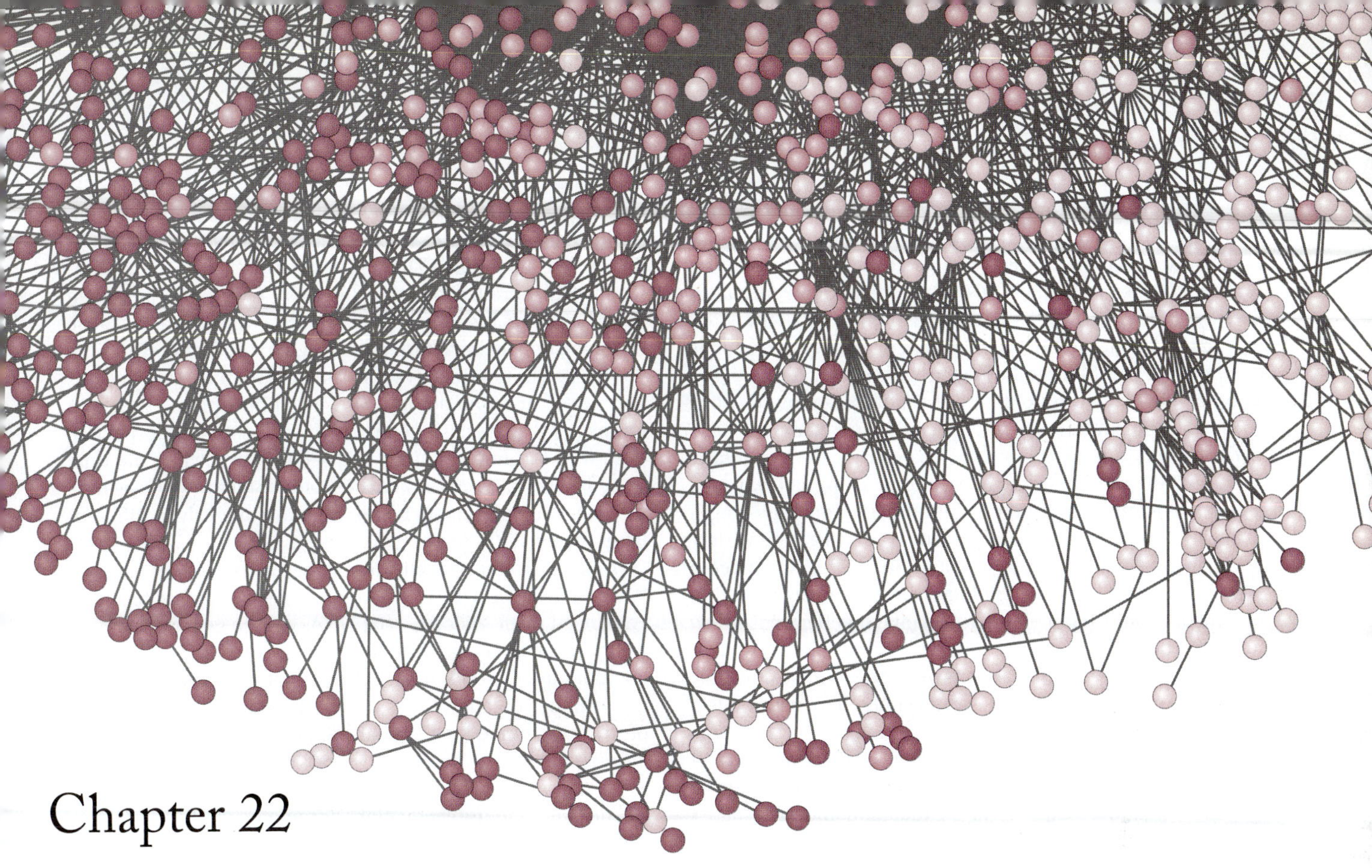

Chapter 22

DNA Repair

22.1 Introduction

Cells constantly experience massive attack from endogenous and environmental factors. It has been estimated that each cell of the human body receives tens of thousands of DNA lesions per day. To take an extreme example, the exposure of skin cells to the UV light from the sun can induce ~100,000 lesions per cell per hour. Lesions in DNA can have deleterious effects on the functioning of the cell, blocking replication and transcription. If these lesions are not repaired, or are repaired in a way that does not restore the original DNA molecule, the consequences to the cell or to the organism can be dire. Mutations accumulate and the genome as a whole becomes very unstable, with numerous changes in chromosome structure. Chromosome translocations and other genome rearrangements occur that will eventually lead to malignant transformation or cell death.

In this chapter, we describe the kinds of lesions that are known to occur in DNA and the best-understood DNA repair pathways. A brief history of the development of this fascinating field of study is presented in **Box 22.1**. These pathways do not act in isolation; they are extensively interconnected in several ways. First, the same type of lesion may be subject to repair by a number of alternative pathways. Second, the pathways often share the same signaling mechanisms; that is, there are signaling molecules that serve to recognize damage and to create a signal that initiates repair, and the same signal may activate several pathways. Third, there are numerous protein molecules that participate in more than one pathway, performing similar functions. Finally, some authors believe that the accepted repair pathway classifications serve more to organize our thoughts than to reflect the biological reality. These authors consider flexibility as the repair systems' most important characteristic; it is often the case that the actual pathway is selected almost at random, depending on the availability of repair factors and the order in which they work.

We begin with a brief, and by no means comprehensive, description of the most common types of lesions in DNA.

Box 22.1 A brief history of the early years in the DNA repair field The remarkable thing about genetic or DNA repair is that the phenomenon was recognized long before it was realized that it was actually DNA that was being repaired, or even that DNA had much biological significance. The story can be traced back to the classical genetic studies of fruit flies by Thomas Hunt Morgan in the early twentieth century (see Chapter 2). This work pointed out the importance of spontaneous gene mutations. A major advance came in 1927, when Hermann Muller found that exposure to X-rays greatly increased the mutation rate in fruit flies. Within a few years, a number of other laboratories found that short-wavelength ultraviolet light could also induce mutations in a wide variety of organisms, including bacteria and fungi. This instigated a new field, called radiation biology. The slow realization that organisms could repair radiation damage generated a new field of study, ultimately termed DNA repair. The early history of this field is rich in incidents of unexpected discoveries, and sometimes more than one group claimed to have made a discovery first (**Figure 1**).

In 1935, Alexander Hollaender, then at the University of Wisconsin, made a most peculiar and puzzling observation. If *E. coli* bacteria were exposed to UV radiation and then plated onto nutrient agar, some colonies appeared only after a considerable delay. One explanation was that the damage to the genes caused by the UV irradiation was somehow being spontaneously repaired. However, these holding recovery experiments were hard to reproduce, and the phenomenon remained an obscure puzzle for many years.

In 1948, Albert Kelner was a young man just beginning his scientific career in the Cold Spring Harbor Laboratory on Long Island, NY. He decided to reinvestigate holding recovery on UV-irradiated *E. coli*. The results from one experiment to another seemed maddeningly inconsistent. Looking at the phenomenon in the fungus *Actinomyces* rather than *E. coli* did not help, nor did careful control of the temperature of incubation. But one day in September of 1948, Kelner noted a most curious correlation: whenever the UV-irradiated samples were subsequently held in full daylight, recovery was remarkably stronger. A series of light-controlled experiments quickly confirmed that recovery from UV damage was somehow stimulated by light in the visible region of the spectrum. Remarkably, at about the same time, Renato Dulbecco, a researcher in the laboratory of Salvatore Luria at the University of Indiana, stumbled upon the same discovery. In Dulbecco's case, it was because he forgot, one night, to turn out the lights in the laboratory. The discovery of photoreactivation, as it came to be called, illustrates two important points about scientific progress: first, that important discoveries often come from the most unexpected, even weird-seeming results; and second, that when the time has come for a discovery, because all of the preliminary data are in, it can often happen almost simultaneously in different laboratories.

Note that photoreactivation was discovered before the Watson–Crick structure and the Hershey–Chase experiment convinced most scientists that DNA was the genetic material. Neither the target nor the mechanism of photoreactivation could be understood in 1948. It was not until 1961 that work in a number of laboratories demonstrated that the formation of thymine dimers and similar molecular species is the DNA lesion repaired by photoreactivation. The photoactivatable enzyme responsible was not isolated until 1983.

Although photoreactivation dominated early interest in the repair of radiation damage, it soon became obvious that other mechanisms were at work. Experiments using chemical mutagens, as well as radiation studies in which light was excluded after treatment, also showed the steady accumulation of viable cell colonies with time of storage. Such liquid storage experiments were conducted in a number of laboratories in the 1950s. The mechanisms for nonradiative repair remained obscure until a crucial experiment was performed in 1963 by Richard Setlow, a researcher at the Oak Ridge National Laboratory in Tennessee. By this time, thymine dimers were recognized as a major product of UV damage, and their cleavage by the photosensitive reactivating enzyme was understood. Setlow used an *E. coli* mutant that was resistant to photoreactivation but would still reactivate in the dark. The presumption was that some other process was cleaving the dimers. Yet analysis of the reactivated cells indicated that the dimers were still there.

1927 — **H. Muller:** X-rays induce mutations
1928 — **F. Gates:** UV at DNA absorption maximum most lethal for bacteria
1935 — **A. Hollaender and J. Curtis:** Recovery after UV irradiation
1941 — **A. Hollaender and J. Emmons:** Action spectrum = UV absorption spectrum
1944 — **O. Avery, C. MacLeod, and M. McCarty:** DNA is the genetic material
1949 — **A. Kelner:** Photoreactivation
R. Dulbecco: Photoreactivation
1953 — **J. Watson and F. Crick:** Structure of B-DNA
J. Weigle: Reactivation of phage λ: basis for Salt Overly Sensitive (SOS) model
1956 — **S. Goodgal:** Evidence for role of enzymes in photoreactivation
1961 — **R. Setlow and J. Setlow:** Thymine dimers shown to be DNA lesions
1964 — **R. Setlow and W. Carrier:** Excision repair
R. Boyce and P. Howard-Flanders: Excision repair
1974 — **M. Radman:** Full description of SOS pathway
T. Lindahl: Base excision repair
1975 — **J. Wildenberg and M. Meselson:** Mismatch repair

Figure 1 Major events in the first 50 years of DNA repair research.

The key to the puzzle came when the DNA from the recovered strains was fractionated into high-molecular-weight, acid-insoluble and low-molecular-weight, acid-soluble oligonucleotides. The thymine dimers were found only in the latter fraction, indicating that small oligonucleotide fragments containing the thymine dimers had been excised from the genomic DNA. Thus, the idea of excision repair was born. Like photoreactivation, excision repair was discovered almost simultaneously in two different labs, by Setlow and by Richard Boyce and Paul Howard-Flanders at Yale. After a hiatus of about 10 years, the repair field saw another burst of fundamental advances in the mid-1970s. In 1974, Tomas Lindahl, of the Karolinska Institute in Stockholm, elucidated the mechanism of base excision repair. In the same year, Miroslav Radman at Harvard provided an overall view of the salt overly sensitive or SOS response, first uncovered by Jean Weigle in collaboration with Matthew Meselson at CalTech in 1953. In the following year, Judith Wildenberg and Matthew Meselson described the mechanism of mismatch repair.

22.2 Types of lesions in DNA

Natural agents, from both within and outside a cell, can change the information content of DNA

There are numerous agents that can affect the chemical structure of DNA and thus its informational content. In addition to the well-known external agents, physical and chemical, coming from the environment, numerous attacks on DNA may come from within the cell itself. The intracellular culprits include, for example, both oxidative stress and the by-products of errors that occur during scheduled biological processes, such as DNA replication, V(D)J recombination in immunoglobulin genes, and meiotic recombination.

Lesions can come in a variety of shapes and forms (**Figure 22.1**). The DNA backbone may be broken, giving rise to single-strand or double-strand breaks: the latter usually appear if two single-strand breaks occur in close proximity, within about one helical turn of the DNA molecule, on opposite DNA strands. The bases themselves can undergo chemical changes, such as **deamination**, oxidation, and **alkylation**; if unrepaired, these lesions lead to mutations that become fixed during DNA replication. Intra- and interstrand cross-links are also common forms of DNA lesion. Figure 22.1 also illustrates the intrastrand cross-links caused by UV light. Interstrand cross-linking also occurs frequently, sometimes as a result of a genetic disorder, and if unrepaired leads to severe disease conditions such as Fanconi anemia (**Box 22.2**). Another major lesion is depurination, when a purine base is cleaved off the backbone. This type of damage can also give rise to mutations, unless the repair machinery finds guidance as to what base, in the context of a dNTP, should be incorporated across from the missing base.

22.3 Pathways and mechanisms of DNA repair

DNA lesions are countered by a number of mechanisms of repair

DNA repair is a collective term that encompasses all biological processes during which alterations in the chemistry of DNA are removed and the integrity of the genome is restored. Many different, but still highly integrated, processes are used by cells to repair DNA damage. They are usually damage-specific (**Figure 22.2**) and some may be biological species-specific. Here, as an overview, we mention only the main characteristic features of each pathway.

- **Direct repair**: damaged DNA base, O^6-alkylguanine or a cyclobutane pyrimidine dimer, undergoes a chemical reaction to restore the original structure.
- **Nucleotide excision repair**, **NER**: repairs helix-distorting base lesions. The mechanism involves excision of a 22–30-nt fragment that contains the damage and use of the resulting ssDNA strand as a template for DNA polymerase action, followed by ligation. There are several subpathways, which will be described in detail. The process uses different mechanisms in bacteria and in eukaryotes.
- **Base excision repair**, **BER**: removes abnormal bases that result from chemical alterations in DNA bases. The first step involves cleavage of the glycosidic

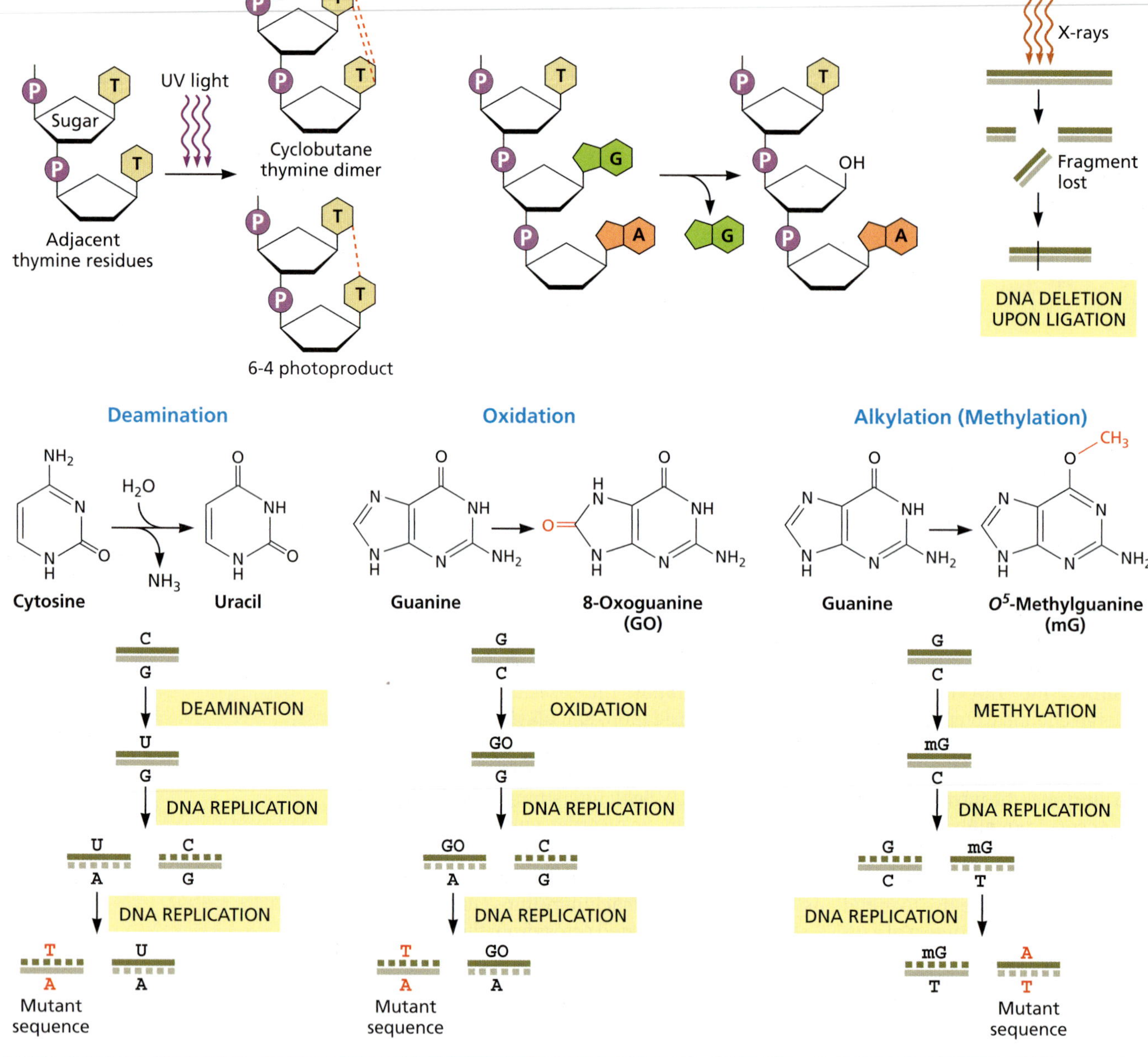

Figure 22.1 Changes that elicit DNA repair responses. These include changes in the chemical structure of DNA and the introduction of double-strand DNA breaks. If not repaired, most of these changes can affect the genetic information stored in DNA, after a couple of replication cycles.

bond connecting a damaged base to the DNA sugar-phosphate backbone, with the removal of the base; then nucleases, polymerases, and ligases come into play.

- **Mismatch repair, MMR**: detection of mismatches and insertion or deletion loops triggers an incision of one of the DNA strands, which is further processed by nucleases, polymerases, and ligases. The proteins involved differ between bacteria and eukaryotes. In *Escherichia coli*, the process is directed by DNA methylation; the mechanism in eukaryotes is unknown.
- **Homologous recombination repair, HR repair**, and **nonhomologous end-joining, NHEJ**: error-free and error-prone mechanisms, respectively, to repair double-strand breaks, DSBs, and other lesions. HR can take place only in the late S and G_2 phases of the cell cycle, whereas NHEJ is active throughout the cell cycle. This distinction is due to the fact that HR needs a homologous intact

Box 22.2 Do defects in the response to DNA damage contribute to aging? Aging is a complex set of processes in metazoans that leads to progressive functional decline and eventual death. It has been thought for years that organisms age as a consequence of stochastic deterioration of biomolecules, caused by oxygen radicals and other endogenous and exogenous harmful compounds or exposure to physical agents such as UV light and X-radiation. The past 20 years have seen a paradigm shift resulting from the experimental identification of life-span-extending loss-of-function mutations, initially in the model nematode *Caenorhabditis elegans*. Mutations in insulin or insulin-like growth factor signaling were found to extend the nematodes' life span and to regulate entry into a life stage that larvae enter when food is scarce. In this stage, called dauer, which is German for lasting, metabolism is severely restricted and the larvae can survive for many weeks. Once conditions improve and food becomes available, the larvae switch back to normal development. Similar effects involving suppression of insulin/insulin-like growth factor signaling have been observed in *Drosophila* and mammals. The popular calorie restriction as a way to prolong life mimics the slow metabolism occurring as part of the natural stress response.

The connection between the insulin-like growth factor pathway and resistance to stress is very complex. Earlier in life, while the organism is developing and reproducing, the pathway is very active and beneficial for overall development at the expense of stress resistance; later in life, this may be detrimental. Thus, mutations that suppress this pathway may be beneficial in later years when development and reproduction are no longer a requirement, but stress resistance is a must.

Aging has been very difficult to study at the molecular level because of its complexity and the absence of meaningful leads to follow. A major effort has been aimed at understanding the causes of human diseases that are characterized by premature aging, known as progeroid syndromes. Table 1 lists some of the most prevalent of these conditions with their symptoms, the mutated repair genes, and the DNA repair pathways that are affected. The study of these conditions necessitated the creation of mouse models with phenotypes that closely mimic those of the respective human syndrome. Collectively, the studies on the human conditions and on the mouse models reveal that premature-aging syndromes are caused by defects in the ability to maintain genome integrity. Thus, the accumulation of unrepaired DNA damage might be the driving force behind aging and age-related pathology.

Table 1 Most prevalent premature-aging syndromes and the gene and repair pathway affected. (Adapted from Schumacher B, Garinis GA & Hoeijmakers JHJ [2008] *Trends Genet* 24:77–85.)

Syndrome	Clinical features	Mutated genes	Repair processes affected
Cockayne syndrome (CS)	neuronal degeneration, loss of retinal cells, cachexia or wasting syndrome: loss of appetite and weight, muscle atrophy, fatigue, weakness	*CSA, CSB*	transcription-coupled NER[a]
trichothiodystrophy	neurological and skeletal degeneration, cachexia, ichthyosis or dry, rough, scaly skin, characteristic brittle hair and nails	*XPB, XPD, TTDA*	transcription-coupled NER
xeroderma pigmentosum (XP)	hypersensitivity to sun exposure, pigmentary alterations and premalignant lesions in sun-exposed skin areas, extremely high incidence of skin cancer	*XPA–D, XPF, XPG*	NER
Fanconi anemia (FA)	pancytopenia or low number of blood cells, bone marrow failure and renal dysfunction, abnormal pigmentation, short stature, cancer	*FANC, BRCA2*	DNA-cross-link repair
Nijmegen breakage syndrome (NBS)	immunodeficiency, increased cancer risk, growth retardation	*NBS1*	DSB repair; telomere instability
Bloom syndrome (BLS)	immunodeficiency, growth retardation, genomic instability, cancer	*BLM helicase*	mitotic recombination
Werner syndrome (WS)	atrophic skin, thin gray hair, osteoporosis, type II diabetes, cataracts, arteriosclerosis, cancer	*WRN helicase*	DNA recombination, telomere maintenance
Rothmund–Thomson syndrome (RTS)	growth deficiency, graying hair, juvenile cataracts, skin and skeletal abnormalities, osteosarcoma, skin cancers	*RECQL4 helicase*	repair of oxidative DNA damage
ataxia telangiectasia (AT)	progressive cerebellar degeneration leading to severe ataxia or lack of voluntary coordination of muscle movements, telangiectasia or dilated blood vessels, immunologic defects, cancer	*ATM*	DSB signaling response

[a]NER, nucleotide excision repair.

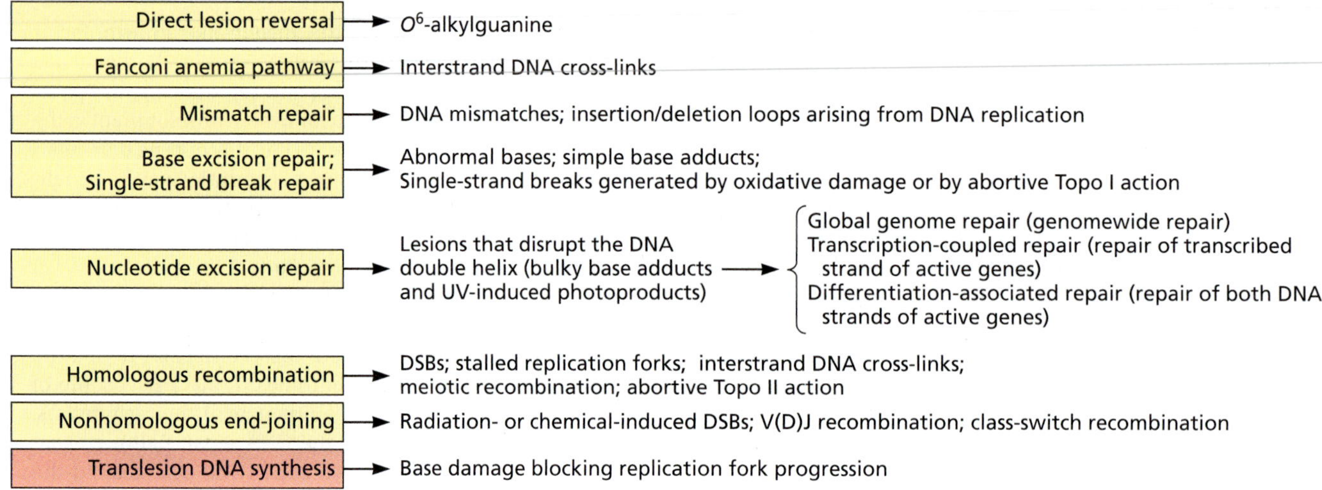

Figure 22.2 Various DNA repair pathways in eukaryotes and the main lesions that they repair. All classical repair pathways that lead to the repair of a lesion, with or without errors, are marked in yellow boxes. There is another, less desirable mechanism for dealing with lesions, known as translesion DNA synthesis, shown in the red box. Rather than repairing the lesion, this mechanism allows it to be bypassed during DNA replication. In addition to repairing harmful lesions in the DNA molecule, some of the pathways are responsible for normal physiological processes, such as meiotic recombination and rearrangement of immunoglobulin genes to create the repertoire of immunoglobulin molecules.

DNA duplex, from a sister chromatid, as a template to fix the damage in an error-free way; such homologous duplexes exist only following DNA replication. On the other hand, NHEJ uses no template for the repair process and is thus error-prone and cell-cycle-independent. Numerous and different proteins are involved in each pathway. HR is most frequently initiated by generation of

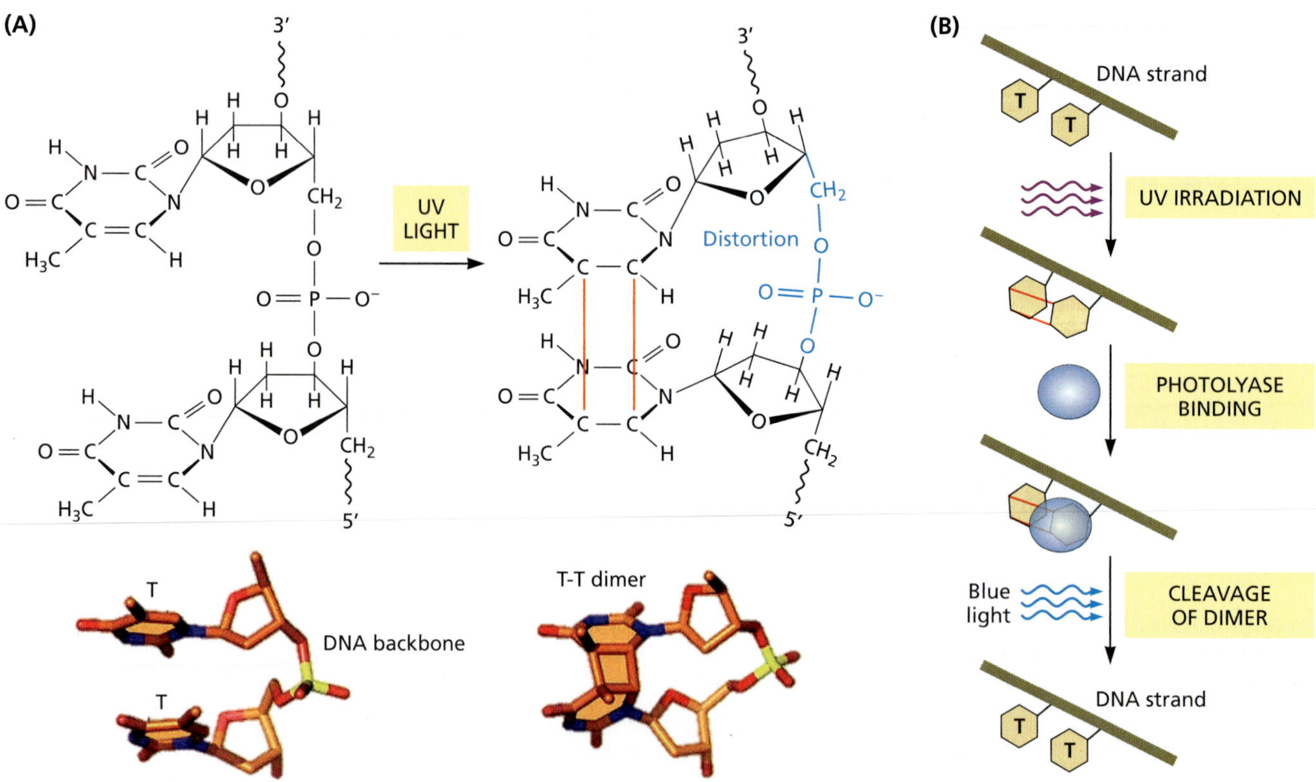

Figure 22.3 Direct repair of thymine dimers by DNA photolyase. This enzyme is also known as photoreactivating enzyme. (A) UV light causes photodimerization of adjacent thymidylate residues, causing distortion of the DNA backbone. (B) Schematic of DNA-mediated direct repair of thymine dimers.

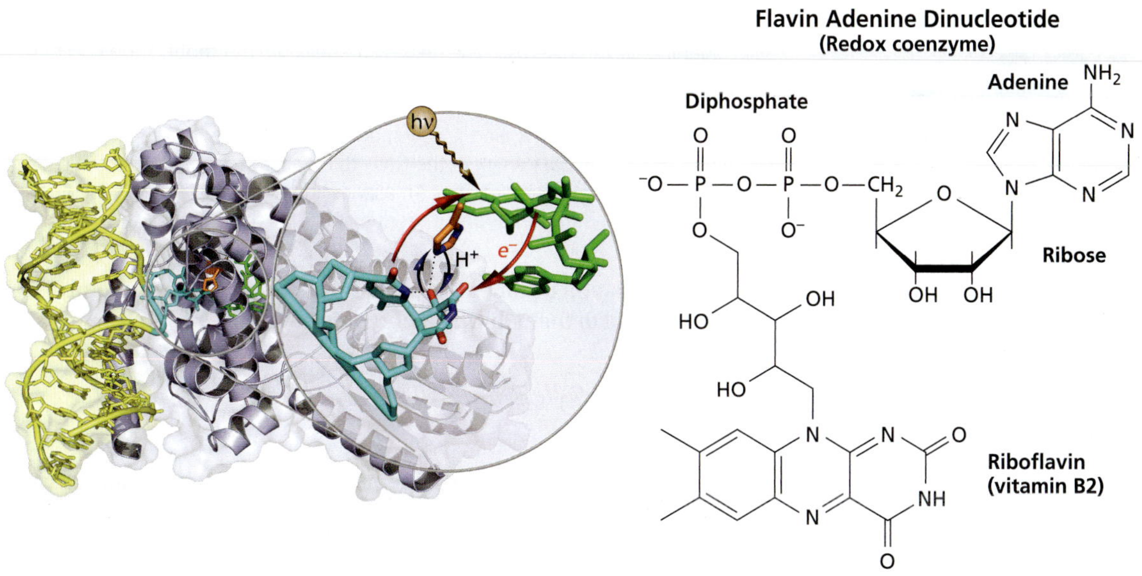

Flavin Adenine Dinucleotide (Redox coenzyme)

Figure 22.4 Action of DNA photolyase. Photolyase is a 471-amino-acid protein that acts through two chromophore co-factors, $FADH^-$, flavin adenine dinucleotide, and MTHF, 5,10-methenyltetrahydrofolylpolyglutamate. The reaction involves the following steps. (Step 1) In a light-independent step, the enzyme binds to the DNA section that contains the lesions. The discrimination ratio between damaged and intact DNA is very high, 10^5. The thymine dimer to be cleaved, shown in cyan, flips out of the DNA helix, shown in yellow. (Step 2) Blue light is absorbed by MTHF, followed by radiationless excitation energy transfer from MTHF* to $FADH^-$. (Step 3) An electron is transferred from the excited $FADH^{-*}$ to the cyclobutane thymine dimer in the bound DNA; this leads to splitting of the dimer to produce intact DNA. (Step 4) Electron transfer from the repaired DNA back to the FADH completes the catalytic cycle. (Left, from Li J, Liu Z, Tan C et al. [2010] *Nature* 466:887–890. With permission from Macmillan Publishers, Ltd.)

a DSB, followed by numerous steps that lead to the formation of crossover or noncrossover products (see Chapter 21). In NHEJ, existing DSBs are detected and bound by a damage sensor, which in turn leads to recruitment of other proteins.

There is still another, relatively recently recognized mechanism that allows cells to continue functioning in the presence of lesions that would normally block DNA replication. This mechanism allows the replication machinery to bypass the lesion without repairing it and depends on the activity of special, low-fidelity polymerases. This pathway is appropriately termed translesion DNA synthesis.

Thymine dimers are directly repaired by DNA photolyase

Short-wavelength UV radiation (see Chapter 1) can induce the formation of covalent dimers between pyrimidine bases, most often thymines, in the DNA chain. Remarkably, **DNA photolyase**, also known as photoreactivating enzyme, repairs cyclobutane pyrimidine dimers when activated by the absorbance of visible light, ~370 nm (**Figure 22.3** and **Figure 22.4**). The enzyme binds to the dimers in a light-independent fashion. In the presence of visible light, the bonds linking the adjacent pyrimidine rings are broken; the enzyme can then dissociate in the dark.

DNA photolyase enzyme contains two chromophores, each of which absorbs light of characteristic wavelength, depending on its structure (see Figure 22.4). Blue light from the visible spectrum is first absorbed by MTHF, 5,10-methenyltetrahydrofolylpolyglutamate, and the energy is then transferred to the other co-factor, flavin adenine dinucleotide or $FADH^-$. The excited $FADH^-$ transfers an electron to the thymine dimer, which cleaves the covalent bonds between the two bases, restoring the original structure. Electron transfer from the repaired DNA back to the FADH completes the catalytic cycle.

Photolyase exists in bacteria and many eukaryotes but not in placental mammals, including humans, where thymine dimers are repaired by a different pathway, NER. A curious fact: thinning of the ozone layer, leading to increased UV irradiation, has been

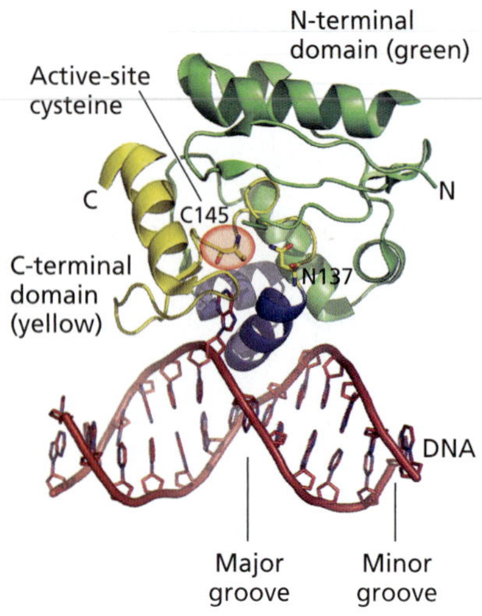
N-terminal domain (green)
Active-site cysteine
C
C145
N
C-terminal domain (yellow)
N137
DNA
Major groove
Minor groove

Figure 22.5 Structure of human AGT bound to an alkylated lesion in DNA. AGT is O^6-alkylguanine alkyltransferase. The helix–turn–helix or HTH motif, shown in blue, binds to the minor groove of the DNA. Well-conserved, small hydrophobic helix recognition residues, not labeled here, allow the helix to pack closely within the DNA minor groove. This minimizes sequence-specific interactions, and is advantageous for nucleotide flipping: the alkylated base has to flip out from the base stack in the DNA to be able to interact with the cysteine, labeled C145, in the active site of the enzyme. An asparagine hinge, labeled N137, couples the HTH with the active-site motifs. An arginine residue stabilizes the extrahelical conformation.

blamed for population declines in certain frog species that lack photolyase. UV light is especially harmful to them during embryonic development in clear lakes.

The enzyme O⁶-alkylguanine alkyltransferase is involved in the repair of alkylated bases

Treatment of DNA with alkylating agents results in the formation of various modified DNA bases. If not repaired, some of these alkylated bases are mutagenic (see Figure 22.1B) and some are lethal to the cell. Some alkylating agents are used in cancer chemotherapy because alkylated DNA bases block DNA replication, but the activity of the ***O*^{6}-alkylguanine alkyltransferase** or **AGT** repair pathway limits the effectiveness of alkylating chemotherapies.

Repair of this type of damage involves AGT (**Figure 22.5**), an unusual enzyme that performs a stoichiometric and irreversible transfer of the alkyl group from the O^6-alkyl adduct to a cysteine in its own active site, without inducing DNA breaks. Thus, AGT is both a transferase and an acceptor of the alkyl group. Having become alkylated, AGT can no longer remove the alkyl group. The enzyme is degraded following the transfer; it might be called a one-time enzyme and its activity a suicide reaction. The term enzyme is, however, a misnomer in this case: at the end of the reaction, a true enzyme is exactly the same as at the beginning, which is evidently not the case here. Examples of AGTs are known from all three domains of life: Bacteria, Archaea, and Eukaryota. However, some organisms, including plants and *Schizosaccharomyces pombe,* lack an enzyme with such functionality.

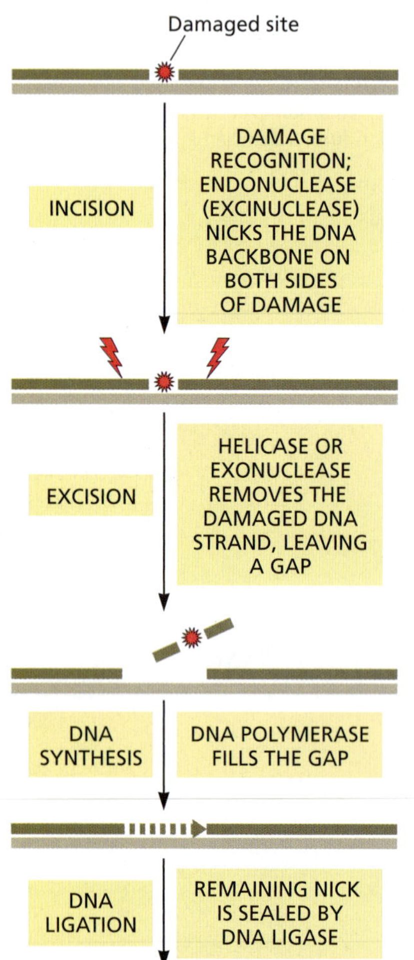
Damaged site
INCISION
DAMAGE RECOGNITION; ENDONUCLEASE (EXCINUCLEASE) NICKS THE DNA BACKBONE ON BOTH SIDES OF DAMAGE
EXCISION
HELICASE OR EXONUCLEASE REMOVES THE DAMAGED DNA STRAND, LEAVING A GAP
DNA SYNTHESIS
DNA POLYMERASE FILLS THE GAP
DNA LIGATION
REMAINING NICK IS SEALED BY DNA LIGASE

Figure 22.6 Nucleotide excision repair. The main pathway in both bacteria and eukaryotes involves four consecutive steps—incision, excision, DNA synthesis, and DNA ligation—which are performed by specific enzymes.

Nucleotide excision repair is active on helix-distorting lesions

Nucleotide excision repair, NER, is the most versatile DNA repair system. It takes care of a wide array of DNA damage, from UV-induced lesions, to intrastrand cross-links and bulky chemical adducts, to lesions produced by reactive oxygen species, such as 8-oxoguanine. The unifying feature of these lesions is that they all disrupt the double-helical DNA structure. Specific proteins act to recognize these lesions. The main NER pathway is the same in both bacteria and eukaryotes in that it involves four steps following lesion recognition (**Figure 22.6**):

(1) incision

(2) excision of a short single-stranded segment of DNA spanning the lesion

(3) DNA synthesis using the intact complementary strand as a template

(4) DNA ligation

In bacteria, a complex of UvrA and UvrB proteins tracks along DNA until it reaches a thymine dimer or other helix-distorting lesion, where it halts and forces the DNA to bend (**Figure 22.7A**). UvrA then dissociates, allowing UvrC, a nuclease, to bind. The UvrBC complex cuts on both sides of the dimer, in a somewhat asymmetric way. After the UvrBC complex leaves, helicase D unwinds the DNA to release the strand containing the lesion; then DNA polymerase and ligase act to seal the nick. The structure of the damage sensor complex $UvrA_2B$ is presented in **Figure 22.7B**.

The process in eukaryotes is more complex and involves numerous proteins. We distinguish between **global genome repair** or **GGR** and **transcription-coupled repair** or **TCR**, two pathways that target the same lesions, either in the entire genome or in

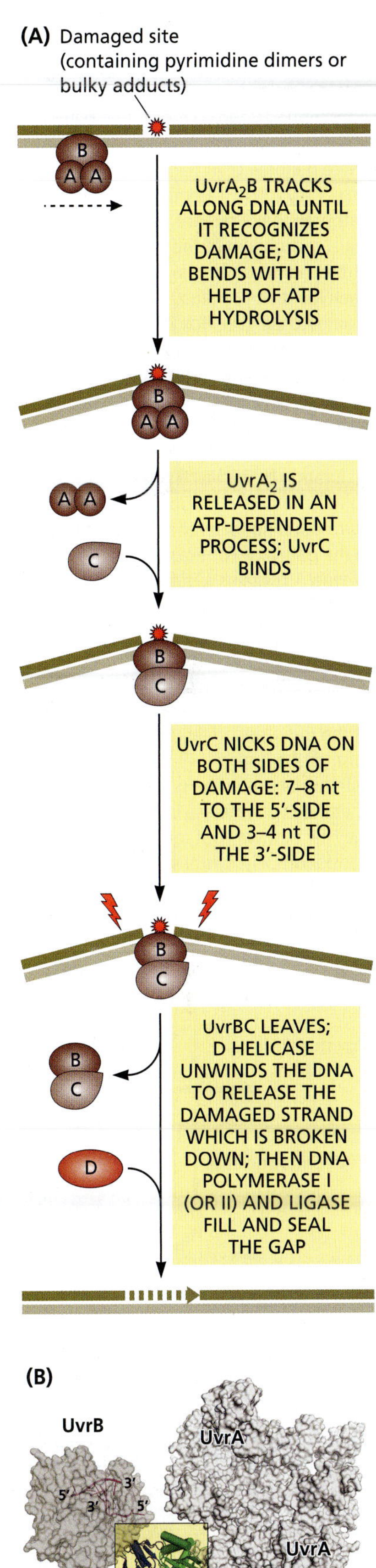
(A) Damaged site
(containing pyrimidine dimers or bulky adducts)
B
A A
UvrA$_2$B TRACKS ALONG DNA UNTIL IT RECOGNIZES DAMAGE; DNA BENDS WITH THE HELP OF ATP HYDROLYSIS
UvrA$_2$ IS RELEASED IN AN ATP-DEPENDENT PROCESS; UvrC BINDS
C
UvrC NICKS DNA ON BOTH SIDES OF DAMAGE: 7–8 nt TO THE 5′-SIDE AND 3–4 nt TO THE 3′-SIDE
UvrBC LEAVES; D HELICASE UNWINDS THE DNA TO RELEASE THE DAMAGED STRAND WHICH IS BROKEN DOWN; THEN DNA POLYMERASE I (OR II) AND LIGASE FILL AND SEAL THE GAP
D
(B)
UvrB
UvrA
UvrA
5′
3′
3′
5′

Figure 22.7 A more detailed look at nucleotide excision repair in bacteria. (A) NER pathway, with participating proteins. (B) Model of the damage sensor UvrA$_2$B, which has a flat and open structure. Note that the DNA-binding path on UvrA, shown in pink, as proposed on the basis of site-directed mutagenesis studies, neatly aligns with the crystallographic position of DNA on UvrB. The model is based on the experimentally determined high-resolution structure of the interaction domains of UvrA, aa 131–245, and UvrB, aa 157–250, boxed in yellow. Understanding the exact mechanism of damage recognition awaits further studies.

transcribed genes only. TCR occurs also in bacteria. Global repair is essential in proliferating cells because it ensures the integrity of the genome, which must be faithfully replicated and transcribed. The situation in terminally differentiated cells is different because the DNA in these cells is never replicated; these cells could arguably accumulate numerous lesions in the genome, as long as they are able to maintain the integrity of the genes needed for viable cell function. These genes are transcriptionally active; thus, they use a specialized pathway, TCR, which targets several repair systems to transcribed genes. Moreover, the transcribed strand is the one that is preferentially repaired. We now know that the other strand is also regularly checked, as its integrity is needed for its role as a template strand in DNA repair. The selective repair of transcriptionally active genes is believed to relieve post-mitotic cells from the burden of having to repair their entire genomes.

The presence of unrepaired damage in transcribing genes has two major consequences. (1) If a lesion creates miscoding or noncoding codons, a mutant, most often nonfunctional protein will be produced. (2) RNA polymerase II may stall at the lesion, preventing the production of RNA and protein products; moreover, polymerase stalling withdraws a polymerase molecule from the pool of active polymerases. The stalled polymerase can be degraded via the ubiquitin pathway (see Chapter 18); alternatively, the stalled polymerase may serve as a signal for TCR to occur.

As **Figure 22.8** depicts, GGR and TCR use practically the same pathway, the difference being in the initial recognition of the damage. The pathway is well understood in humans because of extensive analyses of gene mutations that cause the disease xeroderma pigmentosum (see Figure 22.8 and Box 22.2). Another well-studied example of TCR is found in human neurons, in which UV-induced lesions are proficiently and selectively removed from active genes, even though the bulk of the genome, including silent genes, is not efficiently repaired.

Base excision repair corrects damaged bases

Base excision repair, BER, is the pathway that corrects for the presence of uracil, resulting from hydrolytic deamination of cytosine; methylated bases; and oxidized bases such as 8-oxoguanine. It also repairs some single-strand breaks generated by oxidative damage or by abortive topo I action. The active-site tyrosine of topo I forms a transient covalent bond with the 5′-phosphate at the single-strand break it initially creates (see Chapter 4). If, for some reason, the enzyme cannot proceed further and becomes trapped in the bonded state, then the single-strand break persists and needs to be repaired.

Base excision repair begins with cleavage of the glycosidic bond between the damaged base and deoxyribose by a specialized enzyme, **N-glycosylase** (**Figure 22.9**). The apurinic/apyrimidinic or AP site so formed is recognized by an AP endonuclease, which cleaves on the 5′-side of the AP site. This intermediate is then processed by two different pathways that involve different enzymes. Depending on the length of the stretch being replaced in the process, we distinguish between short patch repair, which occurs in 99% of cases, where only one nucleotide is replaced, and long patch repair, used in 1% of cases, where two or more nucleotides are replaced. What determines the use of one over the other is unclear.

Most cells contain several N-glycosylases that are specific for the alkylated bases N-methyladenine, 3-methyladenine, and 7-methylguanine. Another enzyme, **uracil-DNA glycosylase** or **UNG**, excises unwanted uracil bases in DNA by an extrahelical

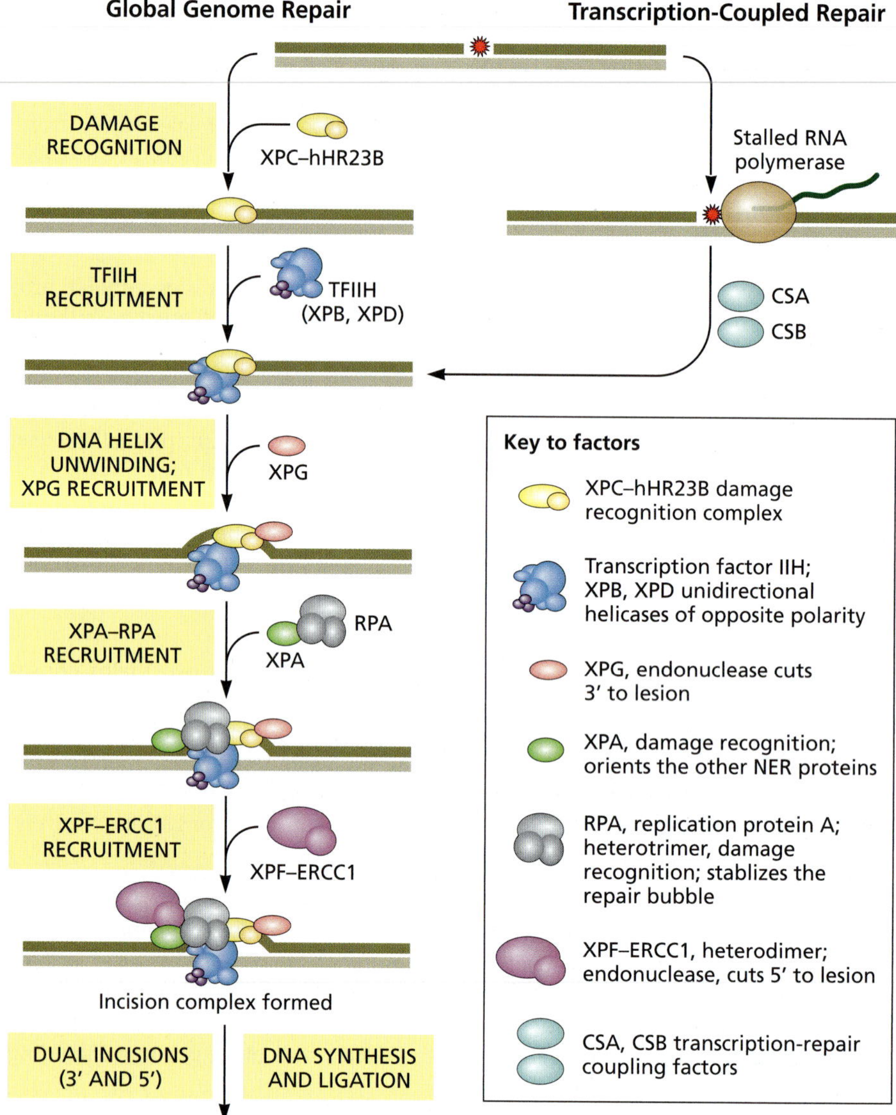

Figure 22.8 Human NER excision complex assembly. We distinguish between global genome repair and transcription-coupled repair, two pathways that target the same lesions, either in the entire genome or in transcribed genes only. They follow essentially the same pathway, the difference being in the initial recognition of the damage. Roles of the individual protein factors have been identified by studying xeroderma pigmentosum (see Box 22.2) patients and cells derived from them, identifying the genes affected by mutations, and then performing genetic complementation tests to see whether the mutated recessive genes can complement each other, with restoration of wild-type phenotypes. Seven complementation groups were identified, corresponding to seven genes participating in the pathway. For example, XP group or type A contains mutations in the gene coding for XPA, which participates in damage recognition and proper orientation of the other proteins in the complex; type B mutations affect the *XPB* gene, which encodes a helicase subunit of the transcription factor TFIIH. A variant form of the disease, XP-V, is associated with mutations in the gene for polymerase η, eta; this polymerase is not active in the NER pathway but rather performs translesion DNA synthesis.

base recognition mechanism (**Figure 22.10**). A key process is the singling out of rare uracil-containing base pairs, U-A and U-G, in a background of ~10^9 T-A or C-G base pairs in the human genome. The enzymatic discrimination of thymine and uracil begins with thermal or spontaneous melting of T-A and U-A base pairs and not with the active participation of the enzyme; thus, base-pair dynamics plays a crucial role in the genomewide search for uracil. Uracil emerging from the DNA bases stack is trapped by the enzyme in a very unstable intermediate, which then transitions to a stable, fully flipped state in which uracil can be removed from the DNA by the enzyme.

Mismatch repair corrects errors in base pairing

Despite careful proofreading, mismatched base pairs and insertion/deletion loops do arise during DNA replication (see Chapters 19 and 20). These errors are repaired promptly by a process that involves a single-strand incision. Mismatch repair, MMR, increases the overall fidelity of replication by 2–3 orders of magnitude. In both bacteria

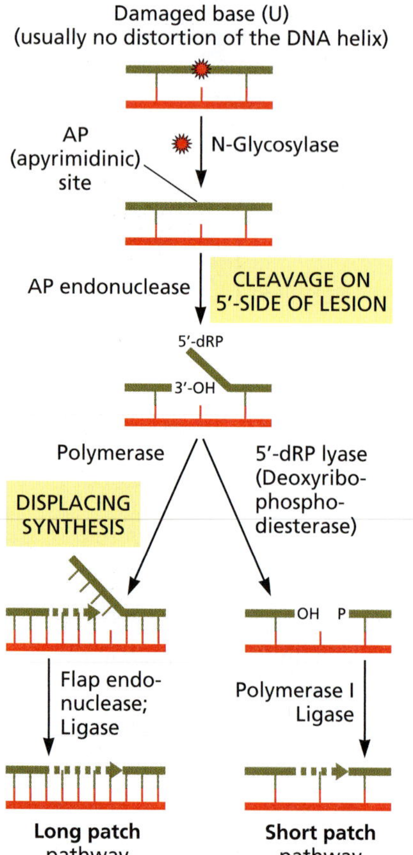

Figure 22.9 Base excision repair. The process begins with cleavage of the glycosidic bond between the damaged base, U, and deoxyribose; the apyrimidinic or AP site created is recognized by an AP endonuclease, which cleaves on the 5′-side of the site. Some enzymes have both glycosylase and AP endonuclease activities in a single polypeptide chain. Furthermore, two pathways—long patch and short patch, depending on the enzymes involved—can then complete the repair.

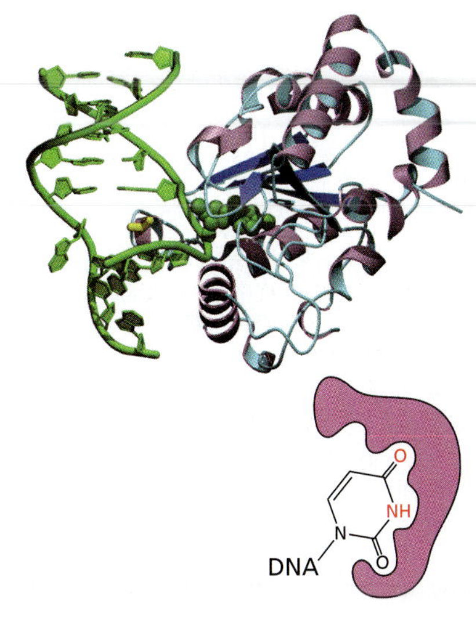

Figure 22.10 Structure of uracil-DNA glycosylase, UNG. View of the extrahelical uracil, represented as green spheres in otherwise double-helical DNA, in the fully flipped state trapped by the human enzyme. The boxed schematic helps in understanding the actual structure. (Adapted from Parker JB, Bianchet MA, Krosky DJ et al. [2007] *Nature* 449:433–437. With permission from Macmillan Publishers, Ltd.)

and eukaryotes, MMR requires Mut proteins; the genes encoding these proteins were identified by mutations in them that have a mutator phenotype, that is, increased frequency of spontaneous mutations.

Methyl-directed mismatch repair in bacteria uses methylation on adenines as a guide

MMR confronts a serious problem in selecting the base that needs to be repaired. In other words, how does the system know which base in a mismatched base pair is wrong and should be removed, and which to keep? In bacteria, the problem is solved by using methylation on adenines as a guide (Figure 22.11).

The sequence GATC is always marked with a methyl group, on the A, on both parental strands, but methyl groups are not added immediately to newly synthesized strands: there is a lag of around 2 minutes before the new strand is methylated. Methylation of the new strand requires the activity of Dam methyltransferase, whose target is the adenine in the frequently occurring GATC tetranucleotide, which occurs statistically once about every 256 bp. The enzyme is a hemimethylase, as it uses methylation on the old strand to target methylation on the new strand. Recall that the corresponding eukaryotic enzyme that methylates the newly synthesized DNA strand is Dnmt1 and that it works on CpG dinucleotides; however, it does not participate in MMR. Thus, the temporary lack of ^{me}A on one of the strands marks this strand as the newly synthesized strand that must contain the erroneous base. This mark allows the repair enzymes to discriminate between a correct base on the old strand and an erroneous base on the newly synthesized strand.

The steps in the process as it occurs in *E. coli* are illustrated in Figure 22.12. MMR is initiated when a complex of MutS–MutL binds to mismatched DNA. Our present, more detailed understanding of how the MutS sliding clamp is activated and works is presented in Figure 22.13 and Figure 22.14. MutS interacts with the β clamp accessory protein that is required for processive DNA replication (see Chapter 19), and it may help to deliver MutS to mismatches. ATP binding releases the complex from the mismatch-containing site so that it is free to translocate along the DNA. When the sliding clamp reaches the protein MutH, which is prebound to CTAG in the new unmethylated strand, it activates the MutH nuclease activity. MutH is a member of the type II family of restriction endonucleases. It introduces a nick in the newly synthesized, temporarily unmethylated strand at hemimethylated sites located within about ~1 kb of the error. Type II restrictases cleave the DNA on both strands (see Chapter 5), so it is still a mystery how MutH acts here to nick just a single strand. The resulting nick, which can be either 3′ or 5′ to the mismatch, is the entry point for MutL-dependent loading of DNA helicase II and binding of single-strand DNA-binding or SSB proteins. Working together, these proteins generate single-stranded DNA, ssDNA, that is digested by either 3′- or 5′-exonucleases, depending on the location of the nick relative to the mismatch. The stretch of ssDNA that is excised is quite extensive, sometimes reaching more than 1 kb away from the methylation site. Having identified the unmethylated new strand, the machinery must then excise all nucleotides that separate the methylation site from the actual error and beyond. This excision removes the error and allows highly accurate DNA polymerase III to resynthesize the strand correctly. DNA ligase seals the nick to complete MMR.

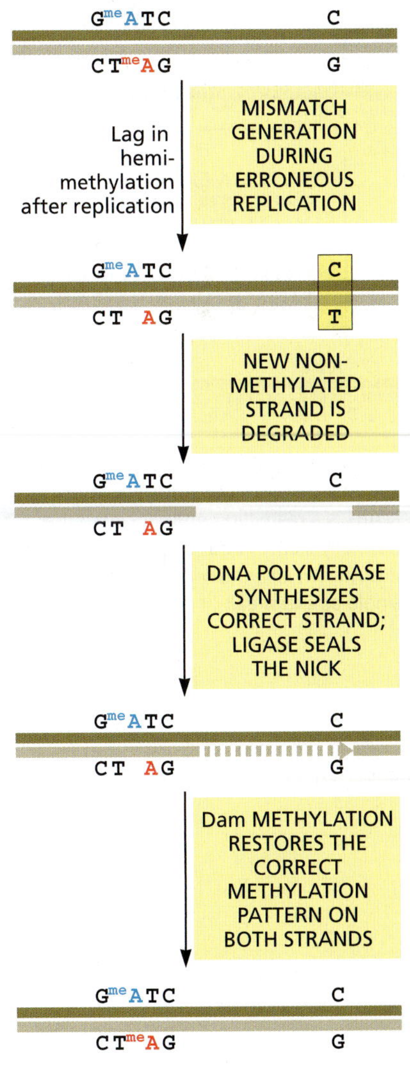

Figure 22.11 Mismatch repair during bacterial replication: Selection of the strand to be repaired. In the bacterial genome, the sequence GATC is always methylated on the A on both DNA strands. The lag in addition of new methyl groups to the newly synthesized strand during replication allows the repair enzymes to discriminate between the old and new strand, and hence to repair the mismatched base in the new strand.

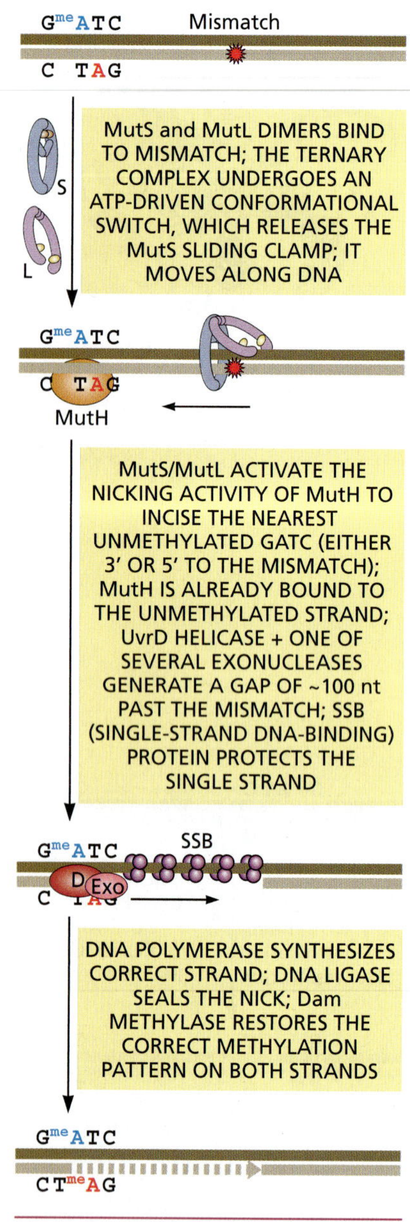

Figure 22.12 Mechanism of mismatch repair during bacterial replication. MutH is a restriction endonuclease that normally cleaves both strands of DNA, so it is not yet clear how it introduces only a single-strand break or nick during MMR.

Mismatch repair pathways in eukaryotes may be directed by strand breaks during DNA replication

The choice of which strand is repaired is clearly methyl-directed in bacteria, but the situation in eukaryotes is much less well understood. MutS and MutL are highly conserved but MutH is found only in Gram-negative bacteria; no functional homolog has been identified in other organisms. This fact led to the suggestion that mismatch processing in eukaryotes is directed by strand breaks that occur during replication, such as the 3′-end of the leading strand or the ends of Okazaki fragments (**Figure 22.15**). Repair is initiated when complexes of MutS homologs, MSHs, either MSH2–MSH6, known as MutSα, or MSH2–MSH3, known as MutSβ, bind to a mismatch. The process presented in **Figure 22.16** has been proposed on the basis of studies in which the entire mechanism was reconstituted from recombinant proteins: either MutSα (**Figure 22.17**) or MutSβ, plus MutLα; replication protein A, RPA; exonuclease 1, EXO1; proliferating cell nuclear antigen, PCNA; replication factor C, RFC; DNA polymerase δ, Pol δ; and DNA ligase. To date, no eukaryotic DNA helicase has been shown to participate in the repair of replication errors. As in *E. coli*, more than one eukaryotic exonuclease has been implicated in MMR, and several other proteins are also required. DNA resynthesis is catalyzed by an aphidicolin-sensitive polymerase, probably DNA polymerase δ.

Repair of double-strand breaks can be error-free or error-prone

Double-strand breaks, DSBs, are probably the most harmful of DNA lesions. The severe consequences of DSB lesions can be explained by two major factors: first, DSBs are difficult to repair without introducing errors or mutations in the DNA sequence, and second, the disruption of continuity in the molecule leads to numerous chromosomal translocations and other rearrangements, which pose a threat to the genomic integrity of a cell. Cells have developed highly sophisticated mechanisms to repair such damage, collectively known as the **DNA damage response**, **DDR**. DDR is also elicited in response to long stretches of ssDNA.

DDR is a major factor that controls the functioning of **cell-cycle checkpoints**. Genetic and biochemical studies have defined points in the cell cycle at which the cell stops temporarily to check whether conditions are ripe for the cell to transit to the next phase of the cell cycle. A general overview of the cell cycle and the related field of cell-cycle regulation can be found in Box 2.2 and Box 20.2. These conditions include the overall metabolic status of the cell; the presence of the correct molecules, such as the deoxyribonucleoside triphosphates needed as precursors for DNA replication, in the needed amounts; and last but not least, the integrity of the DNA molecules. When DNA lesions, especially DSBs, are detected, the cell cycle is halted until the damage is repaired.

Two major pathways are used to repair DSBs: homologous recombination, HR, and nonhomologous end-joining, NHEJ. The key proteins involved in each of these pathways are listed in **Table 22.1**.

Homologous recombination repairs double-strand breaks faithfully

Homologous recombination, HR, depends on the availability of intact DNA sequences in sister chromatids that can be used as templates for faithful repair. Thus, it can only occur in S and G_2 phases. In addition to DSBs, this pathway takes care of stalled replication forks and interstrand DNA cross-links. HR is also involved in physiologically relevant processes such as meiotic recombination and the formation of abortive topo II intermediates. We presented the general features of HR and its roles in such processes in Chapter 21. Here, we introduce another important player that was not discussed previously, the **Mre11–Rad50–NBS1 complex**, abbreviated **MRN**.

The importance of the MRN complex in the cellular response to DSBs was initially recognized in the study of two human conditions: ataxia telangiectasia-like disorder, caused by mutations in *MRE11*, the meiotic recombination 11 gene, and Nijmegen breakage syndrome, caused by mutations in the *NBS* gene; the yeast homolog of

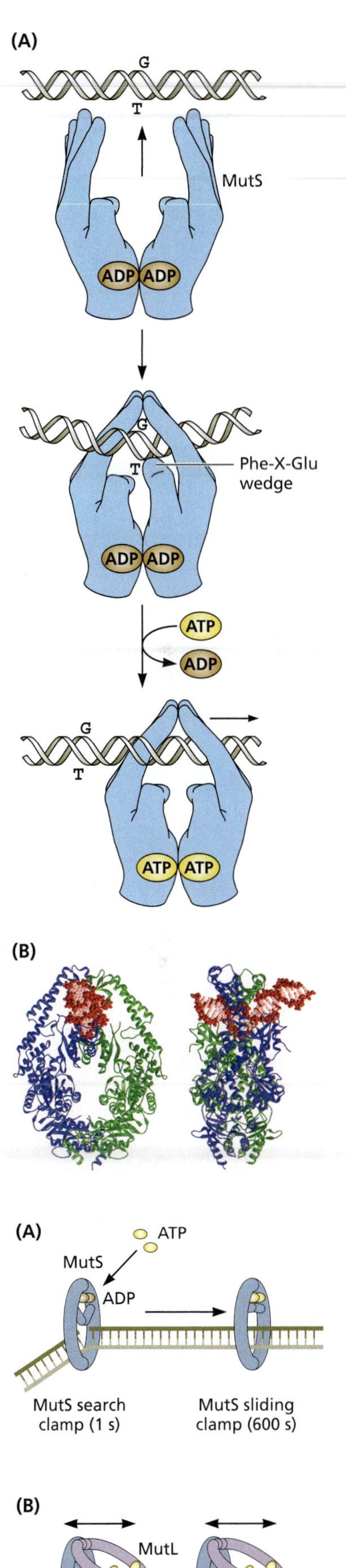

Figure 22.13 MutS sliding clamp and its activation. (A) When not associated with DNA, the finger domains of MutS are unstructured and open, and the ATP-binding sites are dimerized. In the presence of mismatched DNA, the ADP-bound form of MutS wraps around the DNA like a pair of praying hands and is anchored at the mismatch site, G-T, by a Phe-X-Glu wedge, the thumb of one hand, inserted into the DNA minor groove. ADP/ATP exchange brings about a conformational change that releases the thumb from the mismatch site but leaves the fingers closed around the duplex: the clamp is now free to translocate along the DNA in either direction. (B) Structure of MutS–DNA complex: the MutS dimer embraces the DNA, inducing a pronounced kink; MutS itself also undergoes conformational change. The role of ATP hydrolysis is unknown; it may be needed for translocation. (A, adapted from Jiricny J [2006] *Nat Rev Mol Cell Biol* 7:335–346. With permission from Macmillan Publishers, Ltd. B, from Obmolova G, Ban C, Hsieh P & Yang W [2000] *Nature* 407:703–710. With permission from Macmillan Publishers, Ltd.)

this gene is *Xrs2* (Box 22.3). The MRN complex is used in both HR and NHEJ, despite their different mechanisms. This is the complex that initially senses the damage and interacts with the broken DNA ends in such a way as to keep the ends in close proximity with each other to allow repair to occur.

In addition to this purely structural role, the two major subunits of the MRN complex, Mre11 and Rad50, possess enzymatic activities that are essential to its function; the role of NBS1 is much less clear. Mre11 is a multifunctional protein that exhibits numerous activities *in vitro*: it is a nuclease that also has strand-dissociation and strand-annealing activities. The nuclease activity is highly regulated by the other two protein subunits, by ATP, and by sequence homology in the DNA substrate. Exactly how these activities are modulated and coordinated *in vivo* remains to be elucidated. It is clear that Mre11 participates in the initial processing of DNA ends, which contain adducts that could interfere with further processing, and in resolving possible secondary structures of DNA ends.

Rad50 binds and hydrolyzes ATP. The ATP-binding motifs at both ends of the Rad50 polypeptide chain are crucial to its function, as mutations in these motifs result in a null phenotype in yeast and in partial loss of nuclease activity in the human complex *in vitro*. The domain and crystal structures of subunits of the MRN complex, and the conformational transitions that occur upon DNA binding to ensure DNA tethering, are presented in Figure 22.18, Figure 22.19, and Figure 22.20.

The central role of MRN in both DSB repair pathways, as well as in the early steps of meiotic recombination and in maintenance of telomeres, warrants further studies of the complex. Understanding the exact mechanism of MRN action might also lead to important clues for the treatment and eventual cure of the genetic diseases associated with its malfunctions.

Nonhomologous end-joining restores the continuity of the DNA double helix in an error-prone process

Nonhomologous end-joining, NHEJ, is the major pathway that repairs DSBs. It simply restores the integrity of the DNA molecule by joining the two ends in an

Figure 22.14 Dynamics of MutS sliding clamp action as deduced from single-molecule spFRET studies. (A) MutS forms a transient clamp, associated with DNA for about 1 s, that scans duplex DNA for mismatched nucleotides, moving ~700 bp by one-dimensional rotational diffusion. The identification of a mismatch provokes ATP binding, which induces the formation of MutS sliding clamps. The clamps have unusual stability on the DNA, ~600 s; this is the length of time needed for the entire process of MMR to occur. Thus, ATP binding may transform short-lived MutS lesion-scanning clamps into highly stable clamps that are capable of competing with DNA-binding proteins, or chromatin in eukaryotes, and recruiting the mismatch repair machinery. The stability of the clamp on the DNA is independent of ATP hydrolysis, the role of which remains unclear. (B) Short ssDNA excision tract associated with MMR provokes the release of the complex from the DNA; the released complex can then be recycled. (Adapted from Jeong C, Cho W-K, Song K-M et al. [2011] *Nat Struct Mol Biol* 18:379–385. With permission from Macmillan Publishers, Ltd.)

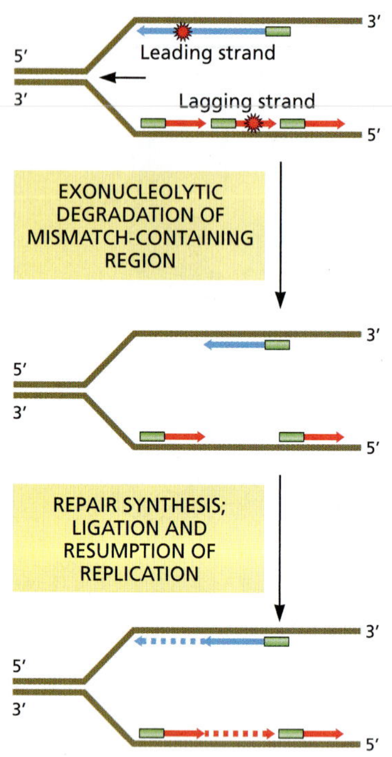

Figure 22.15 Model for mismatch repair in eukaryotes. Mismatch processing in eukaryotes is probably directed by the presence of ssDNA at strand breaks that occur during replication, at the 3′-end of the leading strand or at the ends of Okazaki fragments. The schematic shows two separate lesions, on the leading and lagging strands. Repair in the leading strand might begin at the 3′-terminus, with the polymerase resynthesizing the region; in the lagging strand, an entire Okazaki fragment may be removed, with degradation commencing at either end; extension of the fragment closest to the replication fork would replace the degraded one. (Adapted from Jiricny J [2006] *Nat Rev Mol Cell Biol* 7:335–346. With permission from Macmillan Publishers, Ltd.)

error-prone process. The errors in the DNA sequence of the restored DNA molecule result from the fact that the DSBs have very heterogeneous DNA ends, with little or no terminal microhomology, and therefore are mostly not compatible with ligase enzymes. For NHEJ to occur, ligase-compatible ends must first be produced. There are numerous possible outcomes of the repair process, in the hundreds even for one pair of starting DNA ends, depending on how the ligase-compatible ends are generated. There is often nucleotide loss or addition at the rejoining site. Thus, the process leaves "information scars," to use Michael Lieber's terminology, at most sites of repair. With time, the occurrence of many repairs will lead to accumulation of randomly located mutations over the entire genome in each somatic cell. The positive aspect of NHEJ that compensates for these sequence errors is that it quickly

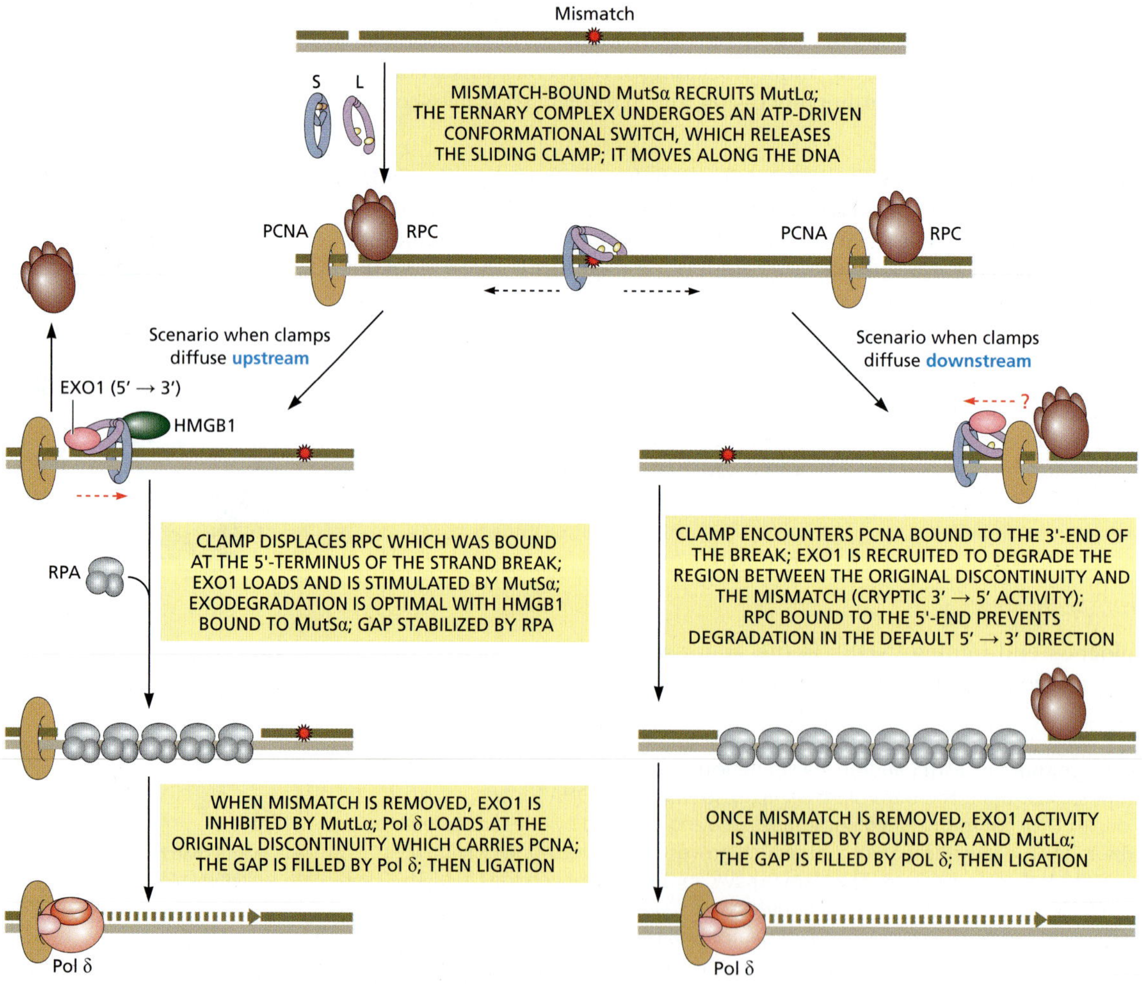

Figure 22.16 Proposed mechanism of mismatch repair during DNA replication in eukaryotes. Note that a cryptic activity of EXO1 in the downstream scenario does not need to be invoked; maybe Mre11 or MutSα, RFC, and PCNA activate a latent MutLα endonuclease activity in an ATP- and mismatch-dependent manner, as suggested by Paul Modrich. (Adapted from Jiricny J [2006] *Nat Rev Mol Cell Biol* 7:335–346. With permission from Macmillan Publishers, Ltd.)

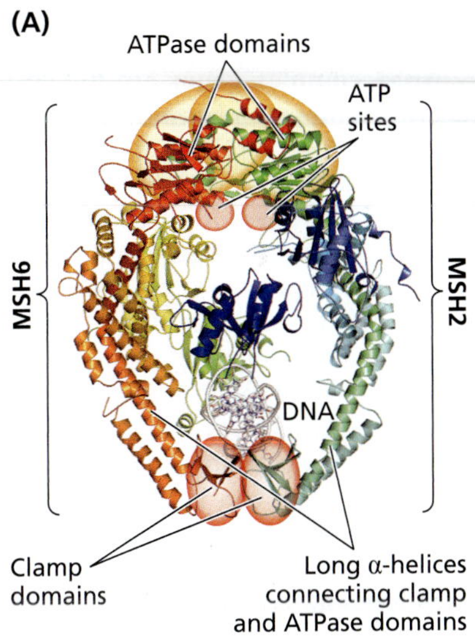

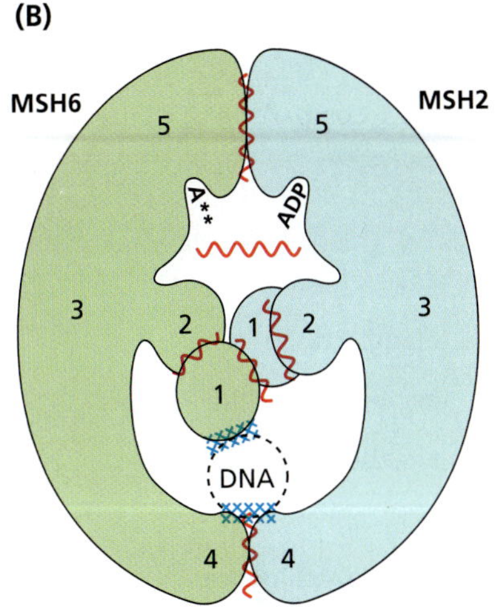

Figure 22.17 Structure of human MutSα–ADP–G·T mispair complex. (A) Crystal structure at 3 Å resolution. MutSα is an asymmetric dimer of MSH2 and MSH6. The small, largely β-strand clamp domains (encircled in red) contact the DNA around the mismatch. MSH6 makes extensive contacts along a 6-bp stretch of essentially B-form DNA on one side of the mispair, whereas MSH2 contacts bases on both sides of the mispair. Long α-helices connect the clamp and ATPase domains of MutSα to execute the cross-talk between the DNA substrate binding site and the ATP-binding site. ATP binding leads to dissociation of the protein from the mispair and its movement along the helix. (B) Schematic based on the crystal structure. Individual domains in both proteins are numbered. Blue hatching indicates protein–DNA interactions; red lines indicate the positions of likely conformationally responsive interdomain interfaces. (B, adapted from Warren JJ, Pohlhaus TJ, Changela A et al. [2007] *Mol Cell* 26:579–592. With permission from Elsevier.)

restores the structural integrity of the DNA molecule, and thus the chromosome, at sites where the breakage would otherwise result in loss of large segments of the genome.

Like other repair processes, the NHEJ pathway begins with recognition of the damage. NHEJ uses the highly abundant **Ku70–Ku80 dimer**, ~300,000 molecules per cell in humans, for this task. Ku was discovered as an autoantigenic protein and was named after a scleroderma patient with initials K.U. Ku can only load and unload onto DNA at DNA ends, and two Ku complexes form during repair, one at each of the two DNA ends to be joined. Ku is believed to be able to recruit the other players that have a role in NHEJ—nuclease, polymerase, and ligase—in any order. This flexibility contributes to the diverse array of outcomes that can arise from identical starting ends.

The other important players in NHEJ are:

- The catalytic subunit of DNA-dependent protein kinase, DNA-PKcs, whose kinase activity is activated upon binding to the ends of the severed DNA molecule. DNA-PKcs phosphorylates a number of proteins involved in this pathway—RPA, DNA ligase IV or LigIV, and its partners XRCC4 and XLF/Cernunnos—as well as itself. It also binds to and regulates the activity of Artemis.
- Artemis, an endonuclease believed to be involved in the preligation processing of DNA ends. Other enzymes such as TdT, terminal deoxynucleotidyltransferase, and PNK, polynucleotide kinase, are also involved in DNA end processing.
- Translesion DNA polymerases, Pol μ and Pol λ.
- LigIV-XRCC4-XLF complex, which is specifically involved in this pathway.

Table 22.1 Major proteins that participate in the two major pathways for DNA damage repair. (Adapted from Mladenov E & Iliakis G [2011] *Mutat Res* 711:61–72.)

Function	Homologous Recombination, HR	Nonhomologous End-Joining, NHEJ
Single-strand break sensor molecules	MRN[a]	Ku70–Ku80
DNA end-processing enzymes	MRN, CtIP, Exo1, Dna2	Artemis, TdT[a], PNK[a]
recombinases	Rad51	
DNA repair mediators	Rad52, BRCA2, Rad51 paralogs	DNA-PKcs
polymerases	Pol δ, Pol ε	Pol μ, Pol λ
ligases	ligase I	ligase IV
ligase-promoting factors	PCNA?	XRCC4, XLF–Cernunnos

[a]MRN, Mre11–Rad50–Nbs1 complex; TdT, terminal deoxynucleotidyltransferase; PNK, polynucleotide kinase; DNA-PKcs, DNA-dependent protein kinase catalytic subunit.

Box 22.3 A Closer Look: Mutations in the MRE11–RAD50–NBS1 complex genes are linked to genetic diseases The MRN complex and its individual components are involved in a wide variety of cellular responses to DNA damage, including that induced by ionizing radiation and radiomimetic drugs. Mutations in two of the components, MRE11 and NBS1, are linked to two rare genetic disorders, ataxia telangiectasia-like disorders or ATLD and Nijmegen breakage syndrome or NBS, respectively. The clinical features of these two disorders have many features in common with those of another prominent genomic-instability disease, ataxia telangiectasia or AT (see Box 22.2). Classical AT is caused by biallelic truncating mutations that cause total absence of the ATM protein, one of the two protein kinases responsible for the sensing and initial activation of the complex cellular response to DSBs. AT is a severe condition that can be recognized in very young children by lack of coordination of motion, known as ataxia, from the Greek a taxis, meaning without order or coordination, and by dilated blood vessels, known as telangiectasia, usually in the eyes. The other major clinical features of AT are genome instability, immunodeficiency, predisposition to lymphoid and other cancers, and sensitivity to ionizing radiation. ATLD and NBS are milder conditions, with slower progression and a later onset of neurological features (Table 1).

All three disorders exhibit very similar cellular phenotypes: hypersensitivity to ionizing radiation, IR, and failure to induce stress-activated protein kinases following exposure to IR. They also show a phenomenon called radioresistant DNA synthesis or RDS, in which irradiated cells fail to elicit an S-phase checkpoint response and the cells continue with DNA replication despite the presence of DSBs. All three disorders are typified by a chromosome translocation between chromosomes 7 and 14. Nevertheless, the clinical symptoms of the three disorders are quite different, with both shared and unique characteristics. This makes it probable that the two MRN subunits affected by mutations that are associated with ATLD and NBS have other yet-to-be-determined roles in addition to functioning in the complex. Nevertheless, the clinical characteristics that ATLD and NBS share with AT point to the involvement of complex networks in the DNA damage response, with numerous interacting pathways.

Table 1 Comparison of clinical features of AT, ATLD, and NBS.

Clinical feature	Ataxia telangiectasia, AT	Ataxia telangiectasia-like disorders, ATLD	Nijmegen breakage syndrome, NBS
ataxia	+	+	–
telangiectasia	+	–	–
lymphoid and other tumors	+	not known	+
skin abnormalities	+	not known	+
microcephaly	–	–	+
normal intelligence	+	+	+/–
congenital malformations	–	–	+
7/14 chromosome translocations	+	+	+
reduced Ig levels	+	only specific antibodies	+

Plus sign means feature affected, minus sign means feature not affected.

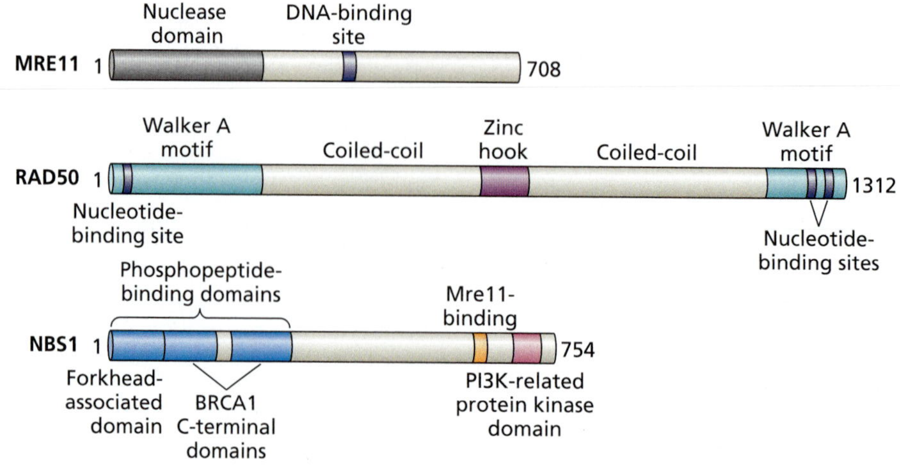

Figure 22.18 Domain structure of the three proteins of human MRN complex. MRN complex is MRE11–RAD50–NBS1. RAD50 and MRE11 are highly conserved and their roles have been, for the most part, determined on the basis of genetic and biochemical studies. NBS1, on the other hand, shows a high degree of sequence variability among species and its role remains enigmatic. The importance of the complex is evidenced by the fact that null mutations in any of these proteins in mice and humans are embryonically lethal.

Figure 22.19 Crystal structure of archaean Mre11–Rad50–ATPγS. This complex is from *Methanococcus jannaschii*. The complex consists of a head and two arms; the head is composed of core domains of both proteins, and the arms are formed by the coiled-coil domain of Rad50 and the C-terminal domain of Mre11. The two proteins regulate each other through extensive interactions. Mre11 brings two Rad50 molecules into close proximity and promotes their ATPase activity. The C-terminal domain of Mre11 holds the coiled-coil arms of Rad50; the capping domain of Mre11 stabilizes portions of Rad50; and finally, Mre11 dimerizes through its nuclease domains. When ATP is bound to Rad50, the nuclease activity of Mre11 is negatively regulated through blocking of the active site. ATP hydrolysis promotes substantial conformation change in the flexible C-linker, which leads to unmasking of the nuclease active site.

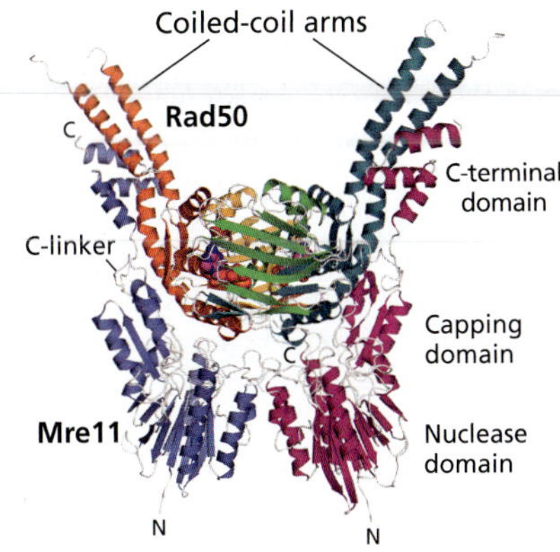

The steps in the classical NHEJ process—the alternative term canonical is also used—and the proteins involved are depicted in detail in **Figure 22.21**, **Figure 22.22**, **Figure 22.23**, and **Figure 22.24**.

Some researchers define this classical pathway as D-NHEJ to emphasize its dependence on DNA-PKcs, an evolutionarily new component that is absent from bacteria and lower eukaryotes. Early studies using SV40 viral DNA substrates in cultured monkey cells and *in vitro* experimentation showed the existence of a pathway that is independent of DNA-PKcs, Ku, or LigIV. This backup pathway, designated by some as B-NHEJ but also known as alternative NHEJ, microhomology-mediated end-joining, Ku-independent end-joining, or LigIV-independent end-joining, uses PARP-1 as the DNA damage sensor, MRN and CtIP as end-processing enzymes, and histone H1 and Werner helicase as mediators in the process; see Table 22.1 for the protein functional categories in HR and D-NHEJ. Finally, the involvement of Pol β, ligase III, and XRCC1 distinguishes this pathway from HR and D-NHEJ. The backup pathway may lead to a higher incidence of cancers than the other two and thus deserves the attention of the scientific community.

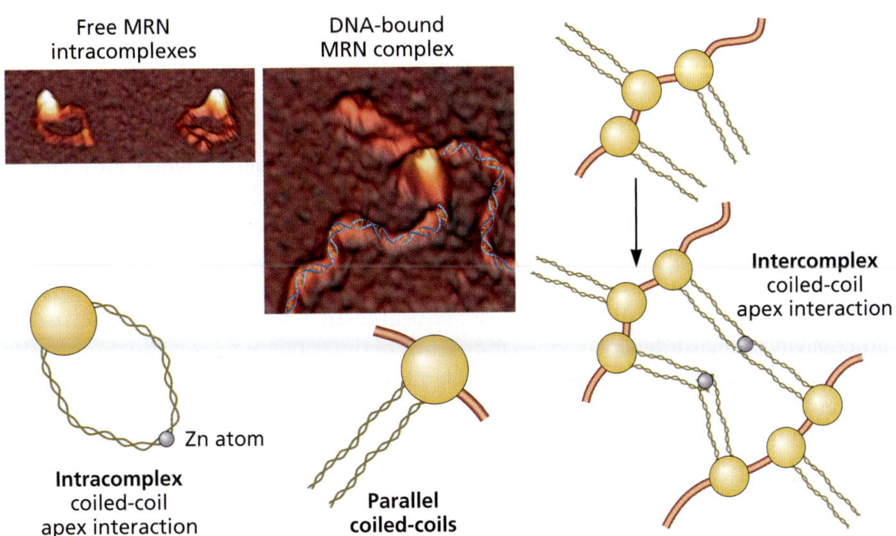

Figure 22.20 Conformational change in MRN complex upon DNA binding. The conformational change was deduced from atomic force microscopy or AFM imaging as shown here. The complex consists of a head and two arms; the arms are highly flexible and dynamically fluctuate between open and closed conformations, resulting from the ability of the apices to self-associate through binding of a Zn atom. This dynamic architecture is markedly affected by DNA binding, which leads to different orientation of the coiled coils. Conformational change is believed to be instrumental in the ability of MRN to keep broken DNA partners in close proximity, thus facilitating their repair. On linear DNA, the complex oligomerizes near DNA ends; these oligomers can now tether DNA molecules together by intermolecular interaction of the apices. (Adapted from Moreno-Herrero F, de Jager M, Dekker NH et al. [2005] *Nature* 437:440–443. With permission from Macmillan Publishers, Ltd. Top left, courtesy of Cees Dekker, Delft University of Technology.)

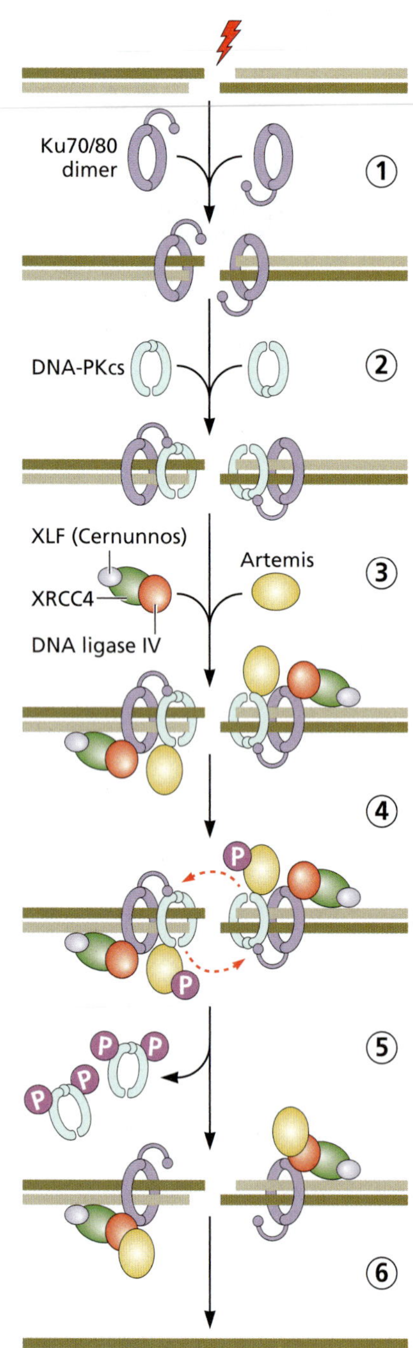

Figure 22.21 Events leading to NHEJ of an ionizing radiation-induced DSB. Ionizing radiation, IR, can induce single-strand breaks on opposite DNA strands, resulting in the formation of a DSB with short overhanging ends. The steps in the repair process are as follows: (Steps 1 and 2) The DSB is detected by Ku70–Ku80 heterodimer, which binds to the DNA ends and recruits DNA-PKcs through its flexible C-terminal region. Binding of DNA-PKcs induces inward translocation of the Ku dimer by about one helical turn, presumably to facilitate access of further repair proteins to the business end of the break, and mediates synapsis of the ends. (Step 3) DNA ends are processed by one or more enzymes that may include the following: (a) Artemis, a versatile nuclease capable of cleaving a wide variety of DNA structures. (b) DNA polymerases, not shown here: translesion polymerases Pol λ and Pol μ; and TdT, terminal deoxynucleotidyltransferase. Each of these polymerases can be used, depending on the structure of the ends to be repaired: 3′- or 5′-overhangs, blunt ends, or small single-stranded gaps. TdT can add untemplated nucleotides to DNA ends, whereas the translesion polymerases can fill gaps. (c) XRCC4–XLF (Cernunnos)–DNA ligase IV complex. DNA ligase IV is specific for this pathway and acts in a complex with two other proteins: XRCC4, which stabilizes the ligase and binds to DNA, and XLF, or XRCC4-like, whose function is still not well-defined but is important, because cells carrying mutations in *XLF* are radiosensitive and deficient in DSB repair. (Steps 4 and 5) Threading of ssDNA ends into cavities in the DNA-PKcs molecule activates the kinase. DNA-PKcs undergoes autophosphorylation, which opens the central DNA-holding cavity, releasing the protein from the DNA ends. This may provide preferential access to XRCC4–XLF–DNA ligase IV. DNA-PKcs phosphorylates Artemis, and possibly other proteins including Ku, activating its endonuclease activity. (Step 6) If the DNA ends are compatible, ligation occurs immediately. If the ends are not compatible, XRCC4–DNA ligase IV remains in the synaptic complex, while its polymerase and nuclease activities process the ends. As soon as the ends are processed into a compatible substrate, XRCC4–DNA ligase IV completes the joining reaction. (Adapted from Dobbs TA, Tainer JA & Lees-Miller SP [2010] *DNA Repair* 9:1307–1314. With permission from Elsevier.)

22.4 Translesion synthesis

Both endogenous and environmentally induced damage to the genome are primarily removed by DNA repair mechanisms, but the damage that remains blocks the progression of the replication fork. The activity of specialized, low-fidelity, low-processivity DNA polymerases enables cells to tolerate damage by replicating lesion-containing DNA without removing the lesion. The process has been termed translesion synthesis, TLS, and the many polymerases capable of TLS are called **translesion synthesis or bypass polymerases**.

Mammalian genomes encode 15 different DNA polymerases, most of which are involved in TLS. Normal replication is very fast—in bacteria ~1000 nucleotides are added to the nascent chain per second—and very accurate, with ~1 error per 10^6 incorporation steps (see Chapters 19 and 20). On the leading strand at least, it is highly processive. TLS enzymes, on the other hand, incorporate only a few nucleotides before they dissociate from the template. Interestingly, the presence of accessory proteins increases the *in vivo* processivity of high-fidelity enzymes thousands of times but increases the processivity of lesion-bypass polymerases only very slightly. Moreover, the low fidelity of lesion-bypass polymerases leads to the creation of numerous mutations in the DNA across the lesion they bypass. Even when they copy normal, undamaged DNA, their intrinsic error frequency is in the range of $1–10^{-3}$. How polymerases process lesions and whether they induce mutations are dictated by the type of lesions, as the bypass polymerases are lesion-specific; the sequence context; and the specific polymerase involved.

DNA polymerases are categorized into several families based on sequence similarity: A–D, X, Y, and RT. High-fidelity polymerases are members of the A and B families, whereas lesion-bypass polymerases belong primarily, but not exclusively, to the Y family. The TLS process itself involves switching from the classical polymerase, which will have stalled at a lesion, to a lesion-specific bypass enzyme, which extends the new

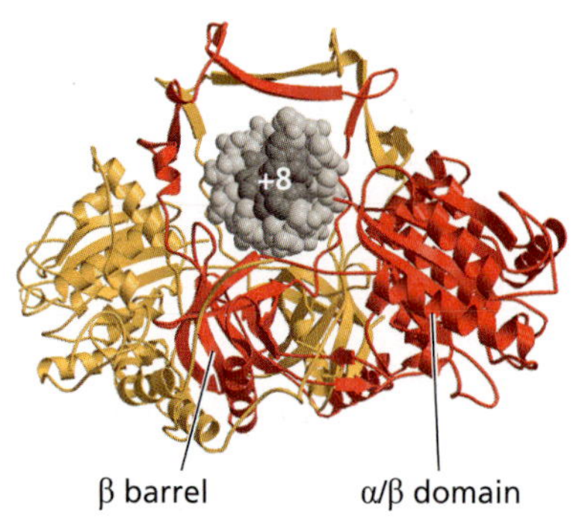

Figure 22.22 Structure of Ku70–Ku80 dimer bound to DNA. The view down the DNA helix is depicted. Ku70 is colored red and Ku80 is orange. A 14-bp duplex portion of DNA is shown; the terminal base pair at the broken DNA is numbered +8. The sugar–phosphate backbone is colored light gray and the bases are dark gray. (From Walker JR, Corpina RA & Goldberg J [2001] *Nature* 412:607–614. With permission from Macmillan Publishers, Ltd.)

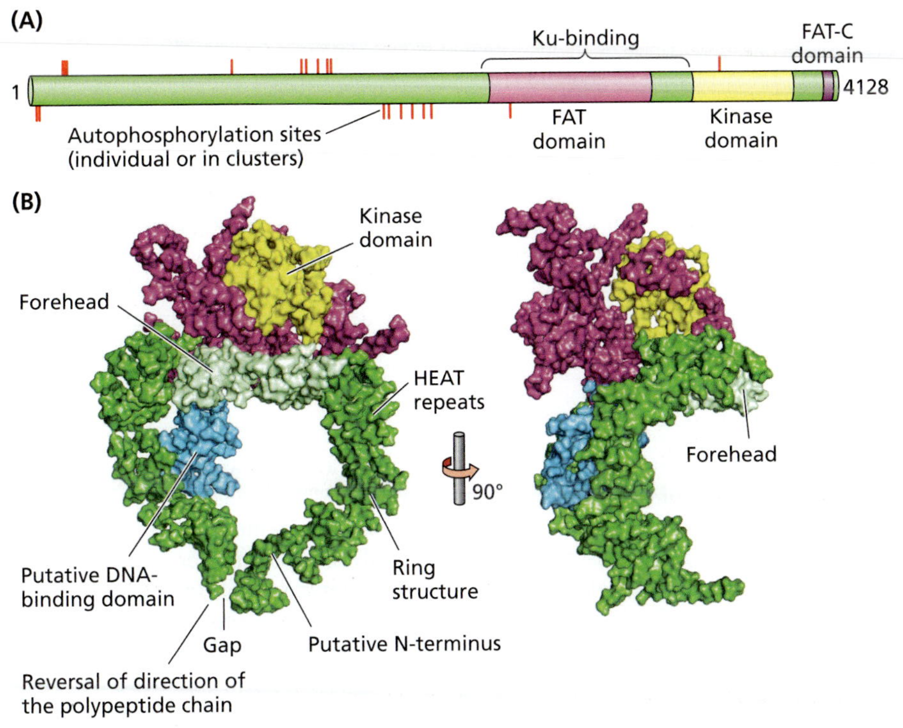

Figure 22.23 Structure of human DNA-PKcs. (A) Domain structure. (B) Crystal structure at 6.6 Å resolution. The many α-helical HEAT repeats or helix–turn–helix motifs facilitate bending of the polypeptide chain into a hollow circular structure, which has a concave shape rather like a cradle when viewed from the side. A conformational change triggered by phosphorylation—red bars indicate phosphorylation sites in part A—would widen the gap in a movement resembling bent arms swinging apart, so that DNA-PKcs can be released from the DNA. The carboxy-terminal kinase domain is located at the top of this structure, and the DNA-binding domain is inside. Finally, the FAT domain, which is present only in the phosphatidylinositol 3-kinase family, and the highly conserved small C-terminal domain FAT-C are positioned around the kinase domain. The three domains that form the head–crown structure were identified from EM studies. (B, from Sibanda BL, Chirgadze DY & Blundell TL [2010] *Nature* 463:118–121. With permission from Macmillan Publishers, Ltd.)

chain only by a couple of nucleotides, and a further switch to a second bypass polymerase, usually Pol ζ, which belongs to the high-fidelity B family (**Figure 22.25**). This second switch is needed to extend the DNA molecule to a sufficient length beyond the lesion, so that the lesion-induced distortions of the DNA structure are no longer in the way of a classical polymerase. At this point, a third polymerase switch takes place, with the re-recruitment of the classical enzyme. This highly orchestrated process must involve many control steps, especially in determining the choice of the polymerase capable of bypassing the particular lesion at hand. For example, of the three bypass polymerases that belong to the X family, Pol β bypasses 1-bp gaps with 3′-OH and 5′-PO_4, Pol λ bypasses the same lesion or a discontinuous template with paired termini, and Pol μ bypasses recessed DNA ends or ends where no complementarity of bases exists whatsoever. It can also perform template-independent DNA synthesis. Note from Table 22.1 that Pol λ and Pol μ are also the polymerases involved in NHEJ, reinforcing the point that DNA repair processes form a complex network in the cell.

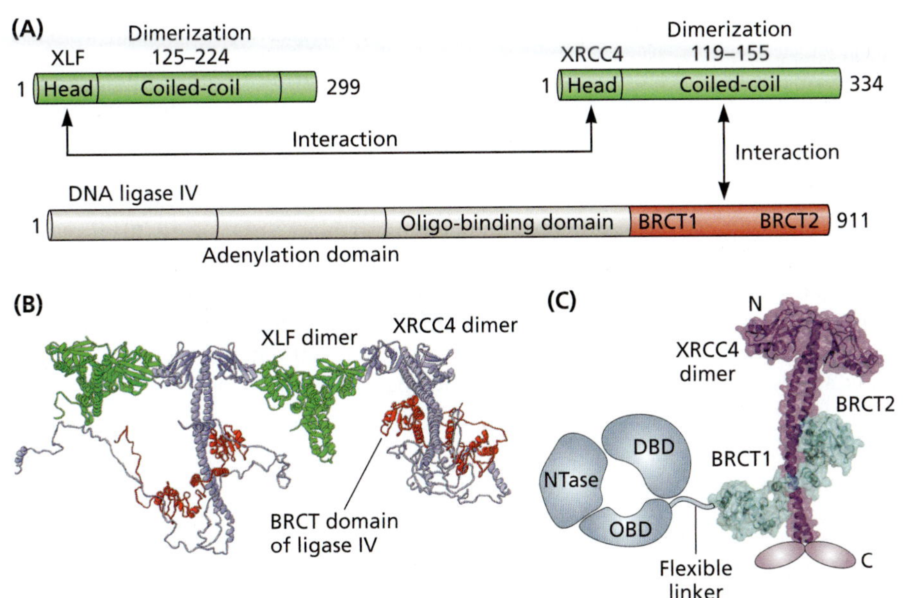

Figure 22.24 XLF–XRCC4–LigIV complex forms a filament that facilitates DNA binding and ligation. (A) Domain structure of XLF, XRCC4, and DNA ligase IV. Binding of XLF and XRCC4 is mediated by head-to-head interactions. XRCC4 also interacts with the tandem C-terminal BRCT domains of DNA ligase IV. (B) In the filament formed by all three proteins, alternating repeating units of XLF and XRCC4 place the BRCT domain of LigIV on one side of the filament. (C) Tandem BRCT domains are connected through a flexible linker to the catalytic nucleotidyltransferase, NTase, and DNA-binding domains, DBD, of the ligase, thus extending away from XRCC4. This flexible connection allows DNA end ligation without the need for LigIV to dissociate from the filament. (B, from Hammel M, Yu Y, Fang S et al. [2010] *Structure* 18:1431–1442. With permission from Elsevier. C, adapted from Perry JJP, Cotner-Gohara E, Ellenberger T & Tainer JA [2010] *Curr Opin Struct Biol* 20:283–294. With permission from Elsevier.)

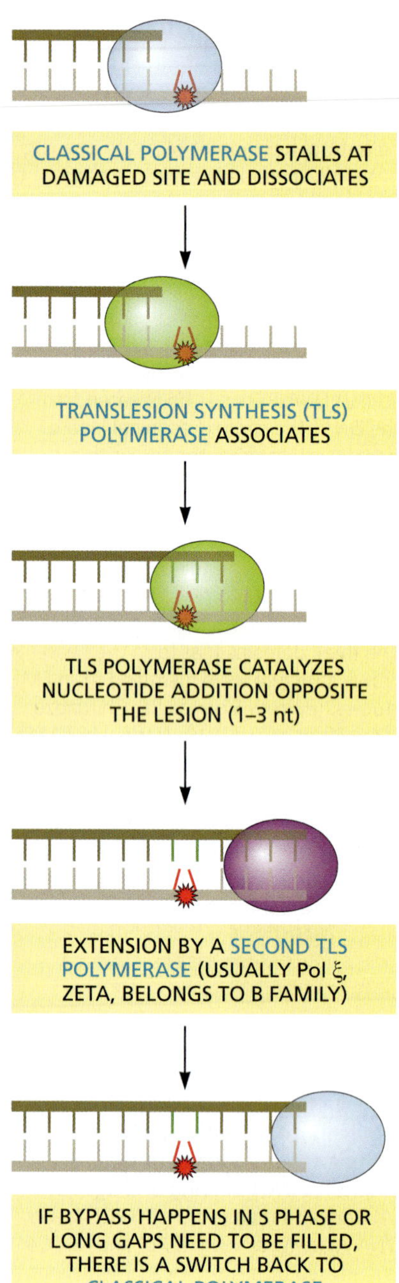
CLASSICAL POLYMERASE STALLS AT DAMAGED SITE AND DISSOCIATES
TRANSLESION SYNTHESIS (TLS) POLYMERASE ASSOCIATES
TLS POLYMERASE CATALYZES NUCLEOTIDE ADDITION OPPOSITE THE LESION (1–3 nt)
EXTENSION BY A SECOND TLS POLYMERASE (USUALLY Pol ξ, ZETA, BELONGS TO B FAMILY)
IF BYPASS HAPPENS IN S PHASE OR LONG GAPS NEED TO BE FILLED, THERE IS A SWITCH BACK TO CLASSICAL POLYMERASE

Figure 22.25 Polymerase switching in translesion synthesis. Switching back to replicative DNA polymerase may involve deubiquitylation of PCNA. (Adapted from Washington MT, Carlson KD, Freudenthal BD & Pryor JM [2010] *Biochim Biophys Acta* 1804:1113–1123. With permission from Elsevier.)

From a structural point of view, all known DNA and RNA polymerases contain a core domain consisting of a palm, fingers, and thumb; the active sites are formed by the palm and finger domains. The Y-family lesion-bypass polymerases contain an additional little finger (**Figure 22.26** and **Figure 22.27**), which makes the active site remarkably solvent-exposed and flexible, allowing it to accommodate erroneous bases or other lesions.

Many repair pathways utilize RecQ helicases

The **chicken-foot model** has been postulated to describe replication restart, a process involving the action of **RecQ helicases**, a subfamily of DNA helicases that are highly evolutionarily conserved (**Figure 22.28**). The family was named after the *recQ* gene in *E. coli*, and RecQ is the sole member of the family in *E. coli*. Lower eukaryotes also have only one RecQ-type helicase: Sgs1 in *Saccharomyces cerevisiae* and Rqh1 in *S. pombe*. Humans have five members of the RecQ family, and genetic defects in three of them lead to severe diseases (**Box 22.4**). Loss of RecQ helicase function leads to loss of genome integrity, resulting from hyperrecombination in particular. RecQ helicases participate in many repair pathways, including double-strand repair through HR. We mentioned Werner and Bloom helicases, members of the RecQ helicase subfamily, in connection with ALT, alternative lengthening of telomeres (see Chapter 20); they are discussed further in **Box 22.5**.

22.5 Chromatin as an active player in DNA repair

DNA repair in eukaryotic cells takes place in chromatin, so the repair machinery must deal with the presence of nucleosomes and higher-order chromatin structure (see Chapter 8). That chromatin plays a role in determining the distribution of DNA damage and the extent and rate of DNA repair has been known for some time. *In vitro* experiments in human cell extracts, using individual nucleosomes or nucleosomal arrays reconstituted on plasmid DNA as substrates for repair, have clearly indicated that repair of naked DNA is more efficient than repair of chromatin DNA, and linker DNA in chromatin is more accessible to repair than the DNA on nucleosomal core particles. It became clear that chromatin rearrangement is often required to allow access of repair proteins to DNA damage sites. This led to the formulation, in the early 1990s, of the access, repair, and restore model or ARR of NER.

Today, we know that chromatin not only is a passive barrier to repair but should be considered as an integral player in DDR. To accommodate this new information about the active participation of chromatin in the repair pathways, the ARR model is being replaced by the prime, repair, and restore model or PRR. Three major components of the chromatin toolbox are utilized in repair: (1) histone variants and the enzymes that place or remove postsynthetic modifications on them (**Table 22.2**); (2) chromatin remodelers; and (3) chromatin chaperones. The interplay among these participants and repair depends on the context and specific repair pathway and is extremely complex. It is also very poorly understood. For that reason, we focus on a couple of examples that illustrate the active participation of chromatin structure and dynamics in the three stages of DNA repair: initiation, actual repair, and termination.

Histone variants and their post-translational modifications are specifically involved in DNA repair

To date, we know of three H2A variants that are actively involved in repair: H2A.X, H2A.Z, and macroH2A (see Chapters 8 and 12). Here we focus on **H2A.X** and its phosphorylated form, γH2A.X.

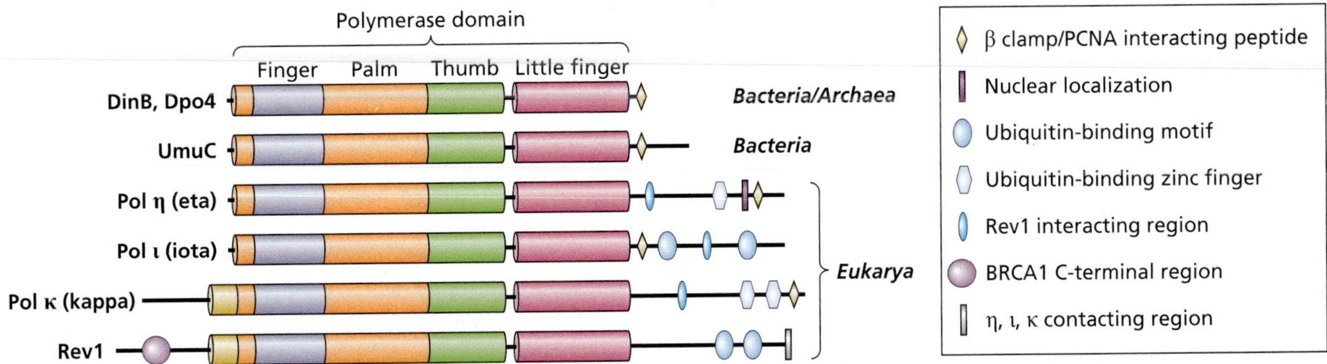

Figure 22.26 Structural domains of Y-family translesion DNA polymerases in the three domains of life. Palm, thumb, and finger subdomains are found in the polymerase domains of all known DNA and RNA polymerases. Uniquely in Y-family polymerases, the C-terminus of the catalytic core also contains a little finger domain, also called the polymerase-associated domain or PAD, of ~100 residues. Two of the polymerases shown have an additional N-terminal domain, depicted in gold. The regulatory sequences are represented, as indicated in the key. Mouse and human Rev1 are unique in possessing an additional N-terminal BRCT domain. Defects in the gene encoding human Pol η cause the sunlight-sensitive and cancer-prone xeroderma pigmentosum variant syndrome, XP-V. (Adapted from Yang W & Woodgate R [2007] *Proc Natl Acad Sci USA* 104:15591–15598. With permission from National Academy of Sciences.)

Phosphorylation of the C-terminus of histone variant H2A.X is one of the first events to occur during the repair of DSBs. H2A.X contains a core sequence, which is highly conserved with other H2A variants, and a short, four-amino-acid conserved tail that is connected to the core by a linker of variable length (Figure 22.29A). The conserved C-terminal tail protrudes from the nucleosome core particle near the entry/exit point of the DNA (see Chapter 8). In mammals, the relative content of H2A.X varies among cell types, but in the chromatin 30-nm fiber, one H2A.X molecule can be present as often as every fifth nucleosome. In yeast, H2A.X is the main H2A variant and constitutes about 90% of the total H2A. Phosphorylation of serine in the tail gives rise to

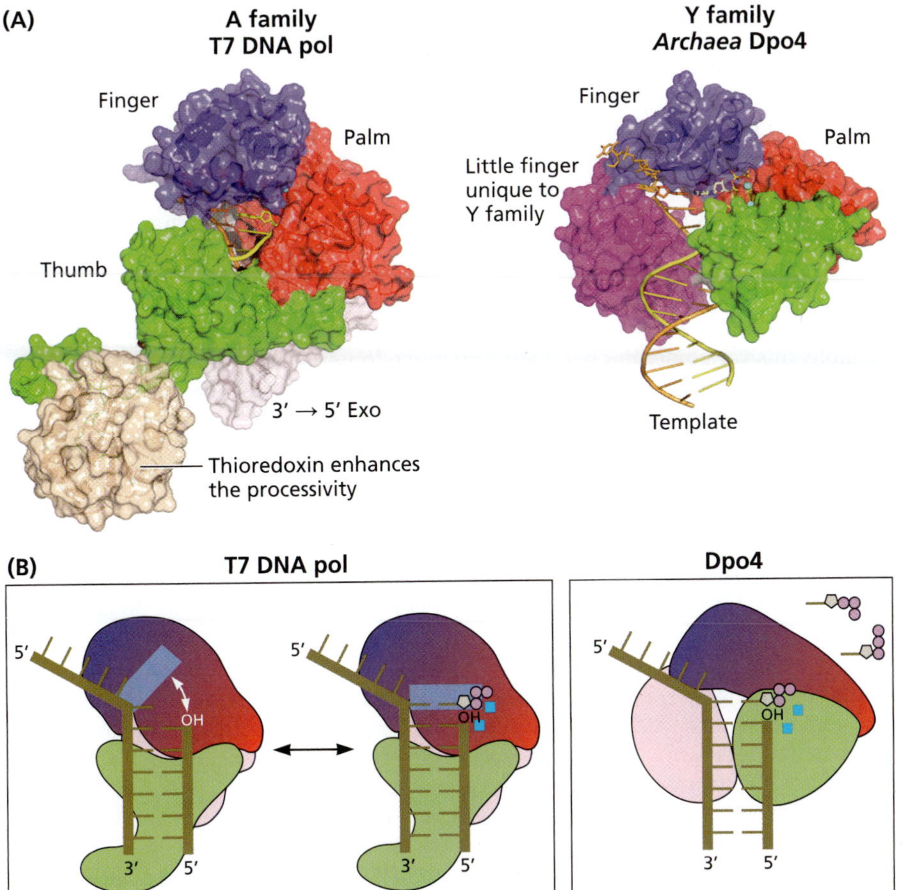

Figure 22.27 Structural comparison of phage T7 DNA pol and translesion DNA polymerase Dpo4. (A) Crystal structures. The active site of Y-family polymerase is pre-formed before substrate binding, in contrast to those of A, B, and RT polymerases. The active site is remarkably solvent-exposed and not as constrained to reject non-Watson–Crick base pairs, allowing erroneous base pairing and accommodation of bulky lesions; the incoming nucleotide is visible only in Dpo4, shown as silver and multicolored sticks. (B) Schematic representation of conformational changes that occur in either polymerase. A helix, shown by a solid blue rectangle, in the finger domain of T7 Pol changes conformation upon binding of the correct dNTP, bringing its α-phosphate into correct orientation with respect to its partner, the 3′-OH of a primer strand; this closes the active site for the polymerization reaction to occur. Movement of the helix is indicated by a double-headed arrow. By contrast, the spacious and open active site of Dpo4 allows several different dNTP conformations, as diagrammed in the upper right corner, and makes it difficult to correctly align the 3′-OH of the primer strand, dNTP, and metal ions with the catalytic site. This contributes to the inefficiency of these enzymes. (Adapted from Yang W & Woodgate R [2007] *Proc Natl Acad Sci USA* 104:15591–15598. With permission from National Academy of Sciences.)

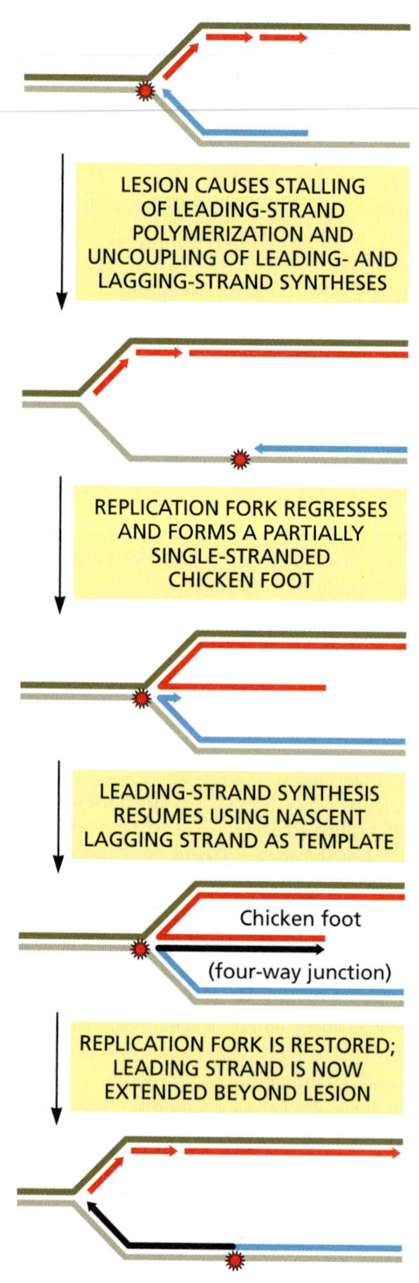
LESION CAUSES STALLING OF LEADING-STRAND POLYMERIZATION AND UNCOUPLING OF LEADING- AND LAGGING-STRAND SYNTHESES
REPLICATION FORK REGRESSES AND FORMS A PARTIALLY SINGLE-STRANDED CHICKEN FOOT
LEADING-STRAND SYNTHESIS RESUMES USING NASCENT LAGGING STRAND AS TEMPLATE
Chicken foot
(four-way junction)
REPLICATION FORK IS RESTORED; LEADING STRAND IS NOW EXTENDED BEYOND LESION

Figure 22.28 Chicken-foot model for replication restart. A lesion in the leading-strand template, shown as a red starburst, blocks leading-strand synthesis. Lagging-strand synthesis continues for a short period. The fork regresses, allowing the nascent strands to anneal and creating a chicken-foot structure or four-way junction. The leading strand is then extended by use of the longer lagging strand as a template. The fork is reset, presumably by a RecQ helicase, via branch migration of the four-way junction, and now the leading strand has been extended beyond the lesion. Replication can recommence. (Adapted from Bachrati CZ & Hickson ID [2003] *Biochem J* 374:577–606. With permission from Biochemical Society.)

γH2A.X. γH2A.X is not essential to DSB repair, as H2A.X-null mice are viable, but its presence accelerates the initial rate of repair and may increase its fidelity.

Here, we consider the situation in mammalian cells, as there are rather significant differences among species, both in the enzymes that participate in phosphorylation/dephosphorylation events and in the protein components of the mature repair foci. Immediately following DSB formation, a large number of the resident H2A.X molecules undergo phosphorylation to form γH2A.X foci around the break (Figure 22.29B). The initial small γH2A.X focus quickly spreads to surrounding regions, and at the end of the repair process it disappears. Increased radiation dose leads to a linear increase in the number of γH2A.X foci. The kinases responsible for the modification are the damage-signaling kinases ATM, ataxia telangiectasia mutated (see Box 22.3); ATR, ATM related; and DNA-PK, DNA-dependent protein kinase. The focus is a very large structure that forms within minutes of exposure to ionizing radiation at each DSB site (Figure 22.30A). In a typical human cell, H2A.X is ~10% of total H2A; thus a 40 Mbp locus will contain about 40,000 H2A.X molecules, ~10% of which appear to be phosphorylated at any one time. In theory, about 200 foci would cover the entire human genome.

The focus also contains a myriad of DNA repair proteins and chromatin remodeling factors that are recruited to the site by direct or indirect interactions with the phosphorylated γ epitope. The focus possesses a defined microstructure (Figure 22.30B and Figure 22.30C), the center of which contains the resected ssDNA regions and the proteins that bind to them. Examples of these are Rad51 and Rad52, the ATR kinase and its interacting partners ATRIP and RPA, Mre11, NBS1, and 53BP1. The flanking regions contain a distinct set of proteins that spread from the initial small focus, thus amplifying the DDR. The MRN sensor complex, the mediator MDC1, 53BP1, and BRCA1 are found in the central region but also spread up to a megabase away from the physical break. The proteins at the foci show different behavior during the cell cycle; some, including Mre11, NBS1, and 53BP1, dissociate during G_2 and are almost absent in metaphase; they then rebind in G_1. Other proteins, such as MDC1 and BLM, remain in the foci throughout the cell cycle.

Importantly, γH2A.X foci also form at damaged or eroded telomeres and in cancer cells (see Figure 22.29B). The number and size of γH2A.X foci in cancer cells are highly variable; this variability has been attributed to telomeric foci. There is also variability among cells in the same culture, which is also primarily due to telomeric foci.

Thus, γH2A.X can be viewed as a first responder, providing a platform for assembly of the functional repair complex and the chromatin remodelers that are needed to open up chromatin structure for the repair to occur.

The recruitment of γH2A.X and its further role as a platform for the assembly of numerous repair factors and chromatin remodelers is presented in schematic form in Figure 22.31, which summarizes the sequence of events that occurs on chromatin

Box 22.4 RecQ helicases, DNA repair, and human disease Helicases are enzymes that unwind the complementary strands in duplex DNA, RNA, or DNA–RNA heteroduplexes. They participate in numerous processes in which the two strands of the nucleic acid need to be separated, including DNA replication, DNA recombination, and DNA repair. Each organism possesses a myriad of helicases that are classified according to their characteristic features and the number of domains that perform strand separation. In addition, helicases can be monomeric, dimeric, or hexameric and differ in their mechanism of action.

Curiously, the hexameric helicase complexes are ring structures that possess helicase domains and can translocate on ssDNA or dsDNA in an ATP-dependent manner, but they do so without actually performing strand separation. The bacterial DnaB protein is an example of such a hexameric helicase, which participates in the initiation of DNA replication; it translocates along ssDNA which is threaded through the hole of the ring (see Figure 19.23). The *E. coli* RuvB protein, which participates in branch migration during homologous recombination, encircles both strands of the duplex (see Figure 21.9 and Figure 21.11). Some chromatin remodeling complexes, such as SWI/SNF (see Chapter 12), also belong to this group. These translocases couple ATP hydrolysis to motion without performing strand separation and should probably not be labeled as helicases, despite the presence of helicase domains.

The RecQ helicase family has attracted considerable attention because of a disease connection in humans. This family is highly conserved in evolution (**Figure 1**) and is named after its founding and best-studied member RecQ in *E. coli*. Bacteria and yeast have only one family member but humans have five, and mutational defects in three of them give rise to three genetic disorders—Bloom, Werner, and Rothmund–Thomson syndromes—associated with predisposition to cancer and/or premature aging (see Box 22.2 and Box 22.5).

RecQ helicases have earned the designation "caretakers of the genome," in the words of Ian Hickson. This is mainly because they participate in HR by helping to resolve Rad51-mediated recombination events at the four-way junction stage. RecQ helicases are also involved in suppression of unwanted recombination events by disrupting Rad51–RecA filaments; in this role they act as anti-recombinases. They are also involved in at least three other steps of HR.

Finally, the RecQ helicases are structure-specific and can act on a variety of intermediates in a number of processes. Thus, they can resolve, or unwind, replication or repair intermediates such as 3′-tailed structures, replication forks, duplex forks, bubbles, and hairpins; recombination intermediates such as three- and four-way junctions, D-loops, and triple helices; and the G-quadruplex structures that form at telomeres (see Chapter 8). All of these activities are needed to avoid structural impediments to normal DNA transactions.

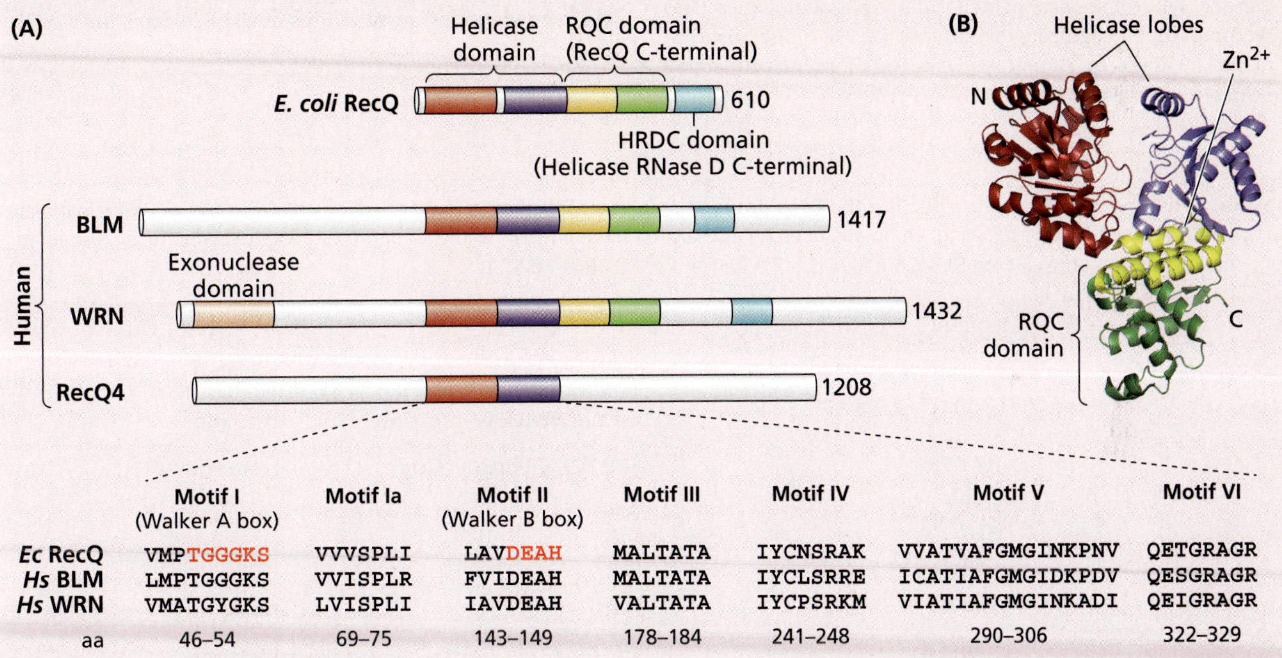

Figure 1 RecQ DNA helicases. (A) Domain structure of selected members of the RecQ family: *E. coli* RecQ and three human proteins whose genes are mutated in Bloom, Werner, and Rothmund–Thomson syndromes, BLM, WRN, and RecQ4, respectively. The proteins are aligned by their conserved helicase domains. In addition, these members of the RecQ family contain two additional highly conserved regions. The RQC domain is unique to RecQ helicases and probably participates in the numerous protein–protein interactions characteristic of these helicases; the HRDC domain seems to be involved in binding to DNA substrates. WRN also has an additional exonuclease domain with 3′ → 5′ exonucleolytic activity. The sequence below the schematic shows the seven helicase motifs typically present in superfamily 2, to which RecQ helicases belong. Motif I, or Walker A box, interacts with the phosphate moieties of NTPs via the amino group of lysine. Motif II, or Walker B box, contains the acidic tetrapeptide DExx; the glutamate, E, is part of the catalytic center. Both motifs participate in coordination of the catalytic Mg^{2+}. Motifs Ia, IV, and V probably serve to bind the DNA structures. (B) Crystal structure of the catalytic core RecQΔC: residues 1–208 are shown in red, 209–340 in violet, 341–406 in yellow, and 407–516 in green, corresponding to color coding in part A. Helicase motifs I, Ia, II, and III are in the red domain; motifs IV, V, and VI are in the violet domain. The yellow subdomain binds Zn; two of the known BLM mutations disrupt Zn binding and the activity of the enzyme. Finally, the green subdomain is a specialized helix–turn–helix fold called a winged-helix subdomain that makes contacts with both the minor and the major groove. It has been proposed that the Zn-binding region, together with the winged-helix domain, may form extended DNA- and protein-binding sites. Such extensive interaction sites are needed for recognition of and binding to numerous DNA structural intermediates in DNA replication, recombination, and repair. (A, top, adapted from Bachrati CZ & Hickson ID [2003] *Biochem J* 374:577–606. With permission from Biochemical Society. B, courtesy of James Keck, University of Wisconsin.)

Box 22.5 Human diseases linked to RecQ helicase mutations
Diseases caused by mutations in the RecQ helicase genes are rare autosomal recessive genetic disorders, occurring mostly through consanguineous marriages. All of these disorders show genomic instability associated with cancer predisposition. Expression of the respective helicase is highly up-regulated in rapidly growing or immortal cells. The clinical features of three major disorders and some facts about the protein defects are presented here.

Bloom syndrome
The disease was first described by dermatologist David Bloom in 1954. It is characterized by proportional dwarfism, narrow face and prominent nose and ears, male infertility and female subfertility, frequent infections, sun-induced erythema, type II diabetes, and predisposition to a wide range of cancers such as non-Hodgkin's lymphoma, leukemias, and carcinomas of the breast, gut, and skin, with mean age of onset of 24. The mutated protein is RecQ2 or BLM.

Werner syndrome
Werner syndrome was named after ophthalmologist Otto Werner, who recognized the disease in 1904. The disorder manifests itself in numerous features of premature aging and sequential appearance of age-related diseases, retarded growth, hypogonadism, immunodeficiency, type II diabetes, cancers of mesenchymal origin, and soft-tissue and osteogenic sarcomas; in the general population carcinomas to sarcomas are 10:1, while in Werner syndrome it is 1:1.

The gene affected in Werner syndrome patients encodes protein RecQ3 or WRN. Some WRN mutants are truncated versions of the protein that lack the C-terminal NLS, nuclear localization signal, and cannot migrate to the nucleus. Unfortunately, there is no good mouse model of the disease, since the mouse protein lacks the NoLS, nucleolar localization signal, adjacent to the NLS, and is not localized in the nucleolus, while the human WRN can traffic between the nucleoplasm and nucleolus.

Rothmund–Thomson syndrome
The syndrome was first described by ophthalmologist August von Rothmund in 1868. Patients affected exhibit short stature, skeletal abnormalities, skin pigmentation changes, skin atrophy, congenital cataracts and bone defects, premature aging, and cancers, primarily osteogenic sarcoma. The genetic defect has been identified as mutations in a single gene, the *RecQL4* gene.

Table 22.2 Histone modifications affecting the DNA damage response. Modifications are classified according to their participation in the consecutive steps in DDR: signaling, chromatin opening at the beginning of repair, and chromatin restoring following repair. (Adapted from Rossetto D, Truman AW, Kron SJ & Côté J [2010] *Clin Cancer Res* 16:4543–4552.)

Type of modification	Residue modified (human)	Enzyme responsible	Step in DDR affected
acetylation	H4/H2A.X	Tip60/yNuA4	opening
	H3 K9	Gcn5, CBP/p300	opening
	H3 K56	yRtt109, CBP/p300, Gcn5	restoring
	H3 K14, K23	Gcn5	restoring
	H4 K5, K12	Hat1	restoring
	H4 K91	Hat1	restoring
deacetylation	H3/H4 K	Sin3/Rpd3, Sir2, Hst1/3/4	restoring
methylation	H4 K20	Set8/Suv4–20	signaling
	H3 K79	Dot1	signaling
phosphorylation	H2A.X S139	ATM/ATR, DNA-PK	signaling
	H4 S1	casein kinase 2	restoring
	H2B S14	Ste20	restoring
dephosphorylation	H2A.X Y142	EYA1	signaling
	H2A.X S139	yPph3/hPP4, PP2A, PP6, Wip1	restoring
ubiquitylation	H2A/H2A.X	RNF8/RNF168	signaling
	H4 K91	BBAP	signaling
	H2A K119	Ring1b/Ring2	restoring
sumoylation	H2A.Z K126/133	?	signaling

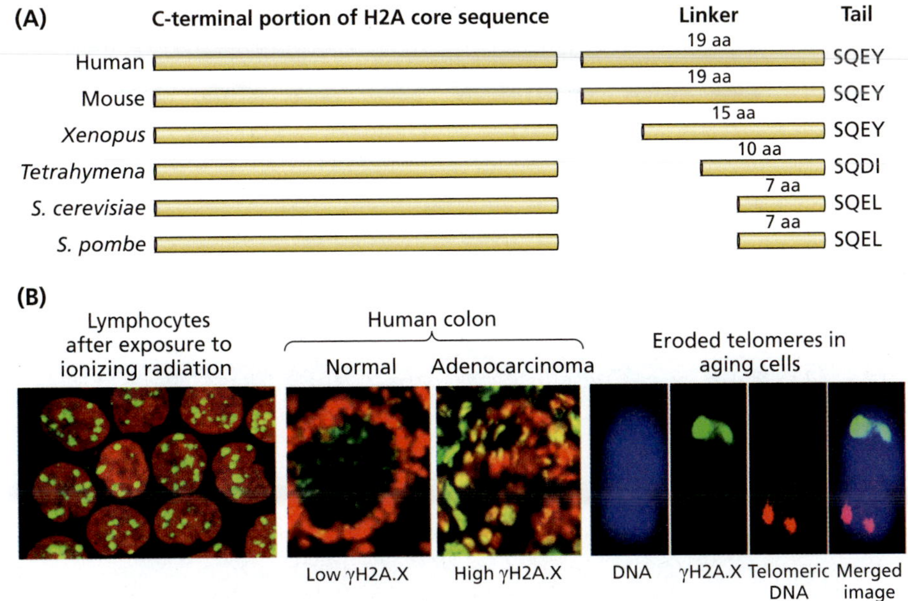

Figure 22.29 Histone variant γH2A.X forms repair foci around DSBs. (A) Structure of H2A.X variants with phosphorylatable tail. (B) Cytological view of foci formed in interphase chromatin or mitotic chromosomes following ionizing irradiation of cells in culture. DNA is stained with red (left and middle) or blue (right) fluorescent dye; γH2A.X is stained with a green fluorescent dye. (Left) Cells 30 min after exposure to ionizing radiation. (Middle) Frozen sections of normal and malignant human colon. Note the presence of numerous foci in the adenocarcinoma. (Right) The eroded telomere in the imaged chromosome from an aging cell does form a γH2A.X focus but is too short to bind to the telomeric DNA probe; the putative functional telomere binds the DNA probe and lacks a γH2A.X focus. (B, courtesy of William Bonner, National Institutes of Health. B, middle, from Bonner WM, Redon CE, Dickey JS et al. [2008] *Nat Rev Cancer* 8:957–967. With permission from Macmillan Publishers, Ltd.)

during DSB repair. Recently, the highly reproducible formation of γH2A.X foci during DNA repair is finding useful application in the clinic (**Box 22.6**).

The newest experimental evidence points to the ways in which the repair process is terminated. It is clear that the phosphorylated form of H2A.X must be removed as a

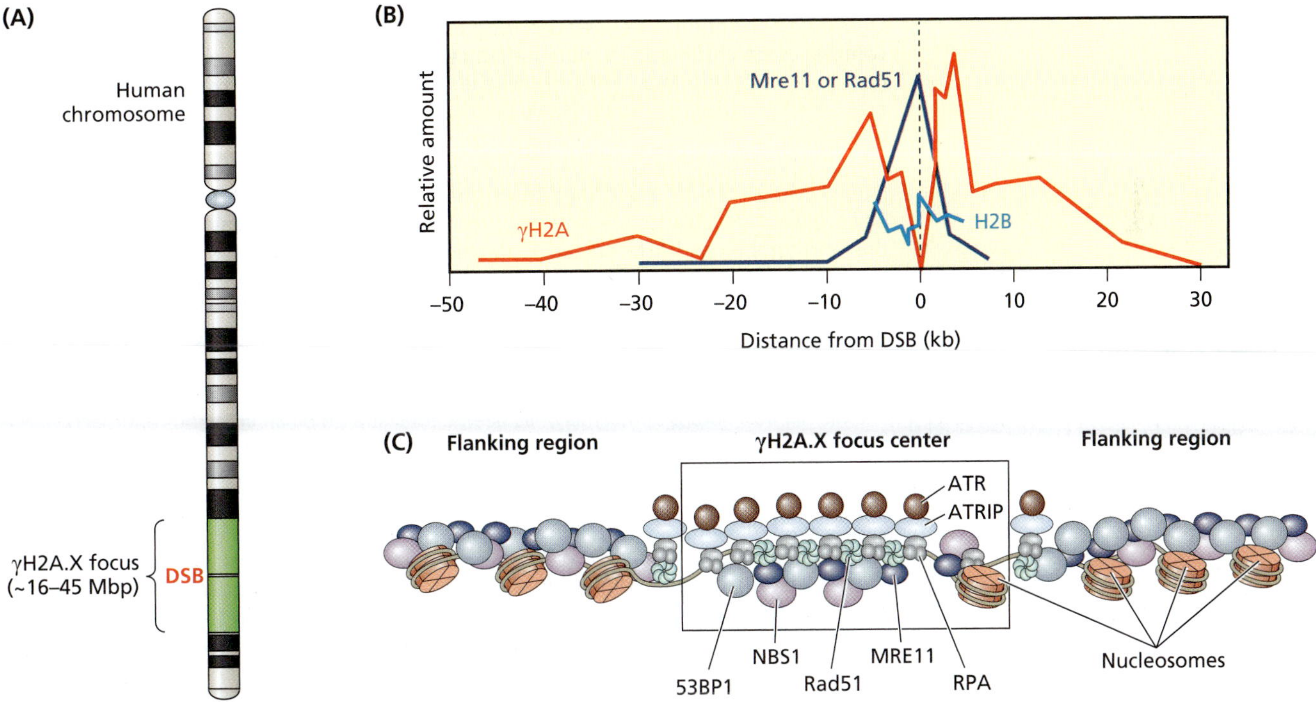

Figure 22.30 Repair foci around DSBs. (A) A single γH2A.X focus in the context of a sample human chromosome. (B) Chromatin immunoprecipitation studies performed in yeast show that histone H2A, the yeast functional analog of mammalian H2A.X, is not phosphorylated near the site of DSBs. The level of histone H2B shows the presence of nucleosomes around the break point. Finally, Mre11 and Rad51 are present at high levels around the break. (C) Similar studies performed on other participants in the repair process have provided an insight into the fine structure or microstructure of a γH2A.X focus, as the one presented here in schematic form. The γH2A.X focus center contains resected ssDNA; the proteins at the center include signaling molecules and factors that restore chromatin structure following repair. Regions that flank the focal center contain proteins that spread and amplify the DNA damage response. (A–B, adapted, courtesy of William Bonner, National Institutes of Health. C, adapted from Misteli T & Soutoglou E [2009] *Nat Rev Mol Cell Biol* 10:243–254. With permission from Macmillan Publishers, Ltd.)

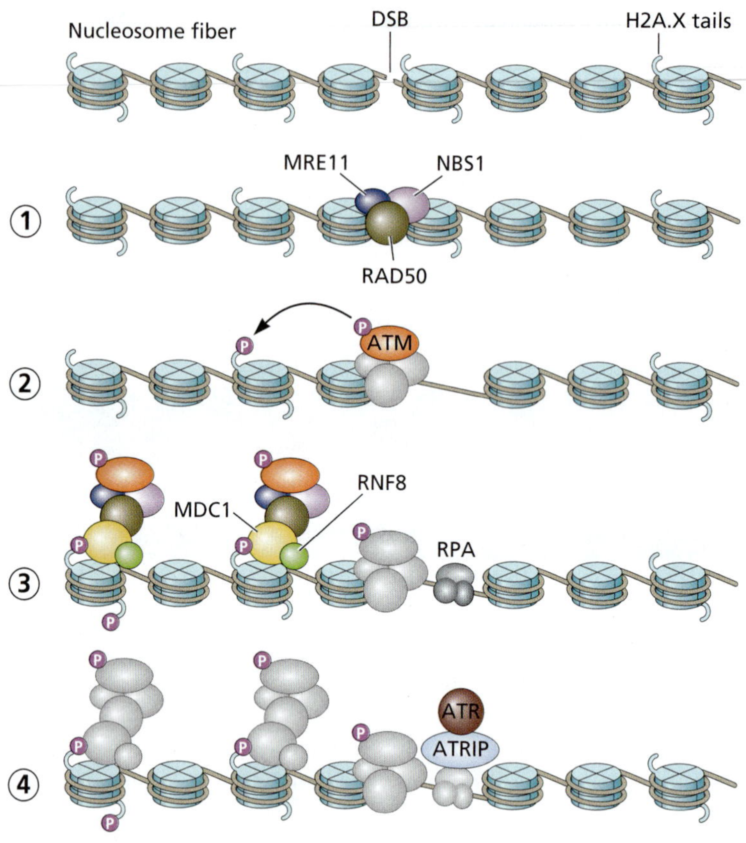

Figure 22.31 Sequence of events during assembly of DNA damage response, DDR, complexes on chromatin in mammalian cells. Several major steps are recognized. For clarity, we have used color to depict only the newly added components; the ones bound in previous steps are in gray. (Step 1) The presence of a DSB in the chromatin fiber is sensed by the MRN complex. (Step 2) MRN activates the signal transducer kinase ATM, which phosphorylates histone H2A.X, creating γH2A.X, in the nucleosomes that flank the break. (Step 3) MDC1, mediator of DNA damage checkpoint 1, has high affinity for γH2A.X; it binds and acts as a platform for the entire complex. MDC1 recruits additional copies of the MRN complex and ATM, spreading the focus along the chromosome. It also recruits chromatin remodelers, not shown here, and histone modification complexes. At this step, the single-stranded portions of DNA are covered by replication protein A, RPA. (Step 4) Ubiquitylated histones recruit the downstream checkpoint mediators 53BP1, p53 binding protein 1, and BRCA1, breast cancer susceptibility protein 1, not shown here. RPA recruits another downstream signal-transducing kinase, ATR, through its interaction partner ATRIP. The hierarchical nature of assembly has been demonstrated by an ingenious approach: rather than performing laborious order-of-addition experiments, researchers carried out tethering experiments. In these experiments, the protein of interest is directly tethered to chromatin as a fusion protein with the bacterial Lac repressor; the fusion protein binds to a *lac* operator incorporated into the eukaryotic genome. In this way, the protein of interest becomes bound to chromatin, avoiding the need for upstream factors. Such experiments showed that the situation is more complex: the simple hierarchical organization of the repair complex is complicated by numerous instances of cross-talk between factors. For example, MDC1 tethering leads to recruitment of the upstream factors MRE11 and NBS1 in a γH2A.X-independent fashion, indicating that the ability of downstream factors to recruit upstream factors is probably a way to amplify the damage signal. (Adapted from Misteli T & Soutoglou E [2009] *Nat Rev Mol Cell Biol* 10:243–254. With permission from Macmillan Publishers, Ltd.)

signal for resumption of the cell cycle following successful repair. This occurs in two major ways: either H2A.X is dephosphorylated or it is replaced by the canonical H2A variant. How is this achieved? First, several phosphatases have been described, in both yeast and mammals, that directly remove the phosphate group, thus negatively

Box 22.6 Histone γH2A.X foci monitoring and clinical practice γH2A.X foci can be used in the clinic to evaluate doses of ionizing radiation, to detect precancerous cells, to stage cancers, or to monitor the effectiveness of treatment. As mentioned above, increased radiation dose leads to a linear increase in the number of γH2A.X foci. The relationship is so highly reproducible and robust, both in cultured cells and in whole organisms, that researchers have suggested using measurements of these parameters as a radiation dosimeter. Methods have been developed to do the analysis on plucked hair, so as to avoid withdrawal of blood or painful skin biopsies (**Figure 1**). To date, the γH2A.X locus remains the most sensitive way to detect a DSB.

Finally, there seems to be a connection between γH2A.X function and cancer. It has been shown consistently in laboratory settings that *H2A.X*$^{-/-}$ mice survive well in unstressed conditions but are less efficient in DSB repair, leading to an increased incidence of chromosomal abnormalities. These mice are prone to developing tumors in a p53-null background. Thus, γH2A.X may be considered a tumor suppressor.

Certain cancers are often characterized by loss of the chromosomal region that contains the *H2A.X* gene, but at least one is known with an amplification of the *H2A.X* gene. Studies have looked into the possibility of using γH2A.X levels in patients to follow treatment effectiveness. These studies looked at γH2A.X levels in easily obtainable normal surrogate cells from patients, such as skin and leukocytes, 1–2 days after drug administration, when the drug is most active. The usefulness of this technique will depend on whether this early drug response in surrogate cells correlates with the response of the tumor several months later. If so, procedures and drugs could be optimized for the most favorable outcome.

The detection and quantitation of γH2A.X foci can be successfully used in assessing overall DNA damage and repair caused by the environment. Studies have used this criterion to examine the effect of mobile phone use and eating organic versus regular apples, for example. There is little doubt that these methods will be further developed for broader assessment of environmentally caused organismal stress.

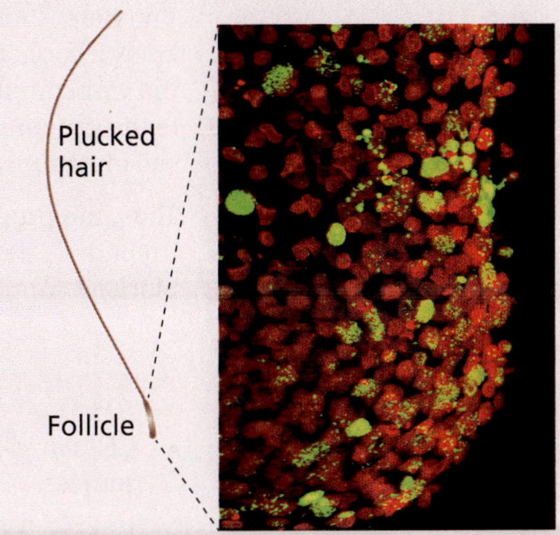

Figure 1 Measuring radiation exposure in follicle cells of plucked hair by quantifying γH2A.X loci by immunofluorescence. The image of a group of follicle cells was taken 30 min following irradiation. Note the good H2A.X immunofluorescence signal at the follicle base. The strength of the signal is dose-dependent, making it a reliable measure of radiation exposure. (Courtesy of William Bonner, National Institutes of Health.)

regulating the repair function, that is, terminating repair. Second, chromatin remodelers and/or histone chaperones are involved. Thus, the histone acetylase complex TIP60, Tat-interactive protein 60, in *Drosophila* and humans acetylates γH2A.X and promotes its replacement by the unmodified variant. The histone chaperone FACT, facilitates chromatin transcription (see Chapters 12 and 20), is poly(ADP)ribosylated after genotoxic stress; this modification disrupts its association with nucleosomes and its role as an exchange factor for H2A-H2B dimers. INO80, inositol-requiring 80, is another remodeler that counteracts the turnover of H2A.X in yeast. Thus, a complex network of factors controls γH2A.X dynamics, and thus DNA repair and cell-cycle checkpoints.

Some of the H3 variants, and the specific chaperones involved in their deposition on chromatin, are also involved in chromatin dynamics during repair. The major histone subtype of a specific histone type is synthesized only during DNA replication, taking part in the replication of chromatin (see Chapter 8). Minor variants are synthesized throughout the cell cycle to perform specific functions that require replacing one histone subtype with another in a preexisting chromatin fiber. One of the two H3 replicative variants, H3.1 or H3.2, is deposited at both UVC- and laser-induced damage sites by the histone chaperone **CAF1, chromatin assembly factor 1**. Thus, H3.1 deposition is not restricted to replicating DNA but also takes place during DNA repair. Such a deposition could provide a window of opportunity to replace resident histones, carrying their respective modifications, with new unmarked histones from the soluble nucleoplasmic pool. Such a replacement may have far-reaching consequences on a variety of processes, including transcription, because it disrupts the parental histone marking. Data exist to support such a notion.

22.6 Overview: the role of DNA repair in life

It should be apparent from this chapter that organisms make great effort to minimize damage to their DNA and to repair that damage once it occurs. This should not be surprising, as the major lesson of our whole study has been that the DNA blueprint specifies, directly or indirectly, almost everything that happens during the growth and development of cells and organisms. If the blueprint is damaged, the resulting structures or processes must, to some extent, be flawed.

But there is another side to the coin. If repair of DNA lesions were perfect, life as we know it could not exist. The process of evolution, from the simplest proto-organism to the vast complexity and diversity we see today, has depended on unrepaired lesions in the genome. We are what we are because of a long history of uncorrected errors we call mutations.

Thus, the proper perspective is to see that repair must be good enough to prevent irreparable damage to each individual organism, but it must not be too good, so as to block evolution. The balance has somehow been maintained over the ages.

Key concepts

- Cellular DNA may suffer damage from either environmental or endogenous sources.
- Lesions occur in many different forms, including single-strand breaks, double-strand breaks, chemical damage to bases, and intra- or interstrand cross-linking.
- Repair can be direct, reversing the lesion reaction, or may involve excision of one or a number of nucleotides surrounding the lesion, removal of a base, correction of mismatches, and either homologous or nonhomologous recombination. Sometimes the lesion is just ignored in replication.
- Direct repair occurs, for example, in photolyase-mediated reversal of pyrimidine dimers, or repair of alkylated sites on bases.
- Nucleotide excision repair, NER, involves cutting out a single-stranded oligonucleotide containing the lesion, resynthesis of the removed strand, and ligation. In eukaryotes, such repair is often focused on transcribing regions.
- Base excision repair, BER, corrects for incorrect or modified bases resulting from chemical insult or conversion, such as C/U transformation. The errant base is simply cut out and replaced with the correct base; in an alternative mechanism, an entire patch surrounding the lesion is replaced.
- Mismatch repair, MMR, corrects for erroneous base pairings that are produced during replication. The main difficulty here is in recognizing which strand is wrong. In bacteria, the methylation marking of GATC sequences, which does not occur immediately after replication, distinguishes the new, that is, error-containing, strand from the template. The corresponding process in eukaryotes is not understood.
- Double-strand breaks, DSBs, are very harmful and difficult for the cell to deal with. There are two general mechanisms. Homologous recombination, HR, depends on the existence of a sister chromatid, which can be used to align the ends to join and is considered essentially error-free. This is possible only in the S or G_2 phases of the cell cycle. Nonhomologous end-joining, NHEJ, restores the integrity of the chromosome by more or less random rejoining. Clearly this can lead to the loss or gain of genetic information and is thus error-prone.
- Sometimes the cell simply ignores a lesion and replicates the DNA that includes it in a mechanism called translesion synthesis, TLS. This is accomplished by a gallery of minor polymerases, 15 in mammals, that can carry out error-prone replication.

- In addition to all the complexities mentioned above, DNA repair in eukaryotes must deal with nucleosomes and the structure of chromatin fibers. It is now clear that for repair to occur, histone modification and probably chromatin remodeling must take place. Although the full picture is yet unclear, there is good evidence that certain histone variants, and chemical modification of those variants, play an important role.

Further reading

Books

Friedberg EC, Walker GC, Siede W et al. (2006) DNA Repair and Mutagenesis, 2nd ed. ASM Press.

Kelley MR (2011) DNA Repair in Cancer Therapy: Molecular Targets and Clinical Applications. Academic Press, Elsevier.

Wei Q, Li L & Chen DJ (2007) DNA Repair, Genetic Instability, and Cancer. World Scientific Publishing Company.

Reviews

Bonner WM, Redon CE, Dickey JS et al. (2008) γH2AX and cancer. *Nat Rev Cancer* 8:957–967.

Broyde S, Wang L, Rechkoblit O et al. (2008) Lesion processing: High-fidelity versus lesion-bypass DNA polymerases. *Trends Biochem Sci* 33:209–219.

Chapman JR, Taylor MRG & Boulton SJ (2012) Playing the end game: DNA double-strand break repair pathway choice. *Mol Cell* 47:497–510.

D'Amours D & Jackson SP (2002) The MRE11 complex: At the crossroads of DNA repair and checkpoint signalling. *Nat Rev Mol Cell Biol* 3:317–327.

Deem AK, Li X & Tyler JK (2012) Epigenetic regulation of genomic integrity. *Chromosoma* 121:131–151.

Friedberg EC (2008) A brief history of the DNA repair field. *Cell Res* 18:3–7.

Hanawalt PC (2002) Subpathways of nucleotide excision repair and their regulation. *Oncogene* 21:8949–8956.

Jackson SP & Bartek J (2009) The DNA-damage response in human biology and disease. *Nature* 461:1071–1078.

Jiricny J (2006) The multifaceted mismatch-repair system. *Nat Rev Mol Cell Biol* 7:335–346.

Lieber MR (2010) The mechanism of double-strand DNA break repair by the nonhomologous DNA end-joining pathway. *Annu Rev Biochem* 79:181–211.

Misteli T & Soutoglou E (2009) The emerging role of nuclear architecture in DNA repair and genome maintenance. *Nat Rev Mol Cell Biol* 10:243–254.

Mladenov E & Iliakis G (2011) Induction and repair of DNA double strand breaks: The increasing spectrum of non-homologous end joining pathways. *Mutat Res* 711:61–72.

Moses RE & O'Malley BW (2012) DNA transcription and repair: A confluence. *J Biol Chem* 287:23266–23270.

Nag R & Smerdon MJ (2009) Altering the chromatin landscape for nucleotide excision repair. *Mutat Res* 682:13–20.

Ouyang KJ, Woo LL & Ellis NA (2008) Homologous recombination and maintenance of genome integrity: Cancer and aging through the prism of human RecQ helicases. *Mech Ageing Dev* 129:425–440.

Rossetto D, Truman AW, Kron SJ & Côté J (2010) Epigenetic modifications in double-strand break DNA damage signaling and repair. *Clin Cancer Res* 16:4543–4552.

Sancar A (2008) Structure and function of photolyase and *in vivo* enzymology: 50th anniversary. *J Biol Chem* 283:32153–32157.

Schumacher B, Garinis GA & Hoeijmakers JHJ (2008) Age to survive: DNA damage and aging. *Trends Genet* 24:77–85.

Soria G, Polo SE & Almouzni G (2012) Prime, repair, restore: The active role of chromatin in the DNA damage response. *Mol Cell* 46:722–734.

Stracker TH & Petrini JHJ (2011) The MRE11 complex: Starting from the ends. *Nat Rev Mol Cell Biol* 12:90–103.

Sung P & Klein H (2006) Mechanism of homologous recombination: Mediators and helicases take on regulatory functions. *Nat Rev Mol Cell Biol* 7:739–750.

Svejstrup JQ (2010) The interface between transcription and mechanisms maintaining genome integrity. *Trends Biochem Sci* 35:333–338.

Wood RD (2010) Mammalian nucleotide excision repair proteins and interstrand crosslink repair. *Environ Mol Mutagen* 51:520–526.

Wu L & Hickson ID (2006) DNA helicases required for homologous recombination and repair of damaged replication forks. *Annu Rev Genet* 40:279–306.

Wyman C & Kanaar R (2006) DNA double-strand break repair: All's well that ends well. *Annu Rev Genet* 40:363–383.

Yang W & Woodgate R (2007) What a difference a decade makes: Insights into translesion DNA synthesis. *Proc Natl Acad Sci USA* 104:15591–15598.

Yoshida K & Miki Y (2004) Role of BRCA1 and BRCA2 as regulators of DNA repair, transcription, and cell cycle in response to DNA damage. *Cancer Sci* 95:866–871.

Experimental papers

Lammens K, Bemeleit DJ, Möckel C et al. (2011) The Mre11:Rad50 structure shows an ATP-dependent molecular clamp in DNA double-strand break repair. *Cell* 145:54–66.

Moreno-Herrero F, de Jager M, Dekker NH et al. (2005) Mesoscale conformational changes in the DNA-repair complex Rad50/Mre11/Nbs1 upon binding DNA. *Nature* 437:440–443.

Volker M, Moné MJ, Karmakar P et al. (2001) Sequential assembly of the nucleotide excision repair factors *in vivo*. *Mol Cell* 8:213–224.

Glossary

-10 box Consensus sequence, TATAAAT, closest to the transcription start site in bacterial genes, separated from it by ~10 nucleotides. Eukaryotes possess a similar sequence. Also known as **TATA box**.

-35 region Second consensus sequence, TTGACA, in bacterial promoters, situated further upstream of the transcription start site, at a distance of ~35 nucleotides.

3′ → 5′ exonuclease Exonuclease that degrades a polynucleotide chain starting at the 3′-end.

5′ → 3′ exonuclease Exonuclease that degrades a polynucleotide chain starting at the 5′-end.

4C technology Circularized chromosome conformation capture, used to identify DNA regions on the same or different chromosomes that interact with each other *in vivo*. It is an extension of 3C technology; the identified interacting regions are further analyzed by microarrays or DNA sequencing.

5C technology Chromosome conformation capture carbon copy: Expansion of 3C technology, which converts physical chromatin interactions into specific ligation products that can be quantified by polymerase chain reaction-based methods. Complex 3C libraries produced as the end product of 3C technology are amplified by a highly multiplexed method; then the amplified libraries are analyzed by microarrays or DNA sequencing.

43S pre-initiation complex In eukaryotic translation, the ternary complex of charged initiator methionine-tRNA, initiation factor eIF2, and GTP bound to the small ribosomal subunit and other initiation factors.

48S closed complex In eukaryotic translation, this complex is in equilibrium with the **48S open scanning complex**. It investigates the identity of the codon present in the peptidyl half-site of the small subunit. Upon identification of the start codon, the phosphate product of GTP hydrolysis is released; this ensures the irreversibility of initiation and transition to elongation.

48S open scanning pre-initiation complex In eukaryotic translation, complex formed by binding of additional factors to the **43S pre-initiation complex** in cases where the mRNA molecules contain stable secondary structures. The 48S complex removes these secondary structures.

β clamp Ringlike structure of subunits of DNA polymerase III that clamps about the DNA to hold the polymerase and assure processivity.

A-form DNA Structural variant of the DNA double helix that exists under conditions of low humidity. The helix is right-handed and contains ~11 bp/turn; it is wider than the canonical Watson–Crick B-form. This structure is also found in duplex RNA.

abortive transcription Transcription that prematurely terminates after a few ribonucleotides are joined together.

acceptor stem Portion of the tertiary structure of a tRNA molecule formed by base pairing between sequences at the 5′-end and the 3′-end. It is extended by a universal trinucleotide motif, CCA, which serves as the site of amino acid attachment.

accommodation Conformational relaxation in an aminoacyl-tRNA, which when bound to the elongation factor EF-Tu is distorted; the relaxation facilitates further steps in translational elongation.

activator Protein that binds to DNA, interacting with promoter regions to stimulate gene transcription.

active site Portion of an enzyme molecule that binds the substrate and catalyzes conversion of the substrate into product. It is usually located in clefts between individually folded protein domains, or between subunits in a multisubunit enzyme, or in pockets on the protein surface.

adaptor molecules Molecules postulated by Francis Crick as intermediates to match amino acids with DNA triplet nucleotide code. Later identified as transfer RNAs.

adenine (A) Purine base found in DNA and RNA .

ADP(ribosyl)ation Enzymatic addition of ADP-ribose units to a protein or an existing ADP-ribose chain.

AIDS (acquired immunodeficiency syndrome) Serious disease transmitted by the HIV retrovirus. It has been the subject of intensive research, resulting in considerable advances in prevention and treatment.

alarmone Molecule, for example, ppGpp, that mediates the stringent response of bacteria to stress.

alkylation Addition of alkyl groups.

allele Particular version of a gene that occurs in the genome. In most cases, each gene in a somatic, diploid cell has two alleles, one inherited from the mother and the other one from the father. One gene may also have multiple alleles that arise through a variety of mechanisms not directly related to fertilization. The most frequent allele in a population is called the wild type.

allele frequency Frequency at which a given allele is represented in a population, that is, the number of copies of a particular allele divided by the number of copies of all alleles of that gene.

allosteric modulator Molecule that binds to a regulatory site or subunit of an allosteric protein to modulate its conformation, and thus its activity. Also known as **effector**.

allostery Intramolecular interactions in which the binding of one ligand to a macromolecule can influence the macromolecule's affinity for others. If one substance influences its own binding, the interaction is termed homeotropic; if one substance modifies the binding of others, the interaction is heterotropic.

alpha-helix (α-helix) One of two main types of secondary structure in a protein molecule. The helix is right-handed, with a rise of 0.15 nm/amino acid residue and 3.6 residues/turn. The helix is stabilized by hydrogen bonds between the carbonyl oxygen of each residue and the amide hydrogen of the fourth downstream residue, counted from N- to C-terminus.

alternative DNA structures Locally denatured regions, cruciforms, Z-DNA, or triplex DNA that form under negative supercoiling stress. They can regulate transcription either positively or negatively.

alternative splicing **Splicing** that involves different choices of exons to be represented in mRNA. Some forms of this kind of splicing may result in inclusion of introns into the mRNA or produce mRNAs having alternative 5′- or 3′-ends.

amino acid Building block of polypeptide chains. Consists of an asymmetric carbon atom to which four different groups are attached: a carboxyl group, an amino group, a hydrogen atom, and a specific side chain, termed R group. The building blocks, 22 types in proteins, differ in the chemical nature of the R group.

amino acid residue Portion of an amino acid that is incorporated into a polypeptide chain.

aminoacyl-tRNA tRNA linked to the cognate amino acid; also known as charged or aminoacylated tRNA.

aminoacyl-tRNA synthetase Enzyme that links a tRNA to the cognate amino acid to form the aminoacyl-tRNA.

aminoacyl site (A site) One of three distinct sites on a ribosome occupied during protein synthesis by different forms of tRNAs. This is the entry site for an incoming aminoacyl-tRNA.

Angelman syndrome Imprinted genetic disorder characterized by mental retardation, seizures, and other neurological characteristics. In two specific regions of the brain, the paternal gene for the E3 HECT-domain ubiquitin ligase is silenced and the maternal allele is transcriptionally active. If the maternal allele is lost by deletion or mutated, the syndrome develops.

antibody Protein produced by the immune system to specifically bind to and neutralize foreign substances.

anticodon Sequence of three nucleotide residues that base-pairs with a codon in translation.

antigen Substance or portion of a molecule that binds to a specific antibody molecule. Large biological polymers usually contain several such portions, termed antigenic determinants or epitopes. An immunological response usually results in the production of numerous diverse antibody molecules, each of which recognizes and neutralizes a specific epitope.

antigens H, A, and B Oligosaccharides of differing chemical nature that exist on the surface of red blood cells; their presence defines the various blood groups in humans and the rules for blood transfusion.

apoptosis Programmed cell death in multicellular organisms. The process requires the activity of nucleolytic and proteolytic enzymes that destroy structures of unneeded or diseased cells, as, for example in eukaryotic development or cancer cells.

archaea In the three-domain classification of living organisms, a coherent group of unicellular organisms. Archaea resemble bacteria morphologically but are much closer to eukaryotes in several metabolic pathways, notably those involved in transcription and translation.

Argonaute Family of enzyme proteins that plays a major role in RNA interference, a process in which small noncoding RNA molecules bind to mRNA molecules to inhibit gene expression. The small RNA molecules guide these proteins to their specific targets in the mRNA through base pairing; this leads to mRNA cleavage or translation inhibition.

assembly maps for ribosome biogenesis Derived from *in vitro* ribosome assembly studies with purified components; show the sequence and interdependency of addition of protein components to the ribosomal subunits.

atomic force microscope (AFM) Instrument that can image microscopic objects, presumably at atomic resolution, by passing a fine-tipped sensor across the objects immobilized on a smooth surface. The instrument senses attractive or repulsive interaction forces between the tip of the sensor and molecules on the surface. In certain applications, it can also be used to micromanipulate molecules.

ATP (adenosine triphosphate) Ribonucleoside in which three phosphoryl groups are linked to the 5′-oxygen atom of the ribose. It is the main source for energy in the cell, since the phosphoanhydride linkages contain considerable chemical potential energy: hydrolysis of these linkages releases usable energy to power biochemical reactions and processes.

attenuation Regulation of transcription via premature termination.

***Att* sites** Sites on phage and bacterial genomes that participate in integration of temperate phage DNA into host DNA.

autonomous replicating sequence (ARS) Nucleotide sequences in yeast chromosomes that contain origins of DNA replication. When such sequences are incorporated into plasmid molecules, they confer upon these plasmids the ability to replicate independently from the replication of the yeast genome.

autonomous retrotransposon **Retrotransposon** that carries all genetic elements necessary for transposition including a gene for reverse transcriptase.

B-form DNA One of two major double-helical structures of DNA; the one favored under aqueous conditions and thus *in vivo*. It approximates the structure first proposed by Watson and Crick.

backtracking Reversal of an RNA polymerase movement on a DNA template; usually for short distances only.

bacteria In the three-domain classification, all microbes except **archaea**.

bacterial artificial chromosome (BAC) Cloning vector that is an *Escherichia coli* plasmid capable of carrying large inserts and controlled replication.

Barr body-deficient histone H2A (H2A.Bbd) Nonallelic replacement variant of histone H2A that is largely excluded from the inactive X chromosome, also known as Barr body. It overlaps regions of histone H4 acetylation in transcriptionally active regions.

barrier activity Types of insulators that physically prevent the spread of repressive heterochromatin structures.

basal transcription factors Eukaryotic RNA polymerases are not capable of recognizing promoter sequences; promoter recognition occurs by binding of specialized protein factors, which then recruit the polymerase. Also known as **general transcription factors**.

base excision repair (BER) DNA repair by directly removing and replacing an incorrect base.

beta-sheet (β-sheet) One of two main types of secondary structure in a protein molecule. Two or more chains lie side-by-side, parallel or anti-parallel, connected by hydrogen bonds between amide hydrogens and carbonyl oxygens on adjacent chains.

bidirectional transcription Transcription that occurs on both strands of DNA. The progression of RNA polymerases along each strand can be divergent, where the polymerases move away from each other to produce non-overlapping transcripts, or convergent, where the polymerases move towards each other, producing partially complimentary transcripts.

bivalents In meiosis, two homologous chromosomes linked by the synaptonemal complex along their entire length.

blunt end DNA terminus in which the two strands end in parallel, so there are no overhangs on either strand. Also known as flush end.

branch migration In recombination, sliding of the crossover point, which takes place on 4-way junctions.

branch site Site within an intron, usually A, that participates in splicing.

breast cancer susceptibility genes (BRCA1, BRCA2) Confer increased tendency to development of breast cancer through a multitude of molecular mechanisms.

breathing transition Conformational change in a nucleosome that involves unwrapping of only a few base pairs of DNA from the histone core. *See* **opening transition**.

bridge helix α-Helix that connects the two large subunits of eukaryotic RNA polymerase II; implicated in translocation motions of the enzyme along the DNA during elongation.

bromodomain Protein domain that favors transcription factor binding to acetylated lysines in histones, for example.

Brownian ratcheting Mechanism for one-directional advancement of a molecular process. The process takes a step driven by thermal motion but is prevented from reversing this step because of an irreversible block, often involving nucleotide hydrolysis.

C-terminal domain (CTD) Domain of Rpb1, the largest subunit of RNA polymerase II, which contains the repeated heptapeptide sequence YSPTSPS. Post-translational modifications of the heptapeptide create binding platforms for specific regulatory proteins.

C-terminus End of a polypeptide chain that carries the unreacted carboxylate group. Contrast with N-terminus.

canonical histones Set of histones that form the classical histone core of the nucleosome; to be distinguished from nonallelic variant histones, also known as replacement variants.

canonical nucleosome Classical DNA–histone octamer particle, containing only canonical histones.

cap analysis of gene expression (CAGE) Technique for isolating 5′-capped mRNA followed by sequencing. Provides information on the expression of protein-coding genes in different cell types under a defined set of conditions.

capsid Protein shell or container of the genetic material of a virus.

caspases In **apoptosis**, two classes of proteolytic enzymes, initiators and effectors, that are activated in succession. These contain cysteine in the catalytic center and have an unusual stringent specificity for cleavage only following Asp residues. Thus, they do not cause total protein degradation but rather partial chain fragmentation.

catabolite repression Repression by glucose or its breakdown products of bacterial genes responsible for catabolism of substrates other than glucose, which is the preferred source of energy.

CATH Semiautomatic classification of protein domains according to categories of class, architecture, topology, and homologous group.

caveolin-1 Critical protein component of caveolae, invaginations in the plasma membrane that contain specific lipids with signaling functions. Involved in insulin signaling.

CCCTC-binding factor (CTCF) Ubiquitous Zn-finger protein with numerous functions. It is considered a master organizer of the genome because of its ability to form chromatin loops between DNA sites on the same chromosome or to bridge DNA sequences located on different chromosomes. Thousands of CTCF binding sites exist throughout eukaryotic genomes.

cDNA library Collection of cloned DNAs that corresponds to the set of expressed messenger RNAs of a cell type or organism.

cell cycle Cyclic temporal program of processes during eukaryotic cell division.

cell-cycle checkpoints Defined points in the cell cycle at which the cell stops temporarily to check whether conditions are ripe for the cell to transit to the next phase of the cell cycle.

cell division cycle mutants (*cdc*) Yeast mutants with modifications, such as blocks or premature transitions, in the cell cycle. Used to study regulation of the cell cycle.

cell sorting Preparative variant of flow cytometry, which allows the fractionation of cellular populations into fractions highly enriched in cells with particular properties.

central dogma Name given to Francis Crick's hypothesis that information flows only from DNA to protein, not vice versa.

centromere Region on a chromosome where spindle fibers attach during cell division.

chain termination method Technique for DNA sequencing that involves termination of growing polynucleotide chains at selected residue types. Chains are polymerized on a DNA template whose sequence is being determined.

chameleon sequences in proteins Relatively short amino acid sequences in proteins that exist in different conformations in different environments.

chaperone Macromolecular complex that aids in the correct folding of another macromolecule. There are macromolecular chaperone complexes for proteins and for nucleic acids.

Chargaff's rules Empirical rules describing stoichiometric equivalencies in base compositions of natural DNAs; for example, G = C, A = T.

chemical sequencing DNA sequencing method employed for the first sequence determinations. The method involves four different reactions of random chemical cleavage of the polynucleotide chain, each specific for one of the four bases.

chiasmata Crossover sites in meiotic chromosomes where exchange of genetic material takes place between two homologous non-sister chromatids.

chicken-foot model Describes replication restart after a lesion in the leading-strand template blocks synthesis of leading-stand DNA. The process involves the action of **RecQ helicases**.

cholesterol Sterol, prominent as a membrane constituent. It tends to give rigidity to membrane structure.

Christmas tree Microscopic appearance of ribosomal genes transcribed at high frequency. Numerous RNA polymerases move along the DNA template simultaneously, with the nascent RNA transcripts extending from the DNA in an almost perpendicular direction. The transcripts have a knobby appearance due to the presence of proteins that bind to the nascent RNA chain.

chromatin DNA–histone complex that constitutes mitotic and interphase chromosomes. *See also* **histones**.

chromatin assembly factor 1 (CAF1) Histone chaperone that deposits histones H3 and H4 onto replicating chromatin.

chromatin hub Dynamic spatial organization of genes and chromatin regions that affect the transcriptional activity of resident genes. A well-studied example is the β-globin locus in the mouse, which contains four globin genes expressed at different stages of erythroid development.

chromatin immunoprecipitation (ChIP) Method for isolating chromatin–protein complexes by immunoprecipitation with antibodies to the protein.

chromatin remodeling Rearrangement of chromatin structure and/or nucleosome positioning by dedicated enzyme complexes.

chromatosome Nucleosome plus **linker DNA** and a molecule of **linker histone**.

chromodomain Domain in nuclear proteins that binds to methylated lysine and arginine residues in proteins.

chromosome Structure composed of a very long DNA molecule and associated proteins, which carries part or all of the hereditary information of an organism. Especially evident in plant and animal cells undergoing mitosis or meiosis, during which each chromosome in interphase becomes condensed into a compact rodlike structure visible in the light microscope following staining with specific stains. From the Greek chroma, color, and soma, body; that is, colored body.

chromosome territories Regions in the nucleus, each occupied by a specific interphase chromosome.

circular permutation Differing ways of ordering elements in a circular structure, such as a circular polypeptide.

***cis* interactions** Interactions between *cis*-regulatory DNA sequence elements, that is, elements in the vicinity of a gene in bacteria or on the same chromosome in eukaryotes. Contrast with ***trans* interactions**.

***cis*-regulatory lexicon** Entire set of *cis*-regulatory elements in the human genome, as defined by the **ENCODE project**. This set has co-evolved with the set of transcription factors that bind to regulatory sequences.

***cis–trans* interactions** Interactions between *cis*-regulatory DNA sequence elements and *trans*-acting protein factors, encoded elsewhere in the genome.

clamp loader Macromolecular complex that functions to load **β clamps** onto the leading and lagging strands in DNA replication. The loader also functions to organize the multiple elements of the replisome.

clone Group of cells, organisms, or DNA sequences that are genetically identical because they are derived from a single ancestor.

cloning Production of multiple identical copies or clones of an entity: in molecular biology, most often a DNA sequence; in biology, often an entire organism.

closed complex In transcription, the RNA polymerase–DNA complex first formed, before the transcription bubble opens to form an **open complex**.

cloverleaf structure Two-dimensional structure of a generalized tRNA molecule that contains four different arms formed by intramolecular base pairing. Anticodon and acceptor arms are named to reflect their function; the TψC arm and the D arm are named to reflect the high content of modified bases.

co-dominance Genetic condition where dominance is shared by two or more alleles; the progeny exhibit a combination of phenotypical parental traits.

co-immunoprecipitation (co-IP) Method for identification of interacting proteins: immunoprecipitation of one protein can result in precipitation of strongly interacting protein partners. Also known as pulldown.

coding strand In transcription, DNA strand that has the same sequence as the RNA that is transcribed; also called the sense strand or the nontemplate strand.

codon Sequence of three nucleotide residues in DNA that specifies a particular amino acid residue in the genetic code.

cognate amino acid Correct amino acid attached to a tRNA that carries the anticodon for that amino acid.

cohesin Complex of proteins containing **SMC proteins** that holds sister chromatids together along their length before their separation during the final stage of mitosis.

composite transposons In bacteria, transposons that carry, in addition to sequences and genes essential for transposition, extra genes; these are flanked on each side by IS elements and often confer antibiotic resistance.

condensin Complex of proteins containing **SMC proteins** that plays a role in chromosome condensation and segregation during mitosis and meiosis.

conjugation Process of transfer of genetic information between two bacteria of opposite sexes; the process involves direct cell-to-cell contact. This is one mechanism of horizontal gene transfer. The other two mechanisms, **transformation** and **transduction**, do not involve cell-to-cell contact.

consensus sequence For a group of nucleotide or amino acid sequences that show close similarity but are not identical, this is made up of the most frequently utilized residue at each position on the sequence. An example for which a consensus sequence has been derived is the promoter elements upstream of bacterial protein-coding genes; the individual promoter elements bind to RNA polymerase holoenzyme with slightly different affinities.

constitutive Denotes a general process or structure, not regulated by external stimuli.

constitutive heterochromatin Regions of interphase chromatin that are highly compacted and transcriptionally silent under all conditions for all cells of a given species. This state is irreversible; heterochromatic genes are never converted to euchromatin. Typical of centromeres, telomeres, and repetitive sequences.

constitutive transcription Transcription of genes whose products are essential to the vital functions of any cell, termed **housekeeping genes**, transcribed at all times. The level of expression of housekeeping genes may vary under different growth conditions.

copy DNA (cDNA) Double-stranded DNA molecules produced in laboratory setting transcribed from RNA molecules, by use of **reverse transcriptase**.

core histones Four histone types, H2A, H2B, H4, and H3, that make up the protein core of the nucleosome.

core promoter In eukaryotic genes, the region of the promoter that is located immediately upstream of the transcription start site; it is the docking site for general transcription factors.

core RNA polymerase Portion of the multisubunit enzyme containing only those subunits essential for transcription.

CpG islands Genomic regions with high density of the dinucleotide CpG; despite the enrichment in CpGs, which are the best substrates for cytosine methylation, these regions are undermethylated when compared to bulk DNA.

cross-talk in *cis* Mechanism of post-translational histone modification where modification of one amino acid residue affects the modification of another residue in the same histone molecule.

cross-talk in *trans* Mechanism of post-translational histone modification where a modification of one amino acid residue affects the modification of another residue in another histone molecule, residing either in the same nucleosome or in a different one.

crossover (or crossing over) Point in recombining DNA sequences where physical exchange between homologous DNA sequences occurs. These are recognized as microscopically observable **chiasmata**.

crossover hotspot instigator (*Chi*) In homologous recombination, binding of RecC to the specific nucleotide sequence *Chi* pauses the RecBCD complex and prevents further digestion of the 3′-tail of the DNA strand.

crown gall Plant tumor produced in response to bacterial infection by *Agrobacterium tumefaciens*.

cruciform Cross-shaped; used to refer to polynucleotide structures.

cryo-electron microscopy (cryo-EM) Version of transmission electron microscopy in which samples are quick-frozen in vitrified water. It allows observation of particles trapped in the water at varied orientations and hence reconstruction of three-dimensional structure.

cyclins Proteins involved in regulation of the cell cycle in eukaryotes. They vary in amount throughout the cycle and may stimulate or block transitions from one cell cycle phase to the next.

cytoplasmic polyadenylation element (CPE) Specific nucleotide sequence that may be present in the 3′-untranslated region of mRNA; responsible for lengthening of already shortened poly(A) tails to further stabilize the message.

cytosine (C) Pyrimidine base found in DNA and RNA.

Dcp2 Decapping enzyme in the 5′ → 3′ pathway of degrading mRNA.

deadenylation Process of removing the poly(A) tail from nonfaulty mRNA molecules as a first step in their degradation.

deamination Removal of amino groups.

decoding center (DC) Site on the small subunit of the ribosome where the codon on the mRNA pairs with the anticodon on the appropriate aminoacyl-tRNA.

degenerate Used to describe an essential property of the genetic code: most amino acids are specified by more than one codon. In practical terms, the first two letters of codons are enough to specify one amino acid; the third letter may differ.

denature Change a polypeptide or polynucleotide from its native conformation. DNA denaturation involves separation of the two complementary strands; protein denaturation disrupts the native folding of the polypeptide chain without breaking covalent bonds.

Deoxyribonucleic acid (DNA) Polynucleotide formed from covalently linked deoxyribonucleotide units. The store of hereditary information within a cell and the carrier of this information from generation to generation.

Dicer Ribonuclease that cleaves double-stranded RNA or pre-microRNA into short double-stranded RNA fragments, known as small interfering RNA or microRNA, respectively. Also part of a complex that degrades mRNA.

diffraction pattern Pattern produced when a beam of electromagnetic radiation passes through a periodic structure with spacings comparable to the wavelength of the radiation.

diploid Genetic state in which two copies of each chromosome are present in a eukaryotic cell.

direct repair Repair by chemical reversal of a DNA lesion, as in photolytic repair of thymidine dimers.

dispersed promoters One of two broad types of eukaryotic promoters, containing multiple weak transcription start sites over a region of 50–100 nucleotides. Contrast with **focused promoters**. Also known as broad promoters.

DnaA boxes Multiple 9-bp repeat elements in the *Escherichia coli* origin of replication, to which **initiator protein DnaA** binds to form **primosomes**.

DNA binding domain (DBD) Distinct domain in transcriptional activators that binds to specific nucleotide sequences in regulatory gene regions.

DNA damage response (DDR) Set of sophisticated mechanisms to repair double-strand breaks; also elicited in response to the presence of long stretches of single-stranded DNA.

DNA gyrase Topoisomerase that catalyzes the incorporation of negative superhelical turns in DNA.

DNA helicase Enzyme that unwinds double-stranded DNA. Essential in DNA replication.

DNA ladders Electrophoretic patterns of DNA fragments resulting from partial digestion of chromatin with **micrococcal nuclease**, followed by DNA purification and electrophoretic analysis in agarose gels. The DNA fragments obtained are multiples of a constant length, termed **repeat length**.

DNA ligase Any of a class of enzymes that catalyze the covalent connection of DNA molecules.

DNA linkers Small oligonucleotides that can be chemically linked to DNA blunt ends; they carry restriction sites that, when cleaved by the respective restriction endonuclease, give rise to sticky ends for efficient ligation of DNA fragments.

DNA methyltransferase (DNMT) Any of a number of enzymes that catalyze the methylation of DNA bases.

DNA methyltransferase 1 (DNMT1) Methylation enzyme that recognizes and methylates cytosine residues in CpG dinucleotides in the newly replicated DNA strand, thus preserving the methylation pattern of the genome on both strands.

DNA photolyase Enzyme that uses visible light as an energy source to repair thymidine dimer lesions in DNA. Also called photoreactivating enzyme.

DNA protection during starvation protein (Dps) Protein that accumulates in starved bacteria, forming liquidcrystal co-structures with the DNA that protect the genome from degradation.

DNA recombination Re-sorting of genetic segments by exchanges of stretches of duplex DNAs.

DNA tiling array Kind of microarray in which short oligonucleotide probes called targets cover the genomic region in an overlapping way. Such a design allows each entity, either protein or nucleic acid fragment, that is interrogated for interactions in the genomic region of interest to be detected in several neighboring spots, thus increasing the accuracy of the results.

DNase I Endonuclease that makes only single-strand cuts in DNA; used to study the internal organization of nucleosomes and to identify regulatory regions in the genome.

DNMT3a and DNMT3b Two enzymes responsible for introducing new methyl groups into previously unmethylated CpG dinucleotides; involved in global *de novo* methylation during embryonic development.

domains Portions of a macromolecule that exhibit a folded structure that is distinguishable from other parts of the molecule and give evidence of independent folding. Often, these structures in a protein molecule carry out distinct functions.

double-sieve mechanism Mechanism used by aminoacyl-tRNA synthetases to ensure fidelity of the aminoacylation process. A coarse sieve excludes the activation of amino acids larger than the cognate amino acid; a second, finer sieve would hydrolyze amino acids smaller than the cognate one.

downstream In the 5′ → 3′ direction from a given DNA position. Contrast with upstream.

duplex In molecular biology, any two-stranded DNA or RNA structure.

***Eco*RI** Widely used restriction endonuclease from *Escherichia coli*, useful in cloning because it produces **sticky ends**.

editing In molecular biology, correction of mistakes made in a replicative process or during aminoacylation of tRNAs. *See also* **double-sieve mechanism**.

EF-G Factor in bacterial translational elongation that, along with GTP hydrolysis, promotes **translocation**.

eIF2 Together with GTP and methionine-charged initiator tRNA, forms the ternary complex that delivers initiator tRNA to the ribosome during initiation.

electron microscope (EM) Microscope that uses a focused beam of electrons instead of light rays to image objects.

electrophoresis Migration of macromolecules in an electric field. It is widely used, in many variants, for separating or analyzing mixtures of proteins or nucleic acids.

electroporation Technique of inserting DNA into cells by strong electric shock.

electrospray ionization (ESI) Method used in mass spectrometry to introduce molecules, often large ones, into the spectrometer for analysis. One or a few molecules are present in each tiny drop.

elongation In transcription, translation, and DNA replication, the stage in the process in which the product polymer chain is elongated. Distinguished from **initiation** and **termination**.

elongation complex Complex between RNA polymerase and DNA during elongation of transcription. The transition from initiation to elongation is accompanied by substantial movements of some portions of the polypeptide chain to ensure the firm grip on the DNA needed for processivity.

elongation factor Tu (EF-Tu) Factor in bacterial translation that, along with GTP, carries aminoacyl-tRNAs to the A site of the ribosome during elongation.

embryonic stem (ES) cell Cell from a eukaryotic embryo that has the potential to differentiate into a variety of somatic cells.

Encyclopedia of DNA Elements (ENCODE) Multi-laboratory international project to explore the structure and expression of the human genome.

endonuclease Enzyme that cuts phosphodiester bonds within a nucleic acid molecule.

***endo* nucleoside** Nucleoside with the sugar in the *endo* conformation.

energy charge Measure of the energy available to a cell from nucleoside phosphate hydrolysis.

enhanceosome Higher-order protein complex assembled at gene enhancer elements to promote transcription; also used for the particular structure that is created at RNA I polymerase promoters to initiate transcription. In this structure, a dimer of the upstream binding protein, UBP, interacts with two regions in the gene promoter, causing DNA looping.

enhancer Genetic element, sometimes remote, that increases a gene's expression through protein binding. DNA looping caused by interactions between the enhancer-bound factors and promoter-bound factors may be involved in the mechanism of transcriptional stimulation. Contrast with silencer.

enhancer-blocking activity Types of insulators that block communications between enhancers and promoters to repress transcription.

enzymatic sequencing Another name for the chain termination method, so called because it utilizes enzymatic copying of the nucleic acid, periodically blocked by withholding a monomer type.

epigenetic Changes in the informational content of the genome in a cell that are outside of DNA; not heritable through the sexual reproduction cycle.

epigenetic regulation Regulation of gene activity through heritable molecular events that cannot be explained by genetic principles, that is, by mutations in the DNA. Today a broader definition is used, referring to all post-translational modifications of histones that affect gene expression, whether heritable or not. From the Greek epi, for on, upon, over; and genetics.

epistasis Genetic condition in which one gene masks the phenotypic expression of another.

epitope Portion of an antigen molecule that is actually recognized by the antibody. Also called antigenic determinant.

erasers Enzymes that remove post-translational marks from histones.

euchromatin Less-condensed portion of chromatin in the nucleus. More often involved in transcription than heterochromatin.

eukaryotes Organisms that possess nucleated cells. Distinguished from bacteria and archaea, which lack nuclei and other organelles.

eukaryotic initiation factor 4F (eIF4F) Complex of three proteins, E, G, and A, that binds to the 5′-cap structure in mRNA and is required for initiation. eIF4E is the cap-binding protein, eIF4A is an RNA helicase, and eIF4G is a scaffolding protein that, by interacting with other proteins, circularizes the mRNA for efficient translation.

exit site (E site) One of three sites on the ribosome occupied by tRNA. This site is occupied by the tRNA that has been deacylated and is about to leave the ribosome.

exon Segment of a protein-coding gene that is retained in the mature mRNA and is translated into protein. Contrast with intron.

exon definition In pre-mRNA containing multiple exons and introns, the initial interaction between the protein factors bound at the splice sites occur across the exon. Contrast to intron definition, where the interactions required for splicing occur across the intron.

exon-junction complexes (EJC) Protein complexes that are bound to exon–exon junctions, ~20–24 nucleotides upstream of the junctions. During normal translation, the ribosome proceeds along the polynucleotide chain, displacing EJC. Involved in the process of mammalian nonsense-mediated decay, which degrades mRNA molecules containing premature stop codons.

exonic splicing enhancers and silencers (ESE and ESS) Regions of exons that can enhance or repress splicing by binding splice regulatory proteins.

exonuclease Enzyme that degrades nucleic acids from the end(s), removing one nucleotide at a time.

***exo* nucleoside** Nucleoside with the sugar in the *exo* conformation.

exosome complexes Protein complexes in eukaryotic cells and archaea with specific ribonuclease 3′ → 5′ activities. Exosome substrates include mRNA, rRNA, and a variety of small RNAs. Bacteria have a simple complex called the degradosome.

expression vector **Vector** that allows expression of genes within host cells; this requires that they contain the proper regulatory sequences for transcription and translation.

expressivity Refers to the degree or intensity with which a particular genotype is expressed in a phenotype.

extrinsic and intrinsic pathways of apoptosis Two major pathways that trigger and execute **apoptosis**, depending on the nature of the death signal. Signals from outside the cell activate the extrinsic pathway; stress conditions inside the cell activate the intrinsic pathway. The two pathways converge at the last executional stage of total cell degradation.

facilitates chromatin transcription complex (FACT) Eukaryotic protein complex that aids transcriptional elongation by interaction with DNA and histones H2A and H2B, facilitating their release from nucleosomal particles in front of the moving polymerase. It also helps subsequent rebinding of these histones once the polymerase has passed.

factor for inversion stimulation (Fis) Transcriptional activator protein in bacteria that binds upstream from a promoter site and interacts with the C-terminal tail of RNA polymerase to recruit it to the promoter; in addition, it helps to compact DNA by bending it.

facultative Process or structure dependent on circumstances.

Ferguson plot Graph of gel electrophoretic data in which the logarithm of the relative mobility is plotted versus the concentration c of the gel. The slope depends on molecular size, and the intercept at $c = 0$ corresponds to the free mobility of the molecule.

flippase Enzyme that promotes end-to-end rotation of proteins in membranes.

flow cytometry Technique for either counting or separating cell types in a mixture. Separation may be on the basis of cell size, internal structure, or content of a fluorescently labeled component like DNA.

fluorescence Re-radiation of light absorbed by a molecule at lower frequency after energy loss in the absorber.

fluorescence imaging with one-nanometer accuracy (FIONA) Fluorescence microscopy technique utilizing photon counting over time to refine the position of a molecule or fluorophore in a structure.

fluorescence resonance energy transfer (FRET) Method to measure the distance between two fluorophores by observing energy transfer from one to the other. The efficiency of such transfer decreases rapidly with increasing distance between the two fluorophores.

focused promoters One of two broad types of eukaryotic promoters, containing a single or several tightly clustered transcription start sites. Contrast with **dispersed promoters**. Also known as sharp promoters.

fork trap DNA element that halts progression of the replication fork. In a circular chromosome, there will be fork traps functioning for each direction of replication.

formaldehyde-assisted isolation of regulatory elements (FAIRE) Isolation of nucleosome-depleted genomic regions by exploiting the difference in cross-linking between nucleosomal histones and DNA, which is of high efficiency, and cross-linking between sequence-specific regulatory factors and DNA, which is of low efficiency. Cross-linking is followed by phenol extraction and sequencing of DNA fragments in the aqueous phase.

four-way junction (4WJ) Branched structure involved in strand exchange during DNA recombination. It is stabilized by specific proteins and is capable of migration along the recombining duplexes. Also known as a Holliday junction.

frameshift Mutation that shifts the reading frame of a gene; that is, deletion or insertion of any number of nucleotides other than a multiple of three. Frameshift mutations lead to the production of a protein whose sequence is totally changed 3′ to the mutation.

functional element of DNA Term introduced by the ENCODE project to define a discrete genome segment that encodes a defined product, such as RNA or protein, or displays a reproducible biochemical signature, such as protein binding or a specific chromatin structure.

fusion proteins Constructed proteins in which two protein sequences are fused together. These proteins can be produced by the cloning of fused DNA sequences in an **expression vector.**

G-quadruplex Four-strand DNA conformation found in **telomeres.**

gel electrophoresis **Electrophoresis** in a gel matrix. Agarose and polyacrylamide are the most commonly used matrix formers.

gene addition Modification of the genome of an organism by the addition of a gene, even though a native form of the gene is already present.

gene gun Device for shooting vector-containing particles into host cells.

gene replacement Modification of the genome of an organism by the addition of a gene, with concomitant deletion of the native form of the gene.

gene silencing Permanent, irreversible loss of the ability of a gene to be transcribed. When this loss occurs, genes are organized as condensed constitutive **heterochromatin**.

genome Entire genetic content of a cell or a virus.

genome segmentations or major classes of genome states Major classes of genome transcriptional states as defined by the **ENCODE project**. These are transcribed genes, their transcription start sites and promoters, strong and weak enhancers, CTCF-binding sites, and transcriptionally repressed regions.

genomic libraries Collections of cloned DNA molecules representing entire genomes.

GINS complex Complex of four proteins that helps to form the pre-initiation complex in eukaryotic DNA replication.

global controls of overall expression patterns Global level of **DNA supercoiling** is one example of global control.

global genome repair (GGR) Pathway of nucleotide **excision repair** active in the entire genome.

global regulators Small number, around 10, of transcription factors that control very large sets of genes.

global supercoiling Overall degree of supercoiling of the bacterial chromosome. The activity of the enzyme **gyrase** that introduces negative superhelical stress into DNA is dependent on the level of ATP; thus the overall cellular energetics play a role in determining this level of supercoiling.

glycoproteins Proteins to which glycosyl groups are covalently attached. These may be individual sugar molecules or carbohydrate chains.

glycosylation Addition of glycosyl group(s) to a molecule such as a protein. Attachment to proteins may be N-linked, for example, through asparagine, or O-linked, as through threonine or other amino acids containing hydroxyl groups.

golden rice Genetically modified variety of rice engineered to produce β-carotene, a precursor of vitamin A, in the grains; the increased content of β-carotene gives the golden color of the rice grains.

GroEL/GroES **Chaperone** in bacterial cells; a huge multisubunit complex that contains a central hydrophobic cage, which can accommodate a protein and help it to fold correctly.

group I introns Class of self-splicing introns that excise themselves from mRNA, tRNA, and rRNA precursors; use a guanosine molecule as a cofactor. Very abundant in bacteria and lower eukaryotes, rare in animals.

group II introns Class of self-splicing introns that excise themselves from mRNA, tRNA, and rRNA precursors in organelles of fungi, protists, and plants; use an internal adenosine as a cofactor.

guanine (G) Purine base found in DNA and RNA.

H-DNA *See* **triple helix.**

H2A.X Nonallelic histone replacement variant that contains a short conserved C-terminal tail; phosphorylation of a serine residue in the tail produces γH2A.X. Phosphorylation of a large number of this variant occurs immediately following double-strand break formation; the phosphorylated serine then recruits a myriad of DNA repair proteins.

***Hae*III** Widely used restriction endonuclease from *Haemophilus aegyptius.* Produces blunt ends but is useful as a frequent cutter for DNA degradation, with a recognition site size of 4 nucleotides.

hairpin Polynucleotide conformation, usually RNA, in which a self-complementary chain folds back and hydrogen-bonds to itself to form a hairpin-like structure.

handshake motif Structure that defines histone–histone interactions between histone folds, the tertiary motifs common to all four **core histones.**

haploid Having only one duplex copy of the genome. Gametes, eggs and sperm, are haploid cells that combine their genomes during fertilization to form diploid somatic cells.

heat unstable (HU) protein Bacterial protein that binds to the DNA minor groove, thus bending the DNA; major component of the bacterial nucleoid.

helicase DnaB Enzyme absolutely required for initiation of DNA synthesis at *Escherichia coli* origin of replication. Two helicase molecules bind to opposite DNA strands to unwind the DNA in opposite directions.

helicase loader DnaC Protein that loads the helicase DnaB onto the *Escherichia coli* origin of replication; an ATPase.

helicases Large class of essential enzymes that move directionally along the phosphodiester backbone of double-stranded nucleic acids to separate the two strands. Such separation is essential for all DNA transactions, such as replication, recombination, repair, and transcription.

helix-destabilizing proteins Proteins that bind to single-stranded portions of denatured DNA to temporarily maintain the denatured state during DNA replication, recombination, and repair. Also known as single-strand binding proteins.

helix–turn–helix Common protein structural motif involved in binding to DNA.

hemisome Nucleosome particle containing only one tetramer of the core histones CENP-A, which is a replacement variant of histone H3, H4, H2A, and H2B.

heterochromatin Condensed, transcriptionally inactive chromatin regions. Contrast with **euchromatin**.

heterochromatin protein 1 (HP1) Protein involved in gene silencing in organisms other than budding yeast. One of its domains recognizes methylated lysines in histone H3; a closely related domain interacts with other proteins, including histone methylation enzymes.

heterogeneous nuclear ribonucleoproteins (hnRNPs) Structurally diverse group of proteins that affect splicing by binding to splicing silencers. They are usually complexed with small RNA molecules, hence the name.

heterophagy Process in which **lysosomes** digest exogenous proteins brought to the lysosome via receptor-mediated endocytosis or pinocytosis; it also refers to degradation of exogenous particles internalized through phagocytosis.

high-frequency recombinants (Hfr) Bacterial strains that have a high frequency of conjugation and thus DNA recombination.

histone chaperones Proteins that facilitate the deposition of histones onto DNA during nucleosome formation on newly replicated DNA or during histone exchange in pre-formed nucleosome particles.

histone fold Three-dimensional fold, HTHTH, common to the histones in the nucleosome core. Individual histone molecules interact through the middle helix of the fold to form H2A-H2B and H3-H4 dimers via the **handshake motif**.

histone variant Nonallelic variants of certain histones that replace the **canonical histones** in some nucleosomes. In some cases, like H2A.Z and H3.3, the variants are concentrated in nucleosomes in transcribed regions of chromatin. Other variants, such as macroH2A, are located in transcriptionally inactive chromatin regions.

histone H2A.Z **Nonallelic replacement variant** of histone H2A that is significantly enriched in promoter regions genomewide. Two nucleosomes containing these variants flank **nucleosome-free regions** at transcription start sites. The presence of this variant correlates with transcription in humans but anticorrelates with transcription in yeast.

histone H3.3 **Nonallelic replacement variant** of histone H3 that marks actively transcribed gene regions, where nucleosomes are constantly disassembled and reassembled. The two nucleosomes flanking the nucleosome-free region in promoters contain this variant as well as H2A.Z.

Histone-like nucleoid structuring protein H-NS Bacterial protein that forms bridges between stretches of DNA, thus helping to condense the bacterial chromosome.

histones Small, evolutionarily conserved basic proteins that constitute the protein basis of the structure of eukaryotic chromatin.

Holliday junction (HJ) Branched structure involved in strand exchange during DNA recombination. It is stabilized by specific proteins and is capable of migration along the recombining duplexes. Also known as a four-way junction.

holoenzyme Complete, functional enzyme complex, containing all essential subunits.

homeobox Region of about 200 base pairs found in genes that control development, known as homeotic genes; codes for helix–loop–helix DNA-binding motifs in sequence-specific protein factors.

homologous recombination (HR) Recombination that requires strong sequence homology between the recombining DNA sequences.

homologous recombination repair (HR repair) DNA repair accomplished via homologous recombination.

horizontal gene transfer Extracellular transfer of DNA from one bacterium to another. Also termed lateral gene transfer, to distinguish it from vertical gene transfer, the transmission of genes from parents to progeny.

host restriction Ability of some bacteria to restrict the growth of certain bacteriophage. It was explained by the existence of restriction endonucleases and DNA methyl transferases. Methylation protects the host DNA from the endonuclease, which cleaves only the unmethylated sequences of the invading phage.

housekeeping genes Those genes essential for cell survival; active most of the time.

Hsp90 Protein chaperone with multiple roles in both protein folding and protein degradation.

human immunodeficiency virus (HIV) Retrovirus responsible for **AIDS**.

hybrid states On the ribosome, intermediate states in the elongation cycle in which a given aminoacyl-tRNA is bound between two different half-sites, one on the large subunit and the other on the small subunit.

hydrolysis Cleavage of a covalent bond with the introduction of a water molecule to carry out the reaction. The hydrogen and hydroxyl from the water molecule are attached to the two fragments produced.

hydroxyurea (HU) Chemical widely used to block DNA replication in both bacterial and eukaryotic cells. It target ribonucleotide reductase, the essential enzyme that reduces ribonucleotides to deoxyribonucleotides, thus depleting cells of DNA precursors. Used in clinical practice to treat various diseases, including cancer.

identity elements in tRNAs Sequence elements in any of the four tRNA arms, which ensure that the correct tRNA is presented to the correct aminoacyl-tRNA synthetase, thus ensuring fidelity of charging the tRNA with its cognate amino acid.

imitation switch (ISWI) Family of chromatin remodeling complexes found in eukaryotic organisms ranging from yeast to humans.

immunoglobulins Class of proteins that function as antibodies in the immune response.

immunoprecipitation (IP) Method in which antibodies are used to purify proteins out of complex biological samples such as cell extracts or bodily fluids.

imprinted gene Gene whose alleles are expressed in a parent-of-origin-specific manner. The allele that is silenced may come from either parent and is marked for silencing by DNA methylation during development of the germline. These marks are maintained through all mitotic cell divisions.

incomplete dominance Genetic state where the phenotype of the heterozygous genotype is hybrid or intermediate between the phenotypes of the homozygous parents.

indirect end-labeling Method to determine nucleosome positions in a defined genomic region. DNA purified following **micrococcal nuclease** digestion of chromatin is further cleaved by a pair of restriction endonucleases that cut at defined sites bordering the region of interest. The lengths of the resulting DNA fragments are used to infer the position of the nucleosome.

inducer Low molecular weight substance that promotes expression of a bacterial operon by binding to a repressor molecule, causing it to dissociate from the gene promoter.

initiation In transcription, translation, and DNA replication, the beginning of the process of polymerization. It may require special initiation factors that are different from factors needed in **elongation** or **termination**.

initiation complex Complex between RNA polymerase and DNA during initiation of transcription.

initiation factor Protein factor needed for promoter recognition during initiation of transcription in bacteria; the initiation factor belongs to a family of proteins called Σ factors.

initiator protein DnaA Protein that binds to multiple 9-bp repeat elements called DnaA boxes in the *Escherichia coli* origin of replication; an ATPase.

initiator tRNA tRNA that carries the amino acid corresponding to the N-terminal residue of the polypeptide chain. This will be fMet-tRNA in bacteria and a special Met-tRNA in eukaryotes.

insertion sequences (IS) Bacterial transposon containing inverted terminal repeats and genes essential for transposition.

insertion vector Bacteriophage vector in which a recombinant fragment has been inserted into the phage genome without removing phage DNA, in contrast with **replacement vector**.

insulator Genetic element that separates an enhancer from a promoter; insulator activity occurs primarily through DNA looping mediated by the protein CTCF.

integrase Any enzyme that catalyzes the integration of one piece of DNA into another. Essential, for example, for the lysogenic life style of **temperate phage**.

integration host factor (IHF) Bacterial protein that binds to the DNA minor groove, thus bending the DNA; major component of the bacterial nucleoid.

interactome maps Maps summarizing the interactions among a group, often large, of substances like proteins.

internal ribosome entry sites (IRES) Sites in the 5′-end of mRNA molecules used to initiate translation under conditions in which normal cap-dependent initiation is impaired. IRES function does not require the cap structure or any factor needed to remove secondary structures from the mRNA.

intertelomeric D-loop Structure involved in mechanisms of alternative lengthening of telomeres or ALT pathways. The 3′-overhang of one telomere invades the telomeric duplex of another chromosome; telomere lengthening occurs by use of the sequence information in the second chromosome.

intrinsically disordered proteins or regions Protein molecules, or portions thereof, that exhibit no defined secondary or tertiary structure. Such proteins or regions will behave essentially as **random coil**, with enough flexibility to interact with numerous partners.

intron Segment of a protein-coding gene that is not retained in the mature mRNA and is not expressed as protein. Contrast with exon.

intron definition Splicing defined by interaction of the spliceosome with the two ends of the intron to be spliced out; contrast to **exon definition**.

intronic splicing enhancers and silencers (ISE and ISS) Regions of introns that can enhance or repress splicing by binding splice regulatory proteins.

isoacceptor tRNAs tRNA molecules containing different anticodons that specify the same amino acid.

isochores Large regions, >300 kilobase pairs, in mammalian genomes that contain DNA with homogeneous, either high or low, GC content and gene density. Homogeneity of GC content contrasts with heterogeneity over the entire genome. GC-rich, gene-rich isochores are replicated early in S phase, whereas those that are GC-poor and gene-poor replicate late. From the Greek chore, combining form, and iso, equal or identical.

isoelectric focusing Electrophoresis carried out in a pH gradient. Each kind of protein migrates to the position in the gradient corresponding to its **isoelectric point**.

isoelectric point pH at which a protein carries zero net charge.

isopeptide bond Covalent bond between one protein and a side-chain residue on another. The best-known example is conjugation of ubiquitin to target proteins.

isoschizomers Different restrictases that recognize the same target sequence. They may not necessarily cut in the same manner.

junk DNA Term formerly applied to noncoding DNA in the genome that was thought to have no useful function. We now know that most or all DNA is transcribed and thus may function in certain ways, so the term is now rarely longer used.

kinases Large group of enzymes that add a phosphate group to serine or threonine residues in proteins. A small percentage of phosphorylation reactions also occur on tyrosines and histidines.

kinetic proofreading Mode of proofreading during protein synthesis that depends upon the slow rate at a particular step of translation to increase chances to detect and correct errors.

Klenow fragment Proteolytic fragment of DNA Pol I lacking the 5′ → 3′ exonuclease activity of the holoenzyme; it retains the other two enzymatic activities, 5′ → 3′ polymerase and 3′ → 5′ exonuclease. Widely used in laboratory research.

knock-downs Genetically modified organisms in which one or more genes have been decreased in expression by modification of regulators.

knock-ins Genetically modified organisms in which one or more genes have been inserted.

knockouts Genetically modified organisms in which one or more genes have been eliminated.

Ku70–Ku80 dimer Highly abundant protein dimer that initiates **nonhomologous end joining**. Two dimers bind to the two DNA ends to be joined and recruit the other players, nucleases, polymerases, and ligases, in the process.

***lac* operon** Group of three contiguous genes required for lactose metabolism in bacteria; the three genes are regulated as a whole.

lagging strand DNA strand that is synthesized discontinuously during replication.

lamina-associated domains Portions of the chromatin fiber that are associated with the nuclear lamina structure and are transcriptionally inactive.

leading strand DNA strand that is synthesized continuously during replication.

leucine zipper Common three-dimensional α-helical structural motif, mainly in DNA binding proteins, that allows protein–protein interactions. The primary structure of such motifs shows a periodic repetition of leucine residues at roughly every seventh position; when these polypeptide segments form an α-helix, the leucine residues from one α-helix can interdigitate with those from a similar α-helix, belonging either to the same polypeptide chain or to a different one. This formation facilitates protein dimerization, often needed for the activity of transcription factors.

ligand In biochemistry, small molecule that binds to a macromolecule.

linkage In genetics, coordinated transfer of two genes during meiosis. Genes that are nearer to each other on a chromosome have a lower likelihood to be separated during genetic recombination. The frequency of this transfer can be used to map genes on chromosomes.

linker DNA In chromatin, the DNA between nucleosomes. Also known as spacer DNA.

linker histones Histones of the H1 family that are found on linker DNA.

linking number Number of times the two strands of a closed circular DNA are interlinked; it is equal to the number of times the strands cross each other by either twist or writhe.

lipofection Insertion of recombinant DNA into a cell via a liposome.

liposome Lipid micelle that can transport foreign molecules, such as DNA, into cells. Used for **lipofection**.

locus control region (LCR) Chromatin region that controls developmentally regulated expression of individual genes within a gene cluster; an example is the β-globin locus.

long interspersed nucleotide element 1 (LINE-1, L1) Very abundant class of non-LTR retrotransposons. The only currently active transposon that can transpose itself and all other elements.

long noncoding RNAs (lncRNAs) Long RNAs that are expressed but code for no protein; they have some function in gene regulation.

long terminal repeats (LTRs) Repetitious DNA sequences, several hundred base pairs in length, found at the ends of retrovirus proviruses or retrotransposons. These sequences are required for the virus to insert its genetic material into the host genome.

looping model of enhancer or silencer action Model for mechanism of action of distant transcription control elements: transcription factors that bind to specific sequences in the enhancer or silencer interact with protein factors bound to the promoter, causing the intervening DNA to loop.

LTR retrotransposons **Autonomous retrotransposons** with long terminal repeats at both ends. To be contrasted with non-LTR transposons, which do not carry such terminal repeats.

lysosome Eukaryotic organelle that contains more than 50 different hydrolytic enzymes used to break down biomolecules and cellular debris.

macroH2A **Nonallelic replacement variant** of histone H2A that is present only in vertebrates; its presence leads to transcriptional silencing. Contains a long C-terminal non-histone region termed the macrodomain.

macromolecules Long chains of covalently linked, small organic or carbon-containing molecules. The principal building blocks from which a cell is constructed, which confer the most distinctive properties of living things.

matrix-assisted laser desorption and ionization (MALDI) Technique for introducing large molecules into a mass spectrometer. The desired substance is trapped in an inert matrix that is volatilized by a laser beam, freeing the molecules.

matrix attachment region (MAR) DNA region of 1–2 kilobases, highly enriched in AT base pairs, that preferentially attaches to the nuclear matrix in interphase nuclei.

Mcm2–7 complexes In eukaryotic replication, hexameric helicases that unwind DNA in opposite directions in two divergent replication forks.

Mediator Large protein complex that is essential for eukaryotic transcription; thus can be considered a general transcription factor. In addition, it serves as a bridge between Pol II and the sequence-specific proteins that recognize regulatory elements surrounding the gene, thus acting as a co-activator or co-repressor.

meiosis Cell division to produce haploid gametes, each with only one copy of the genome.

meiotic bouquet In meiosis, structure formed by clustering of telomeres at one end of the nucleus through attachment to the nuclear envelope. Chromosomes resemble a bouquet of flowers with gathered stems, hence the name.

messenger RNA (mRNA) RNA that is complementary to one strand of a protein-coding DNA gene and carries this sequence to the ribosome to direct protein synthesis.

methyl-CpG-binding proteins Various proteins that recognize methylated genome regions; these protein share a common methyl-CpG-binding domain or MBD.

methylation-sensitive isoschizomers Pairs of **isoschizomers** that differ in their response to methylation of the DNA substrate.

micro- and macroautophagy Processes in which **lysosomes** digest endogenous cellular proteins and organelles, respectively.

micrococcal nuclease (MNase) Endonuclease from *Staphylococcus aureus* that makes double-strand cuts in DNA. It plays an important part in chromatin studies by selectively cleaving in linker DNA between nucleosome core particles.

microRNAs (miRNAs) Type of small silencing RNAs very abundant in animals. Interact with target mRNA sequences by base pairing, the extent of which determines the specific mechanism of the silencing reaction.

miRNA-induced silencing complex (miRISC) Molecular complex containing miRNA and **Argonaute protein** that mediates regulation of translation.

mismatch repair (MMR) Repair of DNA duplexes in which mismatches have occurred during DNA replication. MMR also repairs small insertion or deletion loops arising from faulty DNA replication.

mitosis Cell division in which two double-stranded copies of parental DNA are produced and the resulting daughter cells contain a diploid number of chromosomes. Typical of eukaryotic somatic cells.

mitotic chromosome Compacted eukaryotic chromosome at mitosis.

mobile clamp Molecular clamp that can enclose a polynucleotide chain to assure processivity of a process, such as DNA replication.

modifier gene Gene that has small quantitative effects on the level of expression of a second gene.

monoallelic gene expression Preferential expression of only one of the usual two alleles in a somatic cell. This phenomenon, also known as allelic exclusion, is rare; in most cases, the two alleles of a gene are expressed.

monocistronic Term used to denote mRNA molecules that encode only one single polypeptide. Contrast with **polycistronic**.

monoclonal antibody Antibody produced by a single B cell or clone thereof. These antibodies will bind a single antigenic determinant or epitope present in a biological macromolecule. The normal immunological response involves production of polyclonal antibodies, that is, a large variety of antibodies, each recognizing a specific epitope.

Mre11–Rad50–Nbs1 complex (MRN) Eukaryotic complex of three proteins involved in the initial steps of double-strand break repair, by both **homologous recombination** and **non-homologous end joining**. Senses damage and interacts with broken DNA ends to keep them in close proximity for repair to occur. Mutations in *MRE11* and *NBS* genes in humans lead to genetic diseases.

multiprotein complex Term usually reserved for very large complexes involving a number of different kinds of proteins. Usually these act as protein machines to carry out or regulate vital cellular functions.

N-end rule Rule that defines the mechanisms by which proteins to be degraded by the proteasome are recognized by the ubiquitylation enzymatic machinery. The rule states that the nature of the N-terminal amino acid residue in a protein channels it for degradation. Some amino acids can be directly recognized by a RING-finger E3 ligase; others must first have an arginine residue added; Asn or Gln must first undergo deamidation reactions.

N-glycosylase Enzyme that cleaves the glycosidic bond between the damaged base and deoxyribose in the first step of **base excision repair**.

N-terminus End of a polypeptide chain at which the amino group is unreacted. Contrast with C-terminus.

near-cognate amino acids Amino acids so similar in structure to the cognate amino acid that they can be misidentified by the aminoacyl-tRNA synthetase as the cognate amino acid in loading to tRNA. Contrast with noncognate amino acids, whose structure differs significantly from that of the cognate amino acids and are thus not likely to be misidentified by the enzyme.

nick As applied to double-stranded polynucleotides, a cut in one strand only.

no-go decay (NGD) Mode of mRNA decay induced by prolonged stalling of a ribosome on the mRNA molecule.

non-stop decay (NSD) Mode of mRNA decay induced by lack of an appropriate stop codon.

nonautonomous retrotransposon **Retrotransposon** that requires additional genetic elements for transposition.

noncoding RNAs (ncRNAs) RNA molecules that are the final products of a gene and do not code for protein. These RNAs serve as enzymatic, structural, and regulatory components for a wide variety of processes in the cell.

nonhomologous end-joining (NHEJ) Repair of DNA double-strand breaks without requirement for homology at the severed ends. This mode of repair is error-prone and cell-cycle independent.

nonhomologous recombination Recombination in which the exchanging DNA segments are not homologous in nucleotide sequence.

nonsense-mediated decay (NMD) Mode of mRNA decay induced by encounter of the ribosome with a premature stop codon.

nontemplate strand Same as **coding strand**.

nonviral vectors Alternative ways of introducing foreign DNA into cells developed to overcome safety issues of viral vectors. These include liposomes, artificial lipid spheres containing aqueous solutions of DNA, and artificial chromosomes, large recombinant constructs that contain centromeres and telomeres of native chromosomes to ensure stable reproduction of the foreign DNA.

nuclear matrix Highly dynamic network of insoluble protein fibers found throughout the volume of the interphase cell nucleus; helps in organizing chromatin fibers into individual loop domains of varying functionalities.

nuclease-hypersensitive sites DNA regions in eukaryotic chromatin that are nucleosome-free and thus highly sensitive to cleavage by the enzyme DNase I.

nucleic acids Polynucleotides; polymers of nucleoside monophosphates; DNA and RNA.

nucleoid Protein–DNA complex found in the bacterial cytoplasm. It serves some of the same functions as eukaryotic chromatin, such as DNA compaction and organization of DNA in supercoiled loops.

nucleolus Nuclear substructure, plural nucleoli, formed by intra- and interchromosomal interactions that contain tandemly repeated ribosomal genes. Site of ribosome synthesis and assembly.

nucleoporins Class of structural proteins of nuclear pores.

nucleoside Molecule in which a purine or pyrimidine base is linked to the 1′-position of a ribose or deoxyribose moiety.

nucleosome General term for DNA–histone octamer particles that are major repeating units of eukaryotic chromatin. *See also* **nucleosome core particle** and **octasome**.

nucleosome core particle Specific structure with 147 base pairs of DNA wrapped about an octamer of histones.

nucleosome-free regions (NFR) Regions in eukaryotic promoters that are permanently devoid of nucleosomes, regardless of the transcriptional state of the linked gene.

nucleosome positioning Placing of nucleosomes on specific DNA sequences.

nucleotide Nucleoside with one, two, or three phosphate groups bound at the 5′-hydroxyl.

nucleotide excision repair (NER) DNA repair of lesions such as bulky adducts and UV-induced photoproducts that disrupt the DNA double helix; accomplished by removing a number of nucleotides from the damaged strand, followed by filling in.

***O*^{6}-alkylguanine alkyltransferase (AGT)** Unusual enzyme involved in the repair of alkylated bases in the DNA. Catalyzes irreversible transfer of the alkyl group to a cysteine in its own active site; the enzyme is degraded following the transfer in a suicide reaction. Thus, it is not a true enzyme.

octasome Nucleosomal particle with an octamer of histones binding 147 base pairs of DNA; also known as **nucleosomal core particle**.

Okazaki fragments Short pieces of DNA strand synthesized on the **lagging strand** template in replication of double-stranded DNA.

open complex In transcription, the complex in which the DNA strands have been separated to form the **transcription bubble** within the RNA polymerase.

opening transition Structural change in a nucleosome where a major portion of the DNA unpeels from the histone core. Much less frequent than **breathing transitions**. Also known as unwrapping transition.

operator In bacteria, a genomic region that binds a protein and thereby regulates an **operon**.

operon In bacteria, a group of contiguous genes of related function, regulated as a group.

origin of replication Specific nucleotide sequence on DNA where replication can begin.

origin replication complex (ORC) Multisubunit protein complex that assembles at replication origins and is required for replication initiation.

overhang At a DNA end, extension of one strand beyond the other.

paired-end tags (PET) Short, ~13 base pair sequences at the two ends of a DNA fragment that are unique as a pair to exist only once in a genome, allowing identification of the

DNA sequence between them by computational methods applied to whole genome sequences.

palindrome Sequence of letters, words, or symbols that reads the same forward or backward. When palindromes occur in single-strand nucleic acids, they are prone to form hairpin structures.

pausing Short or longer temporal stops during the process of transcription. Resumption of elongation is spontaneous and fast following short pauses; in contrast, long pauses can be overcome only by endonucleolytic activity of the polymerase, which cleaves of a portion of the nascent transcript, in conjunction with specific protein elongation factors: GreA and GreB in bacteria, TFIIS in eukaryotes. Contrast with **stalling**.

pBluescript Commercially available phagemid cloning vector that includes a β-galactosidase gene, allowing a color reaction for occur.

pBluescript II Commonly used phagemid cloning vector that combines the features of phages and plasmids. It is also an expression vector because it contains certain promoter sequences that allow the expression of the inserted DNA sequence.

penetrance Measure of how many members of the population exhibit the phenotype expected from the genotype.

penicillin Antibiotic that binds to, and irreversibly inhibits, one of the enzymes that participates in biosynthesis of **peptidoglycans**. Prevents further bacterial growth and proliferation. Widely used in clinical practice.

peptide Small oligomer of amino acids connected by peptide bonds.

peptide bond Covalent bond formed by elimination of a water molecule between the amino group of one amino acid and the carboxyl of another.

peptidoglycans Compounds in which polysaccharide chains are linked to small peptides, usually forming large cross-linked complexes. They are found in bacterial cell walls. Peptidoglycans have different organization in the two major classes of bacteria, Gram-negative, and Gram-positive.

peptidyl site (P site) Site on the ribosome that holds the peptidyl–tRNA, carrying the growing polypeptide chain. A new charged tRNA can enter the neighboring A site; the peptidyl-tRNA in the P site and the aminoacyl-tRNA in the A site are now positioned to form of a new peptide bond.

peptidyl transferase center (PTC) Center in the large ribosomal subunit where actual formation of the peptide bond between the aminoacyl-tRNA in the A site and the peptidyl-tRNA in the P site occurs.

persistence length Measure of the stiffness of a polymer molecule; a stiffer molecule has a longer persistence length.

phage display Expression of cloned genes as fusion proteins on the surface of an M13 bacteriophage vector.

phase variation Reversible regulated generation of variant bacteria that differ in the presence or absence of specific surface antigens. A well-studied example is the synthesis of pyelonephritis-associated pili in uropathogenic *Escherichia coli*.

phenocopying Describes situations in which environmental agents cause changes in phenotype that mimic effects of gene mutations.

phenotype Observable morphological or biochemical characteristics of an organism.

phosphatase Any of a class of enzymes catalyzing the hydrolysis of phosphate groups on a substrate.

Piwi-interacting RNAs (piRNAs) Type of small noncoding RNAs made in the germline that, in complex with Piwi proteins, keep in check the movement of transposable elements by transcriptionally silencing transposon genes and destroying RNAs produced by them. They originate from extremely long single-stranded RNA transcripts, usually antisense, and do not require Dicer for biogenesis.

plasmids Small, extrachromosomal DNA molecules found in most bacterial cells; they can replicate independently of the cell DNA. These molecules were the first cloning vectors and are still often used.

pleiotropy In genetics, the situation where one gene affects multiple, seemingly unrelated phenotypical characteristics.

pluripotency With respect to cells, ability to differentiate into multiple cell types. Completely pluripotent cells, such as embryonic stem cells, can form every cell of an organism. Multipotent cells can differentiate into limited cell types; for example, hematopoietic cells can differentiate into lymphocytes, monocytes, erythrocytes, and other blood cell types but cannot give rise to, for example, brain cells.

point centromeres Localized, well-defined **centromeres**.

Pol II core promoter DNA nucleotide sequence around the transcription start site; it binds RNA polymerase and directs initiation of transcription.

poly(A)-binding protein (PABP) Protein of various functions that recognizes and bind to the poly(A) tail in mRNAs.

poly(ADP-ribose) polymerase 1 (PARP-1) Extremely abundant eukaryotic enzyme that performs poly(ADP-ribosyl)ation of itself, known as automodification, and of numerous other proteins, known as heteromodification.

poly(ADP-ribosyl)ation (PARylation) Enzymatic addition of ADP-ribose units or chains of such units to proteins.

polyadenylated Messenger RNAs that have poly(A) tails attached at the 3′-end.

polycistronic Refers to DNA or RNA sequences that contain several genes or their transcripts in tandem. Common in bacteria, rare in eukaryotes.

polyclonal antibody Antibodies produced in the course of normal immunological response; will usually respond to a variety of epitopes, in contrast to a **monoclonal antibody**.

polycomb repressive complexes (PRC1 and PRC2) Complexes that silence hundreds of genes that encode developmental regulators in plant and animals. Classic examples of such genes are the **homeobox** genes that are active in early development but are irreversibly repressed when they are no longer needed later in development. Both complexes contain subunits with specific histone-modifying activities.

polymerase Any of a large group of enzymes that catalyze the polymerization of nucleic acid monomers. There are RNA polymerases and DNA polymerases. All add monomers in the 5′ → 3′ direction. Most, but not all, require a template DNA or RNA.

polymerase chain reaction (PCR) *In vitro* technique for greatly increasing the amount of a desired DNA fragment by repeated cycles of copying strands of denatured DNA templates starting from primer oligonucleotides. *See also* **real-time PCR**.

polynucleotides Polymers of nucleotide residues; nucleic acids.

polypeptide Polymer formed by the connection of a number of amino acids through peptide bonds.

polyribosomes Structures reflecting the simultaneous translation of large mRNA molecules by numerous ribosomes moving along the molecule.

polytene chromosome Multiple copies of a chromosome packed in parallel, as found in some insect cells, for example, in the salivary glands of *Drosophila melanogaster*.

Pol α In eukaryotes, DNA polymerase involved in replication initiation. Once **primase** has synthesized the RNA primer, this polymerase elongates it with ~20 nucleotides. It has limited processivity and cannot proofread errors, thus it is not suited for synthesizing long stretches of DNA strands.

Pol Δ In eukaryotes, DNA polymerase that synthesizes the lagging DNA strand.

Pol ε In eukaryotes, DNA polymerase that synthesizes the leading DNA strand.

pre-initiation complex (PIC) In any process of transcription, translation, or replication, a specific protein complex that forms about the origin site, to prepare the first steps of polymerization.

pre-messenger RNA (pre-mRNA) RNA produced in the eukaryotic nucleus as a primary transcript from a protein-coding gene that has not been spliced or otherwise processed to yield mature messenger RNA.

pre-replication complexes (pre-RCs) Protein complexes assembled on origin of replication sequences during M and G_1 phases; these are then activated at various times during S phase.

primary structure Sequence or order of amino acid residues in a protein molecule.

primase Enzyme that synthesizes primers on DNA segments to be replicated.

primer In replication, short oligonucleotide that base-pairs with the DNA template strand and provides a free 3′-OH group for primer extension by polymerase from this point. Also refers to DNA primers used in **polymerase chain reaction**.

primer extension Method usually used to determine transcription start sites. A radioactive oligonucleotide or primer is annealed to a region near the 3′-end of the mRNA and then extended by **reverse transcriptase** to the 5′-end of the RNA. The extended primer is analyzed for length to pinpoint the transcription start site. A modification of the method can be applied to a **micrococcal nuclease** digest of chromatin, using primers complementary to ends of the resulting purified DNA fragments and DNA polymerase.

priming Use of a primer to initiate polymerization of a new DNA strand.

primosome In *Escherichia coli*, a multiprotein complex that contains the primase and is essential for primer formation in DNA replication.

principle of genetic equivalence In cell biology, refers to the fact that all cells of an adult organism are genetically identical and differ only in the subsets of genes that are expressed, that is, transcribed and translated. Also known as principle of nuclear equivalence.

processed pseudogene Pseudogene that has been produced by retrosynthesis of a normal transcript, followed by insertion into the genome. They lack functional promoter elements and most are transcriptionally inactive; sometimes they co-opt nearby promoters and can be expressed.

processing bodies (P bodies) Cytoplasmic protein aggregates that incorporate unused mRNA for either degradation or recycling through **stress granules**.

processive Copying processes such as transcription, translation, and replication are said to be processive if they can proceed for many steps without interruption, thus synthesizing long stretches of the respective polymer.

prokaryotes Single-celled microorganisms whose cells lack a well-defined, membrane-enclosed nucleus. Either bacteria or archaea.

proliferating cell nuclear antigen (PCNA) In eukaryotes, processivity factor in DNA replication. Counterpart of bacterial β-clamp.

promoter Genetic element that is essential for the expression of a gene. They usually lie near transcription start sites, and bind RNA polymerase and initiation factors.

promoter escape Step in transcription in which the RNA polymerase starts moving along the DNA template strand, leaving the promoter and proceeding into the gene body. Also known as promoter clearance.

promyelocytic leukemia bodies (PML) Cytologically recognizable structures where alternative lengthening of telomeres occurs.

proofreading In molecular biology, examining a copy of a sequence for mistakes. It is usually combined with **editing** to correct the copy.

prophage With **temperate phage**; the phage genome that has been inserted into the host genome.

proteasome Multiprotein complex that catalyzes the proteolysis of malformed or obsolete proteins in the cell. The proteins have been marked for destruction by **ubiquitin.**

protein splicing Process in which different portions of a polypeptide chain are cut and then spliced together to form a new molecule. Performed by proteolytic enzymes that perform the hydrolysis reaction in reverse: they catalyze peptide-bond formation using energy provided by simultaneous proteolysis of a nearby peptide bond.

proteoglycans Peptide–carbohydrate adducts that are a major component of the extracellular matrix, the substance between cells in a tissue.

proteolysis **Hydrolysis** of proteins.

proteome Ideally, the complete list of proteins and their interactions in an organism.

protomer One unit in a multisubunit protein or in any protein with quaternary structure. They may be of one type or many.

protoplast Cell, usually plant or yeast, from which a hard polysaccharide coat has been removed by enzymatic digestion to expose the outer plasma membrane.

proximal promoter In eukaryotic gene promoters, which consist of proximal and **core promoters**, these lie just upstream of the core promoter and serve to regulate the activity of the latter through binding of sequence-specific transcription factors.

pseudogene Genetic element that is very similar or identical to a protein gene but cannot be expressed, usually for lack of a functional promoter.

quantitative PCR (qPCR) Variant of **polymerase chain reaction** in which the quantity of product is monitored throughout multiple reaction cycles. Also known as real-time PCR.

quaternary structure Level of protein structure produced by association, usually noncovalent, between protomers.

Rabl configuration Spatial arrangement of interphase chromosomes in nuclei such that centromeres are clustered at one end of the nucleus and telomeres at the other; thus the chromosome arms extend from the centromeres to the telomeres throughout the nuclear volume. Such general organization facilitates homologous recombination by positioning homologues closer together than they would be in a totally random coil of chromosomes.

Rad51 Eukaryotic recombinase, counterpart of bacterial RecA.

radial loop model Model for mitotic chromosome structure in which loops of chromatin extend from a proteinaceous core or **scaffold**.

random coil Refers to a linear polymer that has no secondary or tertiary structure but instead is wholly flexible with a randomly varying geometry; this is the state of a denatured protein or nucleic acid.

rapid amplification of cDNA ends (RACE) Method for obtaining the full-length sequence of an RNA transcript. Based on reverse transcription polymerase chain reaction, RT-PCR, which allows the production of amplified cDNA copies of the RNA transcript, which are then sequenced.

readers Proteins that interpret epigenetic marks on chromatin by binding to such marks.

reading frame In a nonpunctuated code, codons of length n can be read in n different ways or frames, depending on where reading begins. The triplet genetic code can be read in three different frames, giving rise to as many as three different polypeptide chains.

real-time PCR Variant of **polymerase chain reaction** in which the quantity of product is monitored throughout multiple reaction cycles. Also known as quantitative PCR.

RecA In homologous recombination, single-strand binding protein that forms a right-handed helical filament around the single-stranded DNA regions formed by the action of **RecBCD**. It mediates the strand-exchange reaction by invading the double-helical DNA in search of homologous sequences.

RecBCD Three-subunit protein complex that creates single-stranded regions around double-strand breaks in DNA; these regions serve as binding sites for RecA, to initiate strand exchange in homologous recombination. RecB is a $3' \rightarrow 5'$ helicase and a nuclease; **RecD** is a $5' \rightarrow 3'$ helicase; RecC provides a scaffold for organizing the entire complex and recognizes and binds to a specific sequence, **crossover hotspot instigator** or *Chi*.

recognition motifs Protein structural motif involved in locating specific nucleic acid sites. Examples include helix–turn–helix motif, leucine zipper, and zinc finger.

recombinant DNA technology Procedures in which DNA segments from different molecules are joined together to form a new contiguous DNA molecule in either research or industry.

recombinational hotspots Nonrandom distribution of genomic regions containing clusters of double-strand breaks in DNA; these are regions of active homologous recombination. Double-strand breaks in non-hotspot regions are recombinationally silent and are repaired as normal DNA damage.

RecQ helicases Helicase family in both bacteria and eukaryotes; includes **Werner and Bloom helicases**.

recruitment model for transcriptional activation Model stating that transcriptional activators act by merely recruiting the transcriptional machinery to the promoter.

reduced representation bisulfite sequencing (RRBS) Method for determining the methylation status of individual cytosines in small CpG portions of the genome in a quantitative manner. Unmethylated cytosines in a DNA sequence are converted to uracil by treatment with bisulfite, followed by sequencing. The assay is preceded by enrichment for CpG-rich genomic regions by restriction endonuclease treatment with enzymes that cut around CpG dinucleotides.

regional centromeres Centromeres that are not highly localized with respect to the underlying DNA sequence, in contrast to **point centromeres**.

regulated transcription Transcription of genes that are expressed only under certain conditions. Transcriptionally-regulated genes are either ON or OFF. An additional level of regulation in the ON state allows the production of products at different levels.

regulon In bacterial transcription, a group of unlinked, nonadjacent genes that are regulated as a group; an order of regulation higher than operons.

release factor (RF) Protein that facilitates termination of translation when a stop codon enters the A site of the ribosome.

repeat length In chromatin, the average center-to-center distance, in base pairs, between adjacent nucleosomes as determined by analysis of **electrophoretic DNA ladders**.

replacement variants Histone variants that replace **canonical histones** in some nucleosomes. Replacement can occur at any phase of the cell cycle; by contrast, canonical histones are incorporated into chromatin only during S-phase.

replacement λ based vectors Cloning vectors based on the fact that about one-third of the phage λ genome is not necessary for replication and can be replaced by any desired sequence without affecting the infectivity of the recombinant phage.

replication factor C (RFC) In eukaryotes, clamp loader in DNA replication. Counterpart of bacterial clamp loader DnaX.

replication fork Forklike structure produced during the replication of one DNA duplex to form two daughter duplexes. The fork travels from the replication origin to the site of termination.

replication origin (*oriC*) Nucleotide sequences on DNA where replication can begin. The origin of replication in *Escherichia coli*.

replication protein A (RPA) In eukaryotes, heterotrimeric protein complex that serves as single-stranded DNA binding protein in DNA replication. Counterpart of bacterial SSB protein.

replicon Stretch of DNA replicated from one origin.

replisome Multiprotein complex that replicates DNA.

repressor Protein that represses transcription by binding to specific nucleotide sequences, usually located in regulatory regions in or near a gene.

residue Portion of a monomer unit that remains in a polymer after polymerization.

restriction endonuclease (restrictase) Enzyme that cleaves double-stranded DNA following recognition and binding to a specific nucleotide sequence. These enzymes are part of the bacterial restriction system, destroying foreign DNA invading the cell. Also known as restrictase.

retrotransposon In eukaryotes, transposon that uses a retrovirus-like mechanism for transposition, involving reverse transcription of RNA.

retrovirus Any RNA virus in which the RNA template that enters the cell is copied into a DNA duplex; this can be integrated into the cell DNA. This process requires **reverse transcriptase**.

reverse transcriptase (RT) Enzyme, first discovered in retroviruses that catalyzes the formation of double-stranded DNA by use of an RNA molecule as template. Also known as **RNA-dependent DNA polymerase**.

reverse transcription PCR Variant of **polymerase chain reaction** in which multiple copies of DNA are produced from an RNA template. The RNA is first transcribed by reverse transcriptase into copy DNA, which then serves as template for polymerase chain reaction.

rheostat control Regulation of biological processes that is graduated, not on/off.

Rho factor Protein factor functioning in one mode of transcription termination in bacteria.

Rho utilization (*rut*) site G-rich sequence in bacterial nascent transcripts that serves as a binding site for the **Rho termination factor**.

ribonucleic acid (RNA) Polymer of ribonucleotide residues.

ribosomal protein (r-protein) One of numerous proteins that, together with rRNA, form the ribosome.

ribosome Subcellular particle, composed of RNA and proteins, which is the site of protein synthesis.

ribosome releasing (or recycling) factor (RRF) Protein factor that facilitates release of the ribosome from mRNA at the termination of translation. This allows recycling of ribosome subunits, hence this factor is also known as ribosome recycling factor.

riboswitch Sequence in some mRNAs that will change in conformation in response to some external stimulus, such as temperature or binding of a metabolite, and thereby switch on or off the translation of that mRNA.

ribozyme RNA molecule that possesses catalytic functions like a protein enzyme.

Ring1b–Bmi1 Two subunits of **polycomb repressor complex 1**, which place a monoubiquitylation mark on Lys119 of histone H2A, as part of its repressive function. Both subunits contain RING-finger domains, which extensively interact with each other, although only the RING-finger domain of Ring1b has a catalytic function.

RNA-dependent DNA polymerase Enzyme that catalyzes the formation of double-stranded DNA by use of an RNA molecule as template. Also known as reverse transcriptase.

RNA polymerase (RNAP) Enzyme complex that catalyzes the polymerization of an RNA strand complementary to a DNA template strand.

RNA processing Various processes that lead to transformation of primary transcripts into mature functional RNA molecules. Includes splicing, editing, polyadenylation, etc.

RNase P Enzyme that creates the mature 5′-end of all tRNAs, consisting of a highly structured RNA molecule and a small protein. Catalysis is carried out by the RNA component; thus the enzyme is a ribozyme.

rolling circle replication Mode of replication of some viruses in which a polymerase may make many passes around a circular genome, producing multiple tandem transcripts. In this mechanism, one DNA strand remains intact to serve as a template for DNA synthesis of the other strand.

RuvA–RuvB complex In **branch migration**, RuvA recognizes and binds to 4-way DNA junctions. RuvB binds next and serves as the molecular motor that rotates two of the four DNA arms in the 4-way junction in opposite directions; this results in rotational movement of the other two strands that are thus being drawn into the junction.

RuvC Resolves 4-way junctions in the final step of **branch migration**. Cleaves DNA at symmetrical positions across the junction, creating the final products of recombination.

S1 nuclease protection One of two classical methods for mapping the position of transcription start sites. The method is based on treating DNA–RNA heteroduplexes between 5′-labeled fragments of a gene and total RNA from cells expressing the gene with the fungal enzyme S1 nuclease. S1 specifically cleaves single-stranded nucleic acids, until it reaches the boundaries of the duplex region. The length of the protected region is determined by high-resolution electrophoresis.

satellite DNAs Small DNA segments with density quite different from bulk DNA; they will band separately, as satellite bands, in a density gradient in the ultracentrifuge.

scaffold Proteinaceous core of a mitotic chromosome.

scaffold attachment regions (SARs) DNA sequences that attach to the **chromosome scaffold**. They are the analog of matrix attachment regions in the interphase nucleus.

scanning or tracking mechanisms of enhancer action Alternative mechanism to the **looping model**. Activator proteins originally bound to enhancers move continuously along DNA until they encounter proteins that are already bound to the promoter. Tracking involves moving along the backbone of the double-helical DNA, thus affecting superhelical tension in the DNA template and the integrity of nucleosomes.

SDS gel electrophoresis Gel electrophoresis of proteins carried out in the presence of non-ionic detergent sodium dodecyl sulfate, SDS. Under these conditions, proteins unfold and can be separated according to chain length.

secondary structure Portions of regular, repeating folded structure in a protein molecule. The α-helix and β-sheet structures are the most important, but certain types of turns may be included, as well as other helix types.

self-assembly principle Hypothesis that proteins can assemble their higher-order structure on the basis of their sequence information, without aid of chaperones. This has proved true for some proteins but not for all.

semiconservative replication Form of DNA replication in which a complimentary copy of each strand is made, so that the daughter DNA molecules are each half new DNA and half old. Of several possible modes of replication, this is the one found in nature.

separase Protease that cleaves one of the non-SMC subunits of **cohesin** to open the cohesin ring and enable sister chromatid separation and segregation at anaphase.

sequence logo Way of representing the relative frequency of residue utilization in nucleotide or amino acid sequences such as promoters or in portions of related polypeptide chains.

shelterin Multiprotein complex that induces formation of a **t-loop** to protect single-stranded telomere overhangs from being mistakenly recognized as damage and repaired.

shelterin complex Protein complex that protects telomeres from degradation or inappropriate DNA repair.

Shine–Dalgarno sequence (SD sequence) Sequence found at the 5′-end of bacterial mRNA, which is complementary to a ribosomal RNA sequence. This allows recruitment of the ribosome to the mRNA and ensures correct alignment of the start codon in the mRNA for initiation of transcription.

short interspersed element (SINE) Non-LTR transposon that does not encode a functional reverse transcriptase and cannot transpose without the help of other mobile elements.

short noncoding RNAs Small noncoding RNA transcripts with regulatory functions in transcription and translation. Several classes are known, including microRNA (miRNA), small interfering RNA (siRNA), small nucleolar RNA (snoRNA), and Piwi-interacting RNA (piRNA), among others.

shotgun sequencing approach Strategy for sequencing large genomes. Involves creation of a large series of recombinant clones, their sequencing, and alignment of clones that contain overlapping sequences. Computational methods are used to reconstruct the sequence of the entire genome.

signal recognition particle (SRP) Membrane-bound protein complex that recognizes signal sequences for insertion.

signal sequence Sequence at the N-terminus of a newly synthesized polypeptide chain that facilitates insertion of the polypeptide into a membrane. Sometimes called leader sequence.

silencers Genetic element, often located far from a gene, that inhibits the transcription of that gene. Contrast with enhancer.

silent information regulator 2 (Sir2) Histone deacetylase involved in silencing of heterochromatic structures at yeast telomeres. It is considered a longevity factor in both yeast and humans.

single-strand binding protein (SSB) Protein that binds preferentially to single-stranded nucleic acids.

site-directed mutagenesis Technique for the production of sequence modifications or mutations at specific desired sites in a cloned portion of the genome; these modifications will be heritable.

site-specific recombination Recombination in which DNA strand exchange occurs between DNA segments that possess only a limited degree of sequence homology.

sliding clamp In DNA replication, a ringlike protein structure that surrounds the DNA and binds the polymerase to prevent its dissociation from the template, thus serving as a factor to promote processivity.

slippage Erroneous movement of DNA polymerase along DNA templates that contain short repeated sequences, creating expansions or deletions of these sequences. This causes a number of genetic diseases.

small interfering RNAs (siRNAs) Short, 21–26 base pair RNAs derived from double-stranded RNAs by the activity of a specialized cytoplasmic RNase called Dicer. They inhibit gene expression by directing destruction of complementary mRNAs.

small ubiquitin-like modifier (SUMO) Small protein with structural resemblance to **ubiquitin** that has a wide variety of cellular functions.

Sm proteins Highly conserved family of proteins that function in numerous aspects of RNA metabolism: RNA processing in the nucleus, mRNA decay in the cytoplasm, etc. Two distinct complexes of seven nonidentical subunits exist that differ in only one subunits; both complexes have a ring structure.

SNARE proteins Protein molecules that aid in attaching vesicles to membranes of target organelles, such as lysosomes.

spinal muscular atrophy (SMA) Genetic disorder characterized by atrophy of the muscles used for walking, sitting up, and controlling head movement. Results from deletion or mutation of the *SMN1* gene, the protein product of which is essential for survival of motor neurons.

spliceosome Nuclear protein–RNA complex upon which splicing occurs.

splicing Nuclear processing of pre-messenger RNA molecules that removes introns and reconnects exons in the uninterrupted coding sequence of mature mRNA.

SR proteins Class of serine/arginine-rich proteins that affect splicing by binding to splicing enhancers or silencers.

***SRY* gene** Gene on the male Y chromosome involved in determining the anatomical sex of males.

stalling Long-duration halting, usually irreversible, in transcription. It causes disintegration of the polymerase complex and release of the nascent transcript.

start signal Triplet that signals the place where translation of mRNA into protein should begin. In protein synthesis, codes for initiation methionine. Also known as start codon.

start site selection Some mRNAs have multiple potential start sites for initiation of translation, which may be selected on the basis of cellular needs.

statistical positioning Positioning of nucleosomes that often follows downstream from a well-positioned nucleosome. Positioning becomes less regular farther from the well-positioned nucleosome.

stem cell Undifferentiated eukaryotic cell that may differentiate to produce cells with differing capabilities.

sticky ends Refers to ends of DNA fragments with **overhangs** that are complementary to overhangs produced by the same enzyme elsewhere. Complementary overhangs will hybridize to each other, favoring recombination.

stimulons Also known as modulons. Sets of regulons that are controlled in coordination. An example is the **CAP** modulon that contains all regulons and operons regulated by cAMP-bound CAP.

stop codon Triplet that signals the place where translation of mRNA into protein should terminate.

stress granules Cytoplasmic aggregates of proteins and RNAs, formed in response to stress, that can store and recycle unused mRNA.

stringency Conditions such as temperature, salt concentration, etc., employed in hybridization reactions, specifically in **polymerase chain reaction**. When stringency is high, very close matching between the sequence to be replicated and the primer is required; low stringency is more forgiving.

stringent response Transcription regulation in bacteria during starvation that involves ligand binding to RNA polymerase. Biochemically, expresses itself as shutting down the synthesis of stable RNA molecules, such as rRNA, in favor of enhancing the synthesis of some amino acid biosynthesis genes.

structural maintenance of chromosomes (SMC) proteins Large family of ATPases that are involved in higher-order chromosome organization in both bacteria and eukaryotes. In eukaryotes, they are part of the protein complexes **cohesin** and **condensin**.

sucrose gradient sedimentation Sedimentation analysis in which separation of components in bands is stabilized by a gradient in sucrose concentration.

supercoil density Measure of the degree of supercoiling of a molecule. The difference in linking number from the relaxed value divided by the relaxed value.

supercoiled domains Loops in the bacterial chromosome under different degrees of superhelical stress. The degree is highly regulated by **topoisomerases** and by structural proteins that bend or twist the DNA; this in turn regulates transcription.

superhelical stress Tension induced in duplex DNA by supercoiling.

supraspliceosome Complex of four closely packed spliceosomes that exists *in vivo* and allows for simultaneous occurrence of four splicing events on long pre-mRNA molecules.

Svedberg (S) Conventional unit for sedimentation coefficients, defined as 10^{-13} s.

switch/sucrose nonfermenting (SWI/SNF) Family of chromatin remodeling complexes; yeast and humans share some subunits, especially the ones that carry out ATP hydrolysis

synapsis Alignment of homologous chromosomes in meiosis.

synaptonemal complex (SC) Protein complex that functions primarily as a scaffold to hold chromosomes together in synapsis.

t-loop Structure formed as a result of interaction of the single-stranded overhang at telomere end with a homologous region further upstream. Its formation is induced by the protein complex **shelterin**, which binds to the overhangs and shelters the structure from unwanted repair and degradation.

tandem chimerism Process of alternative slicing, which leads to creation of intergenic products by combining exons of two or more neighboring genes. It occurs as a result of long-range transcription.

Taq polymerase Heat-stable DNA polymerase from the thermophilic bacterium *Thermus aquaticus*. It is useful at the high temperatures used in **polymerase chain reaction**.

TATA-binding protein (TBP) Protein in the eukaryotic transcription initiation complex that binds to TATA. The other general transcription factors and RNA polymerase II assemble around TBP.

TATA-binding protein associated factors (TAFs) General transcription factors associated with TBP or TATA-binding protein, essential for initiation.

TATA box Oligonucleotide sequence, usually found at transcription promoter sites in eukaryotes. Binds TBP or TATA-binding protein, a general transcription factor essential for correct attachment of RNA polymerase.

telomerase Ribonucleoprotein reverse transcriptase that adds short DNA sequence repeats to telomeres, thus preventing undesirable shortening and gene loss from chromosome ends.

telomeres Structures at chromosome ends, consisting of a number of DNA repeats. Telomeres protect the chromosomal DNA from deleterious end deletion due to nonreplacement of primers during replication.

temperate phage Bacteriophage that can, after infection of a bacterium, either be replicated and lyse the bacterium, in what is known as lytic life style, or become integrated into the bacterial genome, in what is known as lysogenic life style.

template strand Strand of duplex DNA that serves as a template for transcription. It is therefore complementary in sequence to the transcript.

termination In transcription, translation, and DNA replication, the ending of production and release of the polymeric product: RNA, protein, or DNA.

termination signal Signal, usually a nucleotide sequence, that triggers termination of transcription. There are also signals in mRNA to terminate translation and in DNA to terminate replication. Also known as termination sequence.

termination sites (*Ter* sites) Sites where a polymerization process such as DNA, RNA, or protein synthesis is terminated. Together with initiation sites. They specify the length of a biopolymer.

termination utilization substance (Tus) In *Escherichia coli*, protein factor that binds to termination site sequences. These arrest replication fork movement in one direction only; that is, they are polar.

termination zones Rather large regions in the bacterial chromosome where termination of DNA replication occurs.

tertiary structure Three-dimensional folded structure of a protein or RNA molecule.

tetrasome Nucleosome-like structure in which DNA is coiled about a single H3-H4 histone tetramer.

TFIIS Specific eukaryotic elongation factor that acts to overcome transcriptional pausing caused by backtracking of RNA polymerase II along the DNA template. Bacterial counterparts are GreA and GreB.

three-polymerase replisome Replisome carrying an extra core polymerase, so as to facilitate rapid restarting of Okazaki fragments.

threshold cycle That cycle in quantitative or **real-time polymerase chain reaction** at which the desired product first becomes detectable.

thymine (T) Pyrimidine base found in DNA but not in RNA.

toeprinting Primer extension technique that can locate the edge of a protein bound to a nucleic acid; often used to locate stalled ribosomes on mRNA molecules.

topoisomerases Enzymes that can change the linking number of DNA. They accomplish this by cutting and then relaxing or further supercoiling the molecule, and then resealing the DNA. Type 1 topoisomerases cut a single strand, while type 2 topoisomerases cut both strands of the duplex DNA.

topoisomers DNA molecules that differ only in linking number.

topologically constrained Double-stranded polynucleotide being held, by either circular structure or outside constraints, so that the linking number cannot change without cutting and resealing the polynucleotide chain.

***trans* interactions** Interactions between sequence elements on two different chromosomes. Contrast with *cis* **interactions**.

***trans*-splicing** Splicing that connects distant exons; occurs as a consequence of long-range transcription and leads to production of intergenic splicing products.

transactivation domain Protein domain in transcription factors, which, in response to molecular signals, promotes transcription by interacting with protein components of the transcriptional machinery. Also known as transcription activation domain or AD.

transcription Reading of a DNA template strand into an RNA sequence.

transcription bubble Opened region of about 13–14 base pairs in the DNA located within the RNA polymerase. It allows access to DNA bases for initiation and elongation to occur. The bubble moves with the polymerase along the template.

transcription-coupled repair (TCR) Pathway of **nucleotide excision repair** active in transcribed genes only. Moreover, the transcribed DNA strand is preferentially repaired.

transcription factor Protein that binds to a promoter sequence or regulatory regions such as enhancers and silencers, thereby activating or sometimes repressing transcription.

transcription factor IID (TFIID) Multisubunit protein complex that binds to promoter first. Contains the **TATA-binding protein TBP** and **TATA-binding protein associated factors TAFs**.

transcription start site (TSS) Precise position on DNA where transcription will begin.

transcriptional activation domain (AD) Distinct domain in transcriptional activators that carry the activation function. They interact directly with protein components of the transcriptional machinery.

transcriptome Total of transcribed regions of the genome.

transcriptome analysis High-throughput sequencing method that allows measurement of transcriptional activity of thousands of genes at once to create a global picture of cellular function. Can analyze entire sequenced genomes, including the human genome.

transduction Transfer of DNA sequences from one bacterium to another via infection with temperate phage.

transfer RNA (tRNA) Molecule that can bind a specific amino acid and carry it to a **ribosome** where its specific anticodon will match a codon on the messenger RNA. Thus, it makes the connection between RNA and protein sequences.

transformation Any change in an organism's genome caused by introduction of foreign DNA. The term malignant transformation is used to indicate progression to a cancerous state in animal cells, independent of the mechanism.

translation Production of a specific polypeptide or protein in response to a specific mRNA. This process occurs on the **ribosome** and is mediated by tRNAs.

translesion DNA synthesis Synthesis across a DNA lesion without repair of the lesion.

translesion synthesis or bypass polymerases Specialized, low-fidelity, low-processivity DNA polymerases that replicate lesion-containing DNA without removal of the lesion.

translocation In protein synthesis, a complex series of steps on the ribosome after peptide bond formation, by which the mRNA is moved forward by precisely one codon in the $3' \rightarrow 5'$ direction. This movement positions the next codon in the A site of the ribosome.

translocon Multiprotein complex that participates in transport of proteins into or across membranes.

transpeptidation Transfer of an amino acid or a peptide from one peptide chain to another. Protein splicing is one of several such reactions.

transposable element (transposon) DNA element that can move or transpose itself from one place to another in the genome.

transposase Class of enzymes coded in transposons; catalyze transposition.

tripeptide anticodon motif Tripeptide in the **release factor** that recognizes the stop codon and thus facilitates appropriate release of the synthesized polypeptide.

triple helix Helix of three polypeptide strands or three polynucleotide strands.

tRNA nucleotidyltransferase Enzyme that adds the universal triplet CCA to the 3′-end of tRNA precursors already trimmed to their final length by several nucleases.

trombone In DNA replication, a looping out of a portion of the lagging-strand template to allow synchronization of lagging-strand synthesis with leading-strand synthesis.

twin supercoil domain model Model for transcriptional elongation that takes into account differences in supercoil stress in front of and behind an advancing polymerase. The advancing polymerase, as well as any other molecular motor moving along double-stranded topologically constrained DNA, creates domains of positive superhelical stress in front and negative superhelical stress in its wake.

twist In DNA, the number of times the two strands cross over each other, excluding writhing. *See also* **writhe** and **linking number**.

two-dimensional gel electrophoresis Electrophoretic separation of macromolecules in a gel that is carried out sequentially in two directions. The first dimension for protein separation may involve, for example, separation by isoelectric focusing, followed by SDS gel analysis in the second dimension.

two-part primer In eukaryotic DNA replication, primers contain both an RNA and a DNA segment.

ubiquitin Small protein that is often attached *in vivo* to other proteins as a marker. An example is the marking of proteins for destruction by the proteasome.

ubiquitin-activating enzymes (E1) Enzymes that perform the first step in the ubiquitylation reaction, activation of **ubiquitin**, by forming a covalent bond between the C-terminal glycine of ubiquitin and an SH group of the enzyme.

ubiquitin chains Formed by the formation of **isopeptide bonds** between the C-terminal Gly76 of the conjugated ubiquitin and an ε-amino group of a lysine residue in another ubiquitin molecule. Depending on which of the ubiquitin internal lysines are used, the resulting chains have very different structures and functions.

ubiquitin-conjugating enzymes (E2) Enzymes that perform the second step in the ubiquitylation reaction, transfer of the activated ubiquitin to an E2 cysteine.

ubiquitin-fusion degradation pathway (UFD) Second pathway, in yeast, that leads to polyubiquitylation and then proteasomal degradation of protein substrates. Acts on ubiquitin fusion proteins in which the N-terminal ubiquitin cannot be removed by deubiquitylation. The N-terminal ubiquitin is recognized as a degradation signal by specific ubiquitin ligases.

ubiquitin–protein ligase enzymes (E3) Enzymes that perform the third step in the ubiquitylation reaction, transfer of E2-conjugated ubiquitin to protein targets. There are two classes of E3 enzymes, depending on the domain they contain and the mechanism of action: HECT-domain and RING-domains E3s.

unnatural amino acids Amino acids that do not normally exist in nature; they can be produced by *in vitro* techniques and incorporated into proteins, thus changing their structure and functionality. Can be inserted into a polypeptide sequence to create proteins of desired characteristics.

untranslated regions (UTRs) Portions of mRNA, usually at termini, that are not translated into protein.

upstream In the $3' \rightarrow 5'$ direction from a given DNA position. Contrast with downstream.

upstream activating sequence (UAS) In yeast, a single sequence upstream of promoters that replaces proximal promoter elements in higher eukaryotes. It stimulates transcription and is often considered analogous to enhancers in multicellular eukaryotes.

uracil (U) Pyrimidine base found in RNA but not in DNA.

uracil-DNA glycosylase (UNG) Enzyme that excises unwanted uracil bases in DNA in the base excision repair pathway.

variable domain (V) Portion of an antibody molecule that is unique, to specifically recognize only one particular epitope. The remainder constitutes the constant domain, common to all antibodies of a class.

vector DNA used for transmitting other DNA molecules into cells. Plasmids, bacteriophage, and artificial chromosomes all are used as vectors in recombinant DNA technology.

viral vectors Artificial constructs based on viral genomes used for efficient delivery of foreign DNA into cells.

Werner (WRN) and Bloom (BLM) helicases Two helicases that interact physically and functionally with one subunit of **shelterin**. Mutations in the respective genes lead to Werner and Bloom syndromes, characterized by premature aging.

wobble Ability of the third base of a codon to shift so it can pair with alternate bases, thus allowing one tRNA to match with more than one codon, all encoding the same amino acid.

writers Enzymes that place epigenetic marks in chromatin proteins. *See also* **readers**.

writhe With respect to supercoiled DNA, the number of times the helix axis crosses itself. It can be interchanged with twist without affecting linking number. *See also* **linking number** and **twist**.

X-linked mental retardation Genetic neurological disorder characterized by impaired intellectual functioning. Results from deletion of a gene, *VCX-A*, the protein product of which specifically binds the 5′-end of capped mRNA to prevent decapping mediated by **Dcp2**.

yeast artificial chromosome (YAC) High-capacity vector that contains sufficient elements of yeast chromosomes, including telomeres, centromeres, and replication origin, to both accommodate large inserts and replicate in a yeast host.

yeast episomal plasmid vector (YEP) Example of a shuttle cloning vector containing DNA sequences that ensure its replication in both bacteria and yeast.

Z-form DNA Unusual, left-handed DNA conformation that exists under specific ionic conditions *in vitro*. It is found also *in vivo*; its biological significance is uncertain.

zinc finger DNA binding motif in proteins, in which the peptide chain is folded into a finger stabilized by a zinc atom.

Index

3C (chromosome conformation capture) methodology 233, 306
3'→5' mRNA degradation 408, 410, 412, 415
3' end nucleotide addition 316
4C (chromosome conformation capture on chip) methodology 233, 522
4WJ *see* four-way junctions
5C (chromosome conformation capture carbon copy) methodology 305, 306, 307, 313
5'- and 3'-untranslated regions (UTRs), eukaryotic mRNA 153, 358, 360–1, 362, 399, 402–3
5'→3' mRNA degradation 408–10, 412, 415
5' end capping, RNA 316, 318, 319–20
30s ribosomal subunits 361, 362, 363, 364, 365, 366
40s ribosomal subunits 362, 363, 367
48s translation complexes 378–9, 380
50s ribosomal subunits 362, 363, 364, 365
60s ribosomal subunits 362, 363, 367
70s ribosomal particles 362, 363, 364
80s ribosomal particles 362, 363

abortive transcription 196, 206–9
acceptor (A) site, ribosomes 348–9, 365
acceptor stem, tRNA molecules 350, 351
accessible chromatin 298–299, 301–302, 303, 311–312, 313
accommodation, distorted tRNA 381, 384
accuracy of translation 371, 372
acetylation
  histones 272, 273, 274, 275
  proteins 55, 436, 442
activation domains (ADs) 264
activator recruitment model 266
activators, transcription regulation 244, 245, 264–267
activator-repressor interactions 245
active/inactive genes, spatial separation in nucleus 233–4
adaptor hypothesis 147, 149, 349
adenine (A) 66–7, 72, 73
adenosine triphosphate (ATP) 3, 44, 65, 203, 255
  hydrolysis 55, 57–8, 81–2, 183, 185
ADP-ribosylation/ADP-ribose polymerization 279–80, 281
ADs *see* activation domains
AFM *see* atomic force microscopy
A-form DNA 72, 74, 75
African trypanosomiasis 545–546
aggregated structures, proteins 51, 53
aggregation plots
  genomic signals 303
  see also clustered aggregation tool
aging 505, 556, 572, 573
Agrobacterium tumefaciens 119–20
alarmone, gene expression control 243
O6-alkylguanin alkyltransferase (AGT) 556
alkylation, DNA damage 551, 552, 556
alleles 17, 20–1
allosteric control of protein function 248
allosteric feedback control, transcription regulation 250
allosteric modulators, protein structure 44, 47
allostery 47
α-helix 42, 43
α-subunits, RNA polymerase 201, 203, 204
alternative codon usage 152
alternative lengthening of telomeres (ALT) pathway 506–7, 517
alternative splicing 322–8
  cancer 326, 327–8
  diversity of gene products 155–6
  evolution 323
  mechanisms 324–6
  regulation 329–33
  RNA secondary structure effects 332
  tandem chimerism 324, 328
  trans-splicing 328–9
alternative untranslated regions 360
Altman, Sidney 316, 318
amino acids
  classic protein constituents 33–4
  protein structure 33, 35, 43, 46
  tRNA specificity 351–2
aminoacylation, tRNA 351–4, 355–6, 357
aminoacyl-tRNA synthetases 351–2, 353, 355–6, 358
anemia *see* Fanconi anemia; sickle cell anemia
Angelman syndrome 449–50
annealing, PCR 111
antibiotics
  bacterial transcription inhibition 212
  DNA replication inhibition 473, 474
  peptidoglycan synthesis inhibition 439
  resistance 94, 100–2, 117, 374, 535–6
  ribosome targeting 374–5
  species specific 474

antibodies 542–5
anti-cancer agents 81, 83, 212, 556
anticodon arm of tRNA 350, 351
anticodons 84
antigenic determinants (epitopes) 542, 543
antigen switching 545–6
anti-nucleosides versus *syn*-nucleosides 67, 68
apoptosis 433, 437–9
Arabidopsis thaliana 27, 28
ara operon, transcription regulation 252–5
archaea
  classification of life 4–5
  DNA compacting proteins 176
  protein translocation 424, 425
  ribosomes 361, 362
  RNA degradation 339–40
  transcription regulation 241
ARE-mediated RNA decay (AMD) 413
Argonaute (Ago) proteins 342, 343, 344, 345, 406, 408, 409
ARSs *see* autonomous replicating sequences
artificial chromosomes 106–8, 114, 124, 490
ASF1 histone chaperone 501, 502, 503
A site *see* acceptor site
assembly maps for ribosome biogenesis 365, 366
asymmetric DNA Pol III holoenzyme complex 462
ataxia telangiectasia like disorders (ATLD) 564
atomic force microscopy (AFM) 11, 141, 182–3
ATP *see* adenosine triphosphate
attachment sites (*attB*/*attP*), phage integrated 541, 542
attenuation, *trp* operon 251–2, 253
AU-rich elements (AREs) 408, 413
autonomous replicating sequences (ARSs) 488, 490, 493
autonomous retrotransposons 537–8
autophagy 445, 446
auxiliary factors, ribosomes 366

backtracking, in transcription 198, 200, 268
BACs *see* bacterial artificial chromosomes
bacteria
  classification of life 4–5
  conjugation 23, 25, 247–8, 516–17
  constitutive versus regulated transcription 241
  coordinated synthesis of ribosome components 396–7
  DNA cloning 94, 101, 102
  DNA replication 461–78
  environmental responses 243, 395–7
  gene expression control 242–3, 255–8
  gene mapping 25
  genome organization 165–8
  global transcription regulation 244, 255, 256, 258
  Gram staining 436, 439
  heat responses 205–6, 207, 404, 405
  homologous recombination 517–25
  horizontal gene transfer 23, 25, 247–8, 516–17
  integration of phage DNA into genome 540–2
  introducing recombinant DNA 109–10, 119–20
  methyl-directed mismatch repair 559, 560
  mRNA, Shine-Dalgarno sequences 357–8, 361
  nucleoids 166–8
  nutrient deficiency 243
  pathogenicity regulation by phase variation 257
  peptidoglycans in cell walls 436, 439
  pharmaceutical production 116–18
  physiologically important operon regulation 246–55
  protein translocation 423–4, 425, 426
  ribosome number, environmental responses 395–7
  ribosomes 361, 362, 363–6
  riboswitches 400, 404–5, 406
  RNA degradation 339, 340, 341
  RNA processing 315–17
  structural proteins for genome organization 166–8
  structure 4
  transcription 202–12
    aborted 206–9
    elongation 209–10
    inhibition by antibiotics 212
    initiation 202–6, 207
    regulation 241–59
    RNA polymerase subunits and structure 201, 203, 204
    termination 210–11
    translation coupling 371, 373
  transcription initiation factors 205–6, 207
  transcription/translation velocity 371
  transformation and transduction 25, 517
  translation
    elongation 379–88
    GTP hydrolysis 274, 275
    initiation 376, 377–8
    regulation 404–5, 406
    termination 389–90
    transcription coupling 371, 373
  transposons 535–6
  virulence genes 119, 404, 405
  see also antibiotics
bacterial artificial chromosomes (BACs) 106–7, 114
bacteriophages
  bacterial genome integration 540–2
  DNA cloning vectors 103–5, 106
  DNA replication 478–81
  genetic study 26–7
  Hershey-Chase experiment 71
  phage display techniques 104–5, 107
  phage λ
    genetic study 26–7
    lysogenic cycles and DNA recombination 540–2
    recombinant DNA cloning vectors 103–5
  phage M13, recombinant DNA cloning vectors 104, 106, 107, 113
  plasmid combination vectors 105
  RNA polymerases 201, 202
  structure and genome packaging 164
  transduction 25
Baldwin, Robert 355
band shifts, gel electrophoresis 139
Barr body-deficient histone H2A (H2A.Bbd) 281
barrier activity 263, 264
basal transcription 223, 226, 263, 267
base excision repair (BER) 551–2, 554, 557–8

base-pair recognition, protein binding 131
bases 66–7
BER *see* base excision repair
Berg, Paul 355
β-carotene, golden rice 121–122
β-furanose rings, nucleosides 67
β-sheets 42, 43
β-sliding clamp 462–6, 471
β/β'-subunits 201, 203, 204
B-form DNA 69, 71, 72, 73, 74, 75
bidirectional replication 461–2, 472, 479, 481, 486
bidirectional transcription 195, 231
binding immunoglobulin protein (BiP) 425, 431
bioengineering, proteins with unnatural amino acids 355, 359–60
biological membranes (biomembranes) 421–33
see also membrane...
biotin-labeled DNA samples 89
BiP ATPase 425, 431
bivalent chromosomes 531
Blackburn, Elizabeth 505
BLM *see* Bloom helicase
blood groups 444–5
Bloom (BLM) helicase 506
Bloom syndrome 506, 572
blunt (flush) DNA ends 97, 99, 100
botulin neurotoxin 428, 429
branch migration, Holliday junctions 524–5
branch site, spliceosome 321, 322, 324
BRCA genes 451, 526, 527
breast cancer 414, 447, 451
breathing transitions, nucleosome dynamics 176
breeding experiments 16–17, 20, 22
Brenner, Sydney 147, 148, 149, 150
bridge helix, RNA polymerase Pol II 218, 221
broad-type (dispersed) promoters 222
Brownian ratcheting
membrane translocation 425, 431
nuclear export of mRNA 336, 338
ribosome/RNA translocation 384, 386–387
transcription 195, 221
bubble trapping 488
budding yeast *see Saccharomyces cerevisiae*
bypass polymerases 566

Caenorhabditis elegans 27–8, 61, 63, 185, 342–3, 553
CAF1 *see* chromatin assembly factor 1
CAGE *see* cap analysis of gene expression
CAGT *see* clustered aggregation tool
cAMP *see* cyclic AMP
cancer
alternative splicing 326, 327–8
anti-cancer agents 81, 83, 212, 556
BRCA genes 451, 526, 527, 574
cell cycle regulation 490
chromatin remodeling factors 284, 575
CpG methylation patterns 287
E3 ligases 449, 451
premature aging syndromes 553, 572
recombination mediators 527
Sm/LSm1 protein 414
telomeres/telomerase 505, 506, 507, 570
transcription regulation 278, 287, 288, 294
translation factors 372
canonical histones 174, 177
canonical nucleosomes 174, 176, 186, 187
CAP *see* catabolite activator protein
cap analysis of gene expression (CAGE) 235, 236, 312
Cap-binding complex, mRNA 320
Cap-dependent initiation 400–1, 402
capping RNA, 5' ends 316
capsid, virus 164, 165
carbohydrates *see* glycosylated proteins; polysaccharides; sugars
carcinogenesis *see* cancer
β-carotene, golden rice 121–2
caspases, apoptosis 433, 437–9
catabolite activator/regulatory protein (CAP/CRP) 134, 249–50, 251, 256, 257
CATH (Class/Architecture/Topology or fold/Homologous superfamily) algorithm 48–51
CCCTC-binding factor (CTCF) 233, 300, 301
cDNA, reverse transcription of messenger RNA 112, 114, 117
Cech, Thomas R. 317, 318
cell cycle 18, 19, 25–6, 487, 489–90
cell-cycle checkpoints 489, 560, 564, 575
cell division 18–19
see also meiosis; mitosis
cell sorting, flow cytometry 26
cell structure discovery 3–4
cell transformation methods, recombinant DNA 110
central dogma 84, 147, 509
centrifugation 8, 87, 102, 147–8
see also sedimentation...; ultracentrifuge
centromeres 174, 184–7, 189, 520, 521–2, 531
chain-termination method, DNA sequencing 90
Chamberlin, Michael 355
chameleon sequences in proteins 44
chaperones
histones 177–8, 499, 500–2
protein folding 55, 57–8
chaperonin 60/chaperonin 10 55, 58
Chargaff's rules 69, 72
Chase, Martha 71
CHD (chromo-helicase-DNA) binding protein 283–4
checkpoints in cell cycle 489, 560, 564, 575
chemical sequencing, DNA 90
chemiluminescent detection of labeled DNA samples 89
chemotherapy/anti-cancer agents 81, 83, 212, 556
Chi *see* crossover hotspot instigator
chiasmata 517, 530, 531
chicken-foot structure, DNA replication 498, 522, 523, 568, 570
ChIP (chromatin immunoprecipitation) 142–3, 333, 397, 486, 532
ChIP-on-chip 143
chips (DNA microarrays) 143
see also chromosome conformation capture on chip

ChIP-seq 301, 302, 307, 309, 312, 313
chloroplasts
  genome structure 507, 508
  organelle-specific transcription enzymes 216
  ribosomes 362
  transformation 110
cholesterol, membranes 422, 432
chromatids 18
  meiosis 517, 530, 531
  mitosis 182, 183, 184, 185, 506
chromatin 168–82
  centromeres 186
  compaction and decompaction 169, 182
  discovery 168, 169
  DNA methylation effects 312, 313
  DNA-protein interaction 128
  DNA repair 568–70, 572–5
  DNase I sensitivity 298–9, 301–2
  fibers 178–81
  higher-order structures 178–82
  lamina-associated domains 181
  nuclease digestion 172
  nucleosome structures 168–78
  origins of replication 493
  psoralen cross-linking 397–8
  regulatory elements 301–5
  remodeling 216, 282–4
  reproduction of structure 485, 499–504
  splicing regulation 332–3
  study methods 172–3
  transcription regulation 268–71
chromatin assembly factor 1 (CAF1) 575
chromatin hub 264, 265
chromatin immunoprecipitation (ChIP) 142–3, 333, 397, 486, 532
chromatin maps 276
chromatin remodeler superfamilies 282–3
chromatin remodeling 178, 269, 282, 284, 286, 291–2, 333, 570–1, 577
chromatosome structure 179
chromosome conformation capture (3C) methodology 233
chromosome conformation capture carbon copy (5C) methodology 305, 306, 307, 313
chromosome conformation capture on chip (4C) methodology 233, 522
chromosomes
  interphase structure 168, 169, 181
  mapping 22
  mitotic structures 169, 182–90
  number variation among taxa 157
  territories 181, 182, 232–3
circular DNA structures, *see also* DNA loops; protein-induced DNA bending; supercoiled DNA
circular genomes
  bacteria 461, 472, 474, 478
  bacteriophages 480, 481
  mitochondria 507–8
circular permutation assay 141
circular proteins 433, 435, 436
cis-acting elements
  Dcp2 activation 408, 413
  mRNA translation regulation 399
  transcription factor co-evolution 310
cis–trans interactions
  splicing regulation 330–2
  transcription regulation 244
cis and *trans* post-transfer editing, mischarged tRNA 354, 357
clamp loader
  bacterial DNA replication 464–8
  eukaryotic DNA replication 499
class I transposons (retrotransposons) 537–8, 539–40
class II transposons (DNA transposons) 536, 538–9
Class/Architecture/Topology or fold/Homologous superfamily (CATH) algorithm 48–51
classical genetics 15–29
classification of life 4–5
clone-by-clone sequencing *see* hierarchical shotgun sequencing approach
cloning
  animals 124–5
  definitions 94
  see also DNA cloning
closed complex, RNA polymerase–DNA 196, 197
clustered aggregation tool (CAGT) 303, 304, 305
CMG complex (Cdc45/Mcm2-7/GINS) 494
coding strand of DNA *see* nontemplate strand of DNA
co-dominance of genes 20, 21
codon exchange 109
codons, tRNA matching 380
codons/anticodons
  adaptor molecule concept 147, 149
  genetic code elucidation 149–52
  genetic information flow 84
cognate amino acids 351
cognate/noncognate DNA sequences 97, 98
cohesins 183, 184, 185
co-immunoprecipitation (co-IP) 40–1
complementation, multifactorial inheritance 20
complete dominance of genes 20, 21
composite transposons 535–6
condensins 183, 184, 185
confocal fluorescence microscopy 9–10, 209, 486
conjugation
  bacteria 23, 25, 247–8, 516–17
  proteins 52, 445, 446, 453, 454
consensus sequences 202–4, 205
constant domains, immunoglobulins 542, 543, 544
constitutive heterochromatin 275–6, 277
constitutive origins of replication, eukaryotic DNA 486
constitutive transcription 241, 247
continuously varying polygenic traits 21
copy-and-paste (replicative) transposon movement 538–9
core promoters, eukaryotic transcription 222, 226, 227, 262
core RNA polymerase subunits 201, 202, 203
COs *see* crossovers
cosmids 105
co-transcriptional RNA processing 317, 319, 332, 337
co-translational translocation of proteins 424, 425–6, 429, 431
covalently closed circles (CCC) 102

CPE *see* cytoplasmic polyadenylation element
CpG dinucleotides 285–6
CpG islands 286
Crick, Francis 69–70, 72–3
adaptor hypothesis 349
discovery of gene translation processes 147, 148, 149, 150, 151
wobble hypothesis 350, 351, 352
cross-links, DNA damage 551, 552
crossover hotspot instigator (Chi) 519, 520
crossovers (COs), homologous recombination 517, 518, 524, 529, 530
CRP (catabolite regulatory protein) *see* catabolite activator protein
cruciforms (double hairpins), DNA 75
cryo-electron microscopy (cryo-EM) 7, 10–11, 373–4, 377–8
cryptic unstable transcripts (CUTs) 231
crystal structures
2D proteins on lipid layers 217
nucleosomes 171
RNA polymerase Pol II 217–18
self-assembly DNA 76
see also X-ray crystallography
CTCF *see* CCCTC-binding factor
CTCF-binding sites 300, 301
C-terminal domain (CTD) 218–19, 220
cut-and-paste (nonreplicative) transposon movement 538–9
CUTs *see* cryptic unstable transcripts
cyclic AMP (cAMP), CAP protein 249, 250, 254, 256, 257
cytoplasmic mRNA 147–8, 415, 416
cytoplasmic polyadenylation element (CPE) 360–1
cytosine (C) 66–7, 72, 73

D-amino acids 33
D arm, tRNA molecules 350
DBs *see* DNA-binding domains
DC *see* decoding center
DDR *see* DNA damage response
deamination, DNA damage 551, 552
decapping *see* mRNA decapping
decapping enzyme Dcp2 408, 412, 413, 415
decapping enzyme DcpS 408, 410, 412, 415
decoding center (DC) 361, 376
decoding process 380
degenerate genetic code 151
demethylation of DNA 257, 287, 288, 289
denaturation (melting) of DNA 71, 75
DNA replication 460, 470
PCR 111
SSB proteins 128–9, 130
transcription 196, 197, 199
density gradient equilibrium centrifugation 87
density gradient velocity sedimentation 87, 147–8
see also sedimentation analysis
deoxyribonucleic acid *see* DNA
deoxyribonuclease I *see* DNase I
deoxyribose sugar 66–7
depurination, DNA damage 551, 552
designer proteins 355, 359–60
development
Caenorhabditis elegans 28
DNA methylation 285, 287, 288
embryonic stem cells 19
expression regulation 264, 265, 276–7
genome reorganization 534
DHSs *see* DNase I hypersensitive sites
diabetes 117, 407, 410–11
diagnostics, recombinant DNA 118
differential sedimentation *see* sedimentation...
digoxigenin-labeled DNA 89
dimeric proteins, leucine zippers 135, 136
dimer removal/replacement 284
diploid cells 17, 18
direct repair of DNA damage 551
discontinuous DNA replication 458–9, 460
disordered proteins/regions 49–51, 52
disordered RNA structures 83
dispersed core promoters 222
distal regulatory elements 262, 263, 305
D-loops, telomeres 186, 506
Dmc1 recombinase 526, 527, 532, 534
DNA (deoxyribonucleic acid)
alternative forms 71, 72, 74, 75
base sequence effects physical properties 74–5
chemical structure 65, 66, 129–30
duplex forms 70, 71
evidence of information carrying 70, 71
flexibility and form 74–5
historical research 69–71, 72–3
melting temperatures 71, 75
physical structure 69–82
recombination 21, 22, 515–48
Sanger sequencing reaction 90, 91
sequencing methods 90–1
site-specific protein binding 129–30
structure, Watson–Crick model 25
tertiary structures 75–6
DnaA boxes (binding sites) 474, 475, 477
DnaA initiation protein 474–7
DNA-affinity chromatography 129
DNA amplification by PCR 110–12
DNA bending *see* protein-induced DNA bending
DnaB helicase 460, 465, 471, 474–5, 476
DNA-binding domains (DBDs) 264, 266, 527
DNA-binding proteins 129, 132, 134–6
DNA breakage
topoisomerases 79, 80
see also double-strand breaks; single-strand breaks
DNA-bridging proteins 166
DnaC helicase loader 474–6
DNA circularization assay 141, 142
DNA cloning 94–5, 100–8
DNA cloning vectors 94, 101, 102–8
DNA compaction 163–91
DNA damage
aging and disease 553
lesion types 551, 552
repair 517, 551–77
DNA damage response (DDR) 560, 568, 570, 572, 574
DNA denaturation *see* denaturation (melting) of DNA

DNA-dependent protein kinase subunit (DNA-PKcs) 563, 565, 566, 567
DNA-dependent RNA polymerases 201, 216–17
DNA enzyme types 95–6
DNA footprinting methods 140
DNA-gyrase 81
DNA hybridization, labeling genes 95
DNA ligases, cloning 95–6, 98–9
DNA linkers, joining blunt ends 99, 100
DNA looping 133–4, 141, 245, 246, 253–4, 263
DNA melting *see* denaturation (melting) of DNA
DNA methylation
  carcinogenesis 287
  chromatin accessibility/transcription relationships 311–12, 313
  ENCODE project 308–9, 313
  enzymatic machinery 288–9
  epigenetics 271, 285–9
  imprinted genes 263–4, 287
  methyl-CpG-binding proteins 289, 290, 291
  Rett syndrome 289, 291
  transcription regulation 255–6, 285–9
DNA methyltransferase (DNMT) 285–9
DNA microarrays (chips) 143
DNA photolyase 554, 555–6
DNA-PKcs *see* DNA-dependent protein kinase subunit
DNA polymerases
  bacterial
    need for primer 459
    Pol I 462, 463, 468–70
    Pol III 462, 463, 468–9, 471
    proofreading 462, 468–9, 470
    two complexes 459
  DNA replication 459
  eukaryotes 495–6
    Pol α 495, 496, 497
    Pol α 496
    Pol α 496, 497
  mitochondria, Pol α 508
  switching 566–7, 568
  translesion synthesis 566–8
DNA protection during starvation protein (Dps) 167, 168
DNA-protein interactions *see* protein-DNA interactions
DNA repair 517, 549–77
DNA replication 25, 457–514
  all organisms 457–61
  bacteria 461–82
  drug inhibition 473–4
  eukaryotes 485–514
    chromatin replication 499–504
    chromosome ends 504–7
    elongation 495–9, 500, 501
    initiation 485–95
    slippage errors 497–8
    timing regulation 491–2
  mitochondria 507–9
  requirements 459
  retroviruses 509–12
DNase I footprinting method 140
DNase I footprints, regulatory elements 302–3, 306, 308, 309, 310, 311
DNase I hypersensitivity 298–9, 301–2, 303, 305–7, 313
DNase-seq 301, 313
DNA supercoiling *see* supercoiled DNA
DNA transposons (class II transposons) 536, 538–9
DNMT *see* DNA methyltransferase
domains, classification of life 4–5, 6
dominance of genes 17, 20, 21
dormant origins of replication 486
double hairpin *see* cruciforms; hairpin structures
double helix DNA structure 70, 72–3, 74, 75
double-sieve proofreading 353, 355–6
double-strand breaks (DSBs)
  DNA damage 551, 552
  end resection by RecBCD complex 518–20
  error free versus error prone repair 560–1
  meiotic recombination 532
  programmed 531, 532, 533
  repair 517, 552, 555, 560–3, 565–6
double-stranded RNA-binding domain (dsRBM) 137
downstream, RNA molecules 195
Dps *see* DNA protection during starvation protein
Drosophila melanogaster
  alternative splicing 326
  chromosome mapping 22
  genetic research 21–2
  heat-shock genes/transcription stalling 267, 269, 270
  as model organism 27
  polytene chromosomes 22, 23
  small silencing RNAs 343, 344, 345
drug inhibition of DNA replication 473–4
drug production, recombinant DNA 116–19
drug resistance 94, 100–2, 117, 374, 535–6
DSBs *see* double-strand breaks
dsRBM *see* double-stranded RNA-binding domain

E1 *see* ubiquitin-activating enzyme
E2 *see* ubiquitin-conjugating enzyme
E3 *see* ubiquitin-protein ligase enzyme
EcoRI restriction endonuclease 96, 100, 101
effectors, protein structure 44, 47
eIFs *see* eukaryotic initiation factors
electromagnetic radiation 2–3
  see also UV radiation; X-rays
electron microscopy (EM) 6–8, 10–11
  cryo-EM 7, 10–11, 373–4, 377–8
  DNA looping assay 141
  mitotic chromosomes 182
  transcription visualization 198, 201
electrophoresis
  general principles 37
  see also gel electrophoresis
electrophoretic mobility shift assay (EMSA/gel shift assay) 139
electroporation 110
electrospray ionization (ESI) 56
ELISA *see* enzyme-linked immunosorbent assay
elongation complex, RNA polymerase 201, 202, 204
elongation factors 268

elongation factor Tu (EF-Tu)-GTP complex 380, 381
elongation phase
  DNA replication 495–9, 500, 501
  transcription 196, 197, 198
    bacterial 209–10
    eukaryotes 216, 218, 219–23, 267–8, 270
  translation 348–9, 379–88
EM *see* electron microscopy
embryonic stem (ES) cells 19, 62, 116, 286, 491
EMSA *see* electrophoretic mobility shift assay
Encyclopedia of DNA elements (ENCODE) project
  chromatin structure 301–5
  consortium 153–4
  gene regulation insights 305–12
  methods 298–9, 306, 308–9, 312–13
  transcription of genome 228–31
  transcription regulation 297–314
end modifications, RNA 316, 318–21
endogenous retroviruses 537
endonucleases 96
  see also restriction endonucleases
endonucleolytic activity, RNA polymerase complex 198
endonucleosides 68
endoplasmic reticulum (ER) 424–5, 429, 443
end resection, homologous recombination 518–20
end structures of genes 234, 235
enhanceosome, transcription by Pol I 226, 227, 228
enhancers
  gene translation 153
  genome segmentation by transcription state 300–1
  splicing regulation 330
  transcription regulation in eukaryotes 262, 263
enveloped viruses 165
environmental responses
  bacteria 241–3, 395–7
  mimicking mutations 21
  ribosome number 395–7
  transcription regulation 241–2
enzymatic sequencing, DNA 90, 91
enzyme-bridging model 79, 80
enzyme-linked immunosorbent assay (ELISA) 40
epigenetics 271, 485, 503–4
epistasis 20
epitopes (antigenic determinants) 542, 543
ER *see* endoplasmic reticulum
erasers, histone modification removal 272
erythroid cell differentiation 264, 265
ES *see* embryonic stem cells
Escherichia coli
  as model organism 27
  pathogenicity 257
  SSB proteins 129, 130
  transcription regulation 244, 246–50, 257–8
ESE *see* exonic splicing enhancers
ESI *see* electrospray ionization
ESS *see* exonic splicing silencers
Eubacteria
  classification of life 4–5
  see also bacteria
euchromatin 181, 275, 277, 279
eukaryotes
  chromatin 168–82
  classification of life 4–5
  complexity 261
  co-translational translocation of proteins 424, 425–6, 429, 431
  DNA replication inhibition by hydroxyurea 474
  gene nomenclature 261
  genetic manipulation 115–24
  homologous recombination 517, 525–32
  internal structure 3–4
  introns and gene splicing 152–3
  mismatch repair 560, 562, 563
  mitotic chromosomes 182–90
  mRNA untranslated regions 358, 360, 362
  post-translational protein translocation 425–6, 431
  recombinant DNA 110
  ribosomal component synthesis regulation 397–9
  ribosomes 361–2, 366–7
  RNA degradation 339, 340, 341
  RNA processing 315–46
  transcription regulation 261–96
  translation 377, 378–9, 380, 388, 389, 392
  transposons 536–40
  vesicle protein transport 428–9, 432
eukaryotic artificial chromosomes 107–8, 124, 490
eukaryotic initiation factors (eIFs) 378, 400–2
evolution
  alternative splicing 323
  conservation of RNA polymerase Pol II structure 219
  conserved/varied protein translocation pathways 425–6
  gene diversity mechanisms 154–6
  genome comparisons 156–60
  transcription factors/binding sites 309–11
  transposable DNA elements 533–4
excision repair 551
exit tunnel/site, ribosome 348, 349, 365, 387, 388
exon-definition model of splicing 329–30
exonic splicing enhancers (ESE) 330
exonic splicing silencers (ESS) 330, 332
exon–intron architecture 329–30, 329030
exons 152–3
  see also splicing...
exonuclease activity of DNA polymerases 462, 463, 468, 469, 496
exonucleases
  definition 96
  DNA footprinting method 140
  mRNA degradation 408
exonucleosides 68
exosome complexes 317, 339–40
expression regulation
  DNA sequences 153
  ENCODE project insights 305–12
  eukaryotes 215
  protein–nucleic acid complexes 163
  small silencing RNAs 342–4
  see also transcription regulation; translation regulation

expression of transgenes 122, 123
expression vectors for recombinant genes 108–9
expressivity of genes 21
FACT (facilitates chromatin transcription) complex 271
factor-dependent termination of transcription 210–11
Factor for inversion stimulation (Fis) 166, 167, 168
facultative heterochromatin 275, 277
FAIRE *see* formaldehyde-assisted isolation of regulatory elements
FAIRE-seq, ENCODE project 313
Fanconi anemia 527, 551, 553, 554
faulty mRNA destruction 416–18
feedback control
  transcription 250–1, 255
  translation 396
Ferguson plots, electrophoresis 37, 38
Fersht, Alan 355, 356
F factors (functional fertility plasmids) 106
filament quaternary structures of proteins 53
filter binding assay 128
FIONA *see* fluorescence imaging with one-nanometer accuracy
Fis *see* Factor for inversion stimulation
Fisher, Edmond 440
fission yeast *see Schizosaccharomyces pombe*
flexible origins of replication 486–487
flippases 443
flow cytometry cell sorting 26
fluorescence imaging with one-nanometer accuracy (FIONA) 10
fluorescence resonance energy transfer (FRET) 10, 208, 209, 376–7, 561
fluorescence techniques 8–10, 90
fluoroquinolones 473
flush (blunt) DNA ends 97, 99, 100
focused (sharp-type) promoters 222
folded fiber model, chromosome 187–8
forensic DNA techniques 112, 160
fork *see* replication fork
formaldehyde-assisted isolation of regulatory elements (FAIRE) 298–9
N-formylmethionine, bacterial translation initiation 277
four-arm cloverleaf structure, tRNA 350
four-way junctions (4WJ/Holliday junctions) 517–18, 523–5, 541–2
frameshift mutations 149, 150, 151, 154
Franklin, Rosalind 70, 72, 73
free energy, nucleoside triphosphate cleavage 195
FRET *see* fluorescence resonance energy transfer
frogs 27, 124, 533, 556
functional elements of DNA 154, 297–9
functional fertility plasmids (F factors) 106
  see also conjugation
β-furanose rings, nucleosides 67
fusion proteins 109

gametes 17, 18
gametogenesis 288
Gamow, George 349
gel electrophoresis
  circular/coiled nucleic acids 75, 85–6
  DNA-protein interactions 139–42
  general principles 37
  linear nucleic acids 85
  mapping origins of replication in eukaryotic DNA 487
  proteins 37–9
  two-dimensional 7, 10, 12, 39, 45, 86, 186, 217–8
gel shift assay/gel retardation assay/electrophoretic mobility shift assay 139
gene addition therapy 122
gene conversion, antigen switching 546
gene definitions 23, 145–7, 153–4
gene diversity 154–6
gene duplication 155
gene end structure analysis 234, 235
gene guns 110
gene mapping 22, 25, 231
gene-poor chromosomes 181, 182
general transcription factors (GTFs) 223, 226
gene replacement therapy 122
gene-rich chromosomes 181, 182
gene silencing 275–7, 406, 408, 410, 413
gene splicing *see* exons; introns; splicing...
gene therapy 122–4
genetic code 84
  elucidation 149–52
  properties 149, 150, 151
genetic diseases
  understanding of genetics 21, 23, 24
  see also human disease
genetic engineering
  plants 118–22
  see also recombinant DNA; transgenic animals
genetic equivalence principle 124
genetic information flow 84, 147, 509
genetic manipulation, eukaryotic organisms 115–24
genome, global repair 556–7
genome organization
  Archaea 164, 176
  bacteria 165–8
  eukaryotes 168–90
  viruses 164–5
genome packaging 163–91
genome segmentation by transcription state 300–1
genome sequences
  evolution studies 156–60
  gene concept 153–4
genome size, variation among taxa 156–7
genome-wide bubble trapping protocol 488
genome-wide localization analysis (GWLA) 143
genome-wide transcriptome analysis 235, 237–8
genomic CpG methylation patterns 286–8
genomic libraries 112
GINS (go/ichi/nii/san) complex 494
global genome repair (GGR) 556–7
global transcription regulators 244, 255–6, 258
globin genes 155, 264–5

globin protein structures 13, 36, 44–5, 52–3
glycoproteins 441–2
glycosidic bonds 66–7
N-glycosylase 557–8
N-glycans 442–3
glycosylated proteins 55, 436, 439, 441–5
golden rice 121–2
Golgi apparatus 425, 427–8, 443, 446
Golgi, Camillo 427
golgins, *trans*-Golgi network 428
G-quadruplexes, telomeres 187
Gram, Hans Christian 436
Gram staining of bacteria 436, 439
GreA/GreB protein elongation factors 198
GRFs *see* general transcription factors
Grieder, Carol 505
Griffith, Frederick 70
GroEL/GroES chaperone system 55, 58
group II introns, self-splicing RNA 334, 336
group I introns, self-splicing RNA 334–5
GTP *see* guanosine triphosphate
guanine (G) 66–7, 72–3
guanosine triphosphate (GTP) 65, 203
GW182 protein 406, 408, 410
GWLA *see* genome-wide localization analysis

H2A.Bbd *see* Barr body-deficient histone
H2A.X/γH2A.X, histone phosphorylation 568–70, 573–5
hairpin structures
  DNA cruciforms 75
  RNA 83, 251–2, 253, 356, 362
half-cruciform structures 83
haploid gametes 17, 18
Hartwell, Leland 489
heat shock
  bacterial transcription initiation factors 205–6, 207
  stimulating recombinant DNA uptake 109
heat-shock genes
  activation 331
  thermosensing riboswitches 404
  transcription stalling 267, 269–70
heat-shock protein 90 (hsp90) 57, 59, 266, 331
heat stress, splicing inhibition 331
heat unstable (HU) protein 166, 167, 476
heavy chains, immunoglobulins 542, 543–4
HECT-domains 447–51, 453
helical coiling of radial loop model of chromosome structure 187–8
helicase-primase complex 471–2, 474
helicases
  bacteria 460, 464–5, 475–6
  double-strand break end resection 518–19
  eukaryotes 493–4, 499, 501, 506
  unwinding DNA for SSB protein attachment 129
  see also DnaB; Mcm2-7 complexes
α-helix, protein secondary structure 42–3
helix-destabilizing proteins *see* single-strand DNA-binding proteins
helix-turn-helix (HTH) motif 132, 134–5, 250–2, 556
hemisomes (half nucleosomes) 177, 186–7
hemoglobin 45, 52–3
hemophilia 257, 528–9
hepatitis B vaccine production 117–19
Hershey, Alfred 71
Hershey-Chase experiment 71
Hertwig, Otto 169
heterochromatic protein 1 (HP1)-mediated gene repression 277–9
heterochromatin 181, 263, 275–6, 277
heterogeneous nuclear ribonucleoproteins (hnRNPs) 331
heteroheptameric ring complexes 409, 413–4
heterophagy 445
heterozygous genotypes/phenotypes 17, 21, 24
Hfr *see* high frequency recombinants
hierarchical helical coiling model of chromosome structure 187–8, 190
hierarchical shotgun sequencing approach 113–4
high frequency recombinants (Hfr) 247–8
high-resolution/high-density DNA microarrays (chips) 143
high-throughput sequencing 298–9
histone-like nucleoid structuring protein (H-NS) 166, 167, 168
histone modification 175–6, 398, 568–70, 572–5
  epigenetic regulation 271–85
  gene silencing 275–7
  genome segmentation 301
  methylation regulating DNA replication 495
histone mRNAs degradation 409, 413–4
histones
  chaperones 177–8, 499, 500–2
  discovery 169
  DNA/chromatin replication 502
  H1 502, 503
  H2A-H2B dimers 500–1, 502
  H2A.Z/H3.3/H2A.Bbd 280–2
  H3-H4 dimers 501–3
  nonallelic variants 171, 174–6
  nucleosome structure 170–8
  types 168–9, 170
  variants 280–2, 568–70, 572–5
history of discovery and understanding
  cell cycle regulation 489
  chromatin 168–70
  discontinuous DNA replication 460
  DNA polymerase 468
  DNA recombination 515–6
  DNA repair 550–1
  first recombinant DNA 101
  gene concept 23, 145–7, 153–4
  gene expression in *Escherichia coli* 247–8
  genetic code 149–52
  Golgi apparatus 427
  insulin 117
  proofreading amino acid selection 355–6
  restriction endonuclease 96
  retroviruses 509–10
  ribosomes 362–3
  RNA structures and functions 147–9
  semiconservative DNA replication 458
  telomeres 505
  translation 349

HIV (human immunodeficiency virus) 118, 165, 509, 512
hnRNPs *see* heterogeneous nuclear ribonucleoproteins
H-NS *see* histone-like nucleoid structuring protein
Holley, Robert 148, 150
Holliday junctions (HJ/four-way junctions/4WJ) 517–18, 523–5, 541–2
holoenzymes 202–6
homologous recombination (HR) 515–32
- alternative lengthening of telomeres pathway 506–7, 517
- bacteria 516–25
- DNA repair 517, 552, 554–5, 560–1, 563, 565, 571
- double-strand break end resection 518–20
- eukaryotes 517, 525–32
- horizontal gene transfer in bacteria 516–17
- malfunctions and human disease 527–9
- meiosis 529–32, 533–4
- nucleus organization 520, 521–2
- strand invasion and strand exchange 518–20, 523
- transgenic organisms 115–16
- *Trypanosoma brucei* antigen switching 546

Hoogsteen pairing of DNA 75, 76
horizontal gene transfer, bacteria 23, 25, 247–8, 516–17
host restriction 96
housekeeping genes 241
HP1 *see* heterochromatic protein 1
HR *see* homologous recombination
Hsp90 (heat-shock protein 90) 57, 59
HTH *see* helix-turn-helix motif
HU *see* heat unstable protein; hydroxyurea
human artificial chromosomes 124
human disease
- African trypanosomiasis 545–6
- E3 ubiquitin ligases 449–451
- gene regulatory and transcriptional component mutations 267, 269
- methyl-CpG-binding protein mutations 291
- Mre11–Rad50–NBS1 complex mutations 564
- premature aging syndromes 553
- RecQ helicase mutations 572
- translational fidelity 358

human genome
- elements 154
- ENCODE project 228–31
- gene functions 230
- genome segmentation by transcription state 300–1
- genomic libraries 112
- hierarchical shotgun sequencing approach 114
- transcription regulation 297–314

Human *Hsp70* gene, promoter region 262
human immunodeficiency virus (HIV) 118, 165, 509, 512
human protein interactome 63
Hunt, Timothy 489
hybridization, nucleic acids 88–90
hybridoma technology 543
hybrid state formation in ribosome 383–5
hydrogen bonds
- DNA 70, 73, 75
- DNA–protein 130, 131, 147
- proteins 39, 42–3
- tRNA 148, 350–1

hydrolysis of proteins 35, 430–5, 445, 447
hydrophilic/hydrophobic residues, protein structure 33, 43, 46
hydroxyurea (HU) 474
hypermethylation of CpG islands 287
hypersensitivity to DNase I 298–9, 301–2
hypomethylation of genome, cancer 287
ID *see* intrinsically disordered proteins/regions
identity elements, tRNA 352–4
IFs *see* initiation factors
Ig *see* immunoglobulins
IHF *see* integration host factor
imitation switch (ISWI) 283–4
immune system 122–3, 542–6
immunoblotting techniques 40–1
immunoglobulins (Ig) 40, 48, 542–3, 544–5
immunological methods, proteins 40–1
immunoprecipitation (IP) 40–1
- *see also* chromatin immunoprecipitation

imprinted genes 263–4, 287, 449–50
inactive (dormant) origins of replication 486
inchworming model of abortive transcription 208
incomplete dominance of genes 20, 21
independent assortment of alleles 20
index nucleosomes 178
indirect end-labeling 173
inducers, transcription 245, 246–8, 249, 250
information flow 84, 147, 509
inheritance patterns, Mendelian 16–17, 20
inhibition of transcription 244, 245, 246
initiation complexes
- DNA replication 493–5
- RNA polymerase 201, 202–6

initiation of DNA replication 472–6, 485–95
initiation factors (IFs), translation 378
initiation factor (σ), RNA polymerase 201, 202, 203, 204–6, 207, 209, 242
initiation of transcription 196, 202–6, 222–4, 226–9
initiation of translation 348, 378
initiator tRNA (tRNAi) 348, 377
INO80/SRCAP (inositol-requiring/SNF-2-related CREB-binding activator protein) 384
insertion sequences, transposons 535–6
insertion vectors, DNA cloning 103
insulators, transcription regulation 262–3
insulin 117, 430–1, 433
insulin-like growth factor 553
integral membrane proteins 422–6
integrase 541
integration host factor (IHF) 166, 167, 541–2
interactome maps 61–3
internal ribosome entry sites (IRES) 360, 400, 401–2, 403
internal structure of cells 3–4
interphase nucleus
- chromosome structure 168–9, 181
- homologous recombination 520, 521–2

interstrand cross-links 551, 554
intertelomeric D-loop 506

intrastrand cross-links 551, 552
intrinsically disordered (ID) proteins/regions 49–51, 52
intrinsic curvature localization in DNA fragments 141
intrinsic termination of transcription 210
intron-definition model of splicing 329–30
intron–exon architecture 329–30
intronic splicing enhancers (ISE) 330
intronic splicing silencers (ISS) 330
introns
  discovery 152–3
  removal for maximizing recombinant gene expression 109
  see also splicing of RNA
ionizing radiation 550, 564, 566
IP *see* immunoprecipitation
IRES *see* internal ribosome entry sites
ISE *see* intronic splicing enhancers
isoacceptor tRNA molecules 350
isochores, mammalian genome 492
isoelectric focusing 39
isopeptide bonds 445–6, 447–8
isoschizomers 97
ISS *see* intronic splicing silencers
ISWI *see* imitation switch

Jacob, François 147, 148, 149, 247–8
junk DNA, use of term 159, 228, 231, 334, 533

Kendrew , John 45
K-homology (KH) motif 137
Khorana, Gobind 149, 150
kinases 435, 440, 441
  see also DNA-dependent protein kinase subunit
kinetic model, splicing regulation 332–3
kinetic proofreading 385
kinetochore 184, 186–7
kingdoms, classification of life 4–5
Klenow fragments 468–9
knock-down/knock-in/knockout organisms 116
Kornberg, Arthur 460, 468
Kornberg, Roger 217
Krebs, Edwin 440
Ku70–Ku80 dimer 563, 566

L1 *see* long interspersed nucleotide element 1
labeling DNA 89, 95, 470
Lac operon transcription 246–50, 251
Lac repressor 132, 133, 246, 249, 250
Laemmli, Ulrich 188–9
lagging strand, DNA replication 459, 460–1, 468, 496
lamina-associated domains, chromatin 181
L-amino acids 33
Landsteiner, Karl 444
La proteins, *trans*-acting elements 399
LCRs *see* locus control regions
leading strand, DNA replication 459, 496
lesion types, DNA damage 551, 552
leucine zippers 135–6
licensing for initiation, DNA replication 493, 495
ligation, DNA ligase 98–9, 100
light chains, immunoglobulins 542–4
light microscopy 2–4, 8–10
LINE-1 *see* long interspersed nucleotide element 1
linkage of genes 20, 22
linker histones 179–80
linking number, circular DNA structures 77, 78
lipid bilayers 422, 423
liposome-mediated gene transfer (lipofection) 110, 124
Listeria, virulence genes 404, 405
lncRNAs *see* long noncoding RNAs
locus control regions (LCRs) 262, 264
long-flap pathway, Okazaki fragment maturation 499, 500, 501
long interspersed nucleotide element 1 (LINE-1/L1) 538, 539
long noncoding RNAs (lncRNAs) 154, 231, 232, 289–90, 292–3, 294
long pauses in transcription 198, 200
long terminal repeats (LTRs) 510–1, 537
loop formation (trombone), replisomes 468
looping models, enhancer and silencer action 263
LSm1-7 protein complexes 407, 413–14
LTRs *see* long terminal repeats
lysogenic/lytic modes, bacteriophages 25, 541–2
lysosome 445

M13 bacteriophage 106–7
MAbs *see* monoclonal antibodies
McClintock, Barbara 533–4
Mcm2–7 complexes 493–5, 499
macroautophagy 445
macroH2A histone variant 282
magnetic resonance imaging/spectroscopy (MRI/MRS) 46–7
magnetic tweezers (MT) 199, 200, 208, 209
maize 121, 533, 534–5
major classes of genome states 300–1
major groove of DNA 73, 75
  protein binding 129–30, 132, 133, 134–5, 250, 251
MALDI *see* matrix-assisted laser desorption and ionization
mammals
  cloning 124–5
  isochores and DNA replication 492
  ribosomal protein regulation 399
  transgenic animals 115–16
marker genes
  plant genetic engineering 120
  transgenic mice 115
  vectors 94, 100, 103, 108, 109
MARs *see* matrix attachment regions
mass spectrometry (MS) 56
matrix-assisted laser desorption and ionization (MALDI) 56
matrix attachment regions (MARs) 189
MBD *see* methyl-CpG-binding domain
mediator protein complex 224, 225
meiosis 18, 529–32, 533, 534
meiotic bouquet 531, 533
membrane proteins 422–3, 424, 425, 426
membrane structures 422–3
membrane transport of proteins through 423–9, 430, 431, 432
  bacteria 423–4, 425, 426
  eukaryotes 424, 425–6, 428–9, 431, 432

Mendel, Gregor 15–17, 20
Meselson, Matthew 458
Meselson-Stahl experiment 457, 458
Meselson-Weigle experiment 516
messenger RNA *see* mRNA
metabolites
  riboswitch responses 404–5, 406
  transcription regulation 284–5
methionine, tRNAs for start and internal codons 277, 278
methylated guanosine monophosphate 319–20
methylation
  DNA damage 552
  DNA resistance to degradation 96
  histones 272, 274, 275, 276–8, 495
  proteins 55
  type I restrictases 98, 99
  see also DNA methylation
methylation-sensitive isoschizomers 97, 98
methyl-CpG-binding domain (MBD) 289, 290
methyl-CpG-binding proteins 289, 290, 291
5-methylcytosine (m5C) on CpG dinucleotide 285–6
methyl-directed mismatch repair 559, 560
microautophagy 445
microRNA-induced silencing complex (miRISC) 406, 408, 410
microRNAs (miRNAs) 342, 343, 361, 405–7, 408–9, 410–11
microsatellites 160
microscopy *see* atomic force microscopy; electron microscopy; light microscopy
Miescher, Friedrich 168, 169
mini-satellites 160
minor groove of DNA 73, 75
  protein binding 129–30, 133, 166, 223, 224
miRNAs *see* micro-RNAs
misfolded proteins 54–5
  destruction 55, 57, 58
  translation fidelity 358
mismatch repair (MMR) 552, 554, 558–60, 561, 562, 563
mitochondria 216, 362, 507–9
mitosis 18–19, 183, 184, 185
  chromosome structure 182–90
  homologous recombination 517
MMR *see* mismatch repair
mobile clamp domain 218
model organisms 26–8
modified nucleosides, tRNA 351
modifier genes 21
modulons *see* stimulons
molecular motions 195
molecular theory of genetics 23
monoallelic gene expression 20
monocistronic mRNAs 373
monoclonal antibodies (MAbs) 543
Monod, Jacques 147, 247–8
Morgan, Thomas Hunt 16, 21–2
mouse (*Mus musculus*)
  DNA methylation patterns during development 287, 399
  experimental organisms 27–8, 70
  homologous chromosome regions with humans 158
  monoclonal antibody production 543
  transgenic organisms 115–16, 291, 505
Mre11-Rad50-NBS1 complex (MRN) 560–1, 563, 564, 565
MRI *see* magnetic resonance imaging
MRN *see* Mre11-Rad50-NBS1 complex
mRNA, 5' capping 319–20
mRNA 84, 356–61, 362
  alternative splicing 155–6, 322–8
  cDNA libraries 114
  discovery 148–9
  eukaryotic 3'-/5'-untranslated regions 358, 360, 362
  export from nucleus 321, 335–6, 338
  hairpin formation 83, 251–2, 253, 356, 362
  introns/exons 152–3
  life history 335–8
  monocistronic/polycistronic 373
  multiple translations 372–3
  polyadenylation of 3' end 320–1
  riboswitches 404
  sequestration in P bodies and stress granules 412–13, 415–16
  Shine-Delgarno sequences 357–8, 361
  stability 373, 407–18
  structure 356–7, 361, 362, 404
  translocation in translation 379, 382, 383–5
mRNA decapping
  3'→5' mRNA degradation pathway 408, 410, 412, 415
  5'→3' mRNA degradation pathway 408–9, 412, 413, 414
  diseases 415
mRNA degradation
  3'→5' pathway 408, 410, 412, 415
  5'→3' pathway 408, 412, 413, 415
  deadenylation 407–8, 412
  destruction of faulty molecules 416–18
  no-go decay 417, 418
  nonsense-mediated decay 413, 416–17
  non-stop decay 417–18
  processing bodies 412–13, 415–16
mRNP states, in cytoplasm/P bodies and stress granules 416
MRS *see* magnetic resonance spectroscopy
MS *see* mass spectrometry
MT *see* magnetic tweezers
Muller, Hermann Joseph 16, 22
multidomain protein structures 44, 49
multifactorial inheritance 20, 21
multigene family evolution 155
multiprotein complex structures 53
Mus musculus see mouse
mutations
  evolution 154–5
  frameshift 149, 150, 151, 154
  gene regulatory and transcriptional components 267, 269
  methyl-CpG-binding proteins 291
  Mre11-Rad50-NBS1 complex 564
  phenotype effects 20, 21
  RecQ helicase 572
MutH, mismatch repair 559, 560
MutL, mismatch repair 559, 560, 562
MutS sliding clamp 559, 560, 561, 562, 563
myoglobin 13, 36, 44, 45, 52–3
myohemerythrin 36–7

NAD (nicotinamide adenine dinucleotide)-dependent histone deacetylators 277
Nanog protein interaction network 62
NARL transcription factor 244
NBS *see* Nijmegen breakage syndrome
NCOs *see* noncrossovers
near-cognate amino acids 353, 357
NER *see* nucleotide excision repair
neurodevelopmental disorders 289, 291
neurotoxins 428, 429
NFRs *see* nucleosome free regions
NGD *see* no-go decay
N-glycosylase 557, 558
NHEJ *see* nonhomologous end-joining
nick translation 468, 470
nicotinamide adenine dinucleotide (NAD) 277
Nijmegen breakage syndrome (NBS) 564
Nirenberg, Marshall 149, 150
N-linked glycosylation 442, 443
NMD *see* nonsense-mediated decay
NMR *see* nuclear magnetic resonance
no-go decay (NGD) 341, 417, 418
nonautonomous retrotransposons 537, 538
noncanonical amino acids 354–5, 359
noncoding RNAs (ncRNAs) 84, 154, 231, 232, 289–90
noncovalent bonds, protein–nucleic acid interactions 127
noncrossovers (NCOs), homologous recombination 518, 529, 530
non-enveloped viruses 164–5
nonhomologous end-joining (NHEJ) 552, 554–5, 561–3, 565–7
nonhomologous recombination 532–40
non-LTR retrotransposons 538
non-overlapping triplet code 149, 151
non-protein coding DNA transcription 228–31
nonreplicative transposon movement 538–9
nonsense-mediated decay (NMD) 341, 413, 416–17
nonspecific protein binding to DNA 128–9, 130–1
non-stop decay (NSD) 341, 417–18
nontemplate strand of DNA 193–4
nonviral gene therapy vectors 124
NPCs *see* nuclear pore complexes
N-terminal signal sequences, protein translocation 424, 425, 426, 429–30
NTPs *see* nucleoside triphosphates
nuclear magnetic resonance (NMR) 12, 46, 139
nuclear matrix, genome organization 181
nuclear pore complexes (NPCs) 321, 335–6, 337–8
nuclease-hypersensitive sites 178
nucleases, chromatin digestion 172
nucleic acids 65–92
  chemical structure 65–9
  hybridization 88–90
  information flow 84–91
  labeling for hybridization study 89
  physical structures 69–83
  study methods 84–91
  see also DNA; RNA
nucleoids/nucleoid associated proteins 166–8
nucleoli 181, 225, 367
nucleoside 5'phosphates 67
  see also nucleotides
nucleoside diphosphates 66
nucleoside monophosphates 65, 66
  see also nucleotides
nucleoside phosphates 65–7
nucleosides 65–7
  RNA versus DNA 67
nucleoside triphosphates (NTPs) 65, 66, 67
  cleavage energy 195, 221
  RNA synthesis 194, 195, 203
nucleosome free regions (NFRs) 178, 262, 269, 270, 276, 493
nucleosomes 168–78
  assembly by histone chaperones 177–8
  canonical versus non-canonical structures 176–7, 186–7
  centromere 186, 187
  chromatin remodeling models 284
  chromatin replication 502
  chromatin structure during transcription 268–71
  crystal structure 171
  dynamic transitions 176–7
  elucidation of chromosome structures 170
  positioning along chromatin fibers 172–3, 178, 179
  positioning patterns at promoter sites 303
  structural types and nomenclature 176–7
nucleotide addition cycle 219–23
nucleotide excision repair (NER) 551, 553, 554, 556–7
nucleotides 65–6
  components 66–8
  conformations of components 67–8
  polymerization 68–9
Nurse, Paul 489
nutrient deficient bacteria, gene expression 243

O^6-alkylguanine alkyltransferase (AGT) 556
OB-fold *see* oligonucleotide/oligosaccharide-binding motif
octamer eviction, chromatin 284
Okazaki fragments
  DNA replication 459, 460–1, 463, 468–70
    bacteria 459, 463, 468
    bacteriophages 479
    eukaryotes 496, 499, 500, 501
Okazaki, Reiji 460–1
oligonucleotide/oligosaccharide-binding motif (OB-fold) 100, 130
oligosaccharides, protein links 442, 443
oligo(U) tracts 408, 413
O-linked glycosylation 441, 442
Ω-subunits, RNA polymerase 201, 203, 204
one-dimensional searching, protein–DNA interactions 195, 196
open chromatin regions, DNase I hypersensitivity 298–9, 301–2, 303, 313
open circles (OC), plasmids 102
open complex, RNA polymerase–DNA 196, 197
opening transitions, nucleosome dynamics 176
open reading frames (ORFs) 143, 282, 401, 403
operons 244, 246–55
optical tweezers (OT) 199, 200
ORC *see* origin recognition complex
organelle-specific transcription enzymes 216

orientation of genes, direction of transcription 193–4
origin recognition complex (ORC) 490, 493, 494, 495
origins of replication 461, 472, 485–8, 490–1, 493, 495
OT *see* optical tweezers
overhangs (sticky DNA ends) 96, 97, 99, 187
oxidation, DNA damage 551, 552
oxidative stress 413, 416

PABP *see* poly(A)-binding protein
PAbs *see* polyclonal antibodies
packaging genomes 163–91
pair-end di-tag analysis (PET) 235, 236
PaJaMo experiment 247–8
palindrome sequences 75–6, 83
pancreatic deoxyribonuclease *see* DNase I
pap operon, transcription regulation 257
paracodons 352
parasites
  malaria 24
  Trypanosoma brucei 545–6
parasitic (selfish) DNA concept 334, 533
Pardee, Arthur 247–8
PARP1 *see* poly(ADP-ribose)polymerase 1
PARylation *see* poly(ADP)ribosylation
pathogenicity in bacteria 119, 257, 404, 405
Pauling-form DNA 73
Pauling, Linus 42–3
pauses in transcription 198, 200, 268
PAZ domain, argonaute proteins 342
PCNA *see* proliferating cell nuclear antigen
PCR *see* polymerase chain reaction
penetrance of genes 21
penicillin 439
peptide bonds 34–5, 381–3, 384, 385
peptidoglycans 432, 436, 439
peptidyl (P) site, ribosomes 348, 349, 365
peptidyl transfer 382
peptidyl transferase center (PTC) 361, 376, 381, 382, 385
peripheral membrane proteins 422, 424
Perutz , Max 45
PET *see* pair-end di-tag analysis
PF *see* promoter flanking regions
phage display techniques 104–5, 107
phagemids 105
phage *see* bacteriophage
pharmaceuticals
  recombinant DNA 116–19
  see also drug...
phase variation, bacterial pathogenicity regulation 257
phenocopying 21
phenotype inheritance patterns 16–17, 20–1
phosphatases 435–6, 441
phosphodiester bonds 65–6, 68, 194, 206
phospholipids 422, 423
phosphorylation of proteins
  C-terminal domain of Rpb1 219, 220
  histones 272, 275
  post-translational processing 55, 435–6, 440, 441
photoreactivation 550, 554, 555–6
Piwi domain, argonaute proteins 342
Piwi-interacting RNAs (PiRNAs) 342, 344, 345
plants
  cell structure 4
  genetic engineering 118–22
  hepatitis B vaccine production 118–19
  introducing recombinant DNA into cells 110
plasmid/phage combinations 105
plasmids
  Agrobacterium tumefaciens 119–21
  bacterial conjugation 23
  DNA cloning vectors 94, 101, 102–3
  DNA replication 478, 481
  isolation from bacteria 102
  recombinant insulin production 117
  shuttle vectors 109
plasmids with *cos* sites *see* cosmids
plectonemes, supercoiled DNA 78–9
pleiotropy 20
PML *see* promyelocytic leukemia bodies
point centromeres 185
Pol I/Pol II/Pol III *see* RNA polymerases
poly(A)-binding protein (PABP) 321, 406, 408, 410
polyadenylation of RNA transcript 224–5, 320–1
poly(ADP-ribose)polymerase 1 (PARP-1) 279–80, 281
poly(ADP)ribosylation (PARylation) of proteins 279–80, 281
poly(A) tails on mRNA
  cytoplasmic polyadenylation element 360–1
  length and functions 320–1
  translational regulation role 407–8, 412
  see also decapping
polycistronic mRNAs 373
polyclonal antibodies (PAbs) 543
polycomb repressive complexes (PRC1/PRC2) 276–7
polygenic traits 21
polymerase, *see also* DNA polymerase; RNA polymerase
polymerase chain reaction (PCR) 110–12, 118, 235, 306, 399
polynucleotides *see* nucleic acids
polypeptides 34–7
  see also proteins
polyribosomes 372, 373
polysaccharides
  cell walls/capsules 7, 70, 110, 436
  see also glycosylated proteins
polytene chromosomes 22, 23
polyubiquitylation of proteins 448, 449, 451–2, 453
post-transcriptional processing 315–46
post-transfer quality control mechanisms, tRNA aminoacylation 353–4, 357
post-translational modifications (PTMs) 55, 429–56
  acetylation 436, 442
  chemical modification 433–6, 439–55
  circularization 433, 435, 436
  epigenetic regulation 271–85
  glycosylation 436, 439, 441, 442, 444–5
  Golgi 427
  histones 175–6, 271–85
  phosphorylation 435–6, 440, 441

proteolytic 429–33, 434, 435
transcription factors 244
post-translational translocation of proteins 424, 425–6, 431
PRC1/PRC2 *see* polycomb repressive complexes
predicted enhancers, genome segmentation by transcription state 300–1
pre-initiation complexes
DNA replication 493, 494
transcription 223–4, 228
premature aging syndromes 506, 553, 572
premature stop codons 416–17
pre-messenger RNA 153
pre-replication/pre-replicative complexes (pre-RCs) 493, 494, 495
pre-transfer quality control mechanisms 353, 357
primary structure of proteins 35–7
primase release 471
primer extension, mapping method 173, 234, 469
primer extension/elongation, PCR 111
primer extension inhibition by ribosomal-mRNA complexes 379, 381
primers
bacterial DNA replication 459, 460, 466, 468, 471
eukaryotic DNA replication 495, 496, 504
priming-loop formation 471
primosome 496, 497
principle of genetic equivalence 124
processed pseudogenes 538
processing bodies (P bodies) 412–13, 415–16
processivity factors 462–4, 499
programmed cell death (apoptosis) 433, 437–9
programmed double-strand breaks 531, 532, 533
proinsulin 430–1, 433
prokaryotes
classification of life 4–5
see also bacteria
proliferating cell nuclear antigen (PCNA) 499
promoter escape (clearance) 196
promoter flanking regions (PF) 300, 307
promoter-proximal stalling 267–8
promoters
bacteria 202–4, 205, 242–6, 255
binding study 199
DNase I sensitivity signatures 302
eukaryotes 222, 226, 227, 231, 232
gene expression 153, 155
transcription 196, 197
promoter–distal DNase I hypersensitive site connectivity 305, 307
promyelocytic leukemia (PML) bodies 506, 507
proofreading
bacterial DNA replication 462, 468–9, 470
eukaryotic DNA replication 496–7
translation 385
tRNA aminoacylation 353–4, 355–6, 357, 358
prophages 541
prostate cancer 327, 328, 414
proteases
caspases in apoptosis 433, 437–9
neurotoxins acting on SNARE proteins 428, 429
protein splicing 431–3
proteasome
protein destruction 57, 58–60, 433, 439, 451–2, 453
protein splicing 432, 434
see also ubiquitin-proteasome system
protein-coding genes
complex structure 153
numbers 154, 157
protein degradation
apoptosis 433, 437–9
ubiquitin-proteasome system 433, 439, 445–7, 451–2, 453
protein–DNA interactions 128–36
affinity chromatography 129
bacteria 166–8
bending and looping 132–4, 141, 142, 166
DNA footprinting 140
electrophoresis 139–42
mitotic chromosomes 183–90
nonspecific binding 128–9
nucleosomal structure of chromatin 168–78
site-specific binding 129–36
study methods 128, 129, 139–43
viruses 165
protein elongation factors 198
protein engineering 355, 359–60
protein-induced DNA bending
free energy price 132
genome organization 166
looping 133–4
minor-groove binding 129–30, 133, 166, 223, 224
study methods 141–2
protein–nucleic acid interactions 127–144
complexes with structural/regulatory functions 163
study methods 128, 129, 139–143
see also protein–DNA interactions; protein–RNA interactions; transcription factors
protein–RNA interactions 136–44
RNA binding 136–8
RNA polymerase complexes 201–2
proteins 31–64
algorithms to identify and classify domains from sequence 45, 47, 48–9
allostery 44, 47, 248
chaperones to assist folding 55, 57–8
chemical modification of side chains 433–55
cleavage 430–1, 433
degradation/destruction 35, 57, 58–60, 432, 433
DNA-binding 129, 132, 134–6
domain types 44–5, 48
early understanding of gene function 23, 24
electrophoresis 37–9
fate after synthesis 57
filaments 53
folding 39, 42, 53–5, 57
functions 31–3
interaction networks 61–3
intrinsically disordered structures 49–51, 52
mature proteins from precursors 430–1

proteins (*continued*)
membrane associated 422–3, 424
membrane insertion 425, 426
molecular interactions determining structure 42
N-terminal signal sequences 424, 425, 426, 429, 430
postsynthetic modifications 55
primary structure 35–7
processing and modification 421–56
proteolytic processing 429–33, 434, 435
purification and identification techniques 40–1
quaternary structure 51, 53
secondary structure 39, 42, 43
side chain modification 433–55
signal peptide removal 425, 426, 430
splicing 431–3, 434, 435
structural classification by CATH 48–51
structural motifs
helix-turn-helix 132, 134–5, 250, 251, 252, 556
leucine zippers 135, 136
oligonucleotide/oligosaccharide-binding motif/ OB-fold 100, 130
zinc fingers 135, 136, 137, 138
structure 33–58
tertiary structure 42–4
translocation through biomembranes 423–9, 430, 431, 432
types 32–3
X-ray crystallography 45, 46
see also enzyme...; histone...
protein scaffold, mitotic chromosome assembly 189, 190
protein targeting cycle 429
proteoglycans 436, 442
proteolysis 35, 430–435, 445, 447
the proteome 60–61
protomers 51
protoplasts 110
provitamin A *see* β-carotene
proximal promoters 262
pseudogenes 154, 155, 538
P site *see* peptidyl site
psoralen cross-linking 397–398
PTC *see* peptidyltransferase center
PTMs *see* post-translational modifications
pulldown *see* co-immunoprecipitation
purines 66–7, 69
pyrimidines 66–7, 70

Q-SNAREs 428–9, 432
quality control
RNA 339–41
tRNA charging with cognate amino acids 353–4, 355–6, 357
see also proofreading
quantitative PCR (real-time PCR) 111–112, 143
quantitative traits 21
quaternary structure of proteins 51–53

Rabl configuration of interphase chromosomes 520
RACE *see* rapid amplification of cDNA ends
Rad51 recombinase 520, 521, 525, 526, 527, 531–532
Rad52, recombination mediator 526–527
RAD52 epistasis group 525
radiation *see* electromagnetic radiation; ionizing radiation; UV radiation; X-rays
radial loop model of mitotic chromosome structure 187–8, 189, 190
Ramachandran plots 43
random coil RNA structures 83
rapid amplification of cDNA ends (RACE) 235
ratchet mechanisms *see* Brownian ratcheting
readers of chromosome marks 272, 273–5, 289
reading frames 149, 150, 151
real-time PCR (quantitative PCR) 111–112
RecA, homologous recombination 519, 520, 521–3, 525
RecBCD complex, end resection 518–20
reciprocal exchange, antigen switching 546
recombinant DNA 93–126
gene expression 108–9
human insulin 117
introducing into host cells 109–10
pharmaceutical applications 116–19
plant genetic engineering 118–22
practical applications 116–25
sequences, genomic libraries 112
technology 93–126
transgenic animals 115–16
uses 93
recombinases, *see also* Rad51
recombination 21, 22, 515–48
homologous 516–32
nonhomologous 532–40
site-specific 540–6
recombinational hotspots 532
recombination mediators 526–527
RecQ helicases, DNA repair 568, 570–572
recruitment models 266, 332–3
reduced representation bisulfite sequencing (RRBS) 308, 309, 312, 313
regulated versus basal transcription 262
regulated versus constitutive transcription 241
regulatory proteins
connection to Pol II 224
splicing regulation 330–331
regulons 256
re-initiation of transcription cycle 227
relaxed plasmids 102
release factors (RFs) 389–90
replacement variants, histones 174
replacement vectors 104
replication factor C (RFC) 499
replication foci 485, 486, 488, 490, 492
replication fork 457–459, 460
collapse rescue mechanisms 517
core bacterial proteins 466
helicase activation 494
progression 471–2, 485, 486
replication origin *see* origins of replication
replication protein A (RPA) 499
replicative transposon movement 538–9

replisomes
  bacterial DNA replication 461, 462, 464–5, 468, 470–2, 473
  eukaryotic versus bacterial 495–7, 499
  pausing 471
repressed regions of genome 300, 301
repressors
  displaced from DNA by inducers 248, 249, 250
  feedback from end product 250–1, 255
  transcription regulation 244, 245, 246, 264–267
re-replication of DNA prevention 495
restriction endonucleases (restrictases) 94–9
retinoblastoma 21
retrotransposons 537–8, 539–40
retroviruses 509–12
Rett syndrome 289, 291
reverse transcriptase (RT) 46, 469, 509–12, 537
reverse transcription
  cDNA 112, 114, 117
  retrotransposons 340, 537–8, 539, 540
  retroviruses 509–12
  telomeres 505, 506
reverse transcription polymerase chain reaction (RT-PCR) 112, 118, 235
RFA *see* replication factor A
RFC *see* replication factor C
RFs *see* release factors
RF-Tu *see* elongation factor Tu
rheostat control of transcription 284–285
Rho termination factor 210–211
Rho utilization site (rut site) 211
ribonucleic acid *see* RNA
ribonucleoprotein (RNP) particles 361–362
ribose sugar 66–7
ribosomal proteins (r-proteins) 396–7, 398–9
ribosomal releasing (recycling) factor (RRF) 389, 391
ribosomal RNA 5, 225–7, 228–9, 316–17
ribosomes 84, 361–7
  antibiotic effects 374–5374–5
  archaea 361, 362
  assembly 365–366, 367
  bacteria 361, 362, 363–366
  bacterial transcription-translation coupling 251–252, 253
  biosynthesis regulation 396–399
  coordinated synthesis of components 396–397
  decoding center 361
  discovery and function elucidation 147–9
  eukaryotes 361–2, 366–7
  in vitro assembly process 365, 366
  in vivo assembly process 366, 367
  mRNA alignment 357–358, 361
  number regulation 395–9
  organelles 362
  peptidyltransferase center 361
  recycling 390–1
  subunit structure 361–4, 365
  translation 348–9, 364–5
  translation initiation on small subunit 377
  translocation of mRNA during translation 379, 384, 386–7
  tRNA binding sites 348–9, 365
riboswitches 400, 404–405, 406
ribozymes 83, 316, 317–318, 334
rice, genetic engineering 121–122
RING-finger domains 447–53
RNA 66–7
  amplification by RT-PCR 112
  chemical structure 65
  degradation 339–41
  functions 83, 84, 147–9
  physical structure 82–3, 148, 404
  quality control 339–41
  replication primers 459, 460, 466
  retrotransposons 537, 539–40
  reverse transcriptase 46, 469, 509–12, 537
  separation by size 147–8
  stability 320–1
  synthesis 194, 196, 197, 206
  see also individual RNA types; RNA polymerases; transcription
RNA-dependent RNA polymerases 201
RNA enzymes *see* RNase P; ribozymes
RNA-PET, ENCODE project 313
RNA polymerases (RNAPs)
  bacterial 201–6
  bacterial subunits 201, 203, 204
  discovery 203
  eukaryotic 201, 216–17
  initiation complex 201, 202–6
  molecular motions 195
  organelle-specific enzymes 216
  Pol I 216–17, 225–7, 228
  Pol II 216–25
  Pol III 216–17, 227, 229
  promoter-proximal stalling 267–8
  recruitment by transcription factors, bacteria 245
  removal of nucleosomal barrier 270
  transcription 194, 195–6, 197, 198, 201–2
  viral/phage 201, 202
RNA processing 315–46
  bacteria 315–17
  categories 315
  end modifications 316, 318–21
  life history of mRNA molecule 335–8
  post-transcriptional sequence editing 336, 338
  quality control 339–41
  small silencing RNAs 342–4, 345
  splicing 321–33
  tRNA/rRNA 316–17
RNA-protein interactions *see* protein-RNA interactions
RNAPs *see* RNA polymerases
RNA-recognition motif (RRM) 137, 138
RNase P, 5' end capping of tRNA 316
RNA-seq, ENCODE project 312
RNA Tie Club 349
RNP *see* ribonucleoprotein
rolling circle replication 480, 481
Rothman, James 427
Rothmund-Thompson syndrome 572
Rous, Peyton 509

Rpb1 subunit of RNA polymerase Pol II 218–19, 224
Rpb4/7 subcomplex of RNA polymerases 216, 217, 224
RRF *see* ribosomal releasing (recycling) factor
RRM *see* RNA-recognition motif
rRNA genes (rDNA) 397–8
R-SNAREs 428, 432
RT *see* reverse transcriptase
rut site *see* Rho utilization site
RuvA-RuvB complex 524–5

S1 nuclease protection 234
Saccharomyces cerevisiae
 cell-cycle regulation study 489
 chromosome maps 22
 histone modification enzymes 273
 homologous recombination factors 526
 as model organism 27
 origins of DNA replication 488, 493
 replicative polymerases 496
 RNA polymerase structure 217–219, 220
 ubiquitin system 449
 see also yeast
Sanger sequencing reaction 90, 91
SARs *see* scaffold attachment regions
satellite DNA 159, 160
SC *see* synaptonemal complex
scaffold, mitotic chromosome assembly 189, 190
scaffold attachment regions (SARs) 189
scanning/tracking mechanism model, enhancer action 263
Schizosaccharomyces pombe 27, 489, 490
Schoenheimer, Rudolf 445
scrunching model of abortive transcription 208, 209
SD *see* Shine–Dalgarno sequences; strand-displacement model
SDS *see* sodium dodecyl sulfate
SDSA *see* synthesis-dependent strand annealing
SecA, bacterial protein translocation 424, 426
secondary structures
 nucleic acids 165, 166–168, 361
 proteins 39, 42
second law of thermodynamics 195
second-round pre-RC formation prevention 495
secretory protein translocation 424
SecYEG complex 425
sedimentation analysis of nucleic acids 87, 147–8, 160, 460, 492, 516
sedimentation coefficient of proteins 36, 362–3
seed regions, microRNAs pairing with mRNA sequences 406, 407, 408, 409, 411
self-assembly principle, protein folding 54
self-cleavage reactions, ribozymes 317–18
selfish DNA concept 334, 533
 see also junk DNA
self-splicing RNA 334–5
semiconservative replication of DNA 70, 73, 74, 457, 458
sense (nontemplate) strand of DNA 193–194
separase, cohesin cleavage 183, 185
sequence logos 251
sequencing entire genomes 112–14
sequencing methods for DNA 90–91
Ser-Arg (SR) proteins 330–331
sexual reproduction 18
sharp-type promoters 222
β-sheets 42, 43
shelterin complex 187, 506, 507
Shine-Dalgarno (SD) sequences 357–8, 361
short-flap pathway, Okazaki fragment maturation 499, 500, 501
short interspersed nucleotide elements (SINEs) 538
short noncoding RNAs 231
short pauses in transcription 198
short tandem repeats 159, 160
shuttle vectors, recombinant DNA 109
sickle cell anemia (SCA) 23, 24
α-initiation factors, RNA polymerase 201, 202–206, 207, 209, 242
signaling
 intrinsically disordered proteins 51
 membrane protein receptors 423
 protein phosphorylation 435–6
signal peptidase, N-terminal signal sequence removal 425
signal recognition particle (SRP) 425, 429, 430
signal sequences, co-translational protein translocation 424, 425, 426, 429, 430
silencers
 splicing regulation 330
 transcription regulation 153, 262, 263
 see also small silencing RNAs
silencing of genes, histone modification 275–6
silent information regulator 2 (Sir2) 277
SINEs *see* short interspersed nucleotide elements
single-gene inheritance 20, 21
single-molecule cryo-electron microscopy 10–11
single-molecule techniques, transcription study 195, 198, 199–200
single-pair fluorescence resonance energy transfer (spFRET) 209, 376–7, 561
single-strand breaks, DNA damage/repair 551, 554, 556
single-strand DNA-binding protein (SSB) 128–129, 130, 461, 462, 466, 499, 522
single-strand DNA (ssDNA), bacteriophages 106, 480, 481
Sir2 *see* silent information regulator 2
siRNAs *see* small interfering RNAs
site-directed mutagenesis 112, 113, 142
site-specific protein binding to DNA 129–133
site-specific recombination 540–546
sleeping sickness 545–546
sliding action, chromatin remodelers 284
sliding clamps
 DNA repair 559, 560, 561, 563
 DNA replication 461, 462–466, 471, 499
 see also proliferating cell nuclear antigen
slippage mechanism, telomeres 505
small interfering RNAs (siRNAs) 342, 344
small nuclear RNAs (snRNAs) 322
small ribosomal subunits 361–6, 377
small RNAs 154, 227–8, 232

small silencing RNAs (ssRNAs) 342–4, 345
small ubiquitin-like modifier (SUMO) 452–455
SMC *see* structural maintenance of chromosomes proteins
Sm proteins, RNA metabolism 409, 414
SNARE proteins 428–429, 432
snRNAs *see* small nuclear RNAs
solenoidal (toroidal) DNA 78–9
spatial organization, transcription 232–4
spFRET *see* single-pair fluorescence resonance energy transfer
S phase of cell cycle 174, 488, 490
spinal muscular atrophy (SMA) 415
spliceosomes 321–322, 323, 324
splicing regulation 329–33
  cis-trans interactions 330–2
  exon–intron architecture 329–30
  heat-stress inhibition 331
  regulatory proteins 330–1
  r-protein synthesis control 399
splicing of RNA 321–33
  self-splicing 334–5
  see also alternative splicing
SR *see* Ser-Arg proteins
S RNA, signal recognition particles 425
SRp38 protein 331
SRP *see* signal recognition particle
SRY gene, sex determination 158–9
SSB proteins *see* single-strand DNA-binding proteins
ssDNA *see* single-stranded DNA
ssRNAs *see* small silencing RNAs
stable unannotated transcripts (SUTs) 231
Stahl, Franklin 458
start codon, initiator tRNA 277
start site selection, translation 378–9
steric effects 35, 42, 245, 246, 355, 356
steroid receptors 135, 136
sticky ends (DNA overhangs) 96, 97, 99, 187
stimulation of transcription 244, 245
stimulons 256
stop codons 148, 151
strain in circular DNA 77
strand-displacement (SD) model 508, 509
strand exchange, homologous recombination 518, 520, 521, 523, 524
strand invasion, homologous recombination 518, 519, 520, 523
strength of splice sites 329
stress, superhelical DNA 209–210, 270
stress granules 412–13, 415–16
stress responses
  insulin-like growth factor 553
  RNA splicing inhibition 331
  stringent control of gene expression in bacteria 243
stringency of conditions, PCR 111
stringent plasmids 102
stringent response, transcription regulation 243
strong enhancers 300–1
structural maintenance of chromosomes (SMC) proteins
  alternative lengthening of telomeres pathway 506, 507
  bacterial genome organization 166, 167, 168
  cohesins/condensins in mitotic chromosomes 183, 184
structural models, mitotic chromosome structure 187–188, 190
structural proteins
  bacterial genome organization 166
  mitotic chromosome scaffold 189, 190
  see also histones
Sturtevant, Alfred 16, 22
sugars
  glycosylated proteins 55, 436, 439, 441–5
  metabolism in bacteria 246–50, 252–3
  ribose/deoxyribose 66–7
SUMO *see* small ubiquitin-like modifier
sumoylation of proteins 452–455
supercoiled DNA
  bacterial expression regulation 255–256
  bacterial genome organization 166
  bacterial replication 476
  coiling processes 76–81
  distinguishing positive and negative coiling 85–6
  gel electrophoresis 85–6
  plectonemes 78–9
  superhelical stress 270
  topoisomerases 79–81, 83
  topoisomers 85–6
  topologically constrained loops 76
  transcription 199, 209–10, 255–6, 270
superhelical density 78
superhelical stress, transcription 209–10, 270
supraspliceosome particles 322
SUTs *see* stable unannotated transcripts
Svedberg (S), molecular sizes 36
SWI/SNF (switch/sucrose nonfermenting) chromatin remodeler superfamily 283
synapsis 530, 531
synaptonemal complex (SC) 530–1, 532
synthesis dependent strand annealing (SDSA) 518, 529, 530
syn-versus *anti*-nucleosides 67, 68
systemic lupus erythematosus (SLE) 413–14

TAFs *see* TATA-binding protein associated factors
tandem chimerism 324, 328
Taq polymerase 111
target-primed reverse transcription (TPRT) 539–540
TATA-binding protein associated factors (TAFs) 224
TATA-binding protein (TBP) 224, 226
TATA box 222, 223–4
TCR *see* transcription-coupled repair
telomerase 504–6
telomeres 184–7, 189, 277, 504–7
Temin, Howard 509
temperate phages 25, 540–2
  see also bacteriophages, phage
template, DNA replication 459
template strand of DNA, transcription 193, 194
5'-terminal oligopyrimidine (TOP) sequences 399
termination of replication 476–8
termination of transcription 196, 198, 210–211, 224–225
termination utilization substance (Tus) 477–478

tertiary structure of DNA 75–76
tertiary structure of proteins 42–51
tetanus neurotoxin 428, 429
TFs *see* transcription factors
TGN *see trans*-Golgi network
thale cress *(Arabidopsis thaliana)* 27, 28
thermosensing riboswitches 404, 405
three-polymerase replisome 472
threshold cycle (CT) 111–112
thymine dimer repair, DNA photolyase 554, 555–6
thymine (T) 66–7, 72, 73, 193–4
time-of-flight mass spectrometry 56
t-loops, telomeres 186, 187
toeprinting 379, 381
topoisomerases
  bacterial genome organization 166
  DNA replication 460–1, 466, 478
  inhibitors as anti-cancer agents 83
  supercoiled DNA 79–81, 83
  type I/II 79, 80, 81
topoisomerase/topo II catalytic cycle 79, 82
topoisomers, supercoiled DNA 85–6
topologically constrained loops 76, 209–10
toroidal (solenoidal) DNA 78–9
TPRT *see* target-primed reverse transcription
TψC arm, tRNA molecules 350
tracking/scanning mechanism model 263
trans-acting elements 399
trans–cis interactions *see cis–trans* interactions
transcribed genes, genome segmentation 300
transcription 193–239
  aborted 196, 206–9
  backtracking 198, 200
  bacteria 202–212, 241–259
  bidirectional 195
  common mechanisms 193–202
  direction 193–4
  DNA rotation 197, 200, 209
  DNA-RNA 82, 84
  electron microscopy 198, 201
  elongation 196, 197, 198
    bacteria 209–10
    eukaryotes 216, 218, 219–23, 267–8, 270
  eukaryotic 215–239
    organization and function 228–234
    Pol I 216–17, 225–7, 228
    Pol II 216–25
    Pol III 216–17, 227, 229
    regulation 161–96
    study methods 234–8
  fidelity 221
  initiation 196, 202–206, 222–224, 262–267
  interruption and pauses 197–198, 200, 268
  nontemplate strand of DNA 193–4
  ribosomal RNA precursors 225–227
  study methods 198–201, 234–238
  template strand of DNA 193, 194
  termination 196, 198, 210–211, 224–225
  see also transcription regulation
transcriptionally repressed regions (R) 300, 301
transcription bubble 194, 196, 197, 198
transcription-coupled repair (TCR) 556–7
transcription factor binding sites (TF-binding sites) 264, 266, 302–5, 307
transcription factories 232–3
transcription factors (TFs)
  acting together 244–5
  activators versus repressors 244, 245
  binding defines structure and function of regulatory regions 305–9
  combinatorial interactions 267
  eukaryotic transcription initiation 223–4, 226
  general 223, 226
  gene transcription regulation 264–267
  interaction networks 307–9
  networks 257–258
  proteins that are both activators and repressors 252–255
  regulation in bacteria 244–6
  structure/binding site co-evolution 309–11
  synergistic interaction 244, 262
  TFIIB/E/F/H/J 223, 224, 225
  TFIID 223, 224, 266, 268
  TFIIF-like subcomplex of RNA polymerases 217, 222, 223, 224, 225
  TFIIIA 135, 229
  TFIIIB/TFIIIC 185, 229
  TFIIS protein elongation factor 198, 268
transcription rate, splicing regulation 332–333
transcription regulation
  bacteria 241–59
    ara operon 252–5
    coordination of gene expression 256–8
    DNA modification 255–6
    global control of expression patterns 244, 255, 256, 258
    history 247–8
    Lac operon 249–50, 251
    promoters 242
    specific gene regulation 244–55
    stringent control 243
    trp operon 250–2, 253
  eukaryotes 261–96
    chromatin structure 268–71
    DNA methylation 285–9, 291
    elongation 267–8
    epigenetics 271
    histone modification 271–85
    initiation 262–7
    long noncoding RNAs 289–90, 292–3, 294
    measuring activity of regulatory elements 293
    metabolites 284–5
    polytene chromosomes 23
    rheostat control 284–5
    ribosomal protein synthesis 399
  human genome 297–314
    chromatin structure 301–5
    DNA methylation 311–12
    ENCODE project 297–314
    segmentation by transcription state 300–1

transcription factor binding 305–11
transcription start sites (TSS)
bacterial 202–203
bidirectional transcription 231
eukaryotic 222, 231
general 194, 196
human 300, 302–303, 304
mapping methods 234, 235, 236
transcriptome analysis 235, 237–8
transduction of bacteria 25
transesterification of RNA 321, 322
transfection efficiency 123
transfer RNA *see* tRNA
transformation 25, 110, 517
transgenic organisms 115–116, 291, 505
trans-Golgi network (TGN) 428
transient excursion model of abortive transcription 208
translation 347–420
coupled to transcription in bacteria 251–252, 253
decoding process 380
efficiency 360–1
elongation 348–349, 379–388
fidelity 353–354, 355–356, 357, 358
GTP hydrolysis 274, 275
history 349
initiation 348
bacteria 376, 377–378
eukaryotes 377, 378–379, 380, 400–402
regulation 400–407
overview 347–349
participants 347–69
process 371–93
RNA function elucidation 147–9
RNA-protein 84
speed and accuracy 371–3
study methods 373–7
termination 348, 349, 388–392
translocation association 423–424, 425–426, 429, 431
translation factors, potential oncoproteins 372
translation regulation 395–420
human disease 407, 409, 410–11, 413–14, 415
initiation 400–7
metabolite responses 404–5, 406
microRNA-induced silencing complex 406, 408, 410
microRNAs 405–7, 408, 409, 410–11
mRNA sequestration in P bodies and stress granules 412–13, 415–16
need for 395
ribosomal protein synthesis in mammals 399
ribosome number 395–9
riboswitches 404–5, 406
temperature responses 404, 405
translesion synthesis, DNA repair 522, 554, 555, 566–8
translocation
DNA in eukaryotic transcription 219, 220–221
mRNA with respect to ribosomes in translation 379, 382, 383–385, 386–387
retroviral DNA 511–512
translocons, membrane transport 424, 425
transmembrane proteins 422–3, 424, 425, 426
transmission electron microscopes 7
transposable elements (transposons) 160, 532–540
bacterial 535–536
class I-retrotransposons 537–538, 539–540
class II-DNA transposons 536, 538–9
eukaryotic 536–540
repression 344
transposase 535, 539
trans post-transfer editing 354, 357
trans-splicing 328–9
TRAP *see* tryptophan RNA-binding attenuation protein
tripeptide anticodon motifs 390
triplet code concept 148, 149
triplex (triple helix) DNA 75, 76
tRNA 84, 349–56, 357, 358
aminoacylation 351–4
codon matching 380
discovery 147–148
movement through ribosome 348–349
noncanonical amino acid binding 354–5, 359
processing 316
proofreading 353–4, 355–6, 357
structure 148, 350–1
transcription by Pol III 227
see also aminoacyl-tRNA synthetases
tRNAi (initiator tRNA) 348, 377
tRNA nucleotidyltransferase 316
trombone (loop formation) 468
trp operon 250–2, 253
Trypanosoma brucei 545–6
tryptophan RNA-binding attenuation protein (TRAP) 138
tryptophan synthesis 250–2, 253
TSS *see* transcription start site
tumor-inducing (Ti) plasmid 119–20
tumor suppressor proteins/genes
ATBF1 327–8
BRCA1 451
BRCA2 527
candidate genes 284
E3 ligases 451
gene silencing/reactivation 287
H2A.X 575
p53 49–50, 52, 294, 403, 449
Tus (termination utilization substance) 477–478
twist, strained circular DNA 77, 78
two-dimensional gel electrophoresis 39
two-gate model, DNA transport 79, 82
two-hit hypothesis 21
type I restrictases 96, 97, 98, 99
type II restrictases 96, 97, 101
type III restrictases 97, 98
type I/II topoisomerases 79, 80, 81

UBF *see* upstream binding factor
ubiquitin, *see also* small ubiquitin-like modifier
ubiquitin-activating enzyme (E1) 447, 448, 449, 453, 454
ubiquitin-conjugating enzyme (E2) 447, 448, 449, 450–1, 453, 454

ubiquitin-proteasome system 433, 439, 445–7, 452
ubiquitin-protein ligase enzyme (E3) 447, 448, 449–51, 453, 454
ubiquitylation
  histones 275, 276–277
  post-translational processing 445–452, 453, 454
  proteins marked for destruction 55, 57, 58–59
UCE *see* upstream control element
ultracentrifugation 36, 147–8, 160, 362, 548
UNG *see* uracil-DNA glycosylase
unidirectional motion of RNA polymerase 195, 196
universality of genetic code 151–2
unnatural amino acids 355, 359–60
unpunctuated triplet code 149, 151
5'-3'-untranslated regions (UTRs) 153, 358, 360–1, 362, 399, 402–3
unusual amino acids *see* noncanonical amino acids; unnatural amino acids
upstream, RNA molecules 195
upstream activating sequence (UAS) 262
upstream binding factor (UBF) 226, 227
upstream control element (UCE) 226, 227
uracil-DNA glycosylase (UNG) 557–558, 559
uracil (U) 66–67, 193–194
UTRs *see* 5'-3'-untranslated regions
UV radiation 3, 549, 550, 552, 554, 555–6

vaccine production 117–19
variable domains, immunoglobulins 542, 543, 544
variable surface glycoprotein (VSG) 545–546
vectors
  artificial chromosomes 106–8, 114, 124, 490
  bacteriophage 103–5, 106, 107, 113
  DNA cloning 94, 101, 102–8
  gene therapy 123–4
  plant genetic engineering 119–20
  shuttle vectors 109
Vibrionaceae specific DNA replication inhibition 474
viral RNA, self cleavage 318
viral vectors, gene therapy 123–4
virulence genes 119, 404, 405
viruses
  electron microscopy 7, 8
  genome organization 164–5
  HIV 118, 165, 509, 512
  retroviruses 509–12
  RNA polymerases 201, 202
  see also bacteriophages
vitamin A deficiency 121–122
VSG *see* variable surface glycoprotein

Watson, James 69–70, 72–73, 349
weak enhancers (WE) 300–301
Werner syndrome 506, 572
Werner (WRN) helicase 506
Western blotting 40, 41
whole-animal cloning 124–125
whole-genome sequencing 298–299
whole-genome shotgun approach 113
wild-type alleles 20, 21
wild-type phenotype reversal with complementation 20
Wilkins, Maurice 70, 72, 73
wobble hypothesis/rules 151, 350, 351, 352
writers, histone modification marks 272–3
writhe, strained circular DNA 77, 78
WRN *see* Werner helicase

X-linked mental retardation 415
X-ray crystallography 45, 46, 139, 374–6
X-ray diffraction 11–12, 13, 70–1, 72, 217–18
X-rays 3, 525, 550, 552

YACs *see* yeast artificial chromosomes
Y chromosome 157–159
yeast
  co-translational versus post-translational translocation 426
  phosphorylation and dephosphorylation events 435, 441
  protein interaction maps 62
  ribosomal protein synthesis regulation 399
  RNA polymerase II structure 217–219
  ubiquitin system 447, 449, 552
  see also Saccharomyces cerevisiae; Schizosaccharomyces pombe
yeast artificial chromosomes (YACs) 107–8, 490
yeast episomal plasmid vector (Yep) 109

Zea mays (maize) 121, 533, 534–5
Z-form DNA 72, 74, 75
zinc fingers (ZnFs) 135, 136, 137, 138
zygotes 17, 18

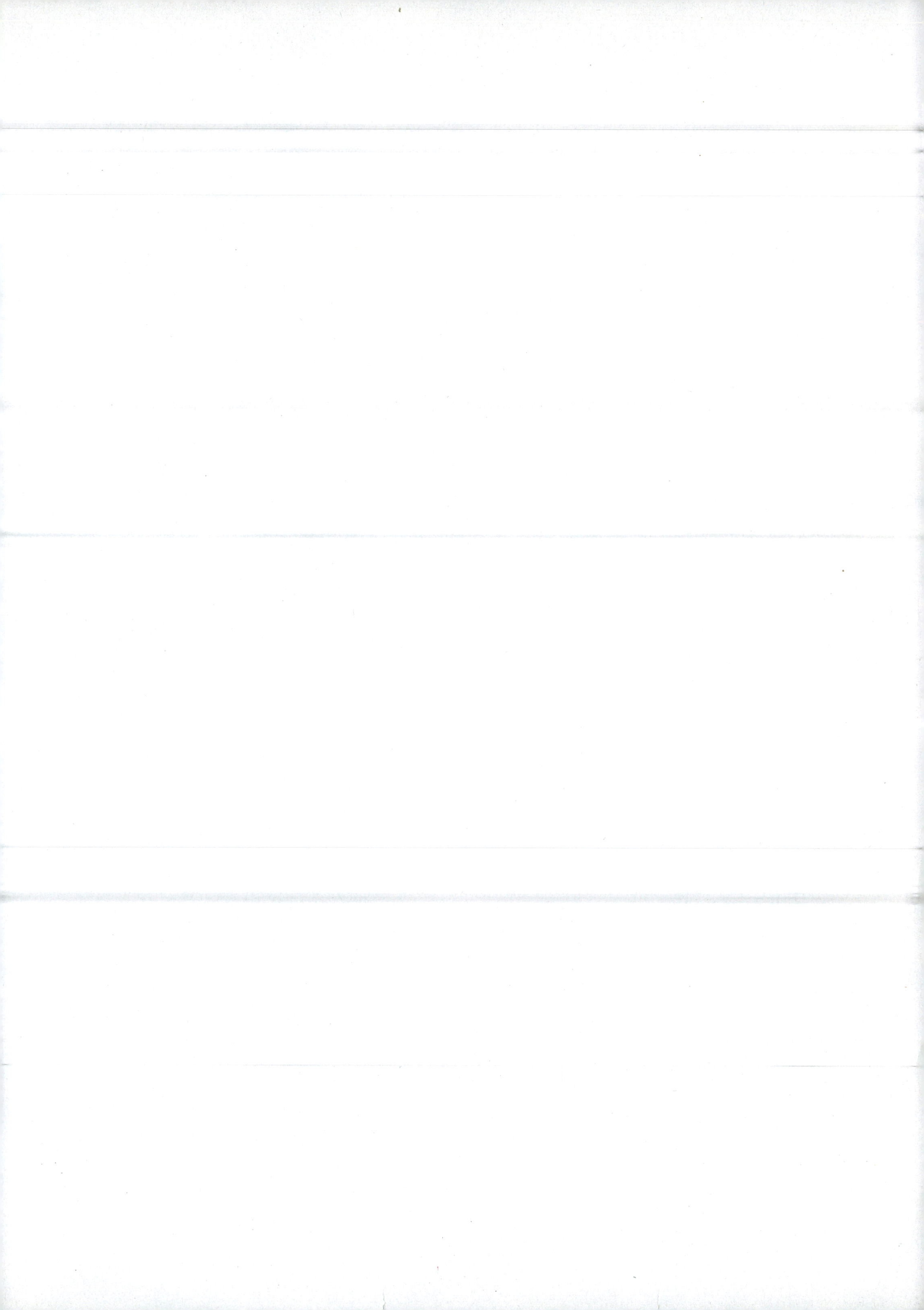